Birkhäuser

Studies in Universal Logic

This series is devoted to the universal approach to logic and the development of a general theory of logics. It covers topics such as global set-ups for fundamental theorems of logic and frameworks for the study of logics, in particular logical matrices, Kripke structures, combination of logics, categorical logic, abstract proof theory, consequence operators, and algebraic logic. It includes also books with historical and philosophical discussions about the nature and scope of logic. Three types of books will appear in the series: graduate textbooks, research monographs, and volumes with contributed papers.

More information about this series at http://www.springer.com/series/7391

Ángel Garrido • Urszula Wybraniec-Skardowska
Editors

The Lvov-Warsaw School. Past and Present

Birkhäuser

Editors
Ángel Garrido
UNED
Department of Fundamental Mathematics
Madrid, Spain

Urszula Wybraniec-Skardowska
Cardinal Stefan Wyszyński University
in Warsaw
Warsaw, Poland

ISSN 2297-0282 ISSN 2297-0290 (electronic)
Studies in Universal Logic
ISBN 978-3-030-09740-0 ISBN 978-3-319-65430-0 (eBook)
https://doi.org/10.1007/978-3-319-65430-0

Mathematics Subject Classification (2010): 01A72, 03-03

Softcover re-print of the Hardcover 1st edition 2018

Printed on acid-free paper

This book is published under the trade name Birkhäuser, www.birkhauser-science.com by the registered company Springer International Publishing AG part of Springer Nature.
The registered company address is: Gewerbestrasse 11, 6330 Cham, Switzerland

Preface

This book is about the Lvov-Warsaw School. I did not "order" this book, but I am very glad to have received this proposal and the book to be published in a book series I have created inspired by this school.

The Lvov-Warsaw School is one of the most important schools in the history of logic but is still not very well known outside of Poland and a circle of aficionados. This book with contributions about the main members of this school and their works will certainly help to fill the gap, reinforcing the already existing attraction and promoting new interests for this school.

I would like to thank the editors of this book, Urszula and Angel, for their considerable efforts to gather all the papers included in this big volume, as well as all the contributors of this book and Birkhäuser staff in Basel.

Editor-in-Chief
Studies in Universal Logic
Rio de Janeiro, Brazil
December 19, 2017

Jean-Yves Beziau

Contents

Part I
Twardowski's School: The Period of Crystallization of LWS

Introduction. The School: Its Genesis, Development and Significance

Urszula Wybraniec-Skardowska

Abstract The Introduction outlines, in a concise way, the history of the Lvov-Warsaw School—a most unique Polish school of worldwide renown, which pioneered trends combining philosophy, logic, mathematics and language. The author accepts that the beginnings of the School fall on the year 1895, when its founder Kazimierz Twardowski, a disciple of Franz Brentano, came to Lvov on his mission to organize a scientific circle. Soon, among the characteristic features of the School was its serious approach towards philosophical studies and teaching of philosophy, dealing with philosophy and propagation of it as an intellectual and moral mission, passion for clarity and precision, as well as exchange of thoughts, and cooperation with representatives of other disciplines. The genesis is followed by a chronological presentation of the development of the School in the successive years. The author mentions all the key representatives of the School (among others, Ajdukiewicz, Leśniewski, Łukasiewicz, Tarski), accompanying the names with short descriptions of their achievements. The development of the School after Poland's regaining independence in 1918 meant part of the members moving from Lvov to Warsaw, thus providing the other segment to the name—Warsaw School of Logic. The author dwells longer on the activity of the School during the Interwar period—the time of its greatest prosperity, which ended along with the outbreak of World War 2. Attempts made after the War to recreate the spirit of the School are also outlined and the names of continuators are listed accordingly. The presentation ends with some concluding remarks on the contribution of the School to contemporary developments in the fields of philosophy, mathematical logic or computer science in Poland.

Keywords Lvov-Warsaw School · Twardowski · Warsaw School of Logic · Łukasiewicz · Leśniewski · Tarski · Lvov-Warsaw School Phenomenon · Polish logic after the War · Mostowski · Rasiowa · Grzegorczyk · Suszko

Mathematics Subject Classification (2000) Primary 01-02, 03-03

Á. Garrido, U. Wybraniec-Skardowska (eds.), *The Lvov-Warsaw School. Past and Present*, Studies in Universal Logic,
https://doi.org/10.1007/978-3-319-65430-0_1

1 A Short History and Influence of the School[1]

1.1 Twardowski's School: The Period of Crystallization of the L-WS

The beginnings of the Lvov-Warsaw School (L-WS by acronym) date back to the end of the nineteenth century in Lvov, precisely to 1895, when Kazimierz Twardowski (1866–1938), who, at the age of 29 (in 1895), was appointed professor of philosophy in Lvov, then an Austrian town (Fig. 1).

Kazimierz Twardowski took the chair of Philosophy at Lvov University.

He came from Vienna, where he belonged to the last group of students of Franz Brentano. His main aim was to build and organize a strong philosophical circle in Poland, developing the scientific philosophy and methodology in the Brentano's spirit. Poland was at that time partitioned between Austro-Hungary, Germany and Russia; and Lvov belonged to the Austro-Hungarian Empire. Organizing regular and modern philosophical studies was not easy. In Lvov at that time there was no relevant library, or respectable philosophical studies; neither were there any philosophical seminars. Twardowski himself organized such a seminar and a library, donating his own book collection. His incredibly serious treatment of philosophical investigations and philosophy teachings consisting in promoting a clear and critical way of thinking and cultivating scientific, objective truth, his moral attitude towards philosophy in connection with his amazing charisma as a teacher and tutor, and friendly relations with students, were the main reasons of his becoming highly recognized by both students and reputable representatives of other fields of science, not only Lvov philosophy. Also his scientific-organizational achievements and ventures which he accomplished played a significant role. All those features and abilities drew many young people towards philosophy, and their numbers began to grow quickly. Many

Fig. 1 Kazimierz Twardowski

[1] Based on the entry from Stanford Encyclopedia of Philosophy 'Lvov-Warsaw School' elaborated by Jan Woleński [21], his famous book [20] and his article [22] and also the book edited by Jacek Jadacki and Jacek Paśniczek [7], the book by Jacek Jadacki [6] and the essay by Urszula Wybraniec-Skardowska [23].

of them became later Twardowski's devoted disciples. Twardowski promoted many PhD students, who later earned their professorships not only in Poland, but also in various countries around the world, cultivating the tradition and the program of the School.

Among the characteristic features of the School was its serious approach towards philosophical studies and teaching of philosophy, dealing with philosophy and propagation of it as an intellectual and moral mission, passion for clarity and precision, as well as exchange of thoughts, cooperation with representatives of other disciplines at home and abroad, and also fruitful collaboration with mathematicians. The L-WS found its own scientific style of philosophizing and met international standards of training, rigor, professionalism and specialization. The Lvov-Warsaw School was the first of its kind in Poland. At the same time, its community managed to establish contacts between its philosophers and world philosophy.

In this philosophy program, logic—in its broader sense (formal logic, semantics and methodology of sciences)—was a meaningful element and although Twardowski was not a logician and was never a devotee of mathematical logic, he did lecture logic as well (see Fig. 2).

He credited Stanisław Piątkiewicz with forming the foundations of formal logic (then called logistics) in Poland, and recognized Lvov as the place of its birth.

The first students of Twardowski were Władysław Witwicki (1878–1948) and Jan Łukasiewicz (1878–1956). Both earned their Doctorate degrees under his supervision. Witwicki specialized in psychology. Łukasiewicz was Twardowski's first student to

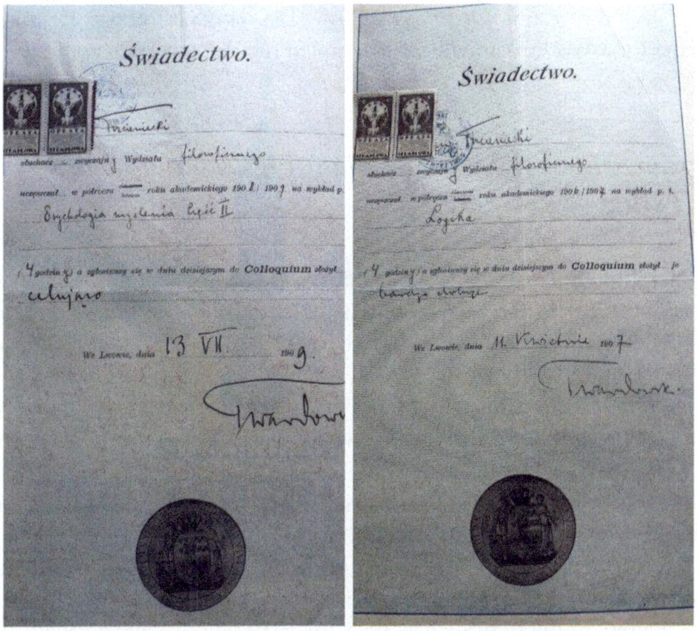

Fig. 2 Student's certificates of Jan Trzciniecki from Lvov University Faculty of Philosophy, lectures in Psychology of Thinking and Logic, signed by Twardowski (photos from the family archives of Irena Trzciniecka-Schneider)

be interested in logic. After obtaining his PhD (1902) and the habilitation (1906), he systematically lectured logic (since 1906). The list of other, trained mainly by Twardowski, eminent individuals strongly interested in logic, who also attended courses conducted by Łukasiewicz include (in alphabetical order): Kazimierz Ajdukiewicz (1882–1963), Tadeusz Czeżowski (1889–1981), Tadeusz Kotarbiński (1886–1981), Zygmunt Zawirski (1882–1948) and Stanisław Leśniewski (1886–1939), who joined the group later, in 1910.

Twardowski promoted 30 full professors of different domains of science; he graduated many more PhD students.

Ajdukiewicz played an important role in logic at Lvov University. He also lectured for mathematicians (before Leon Chwistek got professorship in mathematical logic). Ajdukiewicz gave philosophy a clearly logicizing character.

Twardowski and Ajdukiewicz trained a group which included three women: Izydora Dąmbska (1904–1983), Seweryna Rohman (1904–1978), Maria Kokoszyńska (1905–1981); Henryk Mehlberg (1904–1978) and Zygmunt Schierer (1900?–1943) also were a part of the group. One of the prominent listeners of Twardowski's and Ajdukiewicz's lectures was Eugenia Ginsberg-Blaustein (1905–1944), together with her husband Leopold Blaustein (1905–1944). They all belonged to the second generation of Twardowski's School. So did Janina Hosassion-Lindenbaum (later Mrs. Lindenbaum; 1899–1942), who was a philosopher, and a member of Kotarbiński's strong philosophical group.

The L-WS was also joined by a group of catholic philosophers, including Father Innocenty (Józef Maria) Bocheński (1902–1995) and Father Jan Salamucha (1904–1944).

All the representatives of the L-WS mentioned here are, in this book, classified to belong to the period of *Crystallization* (1900–1918). This period was terminated by the outbreak of the First World War, which interrupted regular scientific activities.

1.2 Warsaw School of Logic, Its Main Figures and Ideas: The Period of Prosperity

The third generation of the students who later reached a great significance in developing the teachings and ideas of the L-WS consists essentially of well-known representatives of the so-called Warsaw School of Logic (by acronym—WSL), acting in the interwar period (1918–1939). In this book, the period is called the *Prosperity* period of the L-WS. It began when in 1916 the University of Warsaw was reactivated and Poland regained its independence in 1918. Kotarbiński and Wytwicki were offered positions as professors at the University of Warsaw at the time.

Many scholars from Lvov moved to Warsaw to organize and build national academic life. Twardowski's School ceased to be a school of Lvov and became a nationwide school. During the interwar period, his former students held chairs in philosophy departments at all of the Polish universities, with the exception of the Catholic University of Lublin.

The main centers of the L-WS since 1918 till 1939 were Lvov and Warsaw. It should be added that Lvov and Warsaw were also the two centers of the Polish mathematical school, which acted in parallel, but not independently of the L-WS. The cooperation of the Lvov-Warsaw philosophers with mathematicians evolved into establishing a logical school based in Warsaw (WSL).

Lvov was still a strong philosophical center, where under the influence of Twardowski, Ajdukiewcz and their disciples there was an exchange of ideas by representatives of various generations of the L-WS. Among the philosophical specialties there prevailed methodology of science and philosophy of language. Kotarbiński, who moved from Lvov to Warsaw in 1919, for the professorship, created a strong group of scholars working mainly in philosophy of science. Kotarbiński educated a whole pleiad of disciples in Warsaw, among others, Dina Sztejnbarg (later Mrs. Janina Kotarbińska; 1901–1997), Edward Poznański (1901–1976), and Aleksander Wundheiler (1902–1957) (the last two of them were the authors of the famous article on the concept of truth in physics).

Warsaw was mainly the center of logic. The development of logic in Warsaw was to a large extent the merit of the mathematician Zygmunt Janiszewski (1888–1920), belonging to the first generation of Twardowski's students, also vitally interested in logic and foundations of mathematics. He was one of the main initiators of the mutual cooperation between mathematicians and philosophers of Twardowski's School.

According to Janiszewski's program for development of mathematics, mathematical logic and foundations of mathematics had a great importance and played a special role. Logic in Lvov did not play such a role as it did in Warsaw. This happened mainly due to the fact that two philosophers, namely Łukasiewicz and Leśniewski, became professors of the University of Warsaw at the Faculty of Mathematical and Natural Sciences. They were put in charge of chairs in the mathematical environment and community at the Faculty (Łukasiewicz since 1915, Leśniewski since 1919).

The Warsaw School of Logic (WSL) grew out of Twardowski's School. The WSL had double roots: philosophical and mathematical. The combination of logic and philosophy in the history of the WSL did not limit itself to the fact that it was genetically related to philosophy. There was a close contact between its logicians and philosophers, in particular with Tadeusz Kotarbiński.

Stanisław Leśniewski and Jan Łukasiewicz, the WSL's two founders, and their student, Alfred Tarski, who obtained his doctor's degree in 1924 with an impressive scientific output, are the three outstanding representatives of the Warsaw School (Fig. 3):

Fig. 3 Jan Łukasiewicz, Stanisław Leśniewski and Alfred Tarski

Łukasiewicz and Leśniewski were logicians and philosophers graduated at Lvov. They continued the philosophical thoughts, teaching methods and scientific research organization of Kazimierz Twardowski, but were protagonists of formal techniques in science, similar to those applied in mathematics. And the connection between philosophers and mathematics, initiated by Janiszewski and his program for mathematical research, manifested itself, among others, in the fact that Warsaw leading mathematicians (Wacław Sierpiński, Stefan Mazurkiewicz, Kazimierz Kuratowski) assigned a considerable role of logic to Jan Łukasiewicz and Stanisław Leśniewski, philosophers by education. Both began teaching mathematical logic not only in the mathematical environment, but also in the philosophical one. Logic became an attractive subject to study in Warsaw. And even though the WSL was a joint creation of philosophers and mathematicians, logic—as a subject of its research—was not regarded as a part of mathematics or philosophy, but as an autonomous science.

Well-known representatives of WSL (in alphabetical order) were: Stanisław Jaśkowski (1906–1965), Adolf Lindenbaum (1904–1941?), Czesław Lejewski (1913–2001), Andrzej Mostowski (1913–1975), Moses Presburger (1904?–1943), Jerzy Słupecki (1904–1987), Bolesław Sobociński (1904–1980), and Mordechaj Wajsberg (1902–1942?). Most were mathematicians, with the exception of Sobociński, who graduated in philosophy, and Lejewski, who studied classical philology.

In the WSL, unprecedented results were achieved. Heinrich Scholz from Münster said:

> Warsaw became the main centre of logical studies.

Thus, in the lifetime of one generation, Polish logic grew from ground level to the acme of international acclaim. In the well-known book of A. Fraenkel, Y. Bar-Hillel, and A. Levy [5, p. 200], it is stated that:

> Probably no other country, taking into account the size of its population, has contributed so greatly to the development of mathematical logic and foundations of mathematics as Poland.

and that:

> this curious fact should be explained sociologically.

The masters of WSL—Leśniewski, Łukasiewicz and Tarski—were individualists with different personalities, who shaped their students in many ways. Leśniewski gave the WSL its theme of synthesis, Łukasiewicz its dynamics, and Tarski the contact with mathematics (see Woleński [20]). As Jan Woleński writes [20], there were no divisions in the School into 'old' and 'young', 'beginner' and 'advanced'. The School had a strong emphasis on cooperation, irrespective of social position, views, or character of members. Thus, it combined scientists active in public and academic life. Members also differed in socio-political and religious views, or their personalities: Łukasiewicz and Sobociński were conservatives; Lindenbaum and Presburger leaned towards communism. Some were devout Catholics, others were followers of Judaism, some were atheists. Except Leśniewski, who was quite reserved, most were persons of a rather friendly nature. Regardless of their differences, they united around a shared scientific idea, the charisma of their teachers, awareness of their exceptionality, and their position in the development of logic in the world.

The problems which the WSL dealt with belonged mostly to mathematical logic. Still, it all started in philosophy. Both Łukasiewicz and Leśniewski earned their Doctor's degrees in philosophy at the University of Lvov: the first one—in 1902, and the other one—in 1912; both under the supervision of Kazimierz Twardowski. Łukasiewicz started with methodology of empirical sciences. But his monograph *About the Principle of Contradiction in Aristotle* (published in Polish in 1910, see [11]) adds a short lecture on 'algebraic logic'. The works by Łukasiewicz, besides those of Jan Śleszyński, are the first in Poland on mathematical logic. Łukasiewicz never resumed his research into the methodology of science. Leśniewski's studies in the time before the First World War concerned primarily problems of semantics of colloquial language and antinomies.

The philosophical education of the WSL founders strongly influenced their disciples and their output in mathematical logic, inducing great care for intuitive value. The representatives of the WSL connected philosophical questions to those of formal logic, solving classical problems of philosophy by its means. For instance, Łukasiewicz was convinced that the three-valued logic he had created in 1920 (see [12]) indeed cast new light on the problem of determinism. And of course, in the famous work [18] (see [19]) on the concept of truth (published in Polish in 1933 and translated into many languages), Tarski solved one of the fundamental questions of the theory of knowledge in such an undisputable manner, that probably no other account could claim. As to the self-image, Leśniewski called himself a 'philosopher-apostate', while Łukasiewicz considered himself a philosopher. Still, Łukasiewicz also did purely formal work, while to Leśniewski logic was always a tool for philosophical questions.

Another feature of the WSL was a drive for full, precise and simplest solutions to problems. This 'perfectionism' (Łukasiewicz) caused the Warsaw logicians to often release results with a delay, even at the risk of losing priority. They delighted in formally perfecting systems, simplifying axioms several times. A peak achievement was reduction of axioms to only one, as short as possible in terms of symbols. The most surprising results in this area were achieved by Łukasiewicz and Sobociński, and it is worth recalling that Łukasiewicz and Leśniewski created two types of original and inventive logical symbolism.

We will not describe concrete achievements in detail, since there are many sources. A core subject in the WSL was methodology of the propositional calculus, initiated by Łukasiewicz. Tarski's results on the notion of truth are the most outstanding. Famous other contributions were by Lindenbaum, Wajsberg on intuitionism, or Jaśkowski's natural deduction system. Tarski's subsequent accomplishments extended to the methodology of all deductive systems. Starting from 1930, he initiated the abstract study of axiomatic systems, but also a standard semantic viewpoint in his paper on the notion of truth [18, 19]. Tarski led a very busy didactic activity, and his students who obtained outstanding results already before the World War 2 included Andrzej Mostowski and Wanda Szmielew.

A highly original strand in the Warsaw School was Leśniewski's creation of the systems of '*prototetics*', '*ontology*' and '*mereology*' in 1929–1930 (see [8, 9]), trying to improve on the mathematical foundations of Russell and Whitehead. Especially,

Leśniewski's ontology has continued to attract interest, for instance from Jerzy Słupecki, Andrzej Grzegorczyk and Bar-Hillel, and in the early 1970s Boguslaw Iwanuś. But perhaps his most famous system is the mereology, studied by Tarski and many others, that still finds applications in geometry, biology, and linguistics today.

The notion of a scientific school is a complex one. Members should address a common problem with shared methods of investigation. The WSL satisfied this to a high degree. Also, a true school should have results that create a valuable whole.

What was the lasting contribution of the WSL? The 400-page volume *Polish Logic 1920–1939* (edited by S. McCall [16] in 1967), which has translations of 17 articles by Polish logicians. All, except two, are from the WSL. Also, during that period, the *Selected Works* of Łukasiewicz [14] and [15] appeared; the first edited by J. Słupecki, the second by L. Borkowski. Half of them were papers from the interwar period. Also there was a monograph by E.C. Luschei on *The Logical Systems of Leśniewski* [10]. A wide selection of Tarski's pre-war articles was *Logic, Semantics, Mathematics* [19].

A scientific school should also be characterized by personal contacts of its members, an atmosphere of constant discussion and exchange of thoughts. The following excerpt from Łukasiewicz's preface to his *Elements of Mathematical Logic* [13] illustrates what the cooperation at Warsaw University looked like:

> I owe the most to the scientific atmosphere created at Warsaw University in the field of mathematical logic. It is in discussions with my colleagues, mainly Professor Leśniewski and Assistant Professor Tarski, and often also with students of theirs and mine that I had a chance to comprehend many a notion, absorb new ways of expressing myself, and learning many a new result whatever their authors were like.

The second period of the L-WS was a phase of splendid creative prosperity: talents that crystalized during the first period brought results in the form of original scientific work—ideas, conceptions and scientific systems. It needs to be emphasized that the L-WS, although known first of all for its achievements in the field of logic, presented a most pluralistic character with regards to interests and views.

1.3 The War and Post-War Period

1.3.1 Annihilation

1939 is regarded as the last year of the WSL. Twardowski and Leśniewski died before 1939. The invasion of Poland by Germany on 1 September 1939 and by the Soviet Union on 17 September 1939 began the Second World War. Poland was again divided, under the German-Soviet pact of 23 August 1939. Warsaw was completely destroyed. Many key members of the L-WS were forced to leave Warsaw (Fig. 4).

Fig. 4 Warsaw after bombardment

A period of unprecedented persecution and extermination of Polish citizens (especially Polish Jews) began, as well as pillage on a massive scale. Most people mentioned above lost their lives: the Lindenbaums, Presburger, Salamucha, Schmierer and Wajsberg. The Polish Government-in-Exile was constituted.

Numerous representatives of the L-WS mentioned above emigrated from Poland during the Second World War or shortly after it: Łukasiewicz (Dublin), Tarski (Berkeley), Lejewski (Manchester), Mehlberg (Toronto, Chicago), Sobociński (Notre Dame), Wundeheiler (New York), Bocheński (Fribourge) and Poznański (Jerusalem, before 1939). Zawirski died in 1947.

On the home front in Poland itself, the resistance movement gradually evolved into an Underground State. Although teaching was conducted in secret, still scientific and cultural activity was interrupted.

1.3.2 WW2 Impact

Polish Logic after the Second World War, since 1945, never regained the renown of the WSL, not so much by losses in human resources as in rhythms of scientific activity, and losses to libraries and manuscripts. The new academic life, and lack of freedom in expressing thoughts, did not favor logic either. Lvov found itself outside Poland (it has been in the Ukraine since the War). Polish country was exposed to the darkest many years of communist terror. In the new political situation many scholars of the L-WS continued teaching and working.

Philosophers and logicians from Warsaw, who did not leave Poland, went elsewhere. For instance, Stanisław Jaśkowski found himself in Toruń (the city of Nicholas Copernicus) in 1945, continuing his scientific and didactic activity in logic, but also in general mathematics.

Jerzy Słupecki was a typical illustration of how the Warsaw School spread in new ways. After the War, he found himself in Lublin and then in Wrocław (in 1948). Wrocław was then a home of displaced people from all over Poland, and expatriates from Lvov.[2]

Słupecki as his main scientific goal chose to continue, popularize and extend the studies of the WSL (in particular the outputs of Łukasiewicz, Tarski and Leśniewski). Słupecki also started working for the Opole Teacher's Training College, 80 km to the east of Wrocław, where he founded his own center of logic.

But after the War, a new Warsaw center of mathematical logic and foundations was started by Andrzej Mostowski, who stayed in Warsaw at the time when the situation of logic was difficult.

Mostowski kept close contacts with mathematicians like Kazimierz Kuratowski and Wacław Sierpiński. Continuing Tarski's work, his center studied set theory, model theory, decidability, algebraic and topological methods in logic. One persistent topic in his center has been generalized quantifiers, introduced by him in 1957 [17]. Famous names from this center are Helena Rasiowa (1917–1994) and Andrzej Grzegorczyk (1922–2014), who continued—he is deceased the pre-war traditions as a logician, mathematician, philosopher and ethicist.

Mostowski and Rasiowa educated a numerous group of logicians and mathematicians, who are classified to a new generation of the L-WS, and who kept up its renown.

Polish logic after the Second World War owed much to Tadeusz Kotarbiński and Kazimierz Ajdukiewicz, who still taught actively. Kotarbiński's work initiated the new discipline—praxeology. Ajdukiewicz's early ideas opened a way to study questions and answers in modern logic [1], the categorial grammar [4], and the notion of meaning [2, 3]. It should be added that in 1955 Ajdukiewicz founded the journal *Studia Logica*; *Studia Logica* is now a significant international journal (incidentally, one of many founded by Polish logicians before the War).

The figures of the Warsaw post-war logic should also include Roman Suszko (1919–1978)—a logician and philosopher (a disciple of Zawirski and Ajdukiewicz) who, together with Ryszard Wójcicki, created in the late 1960s an active logical center at Polish Academy of Science.

[2]A strong group of logicians and philosophers gathered in Wrocław: mathematicians, philosophers and logicians. After the War, Wrocław became perhaps one of the leading mathematical and logical centers in Poland; two famous names are Czesław Ryll-Nardzewski and Jerzy Łoś—prominent logicians and mathematicians.

1.3.3 Polish Logic Today

Polish logic and philosophy still function with vigor today. We have already mentioned the strong ongoing tradition of mathematical logic, emanating eventually from the WSL described in the above. As for formal and philosophical logic, a continued influence of the L-WS is described by Jacek Jadacki in the book [7] edited by himself and by Jacek Paśniczek.

It should also be mentioned here that many Polish logicians of the new generation found a new scope for their talents, namely, in the field of computer science (that did not exist at all in the pre-War period) and many of them are internationally famous.

2 On the Structure and Contents of the Anthology

The book is divided into three parts related to each other. After Introduction the next three parts are in accordance with Sects. 1.1–1.3. Thus, the main contents of the anthology are included into the following parts:

Part I Twardowski's School: The Period of Crystallization of L-WS,
Part II Warsaw School of Logic, Its Main Figures and Ideas: The Period of Prosperity,
Part III The War and Post-War Period.

Every part consists of biographies of almost all the leaders of the L-WS, mentioned in particular sections, respectively. They will be followed most often by articles related to their outputs. All the biographies and articles were elaborated by Polish and foreign experts on the achievements of the L-WS representatives.

In Part II we omit biographies of Łukasiewicz and Leśniewski—the founders of Warsaw School of Logic—which are presented in Part I.

Acknowledgements First of all, the editors would like to thank Professors Jacek J. Jadacki, Jan Woleński and Andrzej Skowron for their kind support and advice, which helped to give the anthology its present form. We would like to express our utmost gratitude to all the authors of this volume, who belong to a new generation of L-WS, for their input and unrelenting efforts to accomplish sometimes very demanding tasks. Their contributions will certainly help to preserve the history of Lvov-Warsaw School for future generations.

Individual thanks are due to Iwona Bajek for the English-language verification of a number of texts in this book, and Dr. Zbigniew Bonikowski for the preparation of many works for publication in LaTeX and the style of Springer.

References

1. Ajdukiewicz, K.: Analiza semantyczna zdania pytajnego [A semantic analysis of a question sentence]. Ruch Filozoficzny **X**, 194–195 (1926)
2. Ajdukiewicz, K.: O znaczeniu wyrażeń [On meaning of expressions]. In: Księga Pamiątkowa Polskiego Towarzystwa Filozoficznego we Lwowie, Lvov (1931)
3. Ajdukiewicz, K.: Sprache und Sinn. Erkenntnis **IV**, 100–138 (1934)

4. Ajdukiewicz, K.: Die syntaktische Konnexität. Studia Philosophica **1**, 1–27 (1935) (English translation: Syntactic connection. In: McCall, S. (ed.) Polish Logic in 1920–1939, pp. 202–231. Clarendon Press, Oxford (1967))
5. Fraenkel, A., Bar-Hillel, Y., Levy, A.: Foundations of Set Theory. North-Holland Publishing, Amsterdam (1958)
6. Jadacki, J.: Polish Analytical Philosophy. Semper, Warsaw (2009)
7. Jadacki, J., Paśniczek, J. (eds.): The Lvov-Warsaw School — The New Generation. Poznań Studies in the Philosophy of Sciences and Humanities, vol. 89. Rodopi, Amsterdam/New York (2006)
8. Leśniewski, S.: Grundzüge eines neuen Systems der Grundlagen der Mathematik. Fundam. Math. **14**, 1–81 (1929)
9. Leśniewski, S.: Über die Grundlagen der Ontologie. Comptesrendus des séances de la Société des Sciences et des Lettres de Varsovie, Classe II **23**, 111–132 (1930)
10. Luschei E.C.: The Logical Systems of Leśniewski. North-Holland Publishing, Amsterdam (1962)
11. Łukasiewicz, J.: O zasadzie sprzeczności u Arystotelesa. Studium krytyczne [About the Principle of Contradiction in Aristotle. Critical study]. PAU, Kraków (1910)
12. Łukasiewicz, J.: O logice trójwartościowej [About the three-valued logic]. Ruch Filozoficzny **5**, 169–171 (1920)
13. Łukasiewicz, J.: Elementy logiki matematycznej [Elements of Mathematical Logic]. Nakładem Komisji Wydawniczej Koła Matematyczno-Fizycznego Słuchaczów Uniwersytetu Warszawskiego, Warszawa (1929). Presburger, M. (ed.) (2nd edn. PWN, Warszawa (1958))
14. Łukasiewicz, J.: Z zagadnień logiki i filozofii. Pisma wybrane [From Issues of Logic and Philosophy. Selected Works]. PWN, Warszawa (1961). Selection, introduction and notes by J. Słupecki
15. Łukasiewicz, J.: In: Borkowski, L. (ed.) Selected Works. North-Holland Publishing, Amsterdam (1970)
16. McCall, S. (ed.): Polish Logic in 1920–1939. Clarendon Press, Oxford (1967)
17. Mostowski, A.: On a generalization of quantifiers. Fundam. Math. **XLIV**, 12–36 (1957)
18. Tarski, A.: Pojęcie prawdy w językach nauk dedukcyjnych [The Concept of Truth in Languages of Deductive Sciences]. Towarzystwo Naukowe Warszawskie, Warszawa (1933) (German version: Der Warhheitsbegriff in den formalisierten Sprachen. Studia Philosophica **I**, 261–405 (1936); English translation: The Concept of Truth in Formalized Languages. In: Tarski, A. Logic, Semantics, Metamathematics: Papers from 1926 to 1938, pp. 151–278. Clarendon Press, Oxford (1956))
19. Tarski, A.: Logic, Semantics, Metamathematics: Papers from 1926 to 1938. Clarendon Press, Oxford (1956). Trans. J.H. Woodger (2nd edn edited and introduced by John Corcoran, Hackett Publishing Company, Indianapolis (1983))
20. Woleński, J.: Logic and Philosophy in the Lvov-Warsaw School. Kluwer Academic Publishers, Dordrecht (1989)
21. Woleński, J.: Lvov-Warsaw School. In: Zalta, E.N. (ed.) The Stanford Encyclopedia of Philosophy (Winter 2015 edn.). https://plato.stanford.edu/archives/win2015/entries/lvov-warsaw/
22. Woleński, J.: Logic and the foundations of mathematics in Lvov. Banach Cent. Publ. **87**(1), 27–44 (2009)
23. Wybraniec-Skardowska, U.: Polish Logic, Some Lines from a Personal Perspective. Publications of Institute for Logic, Language and Computation, Amsterdam (2009)

U. Wybraniec-Skardowska (✉)
Cardinal Stefan Wyszyński University in Warsaw, Warsaw, Poland
e-mail: uws@uni.opole.pl

Kazimierz Twardowski: A Great Teacher of Great Philosophers

Anna Brożek

Who really is a patriot, is not hungry for fame.
His inner conviction is enough for him,
that he makes what he should make,
and none of his effort will be wasted;
his sweat and — when it is needed — his blood
produces the most beautiful and most wonderful seeds:
the greatness and happiness of the Fatherland.
On patriotism (Twardowski [7, p. 454])

Abstract Kazimierz Twardowski was the founder of the Lvov-Warsaw School, one of the greatest phenomena in the European culture. The School had its representatives in all scientific disciplines, logic and mathematics including. Among his pupils are such great figures of European philosophy like: Jan Łukasiewicz, Stanisław Leśniewski, Tadeusz Kotarbiński, Władysław Tatarkiewicz and Kazimierz Ajdukiewicz. The paper presents his life and various fields of his rich activity, as well as the list of his main works and the greatest achievements in philosophy. He is best known in the world as the author of the distinction presentation-object-content, act-product, as well as the supporter of the postulate of clarity of thought and speech and the critic of psychologism, relativity of truth, symbolomania and pragmatophobia in science.

Keywords Act-product · Clarity · Kazimierz Twardowski · Lvov-Warsaw School · Pragmatophobia · Presentation · Psychologism · Relativity · Symbolomania

Mathematics Subject Classification (2000) Primary 01A70, 03A05; Secondary 03B65

1 Life

He was born on October 20, 1886, in Vienna; he died on February 11, 1938 in Lvov.

After several years of education at home and in public school, Twardowski, through the efforts of his father, obtained a place in a prestigious high school in Vienna—*Theresianum*. This school provided him with a comprehensive classical education;

Á. Garrido, U. Wybraniec-Skardowska (eds.), *The Lvov-Warsaw School. Past and Present*, Studies in Universal Logic,
https://doi.org/10.1007/978-3-319-65430-0_2

in *Theresianum*, he was also familiar with philosophical problems. Thanks the iron discipline prevailing in the school, Twardowski got accustomed to systematic, persistent work—and thanks to the example instilled in him at home, from an early age he wanted to serve Poland and the Poles.

After graduating from *Theresianum* Twardowski began his studies at the University of Vienna, initially at the Faculty of Law. For some time he lived at the property of Earl Wojciech Dzieduszycki in Jezupol where he was teaching the children of the earl. Perhaps partly due to Dzieduszycki—a philosopher, a politician and a writer—with whom he had cordial relations, Twardowski moved to the Philosophical Faculty. Studying philosophy, he was influenced to the greatest degree by Franz Brentano—who became for Twardowski the model of a teacher and a researcher.

After graduating Twardowski performed annual military service. The doctoral degree he earned in 1891 was based on a dissertation on the views of Descartes. The official promoter of this work was Robert Zimmermann (at this time, Brentano could not officially be a promoter of any doctorate).

In 1892, Twardowski married Kazimiera Kołodziejska, with whom he had three daughters: Helena, Aniela and Maria. All were well-educated, and the youngest of them married Kazimierz Ajdukiewicz, one of Twardowski's closest disciples.

Thanks to a scholarship from the Austrian Government, in 1892 Twardowski traveled to Leipzig and Munich. After returning to Vienna, he took a job as an official, so that he could earn money as a tutor while preparing his habilitation. He received it in 1894 for the thesis *Zur Lehre und Gegenstand der Inhalt der Vorstellung*. This dissertation was, until recently, the most famous of Twardowski's works, since it was published in German. After obtaining habilitation, in the academic year 1894/1895, Twardowski taught in Vienna as a Privatdozent.

As early as the next year, thanks to a happy coincidence, as the only 29-year-old scholar, Twardowski took over the chair of philosophy at the University of Lvov. Twardowski quickly proved to be an equally charismatic teacher, like his Viennese master. Twardowski remained faithful to the University of Lvov, working there until retirement, i.e. until 1930, and after that he continued to take part in its life.

2 Personality

He was one of the most important and most versatile characters of Polish culture of the first half of the twentieth century.

As a scholar, he was the author of excellent work from all the areas of philosophy—from logic, through ontology and epistemology, to aesthetics and ethics—and also works on its history, on psychology and on didactics; he also composed songs full of expression, most of which were unfortunately lost during the turmoil of the wars.

As a teacher, he was a master of several generations of Polish philosophers and, more broadly, humanists.

As a citizen, he set the tone of the teaching environment, presiding, in the first decade of the twentieth century, over the influential Society of Teachers of Higher Education and leading an extensive lifelong correspondence with many opinion-forming representatives of the country. Thanks to cautious, but firm and persistent action, he led, among other things, to fully repolonizing education in the Austrian partition of Poland.

For the elementary component of citizenship he felt patriotism. The latter—understood as an actually active attitude—Twardowski distinguished from ceremonial patriotism. A ceremonial patriot has a sentimental attitude to his country and nation, gladly manifesting this attitude on certain occasions. However, he can not afford to sacrifice his own good for the good of his countrymen.

3 Writings

The most important of Twardowski's book publications published in his lifetime—are: *Zur Lehre vom Inhalt und Gegenstand der Vorstellungen* [1], *Imageries and concepts* [3], *The essential concepts of didactics and logic* [4], *Six lectures on medieval philosophy* [5], *Addresses and papers* [6] and *Philosophical dissertations and papers* [9].

After his death the following appeared: *Selected philosophical writings* [10], *Collected psychological and pedagogical writings* [11], *Ethics* [12], *Diaries*. Vol. I–II [13], *Philosophy and music* [15], *Youthful diaries* [16] and *Thought, speech and action*. Vol. I–II [17].

In English, the following of Twardowski's books were published, i.a.: *On the content and object of presentations* [1], *On actions, products and other topics in philosophy* [14] and *On prejudices, judgments and other topics in philosophy* [18].

4 Views

It is often said that the greatness of Twardowski lies exclusively in his organizational and pedagogical activities: in particular, that he was the founder of the Lvov-Warsaw School. Such an assessment is given even by many representatives of the School itself.

However, it should be kept in mind that only some of the philosophical views of Twardowski were expressed in print during his lifetime. In this case, Twardowski was like his master—Franz Brentano.

This involved, among others, the fact that Twardowski not only preached, but also realized, to the highest degree, the postulate of accuracy and precision. Anyone who obeys this postulate—knows how difficult it is to give final form to our own thoughts. Twardowski severely judged those who lacked awareness of this situation:

> How many authors publish philosophical papers, essays, and even extensive volumes, which directly swarming with equivocations, incomprehensibilities, thoughts, hopes, inconsistencies, vague expressions and paralogisms, and who, together with their publishers, loses nothing, provoking sound and often admiration in wide reading circles! I could envy them the ease of their literary production; but, in fact, I underestimate the products of their pens and I resent their behavior which is highly detrimental to the logical development of philosophical thinking [8, p. 32].

Meanwhile, an analysis of even only the printed legacy of Twardowski shows that he has won important theoretical results in all disciplines of philosophy. Here we have an outline of these results.

4.1 Metaphilosophy

4.1.1 Philosophy and Worldview

Twardowski regarded philosophy as a conglomerate of disciplines, which consists of logic, psychology, metaphysics, ethics, aesthetics and the history of philosophy. He pointed also to a particular trait of all philosophical disciplines: the objects of philosophical research are given in internal experience (introspection).

The word "philosophy" is sometimes used as a synonym for the word "worldview". Twardowski's postulate was separating philosophical problems, cultivated with scientific methods, from the problems of a worldview, whose field is one of personal beliefs; giving the latter the form of a theory which he considered pure speculation.

In Twardowski's relation to faith—we need to distinguish its relationship to the Catholic dogma, to Catholic ethics and Catholic ceremony. As for the first and second case, Twardowski wanted, in Christianity, a dominance of the ethical dimension over—let us call it so—the dogmatic.

4.1.2 Brightness of Language

One of the main causes of errors in philosophy, he recognized, is the lack of clarity of thought, coupled closely with the lack of clarity of speech; the lack of clarity of speech can at the same time be regarded as a symptom of the lack of clarity of thought, as discursive thinking can not be extralinguistic.

Accordingly, philosophical terminology required reconstruction. In this reconstruction, he recommended using the analytical method (which was also used by him). Such a reconstruction is carried out in two stages. The first stage is to examine the essential properties of several typical designates of a given term and to formulate its analytical definition. In the second stage this definition is adjusted by checking whether the analytical theses implicated by this definition are true for other designata of the defined term.

Twardowski advocated the careful use of formal logic in the analysis of philosophical problems: he was the enemy of (as he expressed it) symbolomania, i.e. abusing logical symbolism, and of pragmatophobia, related to the first, i.e. focusing our attention on the syntactic aspects of this symbolism. For logical structures—including those that are formally flawless—are not always based on well defined intuitions, and these structure do not often have any model in the field of philosophical studies.

4.1.3 Independence of Thought

Twardowski was a firm defender of the independence of thought.

In his opinion, our beliefs are often dependent on: those deemed by us to be authorities; the environment in which we live; the language we use; our mental organization and finally, our emotions.

To the question of whether we can think independently, Twardowski responds in a complex way. There are things or domains, in which we all are doomed to be—more or less—dependent in thinking; however, some of us, in some cases, can think independently. Independence of thought is therefore an ideal which—according to Twardowski—we ought to and, which is important, we are able to pursue successfully.

4.1.4 Anti-psychologism

Twardowski regarded logic and psychology to be propaedeutic disciplines of philosophy. Earlier, however, he had rejected the psychologism prevailing at the time of his youth, i.e. the view, according to which, logic is a part of psychology or should rely upon it.

Psychologism was, according to Twardowski, untenable for the following reasons: (1) logic was created and developed independently of psychology; (2) theses in psychology (which are generalizations of experiential data) are only probable, whereas theses in logic are reliable (and, consequently, unquestionable), and being reliable, they can not follow psychological theses; (3) psychology is the theory of the actual activity of thinking—whereas logic deals with evaluating the typical forms of the products of thinking (i.e. thoughts) in terms of their accuracy.

4.2 Metaphysics

Twardowski never disavowed metaphysics as such, but he had a very rigorous view of the conditions that must be met by philosophical research, if their results could be included in scientific knowledge. However, he did not share the (positivist) belief that since the object of study of metaphysics—and wider, in philosophy—is neither the world of sensual phenomena, nor the world of mental phenomena, it is not science. The idea is that outside these phenomena there are still objects of another type (e.g. relationships)—and they are examined in, among other disciplines, metaphysics in a manner adopted in other sciences, in particular, by describing these objects and using inductive and deductive reasoning to define and justify statements about these objects, just as in the natural sciences.

At the same time, Twardowski was a supporter of the cumulative conception of philosophy. The long-term goal of philosophy is to construct a synthesis of the scientific theories which would be the theory of all objects (and not only a theory of sensual or mental phenomena); such a synthesis has still not been achieved, but its elements already exist and we should patiently seek others. For, according to Twardowski, it is better to accept no synthesis at all—than to accept a fallacious one.

4.2.1 Anti-psychologism

We owe to Twardowski a developed theory of objects, formulated, by the way, before Alexius Meinong.

According to Twardowski, each object—regardless of its existential and metaphysical category—is a homogeneous whole, composed of various properties. The components of objects are whatever can be distinguished in these objects: a particular component is something that can be distinguished in fact; a component of the abstract is something that can be distinguished only intentionally. All properties of objects and all relationships between the parts of these objects are abstract components. For example, the stem and thorns of a certain rose are its concrete components, whereas its redness and its being-a-plant are its abstract components.

4.2.2 Categorization of Objects

As one of their tasks, metaphysicians always sought to make a categorization of objects. However, few of them realized this goal in a logically satisfactory way, such that these categorizations took the form of logically correct classifications; fortunately Twardowski belonged to this smaller group.

He carried out logically correct classifications of objects based on different existential aspects and different ontic aspects. In the first case, he found three dichotomous classifications: into possible and impossible objects, into existing and non-existing objects, and into real and unreal objects. In the second case: into individual and general objects, into simple and complex objects, as well as into corporal and spiritual objects. This made it possible to recognize the traditional categorizations as the result of not fully intentionally crossing the classifications made by Twardowski.

4.2.3 Actions and Products

An important step on the way to Twardowski's rejecting psychologism, was his analysis of actions and the products of actions.

Actions are states of a special type. They are associated with certain specific objects, which he called "products": and so a picture is the product of painting (or drawing), an inscription is the product of writing, a thought is the product of thinking, etc. Among the artifacts, there are relatively unstable products, which can be distinguished from the corresponding actions only mentally (*scil.* by abstraction), and relatively stable products. A jump is an unstable product of jumping and a dance is an unstable product of dancing. A picture as a product of painting, a hole as a product of digging—are relatively stable products of their relevant actions.

Products of physical actions—i.e. physical products—are either unstable (e.g. a cry as a product of crying, a jump as a product of jumping, a rotation as a product of rotating) or stable (e.g. a print as a product of printing, a braid as a product of braiding). Stable products of physical actions exist more than the actions by which they arise. However, all the products of mental actions—i.e. mental products—are unstable (e.g. a thought as a product of thinking, an experience as a product of experiencing, a decision as the product of deciding).

Some actions are directed at certain objects. Objects—especially things—at which physical actions are directed, create the material (*scil.* basis) of these actions (e.g. sand, which reflects the rate of person walking along this sand).

The product of a mental action directed at a certain material is not a material itself, but is a new structure of this material (created by this action). The object, at which a certain mental action is directed, is the object of this action (e.g. a landscape imagined by someone). Actions that are directed at something—are intentional actions.

Certain properties of products are not properties of the actions that create these products. For example—it happens that our dream is false, but action of dreaming itself is not false; a question may be confusing, but it cannot be the confusing action of asking a question.

Twardowski made a distinction of actions and the contents of presentations in terms of actions and products. He considered stable products to be psychophysical objects; he identified cultural artifacts just with such objects.

4.2.4 The Soul

Twardowski proposed an original approach to the problem of the ontological status of the soul.[1]

According to Twardowski, the thesis about the existence of the soul—as a substratum of experiences—is an obvious thesis and as such it does not need any proof. An opponent to this thesis—an asubstantialist—could not use the pronoun "I" in the ordinary sense of this pronoun. The so-called group of mental phenomena, which is identified by asubstantialists with "I", would not be able to determine that a given experience belongs just to it (a similar difficulty would arise with any internal perception); for a similar reason, this group would not say of itself that it knows something.

Twardowski justified the thesis of the simplicity of the soul in such a way. If a substratum of experiences consisted of parts, and some two experiences (e.g. a visual and an auditory impression) were located in two different parts of the substratum, it would not be possible to compare them; meanwhile, such a comparison is a fact. Since the soul is simple (*scil.* it is a mental atom), it is eternal, because its possible creation or annihilation cannot be naturally explained. It remains to be assumed that the natural eternity of souls-atoms goes hand in hand with the fact that they are created by God (creating any objects by an eternal God does not require that that they occurred later in time than God—and thus were non-eternal) and at one moment (e.g. at the moment of the birth of a human being) are revealed in the temporal-spatial world.

The thesis of the simplicity of the soul is one of the premises of reasoning justifying the thesis of the immortality of the soul. The second premise is the principle of conservation of energy—taken from the natural sciences. While Twardowski was aware of certain imperfections of his reasoning, he considered it to be conclusive.

4.3 Psychology and Epistemology

Standing on the position of anti-psychologism—Twardowski devoted much of his attention to psychological and meta-psychological considerations, with respect to considerations bordering on psychology, epistemology and methodology. Until the end of his life, he believed that psychology is the basic philosophical discipline due to the fact that it is the theory of (real) thinking.

---

[1]Twardowski's views concerning this subject are little known. Only recently was his extensive German manuscript from 1895 published. Cf. [1].

Twardowski practiced empirical psychology, i.e. psychology that justifies its statements on the basis of experience—unlike, for example, political history, whose tested facts (as past) need to be reconstructed on the basis of indirect data, i.e. testimonies, and «natural» history, when investigating the history of certain fragments of nature. At the same time, empirical psychology appeals to external experience (*scil.* extraspection) and internal experience (*scil.* introspection), introspection being its ultimate foundation. So psychology may be considered a quasi-historical science: lying on the border of empirical and historical disciplines—wherein some fields of psychology are closer to the first (as e.g. examining the psyche of healthy people), and some to the others (as e.g. psychiatry).

4.3.1 Components of Consciousness

Experiential components of consciousness—mental facts—are spiritual actions (which can be only mentally separated from each other) and their products. The source of knowledge concerning psychic facts cannot be located in the senses, but rather self-consciousness. In such a direct way, only states of our own consciousness are knowable.

The basic kind of spiritual action is presenting something for the substratum of these actions. It is a necessary condition for all other—secondary—types of spiritual actions, in particular: judging, reasoning (or "pondering, hesitating, comparing, devising, synthesizing, distinguishing etc."), feeling and deciding. On the other hand, judging is a necessary condition of feeling—in any case convictional feeling—and deciding.

Twardowski thought, at the same time, that presenting and judging are kinds of thinking; this is suggested, among other things, by the fact, that there are two contexts of the word "to think", namely: "to think of *X*", and "to think that *p*". To think of *X*—is the same as—to present *X* (*scil.* to imagine or comprehend), and to think that *p*—is the same as—to judge that *p*. One of the kinds of thinking is also reasoning, which—according to Twardowski—is judging about judgments.

Both primary and secondary spiritual actions are intentional actions: they require an object.

The spiritual conditions of spiritual actions are dispositions. Dispositions are only hypothetical components of consciousness. According to the law of practice, it is assumed that the occurrence of any mental fact leaves a disposition to reproduce it.

4.3.2 Acts, Contents and Objects of Presentations

The legitimacy, which Twardowski permanently secured in the history of European philosophy, was in carrying a precise delimitation of the three aspects of presentation: act, content (i.e.—as he put it—the product of action) and object. Twardowski accepted this distinction throughout all his scientific activity, but he made some modifications to the original conception of this distinction.

Act and content of presentation are the metaphysical (*scil.* abstract) parts of presentation. In languages, presentations correspond to names; the content of presentations corresponds to the meaning of names, and the object of presentations—to what names designate.

The object of a given presentation is different from the content of this presentation, because: (a) when we perform a negative judgment of the type "*A*-being-*B* does not exist", then we deny the existence of the object A-being-B given in presentation, which forms the basis of the performed judgment, but, at the same time, we realize that the content of this presentation contains something (namely B), which means that the object of this presentation does not exist in reality; (b) there are various presentations of the same object (i.e. we can imagine the same A, in one time, as P, and in another time, as Q). The expression "presented object" is ambiguous; it may refer (in the determining meaning) to a real object, which is presented by someone—or (in the modifying meaning) to the mental presentation of this real object.

Twardowski divided presentations into images (*scil.* concrete, pictorial presentations) and concepts (*scil.* abstract, non-pictorial presentations). At the same time, he distinguished perceptive images—from reconstructive (memory-based) and productive (based on imagination) images. The difference between them is qualitative rather than only quantitative. Perceptive presentations—unlike reconstructive and productive ones—are more vivid and independent of our will.

Twardowski defined concepts as presentations consisting of a vague underlying image and presented judgments which assign to the object of this image properties constituting the connotation of the expression with which this concept is linked.

By virtue of their object—concepts can be divided into synthetic and analytic. Objects of synthetic concepts are things or persons (e.g. the concept of a tree, of a kingdom, of God). The underlying image of a synthetic concept is the (reconstructive or productive) image of an object similar to the one that has to be conceptualized. One of the components creating the content of a given concept is an image of the statement stating that the object of the underlying image has properties, which de facto are not possessed by this object.

Both types of concepts—i.e. synthetic and analytic—have two forms: proper and shortened, *scil.* hemisymbolic or symbolic. In the hemisymbolic form, the act of conceptualizing contains the image of the name designating the object of this concept and the accurate underlying image. In the symbolic form, there is only the first component (*scil.* the image of a name).

4.3.3 Judgments

The starting point in the theory of judgments was to Twardowski the idiogenic conception of judgment. According to this conception, judging is recognizing or rejecting the existence of a certain object.[2] Just as within the domain of presentations, Twardowski distinguished, in the case of judgments, content and the object of them. The object of a judgment—is what we recognize as existing or reject as non-existing. The content of a

[2]In Twardowski, we find a characteristic hesitation, which will be also visible in many of his disciples: either a logical value should be considered a defining property of "judgment", or "judgment" should be defined without using the notion of truth with respect to falsehood. Here we are dealing with the second solution. But sometimes Twardowski called "judgment" a mental action, "in which truth or falsehood is contained". This involves "truth" and "falsehood" in the proper sense. The external criterion of truthfulness understood in such a manner is conformity with the truth, and of falsehood—fallacy. The internal criterion is whether "true" and "false" are adjectives determining nouns (and thus they are determinators), where they stand—or adjectives which retain the changed sense of those nouns (and thus—in the latter case—they are modifiers).

positive judgment is the existence of the object, the content of an appropriate negative judgment is the non-existence of the object. Each judgment is a superstructure on an underlying presentation. The underlying presentations of judgments have an existential paraphrase as in the judgments "There is a city where Kazimierz Twardowski was born" and "There is a city where the peace conference was held after the Napoleonic wars", which have the same object: the city of Vienna. These judgments are different because they are different as to the content of their underlying presentation: in the first case, Vienna is presented as a city which is the birthplace of Kazimierz Twardowski; in the second case—as a city where the peace conference was held after the Napoleonic wars.

Twardowski noticed, however, that the traditional idiogenic theory encounters some difficulties—especially, e.g., in the case of judgments on past objects as well as in the case of judgments about relationships.

The problem of judgments about a future object was reconstructed by Twardowski as follows. Consider the judgment "There was a king". Brentano would interpret this judgment as acceptance of the existence of a past king. However, if "to be" means the same as "to be now", and "past king" designates such a king who was formerly and now is gone—then we are in contradiction: because we receive the judgment "There exists something that once existed, but no longer exists".

Twardowski sketched therefore a modification of the theory judgment, avoiding the identified difficulties. He contrasts in it two types of judgments: the existential and the relational, the term "existential judgment" coming from the content, and the term "relational judgment" coming from the object of presentation. The content of existential judgment is namely existence, and the object of relational judgment—wherein the term "relationship" was used by Twardowski in the meaning similar to that, in which the term "state of affairs" is used today. In relational judgment, a certain relationship is accepted or rejected, and the content of such a judgment is subsistence (Germ. *vorhandensein Bestehen*). In this case, the overall form of judgment is: $\pm bA$, where '$\pm$' means the acceptance or the rejection, 'b'—subsistence, and 'A'—an appropriate relationship. The characteristic fact here is the presence of the difference between the judgments: "God exists" and "There was a king". While the content of the first judgment is the existence of God, the content of the second is the subsistence of a relationship of possession taking place between a certain king and a preceding period (or otherwise: the subsistence of a relationship of possession occurring between a certain king and the property of being contemporary with some past period).

4.3.4 Truth and Error

Twardowski gave convincing arguments in favor of alethic absolutism, showing that relativists, supposing that the logical value of judgments may vary in time and space, have in mind de facto not judgments, but rather their imperfect language expressions which are rarely reliable sentences. For example, maintaining that the judgment that it is raining is relatively true, since it is true only in certain circumstances—is a misunderstanding. When we say "It is raining"—we express the judgment that it is raining at such and not another place and at such and not another time, and the judgment that it is raining in these and not other circumstances is absolutely true or absolutely false.

Twardowski devoted a lot of attention to the analysis of prejudices.

Prejudice, according to Twardowski, is a false belief, asserting the occurrence of a certain regularity—a belief which is supported by someone without sufficient justification. Twardowski distinguished two particular kinds of prejudices: superstitions and survivals. A superstition is a prejudice concerning the accuracy occurring between natural and supranatural phenomena. In contrast, survival is a prejudice which was once common but its detailing is not a prejudice.

4.4 *Semiotics*

4.4.1 Functions of Speech

According to Twardowski—language is a double instrument by the fact that particular language signs perform a dual representative function (*scil.* functions of expressing) in the face of spiritual actions (primarily, in the face of emotional-volitional actions; secondarily, in the face of intellectual actions): objective and subjective functions.

Objective functions—significative and denotative ones—rely in turn on the fact that a language sign signifies a product of represented spiritual actions and designates the object of these actions. Subjective functions—indicative and evocative ones—rely in turn on the fact that a language sign indicates an action on the part of the sender of this sign and inspires action in its recipient; the matter here is about actions, the products of which are meant by this sign. It is thanks to the indicative function that the language signs (of natural speech) may indicate, among other things, emotional shades (*scil.* emotional moments), marked not only in the choice of vocabulary (cf. e.g. a pair of words: "nag"—"horse", "to give up the ghost"—"to die",[3] the first positions of which are occupied by the words of a specific color), but also in the manner of pronouncing words. On the other hand, it is thanks to the evocative function, that language signs can be understand at all.

Twardowski notes that language signs are usually ambiguous. The same kind of sign can have several logical meanings (e.g.: "my portrait" may mean the same as "a portrait done by myself", "a portrait, which belongs to me" or "portrait, which presents me"). Speech features also multinominality. One type of thinking corresponds sometimes with many kinds of words of speech.

Ambiguity and multinominality speak against a parallelism between speaking and thinking; and the fact that we can speak without thinking and we can think without speaking (also continuously—as it is, for example, in the case of dumb men) speaks against the identity of speaking and thinking. But there is a strong correlation between speaking and thinking. Firstly, signs of speech—as symbols of corresponding thoughts—remain in a causal relationship with them. Secondly, in thinking—especially more abstract thinking—there are numerous «shortened» concepts (*scil.* hemisymbolic and symbolic concepts, as mentioned above). Thirdly, the relationship between thinking and speaking means that people thinking vaguely cannot clearly express their thoughts.

The set of language signs has two subsets: sentences and parts of sentences. The latter can in turn be presenting or cathegorematic signs, or connecting or syncathe-

[3]In this paper, all examples come from Twardowski himself.

gorematic signs. Cathegorematic signs perform representative functions in relation to presentations—they mean something; syncathegorematic signs are only co-meaning. Typical cathegoremata are names; typical syncathegoremata are connectors.

4.4.2 Names

Of the grammatical parts of speech, names tend to generally be proper nouns (e.g.: "Lvov", "Stanislaus Augustus") and common nouns (e.g.: "father", "soul"), pronouns (e.g.: "he", "we", "something"), adjectives (e.g.: "white", "gloomy"), numerals (e.g.: "two," "tenth", "fifty seven and a half") and verbs (e.g.: "to run", "to learn", "to expire")—namely those which are grammatical subjects (except sentences deprived of a predicate), predicates (except sentences deprived of a subject), predicatives or complements of simple sentences. Names may be not only single nouns, but also whole noun phrases, and therefore fusions of nouns with other expressions (e.g.: "human eye", "this man", "a man", "the second son", "the highest mountain in Europe", "the son who insulted his father").

According to Twardowski—as there are no objectless presentations, so too are there no empty names: each name designates a certain entity—and exactly one entity (even a complex name). On the other hand, the fact that a given object is designated does not imply the existence of this object.

4.4.3 Declaratives

Among sentences, Twardowski distinguished not only expressions representing judgments, but also those representing orders, wishes, requests, questions or curses.

He called sentences representing judgments—"declaratives". A declarative means the frame of a given judgment (*scil.* existence); it designates the object of this judgment; it indicates a judging performed by the speaker and raises performing—or only presenting (and thus understanding)—the judgment by the listener. Moreover, not every declarative can be described as a "sentence" in the grammatical sense (cf. e.g. the word "Fire!", representing sometimes the judgment that it is burning).

A special type of declaratives are definitional formulas of the structure "*A* is (identical with) *B*" (e.g. "*A* state is a public community which includes a sedentary population within a given territory as a community of rulers and the ruled"). These declaratives do not de facto represent the judgments that *A* is (identical with) *B* (especially in the example, given above, we claim nothing about a state), but rather judgments, that '*A*' designates *B* (in particular we claims in this definition that by the word "state" we designate a certain object, whose representation has been constructed in this way).

Truthfulness, as well as evidence and probability—and their inverses—could be assigned to declaratives only indirectly: due to judgments represented by them. Absolute truthfulness is namely possessed by a declarative representing only true judgments. Such declaratives should be complete ones. About a declarative which sometimes represents true judgment and sometimes wrong, we can say that it is relatively true. Among such declaratives, there are elliptical declaratives (e.g.: "It is raining", "A cold bath is a healthy thing", "Apartments in Lvov are expensive"). Similarly, if talking about synonymy in relation to judgments is ridiculous, so much is it so in relation to declaratives. Namely, two declaratives are synonymous when they represent the same judgment.

4.4.4 Connectors

According to Twardowski, among *lato sensu* connectors, a special place is occupied by determinators, i.e. nouns, pronouns, adjectives, adverbs and numerals—and the phrases equivalent to them—which occur in attributive contexts (e.g.: “speed OF MOTION”, the “father OF SOCRATES”, “the gable wall OF A HOUSE”; “THIS man”, “A man”, “NO man”). In predicative contexts, they are cathegoremata.

A determinator—together with the name for which it stands—co-represents the presentation of a judgment relating to the object designated by the determined expression. Depending on the subject of this judgment, an ascriptive function is performed by a determinator in relation to a given determined expression, consisting in determination, abolution, confirmation or modification (i.e. both abolution and determination). The determining and abolutioning function consists in changing the meaning and, in particular, either in its enrichment (cf. “good man”), or its depletion (cf. “alleged shape”). The confirming function consists in strengthening or restoring some of its components (cf. “actual fact”). The determining, abolutioning and confirmating functions are simple ones. On the other hand, the modifying function is complex, consisting in simultaneous enrichment and depletion of meaning (cf. “the former minister”).

Stricto sensu connectors contain conjunctions which may occur either in a nominal context (e.g. “non-”) or in a propositional context (e.g. “no”, “or”, “if”).

The conjunction “non-” in the nominal context (e.g. “non-Greek”) performs a specific modifying function in relation to the meaning of its nominal element: the infinitive function namely. It changes the meaning of its argument so that the meaning of the whole context becomes identical with the meaning of a generic name (*genus proximum*) superior in relation to this element (here: the meaning of the name “man”), enriched with a presentation of the judgment refusing to the object of this name—the specific features of the object of the negated name (here: the name “Greek”). With all this, the rule of infinitation is in force, allowing the addition of “non” only to such names which are subordinate a certain name. This rule is violated by the expression “non-entity”, because for the word “entity” we can not establish a *genus*. Thus “non-entity” is a nonsense term. Therefore, if we considered the noun “nothing” as synonymous with the expression “a non-entity”, we would also consider “nothing” to be a nonsense term. In fact, “nothing” is a syncathegorematic element of negative sentences: it is either a propositional connector or it may be the negation of an existential quantifier. The phrase “Nothing is eternal” therefore means the same as “There is no entity which would be eternal”.

As for conjunctions in propositional contexts, e.g. the conjunction “or”, they fulfil the modifying function in relation to sentences which are combined with the aid of them: these sentences do not represent in this context performed judgments but presented judgments, and a whole alternative sentence expresses the probability of these presented judgments. The degree of this probability remains in inverse proportion to the number of main elements of this context. A similar modifying function—at least in certain contexts—is fulfilled by the connective “if” in relation to sentences-elements (*scil.* predecessor and successor). The whole of such a context expresses the judgment that the logical relation of consequence occurs between the presented judgments. We are talking here about so-called formal (*scil.* logical) truths—i.e. the judgments considered consequences of the relevant reasons—that is the "truth" is opposed sometimes to material truths. However, if these "truths" are true, it is precisely in the same sense in which any materially true judgment is true. Thus, formal truths are a kind of material truths, i.e. judgments which state what is or deny what is not.

4.5 Ethics

Twardowski was a forerunner (anyway, in Poland) of the program of so-called independent ethics, i.e. an ethics free from ideological assumptions. Among the traditional ethical problems, he separated problems which, according to him, could be considered scientifically. He claimed that the task of scientific ethics consisted in a description of the conditions in which the aspirations and actions of individual people and communities could be reconciled to the greatest extent.

4.5.1 Ethical Absolutism

Just as in the theory of truth, he also presented in ethics a profound critique of axiological relativism and skepticism.

In support of ethical agnosticism, the following arguments were formulated, among others: (a) there are no absolutely obligatory ethical norms (*resp.* criteria); (b) there are no commonly obligatory norms (at all times and in all societies); (c) the terms "good" and "bad" have different meanings in different times and places. According to Twardowski—these must be rejected because: (a) when we precisely formulate norms which are considered general judgments, they appear to be norms restricted to a certain domain; (b) the fact that, at certain times or in certain communities, certain norms are not obligatory, should be understood in such a way that either they are not (at that time or in these communities) recognized as obligatory, and they can be mistakenly recognized as such, or that they (despite the general formulation) do not apply de facto to all cases—i.e. they are not applicable in all of them; (c) the variability of notions does not imply the variability of norms.

Among the arguments in favor of ethical relativism, sometimes the consequences that the theory of evolution entails to ethics are mentioned. Twardowski—without ruling on the merits of the latter—carried out a detailed analysis of these alleged consequences.

He showed that the spread of the theory of evolution caused the appearance of the ethical conception, according to which ethical convictions were the result of natural selection. Moreover, some people began to regard as moral every, and only such, an action, which contributes to maintaining our own life and the survival of the species. This leads to changing ethical beliefs along with changing living conditions and, consequently, to—as it was called by Twardowski—"ethical anarchy".

It is hard—Twardowski emphasized—not to acknowledge the fact that people differ in their assessments as to the validity of various things: in the assessments of ethical actions (performed "on the basis of conscience"), in aesthetic assessments (performed "on the basis of taste"), as well as in logical assessments, i.e. assessments of beliefs in terms of truth and falsehood (performed "on the basis of reason"). However, this does not imply that there are no generally obligatory norms in matters of conscience, taste and reason. Differences of opinions in these matters may have a source in the fact that we do not have a theory to order these matters satisfactorily.

These are not "moral truths"—as Twardowski termed them—developed in the course of evolution, but the development of humanity itself, improving in terms of "reason, conscience and aesthetic taste."

4.5.2 Freedom of Will

The question of free will is linked with determinism, when the freedom of will of *X* is recognized as a lack of causes for the decision-making acts (*scil.* resolutions) of *X*. The motives and personality of *X* (*scil.* the totality of *X*'s dispositions) could be such causes.

Twardowski was interested not in resolving the controversy of determinism-indeterminism—although he was convinced that determinism is the most probable of all possible positions in this issue. He wanted to determine whether determinism entails such ominous effects for ethics as anti-determinists sometimes thought. Twardowski's answer was negative: if resolutions had in fact causes-motives and the personality of the acting person influenced them, then nothing stood in the way of assessing ethical actions just because of these motives and this personality. Because, generally speaking, "never and nowhere is the necessity of a certain phenomenon an obstacle to evaluating it".

But the question arises as to whether *X* is responsible for the actions of *X* taken under the resolution of *X*, determined by the motives and personality of *X*. According to Twardowski, if a certain action of *X* is taken in such circumstances—that can be (as Twardowski used to say) "accounted" as *X*'s action, and therefore *X* is a perpetrator of this action—then *X* is responsible for this action to *Y* (if, in particular, *Y* has the right to obtain compensation): this action is his merit (if good) or his guilt (if bad), and therefore it is worthy of reward or punishment.

Twardowski thought that although the existence of free will was not necessary for moral improvement, for some people, believing (alternatively: instinctive believing) in its existence was needed. If the thesis of determinism was true, then the reason for the existence of practical ethics (*eo ipso* for formulating moral norms) could be that some people carried about "the positive value of resolutions".

4.6 Aesthetics

In the field of aesthetics—Twardowski conducted, among other things, an analysis of two questions examined in the context of experimental, and therefore scientific aesthetics: the question of the criterion of beauty in art, and the question of how music is able to perform its evocative function.

Experimental aesthetics is practiced by a person, who verifies, with the aid of carefully selected experiments, hypotheses posed by aestheticians and concerning aesthetic assessments and preferences. One such hypothesis is that "only sizes of figures based on the line of the so-called golden ratio could be considered beautiful". As has been shown by experimental research, this hypothesis—according to Twardowski—is verified by "aesthetically educated minds", and is falsified by "aesthetically uneducated ones".

Twardowski posed the question of how music evokes musical experiences in the audience. He answered that question as follows: pleasure arises as a result of listening to music (irrespective of any text to which it is associated), because: (a) this listening does not require any effort; (b) sounds of music themselves are "pleasant"; (c) listening to music, we have a "sense of both diversity and homogeneity of a certain number of impressions". But a musical composition which is listened to also evokes experience with an analogical structure—with its own (melodic, rhythmic, chordal) structure.

5 Philosophical School

Twardowski conducted a de facto reorganization of the philosophical life in Lvov. He led to the origin of modern philosophical studies, to the establishment of various philosophical institutions—and he instilled in his students and all the people affected by him—the love of truth and diligent work. He instilled in his pupils not so much individual ideas, but rather certain methodological assumptions.

In terms of philosophical and ideological beliefs, Twardowski's students differed very much: among them were ontological dualists and monists; axiological absolutists and relativists; opponents and proponents of multi-valued logic; conservatives and socialists; priests and atheists. What they shared was the above-mentioned precision in expressing thoughts and attention to properly justifying accepted theses—and, associated with those qualities, carefulness in solving problems and a critical attitude in relation to results: primarily one's own—but subsequently, also others.

The didactic activity of Twardowski had a lot of components.

First, he delivered cycles of so-called introductory lectures concerning in fact—successively—all the systematic disciplines of philosophy and the history of philosophy. Secondly, he had lessons for a narrower group of students, forming in them the ability of independent philosophical thinking. Thirdly, he delivered lectures to a wider audience: within the university, designed for all faculties—and outside the university, including non-academic groups. Fifthly, he participated in the discussions of the Philosophical Circle, existing at Lvov University before his arrival in Lvov, and then in the Polish Philosophical Society, created by him in 1904. Sixthly, the program to create a serious study of philosophy in Poland also involved creating the magazine *Philosophical Movement* (1911) and involving in the organization of the existing *Philosophical Review*.

Twardowski's program has brought great results. After World War I, Twardowski's School spread to other Polish philosophical centers: the Lvovian wing was, first of all, supplemented by Warsawian one. This happened because chairs of philosophy and logic in Warsaw were taken by some of the first generation of Twardowski's pupils. Thus it came to the formation of the Lvov-Warsaw School, the largest, most influential and most famous in the world—so far—Polish school of philosophy. It included philosophers of the caliber of Jan Łukasiewicz, Władysław Witwicki, Zygmunt Zawirski, Stefan

Baley, Stanisław Leśniewski, Tadeusz Kotarbiński, Władysław Tatarkiewicz, Tadeusz Czeżowski, Kazimierz Ajdukiewicz and Izydora Dąmbska.

The value of Twardowski's work is shown in the fact that at the time of his death—his immediate disciples held chairs in philosophical domains (and many of Twardowski's pupils also held chairs in other fields) at almost all of the Polish universities (i.e. in Warsaw, Poznań, Cracow, Lvov and Vilna).

6 Conclusions

The basis of Twardowski's philosophical views is a pluralistic conception of being, a bipolar conception of actions, an intentional conception of consciousness, a triadic conception of presentations, a presentative conception of images, an allogenic conception of concepts, a constructivist conception of universals, a gradational conception of the accuracy of concepts, and an idiogenic conception of judgments.

These views form the metaphysical basis of his descriptive semiotics. This is an abstract theory of language phenomena, i.e. a system of definitions and their consequences—built with the method of logical analyses, preceded by an inventory and classification system supplementing these phenomena. The core of this theory is a psycho-physical conception of signs and a functional conception of expressions.

In semiotics, Twardowski is to be respected fourfold: as a critic of other people's conceptions, a creator of his own analyses, refinements and theses, a precursor of new ideas and methods, and an impeller of later polemics.

It was under Twardowski's influence that Jan Łukasiewicz became an anti-psychologist. As a result of his criticism of relativism, the classical theory of truth was adopted widely in Poland and became the background of the semantics of Alfred Tarski. Distinguishing content and the object of presentation in the form proposed by Twardowski has been assimilated and further established by Meinong. The criticism of idealism, based on these distinctions, was continued later, fruitfully, by Ajdukiewicz. The view that probability is possessed by presentations of judgments and not by judgments themselves, was taken over from Twardowski, i.a. by Łukasiewicz. Coming from Twardowski's idea of the relationship of action-product—as a relationship of causation different from causal connection—resulted in the future praxeology of Kotarbiński. The idea of examining objects as correlates of mental actions, i.e. as existentially neutral objects—was also present in the ontology of Meinong and in the phenomenology of Edmund Husserl—and was revived later in Saul Kripke's semantics of possible worlds. The idea of images—as a possible underlying presentations of concepts—has allowed the extension of normal defining beyond the classic formula (*per genus*). The idea of concepts as presentations containing, i.a. presented judgments in its content—can be regarded as the announcement of a Russellian reduction of notions to propositional functions. The idea of presented judgment as a presentation of performed judgment resulted in the Ingardenian conception of quasi-judgments. The idea of a work of art as a product of an artist's actions, different from the material of these actions—was developed in detail by Roman Ingarden in his intentional aesthetics. In Twardowski, we should look for the prototypes of Czeżowski's method of analytical description and the Ajdukiewiczian method of semantic translation (of interpreted texts).

All this is enough to consider Twardowski as, not only an extraordinary teacher, but also an excellent scholar.

Kazimierz Twardowski's Works Mentioned in the Paper

1. Zur Lehre vom Inhalt und Gegenstand der Vorstellungen. Eine Psychologische Untersuchung. Wien, Hölder (1894) (English translation: On the content and object of presentation. A psychological investigation. Martinus Nijhof, The Hague (1977))
2. Die Unsterblichkeitsfrage. Wydawnictwo WFiS UW, Warszawa (2009) (Manuscript 1895)
3. Wyobrażenia i pojęcia [Images and concepts]. In: Wybrane pisma filozoficzne [Selected philosophical writings]. PWN, Warszawa (1965), pp. 114–197 (Original work published 1898. English translation: Imageries and concepts. Axiomathes, vol. I, 79–104 (1995))
4. Zasadnicze pojęcia dydaktyki i logiki [The essential concepts of didactics and logic]. Nakładem Towarzystwa Pedagogicznego, Lwów (1901)
5. O filozofii średniowiecznej wykładów sześć [Six lectures on medieval philosophy]. H. Altenberg, Warszawa (1910)
6. Mowy i rozprawy z okresu działalności w Towarzystwie Nauczycieli Szkół Wyższych [Addresses and papers from the period of activity in the Association of Teachers in Higher Education]. Nakładem TNSW, Lwów (1912)
7. O patriotyzmie [On patriotism]. In: Myśl, mowa i czyn [Thought, Speech and Actions]. Vols. I–II. Copernicus Center Press & Wydawnictwo Naukowe Semper, Kraków & Warszawa (2013–2014), vol. I, pp. 436–455 (Original work published 1919)
8. Autobiografia filozoficzna [Philosophical autobiography]. In: Przegląd Filozoficzny vol. I(1), pp. 19–33 (1992)(Original work published 1926)
9. Rozprawy i artykuły filozoficzne [Philosophical dissertations and papers]. Księgarnia SA Książnica-Atlas TNSW, Lwów (1927)
10. Wybrane pisma filozoficzne [Selected philosophical writings]. PWN, Warszawa (1965)
11. Wybór pism psychologicznych i pedagogicznych [Collected psychological and pedagogical writings]. WSiP, Warszawa (1992)
12. Etyka [Ethics]. Wydawnictwo Adam Marszałek, Toruń (1994)
13. Dzienniki [Diaries]. Vols. I–II. Wydawnictwo Adam Marszałek, Toruń (1997)
14. Twardowski, K.: On Actions, Products and Other Topics in Philosophy. Rodopi, Amsterdam (1999)
15. Filozofia i muzyka [Philosophy and music]. Wydawnictwo Naukowe Semper, Warszawa (1995)
16. Dzienniki młodzieńcze [Youthful diaries]. Wydawnictwo Naukowe Semper, Warszawa (2013)
17. Myśl, mowa i czyn [Thought, Speech and Actions]. Vols. I–II. Copernicus Center Press & Wydawnictwo Naukowe Semper, Kraków & Warszawa (2013–2014)
18. Brozek, A., Jadacki, J.: On Prejudices, Judgments and Other Topics in Philosophy. Rodopi, Amsterdam (2014)

A. Brożek (✉)
Institute of Philosophy, University of Warsaw, Warsaw, Poland
e-mail: abrozek@uw.edu.pl

Jan Łukasiewicz: A Creator of New Ideas in Logic and a Reinterpreter of Its History

Jacek Jadacki

Poetic works do not differ from scientific in more imaginative fantasy. [...] However, scientists differ from poets in this respect, that the former REASON *always and everywhere.*
About creativity in science (Łukasiewicz [3, p. 32])

Abstract Jan Łukasiewicz was a leading figure of the Warsaw School of Logic—one of the branches of the Lvov-Warsaw School. The paper presents his personality, life and didactic activity, as well as the list of his main works and the greatest achievements in logic, its history and philosophy. In propositional logic, he invented the bracketless symbolism and constructed many systems of this logic. One of his greatest achievements was discovering three-valued logic. He also reconstruct the history of logic with the apparatus of modern logic. In ontology, Łukasiewicz made a logical analysis of the notion of causality and determinism. In epistemology, he established the precise formulation of the paradox of the liar.

Keywords Determinism · History of logic · Jan Łukasiewicz · Lvov-Warsaw School · Many-valued logic · Propositional logic

Mathematics Subject Classification (2000) Primary 01A70, 03B05; Secondary 03B50

1 Personality

People dabbling in science are eminent, if they see fundamental questions in their discipline and find original responses to these questions. If the theoretical construction proposed by an eminent scholar—in particular, an eminent logician—also has the value of simplicity and beauty, one can say about him that he is a genius. Łukasiewicz was certainly a genius in this sense.

It was said of him that he was shy, sensitive and irritable. He was sensitive to how others judged him—and whether he was appreciated by them. He could not hide the fact that he cared about his recognition.

Á. Garrido, U. Wybraniec-Skardowska (eds.), *The Lvov-Warsaw School. Past and Present*, Studies in Universal Logic,
https://doi.org/10.1007/978-3-319-65430-0_3

Sometimes, talents are born—as in the Polish phrase—*on the stone*; but a talent may not generally develop so as to become a genius: to do this, one needs the appropriate soil. This soil—in the case of Łukasiewicz—was the mental environment brought to life by Kazimierz Twardowski, that is, in short, the philosophical Lvov-Warsaw School. Łukasiewicz grew in this environment—and then he co-created this environment, not without mutual theoretical interaction also with his own colleagues and students.

What was the relationship of Łukasiewicz to that environment?

He identified himself with the Brentanian roots of the School—but not with all of them. Twardowski was highly estimated by him—but not for everything. He appreciated Stanisław Leśniewski—as a logician—at least initially, higher than himself.

Three women played a great role in the life of Łukasiewicz: firstly—his mother, Leopoldyna née Holtzer; then—princess Maria Józefina Sapieżanka, who was the object of his great love, but without reciprocity; in the end—his wife, Regina née Barwińska, who was his bedrock especially in the last years of his life (even though—let us add—a sometimes troublesome bedrock due to her antagonistic character).

He was a great Polish patriot, but his ancestors were Ruthenian (paternal) and Tyrolean (maternal). To strangers, with whom he came into contact, he referred differently. He had friends among the Germans—but he did not feel good in Germany. He treated Ukrainians with sympathy—but he was, i.a., an opponent of ukrainizing the Lvov University. He felt aversion to some Poles of Jewish origin—but he fiercely opposed anti-Jewish movements in the academic circles. In the last decade of the life, Ireland became his second homeland—but his attitude to the Irish people was rather (unfairly!) dry.

On the one hand he was a man of deep faith: in particular, a Catholic and a practicing Catholic. On the other hand, he avoided the public «confession of faith». He also avoided political declarations—which does not mean that he did not have an explicit (conservative) view on these issues.

2 Life

He was born on December 21, 1878, in Lvov; he died on February 13, 1956 in Dublin.

He studied law and later philosophy with Twardowski at the Lvov University—and at the universities of Berlin and Louvain. After his doctorate (1902) and habilitation (1909), he was a participant in the seminar of Alexius Meinong in Graz (1910). From the years 1911–1915, he was a professor of philosophy and logic at Lvov University; from the years 1915–1939 at the Warsaw University (with breaks from 1918–1920 and 1924–1929), where he served twice as the rector (1922/1923 and 1931/1932). In 1919 he was the Minister of Religious Denominations and Public Education. During one of the German raids at Warsaw in September 1939, his library collections and rich manuscript legacy burned totally. In 1944, as an avowed anti-communist—in the face of the approaching front of the Soviet troops, he decided to go to Switzerland, but because of the tense political situation after the assassination attempt on Adolf Hitler, he had to stop in Münster (under the supervision of Heinrich Scholz). After the war, he settled first in Brussels and then in Dublin, where he was professor of logic at the Royal Irish Academy.

He was a member of the Polish Academy of Sciences, the Polish Scientific Society in Exile, and received honorary doctorates from universities in Münster and Dublin. He was a leading figure in the Warsaw School of Logic, being an essential component of the philosophical Lvov-Warsaw School. The spirit of the School found in his philosophical and logical works its most perfect incarnation.

3 Works

Among the most important publications of Łukasiewicz, there are the following books: *On the principle of contradiction in Aristotle* [2], *Die Grundlagen der logischen Wahrscheinlichkeitsrechnung* [5] and *Aristotle's syllogistic from the standpoint of modern formal logic* [11].[1]

Most of his papers were collected in two Polish volumes: *Selected problems of logic and philosophy* [13] and *Logic and metaphysics. Miscellanea* [19].

Many of his works appeared also in translation into other languages, including English [13], French [16, 20, 22], Spanish [17], Japanese [18] and Russian [12, 21].

Łukasiewiczian *Elements of mathematical logic* [8] is the model of a manual. This work includes a lecture on the axiomatic system of classical propositional calculus, propositional calculus with quantifiers and a fragment of the calculus of names (Aristotelian syllogistic included). An unusual source for the history of Polish culture of the twentieth century is Łukasiewicz's *Diary* [23].

The most important scientific achievements were Łukasiewiczian logic, the philosophy and history—as well as ontology and methodology of philosophy.

4 Metaphysics

4.1 Ontology

Of ontological issues, Łukasiewicz was especially interested in two problems (though interrelated in many ways): what is a causal relationship and what are the reasons for determinism.

4.1.1 Causal Relationship

An attempt to define "causal relationship" was made by Łukasiewicz in his classic dissertation "Analysis and construction of the concept of cause" [1]. The analytical part of this dissertation has little equal in philosophical literature—and not only in Polish writing.

[1]This work is a continuation of the Polish monograph, prepared in 1939, which unfortunately burned during World War 2.

In the constructive part, Łukasiewicz proposed reducing the notion of a causal relationship to the notion of necessity. This reduction could be simplified in such a way: The fact that object P_1 has feature c_1 is the reason for the fact that object P_2 has feature c_2—when if object P_1 has feature c_1, then object P_2 must have feature c_2, where P_2 must have feature c_2, when if object P_2 did not have feature c_2, then object P_2 would be an internally contradictory object.

Łukasiewicz initially thought that he gave, in this way, an equivalent definition of "causal relationship". But it finally turned out, that the necessity of the relationship between the two states of affairs does not settle the fact that this is a causal relationship, since the impossibility of the occurring state of affairs S_2 without the occurring state of affairs S_1 can take place, i.a. in the case where the occurring state of affairs S_1 is later than the occurring state of affairs S_2, or when the two states of affairs are timeless (like, e.g., the fact that a certain figure is a square necessarily involves the fact that this figure is a rectangle)—whereas it is assumed that every state of affairs is later than its cause.

Consequently, the definition proposed by Łukasiewicz can be treated at most as an inclusive definition, indicating only a necessary condition for the occurrence of a causal connection.

4.1.2 Determinism

The ontological thesis of determinism (in one of its versions) holds that each state of affairs occurring in the real world is uniquely determined by causes preceding it.

Łukasiewicz pointed out that among the premises forming the basis of the thesis of determinism there are two principles: the principle of the excluded middle and the principle of causality [6]. The first—let us remember—states (in a certain version) that for any state of affairs, this state of affairs occurs or does not occur. The second one states (in a certain version) that every state of affairs has a cause in some previous state of affairs. Following the principle of the excluded middle applied to the future, we must recognize that the occurrence of any future state of affairs has been already determined; but—as Łukasiewicz says—there is no compulsion to accept the principle of the excluded middle (on the grounds of, e.g., trivalent logic, this rule does not apply). Supposing the principle of causality, we must accept the fact that causal-effectual chains are infinite «in the past» (i.e., they are eternal); but—as Łukasiewicz says—one can accept the hypothesis of infinite causal-effectual chains and at the same time recognize that in a certain—distant enough—moment, there is «already» no reason for a given state of affairs, if only one assumes that time intervals between successive links in the causal-effectual chain decrease «back» unlimitedly; under such a condition, adopting the principle of causality is compatible with indeterminism.

4.2 *Epistemology*

Łukasiewicz's attitude to epistemology was more than critical: he considered most epistemological problems to be apparent ones.

The problem of truth was one of the real problems to which he devoted a lot of attention; it was understandable: truth is one of the logical values, and Łukasiewicz worked on constructing logical systems in which more than two logical values could be admissible, and not only truth and falsehood.

By "truth"—or more precisely: a "true sentence"—Łukasiewicz understood a sentence that "admits this property to an object, which is really possessed by this object, or that refuses this property, which is not really possessed" [5, p. 55]. We use different criteria for the truth of sentences, but we are not able to justify that these or other criteria are valid; an attempt to give such a justification always ends in either a vicious circle or *regressus in infinitum*.

Łukasiewicz was probably the first to establish the precise formulation of the so-called paradox of the liar. As the source of the paradox, he indicated a sentence of the type "Sentence S is false": such that, if sentence Z, that is, the sentence "Sentence S if false", is true, it is just as it says, so it is false; if sentence S is false, then it is just not the case, so sentence S is not false—or it is true. This formulation became an inspiration for the semantic conception of truth, proposed by the student of Łukasiewicz, Alfred Tarski.

5 Logic

5.1 Propositional Logic

A particular subject of Łukasiewicz's interest was a classic propositional calculus, i.e. a logical theory which reports the meaning of conjunctions (functors) connecting sentences—such as "and", "or", "if …, then", "always and only if"—and the negation "it is not the case that".

Let us start with the fact that Łukasiewicz invented for this calculus a special symbolism, called "bracketless symbolism" and later "Polish symbolism". It consists in recording complex sentences in such a way that at first we give a functor, followed by sentences «bonded» by this functor. For example, the formula "If p, then (q and r)" is recorded in such a symbolism as: CpKqr (where 'C' signifies functor "if …, then", and 'K'—functor "and"). In the parenthetical symbolism, this formula has the form: $p \rightarrow (q \wedge r)$. Note that the bracketless formula, mentioned above, has two symbols less than its equivalent parenthetical formula.

Now: propositional calculus in the form of axiomatic theory is a set of statements about these conjunctions, each of which is either an axiom (and therefore the claim accepted without proof), or can be derived from axioms using determined rules called "rules of inference". Łukasiewicz constructed many versions of such propositional calculus—differing, among other, as to adopted axioms and their number, rules of inference and their number, as well as which functors are considered primary and which secondary, i.e. definable with the use of the former. Among those versions, one was considered a classic one: it is a system which operates one functor (i.e. the functor "always if …"—or the functor of implication), universal quantifier ("every …") binding propositional variables, four rules of inference, symbols of truth and falsehood, and symbols of accepting and rejecting sentences.

Łukasiewicz's desire was to create a system which would contain as little as possible axioms, rules of inference etc. He succeeded in this respect, firstly, in inventing a system of implicational propositional calculus (or a calculus operating with only one functor: the functor of implication) based on one axiom numbering—in the bracketless symbolism—13 symbols (CCC*pqr*CC*rp*C*sp*), and then to prove that it is the shortest axiom of that implicational calculus. Secondly, Łukasiewicz honed the rules of inference. He invented, among others, a useful version of the rule of substitution, i.e. the rule according to which in a sentence containing variables we are free to insert in their place other variables or constants, as long as we are doing it consequently (i.e. to insert these variables or constants into the places of each appearance of a substituted variable). He developed a more precise reasoning called "generalizing deduction", i.e. the method of proving general statements on the basis of their particular cases. He codified the matrix characteristics of the functor of propositional calculus and the method examining the tautologicality of formulas of that calculus, involving the compilation of possible combinations of substituting propositional variables in these formulas by symbols of truth and falsehood.

5.2 Propositional Logic with Quantifiers

In turn, the system of propositional calculus with quantifiers was based by Łukasiewicz on: (a) the universal quantifier ($\prod$) and the functor of implication (C) as primary concepts; (b) the three axioms (Tarski and Paul Bernays's ones: C*q*C*pq*, CCC*pqpp* and CC*pq*QQ*qr*C*pr*); (c) the definition of negation ($\mathrm{N}p = \mathrm{C}p \prod pp$) and (d) five rules of inference: the rule of substitution (modified in comparison to the analogical rule in propositional calculus without quantifiers), the rule modus ponens, the rule of replacement, the rule of combining, and the rule of skipping quantifiers. In this system, Łukasiewicz presented proofs of 19 theorems (he left five theorems without proof), including the proofs of three axioms of the system of propositional logic without quantifiers, described above.

5.3 Reconstruction of Syllogisms

Łukasiewicz has reconstructed the most significant part of the logic of Aristotle, i.e. "assertoric" syllogistic (*scil.* non-modal) [11]. His intention was to make this reconstruction: on one hand—according to the intentions of the great philosopher; on the other hand—developed in the spirit of modern logic.

According Łukasiewicz—Aristotle's syllogistic is a part of the logic of names, namely a formal theory of three constants: "all … are …", "none … are …", "some … are …" and "some … are not …", where the values of the variables representing arguments of those functors of two arguments are adopted only in general terms (in particular, with the exclusion of empty and negative names). The so constructed syllogistic is superstructured over propositional logic; in particular, it contains the following constants of this logic: "if …, then …", "… and …" and (in some proofs), "it is not the case that … ."

Aristotle tried to axiomatize syllogistic, taking as its basis four modes of the first syllogistic figure, eventually reducing finally to two of them (*Barbara* and *Celarent*). It turned out, however, that we need to add to them two laws of conversion, and (in some cases) two laws of identity. The simplest axiomatic base contains as primary terms—the constants "all ... are ..." and " ... some ... are ..." (the rest can be defined with their help and the negation of sentences), and as axioms—two laws of identity and the moods *Barbara* and *Datani* (or: *Barbara* and *Dimaris*).

Reduction of imperfect syllogisms to perfect ones, postulated by Aristotle, was interpreted by Łuksiewicz as the proof of theorems of the system (*scil.* deriving them from axioms). According to Łukasiewicz, Aristotle gave not only proofs of the true syllogistic formulas, but he also tried to show that all the others formulas are wrong, and as such should be rejected. He rejected inconclusive formulas, usually using the method of exemplification by means of appropriate concrete terms (which satisfy "premises", but do not satisfy—"conclusions").

5.4 Many-Valued Logic

One of Łukasiewicz's most important achievements was discovering three-valued logic[2]—and more generally: logics more than two-valued; the philosophical importance of the three-valued logic was compared, by Łukasiewicz himself, to the importance of non-Euclidean geometry in mathematics.

Existing logical calculi based on the principle of bivalence—i.e. on the assumption that every sentence has exactly one logical value: it is either true or false: tertium non datur. Łukasiewicz generalized the concept of logical value in such a manner that he allowed the existence of "intermediate" values between truth and falsehood: one—in the case of the three-valued logic, two—in the case of the four-valued logic, or more—up to the infinitely-many-valued (in short: $\aleph_0$-valued) logic.

Behind the idea of the three-valued logic there were the following insights. There are sentences which at the given moment can not be principally (and not, e.g., because of someone's ignorance) determined to be true or false. These are sentences about future events, which at present are not determined (i.e. those that there are currently neither the cause of their occurrence, nor the cause of their non-occurrence—or sentences about past events, the effects of which have completely "expired" (i.e. those for which there are currently no effects of their occurrence). These are just sentences having a third logical value. So sentences can be not only true or false, but also undetermined.[3]

An example of intuitive interpretation of the four-valued logic is its interpretation in terms of modal logic, that is, one in which there are functors of the type "it is necessary

[2]The first outline of non-Chrisipian logic dates back to 1917, and the first system of it—to 1920.

[3]It may be worth noting that one of the consequences of adopting the so characterized indefiniteness as a third value in the three-valued logic is that we should to accept the view that sentences change their logical value in time: e.g. sentence undetermined at a certain moment may come to be true (or false) at some later moment, namely, one in which adequate causes have already occurred. Someone who would not want to agree with such a consequence, could not also agree with the sketched interpretation of the third value. Questioning the interpretation does not involve, of course, questioning the interpreted calculus.

that ..." and "it is possible that ...." Łukasiewicz made this interpretation when it turned out that the characteristics of propositional functors in the conceptual apparatus of the three-valued logic is unsatisfactory.

As regards the infinitely-many-valued logic, Łukasiewicz claims (at least initially) that logical values present in this logic can be identified with degrees of probability.

5.5 *Metalogic*

Among Łukasiewicz's great achievements, there were the results of his metalogical analysis, and in particular: his results of research on consistency, independence and the completeness of axioms of the propositional calculus. The issue at stake was to determine whether or not the constructed sets of axioms contained axioms negative to each other such that some of them resulted in others, and finally, whether these axioms are sufficient to prove all the theses of the system, which appears to be true.

It was also demonstrated here that classical propositional calculus is a fragment of intuitionistic logic. If one considers that at the core of the latter belief there is the conviction that only proved theorems can be considered as theorems of a certain mathematical theory—and therefore among its theorems there is not, e.g., the law of excluded middle ("*p* or it is not the case that *p*")—then Łukasiewicz's result sounds surprising (removing this impression requires suitably enriching intuitionistic logic itself).

5.6 *Philosophy of Logic*

5.6.1 Anti-psychologism

In the second half of the nineteenth century—not without relation to the rapid development of psychological research—tendencies appeared to reformulate all philosophical issues, including problems of logic, in such a way that every philosophical (and logical) problem was replaced by its psychological paraphrase.

Łukasiewicz was one of those people who revolted earliest and most strongly against such a psychologization—especially the psychologization of logic. His anti-psychologism was reflected, among other things, in the fact that he precisely contrasted laws of logic with their alleged equivalents in the form of psychological laws, and—let us add—ontological ones. He stressed, e.g., the difference between logical, psychological and ontological principles of contradiction: the first principle (in one of its versions) is the law of propositional logic, according to which it is not the case that both *p* and not-*p*; the second principle (in one of its versions) claims that in reality is never so that a certain state of affairs occurs and does not occur at the same time; the third principle (in one of its versions) is the view that no one alive can hold a certain conviction and its negation at the same time.

Łukasiewicz—analyzing these versions of the principle of contradiction—noted that an experience is not able to confirm the ontological version of these principle. Here are excerpts of this argument—a beautiful example of Łukasiewicz's philosophical prose:

> Any movement [...] takes place in such a way that the changing object loses some features that it possessed, and acquires new ones that it did not possess. In both cases, contradiction would arise, if there were not different temporal determinations.
>
> If the change is CONTINUOUS, e.g. the movement of an arrow released from a bow, [...] then in every smallest interval of time, the changing object loses in turn some features, and purchases second ones. The moving arrow is in any two moments of time in DIFFERENT places. [...] What [...] happens when this distance decreases to zero when we shall consider only ONE moment as unextended on timeline?
>
> Once, we heard the fable that when a princess pricked her little finger on a spindle, she fell immediately into a hundred-year's deep dream, and all life around her slept as well. In such a way, the legendary Popiel's court froze in the blink of an eye, enchanted by Rzepicha in the songs of *King of Spirit*. Suppose that what is only a poetic fantasy has become reality. [...] The arrow would [then] rest motionless in a certain place. But how do we know that it would be only in ONE place? Why, in an unextended moment, in a temporal point of the section, could it not be in at least two different places and thus be in a certain place and not be there at the same time? [...] Experience is silent on this issue. [...] All the perceived phenomena LAST shorter or longer and SHOULD last for a minimum period of time to be noticed by us at all. We do not know what happens in an unextended moment. However the principle of contradiction applies to such a moment; because if we say that the arrow cannot AT THE SAME TIME be and not be at the same place, the phrase "at the same time" refers to THE SAME, so the only ONE, unextended moment. [2, pp. 136–138]

So much for the ontological version of the principle of contradiction. As for the empirical justification of the psychological version, it would require painstaking research, which has not yet been undertaken and whose desirability at all is in question in light of the statements of some people (including philosophers and mystics) that they entertain contradictory beliefs. Finally—the only justification for accepting the logical version of the principle of contradiction is that it makes it possible to prove with a certainly unattainable in other cases that someone is mistaken or lying; because we recognize (assuming the principle of contradiction) a conviction entailing a contradiction as mistaken—and we have (assuming this principle) the basis to believe that someone is lying when that person has once said, that p, and some other time, that not-p; the inability to prove that someone is wrong or lying, would have far-reaching—negative—consequences for social life, because, among other things, it would make it difficult, and in some cases even prevent issuing of righteous judgments by the courts.

5.6.2 Metaphysical Neutrality

Łukasiewicz argued not only for depsychologizing logic, but also for—so to speak—the metaphysical neutralizing of it. The idea was that—already in the twentieth century—there were tendencies to treat calculi of logical systems as systems implying one or another ontological or epistemological positions. Additionally, to show the groundlessness of such tendencies—it is enough to exactly distinguish the logical system from its permissible interpretation—and from metaphysical positions occupied by a logician who is the creator of this system or its interpreter. Logicians tend to be of defined metaphysical views—or of a defined worldview—but they feed these views as metaphysicians or members of a particular philosophical orientation, not as logicians: logical systems, as such, neither assume nor imply any metaphysical theses or any theses of faith or unbelief. Moreover, practicing formal logic does not require, e.g., accepting nominalism—or the view that logical systems are systems of unretrievable, «senseless» (i.e. having no reference to reality) symbols-strings. If it were so, logical systems would be always sets of a finite number of theses—because we can not «produce» (or build) an infinite number of symbols-strings. Meanwhile, these systems are infinite sets (it is enough to note that if the statement "If p, then p" is a thesis of the propositional calculus, then the statements "If *p*, then (*p* or *q*)", "If *p*, then [*p* or (*q* or *r*)]" etc. *ad infinitum* are such theses also. Similarly, e.g., the existence of many logical systems does not support relativism and conventionalism—or the view that our images of the world are determined by the freely accepted conceptual apparatus, so that none of them can reasonably be considered as a «true» image; because in this case, the issue is decided by experience, providing us knowledge about what the world «truly» is—not the mere fact of the existence (*scil.* constructability) of various its images.

5.7 *History of Logic*

Łukasiewicz's basic postulate in relation to historical research on logic (and more generally—philosophy) was the postulate to reconstruct history with the apparatus of contemporary logic; the first work realizing this postulate was the dissertation *On the principle of contradiction in Aristotle* [2], although the postulate itself was explicitly formulated much later [9].

According to this postulate—Łukasiewicz analyzed two great logical systems of antiquity: Aristotle's syllogistic and Stoic logic. Results of this analysis were surprising—and they questioned existing views on both systems.

It turned out, firstly, that Aristotle's syllogistic is not a theory of inference (as was previously thought), but a calculus of names: in particular, Aristotle's syllogisms are not rules of inference (of the type: If we accept the premise "Each *A* is *B*" and the premise "Each *B* is *C*", we must accept the conclusion "Each *A* is *C*"); they are theses of the calculus of names (of the type: If each *A* and each *B* is *C*, then each *A* is *C*). By the way, it turned out that Aristotle is the inventor of nominal variables and his syllogistic was the first (admittedly imperfect) axiomatized system in the history.

Secondly, Łukasiewicz showed that the Stoic logic is not (as previously thought) a calculus of names, but historically the first the system of propositional logic having its extension in the Middle Ages under the name of the theory consequences: in opposition to Aristotelian syllogisms—Stoic syllogisms are the rules of inference.

6 Methodology

In methodology—Łukasiewicz proposed an original classification of reasoning and reinterpreted the notion of inductive reasoning, the notion of probability and the notion of magnitude.

6.1 Reasoning

According Łukasiewicz—to reason (let us add: validly) is the same as to select for a certain sentence *S* a reason (or a sentence which is followed by sentence *S*) or a consequence (i.e., a sentence which follows sentence *S*); in the first case we are talking about reduction; in the second case—about deduction. Sentences—given and selected—may be accepted or not accepted (as true sentences) by the reasoner before beginning to reason. Depending on which of these two cases takes place—reduction is either an explanation (when we select an unaccepted as yet reason for an accepted sentence) or proof (when we select an accepted reason for an unaccepted as yet sentence); on the other hand deduction is inference (when we select an unaccepted as yet consequence to an accepted sentence) or confirmation (when we select an accepted consequence to an unaccepted as yet sentence).

The reasoning consisting in the fact that a certain law is accepted—therefore a general statement stating occurrence of certain general regularities—on the basis of earlier accepted individual sentences stating occurrence of a certain number of cases of this regularity is called "inductive reasoning" (more accurately: "incomplete enumerating induction"). Before Łukasiewicz induction was believed to be a kind of deduction. In light of Łukasiewiczian conception of reasoning, this is a misconception—inductive reasoning is a kind of explanation (in which a reason is selected to individual sentences), and thus it is a kind of reductive, and not deductive reasoning.

6.2 Probability

In the traditional interpretation of probability, the probability of the occurrence of a certain events is discussed, e.g. about the probability that picking up a random a ball from a face-down box in which there are four balls: white, black, red and blue—I shall take up the white ball. Łukasiewicz proposed a «propositional» interpretation of probability. Certain probability is entitled not to the fact that (e.g.) I shall take up the white ball from the box, but to the sentence "I shall take up the white ball from the box". This sentence is a substitution of the formula "I shall take up ball *x* from the box". In this situation we can insert in place '*x*' a name of one of four colors: white, black, red or blue—but only in the case of one substitution, will the formula be transformed into a true sentence. Therefore, we can say that the probability of the sentence "I shall take up the white ball from the box" is equal to 1/4 in this case.

6.3 Magnitude

Łukasiewicz proposed—in the place of a very complicated «classic» (but logically flawed) definition of "magnitude" given by the mathematician Stanisław Zaremba—a very simple definition, according to which magnitude is the same as an element of a certain well-ordered set, or such a set, whose elements are «arranged» sequentially one after another, because it is determined that a suitable relationship for them is both asymmetrical, transitive and consistent (i.a. the relation of being-greater-than is such an ordering relation in the class of natural numbers).

6.4 Methodology of Philosophy

The methodology of science can be regarded as a description of how—and, in particular, by means of which research methods—science is actually done. We can also treat it as a set of rules defining how science should be done.

Łukasiewicz's contribution to the methodology of philosophy belongs mainly to methodology in this second sense. Łukasiewicz's expectations in this regard were very far-reaching, although he expressed them in a simple postulate:

> Metaphysical problems [have not been] resolved, but I do not think that [they are] unsolvable. We only need to approach them with the scientific method: with the same proven method which is used by a mathematician or a physicist. And above all, we need to learn to think: clearly, logically and strictly. [7, p. 368]

In practice, Łukasiewicz advocated practicing «axiomatic» philosophy, i.e. a philosophy with a structure similar to axiomatic logical systems.

The basis of such a philosophy should be its axioms. Łukasiewicz wrote:

> Every sentence, especially those that will be the basis of philosophy, [...] [should be] formulated as precisely as possible because only then can we duly justify or know the direct evidence of these sentences. [10, p. 372]

Evaluation of the current—especially so-called modern—philosophy was negative:

> The state of pre-Kantian philosophy: on the one hand, fantastic dreams, not withstanding scientific criticism; on the other hand, radical, dogmatic, unfounded skepticism. [But then also:] as we approach Kantian philosophy with the requirements of scientific criticism, its construction collapses like a house of cards. At every step, we have vague notions, incomprehensible sentences, unfounded statements, contradictions and logical errors. [7, p. 368]

The conclusion was crushing:

> Caused by a negligence of logic and by the ensuing mental dressage, the whole of modern philosophy was paralyzed by an impotence of strict and clear thinking. [10, p. 373]

7 Teacher

Even an approximate list of Łukasiewicz's students is not known; neither the audience of his lectures, nor the participants of his seminars. Even the list of masters and doctors promoted by him is incomplete and it raises doubts at various points.

Certainly, his postgraduate students were: Maria Ossowska, in the years 1923–1927, a senior assistant at the Seminary of Philosophy of Łukasiewicz; Mordchaj Wajsberg and Zygmunt Kobrzyński, both of whom died during World War 2—so they did not survive their promoter; finally, Stanisław Jaśkowski and Jerzy Słupecki. It seems that under the guidance of Łukasiewicz a doctorate on the Stoic logic was prepared by Czesław

Lejewski but there was no promotion because of the outbreak of the war. However, it is not clear, e.g., whether Bolesław Sobociński was a postgraduate student of Łukasiewicz or Leśniewski.

Łukasiewicz's assistants—though not postgraduate students—were: Tarski (from 1929) and Henryk Hiż (from 1940 to 1944).

Those who witnessed Łukasiewicz's lectures—colleagues and students—stressed that he was an excellent didactician. This was manifested in the fact that he could make contact with the participants of his lessons, and his lectures were strict and at the same time interesting and affordable, with a clear structure, delivered fluidly and in beautiful language.

Jan Łukasiewicz's Works Mentioned in the Paper

1. Analiza i konstrukcja pojęcia przyczyny [Analysis and construction of the notion of cause]. In: Z zagadnień logiki i filozofii. Pisma wybrane [Selected Problems of Logic and Philosophy]. PWN, Warszawa (1961), pp. 9–62 (Original work published 1906)
2. O zasadzie sprzeczności u Arystotelesa. Studium krytyczne [On the principle of contradiction in Aristotle. A critical study]. Akademia Umiejętności, Kraków (1910)
3. O twórczości w nauce [On creativity in science]. In: Logika i metafizyka [Logic and Metaphysics]. Miscellanea. Wydawnictwo WFiS UW, Warszawa (1998), pp. 9–33 (Original work published 1912)
4. Die logischen Grundlagen der Wahrscheinlichkeitsrechnung. Spółka Wydawnicza Polska, Kraków (1913)
5. Zagadnienia prawdy [Problems of truth]. In: Logika i metafizyka [Logic and metaphysics]. Miscellanea. Wydawnictwo WFiS UW, Warszawa (1998), pp. 55–56 (Original work published 1913)
6. O determinizmie [On determinism]. In: Z zagadnień logiki i filozofii. Pisma wybrane [Selected Problems of Logic and Philosophy]. PWN, Warszawa (1961), pp. 114–126 (Original work published 1922–1923)
7. Kant i filozofia nowożytna [Kant and modern philosophy]. In: Logika i metafizyka [Logic and metaphysics]. Miscellanea. Wydawnictwo WFiS UW, Warszawa (1998), pp. 365–368 (Original work published 1924)
8. Elementy logiki matematycznej [Elements of mathematical logic]. Wydawnictwo Koła Matematyczno-Fizycznego Słuchaczów Uniwersytetu Warszawskiego, Warszawa (1929)
9. Z historii logiki zdań [From the history of propositional logic]. In: Z zagadnień logiki i filozofii. Pisma wybrane [Selected problems of logic and philosophy]. PWN, Warszawa (1961), pp. 178–194 (Original work published 1934)
10. Kartezjusz [Descartes]. In: Logika i metafizyka [Logic and Metaphysics]. Miscellanea. Wydawnictwo WFiS UW, Warszawa (1998), pp. 370–374 (Original work published 1938)
11. Aristotle's Syllogistic from the Standpoint of Modern Formal Logic. Clarendon Press, Oxford (1951)
12. Аристотелевская силлогистика с точки зрения современной формальной логики. Издательство Иностранной Литературы, Москва (1959)
13. Z zagadnień logiki i filozofii. Pisma wybrane [Selected Problems of Logic and Philosophy]. PWN, Warszawa (1961)
14. Elements of Mathematical Logic. Pergamon Press, Oxford (1966)
15. Selected Works. North-Holland Publishing Company & PWN. Amsterdam and Warszawa (1970)
16. La syllogistique d'Aristote. Armand Colin, Paris (1972)
17. Estudios de lógica y filozofía. Biblioteca de la Reista de Occidente, Madrid (1975)
18. Sūri rondigaku genron. Bunka Shobō Hakubunsha, Tokio (1992)
19. Logika i metafizyka [Logic and metaphysics]. Miscellanea. Wydawnictwo WFiS UW, Warszawa (1998)
20. Du principe de contradiction chez Aristote. L'Éclat, Paris (2000)

21. О принципе противоречия у Аристотеля. Критическое исследование. Центр Гуманитарных Инициатив, Москва & Санкт-Петербург (2012)
22. Écrits logiques et philosophique. Vrin, Paris (2013)
23. Pamiętnik [Diary]. Wydawnictwo Naukowe Semper, Warszawa (2013)

J. Jadacki (✉)
Institute of Philosophy, University of Warsaw, Warsaw, Poland
e-mail: j.jadacki@uw.edu.pl

Kazimierz Ajdukiewicz: The Cognitive Role of Language

Anna Jedynak

Abstract Kazimierz Ajdukiewicz was an eminent representative of the Lvov-Warsaw School. His main interest was the cognitive role of language. His radical conventionalism intended to explain rapid and fundamental changes in science. He used the method of philosophical paraphrase to make traditional metaphysical questions decidable. Then he drew metaphysical conclusions from the so called "semantic epistemology" based—according to his programme—on semantics (which played an important role in his research) and formal logic. His categorial grammar aiming to formulate the general criteria of syntactic coherence was the first grammar based exclusively on the structural properties of expressions. He also undertook a number of methodological issues, both general and detailed. He was interested in the theory of definition, the theory of questions, the problem of rationality of fallible inferential methods, the foundation of sentences, classification of reasonings, of sciences and of axiomatic systems, and in the reconstruction and evaluation of scientific procedures.

Keywords Ajdukiewicz · Meaning-rules · Methodology of science · Radical conventionalism · Semantic epistemology · Scientific procedures.

Mathematics Subject Classification (2000) Primary 03A10, 03A99, 03B42, 03B65; Secondary 03B48

1 Life

Kazimierz Ajdukiewicz (1890 Tarnopol–1963 Warsaw) was an eminent representative of the Lvov-Warsaw School. In Lvov, he studied philosophy, mathematics, and physics. He was a disciple (and later on, also a son-in-law) of the founder of the School, Twardowski, but also studied under other teachers, such as Łukasiewicz, Sierpiński, and Smoluchowski. After having obtained his PhD degree in 1912 (*On the Relation between the Apriorism of Space in Kant and the Question of the Genesis of the Representation of Space*), he deepened his studies in Göttingen, where he had the opportunity to attend lectures by Husserl and Hilbert. During World War I, he served in the Austrian army, to

Á. Garrido, U. Wybraniec-Skardowska (eds.), *The Lvov-Warsaw School. Past and Present*, Studies in Universal Logic,
https://doi.org/10.1007/978-3-319-65430-0_4

which he was conscripted, and later, in 1920, during the Bolshevik War, as a volunteer in the Polish army. In 1921, he received his habilitation (*From the Methodology of Deductive Sciences*). In the interwar period, he was first a professor at the University of Warsaw, and then, since 1928, at the University of Lvov. He survived World War II in Lvov, involved in administrative work, clandestine teaching, and—in periods free of the German occupation—lectures in scientific institutes. After the war, he was a professor at the University of Adam Mickiewicz in Poznań (where in the years 1948–1952 he served as rector) and the University of Warsaw (since 1954). He served many functions related to the animation and organization of scientific life. He participated in national and international congresses, organizing some of them; he edited prestigious scientific journals ("Studia Philosophica" and "Studia Logica"), travelled abroad in connection with his scientific activities (Great Britain, Austria, USA), was an active member of the Polish Academy of Sciences and other scientific associations, and established and led the Department of Logic at the University of Warsaw. In 1962, the University of Clérmont-Ferrand awarded him an *honoris causa* doctorate. The University of Adam Mickiewicz in Poznań had a similar intention, which was not carried out because of the death of Ajdukiewicz. In the postwar period, Ajdukiewicz and Kotarbiński were the main representatives of the Lvov-Warsaw School in Poland. They continued the School tradition of practicing broadly conceived logic, which included semiotics and methodology, and of using it to resolve philosophical issues. Ajdukiewicz described himself as a philosopher and a logician (in that order). He initiated and coordinated many research themes and educated students who later continued this work. A strong personality, he was considered a great authority among his colleagues and students.

2 Main Publications

The most important publication of Ajdukiewicz's papers is *Language and Cognition*, comprising a selection of his texts from the years 1920–1939 [2] and 1945–1963 [7]. The essential texts from that publication have also been translated into English and published as *The Scientific World-Perspective and Other Essays*, 1931–1963 [31]. Results of the author's research in methodology were published posthumously as *Pragmatic Logic* [12], conceived as a university textbook. His *Problems and Theories of Philosophy* [11] remained the main philosophical textbook for several generations of Poles. Both have also been translated and published in English.

3 Scientific Activity

3.1 Meaning-Rule Conception and Radical Conventionalism

The driving force of young Ajdukiewicz's philosophical inquiry was his interest in the cognitive role of language [38]. Inspired by the ideas of French conventionalists, especially Le Roy, he wanted to present the latter's free philosophical insights in a strictly scientific manner. As a result, he created his radical conventionalism. First, however, he needed a precise conception of meaning, because the question he investigated was how the choice of language, and the meanings of its expressions in particular, affected the process and outcome of cognition. For this reason, he took up studies in semiotics [23, 40]. He developed the meaning-rule conception of meaning, with language understood as a kind of deductive system governed by what he called meaning-rules [20]. These meaning-rules prohibited the language user to refuse accepting some of its sentences—either unconditionally (axiomatic meaning-rules), or by virtue of prior acceptance of other sentences (deductive meaning-rules), or in the face of certain experiential data (empirical meaning-rules)—otherwise the meaning of expressions of the language would be violated. Sentences accepted under those meaning-rules constituted, respectively, axiomatic, deductive, and empirical theses of the language. Ajdukiewicz considered as synonymous both expressions of the same language, which were interchangeable within its meaning-rules, providing the latter remained intact, as well as expressions from different languages, which occupied an analogous structural position, each within the meaning-rules of their respective languages. In view of this, meaning was the property—defined by abstraction—common to all synonymous expressions.

Ajdukiewicz proceeded to investigate only the languages he called closed and connected, that is, those in which all expressions were interconnected in terms of meaning by way of meaning-rules and to which no new meanings could be added since they already contained all the meanings from the given relationship network. He called the set of meanings of a closed and connected language a conceptual apparatus, and the set of theses of such a language a world-perspective. Then he formulated the thesis of radical conventionalism: a world-perspective depends on the choice of conceptual apparatus [35]. When applied to languages other than closed and connected ones, this conception did not claim anything unusual: only that sentences accepted by virtue of the meaning of their expressions depended on those very meanings. Its originality lay in its application to closed and connected languages and in the assumption that various such languages existed, based on different conceptual apparatuses. Two such languages (since they could not be enriched in a connected way) either used identical conceptual apparatuses and differed only by the sound of their expressions, or used completely different apparatuses that had nothing whatsoever in common. In the second case, those languages were completely mutually untranslatable, that is, no expression of either language had a synonym in the other one. Moreover, a change of meaning of even one expression of a closed and connected language led to a change of the meanings of all its expressions, and so a shift to another apparatus. This in turn meant a transition to a fundamentally different cognitive perspective.

Radical conventionalism was supposed to give an account of the profound transformations in science, later called scientific revolutions. According to Ajdukiewicz, what changed were not the theories as such but something more fundamental, namely, languages and associated conceptual apparatuses by means of which theories were expressed. A shift to another apparatus made the old theory and the new one mutually untranslatable. Science, therefore, did not develop cumulatively. However, Ajdukiewicz pointed out some evolutionary trends of conceptual apparatuses, which allowed progress in science to take place. He abandoned this conception partly due to the shortcomings that Alfred Tarski had found in the underlying meaning-rule conception of meaning, and partly because he had concluded that the notion of a closed and connected language was empty, and as a result, that radical conventionalism itself did not have a real exemplification. He continued promoting the main thesis of radical conventionalism in a weakened version and a different formulation, applying it to ordinary, actually used languages, though as a result, it lost its clarity and originality.

3.2 Radical Empiricism

While radical conventionalism emphasized the cognitive role of language, in later years Ajdukiewicz seemed to shift to the position of radical empiricism, minimizing that role. He pondered whether it was possible to construct a language governed exclusively by empirical rules of meaning. First, he pointed out the possibility of abandoning axiomatic rules [21], and then also deductive ones [26]. He concluded that although it had not been practised before, it was possible to acquire and express knowledge in a language free of all a priori components, in which even logic would be based on experience. He also noticed that constructing such a language required a reconstruction of the notion of meaning, which he managed to sketch only roughly [26]. This evolution of Ajdukiewicz's views can, however, be seen differently and interpreted as a shift from conventionalism not so much to empiricism as to metaconventionalism [37]. This is because we can choose the kind of language we need: is it supposed to be governed by all types of rules, or only by empirical ones? This choice determines not the meanings of expressions but the idea about what meaning actually is, the acceptance not of the sentences in question but of epistemological theses. Ajdukiewicz also drew close to empiricism by demonstrating that sentences traditionally considered to be analytical required resorting to experience, and specifically to existential premises [28]. He thereby initiated a lively debate in Polish philosophical literature on the notion of an analytic sentence.

3.3 Rationalism, Realism, Classical Logic

Of special importance for Ajdukiewicz were rationalist standards of pursuing philosophy. He shared the view, common in the Lvov-Warsaw School, that the necessary condition of cognitive rationality of problems one investigated was their communicability and intersubjective testability. He coined the term *anti-irrationalism*, which meant rationalism in a broad sense, including empirical methods, as opposed to narrow rationalism that excluded empiricism. He consistently espoused the classical notion of truth, the reality of the outside world, and also classical logic, upholding the principle of bivalence and ungradability of truth. He successfully defended the law of non-contradiction against ideologues who, in the postwar period, resorting to marxist principles, argued that the universality of movement and changes in the world led to inevitable contradictions [15].

3.4 *Semantical Epistemology and Metaphysics*

Although the criterion of intersubjective testability resembles the neopositivist criterion of sense, the Lvov-Warsaw School—in contradiction to the Vienna Circle—did not avoid traditional metaphysical problems. Ajdukiewicz reconciled metaphysical aspirations with the criterion of testability by using his method of philosophical paraphrase: his way of explaining philosophical concepts was meant to make notoriously vague metaphysical problems, expressed through those concepts, graspable and decidable. He tried to combine maximalistic aspirations with the "toolbox" of a minimalist. In his view, metaphysics could be based on epistemology (which he did not see as part of metaphysics). He formulated and pursued the following programme of semantic epistemology: all cognition manifested in language, therefore the study of cognition could be brought down to the study of its linguistic results, i.e. the sentences of the language in which they were expressed. Those sentences, their mutual connections and relationships with reality, were the domain of semantics, based on the achievements of logic. And semantics and logic provided well-founded conclusions. According to his method, only after having arrived at such beliefs about the nature of cognition could we draw metaphysical conclusions about the nature of existence.

In the early period of Ajdukiewicz's work semantics was riddled with antinomies. Therefore, the meaning-rule conception was not based on semantic concepts. Instead of the notion of truth, the notion of a thesis of language played an important role in it. However, Ajdukiewicz did not equate these two notions. On the contrary—as soon as semantics had been cleared of antinomies, he carefully differentiated them. Therefore, it is difficult to understand why later he encountered the ideologically motivated charge that he equated these concepts, and that he based metaphysics on views relativized to language, thereby allegedly turning out to be an idealist [39]. This criticism (which Ajdukiewicz did not neglect to respond to [36]) was especially poignant for him because the views imputed to him were not only fundamentally different from his own but also ones he himself argued against.

3.4.1 Polemic with Neo-Kantian Objective Idealism

It was precisely for the purpose of this polemic that Ajdukiewicz made a distinction between the concepts of truth and thesis [32]. The polemic itself was a model example of his application of the paraphrase method and the use of semantical epistemology in metaphysics. In his view, the main thesis of idealism—that the world did not exist independently but was a correlate of an objective spirit or the transcendental subject—was unclear, and its central idea required an explanation. At the time, idealists conceived of this superindividual subject not as some higher self but a system of ideas and judgments dictated by transcendental norms. These norms were supposed to contain the criteria of truth, while the world as a correlate of transcendence was supposed to depend on truth. This established an ontic order completely contrary to Ajdukiewicz's philosophical beliefs. In his opinion, criteria did not define truth and truth did not define the world. The truth was (on the basis of a given language) secondary to events in the world, and the criteria of truth were secondary to truth itself. To make the problem of idealism graspable, Ajdukiewicz interpreted transcendental norms as meaning-rules of language, and judgments constituting the objective spirit as language theses dictated by meaning-rules. If the notion of a thesis did not differ from the notion of truth, idealism paraphrased in such a manner would claim that the set of true sentences depended on linguistic rules. And as a consequence, that the world had to conform to the rules of language. In order to reject this view, Ajdukiewicz had to demonstrate that the notion of a thesis differed from the notion of truth. Then language rules would determine the set of theses but not the set of truths, and thus, not the external world. To do this, Ajdukiewicz resorted to Gödel's theorem, according to which in sufficiently rich languages (and such languages are used in science) one could formulate undecidable problems, that is, such questions that any answer, as well as its negation, were not theses. At the same time, by virtue of the law of excluded middle, one of the two contradictory sentences had to be true. Ajdukiewicz gave an example of an undecidable problem, pointing out that one of the possible answers (though which one, was not known) was true, but it was not a thesis. He concluded that not all truths were theses, and that the directives did not exhaust the richness of the world, therefore idealism was false. If this result may raise doubts, it is only in connection with the accuracy of the paraphrase of objective idealism.

3.4.2 Polemic with Subjective Idealism

By taking it up, Ajdukiewicz explained in more detail the programme of semantical epistemology [17]. Now, in his opinion, metaphysical conclusions could be drawn from reflection on cognition only when the investigated cognitive results were expressed in the language of semantics. Such reflection then addressed the relations of language to the outside world, somehow taking this world into account and offering a cognitive transition to its affairs. This, however, was not possible when such reflection referred only to intralinguistic relations, or possibly also mental phenomena, but did not touch the sphere independent of consciousness. There was then no transition to that sphere and one could not predicate anything about it. According to Ajdukiewicz, this was actually the kind of language, free of object-related expressions, that subjective idealists used: they spoke only about ideas and their perception, and this did not justify any beliefs about the real external world. Moreover, in contradiction to natural language, they redefined certain expressions (e.g. they conceived of the *body* as a system of ideas). The resulting equivocation created the impression that they used objective language, and so their conclusions about existence—although counterintuitive—seemed justified. In actual fact, all the reflections of subjective idealists took place in the sphere of experiences and sensations, without reaching external reality. Therefore, they could not even deny the existence of that reality. If, however, they uttered such sentences, it meant they were trying to restore object language. This would not do them much good, however, because it immediately violated the meaning-rules of the common language. After all, in the face of certain data, those rules led to the acceptance of such sentences as *these are cats*, and as a consequence, deductive rules required the acceptance of such sentences as *cats exist*. And so the idealists either did not get involved in the object language at all, and had no means to unequivocally deny the real existence of things, or did get involved in it but broke its rules by stating their views. Proponents of the existence of intentional entities were in a better position than the idealists [24]. This was because intentional existence was ascertained by empirical criteria (e.g. one had to check what Homer had written about), expressed in object language, the rules of which allowed distinguishing real existence from intentional existence.

3.4.3 Polemic with Reism

Ajdukiewicz's ontological interests also included the issue of universals [25]. Referring to T. Kotarbiński's reism, he noted that the ontological version of that view, according to which things existed but not universals, was incompatible with the semantic version, understood as a programme of not using apparent names, that is, names other than those of specific things [5]. This was because if reists wanted to express the negative part of their ontological argument, they had to use the apparent name *universals*. Granted, they did allow apparent names in sentences, from which those names could be eliminated without a change in meaning, but was this the case here? If the negative argument could be translated into a sentence only about things, it would be at odds with the reists' intentions. And if it could not, the reists would violate their own programme by uttering it. Ajdukiewicz had an impact on the reception of reism in the Lvov-Warsaw School. His critique was resisted only by a weakened version of semantic reism, which came down to the search for reistic substitutes of abstract names.

Moreover, Ajdukiewicz pointed out that natural language did not determine whether all names constituted one semantic category (and accordingly, whether everything existed in the same way) or general names differed categorically from individual ones (and accordingly, whether different entities might exist in different ways). He concluded that the process of language clarification could develop in various directions, thereby leading to different concepts of existence and different answers to the question about the existence of universals. He held against reism not that it had taken avail of one of those possibilities but that it had ascribed an absolute value to it. He himself preferred to use a language that made ontological room for universals, even though reistic language tempted him with the simplicity of some of its solutions. He stressed the necessity of relativizing ontological considerations to the language in which they were conducted. After all, depending on the language rules, the word *existence* could acquire different shades of meaning. Ajdukiewicz had thereby anticipated the ontology later proposed by Quine, which came down to the ontological commitments of language. In this sense, Ajdukiewicz's ontological interests fit into his broadly conceived research on the cognitive role of language.

3.4.4 Logic and Natural Language

Nevertheless, Ajdukiewicz did not see in logic the nostrum for all philosophical problems. He delimited the applicability of pure logic to philosophical issues [22]. For him, this limit was the necessity of validating each time the paraphrase of a philosophical problem in the language of logic. On the other hand, he criticized the view that logic and ordinary language were incompatible, proposing a pragmatic solution to the paradox of material implication [16]. What this paradox comes down to is that certain implicational sentences, true by virtue of the logical interpretation of the implication functor, are unacceptable, which seems to undermine this interpretation. Ajdukiewicz made a distinction between what the sentence asserted and what it expressed. He showed through examples that sometimes a sentence asserted a true state of affairs but expressed such conviction states of the person uttering them, about which we knew from context that he or she could not hold them. In Ajdukiewicz's opinion, that was the reason of the unacceptability of the aforementioned sentences.

3.5 *Investigation of the Syntax and Structure of Expressions*

Ajdukiewicz was the creator of categorial grammar and of the notation specifying the syntactic position of expressions, which he successfully used when addressing semiotic issues.

3.5.1 Categorial Grammar

The resolution of semantic antinomies resulted in Ajdukiewicz's work not only in semantic epistemology but also in taking up by him of the issue of syntactic meaningfulness of expressions. His categorial grammar—historically, the first grammar based exclusively on the structural properties of expressions—formulates the general criteria of syntactic coherence [33]. In other words, it indicates what predetermined steps should be taken with respect to any expression, so that the obtained results automatically show whether the words in this expression are assembled to form a meaningful whole. Ajdukiewicz took the notion of semantic category from Husserl and Leśniewski. He considered a category to consist of expressions that were exchangeable in any meaningful context. He distinguished categories of sentences, names, and functors. He divided functors into subcategories, depending on the category that the expressions those functors built belonged to and on the number and category of arguments those functors operated on. He developed a notation that assigned to each expression an index of its category. To sentences and names, he assigned simple indices (in the form of letters s and n), and to functors, indices resembling fractions, with the numerator containing the index of the expression that the given functor built, and the denominator containing the indices of all the arguments. To investigate the syntactic meaningfulness of any expression, one had to assemble the indices of all its component expressions, and then to "reduce" indices similarly to reducing fractions, i.e. if any index had its counterpart in the denominator of another index, both of those homomorphous symbols should be cancelled. An expression was syntactically coherent if and only if after making all possible cancellations we obtained a single index. It was not coherent if we were left with more than one index after all possible cancellations had been made. Thus, all sentences, names, and functors were meaningful, but not their accidental combinations.

3.5.2 Syntactical Position of Expressions

Apart from categorial indices, Ajdukiewicz eventually proposed another notation, which unequivocally determined the syntactic positions of all component expressions in a compound expression [30] (which was not the case with categorial notation). Syntactic symbols of individual words in a compound expression allowed a recreation of its structure even if those words were given in a random order. They gave the notion of a syntactic position a precise meaning, independent of common intuitions. Ajdukiewicz used this notation, when towards the end of his life he sketched a new conception of meaning. He assumed that the meaning of a compound expression was a relationship that ascribed the denotation of each of its component expressions to its syntactic position within the whole. This was based on a simple idea that to understand an expression, one needed to know which fragments of reality the individual words of this expression related to and in what way they had been connected with one another within it. Another late semiotic idea of his was that synonymous expressions evoked in language speakers thoughts that were identical in some essential respects, like the object reference of an expression or its emotional colouration [12, pp. 7–15]. He did not manage to develop this idea further—exactly which aspects of thought were significant in terms of the meaning of the experiences eliciting them, and which were not, and whether those insignificant ones were incommunicable, and thus non-rational. He also used the notation of syntactic symbols to analyze the relationship between the subject and the predicate of a sentence [34] and for sketching the method of eliminating intensional expressions [6, 18]. His aim was not to remove those expressions from the language but to reinterpret them in a way that would make them extensional. He concluded that the imprecise natural language allowed this reinterpretation, though it did not require it. His concept was to have the functor dissemble the expression not into a few compound arguments, but only into the simplest component expressions. The denotation of a sentence could change when a compound argument of the main functor changed to another one with the same denotation but not when arguments were taken to be only the simplest expressions (including their syntactic symbols within subordinate sentences). An analogous exchange on the lowest structural level retained the meaning of the whole and its denotation.

3.6 Methodology

Ajdukiewicz valued scientific knowledge highly and devoted a great deal of attention to it. His scientism was evidenced by reconstruction of scientific procedures, ideas aimed at improving them, and the postulate to base philosophical methods on scientific ones (which he himself pursued). He distinguished two disciplines dealing with science: metascience and methodology [13]. The first one was an exact theory of deductive systems. The second one belonged to the humanities and dealt with the activities of scientists and the purposes they seemed to pursue. It also attempted to understand the mechanisms of science and its developmental trends.

Science requires precise, specialized terminology and the ability to formulate problems and to find their valid solutions. Hence Ajdukiewicz's interest in the theory of definition, theory of questions, and the issue of justification. His texts on methodology, innovative at the time of publication, do not differ much from the contents of contemporary textbooks of broadly understood logic. It had been due to his contribution, however, that these contents have been accepted, popularized, and developed further.

3.6.1 Theory of Definition

Ajdukiewicz returned to the issue of definition a number of times. He compiled matters relating to definitions, systematized conceptually the foreground of definition theory, and created its foundations. Taking the procedures used in formalized theories as a starting point, he came to the conclusion that not all aspects of the definition theme were reflected there. He distinguished real definitions (that unambiguously characterize an object), nominal definitions (that give translation rules), and arbitrary definitions (or meaning postulates). He pointed out that these three intersecting notions of definitions could not be contained in a uniform general theory, because they were so different in terms of their intension that a superordinate *genus* did not exist for them [1].

It is not possible to determine the general form of even just nominal definitions, as they have to be relativized to their languages, and the latter differ in terms of meaning rules. Only some principles of constructing definitions apply to all languages. Ajdukiewicz distinguished analytic and synthetic definitions. He characterized types of definitions with different structures. He recapitulated and collected the conditions of correctness and usefulness of definitions: formal (e.g. conditions of consistency, non-creativity, existence, and uniqueness), semantic (e.g. adequacy condition), and pragmatic (e.g. avoiding the fallacy of *ignotum per ignotum*). He noted that not all kinds of definitions gave the translation of the defined term directly, but under certain formal conditions they allowed translatability. He showed how to formulate definitions to ensure translatability.

He did not share the view of nominalists who, denying the existence of species, considered all definitions nominal. He believed that such an approach left no room for inquiry into what certain things were, limiting attention to what their names meant. He noticed that at an early stage of development of a scientific discipline, primarily criterial definitions were formulated. Then, as research moved deeper into phenomena, criteriality gave way to grasping the essence of things, understood as those properties of things, from which, based on natural laws, the other properties could be deduced. In this way, the conceptual framework was gradually penetrated by knowledge of empirical origin. This corresponded to Ajdukiewicz's observations that conceptual apparatuses tended towards rationalization (i.e. the process of making originally empirical problems decidable in a purely conceptual sense, by way of language modification) and that empirical components did exist in sentences traditionally considered analytic, in particular in arbitrary definitions.

3.6.2 Theory of Questions

Ajdukiewicz proposed a useful terminology that referred to Twardowski's proposals and proved convenient for the logical analysis of questions and answers [19]. He defined the range of the unknown of a question, positive and negative assumptions of questions, as well as didactic, captious, and suggestive questions. He distinguished decision questions and complementation questions; questions posed properly and improperly, and proper, improper, and correcting answers; partial, complete, and exhaustive answers; and direct and indirect answers.

3.6.3 Rationality of Inferences and Foundation of Sentences

In the time when induction logic was taking its first steps, Ajdukiewicz tackled the problem of the rationality of fallible methods of inference [29]. He considered a fallible inference rational when in the long run, actions based on it brought more profits than losses, that is, the degree of subjective certitude with which the conclusion was accepted did not exceed the degree of reliability of the scheme of the applied inference. He assumed the degree of acceptance of the conclusion could be calculated based on behavioral criteria, and that the degree of infallibility of the scheme was the relative frequency of achieving true conclusions based on it, providing the premises were true (he was aware this explanation was incomplete because in practice the data for calculations was unavailable). He pointed out that whereas previously the rationality of inference had been based on probability theory and had relied only on a sense of obviousness, he demonstrated the correctness of such an approach in view of a general pragmatic theory of rationality, associated with the balance of profits and losses.

However, Ajdukiewicz did not see in this solution a way leading to the formulation of criteria of justifying sentences. On the contrary, in the field of justification he perceived still unsolved problems that others did not. He came to the conclusion that the very notion of justification, central in methodology, was unclear. At the same time he admitted that in science, justification or rather the sense of justification was a fact. He outlined a methodological programme aimed at explicating that notion [27]. In his view, even though methodology had reconstructed various types of induction, we still did not know what distinguished inferences approved by scientists, since after all, they did not consider each generalization justified. This would have to be investigated, and based on this, the notion of justification recreated. As far as deductive sciences were concerned, it would involve determining what conditions imposed upon assumptions and rules of transformation would guarantee justification of secondary theorems. From a broader perspective, such a programme would proceed to a reconstruction of the scientific method and to the determination of its capabilities and limitations. Ajdukiewicz returned to the issue of justification many times, taking up more detailed problems.

3.6.4 Classification of Reasonings

He performed a methodologically useful classification of reasonings, combining their various divisions: into simple and complex ones; into deductive and nondeductive ones, and among the latter, into rational and logically valueless ones; further, into those based on inference and those based only on hypothetical derivation of given sentences from others; into spontaneous and task guided ones, and among the latter, into those directed by decision questions, by complementation questions, and by the task of performing a proof. He took a critical view of the earlier classification elaborated by Łukasiewicz and modified by Czeżowski, which in his opinion missed part of the logical reality [8].

3.6.5 Classification of Sciences and Axiomatic Systems

Regarding types of inferences, Ajdukiewicz distinguished deductive and nondeductive sciences, and regarding ultimate premises that were not subject to justification, he distinguished deductive sciences based on axioms, empirical sciences based additionally on observational sentences, and the humanities, based additionally on interpretations of other people's broadly understood utterances. He specified stages of development of deductive sciences (preaxiomatic, intuitive axiomatic, and formalized axiomatic). Actually, the very beginnings of his activity involved reflection on axiomatic systems, thus paving the way in Poland for the development of deductive disciplines [10]. Only later did he turn from metascience to methodology (and even though he had considerable achievements in the latter, in the last years of his life he expressed regret about not having been involved in it enough earlier on). In the period of his metascientific investigations he formulated structural definitions of the notions of logical proof and logical consequence, and got involved in proofs of consistency and the concept of existence in mathematics; he pointed out how to strengthen the foundations of contemporary predicate calculus in order to derive traditional syllogistic from it; and he investigated the conditions of reversibility of premise and consequence.

Later on, he classified axiomatic systems, combining two of their divisions: into reductive and deductive ones, and into hypothetical (i.e. neutral) and assertive ones [14]. He performed a critical analysis of the ways of justifying axioms in assertive-deductive systems, and presented the possibility of limiting deductive sciences to uninterpreted, formalized neutral systems. He saw hypothetical-reductive systems in empirical sciences that passed from consequences to premises and at the same time aspired to the formulation of precise theories. Such a methodologically uniform classification partially eliminates the gap between deductive and empirical sciences.

3.6.6 Reconstruction and Evaluation of Scientific Procedures

Ajdukiewicz reconstructed such procedures as generalization of observations, testing of hypotheses, formulation of theories, or revision of principles [12]. He distinguished various types of natural laws, according to the dependencies they described. He devoted particular attention to statistical laws. For various kinds of fallible reasonings, he presented a method for calculating the degree to which assumptions increased the probability of the conclusion on the basis of accepted knowledge. He came up with a number of interesting theorems about the impact of evidence on the degree of validation of hypotheses and generalizations. They allowed a comparative evaluation of the degree of validation of laws, but not an absolute evaluation (and here Ajdukiewicz saw the possibility of further research). He proved the rationality of strategies usually used in science for increasing the degree of validation of laws.

He critically examined ways of justification adopted in the empiricist tradition. He considered justification that appealed to observations unscientific because direct observation was neither intersubjective nor reproducible (another observation, although similar, was never the same as the previous one). The scientific method, however, required intersubjectivity and repeatability. Scientificity began with generalizations that—paradoxically—were based on an unscientific foundation. However, Ajdukiewicz did not join the critics of direct justification, who referred to the theory ladenness of observations. On one hand, he minimized the role of terminological conventions, and on the other, he emphasized it. He minimized it by demonstrating that for justifying analytic sentences, conventions required supplementing with existential premises. He emphasized it by demonstrating that problems in science, which were seemingly solved arbitrarily (e.g. *what time measurement is really accurate?*), were based on implicitly accepted conventions that specified the meaning of certain vague expressions (e.g. *a reliable time measurement device*), thereby assuring a precise meaning, graspability, and decidability of certain problems (e.g. *are two given time periods equal?*).

Following Twardowski, Ajdukiewicz understood the humanities antinaturalistically. He categorized them as idiographical ones (e.g. history), nomothetical ones (e.g. psychology), and evaluative ones (e.g. cultural studies). The evaluative ones were supposed to reveal the values being the goal of human actions, and to evaluate the effectiveness of those actions. In psychology, contrary to the neo-positivists, he rejected physicalism and psychophysical reductionism but accepted introspection.

3.7 Axiology

Ajdukiewicz did not focus on axiological problems only in the context of methodology of the humanities. His metaethical views were close to intuitionism, although he considered the concept of intuition vague and preferred to talk about feeling or conscience [4]. He considered values objective and knowable through feelings. He objected to defining values in terms of feelings. This, he wrote, would resemble defining the properties of things in terms of sense impressions. In his view, empirical properties and values were primary, and the reception of both secondary. He saw also other analogies between axiological and empirical cognition: in both cases one began with direct experience and in both cases the cognitive faculties—whether feeling or the physical senses—were sometimes fallible. Then experience was generalized and principles (moral or empirical) formulated. After that, in both areas individual sentences (resp. judgments or descriptions) could be justified directly or indirectly, by inferring them from principles.

Ajdukiewicz was interested in psychology and human behaviour, and he characterized a number of important concepts in that area. Prudence and the ability to reflect were the traits in people he valued more than spontaneous abandon. In terms of seeking satisfaction in life, he encouraged keeping a healthy balance between the pursuit of specific goals and the enjoyment of one's current activity [9]. He performed a thorough examination of the notion of justice [3].

One of his concerns was education in logic and philosophy. He called for the promulgation of logical culture and presented concrete ideas how to go about it. He authored a few logic and philosophy textbooks designed for students of various levels.

4 Disciples and Continuators

Ajdukiewicz was a great animator of scientific life. He initiated many kinds of research, sketching programmes of their further development. Some of the problems he had addressed were later continued by his disciples, many of them prominent in their own right. He supervised Master's or PhD dissertations of the following persons: Zygmunt Schmierer (who did not survive the war and could not develop his interests), Stefan Swieżawski (metaphysics, history of philosophy), Jerzy Giedymin (methodology of history and social sciences, conventionalism), Roman Suszko (logic, epistemology) Henryk Skolimowski (analytic philosophy, axiology), Adam Nowaczyk (logical foundations of language and cognition). He also had an impact on the scientific development of such philosophers as Henryk Mehlberg (philosophy of mathematics and of empirical sciences), Maria Kokoszyńska-Lutmanowa (philosophy of science, methodology), Izydora Dąmbska (semiotics, history of philosophy), Seweryna Łuszczewska-Rohmanowa (theory of knowledge, language of science), Janina Hosiasson-Lindenbaum (probabilistic validation of fallible inferences), Klemens Szaniawski (logic of fallible inferences in the context of decision theory and the theory of rationality), Halina Mortimer (induction logic and its history), Jerzy Pelc (logical semiotics), Witold Marciszewski (pragmatic aspects of cognition, artificial intelligence), Marian Przełęcki (semantic reconstruction of empirical theories, metaethical intuitionism), and Ryszard Wójcicki (pragmatic reconstruction of empirical theories).

References

1. Ajdukiewicz, K.: Three concepts of definition. Logique et Anal. **1**, 115–126 (1958)
2. Ajdukiewicz, K.: Język i poznanie. Tom I. Wybór pism z lat 1920–1939 [Language and Cognition. Vol. 1. Selected Papers 1920–1939]. PWN, Warsaw (1960)
3. Ajdukiewicz, K.: O sprawiedliwości [On justice]. In: Język i poznanie. Tom I. Wybór pism z lat 1920–1939 [Language and Cognition. Vol. 1. Selected Papers 1920–1939]. PWN, Warsaw (1960), pp. 365–376 (Originally published in: Kultura i wychowanie **2**, 109–121 (1939))
4. Ajdukiewicz, K.: Postępowanie człowieka [Human conduct]. In: Język i poznanie. Tom I. Wybór pism z lat 1920–1939 [Language and Cognition. Vol. 1. Selected Papers 1920–1939]. PWN, Warsaw (1960), pp. 317–364 (Originally published in: Propedeutyka filozofii dla liceów ogólnokształcących, pp. 156–214. Książnica-Atlas, Lvov-Warsaw (1938))
5. Ajdukiewicz, K.: Reizm [Reism]. In: Język i poznanie. Tom I. Wybór pism z lat 1920–1939 [Language and Cognition. Vol. 1. Selected Papers 1920–1939]. PWN, Warsaw (1960), pp. 79–101 (Originally published in: Przegląd Filozoficzny **33**, 140–160 (1930))
6. Ajdukiewicz, K.: A method of eliminating intensional sentences and sentential formulae. In: Atti del XII Congresso Internazionale di Filosofia, Venice 1958, vol. 5, pp. 17–24. G.C. Sansoni, Florence (1960)
7. Ajdukiewicz, K.: Język i poznanie. Tom II. Wybór pism z lat 1945–1963 [Language and Cognition. Vol. 2. Selected Papers 1945–1963]. PWN, Warszawa (1965)
8. Ajdukiewicz, K.: Klasyfikacja rozumowań [A classification of reasonings]. In: Język i poznanie. Tom II. Wybór pism z lat 1945–1963 [Language and Cognition. Vol. 2. Selected Papers 1945–1963]. PWN, Warszawa (1965), pp. 206–225 (Originally published in: Stud. Logica **2**, 278–299 (1955))
9. Ajdukiewicz, K.: Pochwała życia pracowitego [Praise of a busy life]. In: Język i poznanie. Tom II. Wybór pism z lat 1945–1963 [Language and Cognition. Vol. 2. Selected Papers 1945–1963]. PWN, Warszawa (1965), pp. 77–82 (Originally published in: Wiedza i życie **17**(4), 307–312 (1948))

10. Ajdukiewicz, K.: From the methodology of deductive sciences. Stud. Logica **19**, 9–46 (1966). Trans. J. Giedymin (Originally published as: Z metodologii nauk dedukcyjnych. Wydawnictwo Polskiego Towarzystwa Filozoficznego, Lvov (1921))
11. Ajdukiewicz, K.: Problems and Theories of Philosophy. Cambridge University Press, Cambridge (1973). Trans. H. Skolimowski and A. Quinton (Originally published as: Zagadnienia i kierunki filozofii. Czytelnik, Warsaw (1949))
12. Ajdukiewicz, K.: Pragmatic Logic. Synthese Library, vol. 62, Reidel. Dordrecht-Boston (1974). Trans. O. Wojtasiewicz (Originally published as: Logika pragmatyczna. PWN, Warszawa (1965))
13. Ajdukiewicz, K.: Methodology and metascience. In: Przełęcki, M., Wójcicki, R. (eds.) Twenty-Five Years of Logical Methodology in Poland. Synthese Library, vol. 87, pp. 1–12. Reidel, Dordrecht-Boston; PWN, Warsaw (1977). Trans. E. Ronowicz (Originally published as: Metodologia i metanauka. życie Nauki **6**, 4–15 (1948))
14. Ajdukiewicz, K.: The axiomatic systems from the methodological point of view. In: Giedymin, J. (ed.) The Scientific World-Perspective and Other Essays, 1931–1963. Synthese Library, vol. 108, pp. 282–294. Springer, Dordrecht/Boston (1978). Trans. J. Giedymin (Originally published as: Systemy aksjomatyczne z metodologicznego punktu widzenia. Stud. Logica **9**, 205–218 (1960))
15. Ajdukiewicz, K.: Change and contradiction. In: Giedymin, J. (ed.) The Scientific World-Perspective and Other Essays, 1931–1963. Synthese Library, vol. 108, pp. 192–208. Springer, Dordrecht/Boston (1978). Trans. J. Giedymin (Originally published as: Sprzeczność i zmiana. Myśl Współczesna **8/9**, 35–52 (1948))
16. Ajdukiewicz, K.: Conditional statement and material implication. In: Giedymin, J. (ed.) The Scientific World-Perspective and Other Essays, 1931–1963. Synthese Library, vol. 108, pp. 222–238. Springer, Dordrecht/Boston (1978). Trans. J. Giedymin (Originally published as: Okres warunkowy i implikacja materialna. Stud. Logica **4**, 117–134 (1956))
17. Ajdukiewicz, K.: Epistemology and semiotics. In: Giedymin, J. (ed.) The Scientific World-Perspective and Other Essays, 1931–1963. Synthese Library, vol. 108, pp. 182–191. Springer, Dordrecht/Boston (1978). Trans. J. Giedymin (Originally published as: Epistemologia i semiotyka. Przegląd Filozoficzny **44**, 336–347 (1948))
18. Ajdukiewicz, K.: Intensional expressions. In: Giedymin, J. (ed.) The Scientific World-Perspective and Other Essays, 1931–1963. Synthese Library, vol. 108, pp. 320–347. Springer, Dordrecht/Boston (1978). Trans. D. Pearce (Originally published in: Stud. Logica **20**, 63–86 (1967). Translated by O. Wojtasiewicz from an unpublished typescript of a lecture in the United States in 1959)
19. Ajdukiewicz, K.: Interrogative sentences. In: Giedymin, J. (ed.) The Scientific World-Perspective and Other Essays, 1931–1963. Synthese Library, vol. 108, pp. 155–164. Springer, Dordrecht/Boston (1978). Trans. J. Giedymin (Originally published as: Zdania pytajne. In: Logiczne podstawy nauczania. Nasza Księgarnia, Warszawa (1934))
20. Ajdukiewicz, K.: Language and meaning. In: Giedymin, J. (ed.) The Scientific World-Perspective and Other Essays, 1931–1963. Synthese Library, vol. 108, pp. 35–66. Springer, Dordrecht/Boston (1978). Translated from German by J. Wilkinson (Originally published as: Sprache und Sinn. Erkenntnis **4**, 100–138 (1934))
21. Ajdukiewicz, K.: Logic and experience. In: Giedymin, J. (ed.) The Scientific World-Perspective and Other Essays, 1931–1963. Synthese Library, vol. 108, pp. 165–181. Springer, Dordrecht/Boston (1978). Trans. J. Giedymin (Originally published as: Logika i doświadczenie. Przegląd Filozoficzny **43**, 3–21 (1947))
22. Ajdukiewicz, K.: On the applicability of pure logic to philosophical problems. In: Giedymin, J. (ed.) The Scientific World-Perspective and Other Essays, 1931–1963. Synthese Library, vol. 108, pp. 90–94. Springer, Dordrecht/Boston (1978). Trans. J. Giedymin (Originally published as: O stosowalności czystej logiki do zagadnień filozoficznych. Przegląd Filozoficzny **37**, 323–327 (1934))
23. Ajdukiewicz, K.: On the meaning of expressions. In: Giedymin, J. (ed.) The Scientific World-Perspective and Other Essays, 1931–1963. Synthese Library, vol. 108, pp. 1–34. Springer, Dordrecht/Boston (1978). Trans. J. Giedymin (Originally published as: O znaczeniu wyrażeń. In: Księga Pamiątkowa Polskiego Towarzystwa Filozoficznego we Lwowie, pp. 31–77 (1931))
24. Ajdukiewicz, K.: On the notion of existence: some remarks connected with the problem of idealism. In: Giedymin, J. (ed.) The Scientific World-Perspective and Other Essays, 1931–1963. Synthese Library, vol. 108, pp. 209–221. Springer, Dordrecht/Boston (1978). Trans. J. Giedymin (Originally published as: W sprawie pojęcia istnienia. Kilka uwag w związku z zagadnieniem idealizmu. Stud. Philos. **4**, 7–22 (1949/50))

25. Ajdukiewicz, K.: On the problem of universals. In: Giedymin, J. (ed.) The Scientific World-Perspective and Other Essays, 1931–1963. Synthese Library, vol. 108, pp. 95–110. Springer, Dordrecht/Boston (1978). Trans. J. Giedymin (Originally published as: W sprawie 'uniwersaliów'. Przegląd Filozoficzny **37**, 219–234 (1934))
26. Ajdukiewicz, K.: The problem of empiricism and the concept of meaning. In: Giedymin, J. (ed.) The Scientific World-Perspective and Other Essays, 1931–1963. Synthese Library, vol. 108, pp. 306–319. Springer, Dordrecht/Boston (1978). Trans. J. Giedymin (Originally published as: Zagadnienie empiryzmu a koncepcja znaczenia. Stud. Filozoficzne **1**, 3–14 (1964))
27. Ajdukiewicz, K.: The problem of foundation. In: Giedymin, J. (ed.) The Scientific World-Perspective and Other Essays, 1931–1963. Synthese Library, vol. 108, pp. 295–305. Springer, Dordrecht/Boston (1978). Trans. J. Giedymin (Originally published as: Zagadnienie uzasadniania. Stud. Filozoficzne **2**, 3–16 (1963))
28. Ajdukiewicz, K.: The problem of the foundation of analytic sentences. In: Giedymin, J. (ed.) The Scientific World-Perspective and Other Essays, 1931–1963. Synthese Library, vol. 108, pp. 254–268. Springer, Dordrecht/Boston (1978). Trans. J. Giedymin (Originally published as: Zagadnienie uzasadnienia zdań analitycznych. Stud. Logica **8**, 259–272 (1958))
29. Ajdukiewicz, K.: The problem of the rationality of non-deductive types of inference. In: Giedymin, J. (ed.) The Scientific World-Perspective and Other Essays, 1931–1963. Synthese Library, vol. 108, pp. 239–253. Springer, Dordrecht/Boston (1978). Trans. D. Pearce (Originally published as: Zagadnienie racjonalności zawodnych sposobów wnioskowania. Stud. Filozoficzne **4**, 14–29 (1958))
30. Ajdukiewicz, K.: Proposition as the connotation of a sentence. In: Giedymin, J. (ed.) The Scientific World-Perspective and Other Essays, 1931–1963. Synthese Library, vol. 108, pp. 348–361. Springer, Dordrecht/Boston (1978). Trans. D. Pearce (Originally published in: Stud. Logica **20**, 87–98 (1967). Translated by O. Wojtasiewicz from an unpublished typescript of a lecture in the United States in 1959)
31. Ajdukiewicz, K.: In: Giedymin, J. (ed.) The Scientific World-Perspective and Other Essays, 1931–1963. Synthese Library, vol. 108. Springer, Dordrecht/Boston (1978)
32. Ajdukiewicz, K.: A semantical version of the problem of transcendental idealism. In: Giedymin, J. (ed.) The Scientific World-Perspective and Other Essays, 1931–1963. Synthese Library, vol. 108, pp. 140–154. Springer, Dordrecht/Boston (1978). Trans. J. Giedymin (Originally published as: Problemat transcendentalnego idealizmu w sformułowaniu semantycznym. Przegląd Filozoficzny **40**, 271–87 (1937))
33. Ajdukiewicz, K.: Syntactic connexion. In: Giedymin, J. (ed.) The Scientific World-Perspective and Other Essays, 1931–1963. Synthese Library, vol. 108, pp. 118–139. Springer, Dordrecht/Boston (1978). Translated from German by H. Weber (Originally published as: Die syntaktische Konnexität. Stud. Philos. **1**, 1–27 (1935))
34. Ajdukiewicz, K.: Syntactical connections between constituents of declarative sentences. In: Giedymin, J. (ed.) The Scientific World-Perspective and Other Essays, 1931–1963. Synthese Library, vol. 108, pp. 269–281. Springer, Dordrecht/Boston (1978). Trans. J. Giedymin (Originally published as: Związki składniowe miedzy członami zdań oznajmujących. Stud. Filozoficzne **6**, 73–86 (1960))
35. Ajdukiewicz, K.: The world-perspective and the conceptual apparatus. In: Giedymin, J. (ed.) The Scientific World-Perspective and Other Essays, 1931–1963. Synthese Library, vol. 108, pp. 67–89. Springer, Dordrecht/Boston (1978). Translated from German by J. Wilkinson (Originally published as: Das Weltbild und die Begriffsapparatur. Erkenntnis **4**, 259–287 (1934))
36. Ajdukiewicz, K.: My philosophical ideas. In: Sinisi, V., Woleński, J. (eds.) The Heritage of Kazimierz Ajdukiewicz. Poznań Studies in the Philosophy of the Sciences and the Humanities 40, pp. 13–33. Rodopi, Amsterdam-Atlanta, GA (1995). Trans. R. Harandon (Extensive excerpts from the original publication: W sprawie artykułu prof. A. Schaffa o moich poglądach filozoficznych [Concerning professor A. Schaff's paper on my philosophical views]. Myśl Filozoficzna **2**, 292–334 (1953))
37. Giedymin, J.: Radical Conventionalism, its Background and Evolution: Poincaré, Le Roy, Ajdukiewicz. Editor's Introduction to Giedymin, J. (ed.) The Scientific World-Perspective and Other Essays, 1931–1963. Synthese Library, vol. 108, pp. xix–liii. Springer, Dordrecht/Boston (1978)
38. Jedynak, A.: Ajdukiewicz. Wiedza Powszechna, Warsaw (2003)
39. Schaff, A.: Poglądy filozoficzne Kazimierza Ajdukiewicza [Philosophical views of Kazimierz Ajdukiewicz]. Myśl Filozoficzna **1**, 209–256 (1952)
40. Wójcicki, R.: Ajdukiewicz: teoria znaczenia [Ajdukiewicz: Theory of meaning]. Pró-szyński i S-ka, Warsaw (1999)

A. Jedynak (✉)
The University of Warsaw, Warszawa, Poland
e-mail: anna.jedynak@uw.edu.pl

On Ajdukiewicz's Project of the Semantic Theory of Knowledge

Adam Olech

Abstract Ajdukiewicz's the project of semantic theory of knowledge comes down to the statement that reflection upon logically understood concepts and judgments—which constitute the logically understood knowledge—is equivalent to the reflection upon terms and sentences whose meanings are those concepts and judgments. The value of this equivalence lies in the fact that it allows to apply the results obtained in metamathematics, treating of expressions and sentences, to epistemology, which pursuant to the mentioned equivalence is also treating of expressions and sentences, and which—similarly to mathematics—is also a metatheoretical discipline. What is more, this equivalence allows to maintain the connection of semantic theory of knowledge, that speaks about expressions and sentences, with traditional theory of knowledge, that speaks about concept and judgments understood in a logical way.

This equivalence was justified by Ajdukiewicz. Yet in his justification he did not refer to any philosophy of a language, including the philosophy of language and knowledge. Not every philosophy of language is a philosophical language reason of this justification, that is not every philosophy of language is a philosophical language justification of this justification. The one which is appropriate for this project is, in my opinion, Edmund Husserl's intentional theory of expressions from his *Logical investigations*. Indicating the appropriateness of this philosophy of language for this project is the main task which I undertake in this paper.

Keywords Ajdukiewicz's metaepistemological opinions · Logical understood concepts and judgments · Linguistic meanings of expressions · Concepts and judgments determined as to their contents · Husserl's intentional theory of meaning

Mathematics Subject Classification (2000) Primary 99Z99; Secondary 00A00

This paper is based on a part of research results, which have been presented in my monograph [39].

Á. Garrido, U. Wybraniec-Skardowska (eds.), *The Lvov-Warsaw School. Past and Present*, Studies in Universal Logic,
https://doi.org/10.1007/978-3-319-65430-0_5

1 Introduction

1.1 The semantic theory of knowledge as the meta-epistemological proposal and as its implementations

The name "semantic theory of knowledge" was introduced by Kazimierz Ajdukiewicz. He used it for the first time in a paper titled *A Semantical Version of the Problem of Transcendental Idealism* [*Problemat transcendentalnego idealizmu w sformułowaniu semantycznym*], in which he wrote that the semantic theory of knowledge is such a manner of practicing the theory of knowledge

> [...] in which epistemological problems are programmatically studied from the perspective of language as a system of expressions endowed with meaning. Its theses are formulated in such a way that they concern expressions, i.e. sentences and terms, but sentences and terms of a definite language which endows them with meaning.[1] [12, p. 142]

That paper is composed of two parts: a lecture explaining what the semantic theory of knowledge is, which lecture is *de facto* a meta-epistemological *credo* of this philosopher, and of an implementation of this theory consisting in analysis of the principal thesis of Heinrich Rickert's transcendental idealism, and in consequence in refuting this thesis as being contradictory to Gödel's first incompleteness theorem. I isolate this lecture from the entire paper, since it enjoys a specific, meta-epistemological autonomy, and I describe it as a "meta-epistemological project of a semantic theory of knowledge". This project (proposal) and its implementations—since apart from the one indicated above, there were others—are one of the most profound developments of the twentieth century analytical philosophy.

The proposal itself, in terms of volume, was presented on *two pages* of the abovementioned paper, and is *de facto* an introduction thereto. That I attach such importance to these two pages and focus my deliberations around them, results from the fact that I do not find in the entire philosophical-analytical literature an argumentation which would as convincingly support the thesis that cognitive questions should be approached from the language perspective.

Those Ajdukiewicz's papers which semiotically analyse traditional philosophical problems, in particular those whose subject is the dispute between epistemological and metaphysical idealism and epistemological and metaphysical realism are the implementations of the abovementioned meta-epistemological proposal. If an epistemological analysis of these problems was described as "semantic theory of knowledge", and exactly this name is used in this case, then the proposal of the semantic theory of knowledge can

[1] This paper is an extended version of his lecture, which he gave at the Third Polish Philosophical Congress in Cracow in 1936. A historical-terminological remark: the concept of "the semantic theory of knowledge" was used for the first time by Ajdukiewicz during that lecture and not in the abovementioned paper. One needs to remember however that using the term "semantics" in the 1930s, Ajdukiewicz understood this term in the same manner in which the term "semiotics" is understood contemporarily. If we therefore abstracted from this historical context and applied the contemporary terminology to Ajdukiewicz's semantic-epistemological studies, then we should describe these studies as "semiotic theory of knowledge", since Ajdukiewicz's semantic theory of knowledge engages all three components of logical semiotics, i.e. syntax, semantics and pragmatics.

be described as "an answer to the question, whether a semantic theory of knowledge is possible, and if yes, under what conditions".

The semantic theory of knowledge analyses traditional philosophical problems approaching them from the language perspective, and does so in a manner that makes it possible to apply logical tools, which is particularly important for a semantic epistemologist. Thus, the proposal of the semantic theory of knowledge is an answer to the question whether it is possible to approach knowledge from the language perspective, and if it is, how to do it, as knowledge understood in a traditional manner is composed of the cognitive subject's cognitive acts and their results. In other words, the proposal of the semantic theory of knowledge is an answer to the question, whether it is possible to translate a cognitive act and its result into a linguistic act and result, and it if its, how to do it, for this translation to be a translation as to the scope, and at the same a translation which perceives the language as a logically analysed medium.[2]

I therefore treat the question about the philosophy of language as interesting, or even as indispensable for the proposal of the semantic theory of knowledge, since the philosophy of language is fundamental for the discussed proposal, as well as for every meta-philosophical declaration characteristic for an analytical philosopher proclaiming that instead of concepts and propositions it is possible and advisable to speak of expressions, i.e. of names and sentences, whose meanings these concepts and propositions are. When presenting his proposal, Ajdukiewicz did not legitimize it from the point of view of philosophy of language and to be precise from the point of view of the philosophy of language and knowledge understood as cognitive acts and results connected with them, i.e. as cognitive creations. And I find this question, the question of philosophical-linguistic legitimisation of the meta-epistemological proposal, to be of key importance, which is reflected in this article. And a as philosophy of language adequate in this respect I indicate the philosophy of Edmund Husserl presented in the second volume of his *Logical Investigations*,[3] which—similarly to the solutions contained in both volumes of *Logical Investigations*—influenced the philosophical convictions of many representatives of the Lvov-Warsaw School, and apart from them—which is obvious—also the most

[2]Ajdukiewicz contends, speaking of the semantic theory of knowledge, that speaking of sentences is equivalent to speaking of propositions, which are the meanings of those sentences—c.f. [12, p. 141]. He therefore contends—I understand his pronouncement this way—that those two ways of speaking are a translation as to the scope. The context in which the equivalence claim appears, is as follows: "So, for example, sentences about the relation of consequence or about the relation of inconsistency, etc. between judgments are equivalent with sentences asserting suitable relations between the sentences whose meanings are those judgments" [12, p. 141]. The question which I am posing at this point, but to which I am not giving a definite answers is as follows: Perhaps we should be speaking in this case not only of a translation as to the scope, but also of a translation as to the content. What motivates me to ask this question is the fact that the philosophy of language underlying Ajdukiewicz's theory of knowledge is—of which I am deeply convinced—the Husserlian intentional theory of expressions. And from the perspective of this theory it is impossible to speak of the meanings of sentences, without at the same time speaking of sentences, and the other way round. What is more, it is impossible to speak of meanings of expressions, i.e. of the contents of acts understood *in specie*, without speaking of these acts, as well as of expressions, in which these acts are involved as their meaning intentions.

[3]In the English language edition of *Logische Untersuchungen Untersuchung* I (Investigation I) titled *Ausdruck und Bedeutung* [*Expression and Meaning*] is included in volume one, whereas in the original it is included in volume two. Referring to the original, I will be thinking of [17].

distinguished Husserl's student, i.e. Roman Ingarden. This is the philosophy that makes it possible to reasonably speak of a close link between thinking and speaking, which I wish to particularly emphasize, and provides thereby a philosophical-linguistic legitimisation for the meta-epistemological proposal. And by legitimizing this proposal, it makes it possible to demonstrate the links between the traditional theory of knowledge with the semantic theory of knowledge.

The fact that I perceive Husserl's philosophy of language as underlying Ajdukiewicz's meta-epistemological proposal and its implementation, i.e. underlying the semantic theory of knowledge, is not only an assumption of mine, but—which I try to demonstrate—an assumption that is fully justified. Although, when justifying his proposal, Ajdukiewicz does not rely on Husserl's philosophy of language, nonetheless when speaking of knowledge determined as to its content and of knowledge undetermined as to its content, and the concepts of knowledge understood in this way are of key importance for this proposal, he makes a reference to the "spirit" of Husserl's deliberations contained in *Logical Investigations*, and, which is symptomatic, Ajdukiewicz makes use of a similar example as was used by Husserl in Investigation V when he spoke of the intentional and semantic essence of an act, which concepts correspond to Ajdukiewicz's concept of "knowledge determined as to its content".[4] Even if in the fragment of his paper which discusses knowledge determined as to its content (and knowledge understood in this manner is based on the close link between thought and language) Ajdukiewicz makes use of Husserl's establishments in an implicit manner, yet 3 years later in a paper titled *Język i znaczenie* [*Language and Meaning*], when writing in another contest about the link between thought and language, he explicitly relies on Husserl, and to be more precise: on Investigation I: *Expression and Meaning* from volume one of *Logical Investigations*.[5]

1.2 The nominalistic guise of the meta-epistemological project

The project of semantic theory of knowledge originated not only in the Husserlian and anti-psychological climate, but also in the nominalist and logical climate. The latter climate in Polish philosophy was created by such philosophers as Jan Łukasiewicz, Stanisław Leśniewski, Tadeusz Kotarbiński, Alfred Tarski. And this other climate also needs to be borne in mind, if ones speaks of the semantic theory of knowledge understood both as a proposal and as implementations of this proposal, and this climate was postulated, in order to present philosophical problems in a nominalist form. This is what

[4]Fragments of Ajdukiewicz's and Husserl's work, which are convergent in this respect, are: [12, pp. 142–143] and [19] (Investigation V: *On intentional experiences and their 'contents'*, §20 *The difference between the quality and the matter of an act* and §21, *The intentional and the semantic essence*), pp. 121–125, as well as [18] (Vol. 2, Part 1 of the German Editions) (Investigation I: *Expression and Meaning*, §28 *Variations in meanings as variations in the act of meaning*), p. 223.

[5]See [8, p. 147] and/or [15, p. 37]. A sentence from this pages which is worth citing is: "The representation [of a sentence—A.O.] enters fully into the judgment-process and, indeed, forms its essential part. This has been convincingly demonstrated by Husserl". This sentence ends in a footnote referring to E. Husserl [17, Investigation I: *Ausdruck und Bedeutung*].

J. Łukasiewicz wrote about the nominalist form of contemporary logic, and what *mutatis mutandis* applies to analytical philosophy pursued in a logical manner:

> Contemporary logistic has a nominalistic guise. It refers not to concepts and judgements, but to terms and propositions, and treats those terms and propositions not as *flatus vocis*, but—having a visual approach—as inscription having certain forms. In accordance with that assumption, logistic strives to formalize all logical deductions, that is, to present them so that their agreement with the rules of inference, i.e., the rules of transforming inscriptions, can be checked without any reference to the meanings of the inscriptions. [27, p. 198] and/or [32, p. 222]

This is what Łukasiewicz wrote, but then he added that the nominalist form is indeed only a form, since the language of logic is impossible to be treated consistently in a nominalist manner, without encountering insurmountable difficulties thereby. Consequent nominalism requires for example that the language be treated finistically, i.e. it allows one to operate on a finite number of inscriptions only, despite the fact that both in case of artificial languages as well as in case of natural languages one needs to take into account an infinite number, in case of the latter group of languages—at least a potentially infinite number. Nominalism requires also—let us add—that expressions be treated only as expressions-specimens, and does not allow expressions-types, yet it is impossible to manage without the latter.

Łukasiewicz, unlike Leśniewski, Kotarbiński and Tarski, was not always a nominalist, after all in the 1930s he no longer was one, nonetheless each of these philosophers and logicians, irrespective of the fact whether he was a factual nominalist, presented philosophical and logical-philosophical questions in a nominalist way.[6] And it is this presentation made in a nominalist way that makes it possible to analyse these issues with the use of the contemporary logical apparatus, in which one speaks of expressions, and not of concepts and propositions, which were handled by traditional logic, and which were and are handled by the traditionally conceived theory of knowledge.

If we were to elaborate on this compactly put postulate, ordering us to present philosophical questions in a nominalist form, and to adjust it to the needs of epistemology understood in an anti-psychological manner, this postulate could be as follows:

> speaking of products of cognitive acts needs to be replaced by speaking of expressions, including of sentences of language L, since the logical-linguistic ways of understanding them are not entangled in disputes, in which products of cognitive acts constituting the meanings of expressions, i.e. logically understood concepts and propositions, are entangled. What is more, such presentation of the issue is heuristically promising, since it provides a possibility to use in epistemology results obtained in meta-mathematics, i.e. in a meta-theoretical discipline analogous to epistemology. And

[6]At some point Łukasiewicz was a nominalist, however his nominalism resulted from his philosophical immaturity. This is what he wrote on nominalism in an article [28] from 1937: "I will quite frankly admit that if someone had asked me not that long ago whether as a logistician [i.e. a logician—A.O.] I professed to nominalism, I would have given an affirmative answer without hesitation, since I had not been giving much thought to the nominalist doctrine itself, and I only focused on logistic practice. And so, logistics aspires to the greatest accuracy possible, and this accuracy can be attained by construction of a language that would be as precise as possible.[...] [Yet] is this strive for language precision and formalisation nominalism already? I do not believe so. Logistic would assume a nominalist standpoint, if it treated names and sentences only as signs of a certain form, without caring whether these signs meant anything. Logistic would then become a study of some ornaments or figures, which we draw and order in accordance with certain rules, playing with them as if with a game of chess. It would be impossible for me to accept this view. [...] I could no longer assume today a nominalist standpoint in logistic. But I am saying this as a philosopher and not as a logician. Logic cannot resolve this problem, since it is not philosophy" ([29, pp. 212–214] and/or [31, pp. 239–241]).

it is justified to use these results, if one demonstrates that the products of cognitive acts, described by non-psychologically understood epistemology, are meanings of expressions of language L, in which the cognitive subject verbalizes his cognitive acts. In such case, speaking of expressions of language L, in which cognitive acts are verbalized, will be tantamount to speaking of products of these acts.[7]

I contend that the proposal of the semantic theory of knowledge was an implementation of the above postulate. What is more, it was the fullest and the most meaningful implementation. Contending the above, I do not forget that Ajdukiewicz was at the same time far from factual nominalism, which is demonstrated by Ajdukiewicz's appreciation that he always had for the Husserlian theory of language and his repeated defence of the universals (Cf. the same author on this: [1, 7, 13]).

2 Presentation of the Meta-Epistemological Project of the Semantic Theory of Knowledge and Analysis Thereof

2.1 On the necessity to approach knowledge from the perspective of language

The theory of knowledge can be pursued as a theory of cognitive acts or a theory of objective cognitive contents. Objective contents can be understood in various ways—Kazimierz Twardowski understood them as products of cognitive acts, Bernard Bolzano—as *Begriffe* and *Sätze 'an sich'*, Gottlob Frege—as *Sinne* and *Gedanken*, Edmund Husserl—as expression meanings (*ausdrückliche Bedeutungen*), and also as meanings 'in themselves' (*Bedeutungen 'an sich'*), and others—as creations of an objective sprit.

If the theory of knowledge is to be a theory of objective and at the same time *determined cognitive contents*, a theorist of knowledge must approach knowledge from the perspective of language, and this means that he needs to treat logical concepts and propositions, which comprise objective cognitive contents, as linguistic meanings of terms and sentences. And the theory of knowledge understood in such way is identical to the semantical theory of knowledge, which does not directly describe such concepts and propositions, but describes them indirectly, since it directly describes expressions, whose logical meanings are those concepts and propositions.[8]

---

[7]To emphasize the significance which I attach to these contentions, I am writing down this postulate using a different font.

[8]The contents described in this paragraph is the matter of the meaning intention act together with the quality of this act. These two components of the act understood in such way constitute its meaning essence, and the meaning essence understood *in specie* is the logical meaning of the expression—the expression in which this act is involved. The meaning essence of an act ascribing meaning to an expression is an immanent content of the act, i.e. the subjective, empirically-real content. Yet this essence understood *in specie*, i.e. the product of its idealising abstraction, gives us the ideal content, i.e. logical, objective content. I will return to the issue which I am signalling right now at the end of this chapter. And the fact that when I speak of cognitive contents I understand them in the Husserlian way is justified by the fact that more or less clearly this is the way they were understood by Ajdukiewicz, as well as by the fact that such understanding of cognitive contents makes it possible to justify the strict connection between

That we need to approach knowledge from the perspective of the language if we want to speak of knowledge determined as to its content, results from the fact that it is impossible to name a certain cognitive content, i.e. some designated logical concept or proposition, other than by characterising this concept or proposition as a linguistic (logical) meaning of a certain term or sentence. Relativization of concepts and propositions understood in such way, results in the fact that the theory of knowledge describing them is not a theory of knowledge is general, i.e. it is not a theory of concepts and propositions 'hanging in the air', as father J.M. Bocheński called them, but is a theory of knowledge determined as to its content. And if knowledge in logical sense is understood as identical to language meanings of expressions, *then each sentence which asserts something about concepts and propositions in a logical sense, corresponds to an equivalent sentence which asserts something about terms and sentences, whose meanings these concepts and propositions are, i.e. each sentence of the traditional theory of knowledge, describing knowledge understood production-wise, corresponds to an equivalent sentence of the semantic theory of knowledge.*[9]

Ajdukiewicz in the following manner justifies the necessity to approach knowledge from the perspective of language, i.e. he justifies in the following manner what I call the "proposal of a semantic theory of knowledge"—below I quote extensive fragments of this justification, which are at the same time the essence of the semantic theory of knowledge, i.e. the thing which is the subject of analysis in this paper:

> [...] It is impossible to name a given concept or judgment except by characterizing them as the meanings of certain terms or sentences. Prima facie, the contrary appears to be the case, namely that one does not have to study concepts and propositions indirectly through language. For to refer to certain determinate cognitions (with definite content) one uses such phrases as, for example, 'the concept of a triangle', 'the proposition that $2 \times 2 = 4$', etc. In such phrases one does not mention—or so it appears—any expressions whose meanings would be those cognitions. This is an illusion, however. In a phrase like 'the concept of a triangle', provided it is to serve as a singular name of a determinate concept, the word 'triangle' is not being used in the usual way as, for example, in the sentence 'a triangle is a plane figure', where 'triangle' denotes a class of geometrical figures; in the phrase 'the concept of a triangle' the word 'triangle' occurs *in suppositione materiali*, i.e. as its own name. For if we were to regard 'the concept of a triangle' as containing 'triangle' in the normal supposition, i.e. as a name of triangle, then that expression would have, from this point of view, the same syntactic structure as, for example, 'John's father'. 'John's father' denotes the only object, which stands in the fatherhood-relation to John. Similarly, other expressions of the same structure, provided they are singular names, denote a unique object which stands in a certain relation to the object denoted by the term in the genitive case. The relation in question is indicated by the noun in the nominative case. (In Russell's logical symbolism, so-called descriptive function '*R*'*x*' corresponds to expressions of this sort.) One may say, therefore, that John's father is identical with the (unique) object which stands in the relation of fatherhood to John. Now if the 'concept of a triangle' were to be regarded as having the same syntactical structure as 'John's father', then one would have to say that 'the concept of a triangle' denotes the only object which stands to triangle

thought and language, i.e. the strict connection between knowledge determined as to its contents with the logical (language) meaning of expressions, which I have already described in the Introduction.

[9]The traditional theory of knowledge is both about cognizing and knowledge, i.e. it is both about cognitive acts, as well as about products of these acts (speaking the language of actions and products of Kazimierz. Twardowski). If we—in the name of anti-psychologism—limited the subject of study of the theory of knowledge to products of cognitive acts, then we would receive the abovementioned equivalence of the traditional and at the same time anti-psychologically oriented theory of knowledge (speaking of concepts and propositions in the logical sense) and the semantic theory of knowledge (speaking of expressions of language *L*, whose meanings these concepts and propositions are).

in the relation of being its concept or—in other words—that the concept of a triangle is identical with the only object which stands to triangle in the relation of being its concept. By analogy, if we consider in a similar fashion 'the concept of a trilateral figure' one would have to say that the concept of a trilateral figure is identical with the only object which stands to the trilateral figure in the relation of being its concept. However, a triangle is the same as a trilateral figure. If, on the other hand, *a* is identical with *b*, then whatever the relation *R*, the only object which stands in the relation *R* to *a* is identical with the only object which stands in the relation *R* to *b*. Therefore, the only object which stands to a triangle in the relation of being its concept, i.e. the concept of a triangle, would be identical with the only object which stands to a trilateral figure in the relation of being its concept, i.e. the concept of a trilateral figure. This is not the case, however; the concept of a triangle is not identical with the concept of a trilateral figure (nota bene, if the expressions used here, 'the concept of a triangle', and 'the concept of trilateral figure' are to serve as singular names of certain determinate concepts and not as universal terms denoting classes of concepts which correspond to the symbolic schema '$\overrightarrow{R}$'x').

In order to talk about a determinate concept of a triangle, we must not use 'the concept of a triangle' as if the term 'triangle' in the normal supposition (as the name of triangle) occurred in it. Rather we ought to understand that term as containing the word 'triangle' in the material supposition, i.e. as the name of itself. Accordingly, expressions like 'the concept of a triangle'—if they are to be singular names of determinate concepts—ought to be used as abbreviations of expressions such as "the concept which constitutes the meaning of the term 'triangle' " i.e. as expressions each characterizing its designatum as an object which is the meaning of the term 'triangle'. If we want to name a determinate concept we ought, therefore, to write either "the concept which constitutes the meaning of the word 'triangle' " or—briefly—"the concept of 'triangle' " keeping in mind that the word 'triangle' is used here in *suppositione materiali*, i.e. as the name of itself. This does not mean that 'the concept of a triangle' with 'triangle' in the normal rather than material supposition, is not a grammatically correct expression. It is; one has to bear in mind though that then it is not a singular name of a determinate concept, but rather a universal term to whose extension belong all concepts which constitute the meanings of the terms co-extensive with the term 'triangle'.

The semantic theory of knowledge is thus intentionally on the same path which had been followed by the epistemologists (without their realizing it) whenever they talked about certain determinate concepts, judgments, etc., or, which should have been followed by them had they expressed their ideas more precisely. [12, pp. 142–143] and/or [5, pp. 266–267]

2.2 The analysis of the necessity to approach knowledge from the perspective of language

Below I present an analysis and a commentary to the above justification, which I will begin with elementary, but—as it will prove in the course of the analysis—indispensable distinctions:

1. The designatum of the *concept* of a triangle is a triangle. The designatum of the *concept* of a trilateral figure is a trilateral figure, and a triangle is the same as a trilateral figure. But:
2. The designatum of the *name* "concept of a triangle" is any concept referring to a triangle, and there are *many* such concepts. One of them is the concept of a triangle constituting the meaning of the name "triangle".
   The designatum of the *name* "concept of a trilateral figure" is any concept referring to a trilateral figure, and there are *many* such concepts. One of them is the concept constituting the meaning of the name "trilateral figure".

Both the first and the second name are general names—each of them has more than one designatum.

3. The designatum of the *name* "concept of 'triangle' " is the concept of a triangle being the meaning of the name "triangle", and there is only *one* such concept.
   The designatum of the *name* "concept of 'trilateral figure' " is the concept of a trilateral figure being the meaning of the name "trilateral figure", and there is only *one* such concept.
   Both the first and the second name are singular names—each of them has one designatum.
   However:
4. The denotation of the name "concept of a triangle" is a set of concepts of the same scope as the name "triangle".
   The denotation of the name "concept of a trilateral figure" is a set of concepts of the same scope as the name "trilateral figure",
   and therefore: these names have the same denotation, which is a multi-component set.
5. Denotation of the name "concept of 'triangle' " is a one-component set, which is the concept of a triangle being the meaning of the name "triangle".
   Denotation of the name "concept of 'trilateral figure' " is a one-component set, which is the concept of a trilateral figure being the meaning of the name "trilateral figure",
   and therefore: these names have different denotations.

I wish to emphasize a material difference between point (1) and points (2) and (3): in point (1) one speaks of the designatum of the *concept* of a triangle, whereas in points (2) and (3) ones speaks of the designata of the *names*: "concept of a triangle" ("concept of a trilateral figure") and "concept of 'triangle' " ("concept of 'trilateral figure' "). There also is a difference between point (2) and point (3), which is demonstrated by different use of the word "triangle": in the first case the word is used in the normal supposition—in point (2), and in the second case it is used in *suppositione materiali*—in point (3). And this makes the names in point (2) general names, i.e. names with multiple designata, and each of the designata is a concept referring to a triangle, but the names in point (3) are singular names—their designata are elementary concepts,[10] each of which is the meaning of the name put in the internal quotation marks. I will return to these distinction is further analyses, for now I will express them in the shortest manner possible:

The name "triangle" designates a triangle, the concept of a triangle also designates a triangle, but the name "concept of a triangle" designates any concept referring to a triangle, and the name "concept of 'triangle' " designates one concept, the concept of a triangle, which is at the same time the meaning of the name "triangle" and which refers to a triangle.

[10]In this case I am speaking of elementary concepts, and not of singular concepts. I call them elementary concepts since they are elements of sets, which in this case are general concepts. And so, for example the concept of a "triangle" being the designatum of the name "concept of 'triangle' ", is an element of a set of concepts with the same scope, i.e. of concepts each of which refers to a triangle. We cannot however say about the concept of a triangle that it is a singular concept. Singularity of a concept means, this is the language usus in this respect, that the concept has one designatum only. Yet, any triangle, and there are many of them, is a designatum of the concept of a "triangle". With respect to the name "concept of 'triangle' " we can indeed say that it is singular, since there exists only one such concept, which is the designatum of the name "concept of 'triangle' ", and which at the same time is the meaning of the name "triangle".

Let us therefore use the above distinctions in commentaries and analyses of the quoted justification: If we are anti-psychologists, then following Ajdukiewicz, we differentiate concepts in the psychological sense from concepts in the logical sense, as well as—propositions in the psychological sense from propositions in the logical sense; and bearing in mind the distinctness of both types of concepts and of both types of propositions, we include both the former and the latter into the scope of epistemological deliberations.[11] The former, i.e. the concepts, are respectively—psychological and logical meanings of nominal expressions; the latter—are psychological and logical meanings of sentences. By restrictive understanding of cognitive acts, i.e. by such understanding of cognitive acts which classifies only verbalized cognitive acts as acts deserving to be called "cognitive" and by acceptance of the Husserlian philosophy of language,[12] concepts in the psychological sense are identical to acts of meaning intention involved in intuitive presentations of nominal expressions, whereas propositions in the psychological sense are identical to acts of meaning intention involved in intuitive presentations of propositions. And consequently, concepts in the logical sense are identical to understood *in specie* intentional essences of acts involved in intuitive presentations of nominal expressions,

[11]The direct subject of deliberations of the semantic theory of knowledge are expressions whose logical meanings are concepts and propositions understood logically, and thus an indirect subject of deliberations of this theory are those propositions and concepts—deliberations concerning these propositions and concepts are equivalent to deliberations concerning expressions. The semantic theory of knowledge programmatically narrows the object of its interest to knowledge understood logically. This does not mean however that the author of the semantic theory of knowledge eliminates from the scope of epistemological deliberations knowledge understood psychologically, i.e. concepts and propositions understood psychologically. He never did so. In any case, it is impossible to speak of logically understood concepts and propositions in abstraction from their psychological counterparts, since they are, these logical units, ideally understood meaning essences of the acts of conceiving and judging involved in nominal expressions and sentences. This is how they were described by the author of *Logical Investigations*, and following this author, i.e. Husserl, Ajdukiewicz understood them in the same way. And the fact that he explicated them in a syntactic-pragmatic manner in his directive concept of knowledge is a separate matter, which is not mutually exclusive with the former. One should add one additional remark to the above: despite the fact that the semantic theory of knowledge directly describes expressions equipped in meaning, which means it is pursued based on language interpreted intentionally (but after the breakthrough caused by the semantic works of Tarski, the author of the semantic theory of knowledge included in a significant manner the semantic aspect of language into his conceptual apparatus), it needs to be born in mind that the author of the semantic theory of knowledge treats linguistic expressions as three-layer objects: the physical sign—the act of meaning intention involved in that act (being the psychological meaning of the expression)—the logical meaning (being the ideal understanding of the intentional essence of a meaning generating act). And since it is impossible to speak of an expression without the meaning intention act, it needs to be stated that this is a subjective-objectivist view of language expressions. The objectivist component is the expressed sign and the logically understood meaning; the subjective component is the act of meaning intention involved in the expression sign.

[12]Speaking of the Husserlian philosophy (concept) of language (expression or meaning) in this article, I always mean the one which this philosopher presented in *Logical Investigations*. One needs also to bear in mind that Ajdukiewicz accepted this philosophy throughout the entire period of his academic activity, and not only at the time when after his doctoral thesis he was going in 1913 for Göttingen, to Husserl and Hilbert, being under the impression of *Logical Investigations*, but also at the time when he was writing his last work [10], an unfinished work, interrupted by the author's death. ([10] was prepared for print by H. Mortimer and K. Szaniawski, *Foreword* was written by K. Szaniawski, the work was published 2 years after the author's death.) I have described the fact that Ajdukiewicz accepted the Husserlian concept of the language also as the author of [10] in my articles, e.g. [35, 40].

whereas propositions in the logical sense are identical to understood *in specie* intentional essences of acts involved in intuitive presentations of sentences.

Bearing in mind the above, and limiting oneself to the example of the *ambiguous* name "concept of a triangle" (written down without caring for potential internal quotation marks), one is able to indicate the following meanings, which are connected with this name:

1. If in the name "concept of a triangle" the word "triangle" is in normal supposition, i.e. if it is not put in quotation marks, then this name has the following form: "concept of a triangle", and as such it is a general name.
2. If in the name "concept of a triangle" the word "triangle" is in *suppositione materiali*, i.e. if this word in put in quotation marks, then this name has the following form: "concept of 'triangle' ", and as such it is a singular name.

Ad (1) Similarly as in the case of other names, also in this case we differentiate the objective references of names: "concept of a triangle" and "concept of 'triangle' ", from the meanings of these names. Both the objective references of these names—and when saying this I mean their designata—as well as their meanings are concepts, however these are not the same concepts.[13] Let us deal with the objective references of these names, since they, and not their meanings, are possible to be taken into account, when we analyse the question of knowledge both determined and undetermined as to its content, i.e. the question whose purpose it is to legitimize the principal thesis of the semantic theory of knowledge. And this thesis proclaims that instead of concepts and propositions understood logically, we can speak of expressions, whose meanings these propositions and concepts are. What is more: we not only can, but we ought to speak of expressions, if we wish to discuss knowledge determined as to its content. And this is tantamount to the assumption that we ought to speak of those concepts and propositions, which are the meanings of these expressions.

The name "concept of a triangle" is a general name, and its designatum is every concept which refers to a triangle, i.e. every concept whose designatum is a triangle; and there are many such concepts. The designatum of the name "concept of a triangle" is therefore: the concept of "triangle", the concept of "trilateral figure", the concept of "polygon with the sum of internal angles equal to 180°" etc. When, however, we speak of a concept, we could have in mind a concept in the psychological sense or a concept in the logical sense; thus the name "concept of a triangle" is not only a general name, but at the same time an ambiguous name—each of its meanings indicates, as its scope, a set of concepts in the psychological sense or a set of concepts in the logical sense.

A concept in the psychological sense is an act of consciousness of specified quality and specified matter, i.e.—in case of the latter—of specified sense of objective understanding. Unity of the both of them, i.e. unity of the quality and of the matter of the act, is the

[13]That concepts are meanings of these names results from the fact that the meaning of each name is a concept. However, the meanings of these names are concepts of a higher tier as compared to concepts being the designata of these names. And so, the name "concept of 'triangle' " has a meaning which is a concept, and it has a designatum which also is a concept. The latter, i.e. the designatum of the "notion of 'triangle' " is the meaning of the name "triangle"; the former is the meaning of the name "notion of 'triangle' ". And thus the meaning of the name "concept of 'triangle' " is a concept of a concept, to be precise: a concept of a concept of a "triangle".

intentional essence of the act, whereas a concept in the logical sense is the intentional essence of the act understood *in specie* or to put it in other words: its ideal abstraction. Speaking of concepts in the psychological sense, I obviously mean acts ascribing meaning to nominal expressions, i.e. acts of meaning intention involved in intuitive presentation of these expressions; the intentional essences of these acts are meaning essences of these acts. These essences understood *in specie*, i.e. the logical meanings of expressions—irrespective of the fact whether this is a nominal or a propositional expression[14]—are described by Husserl with the general name of "expression meanings" (*ausdrückliche Bedeutungen*) and juxtaposed with meanings 'in themselves' (*Bedeutungen 'an sich'*), which have not 'happened' to be meant by any expression.

Let us call the ideal abstractions of meaning essences of acts ascribing meaning to expressions—"ideal contents".[15] We will then say that if we spoke of the name "concept of a triangle", and took into consideration its logical meaning, then the designata of this name would be ideal concepts, i.e. ideal (logical) concepts, such as: concept of a triangle, concept of a trilateral figure, concept of a polygon with the sum of internal angles equal to 180°, etc., each of which refers to a triangle. In other words, speaking of the name "concept of a triangle", we would have in mind the *plurality* of its ideal contents. Thus to the question: what do we have in mind when we speak of the name "concept of a triangle" we should answer: we have in mind the *plurality* of these ideal concepts. And if this is so, then the name "concept of a triangle" *is not determined as to its content* or, to put it in other words, *it is not specified as to its content.*[16]

All of these ideal contents, which comprise the plurality of ideal contents, can be *linked* to a relevant nominal expression, and then the general name "concept of a triangle" will 'break up' into a plurality of singular names, each of which is determined as to its content. This linking will consist in that, that a particular ideal content, *by being linked to a relevant expression, shall become the meaning of this expression.* It therefore turns out that the abovementioned plurality of ideal contents is nothing else than the plurality of potential meanings of expressions determined as to their content.

What has been said so far with respect to this issue, can be expressed in the following manner: If in the name "concept of a triangle" the expression "triangle" is used in normal supposition,[17] then the name "concept of a triangle" will not be a singular name of the

[14]I understand a propositional expression as an expression whose logical (language) meaning is a proposition in the logical sense.

[15]This is exactly how E. Husserl describes them (and he adds that they can be also described as logical or intentional contents, as an intending sense or sense, or simply as meaning), when he juxtaposes them with subjective (immanent, empirically-real) contents, i.e. those on which the idealizing abstraction is carried out—cf. [18, pp. 196, 200, 202]. This immanent content would be meaning in the psychological sense (psychologically understood concept or proposition), whereas the ideal (logical) content would be meaning in the logical sense (a logically understood concept or proposition).

[16]An analogous reasoning can be carried out, if speaking of the name "concept of a triangle" we are taking into consideration its psychological meaning as well. Then instead of concepts in logical sentence, i.e. of ideal contents of cognitive acts, we will be speaking of concepts in the psychological sense, i.e. of immanent contents of these acts.

[17]In [12] which is analysed in this point of the deliberations, Ajdukiewicz uses two types of suppositions: the *suppositione materiali* and the normal supposition, and then—making a reference to the normal supposition—he uses the name "ordinary" and makes no comment on this. I believe that Ajdukiewicz's use of the name "ordinary supposition", in the context of the previously introduced—in the analysed

concept being the meaning of the name "triangle", but will be a general name, denoting an entire class (set)[18] of elementary concepts, each of which is a meaning of the name equivalent to "triangle", such as for example the name: "trilateral figure", "polygon with the sum of internal angles equal to 180°", etc.

Und thus, if an elementary concept being the meaning of the name "triangle" is the designatum of the singular name "concept of 'triangle' ", then the designata of the general name "concept of a triangle" are concepts being the meanings of a *possible* complex name in the form of: $a_1 \wedge a_2 \wedge \ldots \wedge a_n$, where:

1. any a_k is the symbol of a name designating a *partial* content, comprising the full content of the general concept of a triangle;
2. symbol "$\wedge$" is used to denote conjunction understood as a name-forming functor with name arguments;
3. n belongs to the set of natural numbers, whereas k is greater or equal to 1 and smaller or equal to n.

An example of such a complex name is: "a triangle **and** a trilateral figure **and** polygon with the sum of internal angles equal to 180° **and** ...". Account taken of the number of elementary concepts comprising the scope of the name "concept of a triangle" which is nearing infinity, i.e. the full content of the general concept of a triangle, also the number of components of the complex name in the form of: $a_1 \wedge a_2 \wedge \ldots \wedge a_n$ will be nearing infinity.

Conclusion: the name "concept of a triangle" does not mean, unlike the name "concept of 'triangle' ", a concept determined as to its content, that is—in other words—the concept of a triangle is not, unlike the concept of "triangle", a concept determined as to its content. And saying this, we bear in mind that the name "concept of 'triangle' " is an abbreviated form of the name "concept being the meaning of the name 'triangle' ".

Ad (2) The name "concept of 'triangle' " is a singular name. Its sole designatum is the concept being one of the ideal concepts comprising the abovementioned pluralities of ideal concepts—to be precise: it is the ideal content connected with the expression "triangle" and being the meaning of this expression. And since it is linked with the expression "triangle", therefore the fact of this link needs to be marked, by putting this word into special quotation marks, which—being an integral part of this expression—cause this expression to occur in *suppositione materiali*. Therefore, the name "concept of 'triangle' " is *de facto* an abbreviation of the expression "concept being the meaning of the expression 'triangle' ", and the name "concept of 'trilateral figure' " is an abbreviation of the expression "concept being the meaning of the expression 'trilateral figure' ", and the name "concept of 'polygon with the sum of internal angles equal to 180°' " is an abbreviation of the expression "concept being the meaning of the expression 'polygon with the sum of internal angles equal to 180°' ", etc. And thus names: "concept of

paper—distinction into *suppositione materiali* and the normal supposition, means that when he speaks of the normal supposition and he means the one which was described by Piotr Hiszpan as *suppositio personalis*, and which in a different terminology has been described as *suppositio simplex*, i.e. as regular (or ordinary) supposition or as simple supposition.

[18]In this paper I am using these terms interchangeably. Ajdukiewicz speaks in this case of a class of concepts (cf. the extensive fragment of his justification cited above concerning the indispensability of approaching knowledge from the language perspective).

'triangle' ", "concept of 'trilateral figure' " etc. are names whose designata are concepts determined as to their content, i.e. such concepts as: concept of "triangle", concept of "trilateral figure", etc. are concepts determined as to their content, but such names as: concept of a triangle, concept of a trilateral figure are not.

2.3 Husserl and the issue of being determined as to content (definite in content)

There is a passage in *Logical Investigations* pertaining to the issue of being determined as to content, which corresponds to what has been discussed above, but which moreover has a certain more general value—both ontologically and epistemologically. It is worth citing this passage, here it is:

> Everything that is, can be known "in itself". Its being is a being definite in content [*ist inhaltlich bestimmtes Sein*—A.O.], and documented in such and such "truths in themselves". What is, has its intrinsically definite properties and relations [...]. But what is objectively quite definite, must permit objective determination, and what permits objective determination, must, ideally speaking, permit expression through wholly determinate word-meanings. To being-in-itself correspond truths-in-themselves, and, to these last, fixed, unambiguous assertions. Of course, to be able to say all this actually, would require, not merely the necessary number of well-distinguished *verbal signs*, but a corresponding number of *expressions* having precise meanings—in the strict sense of expressions. We must be able to build up all expressions covering all meanings entering into our theory, and to identify or distinguish such meanings with self-evidence. [18, p. 223] and/or [17, p. 90]

This passage can be commented on in the following manner: Also with respect to a general concept, for example with respect to the general concept of a triangle, it is possible to say that it is determined as to its content, however this is a determinateness of another kind, namely a determinateness *in itself* and only *in itself*. On the other hand, with respect to an elementary concept, e.g. with respect to the elementary concept of "triangle" or with respect to the elementary concept of "trilateral figure", we will say that it is a concept specified as to its content for *cognitive subject S speaking language L*, and a cognitive subject always speaks some language. What is more, one is a cognitive subject, only if at the same time one is a linguistic subject—this is what an analytical philosopher would say. And this means that one is a cognitive subject, if his cognitive acts are at the same time acts involved in intuitive presentations of word creations, that is if his cognitive acts are acts ascribing meaning to expressions. This issue was understood in this way by Ajdukiewicz, who at the same time claimed that a convincing, phenomenological description of this issue had been provided by Husserl in the second volume of *Logical Investigations*.[19]

[19]See [14, p. 37, footnote 2]. This footnote refers to the abovementioned Investigation titled *Expression and Meaning* from Husserl's *Logical Investigations*. And this is this fragment from *Language and Meaning*, which ends in a reference to this investigation and which is written entirely in accordance with the spirit and the letter of the *Logical Investigations*: "Articulate judging takes place mostly (if not always) in reference to voiced or unvoiced speaking, viz. reading, writing, listening, etc. That is to say, articulate judging is a composite psychic process in which usually there can be discerned a more or less fragmentary intuitive representation of a word-image. This intuitive representation is then mixed with certain others (without analysis of the distinguishable components) into the unity of the articulate judging.

2.4 *What are the concepts and propositions "in themselves"?*

When an analytical philosopher equalizes concepts and propositions understood in the logical sense with the logical (linguistic) meanings of expressions, what he means are concepts understood as singular (elementary) objects, and not general concepts, and the analytical philosopher does so in the name of the postulate demanding determinateness of knowledge as to its content. This determinateness is assured by the above mentioned equalization, which at the same time is a close link between knowledge and language. Determinateness of knowledge understood in the above way constitutes also the linguistic objectification (in the sense of inter-subjectification) of knowledge, i.e. it makes knowledge universally important.

An analytical philosopher rejects concepts and propositions 'in themselves' since they are not meanings of actual expressions, but are at most meanings of possible expressions—analogous to the aforementioned *possible* complex name in the form of: $a_1 \wedge a_2 \wedge \ldots \wedge a_n$. General concepts understood this way are a class of possible meanings of a thinkable multi-element nominal expression, whose every element—being an argument of the name-forming conjunction functor—means one of many contents comprising a general concept and each of these contents is an elementary concept. Lack of acceptance for concepts 'in themselves' in this case means as much as negating a contention saying that these concepts make up knowledge understood in the logical sense, which at the same time has the value of objectivity in the sense of: *actual* inter-subjectivity, i.e. in the sense of: *actual* determinateness as to its content.

Strictly speaking, an analytical philosopher does not have to reject concepts 'in themselves' just because they are general concepts. He does not have to reject them also because in their content constitution they are transcendent with respect to language, which means that they are transcendent with respect to the subject. Such concepts can occur in an analytical philosopher's ontology, as long as they do not aspire to be actual knowledge; since—in an analytical philosopher's opinion—only knowledge determined as to its content deserves to be described that way.[20]

We consider it fallacious to characterize matters in such a way that in the cases above judging is linked to the sentence-representation simply on the basis of association. The representation enters fully into the judgment-process and, indeed, forms its essential part. This has been convincingly demonstrated by Husserl. An articulate judgment-process whose essential part is the intuitive representation of a sentence we shall call 'verbal judging'. (Here we leave open the question whether there are in general articulate but non-verbal judgings.) Scientific judgment-processes in mature form are always of the verbal sort" [14, p. 37].

The last sentence of this quotation deserves attention, since this sentence says the same thing that has already been said above, namely that a cognitive subject is cognitive in the strict meaning of the work "cognition" ("knowledge") if, and only if the cognitive subject is a linguistic subject. What is also noteworthy is the fact that the justification of this thesis has been taken by the author from the phenomenological linguistic descriptions of expressions and their meanings written by the father of phenomenology, E. Husserl, and yet the representatives of this philosophical belief do no acknowledge this thesis which perceives a cognitive subject as a linguistic subject (I have devoted the final remarks contained in point 7 of this article to this issue).

[20]With respect to the issue of the possibility of existence in an ontology of an analytical philosopher of concepts and propositions not being meanings of expressions, I would like to note the following thing: And so, in an article [6] from 1948 Ajdukiewicz distinguishes three meanings of the term

2.5 The general concepts as classes and as collectives

An analytical philosopher, when speaking of knowledge, can have in mind either possible or actual knowledge, whereby he is willing to ascribe epistemological value only to the latter:

(a) Possible knowledge—means knowledge undetermined as to its content, but which is possible to determine. Such knowledge is composed of logical propositions, which are possible meanings of sentences, and of general concepts composed of elementary concepts (elementary contents), each of which is a possible meaning of some nominal expression.
(b) Actual knowledge—means knowledge determined as to its content. Such knowledge is composed of logical propositions which are actual meanings of sentences and from elementary concepts being meanings of actual expressions, which are the components of general concepts.

Speaking of general and elementary concepts, what I mean are logical concepts, also known as ideal concepts. With respect to general concepts understood in such way I contended that they are a *class* of possible meanings of a thinkable multi-element nominal expression, each element of which—being an argument of the name-forming conjunction functor—means one of many contents *comprising the general concept*, and each of these contents is an elementary concept. Ajdukiewicz described general concepts as *classes* of concepts, and I would like to add that they are classes of elementary concepts.[21]

R. Ingarden, whose philosophy of language also conceives of such concepts, treated them, as one can suspect, as *wholes*, i.e. as *collectives*. He understood meanings of expressions as actualizations of *a part* of ideal senses contained in a general concept.[22] If

"science": the psychological meaning, by which science is composed of concepts and propositions in the psychological sense; the logical meaning, by which science is composed of concepts and propositions in the logical sense, being the logical (ideal) meaning of terms and sentences, and the ideal meaning, by which science is composed of concepts and propositions in the logical sense, to which it could have happened that they never became meanings of any expressions. Thus, the last understanding of the term "science" is composed of meanings 'in themselves' (the Husserlian *Bedeutungen 'an sich'*), to which it could have happened that they never became meanings of any expressions (the Husserlian *ausdrückliche Bedeutungen*). Although when speaking of ideal understanding of "science" Ajdukiewicz speaks only of ideally understood sentences, and not of ideal meanings 'in themselves', nonetheless speaking of such ideal sentences, i.e. *de facto* of sentences 'in themselves', is equivalent to speaking of meanings 'in themselves'; what is sure is that the ontic status of these beings is the same—they are for sure radically atemporal, unlike concepts and propositions understood psychologically, but also unlike concepts and propositions understood logically, being at the same time the logical meanings of expressions (*ausdrückliche Bedeutungen*), which concepts and propositions, being involved in expressions of a particular language L, are thus involved in time, i.e. in the temporality of this language and the history of science pursued in the language.

[21] Cf. [12, p. 143]. Ajdukiewicz writes in there that such expressions as "concept of a triangle" or "concept of a trilateral figure" mean entire classes of concepts. He does not write that they mean the entire classes of elementary concepts. The adjective "elementary" is my qualification in this respect used to designate a concept determined as to its content.

[22] Cf. [25, p. 139] and/or [24, Chapter V § 16]. Strictly speaking, Ingarden does not use the expression "collective", he does use the expression "part", however. And this is the signalled view of Ingarden, which needs to be quoted in full, since the matter of ideal concepts and of the relation between them and

he indeed conceived of ideal concepts as collectives, then I find Ajdukiewicz's, rather than Ingarden's, approach to be right, among others due to the fact that Ajdukiewicz's approach makes it possible to speak of content-deprived general concepts, and I am inclined to consider transcendental concepts to be such content-deprived general concepts.[23] Yet, it is impossible to reasonably speak of an empty whole or an empty collective, since from the perspective of mereology empty sets do not exist.

If one acknowledges that a general concept is a class of elementary concepts, one needs to acknowledge at the same time, that a general concept is an object of a logical type superior towards the elementary concept corresponding thereto, and in consequence, names which mean general concepts (e.g. the name "concept of a triangle") are of a higher logical type (are names of a higher syntactical category), than the names marking elementary concepts (e.g. the name "concept of 'triangle' "). And if this is so, then the conjunction being a name-forming functor with name arguments, as already discussed, needs to be understood in an enumerative sense, and not in a synthesizing sense. We would be dealing with synthesizing sense, if general concepts undetermined as to their content were collectives of elementary concepts determined as to their content.[24]

meanings of expressions, is one of the most important, if not the most important matter in the philosophy of language: "[...] m e a n i n g of the word 'square' within its material content contains c u r r e n t l y only a certain p a r t of what is contained in the c o n c e p t of a square, *resp.* in the idea of a square. On the other hand, the meaning of the expression 'rectangular equilateral parallelogram' contains c u r r e n t l y a n o t h e r part of the content of t h e s a m e concept, and namely the part thanks to which its object gets constituted through the multiplicity of ideal qualities equivalent to squareness" [25, p. 139, translated by A.O.].

[23] I write about transcendental concepts, as content-deprived general concepts in an article [34]. On this topic see also my article [37]. When I speak of the content emptiness of transcendental concepts, what I mean, which I have expressed in the abovementioned articles, is the objectivist understanding of content, i.e. such understanding by which the content of a given concept is composed of features inherent to all designata of a given concept, i.e. the characteristic features of the designata of that concept. Not going into details connected with this issue, in this footnote I will only state that account taken of all designata of the transcendent concept of "being", we will not demonstrate any such feature C, which could be common for all designata of the distributively understood concept of "being". And this means that the concept of "being" is content-deprived. Yet, the content described in these deliberations is the content connected with the act of consciousness, and to be more precise: the immanent content of the act understood *in specie*, or in other words, this is the content being the result of idealising abstraction carried out on the immanent content—and thus an ideal, objective content, which is also described as "the sense of the objective understanding" or in short as "sense". If however the transcendent concept of "being" is content-deprived, within the aforementioned meaning, i.e. such meaning where content is defined in categories of features, then it is also content-deprived within the second meaning of the term "part", in case of which speaking of content one means *the sense of objective understanding*. This is so, because features are also objects (but objects of a different tier than the objects they are inherent to), and being objects, they are thus objects of objective understanding, which is inherent to sense. Absence of a feature common for all designata of the transcendent concept of "being" entails thus the absence of the sense of objective understanding, *ergo*: the absence of logically understood content. At a side, the same feature C can be, as any other object, presented with the use of acts of various content, and thus various senses can refer to the same feature C—this is the main reason why we should differentiate between content specified in categories of features from contents understood logically, that is, in other words—from content understood as intentions of expressions. This issue, only signalled at this point, deserves to be treated separately.

[24] Conjunction understood as a name-forming functor with name arguments, in an expression "A and B are C" has: (1) enumerative sense, when both the object marked with "A" as well as the object marked

2.6 *Ajdukiewicz's expression meanings as Husserl's ausdrückliche Bedeutungen*

This *content*, being the focus of our deliberations in this paper, whose central theme is knowledge determined as to its content, is the matter of the meaning intention act stemming from *Logical Investigations*, which, as a dependent moment of the act, coexists together with another moment, inseparably connected with the latter and as dependant, i.e. with the quality of the act, described several years later by Husserl in *Ideas* as the "theoretical character of the act". Unity of them both, being an immanent content of an act, constitutes the intentional essence of the act. And the latter understood *in specie* is the objective, i.e. ideal content of the act.

Bearing in mind the influence of Husserl's philosophy's of language on the philosophy of language of the author of the semantic theory of knowledge, one must—speaking of these matters—use Husserlian terminology, and moreover one must, in relation to Sect. 2.1. of this paper and the footnote to that subparagraph, describing the subjectively and objectively understood cognitive content, refer to a relevant passage from *Logical Investigations*. This passage constitutes a philosophical-linguistic background and a philosophical-linguistic legitimization of the semiotic justification of the thesis, quoted at the beginning of this chapter, that if one wishes to speak of knowledge determined as to its content, one must necessarily approach it from the perspective of the language.

> Our distinction posited two sides in every acts: its quality, which stamped it as, e.g., presentation or judgement, and its matter, that lent it direction to an object, which made a presentation, e.g., present *this* object and no object. [...] One can readily see, In fact, that *even if quality and objective direction are both fixed at the same time, certain variations remain possible*. Two identically qualified acts, e.g. two presentations, may appear directed, to the same object, without full agreement in intentional essence. The ideas *equilateral triangle* and *equiangular triangle* differ in content, though both are directed, and evidently directed, to the same object: they present the same object, although 'in a different fashion'. [...] The matter, therefore, must be that element in an act which first gives it reference to an object, and reference so wholly definite that it not merely fixes the object meant in a general way, but also the precise way in which it is meant. [...] It is *the objective, the interpretative sense (Sinn der gegenständlichen Auffasung, Auffasungsinn)* [...]. In so far as quality and matter now count for us(as will be shown later) as the wholly essential, and so never to be dispensed with, constituents of an act, it would be suitable to call the union of both, forming one part of the complete act, the act's *intentional essence*. To pin down this term, and the conception of the matter it goes with, we simultaneously introduce a second term. To the extent that we deal with acts, functioning in expressions in sense-giving fashion, or capable of so functioning—whether all acts are so capable must be considered later—we shall speak more specifically of the *semantic essence* of the act. The ideational abstraction of this essence yields a 'meaning' in our ideal sense.[25] [19, pp. 121–123]

---

with "B" belong to the class of objects marked with "C" and when at the same time A and B are different objects; (2) synthesizing sense, if the objects marked with "A" and with "B" taken together create a new, different object marked with "C".

[25] During the Lectures on Logic Semiotic, which Ajdukiewicz gave at the John Casimir University of Lvov in Autumn 1930, he spoke in the same manner of the logical meaning of expressions. In the last lecture from that cycle, i.e. in Lecture XVIII given on 9 December 1930, he repeated a fragment of Investigation V from the second volume of *Logical Investigations*, which I have already quoted above. Lecture XVIII, being the last of the cycle, was a summary of that cycle, it therefore presented conclusions which Ajdukiewicz accepted, and one of the main topics, which these lectures were devoted to, was the

Meaning-generating acts, as mentioned above, are described by Husserl in *Logical Investigations* also as die Bedeuten or as "meaning intentions" (*die Bedeutungsintentionen*). The ideal content of these acts, obtained as a result of the abovementioned abstraction, is the expression meaning (ausdrückliche Bedeutung), meaning connected with the language, unlike the meaning 'in itself' (*Bedeutung 'an sich'*), which is extra-linguistic. Ajdukiewicz explicated these expression meanings in a semiotic manner and in accordance with the spirit of Hilbert,[26] as common properties of synonymous expressions in a given language L. Yet he did so not with the use of syntactic categories, but with the help of syntactic-pragmatic categories, since the latter, the pragmatic categories—describing acts of understanding of expressions or acts of acceptation of sentences—are capable of coping with explication of quality and matter of a meaning-generating act. Therefore, expression meanings—as described by Husserl in the above mentioned work, and as described by Ajdukiewicz after him—are understood *in specie* essences of meaning intention acts, i.e. they are—roughly speaking—meaning acts *in specie*. Individual meaning intentions fall under an ideal *species*, and this *species* is something that—as in Aristotelian conceptual realism—exists *in individuo* (cf. [21, p. 438]) within the scope of meaning-generating acts, i.e. within the scope of meaning intentions.

Expression meanings are those meanings 'in themselves' that 'happened' to be meant by a given expression, but apart from them there exist also such meanings 'in themselves' (*Bedeutungen 'an sich'*), which even due to our limited cognitive capabilities will never be expressed in meaning-generating acts, i.e. they will never

> [...] become real in human mental life. [18, p. 233]

This is because acts of meaning intention are their realization and not their *quasi*-realisation, as it is the case with the already mentioned views of R. Ingarden in this respect. What is ideal, objective and transcendent through operation of an intending act

---

issue of meaning of expressions. I edited and published Lecture XVIII as a component of my paper [38]. I quoted fragments of Lecture XVIII in article [40]. As to the answer to the question, what the logically (linguistically) understood meaning of expressions is, I would like to cite the first and the last sentence of this Lecture—here they are: "One of the best solutions in this case is what was done by Husserl ..." [38, p. 171]. Then Ajdukiewicz present the Husserlian understanding of the meaning of expressions, repeating what Husserl said in the above quotation from Investigation V, and this is followed by a conclusion, constituting the last sentence of this lecture, and at the same time the last sentence of the entire cycle of lectures on logical semiotic, which is as follows: "We can say now that the meaning of a word in such and such form is the meaning essence of thoughts, which must be involved in this word for this word to be used as a word of such and such language" [38, p. 172]. If one wanted to correct Ajdukiewicz, and it is necessary to do this at this point, then one would need to say that the logical meaning of an expression (and this is the understanding of meaning that he has in mind) is not the meaning essence of an act of meaning intention involved in this expression, but it is this essence understood *in specie* or, in other words, the ideal abstraction of this essence. This is so, because the essence of an act of meaning intention is the quality and the matter or this act, and those comprise what you can call the immanent content of the act, and the objective (ideal) content is the immanent content understood *in specie*.

[26]During the session of the International Philosophical Congress in Prague in 1934, Ajdukiewicz presented his most important work to that date in the following manner: "As the first person in Poland—as it seems—he formulated (under the influence of Hilbert) an idea of strictly formalized deductive study of structural (and thus abstracting from the meaning of words) directives of reasoning. Following this idea, he tried to define the meaning of words as a logical structure of certain relations existing between expressions of a given language, which apart from the sound of the expressions is necessary for a characteristic of the language" [2, p. 405, translated by A.O.].

falls within the scope of this act. And since without this act there is no expression, since the act makes an expression an expression, finding its support in a physical expression sign, then an expression is a physical-psychological-logical creation. Such understanding of expressions cannot affect the proposal of semantic theory of knowledge claiming that speaking of expressions is tantamount to speaking of logically understood concepts and propositions, which are the meanings of expressions.

The psychological components of these creations—are the contents of the meaning-generating acts, which are subjective (immanent) cognitive contents. Those, in turn, understood *in specie*, are objective cognitive contents and are identical to the logical meanings of expressions, and logical meanings are identical to linguistic meanings.[27] And so, by the aforementioned determination of cognitive acts in terms of acts of meaning intentions involved in expressions, the cognitive subject is a linguistic subject, to be precise: it is a part thereof, it is not identical therewith, since not every linguistic activity is a cognitive activity, but every cognitive activity is at the same time a linguistic activity.

The last sentence requires a commentary. An analytical philosopher, and I mean in this a case a philosopher who accepts the Husserlian theory of expressions, does not find cognitive acts to be identical to meaning intention acts, but for certain agrees with the thesis that each cognitive act is *de facto* an act of meaning intention involved in a sentence, which is a verbalization of an act of judgement. This is so, since only verbalized cognitive acts—and what is more only literally verbalized acts, which will be discussed below—deserve, in the opinion of an analyst, to be called cognitive acts, and a verbalized cognitive act is an act whose component is a more or less exact visual presentation of a word creation. And therefore, to be precise, the abovementioned thesis has—in another, equivalent wording—the following form: the scope of the concept "cognitive act" is subordinate to the scope of the concept of "linguistic act" or, to put

[27] I demonstrate the fact that Ajdukiewicz's directive theory of meaning defining meaning of expressions as a class of abstraction (common feature) of synonymous expressions is an explication of Husserl's intentional theory of meanings of expressions in an article [35] and [40]. The main topic of this article is to present the explicating sequence from the act of meaning expression *in specie* (*ausdrückliche Bedeutung*) up to the language expression as a common feature of synonymic expressions. The conclusion of this article—which I quote literally—is as follows: Ajdukiewicz's *explicatum* for the Husserlian *ausdrückliche Bedeutung* is a shared property of synonymous expressions, hence identical ones. Husserlian identity of species is Ajdukiewicz's synonymy. The possible sequence of expressions demonstrating that Ajdukiewicz's *explicatum* is translatable with no change of sense to Husserlian categories (and conversely, too, of course) would be as follows:

1. linguistic meaning—the same as shared property of expressions (The precise shape of this sentence would be this: the meaning of expression E in language L is the same as the shared property of synonymous expressions with this expression in language L. The remaining sentences of the sequence ought to be made more precise in a similar fashion, which I have given up for stylistic considerations), (if an expression—the same as to be used as an expression, then)
2. linguistic meaning—the same as shared property of the uses as expressions of language, (if used as an expression of language—the same as to be understood, then)
3. linguistic meaning—the same as shared property of the acts of understanding, (if an act of understanding—the same as an act of meaning-intention, then)
4. linguistic meaning—the same shared property of acts of meaning-intentions, (if a shared property of acts of meaning-intentions—the same as their essence, that is species, then)
5. linguistic meaning—the same as an act of meaning-intention in specie [40, p. 151].

it in other words, the scope of the concept "cognitive activity" is subordinate with respect to the scope of the concept "linguistic activity". Therefore, consequently, it needs to be said with respect to the cognitive subject that it is not identical to the linguistic subject, but is 'in the power' of the linguistic subject, or in other words, that the cognitive subject is a component of the linguistic subject.

2.7 *On relation between concepts: "cognitive act" and "linguistic act"*

When discussing these things, it is impossible not to mention the phenomenological standpoint in this respect—the thesis of a phenomenologist concerning the issue of the relation between the scopes of the concepts: cognitive act and linguistic act is that the scopes of these concepts intersect, which means that *only some* of the cognitive acts are linguistic acts, i.e. some cognitive acts are linguistic acts and some are not. Crossing of the scopes of these concepts, and to be precise their independence,[28] results from the fact that a phenomenologist believes, contrary to an analytical philosopher, that although some cognitive acts and their results are not verbalized, yet they have every right to be classified as cognitive. Cognitive acts are the acts and results of direct knowledge, specific for this philosophical belief.[29]

In this situation it could seem that the standpoint of an analyst and the standpoint of a phenomenologist are contradictory, since the following categorical sentence: Every *C* is *L* (where *C* symbolizes cognitive acts, and *L*—linguistic acts), expressing the thesis of an analyst, and the complex sentence: Some *C* are *L* and some *C* are not *L*, expressing the thesis of a phenomenologist, seem to be in a relation of contradiction, i.e. seem to be disjunctive. It turns out, however, that different ways of understanding of the cognitive act by the analyst and the phenomenologist result in the fact that this dispute is an apparent one, since each of them substitutes a different value for the name variable *C*. Distinctness of these values results from the different answers given to the following questions: What deserves to be called "a reason legitimizing knowledge"? Should we limit ourselves only to those reasons which result in knowledge having the feature of objectivity, understood as inter-subjective verifiability and communicatibility? Or should we perhaps, at the expense of objectivity understood in such way, valorise in knowledge an objectivity understood otherwise, namely as objective validity? In other words, what should an epistemologist value more: intersubjectivity of knowledge at the expense of a smaller number of truths, or should he valorise a greater number of truths at the expense of the intersubjectivity of knowledge? As it is easily noticeable, answers to these questions are not axiologically

[28] If scopes of two names (concepts) are crossing, and the logical sum of those scopes is properly contained in the class, within which their mutual relation is considered, then such crossing of scopes is referred to as "independence of names (concepts)". If however a logical sum of the crossing scopes of names (concepts) is equal to the class, within which their mutual relation is considered, then this relation is described as "sub-opposition of names (concepts)"—on this cf. [26, pp. 98–101].

[29] As to the non-verbality of cognitive acts and results in phenomenological research—see e.g. [20]. What deserves particular attention in this respect are also the lectures of R. Ingarden from 1948/1949 given at the Jagiellonian University, concerning the role of language in science (in knowledge)—see [23, in particular p. 95 et seq.].

neutral, and therefore the dispute between an analyst and a phenomenologist with respect to the role of the language in knowledge is not a dispute in the strict sense of the word, since, first, they understand the term "knowledge" differently, and second, the dispute as to the manner of understanding the term "knowledge" is not a dispute in the strict meaning of the word, since it is axiologically involved. A closer examination of this dispute would lead us to the conclusion, that each of them, the analyst and the phenomenologist, differently understands not only the cognitive actions and results, but also the issue of verbalization of cognitive actions and results, i.e. each of them differently understands the term "linguistic actions", which is involved in this dispute—for the phenomenologist these can be a kind of linguistic actions that are unacceptable for the analyst, since the analyst does not accept for example the language of comparisons and metaphors—this issue is tackled in the fragment from Ajdukiewicz's speech at the 1935 Paris Philosophical Congress, which I present below.

As to the aforementioned dispute in the strict meaning of the word, we would be dealing with it, if the standpoints of the disagreeing parties were possible to express with sentences in the logical sense and if these sentences were in the relation of contradiction or in the relation of exclusive or ordinary alternative or in the relation of disjunction.[30]

2.8 *Ajdukiewicz's and Husserl's attitude towards the knowledge definite in content*

One should bear in mind the differences between the standpoint of an analytical philosopher (Ajdukiewicz) and of a phenomenologist (Husserl or Ingarden) with respect to the question of the relation: cognitive acts and results vs. linguistic acts and results (and therefore, the differences between their standpoints with respect to the question of the relation: cognitive subject vs. linguistic subject), when one speaks of the fact that the intentional theory of expressing and meaning, presented in the second volume of *Logical Investigations*, was used by Ajdukiewicz to present verbalized cognitive acts and results connected therewith in the Husserlian way, as well as, which is connected with the above, to understand the linguistic meaning of expressions in the Husserlian way, and in consequence, to understand in this way knowledge determined as to its content. Ajdukiewicz did not acknowledge however, and this needs to be borne in mind, the sources of direct knowledge which are characteristic for phenomenology. And, for example, in the course of a discussion with R. Ingarden, after his congress paper, he concluded:

> If some concept of proposition (in the logical sense) does not constitute a meaning of any expression, then it is impossible to say anything about it, which would pertain to its content. All that pertains to such concepts and propositions would therefore be inexpressionable, and therefore it could not belong to any science, if science is understood as something that is publically (inter-individually) available. This remark pertains in particular to the alleged theory of indirect knowledge. [4, p. 338]

[30] I write more about the dispute between the analyst and the phenomenologist on the role of language in knowledge an article [36].

He expressed similar views at the aforementioned Paris International Philosophy Congress. Discussing the meta-philosophy accepted in the Lvov-Warsaw School, he said that:

> In order to make science, it is not enough to observe the rules of intellectual honesty, i.e. to allow oneself to be guided by expressing opinions by nothing else than an honest conviction rooted in deep reflection. It is also necessary to express oneself in an intersubjectively communicative language, and claim only that which one is able to determine and justify, being certain of such justification.
>
> Further, for a linguistic utterance to be intersubjectively communicative, it is not enough for such utterance to be easily understandable for others—it is also necessary that one is able to make sure that the utterance has been understood in its correct sense. We consider a justification to be sure, if it is controllable by others, who could verify and repeat it. Intellectual work, which were unable to meet the presented requirements, could never become an object of cooperation and would not deserve the right to be called science.
>
> Observing these two postulates determines the method and the language, limiting at the same time the field of our philosophical activity. If we want to satisfy both of the requirements we are discussing, we can make use neither of the Bergsonian intuition nor of Husserl's "Wesenschau" (with all due respect for their significance in the fields relevant for them), nor of any other similar methods, until the results, to which they lead, can be formulated in an intersubjectively communicative manner and can be verified in a manner which will allow us to be sure. Although theses arrived at with the use of these methods can be verbally communicative, yet use of words in this case is only suggestive, which means that words are used only to invoke desired mental reactions of the listener. One therefore expresses these theses in a metaphorical manner, and uses thereby comparisons and examples, one is unable however to formulate them in expressions of direct character, i.e. such expressions which to be received correctly need only to be understood literally.[31] [16, pp. 123–124]

3 Conclusion

Semantic theory of knowledge grew out of the spirit of the anti-psychological breakthrough, and therefore—according to this theory—if we want to speak of logically (linguistically) understood knowledge determined as to its content, then instead of logical concepts and propositions, comprising knowledge understood in the logical way, we should speak of expressions, whose linguistic meanings these concepts and propositions are—and this will be semiotically substantiated. And moreover, what I have been trying to demonstrate, this will be also substantiated from the perspective of the philosophy of language, if this philosophy is the philosophy presented in the *Logical Investigations*. What is more, by proceeding this way, we can apply, in the theory of knowledge conceived of in such manner, any results obtained in meta-mathematics, which is a meta-theoretical discipline analogous to epistemology. Logically understood concepts and propositions, being the linguistic meanings of expressions, are their logical meanings, and being at

[31] During that speech Ajdukiewicz used the name "the Lvov-Warsaw School"—perhaps this was the first time this name was used for the first time.

the same time intensions[32] of expressions, are identical to the contents of mental acts understood *in specie*—the acts which are expressed by these expressions.

References

1. Ajdukiewicz, K.: W obronie uniwersaliów (On Defense of Universals). Ruch Filozoficzny **13**, 40b–41b (1932)
2. Ajdukiewicz, K.: Logistyczny antyirracjonalizm w Polsce [Logistic Anti-irrationalism in Poland]. Przegląd Filozoficzny **37**(4), 399–408 (1934) (Reprint in German as Ajdukiewicz, K. (ed.) Der logistische Antiirrationalismus in Polen. Erkenntnis **5**, Leipzig, 151–164 (1935))
3. Ajdukiewicz, K.: Der logistische Antiirrationalismus in Polen. Erkenntnis **5**, Leipzig, 151–164 (1935)
4. Ajdukiewicz, K.: Zagadnienie idealizmu w sformułowaniu semantycznym [The Issue of Idealism in the Semantical Formulation]. Przegląd Filozoficzny **39**(4), 334–340 (1936). Księga Pamiątkowa III Polskiego Zjazdu Filozoficznego. Kraków 1936 [Commemorative Volume of the Third Polish Philosophical Congress. Cracow 1936]
5. Ajdukiewicz, K.: Problemat transcendentalnego idealizmu w sformułowaniu semantycznym. Przegląd Filozoficzny **40(5)**, 271–287 (1937) (Reprint in Polish in Ajdukiewicz, K. (ed.) Język i poznanie. Wybór pism z lat 1920–1939 [Language and Knowledge. Selection of Papers from Years 1920–1939], vol. 1, pp. 264–277. Państwowe Wydawnictwo Naukowe, Warszawa (1960). Reprint in English as Ajdukiewicz, K. (ed.) A Semantical Version of the Problem of Transcendental Idealism. In: Introduction by J. Giedymin (ed.) The Scientific World-Perspectives and Other Essays 1931–1963, pp. 140–154. D. Reidel Publishing Company, Dordrecht/Holland/Boston (1978))
6. Ajdukiewicz, K.: Metodologia i metanauka [Methodology and Meta-science]. życie Nauki **6**(31–32), 4–15 (1948) (Reprint in Polish in Ajdukiewicz, K. (ed.) Język i poznanie. Wybór pism z lat 1945–1963 [Language and Knowledge. Selection of Papers from Years 1945–1963], vol. 2, pp. 117–126. Państwowe Wydawnictwo Naukowe, Warszawa (1965))
7. Ajdukiewicz, K.: Trzy pojęcia definicji (Three Concepts of Definitions). Studia Filozoficzne **5**(8), 3–16 (1958) (Reprint in Polish in Ajdukiewicz, K. (ed.) Język i poznanie. Wybór pism z lat 1945–1963 [Language and Knowledge. Selection of Papers from Years 1945–1963], vol. 2. Państwowe Wydawnictwo Naukowe, Warszawa (1965))
8. Ajdukiewicz, K.: Język i poznanie. Wybór pism z lat 1920–1939 [Language and Knowledge. Selection of Papers from Years 1920–1939], vol. 1. Państwowe Wydawnictwo Naukowe, Warszawa (1960)
9. Ajdukiewicz, K.: Język i znaczenie. In: Ajdukiewicz, K. (ed.) Język i poznanie. Wybór pism z lat 1920–1939 [Language and Knowledge. Selection of Papers from Years 1920–1939], vol. 1, pp.145–174. Państwowe Wydawnictwo Naukowe, Warszawa (1960). Trans. from German into Polish by F. Zeidler (Reprint in English as Ajdukiewicz, K. (ed.) Language and Meaning. In: Introduction by J. Giedymin (ed.) The Scientific World-Perspectives and Other Essays 1931–1963, pp. 35–66. D. Reidel Publishing Company, Dordrecht/Holland/Boston (1978). The orig.: *Sprache und Sinn.* Erkenntnis 4, Leipzig, 100–138 (1934))
10. Ajdukiewicz, K.: In: Mortimer, H., Szaniawski, K. (ed.) Logika pragmatyczna [Pragmatic Logic]. Państwowe Wydawnictwo Naukowe, Warszawa (1965). Wstęp [Foreword] by K. Szaniawski

[32]In the 1930s Ajdukiewicz did not deal with the question of intensionality. The first work in which the concept of "intensionality" was described more extensively is Ajdukiewicz's article [9] from 1958 in which—additionally to the topic indicated in the title—he argues that due to the intensional character of certain expressions, it is impossible to eliminate real definitions from the general theory of definitions and to replace them with nominal definitions in objective stylistics, without encountering insurmountable difficulties. The need for such replacement is motivated by some philosophers with their nominalist views, since the universale is the thing to which the real definition refers to.

11. Ajdukiewicz, K.: Język i poznanie. Wybór pism z lat 1945–1963 [Language and Knowledge. Selection of Papers from Years 1945–1963], vol. 2. Państwowe Wydawnictwo Naukowe, Warszawa (1965)
12. Ajdukiewicz, K.: A Semantical Version of the Problem of Transcendental Idealism. In: Introduction by J. Giedymin (ed.) The Scientific World-Perspectives and Other Essays 1931–1963, pp. 140–154. D. Reidel Publishing Company, Dordrecht/Holland/Boston (1978)
13. Ajdukiewicz, K.: On the problem of universals. In: Introduction by J. Giedymin (ed.) The Scientific World-Perspectives and Other Essays 1931–1963, pp. 95–110. D. Reidel Publishing Company, Dordrecht/Holland/Boston (1978)
14. Ajdukiewicz, K.: Language and Meaning. In: Introduction by J. Giedymin (ed.) The Scientific World-Perspectives and Other Essays 1931–1963, pp. 35–66. D. Reidel Publishing Company, Dordrecht/Holland/Boston (1978)
15. Ajdukiewicz, K.: In: Introduction by J. Giedymin (ed.) The Scientific World-Perspectives and Other Essays 1931–1963. D. Reidel Publishing Company, Dordrecht/Holland/Boston (1978)
16. Ajdukiewicz, K.: Przemówienie powitalne delegacji polskiej na Międzynarodowym Kongresie Filozofii Naukowej na Sorbonie w roku 1935 [Welcoming Speech of the Polish Delegation at the International Scientific Philosophy Congress in Sorbonne]. Principia. Pisma koncepcyjne z filozofii i socjologii teoretycznej **10–11**, 123–125 (1994). Trans. from French into Polish by J. Hartman (The orig. in: Actes du Congrés international de philosophie scientifique, vol. **1**. Philosophie scentifique et empirisme logique, Sorbone, Paris (1935))
17. Husserl, E.: Logische Untersuchungen. Untersuchungen zur Phänomenologie und Theorie der Erkenntnis. Bd. 2, Teil 1. Max Niemeyer, Halle (1928)
18. Husserl, E.: Logical Investigations, vol. 1. Routledge, London and New York (2001) (J.N. Findlay, trans. from the 2nd German edition of Logische Untersuchungen)
19. Husserl, E.: Logical Investigations, vol. 2. Routledge, London and New York (2001) (J.N. Findlay, trans. from the 2nd German edition of Logische Untersuchungen)
20. Ingarden, R.: Dążenia fenomenologów [The endeavours of phenomenologists]. Przegląd Filozoficzny **22**(3), 118–156, and **22**(4), 315–351 (1919) (Reprint in Ingarden, R. (ed.) Z badań nad filozofią współczesną [From Studies of Contemporary Philosophy], pp. 269–379. Państwowe Wydawnictwo Naukowe, Warszawa (1963))
21. Ingarden, R.: Główne fazy rozwoju filozofii E. Husserla [The Main Gestation Periods of Husserl's Philosophy]. In: Ingarden, R. (ed.) Z badań nad filozofią współczesną [From Studies of Contemporary Philosophy], pp. 383–452. Państwowe Wydawnictwo Naukowe, Warszawa (1963). (the modified reprint from: idem, Główne linie rozwoju poglądów filozoficznych Edmunda Husserla [The Main Lines of Development of Husserl's Philosophical Ideas]. Przegląd Filozoficzny **42**(2), 25–76 (1939–1945))
22. Ingarden, R.: Z badań nad filozofią współczesną [From Studies of Contemporary Philosophy]. Państwowe Wydawnictwo Naukowe, Warszawa (1963)
23. Ingarden, R.: Z teorii języka i filozoficznych podstaw logiki [From the Theory of Language and the Philosophical Grounds of Logic]. Państwowe Wydawnictwo Naukowe, Warszawa (1972)
24. Ingarden, R.: The Literary Work of Art. Northwestern University Press, Evanston, IL (1973). Trans. by George G. Grabowicz
25. Ingarden, R.: O dziele literackim. Badania z pogranicza ontologii, teorii języka i filozofii literatury, 2nd edn. Państwowe Wydawnictwo Naukowe, Warszawa (1988). Trans. from German into Polish by M. Turowicz (The orig.: Das literarische Kunstwerk. Eine Untersuchung aus dem Grenzgebiet der Ontologie, Logik und Literaturwissenschaft. Niemeyer, Halle (1931). Reprint in English as The Literary Work of Art. Northwestern University Press, Evanston, IL (1973). Trans. by George G. Grabowicz)
26. Jadacki, J.J.: Spór o granice języka [Dispute on the Limits of Language]. Wydawnictwo Naukowe Semper, Warszawa (2002)
27. Łukasiewicz, J.: Logistyka a filozofia [Logistic and Philosophy]. Przegląd Filozoficzny **39**(2), 115–131 (1936) (Reprint in Polish in Łukasiewicz, J. (ed.) Z zagadnień logiki i filozofii. Pisma wybrane [Topics in Logic and Philosophy. Selected Writings], pp. 195–209. Państwowe Wydawnictwo Naukowe, Warszawa (1961). Reprint in English as Łukasiewicz, J. (ed.) Logistics and Philosophy. In: Borkowski, L. (ed.) Selected Works, pp. 218–235. North-Holland Publishing Company, Amsterdam/London (1970). PWN — Polish Scientific Publishers, Warszawa (1970))

28. Łukasiewicz, J.: W obronie logistyki. Myśl katolicka wobec logiki współczesnej [In Defence of Logic. Catholic Thought in Comparison with Contemporary Logic]. Studia Gnesnensia **15** (1937) (Reprint in Polish as Łukasiewicz, J. (ed.) W obronie logistyki [In defense of logistic]. In: Z zagadnień logiki i filozofii. Pisma wybrane [Topics in Logic and Philosophy. Selected Writings], pp. 210–219. Państwowe Wydawnictwo Naukowe, Warszawa (1961). Reprint in English as Łukasiewicz, J. (ed.) In defense of logistic. In: Borkowski, L. (ed.) Selected Works, pp. 236–249. North-Holland Publishing Company, Amsterdam/London (1970). PWN — Polish Scientific Publishers, Warszawa (1970))
29. Łukasiewicz, J.: W obronie logistyki [In defense of logistic]. In: Łukasiewicz, J. (ed.) Z zagadnień logiki i filozofii. Pisma wybrane [Topics in Logic and Philosophy. Selected Writings], pp. 210–219. Państwowe Wydawnictwo Naukowe, Warszawa (1961)
30. Łukasiewicz, J.: Z zagadnień logiki i filozofii. Pisma wybrane [Topics in Logic and Philosophy. Selected Writings]. Państwowe Wydawnictwo Naukowe, Warszawa (1961). Selected by and provided with an introduction and footnotes by J. Słupecki
31. Łukasiewicz, J.: In defense of logistic. In: Łukasiewicz, J.: In: Borkowski, L. (ed.) Selected Works. North-Holland Publishing Company, Amsterdam/London (1970). PWN — Polish Scientific Publishers, Warszawa (1970)
32. Łukasiewicz, J.: Logistics and Philosophy. In: Borkowski, L. (ed.) Selected Works, pp. 218–235. North-Holland Publishing Company, Amsterdam/London (1970). PWN — Polish Scientific Publishers, Warszawa (1970)
33. Łukasiewicz, J.: In: Borkowski, L. (ed.) Selected Works. North-Holland Publishing Company, Amsterdam/London (1970). PWN — Polish Scientific Publishers, Warszawa (1970)
34. Olech, A.: W sprawie pojęć uniwersalnych i transcendentalnych [On Universal and Transcendental Concepts]. Filozofia Nauki **6**(2/22), 35–49 (1998)
35. Olech, A.: Ajdukiewicz a Husserl wobec kwestii znaczenia wyrażeń [Ajdukiewicz and Husserl on the Issue of the Meaning of Expressions]. Studia Semiotyczne **24**, 141–161 (2001) (Reprint in English as Olech, A. (ed.) Ajdukiewicz and Husserl on the issue of the meaning of expressions. Stud. Semiotyczne — English Supplement **24**, 130–154 (2015/2001). http://studiaes.pts.edu.pl/volume-xxiv.html)
36. Olech, A.: W sprawie sporu o rolę języka w poznaniu [On the Dispute Over the Role of Language in Knowledge]. Przegląd Filozoficzny — Nowa Seria **14**(2/54), 221–229 (2005) (Reprint in: Centrum Badań Filozoficznych im. R. Ingardena; http://www.roman-ingarden.phils.uj.edu.pl/artykuly.php)
37. Olech, A.: O pojęciach uniwersalnych i transcendentalnych [Concerning Universal and Transcendental Concepts]. In: Żurkowska, G., Blandzi, S. (eds.) Rezonujący rozum nauki a rozumność intuicji. Księga dedykowana Profesor Alinie Motyckiej [The Resonating Reason of Science and the Rationality of Intuition. A Commemorative Volume Dedicated to Professor Alina Motycka], pp. 364–378. Wydawnictwo IFiS PAN, Warszawa (2009)
38. Olech, A.: O Ajdukiewiczu i Ingardenie – uczniach Husserla. W setną rocznicę ukazania się Księgi I Idei (On Ajdukiewicz and Husserl – the Husserl's Pupils. On the Hundredth Anniversary of the First Edition of Book I of Husserl's Ideen). In: Woźniczka, M. (ed.) Filozofia polska na tle filozofii europejskiej (Polish Philosophy on the Background of European Philosophy), pp. 159–176. Wydawnictwo Akademii im. Jana Długosza w Częstochowie, Częstochowa (2014)
39. Olech, A.: Semantyczna teoria poznania [Semantic Theory of Knowledge]. Wydawnictwo im. S. Podobińskiego Akademii im. Jana Długosza w Częstochowie, Częstochowa (2014)
40. Olech, A.: Ajdukiewicz and Husserl on the issue of the meaning of expressions. Stud. Semiotyczne — English Supplement **24**, 130–154 (2015/2001). http://studiaes.pts.edu.pl/volume-xxiv.html

A. Olech (✉)
Institute of Philosophy, Jan Dlugosz University in Częstochowa, Częstochowa, Poland
e-mail: adamolech@wp.pl

Categorial Grammars and Their Logics

Wojciech Buszkowski

Abstract This paper surveys the development of categorial grammars (also called type grammars) with emphasis on type logics (i.e. logical calculi underlying these grammars) and their relation to the origin in Ajdukiewicz (Studia Philosophica 1:1–27, 1935).

Keywords Categorial grammar · Type · Lambek calculus · Substructural logic · Parsing

Mathematics Subject Classification (2000) Primary 03B65; Secondary 68T50

1 Introduction

In the modern literature Kazimierz Ajdukiewicz is commonly accepted as the father of categorial grammars: formal grammars assigning logical types to expressions. His seminal paper [4] provided a clear idea of these grammars. Ajdukiewicz acknowledged an impact of Husserl and Leśniewski. From Husserl [27] he took the idea of semantical categories which can be defined in terms of mutual substitution of expressions in meaningful or sentential contexts. From Leśniewski he took a classification of categories in basic categories and functor categories. Here I cannot cite any single publication of Leśniewski, since in his works he never wrote a longer passage on this matter. Ajdukiewicz and other authors cite [41], but this paper on new foundations of mathematics merely contains short notes on 'the theory of semantical categories'; different logical symbols are characterized as functors of a particular category (no special symbols for categories are introduced). It seems that more elaborated considerations only appeared in Leśniewski's oral lectures.

Ajdukiewicz introduced a system of indices for categories. This system is employed in his procedure for verifying the 'syntactic connexion' of expressions. In [4] he writes: "We shall base our work here on the relevant results of Leśniewski, adding on our part a symbolism, in principle applicable to almost all languages [and enabling us to build a calculus], which makes it possible to formally define and examine the syntactic connexion

Á. Garrido, U. Wybraniec-Skardowska (eds.), *The Lvov-Warsaw School. Past and Present*, Studies in Universal Logic,
https://doi.org/10.1007/978-3-319-65430-0_6

of a word pattern."[1] (The passage in brackets has been omitted in the English translation in [6]; I add it, since the word 'calculus' is quite important.) In fact, the indices for categories were introduced in Ajdukiewicz's earlier paper [3], where they were used in a semantical analysis of the problem of universals.

Let me briefly comment on terminology. The indices for categories will be called *types*, according to modern standards in logic. Categories can be understood as some sets of expressions or sets of ontological objects (having a common type). Although Leśniewski and Ajdukiewicz use the term 'semantical category' (after Husserl), the term 'syntactic(al) category' is more appropriate. This was noticed by Ajdukiewicz [5]: "The concept of semantical categories must be clearly distinguished from the concept of syntactical categories. The term 'semantical category' was introduced for the first time by Husserl; however, the concept he associated with it would correspond better to the term 'syntactical category'. For Husserl pointed out that the expressions of a language may be classified according to the role they can play within a sentence. He defined, therefore, the categories from the syntactical viewpoint." Ajdukiewicz [5] outlined a theory of semantical categories: the type of an expression is determined by the ontological type of the denotation of this expression. The first semantical interpretation of [4] is due to Bocheński [9].

The connections of syntactic (semantic) types with categories defined by mutual substitution are by no way obvious, nor simple. They are quite tight for deterministic (or: rigid) grammars which assign at most one type to one expression, but become less regular for categorially ambiguous grammars. A thorough discussion of this topic can be found in [15].

The present paper focuses on categorial grammars: how they developed from the origin in [4] to their modern forms. Categorial grammars are also called 'type grammars', and the latter term seems better. The classification of expressions in categories appears in different grammar formalisms (e.g. phrase structure grammars), whereas logical types assigned to expressions are characteristic of the grammars considered here. To emphasize this Morrill [47] and others use an even more explicit term 'type logical grammar'. In the present paper both terms are used: the former in traditional names of grammars, the latter in general considerations.

The impact of [4] can be seen in several areas of formal linguistics and logical philosophy of language. Type grammars belong to formal linguistics, since their main intention is to describe natural language. They are closely related to type-theoretic semantics of natural language, initiated by Monatgue [42], with an explicit reference to [4], and extensively studied by many authors as Montague Grammar. Some other works may be counted to the logical turn: they study types and categories in formal languages of logic and mathematics. This direction was represented in Poland by Suszko [58, 59], Wybraniec-Skardowska [67] and others. Suszko elaborated a formal framework for syntax and semantics of higher-order languages. Wybraniec-Skardowska presented a general theory of 'categorial languages' with a distinction between expression-tokens and abstract expressions (this theory, however, does not directly address natural language). Tałasiewicz [60] provided a philosophical analysis of Ajdukiewicz's approach, applied to natural language, with an interpretation in terms of situation semantics,

[1] All citations from Ajdukiewicz are based on the English translations of Ajdukiewicz's original papers, collected in [6].

It is a bit surprising that just the linguistic turn leads to new logical calculi (Lambek logics) and models (residuated algebras), whereas the logical turn usually focuses on some standard logical systems (higher-order logics, type theories). To keep this survey in a reasonable size I will mainly write on the new logics elaborated for type grammars and only briefly note some links with other developments. Montague Grammar and its descendants cannot be discussed in detail; the reader is referred to [65] and the items cited there.

This survey is addressed to a wide community, not necessarily experts in type grammars. Therefore I omit mathematical subtleties. I do not even discuss all important mathematical results in this area; I only briefly note some of them to clarify the main ideas. Nonetheless an acquaintance with general logic and formal linguistics may help the reader to follow the text. I provide a couple of linguistic examples, but all are quite simple. The reader is referred to [40, 45–48] for a more advanced linguistic material.

Section 2 is concerned with basic categorial grammars, a framework directly related to Ajdukiewicz's proposal (modified in [8]). Section 3 discusses the Lambek calculus and several analogous systems with a particular emphasis on their role in type grammars. At the end, I defend the view that Lambek logics are important, general logics of syntactic and semantic types (besides other applications), but not as good for efficient parsing. An optimal strategy seems the following: (1) to apply Lambek logics in metatheory and on the lexical level, (2) to preserve the Ajdukiewicz system as a parsing procedure for compound expressions.

2 Basic Categorial Grammars

Ajdukiewicz (following Leśniewski) distinguishes two basic categories: sentence (type s) and name (type n), but stipulates that in general "nothing could be decided about the number and kind of basic semantic [categories] and functor categories, since these may vary in different languages." The types of functor categories have the form of fractions:

$$\frac{\alpha}{\beta_1 \ldots \beta_n};$$

an expression of this type with arguments of type $\beta_1, \ldots, \beta_n$ forms a compound expression of type α. For example, an intransitive verb is of type $\frac{s}{n}$, a transitive verb of type $\frac{s}{n\,n}$, a sentential connective of type $\frac{s}{s\,s}$, an adverb of type $\frac{\alpha}{\alpha}$ for $\alpha = \frac{s}{n}$, and so on.

The procedure of checking the 'syntactic connexion' of a compound expression is designed as follows. First, the expression is rewritten in prefix notation: each functor directly precedes its arguments. So one writes `likes John wine` instead of `John likes wine` and `hardly works John` instead of `John works hardly` (my examples). Second, one considers the sequence of types corresponding to the words of the rearranged expression. For these two examples one obtains the sequences:

$$\frac{s}{n\,n},\ n,\ n \text{ and } \frac{\frac{s}{n}}{\frac{s}{n}},\ \frac{s}{n},\ n.$$

Third, one reduces a block of adjacent types:

$$\frac{\alpha}{\beta_1 \ldots \beta_n},\ \beta_1, \ldots, \beta_n$$

to α and repeats this step as many times, as possible. If this reduction ends in a single type, the expression is qualified to be 'syntactically connected' and assigned the resulting type. The Ajdukiewicz reduction procedure applied to our examples yields:

$$\frac{s}{n\,n},\ n,\ n\ \Rightarrow\ s \text{ in one step,}$$

$$\frac{\frac{s}{n}}{\frac{s}{n}},\ \frac{s}{n},\ n\ \Rightarrow\ \frac{s}{n},\ n\ \Rightarrow\ s \text{ in two steps.}$$

So both expressions are syntactically connected (of type s). In fact, Ajdukiewicz's original procedure was more restrictive: at each step one reduces the left-most occurrence of a reducible pattern, but this constraint narrows its applications [13].

This approach reveals two characteristic components of modern type grammars: (1) *the type lexicon*, i.e. an assignment of types to words, (2) *the type processing machinery*, i.e. a procedure of checking the grammatical correctness of arbitrary expressions and at the same time deriving types of them. In terms of contemporary computational linguistics, (2) is *a parsing procedure*. Ajdukiewicz was the first who clearly formulated the problem of parsing and proposed a parsing algorithm (20 years before mathematical linguistics was founded by Noam Chomsky).

The Ajdukiewicz procedure requires the rewriting of the parsed expression in prefix notation. In practice this restricts its applications to some formal languages. In fact Ajdukiewicz acknowledged that one of his goals was a generalization of the parenthesis-free notation, elaborated by J. Łukasiewicz for propositional logics, toward richer formal languages. On the other hand, his examples came from natural languages, and he expected a wide applicability of his method. Probably he admitted various modifications of the original procedure, when applied in practice.

Bar-Hillel [7] adjusted this approach to natural language. He introduced *directional* types of the form:

$$\frac{\alpha}{\beta_1 \ldots \beta_m;\, \gamma_1 \ldots \gamma_n}$$

and the reduction procedure based on the rule:

$$\beta_1, \ldots, \beta_m,\ \frac{\alpha}{\beta_1 \ldots \beta_m;\, \gamma_1 \ldots \gamma_n},\ \gamma_1, \ldots, \gamma_n \Rightarrow \alpha.$$

Now transitive verbs are assigned type $\frac{s}{n;n}$, and `John likes Mary` is parsed as:

$$n,\ \frac{s}{n;\,n},\ n \Rightarrow s \text{ in one step.}$$

In [8], this approach was modified. After Lambek [36], functor types were restricted to $\alpha\backslash\beta$ and α/β. An expression of type $\alpha\backslash\beta$ (resp. β/α) with an argument of type α on the left (resp. on the right) forms a compound expression of type β. So $\alpha\backslash\beta$ corresponds to $\frac{\beta}{\alpha;}$ in the former notation, α/β to $\frac{\alpha}{;\beta}$, and the fraction $\frac{\alpha}{\beta;\gamma}$ is represented as $\beta\backslash(\alpha/\gamma)$ or $(\beta\backslash\alpha)/\gamma$. The representation of many-argument types by (nested) one-argument types is closely related to 'currying', i.e. the representation of many-argument functions by one-argument functions of higher order, a routine in modern type theories.

The reduction procedure is based on two rules:

$$\text{(RED.1)}\ \alpha, \alpha\backslash\beta \Rightarrow \beta,\ \ \text{(RED.2)}\ \alpha/\beta, \beta \Rightarrow \alpha\,.$$

In [8], *a categorial grammar* is formally defined as a triple $G = (\Sigma, I, s)$ such that Σ is a nonempty finite set, I is a finite relation between elements of Σ and types, and s is an atomic type. The elements of Σ are interpreted as the words of a natural language (then Σ is referred to as *the lexicon*) or symbols of a formal language (then Σ is referred to as *the alphabet*). Nowadays I is called *the type lexicon* or *the initial type assignment*. Often I is represented as a map which assigns finite sets of types to elements of Σ. In examples we write $v : \alpha$ for $\alpha \in I(v)$. One refers to s as *the designated type*. One admits an arbitrary finite set of atomic types.

Finite sequences of elements of Σ are called *strings* (on Σ). The empty string is denoted by ϵ. The string $(v_1, \ldots, v_n)$ is usually written as $v_1 \ldots v_n$. One says that G *assigns* type α to the string $v_1 \ldots v_n$, if there exist types $\alpha_1, \ldots, \alpha_n$, belonging to $I(v_1), \ldots, I(v_n)$, respectively, such that the sequence $\alpha_1, \ldots, \alpha_n$ reduces to α by finitely many applications of rules (RED.1), (RED.2). *The language of* G consists of all strings on Σ which are assigned type s by G.

In the modern literature, categorial grammars in the sense of [8] are called *basic categorial grammars* (BCGs) or: classical categorial grammars, AB-grammars (a credit to Ajdukiewicz and Bar-Hillel).

The main mathematical theorem of [8] establishes the weak equivalence of BCGs and Chomsky's (ϵ-free) context-free grammars (CFGs). Recall that a CFG is defined as a quadruple $G = (\Sigma, N, s, P)$ such that Σ and N are disjoint finite sets (whose elements are treated as simple symbols), $s \in N$, and P is a finite set of pairs (a, x), where $a \in N$ and x is a string on $\Sigma \cup N$. The elements of Σ (resp. N) are called *terminal symbols* (resp. *nonterminal symbols* or *variables*), and s is called *the start symbol*. The pairs in P are called *production rules*. one writes $a \mapsto x$ for (a, x) and interprets it as a rewriting rule: the string yaz can be rewritten as yxz according to this rule (by xy one denotes the concatenation of x and y). *The language of* G (or: generated by G) consists of all strings on Σ which can be derived from s by finitely many applications of the production rules. A CFG is ϵ-*free*, if it contains no nullary rule of the form $a \mapsto \epsilon$. The equivalence theorem states that BCGs and ϵ-free CFGs generate the same class of languages. More precisely, for any BCG G there exists an ϵ-free CFG G' such that $L(G) = L(G')$, and conversely.

The first part of this theorem can easily be proved: a BCG $G = (\Sigma, I, s)$ generates the same language as the CFG with the terminal alphabet Σ, the nonterminal alphabet consisting of all types involved in G and their subtypes (i.e. subterms), the start symbol s, and the production rules reversing (RED.1), (RED.2) (restricted to the types in the nonterminal alphabet) plus the lexical rules $\alpha \mapsto v$, for $\alpha \in I(v)$. The second part is more difficult; the proof in [8] yields, in fact, the Greibach normal form theorem for CFGs (independently proved a few years later).

In opposition to CFGs, BCGs are *lexical*: the whole information on the described language is contained in the type lexicon, whereas the parsing procedure is independent of this particular language (it employs the language-independent rules (RED.1), (RED.2)). This is not the case for CFGs. For instance, a standard non-lexical rule for English is

$s \mapsto np, vp$ (a sentence consists of a noun phrase and a verb phrase). The lexicality is a characteristic feature of all type grammars, considered nowadays. Sometimes it is convenient to admit certain simple non-lexical rules, e.g. $pn \Rightarrow np$ (a proper noun is a noun phrase), but one tends to eliminate them, whenever possible.

The nonterminal symbols of a CFG can be interpreted as names of syntactic categories, like types in a BCG. Types, however, can be compound terms, not just simple symbols. This is significant for lexicality and makes it possible to study logics of types, expressing deeper relations between types.

Although CFGs are weakly equivalent to BCGs, the strong equivalence does not hold; this means that the structured languages differ for the two classes of grammars. For a CFG, each derivation of a string from a nonterminal symbol determines a unique *phrase structure* of this string. For instance, the grammar with production rules:

$$s \mapsto np, vp \quad vp \mapsto tv, np$$

$$np \mapsto \text{John} \;\; np \mapsto \text{tee} \;\; tv \mapsto \text{drinks}$$

admits the derivation:

$$s \Rightarrow np, vp \Rightarrow np, tv, np \Rightarrow \cdots \Rightarrow \text{John, drinks, tee}$$

which yields the phrase structure (John (drinks tee)) or, more explicitly, (John$_{np}$ (drinks$_{tv}$ tee$_{np}$)$_{vp}$)$_s$. These phrase structures can be depicted as binary trees; see Fig. 1.

Similarly, each reduction in a BCG gives rise to a unique phrase structure of the input string. With the type lexicon:

$$\text{John} : np \;\; \text{drinks} : (np\backslash s)/np \;\; \text{tee} : np$$

one obtains the reduction:

$$np,\ (np\backslash s)/np,\ np \Rightarrow np,\ np\backslash s \Rightarrow s\ ,$$

which yields the same phrase structure (John (drinks tee)). We need an auxiliary notion. *The degree* of type α, denoted by $d(\alpha)$, is defined as follows: $d(\alpha) = 0$ if α is atomic, $d(\alpha\backslash\beta) = d(\beta/\alpha) = d(\beta) + 1$. For any phrase structure generated by a BCG G, depicted as a tree, and for any node of this tree, the length of shortest paths from this node to a leaf is not greater than the maximal degree of types involved in G. Therefore a BCG cannot

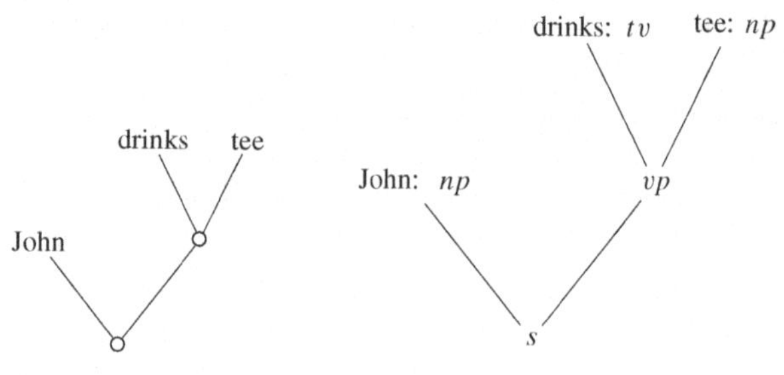

Fig. 1 Derivation trees for a CFG

generate languages of phrase structures with arbitrarily long shortest paths from a node to a leaf. On the contrary, a CFG can generate such languages.

For instance, the CFG with rules $s \mapsto s, s$ and $s \mapsto 0$ generates all possible phrase structures on the alphabet $\{0\}$. The BCG with the type lexicon $0 : s/s$, $0 : s$ generates the same language of strings, which consists of all nonempty strings on $\{0\}$, but not the same language of phrase structures. One only gets the phrase structures: 0, (00), (0(00)), (0(0(00))), and so on, but not ((00)0).

For BCGs, one also considers *functor-argument structures* (fa-structures), i.e. phrase structures augmented with functor markers. For the BCG considered above, the phrase structure (John (drinks tee)) can be refined to the fa-structure (John (drinks tee)$_1$)$_2$, which means that (drinks tee) is the functor in the whole structure and `drinks` is the functor in (drinks tee). Every reduction in a BCG determines a unique fa-structure of the recognized string. The languages of fa-structures and phrase structures take an essential part in the theory of BCGs; see [11, 15]. In particular, syntactic categories can be defined as certain sets of fa-structures rather than strings, which results in a more elegant theory.

The type lexicon of a BCG can assign several types to one word. This reflects *the syntactic ambiguity* of words in natural language. For instance, `and` appears as a sentential connective, but also as a noun connective, verb connective, adverb connective, and others. As a rule, in logical and mathematical formalisms one symbol can be assigned a unique type, which completely characterizes the syntactic role of this symbol. These languages can be described by *rigid* (or: deterministic) BCGs (I is a function from Σ to the set of types).

Worthy of noting, not all languages of formal logic can be described by rigid BCGs. The standard example is the language of (type-free) lambda calculus. Also in the language of first-order logic, a unary function symbol f requires two types t/iv, t/t, where iv is the type of individual variables and t of terms (quantifiers are typed $(s/s)/iv$, where s is the type of formulas). Alternatively, one can assign only t/t to f and admit a non-lexical rule $iv \Rightarrow t$.

The type of quantifiers, given above, adequately characterizes their role in the syntax of first-order logic (in modern setting): the quantifier followed by a variable, next by a formula, yields a formula. It is also fully compatible with Tarskian semantics for this logic. It, however, does not express the variable-binding role of quantifiers. The final part of [4] is devoted to the special status of variable-binding operators, and several authors continue this issue; see [49, 58, 59, 67]. I do not discuss this matter here, since it goes too far from the main topics of this paper.

One of the leitmotives of type grammars is a close relationship between syntax and semantics (the dictum *syntax mirrors ontology*). I have already noted that Bocheński [9] proposed the semantical interpretation of the theory of Leśniewski and Ajdukiewicz, and this turn was adopted by Ajdukiewicz [5]. According to the latter, the basic types are i (individual) and w (truth value; 'value' corresponds to Polish 'wartość' and German 'Wert'). Intransitive verbs are typed $\frac{w}{i}$, as they denote functions from the set of individuals to the set of truth values, transitive verbs $\frac{w}{i\,i}$, as they denote two-argument functions of this kind, (binary) sentential connectives $\frac{w}{w\,w}$, as they denote binary truth-value functions, and so on. Ajdukiewicz [5] brings a radical idea of a *purely flectional* language: the types of words only account for semantical categories of these words (i.e. the ontological status of their denotations), whereas their syntactic roles are described by certain new indices, indicating the position of these words in syntactic trees (some representations of fa-structures). This idea seems very interesting, but the symbolism, proposed in [5], has a limited value, since the new indices show the positions of words in one particular tree, not in any well-formed syntactic tree, containing the given word.

On the other hand, the semantical interpretation is quite fundamental and—modulo terminology and notation—has been commonly adopted in modern type-theoretic semantics. Syntactic types are translated into *semantic types*: atomic types and compound types $\alpha \rightarrow \beta$, where α, β are simpler types. Each atomic type p corresponds to a semantic domain (or: ontological category) D_p. One defines $D_{\alpha\rightarrow\beta}$ as the set of all functions from D_α to D_β. For instance, s is translated into t (the type of truth values) and n into e (the type of entities). Let $\alpha^\bullet$ denote the translation of α. One recursively defines:

$$(\alpha\backslash\beta)^\bullet = (\beta/\alpha)^\bullet = \alpha^\bullet \rightarrow \beta^\bullet.$$

So $n\backslash s$ is translated into $e \rightarrow t$ (the type of sets of entities, identified with their characteristic functions), $s/(n\backslash s)$ into $(e \rightarrow t) \rightarrow t$ (the type of families of sets of entities), and so on. The latter agrees with the interpretation of complete noun phrases as generalized quantifiers; see van Benthem [62].

In semantics, the reduction rules (RED.1), (RED.2) can be interpreted as the application of a function $f \in D_{\alpha^\bullet\rightarrow\beta^\bullet}$ to an argument $a \in D_{\alpha^\bullet}$, which yields $f(a) \in D_{\beta^\bullet}$. Thus, given some fixed denotations of all words of the parsed expression, whose semantic types correspond to their syntactic types, as above, one can determine the denotation of this expression by the (iterated) application of functions to their arguments, following the syntactic reduction procedure. This fully agrees with *the principle of compositionality*, a central idea of logical semantics.

3 Lambek Calculus

3.1 Basic Systems

An essential refinement of BCGs is due to Lambek [36]. His Syntactic Calculus, nowadays called Lambek Calculus and denoted by **L**, is regarded as a basic type logic. Lambek presented his system as an improvement of BCGs: "[...] this paper is concerned with a development of the technique of Ajdukiewicz and Bar-Hillel in a mathematical direction. We introduce a calculus of types, which is related to the well-known calculus of residuals. The decision procedure for this system is solved affirmatively, following a procedure first proposed by Gentzen for the intuitionistic propositional calculus."

Types are built from atomic types by \, / and · (product; some authors write ⊗). An axiomatization of **L** employs *simple sequents* of the form $\alpha \Rightarrow \beta$, where α, β are types. **L** admits the following axioms and inference rules.

$$\text{(Id)}\ \alpha \Rightarrow \alpha$$

$$\text{(A.1)}\ (\alpha \cdot \beta) \cdot \gamma \Rightarrow \alpha \cdot (\beta \cdot \gamma) \quad \text{(A.2)}\ \alpha \cdot (\beta \cdot \gamma) \Rightarrow (\alpha \cdot \beta) \cdot \gamma$$

$$\text{(Res.1)}\ \frac{\alpha \cdot \beta \Rightarrow \gamma}{\beta \Rightarrow \alpha\backslash\gamma} \quad \text{(Res.2)}\ \frac{\alpha \cdot \beta \Rightarrow \gamma}{\alpha \Rightarrow \gamma/\beta}$$

$$\text{(Cut.1)}\ \frac{\alpha \Rightarrow \beta \quad \beta \Rightarrow \gamma}{\alpha \Rightarrow \gamma}$$

The double line in (Res.1), (Res.2) means that these rules can be used in both directions: top-down and bottom-up.

By dropping the associativity axioms (A.1), (A.2), one obtains Nonassociative Lambek Calculus (**NL**), due to Lambek [37]. The counterparts of (RED.1), (RED.2):

$$\text{(Red.1)}\ \alpha \cdot (\alpha\backslash\beta) \Rightarrow \beta\,, \quad \text{(Red.2)}\ (\alpha/\beta) \cdot \beta \Rightarrow \alpha$$

are provable in **NL**, using (Id), (Res.1), (Res.2). We, however, obtain (infinitely) many other laws. Here are some examples.

- (L1) $\alpha \Rightarrow (\beta/\alpha)\backslash\beta$ and $\alpha \Rightarrow \beta/(\alpha\backslash\beta)$,
- (L2) $\alpha \Rightarrow \beta\backslash(\beta \cdot \alpha)$ and $\alpha \Rightarrow (\alpha \cdot \beta)/\beta$,
- (L3) $(\alpha\backslash\beta) \cdot (\beta\backslash\gamma) \Rightarrow \alpha\backslash\gamma$ and $(\alpha/\beta) \cdot (\beta/\gamma) \Rightarrow \alpha/\gamma$,
- (L4) $\alpha\backslash\beta \Rightarrow (\gamma\backslash\beta)\backslash(\gamma\backslash\alpha)$ and $\alpha/\beta \Rightarrow (\alpha/\gamma)/(\beta/\gamma)$,
- (L5) $(\alpha\backslash\beta)/\gamma \Leftrightarrow \alpha\backslash(\beta/\gamma)$ ($\Leftrightarrow$ stands for both $\Rightarrow$ and $\Leftarrow$).

(L1), (L2) are provable in **NL**, but (L3), (L4), (L5) in **L** only. Other laws can be obtained, by using the monotonicity rules: from $\alpha \Rightarrow \beta$ infer $\gamma \cdot \alpha \Rightarrow \gamma \cdot \beta$, $\alpha \cdot \gamma \Rightarrow \beta \cdot \gamma$, $\gamma\backslash\alpha \Rightarrow \gamma\backslash\beta$, $\beta\backslash\gamma \Rightarrow \alpha\backslash\gamma$, $\alpha/\gamma \Rightarrow \beta/\gamma$, $\gamma/\beta \Rightarrow \gamma/\alpha$, which are derivable in both systems.

The most general algebraic models of **NL** are *residuated groupoids*, i.e. ordered algebras $(A, \cdot, \backslash, /, \leq)$ such that $(A, \leq)$ is a partially ordered set, and $\cdot, \backslash, /$ are binary operations on A, satisfying *the residuation laws*:

$$\text{(RES)}\ a \cdot b \leq c \text{ iff } b \leq a\backslash c \text{ iff } a \leq c/b, \text{ for all } a, b, c \in A\,.$$

The operations $\backslash$, $/$ are called *the residual operations* for product. *Residuated semigroups* are residuated groupoids such that $\cdot$ is associative; they are models for **L**. Both systems are *strongly complete* with respect to the corresponding models: the sequents provable in the system from a set of nonlogical hypotheses are precisely those sequents which are true in all models, for all valuations μ, satisfying the hypotheses. $\alpha \Rightarrow \beta$ is *true* for μ, if $\mu(\alpha) \leq \mu(\beta)$.

According to Lambek [36], the intended models for **L** are *language models*, i.e. some algebras of languages (by a language one means a set of strings). By Σ^+ we denote the set of all nonempty strings on Σ. For $L_1, L_2 \subseteq \Sigma^+$, one defines:

$$L_1 \cdot L_2 = \{xy : x \in L_1,\ y \in L_2\}\,,$$

$$L_1\backslash L_2 = \{y \in \Sigma^+ : xy \in L_2 \text{ for any } x \in L_1\},$$

$$L_1/L_2 = \{x \in \Sigma^+ : xy \in L_1 \text{ for any } y \in L_2\},$$

where xy denotes the concatenation of strings x and y. It is easy to show that the powerset of Σ^+ with $\cdot, \backslash, /$ defined as above and inclusion as the order, is a residuated semigroup. $\alpha \Rightarrow \beta$ is true for μ in this model if and only if $\mu(\alpha) \subseteq \mu(\beta)$ (equivalently: every string of type α is of type β). The term 'language model' is due to Pentus [52]; this paper shows *the weak completeness* of **L** with respect to language models (the sequents provable in **L** are precisely those which are valid in all language models). The strong completeness does not hold, but it holds for the product-free **L** [10].

Analogously, the intended models for **NL** are algebras of languages consisting of phrase structures. Let Σ^P denote the set of all phrase structures on Σ. On the powerset of Σ^P one defines $\cdot, \backslash, /$ as above except that Σ^+ is replaced by Σ^P and xy by (x, y). The weak and the strong completeness (with respect to these models) hold for the product-free fragment of **NL** only [23, 33].

The intended models exhibit Lambek's interpretation of categories, which is not the same as in BCGs. For a BCG, the category of type α consists of all (structured) expressions which are assigned this type by the grammar. According to Lambek, the basic categories, i.e. those which are assigned atomic types, generate all other categories by operations $\cdot, \backslash, /$, interpreted in the algebra of languages. In particular, if y is of type $\alpha\backslash\beta$ (resp. β/α), then, for any x of type α, xy (resp. yx) is of type β in a BCG. Lambek replaces 'if ... then' by 'if and only if'. This is an essential difference; it leads to new reduction patterns, like (L1)–(L5), not admitted in BCGs.

This novel understanding of types caused, probably, a relatively small impact of Lambek's approach on his contemporaries. Only in the 1980s there began more systematic studies in Lambek calculi and their role in type grammars and type-theoretic semantics, initiated by W. Zielonka and the present author in Poznań and J. van Benthem and his students (especially M. Moortgat) in Amsterdam. This research was reported in two collection volumes [17, 50]; the second one also contains reprints of some earlier papers. The books [11, 43] elaborate on logical and algebraic properties of Lambek calculi and grammars.

Lambek grammars are defined like BCGs except that the reduction procedure is replaced with the provability in **L**, **NL** or a related system. One employs sequents of the form $\alpha_1, \ldots, \alpha_n \Rightarrow \beta$; in algebras, each comma is interpreted as product. For nonassociative systems, the antecedents of sequents take the form of bracketed sequences, e.g. $(\alpha, (\beta, \gamma))$, which is different from $((\alpha, \beta), \gamma)$. So (Red.1), (Red.2) can be written as (RED.1), (RED.2), and similarly for other laws. WARNING: (RED.1), (RED.2) have been called reduction rules in Sect. 2, but now the term 'rule' is reserved for inference rules of type logics, e.g. (Res.1), (Res.2), (Cut.1), whereas the provable sequents are referred to as laws.

Both **L** and **NL** can be presented as sequent systems [36, 37]. For **L**, the axioms are (Id) and the inference rules are as follows (Γ and Δ stand for finite, possibly empty, sequences of types).

$$(\cdot \Rightarrow)\frac{\Gamma, \alpha, \beta, \Gamma' \Rightarrow \gamma}{\Gamma, \alpha \cdot \beta, \Gamma' \Rightarrow \gamma} \quad (\Rightarrow \cdot)\frac{\Gamma \Rightarrow \alpha \quad \Delta \Rightarrow \beta}{\Gamma, \Delta \Rightarrow \alpha \cdot \beta}$$

$$(\backslash \Rightarrow)\frac{\Gamma, \beta, \Gamma' \Rightarrow \gamma \quad \Delta \Rightarrow \alpha}{\Gamma, \Delta, \alpha\backslash\beta, \Gamma' \Rightarrow \gamma} \quad (\Rightarrow \backslash)\frac{\alpha, \Gamma \Rightarrow \beta}{\Gamma \Rightarrow \alpha\backslash\beta}$$

$$(/\Rightarrow)\frac{\Gamma,\alpha,\Gamma'\Rightarrow\gamma\quad\Delta\Rightarrow\beta}{\Gamma,\alpha/\beta,\Delta,\Gamma'\Rightarrow\gamma}\qquad(\Rightarrow/)\frac{\Gamma,\beta\Rightarrow\alpha}{\Gamma\Rightarrow\alpha/\beta}$$

$$(\mathrm{Cut})\ \frac{\Gamma,\alpha,\Gamma'\Rightarrow\beta\quad\Delta\Rightarrow\alpha}{\Gamma,\Delta,\Gamma'\Rightarrow\beta}$$

One assumes that Γ is nonempty in $(\Rightarrow\ \backslash)$, $(\Rightarrow\ /)$. In sequents, one omits outer parentheses of antecedent sequences and writes Γ, Δ for the concatenation of Γ and Δ.

The sequent system for **NL** is similar. The antecedents of sequents are bracketed sequences of types, hence all rules look a bit differently; see [11, 45].

Clearly these systems are certain intuitionistic sequent systems, types play the role of formulas, and atomic types of variables (or nonlogical constants). The rules $(\cdot\Rightarrow)$-$(\Rightarrow /)$ are *the introduction rules* for connectives, and (Cut) is *the cut rule*.

By dropping (Cut), one obtains the cut-free (sequent system for) **L**. Lambek [36] proved *the cut elimination theorems* for **L** (and in [37] for **NL**): every provable sequent is provable in the cut-free system. As a consequence, both systems possess *the subformula property*: every provable sequent possesses a proof such that each formula appearing in this proof is a subformula of a formula occurring in this sequent. Since, additionally, each introduction rule increases the size of sequents, then the provability in either system is decidable. It is easy to extract language-restricted fragments. For instance, the product-free fragment admits product-free formulas only and drops rules $(\cdot\Rightarrow)$, $(\Rightarrow\cdot)$. **L** is a conservative extension of its language-restricted fragments, and similarly for **NL**.

The product-free **L**, restricted to (Id), $(\backslash\Rightarrow)$ and $(/\Rightarrow)$ ((Cut) is admissible), yields precisely the correct reduction patterns of BCGs. This system is sometimes denoted by **AB**. **L** is much stronger than **AB**, but both systems coincide for sequents of the form $\alpha_1,\dots,\alpha_n\Rightarrow p$ such that p is an atom and no α_i contains a compound type on the argument place. In other words, the order of each α_i is at most 1. *The order* of α, denoted by $o(\alpha)$, is recursively defined as follows: $o(p)=0$ for atomic p,

$$o(\alpha\backslash\beta)=o(\beta/\alpha)=\max(o(\beta),o(\alpha)+1),\ o(\alpha\cdot\beta)=\max(o(\alpha),o(\beta)).$$

For example, $p\backslash q$, $p\backslash(q\backslash r)$, $(p\backslash q)/r$ are of order 1, $p/(q\backslash p)$ is of order 2, and so on (p,q,r are atoms). The product-free **L** is stronger than any extension of **AB** by finitely many new reduction patterns, provable in **L** [69].

Bar-Hillel et al. [8] shows that every ϵ-free CFG G is equivalent to a BCG G' with all types of order at most 1. By the above, the language of G' does not change, if one replaces **AB** by **L**. Consequently, every ϵ-free CFG is equivalent to a Lambek grammar. The converse holds as well [51]. Analogous results for **NL** were obtained in [11, 34].

Due to new laws, Lambek grammars provide a more flexible description of natural language. We consider atomic types s, n, as above, and n^* for plural nouns. In BCGs we get:

1. John likes Jane. $n,\ (n\backslash s)/n,\ n\Rightarrow s$.
2. John works here. $n,\ n\backslash s,\ s\backslash s\Rightarrow s$.
3. John never works. $n,\ (n\backslash s)/(n\backslash s),\ n\backslash s\Rightarrow s$.
4. John works for Jane. $n,\ n\backslash s,\ (s\backslash s)/n,\ n\Rightarrow s$.
5. John works and Jane rests. $n,\ n\backslash s,\ (s\backslash s)/s,\ n,\ n\backslash s\Rightarrow s$.

6. men work. $n^*,\ n^*\backslash s \Rightarrow s$.
7. poor men work. $n^*/n^*,\ n^*,\ n^*\backslash s \Rightarrow s$.
8. men works. $n^*,\ n\backslash s \not\Rightarrow s$.
9. John work. $n,\ n^*\backslash s \not\Rightarrow s$.

Here $\not\Rightarrow$ means that the sequent is not provable in **AB**. The sequents in 8, 9 are unprovable in **L**, either.

Now assign $s/(n\backslash s)$ to `he` and $(s/n)\backslash s$ to `her`; We abbreviate these types as np_s and np_o, respectively, since they correspond to (singular) noun phrase as subject and noun phrase as object.

10. he likes Jane. $s/(n\backslash s),\ (n\backslash s)/n,\ n \Rightarrow s$.
11. John likes her. $n,\ n\backslash(s/n),\ (s/n)\backslash s \Rightarrow s$.
12. he likes her. $s/(n\backslash s),\ (n\backslash s)/n,\ (s/n)\backslash s \not\Rightarrow s$.
13. John works for her. $n,\ n\backslash s,\ (s\backslash s)/n,\ (s/n)\backslash s \not\Rightarrow s$.

The sequent in 12 remains unprovable in **AB**, if even one replaces $(n\backslash s)/n$ by $n\backslash(s/n)$. In **L**, these two types are equivalent, by (L5), and this sequent is provable: use (Red.1) $s/n,\ (s/n)\backslash s \Rightarrow s$, (L3) (to the first and the second type of 12) and (Cut). Also the sequent in 13 is provable in **L**. Notice that `the student follows the teacher` can be parsed like 12 and `John works for a friend` like 13 (assign n_c to common nouns and np_s/n_c, np_o/n_c to articles).

These examples, similar to those in [36], well illustrate the power of Lambek grammars. In a BCG we need at least two types of `likes` (see 10, 11); they are equivalent in **L**, hence only one of them is sufficient. To parse 12 in a BCG we need additional types of words, e.g. $(s/n)/((n\backslash s)/n)$ of `he`; $s/(n\backslash s) \Rightarrow (s/n)/((n\backslash s)/n)$ is an instance of (L4), hence $s/(n\backslash s)$ is sufficient in a Lambek grammar. Even in **NL** one proves $n \Rightarrow np_s$, $n \Rightarrow np_o$ as instances of (L1). This shows that **L** provides some logical transformations of types and explains certain syntactic ambiguities of expressions. Of course, not all; we still need $n\backslash s$ and $n^*\backslash s$ for `worked`, n/n and n^*/n^* for `poor`.

Only four atomic types appear in these examples. Realistic grammars for a natural language employ much more atoms. Lambek [40] uses 33 atomic types for a fragment of English, described by a pregroup grammar (see Sect. 3.2.4). We list some of them.

$\pi =$ subject
$\pi_1 =$ first person singular subject
$\pi_2 =$ second person singular and any plural personal subject
$\pi_3 =$ third person singular subject
$s =$ statement (declarative sentence)
$s_1 =$ statement in present tense
$s_2 =$ statement in past tense
$\bar{q} =$ question
$q =$ yes-or-no question
$q_1 =$ yes-or-no question in present tense
$q_2 =$ yes-or-no question in past tense
$i =$ infinitive of transitive verb
$j =$ infinitive of complete verb phrase
$\bar{j} =$ complete infinitive with `to`
$o =$ direct object

$n =$ name
$n_0 =$ mass noun
$n_1 =$ count noun
$n_2 =$ plural noun
$\bar{n} =$ complete noun phrase
$p_1 =$ present participle
$p_2 =$ past participle

For semantic considerations, however, it is more natural to reduce the number of atomic types. Lambek's $\bar{n}$ can be defined as $s/(n\backslash s)$ or $(s/n)\backslash s$, depending on the role in a sentence (subject or object). Some authors choose np (noun phrase) as an atom and assign $(np\backslash s)/np$ to transitive verbs, instead of $(n\backslash s)/n$ (this neglects tense and number).

Every proof of $\Gamma \Rightarrow \alpha$ in the sequent system of **L** determines a unique bracketing of Γ, and similarly for **NL**. This induces a unique phrase structure of the parsed expression. Due to associativity, **L** is 'structurally omnipotent': every possible bracketing of Γ comes from some proof of $\Gamma \Rightarrow \alpha$ (if (Cut) can be used). Consequently, Lambek grammars based on **L** are not sensitive to phrase structures; they describe languages of strings.

On the contrary, Lambek grammars based on **NL** naturally describe languages of phrase structures. Kandulski [32] shows the strong equivalence of these grammars and BCGs. Therefore some linguists prefer this weaker logic. It is quite weak, indeed; neither 12, nor 13 can be parsed in **NL**, if the same types are used. Worthy of notice, with **NL** one can interchange the roles of functors and arguments. From $x : \alpha$ and $y : \alpha\backslash\beta$ we infer $(x, y)_2 : \beta$, but, using (L1), we obtain $x : \beta/(\alpha\backslash\beta)$, hence also $(x, y)_1 : \beta$.

3.2 Extensions

3.2.1 Multi-Modal Systems

To make it more flexible Moortgat [45] and other authors extend **NL** in different ways: admit several products $\otimes_i$ with residuals $\backslash_i$, $/_i$ and unary modalities $\Diamond_i$, $\Box_i^{\downarrow}$, which form a residuation pair ($\Diamond_i\alpha \Rightarrow \beta$ and $\alpha \Rightarrow \Box_i^{\downarrow}\beta$ are equivalent in models and derivable from each other in the formal system). From $\Diamond_i\alpha \Rightarrow \Diamond_i\alpha$ we obtain $\alpha \Rightarrow \Box_i^{\downarrow}\Diamond_i\alpha$, and from $\Box_i^{\downarrow}\alpha \Rightarrow \Box_i^{\downarrow}\alpha$ we obtain $\Diamond_i\Box_i^{\downarrow}\alpha \Rightarrow \alpha$.

Let us consider an example from [46]. np can be lifted up to both $\Box_n^{\downarrow}\Diamond_n np$ and $\Box_a^{\downarrow}\Diamond_a np$, where the subscripts abbreviate nominative and accusative. We assign np to `John`, `Mary`, $\Box_n^{\downarrow}\Diamond_n np$ to `he`, `she`, $\Box_a^{\downarrow}\Diamond_a np$ to `him`, `her`, and $(\Box_n^{\downarrow}\Diamond_n np\backslash s)/\Box_a^{\downarrow}\Diamond_a np$ to `likes`. The resulting grammar assigns s to `John likes Mary`, `he likes her`, but not to `her likes Mary`.

Another example comes from [45]. Let r be the type of relative clause, e.g. `that Kazimierz wrote` (in `the book that Kazimierz wrote`). With **L** we can assign $r/(s/np)$ to `that`, which yields `that Kazimierz wrote`: r, since

$$r/(s/np),\ np,\ (np\backslash s)/np \Rightarrow r$$

is provable. This sequent, however, is not provable in **NL** (with any bracketing). We assign $r/(s/\Diamond_a\Box_a^{\downarrow}np)$ to `that` and admit the weak associativity axiom:

$$(\alpha \cdot \beta) \cdot \Diamond_a\gamma \Rightarrow \alpha \cdot (\beta \cdot \Diamond_a\gamma).$$

In **NL** we prove $(np, ((np\backslash s)/np, np)) \Rightarrow s$, and consequently,

$$(np, ((np\backslash s)/np, \Diamond_a\Box_a^{\downarrow}np)) \Rightarrow s.$$

This yields $((np, (np\backslash s)/np), \Diamond_a\Box_a^{\downarrow}np) \Rightarrow s$, by the new axiom, hence

$$(np, (np\backslash s)/np) \Rightarrow s/\Diamond_a\Box_a^{\downarrow}np,$$

by $(\Rightarrow /)$. Thus, `Kazimierz wrote`: $s/\Diamond_a\Box_a^{\downarrow}np$ and `that Kazimierz wrote`: r.

These examples show the spirit of multi-modal Lambek grammars, extensively studied by a group of contemporary linguists. The unary modalities are used to construct subtypes and super-types of some types and to restrict associativity and commutativity to some special cases. (By a subtype of α we mean a type β such that $\beta \Rightarrow \alpha$ is true, not a subformula of α.) This resembles the usage of exponentials !, ? in linear logic, where structural rules (weakening, contraction) are limited to formulas $!\alpha$, $?\alpha$. More information on the multi-modal framework can be found in [44–47].

Morrill [48] elaborated Discontinuous Lambek Calculus, a special multi-modal and multi-sorted logic, intended to process types of discontinuous expressions. In language models, one admits strings with some occurrences of | (separator); $\alpha \otimes_i \beta$ denotes the substitution of β for the i-th separator in α. The interpretation of product as substitution also appeared in [11].

3.2.2 Substructural Logics

One can add lattice connectives $\wedge$, $\vee$, satisfying the lattice laws. It suffices to add:

$$\alpha \wedge \beta \Rightarrow \alpha \quad \alpha \wedge \beta \Rightarrow \beta \quad \frac{\alpha \Rightarrow \beta \quad \alpha \Rightarrow \gamma}{\alpha \Rightarrow \beta \wedge \gamma}$$

$$\alpha \Rightarrow \alpha \vee \beta \quad \beta \Rightarrow \alpha \vee \beta \quad \frac{\alpha \Rightarrow \gamma \quad \beta \Rightarrow \gamma}{\alpha \vee \beta \Rightarrow \gamma}$$

to the first axiomatization of **L** or **NL**. The corresponding sequent systems (we skip details) admit cut elimination, hence both logics are decidable. Here are the distributive laws, provable in **NL** with $\wedge$, $\vee$.

$$\alpha \cdot (\beta \vee \gamma) \Leftrightarrow (\alpha \cdot \beta) \vee (\alpha \cdot \gamma) \quad (\alpha \vee \beta) \cdot \gamma \Leftrightarrow (\alpha \cdot \gamma) \vee (\beta \cdot \gamma)$$

$$\alpha\backslash(\beta \wedge \gamma) \Leftrightarrow (\alpha\backslash\beta) \wedge (\alpha\backslash\gamma) \quad (\alpha \wedge \beta)/\gamma \Leftrightarrow (\alpha/\gamma) \wedge (\beta/\gamma)$$

$$(\alpha \vee \beta)\backslash\gamma \Leftrightarrow (\alpha\backslash\gamma) \wedge (\beta\backslash\gamma) \quad \alpha/(\beta \vee \gamma) \Leftrightarrow (\alpha/\beta) \wedge (\alpha/\gamma)$$

Let us note some simple applications of types with $\wedge$, $\vee$ in type grammars. Lambek [37] noticed that a type assignment $I(v) = \{\alpha_1, \ldots, \alpha_n\}$ could be replaced with the rigid type assignment $I(v) = \alpha_1 \wedge \cdots \wedge \alpha_n$. Another application concerns subtypes. Lambek [40] needs nonlogical assumptions $\pi_i \Rightarrow \pi$, $s_j \Rightarrow s$, for $i = 1, 2, 3$, $j = 1, 2$. Instead one can define $s = s_1 \vee s_2$, $\pi = \pi_1 \vee \pi_2 \vee \pi_3$ and apply a pure logic with $\vee$ but no nonlogical assumptions (according to the paradigm of lexicality). Kanazawa [31] proposed feature-decomposition types: `works` is of type $(np \wedge sg)\backslash s$, `work` of type $(np \wedge pl)\backslash s$, `worked` of type $np\backslash s$, and `became` of type $((np\backslash s)/(np \vee ad)$, where np, sg, pl, ad are types of noun phrase, singular, plural, and adjective, respectively.

By **L1** we denote **L** with constant 1 and the axioms:

$$1 \cdot \alpha \Leftrightarrow \alpha \quad \alpha \cdot 1 \Leftrightarrow \alpha\,.$$

L1 is strongly complete with respect to *residuated monoids*, i.e. residuated semigroups with an element 1 (the unit for product). The sequent system is obtained from that for **L** by admitting sequents $\Rightarrow \alpha$ (interpreted in algebras as $1 \leq \mu(\alpha)$), allowing Γ to be empty in rules $(\Rightarrow \backslash)$, $(\Rightarrow /)$, and adding:

$$(1 \Rightarrow)\frac{\Gamma, \Gamma' \Rightarrow \alpha}{\Gamma, 1, \Gamma' \Rightarrow \alpha} \quad (\Rightarrow 1) \ \Rightarrow 1\,.$$

NL1 can be presented in a similar way. Notice that in **L1** one proves new laws, not containing 1, nor the empty antecedent, e.g. $\alpha/(\alpha\backslash\alpha) \Rightarrow \alpha$; $\Leftarrow$ is provable in **L**. To prove the former, from $\alpha \Rightarrow \alpha$ infer $\Rightarrow \alpha\backslash\alpha$, by $(\Rightarrow \backslash)$, then apply $(/ \Rightarrow)$. This proof works in **NL1** as well.

The language models for **L1** are algebras of subsets of $\Sigma^* = \Sigma^+ \cup \{\epsilon\}$; the operations $\cdot, \backslash, /$ for languages are defined as above except that Σ^+ is replaced with Σ^*. The language $\{\epsilon\}$ is the unit for product. The intended models for **NL1** employ languages of phrase structures, now enriched with the empty structure ϵ such that $(\epsilon, x) = (x, \epsilon) = x$ for any phrase structure x.

Some linguists object the suitability of **L1** as a logic for type grammars. $\alpha/\alpha \Leftrightarrow (\alpha/\alpha)/(\alpha/\alpha)$ is provable in **L1**, hence n_c/n_c and $(n_c/n_c)/(n_c/n_c)$ are equivalent, but the former is a natural type of adjectives and the latter of adverbs.

On the other hand, logicians prefer **L1** and its extensions. In these systems, some formulas are provable (a formula α is said to be *provable*, if $\Rightarrow \alpha$ is provable); for example, $\alpha\backslash\alpha$ in **NL1** and $(\alpha\backslash\beta)\backslash((\gamma\backslash\alpha)\backslash(\gamma\backslash\beta))$ in **L1**. Furthermore, every sequent is deductively equivalent to a formula, e.g. $\alpha, \beta \Rightarrow \gamma$ to $\beta\backslash(\alpha\backslash\gamma)$. Accordingly, these logics can be presented in the form of Hilbert style systems and more easily compared with other nonclassical logics. For example, the product-free **L1** can be axiomatized as a Hilbert style system with the following axioms and rules.

$$\text{(a.1) } 1 \quad \text{(a.2) } 1\backslash(\alpha\backslash\alpha) \quad \text{(a.3) } ((\alpha\backslash\beta)/\gamma)\backslash(\alpha\backslash(\beta/\gamma)) \quad \text{(a.4) } (\alpha\backslash(\beta/\gamma))\backslash((\alpha\backslash\beta)/\gamma)$$

$$\text{(a.5) } (\alpha\backslash\beta)\backslash((\gamma\backslash\alpha)\backslash(\gamma\backslash\beta)) \quad \text{(a.6) } ((\alpha/\gamma)/(\beta/\gamma))/(\alpha/\beta)$$

$$(mp\backslash)\frac{\alpha \quad \alpha\backslash\beta}{\beta} \quad (\backslash - /)\frac{\alpha\backslash\beta}{\beta/\alpha}$$

For the 1-free fragment, (a.1), (a.2) are replaced by (id) $\alpha\backslash\alpha$. Other axiom systems can be found in [70] and for richer logics in [24].

L1 with $\wedge$, $\vee$ is called Full Lambek Calculus (**FL**) and regarded as a basic substructural logic [24]. Substructural logics can be defined as axiom and rule extensions of **FL**. They correspond to some classes (usually varieties or quasi-varieties) of *residuated lattices*, i.e. lattice-ordered residuated monoids. One often adds a new constant 0 and defines *negations*: $\sim\alpha = \alpha\backslash 0$, $-\alpha = 0/\alpha$ (0 is interpreted as an arbitrary element of the residuated lattice).

The term 'substructural logics' refers to the fact that sequent systems for these logics lack some structural rules, characteristic of the Gentzen system for intuitionistic logic: exchange (e), contraction (c), left weakening or integrality (i), right weakening (o). The first three rules have the following forms.

$$(e)\frac{\Gamma, \alpha, \beta, \Gamma' \Rightarrow \gamma}{\Gamma, \beta, \alpha, \Gamma' \Rightarrow \gamma} \quad (c)\frac{\Gamma, \Delta, \Delta, \Gamma' \Rightarrow \alpha}{\Gamma, \Delta, \Gamma' \Rightarrow \alpha}$$

$$(i)\frac{\Gamma, \Gamma' \Rightarrow \beta}{\Gamma, \alpha, \Gamma' \Rightarrow \beta}$$

They express some algebraic properties of product: (e) $a \cdot b = b \cdot a$, (c) $a \leq a \cdot a$, (i) $a \cdot b \leq a$, $a \cdot b \leq b$ (with 1 this amounts to $a \leq 1$). (o) involves sequents of the form $\Gamma \Rightarrow$ (interpreted as $\mu(\Gamma) \leq 0$). They can be eliminated, and (o) can be replaced by the axiom $\Gamma, 0, \Gamma' \Rightarrow \alpha$.

For logics with (e), corresponding to commutative algebras, $\alpha\backslash\beta$ is equivalent to β/α (in algebras $a\backslash b = b/a$), and one writes $\alpha \rightarrow \beta$ for both. The sequent systems are simpler (we omit details). Also $\sim\alpha$ is equivalent to $-\alpha$, and one writes $\neg\alpha$ for both.

From the algebraic point of view, substructural logics treat implication(s) as residual(s) of the product operation; the latter usually differs from the lattice meet. Many well-known nonclassical logics belong to this family: relevant logics, many-valued logics, fuzzy logics, and intuitionistic and classical logics as the limit cases. For instance, intuitionistic logic amounts to **FL** with (e), (c), (i), (o) (in fact, (e) is derivable with (c), (i)), and Łukasiewicz infinitely valued logic to **FL** with (e), (i), (o) and the axiom $\alpha \vee \beta \Leftrightarrow (\alpha \rightarrow \beta) \rightarrow \beta$ [24].

Linear logic of Girard [26] can be presented as **FL** with 0, (e) and the double negation axiom $\neg\neg\alpha \Rightarrow \alpha$ ($\Leftarrow$ is provable); we neglect exponentials !, ?. Noncommutative versions are due to Yetter [68] and Abrusci [1]. The former can be presented as **FL** with 0 and the axioms $\alpha\backslash 0 \Leftrightarrow 0/\alpha$ (hence $\sim\alpha$ and $-\alpha$ collapse in $\neg\alpha$) and $\neg\neg\alpha \Rightarrow \alpha$; the latter as **FL** with 0 and the axioms $\sim -\alpha \Rightarrow \alpha$, $- \sim \alpha \Rightarrow \alpha$ (again $\Leftarrow$ are provable). Both logics are conservative extensions of **FL** without 0 [2]. In [24] they are called Cyclic Involutive Full Lambek Calculus (**CyInFL**) and Involutive Full Lambek Calculus (**InFL**), respectively.

Cut-free sequent systems of linear logics look differently. Formulas are built from atoms by $\otimes$, $\oplus$ and negation(s), where $\otimes$ stands for product and $\oplus$ for the dual product ('par'). In algebras, $a \oplus b = \neg(\neg b \otimes \neg a)$ for Girard's logic and **CyInFL** and $a \oplus b = \sim(-b \otimes -a)$ for **InFL**. One employs classical sequents $\Gamma \Rightarrow \Delta$ or one-sided sequents only: either $\Rightarrow \Delta$ (Schütte style), or $\Gamma \Rightarrow$ (dual Schütte style). Each comma in Γ is interpreted as product and in Δ as dual product. In **InFL**, presented in this way, one can define $\backslash$, $/$ as follows: $\alpha\backslash\beta = \sim\alpha \oplus \beta$, $\alpha/\beta = \alpha \oplus -\beta$.

In the literature on linear logics, $\otimes, \oplus, \backslash, 1, 0$ and negation(s) are referred to as *multiplicatives* and $\wedge, \vee$ (also constants $\top, \bot$, interpreted as the greatest and the least element) as *additives*. According to a different tradition, they are intensional and extensional connectives and constants, respectively. **L1** is often characterized as the intuitionistic fragment of multiplicative linear logic.

Type grammars usually employ basic intuitionistic substructural logics, often not admitting empty antecedents of sequents and being restricted to multiplicative connectives (also multi-modal). Nonetheless the impact of linear logics (which are 'classical') can be seen in current developments. I have already noted an analogy between modalities in type grammars and exponentials of Girard [26]. Also *proof nets*, i.e. a representation of proofs in multiplicative linear logics by means of some graphs of links between formulas, are used as representations of syntactic structures in type grammars, either directly, or in a form suitable for intuitionistic fragments. We cannot discuss this matter here; the reader is referred to [46].

In language models, $\wedge, \vee$ can be interpreted as the set theoretic intersection and union of languages. Then, we obtain a distributive lattice. The distributive laws for $\wedge, \vee$ are not provable in **FL**, nor other logics, discussed above. One can add them as new axioms; it suffices to add:

$$(\mathrm{D})\ \alpha \wedge (\beta \vee \gamma) \Rightarrow (\alpha \wedge \beta) \vee (\alpha \wedge \gamma)\,.$$

Nevertheless, some interesting linguistic interpretations of logics without (D) are possible. Clark [18] introduced *syntactic concept lattices* as a special kind of concept lattices from lattice theory. Let $L_0 \subseteq \Sigma^*$ be a fixed language. Pairs (x, y), for $x, y \in \Sigma^*$, are called *contexts*. For a set of contexts X, one defines $X^{\lhd}$ as the set of all $z \in \Sigma^*$ such that $xzy \in L_0$, for all $(x, y) \in X$. The sets of the form $X^{\lhd}$ are called *syntactic concepts* for L_0. They can be interpreted as the syntactic categories determined by L_0, a reasonable generalization of Husserl's idea, followed by Ajdukiewicz. Since **L** provides nontrivial laws $\alpha \Rightarrow \beta$, syntactic categories in Lambek grammars cannot be equivalence classes (substitution classes). The family of syntactic concepts for L_0 is a complete residuated lattice with operations: $X \wedge Y = X \cap Y$, $X \vee Y$ (resp. $X \otimes Y$, 1) equal to the smallest concept containing $X \cup Y$ (resp. $X \cdot Y$, $\{\epsilon\}$), and $\backslash, /$ defined as for languages.

3.2.3 Semantic Types

The product-free **L** with (e) was studied by van Benthem [62, 63] as a logic of semantic types; we call this logic the Lambek-van Benthem calculus (**LB**). Proofs in a natural deduction system (ND-system) for **LB** can be encoded by some terms of typed lambda calculus, namely linear terms (i.e. every λ binds exactly one occurrence of a variable), satisfying the additional constraint: no subterm is closed. This is an adaptation of the 'Curry-Howard isomorphism' between ND-proofs and lambda terms [55]. Since every ND-proof in **L** can be translated into an ND-proof in **LB**, the former determines a unique lambda term; this lambda term, interpreted in a standard type-theoretic model (see Sect. 2), denotes a semantic transformation corresponding to the syntactic parsing in the grammar. Size limits do not allow us to discuss this framework in detail. Let us consider one example. Recall that the characteristic inference rules of ND-systems are the introduction rules and the elimination rules for connectives.

From $n \Rightarrow n$ and $n\backslash s \Rightarrow n\backslash s$ we get $n, n\backslash s \Rightarrow s$, by the $\backslash$-elimination rule: from $\Gamma \Rightarrow \alpha$ and $\Delta \Rightarrow \alpha\backslash\beta$ infer $\Gamma, \Delta \Rightarrow \beta$ (in an ND-system for **L**). This is translated in **LB** as: from $e \to t \Rightarrow e \to t$ and $e \Rightarrow e$ infer $e \to t, e \Rightarrow t$, by the $\to$-elimination rule (in an ND-system for **LB**). In **L** we obtain $n \Rightarrow s/(n\backslash s)$, by the /-introduction rule, which is translated into $e \Rightarrow (e \to t) \to t$ in **LB**. The ND-proof in **LB** is encoded by the term:

$$\lambda y^{e\to t}.y^{e\to t}x^{e}.$$

This is a linear term, satisfying the additional constraint. In a type-theoretic model, if x^e is valuated as $a \in D_e$ (an individual), then this term denotes the family of all (characteristic functions of) $X \subseteq D_e$ such that $a \in X$. Thus, the syntactic law $n \Rightarrow s/(n\backslash s)$ corresponds to the semantic transformation which sends an individual into the family of all properties (interpreted extensionally) of this individual. In particular, any proper noun (denoting an individual) can be treated as a noun phrase (denoting a generalized quantifier, i.e. a family of sets of individuals).

We have explained a semantic interpretation of (L1). Similarly, (L3) correspond to the composition of functions, (L5) to the interchange of arguments, and so on. An interesting theorem of [63] shows that every provable sequent of **LB** admits only finitely many different proofs up to the equality in lambda calculus; so every expression possesses only finitely many 'semantic readings'. Further studies on this topic can be found in [43, 45].

Analogous correspondences were elaborated for richer logics [66]. Lambek [38] studied category-theoretic interpretations of **L** and its extensions. Abstract Categorial Grammars, introduced by de Groote [20], employ linear lambda-terms as representations of both syntactic structures and semantic structures (logical forms) of expressions in natural language with certain homomorphisms between them.

3.2.4 Pregroup Grammars

Lambek [39] proposed another extension of **L1**, called compact bilinear logic (**CBL**). It corresponds to *pregroups*, i.e. ordered algebras $(A, \cdot, ^r, ^l, 1, \leq)$ such that $(A, \cdot, 1, \leq)$ is a partially ordered monoid and $^r, ^l$ are unary operations, satisfying *the adjoint laws*:

$$a^l \cdot a \leq 1 \leq a \cdot a^l \text{ and } a \cdot a^r \leq 1 \leq a^r \cdot a\,,$$

for any $a \in A$. a^r (resp. a^l) is called the *right* (resp. *left*) *adjoint* of a. (This terminology is transferred from category theory.) Pregroups coincide with the algebras for the multiplicative fragment of noncommutative linear logic of Abrusci [1] such that $\otimes$ equals $\oplus$ and $1 = 0$. The residuals of product are defined by: $a\backslash b = a^r \cdot b$, $a/b = a \cdot b^l$. We have: $a^{rl} = a^{lr} = a$, $(a \cdot b)^r = b^r \cdot a^r$, and similarly for ^{l}. Adjoints reverse the ordering: if $a \leq b$ then $b^r \leq a^r$ and $b^l \leq a^l$.

CBL is a logic of free pregroups. From atoms $p, q, r, \ldots$ one builds *simple types* $p^{(n)}$, where n is an integer. $p^{(0)}$ is interpreted as p, $p^{(n)}$, $n > 0$, as $p^{r \ldots r}$ (n times), and $p^{(n)}$, $n < 0$, as $p^{l \ldots l}$ ($|n|$ times). *Pregroup types* are finite strings of simple types.

One also assumes that the set of atoms is partially ordered by a relation $\preceq$. The relation $\Rightarrow$, between pregroup types, is defined by the following rewriting rules:

- (Contraction) $X, p^{(n)}, p^{(n+1)}, Y \Rightarrow X, Y$,
- (Expansion) $X, Y \Rightarrow X, p^{(n+1)}, p^{(n)}, Y$,
- (Induced Step) $X, p^{(n)}, Y \Rightarrow X, q^{(n)}, Y$, if either $p \preceq q$ and n is even, or $q \preceq p$ and n is odd.

$U \Rightarrow V$ holds, if U can be transformed into V by finitely many applications of these rules.

Pregroup grammars are defined as Lambek grammars except that **L** is replaced by **CBL** (and types of **L** by pregroup types). Lambek [39] shows that (Expansion) can be eliminated from proofs of $X \Rightarrow p$, where p is an atom. Furthermore, (Induced Step) and (Contraction) can be collapsed with one rule of generalized contraction:

- (GCON) $X, p^{(n)}, q^{(n+1)}, Y \Rightarrow X, Y$, with the same condition as in (Induced Step).

If $\alpha_1, \ldots, \alpha_n$ are assigned to $v_1, \ldots, v_n$, respectively, then the grammar assigns p to $v_1 \ldots v_n$, if the concatenation $\alpha_1 \cdots \alpha_n$ reduces to p by a finite number of applications of (GCON) and, possibly, (Induced Step) at the end of the reduction. Such derivations can be presented by means of links, joining the reduced types of (GCON).

For example, we assume `she`: π_3, `will`: $\pi^r s_1 j^l$, `see`: io^l and `him`: o, where i, j are types of infinitive of intransitive verb and infinitive of any complex verb phrase, and π, π_3, o are understood as above. We also assume $\pi_3 \preceq \pi$, $i \preceq j$. Then, `she will see him` is assigned s_1, since:

$$\pi_3,\ \pi^r s_1 j^l,\ io^l,\ o \Rightarrow s_1\,.$$

The reduction can be depicted as follows:

$$\pi_3,\ \pi^r\, s_1\, j^l,\ i\, o^l,\ o$$

where each link corresponds to one application of (GCON).

With `man`: n_1 (count noun), `whom`: $n_1^r n_1 o^{ll} s^l$, `saw`: $\pi^r s_2 o^l$ and $s_2 \preceq s$, one assigns n_1 to `man whom she saw`, by the reduction:

$$n_1,\ n_1^r n_1\, o^{ll}\, s^l,\ \pi_3,\ \pi^r\, s^2\, o^l\,.$$

These examples come from [40] (up to minor changes), where Lambek analyzed many basic grammatical constructions of English within the pregroup framework. In other publications he and his collaborators applied this approach to several languages: German, French, Italian, Polish and some non-European languages; see [40] for references.

Parsing by pregroups is computationally simple; it runs in polynomial time [14], whereas **L** is NP-complete [53]. **CBL** is stronger than **L1**: $(p/((p/p)/p))/p \Rightarrow p$ is provable in **CBL** (define / as above), but not in **L1**. The logical meaning of the new laws is not clear; the latter does not hold even in classical logic (interpret / as implication with the antecedent on the right). No type-theoretic semantics for pregroup grammars is known.

It seems that **CBL** is an algebraic calculus rather than a genuine logic. This opinion is confirmed by the fact that bounded pregroups are trivial (one-element) algebras, hence **CBL** with $\top$ is inconsistent [14]. (The latter paper shows that pregroup grammars are equivalent to CFGs.)

On the other hand, all linguistic examples, analyzed by Lambek and other authors by means of pregroups, can easily be parsed with **L**. We return to `man whom she saw`. The pregroup types, given above, are translations of **L**-types; e.g. `whom`: $(n_1\backslash n_1)/(s_2/o)$, `saw`: $(\pi\backslash s_2)/o$. The sequent:

$$n_1,\ (n_1\backslash n_1)/(s/o),\ \pi_3,\ (\pi\backslash s_2)/o \Rightarrow n_1$$

is provable in **L** augmented with $s_2 \Rightarrow s$, $\pi_3 \Rightarrow \pi$. Therefore, the semantics for these examples can be transferred from **L**.

3.2.5 Modal Logics

At the end, we consider other modal logics, extending **L** and **NL**. Buszkowski and Farulewski [16] studied **NL** with $\wedge, \vee$, which satisfy the laws of a distributive lattice, and its extensions with either classical negation (**BFNL**), or intuitionistic implication and $\top, \bot$ (**HFNL**); these logics were presented as sequent systems with cut. Hilbert-style systems for the latter logics, denoted by **NLC** and **NLI**, were studied in [29, 30]. The connectives are $\wedge, \vee, \Rightarrow, \neg$ (now $\Rightarrow$ stands for the classical or intuitionistic implication, and $\neg$ for the classical or intuitionistic negation) and Lambek connectives $\cdot, \backslash, /$. Lambek's sequents $\alpha \Rightarrow \beta$ are treated as conditionals. **NLC** (resp. **NLI**) can be axiomatized by all tautologies of classical (resp. intuitionistic) propositional logic in the extended language and the rules: *modus ponens* for $\Rightarrow$, (Res.1), (Res.2). In the associative versions **LC**, **LI** one adds axioms (A.1), (A.2). The following formulas, similar to the modal axiom (K), are provable in **NLI**, hence also in **NLC**, **LI**, **LC** (we assume that $\backslash, /$ bind stronger than $\Rightarrow$).

$$\gamma\backslash(\alpha \Rightarrow \beta) \Rightarrow (\gamma\backslash\alpha \Rightarrow \gamma\backslash\beta) \quad (\alpha \Rightarrow \beta)/\gamma \Rightarrow (\alpha/\gamma \Rightarrow \beta/\gamma)$$

It should be emphasized that the theorems (i.e. provable formulas) of these systems are $\top$-theorems: they satisfy $\mu(\alpha) = \top$ in algebras. In substructural logics one usually considers 1-theorems ($1 \leq \mu(\alpha)$ in algebras). Both notions collapse for substructural logics with (i). **LC** is a conservative extension of **L**, and **NLC** of **NL**.

NLC, **LC** and **NLI**, **LI** are, in fact, some classical and intuitionistic multi-modal logics; product and its residuals are binary modalities. This perspective was already admitted in Arrow Logic of van Benthem [64] and multi-modal versions of Lambek calculi. Kaminski and Francez [29, 30] study relational frames for **NLC**, **LC**, **NLI**, **LI**, proving some completeness and decidability results. Interestingly, the undecidability of **LC** follows from some results of [35], whereas **LI** is decidable [30]. For **NLC**, **NLI** even the consequence relations are decidable [16].

3.3 Lambek Versus Ajdukiewicz

Although Lambek logics are much stronger than **AB**, the parsing procedure in Lambek grammars can be carried out in a similar way as in BCGs. The action of **L** and related systems can be reduced to the lexical level: the type lexicon is extended by new types, derivable from the initial types in the system.

For example, if $\alpha, \beta, \gamma \Rightarrow \delta$ is provable in **L**, then $\alpha \Rightarrow (\delta/\gamma)/\beta$ is provable, by $(\Rightarrow /)$, and the sequent:

$$(\delta/\gamma)/\beta, \beta, \gamma \Rightarrow \delta$$

is provable in **AB**. For **NL**, if $\alpha, (\beta, \gamma) \Rightarrow \delta$ is provable, then $\alpha \Rightarrow \delta/(\beta \cdot \gamma)$ is provable, and $\delta/(\beta \cdot \gamma), (\beta, \gamma) \Rightarrow \delta$ is provable in **AB** (with product; see [32]). To eliminate product, one can use $\beta \Rightarrow (\alpha\backslash\delta)/\gamma$ and prove in **AB**:

$$\alpha, (\alpha\backslash\delta)/\gamma, \gamma \Rightarrow \delta\,.$$

For $v_i : \alpha_i, i = 1, \ldots, n$, a successful parsing can be arranged as in Fig. 2 ($\alpha_i \Rightarrow \beta_i$ is provable in a Lambek logic).

In the same way one can arrange semantic derivations: the semantic transformations, definable in (a fragment of) lambda calculus, can be performed on the initial denotations of words in a type-theoretic model, and the denotations of compound expressions are obtained by the (iterated) application of functions to their arguments.

Such laws as (L1), (L2), (L4) produce infinitely many types β derivable from a single type α. For instance, starting from n, one derives:

$$n \Rightarrow s/(n\backslash s) \Rightarrow s/((s/(n\backslash s))\backslash s) \Rightarrow \cdots\,.$$

Nonetheless only finitely many of them are really needed to parse any expression in a particular grammar. Buszkowski [12] shows that every type grammar G, based on **L**, is equivalent to a BCG G' whose type lexicon extends that of G by finitely many new types, derivable in **L** from those in the type lexicon of G. The same was earlier shown for **NL** in [32].

These results seem to support the opinion that Lambek logics can be regarded as general logics of syntactic or semantic types rather than type processing systems in type grammars. The former explain deeper reasons for syntactic ambiguities of expressions and guide our choice of lexical types. On the other hand, parsing can be based on the classical type reduction procedure, proposed by Ajdukiewicz, with necessary modifications.

Fig. 2 Reducing **L**-parsing to **AB**-parsing

$$\begin{array}{ccccc} v_1 & \ldots & v_n & & \\ \alpha_1 & \ldots & \alpha_n & \Rightarrow & s \\ \Downarrow & \ldots & \Downarrow & & \\ \beta_1 & \ldots & \beta_n & \overset{\mathbf{AB}}{\Rightarrow} & s \end{array}$$

This opinion is non-orthodox. Many authors maintain the priority of Lambek logics, directly applied in grammars, according to the general paradigm of *parsing as deduction*. They, however, usually ignore the problems of efficiency. Parsers for BCGs can be designed like for CFGs; they run in cubic time in the length of the parsed expression. This is impossible for type grammars based on **L**, which is NP-complete [53]. Type grammars with **NL** remain polynomial [21], but parsers are not as simple as for BCGs.

At the end of this subsection, let me mention some developments in type grammars, which are closer to Ajdukiewicz.

Combinatory Categorial Grammars (CCGs), developed by M. Steedman, A. Szabolcsi and others, enrich **AB** with finitely many new reduction patterns, semantically corresponding to some combinators, i.e. closed lambda-terms; see [57] for an overview. This direction continues certain ideas of Curry [19] and Shaumyan [54]. Some of the new patterns are provable in **L**, but others require a stronger logic (some instances of exchange and contraction). The Ajdukiewicz procedure enriched with composition laws (similar to (L3), (L4)) was earlier proposed by Geach [25].

Categorial Unification Grammars (CUGs), studied by Uszkoreit [61], admit polymorphic types, containing variables, which range over a family of types. The simplest example is $(x\backslash x)/x$ as the type of and. In the course of parsing, one applies the reduction rules of BCGs and a unification algorithm. For instance, $\alpha, \beta\backslash\gamma \Rightarrow \sigma(\gamma)$, where σ is a substitution such that $\sigma(\alpha) = \sigma(\beta)$.

L with $\wedge, \vee$ can generate some non-context-free languages, e.g. the intersection of two context-free languages [31]. This also holds for grammars based on **AB** with $\wedge, \vee$. Other frameworks going beyond the context-free world are Tupled Pregroup Grammars [56] and Categorial Dependency Grammars [22]. Both approaches employ very restricted types only; the resulting grammars might be presented as BCGs with all types of order at most 1 and certain constraints imposed on reductions.

References

1. Abrusci, V.M.: Phase semantics and sequent calculus for pure noncommutative classical linear logic. J. Symb. Log. **56**, 1403–1451 (1991)
2. Abrusci, V.M.: Classical conservative extensions of Lambek calculus. Studia Logica **71**, 277–314 (2002)
3. Ajdukiewicz, K.: W sprawie 'uniwersaliów' [On the problem of 'universals']. Przegląd Filozoficzny **37**, 219–234 (1934)
4. Ajdukiewicz, K.: Die syntaktische Konnexität (Syntactic connexion). Studia Philosophica **1**, 1–27 (1935)
5. Ajdukiewicz, K.: Związki składniowe między członami zdań oznajmujących (Syntactic connections between constituents of declarative sentences). Studia Filozoficzne **6**(21), 73–86 (1960)
6. Ajdukiewicz, K.: Giedymin, J. (ed.) The Scientific World-Perspective and Other Essays, pp. 1931–1963. D. Reidel, Dordrecht (1978)
7. Bar-Hillel, Y.: A quasi-arithmetical notation for syntactic description. Language **29**, 47–58 (1953)
8. Bar-Hillel, Y., Gaifman, C., Shamir, E.: On categorial and phrase structure grammars. Bull. Res. Counc. Isr. **F9**, 155–166 (1960)
9. Bocheński, J.: On the syntactical categories. New Scholasticism **23**, 257–280 (1949)
10. Buszkowski, W.: Some decision problems in the theory of syntactic categories. Zeitschrift f. mathematische Logik und Grundlagen der Math. **28**, 539–548 (1982)

11. Buszkowski, W.: Logiczne podstawy gramatyk kategorialnych Ajdukiewicza-Lambeka [Logical foundations of Ajdukiewicz-Lambek categorial grammars]. Państwowe Wydawnictwo Naukowe (1989)
12. Buszkowski, W.: Extending Lambek grammars to basic categorial grammars. J. Log. Lang. Inf. **5**, 279–295 (1996)
13. Buszkowski, W.: The Ajdukiewicz calculus, Polish notation and Hilbert-style proofs. In: Woleński, J. (ed.) The Lvov-Warsaw School and Contemporary Philosophy, pp. 241–252. Springer Netherlands, Dordrecht (1998)
14. Buszkowski, W.: Lambek grammars based on pregroups. In: Logical Aspects of Computational Linguistics. Lecture Notes in Computer Science, vol. 2099, pp. 95–109. Springer, Berlin (2001)
15. Buszkowski, W.: Syntactic Categories and Types: Ajdukiewicz and Modern Categorial Grammars. In: Brożek, A., et al. (eds.) Tradition of the Lvov-Warsaw School: Ideas and Continuations, pp. 35–71. Leiden, Boston, Brill-Rodopi (2016)
16. Buszkowski, W., Farulewski, M.: Nonassociative Lambek calculus with additives and context-free languages. In: Languages: From Formal to Natural. Lecture Notes in Computer Science, vol. 5533, pp. 45–58. Springer, Berlin (2008)
17. Buszkowski, W., Marciszewski, W., van Benthem, J. (eds.): Categorial Grammar. John Benjamins, Amsterdam (1988)
18. Clark, A.: A learnable representation of syntax using residuated lattices. In: Formal Grammar. Lecture Notes in Computer Science, vol. 5591, pp. 183–198. Springer, Berlin (2011)
19. Curry, H.B.: Some logical aspects of grammatical structure. In: Jakobson, R. (ed.) Structure of Language and Its Mathematical Aspects. Proceedings of Symposium in Applied Mathematics, pp. 56–68. American Mathematical Society, Providence, RI (1961)
20. de Groote, Ph.: Towards abstract categorial grammars. In: Proceedings of 39th Annual Meeting and 10th Conference of the European Chapter of ACL, pp. 148–155. Association of Computational Linguistics (2001)
21. de Groote, Ph., Lamarche, F.: Classical Non-associative Lambek calculus. Studia Logica **71**, 355–388 (2002)
22. Dekhtyar, M., Dikovsky, A., Karlov, B.: Categorial dependency grammars. Theor. Comput. Sci. **579**, 33–63 (2015)
23. Došen, K.: A brief survey of frames for the Lambek calculus. Zeitschrift f. mathematische Logik und Grundlagen der Math. **38**, 179–187 (1992)
24. Galatos, N., Jipsen, P., Kowalski, T., Ono, H.: Residuated Lattices: An Algebraic Glimpse at Substructural Logics. Elsevier, Amsterdam (2007)
25. Geach, P.T.: A program for syntax. Synthese **22**, 3–17 (1971)
26. Girard, J.-Y.: Linear logic. Theor. Comput. Sci. **50**, 1–102 (1987)
27. Husserl, E.: Logische Untersuchungen Halle (1900–1901)
28. Jakobson, R. (ed.): Structure of Language and Its Mathematical Aspects. In: Proceedings of Symposium in Applied Mathematics. American Mathematical Society, Providence, RI (1961)
29. Kaminski, M., Francez, N.: Relational semantics of the Lambek calculus extended with classical propositional logic. Studia Logica **102**, 479–497 (2014)
30. Kaminski, M., Francez, N.: The Lambek calculus extended with intuitionistic propositional logic. Studia Logica **104**, 1051–1082 (2016)
31. Kanazawa, M.: The Lambek calculus enriched with additional connectives. J. Log. Lang. Inf. **1**, 141–171 (1992)
32. Kandulski, M.: Phrase-structure languages generated by categorial grammars with product. Zeitschrift f. mathematische Logik und Grundlagen der Math. **34**, 373–383 (1988)
33. Kandulski, M.: The non-associative Lambek calculus. In: Buszkowski, W., Marciszewski, W., van Benthem, J. (eds.) Categorial Grammar, pp. 141–151. John Benjamins, Amsterdam (1988)
34. Kandulski, M.: The equivalence of nonassociative Lambek categorial grammars and context-free grammars. Zeitschrift f. mathematische Logik und Grundlagen der Math. **34**, 41–52 (1988)
35. Kurucz, A., Németi, I., Sain, I., Simon, A.: Decidable and undecidable logics with a binary modality. J. Log. Lang. Inf. **4**, 191–206 (1995)

36. Lambek, J.: The mathematics of sentence structure. Am. Math. Mon. **65**,154–170 (1958)
37. Lambek, J.: On the calculus of syntactic types. In: Jakobson, R. (ed.) Structure of Language and Its Mathematical Aspects. In: Proceedings of Symposium in Applied Mathematics, pp. 166–178. American Mathematical Society, Providence, RI (1961)
38. Lambek, J.: Deductive systems and categories I. J. Math. Syst. Theory **2**, 287–318 (1968)
39. Lambek, J.: Type grammars revisited. In: Logical Aspects of Computational Linguistics. Lecture Notes in Computer Science, vol. 1582, pp. 1–27. Springer, Berlin (1999)
40. Lambek, J.: From Word to Sentence: A Computational Algebraic Approach to Grammar. Polimetrica, Milano (2008)
41. Leśniewski, S.: Grundzüge eines neuen System der Grundlagen der Mathematik. Fundam. Math. **14**, 1–81 (1929)
42. Montague, R.: English as a formal language. In: Thomason, R. (ed.) Formal Philosophy: Selected Papers of Richard Montague, pp. 188–221. Yale University Press, New Haven (1974)
43. Moortgat, M.: Categorial Investigations. Logical and Linguistic Aspects of the Lambek Calculus. Foris, Dordrecht (1988)
44. Moortgat, M.: Multimodal linguistic inference. J. Log. Lang. Inf. **5**, 349–385 (1996)
45. Moortgat, M.: Categorial type logics. In: van Benthem, J., ter Meulen. A. (eds.) Handbook of Logic and Language, pp. 93–177. Elsevier/MIT, Amsterdam/Cambridge (1997)
46. Moot, R., Retoré, C.: The Logic of Categorial Grammars: A Deductive Account of Natural Language Syntax and Semantics. Lecture Notes in Computer Science, vol. 6850. Springer, Berlin (2012)
47. Morrill, G.: Type-Logical Grammar: Categorial Logic of Signs. Kluwer, Dordrecht (1994)
48. Morrill, G.: Categorial Grammar: Logical Syntax, Semantics, and Processing. Oxford University Press, Oxford (2011)
49. Nowaczyk, A.: Categorial languages and variable-binding operators. Studia Logica **37**, 27–39 (1978)
50. Oehrle, R.T., Bach, E., Wheeler, D. (eds.): Categorial Grammars and Natural Language Structures. D. Reidel, Dordrecht (1988)
51. Pentus, M.: Lambek grammars are context-free. In: Proceedings of 8th IEEE Symposium on Logic in Computer Science, pp. 429–433 (1993)
52. Pentus, M.: Models for the Lambek calculus. Ann. Pure Appl. Logic **75**, 179–213 (1995)
53. Pentus, M.: Lambek calculus is NP-complete. Theor. Comput. Sci. **357**, 186–201 (2006)
54. Shaumyan, S.: Applicational Grammar as a Semantic Theory of Natural Language. Edinburgh University Press, Edinburgh (1977)
55. Sørensen, M.H., Urzyczyn, P.: Lectures on the Curry-Howard Isomorphism. Elsevier, Amsterdam (2006)
56. Stabler, E.: Tupled pregroup grammars. In: Casadio, C., Lambek, J. (eds.) Computational Algebraic Approaches to Natural Language, pp. 23–52. Polimetrica, Milano (2008)
57. Steedman, M.: Categorial grammar. Lingua **90**, 221–258 (1993)
58. Suszko, R.: Syntactic structure and semantical reference I. Studia Logica **8**, 213–244 (1958)
59. Suszko, R.: Syntactic structure and semantical reference II. Studia Logica **9**, 63–91 (1960)
60. Tałasiewicz, M.: Philosophy of Syntax. Foundational Topics. Trends in Logic, vol. 29. Springer, Berlin (2010)
61. Uszkoreit, H.: Categorial unification grammar. In: Proceedings of 11th International Conference on Computational Linguistics, pp. 187–194 (1986)
62. van Benthem, J.: Essays in Logical Semantics. D. Reidel, Dordrecht (1986)
63. van Benthem, J.: The semantics of variety in categorial grammar. In: Buszkowski, W., Marciszewski, W., van Benthem, J. (eds.) Categorial Grammar, pp. 37–55. John Benjamins, Amsterdam (1988)
64. van Benthem, J.: Language in Action. Categories, Lambdas and Dynamic Logic. North-Holland, Amsterdam (1991)
65. van Benthem, J., ter Meulen. A. (eds.): Handbook of Logic and Language. Elsevier/MIT, Amsterdam/Cambridge (1997)
66. Wansing, H.: The Logic of Information Structures. Ph.D. Thesis, University of Amsterdam (1992)
67. Wybraniec-Skardowska, U.: Theory of Language Syntax. Kluwer, Dordrecht (1991)
68. Yetter, D.N.: Quantales and (non-commutative) linear logic. J. Symb. Log. **55**, 41–64 (1990)

69. Zielonka, W.: Axiomatizability of Ajdukiewicz-Lambek calculus by means of cancellation schemes. Zeitschrift f. mathematische Logik und Grundlagen der Math. **27**, 215–224 (1981)
70. Zielonka, W.: Weak implicational logics related to the Lambek calculus - Gentzen versus Hilbert formalisms. In: Towards Mathematical Philosophy. Trends in Logic, vol. 28, pp. 201–212. Springer, Berlin (2009)

W. Buszkowski (✉)
Adam Mickiewicz University, Poznań, Poland
e-mail: buszko@amu.edu.pl

Material Implication and Conversational Implicature in Lvov-Warsaw School

Rafal Urbaniak and Michał Tomasz Godziszewski

Abstract The relation between indicative conditionals in natural language and material implication wasn't a major topic in the Lvov-Warsaw school. However, a major defense of the claim that the truth conditions of these two are the same has been developed by Ajdukiewicz (Studia Logica IV:117–134, 1956). The first major goal of this paper is to present, assess, and improve his strategy. It turns out that it is quite similar to the approach developed by Grice (Studies in the Way of Words, Harvard University Press, Cambridge, MA, 1991), so our second goal is to compare these two and to argue that the accuracy of Ajdukiewicz's explanation is less dependent on controversial properties of a systematic but convoluted general theory of cooperative communicative behavior. In Lvov-Warsaw school the relation between material implication and indicative conditionals was also discussed by Gołąb (Matematyka 3(5):27–29, 1949) and Słupecki (Matematyka 4(6):32–35, 1949), so the third part of our paper is devoted to their discussion and relating it to Ajdukiewicz's views.

Keywords Material implication · Counterfactuals · Conversational implicature · Natural language · Assertibility · Pragmatics · Cancellability

Mathematics Subject Classification (2000) Primary 03A05, 03B05; Secondary 09B65, 94A99

1 Ajdukiewicz's Equivalence Argument

Ajdukiewicz meant to establish that any conditional of the form:

$$\text{If } p, \text{ then } q$$

has the same truth conditions as the material implication

$$p \to q$$

Á. Garrido, U. Wybraniec-Skardowska (eds.), *The Lvov-Warsaw School. Past and Present*, Studies in Universal Logic,
https://doi.org/10.1007/978-3-319-65430-0_7

(which is false just in case p is true and q false, and true otherwise). His argument has two main stages:

STAGE 1 Showing that natural language disjunction—*not-p or q*—has the same truth conditions as its corresponding natural language conditional—*if p, then q*.

STAGE 2 Showing that this natural language disjunction has the same truth conditions as its corresponding classical disjunction—$p \vee q$.

If we mark having the same truth conditions by '$\Leftrightarrow$', the structure of the argument is this (we make no distinction between "either ... or ..." and "... or ..."):

$$[\text{If } p, \text{ then } q] \Leftrightarrow [\text{Either not-}p \text{ or } q] \text{ (by Stage 1)} \tag{1}$$

$$[\text{Either not-}p \text{ or } q] \Leftrightarrow [\neg p \vee q] \text{ (by Stage 2)} \tag{2}$$

$$[\neg p \vee q] \Leftrightarrow [p \rightarrow q] \text{ (by logic)} \tag{3}$$

$$[\text{If } p, \text{ then } q] \Leftrightarrow [p \rightarrow q] (\text{ by transitivity of } \Leftrightarrow) \tag{4}$$

Let's focus on STAGE 1 now (we'll get to STAGE 2 in Sect. 3). Ajdukiewicz gives two separate arguments for two directions of the equivalence.

$\Leftarrow$:

w.t.s.:	[Either not-p or q] $\Rightarrow$ [If p, then q]
Step 1	(suppose) Either not-p or q.
Step 2	(from Step 1 by disjunctive syllogism) If it is not the case that not-p, then q.
Step 3	(from Step 2 by double negation elimination in the antecedent) If p, then q.

$\Rightarrow$:

w.t.s.:	[If p, then q] $\Rightarrow$ [Either not-p or q]
Step 1	(suppose for contraposition) It is not the case that [either not-p or q].
Step 2	(from Step 1 by De Morgan's law) Neither (not-p) nor q.
Step 3	(from Step 2, assuming conditionals with true antecedents and false consequents are false) It is not the case that [if p, then q].

2 Jackson's Argument

Before we continue with the discussion of Ajdukiewicz's approach to conditionals, let's pause to observe that Jackson [4, pp. 4-6], without reference to Ajdukiewicz (and, we take it, quite independently) formulated a very similar argument. He starts with the following principles:

Truth-functionality (TF)	It is correct to represent English 'not', 'or' and 'and' with $\neg, \vee, \wedge$.
Uncontested principle (UP)	For an indicative conditional to be false it is sufficient that its antecedent is true and the consequent false.
Passage principle(s) (PP)	The following reasoning patterns are valid: p or q. Therefore if not-p, then q. Not-(p and q). Therefore if p, then not-q.

To pass from the material conditional to the indicative conditional, Jackson relies on PP and gives two variants of the argument. The first variant is pretty much the same as in Ajdukiewicz's argument:

w.t.s.:	[Either not- p or q] $\Rightarrow$ [If p, then q]
Step 1	(suppose) Either not- p or q.
Step 2	(from Step 1 by first PP) If it is not the case that not- p, then q.
Step 3	(from Step 2 by double negation elimination) If p, then q.

The second variant is a bit different. First we notice that the material implication has the same truth conditions as not-(p and not-q). Then we argue:

w.t.s.:	Not (p and not-q) $\Rightarrow$ [If p, then q]
Step 1	(suppose) Not (p and not- q).
Step 2	(from Step 1 by second PP) If p, then not-(not- q).
Step 3	(from Step 2 by double negation elimination) If p, then q.

For the other direction, Jackson observes that the uncontested principle allows one to pass from $\neg(p \rightarrow q)$ to not-(if p, then q) and applies contraposition. This results in a move very similar to Ajdukiewicz's argument for $\Rightarrow$.

3 Ajdukiewicz on Disjunction

Let's turn to Ajdukiewicz's STAGE 2. The correspondence between classical disjunction and natural language 'or' isn't terribly problematic. Yet, Ajdukiewicz's way of handling it is quite interesting, because while dealing with it he is forced to make a distinction between truth and assertibility. Let's start with quoting Ajdukiewicz *in extenso*. First,

Ajdukiewicz makes the audience agree that the truth conditions of the relevant sentences are the same:

> I take two pieces of chalk and hide them in my hands, so that the audience doesn't see in which hand I placed each piece: both in my right hand, both in my left hand, or one in my left hand and one in my right hand. Having done this I ask the audience: "am I holding chalk in my left or right hand?" and the response is unanimous and positive. I point out to the audience that there are three possibilities (chalk only in my left hand, chalk only in my right hand, chalk in both hands) and I ask whether no matter which of these is the case they still sustain their agreement with the disjunction: "[a piece of] chalk is in my right or left hand" and whether the only case in which they would consider the disjunction to be false would be if I didn't have chalk in any of my hands—and again, the response is positive. In this way I once again obtain agreement that disjunction in natural language is true if at least one of its arguments is true and false if none of them is. [1, p. 252]

Next, he points out that agreement doesn't have to go hand in hand with truth:

> Now I open my hands and show the audience that both pieces of chalk are, say, in my left hand. In this situation I ask again the same question: am I holding chalk in my left or right hand? This time, the audience is far from unanimosity, and the most popular opinion is that once it is known that chalk is in my left hand, one cannot give a positive response to the disjunctive question. Some would even say that in such a situation one can give neither a positive nor a negative response. This semantic experiment teaches us that we accept a disjunction as long as we know that one of the disjuncts is true, but we don't know which one. Once we find out which of the disjuncts is true, we no longer accept the disjunction. [1, p. 252]

Ajdukiewicz argues that our reluctance to accept the disjunction notwithstanding, it is still true. The disjunction is about pieces of chalks and hands, not about anyone's knowledge, and since the states of affairs involving these pieces of chalk and Ajdukiewicz's hands haven't changed, nor did the truth-value of the disjunction.

Indeed, he points out that suggesting otherwise and claiming that "p or q" really means "At least one of p and q is true and I don't know which" quickly leads to absurdity. For instance:

> For any natural number x, x is even, or x is odd.

is true, whereas:

> For any natural number x, at least one of "x is even" and "x is odd" is true, but I don't know which.

is plainly false, since it implies, among other things:

> at least one of "2 is even" and "2 is odd" is true, but I don't know which.

So, it seems, we refuse to assert the disjunction even though it is true, and more is required for a sentence to be assertible than its truth. What would be the principles governing assertibility resulting in this disparity? Ajdukiewcz starts with a discussion of Quine's proposal.

4 Ajdukiewicz vs. Quine on Assertibility

Ajdukiewicz doesn't think he is original and attributes the distinction between acceptability and truth to Quine.[1] He follows Quine in bringing up two examples:

Example 1 Once we know France is in Europe, we don't utter "France is in Europe or the sea is sweet."

Example 2 Once we know that every S is P (and that there are S), we don't say "at least some S are P."

Quine explains the examples by pointing out that in such contexts the longer sentences are less informative and that shorter and more informative sentences are usually more likely to be uttered for pragmatic reasons.

Yet, Ajdukiewicz is quite unhappy about the use of the notion of informativeness in this account—because he doesn't find this notion clear enough. He claims the only sensible explication of this notion as used by Quine is that p is more informative than q iff q follows logically from p but p doesn't follow from q.

While Ajdukiewicz finds this definition clear enough, on this account, he claims, it is false that if one wants to be as helpful and honest in a conversation as possible, one should always prefer the shorter and more informative claim. For instance, he insists, when we present a reasoning, we often utter statements weaker than the premises in the process, and this would be prohibited by the general principle suggested by Quine. Thus, we might say that Ajdukiewicz discards Quine's general principle by arguing that following it would prohibit us from uttering our arguments.

In the context of our discussion of Ajdukiewicz'a ideas it is worthwhile to clarify Quine's views on the relation between material implication and natural language conditional.

In *Methods of Logic*, [6] discusses briefly whether material implication conforms to the ordinary indicative conditional 'if-then'. He first states that restricting our attention to simple indicative conditionals is justified:

> the material conditional '$p \rightarrow q$' is put forward not as an analysis of general conditionals such as (1) ['If anything is a vertebrate, it has a heart'], nor as an analysis of contrafactual conditionals such as (4) ['If Eisenhower had run, Truman would have lost.'], but, at most, as an analysis of the ordinary singular conditional in the indicative mood... [6, p. 15]

arguing that (1) *must be viewed as affirming a bundle of individual conditionals* which in this context means that it shall be analyzed in terms of first-order logic with quantifiers instead of propositional calculus only, and that...

> ...any adequate analysis of the contrafactual conditional must go beyond mere truth values and consider causal connections, or kindred relationships, between matters spoken of in the antecedent of the conditional and matters spoken of in the consequent... [6, p. 14]

Quine observes that in natural language some counterfactuals with false antecedents and false consequents may be true while other counterfactuals with false antecedents and false consequents can be false. From this, he concludes that the semantics of

[1] He does so without giving any references, but Quine's discussion of these issues can be found in [7] and [6].

counterfactuals is definitely not truth functional, whatever the adequate analysis of their meaning is.

After settling that the material implication may be an analysis only of indicative conditionals, he claims that analyzing conditionals as material implications may be unnatural when there is no relevance between the antecedent and the consequent:

> Even as an analysis of such conditionals, the version '$p \rightarrow q$' is sometimes felt to be unnatural, for it directs us to construe a conditional as true no matter how irrelevant its antecedent may be to its consequent, so long as it is not the case that the antecedent is true and the consequent false. [6, p. 15]

Quine argues that if we have a conditional in which the compounds are irrelevant to each other, it is equally strange to consider it true and to consider it false, irrespective of the truth values of the antecedent the and consequent. He claims that the strangeness of conditionals such as '*If France is in Australia then the sea is sweet*' stems from the fact that linguistic practice usually allows for forming conditionals (of the form 'if p, then q') out of compounds (p and q) which have unknown truth values:

> ...for it is not usual in practice to form conditionals out of component statements whose truth or falsity is already known unconditionally... In practice, one who affirms 'If p then q' is ordinarily uncertain as to the truth or falsehood individually of 'p' and of 'q' but has some reason merely for disbelieving the combination 'p and not q' as a whole. [6, p. 15]

Thus, the question of the nature of the reason to disbelieve 'p and not q' rises. Quine replies, as mentioned above, that the reason is to be provided by some kind of relevance between the compounds of the conditional. However, he claims that this relevance does not have an influence on the *meaning* of the conditional, but rather that *useful applications* of a conditional in linguistic practice are dependent on it:

> Only those conditionals are worth affirming which follow from some manner of relevance between antecedent and consequent—some law, perhaps, connecting the matters which these two component statements describe. But such connection underlies the useful application of the conditional without needing to participate in its meaning. Such connection underlies the useful application of the conditional even though the meaning of the conditional be understood precisely as '$\neg(p \wedge q)$'. [6, p. 16]

In *Mathematical Logic*, [7] presents essentially the same view, adding a remark on the particular role of the truth table for material implication in assessing conditionals in ordinary practice:

> What the truth table adds, in thus deciding the cases beyond the range of ordinary usage, is essentially theoretical; no supplementary practical use of 'if-then' is thereby prescribed. In practice, even in the light of the truth table, one would naturally not bother to affirm a conditional if he were in position to affirm the consequent outright or to deny the antecedent—any more than one bothers to affirm an alternation when he knows which component is true. [7, p. 17]

Hence, according to Quine, there is a difference between the meaning (semantics) of an expression and its useful applications (pragmatics)—the meaning of a conditional is exhausted by the truth tables of material implication, whereas practical useful application of a conditional demands there to be some kind of (e.g. causal) connection or relevance between the matters expressed in the compounds of a given conditional. And, most importantly, the rule is not to assert a conditional if we are already certain of the truth values of its compounds. Consequently, the general rule of reasoning that Quine suggests and the one he takes as an explanation of the strangeness of conditional with mutually irrelevant compounds is that in practice communicating agents choose to assert statements that are as short and as strong as possible. So, when we know the truth values of 'p' and of 'q, it is useless to use the conditional '*if* p *then* q'' in our reasoning:

> Why affirm a long statement like 'If France is in Europe then the sea is salt.' or 'If France is in Australia then the sea is salt.' when we are in position to affirm the shorter and stronger statement 'The sea is salt'? And why affirm a long statement like 'If France is in Australia then the sea is salt.' or 'If France is in Australia then the sea is sweet.' when we are in position to affirm the shorter and stronger statement 'France is not in Australia'? [6, p. 15]

Instead of following Quine, Ajdukiewicz, having found Quine's general principle too strong, and lacking a more successful general principle, settles tentatively with acknowledging the following phenomenon:

DISJUNCTION One normally doesn't utter a disjunction if one knows which of the disjuncts is true.

and goes on to find a more plausible explanation for DISJUNCTION.

5 Expressing vs. Stating

At this point, to provide a more general account of why such a principle should hold, Ajdukiewicz makes a distinction between what a sentence *states* (what has to be the case for it to be true) and what it *expresses*:

> To say that an utterance W of a person O expresses, given the linguistic habits, his state S, is the same as saying that W uttered by O is for the audience (who know of those [linguistic] habits) a sign of O's being in state S, or that uttering W by O allows the audience familiar with those habits to figure out that O is in state S. [1, p. 255]

On this approach, by uttering something, speakers not only state the fact required for the truth of the statement, but also express their states, associated with particular types of utterances by linguistic habits of a given linguistic community. While the statement might be *true* or *false* in virtue of whether what it states obtains, it is *proper* or *improper* in virtue of whether the utterer is in the state expressed by the statement.

Another important aspect of the distinction is that while to come to believe what is stated, one has to believe the statement to be true, while to come to believe what is expressed, one doesn't have to accept the statement itself. It is enough to understand it and to know the relevant linguistic habits.

Now we have reached Ajdukiewicz's general principle:

AJDUKIEWICZ One is unwilling to accept improper claims, even if they are true.

Ajdukiewicz doesn't say anything about this issue, but notice that Ajdukiewicz's criticism of Quine's solution doesn't apply to his own approach. After all, there are many cases of deductions which can be presented properly, because the inference steps do not express anything that contradicts what the premises express.

6 Ajdukiewicz on the Diagnostics of Improper Use

One way of diagnosing our reasons to not assert a particular utterance p suggested by Ajdukiewicz is to ask ourselves: are we willing to accept $\neg p$?[2] If we refuse to assert p because we think it's false, our answer should be positive:

> Whether the refusal to accept a certain sentence is motivated by the unwillingness to accept something false, or the unwillingness to use an expression improperly, can be recognized, among other facts, by the fact, that in the first case the refusal is accompanied by the readiness to accept the negation [of the sentence], while in the second case such a readiness is missing. [1, p. 256]

Now, this seems a bit too hasty. After all, if I have no information about some p (say "aliens exist"), I will refuse to accept it, and I will refuse to accept $\neg p$. By Ajdukiewicz's criterion, this would mean that my refusal to accept p is not motivated by the unwillingness to accept something false. But this doesn't sound right—the main reason why I don't accept sentences about whose truth values I have no information is because as far as I know they might be false, and I wouldn't want to accept a false sentence.

In all fairness, however, when I refuse to assert "aliens exist", it's not only because I don't want to accept a (potentially) false statement, but also because this would suggest to the audience that I do think that I know that aliens exist, and that would be false. But this holds for any sentence which we don't know to be true, including those sentences that we know to be false. So, for instance, I'm unwilling to assert "2+2=5", and one reason for this is that this would express the claim that I know that 2+2=5, which is false. But at the same time, I am willing to assert "$2 + 2 \neq 5$", and so, by Ajdukiewicz's criterion, I am refusing to accept "2+2=5" for truth-related reasons, and not just because it would be simply improper to assert it.

So, we submit, Ajdukiewicz's account of this diagnostics should be revised. First of all, refusing to assert a sentence because it would be improper (because it would express something false) doesn't exclude refusing to assert it because it's false. If we believe that a sentence is false, we should also believe that asserting it would be improper, because the assertion would express our knowledge that it is true (and we don't have that knowledge).

Second, as our example with aliens shows, refusing to assert a sentence because it would be improper (because it would express something false) doesn't exclude refusing to assert it because we wouldn't want to accept something false, even if we don't know the sentence to be false.

Third, the distinction should be rather between sentences that we believe to be false and refuse to accept, and sentences that we don't believe to be false and refuse to accept. To this distinction Ajdukiewicz's diagnostic criterion correctly applies: if I believe something to be false, I will assert its negation, and if I don't believe it to be false, I won't.

The problem is, however, that now the criterion fails to divide sentences where Ajdukiewicz would like it to. After all, if I refuse to accept a disjunction while refusing to accept its negation, all I know is that I don't believe it to be false. Whether it is further the case that I only refuse to accept it because I believe it is true, but it would be improper

[2]For the purposes of this paper we follow Ajdukiewicz in taking acts of acceptance to be public, and so we ignore the distinction between accepting and asserting.

to assert it, or whether I also think I have no good reasons to think it's true is a separate issue that needs to be separately discussed.

There is a charitable and instructive way of reading Ajdukiewicz's remarks about the diagnostic method, though. Consider the example with pieces of chalk. The audience agreed to "p or q". Upon discovering that p is true and q is false, they cease to assert this disjunction. Now, if they thought the fact that p is true and q is false made the disjunction false, they would not only refrain from asserting it, but also were ready to assert its negation. Hence, it seems, Ajdukiewicz is still right that this shift doesn't have much to do with truth conditions of disjunctions.

7 Generalization, Weakening, Moore's Paradox

One remark is in place here. Ajdukiewicz suggests that by uttering a disjunction we express the fact that we know it to be true (and, that we don't know which disjunct is true). And in general, one might think that when uttering any sentence meant to state a fact, in standard contexts we express that we know that the sentence is true.

Both the suggestion about disjunction, and its generalization seem a bit hasty, though. For Ajdukiewicz would like any assertion which expresses a condition which is not satisfied to be improper. But there seems to be nothing improper to assert a true disjunction I *think* I know to be true which I don't really know to be true (and of which I *think* I don't know which disjunct is true), Gettier-like cases abound. And so, we submit, it would be sensible, and it would not damage Ajdukiewicz's approach to conditionals, to weaken the claim to saying that when one, in a standard context, asserts a disjunction (or any sentence meant to state a fact), one expresses the fact that one *believes* (not: *knows*) that disjunction (sentence) to be true. In what follows, we'll keep this remark in mind, but we'll still formulate arguments mostly in terms of knowledge, assuming that unless stated otherwise, considerations still apply after replacing knowledge with belief.

On the other hand, this easy generalization (and, at the same time, weakening) of Ajdukiewicz's suggestions bears directly on what, we think, Ajdukiewicz would say about Moore's paradox, where one says things like "p, but I don't believe that p". Namely, Ajdukiewicz would point out that by asserting a conjunction one expresses one's belief in both conjuncts, and so one expresses one's belief in p, and one's belief in not believing that p. Observe further that assuming introspection, this leads to straightforward contradiction in expressed judgments. The idea is that the expressed content is $Bp \wedge B\neg Bp$. The first conjunct, by introspection yields BBp, and so the whole thing together entails that the belief set is inconsistent.

8 Apparent Connection Between Disjuncts

To use AJDUKIEWCZ to explain DISJUNCTION, Ajdukiewicz observes that a disjunction, apart from *stating* that at least one of the disjuncts is true, *expresses* the fact that the speaker knows that at least one of the disjuncts is true, and that the speaker is ignorant as to which of them it is. For this reason, he suggests, once we know which disjunct is true, we refuse to utter a disjunction.

Ajdukiewcz asks: how can we know that a disjunction of the form "p or q" is true? Well, one option is that we know p, or we know q. In such a case, asserting "p or q" would be improper. Another option is that we're ready to assert q if we find out that p is false: that is, we're ready to infer q from $\neg p$.

So, on this view, in the context of proper assertion, expresses speaker's readiness to assert q if they find out that p is false: to infer q from $\neg p$. Where does this readiness comes from, though? Isn't it supported by some connection between the disjuncts, which is not only expressed, but also stated by a disjunction? Ajdukiewicz disagrees.

What sort of connection would that be? One option would be that it would be psychological: on this approach, the connection is that a disjunction makes one willing or ready to infer one disjunct from the negation of the other one. Ajdukiewicz observes that this notion would relativize truth-conditions of a disjunction to the speaker, and since he finds the idea of a disjunction being true for one person but false for another unpalatable, he rejects this account.

Another idea would be that the connection consists in making the inference under discussion legitimate. But in what sense? If all that is meant here is that if one of the disjuncts is false, the other one is true, then Ajdukiewicz agrees—but this is exactly what a disjunction states, and what is captured by the standard truth table for classical disjunction.

On a stronger interpretation, the claim is that the negation of one disjunct logically entails the other disjunct. But this, Ajdukiewicz observes, doesn't seem to hold for natural language disjunctions. He uses the following example: *I will die on a day with an even date, or on a day with an odd date*. This, he holds, is clearly true.[3] Yet, without additional premises that he will die some day and that each day has either an odd date or an even date, the negation of one of the disjuncts doesn't logically entail the other disjunct.

Could the second proposal be fixed to avoid Ajdukiewcz's criticism by saying that the negation of one of the disjuncts should entail the other disjunct with some additional premises? Not easily, for reasons similar to those for which the cotenability approach to conditionals is unsuccessful. For one would have to specify which additional premises can be used. If any true premises can be used, no interesting connection between disjuncts is required, it is just enough that one of them really is false. Once p is false, $\neg p$ is true, and p *or* q with $\neg p$ (a true sentence after all) logically entails q. One might be tempted to avoid this by saying that only those extra premises can be used to infer q from p *or* q and $\neg p$ which are connected with q, but this would make the account circular.

Let's observe, however, that Ajdukiewicz's criticism isn't lethal. He lists three interpretations of what it would mean for there to be a connection between the relevant sentences, excludes two of those options, and shows that the third one is exactly the one he proposes. What is missing is an argument to the effect that this is a complete list of sensible interpretations. But, absent other interpretations that would avoid criticism, this should be good enough for now.

[3] In fact, he died on April 12, 1963, and so the first disjunct is true.

9 Ajdukiewicz's Diagnostics and Grice's Cancellability

Now that we have presented Ajdukiewicz's defense of the material implication account of indicative conditionals, let's compare it to a much better known defense of the same claim, due to Grice. While analyzing the problem of material implication in his *Logic and Conversation* [3, pp. 3–143], P. Grice considers the so-called Indirectness Condition (IC) associated with the conditional. According to IC, there should be non-truth-functional grounds for accepting $p \rightarrow q$ as the meaning of 'if p then q'. That is, if a subject asserts 'if p then q', they are conventionally committed both to the proposition $p \rightarrow q$ and to IC, which amounts to claiming some causal (or other, but still non-truth-functional) link between p and q.

Grice's analysis of conditionals lies in the scope of his theory of maxims of conversation and of conversational implicature. The main idea is that rational communicative interaction is governed by certain principles and maxims, which despite their prescriptive phrasing actually *describe* how agents behave in order to achieve effective communication in conversation. The most general rule is **the cooperative principle** (CP) which says:

> Make your contribution as is required, when it is required, by the conversation in which you are engaged.

According to Grice, the descriptive content of CP consists in the fact that speakers (generally) observe the cooperative principle, and listeners (generally) assume that speakers are observing it. Fulfilling CP consists then in obeying the so-called maxims of conversation. This means that the requirements of CP are explicated by the following rules:

Maxim of Quality: Contribute only what you know to be true. Do not say false things. Do not say things for which you lack evidence.
Maxim of Quantity: Make your contribution as informative as is required. Do not say more than is required.
Maxim of Relevance: Make your contribution relevant.
Maxim of Manner: Avoid obscurity, avoid ambiguity, be brief and be orderly.

Grice claims that certain utterances during a conversation convey meanings that are not explicitly expressed nor logically implied (entailed) in what is said, but nevertheless can, in some sense, be inferred for pragmatic reasons. Such meanings (or pragmatic inferences) are called *conversational implicatures*:

> I am now in a position to characterize the notion of conversational implicature. A man who, by (in, when) saying (or making as if to say) that p has implicated that q, may be said to have conversationally implicated that q, provided that (1) he is to be presumed to be observing the conversational maxims, or at least the cooperative principle; (2) the supposition that he is aware that, or thinks that, q is required in order to make his saying or making as if to say p (or doing so in those terms) consistent with this presumption; and (3) the speaker thinks (and would expect the hearer to think that the speaker thinks) that it is within the competence of the hearer to work out, or grasp intuitively, that the supposition mentioned in (2) is required. [3, pp. 30–31].

The idea is that a conversational implicature is a pragmatic inference that the listener has to make if she is about to maintain that the speaker is cooperative. More precisely, following [5, p. 113] we might say that the speaker S's saying that p **conversationally implicates** q if:

1. S is presumed to be observing the maxims, or at least (in the case of floutings) the cooperative principle (cooperative presumption).
2. In order to maintain this assumption it must be supposed that S thinks that q (determinacy).
3. S thinks that both S and the addressee H mutually know that H can work out that to preserve the assumption in (1), q is in fact required (mutual knowledge).

One of the most important features of conversational implicatures[4] is cancellability and Grice even claims that conversational implicatures differ from semantic implicatures exactly in being cancellable. This means that they can be consistently dismissed by the speaker or in light of the context. For instance, the speaker might consistently add to the conversation some content that entails the negation of an already introduced implicature—assume that the speaker said 'Some of the students passed the test.' This implicates that not all of the students passed the test. The implicature might be canceled then by uttering: 'Indeed, all of the students passed the test.' Another example is when the speaker adds to the conversation some content that indicates that she is not committed to the implicature or its negation, as in 'Some, maybe all, of the students passed the test.'

Grice also makes one distinction that will be relevant to our concerns. He claims that there is a difference between *particularized* and *generalized* conversational implicatures:

> I have so far considered only cases of what I might call particularized conversational implicature—that is to say, cases in which an implicature is carried by saying that p on a particular occasion in virtue of special features of the context, cases in which there is no room for the idea that an implicature of this sort is NORMALLY carried by saying that p. But there are cases of generalized conversational implicature. Sometimes one can say that the use of a certain form of words in an utterance would normally (in the ABSENCE of special circumstances) carry such-and-such an implicature or type of implicature. Noncontroversial examples are perhaps hard to find, since it is all too easy to treat a generalized conversational implicature as if it were a conventional implicature. [3, p. 37]

A generalized conversational implicature is one which does not depend on particular features of the context, but is instead typically associated with the proposition expressed. Not surprisingly, a particularized conversational implicature is one which depends on particular features of the context.

For instance, imagine a following conversation:

A: Will Sally be at the meeting?
B: Her car broke down.

In this situation, it is implicated by B that Sally will not be at the meeting, but the implicature is particularized, as there is nothing in the content of the expressed proposition that would suggest such an inference. Consider, however, another example:

Mary has 3 children.

---

[4]Among others, such as: calculability, non-conventionality, non-detachability, indeterminacy or reinforcerability. An interested Reader can consult e.g. Levinson [5].

When such a sentence is uttered in a conversation, it is implicated that Mary has no more than 3 children and the implicature is generalized—we can associate it with the proposition expressed, irrespective of the context.

Grice defends the view that the meaning of a conditional on any particular occasion of utterance is simply equivalent to the meaning of material implication. He claims that IC is a generalized conversational implicature. Just as all generalized conversational implicatures, it can be cancelled without contradiction, either by circumstances in the context or by explicit denials. Grice considers two examples of assertions of conditionals which carry no implicature of IC. These are the following:

If I have a red king, I also have a black king

uttered during a card game, and

If Mr Jones has black pieces, Mrs Jones has too

uttered during the run of a particular logical puzzle where the participants are about to guess the identity of the characters in the game and are provided with a piece of information, such as the one given in the conditional above. Grice claims that the total contents of the utterances above is just $p \rightarrow q$.

One of the maxims of Girce's pragmatics is the one of Quantity—it dictates to make, in a given context, the most informative statements of interest as possible. Grice calls the pragmatic virtue of this maxim 'conversational helpfulness'. Its point is to make natural language communication as efficient as possible. This resembles Quine's suggestion that we should utter stronger rather than weaker claims However, as Grice puts it:

> An infringement of the first maxim of Quantity, given the assumption that the principle of conversational helpfulness is being observed, is most naturally explained by the supposition of a clash with the second maxim of Quality ("Have adequate evidence for what you say"), so it is natural to assume that the speaker regards himself as having evidence only for the less informative statement ... [3, p. 33]

The above amounts to the claim that any utterance of 'if p then q' will, unless prevented by context, give rise to the implicature that the speaker does not have definite information about the truth values of p and q. According to Grice, some utterances of 'if p then q' might implicate a stronger condition than the one provided by the truth-tables of $p \rightarrow q$. However, instead of being a part of the meaning, what is implicated then is founded on the Cooperative Principle and the maxims of conversational implicature. Conditionals, then, play, apart from their semantics (given by the truth-table for $\rightarrow$), also roles of pragmatic nature: they enable people to ponder the consequences of certain choices during a conversation. In the light of Gricean theory one might say that it would simply be irrational to use a conditional in certain contexts, for instance when there is no doubt about the truth of the antecedent.

Interestingly, the very same analysis also applies, according to Grice, to disjunctions: a natural-language disjunction of the form 'p or q' shares the logical meaning of '$p \vee q$', but in addition carries a generalized implicature that they are not both true. If the speaker was in a position to offer the more informative form 'p and q', then it would be conversationally more helpful to do so.

Prima facie, both Grice and Ajdukiewicz achieve seemingly equivalent results—the former's conversational implicature and the latter's expression of a mental state are quite similar to each other. It needs however to be noted that Grice, as Quine, has a general principle from which the Maxim of Quantity follows, whereas Ajdukiewicz gets to his result in a different manner. He first acknowledged a simple linguistic phenomenon (DISJUNCTION—see above) and found a relatively uniform yet simple explanation for it. Without assuming generalities of logical (as Quine) or of pragmatic (as Grice) sort, he simply associated effective utterances (i.e. the ones that, during a conversation, successfully deliver the content intended by the speaker) of a given expression with the possession of a certain epistemic state. In contrast with Grice's theory, the accuracy of his explanation does not depend on strongly controversial properties of a systematic but convoluted general theory of cooperative behaviour. Secondly, the nature of the criterion needs to be taken into account—for Grice it seems to be purely pragmatic, conversational and practical; for Ajdukiewicz—it is, in a sense, doxastic—it pertains to expressing the agent's knowledge or belief.

However, it actually might be questioned if Ajdukiewicz's account allows for expressing the content of the mental states only. As he puts it himself, conditionals express certain type of a lack of knowledge of the agent who makes the assertion (that she does not know that the antecedent is true and she does not know that the consequent is false). Moreover, conditionals express that the agent is ready to infer the consequent from the antecedent. As for the former condition, it is rather clearly doxastic or epistemic. But as for the latter, one might ask: what does it exactly mean to express a disposition to make a certain type of inference? It is not completely clear whether one should understand it as expressing a mental state or rather just a disposition to be in a certain type of mental state a disposition to perform a particular type of behaviour.

10 Gołąb on Indicative Conditionals

Material implication as an interpretation of natural language indicative conditionals was also discussed in Lvov-Warsaw school by Gołąb [2] and Słupecki [8]. The former proposed a difficulty to the interpretation and the latter replied. Gołąb complains that when he teaches introductory logic, the claim that a conditional whose antecedent is false and consequent true is true usually startles the audience, because they have strong intuitions there are false conditionals which satisfy these conditions. Gołąb's example is:

If today is Monday, tomorrow is Wednesday.

The sentence, uttered on Tuesday, should be true, if the material reading is adequate, but people usually don't have this intuition.

Another problem brought up by Gołąb is with contraposition—he credits prof. Harassek from Lublin with the following counterexample:

$$\frac{\text{If I'm hungry, I won't swallow a needle.}}{\text{If I swallow a needle, I won't be hungry.}}$$

11 Słupecki's Reply to Gołąb

As for contraposition, Słupecki suggests that the conditionals formulated in the counterexample don't express what is really being claimed. The actual and correct premise of the reasoning should be *Even if I'm hungry, I won't swallow a needle.* But if that's the case, the problem doesn't arise, Słupecki claims, because contraposition doesn't work for *even ... if* conditionals. He claims that for such conditionals there is no rule analogous to the principle of contraposition.

Słupecki's response to Gołąb's first worry is somewhat dismissive. He insists that the sentences in question are true, but it's simply that people wouldn't normally utter them due to some unspoken principles of parsimony, which prevent them from stating useless sentences not worth of interest:

> I suppose that the source of these intuitions is a kind of unformulated and not fully recognized principle of economy which we also follow when we classify the following sentence as nonsense:
>
> $$2+2=4 \text{ and } 2+2=4 \text{ and } 2+2=4$$
>
> —although in logical terms (logically) it is completely correct (sound)—or when we are not willing to accept (assert) compound sentences, the components of which are not materially (factually, objectively) linked or when such a link is bizarre (fanciful, odd, singular). Perhaps we would not be always inclined to count such sentences as false, but rather as superfluous, inexpedient or not worth of serious attention.

Słupecki also suggests a certain teaching trick that is meant to facilitate the digestion of the truth conditions of material implication. The trick consists in asking the students to describe a state of affairs in which a given conditional would be false. For instance

> If it rains, John is not having a walk

The trick is to ask when this sentence would become false. The expected (obvious) answer should be, and usually is: if and only if it rains and John **is** having a walk. Then, it should be an easy way from this point for the students to see that if the above is not the case, the entire conditional must be true.

The problem with the trick is that it doesn't seem to overcome the difficulties that give rise to the questions concerning the semantics of conditionals. It is not too surprising that such heuristics works for sentences in which the antecedent and the consequent are causally related. It is, however, doubtful whether this teaching method would work for problematic examples such as *if today is Monday, tomorrow is Wednesday* or *if it is raining and not raining, then the Moon is made of green cheese.*

12 Ajdukiewicz's Solution and Gołąb's Problem

Obviously, Ajdukiewicz's approach was not available at the time of Gołąb's and Słupecki's discussion, but we can ask whether the solution he proposed successfully handles the case Gołąb brought up (Ajdukiewicz doesn't seem to be aware of Gołąb's and Słupecki's papers on the topic).

A natural thing to say about the conditional *if today is Monday, tomorrow is Wednesday* from Ajdukiewicz's perspective is that when we utter it on Tuesday, we know that the consequent is false, and yet the conditional expresses the claim that we don't have such knowledge.

But what if this reason for the utterance being improper is removed? Suppose it is Tuesday, but neither the speaker nor the audience know what day of the week it is (nor do they have any beliefs about it). The implication in question expresses that the speaker doesn't know which of the disjuncts is true, which is the case. Isn't this a problem for Ajdukiewicz's account? Not really—for there is another reason why an utterance of this conditional in such situation would be improper, stemming from the discussion in Sect. 8.

How would the speaker come to know the conditional? Well, one way is by knowing either in the consequent, or in the negation of the antecedent (this is, by the way, quite a useless way of knowing a conditional, because a conditional known this way cannot be properly uttered). But the speaker doesn't know what day of the week it is, so this way of knowing it is not an option. Another would be, perhaps, to know of a reason why today's being Monday would allow the speaker to infer that tomorrow is Wednesday. But there can't be such a reason, and so the speaker can't also come to know the conditional in this manner. Thus, the speaker is never in position to know the conditional, and consequently, never in position to utter it properly, for the expressed claim that the speaker knows the conditional to be true would always be false.

Acknowledgment This work was completed with the support of the Research Foundation Flanders and Polish National Science Centre (grant DEC-2016/22/E/HS1/00304)

References

1. Ajdukiewicz, K.: Okres warunkowy a implikacja materialna [Conditional sentence and material implication]. Studia Log. **IV**, 117–134 (1956)
2. Gołąb, S.: O kilku trudnościach związanych z prostymi kwestiami logicznymi [On a few difficulties connected to simple logical issues]. Matematyka **3**(5), 27–29 (1949)
3. Grice, P.: Studies in the Way of Words. Harvard University Press, Cambridge (1991)
4. Jackson, F.: Conditionals. Blackwell, Oxford (1987)
5. Levinson, S.C.: Pragmatics. Cambridge University Press, Cambridge (1983)
6. Quine, W.V.: Methods of Logic, 2nd edn. Holt, Rinehart and Winston, Austin (1966)
7. Quine, W.V.: Mathematical Logic, Revised edn. Harvard University Press, Cambridge (1981)
8. Słupecki, J.: O własnościach okresu warunkowego [On the properties of a conditional]. Matematyka **4**(6), 32–35 (1949)

R. Urbaniak (✉)
Institute of Philosophy, Sociology and Journalism, University of Gdańsk, Gdańsk, Poland

M. T. Godziszewski
Department of Logic, Institute of Philosophy, University of Warsaw, Warsaw, Poland

Tadeusz Czeżowski

Dariusz Łukasiewicz

Abstract The aim of this paper is to present life, main works and ideas of Tadeusz Czeżowski, who was a Polish logician, philosopher and an eminent representative of the Lvov-Warsaw School. Czeżowski's contribution to the contemporary philosophy and logic is of universal value and includes such achievements as the classification of reasoning types, the analysis of singular propositions, the analysis of Aristotle's modal logic from the point of view of modern logic, the reconstruction of Brentano's syllogistic and the theory of transcendental concepts.

Keywords Logic · Philosophy · Classification of reasoning types · Kazimierz Twardowski

Mathematics Subject Classification (2000) Primary 01A60, 01A5; Secondary 03B42

1 Life

Tadeusz Czeżowski was born in Vienna on July 26, 1889 in a middle class family. His father was Prefect and later Counsellor of the Governorate of Galicia and was transferred to Lvov in 1899; his mother, Helena Kusché, belonged to the *petite bourgeoisie* of Lvov. He died in Toruń on February 28, 1981.

In 1907 he enrolled at the Faculty of Philosophy of the University of Lvov to study philosophy, mathematics and physics. In 1912 he qualified as a teacher of mathematics and physics and taught in a Lvov grammar school from 1912 to 1914. Along with teaching duties, Czeżowski continued his scientific activity. Under Twardowski's supervision he wrote a dissertation on *The theory of Classes* (*Teoria klas*), for which he obtained his doctoral degree in 1914. At the outbreak of the First World War in 1914, he arrived in Vienna and was appointed by the Rector of the University of Lvov, Kazimierz Twardowski as responsible for the accommodation of war refugees who wished to continue their studies there. After resuming his university studies he returned to Lvov, where he held the position of director of University Chancellor's Office for 3 years (1915–1918). In 1918, Czeżowski was given a post at the newly established Polish Republic Ministry for Religious Confessions and Public Education, where he worked until 1923, first as

Á. Garrido, U. Wybraniec-Skardowska (eds.), *The Lvov-Warsaw School. Past and Present*, Studies in Universal Logic,
https://doi.org/10.1007/978-3-319-65430-0_8

a clerk and then as Director of the Department of Science and Secondary Education. In 1919, he was sent to Vilnius to help in reorganizing the administration of Stefan Batory University. In 1920, Czeżowski obtained his habilitation degree for the work *Variables and Functions* (*Zmiennie i funkcje*), thanks to which he became Assistant Professor. During the Polish-Soviet war in 1920 he was a soldier and was awarded for courage.

In 1923 Czeżowski was offered the Chair of Philosophy at the University of Vilnius when it became vacant. From 1933 to 1935 he was Vice Rector there and, from 1935 to 1937, Dean of the Faculty of Humanities. He was awarded the title of Full Professor in 1936. In 1928 he founded Vilnius Philosophical Society. During all those years in Vilnius, he was a correspondent of *Przegląd Filozoficzny* (*Philosophical Review*), one of the leading Polish philosophical journals, and he collaborated with *Ruch Filozoficzny* (*Philosophical Movement*), another important Polish philosophical journal, founded and edited by Kazimierz Twardowski.

When the University was closed during the Second World War, Czeżowski continued to teach clandestinely, holding as many as 143 lectures, mostly on ethical subjects. During the war Czeżowski was imprisoned twice. He saved the lives of many Jewish people, for which, in 1963, he was awarded the title of *Righteous among the Nations* by Yad Vashem Institute, together with his wife Antonina and his daughter Teresa, and, in 2012, he posthumously obtained the honorary citizenship of Israel.

After the war, Czeżowski had to leave the University of Vilnius and, like other professors of that university, he was moved to a newly established University of Toruń (Nicolaus Copernicus University), becoming one of its main organizers. He held the Chair of Philosophy there till 1951 and then the Chair of Logic. In 1946 he founded Toruń Philosophical Society and, for many years, he was elected the chairman thereof. In 1948 he became the editor-in chief of *Ruch Filozoficzny*. He was also the initiator and organizer of a series of conferences on the history of logic, still regularly taking place in Kraków. Czeżowski retired in 1960. After retirement he continued to teach students and young scholars of Nicolaus Copernicus University interested in philosophy, holding as many as 173 seminars.

In total, he was the author of some 190 scientific publications, including his philosophical books (*Odczyty Filozoficzne*, *Filozofia na rozdrożu*, *Główne zasady nauk filozoficznych*), textbooks on logic, numerous articles and reviews. He also gave some interviews to magazines popularizing science. It is worth noting that during the years of communist dictatorship, which started in Poland immediately after the war, Czeżowski till his death refused any awards offered by the communist state.

Regarding his private life, in 1929 he married Antonina Packiewicz, who had attended his lectures as an external student. The couple had two children: the son who died soon after birth and the daughter Teresa; she died at the age of twenty. After the war, they adopted another girl, Eleonora. In private life, Czeżowski had a true passion for mountains; he was an experienced mountaineer and skier. He was also a lover of music and dance.

2 Main Papers

- O metafizyce, jej kierunkach i zagadnieniach [On metaphysics, its trends and problems] 1948
- Główne zasady nauk filozoficznych [Main principles of philosophical sciences] 1959
- Filozofia na rozdrożu [Philosophy at a crossroads] 1965
- Odczyty Filozoficzne [Philosophical lectures] 1969
- Transcendtentalia – przyczynek do ontologii [Transcendentals and ontology] 1977

3 Views

In philosophy and methodology of science, he accepted holism, empiricism, epistemic individualism and fallibilism (in principle). In philosophy of language, Czeżowski argued for the conception of semantic transparency of the sign. His major achievements in logic include: establishing the number of relations between the scope of the terms of the categorical propositions, an extended version of Aristotle's syllogistic, the analysis of singular propositions, an original reconstruction of Aristotle's theory of modal propositions. Czeżowski's idea was that Aristotle's calculus of modal propositions is dual with calculus of categorical propositions. In metaphysics, Czeżowski elaborated an original theory of transcendental concepts such as goodness, beauty and truth. He assumed that they should be interpreted as *modi essendi* rather than as properties of objects. In meta-ethics, he defended anti-naturalism, cognitivism and infallibility of basic moral principles. In the domain of applied ethics, Czeżowski advocated the egalitarian ethics based on the principle of equal measure, i.e. that human should not value their own good higher than the good of other people.

4 Resonance

The most influential ideas of Czeżowski are his theory of syllogism, theory of ethics as empirical and inductive science and his doctrine of transcendentals.

5 Pupils

To the Vilnius group of his pupils belonged: Benedykt Woyczyński, philosphers of law Józef Zajkowski and Jan Rutski, Saul Sarnaker—a physician interested in philosophy, Aleksandra Zajkowska-Znamierowska, Józef Reutt, Edward Csató, Stefan Buhardt, Abraham Fessel, Sawa Frydmann, Maria Renata Mayenowa, Barbara Skarga, Stefan Wołoszyn, Lidia Wołoszyn. To the Toruń group of Czeżowski's disciples belonged Valdemar Voisé, Leon Gumański, Wacław Kubik, Henryk Moese, Janusz Skarbek, Bogusław Wolniewicz and Zbigniew Zwinogrodzki.

6 Summary

Czeżowski's contribution to the contemporary philosophy and logic is of universal value and includes such achievements as the classification of reasoning types, the analysis of singular propositions, the analysis of Aristotle's modal logic from the point of view of modern logic, the reconstruction of Brentano's syllogistic and the theory of transcendental concepts.

Bibliography

1. Czeżowski, T.: O metafizyce, jej kierunkach i zagadnieniach, 1st edn. Wydawnictwo Uniwersytetu Mikołaja Kopernika, T. Szczęsny, Toruń (1948)
2. Czeżowski, T.: Główne zasady nauk filozoficznych, 3rd edn. Zakład Narodowy im. Ossolińskich, Wrocław (1959)
3. Czeżowski, T.: Filozofia na rozdrożu. PWN, Warszawa (1965)
4. Czeżowski, T.: Odczyty Filozoficzne, 2nd edn. Towarzystwo Naukowe w Toruniu, Toruń (1969)
5. Czeżowski, T.: Transcendtentalia – przyczynek do ontologii. Ruch Filozoficzny **1–2**, 54–56 (1977)

D. Łukasiewicz (✉)
Kazimierz Wielki University, Bydgoszcz, Poland
e-mail: dlukas@ukw.edu.pl

Tadeusz Czeżowski's Theory of Knowledge and Beliefs

Dariusz Łukasiewicz

Abstract The aim of the paper is to present and critically evaluate Tadeusz Czeżowski's views concerning knowledge and belief in the context of some non-Brentanian epistemological concepts and theories. Czeżowski belonged to the most eminent representatives of the Lvov-Warsaw School, which was generally under a strong and many-sided influence of Franz Brentano's philosophy. Czeżowski himself was a disciple of Kazimierz Twardowski, the founder of the Lvov-Warsaw School and one of the most prominent of Brentano's students. That is why contemporary discussions about the philosophy of the Lvov-Warsaw School, and about Czeżowski's philosophy in particular, are often set against Brentanian ideas. However, I would like to consider Czeżowski's epistemology in a historical context which is broader than the Brentanian philosophy only. In fact, it is a context to which also the latter belongs; I mean the epistemological doctrines of Descartes, Locke, Hume, Reid and the contemporary views which refer to them.

Keywords Belief · Epistemological individualism · Faith · Fallibilism · Knowledge · Truth

Mathematics Subject Classification (2000) Primary 01A60, 03B42; Secondary 01A70

1 Introduction

The aim of the paper is to present and critically evaluate Tadeusz Czeżowski's views concerning knowledge and belief in the context of some non-Brentanian epistemological concepts and theories. Czeżowski belonged to the most eminent representatives of the Lvov-Warsaw School, which was generally under a strong and many-sided influence of Franz Brentano's philosophy. Czeżowski himself was a disciple of Kazimierz Twardowski, the founder of the Lvov-Warsaw School and one of the most prominent of Brentano's students. That is why contemporary discussions about the philosophy of the Lvov-Warsaw School, and about Czeżowski's philosophy in particular, are often set against Brentanian ideas. However, I would like to consider Czeżowski's epistemology in a historical context which is broader than the Brentanian philosophy only. In fact, it is a context to which also the latter belongs; I mean the epistemological doctrines of

Á. Garrido, U. Wybraniec-Skardowska (eds.), *The Lvov-Warsaw School. Past and Present*, Studies in Universal Logic,
https://doi.org/10.1007/978-3-319-65430-0_9

Descartes, Locke, Hume, Reid and the contemporary views which refer to them.[1] By doing this I do not want to undermine the great importance of the Brentanian heritage for Polish analytic tradition, but I just aim to broaden a historical context into which that philosophy, as I believe, can be rightly included. To achieve this aim, in the first part of the paper I will present some major ideas of early modern and contemporary epistemology which are relevant to our further considerations and, next, in that context I will discuss Czeżowski's views on knowledge and belief.

In contemporary discussions concerning knowledge, belief and justification in general, there are some concepts and dichotomies which outline the mainstream of the debate, such as epistemological individualism vs. collectivism, methodism vs. particularism, internalism vs. externalism, foundationalism vs. fallibilism or Cartesian vs. Spinozan belief-formation models. All these topics have been widely discussed in the epistemological literature, but I decided to recapitulate them briefly below because they are rather absent from works on the Lvov-Warsaw School. Neither were these concepts used by the philosophers of the School; some of them could not be used simply because they were not formulated yet at the time when the Lvov-Warsaw School developed.

2 Modern Epistemological Individualism: Faith and Testimony in Epistemological Framework

Let us begin with the clarification of the doctrine of epistemological individualism since, as it will become clear below, it plays a particularly significant role in Czeżowski's theory of knowledge. The idea of epistemological individualism can be elucidated by quoting Antoine Arnauld and Pierre Nicole of the Port Royal School, who themselves were not proponents of that view, but made useful conceptual distinctions. They write as follows:

> For there are two general paths that lead us to believe that something is true. The first is knowledge we have of it ourselves, from having recognized and examined the truth either by the senses or by reason. This can generally be called *reason*, because the senses themselves depend on a judgment by reason, [. . .]. The other path is the authority of persons worthy of credence who assure us that a certain thing exists, although by ourselves we know nothing about it. This is called faith or belief, following the saying of St. Augustine: *Quod scimus, debemus rationi, quod credimus, autoritati* (What we know we owe to reason, what we believe, to authority). [2, p. 260]

Similar ideas can be traced in Thomas Aquinas' views, according to which:

> [. . .]it is needful that he [man] be able to stand with as much certainty on what another knows but of which he himself is ignorant, as upon the truths which he himself knows. Hence it is that in human society faith is necessary in order that one man give credence to the words of another, and this is the foundation of justice [. . .]. [1, III, art. i. 3]

[1]It is worth mentioning here Jacek Jadacki's remark: "during the last 50 years traditional epistemology has not been cultivated, in principle, in the Lvov-Warsaw School. It was Ajdukiewicz's principle which was decisive here: according to him, epistemological problems could be studied only after a suitable paraphrase and after such a paraphrase they become indistinguishable from respective methodological problems" [14, p. 97]. In my paper I would like to discuss Czeżowski's epistemology just in the context of "traditional epistemology", whose problems are today probably more appealing and popular among epistemologists than the formal approach preferred by the followers of Ajudukiewicz's paradigm.

In both quotations we find a sharp distinction between beliefs acquired by reason and beliefs acquired by faith. The former are beliefs based on perception or inference and the latter include testimonial beliefs which we owe to someone else's authority. The concept of authority used above is broad and embraces individual agents as well as institutional bodies. A crucial presupposition of the distinction between faith and reason is that individuals are unable to acquire all beliefs which are worthy of holding by themselves. Both Thomas Aquinas (and surely other medieval thinkers as well) and the two authors of the *Port Royal Logic* claim that testimonial beliefs, or simply "faith", make a valuable complement to perceptual and inferential beliefs, or simply, to "reason". What is important here is that beliefs belonging to the body of faith can be true; and truth is a great value in itself. That is why faith is a valuable and legitimate complement to reason.

Contrary to that tradition, Rene Descartes stressed very firmly and clearly that beliefs should be acquired by reason alone:

> [...] hence I thought it virtually impossible that our judgments should be as unclouded and firm as they would have been if we had had the full use of our reason from the moment of our birth, and if we had always been guided by reason alone. [9, p. 117]

Furthermore, beliefs held by others are worthless in our gaining knowledge:

> [...] and yet a majority vote is worthless as a proof of truths that are at all difficult to discover; for a single man is much more likely to hit upon them than a group of people. [9, p. 119]

The same approval of an individual search for knowledge one may find in John Locke's seminal *Essay*:

> [...] we should make greater progress in the discovery of rational and contemplative Knowledge, if we sought it in the Fountain, in the consideration of Things themselves; and made use rather of our own Thoughts, than other Men's to find it. For, I think, we may as rationally hope to see with other Men's Eyes, as to know by other Men's Understandings. So much as we our selves consider and comprehend of Truth and Reason, so much we possess of real and true Knowledge. The floating of other Men's Opinions in our brains makes us not one jot the more knowing, though they happen to be true. What in them was Science, is in us but Opiniatry, whilst we give up our Assent only to reverend Names, and do not, as they did, employ our own Reason to understand those Truths which gave them reputation [...] In the Sciences, every one has so much, as he really knows and comprehends: What he believes only, and takes upon trust, are but shreds; which however well in the whole piece, make no considerable addition to his stock, who gathers them. [17, p. 58]

Thus both Descartes and Locke claim that only beliefs acquired by reason can be regarded as knowledge and can be accepted as legitimate. Since beliefs acquired by reason are beliefs which an individual epistemic agent can gain solely by her own efforts, that position is called epistemological individualism. It is clear that epistemological individualism is a view which was not quite shared by the medieval scholars and their early modern followers (the Port Royal School), who secured some room in the epistemic framework for authority-based beliefs.

It is worth mentioning that epistemological individualism was strongly defended also by the ancient Greeks: Socrates, Plato and Aristotle. They all subscribed to the view that there is a distinction between two kinds of knowledge, *doxa* and *episteme*. *Doxa* concerns perishable things given in perception and includes testimonial beliefs (opinions). *Episteme* concerns eternal forms and ideas; it is necessary and self-evident. Knowledge about eternal forms of perishable things—defined by Plato as justified and true belief—

can only be acquired by inference or a specific kind of transsensorial perception called *noesis*.

Importantly, epistemological individualism has some ethical implications, perhaps most clearly formulated by Locke. If knowledge is of great value for humans and can be gained only by an individual effort, there are some epistemic duties and moral responsibility resulting from this fact.

However, this strong epistemic individualism is not the only approach to knowledge that we can find among early modern epistemologists. Apart from the Port Royal School, which maintained the medieval divide between reason and faith, we find some insights appreciating the role of faith in human life in David Hume's philosophy. In his *Enquiry Concerning Human Understanding*, Hume writes as follows:

> [...] there is no species of reasoning more common, more useful, and even necessary in human life, than that which is derived from the testimony of men, and the reports of eye-witnesses and spectators. [12, p. 111]

The idea defended by Hume is that if we strictly followed Descartes's and Locke's epistemological principles, we would have significantly less true beliefs than in a situation when we do not base all our beliefs on perception and reasoning only. Needless to say, if one consequently adhered to the Cartesian and Lockean ideals, this would result in one's severe cognitive deprivation in everyday life.

However, it was not Hume but Thomas Reid who most vehemently and convincingly defended the status of beliefs based on the word of others, i.e. the beliefs which Arnauld and Nicole had called "faith". Reid claims that it is futile to try to justify the immense body of true beliefs that we owe to others and their reports by consulting our own intellectual resources only—in that case we would be left with little knowledge indeed [19]. In Reid's philosophy of common sense, there are two principles which are assumed to guide our cognitive behavior and which are deeply rooted in human nature; the principle of veracity and the principle of credulity [20, pp. 193–194]. The first principle is "a propensity to speak truth, and to use the signs of language so as to convey our real sentiments". The principle of credulity, in turn, says that humans quite naturally tend to believe what they are told. Reid's position regarding the nature of testimonial beliefs (authority-based beliefs included) is, however, more radical than the views defended by Thomas Aquinas and the Port Royal School. On his view, testimonial beliefs are part of our *knowledge*, not faith, and the act of testimony itself counts as legitimate and sufficient justification for them. Reid even claims that testimonial beliefs have the same epistemological status as perceptual and inferential beliefs. In other words, beliefs based on testimony are as much justified as beliefs based on perception (introspection and memory included) and reasoning. Reidian bold epistemic ideas are grounded in a more general doctrine of human nature. According to this doctrine, human beings are sincere, truthful and willing to rely on the word of others. Sincerity and trust are imposed by the requirements of our everyday life and cognitive behavior, which are inescapably social, not solitary, in nature.

3 Methodism, Particularism and Belief-Formation Process

Connected with the question of what may count as a proper justification for our beliefs (and which beliefs may in effect constitute knowledge) is the divide between epistemological particularism and methodism.[2] Richard Fumerton spelled out the core of the distinction between the two positions as follows:

> Does one first decide what one knows and then try to learn from paradigmatic examples of knowledge the necessary and sufficient conditions for knowledge? Or does one discover first the necessary and sufficient conditions for knowledge and apply what one learns do discover what one knows? [10, p. 36].

Epistemological methodism, summarized in the second of the above questions, is an approach typical of Cartesian-style philosophizing; we define first what knowledge is and then we apply the adopted criteria to particular beliefs to check if they meet those criteria and may be considered knowledge. The particularist understanding of knowledge, articulated in the former question and consistent with Reid's approach, recommends a reversed procedure. At first we accept some self-evident, common sense beliefs as indubitable truths that we know and on that basis we proceed to define what knowledge is. Needless to say, since particularism takes our common-sense beliefs as the starting point in defining knowledge, it offers a broader and more inclusive concept thereof—in contrast to methodism whose criteria-first approach typically results in a more restrictive understanding of what it means to know something.

A very interesting epistemological question, much related to the divide between methodism and particularism, is the following: how do we come to believe that a given proposition is true? In other words, what does a paradigmatic belief-formation process involve? Again, there are two competing approaches to the problem of how our beliefs are formed—named after their famous proponents the Cartesian and Spinozan models, respectively. On Descartes's view, in order to believe something, we must first comprehend the propositional content of a (potential) belief which is given to the mind and only subsequently, if the content is clear enough and some sufficient evidence supports it, can we accept the content as true belief; or we can reject it as false if the evidence we possess is unfavorable. On the Spinozan view, our comprehension of belief's content is simultaneous with our initial acceptance of it; we accept any incoming belief as true, which act is inseparable from our understanding thereof, and only later can the proposition be rejected if we realize that it is incompatible with some other beliefs we hold [24, Part 2, 49].[3] Obviously, the Spinozan account of how we must initially accept beliefs in order to understand them is absolutely congruent with the workings of Reid's credulity principle; human beings have a strong, psychologically grounded tendency to *believe*.

---

[2]The divide concerns the so-called Problem of the Criterion famously formulated anew and brought to the fore by Roderick Chisholm [3].

[3]As pointed to by Gilbert [11, p. 108], Baruch Spinoza was the first thinker to question the Cartesian divide into the passive (comprehension) and active (assessment) powers of the mind, therefore the second of the above-mentioned models of belief formation processes can be called Spinozan. For more on the two models of belief formation, see also Lechniak [16, pp. 73–115].

4 What Can Justify Beliefs: Basic Beliefs, Internalism and Externalism

Another pair of vitally important epistemological concepts which we will refer to when discussing Czeżowski's views on knowledge and belief are internalism and externalism. These two concepts recur in any contemporary discussion on epistemic justification, therefore it will suffice here to mention only the major claims of both. Epistemological internalism is a fairly traditional view connecting belief's justification with an individual's awareness of it and epistemic responsibility. A particular belief is justified only if the epistemic agent holding that belief has mental access to and is aware of what confers justification on that belief. Epistemological externalism, which is a more recent view, rejects the foregoing claim that in order to have justified belief, the epistemic agent must be aware of or must have potential access to what justifies that belief. Externalist reliabilists, for example, maintain that a belief can be sufficiently justified solely by the fact of being generated by a reliable, truth-conducive cognitive process, of which fact the epistemic agent need not be aware whatsoever.

Finally, in our further discussion of Czeżowski's views, we will refer to the concept of basic beliefs. Let us remember that in order to have the status of being basic (or foundational), the belief in question is to be justified non-inferentially, by immediate experience—be it introspection, sensory perception or a kind of intellectual intuition (*noesis*). In other words, a basic belief cannot be justified by itself or by any other belief because then we would encounter the infinite regress problem. Since basic beliefs do not require justification from other beliefs but they provide justification for other, non-basic beliefs, they make a foundation for the whole body of our knowledge—that is the main principle of foundationalism of any strand. In its classical version—defended by Descartes—foundationalism requires that basic beliefs be infallible; they cannot be refuted by any future experimental data. Therefore, knowledge built on them is certain and indubitable. A moderate foundationalism (fallibilism) is a less restrictive view; it holds that there exist basic beliefs but they need not be infallible, i.e. it is possible that they will be refuted by some experimental data in the future. According to moderate foundationalism, for example, perception-based beliefs can be regarded as properly basic even though they are not infallible. There has been much discussion and critique in the contemporary epistemology regarding the very notion of basic beliefs, what justifies them, and the possibility of non-inferential justification of knowledge. As argued by Sellars [23] and many others, the main problem is whether basic beliefs as such are possible at all, i.e. whether any belief can be justified by non-propositional and non-judgmental, purely sensoric experiential data. Another set of problems that any foundationalism faces is related to how a set of basic beliefs—rather limited in number—is to confer justification on the whole body of empirical knowledge that we claim to have. It is not necessary and rather impossible to dwell here on the numerous controversies generated by the concept of basic beliefs; clearly the problems involved are much deeper than the sheer question of whether a basic belief may be fallible or infallible.

5 Czeżowski's Theory of Knowledge and Epistemological Individualism

Having introduced some main ideas and dichotomies of the early modern and contemporary theory of knowledge, I would like to defend the thesis that Tadeusz Czeżowski's epistemological views could be characterized as epistemic individualism, methodism, internalism and fallibilism (to a certain degree), and that Czeżowski assumed the Cartesian model of belief-formation, not the Spinozan one. It is also worth noting that his theory of knowledge may be considered, perhaps with some minor reservations, to represent the majority view in the Polish analytic camp in the twentieth century. There are, however, some problems arising when Czeżowski's theory of knowledge is analyzed in detail; we will discuss some of them below.

Let us begin with epistemological individualism. Undoubtedly, Czeżowski sharply distinguished reason (knowledge) and faith [7, p. 56]. In his view, a belief which is regarded as part of knowledge is self-evident, obvious and firm; its object is comprehended in a clear and distinct way. A belief which is part of faith is not self-evident but vague, shaky and somewhat sketchy. It is typically grounded in emotions, desires or authority. In Czeżowski's theory of knowledge, a crucial role is played by the concept of self-evidence (obviousness). There are four kinds of self-evidence and self-evident beliefs: apodictic, introspective, perceptual and inferential ("demonstrative"). Apodictic self-evidence is predicated of beliefs whose content are basic ontological principles such as the principle of non-contradiction, the axioms of deductive systems or the principle of causality [4, 8]. According to Czeżowski, it is simply the *content* of an apodictically self-evident belief which is responsible for its self-evidence. Introspective self-evidence is predicated of beliefs whose objects are mental states of a conscious subject. The introspective self-evidence is not apodictic because, as Czeżowski writes, mental states are "ontologically contingent" [7, p. 56]. Thus, we may infer—Czeżowski himself does not say it explicitly—that ontological principles (the principle of non-contradiction included) are ontologically necessary. Perceptual self-evidence, in turn, is predicated of perceptual beliefs. It is vividness and persistence of perception which are responsible for self-evidence of perceptual beliefs. These qualities of a perception act can be saved in memory and as long as memory-based beliefs are vivid and persistent, they can be counted as self-evident. Inferential self-evidence is predicated of beliefs which have been logically deduced from self-evident beliefs (premises); an inferential chain transfers and preserves self-evidence of its premises. These four kinds of self-evidence correspond with three kinds of propositions: analytic propositions, empirical propositions (perceptual and introspective propositions included) and inferential propositions.[4] Czeżowski clearly

[4] Czeżowski writes: "Cztery epistemologicznie rozróżnione rodzaje zdań odpowiadające wymienionym czterem rodzajom przekonań oczywistych, mianowicie zdania analityczne, zdania empiryczne doświadczania wewnętrznego i doświadczenia zmysłowego oraz zdania uzyskane drogą wynikania logicznego ze zdań pierwszych trzech rodzajów *wyczerpują* zakres zdań wchodzących w skład wiedzy. Wiara nie wymaga oczywistości." [7, p. 56].

states that our *knowledge* is composed only of the foregoing analytic, empirical and inferential propositions.[5]

I would like to make three remarks concerning Czeżowski's views on self-evidence and knowledge. Firstly, we find in Czeżowski's philosophy two different concepts of empirical self-evidence and empirical propositions. The first one is narrow and embraces only sense perception and introspection. The second one is broad and includes also axiological perception (moral and aesthetic perception of values) and intellectual intuition (*noesis*). In 1945, when Czeżowski was writing his paper in which he provided the foregoing classification of propositions, he held a narrow concept of perception.[6] However, later on he changed his mind and classified axiological insights as axiological self-evidence. It was not meant as a new kind of self-evidence but simply as a component of perceptual knowledge.

Secondly, Czeżowski considered ontological principles to be analytic, but they could be regarded as synthetic if we assume the existence of an ontological perception (*noesis*) in which the basic ontological principles can be given. This interpretation could be supported by Czeżowski's metaphysical theory of transcendentals: truth, goodness and beauty [6]. If we can "perceive" goodness and beauty in axiological insights—as Czeżowski decisively claims—why not assume that we are capable of "perceiving" the truth of ontological principles?

The third point is that, in Czeżowski's framework, beliefs/propositions based on introspection and perception conceived of in a broad sense are directly justified and inferential propositions are justified indirectly [7, p. 56]. This means that a belief can be justified by perception, introspection, memory or inference, or, if it is an analytic belief, it is justified directly by its content, and only beliefs justified in that way can be regarded as knowledge. Testimonial beliefs cannot be counted as knowledge and they should be taken as a matter of faith only. For this reason, Czeżowski's theory of knowledge is an exemplification of epistemological individualism, paradigmatically represented in the early modern philosophy by Descartes.

There is another important and philosophically interesting point in Czeżowski's philosophy which should be mentioned while discussing the Cartesian inspirations in his theory of knowledge. It is a relation between self-evidence and truth. Czeżowski claims that every true belief is self-evident but not conversely, i.e. not every self-evident belief is true [7, p. 57].[7] But let us note that if every true belief is self-evident and every self-evident belief is justified, then it follows that every true belief (proposition) is justified. But this conclusion seems to be simply false and a bit surprising. Surely, not all true beliefs are epistemically justified; there are true beliefs acquired by wishful thinking, guess or chance. A correlation between truth and self-evidence suggested by Czeżowski

[5]I do not want to go here into a detailed discussion on the structure of belief and the relation between proposition and belief. Suffice it to say that in Czeżowski's case a proposition is the content of a belief and a state of affairs is an object of a belief [18].

[6]I mean the paper published first in French, "Quelques problems ansciens sous la forme modern" (Studia Philosophica 1939/1946, 101–113), and next in Polish as "Niektóre dawne zagadnienia w nowoczesnej postaci", the Polish version was published first in *Odczyty filozoficzne* in 1958 and later in the second edition of *Odczyty filozoficzne* in 1969 [7].

[7]This is contrary to later Brentano's view which claims that every self-evident belief is true. Artur Rojszczak [21] discussed Brentano's conception of truth in more detail.

is not quite Cartesian, but it is very strong, though asymmetrical. That is why Czeżowski's theory of knowledge leads to a false conclusion.

There is another striking point in Czeżowski's theory of knowledge which might be elucidated by the aforementioned strong but asymmetrical correlation between truth and self-evidence. The definition of knowledge which Czeżowski used says simply that knowledge is a set of true beliefs/propositions [5, pp. 8, 57]. Therefore, Czeżowski's definition of knowledge is clearly less demanding than the classical definition thereof, formulated by Plato, which defines knowledge as true *and* justified belief. However, if we take into account that every true belief is self-evident and every self-evident belief is justified (by perception, introspection, memory or inference), then every true belief belongs to the body of our knowledge. One can say that all true beliefs are justified and belong to our knowledge. Needless to say, two propositions: every true belief is justified and every true and justified belief is knowledge are not logically equivalent. The first of them is simply false. I suggest that we should treat Czeżowski's definition of knowledge as a semi-Cartesian component of his epistemology.

As argued above, Czeżowski's stance in epistemology is epistemic individualism. However, in his writings there are some remarks and comments which might undermine this interpretation and therefore they deserve our attention.[8] In 1959, Czeżowski states that we owe only a small portion of all our knowledge to our own experience and reasoning [5, p. 43].[9] Most of our beliefs are "second-hand beliefs", which we owe to what other people tell us, what we are taught at different levels of education, what is said in the media, etc. The problem is that if a belief is based on the authority of teachers or books or it is based on other people's reports, then it should be counted as faith rather than knowledge. It seems that Czeżowski's position might be defended as coherent and tenable because, as he claims, we accept belief *p* on the basis of our perception and understanding of its content, and not simply because someone else (agent *O*) informed us about *p*. Czeżowski assumes that, having perceived visual or auditory signs and having understood the content these signs stand for, we ourselves evaluate their cognitive and epistemic value and finally accept the belief as true or reject it as false. However, there arises a problem here. Given Czeżowski's asymmetrical correlation between truth and self-evidence, if *p* is true, then there must be some relevant self-evidence corresponding with *p*. But if *p* says, for example, that Napoleon was defeated at Waterloo in 1815, what kind of self-evidence corresponds with *p*? Surely, it cannot be the mere perception of visual or auditory signs and comprehension of the meaning to which they refer. Czeżowski himself points to the mental states of agent *O* who said or wrote that *p*. But what kind of self-evidence *O* could have, if *O* was not an eyewitness of the reported event? Most probably *O* referred

[8]A reflection about epistemological individualism in Czeżowski's philosophy is not only a matter of historical reconstruction but it can have more general philosophical significance.

[9]Czeżowski writes: Jednakże we współczesnych warunkach życia tylko nieznaczny ułamek posiadanej wiedzy zawdzięcza każdy z nas *własnemu doświadczeniu i własnej myśli*. Większość przekonań nabywamy niejako z drugiej ręki, przejmując gotowe cudze przekonania drogą informacji. Dzieje się to za pośrednictwem słowa mówionego i pisanego, w rozmowie i w listach, przez dzienniki i przez radio, w szkołach przez książki i czasopisma naukowe. Motywem przekonania jest we wszystkich tych przypadkach słuchowe lub wzrokowe spostrzeżenie mowy wraz ze zrozumieniem tego, co się spostrzega; pośrednio jednak trzeba odnieść powstawanie przekonań tą drogą do cudzych przeżyć, znajdujących swój wyraz w spostrzeganych znakach [5, 43].

to the mental states of some eyewitnesses of *p*. But why should we believe that *O* indeed referred to their mental states and that those eyewitnesses were reliable in their reports? We are usually not in a position to check that. Arguably, we accept *p* not because we have any epistemic access to the mental states of other agents, but because we have some perceptual and inferential self-evidence that agent *O* is a competent and sincere person. Therefore, finally we alone are responsible for justifying our belief that *p*, which is a clearly reductionist and individualist stance.

Czeżowski admits that humans have a natural propensity to believe what they are told and this propensity is rooted in their "drive to imitate" [5, p. 43]. However, beliefs acquired and accepted by "the will to believe", to use James's terminology, are not knowledge. Moreover, Czeżowski holds that the will to believe rooted in the imitation drive is responsible for many negative phenomena, such as the spread of rumors and superstitions or giving in to propaganda and other bad epistemic practices—in short, for many false beliefs.[10] That the will to believe and the drive to imitate are assessed negatively is clear from Czeżowski's critical remark describing these propensities as particularly common among young children and primitive tribes, or when mass psychology is at play. This opinion seems a bit strange since the will to believe is grounded simply in human nature and hence it can be attributed to every human, not only to children or primitive tribes. Czeżowski does acknowledge that the will to believe may at times have some positive effects; it may help in someone's recovery, for example. However, there is a huge gap between Czeżowski's noticeably reserved approach to testimonial beliefs and Reid's optimistic epistemological project, in which the human inclination to believe other people is conceived of as our basic cognitive capacity, on a par with perception (cf. Reid's principles of credulity and veracity). For Czeżowski, the natural fact that we tend to rely on the word of others is a kind of epistemic imperfection in us, if not downright impediment in our gaining knowledge—certainly not an epistemic asset. For that reason, testimonial beliefs should be very critically assessed before they are accepted.

But, we may still wonder, if the will to believe others is rooted in human nature, how can we expect an epistemic agent to be ready to asses critically his/her beliefs acquired via testimony and, in this way, to act against his/her natural endowment? I suppose Czeżowski might answer the above question in at least two ways. Firstly, following Aristotle and many others, he might appeal to the great value of knowledge and truth, which surpasses our natural inclinations and imposes some moral obligation on us to do our best in striving to achieve this value [7, pp. 211–212).[11] Secondly, Czeżowski always underlined

---

[10] Czeżowski says: "Jeżeli przekazywanie komuś przekonań innej osoby nie dzieje się w postaci informacji czysto rzeczowej lub nauczania to nazywamy je poddawaniem lub sugerowaniem przekonań, stosowanym lub nadużywanym w reklamie, propagandzie i wszelkiej agitacji, ułatwiającym szerzenie się pogłosek i plotek; ma ono również jednak i dodatnie zastosowania, np. w lecznictwie chorób nerwowych, gdzie poddanie choremu przekonania o możliwości wyzdrowienia jest nieraz warunkiem uleczenia. U jego podstawy leży *naturalna skłonność* do *uwierzenia* w to, co ktoś drugi wyznaje szczerze lub nieszczerze jako swe własne przekonanie; przejawia się w niej właściwy naturze ludzkiej popęd do naśladownictwa, ten sam, który tłumaczy nam zjawiska mody. Ów popęd jest w dziedzinie psychicznej jakby zaraźliwością pewnych zjawisk psychicznych i występuje wyraźnie zwłaszcza tam, gdzie nie dochodzą do głosu indywidualne różnice w rozwoju psychicznym jednostek: u ludów pierwotnych, wśród dzieci, w zjawiskach masowych tzw. psychologii tłumu." [5, p. 43].

[11] Czeżowski writes: "Sądzę bowiem, że także w dzisiejszych czasach zachował swą wartość ideał filozofa starożytnego-mędrca wcielającego zasady teoretyczne" [7, p. 212].

the role of character-making and the importance of moral and cognitive training, which contribute to the development of such intellectual virtues as criticism, exactness, open-mindedness, conscientiousness or reliability [7, p. 212]. However, there is still some tension in Czeżowski's views on knowledge because, as said above, he is explicit that we owe only a small part of what we know to our own experience and reasoning. If knowledge is of such great value to us, which is rather unquestionable, and we owe only a small portion of it to ourselves, this situation can be a bit disturbing for an epistemological individualist for whom an individual search for truth is of great value too.

6 Czeżowski's Internalism and Fallibilism

Regarding the internalism/externalism divide, Czeżowski's position appears to be on the internalist side. If a belief is justified by self-evidence of some kind, be it apodictic, perceptual, introspective or inferential self-evidence, and all justified beliefs are self-evident, then the epistemic agent holding a justified belief clearly has access to the relevant justification [7, p. 56]. However, also here one could trace some tension in Czeżowski's views. Let us remember that when discussing in [5] the problem of what justifies our second-hand knowledge, i.e. the knowledge that we owe to someone else's reports and teachings (significantly, the term *knowledge* is used in that context), Czeżowski points to the mental states of our informers who said or wrote that *p*.[12] The mental state of another person, i.e. his/her perception or reasoning, may provide justification for our belief ("the extendedness hypothesis"), but then it is clearly an externalist position.

It is difficult to speculate how Czeżowski would solve the problem of the criterion, famously reintroduced into the epistemological debate by Roderick Chisholm [3], but most probably he would subscribe to epistemological methodism rather than to particularism. The attention Czeżowski paid to the notion of self-evidence, the types of it and the criteria which beliefs must fulfill to count as knowledge, place him clearly in the Cartesian camp. Let us repeat here the features of beliefs contributing to our knowledge. A belief regarded as part of knowledge must be self-evident, obvious, firm and its object must be comprehended in a clear and distinct way [7, p. 56].[13]

Another question is whether Czeżowski's views on the justificatory role of self-evidence could be regarded as a version of foundationalism. Certainly, it was not Cartesian-style foundationalism, but I think Czeżowski might be understood as representing a bit reformulated version of moderate foundationalism, or fallibilism. He held the view that there are basic beliefs but they are fallible, and even if belief is self-evident, it can be refuted. Every true belief is self-evident but not every self-evident belief is true. According to Czeżowski, every element of a scientific theory, be it an inductive law, deductive axioms, historical hypotheses or a singular proposition based on experiential data, can be rejected; "there are no infallible premises in empirical sciences" [6, p. 88]. When arguing against Cartesian foundationalism, he resorts to the history of science. In

[12] See footnote 9 above.

[13] A method of an analytic description can be used to establish the characteristic of knowledge. This method is very similar to Plato's noeis or Aristotle's induction.

his numerous papers regarding the methodology and philosophy of science, he referred to two—in his view particularly significant—scientific discoveries. The first was the discovery of the non-Euclidean geometries. The non-Euclidean systems of geometry, as Czeżowski argued, undermined Descartes's and Kant's belief in the apparently infallible axioms of the classical geometry. He stressed that there can be various incompatible deductive systems whose axioms are counted as true and, as David Hume had already claimed, that there is no direct and only one relation between formal systems and empirical facts. The second discovery to which Czeżowski referred was made by the French mathematician and logician, Joseph Gergonne (1771–1859). Gergonne argued that the axioms of deductive systems can be regarded as implicit definitions which—similarly to the algebraic equations containing variables—consist of terms whose meaning is constituted by the stipulation that the axioms are true.

Czeżowski observed that not only a priori beliefs and propositions are fallible, but introspective, perceptual and moral beliefs can be fallible as well. But he was not a skeptic. There seems to be no infallible knowledge, but some beliefs are more probable than others and some beliefs are better justified than others—this allows for a progress in science [5, 6]. That progress is possible because there are methods of inferential justification of beliefs by calculations of probabilities. Czeżowski himself subscribed to Hans Reichenbach's logic of probability [7, p. 59].

But it must be stressed here that even despite his firm statements regarding the fallible nature of all scientific knowledge, Czeżowski is clear that there are some infallible beliefs. These are simple moral principles which are universally shared by all people in various cultures and times. Therefore, we are also allowed to assume that they are fully reliable and certain—or simply infallible [7, p. 44].

Another problem is how to interpret the epistemic status of ontological principles such as the principle of non-contradiction or the principle of causality.[14] Czeżowski elaborated a rather sophisticated doctrine of how formal deductive systems are related to the empirical world.[15] From this point of view, ontological principles can be regarded only as "interpretations" of deductive axioms and theorems. If a logical axiom or a theorem is confirmed within the framework of a physical theory, be it the classical Newtonian mechanic, relativistic cosmology or quantum mechanics, it is "materially true", as Czeżowski in a bit old-fashioned style would say. But since we will never be able to provide such a confirmation—taking into account all possible interpretations and all regions of the universe—we will never know whether and which formal and logical truths are always materially true. In other words, we cannot say that a given principle is infallible because we do not know whether every possible empirical and ontological interpretation of a formal axiom or theorem (be it the logical law of non-contradiction or the law of excluded middle) is materially true. On the other hand, Czeżowski, as mentioned above, viewed the ontological principle of non-contradiction as necessary and apodictically self-evident. To sum up Czeżowski's stance concerning the problem

[14] The principle of causality was strongly defended by the members of the Lvov-Warsaw School in face of quantum indeterminacy [25, pp. 249–250].

[15] Surely, Czeżowskian model was much more sophisticated than Russell's model. It was Russell [22, p. 169] who said that "logic is concerned with the real world just as truly as zoology, though with more abstract and general features".

of foundationalism, I think that we could say that he was not a Cartesian foundationalist, but his general adherence to what might be called moderate foundationalism/fallibilism is not unproblematic either since he did assume some infallible moral and ontological basic beliefs.

7 Czeżowski and Some Views on Belief-Formation Processes

Brentano's and Twardowski's heritage is clearly present in Czeżowski's view on belief-formation process, but I wanted to point to the fact that his views in that regard are also a continuation of Descartes's ideas. In his theory of belief formation, Czeżowski directly refers to Brentano's and Twardowski's descriptive psychology.[16] According to the Brentanian approach, a belief (judgment) consists in our acceptance or rejection of the existence of an object given to us in an act called presentation.[17] Certain technical distinctions aside, we can simply say that presentation is a neutral act of comprehension (or understanding) related to a given content.[18] The asymmetrical relation between presentation and belief (judgment) is necessary; there can be no belief without presentation, but presentation does not entail belief. As stated by Czeżowski, in order to assert that "it rains", one must know before what "rain" is, and in order to reject the claim that "ghosts exist", one must know before what the sentence means [5, p. 59]. This is exactly the Cartesian view of how the belief formation proceeds; an act of neutral comprehension of a given content is prior to the act of acceptance or rejection. Though apparently convincing, such a model of belief formation is from psychological perspective rather inadequate, and, consequently, has been criticized by many philosophers and scholars. According to developmental psychology, human beings have a strong natural inclination to accept beliefs right away, without prior considering their content, which view was also admitted by Czeżowski himself (cf. "the drive to imitate"). Thus, beliefs are accepted as if simultaneously with their comprehension, without earlier "presentation", to put it in the Brentanian terminology. Acceptance is psychologically prior to critical assessment and possible rejection. Whatever the evolutionary basis of this mechanism of belief formation may be and its cognitive advantages or disadvantages, we naturally and involuntarily tend to *believe*. As Elżbieta Łukasiewicz describes and summarizes that problem:

> First and foremost, in the mental development of children, the capacities of doubt and disbelief are acquired much later and are deemed more complex than acceptance and trust, which are ontogenetically prior. As Reid's argument has it, if credulity were not a gift of nature and were not primary in humans, but were developed by children on a par with reasoning and other mental

---

[16] Twardowski's theory of judgment is presented and discussed in a comprehensive way by Jacek Jadacki [13, pp. 178–184].

[17] On relations between presentation and judgments see also [18].

[18] There is a distinction between an act (be it presentation or judgement), the content of an act and its object. The distinction in question was forcefully defended by Kazimierz Twardowski and taken over by all his disciples, including Czeżowski. Judgment is an assertive act which can be identified with belief. The characteristic features of judgement (belief) are its propositional structure and being articulated in a language.

> capacities, the attitude of trust should be greater in adolescents and adults than it is in children. Very young children should reveal the most neutral attitude to what they are told compared to other age groups. Since it is exactly the other way round, i.e. credulity is strongest in young children and is lessened with age and experience, then it follows that trust must be the natural endowment of human beings.[19] So children's minds are not Cartesian systems—were they such systems, believing and disbelieving would be equally easy for them and these two abilities would develop at about the same time and rate, which is obviously not the case. [19][20]

Thus, one could say that mere comprehension of a proposition which is free of any assessment is a myth [19]. Also a semantic argument could be raised against the Cartesian and Brentanian models of belief formation. As argued by Johnson-Laird [15, p. 110], in order to understand the meaning of a sentence like *John is a bachelor* one must know the conditions under which the sentence is true (or false). How could we understand the meaning of the above sentence without somehow representing it in the mind as the real state of affairs, i.e. *with* its truth-value attached to it? It is not clear what this truth-neutral representation would represent then [15].

Let us note here that it was Spinoza who was the first to decisively question and reject Descartes's model of belief formation.[21] The objections are serious and cannot be easily dismissed. Yet, I suggest that the Cartesian and Brentanian models of belief formation could be defended by appealing to the distinction between descriptive and normative epistemology. The difference between the two approaches may be expressed in the following two questions. The first one is how *we really form our beliefs*—we know that the Cartesian explanation of that process is at least wanting, if not simply false. The second question is *how we should form our beliefs*, and the Cartesian model appears much more promising when applied to this problem. If knowledge and truth are of great value, then we have epistemic, if not moral, duties to do our best to acquire them. Thus, even if we have a natural tendency to uncritically accept any incoming beliefs in order to understand them, we still have the duty to subsequently evaluate them and we are free to reaffirm or reject them if they do not meet our epistemic standards. By that critical assessment, we preserve the Cartesian and Brentanian character of epistemology whose essential part is an individualist search for truth and knowledge.

In conclusion, we could say that Czeżowski's theory of knowledge and belief is, *via* Brentano's and Twardowski's views, significantly Cartesian. The core idea of this project is that one is obliged to strive for knowledge and one can do it solely by an individual effort and individual cognitive faculties such as perception, introspection, memory and inference. Czeżowski was clearly aware of the role played in our life by others' testimony and he did acknowledge our natural, psychologically grounded inclination to believe others, nevertheless, he was explicit in his writings that beliefs based on authority, emotions or grounded in "the will to believe" do not deserve to be considered part of knowledge; they belong to mere "faith". Contrary to the long medieval tradition, the Port Royal School, David Hume, Thomas Reid and their followers, who considered authority-based faith to be legitimate source of beliefs, Czeżowski showed sympathy with Cartesian and Lockean epistemic individualism and evaluated faith-based beliefs rather negatively.

---

[19] Cf. Reid *Inquiry into the Human Mind*, 1997.

[20] See also Gilbert [11] and Lechniak [16].

[21] See Gilbert [11].

In general, the epistemological program characteristic of almost all representatives of the Lvov-Warsaw School, not only of Czeżowski, was deeply rooted in the Cartesian individualist tradition. Furthermore, it was also under great influence of the philosophy of ancient Greeks, who also greatly valued the ideal of individual search for knowledge. It is true that by downgrading the role of authority and testimony, Czeżowski's theory, like many other epistemological projects, does encounter the problem of testimonial knowledge and lack of proper justification for a huge part of what we claim to know. But, on the other hand, one can find some merits in Czeżowski's epistemic individualism. Undoubtedly, one of its great advantages depends on a very straightforward connection—which in many externalists accounts is missing—between the value of knowledge and its ethical dimension, i.e. the credit we deserve for achieving knowledge if we gain it through our own effort and our epistemic responsibility if we fail to meet sufficient epistemic standards.

Acknowledgements I would like to express my gratitude to Elżbieta Łukasiewicz for both reading and improving the paper.

References

1. Aquinas, T.: Faith, Reason and theology. Questions I–IV of his Commentary on the Trinitate of Boethius. Pontifical Institute of Mediaeval Studies, Toronto (1987). Trans. M. Maurer
2. Arnauld, A., Nicole N.: Logic or the Art of Thinking. Cambridge University Press, Cambridge (1996). Trans. Jill Vance Buroker
3. Chisholm, R.M.: The Problem of the Criterion. Marquette University Press, Milwaukee (1973)
4. Czeżowski, T.: O metafizyce, jej kierunkach i zagadnieniach, 1st edn. Wydawnictwo Uniwersytetu Mikołaja Kopernika, T. Szczęsny, Toruń (1948)
5. Czeżowski, T.: Główne zasady nauk filozoficznych, 3rd edn. Zakład Narodowy im. Ossolińskich, Wrocław (1959)
6. Czeżowski, T.: Filozofia na rozdrożu. PWN, Warszawa (1965)
7. Czeżowski, T.: Odczyty Filozoficzne, 2nd edn. Towarzystwo Naukowe w Toruniu, Toruń (1969)
8. Czeżowski, T.: O metafizyce, jej kierunkach i zagadnieniach, 2nd edn. Wydawnictwo Antyk, Kęty (2004)
9. Eliot, C.W. (ed.): Descartes: Discourse on the Method, of Rightly Conducting Reason and Seeking the Truth in Science. P.F. Collier & Son, New York (1909–1914)
10. Fumerton, R.: The problem of the criterion. In: Greco, J. (ed.) Oxford Handbook of Scepticism, pp. 34–52. Oxford University Press, Oxford (2008)
11. Gilbert, D.T.: How mental systems believe. Am. Psychol. **46**(2), 107–119 (1991)
12. Hume, D.: In: Beauchamp, T.L. (ed.) An Enquiry concerning Human Understanding. Oxford University Press, Oxford/New York (1999)
13. Jadacki, J.: On Kazimierz Twardowski's descriptive semiotics and its metaphysical basis. In: Jadacki, J. (ed.) Polish Analytical Philosophy, pp. 175–199. Wydawnictwo Naukowe Semper, Warszawa (2009)
14. Jadacki, J.: The Lvov-Warsaw School and its influence upon the Polisch Philosophy of the twentieth century. In: Jadacki, J. (ed.) Polish Analytical Philosophy, pp. 77–131. Wydawnictwo Naukowe Semper, Warszawa (2009)
15. Johnson-Laird, P.N.: Computer and the Mind: An Introduction to Cognitive Science. Harvard University Press. Cambridge (1988)
16. Lechniak, M.: Zmiana przekonań. Wydawnictwo Katolickiego Uniwersytetu Lubelskiego, Lublin (2012)

17. Locke, J.: An Essay Concerning Human Understanding. Gateway Editions, Chicago (1956). Preface by Russell Kirk
18. Łukasiewicz, D.: Tadeusz Czeżowski on Existence. In: Chrudzimski, A., Łukasiewicz D. (eds.) Brentano and Polish Philosophy, pp. 183–215. Ontos, Frankfurt (2006)
19. Łukasiewicz, E.: Evidentiality and epistemic justification of beliefs. Kazimierz Wielki Univeristy Press (2018)
20. Reid, T.: In: Brookes, D.R. (ed.) An Inquiry into the Human Mind on the Principles of Common Sense. Pennsylvania State University Press, University Park (1764/1997)
21. Rojszczak, A.: Wahrheit und Urteilsevidenz bei F.Brentano. Brentano-Studien **V**, 187–218 (1994)
22. Russell, B.: Introduction to Mathematical Philosophy. George Allen and Unwin, London (1919)
23. Sellars, W.: Empiricism and abstract entities. In: Schilpp, P.A. (ed.) The Philosophy of Rudolph Carnap, pp. 431–468. Open Court, LaSalle (1963)
24. Spinoza, B.: The Ethics. Princeton University Press, Princeton (1985). Trans. Edwin Curley
25. Woleński, J.: Logic and Philosophy in the Lvov-Warsaw School. Kluwer Academic, Dordrecht (1989)

D. Łukasiewicz (✉)
Kazimierz Wielki University, Bydgoszcz, Poland
e-mail: dlukas@ukw.edu.pl

What Is Reasoning?

Robert Kublikowski

Abstract The usage of language and cognition have perhaps been the oldest specific features of a human mental activity. However, is there anything really exceptional in the fact *that* we use language and *that* we cognise? Is there anything specific in *how* we do it? It seems that reasoning is especially important among various linguistic and cognitive human activities. But what is reasoning? What is human reasoning? Is there anything *specific* about it? Which cases of reasoning are the *correct* ones and *why*? I take as my starting point the views on reasoning presented by the logicians and philosophers belonging to the analytic philosophy of language and epistemology, especially to the Polish analytic philosophy (The Lvov-Warsaw School: Tadeusz Czeżowski, Kazimierz Ajdukiewicz, Janina Kotarbińska and Alfred Tarski). In addressing the above questions I develop an improved view according to which the formal conditions of correctness are based on formal relationships (a consequence etc.). Such conditions depend on informal (material) conditions of correctness which are based on informal (material) relationships (reference, causality, spatial relationships etc.).

Keywords Language · Meaning · Truth · Assertion · Definition · Inference · Reasoning · Formal and informal conditions of correctness · Belief revision · Justification · Knowledge · Cognition

Mathematics Subject Classification (2000) Primary 03B65; Secondary 03B42

1 Introduction

Each of us belongs to the biological species called *Homo Sapiens* in terms of our physical features, especially, in terms of the human nervous system and its main element—the brain. We are, in this respect, a part of nature understood as the material world. But we transcend this material dimension with our minds. In other words, the human mind is something more than a complex of material (physical, chemical or biological) elements and rules according to which such a complex of elements works. Our human mind enables us: to feel emotions (feelings), to decide (the will) and to

Á. Garrido, U. Wybraniec-Skardowska (eds.), *The Lvov-Warsaw School. Past and Present*, Studies in Universal Logic,
https://doi.org/10.1007/978-3-319-65430-0_10

think (the practical and the theoretical reason, the intellect). The intellect is featured by intelligence and an intellectual intuition. An activation of our mental, rational, intellectual, intelligent, intuitional abilities (dispositions) enables us to use language and to cognise.

Cognition may be understood as a process and its result—knowledge. Knowledge is a true and justified belief. A belief is an asserted sentence with its meaning (sense) which is a logical judgment (a proposition). Such a judgment can be justified in a *direct* way—by perception or *indirectly*—by reasoning. However, strictly speaking, three types of a cognitive process (and of knowledge) can be distinguished: the perceptual, the intuitive and the discursive ones. I prefer to qualify these cognitive processes using the term "types" rather than "kinds" as this distinction is not a product of classification but of typology. The boundaries between these cognitive activities (and respectively, their concepts) are vague. So, let us repeat, the following types can be distinguished:

- Perception is a cognitive process obtained by using senses: sight, hearing etc.
- Intuition is a momentous, quick, "without any steps" act of an intellectual seeing (apprehending, grasping, understanding) that something has such and such a feature or something is related to something else.
- A discursive process is "realised in steps": e.g. an analysis, a classification, reasoning, a discussion etc.

These cognitive activities are mutually and closely interrelated. Intuition is engaged in a perceptual act during the process of getting a concept of something which is perceived (conceptualisation). But intuition is also useful or perhaps necessary in the *discursive* process of reasoning:

- to see the *relationship* between a premise and a conclusion (a premise is a sentence which is a starting point and a conclusion—an ending point, a goal of reasoning),
- to grasp that a sentence is true or
- to acknowledge (accept) a sentence as a true premise or a true conclusion of reasoning. Such a pragmatic activity is called assertion.

"A transition from some beliefs to a conclusion counts as inference only if the thinker *takes* his conclusion to be *supported* by the presumed truth of those other beliefs. [...] It's enough that we take our premises to be true, that is, judge them to be true." [6, p. 4]. However, the question arises: "What is it to believe something *because* one *takes* it to be supported by other things one judges to be true? What kind of taking are we talking about?" [6, p. 6]. The notion of "taking" can be explicated in the following way: "(Taking Condition): Inferring necessarily involves the thinker *taking* his premises to support his conclusion and drawing his conclusion *because* of that fact. The intuition behind the Taking Condition is that no causal process counts as inference, unless it consists in an attempt to arrive at a belief by figuring out what, in some suitably broad sense, is supported by other things one believes. In the relevant sense, reasoning is something we *do*, not just something that happens to us. And it is something *we* do, not just something that is done by sub-personal bits of us. And it is something that we do with an *aim*—that of figuring out what follows or is supported by other things one believes." [6, p. 5], [10, 11, 22, 27]. However, what is reasoning? What makes something reasoning?

2 The Definition of Reasoning

Reasoning is a process or an activity concerning sentences or judgments. The process leads to a result: new sentences or judgments. Such a result is a logical structure consisting of judgments which are linked as premises and a conclusion, reasons and a consequence, a starting point and an ending point (a goal) of reasoning [13, p. 119].

Reasoning can be understood in various ways:

- as complex reasoning (something more complicated than inference) or
- as simple reasoning (inference).

Complex reasoning consists of inference and of another procedure (a method), for instance, of questioning. A proof is a kind of complex reasoning.

A proof in a logical sense is a complex composed of a logical judgement to be proved which is linked by a relationship of a consequence with other logical judgements (premises). The conclusion can be deductively derived from the premises [12, pp. 91, 93]. At the starting point we know a conclusion which is the last element of the structure of proof. We ask the guiding question: How do we get to know (prove) that the conclusion is true? After that we try to find the correct premises among judgments we have already accepted. However, the following question arises: "what is it to draw a conclusion from a premise *because* you take the premise to provide support for the conclusion?" [22, p. 389].

If we want to build a proof, we match a true premise with a judgment to be proved. In a case when a proof is deductive reasoning, a premise is a reason and a judgment to be proved (a conclusion) is a consequence. A premise must be connected with a conclusion by a relationship of a *logical consequence*. In other words, a conclusion must follow logically from a premise on the *basis* of a logical law. Such a proof is justifying reasoning because we match a correct premise to support a conclusion. Such a proof is also regressive reasoning because we "*move*" *from* a conclusion *to* a correct premise. So the direction of finding a correct premise is opposite to the direction of the relationship of a logical consequence [12, p. 93]. Yet, what is simple reasoning, that is, inference?

"Inferring is a movement of thought between propositions which may, in special circumstances, result in the thinker coming to judge the proposition inferred to be true." [27, p. 28]. Inference is a way of thinking. But "thinking" is a broader concept than "inference". It means that every inference is a way of thinking, but not every way of thinking is inference.

Inference "involves judging a conclusion to be true because one takes the (presumed) truth of the premise to provide support for that conclusion." [22, p. 389]. The word "*because*" or "'therefore' is used to express or report an inference, it does so by virtue of the contextually salient explanatory relation being precisely the relation that Frege spoke of in describing inferring as believing a conclusion because one takes it to be *justified* by something. It is, in other words, a relation that obtains between some conclusion that is justified, and something else—a reason—that makes it justified. This is not the kind of relation that some epistemologists call 'propositional justification', which is a relation between a person and a proposition that the person is justified in believing, but rather a generalization of the relation that epistemologists call 'doxastic justification', which is a relation between a person's belief and whatever makes that person's belief justified." [22, p. 400]. Thus, some relations can be distinguished:

(a) a relation between something and thinker (a person, a cognitive subject) *S* who believes (judges, accepts, takes etc.) that conclusion (a logical judgment, a proposition etc.) *q* is true.

This something is supposed to justify (support, warrant, entitle etc.) *S* believing that *q* is true. But what is this something? Let us reformulate (a) to see it:

(b) a relation between premise *p* presumed to be true (a reason justifying *q* as true) and *S* who believes that *q* is true because of *p*.

This is a pragmatic relation, that is, a relation between linguistic expressions and language user *S*.

(c) This something can be understood not just as *p*, but as something which justifies *p* as true, something which *p refers to*, something making *p* true (a *truth-maker*).

Such a relation is a semantic relation. But what are the conditions which justify *S* believing that *p* is true? Or when is *p* true? Let us analyse this situation. At the starting point of reasoning (a proof) *S* knows *q* and *S* searches *p* to justify *q* in a logically and materially correct way. In other words, *S* is justified to judge *q* as true if *q* is justified by *p*. And, thinking in a regressive way, that is, "going back", *S* is justified to judge *p* as a true belief and a right reason justifying *q* if *S* is *justified* to judge *p* as a true belief and a right reason justifying *q*. And so on. There appears the infinite regress (or a circularity) in reasoning. However, it is not identical with the infinite regress (or a circularity) in justification. And perhaps a non-inferential justification is accessible. One way of non-inferential justification is to claim that some beliefs—basic beliefs—are justified by intuition (a reflection). Such beliefs must clearly, obviously, surely be true etc. However, the problem with intuition is that a particular *p* is clear etc.—and thus fundamental—for some people, but it is not for others. And it makes the notion of intuition complicated. (I will analyse this problem in the following sections and I will also propose a solution to this problem.)

Inference, or to be more specific inferring, is a mental, *complex* process consisted of judgments. However, inference is not just an accidental collection of judgments. The conclusion is accepted as true on the basis of the acceptance of premises. The crucial issue is the relationship between the acceptance of premises and of a conclusion. Moreover, weaker or stronger acceptance of premises leads to the acceptance of a conclusion which has not been accepted yet or has been accepted with a lower degree of certainty. The degree of acceptance of a conclusion is not higher than the degree of the acceptance of premises [1, p. 107]. Inference is not just "a few sentences". So, what is it that makes inference? Inference consists of sentences. It seems that they are elements of an inferential structure. And yet, what makes them a complex (a structure)? What unites such sentences? What is the basis of the structure? Is it a chain (a sequence) of sentences? What is this chain (this sequence)?

Premises accepted (acknowledged) as true, in other words, asserted premises are the basis of asserting a conclusion. A sentence—a conclusion—is asserted on the basis of a relationship which connects asserted sentences playing the role of premises or a conclusion. But what is this relationship? What relationship justifies the fact that a premise "sends" ("transmits") its meaning—a logical judgment—to a conclusion? [21, p. 260]. The candidates for these relationships are formal and informal relationships [16].

Another question is what makes something a correct, good reasoning? Is incorrect, bad reasoning still reasoning?

3 The Conditions of Correctness of Reasoning

The conditions (rules) allow to judge (evaluate) reasoning in terms of correctness (rightness) or incorrectness (non-rightness). What makes reasoning right? Which premise or conclusion is right? "The premise judgments need to have caused the conclusion judgment 'in the right way.'" [6, p. 3]. It is important to emphasise that an act "might seem right without being right: there is a distinction to be made between seeming right and being right." [11, p. 23]. What is the evidence that such a distinction is useful? "An essential part of the attitude of seeming right is recognizing the possibility of correction. You recognize that a certain sort of challenge to your act is appropriate and may succeed. We may call the challenge 'checking'. When an act seems right to you, relative to a particular rule, you recognize that it might no longer seem right to you if you were later to check what you did. This recognition appears as a disposition. Having the attitude of seeming right involves a disposition to stop having this attitude in particular circumstances, specifically if you were to check the act and it were no longer to seem right. This is a counterfactual disposition, since you might never check. You may not be disposed to check—perhaps because you are confident—but you still have this counterfactual disposition. Checking may consist simply in trying again in the same way to follow the rule, or it may involve something more." [11, p. 22].

Then let us consider that act of checking. Checking involves conditions (rules) of evaluating, in our case, evaluating the correctness of reasoning. Some general conditions of correctness have been distinguished: the *formal* (logical) and *informal* (material) ones: see e.g. [1, pp. 97–181], [12, p. 93].

3.1 The Formal Conditions of Correctness

Formal logic concerns—as the term "formal" suggests—a *form* of used expressions. It does not concern rather a *content* of such expressions. But the opinion that formal logic totally passes over whatever contents would be slightly exaggerated for even shapes p, q, r—representing simple sentences—express the content that p is identical with p, p is not q, p differs from $\rightarrow$ (the symbol of implication) etc.

Sentences which play the roles of premises or a conclusion are linked by *formal* relationships: in such cases a shape, a composition (an order) of expressions and their logical value (truth, falsity) is taken under consideration. Such formal and *inter*-sentential relationships are worked out by a propositional logic and a first-order logic. Sentences consist of names connected by *intra*-sentential relationships which are worked out by a syllogistic logic and a first-order logic [21, p. 261].

It is worthy of distinguishing *implicit* and *explicit* formal conditions of correctness of reasoning [20, p. 234]. Implicit conditions concern the correctness of procedures linked with inference (questions guiding reasoning, definitions of terms used in premises and a conclusion etc.). Explicit conditions fix the correctness of inference itself (e.g. a consequence, truth).

Let us present and analyse some formal implicit conditions:

1. A question which guides inference has to be correct, that is:
   - meaningful (linguistically well constructed),
   - adequate (the question's assumption has to be true),
   - justified (there is a reason to pose such a question),
   - decidable (it is possible to obtain an answer in a finite number of steps),
   - creative (an answer for such a question delivers a new knowledge).
2. Terms used in premises and a conclusion have to be correctly defined:
   - the word defined (*definiendum*) must not be used in the *definiens* in an explicit definition,
   - in the case of a lexical definition the connotation (intension) and the denotation (extension) of the *definiendum* and the *definiens* must be identical, i.e. mutually interchangeable (the condition of adequacy). In other words:
   - the extensions of the *definiendum* and *definiens* must not be mutually exclusive,
   - the extension of the *definiens* must not overlap with the extension of the *definiendum*,
   - the extension of the *definiens* must not be superior to the extension of the *definiendum* (i.e. the definition must not be too broad),
   - the extension of the *definiens* must not be inferior to the extension of the *definiendum* (i.e. the definition must not be too narrow): see e.g. [1, pp. 68–70], [20, pp. 233–234].

   There are also some formal explicit conditions of correctness:
3. It is against the logical law of non-contradiction: : $\neg(p \wedge \neg p)$ to accept contradictory premises.

Therefore, there are a number of formal conditions of correctness of inference. They are based on many different formal relationships. And a logical consequence is not the only formal relationship. However, it is the main one.

4. A logical consequence links the premises and a conclusion in the case of an infallible reasoning: a deduction and a mathematical induction.

The literature concerning a logical consequence is very extensive: see e.g. [2, 14, 15, 23]. Nevertheless, it is not our major topic.

3.2 The Informal (Material) Conditions of Correctness

Not every consequence is a logical one. It is worthy of distinguishing a formal (logical) and an informal (material) consequence. It seems that formal and informal consequences are related to informal conditions of correctness of reasoning.

"We need not treat all correct inferences as correct in virtue of their form, supplying implicit or suppressed premises involving logical vocabulary as needed. Instead, we can treat inferences such as that from 'Pittsburgh is to the west of Philadelphia' to 'Philadelphia is to the east of Pittsburgh,' or from 'It is raining' to 'The streets will be wet,' as *materially* good inferences—that is, inferences that are good because of the content of their *non*logical vocabulary." [9, p. 85], [8, pp. 94–116].

Expressions are connected not only by formal relationships, but also by *informal* ones. Such relationships come out when sentences and names are analysed in terms not only of a logical form, but also of their *contents* [21, p. 262].

There are some informal conditions of correctness:

1. Sentences playing the roles of premises or a conclusion are presumed to be true.
2. Sentences consist of names: individual or general ones. Individual names (e.g. "Christopher Columbus") refer to unique objects. General names refer to classes of objects: names of artefacts (e.g. "a building") and natural kind terms ("gold", "a lemon", "a tiger" etc.).

Such terms refer to natural kinds, i.e. kinds in nature: see e.g. [17]. An ostensive definition is useful for introducing new terms into a language: see e.g. [19].

3. Sentences refer to states of affairs which are linked by many different informal (material) relationships like causality, spatial relationships etc.

So far the following problems (questions) of the paper have been presented: What is reasoning? What is human reasoning? Is there anything specific about it? Which cases of reasoning are the correct ones and why? The definitions of reasoning have been given and some formal and informal conditions of correctness have been shown. However, the main questions have still not been answered in a satisfactory way. So, let us reformulate the questions and consider how logicians can, for instance, know that $((p \rightarrow q) \wedge p) \rightarrow q$. In addition to that, let us express this logical rule (principle) called *modus ponens* (MP or *modus ponendo ponens*—MPP) less symbolically as follows: (if p, then q) & p, so q. And let us analyse the clear physical example of reasoning which fulfills the scheme: *if* water is heated up to the temperature of approximately 100 °C at a standard atmospheric pressure, *then* it boils. In fact, if water is heated up to such a temperature, then it boils. How can logicians know that this scheme is correct?

It is worthy of noticing that "when you reason by modus ponens, you may do some reasoning in identifying what is the antecedent and what is the consequent of the conditional proposition you are reasoning with." [11, pp. 20–21]. Moreover, if "you infer q from p and $p \rightarrow q$, this will seem right relative to the modus ponens rule, but not relative to the rule of inferring a tautology. This means you must have a way of identifying different rules to yourself, which means identifying different dispositions. You must be able to identify some disposition as the modus ponens one, for instance. You can even choose which rule to follow. If you choose the modus ponens rule, this explains why you infer q, rather than something else, from p and $p \rightarrow q$. Choosing a rules mean choosing a disposition. This is not as mysterious as it may sound, and it does not imply you can map out in advance where the disposition will lead in all cases." [11, p. 22].

Perhaps the *machine-like* (syntactic) model of inference is an adequate answer to the question: What is reasoning? It may seem that formal conditions of correctness (a consequence etc.) are more important than informal (semantic) conditions such as truth, reference etc. According to such a model:

- if you apply true premises to a correct scheme of a deductive inference or of a mathematical induction, you will get a true conclusion with an absolute degree of certainty,
- if you apply true premises to a correct scheme of an enumerative induction, you will get a true conclusion with a lower or higher degree of certainty which depends on a quality and a quantity of evidence. Then you will be justified (entitled) to acknowledge a conclusion as true with a lower or higher degree of certainty.

"It is tempting to think that there are two kinds of inference—deductive and inductive. But in what could the difference between these two kinds of inference consist? Of course, in some inferences the premises logically entail the conclusion and in others they merely make the conclusion more probable than it might otherwise be. That means that there are two sets of standards that we can apply to any given inference. But that only gives us two standards that we can apply to an inference, not two different kinds of inference." [6, p. 5].

Does the machine-like model of inference give adequate answers to the questions: What is reasoning? What is human reasoning? Is there anything specific about it? Which cases of reasoning are the correct ones and why? Not quite so.

Thus, let us now notice that "if rule-following is to explain what reasoning is, eventually you must do some rule-following that does not involve reasoning. Moreover, as Boghossian shows clearly, it must not depend on a belief (or any intentional attitude) whose content is that you should act this way. As Boghossian earlier put it, following Wittgenstein, this sort of rule-following must be done 'blindly'" [11, p. 21]. The point is that the *deepest* basis of inference must be non-inferential. It is a *blind* rule-following. How can such a sort of rule-following be blind? In what sense? Is the deepest basis of reasoning blind, that is, accidental? It seems that it is not. Thus how can we explain that an inference, for instance, according to *modus ponens* principle, is not just a matter of good luck? What is the answer to the question: "how could MPP premises warrant MPP conclusions while being blind? Answer: they do, because they are written into the possession conditions for the conditional, and the conditional is a non-defective concept. [...] If we are to make sense of the justified employment of our basic logical methods of inference, we must make sense of [...] *blind but blameless* reasoning—a way of moving between thoughts that is justified even in the absence of any reflectively appreciable support for it." [4, p. 248], [3, 5, 7, 18, 24–26].

To overcome an infinite regress (or a circularity) in reasoning, some conditions in logic have been fixed to be met:

It is important to accept—as a formal foundation of the method of reasoning—a set of rules (*modus ponens*, *modus tollens* etc.) and a set of axioms which are just assumed as clearly, obviously, surely true etc. Such a set of axioms should be:

- independent (the axioms—sentences should not be mutually provable),

- non-contradictory (the axioms and their consequences should not be contradictory),
- complete (every correct sentence or its negation can be proved on the basis of the axioms),
- decidable (there is a method of proving in a finite number of steps whether or not a given sentence belongs to the system) etc.

If you do not assume anything, then you will not prove anything.

It is important also to accept—as an informal (material, empirical) foundation of reasoning—the usefulness of an ostensive definition. If you do not assume it, then you will not get empirical terms, including natural kind terms and empirical sentences which are supposed to deliver an empirical content about some reality.

Perhaps the main claim of this article is also a good candidate for an explanation of why or how we can reason. Namely, it seems that the formal (logical) conditions of correctness of reasoning are based on formal (logical) relationships and these conditions depend on informal (material) conditions of correctness based on informal (material) relationships. The formal relationships are not sufficient for a correct inference and they need to be supported by informal relationships. In other words, "the notion of *formally valid* inferences is definable in a natural way from the notion of *materially correct* ones. [...] the notion of *logically* good inferences is explained in terms of a prior notion of *materially* good ones." [9, pp. 85–86], [8, pp. 104–105].

The formal relationship of the consequence between a sentence which is an antecedent (reason) p and a sentence (the consequent) q in inference seems to depend on the truthfulness of p and q. And the truthfulness of p and q is determined by the relationships of references of p and q to the relevant states of affairs.

Let us analyse the above claims using the example of freezing water. An enumerative induction is used in the context of discovery of regularities concerning such water. Namely, if there is any water at a temperature of about 0 degree Celsius at a standard atmospheric pressure, then the water freezes. It is a complex sentence consisting of two sentences p and q linked by a logical connective of an implication expressing the relationship of a consequence. The implication is true if p and q are true and they are true if p and q—on the one hand—express the given logical judgments and—on the other hand—p refers to the given temperature and atmosphere and q refers to the water which freezes. Water in fact freezes at such temperature and such atmospheric pressure. So, the basis of the whole case is the informal (material) relationship of causality between the states of affairs: the temperature, atmosphere and freezing water. The implication of sentences p and q depends on the (causal) laws of nature.

In the context of justification—however—the empirical claims (conclusions) are supported by deductive inference (precisely speaking, hypothetico—deductive inference). The problem is that deductive inference is infallible if the conditions of its correctness are fulfilled, especially, if premises are true.

But in difficult cognitive cases, that is, if one cannot recognise (know) in a satisfactory, certain way whether empirical premises are true, then one just believes the premises to be true with a lower or higher degree of certainty depending on the quality and quantity of evidence [21, p. 263]. Difficult cognitive situations "disrupt" infallibility of such a deductive "machinery". This is why our minds are able to cognise and express the laws of nature just in a revisable, fallible (tentative) and changeable way.

4 Conclusions

Reasoning is a mental, intellectual activity or its result. At the starting point there are sentences and their meanings (logical judgments) or sentences accepted as true (beliefs). Such sentences play the role of premises or a conclusion. Reasoning is a transition from premises to a conclusion on the basis of formal and informal relationships.

It seems that a good candidate (hypothesis) for an adequate explanation of the human capacity to cognise (to reason) is the *compatibility* of rational (intelligible) nature and of us—humans. Namely, our mental "processes dovetail with the causal structure of the world [...] our processes of belief acquisition are indeed well adapted to providing us with an accurate picture of the world" [17, p. 3]. Our human knowledge of the world is possible on the basis of the assumed hypothetical *fit* between our mental cognitive abilities and the reality, especially the assumed causal structure of the reality which, among others, consists of natural kinds of objects.

This compatibility makes it possible to activate our mental, rational, intellectual, intelligent, intuitive abilities to use language and to cognise (to reason). We are able to reason in a rational way about the material world because—to a certain extent—we are a part of the material world. It suggests the existence of a subtle mental, linguistic and cognitive tuning, by analogy with a discovered subtle cosmological and biological tuning. The explanation of why or how we can reason is a matter of a natural regularity and—to a certain extent—naturally based ability to reason.

However, we humans transcend the material world with our minds as we have such features and functions which do not exist in nature. Such a fundamental feature of the human language and cognition is *normativity*. The conditions and evaluation of the correctness of reasoning: correct (right, valid, sound, good, proper) or incorrect (non-right, invalid, unsound, bad, improper) belong to normative vocabulary and are the evidence of a specifically human normative aspect of reasoning.

References

1. Ajdukiewicz, K.: Pragmatic Logic. PWN – Polish Scientific Publishers, Warszawa and D. Reidel Publishing Company, Dordrecht – Boston (1974) (Originally published in Polish as: Logika pragmatyczna. Państwowe Wydawnictwo Naukowe, Warszawa (1965))
2. Blanchette, P.: Logical consequence. In: Goble, L. (ed.) Blackwell Guide to Philosophical Logic, pp. 115–135. Blackwell Publishers, Oxford (2001)
3. Boghossian, P.: Inference and insight. Philos. Phenomenol. Res. **63**(3), 633–640 (2001)
4. Boghossian, P.: Blind reasoning. Proc. Aristot. Soc. Suppl. Vol. **77**, 225–248 (2003)
5. Boghossian, P.: Inferentialism and the epistemology of logic: reflections on Casalegno and Williamson. Dialectica **66**(2), 221–236 (2012)
6. Boghossian, P.: What is inference? Philos. Stud. **169**, 1–18 (2014)
7. Boghossian, P.: Reasoning and reflection: a reply to Kornblith. Analysis **76**(1), 41–54 (2016)
8. Brandom, B.R.: Making It Explicit. Reasoning, Representing and Discursive Commitment. Harvard University Press, Cambridge, MA, London (1998)
9. Brandom, B.R.: Articulating Reasons. An Introduction to Inferentialism. Harvard University Press, Cambridge, MA, London (2001)
10. Broome, J.: Rationality Through Reasoning. Wiley-Blackwell, Chichester (2013)
11. Broome, J.: Comments on Boghossian. Philos. Stud. **169**, 19–25 (2014)

12. Czeżowski, T.: Proof. In: Czeżowski, T., Knowledge, Science, and Values. A Program for Scientific Philosophy, pp. 91–95. Rodopi, Amsterdam, Atlanta, GA (2000). Ed. L. Gumański (Originally published in Polish as: Dowód. In: Czeżowski, T., Filozofia na rozdrożu (Analizy metodologiczne), pp. 97–105. Państwowe Wydawnictwo Naukowe, Warszawa (1965))
13. Czeżowski, T.: The classification of reasonings and its consequences in the theory of science. In: Czeżowski, T., Knowledge, Science, and Values. A Program for Scientific Philosophy, pp. 119–133. Rodopi, Amsterdam, Atlanta, GA (2000). Ed. L. Gumański (Originally published in Polish as: Klasyfikacja rozumowań i jej konsekwencje w teorii nauki. In: Czeżowski, T., Filozofia na rozdrożu (Analizy metodologiczne), pp. 161–184. Państwowe Wydawnictwo Naukowe, Warszawa (1965))
14. Etchemendy, J.: The Concept of Logical Consequence. Harvard University Press, Cambridge, MA (1990)
15. Haack, S.: Philosophy of Logics. Cambridge University Press, Cambridge (1978)
16. Kamiński, S.: On the classification of reasonings. In: Kamiński, S., Method and Language. Studies in Semiotics and the Methodology of Sciences (in Polish), pp. 231–245. The Learned Society of the Catholic University of Lublin, Lublin (1994). Ed. U. Żegleń
17. Kornblith, H.: Inductive Inference and Its Natural Ground. An Essay in Naturalistic Epistemology. Bradford Book/MIT, Cambridge, MA, London (1995)
18. Kornblith, H.: On Reflection. Oxford University Press, Oxford (2012)
19. Kotarbińska, J.: On ostensive definition. Philos. Sci. **27**(1), 1–22 (1960) (Originally published in Polish as: Tak zwana definicja deiktyczna. In: Fragmenty filozoficzne II. Księga pamiątkowa ku uczczeniu czterdziestolecia pracy nauczycielskiej w Uniwersytecie Warszawskim profesora Tadeusza Kotarbińskiego, pp. 44–74. Państwowe Wydawnictwo Naukowe, Warszawa (1959)). Eds. J. Kotarbińska, M. Ossowska, J. Pelc, M. Przełęcki, K. Szaniawski
20. Kublikowski, R.: Definition within the structure of argumentation. Stud. Log. Gramm. Rhetor. **16**(29), 229–244 (2009)
21. Kublikowski, R.: Meaning, reasoning and normativity in the context of meaning inferentialism of Robert B. Brandom (in Polish). Przegląd Filozoficzny **2**(98), 259–269 (2016)
22. Neta, R.: What is an inference? Philosophical issues. Epistemic Agency **23**, 388–407 (2013)
23. Tarski, A.: On the concept of logical consequence. In: Tarski, A., Logic, Semantics, Metamathematics. Papers from 1923 to 1938, pp. 409–420. Hackett Publishing, Indianapolis, IN (1983). Ed. J. Corcoran (Originally published in Polish as: O pojęciu wynikania logicznego. Przegląd Filozoficzny **39**, 58–68 (1936))
24. Williamson, T.: Understanding and inference. Proc. Aristot. Soc. Suppl. Vol. **77**, 249–293 (2003)
25. Williamson, T.: Reply to Boghossian. Philos. Phenomenol. Res. **82**(2), 498–506 (2011)
26. Williamson, T.: Boghossian and Casalegno on understanding and inference. Dialectica **66**(2), 237–247 (2012)
27. Wright, C.: Comment on Paul Boghossian, "What is inference". Philos. Stud. **169**, 27–37 (2014)

R. Kublikowski (✉)
The John Paul II Catholic University of Lublin, Faculty of Philosophy, Lublin, Poland
e-mail: robertk@kul.lublin.pl

Tadeusz Kotarbiński: Socrates of Warsaw

Jacek Jadacki

The legacy of our millennial past includes treasures mixed with rubble.
Our and your Chopin (Kotarbiński [10, p. 391])

Abstract Tadeusz Kotarbiński was one of the main representatives of the Lvov-Warsaw School. The paper presents his life and didactic activity, as well as the list of his main works and the greatest achievements in philosophy. Kotarbiński supported concretism, radical realism, and the directive of semantic reism. He created praxiology—the theory of effective action. He advocated the system of ethics independent of particular worldview (religious worldview including). Kotarbiński influenced, e.g., Stanisław Leśniewski's ideas of ontology (*scil.* his original logic of names) and of mereology (*scil.* his original theory of collectives).

Keywords Concretism · Tadeusz Kotarbiński · Lvov-Warsaw School · Praxiology · Semantic reism

Mathematics Subject Classification (2000) Primary 01A70, 03A05; Secondary 00A99

1 Life

Tadeusz Kotarbiński was born on March 31, 1886 in Warsaw; he died there on October 3, 1981. In the Kotarbiński family there were many artists: a grandfather of Tadeusz's uncle, Wilhelm, his father, Miłosz, and his brother, Mieczysław—were painters; his uncle, Józef, was an outstanding actor; his mother, Ewa née Koskowska, practiced music.

Even in his school years, in Warsaw, Kotarbiński listened to lectures on philosophy—including the lectures of Stanisław Brzozowski and Adam Mahrburg. In the years 1907–1912, he studied at the University of Lvov with Kazimierz Twardowski (philosophy)—and with Twardowski's most senior disciples: Jan Łukasiewicz (logic) and Władysław Witwicki (psychology), as well as with Stanisław Witkowski (classical philology). After receiving his doctorate in 1912—on the basis of the dissertation *Utilitarianism in the ethics of Mill and Spencer* [1]—he was a teacher of Latin and Greek in Mikołaj Rej Warsaw Grammar School.

Á. Garrido, U. Wybraniec-Skardowska (eds.), *The Lvov-Warsaw School. Past and Present*, Studies in Universal Logic,
https://doi.org/10.1007/978-3-319-65430-0_11

After Poland regained independence in 1918 he was appointed lecturer in philosophy at the University of Warsaw; one year later he became an associate professor, and in 1929—a full professor at this university. He taught there until his retirement in 1961—including from 1939–1944, when the university had to go underground, and in the first years of the communist regime, which, in 1951, deprived him of the chair of philosophy, leaving him only the less "ideological" chair of logic.

In the years 1945–1949, as the rector, he led the organization of the University of Łódź.

Throughout all his life, he took an active part in the philosophical world: in 1915, he belonged to the founding members of the Institute of Philosophy; then he was the president of the Warsaw Philosophical Society (1917–1939) and the Polish Philosophical Society (until 1977). He was a member of the Warsaw Scientific Society (1921–1951) and the Polish Academy of Sciences (1946–1951) until their termination by the communists. He was also a member of the Polish Academy of Sciences since its inception (1952). He received many awards, attained many honors (he was a doctor *honoris causa* of seven universities, a foreign member of five academies, and many scientific societies), was much respected (e.g. in the years 1957–1968 he was a president of the Polish Academy of Sciences, and in the years 1960–1963—a chairman of the *Institut International de Philosophie*)—maybe the most in the postwar period, when the authorities sought to capitalize on his former left-wing sympathies and atheistic and materialistic (but not "dialectical") worldview. Kotarbiński was distinguished in different ways by the authorities, but this did not go hand in hand with his real impact on the main direction of scientific life in the country; however his great intellect and personal integrity and the reliability of the scientific work positively affected the team of disposable "apparatchiks" who had implanted in Poland the "only right" philosophy from the East.

2 Writings

The first book of Kotarbiński was *Practical sketches* [2], in which the main ideas of praxiology appeared. His *Elements of the theory of knowledge, formal logic and methodology of sciences* [3] was a foundation of education for several generations of Polish intelligentsia. The textbook *Logic for lawyers* [4] was studied by many of future Polish lawyers. *Lectures on the history of logic* [7] and *Logic in Poland* [9] constituted an important contribution to the history of logic in general. *Treaty on good work* [5] initiated praxiology as an independent scientific discipline. *Meditations on fair life* [11] gave a summary of his ethical views.

In 1990, the first volume of his *Collected Works* [13] was published. The number of Kotarbiński's works is close to half a thousand.

3 Views

3.1 Semiotics and Methodology

In logic, he dealt creatively mainly with semiotics and methodology.

The central problem of semiotics—the problem of sense—he solved in such a way, that he recognized as making sense all and only sentences not containing onomatoids, or those translatable to such sentences. In the first case, he spoke of basic (literal) sense, and in the second—of a sense of briefly-substituted sense. Onomatoids—or apparent names—he opposed to genuine names, i.e. names of concretes, or precisely speaking—of things, and ultimately—bodies.

He was a supporter of semantic reism, i.e. of the directive to use only language whose sentences are meaningful in the sense explained above. He believed that, in philosophy in particular, using such language—reistic language—protects us against idle disputes, especially ontological ones, consisting e.g. in hypostazing or treating sentences avoiding (reistic) sense or used only in the brief-substitute sense—as sentences saying something about existing objects. Problems, which are objects of such disputes, turn out to be badly formulated.

He accepted the classical conception of truth: a given sentence is true, when it is just as it says. He rejected the nihilistic idea, hence the idea that sentences predicting the truthfulness of a given sentence are synonymous with this sentence itself. He believed that this is so only if the word "true" is used verbally, i.e. if the sentence with truthfulness assigned to it (e.g. the sentence claiming that *a* is *P*), is indicated by putting it in quotation marks (and therefore by using the expression "*a* is *P*"). If, however, the word "truth" is really used, i.e. if the sentence with truthfulness assigned to it is indicated in another way (e.g. by using the expression "the first sentence of this paragraph"), the "truthfulness" can no longer be thus eliminated.

He initially combined the classical conception of truth with logical indeterminism, i.e. with rejecting the principle of the excluded middle, if it would have to proclaim that every statement is either true or false. Although all truths are perpetual, not all of them—are eternal. Truth is perpetual, i.e. if a sentence is now true, it will remain true forever. But there are sentences that are not eternally true, and so being true now, were not always true before. Not being also false, they were undefined in the past. If a sentence is undefined, then it is possible that this sentence or its negation is true. If it is defined, then it is true or false. In turn, if it is true, then it is necessary, and if it is false, it is impossible.

Of the methodological issues—he dealt with, among others, analysis of the reasoning process, which eventually was considered by him as "justifying thinking". Among reasoning, he distinguished—on the one hand—deductive reasoning (i.e. inferring and proving), in which logical consequences are (fully) justified by accepted reason, and reductive reasoning (i.e. testing and explaining), in which logical reasons are justified (only in part) by accepted consequences. On the other hand—he divided reasoning into progressive reasonings (i.e. inferring and testing), i.e. those in which consequences of given reasons are sought, and regressive reasonings (i.e. explaining and proving), i.e. those in which reasons of given consequences are sought.

He indicated a new base of classification of the sciences into the theoretical and the practical. He thought, namely, that both theoretical and practical sciences aim to establish true and justified sentences about the objects of a certain domain, in particular, to establish laws of this domain. Generally, he advocated for criticism, i.e. the postulate of recognizing only sentences justified due to our knowledge. He distinguished practical and theoretical sciences on the ground that practical sciences fixed some truth in order to provide a description of the procedure leading to the formation of certain things, and that in these sciences, handling actions (e.g. experiments) outweigh intellectual actions.

Although Kotarbiński was certainly one of the most eminent Polish philosophers of the twentieth century, he initially proposed to eliminate the word "philosophy" because it does not refer to any well-separated areas of knowledge. Not convincing anyone to abandon this word, he began to introduce at least some conceptual order here. He considered four concepts of philosophy to be the most important ones: practicing philosophy can be identified either with creating a metaphysical worldview, or searching for practical self-knowledge, or indicating an ethical pattern, or building a theory of knowledge. The first understanding of philosophy coincided with what he called the "great philosophy", and more specifically to creating large systems; and the last one—with what he called the "little philosophy", i.e. with philosophical analysis. He himself basically practiced philosophical analysis, but he thought that philosophy does not end on analysis: after conducting analysis, it comes time to create a system.

3.2 Epistemology and Ontology

In the theory of knowledge, Kotarbiński claimed presentationism, called by him "radical realism". The world is directly knowable: objects of knowledge are given to us directly and not through sensual data (contents). He combined presentationism with objectivism: the world exists independently of whether it is or is not known.

He was an empiricist—considering experience to be the ultimate base of knowledge. His empiricism was of an extraspectionistic character. The basis of knowledge is external experience: sentences about introspection are reducible to sentences about extraspection. He proposed that sentences of the type "*A* experiences that *p*" be paraphrased using literally reistic sentences of the type "*A* experiences like this: *p*". These sentences are literally reistic as the onomatoid "that *p*" disappears from them: '*A*' is a genuine name of an "experiencing body", and '*p*' is a description of how—not what—*A* experiences (e.g. seeing something). An introspective sentence can provide, among other things, knowledge about how other people experience—not me alone. And a sentence claiming that someone different from us experiences in such-and-such way, can by understood by us only as an abbreviation of the sentence "*A* experiences as I do, when I would experience like this: *p*". This is therefore the imitationistic conception of introspection. Experiencing here is made independent of the existence of the experienced object: we do not have to accept the existence of that *p*, or some sensory data, in order to experience like someone who experiences in the manner described by the phrase '*p*'.

In theory of objects, Kotarbiński advocated a certain version of nominalism, namely—concretism. According to nominalism, there is exactly one ontological category: the category of individuals. Concretism rolls this out: these individuals are identical with concretes. In other words: every object is a thing. Yet another way: there are only concretes—there are no abstracts. In particular, there are no general objects, i.e. no universals. He argued as follows. Something that would be a universal object (e.g. a man in general) with respect to some individual objects (here: to men) should have all and only properties common to these objects. Suppose that a certain property (e.g. being the author of *The Spirit King*) is the specificity of one of these objects (here: Juliusz Słowacki). According to the (ontological) law of the excluded middle, our universal has this specificity or it does not. It can not, however, have this specificity, because the specificity is not common property of individual objects that fall under this universal. It can not also not have this specificity, because then the universal would have the property of not having this specificity, and this again is not a common property of these objects, because does not reside in this specified object.

Kotarbiński's concretism had a reistic interpretation: every concrete is a thing. The reistic thesis—that there are only things—has become one of the most famous of his views. Ontological reism, in turn, had in Kotarbiński a form of somatism: every thing is a body. So there are only bodies. They are objects being spatio-temporally extended and inert. If we agree, in turn, that the totality of bodies—the collection of all of them—is matter, then somatism is a kind of materialism, because the totality of bodies is also a body.

Initially, materialism was combined with indeterminism, or, strictly speaking, with antipredeterminism. He was an opponent of the thesis that what took place at a certain time was strictly determined by what had happened earlier. It seemed to him that if this were so, as predeterminists proclaim, it would be impossible to act really creatively. Later he departed from this view and concluded that—according to determinism—each event is determined by a set of earlier facts, making together a sufficient condition of this event. He identified the cause of the event with an important component of this condition—i.e. such a component, without which the other components of the team no longer constituted a sufficient condition of a given event.

3.3 Praxiology and Ethics

It is surprising that Kotarbiński, being a reist, devoted so much time to the analysis of action, i.e. to something that did not exist from the point of view of the reistic doctrine, not being a thing. In addition, he analyzed actions by no means in reistic terms. What is more, praxiology—the theory of effective action—is usually tied with the name of Kotarbiński as with reism, and he is rightly regarded as one of main creators of it.

What underlies praxiology is practical realism, i.e. the postulate of sanity in action: reckoning with reality when taking any action—with what was, what is, and what can be.

The most important component of praxiology—as the theory of effective action—is a set of practical directives, i.e. tangible normative sentences of the form: "To achieve this-and-this in such-and-such time, it is good to do that-and-that". The phrase "it is good" is understood in these directives either as a sufficient condition (which is a synonym for "it is enough"), or as a necessary condition (which is synonymous with "it should be"). And to do something—is the same as—to take a sequence of simple actions, involving deliberate and free (i.e. having its source in free impulse) pressure applied to something, causing the creation of something.

Assume that the practical directive specifies that a condition for achieving a given thing in certain circumstances is to do some other thing. The theoretical basis for such a directive is the belief of an overlapping causal link between the occurrence of both of these things. Causing an intended thing requires using specific materials and tools, that is—in general—a specific technical base. On the other hand, doing a thing, which is a condition of causing an intended thing, is a string of deliberate (intentional) actions.

Praxiology provides a theory of the organization of such activities. They are assessed in terms of effectiveness—possibly of varying degrees. An action is effective when it leads to the intended effect; an action blocking this effect is a counter-effective one. Other activities are—from this point of view—ineffective (or in-counter-effective).

A praxiological estimation of actions applies also to their rationality. An action is factually rational when it has a true theoretical basis; it is methodologically rational, when its theoretical base is sufficiently justified.

The efficiency of actions and the rationality of both types are—as we can see—mutually independent.

We should not mix norms (e.g. "Do not do this-and-this!") with normative sentences. Normative statements, regardless of whether they are of the type which has been already mentioned—that is factual—or emotional (of the type "Doing this-and-this is a good/bad thing"), are declarative sentences, so they assessed in terms of truthfulness and validity. While norm—assessed in terms of validity—are devoid of logical value. We do not ask whether they are true or false, but only whether they are or are not valid.

Kotarbiński advocated (meta)ethical absolutism: in spite of the volatility of norms, there are absolute moral criteria which make it possible to decide which norms are valid and which—are not. Ethics is independent of a particular worldview—including a religious worldview. Kotarbiński himself was an atheist, but he distinguished atheism, i.e. the refutation of the existence of God, from blasphemy, i.e. offending the feelings of believers—and certainly he was not a blasphemer.

Justification of moral criteria is provided by elementary moral intuitions, which have their source in conscience, assessing whether human behavior is "venerable" (i.e. morally good), or—"shameful" (morally wrong).

Kotarbiński's chief ethical norm was: Behave as a brave, trustworthy (or responsible) guardian behaves! Be courageous, sacrificial, right and cornered in the fight against the existing evil and prevent evil greater than the existing one from coming to be! This norm also had its freer version—in the form of four recommendations: (1) like doing something, (2) love someone, (3) dont't be a scondrel, and (4) live seriously.

4 Resonance

4.1 Origin

The roots of semantic reism lie in what Kotarbiński heard from the mouth of one of his teachers—Witwicki—and what he himself reported:

> [Witwicki] called for the things themselves, demanding that all abstract ideas should be supported by concretes, any generalization should be illustrated with examples and that these examples should be expressiveliving. [6, p. 81]

A program closer to semantic reism was proclaimed at the time in the Vienna Circle: this was the program of physicalism—especially in the version of Rudolf Carnap. Physicalists conditioned the meaningfulness of sentences on their translatability to a certain distinguished language: the language of things (just like Carnap) or the language of impressions (as with some others neopositivists). There was here, however, an important difference: from the point of view of neopositivism, the thesis of ontological resim must be considered as lacking empirical sense, and so, briefly speaking, as senseless.

In the classification of reasoning, he initially followed Łukasiewicz, but ultimately he accepted understanding of "deduction" and "reduction" offered by Rev. Jan Salamucha.

The direct source of ontological views was logical: namely the calculus of names developed by Leśniewski and called by him "ontology."

The function of the impetus for interest in praxiology was probably fulfilled by Brzozowski's philosophy of working, as well as the analysis of acts and products carried out by Twardowski; without a doubt also, the organizing and pedagogical activity of the latter—activity of rare regularity—became for Kotarbiński a practical verification of his conception of efficient activity. Anticipations of independent ethics can be found in Władysław Biegański.

4.2 *Criticism*

At first, Kotarbiński's logical indeterminism became an object of criticism. Leśniewski convinced him that relativizing the truthfulness to time is inadmissible. The sentences of the form "The sentences "Something is such-and-such" is true in such-and-such time" are meaningful when they can be translated into sentences of the form "The sentence "Something is such-and-such in such-and-such time" is true".

However, destructive criticism touched, above all, Kotarbiński's ontological views. After the Second World War, this came from the side of communist doctrinaires, who came to the (legitimate!) conclusion that reism was not identical with their ideology. But many years before, reism had undergone substantive criticism.

Firstly, it was indicated that reism had significant internal drawbacks.

Thus, Kotarbiński's arguments against abstracts was proved inconclusive. Marian Borowski and Roman Ingarden pointed out that Kotarbiński based the argument against general objects on the mistaken assumption that universals have the same properties (or properties of the same order) as properties which are possessed by individual object falling under these universals.

And Kazimierz Ajdukiewicz showed that reism is an analytical thesis. It can be expressed namely by the statement "Each object is a thing", and this statement is meaningful in reistic language, when the name "object" is a genuine name, that is—is a name of things. Ajdukiewicz noted also that reists, in their own language, can not express the belief in the nonexistence of abstracts, because the sentence of the form "There are no abstracts" can not be considered as a meaningful sentence, even as a sentence in a briefly-substituted sense. The case was put on a knife edge by Stanisław Ignacy Witkiewicz, pointing out that since reism is a convention, it might as well take the view that, e.g., there are only properties or—refuted by reists—contents (sense data).

Secondly, it turned out that reism is not an adequate ontology due to mathematical and physical theories. One of the consequences of adopting the reistic thesis is recognizing as "objectless" a large part of mathematics—including the part of set theory based on the concept of an infinite set. This is because the last concept is in fact a lack of reistic interpretation: no statement of those sets can be translated into an equivalent statement about things.

Klemens Szaniawski and Zdzisław Augustynek raised further the objection that reism—and in any case somatism—comes into conflict with modern physics, which considers not only bodies (corpuscles) real, but also fields (waves). Meanwhile, fields—from the point of view of reism—are abstracts, so they are hypostases.

Already Ajdukiewicz's and Ingarden's criticism caused Kotarbiński to abandon ontological reism for semantic reism—in its liberal version, i.e. not refusing the meaningfulness of unrealistic sentences, but recommending only "as far as possible" avoiding such sentences.

4.3 Continuation

The formulation of the classical conception of truth, made by Kotarbiński, was a reference point of the semantic definition of truth given by Alfred Tarski [1933]. Considerations on absolutism in regard to truthfulness contributed indirectly to Łukasiewicz discovering three-valued logic: he attributed the third value—beyond truthfulness and falsity—just to indeterminate sentences.

Marian Przełęcki came to the conclusion that semantic reism can be defended, provided we find it to be a terminological convention concerning the term "meaningfulness" or a criterion of a minimum ontological commitment of theories.

An interesting attempt to rescue semantic reism was taken by Bogusław Wolniewicz. He suggested that reistic briefly-surrogate senses can be assigned not only to sentences directly translatable to reistic sentences with literal meaning, but also to sentences belonging to any theory for which it is possible to give a set of relevant axioms in reistic language.

Psychologizing imitationism, i.e. supporting the reduction of introspection to extraspection in structural psychology, was postulated by Zbigniew Jordan.

Modification of the original version of ontological reism was made by Janina Kotarbińska. According to her, two meanings of the word "to exist" should be distinguished: the primary and the basic. In the primary meaning—only objects designated by names (apparent names included) exist; in the basic meaning—only things exist. As a result of this distinction, acceptance of the view that every sentence of the form "This-and-this object is that-and-that" implies the existence of this-and-this object but does not force the acceptance of the view that this object exists in the basic meaning of the word "to exist" (i.e. as a thing).

On the other hand, Czesław Lejewski tried to defend the original thesis of ontological reism against charges of tautologicality—considering that, contrary to Ajdukiewicz, the sentence "Each object is a thing" is not an analytical sentence, because the name "object" has a different (poorer) connotation than the name "thing".

It was also attempted to "put into practice" some ideas of praxiology—the slogan of good work, in particular. If these attempts did not bring the expected results, it was probably because the attempts fell on the deaf ears of the statist economy, subordinated entirely to dogmatic ideology.

5 Pupils

As a teacher, Kotarbiński taught, first of all, responsibility for words. As a debater he was a master of what might be called the "idealizing recapitulation". He could so interpret someone else's statement—even a statement of his opponent—that it became BOTH clearer and more justified than in its original form.

Kotarbiński was the second great teacher within the Lvov-Warsaw School: in Warsaw he played the role which was played in Lvov by Twardowski. The list of his pupils starts with such eminent scholars as Maria and Stanisław Ossowskis—and ends with: Jerzy Pelc, Tadeusz Pszczołowski, Szaniawski and Przełęcki.

6 Summary

Karol Irzykowski called him "Socrates of Warsaw"—and this name characterizes Kotarbiński's personality in the most brief way. In logic—his postulate of dehipostazing scientific language made history. In metaphysics—his reism was the most original, but only his imitationistic theory of introspection survived in its original version.

In axiology—the slogan of reliable guardianship turned out to be the most catchy.

Tadeusz Kotarbiński's Works Mentioned in the Paper

1. Utylitaryzm w etyce Milla i Spencera [Utilitarianism in the ethic of Mill and Spencer]. In: Wybór pism [Selected Writings]. Vol. I. Myśli o działaniu [Thoughts on Acting], pp. 215–330. PWN, Warszawa (1957) (Original work published 1912)
2. Szkice praktyczne. Zagadnienia z filozofii czynu [Practical Sketches. Problems from the Philosophy of Act]. In: Wybór pism [Selected Writings]. Vol. I. Myśli o działaniu [Thoughts on Acting], pp. 13–168. PWN, Warszawa (1957) (Original work published 1913)
3. Kotarbinski, T.: Gnosiology. The Scientific Approach to the Theory of Knowledge. Pergamon, Oxford (1966). (Original work in Polish published 1929)
4. Kurs logiki dla prawników [Logic for Lawyers]. Gebethner i Wolff, Warszawa (1951)
5. Traktat o dobrej robocie. Ossolineum (1955) (English translation: Praxiology. An introduction to the sciences of efficient action. PWN – Pergamon Press, Warszawa – Oxford (1965))
6. Nauczyciele sztuki nauczania [Teachers of the Art of Teaching]. In: Studia z zakresu filozofii, etyki i nauk społecznych [Studies of the Domain of Philosophy, Ethics and Social Sciences], pp. 73–83. Ossolineum, Wrocław (1970) (Original work published 1956)
7. Wykłady z dziejów logiki [Lectures on History of Logic]. Ossolineum, Łódź (1957) (French translation: Leçons sur l'histoire de la logique. Presses Universitaires de France, Paris (1964))
8. Wybór pism [Selected Writings]. Vol. I. Myśli o działaniu [Thoughts on Acting]. Vol. II. Myśli o myśleniu [Thoughts on Thinking]. PWN, Warszawa (1957)
9. La logique en Pologne. Son originalité et les influences étrangères. Angelo Signorelli Editore, Roma (1959)
10. Nasz i wasz Szopen [Our and Your Chopin]. In: Studia z zakresu filozofii, etyki i nauk społecznych [Studies of the Domain of Philosophy, Ethics and Social Sciences], pp. 391–395. Ossolineum, Wrocław (1970) (Original work published 1960)
11. Medytacje o życiu godziwym [Meditation of Fair Life]. Wiedza Powszechna, Warszawa (1966)

12. Studia z zakresu filozofii, etyki i nauk społecznych [Studies of the Domain of Philosophy, Ethics and Social Sciences]. Ossolineum, Wrocław (1970)
13. Dzieła wszystkie [Collected Works], vols. I–VI. Ossolineum, Wrocław (1990–2003)

J. Jadacki (✉)
Institute of Philosophy, University of Warsaw, Warsaw, Poland
e-mail: j.jadacki@uw.edu.pl

Agency in a Praxiological Approach

Wojciech W. Gasparski

Abstract The paper presents issues of agency in the approach proposed by Tadeusz Kotarbiński, the Polish philosopher from the Lvov-Warsaw School, author of the philosophy of practicality with its dominant: praxiology. It also outlines a number of other approaches to this notion as well as notions similar and related to the notion of an agent. The conclusion points out that praxiology provides an insight into the reality it studies, but the use to which this is put depends on the users themselves. That use is indirect more often than direct, since it requires reflection on the agent's own practicality and some meta-skills that should characterize reflective practitioners.

Keywords Act · Action · Agent · Change · Goal · Kotarbiński · Practical situation · Praxiology

Mathematics Subject Classification (2000) Primary 00-02

1 Introduction

Agency is one of praxiology's fundamental notions. It was introduced by Tadeusz Kotarbiński, the author of the philosophy of practicality (Gasparski [7]). This philosopher's praxiology triptych consists of three volumes from his complete works, namely *Prakseologia - Część pierwsza* [*Praxiology: Part One*] [12, 462 pp.] containing papers published before World War II and post-war works that featured the expression "dobra robota" ["good work"], *Traktat o dobrej robocie* [literally *A Treatise on Good Work*, published in English as *Praxiology: An Introduction to the Sciences of Efficient Action*, Pergamon Press, New York 1965] [22, 220 pp.] and *Prakseologia – Część druga* [*Praxiology: Part Two*] [23, 699 pp.] containing papers published after World War II. This volume opens with a work called "Abecadło praktyczności" ["ABC of Practicality"] which in concise form presents the main issues of praxiology; the volume also includes meta-praxiological works and papers on science studies. The indexes in the three volumes show that the notions of "agent" and "agency" appear on 91 pages.

Á. Garrido, U. Wybraniec-Skardowska (eds.), *The Lvov-Warsaw School. Past and Present*, Studies in Universal Logic,
https://doi.org/10.1007/978-3-319-65430-0_12

The present paper presents issues of agency in Kotarbiński's approach and further elaboration by his followers. A number of other approaches to this notion as well as notions similar and related to the notion of an agent are also outlined.

2 Agency in Tadeusz Kotarbiński's Approach

2.1 The Beginnings

The term "agent" first appears [13, p. 1] in an essay entitled "Cel czynu a zadanie wykonawcy" ["An Act's Goal versus the Doer's Task"] which Kotarbiński delivered as a lecture in 1910 in Lvov at a session of the Philosophy Club and which was published in the volume *Szkice praktyczne* [*Practical Sketches*] in 1913 [12, pp. 5–85]. This essay [14, pp. 6–19], inaugurating the general theory of action that was yet to be called praxiology, listed its basic notions, among them the "agent, creator or doer" (op. cit., 7). This notion served to define "action" as follows:

> The entirety of work, but exclusively that performed by the doer of the deed, his activities, such as kneading dough and putting it in the oven; walking to the station, buying a ticket and taking one's place in the carriage; manipulating the injection device and a child's skin — let each such entirety be called the 'action' of a given person (op. cit., 8).

Kotarbiński distinguishes action defined as above from an "act", which is:

> The whole process of forces operating, taken together and being the necessary condition of the goal [of a given act] — let this be called an 'act'; hence, for example, the entirety of the baker's muscular effort, the effect of high temperature on the dough, the effect of the yeast's growth power; the entirety of the efforts of the man going to the station, the work of buying a ticket, getting on the train, the work of the train moving; the efforts of the doctor to inject the serum and the effect of different forces of nature in the child's body after the injection, all taken together (op. cit., 8).

The goal of a given act "is the production and existence of the given loaf of bread, the given person's getting to and staying in Kraków, the child's recovery and continued good health", while the goal of an action is "the production, or coming into existence, the lasting of the formed uncooked loaf placed in the oven in the first case, the sitting down and staying in the carriage of the person in the second case, the serum getting under the patient's skin and staying there in the third" (op. cit., 8). Further in the essay the author considers the relations between the two goals, namely: sameness, divergence, opposition, the paradox of acting "in spite of", concurrence, sequential succession, the paradox of indirectness of effect and others. As a result, the goal of an action stops being set in opposition to the goal of an act. Hence:

> We separate [...] from the entirety of the act, as our action [i.e. of us as agents — WWG], part of it on the basis of any of its breakthrough features, meaning those that are materially or methodologically important, which separate all of the activities, including our own, from the rest of the activities which are independent of us. The notion of the goal of the action changes accordingly (op. cit., 12).

An answer is sought to the question: "Where is this directness of our activity [as agents—WG] exerted on external objects (of course as long as we are understood in mental terms, as bodies wanting and acting, and not physically as our bodies)?" (op. cit., 13). After discussing misunderstandings connected with mixing up the two kinds of goal, of an act and of an action (superficial imitation, opposition, primacy and subordination, collaborating counteraction), Kotarbiński proposes: "Let us now assume that there exists such a direct activity, in external work on matter or in internal, purely mental work [...] and let us call it 'absolute action"' (op. cit., 14).

The final issue is the question: "What is the agent of an act morally responsible for?" Kotarbiński offers the following supposition for consideration:

> Perhaps the solution is found in the fact that we are morally responsible for a component of the complex of forces we call an act, one which depends wholly on us and is equal to our effort; [we are responsible] for the component that we sought above in vain, whose existence we assumed and to which we gave the name 'absolute action' (op. cit., 18).

Kotarbiński adds "the shadow of the supposition", as he writes,

> that in each act of a moral nature, this absolute action is always ethical, and only the resultant can be bad; that therefore there is no blame in human acts but only the advantage of alien forces over an absolute action, which always constitutes, even in cases of glaring crime, our ethical merit (ibid.).

2.2 *Further Reflections*

The above was supplemented with some detailed issues discussed in the next two chapters of *Szkice praktyczne*. In the essay "Zagadnienie istnienia przyszłości" ["The Issue of the Future's Existence"] Kotarbiński points to the double-edged nature of being able to do or not to do something. He illustrates this with a rather unpleasant but very instructive example, especially for lawyers:

> It is a certainty that every living being has to die; therefore a killer is not the agent, the author of its death, he is at best only the agent of the kind of death this is and of its acceleration. But whoever punishes a killer with death for a death also is not his death's agent, for the same reason, that's that, if there are no other reasons that would justify condemning this form of retaliation [15, p. 70].

In the next essay, "O rozszerzaniu sfery czynu" ["On Expanding the Sphere of the Act"], we read:

> Who knows if the moment of capacity for performing an act is not always different from the moment of the act itself, just as surely as the moment of that performance is always different from the moment of its product's presence; he is surely not in stark disagreement with common understanding who thinks that before the agent performs the act, he can perform it sooner, and not always just at the moment of the act [16, p. 73].

2.3 *Analytical Definition of Agency*

A rather short paper entitled "Pojęcie zewnętrznej możności działania" ["The Notion of External Possibility of Activity"], published in 1923 in *Przegląd Filozoficzny* (vol. 26, 64–67), is important for the notion of agency. Kotarbiński attached great importance to the analytical definition of agency provided in this paper, as its being quoted in full in [13] testifies. The definition is as follows:

> Due to impulse I of person S, belonging to moment k, S is the agent of fact D from the later moment r and D is the work of S always and only when a set of facts from k containing I defines D and no set of facts from k not containing I defines D [17, p. 104].

This definition is the result of abandoning the indeterminism of "the future's unpreparedness prior to activity" due to the danger of "reduction to a contradiction" [13, p. 2] and of adopting an assumption that is a "demand of determinism" [17, p. 105]. Further on in the treatise Kotarbiński analytically defines the negation of agency and the external possibility of agency, summarizing his thoughts as follows:

> The above reasoning serves to show that it is possible (and how) to select definitions of agency and external possibility of agency, with the help of certain terms, in such a way that they will contain the common meaning of agency and the external possibility of agency and that a certain common supposition will be expressed within determinism. This supposition is that if someone does something, they could have done it and at the same time not have done it, and that abstaining from an act that one could have performed is also an act [17, p. 107].

Kotarbiński points out the discrepancy between the theoretical approach to agency and its common understanding, for example the assertion that a person who had the possibility "to cause what happened later" at the same time "did not have the possibility to prevent it" (op. cit. 107). This is because the common understanding of the possibility of taking action assumes the motive of a "lack of obstacles" and the motive of "sufficient competence", whereas the presented theoretical approach only takes into account the former (op. cit., 108). To conclude:

> ...even just the condition of the possibility of agency on account of a free behaviour from moment k, concerning that very moment k, in combination with the assumption of that behaviour, results in agency (ibid.).

This condition carries with it conditions concerning any moment, which enables the conditions for moments g and n to be drawn as theorems.

In the same year[1] the theory of action was given the name praxiology[2]; this was an initiative that Kotarbiński put forward at the First Polish Philosophy Congress, presenting *Zasady teorii czynu* [*Principles of the Theory of Action*] [18]. In this paper Kotarbiński outlined the programme of praxiology, mentioning the notion of the agent as one of the fundamental praxiological terms. As an example of "defining the meaning" of this notion, he once again defined the relationship of agency, slightly modifying the style of his earlier definitions. This was the new definition:

> ...person S is the agent of fact D, and this fact is that person's doing, always and only if at a moment earlier than the moment of fact D there occurs the fact of a free behaviour of person S, namely impulse I, such that: (1) a certain set of facts contemporaneous with impulse I, containing that impulse, determines fact D due to the inherent law of the sequence of events, (2) no set of facts contemporaneous with impulse I and not containing that impulse determines fact D due to that law (op. cit., 133).

---

[1]The paper was not published until 1927.

[2]Tadeusz Kotarbiński described praxiology as the science of efficient action. Praxiology according to Ludwig von Mises (1987) is the science of means and not goals of action. Both founders of praxiology schools—the Polish and the Austrian school—referred to the French initiator of praxiological research, Alfred Victor Espinas, who focused on analysing the means of achieving goals (realizing intentions). Let us add that in his sociological theory of action, Florian Znaniecki [28] favoured indicating the intention of the acting subject rather than the goal of the activity. See also [3].

The author continued further on:

> ... if person S at the moment of his impulse I and due to that impulse has the external capacity to cause fact D, then: (1) he also has the external capacity to cause the negation of fact D, (2) or he is the agent of fact D (ibid.).

Two years later Kotarbiński published an essay on the agency relationship, illustrating his thoughts with examples serving to highlight the qualities "in which one is usually inclined to see constant traits of an agent" [19, p. 122]. This approach to the problem, which Kotarbiński preferred to a formalized approach, served to prepare the ground, as he said, for presenting an analytical definition of the notion of an agent, adjusted "to the scope of the common understanding of an agent" (op. cit., 128). The definition goes like this:

> John is the agent of a given fact (and that fact is the work of John) means the same as: an earlier free behaviour of John's was an essential component of the all-encompassing complex of contemporaneous facts that creates, in accordance with the laws of nature, the essential condition of that fact (ibid.).

Compared to the earlier approach, the formalized approached is supplemented with the notions of a system and the law of sequence of events. Here is the expanded analytical definition of agency in a formalized version:

> With respect to impulse I of person S, belonging to moment k, S is the agent of fact D from the later moment r, and D is the work of S, always and only if a certain set of facts from k — in system U that includes S — containing I, determines D due to some inherent law of the sequence of facts, and if no set of facts from k, in the same system U, but without S, not containing I, determines D due to such a law [18, p. 128].

The system has to be "sufficiently separate", meaning one "whose parts are not subject to the action of forces from outside it" (op. cit., 129). The law of the sequence of events is meant to "take advantage of the intention contained in the colloquial word 'must' ('... that must have happened after this')" (ibid.). This does not just mean the laws of nature but also "laws" established by people as regulations, but only when acting in accordance with them has become a regularity "of a psychological kind", i.e. "such a disposition has developed in people that they regularly follow this regulation in relevant cases, so a certain regularity of events has developed from this particular structure of people's preferences" (op. cit., 129–130).

To supplement his definitions and disseminate them, the cited author published the entry *Czyn* [*Act*] in volume one of an encyclopaedic outline of contemporary knowledge and culture entitled *Świat i Życie* (*The World and Life*; 1933). In it we read that:

> An acting person is called different things in different cases: an executor, perpetrator, author, agent. We have chosen the last of these words as a technical term of the theory of action, i.e. praxiology. [...] every time one is an intentional agent of something, one is also an unintentional, i.e. involuntary agent of a great many other events [20, p. 135].

The following year saw the publication of a treatise carrying the same title, *Czyn*[3] [*Act*], in which it is stated that "There is no act without an agent. But who do we call an agent? It is the one who made an intentional effort to some aim. He is the agent of everything that occurred as a result" [21, p. 141]. The notion of an agent was extended to include a

[3] Subsequently published as *Czyn* (1934).

set of people and to distinguish "the kind of participation in a collective act when a given participant is an agent of the work accomplished by the act, from participation when he is only a co-agent but not an agent" (op. cit., 144).

2.4 Acting Subject

Kotarbiński also used the term "acting subject"; probably the first time he did so was in the essay "O rozszerzaniu sfery czynu" ["On Expanding the Sphere of Acts"] from 1913, where he indirectly expressed the notion's meaning.

> By the sphere of acts of a given subject I mean the totality of things that he can create, in other words, those that remain in his power, whose existence at a given moment or whose indecision, the beginning of nonexistence or indecision, are — as we say — dependent on that subject, in the sense that the subject can cause an affirmative statement about any of those things to become true at a given time, or can cause it to remain suspended, in that middle that is allegedly excluded from logic, that in other cases the subject can cause that affirmative statement to become false from a given moment, and he can also cause it to remain indecisive, and finally, in some special cases, the subject can cause an affirmative statement about a given thing to become true from a given moment, but can also cause it to become false from that moment [16, pp. 73–74].

Kotarbiński distinguishes between different acting subjects, as follows: (1) an isolated subject: such a creative individual "whose sphere of acts does not overlap with the spheres of acts of others, and thus whose sphere of acts throughout the entire isolation time does not contain objects that would belong to the entirety of things of another's sphere of acts in the time from the moment of isolation" (op. cit., 74); (2) a non-isolated subject: one who "could at a later time have the possibility of action that he or she did not have earlier" (op. cit., 82), but "the power of creating something is weaker in one who shares it with others" (op. cit., 83). The term "acting subject" appears in Kotarbiński's works in the context of collaboration, positive cooperation (organization), negative cooperation (fighting), creating culture, practical mistakes.

2.5 The Agent and the Issue of the Cause of an Effect

Chapter two of *Traktat o dobrej robocie* is devoted to simple acts, the agent and free impulses [22, pp. 15–21]. Using examples, Kotarbiński explains the notion (law) of the inherent sequence of events (e.g. death occurs *after* birth and not simply *later*) essential in order to connect a cause with an effect of action, in other words to determine "a causal link understood in the sense that we will have to consider when defining the notion of the agent of a given outcome" (op. cit., 16).

> Event B is the effect of earlier change A filling moment t, and change A — the cause of event B, always and only if change A is a significant component of the sufficient condition of event B due to moment t and due to the inherent rule of the sequence of events (ibid.).

In this, "any component event of that condition without which the system of the other component events would not be a sufficient condition" (ibid.) is a significant part of the sufficient condition. Thus formulated, this explains a cause in its ordinary, everyday sense,

Kotarbiński writes (e.g. "snow melting was the cause of flooding"). He also notes that we usually see multiple causes "in the group of mutually contemporaneous components of a given sufficient condition for a given effect, and multiple causes of a given effect belonging to its different sufficient conditions, each of which belongs to a different moment" (op. cit., 17).

After this groundwork we can move on to defining the relation of agency, which is the relation between the agent and the work, i.e. the result of activity, meaning an effect that constitutes a change or a state of affairs (being); a cause is always a change. Hence:

> The agent of a given event is the one whose free impulse is the cause of that event (op. cit., 18).

In this, an impulse is a generalized concept of "pressure" understood literally as physical or metaphorically as mental, and "free" means intentional in accordance with the will of the acting subject.

The acting subject is the agent of both what he intended and what he did not intend to do but did do by mistake.

> In the entirety of an event, we are the agent due to the given free impulse, and therefore each such event is our work, its distinguishing feature being that the free impulse was its cause, even if we did not bring about the event intentionally or consciously, and even if we were erroneously convinced at the moment of the impulse that the event would not or could not come about (op. cit., 19).

As a formality, Kotarbiński reiterates that the acting subject "is a flesh-and-blood living person, wanting one thing or another, moving in one way or another, or making a mental effort to achieve what they want" (op. cit., 21). A simple action of the subject thus understood is a single-impulse act (ibid.).

A further chapter of *Traktat* characterizes the notion and types of compound action. A compound action is the fulfilment of the condition that

> a relation of positive or negative cooperation occurs between its components" and when "two acts are linked by such a relation always and only if one of them causes, enables, facilitates, prevents or hinders the other, or when the two have the same impact on a third action. [...] A set of contemporaneous acts (i.e. such that each shares at least part of a moment with another) that form part of a compound action shall be called a chord of actions; a set of consecutive acts (even if parts of their moments overlap) that form a part of a compound action shall be called a series of actions. A series of chords of actions shall be called a complex of actions (op. cit., 48).

The notions of positive cooperation (developed in the chapter on the principles of collaboration) and negative cooperation (developed in the chapter on the theory of struggle) are discussed in the chapter of *Traktat* devoted to collective action. Kotarbiński also discusses the subjectivity of institutions, as follows:

> in our praxiological thinking we will sometimes approach institutions as if they were persons, acting subjects with a special mental and physical structure, even though in our view, no institution is ever, strictly speaking, an acting subject (op. cit., 73).

In the chapter on mental activity, this idea is presented as follows:

> The area of intellectual work is full of reminders of the truth that a team, strictly speaking, is never a subject but only a functional union of subjects, incapable of replacing the subject in acts of learning or acts of decision (op. cit., 191).

2.6 Recapitulation

The notions of an agent and agency are also legal terms, so it is no wonder that Kotarbiński published a recapitulation of the essence of the notion of agency in the law journal *Państwo i Prawo* [23, pp. 42–46]. Let us cite some excerpts from this brief treatise to add point to our report on the history of praxiological analysis of agency in Tadeusz Kotarbiński's approach.

To begin with, we have the notion of a free impulse:

> If someone moved not involuntarily but voluntarily, we shall say that he performed a free impulse; we extend this notion to the sphere of inner efforts by believing that a free impulse was also performed by someone who, for example, curbed the desire to shout out loud or who accomplished a short-term focusing of the attention to remember a forgotten name. [...] Stating the freedom of a given impulse does not mean being grounded in indeterminism, which accepts freedom of will understood as intentional impulses being independent of preceding causes. That an impulse was free only means that someone made the effort that they wanted to make (op. cit., 42).

As for the notion of agent:

> Whoever performed a free impulse at a given moment became the agent of a specific subsequent event always and only if the impulse was a necessary component of a given set of circumstances contemporaneous with it, and that set was a sufficient condition in that moment of the event due to the law of the causal sequence of events. [...] sufficient conditions are sets of events that are mutually contemporaneous. An event is always a change of an object. [...] We accept the deterministic doctrine in strong form. It states that an event has a causal sufficient condition in every earlier moment (op. cit., 43).

Then, characterizing agency in compound actions, Kotarbiński lists examples of compound actions performed by the same acting subject—this is an individual compound action. Next, he discusses a team compound action. He also draws attention to the paradox of co-agency:

> Is it possible [...] that a team is an agent even when there is a lack of agency on the part of all of its members? (op. cit., 45)

The paradox disappears when the free impulse of the other participants in a team action was a part of the sufficient conditions of the compound action. This issue is related to the earlier-discussed question of a team, i.e. two or more acting subjects, as an agent.

> ...this occurs whenever no system of events from a [given] moment containing a free impulse of one of those subjects but not containing the free impulses of all the other subjects, is a sufficient condition of that result, but there exists such a sufficient condition of that result belonging to that moment which contains the whole set of impulses of all those acting subjects as its necessary component (op. cit., 46).

Tadeusz Kotarbiński believed that the problem of agency had not been exhausted yet and thus deserved further consideration. Therefore let us look at some other sources.

3 Other Approaches

3.1 *The Agent and His World According to Jakob Meløe*

The title of this chapter is a reference to the ideas of Norwegian praxiologist Jakob Meløe presented in the work *The Agent and His World* [24]. The notion of "our world" defines the frame of studies on the agent in his world. Our world is the world of practices—the totality of operations performed by one or many agents—in which use is made of what is available, giving it the form of existence. According to Meløe, the basic form of a practical operation is as follows:

> *x operates on y*, where 'x' marks the place of the agent, or the subject of the operation, 'y' the object of the operation, or its target, and where the verb 'operate', or 'operate on', is a stand-in for some suitable verb of action [24, p. 15].

Meløe calls the thing towards which an action is directed a "tautologous object", while the "tautologous subject" is the one who performs a given operation, i.e. the agent.

> To each operation in our world there corresponds a well-defined cut of our world, or a well-ordered niche within it. Without that niche, the operation does not exist as *that* operation. The agent's necessary knowledge of his own operations, or of what he himself is doing, includes necessary knowledge of that niche. That is, the agent necessarily has knowledge of the agent's necessary world.
>
> The agent's necessary world is also the smallest intelligible system within which his operations are intelligible [24, p. 27].

3.2 *The Approach of Evandro Agazzi*

The notion of an operation as a type of action is also used by Evandro Agazzi, who states that "man's actions are always directed by an explicit or implicit confrontation with an 'ought'" [1, p. 107]. An operation is "any human action aimed at the production of a specific and concrete result (in general, an object)", while those actions "for which instead the ideal of perfection concerns the *manner of execution*"—are achievements (examples: language, reasoning, dance etc.). These actions are evaluated on the basis of how they follow the rules of performance. Activities that Agazzi calls pure actions, or simply actions, are activities evaluated not according to their goal but according to an ideal. They are considered right or wrong in themselves (op. cit., 108 onwards).

3.3 *The Approach of Mario Bunge*

Mario Bunge from McGill University in Montreal also defines action by pointing to rationality and morality as dimensions typical of human change-inducing activity.

> The action that one thing or its proxy (the agent) exerts upon another thing (the patient) may be defined as the difference that the former makes to the history of the latter. [...] In human action theory the agent is a human being, or an animal or a machine under his control, and the patient may

be any concrete thing, whether human or not, that coexists with the agent at least during part of the period under consideration. There is interaction if the patient reacts upon the agent, as in the case of work, play, and conversation. And the action is social if both agent and patient are members of the same animal species.

Humans are distinguished from other things not for being doers but for being capable of acting rationally in a morally right or wrong way, i.e. for being able to use knowledge to do good or evil [2, p. 323].

3.4 The Modern Praxiology Approach

Researchers dealing with praxiology[4] propose different factors as the foundation of human activity. Some, as we remember, point to free impulses [12, p. 140], others—to a lack of satisfaction with the state in which someone finds themselves [27, p. 27], others still—to change[5] as the primary goal of every transformation, including transformations caused by humans, i.e. actions [8, p. 18]. What exactly do humans transform? They transform their *practical situations* and/or their contexts from unsatisfactory to satisfactory ones.

What is characteristic of humans, according to modern praxiology [4], is that they act—that each one of us acts—with respect to practical situations of which they—we—are the subjects. Every such situation is an *oikos* of its subject, and the set of these is an *ecology* (*oikos* and *logos*) of practical situations. The practical situation of a given subject is determined by the facts that the subject distinguishes from among other facts due to that subject's values. Values give facts meaning on the basis of which the subject considers them satisfactory or not. If a practical situation is unsatisfactory to the subject, then the subject strives to change the facts in such a way as to achieve a satisfactory situation, But even when the subject considers a situation to be satisfactory, change is still needed. In this case, it is not the kind of "therapeutic" change described above, but a "prophylactic" one serving to prevent any disturbance of the satisfactory situation by natural or artificial (i.e. human-induced) processes. The former type of change applies to the inside of the practical situation, while the latter type concerns the context of the situation—"the rest of the world".

Modern praxiology considers the "existence of action", i.e. the reality connected with activity, in terms—so to speak—of the ontology of practical situations. The practical situation of a subject can be interpreted as a generalization of the concept of personal space introduced by proxemics—a discipline dealing with individual and social space and its perception by humans. Hence, on the one hand praxiology would be a generalization of economics with respect to efficacy-focused behaviours (the "double E" of effectiveness and efficiency), while on the other being a generalization of proxemics with respect to the "bubbles" of practical situations in which each and every one of us is immersed [8].

---

[4]Presented here are excerpts, with minor editing, from the author's earlier works listed in the references [5–11].

[5]Change is also indicated by Mises, who writes that human activity is one of the factors that cause change, it is an element of cosmic activity and becoming; it cannot be reduced to its causes but must be treated as an ultimate given and studied as such [27, p. 32].

Humans behave actively because they have to (e.g. breathing) and because they want to (e.g. driving a car). The former kind is called behaviour while praxiology calls the latter "activity", defining it as a free (i.e. compatible with the acting person's will), conscious human behaviour directed towards a chosen state of things called a target (making it a targeted behaviour). Only a small range of relatively simple actions can be performed by a person—the acting subject—by themselves. These are single-subject actions. All other actions are multiple-subject actions, i.e. actions in which other people (other acting subjects) participate.

Effective accomplishment of intended targets requires the use of appropriate means, and doing this in a way that ensures a surplus of result over cost (economy or efficiency) is the condition of the efficacy of actions. This applies to both single- and multiple-subject actions.

Collaboration of people establishes society as the harmonized activity of many subjects whose actions complement, overlap, support and compete with one another, and also (consciously or not) impede one another, often in brutal and bloody ways. This last element means that actions are considered not only with respect to their praxiological core (subject/subjects and their goal/goals) but also with respect to their context. That context is defined by social consensus subjects (which in itself is a collective activity)—social contract subjects—setting down conditions regarding the targets and means of action which should be met for an activity to win the consent of society. The axiological character of consent is linked to values, i.e. things society treasures above all else. In this sense, praxiology is situated between logic (core) and ethics (context).

Metaphorically speaking, human collaboration from the point of view of praxiology is caused by factors similar to how communicating vessels function: a shortage in one vessel is supplemented with the surplus from another. In activity, some people's lesser dispositional capacity for action is made up for by the greater dispositional capacity of others. To this is added an external factor: unequal distribution of resources causing the necessity for resources to flow from places of surplus to places of shortage. It is thanks to people's conscious efforts to improve unsatisfactory practical situations and/or maintain satisfactory situations, where the conglomerate of these situations is uncountable, that people collaborate. In such collaboration, they achieve the primary targets of organized actions while also meeting their own targets, for which fulfilling the primary target is a means and vice versa [9].

4 Conclusion

Praxiology offers an insight into the reality it studies, but the use that is made of this depends on the users themselves. Such use is more often indirect than direct, since it requires thinking about one's own practicality as an agent, who acts the more efficiently the more of a "reflective practitioner" he is [25, 26]. The meta-skills that reflective practitioners should have for their actions to be suitably efficient for our times are: the ability to gain new skills, the ability to obtain knowledge, the ability to design, and the ability to perform multi-dimensional value judgments within the space defined by the "triple E": effectiveness, efficiency, ethicality [10, p. 35].

Acknowledgements I wish to thank the editors of *Transformacje* annual, in which the Polish version [11] of the present paper is being published, for encouraging me to consider agency in a praxiological approach.

References

1. Agazzi, E.: Dobro, zło i nauka: Etyczny wymiar działalności naukowo-technicznej. Oficyna Akademicka, Warszawa, Poland (1997) (English trans.: Dilworth, C. (ed.) Right, Wrong and Science: The Ethical Dimensions of the Techno-Scientific Enterprise. Rodopi, Amsterdam (2004))
2. Bunge, M.: Treatise on Basic Philosophy. Ethics: The Good and the Right, vol. 8. D. Reidel, Dordrecht (1989)
3. Davidson, D.: Essays on Actions and Events. Clarendon, Oxford (1980)
4. Gasparski, W.: Projektowanie humanistyczne. Prakseologia **2**(78), 27–42 (1981) (English trans.: Humanistic design. In: Jaques, R., Powell, J.A. (eds.) Design: Science: Method, pp. 97–105. Guildford, Westbury House (1981))
5. Gasparski, W.: O pojęciu praktyczności [On the concept of practicality]. Prakseologia **101**, 57–70 (1987/1988)
6. Gasparski, W.: Prakseologia i GOMOPOP [Praxiology and GOMOPOP]. In: Szyszkowska, M. (ed.) Polityka a prawo, gospodarka, moralność [Politics and Law, Economy and Morality], pp. 77–99. Instytut Studiów Politycznych PAN, Warszawa (1992)
7. Gasparski, W.: A Philosophy of Practicality: A Treatise on the Philosophy of Tadeusz Kotarbiński. Societas Philosophica Fennica, Helsinki (1993)
8. Gasparski, W.: Prakseologia [Praxiology]. Wydawnictwo Warszawskiej Szkoły Zarządzania, Warszawa (1999)
9. Gasparski, W.: Działania zbiorowe — ujęcie prakseologiczne [Collective actions — praxiological approach]. In: Matysiak, A. (ed.) Działania zbiorowe — teoria i praktyka [Collective Actions — Theory and Practice], pp. 9–21. Prace Naukowe Akademii Ekonomicznej we Wrocławiu, No. 985, Wydawnictwo Akademii Ekonomicznej we Wrocławiu, Wrocław (2003)
10. Gasparski, W.: Dlaczego ludzie tak działają, jak działają oraz jak działać powinni [Why people act the way they act, and how they shoud act]. Master Bus. Adm. **1**(90), 27–36 (2008)
11. Gasparski, W.: Sprawność w ujęciu prakseologicznym. In: Zacher, L.W. (ed.) Moc sprawcza ludzi i organizacji, pp. 54–76. Poltekst, Warszawa (2016)
12. Kotarbiński, T.: Dzieła wszystkie: Prakseologia Część I [Collected Works: Praxiology Part I]. Ossolineum, Wrocław (1999)
13. Kotarbiński, T.: Od tematu do Traktatu [From the theme to the treatise]. In: Kotarbiński, T. (ed.) Dzieła wszystkie: Prakseologia Część I [Collected Works: Praxiology Part I], pp. 1–4. Ossolineum, Wrocław (1999)
14. Kotarbiński, T.: Cel czynu a zadanie wykonawcy [The goal of an act and the task of the agent]. In: Kotarbiński, T. (ed.) Dzieła wszystkie: Prakseologia Część I [Collected Works: Praxiology Part I], pp. 6–19. Ossolineum, Wrocław (1999)
15. Kotarbiński, T.: Zagadnienie istnienia przyszłości. In: Kotarbiński, T. (ed.) Dzieła wszystkie: Prakseologia Część I [Collected Works: Praxiology Part I], pp. 58–72. Ossolineum, Wrocław (1999) (English transl.: The problem of the existence of the future. Pol. Rev. **13**(3), 7–22 (1968))
16. Kotarbiński, T.: O rozszerzaniu sfery czynu [On the enlargement of the sphere of an act]. In: Kotarbiński, T. (ed.) Dzieła wszystkie: Prakseologia Część I [Collected Works: Praxiology Part I], pp. 72–85. Ossolineum, Wrocław (1999)
17. Kotarbiński, T.: Pojęcie zewnętrznej możliwości działania [On the external possibility to act]. In: Kotarbiński, T. (ed.) Dzieła wszystkie: Prakseologia Część I [Collected Works: Praxiology Part I], pp. 104–108. Ossolineum, Wrocław (1999)
18. Kotarbiński, T.: Zasady teorii czynu [Principles of action theory]. In: Kotarbiński, T. (ed.) Dzieła wszystkie: Prakseologia Część I [Collected Works: Praxiology Part I], pp. 132–133. Ossolineum, Wrocław (1999)

19. Kotarbiński, T.: O stosunku sprawstwa [On the agency relationship]. In: Kotarbiński, T. (ed.) Dzieła wszystkie: Prakseologia Część I [Collected Works: Praxiology Part I], pp. 121–132. Ossolineum, Wrocław (1999)
20. Kotarbiński, T.: Czyn [An act]. In: Kotarbiński, T. (ed.) Dzieła wszystkie: Prakseologia Część I [Collected Works: Praxiology Part I], pp. 134–138. Ossolineum, Wrocław (1999)
21. Kotarbiński, T.: Czyn [An act]. In: Kotarbiński, T. (ed.) Dzieła wszystkie: Prakseologia Część I [Collected Works: Praxiology Part I], pp. 138–152. Ossolineum, Wrocław (1999)
22. Kotarbiński, T.: Dzieła wszystkie: Traktat o dobrej robocie [Collected works, Treatise on good job]. Ossolineum, Wrocław (2000) (English transl. by O. Wojtasiewicz: Praxiology: An Introduction to the Science of Efficient Action. Oxford, Warszawa/ Pergamon Press, PWN (1965))
23. Kotarbiński, T.: Dzieła wszystkie: Prakseologia Część II [Collected Works: Praxiology Part II]. Ossolineum, Wrocław (2003)
24. Meløe, J.: The agent and his world. In: Skirbekk, G. (ed.) Praxeology: An Anthology, pp. 13–29. Universitetsforlaget, Bergen (1983)
25. Schön, D.A.: Educating the Reflective Practitioner. Jossey-Bass, San Francisco (1987)
26. Schön, D.A.: The crisis of professional knowledge and the pursuit of an epistemology of practice. In: Auspitz, J.L., Gasparski, W.W., et al. (eds.) Praxiologies and the Philosophy of Economics, pp. 163–186. Transaction Publishers, New Brunswick (1992)
27. von Mises, L.: Ludzkie działanie: Traktat o ekonomii (fragmenty). Prakseologia **3–4**(128–129), 11–234 (1987) (English trans.: Human Action: A Treatise on Economics. Ludwig von Mises Institute, Auburn (1998))
28. Znaniecki, F.: Wstęp do socjologii [An Introduction to Sociology]. PWN, Warszawa (1988)

W. W. Gasparski (✉)
Koźmiński University, Warsaw, Poland
e-mail: wgaspars@alk.edu.pl

Zygmunt Zawirski: A Creator of New Ideas in Methodology of Science and Scientific Metaphysics

Krzysztof Śleziński

Abstract Zygmunt Zawirski was one of the greatest member of the Lvov-Warsaw School. The paper presents his life and the list of his main works, the greatest achievements in philosophy of nature, methodology of science and metaphysics. He pointed at the possibility of studying the scientific metaphysics with the use of the axiomatic method. His important work was the philosophical issue of time: *The evolution of the notion of time*. Zawirski looked for the justification of the scientific research, listing the problem of induction and deduction, development of the scientific theories and relations of science and metaphysics.

Keywords Methodology of science · Scientific metaphysics · Zygmunt Zawirski · Lvov-Warsaw School · Axiomatic method · Philosophy of nature

Mathematics Subject Classification (2000) Primary 01A70, 03A05; Secondary 00A30

1 Life

Zawirski Zygmunt Michał was born on September 29, 1882, in Podolia in Berezowica Mała next to Zbaraż and died on April 2, 1948, in Końskie.

He has been studying under Kazimierz Twardowski's direction at the Philosophical Faculty of Jan Kazimierz University in Lvov from 1901 till 1906. During his studies he also attended other lectures led, among others, by Mścisław Wartenberg (1868–1938) on issues concerning metaphysics after Kant and Witold Rubczyński (1864–1938) on history of Greek philosophy. He also studied mathematics, physics and philosophy in Berlin (1910) attending lectures led by Carl Stumpf, Georg Simmel and Alois Adolf Riehl and in Paris (1910).

In 1904 Zawirski became one of the founder members of The Polish Philosophical Society, appointed by Kazimierz Twardowski in Lvov. He has achieved a PhD degree in philosophy in July 1906 on the basis of his work, which was written under Twardowski's direction. Zawirski belonged to the first generation of Twardowski's students.

Á. Garrido, U. Wybraniec-Skardowska (eds.), *The Lvov-Warsaw School. Past and Present*, Studies in Universal Logic,
https://doi.org/10.1007/978-3-319-65430-0_13

In 1911 Zawirski started to cooperate with "The Philosophical Movement" and wrote many reports on books and reviews of contents from the French and German philosophical magazines such as "Revue Philosophique de la France et de L'étrange", "Revue de Métaphysique et de Moral" oraz "Archiv für Geschichte der Philosophie". Apart from his reporting activity, he has participated actively in the meetings of the Polish Philosophical Society in Lvov, presenting many papers developed in the form of the scientific articles or more advanced writings. Zawirski achieved the first prize in the 3rd competition of "Philosophical Review" in 1912 on the basis of his work entitled *Causality and Functional Relation. An Epistemological Study* [2]. In this work he demonstrated that it is impossible to reduce completely the notion of causality to the notion of functionality. Since 1915 the subject of his interest included the following problems: hypothesis of constant returns all-matters, inductive metaphysics, relations between metaphysics and science, detailed issues from logic and their significance in mathematical and natural research. The research conducted by him had an impact on development of his opinions.

In 1923 Zawirski presented a paper entitled *Modern Attempts for Axiomatization of Mathematical Nature and their Philosophical Significance* during the 1st Polish Philosophical Convention to share the research conducted and results concerning the implementation of axiomatic method used in philosophy of nature. In the same year he wrote a thesis entitled *Axiomatic Method and Natural Sciences* [5] and presented it to Władysław Heinrich (1869–1957). This work was the basis to initiate the proceedings for the qualification as a university professor at the Jagiellonian University in 1924.

In the period between 1928 and 1936 Zawirski linked with the University of Poznań. Władysław Mieczysław Kozłowski (1858–1935) had retired and Zawirski was appointed as lecturer for theory and methodology of sciences courses at the Humanistic Faculty and since the 1st of August 1929 as associate professor at the Mathematical and History of Science Faculty. The classes led by Zawirski had a good reputation among students. He combined his didactic and scientific activity. He was interested in students' access to the basic philosophical works which were the subjects of his lectures and seminars. Then, he gathered valuable literature in his department, which were destroyed by the Nazis during the World War II. At the time of his lectures, Zawirski focused on the philosophical problems of history of nature, basic problems of mathematics, issues of epistemology and theory of classes and relations.

The period of working for the University of Poznań is the most important stage in his scientific life. He got in the close touch with his master Kazimierz Twardowski. Zawirski sent him reports and reviews of books for "Philosophical Movement" edited by him. He participated in the 7th International Congress of Philosophers in September 1930 in Oxford. Zawirski was awarded with the first and very prestigious prize in Rignan's competition in 1933 announced by the Italian magazine "Scientia" for his work entitled *L'évolution de la notion du temps* [10].

Zawirski was nominated as a full professor in 1934. He stayed 2 years more at the Faculty of Mathematics and Natural Sciences where he was a Dean and a Chair of Theory and Methodological Sciences Department being an active member of scientific life internationally. He participated in the 8th International Philosophical Congress in Prague (1934) and in the 9th Congress in Paris (1934). In the same year he welcomed members of the International Convention of Thomistic Philosophy held in Poznań instead of absent Michał Sobecki, who was the President of Poznań Philosophical Society. Zawirski also participated in the 1st (Paris 1935) and the 2nd (Kopenhaga 1936) Congress of Scientific Philosophy and in the 3rd Polish Philosophical Convention in Cracow in 1936 where he presented a paper entitled *On the matter of Scientific Synthesis* [11].

Near the end of his scientific work at the University in Poznań, he was awarded an honorary doctorate by the University of Poznań and the Faculty of Mathematics and Natural Sciences on the 12th of November in 1936 and accepted it on the 18th of November in 1936 at the hands of president Ignacy Mościcki.

He was asked by Władysław Heinrich in 1935 to chair a faculty after Tadeusz Grabowski (1869–1940). Zawirski accepted it and as a full professor has started his work since the 1st of January in 1937 at the Philosophical Faculty. Later after its division, he worked at the mathematics and Natural Sciences Faculty. In the period between 1938 and 1939 as well as between 1945 and 1946 he was a Dean of this Faculty.

He took over editorial office of "The Philosophical Quarterly" after Władysław Heinrich in 1936. In the period between 1938 and 1945 he was the President of Cracow Philosophical Society and gave papers entitled, among others *On the Scientific Activity of Professor Kazimierz Twardowski* [22] and *The Genesis and Development of Intuitionistic Logic* [21]. In the period between 1938 and 1941 he worked on *Philosophical Dictionary* [23]. Unfortunately, the censorship stopped the printing of dictionary copies in 1948. A manuscript of this dictionary survived in the Polish Academy of Sciences Archives and only some terms were published in 1993.

The Nazis pacifist action "Sonderaktion Krakau" took place on the 6th of November in 1939 against the Polish researchers and scientists. Zawirski was outside Cracow on this very day and due to it he barely avoided a transportation to the Nazis concentration camp in Sachsenhausen. During the World War II he participated in the clandestine academic teaching. After the WW II he was a full professor at the Jagiellonian University. He has been a chairperson of the Cracowian Philosophical Society since 1945.

Zawirski was very active in the scientific life in the period between 1945 and 1948. At that time his works were published as the result of his long standing research. Travelling to Zakopane to attend Philosophical Conference in winter 1947, a luggage with two manuscripts was stolen from him including a methodology manual for natural sciences *On the Scientific Method* [12] and manuscript of Patristic Monography.[1]

Zawirski prepared a written speech for the 10th International Philosophical Congress in Amsterdam in 1948 but unfortunately he did not manage to present it. The work undertaken at this year, and first of all the works of manuscript reconstruction overstrained his organism moving the unexpected catastrophe. Zawirski died suddenly at his son Kazimierz's home in Końskie.

2 Writings

His leading papers include the following: *Causality and Functional Relation* [1], *Philosophical Relativism and the Physical Theory of Relativity* [4], *The Connection between the Principle of Causality and the Principle of Relativity* [6], *Axiomatic Method and*

[1]Roman Ingarden in *Wspomnienia o prof. Zawirskim* (*Recollection of Professor Zawirski* [39, p. 261]) (1948) wrote that the author managed to reconstruct his stolen works.

Natural Sciences [5], *The Eternal Return of the Worlds. Historical-critical Investigations of the Idea of "Eternal Return"* [7], *On Indeterminism in Quantum Physics* [9], *Über das Verhältnis der mehrwertigen Logik zur Wahrscheinlichkeitsrechnung* [19], *L'évolution de la notion du temps* [10], *On the Problem of Scientific Synthesis* [11], *The Importance of Logical and Semantical Investigations for the Theories of Contemporary* [20], *The Genesis and Development of Intuitionistic Logic* [21], *Remarks on the Method for Natural Sciences* [12], *Time: Selections from "The Evolutions of the Notion of Time"* [13], *Science and Metaphysics* [14, 15] and *On the Relation between Metaphysics and Science* [16].

3 Views

We present below only an outline of the whole academic and scientific achievements of Zygmunt Zawirski, grouping it simply into three characteristic areas of research problems originated from philosophy of nature, methodology of science and metaphilosophy. It also includes a discussion on the mutual relations between science and philosophy as well as problems of the scientific metaphysics. Philosophy of nature is one out of many research areas interesting for representatives of Lvov-Warsaw School. Research results from logic, methodology, philosophy of nature and metaphilosophy are commonly discussed. However, little attention is devoted to research from philosophy of nature and scientific metaphysics conducted within this School. The main representative of this trend was Zawirski, who in much more degree used content of the investigated scientific theories than it was observed in philosophy of science.

3.1 Philosophy of Nature

Zawirski was interested in development of natural sciences, first of all the theory of relativity and quantum mechanics. He has left many original works linked with these theories trying to answer the questions stated and philosophical matters discussed at that time. Zawirski presented new philosophical implications derived from these theories in his numerous works and their impact on wider understanding of reality.

One of the first matters out of the natural history raised by Zawirski was a hypothesis on the so called "eternal returns of the worlds". He published a series of three papers on these matters in "The Philosophical Quarterly" in the period between 1927 and 1928. He had already thought these problems over, studied and had been ready for 15 years. Its first versions were presented on the 27th May and 24th June 1911 at the meetings of Polish Philosophical Society in Lvov as well as during the Convention of Polish Doctors and Naturalists in Cracow in the same year.

Zawirski came back to this motive of "eternal returns" in his main work entitled *L'évolution de la notion du temps*, which was published in 1936. His work has been still worth considering due to the clear and profound presentation of history linked with the notion of time, to start with the Pythagoreans and to end with the modern philosophical concepts of his time including H. Bergsona and E. Husserla and the latest relativist and quantum physics. Zawirski, conducting research on nature of time, was interested in the philosophical principles and implications linked with the general theory of relativity in the 1920s of the twentieth century. Only 5 years after Albert Einstein had introduced his theory, which means rather promptly, Zawirski published his paper entitled *Philosophical Reflection on the Theory of Relativity* [3] in 1920. He noticed its crucial importance and explained that it is required to eliminate a number of philosophical premises out of science, which had been accepted earlier by the physical theories and expressed in the notion of absolute time, space and movement. Zawirski paid close attention and followed the development of research undertaken within Einstein's theory. It is confirmed by his numerous reviews of books on this theory published at that time (in total 10 reviews).

Zawirski also investigated problems of casual relation from various perspectives during all his scientific research. He was assured about the importance of this principle for scientific research. In his work entitled *Causality and Functional Relation* [1], he demonstrated that the notion of "casual relation" and "functional relation" differ in terms of content. As the casual relation is the real relation, considering the influences and time relations between reasons and consequences, it cannot be replaced with the functional relation. In his paper entitled *Quantum Theory and the Causality principle* [8] from 1930, he polemicized with Heisenberg's theory that uncertainty principle proves the falsity of causality principle.

Zawirski was a pioneer of works about the implementation of many-valued logics in quantum mechanics and works about the probabilistically based approach to many-valued logics. He linked the system of logic, which can be used in the description of quantum phenomena, with the system of infinite many-values corresponding to the levels of probability. Zawirski designed the system of many-valued logic, which possesses all the advantages of Reichenbach's system developed in 1932. Simultaneously, Zawirski avoids its disadvantages such as Reichenbach's complications unnecessarily introduced for counting implications and equivalence, which leads to an assignment of these logic functions as many values as for the sum and product. Designing the system of many-valued logic, Zawirski used also Jan Łukasiewicz's and Emil Posta's research results. Both of them, independently of each other and almost simultaneously around 1921, generalised of many-valued logic. However, none of them described the relation between logic and theory of probability. Initially, it was achieved only by Hans Reichenbach.

Zawirski pointed at two potential uses of many-valued logic in 1932 in the paper entitled *Les logiques nouvelles et le champ de leur application* in "Revue de Métaphysique et de Morale" [17]. The first one focused on the link between the theory of probability and many-valued logic. The second one focused on the use of Łukasiewicz's three-valued logic in the analysis of wave-particle duality. Nevertheless, Zawirski admitted that this attempt was too early. In further works, our philosopher was for the use of the theory of probability in the description of quantum phenomena. He stated that various degrees of probability can be assigned to quantities of feedback such as time, energy, location and momentum.

3.2 Methodology of Science

Zawirski paid the attention to any signal of philosophical thinking outside the borders of our country. He reacted violently to the methodological research included in Karl R. Popper's work entitled *Logik der Forschung*, published in Vienna in 1934. Zawirski criticised Popper's falsificationism stating that disagreement of some law with one of the recognised and elementary opinions can be regarded as the sufficient reason for the rejection of this law. Zawirski notices that a single empirical law hardly ever is tested perceived separately from the other laws. The whole system of opinions or theory is tested or invalidated. Moreover, a number of laws and independent hypotheses constantly decreases during this test. Then, each general empirical task "takes responsible for itself" for the whole system to which it belongs [20]. In this way development of real sciences constituents a continuation of the theory, which fight with each other and modify constantly. Then their number always decreases.

A problem of testing hypotheses in empirical sciences was also undertaken by Zawirski in the paper entitled *Remarks on the Method of Natural Sciences* [12]. In a similar way to Popper, Zawirski assumes that we derive consequences out of the accepted hypothesis. Then, we test if they agree with the facts concerned with them. The hypothesis is proved correct when there is an agreement between directly observed facts and conclusions derived out of it. If there is a disagreement, the hypothesis should be rejected due to its invalidation. Zawirski in his paper, not mentioning the name of Popper, stills criticised his concept of falsification and impossibility to achieve *experimentum crucis*. Currently, it can be stated that he criticised the so called naive Popper's falsificationism.

Zawirski accepted the present asymmetry between positive and negative result in testing a hypothesis. A question whether the negative result is more sonorous than the positive one remained a problem for him. It is clear for Zawirski that the positive result still does not prove a rightness of the particular hypothesis because it can be changed by the minute. Whereas the negative result does not lead always towards a complete withdrawal of the hypothesis. Invalidation of the hypothesis might be a decisive moment only when none of the notions describing an experiment remained unchanged. Every single change of meaning in the terms used leads to another undertaking of the particular hypothesis, in spite of the fact that it was invalidated. The example of such situation can be earlier rejected wave-particle theory of light after Foucault's experiments and later introduced by the theory of quantum. A similar situation occurs in the following experiments *experimentum crucis* and *instantia crucis* when we select the one which includes some newly-revealed fact out of two competing hypotheses or theories.

Zawirski notices the analogy that is present between a verification of the particular hypothesis and acceptance of the one out of two competing hypothesis. As long as we expect the positive or negative answer in the first case, the positive answer linked with the one hypothesis in the second case is simultaneously the negative answer for the another one. In the situation of *experimentum crucis* it occurs rarely that two competing hypothesis were the opposite opinions. There are usually more complicated hypotheses. Therefore, logical conjunctions of sentences are often checked, whose negation is their alternative. Popper does not notice it in his concept of falsification. Zawirski agrees with Duhem for whom *experimentum crucis* have never invalidated one isolated hypothesis but only the whole theory full of linked tasks.

Comments presented above refer only to the theoretical testing of hypothesis. In practice, the situation does not look as such complicated. Each theory, apart form the formulated laws, includes also numerous terms defined and agreements. The simplest way of testing hypothesis is usually selected. However, more complicated cases occur and may trigger "a revolution in science". This situation occurred during the development of theory of relativity and quantum mechanics when a number of negative experiences increased constantly as well as supporting hypotheses which explained these experiences. After some time, the edifice of classical physics has been reconstructed, which appeared to be the best solution in this difficult situation.

3.3 Metaphilosophy

The scientific interests of Zawirski were link also with more general problems of truth and being. The point of departure for his interests were the profound cognition of classical metaphysics and natural sciences within which more and more often the problems were undertaken previously having been reserved only for philosophers.

In the first decades of the twentieth century Zawirski witnessed an argument and a dispute concerned with the role of natural sciences in the development of general outlook on life. He noticed that both metaphysicians and opponents of metaphysics expressed a need to develop a scientific outlook on life. This fact made him ponder over a possibility to design metaphysics based on experience. He realized very quickly that metaphysics understood in this way would not be able to replace completely classical metaphysics perceived as *scientia entis*. It would not have been balanced only with the synthesis of natural sciences admitted by positivists. However, considering the mutual straggle of the most opponent reasoning movements, he undertook an attempt to develop a middle path leading towards the formation of the scientific metaphysics—critical and open using the results of the empirical experience. Zawirski presented his first ideas about the relation between metaphysics and science on the 5th of May in 1917. It was his lecture entitled *On Relation between Metaphysics and Science* [16] at the meeting of the Philosophical Society in Lvov.

The issue of the relation between metaphysics and science and possibilities of developing a general theory about the reality dominated in Zawirski's research till 1923 when he published his post-doctoral thesis entitled *Axiomatic Method and Natural Sciences* [5]. The work was the summary of the earlier conducted research focusing on the possibility of axiomatization of metaphysical systems. Before 1923 Zawirski had written also two more works at that time on the matter [14, 15] and [16]. Both works remained in the manuscript and only in the period between 1995 and 1996 *Science and Metaphysics* was published in "Philosophy of Science". Then, the second work *On Relation between Metaphysics and Science* was published in 2003.

Zawirski related once again to the problems presented in the post-doctoral thesis in 1936 during the session of the 3rd Polish Philosophical Convention in the paper entitled On the matter of the scientific synthesis [11]. In his presentation he introduced a different argumentation in the defence of scientific metaphysics than the one presented in the work entitled *Axiomatic Method and Natural Sciences*. Following Gödel's statement on the incomplete systems, he argued that we should not refuse the sense of the metaphysical issues only because they are not subject to empirical testing. He also pointed to the tasks of philosophy, stating that it should not undertake the issues for which it is not possible to find methods of their solution.

4 Resonance

Zawirski's works contributed to the development of methodological, logical, ontological and meta-philosophical research. Zawirski introduced the potentials for formalisation of the scientific theories for natural history, the manner of investigating the philosophical assumptions and implications derived from the modern physical theories, the critical analysis of scientific terms including their role in the growth of scientific knowledge as well as the implementation of methodological principles in scientific theories. He showed an opportunity to develop scientific metaphysics using the axiomatic method and opportunity to generalize scientific research philosophically to comprehend the world. He looked for the grounds of scientific research, indicating the methodological problem of induction and deduction, growth of scientific theories, relation between science and philosophy. He pointed at the importance of ontological research linked with time and space.

His suggestions within the methodology of science, which were studied under the impact of critical perception of Popper's work entitled *Logik der Forschung*, appeared to be accurate. It was confirmed by Popper's another work entitled *Conjectures and Refutations* from 1963 where Zawirski's objections were considered (without mentioning his name). Zawirski suggested that induction and deduction are equally impotent in the natural science. He said, that falsification, like verification, has never had the final character of empirical theory, since one does not test one singular theory, but always a group of theories or laws; and also, that the development and changes of science are of the infinite nature. It should be exposed that Zawirski's student named Zygmunt Spira was also interested in Popper's work and developed creatively the issues linked with methodology of natural sciences in the work entitled *Comments on the Methodology and Popper's Theory of Learning* (1946). Zawirski developed a concept of three worlds in the scope of his research such as the world of senses, the psychological world and the world of eternal and ideal being. A concept of three worlds, similar to Zawirski's one, has been developed independently by Karl Popper (*Objective Knowledge* in 1972) and Roger Penrose (*Shadows of the Mind* in 1994).

The research, previously conducted by Zawirski, on the use of the formal logics in the investigation of the physical scientific theories were advanced by his student named Romana Suszko. He developed the theory of models and used it in the investigation of problems beyond mathematics (*Remarks on Sentential Logics* in 1958).

However, Zawirski's main composition entitled *L'évolution de la notion du temps* has been very important scientifically due to the detailed analysis of the time problem. Many issues still discussed currently can be found there. They refer to thermodynamics of irreversible processes, the second law of thermodynamics, theory of dissipative processes (I. Prigogine, I. Stengers, *Order out of Chaos*—1984) and conformal cyclic cosmology (R. Penrose, *Cycles of Time*—2010).

Zawirski's works related to the critical and historical analysis of scientific terms and philosophical reflation over the assumptions for the empirical theories and the investigation of the traditionally philosophical problems entangled into these theories constitute the model prototype of the scientific works from the field of philosophy of nature, natural history and philosophy of nature. The works of this type are still conducted in Cracow background of philosophers such as Józef Życiński (1948–2011) or Michał Heller (born 1936), a director and founder of The Copernicus Centre for Interdisciplinary Studies.

5 Pupils

Zygmunt Zawirski by his didactic and scientific work inspired many Polish logicians and philosophers. The following students, among others wrote their diploma works under his guidance or worked in these fields: Józef Maria Bocheński (1902–1995), Andrzej Grzegorczyk (1922–2014), Zbigniew Jordan (1911–1977), Zygmunt Spira (1911–1942?) and Roman Suszko (1919–1979). His close students are the last three ones.

6 Summary

Zawirski's works and presentations focused on the important research problems in the first part of the twentieth century. In particular, they referred to the philosophy of nature, formal logics, methodology of science and meta-philosophy. He started to cooperate with philosophers, among others, from the Great Britain, France, Germany and Austria. He was active in the scientific work at the international level. He gave lectures and participated in discussions during the International Philosophical Congress including the 7th Congress in Oxford, the 8th in Prague, the 9th in Paris; for the 10th Congress in Amsterdam he had already sent his paper. He was interested in the development and consequences of the general theory of relativity and the quantum theory. From the perspective of developing natural history he investigated questions of the determinism, indeterminism and causality, ontological problems of time and space, methodological problems of the relations between the verification and the falsification, the relations between a hypothesis and the empirical basis, but also the possibility of an *experimentum crucis*. He was interested in the modality of judgements and epistemological principles of the intuitionist logic—Brouwer-Heyting's logic. He tried to applied to the new Łukasiewicz's logic, many-valued logic and quantum mechanics. His metaphilosophical works are also worth paying attention. First of all, the works that focus on the investigation of the relations between science and philosophy are important and the attempt to develop the scientific metaphysics using the axiomatic method.

Zygmunt Zawirski's Works Mentioned in the Paper

1. Przyczynowość a stosunek funkcjonalny [Causality and functional relation]. Przegląd Filozoficzny (PF) [Philos. Rev.] **15**(1), 1–66 (1912)
2. Przyczynowość a stosunek funkcjonalny. Studium z zakresu teorii poznania [Causality and Functional Relation. An Epistemological Study]. Lwów (1912)
3. Refleksje filozoficzne nad teorią względności [Philosophical reflections on the theory of relativity]. PF **23**, 343–366 (1920)
4. Relatywizm filozoficzny a fizykalna teoria względności [Philosophical Relativism and the Theory of Relativity]. Drukarnia Słowa Polskiego, Lwów (1921)
5. Metoda aksjomatyczna a przyrodoznawstwo [The axiomatic method and natural sciences]. Kwartalnik Filozoficzny (KF) [Philos. Q.] **1**(4), 508–545 (1923/1924), **2**(1), 1–58 (1923/1924), **2**(2), 129–157 (1923/1924)
6. Związek zasady przyczynowości z zasadą względności – odczyt habilitacyjny [The connection between the principle of causality and the principle of relativity – postdoctoral thesis]. KF **2**(4), 397–419 (1924)
7. Wieczne powroty światów. Badania historyczno-krytyczne nad doktryną "wiecznego powrotu" [The Eternal Returns of the Worlds. Historical-Critical Investigations of the Idea of "Eternal Returns"]. Uniwersytet Jagielloński, Kraków (1927)
8. Teoria kwantów a zasada przyczynowości [The quantum theory and the principle of causality]. PF **33**(4), 296–301 (1930)
9. W sprawie indeterminizmu fizyki kwantowej [Indeterminism in Quantum Physics]. Książnica Atlas, Lwów (1931)
10. L'évolution de la notion du tamps. Scientia **28**(55), 4, 249–261 (1934)
11. W sprawie syntezy naukowej [On the matter of scientific synthesis]. PF **39**(4), 347–352 (1936)
12. Uwagi o metodzie nauk przyrodniczych [Remarks on the method for natural sciences]. PF **44**(4), 315–318 (1948)
13. Time: selections from "The Evolutions of the Notion of Time". In: Szumilewicz-Lachman, I. (ed.) Zygmunt Zawirski: His Life and Work. Kluwer Academic Publishers, Dordrecht (1994)
14. Nauka i metafizyka (I) [Science and metaphysics I]. Filozofia Nauki [Philos. Sci.] **3**(3), 104–135 (1995)
15. Nauka i metafizyka (II) [Science and metaphysics II]. Filozofia Nauki **4**(1), 131–143 (1996)
16. O stosunku metafizyki do nauki [On the Relation Between Metaphysics and Science]. Wydział Filozofii i Socjologii Uniwersytetu Warszawskiego, Warszawa (2003)
17. Les logiques nouvelles et le champs de leur application. Revue de Métaphysique et de Morale **39**(4), 503–579 (1931)
18. O stosunku logiki wielowartościowej do rachunku prawdopodobieństwa [On the relations between many-valued logic and probability calculus]. In: Pamiętnik XIV Zjazdu Lekarzy i Przyrodników Polskich. Drukarnia P. Mitręgi, Cieszyn (1933)
19. Űber das Verhältnis der mehrwertigen Logik zur Wahrscheinlichkeitsrechnung. Stud. Philos. **1**, 407–442 (1935)
20. Doniosłość badań logicznych i semantycznych dla teorii fizyki współczesnej [The importance of logical and semantical investigations for the theories of contemporary]. PF **41**(1), 25–30 (1938)
21. Geneza i rozwój logiki intuicjonistycznej [The genesis and development of intuitionistic logic]. KF **16**(2–4), 165–222 (1946)
22. O działalności naukowej prof. Kazimierza Twardowskiego [On the scientific activity of professor Kazimierz Twardowski]. RF **15**(3), 157 (1939)

23. Zmarli profesorowie, docenci i asystenci Oddziału Matematyczno-Przyrodniczego Wydziału Filozoficznego [The departed professors, readers and associate professors at Mathematical and Natural History Department at the Faculty of Philosophy]. In: Kronika Uniwersytetu Jagiellońskiego za okres wojny 1939–45 oraz za rok akademicki 1945 [Jagiellonian University Chronicle of the Period Between 1939 and 1945 Plus the Academic Year of 1945]. Kraków (1946)

K. Śleziński (✉)
Institute of Educational Science in Cieszyn, University of Silesia in Katowice, Cieszyn, Poland
e-mail: slez.krzysztof@onet.poczta.pl

Zygmunt Zawirski's Concept of Scientific Metaphysics

Krzysztof Śleziński

Abstract In the thesis *Zygmunt Zawirski's concept of scientific metaphysics* two questions were posed as the means of presenting an explorative issue: Can axiomatic method give rise to philosophy as a science, study of universal "order"? Does scientific metaphysics constitute solid, mathematical representation of our intuitions and instinctive strivings and development of previous attempts to demonstrate the existence of unity among multiplicity of phenomena. The article aims to reconstruct Zygmunt Zawirski's metaphilosophy, among other things: relations between the sciences and philosophy.

Keywords Scientific metaphysics · Zygmunt Zawirski · Relations between the sciences and philosophy · Metaphilosophy

Mathematics Subject Classification (2000) Primary 00A05; Secondary 00A30

1 Introduction

Polish philosophers from the beginning of the twentieth century apart from conducting the detailed research on logics, semiotics, research methodology and ontology, were aware of the general aims headed by philosophy [3, 4]. Their objective constituted the attempt to define the whole view on reality with a particular focus on research method in philosophy, which was discussed at the 2nd Polish Philosophical Meeting in 1927 in Warsaw. At that meeting, Jan Łukaszewicz presented the paper entitled *On Method in Philosophy* [7, 8]. Florian Znaniecki presented his paper entitled *Tasks for Philosophical Synthesis* [25]. In his speech he presented a pluralistic approach to philosophy perceived as the multitude syntheses of knowledge accepted at the separate stages of the development linked with civilisation.

A clear polemical tone is observable in the 1930s, among others in the speeches of Jan Łukasiewicz, Maria Kokoszyńska, Zygmunt Zawirski, Roman Ingarden, Bolesław Gawecki, Tadeusz Garbowski and Joachim Metallmann, on the views of members of Vienna Circle linked with the synthesis of scientific knowledge. The issue has also been undertaken at the 3rd Philosophical Convention in 1936 in Cracow. However, one mutual

Á. Garrido, U. Wybraniec-Skardowska (eds.), *The Lvov-Warsaw School. Past and Present*, Studies in Universal Logic,
https://doi.org/10.1007/978-3-319-65430-0_14

agreement was achieved in terms of developing the general view of realities. Still, the discussion undertaken at this meeting directed the trend of further research on the relations between science and philosophy as well as a study on the scientific metaphysics, as Jan Łukasiewicz appealed to, based on the axiomatic method [8]. It is worth noticing that such attempts had been already undertaken by Benedykt Bornstein [2, 15] and Zygmunt Zawirski, whose suggestions will be discussed below.

2 Relation Between Science and Metaphysics

The departure for Zygmunt Zawirski's study on the scientific metaphysics was the profound comprehension of the classical metaphysics and the scientific theories. In particular, these were the general theory of relativity, thermodynamics and quantum mechanics within which one can understand the problems previously reserved only to philosophers.

Zawirski experienced a dispute related to the role of the natural sciences in the study on the scientific metaphysics in the first decades of the twentieth century. He noticed that both metaphysics and its opponents represent a need for the study on the scientific view of the world. This detail made him started deliberations on the possibility of studying metaphysics based on an experience. He realised very quickly that metaphysics understood in this manner would not be able to replace the classical metaphysics completely perceived as *scientia entis*. It would be an equivalent to it, only to the synthesis of the natural sciences accepted by the Neo-positivists of the Vienna Circle. Taking into consideration, the mutual aspirations for the most opposite directions of thinking, he undertook an attempt to delimit the way in the middle leading towards to the new scientific metaphysics being constituted. The new one was to be critical and open, applying the results of the empirical experience. Zawirski presented the first reflections about the relations between metaphysics and science on the 5th of May in 1917 in his lecture entitled *On the Relation between Metaphysics and Science* [24] presented during a meeting of Philosophical Society in Lvov.

The issues of the relations between metaphysics and science and the possible assumptions accepted in the study on the philosophical theory of reality dominated in Zawirski's research till 1923. This was the year when he published his postdoctoral thesis entitled *The Axiomatic Method and Natural Sciences* [17]. This work has been the summary of the previous research. Moreover, it included the possibility of axiomatization of the metaphysical systems. Before 1923 Zawirski had written also two more works dedicated to the matters undertaken at that time. They were entitled *On the Relation between Metaphysics and Science* (1919), *Science and Metaphysics*—1920 [22, 23]. Both works remained handwritten. It was only between 1995 and 1996 when *Science and Metaphysics* was published in the pages of "Philosophy of Science". Then, *On the Relation between Metaphysics and Science* was published in 2003.

In these works, Zawirski distinguished two areas of empiriologia and meta-empirical knowledge out of the field of metaphysical research heading towards the general knowledge about the whole reality. Empiriologia is for him the science not undertaking the main problems of classical metaphysics while meta-empirical knowledge is the science about of the principle of being. Assigning and giving priority to empiriologia in the research on the whole reality, he did not discredit the need of going in for meta-empirical knowledge. He assumed that the science of empiriologia does not consist of the simple summing up of the results from the detailed sciences and also is not their highest generalisation. Its paragon is the transformation into the axiomatic and deductive system following mathematics.

However, each system of axioms and statements originated out of them tolerate the possibility of many interpretations of their symbols. Then, many interpretations of the results are possible in empiriologia, which occur in the process of meta-empirical knowledge. Following Zawirski, although all interpretations of science in meta-empirical knowledge, leading to various metaphysical systems, possess the same learning value, he followed the idealistic concept criticising the individual metaphysical system. The concept is worth attention and the most credible meta-empirical interpretation of the scientific results defined by empiriologia [24, p. 96].

Empiriologia and meta-empirical knowledge were linked with the use of hypotheses which went beyond experience. These hypotheses in empiriologia contribute to the development of science and serve for the simpler depiction of content in experience. These hypotheses in the meta-empirical knowledge are used in learning about objects and metaphysical problems linked with them, which do not belong to the experience such as the existence of absolute and transcendent being, being after death or deciding about the issue of learning about the world.

Zawirski noticed that the fact that empiriologia existed, metaphysics referring to experience, would allow its many possible interpretations. He also noticed that empiriologia, experiential metaphysics, would never replace the classical metaphysics completely. A possibility of numerous interpretations originates from the fact that universal science about reality is the inductive theory. The authenticity of such theory or any its type cannot be proved and its absolute reliability cannot be shown [22, p. 113]. Symbols of the inductive theory permit the infinite abundance of interpretations. Therefore, none of these interpretations deserves the defined probability ranking. It indicates that inductive metaphysics should not refer to the probability calculus. Taking it into consideration, Zawirski postulates in the studies on the universal theory of reality that empiriologia as the science should link the results of detailed sciences into the system freed from contradictions and remain in such relation to these sciences in which detailed sciences remain in relation towards their particular areas and fields [22, p. 135].

Empiriologia is the scientific metaphysics cannot replace the classical metaphysics. Statements in empiriologia have got the nature of temporary hypothesis and constantly are put into a test of arranging them into the previous experiences in the procedure of *experimentum crucis*. These hypotheses are the important supplements of the scientific picture of the world in empiriologia, which still does not give us its complete picture. There are numerous and important issues for human being, such as ethical postulates, outside the area of the scientific metaphysics.

In the definition of the universal opinion on reality within meta-empirical knowledge, Zawirski achieved an interesting differentiation of *entia rationis*, the world of ideal beings and *entia naturae*, the world of existence. *Entia rationis* exists independently of the learning subject. The world of these beings include among others mathematical objects and content of mathematical natural history. *Entia rationis* is not designed but it is learned and discovered as everything else. The novelty constitutes only their life in our mind as mental existences and as the defined content of our consciousness whereas not as eternal existences beyond time. This world includes also figments of our thoughts, among others metaphysical systems as well as all sorts of logical mistakes having the ideal being and both true or false theories [24, p. 164]. However, should these systems or mistakes exist only in our thoughts, they belong to the world of mental existence.

The world of senses is also the mental construction. Speaking about the mental subjects and similar statements on reality are possible because there is absolute being, over individual one, which following Zawirski's opinion, belongs both to the world of everlasting beings and to the world of existence as well.

The world of *entia rationis* is infinite and inexhaustible, beyond time and endless, unlimited by human thoughts. Human mind should, in Zawirski's opinion, following Ockham's principle, obtain out of this unlimited world of the eternal beings only as much as it is necessary and indispensable to understand the world of existence [24, pp. 159–164]. Moreover, it is forbidden to make a mistake and come down to a blind transfer of the relations linked with into *entia rationis* into the relations linked with *entia naturae*. The relations among *entia rationis* are ideal while the relations among existences are dynamic. Considering an argument that the world of eternal beings incorporates the world of existences, it indicates that all possible relations belong to *entia rationis*. The difference between the one and another relation is that *entia naturae* relations refer to psychological and temporal orders, which cannot be stated about the relations occurring between the eternal beings. Then, the relations which can be implemented in the sense picture of the world define interdependence and temporal order among the empirical data.

Zawirski's research on the scientific metaphysic resulted in the study about the ontological concept of three worlds—the world of senses, the world of psyche and the infinity one. It should be notice that a similar to Zawirski's concept of three worlds was worked out independently by Karl Popper (Objective Knowledge, [12]) and Roger Penrose (*Shadows of the Mind*, [10]). Undoubtedly, there are many similarities among those concepts but also some differences. Each of them was still based on the epistemological and ontological foundation in the analysis of learning and possessing knowledge of the reality.

Zawirski referred to the problems listed in his postdoctoral thesis one more time in 1936 during the meeting of the 3rd Polish Philosophical Society in his paper entitled *On the Matter of Scientific Synthesis* [19]. In his speech he presented a different argumentation in a defence of the scientific metaphysics than the one which was described in his work entitled The Axiomatic Method and Natural Sciences. Following Kurt Gödel's statement about the incomplete systems, he argued that the sense should not be refused in terms of the metaphysical issues only because they are not subjected to the empirical test. He also pointed at the tasks of philosophy, stating that it should not undertake the issues for which a method of its solution could not be appointed.

3 Synthesis of Human Knowledge and Metaphysics

A synthesis of human knowledge was pointed often by the opponents of commonly understood metaphysics in the 1930s of the twentieth century. Their synthetic views of knowledge were anyway a sort of substitute for the acceptance of metaphysical resolutions. The representatives of neo-positivism cited, on the one hand, the representatives of the classical positivism in the process of knowledge unification, for example Comte or Spencer. On the other hand, they tried to refer to Leibniz's concept of *mathesis universalis* [19, p. 347].

The second reference is unfortunate while it would introduce some metaphysical elements into the constructive system of knowledge as it was done by the author of Monadology. Leibniz solved the important issues, linking them more tightly with the elements of his ontological-metaphysical system. The representative of neo-positivism followed Encyclopaedism approach to the presentation of the whole knowledge rather than build a synthesis where non-scientific and metaphysical elements should be introduced.

Zawirski, solving the problem related to the synthesis of human knowledge, was for the impossibility of leading the demarcation line between the scientific synthesis and metaphysics. In Zawirski's opinion, the principal non-settlement of the metaphysical hypotheses could not be accepted. None of the metaphysical or scientific hypotheses could be tested directly. We can only discuss verifiability of the consequences of the accepted hypotheses. The metaphysical and scientific hypotheses are the same on this account. This situation does not change even having accepted Popper's criteria for falsification of the hypotheses. Zawirski agrees that falsification is an effective method of testing hypotheses. However, as he adds, only a few and constructed incorrectly hypotheses are not subjected to this procedure [19, pp. 347–348].

Development a synthesis of human knowledge without a part of science and metaphysics is impossible. Both the scientific terms and the terms of the scientific metaphysics, being under control of experience, are used in the construction of the synthetic view of the knowledge following Zawirski. He defends the scientific metaphysics with terms being under the control of experience and all its theses should fall within a category of testing including their falsification. This range of metaphysics should include the notion of time and space structure plus dynamic reality and the problem of determinism, indeterminism or deliberation [19, p. 348]. In the development of the system for the scientific metaphysics, Zawirski does not eliminate the intuitive learning. However, should the doubts appear or even contradictions of the intuitive details, the intuitive convictions should be consolidated in the axiomatic system. Only at that time one can be sure what s/he wants to say plus what is said will be understood by the others. A depiction of convictions into a system, which includes the appropriate order of sentences together with their consequences, never releases us from a duty of searching for equivalences in an experience for the particular conviction. However, the developed system must be reviewed as well as falsification [19, p. 348]. This approach into developing a system is nothing new. Zawirski notices that in the process of building even the simplest deductive system we deal with the intuitive acceptance of its assumptions.

It is worth mentioning that Zawirski does not remove neither a priori or intuitive elements in the attempt to understand the reality. He demands a depiction od all details first of all into a system, which consequently should be confronted with an experience. In Zawirski's opinion many such systems can be constructed and each of them can be convincing at the equally strong level of obviousness. Zawirski notices that the same set of intuitive details can in the equal degree justify many of the accepted theories. The final decision for which theoretical description of the intuitive details one should support depends on an experience.

The scientific knowledge as well as the scientific metaphysics, in Zawirski's opinion is deprived of Leibniz's illusions, relating to the possibility of settlement for a scientific problem and simultaneously a reasonable one. It can be accepted only in the case of the complete systems that a settlement and sense can coincide with each other. However, one can experience more often the incomplete systems. On the account of this, investigating this type of systems one should consider Gödel's achievements. Still, it should be taken into consideration that within the rich systems, which include arithmetic's, it is possible to construct such sentences that are based on this system and refer to its method legally valid and accepted. These sentences can be neither proved or invalidated. A settlement of the sentences will still take place in the case of enriching a language of the system with the variables of the upper level for logical types. Gödel's achievement reveals a perfection of the modern logics and the results of the research from the field of semantics in the development of language for the system which unifies our knowledge. Following Zawirski, we reach a learning situation when the traditional epistemology must revive through the consideration of research from the semantics and metalogics *The Importance of Logical and Semantical Investigations for the Theories of Contemporary Physics* [21, pp. 25–30]. A similar modernisation should be applied to the traditional metaphysics, which is to form a tighter link with the achievements of the detailed sciences [19, p. 349].

Here one should pay attention to the scientific achievements of Alfred Tarski and Kazimierz Ajdukiewicz, who attempted to bring the notions of the theory of cognition to the sematic notions. In his work entitled *The Notion of Truth in the Languages of the Deductive Sciences* [16], Alfred Tarski demonstrated a possibility of developing a precise and formally correct definition of the right sentence on the basis of the deductive languages, following a classical perception of the truth. As a result of this cognitive detail it became clear that a scope of the term "true sentence" depends on a language, which is the subject of the particular investigation. One can not discuss the meaning of phrases separately from a language where these phrases occur while Ajdukiewicz pointed at the existence of the untranslable. This situation relates to the closed ad coherent languages. It does not refer to the fact that one of them could be too poor [1]. Ajdukiewicz states that the language of the classical and relativistic physics is the example of the untranslatable languages. Whereas in Zawirski's previous theory of physics analysis, following Reichenbach, he assumed that the statements of the classical physics are the border case of the statements attributed to the modern physics. In the perspective of Ajdukiewicz's achievements, this view was taken as a completely wrong one. In Zawirski's opinion, the revealed untranslatability of languages can be a derivative from a differentiation of logics founded by the respective languages. Systems with such qualities of logics are known and one system cannot be translated into the other. Then, a system should be found which could respond to reality [18, pp. 41–42].

3.1 Final Conclusions

Zawirski focused on the investigations of methodological and ontological assumptions in the attempts of developing the scientific metaphysics In the first period of the scientific work. The latter part, investigating the matter he worked on the issues of methodology and semantics.

Zawirski listed the possibilities of formalising the scientific nature theories in the area of the research on studying the scientific metaphysics, the manners of investigating the assumptions and philosophical implications derived from the contemporary physical theories, a critical analysis of these scientific terms including their role in the scientific progress as well as the use of the methodological principles in the scientific theories. He pointed at the possibility of studying the scientific metaphysics with the use of the axiomatic method and the possibility of philosophical generalisation of the scientific research in order to understand the world. He looked for the justification of the scientific research, listing the methodological problem of induction and deduction, development of the scientific theories and relations of science and philosophy.

Zawirski analysing the possibility of describing the scientific metaphysics achieved many valuable results out of the investigations. He stressed the importance of the ontological research related to time and space. On this account, one cannot omit his important work entitled a *L'évolution de la notion du temps*, where we can find many issues which still occur in the scientific discussions [20]. Namely they include thermodynamics of irreversible processes, the second law of thermodynamics, theory of dissipative processes. It should be underlined that the problems undertaken in *L'évolution de la notion du temps* are discussed by I. Prigogin, I. Stengers (Order out of Chaos, [13]) and Roger Penrose (Cycles of Time, [11]).

Zawirski's research are very important for the description of the scientific metaphysics as they relate to the implementation of the formal logics into the natural analyses of the scientific theories. The problems were developed particularly by his student Roman Suszko (1918–1979), who developed the theory of models and implemented it for the investigations of problems outside mathematics (Remarks on Sentential Logics—1958).

A possibility of defying the scientific metaphysics resulted also in the critical and historical analyses of the scientific terms and philosophical reflection over the assumptions of empirical theories [5, 6, 9, 14]. It also led into the investigation of traditionally philosophical problems entangled into natural and mathematical theories. It can be noticed that the Cracow background to which Zawirski was linked in the middle of the twentieth century, continues the research of the notions previously taken by him from the field of natural philosophy, natural history and philosophy of nature. A special attention should be paid to the works of such philosophers as Józef Życiński (1948–2011) or Michał Heller (born 1936), a director and founder of *The Copernicus Centre for Interdisciplinary Studies*.

References

1. Ajdukiewicz, K.: Język i poznanie. Wybór pism z lat 1920–1939 [Language and Cognition. A Selection of Papers from the Period Between 1920 and 1939], vol. I. PWN – Polish Scientific Publishers, Warszawa (1985)
2. Bornstein, B.: Architektonika świata [Architectonic of the World], vols. I–III. Gebethner & Wolff, Warszawa (1934–1936)
3. Heinrich, W.: Filozofia i jej zadania [Philosophy and its tasks]. Kwartalnik Filozoficzny [Philos. Q.] **1**, 1–18 (1922)
4. Heinrich, W.: O zagadnieniach podstawowych filozofji [On the basic philosophical terms]. Kwartalnik Filozoficzny **2**, 269–302 (1924)

5. Heinrich, W.: Teorya poznania [Theory of knowledge]. In: Maczka, J., Janusz, R., Senderecka, M. (eds.). Biblos, Kraków-Tarnów (2005)
6. Kotarbiński, T.: W sprawie pojęcia prawdy [On the matter of the truth notion]. Przegląd Filozoficzny [Philos. Rev.] **37**, 85–91 (1934)
7. Łukasiewicz, J.: O metodę w filozofji [On method in philosophy]. Przegląd Filozoficzny **31**, 3–4 (1928)
8. Łukasiewicz, J.: Co dała filozofii współczesna logika matematyczna? [What was given to philosophy by the modern mathematical logic?]. Przegląd Filozoficzny **39**, 325 (1936)
9. Mehlberg, H.: La théorie causale du temps. Lwów (1935)
10. Penrose, R.: Shadows of the Mind: A Search for the Missing Science of Consciousness. Oxford University Press, Oxford (1994)
11. Penrose, R.: Cycles of Time. An Extraordynary New View of the Universe. Bodley Head, London (2010)
12. Popper, K.R.: Objective Knowledge. Clarendon Press, Oxford (1972)
13. Prigogine, I., Stengers, I. (eds.): Order Out of Chaos: Man's New Dialogue with Nature. Flamingo, London (1984)
14. Śleziński, K.: Dyskusje nad popperyzmem w filozofii polskiej pierwszej połowy XX wieku [Discussions about Popper's ideas in Polish philosophy in the first part of the 20th century]. Studia z Filozofii Polskiej **1**, 243–259 (2006)
15. Śleziński, K.: Benedykta Bornsteina koncepja naukowej metafizyki i jej znaczenie dla badań współczesnych [Benedykt Bornstein's Concept of Scientific Metaphysics and Its Meaning for Contemporary Research]. Wydawnictwo Uniwersytetu Śląskiego, Kraków-Katowice (2009)
16. Tarski, A.: Pojęcie prawdy w językach nauk dedukcyjnych [The notion of the truth in the languages of deductive sciences]. Prace Towarzystwa Naukowego Warszawskiego, Wydział III Nauk Matematyczno-Fizycznych **34**, 1–116 (1933)
17. Zawirski, Z.: Metoda aksjomatyczna a przyrodoznawstwo [The Axiomatic Method and Natural Sciences]. Kwartalnik Filozoficzny, Kraków (1923)
18. Zawirski, Z.: Próby zastosowania logiki wielowartościowej do współczesnego przyrodoznawstwa [Attempts for using many-valued logics in the modern natural sciences]. Sprawozdania Poznańskiego Towarzystwa Przyjaciół Nauk **5**, 41–42 (1931)
19. Zawirski, Z.: W sprawie syntezy naukowej [On the matter of scientific synthesis]. Przegląd Filozoficzny **39**, 347–352 (1936)
20. Zawirski, Z.: Rozwój pojęcia czasu [The Evolution of the Notion of Time]. Uniwersytet Jagielloński, Kraków (1936)
21. Zawirski, Z.: Doniosłość badań logicznych i semantycznych dla teorii fizyki współczesnej [The importance of logical and semantical investigations for the theories of contemporary]. Przegląd Filozoficzny **41**, 25–30 (1938)
22. Zawirski, Z.: Nauka i metafizyka (I) [Science and metaphysics I]. Filozofia Nauki [Philos. Sci.] **3**, 104–135 (1995)
23. Zawirski, Z.: Nauka i metafizyka (II) [Science and metaphysics II]. Filozofia Nauki **4**, 131–143 (1996)
24. Zawirski, Z.: O stosunku metafizyki do nauki [On the Relation Between Metaphysics and Science]. Wydział Filozofii i Socjologii Uniwersytetu Warszawskiego, Warszawa (2003)
25. Znaniecki, F.: Zadania syntezy filozoficznej [Tasks of philosophical synthesis]. Przegląd Filozoficzny **30**, 103–116 (1927)

K. Śleziński (✉)
Institute of Educational Science in Cieszyn, University of Silesia in Katowice, Cieszyn, Poland
e-mail: slez.krzysztof@onet.poczta.pl

Stanisław Leśniewski: Original and Uncompromising Logical Genius

Peter Simons

Leśniewski was notable for the degree of prolixity which he was willing to admit in the interest of complete rigor and precision.
W.V. Quine, *The Journal of Symbolic Logic* 5, (1940), p. 83

Abstract Stanisław Leśniewski was one of the two originators and drivers of the Warsaw School of logic. This article describes his work chronologically, from his early philosophical work in Lvov to his highly original logical systems of protothetic, ontology and mereology. His struggles to overcome logical antinomies, his absolute commitment to logical clarity and precision, and his antipathy towards set theory made his nominalistic approach to logic among the most original of the twentieth century, while his early death and the loss of his papers meant his work was only gradually discovered and appreciated outside Poland.

Keywords Stanisław Leśniewski · Lvov-Warsaw School · Protothetic · Ontology · Mereology · Antinomies

Mathematics Subject Classification (2000) Primary 01A70, 03A05; Secondary 03B05, 03B15, 03B99

1 Life

Stanisław Kazimierz Leśniewski was born in Serpukhov, near Moscow in Russia, on 28 March 1886. His father, Izydor Leśniewski, was an engineer who worked on the construction of the Trans-Siberian Railway. His mother was Helena, née Palczewska. Stanisław was baptised at St. Stanislav's church in St. Petersburg. His mother died when he was young and his father remarried, so Stanisław had several younger step-siblings. He attended classical *Gymnasium* in Irkutsk in Siberia from 1899 to 1903. Between 1904 and 1910 he studied Philosophy and Mathematics at several universities: Leipzig, Zurich, Heidelberg, St. Petersburg, and Munich. In Leipzig, he heard lectures by Wilhelm Wundt, and in Munich by Hans Cornelius, Moritz Geiger and Alexander Pfänder. In 1910, already with a solid philosophical education behind him, he went as a doctoral

Á. Garrido, U. Wybraniec-Skardowska (eds.), *The Lvov-Warsaw School. Past and Present*, Studies in Universal Logic,
https://doi.org/10.1007/978-3-319-65430-0_15

student to Lvov University, then in Austria-Hungary, completing his doctorate under Kazimierz Twardowski in 1912 with a dissertation entitled *Przyczynek do analizy zdań egzystencjalnych* [Contributions to the Analysis of Existential Propositions] [1]. This was also published as an article the previous year in the leading Polish-language philosophy journal *Przegląd Filozoficzny*.

In 1913 he married Zofia Prewysz-Kwintów (1893–1958), from a Lithuanian landowning family. They had no children. Four further publications on the philosophy of logic followed until the outbreak of war, which Leśniewski spent largely in Moscow teaching mathematics at Polish schools. During this period, he developed what he called his theory of collections, which later became known as mereology. After the Bolshevik Revolution in Russia, Leśniewski left Russia for Poland. During the 1919–1921 Polish-Bolshevik War he worked as a codebreaker for the Polish General Staff's Cipher Bureau. He tried unsuccessfully to obtain his habilitation in Lvov, but got it instead in 1918 in the University of Warsaw, where in 1919 he became Extraordinary Professor of the Foundations of Mathematics, a position especially created for him. From then until his final illness Leśniewski lectured regularly on logical and mathematical topics, and built up his logical systems. Leśniewski published nothing between 1916 and 1927. Between 1927 and 1931 he published two major pieces, one a series in Polish on mereology, the other a long article in German on protothetic, his extended propositional logic. Leśniewski was made a full professor in 1936. He contracted thyroid cancer and died on 13 May 1939 at the age of 53 after an unsuccessful operation. He was buried in the historic Old Powązki Cemetery in Warsaw. His papers, entrusted to his student Bolesław Sobociński, included unfinished works on logical antinomies and on many-valued logic. All of these papers were destroyed in Sobociński's apartment during the Warsaw Rising of 1944.

2 Phases of Activity

It is common to divide Leśniewski's work into two phases: an early one (up to 1916) in which he worked informally on questions of the philosophy of logic, and a mature phase (from roughly 1919 to his death) in which he formulated and perfected his formal logical systems. Leśniewski himself gave his imprimatur to the idea that his early work should be clearly separated from his later, since in 1927 he wrote that he formally repudiated all his work before 1914, which he regarded as bankrupt: "philosophical"-grammatical, in his words, with the exception of an argument against Twardowski's conception of general objects, which he still regarded as acceptable.

However, while from the point of view of content there is a clear difference between the work before 1914 and that after, it is something of an oversimplification to divide his activity into just two periods. True, the early papers between 1911 and 1913 are written from the "philosophical"-grammatical point of view that Leśniewski came to despise. Between 1914 and 1916 however, while Leśniewski continued to write in regimented prose, his interests centred around finding a solution to the problem of Russell's Paradox, and mark a transition to his later interests and style. The long period without publication is when Leśniewski was formulating and perfecting his systems. Between 1927 and 1931 there was a burst of publication. Then there was another pause in publishing, when

Leśniewski probably concentrated on writing a large systematic work on antinomies, and on criticising many-valued logics. The year before Leśniewski died, a continuation of the German paper appeared.

So it would be more accurate to say there were five phases: an initial, "philosophical" phase, a transitional phase where Leśniewski was working out how to deal with Russell's Paradox, an intensive developmental phase where his mature systems were formulated and results proved without publication, a concerted publishing phase, with this leading into a second phase of consolidation without publication, the final paper being really an artificially delayed postcript to the earlier burst of publishing. Had Leśniewski survived longer there would doubtless have been another period of publication.

Continuity from 1919 to 1939 was provided both by Leśniewski's efforts to formulate and perfect his systems, and by his unbroken teaching in the University of Warsaw, which centred firmly around the foundations of mathematics and how to approach them properly, being the place where he could try out his ideas and perfect them with the collaboration of students.

3 Early Writings

During his studies in Germany, Leśniewski encountered the philosophy of language of Anton Marty, whose major work *Untersuchungen zur Grundlegung der allgemeinen Grammatik und Sprachphilosophie* (1908), he contemplated translating into Polish. Legend has it that the project foundered after he was unable to find an adequate translation for the second word of the title. It is likely he also encountered Husserl's *Logische Untersuchungen* (1900/1901), and again thought of translating this (somewhat later: 1915/1916), again without outcome. What he later called this "Austrian" approach influenced his early work. This concentrated not on grammar or philosophy of language *per se* but on aspects of the philosophy of logic. The dissertation [1] is dedicated to showing, using notions of denotation and connotation derived from Mill, that existential statements of the form *S exists* are tautological—and false. This barely believable thesis is slightly softened by his claim that the seemingly equivalent sentence *Some object is S* is both non-tautological and may be true. This kind of fine linguistic differentiation was to become the hallmark of his work.

Other writings of this kind from this period were in good part reactions to work by others, but always with an original slant. The paper 'Próba dowodu ontologicznej zasady sprzeczności' [Attempt at a Proof of the Ontological Principle of Contradiction] (1913) [2] is a reaction to aspects of Jan Łukasiewicz's 1910 monograph *O zasadzie sprzeczności u Arystotelesa* [*On the Principle of Contradiction in Aristotle*], a work that was to have far-reaching consequences for Leśniewski's own career. The article 'Czy prawda jest tylko wieczna czy też wieczna i odwieczna?' [Is truth only eternal or both eternal and from eternity?] (1913) [3] criticises an argument by Leśniewski's friend and fellow-student Tadeusz Kotarbiński that statements about future contingents are not true in advance of the outcome, a view later taken up by Łukasiewicz as the motivation for his invention of many-valued logic. Leśniewski upholds the view that statements about future contingents are true or false in advance of the event, and his argument persuaded Kotarbiński to change his mind. The most substantial of Leśniewski's pre-war pieces was 'Krytyka logicznej zasady wyłączonego środk[a]' [Critique of the Logical Principle of Excluded Middle] (1913) [4], again taking on ideas of Łukasiewicz and Twardowski. It contains, among other things, an argument against Twardowski's conception of a general object. Suppose a general object—say the general tomato—is defined as the object having all and only the properties common to all tomatoes. Suppose one tomato weighs 150 g and another tomato weighs 100 g. Then the general tomato has neither the property of weighing 150 g nor the property of not weighing 150 g. Thus, the only way a general object can be non-contradictory is if it is the sole object of its kind, which renders the notion useless. This is the only argument of Leśniewski's early work that he did not "repudiate" in 1927. Another part of the paper discusses the Liar Paradox or Epimenides, the first indication of Leśniewski's long battle against antinomies.

In general, Leśniewski's pre-war papers are characterized by extreme attention to language and how he intends certain words and sentences to be understood, which is not always in the same way as others, and a steely resolve to follow the argument wherever it leads, even if the resulting theses are sometimes counterintuitive. At the same time, his verbal commentary is often colourful and he makes abundant use of scare-quotes. These characteristics were to remain with him throughout his writing career.

4 Antinomies, Classes, Parts

In Łukasiewicz's 1910 monograph on the principle of contradiction in Aristotle, there is an appendix discussing the treatment of contradiction in symbolic logic, employing the notation of Louis Couturat. There is also a discussion of Russell's Paradox of the class of classes which are not members of themselves. Leśniewski read the book in 1911 and was so struck by trying to solve the paradox, which he initially thought was just a trivial puzzle that could be solved in a couple of hours, that he is said to have missed a railway connection from Poland to Russia and endured a long wait. In fact he worked on the problem for 11 years, and never lost his fascination with antinomies. His first published foray into the field was his 1914 paper 'Czy klasa klas, nie podporządkowanych sobie, jest podporządkowana sobie?' [Is the class of classes not subordinate to themselves subordinate to itself?] [5]. The idiosyncratic formulation 'subordinate to itself' rather than 'member of itself' is taken from Łukasiewicz's discussion. While it makes some relatively unimportant assumptions that Leśniewski would later have criticised, the main thrust of the article is that there is no antinomy, because every class (every collection having at least one member) is a member of itself. The reason for this is that Leśniewski understands 'member of *A*' in effect to mean 'part (or whole) of *A*'.

This "solution" was expounded in much greater detail in a forty-page monograph, *Podstawy ogólnej teoryi mnogości I* [Foundations of the General Theory of Collections I] (1916: no further parts appeared) [6] dedicated by Leśniewski to his wife. Consider a line segment *AB*, and suppose it is composed of subsegments *AC*, *CD*, *DE*, *EB*, and that the several segments *AC*, *CD*, *DE* and *EB* are the objects *m*. Then the class of (all the) *m* just is the whole object *AB*, an object that we would now call the mereological sum of the objects *m*. As is clear from Leśniewski's explanations of his terminology, he considered this to be the only sensible way to understand the notions of 'class of *m*' (the sum of all the *m*), 'set of *m*' (a sum of some but not necessarily all the *m*), and 'element of *m*' and 'subset of *m*' (a proper or improper part of the class of the *m*). The monograph declares this to give the true understanding of 'set' (*Menge*) in Cantor, and it is in effect a bid to take over the terminology of the then exploding discipline of set theory and turn it in Leśniewski's direction. That his understanding is clearly *not* that of Cantor and other set theorists is shown by the fact that there are four objects *m* but many more (including *AB* and all its parts) that are elements of *m*, and that a single object such as *AB* can be the class of many distinct groups of objects *n*, even such as have none in common with the *m*, all of which runs counter to Cantor, and standard set theory.

The basic notion of the monograph is that of 'part' (*część*), meaning a proper part, not identical with the whole—Leśniewski's term for a proper-or-improper part is here 'ingredient' (*ingredyens*), but he later preferred the term 'element' for this notion (they are defined differently but are equivalent in the monograph). The development is based on four axioms and three definitions, those for ingredient, set and class. The axioms say that parthood is asymmetric and transitive, and that if there is at least object *m* then there is exactly one class of *m*. Leśniewski later criticised this theory for including definitions at the basic level, and would afterwards always attempt to base his theories on a single notion governed by axioms, before introducing definitions and theorems. On the other hand, the development in the monograph is extremely clearly done and there is very little to criticise from Leśniewski's later formal point of view. The theorems are stated and proved in ordinary Polish augmented with variables, and employing a very restricted vocabulary. They were thus very easy to formalize later.

Leśniewski's concretistic, mereological understanding of the ideas of set theory was destined not to catch on, and led in the 1920s and 1930s to increasingly critical and ultimately personally bitter disagreements with the proponents of what Leśniewski ironically called "official" set theory. To distinguish his views from theirs, in 1927 he renamed his theory 'mereology' [*mereologia*], from the Greek *meros*, part. That is the name that has become standard for formal part-whole theory ever since, and the 1916 monograph was the first formal theory of part and whole to appear in print.

5 Formalization

Upon his appointment to the professorship in Warsaw in 1919, Leśniewski embarked on a 20-year teaching career, the bulk of which was concerned with various aspects of the foundations of mathematics. It was Leśniewski's aim and ambition to provide a rigorous and antinomy-free foundation for mathematics, avoiding the inconsistency of Frege's logical foundation in *Grundgesetze der Mathematik* (1893/1903), the inexactness of Whitehead and Russell's foundation in *Principia Mathematica* (1910–1913), as well as the—to Leśniewski—absurd and counterintuitive set theory developed by Zermelo and others. This was now his life's project.

The initial hurdle to this was the example of Whitehead and Russell, who were notoriously careless about use and mention in the prose passages of *Principia*, a fact which prevented the extremely literal and unsympathetic reader Leśniewski from even understanding their work, and made him shy away from symbolic logic for several years, on the assumption that symbolization itself was problematic. Two factors changed his mind. One was his admiration for the metalogical care and clarity of Frege, notwithstanding the inconsistency of his system, and the other was personal persuasion by Leon Chwistek in 1920.

Leśniewski then set about putting his system into a formal and symbolic guise, starting with the mereology from 1916, which he had continued augmenting and improving in the intervening years, without publication. The way in which Leśniewski came to formalization from already precisely formulated vernacular formulations was somewhat unusual, and explains his attitude to formal logic. It is nowadays generally assumed, in the vein of Hilbert and formalism, that a symbolic system consists of a collection of symbols assembled into formulas by syntactic rules and conventions, given a proof theory by inferences rules and axioms, and endowed with meaning by a formal semantics, often though not always drawing on set theory. In the early twentieth century, several prominent logicians, notable Frege, Whitehead, Russell and Brouwer, took a different view, assuming that the symbols they employed in their logic had a determinate, intuitive meaning, and that axioms and rules were teasing out the logical effects of these intuitive meanings. Leśniewski's sympathies lay with the latter group, since the symbols he started employing from 1920 already inherited their meaning from his prior vernacular formulations and did not need to have it conferred from outside. The point of the axioms and rules was to capture as adequately as possible for logical purposes what these meanings were. This is why, in later writing, Leśniewski ironically describes himself as both a formalist and an intuitionist.

Having dealt with mereology, Leśniewski next turned to the logical apparatus underlying it, in the first place, the logic of names and predicates. Unlike Frege and Russell, Leśniewski was quite happy to allow names to denote more than one individual, following in this regard rather Aristotle, the tradition, and the algebraic logic of Ernst Schröder. He also accepted that a name could be empty, that is, fail to denote anything. The basic notion that he identified as requiring axiomatic determination was that of singular inclusion, in the form 'A is (a) b', for example, 'Socrates is a man'. In Polish, which lacks articles, this is *Sokrates jest człowiekiem*, so logical attention focusses on the single word 'jest'. He then collected some logical truths he considered it to govern, for example 'If A is b, then A is A', and 'If A is B, and B is c, then A is c'. The use of lower and upper case variables is not logically significant, but was merely an informal convention he adopted, taken over from his early writings, that in a subformula which could only be true if the subject was singular, he would use a capital letter for the subject term. For a symbol, Leśniewski took over the Greek lower-case epsilon, 'ε', used by Peano and taken from the Greek *esti*, [is]. In 1920 he managed to come up with a single axiom governing 'is' or 'ε'. Following ideas from Russell's theory of descriptions, according to which '(The) A is a b' means 'There is at least one A, and there is not more than one A, and every A is a b', this was the axiom

> For all A and a: $A\varepsilon a$ if and only if: for some B, $B\varepsilon A$, and for all B and C, if $B\varepsilon A$ and $C\varepsilon A$, then $B\varepsilon C$, and for all B, if $B\varepsilon A$ then $B\varepsilon a$.

Because this primitive enabled Leśniewski to define several other expressions involving uses of the verb *to be*, such as those expressing existence, singular identity, inclusion, and general identity, he came to call the resulting logical system *ontology*. The 1920 axiom was subsequently replaced by shorter ones, and so came to be called the 'long' axiom of ontology, but it remains the most intuitively evident, and embodies a kind of self-definition of 'ε'.

It remained for Leśniewski to formalize the logic of propositional connectives and quantifiers presupposed in ontology and thus in mereology. This proved a little tougher. Leśniewski's view on definitions was that they should be expressed as equivalences, using the connective $\leftrightarrow$ of equivalence, rather than as metalogical abbreviations. He thus wished to base his logic of propositions on material equivalence and universal quantification alone, but could not see how to define conjunction. It was his genial young PhD student Alfred Tarski who found the solution, defining conjunction by quantifying propositional functors, as

$$p \wedge q \leftrightarrow \forall f(p \leftrightarrow (f(p) \leftrightarrow f(q)))$$

so that, by 1923, Leśniewski had a formal system of propositions, propositional functors and quantifiers, which he dubbed *protothetic*. The full system of foundations now comprised protothetic, ontology and mereology, in that order of logical precedence.

6 Consolidation and Publication

Through the early and mid-1920s, Leśniewski did not publish his results, preferring to work on improving them and aspiring to publish a systematic treatise in the style of *Principia Mathematica*. But while he worked on this and discussed his results with others, he increasingly found that results obtained by himself or one or other of his interlocutors were being held back for fear of disputes about priority and responsibility for advances—both Leśniewski and several of his colleagues were notoriously sensitive about attributions of priority. To relieve the difficulty, he decided on a different way of getting his ideas into print, which was a quasi-autobiographical sequence of results in the order he and others had found them. The first result of this was in effect a treatise in eleven sections, 'O podstawach matematyki' [On the foundations of mathematics], published in five parts in *Przegląd Filozoficzny* between 1927 and 1931 [7]. It was affectionately dedicated to "My esteemed and beloved Professor of Philosophy, Dr. Kazimierz Twardowski", from "a philosophical apostate, but a grateful pupil." The paper was a biography of mereology, from 1916 onwards. Like the original, it was expressed in regimented Polish, with a somewhat revised vocabulary, and contained detailed comparisons between the various systems and proofs of over 250 theorems. It also contained a comparison with and criticism of Whitehead's theory of events, to which Tarski had drawn Leśniewski's attention. While Whitehead also used part-whole theory, and had probably developed his ideas at about the same time as Leśniewski, his published formulation was flawed, and Leśniewski mercilessly pointed out the defects.

The other major publication of this time was a long 1929 article, 'Grundzüge eines neuen Systems der Grundlagen der Mathematik' [Fundamentals of a New System of the Foundations of Mathematics] [8], published in *Fundamenta Mathematicae*, also in eleven sections. After a further "autobiographical" account of the development of protothetic, this continued with detailed metalogical specifications, but the development was incomplete at the end of Sect. 11 and it was another 9 years before the series was continued.

In this period Leśniewski also published four short articles, two on single-axiom axiomatizations of group theory and Abelian group theory [9, 10], a short sketch of the metalogic of ontology [11], and an account of principles of definition in propositional logic [12].

7 Final Years

From 1931, Leśniewski's publications faltered again, this time in part because he had fallen out with the other editors of *Fundamenta Mathematicae* over their continued practice of promoting set theory through the journal. A chance to continue the 'Grundzüge' article only came towards the end of the decade, when a new journal, *Collecteana Logica*, was founded, and Leśniewski wrote a long introduction to the continuation noting developments in the intervening years [13], and following it with the completely formal twelfth section, comprising no fewer than 422 theses. But the outbreak of war meant that the journal never appeared. A preprint of the article survived in Harvard, perhaps sent to Quine who then wrote a review for the *Journal of Symbolic Logic* in 1940. Leśniewski was somewhat belatedly promoted to full professor in 1936. At the time of his death he was working on a refutation of many-valued logic, as well as a treatise on antinomies. Neither survived the Warsaw Rising, being lost with his papers and correspondence.

8 Metalogic

Because of his antipathy to set theory, which became the medium of choice for logical semantics, Leśniewski never ventured into semantics, believing it unnecessary for systems whose constants are already meaningful. It was also probably in good part because his former student Tarski employed set theory in his famous paper on truth of 1933 that Leśniewski did not accept Tarski's results, though in the early part of the truth paper Tarski paid generous tribute to Leśniewski for the analysis of the Liar Paradox and the way quotation mark names are used. In his ontological views, Leśniewski was anti-platonistic, and only declined to call himself a nominalist because he thought that some mental phenomena such as after-images defied physicalistic explanation. On the other hand, in his practice of logic, Leśniewski was uncompromisingly nominalistic. In his view, a logical system is a concrete, spatiotemporal collection of marks or inscriptions which can be added to over time by new marks constituting definitions and proofs of theorems. This is in complete contrast to platonistic views, according to which a logic is a system of eternal propositions standing in timeless logical relationships.

Because of this very down-to-earth view of logic, Leśniewski was unable to formulate metalogical principles in what has become the standard way, by a recursive specification, but gave detailed schematic directives on how to extend an existing logical system with new theses. Because the theses are not set out in Plato's heaven, the descriptions of what may count as an acceptable continuation need to be self-adjusting, referring back to expressions already introduced in previous theses. The directives for protothetic include those for substitution, detachment, quantifier distribution, and very importantly, for definitions. Definitions, being concocted like the rest of the system as one went along, were regarded by Leśniewski as object-language equivalences introducing new symbols, effectively new axioms, and they could be creative and non-conservative. Formulating the principles governing what could count as a good definition gave Leśniewski much trouble, and he regarded his directives for adding definitions to protothetic and ontology as his finest work. In protothetic there was only one kind of definition, for sentences and functors, while for ontology there were two kinds, one for adding new sentence-forming functors (predicates of first and higher order), and one for adding names and nominal functors of first and higher order. The directives were expressed in a highly precise way employing a complex regimented vocabulary specified by a series of what Leśniewski called terminological explanations, but which were in effect precise stipulations for the meanings of his metalogical terms.

In the early 1920s, Leśniewski employed a version of Whitehead and Russell's type theory, but he was wary of its apparently inflated ontology and soon reformulated his logical grammar as what was subsequently known as a categorial grammar, according to which any expression is either a sentence or a name or a functor, with precise argument input kinds and output kind, each such kind being what Leśniewski somehat misleading called a *semantic category*. His inspiration for this was Husserl's theory of *Bedeutungskategorien*, though unlike Husserl he understood the categories to be meaningful expressions rather than meanings, and the constraints to be syntactical. The combinatory principles for such a grammar were first set out not by Leśniewski but by Kazimierz Ajdukiewicz. Ajdukiewicz was unable to give satisfactory principles for quantifiers or other variable-binding operators. This mirrored a difficulty faced by Leśniewski, who treated the universal quantifier, binding any finite number of variables from any available categories, as itself syncategorematic, and found himself unable to give directives for adding new binding operators, despite offering students any degree they needed if they could come up with a solution. Leśniewski found his theory of semantic categories to be so natural that he considered a logical system should use it even if there were no threatening antinomies. Tarski initially followed Leśniewski's theory of semantic categories but his later readiness to go beyond the system and allow transfinite types or categories was no doubt another contributory factor to their disagreement.

When Quine met Leśniewski in Warsaw in 1933 they disagreed about whether Leśniewski's liberal practice of allowing quantifiers to bind variables of any category committed him to a correspondingly complex ontology of abstract entities: Quine thought it did; Leśniewski maintained it did not.

9 Teaching and Students

Leśniewski taught in Warsaw from 1919 to 1939, mainly on his own theories, occasionally on work by others, such as Cantor, Zermelo, Peano and Łukasiewicz. Some of his courses extended over 2 academic years, and he used them to trial his own ideas. Lejewski reported that his advanced seminars on directives lasted for three semesters. His style was to work from copious notes, writing theses on the board and asking students for assistance in formulating examples and counterexamples. Quine found it easy to follow despite knowing no Polish. By their nature, his classes did not attract a wide following, and Leśniewski would quietly send away unknowing students, who were just making up their hours, with a positive mark in their student book. Once when unexpectedly many students turned up at the beginning of the class he expressed surprise at their numbers and asked whether they had confused him with Bergson. Those who stayed were dedicated. Leśniewski only ever supervised one doctoral student, Alfred Tarski, who obtained his doctorate in 1923 at the age of 22. Leśniewski would proudly say he had 100% geniuses as doctoral students.

In teaching, as in some of his publications, Leśniewski conducted proofs by a system of natural deduction, which was much easier to follow than full axiomatic proofs. Although one of the first logicians to carry out proofs in this way, he regarded such derivations as mere sketches for "proper" proofs, comparing them with a lounge suit worn in preference to formal evening dress with a stiff shirt and collar. He therefore never formulated exact principles for these derivations, and the impetus for doing so, work pioneered by Stanisław Jaśkowski, came not from Leśniewski but from Łukasiewicz.

10 Colleagues and Personality

Leśniewski remained in various forms of contact with his teacher Twardowski and study colleagues from pre-war Lvov, including Jan Łukasiewicz, Kazimerz Ajdukiewicz and Tadeusz Kotarbiński. Obviously Łukasiewicz was a close colleague in Warsaw, while Tarski soon joined his two teachers as the third major logician of the Warsaw School. There were also his mathematical colleagues in Warsaw, notably Wacław Sierpiński and Kazimierz Kuratowski. His relationships with most of them eventually deteriorated, for a variety of reasons. As the obverse of his obsession with exactness, Leśniewski was a fierce and unsympathetic critic of sloppy thinking and his interventions at seminars in Warsaw put people off from going to speak there. His written criticisms of the ideas of gifted mathematicians such as Hausdorff and von Neumann were often blunt to the point of rudeness, and gained him few friends. His political views shifted rightwards over the years: he supported the authoritarian *Sanacja* regime inspired by Józef Piłsudksi, and became increasingly anti-semitic in his sentiments. At the time of his death his only remaining close friend was the patient and forgiving Kotarbiński, who visited him in hospital in his final days and was with him when he died. On the other hand, he inspired fierce loyalty and admiration among the small group of his closest students, and even his critics had to admit he was one of the most gifted and original of the brilliant interwar generation.

11 Legacy

From the 1930s onwards, a number of factors conspired to keep Leśniewski's work out of the logical mainstream. In part it was his own meticulous but often inconvenient logical practices. In part it was his ideological disagreement over set theory and other platonistic mathematical theories, which meant that he did not participate in the semantic revolution in logic initiated by Tarski. Finally, his tragically early death and the wartime destruction of his papers meant that what passed down was a torso of his work. After the Second World War, his students, particularly Słupecki, Sobociński and Lejewski, made efforts to reconstruct his ideas, recalled from lectures and papers, but most logicians, not least the now famous and influential Tarski, regarded his work as old-fashioned and superseded. Isolated aspects of his work were taken up more widely, in particular mereology, rendered in more conventional logical guise by Henry Leonard and Nelson Goodman, but for the most part his work was regarded as a quaint and ultimately unfruitful branch of modern logic, and largely forgotten. It did not help that most of his work was in Polish, and difficult to access.

From the 1980s things improved somewhat. Translations of rescued lecture notes [14] and English [15] and French [16] translations of his articles brought his primary work to a wider audience. Mereology was instated as a crucial tool of metaphysics. A Polish collected edition, *Pisma Zebrane* [17], a biography by Jacek Jadacki and a commentary monograph by Rafał Urbaniak rounded out the published picture. As the historiography of the Lvov-Warsaw School was written, Leśniewski re-emerged not only as one of the chief drivers of that remarkable school, but more widely as one of the most original and inspirational logicians of the twentieth century.

Stanisław Leśniewski's Works Mentioned in the Paper

1. Przyczynek do analizy zdań egzystencjalnych. Przegląd Filozoficzny **14**, 329–345 (1911) (Cf. [Jadacki, J.J. (ed.): Pisma Zebrane. [Collected Writings], 2 vols, pp. 15–31. Semper, Warsaw (2015)]; English translation: A contribution to the analysis of existential propositions. In: [Surma, S.J., Srzednicki, J.T.J., Barnett, J.D., Rickey, V.F. (eds.): Collected Works, 2 vols, pp. 1–19. Kluwer/Polish Scientific Publishers, Dordrecht/Warsaw (1992), with an annotated bibliography to 1978 by V.F. Rickey])
2. Próba dowodu ontologicznej zasady sprzeczności. Przegląd Filozoficzny **15**, 202–226 (1912) (Cf. [Jadacki, J.J. (ed.): Pisma Zebrane. [Collected Writings], 2 vols, pp. 32–56. Semper, Warsaw (2015)]; English translation: An attempt at a proof of the ontological principle of contradiction. In: [Surma, S.J., Srzednicki, J.T.J., Barnett, J.D., Rickey, V.F. (eds.): Collected Works, 2 vols, pp. 20–46. Kluwer/Polish Scientific Publishers, Dordrecht/Warsaw (1992), with an annotated bibliography to 1978 by V.F. Rickey])
3. Czy prawda jest wieczna, czy też wieczna i odwieczna? Nowe Tory **18**, 493–528 (1913) (Cf. [Jadacki, J.J. (ed.): Pisma Zebrane. [Collected Writings], 2 vols, pp. 147–182. Semper, Warsaw (2015)]; English translation: Is all truth only true eternally, or is it also true without a beginning? In: [Surma, S.J., Srzednicki, J.T.J., Barnett, J.D., Rickey, V.F. (eds.): Collected Works, 2 vols, pp. 86–114. Kluwer/Polish Scientific Publishers, Dordrecht/Warsaw (1992), with an annotated bibliography to 1978 by V.F. Rickey])
4. Krytyka logicznej zasady wyłączonego środka. Przegląd Filozoficzny **16**, 315–352 (1913) (Cf. [Jadacki, J.J. (ed.): Pisma Zebrane. [Collected Writings], 2 vols, pp. 183–222. Semper, Warsaw (2015)]; English translation: The critique of the logical principle of the excluded middle. In: [Surma, S.J., Srzednicki, J.T.J., Barnett, J.D., Rickey, V.F. (eds.): Collected Works, 2 vols, pp. 47–85. Kluwer/Polish Scientific Publishers, Dordrecht/Warsaw (1992), with an annotated bibliography to 1978 by V.F. Rickey])
5. Czy klasa klas nie podporządkowanych sobie jest podporządkowana sobie? Przegląd Filozoficzny **17**, 63–75 (1914) (Cf. [Jadacki, J.J. (ed.): Pisma Zebrane. [Collected Writings], 2 vols, pp. 223–235. Semper, Warsaw (2015)]; English translation: Is the classes of classes not subordinated to themselves, subordinated to itself? In: [Surma, S.J., Srzednicki, J.T.J., Barnett, J.D., Rickey, V.F. (eds.): Collected

Works, 2 vols, pp. 115–128. Kluwer/Polish Scientific Publishers, Dordrecht/Warsaw (1992), with an annotated bibliography to 1978 by V.F. Rickey])
6. Podstawy ogólnej teorii mnogości I. Prace Polskiego Koła Naukowego w Moskwie. Sekcya matematyczno-przyrodnicza, No.2. Popławski, Moscow (1916). No further parts appeared (Cf. [Jadacki, J.J. (ed.): Pisma Zebrane. [Collected Writings], 2 vols, pp. 256–294. Semper, Warsaw (2015)]; English translation: Foundations of the general theory of sets. I. In: [Surma, S.J., Srzednicki, J.T.J., Barnett, J.D., Rickey, V.F. (eds.): Collected Works, 2 vols, pp. 129–173. Kluwer/Polish Scientific Publishers, Dordrecht/Warsaw (1992), with an annotated bibliography to 1978 by V.F. Rickey])
7. O podstawach matematyki, I–V. Przegląd Filozoficzny **30**, 164–206 (1927); **31**, 261–291 (1928); **32**, 60–101 (1929); **33**, 77–105 (1930); **34**, 142–170 (1931) (Cf. [Jadacki, J.J. (ed.): Pisma Zebrane. [Collected Writings], 2 vols, pp. 295–468. Semper, Warsaw (2015)]; English translation: On the foundations of mathematics. In: [Surma, S.J., Srzednicki, J.T.J., Barnett, J.D., Rickey, V.F. (eds.): Collected Works, 2 vols, pp. 174–382. Kluwer/Polish Scientific Publishers, Dordrecht/Warsaw (1992), with an annotated bibliography to 1978 by V.F. Rickey]; French translation: [Sur les fondements de la mathématique. Hermès, Paris (1989)]
8. Grundzüge eines neuen System der Grundlagen der Mathematik. Fundamenta Mathematicae **14**, 1–81 (1929) (Cf. [Jadacki, J.J. (ed.): Pisma Zebrane. [Collected Writings], 2 vols, pp. 489–569. Semper, Warsaw (2015)]; English translation: Fundamentals of a new system of the foundations of mathematics. In: [Surma, S.J., Srzednicki, J.T.J., Barnett, J.D., Rickey, V.F. (eds.): Collected Works, 2 vols, pp. 410–488. Kluwer/Polish Scientific Publishers, Dordrecht/Warsaw (1992), with an annotated bibliography to 1978 by V.F. Rickey])
9. Über Funktionen, deren Felder Gruppen mit Rücksicht auf diese Funktionen sind. Fundamenta Mathematicae **13**, 319–332 (1929) (Cf. [Jadacki, J.J. (ed.): Pisma Zebrane. [Collected Writings], 2 vols, pp. 475–488. Semper, Warsaw (2015)]; English translation: On functions whose fields, with respect to these functions are groups. In: [Surma, S.J., Srzednicki, J.T.J., Barnett, J.D., Rickey, V.F. (eds.): Collected Works, 2 vols, pp. 383–398. Kluwer/Polish Scientific Publishers, Dordrecht/Warsaw (1992), with an annotated bibliography to 1978 by V.F. Rickey])
10. Über Funktionen, deren Felder Abelsche Gruppen in bezug auf diese Funktionen sind. Fundamenta Mathematicae **14**, 242–251 (1929) (Cf. [Jadacki, J.J. (ed.): Pisma Zebrane. [Collected Writings], 2 vols, pp. 714–723. Semper, Warsaw (2015)]; English translation: On functions whose fields, with respect to these functions are abelian groups. In: [Surma, S.J., Srzednicki, J.T.J., Barnett, J.D., Rickey, V.F. (eds.): Collected Works, 2 vols, pp. 399–409. Kluwer/Polish Scientific Publishers, Dordrecht/Warsaw (1992), with an annotated bibliography to 1978 by V.F. Rickey])
11. Über die Grundlagen der Ontologie. Sprawozdania z posiedzeń Towarzystwa Naukowego Warszawskiego, Wydział III [Comtes rendus des séances de la Société des Sciences et des Lettres de Varsovie, Classe III] **22**, 111–132 (1930) (Cf. [Jadacki, J.J. (ed.): Pisma Zebrane. [Collected Writings], 2 vols, pp. 724–745. Semper, Warsaw (2015)]; English translation: On the foundations of ontology. In: [Surma, S.J., Srzednicki, J.T.J., Barnett, J.D., Rickey, V.F. (eds.): Collected Works, 2 vols, pp. 606–628. Kluwer/Polish Scientific Publishers, Dordrecht/Warsaw (1992), with an annotated bibliography to 1978 by V.F. Rickey])
12. Über Definitionen in der sogenannten Theorie der Deduktion. Sprawozdania z posiedzeń Towarzystwa Naukowego Warszawskiego **23**, 289–309 (1930) (Cf. [Jadacki, J.J. (ed.): Pisma Zebrane. [Collected Writings], 2 vols, pp. 746–766. Semper, Warsaw (2015)]; English translation: On definitions in the so-called theory of deduction. In: McCall, S. (ed.) Polish Logic, 1920–1939, pp. 170–187. Clarendon, Oxford (1967). Cf. [Surma, S.J., Srzednicki, J.T.J., Barnett, J.D., Rickey, V.F. (eds.): Collected Works, 2 vols, pp. 629–648. Kluwer/Polish Scientific Publishers, Dordrecht/Warsaw (1992), with an annotated bibliography to 1978 by V.F. Rickey])
13. Einleitende Bemerkungen zur Fortsetzung meiner Mitteilung u. d. T. 'Grundzüge eines neuen System der Grundlagen der Mathematik' [Introductory remarks to the continuation of my article 'Grundzüge eines neuen Systems der Grundlagen der Mathematik']. Collectanea Logica **1**, 1–60 (1938) (Cf. [Jadacki, J.J. (ed.): Pisma Zebrane. [Collected Writings], 2 vols, pp. 570–629. Semper, Warsaw (2015)]; English translation: Introductory remarks to the continuation of my article 'Grundzüge eines neuen Systems der Grundlagen der Mathematik'. In: McCall, S. (ed.) Polish Logic, 1920–1939, pp. 116–169. Clarendon, Oxford (1967). Cf. [Surma, S.J., Srzednicki, J.T.J., Barnett, J.D., Rickey, V.F. (eds.): Collected Works, 2 vols, pp. 649–710. Kluwer/Polish Scientific Publishers, Dordrecht/Warsaw (1992), with an annotated bibliography to 1978 by V.F. Rickey])

14. Grundzüge eines neuen System der Grundlagen der Mathematik (Fortsetzung). Collectanea Logica **1**, 61–144 (Cf. [Jadacki, J.J. (ed.): Pisma Zebrane. [Collected Writings], 2 vols, pp. 630–713. Semper, Warsaw (2015)]; English translation: Fundamentals of a new system of the foundations of mathematics. In: [Surma, S.J., Srzednicki, J.T.J., Barnett, J.D., Rickey, V.F. (eds.): Collected Works, 2 vols, pp. 489–605. Kluwer/Polish Scientific Publishers, Dordrecht/Warsaw (1992), with an annotated bibliography to 1978 by V.F. Rickey])
15. Srzednicki, J.T.J., Stachniak, Z. (eds.): S. Lesniewski's Lecture Notes in Logic. Kluwer, Dordrecht (1988)
16. Surma, S.J., Srzednicki, J.T.J., Barnett, J.D., Rickey, V.F. (eds.): Collected Works, 2 vols. Kluwer/Polish Scientific Publishers, Dordrecht/Warsaw (1992), with an annotated bibliography to 1978 by V.F. Rickey
17. Sur les fondements de la mathématique. Hermès, Paris (1989). Translation of [O podstawach matematyki, I–V. Przegląd Filozoficzny **30**, 164–206 (1927); **31**, 261–291 (1928); **32**, 60–101 (1929); **33**, 77–105 (1930); **34**, 142–170 (1931)] by G. Kalinowski
18. Jadacki, J.J. (ed.): Pisma Zebrane. [Collected Writings], 2 vols. Semper, Warsaw (2015)

P. Simons (✉)
Department of Philosophy, Trinity College Dublin, Dublin 2, Ireland
e-mail: psimons@tcd.ie

Izydora Dąmbska: The First Lady of the Twentieth-Century Polish Philosophy

Anna Brożek and Jacek Jadacki

The main thing is to remain in harmony with ourselves and with the truth, not caring about the rest.
Letter to Maria Obercowa. January 19, 1951 (Dąmbska [5])

Abstract Izydora Dąmbska was one of the most creative women-representatives of the Lvov-Warsaw School. The paper presents her extraordinary personality, life, as well as the list of her main works and the greatest scientific achievements. The main area of her interest was the logic of natural language, methodology and the history of Greek semiotics. She gave meticulous analysis of the relation between conventionalism on one hand, and relativism, scepticism and agnosticism on the other. In semiotics, she proposed new approach to the problem of empty names and material implication, of the correct definition of truth, as well as of pragmatic functions of silence and namelessness. In methodology, her reconstruction of the notion of scientific laws is of great importance.

Keywords Conventionalism · Izydora Dąmbska · Lvov-Warsaw School · Methodology · Pragmatics · Scientific laws · Semiotics

Mathematics Subject Classification (2000) Primary 01A70, 03B42; Secondary 03–03

1 Personality

She was an aristocrat by birth as well as an aristocrat of the spirit and intellect. During World War 2—a soldier of the Polish Underground Army; after the War, she adamantly defended the highest human values and did not enter into any compromise with the communist regime which was ruling Poland. Being the last assistant of Kazimierz Twardowski, the founder of the Lvov-Warsaw School, at the same time, she was one of prominent representatives of the School.

Her philosophical interests were very vast, and her philosophical oeuvre was very rich and creative. This oeuvre included long monographs as well as small—but deep—analytical miniatures.

Á. Garrido, U. Wybraniec-Skardowska (eds.), *The Lvov-Warsaw School. Past and Present*, Studies in Universal Logic,
https://doi.org/10.1007/978-3-319-65430-0_16

Concepts of sign, truth and understanding were the main subjects of her research in semiotics. In ontology, she analyzed both historical theses (irrationalism, conventionalism, relativism, skepticism and agnosticism, determinism and indeterminism) and purely theoretical problems (like forms and the value of instrumental cognition). In methodology, she was occupied by the question of the logical status of scientific laws and of the nature of reasoning by analogy. In light of her origin and attitude to life, it is not surprising that she devoted a lot of her attention to the axiological problem of freedom.

Her work in theoretical philosophy was underpinned by a thorough knowledge of the history of philosophy—especially ancient Greek philosophy.

Dąmbska's method of philosophizing reflected trends prevailing in the Lvov-Warsaw School: she never limited the domain of analysed problems in advance; she was very careful in expressing her opinions and very critical, especially with respect to her own writings.

She accepted—together with the other representatives of the Lvov-Warsaw School—the old principle: *initium doctrinae sit consideratio nominis*. She accepted this in theory—and successfully applied it in practice.

2 Life

She was born on January 3, 1904, in Lvov—and died on June 18, 1983, in Cracow.

Dąmbska studied philosophy at the University of Lvov under Twardowski's guidance (1922–1927); then she was his assistant (1926–1930). Dąmbska complemented her studies in Austria, Germany and France (1930–1931). During the German and later the Russian occupation of Lvov, she was a lecturer of the secret Polish university. After World War 2, threatened with arrest by the Soviet secret service, she moved to Gdańsk. In 1946, Dąmbska received her habilitation at the University of Warsaw, presenting the dissertation *Irrationalism and the scientific cognition* [4]. From the years 1950–1956, and again from 1964 on, she was removed from the university by the communist regime due to political reasons. But she continuously—up to her death—led her *privatissimum* in Cracow.

She maintained close contacts with outstanding representatives of the Lvov-Warsaw School (especially with Władysław Witwicki and Tadeusz Czeżowski) but also with great philosophers from outside of the School, like Henryk Elzenberg and Roman Ingarden—as well as with a great Polish poet, Zbigniew Herbert.

3 Work

The work of Dąmbska contains many original publications, translations of philosophical texts and a huge number of notes on Polish publications in the *Bibliographie de la Philosophie*.

The most important of her books are: *La théorie du jugement de M. Edmond Goblot* [1], *On laws in science* [2], *Outline of the history of Greek philosophy* [3], *Irrationalism and the scientific cognition* [4], *French skepticism of the 16th and 17th centuries* [6], *Two studies from the theory of scientific cognition* [7], *Instruments and object of cognition: from the theory of instrumental cognition. On linguistic philosophy* [8], *Two studies on Plato* [9], *Conventions and conventionalism* [10].

Her selected writings are collected in her *Znaki i mysli. Wybór pism z semiotyki, teorii nauki i historii filozofii.* (*Signs and thoughts. Selected writings in semiotics, epistemology and the history of philosophy*) [11]. Recently, a volume of English translations of her important work was published under the title *Knowledge, language and silence* [14].

4 The Lvov-Warsaw School Versus Neopositivism and Linguistic Philosophy

Dąmbska's philosophical research, which may be called "semiotic-logical analysis", is carried out in the spirit of Twardowski's school. She characterized the scholar spirit of this school in such a way:

> Despite the fact that representatives of the Lvov-Warsaw School understood the scope of philosophy differently, they however agreed to postulate that research should fulfil some definite conditions characteristic for scientific cognition. The most important postulate, consequently realized by members of the School in their works, ordered the application in philosophical research of the method of semantic analysis and logical discourse by appreciating the role of broadly understood intuition in the process of discovering statements; the postulate of clarity, precision and logical correctness in formulating issues, theses and arguments and in defining concepts; finely — the postulate of criticism and antidogmatism in estimation of theoretical assumptions [13, p. 29].

Dąmbska's attitude towards the programs of other currents of analytical movement—like neopositivism and linguistic philosophy—was ambivalent.

She accepted some views of the representatives of the Vienna Circle and the Oxford School, but some their views were rejected by her—like the Carnapian opinion that a system of knowledge is finally reducible to "what is empirically given" or the radical opinion of post-Wittgensteinian philosophers that semiotic analysis is the only method of scientific research in philosophy and, in consequence, that all the main questions of philosophy are incorrectly posed, i.e. are pseudoproblems.

It seems that what unites all these currents of analytic movement is anti-irrationalism. That is why Dąmbska paid considerable attention to the phenomenon of irrationalism.

According to Dąmbska, we should distinguish four main versions of irrationalism: metaphysical, epistemological, logical and psychological.

Metaphysical irrationalism is an opinion that irrationality is an essential property of reality itself; in consequence, using a rational conceptual apparatus to describe such a reality is a kind of deformation; paradoxically, the rational attitude to the irrational reality consists in ... delighted silence. Epistemological irrationalism consists in accepting irrational cognitive methods—like intuition, contemplation, empathy etc.—which are to guarantee the scientific status of logically irrational sentences. It is interesting to note that epistemological irrationalists consider these methods reliable—in opposition to traditional rational methods, i.e. experience and reasoning based on it. In turn, it is considered logically irrational for users of a given language. For all the sentences of this language are either contradictory to laws known to these users, or essentially irresolvable; it is clear that when somebody does not know a certain law, his acceptance of sentences contradictory in fact to this unknown law could not be irrational. Finally, a person who believes in logically irrational sentences—or is ready to use irrational methods of cognition—is psychologically irrational.

According to Dąmbska, there is room for neither epistemological, logical, metaphysical nor psychological irrationalism in science. Science should be rational, i.e. it should use only rational or intersubjective methods of research.

5 Metaphysics

5.1 Ontology

Among ontological questions, the controversies concerning determinism–indeterminism and causalism-acausalism were of special importance for Dąmbska.

In these controversies, she was in favor of determinism and causalism. Accepting the first, she appealed to the fact, that practicing science is rational only on the ground of deterministic hypothesis. Accepting the second, she pointed out that the rule of limited measurability (of the speed and location of physical bodies), supposed in quantum physics, does not speak against the hypothesis of causalism—as some philosophers think; the source of their mistake is confusion of indeterminism with indeterminacy, and the principle of causality with the rule of predictability.

5.2 Epistemology

In the domain of epistemology, two problems especially occupied Dąmbska: the problem of instrumental cognition and the problem of the relation between conventionalism on one hand, and relativism, scepticism and agnosticism on the other.

The problem which arises in face of instrumental cognition, consists in the question of how instruments of such a cognition or—as Dąmbska named them—"cognitive operators" impact the other four elements of cognitive situation, i.e.: the subject of cognition, the process of cognition, the object of cognition and the result of cognition (*scil.* picture of cognized object). As it turns out, instruments affects all these factors, modifying them, and vice versa. Traditional epistemology was not conscious of these complicated interrelations.

In Dąmbska's day, there was a widespread belief that conventionalism was dangerous because of its relativistic, sceptic and agnostic consequences. In light of Dąmbska's meticulous analysis, this belief was based on terminological misunderstandings. Let us note that we have two versions of conventionalism: extreme and moderate. Extreme conventionalism is the view that all scientific laws are conventions (in particular: they are arbitrary definitions) and as such they are not empirically verifiable. Moderate conventionalism simply states that conventions are present in various areas of human life and in various domains of culture. According to Dąmbska, extreme conventionalism is obviously false, whereas moderate conventionalism is obviously true, but both of them imply neither relativism and scepticism nor agnosticism. For instance: relativism assumes that reality itself is contradictory—and conventionalism assumes only that there are different, sometimes contradictory, pictures of reality; scepticism assumes that norms and definitions are accepted purely conventionally—and conventionalism assumes at most that they are accepted with respect to their semantic function in the system of a given theory; agnosticism (considered by Dąmbska to be a radicalized version of scepticism) assumes that the classical conception of truth is binding—and conventionalism accept the classical conception of truth.

6 Semiotics

6.1 Logical and Grammatical Categories

Dąmbska conducted deep criticism of the traditional (grammatical) categorizations of the parts of speech, showing that in fact, it had serious shortcomings.

Here are two examples.

Firstly, she rejected the view that so-called empty names (e.g. "à parca") feature a lack of designata, while proper names (e.g. "Casimir') feature a lack of connotation. According to Dąmbska, all names signify something: possible objects of thought namely; only such an approach is compatible with the common conviction that some sentences containing empty names are true (like e.g. "Morta is a parca"), whereas some of them are false (like e.g. "Morta is a siren"). On the other hand, all names have connotation, proper names included, anyway taken contextually, when their connotation is identical with the connotation of an appropriate description of the individual signified by the proper name (whereas proper names taken acontextually remain variables).

Secondly, she stood in opposition to the idea that logical implications can be considered adequate interpretations of conditional propositions of natural language of the type "If p, then q"; according to Dąmbska the meaning of natural conditionals is the judgement that between what is stated in the antecedent 'p' and what is stated in the consequent 'q', there is a relation of sufficient conditioning. She also proposed a certain modification of Twardowski's conception of the meaning of the conditional proposition in *casus realis* of the type "If it is the case that p, then q". Twardowski identified this meaning with the meaning of the complex of three judgments: (a) a judgement stating the conditioning between 'p' and 'q'; (b) a judgement stating the occurrence of the condition and (c) a judgement stating the occurrence of the conditioned thing. According to Dąmbska, there is only (a) because only (a) is explicitly expressed in this kind of proposition. The remaining two components mentioned by Twardowski—namely (b) and (c)—are what is expressed in this kind of proposition only implicitly.

6.2 Truth Relativized to Language

Dąmbska noted that definitions of truth are rooted in conceptions of language for which the predicate "is true" is defined. She analysed in this respect three conceptions of language which were in circulation in the twentieth century: the correspondent (Alfred Tarski, Rudolf Carnap), the operational (late Ludwig Wittgenstein) and the immanent (Kazimierz Ajdukiewicz). In the first conception, language is treated as a system of signs which refer to a certain objective domain. In the second conception, language is a form of biological and cultural behaviour of a person. In the third conception—language is considered to be a set of signs and directives of creating signs and transforming one sign into another.

Dąmbska shows that definitions of "truth" given in the frame of these conceptions are relativized to them. The classical definition of truth harmonizes with the correspondence conception of language (accepted, by the way, by Dąmbska). She formulates it as follows: An affirmative sentence is true when the state of affairs corresponding to this sentence occurs; a negative sentence is true when a state of affairs corresponding to it does not occur. On the basis of the operational conception, a pragmatic definition of truth is natural. Immanent conception is a basis of the syntactic definition.

According to Dąmbska—all these concepts are derivative with respect to the concept of truthfulness as something that is a feature of judgements in the logical sense (or the logical content of sentences) and what is correlated with some ontical categories—first of all with the concept of existence.

6.3 Understanding

The expression "to understand" occurs, i.a., in contexts such as "*A* understands that *p*" and "*A* understands *X*". The context of the second type has several meanings—to understand *X* is the same as: to know what *X* means, to know what *X* expresses, to know what *X*'s structure is, and in the end—to know what idea is realized by *X*. Dąmbska was convinced that, in all these cases, understanding: (a) concerns objects connected with man's spiritual life; (b) consists in becoming aware of relations which indicate the meanings of these objects, but (c) this becoming aware is repeatable.

Dąmbska was convinced that conditions (a)–(c) may be considered as essential conditions of understanding. With respect to condition (b)—understanding is a fallible cognitive act.

6.4 Silence and Namelessness

Dąmbska's analysis of the semiotic functions of silence has a multidimensional character.

Silence is either a simple lack of speech (not-speaking) or refraining from speaking (signitive silence). Signitive silence analysed as a mark is either a symptom or a signal. Considered as a communicative element of natural language, it is—leaving aside expressive functions—a kind of indexical expression. Besides semantic functions, it performs pragmatic ones, in particular, it is a means of a fight or a way of striving for perfection.

In her semiotic-psychological-cultural research on the concept of namelessness, Dąmbska starts from the ascertainment that, on the one hand, we hold our name in high esteem and, on the other hand, sometimes we pretend to become nameless. Getting rid of a name, changing or hiding it (namelessness) are not indifferent from the psychological point of view (among motives of namelessness there are fear and a need of play) as well as from a sociological point of view (namelessness in action is, first of all, a way of fighting).

7 Methodology

7.1 Justification

To justify a sentence '*p*' is to show that if fulfils sufficient conditions to accept (state, know, suppose, expect) that *p*.

Justification may by direct or indirect (*scil.* by reasoning).

7.1.1 Direct Justification

A sentence is justified directly if we accept it on the basis of our own experiences. Scientific claims should be intersubjectively justified. But, how could we know that the experience of someone else is the same or at least similar to ours—in the face of the same facts?

According to Dąmbska we do not have to know this in order to do science. A hypothesis that other people's experiences are the same or similar to our experience is not—contrary to appearances—a premise of scientific claims, but at the most a metascientific hypothesis. The belief in the similarity of the content of human perception has a similar status as the belief that scientists as such are not liars.

Moreover, if this hypothesis said that scientific theses are theses about intersubjective objects, these objects would not be impressions as such but some relations between impressions. These relations can be intersubjectively cognized. This is certified by the fact that normal people in similar circumstances usually accept the same sentences.

7.1.2 Indirect Justification

Among types of indirect justification, Dąmbska was especially interested in reasoning by analogy.

Dąmbska understood analogy to be "a structural similarity of some sets or systems, i.e. a similarity of relations holding between elements of these sets of parts of these systems, and between the properties determined by these relations".

Reasoning by analogy may take one of two forms:

(a) $[(A : B) :: (C : D)] \Rightarrow [(B : A) :: (D : C)]$.
This is a scheme of, for instance, the following reasoning: If God is for people what a father is for children, then people are for God what children are for a father. Here, the conclusion follows from the premise. This reasoning is deducible (and infallible) and applied for instance in mathematics.

(b) $[(A/B : C/D)) \wedge F(C/D)] \Rightarrow F(A/B)$.
This is a scheme of, for instance, the following reasoning: If God is for people what a father is for children, and a father is the children's just judge, then God is a just judge for people. Here, the conclusion often does not follow from the premises and the premises do not follow from conclusions.

The existence of reasoning of type (b) certifies the fact that the traditional division of processes of reasoning into deductive (in which a conclusion follows from premises) and reductive (in which premises follows from the conclusion) is inadequate.

Reasoning by analogy was considered by Dąmbska to be binding (and *eo ipso* infallible), when the analogy occurring in them is essential, i.e. when the relation "fulfils the conditions which satisfy the same rule or the same law".

Reasoning by analogy often has an insightful character and may serve as a justification of sentences about the future. The last point is based on the assumption of the isomorphism or homomorphism of future events with respect to already given ones.

Appreciating the cognitive value of reasoning by analogy and noting its insightful character, Dąmbska at the same time emphasized the danger connected with making use of analogies in which one element may not be cognized by the rational method in science. Such analogies and reasoning based on them are—according to Dąmbska—irrational.

7.2 *Scientific Laws*

There is no science without laws.

Dąmbska considered scientific laws to be general implications: (a) in which ranges of variables are open classes (infinite or such that we may not decide whether they are finite); (b) concerning a constant connection between phenomena; (c) without any absolute time determination; (d) being an element of a certain science, (e) empirically verified.

Such a concept of law has applications in axiomatized formal disciplines in which theses are tautologies. Here, axioms and some of their consequences—namely those which are of a special importance, for instance, those which are applied in the reasoning of other disciplines or in daily life, or which can simplify these kinds of reasoning—are called "laws". Criteria of this importance are too vague to distinguish laws so understood from a set of tautologies.

This concept has—according to Dąmbska—applications in both the natural and human sciences, especially in history. The peculiarity of history does not consist in the fact that it is an idiographic science (or that it describes individual facts) in opposition to natural sciences which are nomological (or which formulate laws). History differs from natural sciences with respect to the level of complication of the examined facts, which are relatively simple in the case of natural sciences but are very complicated combinations of many phenomena—physical, psychical and sociological ones—in the case of history.

7.3 *Truthfulness of Scientific Laws*

Some scientists claim that truthfulness is neither a necessary nor sufficient condition of scientific claims—in particular regarding physical laws. According to some of them—this is because these laws are in fact arbitrary definitions or their analytical consequences. According to others, laws are only provisory hypotheses, relativized to the changing state of knowledge; if they are approximate hypotheses—they are simply false.

Dąmbska refuted this point of view. Laws of physics would not have any logical value if they could be interpreted as functions whose degree of approximation is undefined (the range of the unknown may be bigger or smaller); but usually in such laws it is (at least provisionally) defined in certain orders. They are not, in Dąmbska's opinion, propositional functions. Even if in physics there are laws in which both antecedent and consequent are false, we may only conclude that such laws are not verifiable. Unverifiability is not the same as falsity.

According to Dąmbska—truthfulness is not a necessary condition of being a law. This is of course classically understood truthfulness—not verifiability (since false sentences may also be verified).

Another thing is that we sometimes simply do not know whether a given law is true. We are inclined to believe in those sentences which are probable.

8 Axiology

In the domain of axiology—Dąmbska accepted absolutism and objectivism.

As opposed to the axiological nihilists, she believed that values exist in reality and are not only purely intentional or fictional objects. As opposed to the relativists—she was convinced that the changeability of conventions "does not testify to the relativism of values, but that they be differently understood" or "that means of realizing them may be chosen differently." As opposed to subjectivists, who wanted to see the source of values in "causative subjects", she wrote:

> The base of establishing [...] the [ethical rules of legal laws] of decisions [...] is created usually by some judgements on values which pretend to be objectively justified; and decision is a choice of a certain norm which — according to the opinion of the decision maker — states an obligation to ways of conduct which aim at realizing or preserving these values [10, p. 114].

8.1 *Man and the World of Values*

The world of man is the world of values. Dąmbska wrote:

> Each of our conscious actions is directed by a desire to realize values giving sense to this action [10, p. 31].

Among sentences which concern the world of values as it is broadly understood—Dąmbska distinguished, i.a., norms, evaluations and axiological sentences *sensu stricto*.

We have already mentioned norms above. Evaluations and axiological sentences *sensu stricto* form a subclass of axiological sentences *sensu largo*. Evaluations—are sentences stating that some objects are valuable (*scil.* that they possess value). Axiological sentences *sensu stricto* are sentences which state what values are and what kinds of them exist.

Among these types of sentences—Dąmbska formulated mostly normative and evaluative sentences. However, she usually supplemented them with axiological sentences *sensu stricto*.

We read in Dąmbska:

> A peculiar feature of metaphysical investigation in the Lvov-Warsaw School is emphasized which is given to axiological moments: moral values which are assumed and produced by making philosophy and to its peculiar ethos, which shapes the life of philosopher [13, p. 29].

There is no doubt that Dąmbska herself had undertaken analysis of "axiological moments" in various domains of her research, first of all in research concerning the theory of science.

She considered accuracy (*scil.* functionality) to be the most important cognitive value of science. Accuracy—as opposed to truthfulness—in a gradable form. The better answers to questions concerning its domain a science gives (the best that can be given under specific conditions of cognition), the more accurate it is.

8.2 The Conception of Freedom

Among the values of human life, Dąmbska analysed i.a. freedom. It was a conscious choice, motivated by the same factors as analyses devoted to silence and namelessness.

Freedom is not always an axiologically positive value. It is so only if freedom is a necessary condition for realising some positive values.

One sometimes distinguishes freedom-from and freedom-to. Dąmbska [12] emphasized the fact that "freedom-from" and "freedom-to" are correlative terms. She wrote:

> Science's freedom from ideological and administrative pressure is freedom towards it inherent function of searching for and delivering truth; freedom of speech is at the same time freedom to publicly proclaim one's views and convictions [12, p. 857].

The correlation occurs that freedom is a necessary (but not sufficient) condition of the possibility of action. And so: freedom from mistake is a condition of morality etc.

8.3 Normative Ethics

In the domain of normative ethics, Dąmbska defended a certain version of perfectionism and intellectualism (and she contrasted the latter to emotionalism). She formulated ethical criterion as follows: Conduct is good if it aims at the perfection of a human; it is bad if it discourages perfection.

9 Significance

The significance of the personality of Dąmbska—especially after World War 2—consisted in the fact that she had the courage to bear witness to the truth in all circumstances of her life.

The significance of the work of Dąmbska—lies in the fact that her work reflects laws of reason and the axiological taste.

That is why she earned the title of the first lady of twentieth-century Polish philosophy.

Izydora Dąmbska's Works Mentioned in the Paper

1. La théorie du jugement de M. Edmond Goblot. Towarzystwo Naukowe we Lwowie, Lwów (1930)
2. O prawach w nauce [On Laws in Science]. Gubrynowicz i Syn, Lwów (1933)
3. Zarys historii filozofii greckiej [Outline of the History of Greek Philosophy]. Filomata, Lwów (1935)
4. Irracjonalizm a poznanie naukowe [Irrationalism and the scientific cognition]. Kwartalnik Filozoficzny **XIV**(i. 2), 83–118 (1937); (i. 3), 185–212 (1938)
5. List do Marii Obercowej [Letter to Maria Obercowa]. In: Perzanowski, J. (ed.) Izydora Dąmbska. 1904–1983, p. 23. Wydawnictwo PAU, Kraków (2001) (Personal communication, 1951)
6. Sceptycyzm francuski XVI i XVII wieku [French Skepticism of 16th and 17th Centuries]. Towarzystwo Naukowe w Toruniu, Torun (1958)
7. Dwa studia z teorii naukowego poznania [The Studies from the Theory of Scientific Cognition]. Towarzystwo Naukowe w Toruniu, Torun (1962)
8. O narzędziach i przedmiotach poznania: z teorii instrumentalnego poznania. O filozofii lingwistycznej [Instruments and Objects of Congition: from the Theory of Instumental Congition. On Linguistic Philosophy]. PWN, Warszawa (1967)
9. Dwa studia o Platonie [Two Studies on Plato]. Ossolineum & PAN, Wroclaw (1972)
10. O konwencjach i konwencjonalizmie [Conventions and Conventionalism]. Ossolineum & PAN, Wroclaw (1975)
11. Znaki i myśli. Wybór pism z semiotyki, teorii nauki i historii filozofii [Signs and Thoughts. Selected Writings in Semiotics, Epistemology and the History of Philosophy]. PWN, Warszawa (1975)
12. Gdy myśle o słowie "wolność" [When I think about the word "freedom"]. Znak **XXXIII**(6), 855–860 (1981)
13. O niektórych koncepcjach metafilozoficznych w Szkole Lwowsko-Warszawskiej [On some metaphilosophical conceptions in the Lvov-Warsaw School]. In: Perzanowski, J. (ed.) Jak filozofowac? Studia z metodologii filozofii [How to Philosophise? Studies from the Methodology of Philosophy], pp. 22–29. PWN, Warszawa (1989)
14. Kowledge, Language and Silence. Selected Papers. Brill/Rodopi, Leiden & Boston (2016)

A. Brożek (✉) • J. Jadacki
Institute of Philosophy, University of Warsaw, Warsaw, Poland
e-mail: abrozek@uw.edu.pl; j.jadacki@uw.edu.pl

Maria Kokoszyńska-Lutmanowa: The Great Polemist

Filip Kawczyński

Abstract The paper concerns life and scientific achievements of Maria Kokoszyńska-Lutmanowa (1906–1981) who was a significant logician, philosopher of language and epistemologist belonging to the second generation of the Lvov-Warsaw School. She is mostly known as the author of the important argumentation against neopositivism of the Vienna Circle as well as one of the main critics of relativistic theories of truth. The article is divided into five sections: 1. Life, 2. Main papers, 3. Views, 4. Pupils, 5. Summary, 6. References.

Keywords Maria Kokoszyńska-Lutmanowa · Vienna Circle · Neopositivism · Justification · Relativism · Absolutism

Mathematics Subject Classification (2000) Primary 01A70; Secondary 03A10

1 Life[1]

Maria Kokoszyńska-Lutmanowa was one of the key figures among the so called second generation of philosophical Lvov-Warsaw School, along with such prominent thinkers as Alfred Tarski, Izydora Dąmbska, Janina Kotarbińska or Henryk Mehlberg. Although the list of Kokoszyńska's philosophical achievements is impressive, for what she is best known is probably her substantial and influential criticism of ideas proposed by the main philosophical force in that time—the Vienna Circle.

Kokoszyńska was born on 06.12.1906 in Bóbrka near Lvov in the family of Eugenia Kokoszyńska (née Sidorowicz) who was a pianist and a music teacher and Stanisław Kokoszyński, a civil servant. She undertook studies in philosophy and mathematics in 1923 at the Jan Kazimierz University in Lvov, where under the supervision of Kazimierz Twardowski—founder of the Lvov-Warsaw School—she wrote a dissertation entitled "Nazwy ogólne i wieloznaczne" (English translation of the title: "General and Ambiguous Names") and completed her PhD in 1928. Apart from Twardowski she has an opportunity

[1]This section of the paper draws on Zygmunt [4].

Á. Garrido, U. Wybraniec-Skardowska (eds.), *The Lvov-Warsaw School. Past and Present*, Studies in Universal Logic,
https://doi.org/10.1007/978-3-319-65430-0_17

to be a student of several other great scholars like Roman Ingarden, Sefan Banach, Hugo Steinhaus or Stanisław Ruziewicz. However, it is Kazimierz Ajdukiewicz who should be considered as the teacher that had the major influence on Kokoszyńska's later development as a philosopher. Besides learning from the superb teachers, during her university years she established long-standing friendships with other Twardowski's students and—as it turned out later—significant philosophers and logicians such as Dąmbska, Seweryna Łuszczewska-Romahnowa or Daniela Gromska.

Between 1930 and 1934 she was an assistant of Ajdukiewicz in the Department of Philosophy of the Jan Kazimierz University. In 1931 she got married to dr. Roman Lutman, polish lawyer, historian and social activist (afterwards the director of Silesian Library and Silesian Institute in Katowice). They had no children.

Between 1934 and 1939, due to being awarded with several scholarships, Kokoszyńska was a visiting scholar in Prague, Vienna, Paris and Cambridge. During her visit in Vienna she was a guest at the seminar organised by the Vienna Circle, where she presented a paper that eventually has been published in the official journal of the *Wiener Kreis*, i.e. in "Erkenntnis" [4]. That allowed her to get in contact with some of the most influential philosophers of that era: Rudolf Carnap, Otto Neurath or Moritz Schlick. In Vienna she also had an opportunity to meet Karl Popper and Kurt Gödel. In 1938 she went to Cambridge where she was supposed to stay for 3 years, however, she got back to Poland in June 1939 because of the war looming. Nevertheless, she managed to attend G.E. Moore's lectures and meet Ludwig Wittgenstein. During the years of travelling around Europe she gave numerous talks at various philosophical conferences, most important of which are: 1st International Congress of Scientific Philosophy (Sorbonne, 1935), 9th International Philosophical Congress (Paris, 1936), 3rd Polish Philosophical Congress (Cracow, 1936), 4th International Congress for the Unity of Science (Cambridge, 1938).

Kokoszyńska spent the war years in Lvov. In 1947 she moved with her husband to Wrocław. The same year she obtained her postdoctoral degree on the basis of her work "W sprawie względności i bezwzględności prawdy" (English translation of the title: "On Relativity and Absoluteness of Truth"). In 1950 she occupied a position of the head of newly-formed department of Logic and Methodology of Science at the University of Wrocław. She governed the department for 26 years and during that period she cooperated with many significant logicians, e.g. Jerzy Słupecki, Ludwik Borkowski, Tadeusz Kubiński, Witold Pogorzelski, Ryszard Wójcicki or Bogusław Iwanuś.

Kokoszyńska became *professor extraordinarius* in 1951 and *professor ordinarius* in 1969. From 1951 to 1954 she was a dean of the Faculty of Philosophy and History of the University of Wrocław and a vice-rector of the university from 01.01.1955 to 30.09.1956.

Kokoszyńska died in Wrocław on 30.06.1981 after struggling with a dread disease.

2 Main Papers

For comprehensive bibliography of Maria Kokoszyńska-Lutmanowa see: Jan Zygmunt, "Bibliografia prac naukowych Marii Kokoszyńskiej-Lutmanowej" *Filozofia Nauki* **46** (2004).

[1] Z semantyki funkcyj zdaniowych [On the Semantics of the Propositional Functions]. In: Księga Pamiątkowa Polskiego Towarzystwa Filozoficznego we Lwowie 12.02.1904–12.02.1929, Lwów (1932)
[2] Syntax, Semantik und Wissenschaftslogik. Actes du Congrés International de Philosophie Scientifique, Paris (1936)
[3] Logiczna składnia języka, semantyka i logika wiedzy [Logical Syntax of Language, Semantics and Logic of Knowledge]. Przegląd Filozoficzny **39**(1), 38–49 (1936)
[4] Über den absoluten Wahrheitsbergiff und einige andere semantische Begriffe. Erkenntnis **6**, 143–165 (1936)
[5] Filozofia nauki w Kole Wiedeńskim [Philosophy of Science in the Vienna Circle]. Kwartalnik Filozoficzny **13**(1), 151–165; (2), 181–194 (1937)
[6] Sur les éléments métaphysiques et empiriques dans la science. Travaus du IX Congrés International de Philosophie, Congrés Descartes, Paris, 108–117 (1937)
[7] Bemerkungen über die Einheitswissenschaft. Ekrenntnis **7**, 325–335 (1938)
[8] W sprawie walki z metafizyką [On Fighting Metaphysics]. Przegląd Filozoficzny **41**, 9–24 (1938)
[9] What Means "Relativity of Truth"? Studia Philosophica **3**, 167–176 (1948)
[10] A Refutation of Relativism of Truth. Studia Philosophica **4**, 93–149 (1951)
[11] (with T. Kubiński and J. Słupecki) Zastosowanie logiki matematycznej do wyjaśniania niektórych pojęć przyrodoznawstwa [Application of Mathematical Logic for Explaining Some Concepts of Natural Sciences]. Studia Logica **4**, 155–211 (1956). Incl. summary in Russian and English
[12] O "dobrej" i "złej" indukcji [On "Good" and "Bad" Induction]. Studia Logica **5**, 43–70 (1957). Incl. summary in English
[13] O stosowalności metody dedykcyjnej w naukach niededukcyjnych [On the Applicability of the Deductive Method in Non-deductive Sciences]. Sprawozdania Wrocławskiego Towarzystwa Naukowego **15** A, 45–49 (1960)
[14] O dwojakim rozumieniu uzasadniania dedukcyjnego [Two Concepts of Deductive Justification]. Studia Logica **13**, 177–196 (1962). Incl. summary in English
[15] O dedukcji [On Deduction]. Studia Filozoficzne **1**, 77–86 (1963). Incl. summary in English
[16] W sprawie koncepcji nauk dedukcyjnych [About the Conception of Deductive Sciences]. Studia Filozoficzne **1**, 57–65 (1967)
[17] Złudzenia aprioryzmu [Illusions of Apriorism]. Ruch Filozoficzny **28**(1–2), 72–78 (1970)
[18] Metodologiczne problemy integracji nauk [Methodological Problems of Integration of Sciences]. Studia Filozoficzne **4**(89), 47–56 (1973)

3 Views

When Kokoszyńska's philosophical career began in the late 1920s, central problems in analytic philosophy concerned language of science, its syntax and semantics. In general, Kokoszyńska shared with other members of Lvov-Warsaw School the view according to which science is thought to be a system of sentences with uniform logical structure and thus it is possible to verify and adjust it by using theoretical tools of logic, methodology or (logical) semantics. Along with e.g. Zawirski, Czeżowski or Mehlberg she paid special attention to the issue of logical form of sentences and their justification. She was an adherent of the view that variables appearing within propositional functions are meaningful and that their meaning is the integral component of the meaning of all substitutions; in other words, it is meaning shared by all these substitutions (see [1]).

When it comes to dealing with the notion of justification of sentences, one of the crucial tasks is defining what does it mean for the sentence to be *true*. According to logical positivists (especially Carnap, but also Neurath or Hempel) truth—as well as some other apparently semantic notions—can be (and should be) defined in terms of syntax and thus, the whole metascientific reflection is reduced to syntactic analysis of the language of science. Kokoszyńska (see [2, 3]) confronted such position with proposals outlined by Tarski and Gödel, and came to the conclusion that positivists' programme is too radical, since—as it stems especially from Tarski's works—semantics is inevitable and necessary for understanding the notion of truth (and thus justification etc).

Kokoszyńska's substantial criticism of the ideas proclaimed by the Vienna Circle—although accompanied by respect and admiration of their contribution to the development of philosophy—did not stop with the above disagreement. In her later works she got back to the caution that the principle of primacy of pure syntax goes too far. One of the main reasons for neopositivists to establish that principle was their idea of uniform scientific language. They believed that such unity can be achieved on that very abstract level of pure syntax. Kokoszyńska pointed (see [5]) that such an attitude is either: (a) futile—as it is merely an assumption that all empirical scientific statements can be expressed in one universal language and as a matter of fact there is no good evidence to accept this assumption as correct; or (b) incorrect—because it has been conclusively shown that logical knowledge cannot be expressed in a single language as semantic statements always demand the other language—namely a metalanguage—to be expressed.

Kokoszyńska attacked also fundamental neopositivist postulate of abandoning metaphysics or strictly speaking—the way in which it has been realised by them. She argued that (see [6, 8]) logical positivists are too hasty in dismissing everything labeled as "metaphysics" and she distinguished two concepts of metaphysics. In the first sense metaphysics is considered as dogmatic and unjustified acceptance of some undetermined propositions. Obviously, Kokoszyńska supports the idea that metaphysics so understood has no entry to the area of serious scientific investigation. In the second sense, however, metaphysics is a set of propositions of some kind, and there are three possible options with regard to what exactly belongs to such a set: (a) determined analytic propositions; (b) undetermined and testable propositions; (c) undetermined and non-testable propositions. As may be expected, Kokoszyńska denies possibility of any scientific interest in the propositions of the last type. However, contrary to members of the Vienna Circle, she is not that certain that propositions of the kinds (a) and (b) also have no *raison d'être*. As a matter of fact, she points out that such statements are indeed present in scientific investigation and work there well as they are general yet still theoretically testable propositions. Worth emphasising, Kokoszyńska believed that propositions of all above types are meaningful; propositions of the type (c) are in fact useless for science, however, there are no reasons to claim that they are senseless.

It may be said, that despite negative assessment of some ways in which neopositivists realised their programme, Kokoszyńska in general shared their main philosophical ideas. Her criticism was aimed not to sink logical positivism but rather to recognise its weak points and thus make it possible to review this theory and eventually make it stronger.

Another field in which Kokoszyńska manifested her polemical edge was the debate concerning relativity of truth. In general, members of Lvov-Warsaw School were absolutists with regard to the issue of truth, and so was Kokoszyńska. Her in-depth and comprehensive analysis of relativistic theories allowed her to provide clear explication of relativity of truth:

> (I) the term 'true' is an incomplete predicate; (II) $\exists X\{Y, Z[(X$ is true with respect to $Y)$ & Non-X is true with respect to $Z)]\}$. [10, p. 94]

She presented a series of arguments against arguments for relativism (see [10]). First, she claimed that some of the arguments for relativism are in fact arguments for a very weak version of relativism, according to which not all theorems are necessary, which is of course the case, but does not entail any interesting theory. Second, she pointed out that many relativists rely on theories from the area of psychology, sociology, pragmatics etc. while in her opinion none of them is appropriate. Third, she showed that relativism often stems from misinterpretation of how deictic expressions function in language. Apart from criticising relativism, Kokoszyńska offered some ideas that should be taken into account when absolutist theory of truth is to be developed. In such a theory it has to be possible to distinguish truth from criteria of truth. The theory has to provide also a method of interpreting all occasional expressions as well as involve the T-convention. Worth mentioning here that Kokoszyńska—who thought that T-convention is a necessary base for any adequate theory of truth—was one of the first philosophers who recognised great value of Tarski's proposal (in one of her letters to him she wrote that even he does not fully appreciate enormous significance of his own theory).

Conflict between absoluteness and relativity affected also Kokoszyńska's late works, concerning deductive and non-deductive justification (see [12–16]). In one of her papers [16] she enters the dispute with Quine over notorious analytic-synthetic distinction. Notion of analyticity was required in her theory to support the distinction for *absolute* deductive justification and *relative* deductive justification. One of the specific features of the former is that it is applicable to analytic statements exclusively. Thus, Kokoszyńska needed a criterion of identifying analytic (*resp.* synthetic) statements and she offered very interesting criterion of that kind, based on the notion of denoting—she claimed that analytic are the statements that are true if and only if appropriate rules of denoting (expressed in metalanguage) fix existing objects as references of terms occurring in a sentence. In other words, the fact that extension in question is not empty, guarantees truth (in case of synthetic statements some further conditions besides existence have to be fulfilled to make a given statement true).

4 Pupils

The most important students of Kokoszyńska are: Tadeusz Kubiński, Ryszard Wójcicki, Witold Adam Pogorzelski, Waleska Rudek, Wanda Charczuk and Leon Gumański.

Worth mentioning that investigations on analyticity conducted by Marian Przełęcki were strongly inspired by works of Kokoszyńska.

5 Summary

Kokoszyńska, although slightly outshined by philosophical giants like Ajdukiewicz or Tarski, stays one of the most significant thinkers of Lvov-Warsaw School. Her works on relativism or analyticity still can be considered as points of reference in a current debates concerning these topics. Especially her theory of analytic-synthetic distinction seems to deserve much more attention than it has received so far. Apart from philosophical achievements Kokoszyńska is also a great source of inspiration as a scholar who was able (and have enough courage) to compete with and argue against some of the most powerful philosophers of her times. Last but not least—being the head of successful department of logic of the University of Wrocław and taking part in numerous philosophical enterprises—she was also a remarkable figure in Polish philosophical society.

References

1. Brożek, A.: O logice i filozofii w Szkole Lwowsko-Warszawskiej. W 30 lat po publikacji "Filozoficznej szkoły lwowsko-warszawskiej". Talk with Jan Woleński [On logic and philosophy in the Lvov-Warsaws School. Thirty years after the publication of "Philosophical Lvov Warsaw School". A talk with Jan Woleński]. In: Brożek, A., Chybińska, A. (eds.) Fenomen Szkoły Lwowsko-Warszawskiej [Phenomenon of the Lvov-Warsaw School]. Wydawnictwo Academicon, Lublin (2016)
2. Jadacki, J.J.: Orientacje i doktryny filozoficzne. Z dziejów myśli polskiej [Philosophical Directions and Doctrines. From the History of Polish Thought]. Wydział Filozofii i Socjologii, Warszawa (1998)
3. Tałasiewicz, M.: Maria Kokoszyńska-Lutmanowa – Methodology, Semantics, Truth. In: Krajewski, W. (ed.) Polish Philosophers of Science and Nature in the 20th Century. Poznań Studies in the Philosophy of the Sciences and the Humanities, pp. 129–133. Rodopi, Amsterdam (2001)
4. Zygmunt, J.: Maria Kokoszyńska-Lutmanowa. In: Kolbuszewski, J. (ed.) Uczeni wrocławscy. Członkowie Wrocławskiego Towarzystwa Naukowego 3 (1953–1996) [Scholars of Wrocław. Members of the Wrocław Scientific Society], pp. 41–44. Wrocławskie Towarzystwo Naukowe, Wrocław (1996)
5. Zygmunt, J.: Bibliografia prac naukowych Marii Kokoszyńskiej-Lutmanowej [Bibliography of scientific works of Maria Kokoszyńska-Lutmanowa]. Filozofia Nauki XII **2**(46), 155–166 (2004)

F. Kawczyński (✉)
Institute of Philosophy, University of Warsaw, Warsaw, Poland
e-mail: f.kawczynski@uw.edu.pl

Seweryna Łuszczewska-Romahnowa

Roman Murawski and Jerzy Pogonowski

Abstract The paper is devoted to the description of life and scientific achievements as well as the influence of Seweryna Łuszczewska-Romahnowa.

Keywords Łuszczewska-Romahnowa · Venn diagrams · Classification · Methodology · Logical pragmatics · Induction

Mathematics Subject Classification (2000) Primary 03-03; Secondary 01A60

1 Life

Seweryna Łuszczewska was born on August 10, 1904 in Mszana near Zborów (Tarnopol Voivodeship, at that time in Kaisertum Österreich). She belonged to a noble aristocratic family: her mother, Maria Antonina was a daughter of Count Wojciech Dzieduszycki (1848–1909), a politician and philosopher. Her father, Konrad Łuszczewski (1876–1937) was a grandson of Minister Jan Paweł Łuszczewski (1764–1812) and himself also a politician. The genealogical tree of Seweryna Łuszczewska-Romahnowa is accessible at:

http://www.sejm-wielki.pl/b/psb.26376.1

She completed her primary and secondary education in Jazłowiec, Jarosław, Kraków and Lvov. She studied at the Lvov University in the years 1922–1928 both: philosophy (under guidance of Kazimierz Twardowski, Kazimierz Ajdukiewicz, and Roman Ingarden) and mathematics (under guidance of Hugon Steinhaus and Stefan Banach). She obtained PhD in philosophy at the Lvov University on the basis of the dissertation *O wyrazach okazjonalnych* (*On occasional terms*). The thesis was written under guidance of Ajdukiewicz, but Twardowski was finally the supervisor. Seweryna Łuszczewska started her academic work at the First Chair of Philosophy (headed by Ajdukiewicz) at the University of Lvov in autumn 1932. Before the World War II she worked also as a teacher in several secondary schools. In 1934 she married Dr Edmund Romahn, a teacher of secondary schools in Lvov and also a philosopher.

Á. Garrido, U. Wybraniec-Skardowska (eds.), *The Lvov-Warsaw School. Past and Present*, Studies in Universal Logic,
https://doi.org/10.1007/978-3-319-65430-0_18

The most tragic period of her life was the wartime 1939–1945. She was expelled from the Lvov University by the authorities of Soviet Ukraine. In the years 1939–1941 (till the seizure of Lvov by Germans) she worked as a teacher in secondary schools in Lvov (lecturing on mathematics and astronomy in Ukrainian) and later as a private tutor. In May 1943 she, together with her husband was arrested by Gestapo and put to prison in Lvov. Then she experienced a martyrdom of Nazi concentration camps: in Majdanek (where her husband was killed in September 1943), Ravensbrück and Leipzig division of the Buchenwald camp. After liberation, she worked shortly as a clerk for UNRRA.

In December 1946 Łuszczewska-Romahnowa came to Poznań and started (January 1, 1947) her work at the Chair of Theory and Methodology of Sciences headed by Kazimierz Ajdukiewicz and converted in 1951 to the Chair of Logic at the Faculty of Mathematics, Physics and Chemistry of the Poznań University. In 1954 she became an associate professor and in 1962 an extraordinary professor. She took the post of the chairman of the Chair of Logic in 1955, after Ajdukiewicz moved to Warsaw (the chair was renamed the Department of Logic and included into the Institute of Mathematics in 1969). She was active at that post up to her retirement in 1974. Most of her academic works were published in that period. Seweryna Łuszczewska-Romahnowa died June 27, 1978 in Poznań.

The main sources of biographical information about Seweryna Łuszczewska-Romahnowa are the papers by Batóg [13, 14] (reprinted also in [16], all in Polish, containing very precise description of the *curriculum vitae* of Seweryna Łuszczewska-Romahnowa) and [15] (in English).

2 Main Works

Publications by Seweryna Łuszczewska-Romahnowa are not numerous. Moreover, most of them are in Polish. Still, they contain original results and reflections, which is witnessed by the fact that some of them were reprinted in the influential collections of papers. One can divide these works into three groups.

2.1 Logical and Mathematical Works

An extensive article [2] from 1953, published in the first volume of *Studia Logica*, is devoted to a graphic method of verifying syllogistic inferences. Her generalization of the method of Venn diagrams is original and makes it possible to avoid certain ambiguities involved in this method. Moreover, the author proposes a discursive equivalent of the method and obtains a decision procedure for the monadic predicate calculus.

Łuszczewska-Romahnowa introduced an important concept of a *natural classification* (cf. [5]). First of all, by an n-level classification of a given set X one understands a sequence of partitions $(F_1, \ldots, F_n)$ of X forming a chain of ever finer partitions of X, where $F_1 = \{X\}$. This means that for each $1 < i \leqslant n$ and every $Y \in F_i$ there exists $Z \in F_{i-1}$ such that $Y \subseteq Z$. Each n-level classification of X is thus a hierarchy of classifications of this set. Any n-level classification $\mathbb{F} = (F_1, \ldots, F_n)$ of X generates a measure of similarity (indistinguishability) between elements of X with respect to $\mathbb{F}$. Namely, let $I_{\mathbb{F}}(x, y)$ (*index of similarity*) be defined as the number of the last level of $\mathbb{F}$ at which objects $x, y \in X$ still remain elements of the same set from the corresponding partition. Let $D_{\mathbb{F}}(x, y) = n - I_{\mathbb{F}}(x, y)$. Then $D_{\mathbb{F}}$ is a quasi-distance function which becomes a standard distance if the last member of $\mathbb{F}$ is a partition of X into one-element sets. Now, it usually happens in empirical sciences that investigated objects are compared in several different aspects and are considered as more or less similar with respect to the criteria taken into account. That is to say, the Nature itself projects similarities between objects. This can be formally expressed by a certain distance function d, established empirically. If at the same time one classifies investigated objects, then one can ask about a correspondence between empirical similarity characterized by d and similarity based on the classifications characterized by $D_{\mathbb{F}}$. Łuszczewska-Romahnowa calls an n-level classification $\mathbb{F}$ *natural*, if $D_{\mathbb{F}}(x, y) < D_{\mathbb{F}}(u, v)$ implies $d(x, y) < d(u, v)$, for all $x, y, u, v \in X$.

Two papers written together with Tadeusz Batóg (cf. [9] and [10]) bring a generalized theory of classifications. The essence of this generalization lies in the possibility of constructing multi-level classifications of a given set such that the number of levels is a transfinite ordinal number. The corresponding definition extends in a natural way the concept of n-level classification mentioned above. The authors provide a few non-trivial examples of such transfinite classificatory systems. Moreover, they obtain an algebraic characterization of the class of all such systems: they prove that it is an arithmetic class in a wider sense. Thus, the introduced concept is indeed a natural one. The authors show also some connections between this concept and Boolean metric spaces.

2.2 Methodological Works

The paper [1] from 1948 deals with ambiguity of terms occurring in the scientific language. According to it, the liberation of the scientific language from all ambiguities does not take place in the research practice but is also in general impossible. Nevertheless, this language functions fairly well as a tool of communication between academicians. This is due to semantic stability of that part of the scientific language which is responsible for logical consequence, truthfulness, empiricalness, etc. Polysemy creates no danger to scientific argumentation as long as the scientists themselves are careful in conducting their argumentation in a logically correct way.

The papers [6] and [8] may be classified as belonging to logical pragmatics. The author tries to formulate anew the classic theory of argumentation errors (which included e.g. *petitio principii*, *non sequitur*, material error). However, she obtains much more than simply a new look at these phenomena. Her most important achievement is, at least in our opinion, a proposal of a precise logical reconstruction of argumentation. It seems likely that she was the first, who suggested an annotated representation of argumentation. Łuszczewska-Romahnowa was aware of the limited scope of application of the proposed formalism (e.g. it does not account for apagogic argumentation). Still, it is a nice piece of work in formal pragmatics, a completely new discipline at that time.

Łuszczewska-Romahnowa criticized the probabilistic approach to the problem of induction in [3] from 1957. After reconstructing the main assumption of that approach she argues that they are not adequate. In her own opinion, even the most sophisticated methods of inductive reasoning hardly enable us to obtain a true general knowledge from true individual knowledge. Hence the probability of the conclusions of inductive reasoning approach zero. This fact is testified by numerous examples from the history of the natural sciences. Theories in such sciences are in a state of permanent change, they undergo correction after correction, in the process of continuous confrontation with the empirical data. The author claims that the real goal at which such sciences are oriented is not the quest for absolute truth but rather to organize our scientific activity in such a way that we can adjust our actions to the conditions imposed by the environment and we can get a general orientation in that environment. Even false theories may have some cognitive value: they pass away, but their life was not meaningless, so to speak. At least, they have provoked us to change our world view, when their predictions collided with the empirical data. To sum up: the goal of induction does not lie in the search of absolute truths, the failure of induction should not be reduced to the fact that its conclusions do not follow logically from its premisses. This view of induction bears thus a pragmatic stigma.

Among the methodological papers of the author one should also mention a short note *Czy filozofia obumiera?* (cf. [7]) which may be considered as a defence of integrity of philosophy. The author criticizes a popular view that philosophy remains unmatured and is unable to solve its own fundamental problems. She points to a dynamic character of philosophical reflection, recognizable for instance in clarification of philosophical language due to the use of tools from logic. As an example, she mentions Ajdukiewicz's analyses of transcendental idealism.

Still another important paper is [11] from 1967 which contains a penetrating analysis of the philosophical ideas of Kazimierz Twardowski. The author reports on Twardowski's program of *scientific philosophy*, discusses its main assumptions and shows how the program in question influenced Polish philosophy.

2.3 Works on History of Logic and Methodology

Professor Łuszczewska-Romahnowa has translated into Polish the famous logical treatise by Arnauld and Nicole *La logique ou l'art de penser*. The translation is accompanied by her introduction containing an analysis of the philosophical thought of the seventeenth century (cf. [4]). In particular, she stresses the differences between Pascal and Descartes concerning the scientific method and the concept of knowledge itself. She has also discussed the role played by the investigations *more geometrico* in European philosophy of the seventeenth century.

Seweryna Łuszczewska-Romahnowa wrote also a short note [12] about the history of logical investigations in Poznań (up to 1973). Our papers [17] and [18] are, in a sense, a continuation of her work.

3 Views

Without any doubt the views of Professor Łuszczewska-Romahnowa were shaped under the influence of the Lvov-Warsaw philosophical school. She was most close in her views to Kazimierz Ajdukiewicz. This can be easily seen from the synthesis of analytical philosophy and logic so characteristic for her style of writing.

According to Tadeusz Batóg, her political views (already before the World War II) were decisively leftist, in spite of her own aristocratic background.

4 Resonance

Professor Łuszczewska-Romahnowa lectured on logic and the methodology of sciences for a few generations of students in Poznań. She was responsible for all examinations in philosophy obligatory for students applying for a PhD at the faculties of mathematics, physics, chemistry, biology and geology of the Poznań University.

5 Pupils

Professor Łuszczewska-Romahnowa was a supervisor of three doctoral dissertations, i.e. those of Tadeusz Batóg, Jerzy Czajsner and Mieczysław Jarosz. In turn, the first of them supervised before his retirement four students (who have their own students) so that Łuszczewska-Romahnowa has totally a dozen of scientific descendants (as of 2016). Tadeusz Batóg was her follower at the post of the chairman of the Department of Mathematical Logic in the years 1974–1996, since 1996 the chairman is Roman Murawski.

6 Summary

The life of Seweryna Łuszczewska-Romahnowa was heroic. Due to the horror of war and Nazi persecution she experienced a tragedy of losing her husband, she was imprisoned in the concentration camps, her health was ruined. Despite all this, she conscientiously fulfilled all her academic duties during the three decades after the war. She is remembered as a very modest person, always helpful and favorably disposed towards others. These qualities of her personality, as well as her scientific achievements gained her a high esteem in the academic community.

Acknowledgements Jerzy Pogonowski acknowledges the financial support of National Science Centre [Narodowe Centrum Nauki] (grant No 2015/17/B/HS1/02232).

References

Selected Works by Seweryna Łuszczewska-Romahnowa

1. Wieloznaczność a język nauki [Polysemy and the language of science]. Kwartalnik Filozoficzny **17**, 47–58 (1948) (Reprinted in: Jerzy Pelc (ed.) Semiotyka polska 1894–1969. Państwowe Wydawnictwo Naukowe, Warszawa (1971))
2. Analiza i uogólnienie metody sprawdzania formuł logicznych przy pomocy diagramów Venna [An Analysis and Generalization of Venn's Diagrammatic Decision Procedure]. Stud. Logica **1**, 185–213 (1953)
3. Indukcja a prawdopodobieństwo [Induction and probability]. Stud. Logica **5**, 71–90 (1957)
4. Logika z Port Royal na tle historii logiki i metodologii XVII wieku [Port Royal logic from the perspective of history of logic and methodology of the 17th century]. Introductory essay in: Arnauld A., Nicole, P., Logika, czyli sztuka myślenia, pp. X–XXVI. Państwowe Wydawnictwo Naukowe, Warszawa (1958)
5. Classification as a kind of distance function. Natural classifications. Stud. Logica **12**, 41–81 (1961) (Reprinted in: Przełęcki, M., Wójcicki, R. (eds.) Twenty-Five Years of Logical Methodology in Poland, pp. 341–373. D. Reidel, Warszawa-Dordrecht (1977))
6. Pewne pojęcie poprawnej inferencji i pragmatyczne pojęcie wynikania [On a certain concept of correct inference and a pragmatic concept of entailment]. Stud. Logica **13**, 203–208 (1962) (Reprinted in: Pawłowski, T. (ed.) Logiczna teoria nauki, pp. 163–167. Państwowe Wydawnictwo Naukowe, Warszawa (1966))
7. Czy filozofia obumiera? [Is philosophy withering?] Stud. Filozoficzne **1**, 187–196 (1962)
8. Z teorii racjonalnej dyskusji [From the theory of rational discussion]. In: Rozprawy logiczne. Księga pamiątkowa ku czci profesora Kazimierza Ajdukiewicza, pp. 163–167. Państwowe Wydawnictwo Naukowe, Warszawa (1964)
9. A generalized theory of classifications I. Stud. Logica **16**, 53–74 (1965). Co-author: Tadeusz Batóg
10. A generalized theory of classifications II. Stud. Logica **17**, 7–30 (1965). Co-author: Tadeusz Batóg
11. Program filozofii naukowej Kazimierza Twardowskiego [Kazimierz Twardowski's program of scientific philosophy]. Stud. Filozoficzne **4**, 154–168 (1967)
12. Logika [Logic]. In: Labuda, G. (ed.) Nauka w Wielkopolsce, pp. 190–198. Wydawnictwo Poznańskie, Poznań (1973)

Other Sources

13. Batóg, T.: Seweryna Łuszczewska-Romahnowa. Stud. Filozoficzne **1**, 189–194 (1979) (Reprinted in: Batóg, T.: Od Kanta do homofonów (Artykuły, autoreferaty, biografie, recenzje) [From Kant to homophones. Articles, abstracts, biographies, reviews], pp. 232–237. Wydawnictwo i drukarnia Uni-Druk s.j., Poznań (2015))
14. Batóg, T.: Romahnowa (Łuszczewska-Romahnowa). Polski Słownik Biograficzny Tom **XXXI**, z. 4, 575–576 (1989) (Reprinted in: Batóg, T.: Od Kanta do homofonów (Artykuły, autoreferaty, biografie, recenzje) [From Kant to homophones. Articles, abstracts, biographies, reviews], pp. 237–239. Wydawnictwo i drukarnia Uni-Druk s.j., Poznań (2015))
15. Batóg, T.: Seweryna Łuszczewska Romahnowa – logic and philosophy of science. In: Krajewski, W. (ed.) Polish Philosophers of Science and Nature in the 20th Century. Poznań Studies in the Philosophy of the Sciences and the Humanities, vol. 74, pp. 113–119. Rodopi, Amsterdam (2001)
16. Batóg, T.: Od Kanta do homofonów (Artykuły, autoreferaty, biografie, recenzje) [From Kant to homophones. Articles, abstracts, biographies, reviews]. Wydawnictwo i drukarnia Uni-Druk s.j., Poznań (2015)

17. Murawski, R., Pogonowski, J.: Badania logiczne na Uniwersytecie Poznańskim w latach 1945–1955 [Logical investigations at the University of Poznań in 1945–1955]. Investigationes Linguisticae **XIV**, 68–77 (2006)
18. Murawski, R., Pogonowski, J.: Logical investigations at the University of Poznań in 1945–1955. In: Bernhard, P., Peckhaus, V. (eds.) Methodisches Denken im Kontext, pp. 239–254. mentis Verlag, Paderborn (2008)

R. Murawski (✉)
Adam Mickiewicz University in Poznań, Faculty of Mathematics and Computer Science, Department of Mathematical Logic, Poznań, Poland
e-mail: rmur@amu.edu.pl

J. Pogonowski
Adam Mickiewicz University in Poznań, Institute of Psychology, Department of Logic and Cognitive Science, Poznań, Poland

On Existential Dependence and Independence in the World of Thoughts and States of Affairs (with Reference to Eugenia Ginsberg-Blaustein's and Roman Ingarden's Analyses)

Urszula M. Żegleń

Abstract The purpose of my paper, which has been inspired by Eugenia Ginsberg-Blaustein's analysis of the concepts of existential dependence and independence, is to shed further light on the concept of existential dependence and connected concepts in their applications to the mental sphere, namely to the analysis of mental content. To realize this purpose, I shall extend Ginsberg-Blaustein's analysis to Roman Ingarden's phenomenological approach with his characterization of the concepts of heteronomous existence, existential dependence and other associated concepts.

Keywords Existential dependence · Thought · Mental content · Judgment · Act of consciousness · Object of act · State of affairs

Mathematics Subject Classification (2000) Primary 03B40, 00A15; Secondary 81P05

1 Introduction

Eugenia Ginsberg-Blaustein's paper On the Concepts of Existential Dependence and Independence was originally published in Polish in 1931 in the anniversary collection of papers presented in the Polish Philosophical Society in Lvov in the period 1904–1929 [2]. After half a century, her paper was translated into English and published in the collection of papers edited by Barry Smith on logic and formal ontology [7]. Since then, the paper has inspired further analyses, the example of which is the formal analysis given by Marek Magdziak [23]. Ginsberg-Blaustein in her critical considerations of the concepts of existential dependence and independence points out difficulties in their precise definitions in the analyses given by Karl Stumpf, Alois Höfler, Kazimierz Twardowski, and especially by Edmund Husserl, with reference to whom she raises serious objections concerning inadequacy of his definitions. In her attempt avoid the difficulties, she gives her own

Á. Garrido, U. Wybraniec-Skardowska (eds.), *The Lvov-Warsaw School. Past and Present*, Studies in Universal Logic,
https://doi.org/10.1007/978-3-319-65430-0_19

analysis, in which she distinguishes between two kinds of dependence and independence: psychological and ontological. The first one is distinguished with regards to the possibility or impossibility of separated presentation (Germ. *Vorstellung*), the second one—with regards to the possibility or impossibility of separated existence. The main idea of her analysis is reference to the concept of state of affairs and its subject and both concepts are treated as primitive. She emphasizes the application of the analysed concepts in many areas of philosophical inquiries, especially (1) in ontology—in the theory of objects (with interesting reference to the theory of objects of higher order) and in the theory of parts and wholes (to which she devotes special attention in her analysis), (2) in epistemology—in the theory of acts and products, and (3) in the philosophy of language—in the theory of meaning. Today the issue of existential dependence and independence belongs to the fundamental topics of analytic metaphysics and phenomenological ontology.

The ontological analysis of dependence and independence and other connected concepts was developed in detail by Roman Ingarden in his *Controversy over the Existence of the World* (in the first two volumes originally published in Polish in 1947, vol. I and in 1948, vol. II).[1] Ingarden provides the most precise characterization of these concepts, which are treated as existential moments and are crucial ontological categories in his inquiries into the controversy between realism and idealism in question of the existence of the world. In Ingarden's analysis, the notions 'dependence' (Pol. 'zależność', Germ. 'Abhändigkeit') and 'independence' (Pol. 'niezależność', Germ. 'Unabhändigkeit') have different meanings than the notions 'samoistność' and 'niesamoistność', which were originally used by Ginsberg-Blaustein in her paper. The translator of her paper justifies his choice of translation and makes some helpful remarks in which he stresses that the Author followed the Husserlian rather than Ingardenian terminology, and applies Ginsberg-Blaustein's own convention of her interpretation of Ingarden's notions.

In my paper I shall extend Ginsberg-Blaustein's analysis to Ingarden's definitions, and use them in my analysis, applicable to the mental sphere. The purpose of my paper is the application of these basic concepts to the analysis of mental content. Today the problem of mental content belongs to the most topical questions in the philosophy of mind. There are many various theories of mental content which are developed in a new context of research provided by cognitive science. I want, however, to make use of these philosophical means of analysis, especially those from Ingarden's theory, in the characterization of mental content considered from the epistemological-psychological point of view.

[1] In spite of the fact that Ingarden as phenomenologist was far from developing philosophy in the manner of the Lvov-Warsaw School he was connected with the Lvov School. He started his philosophical education under Kazimierz Twardowski (who, like Husserl was a student of Franz Brentano), whose classes he attended in the period 1911–1912. His further academic career was associated with Lvov, where in 1924 he received his Habilitation and in 1925 he was employed as Private Docent at Jan Kazimierz University. In 1933 he received the position of Associate ('Extraordinary') Professor and was appointed Head of the Chair of Philosophy, succeeding Kazimierz Twardowski. The Second World War dramatically interrupted educational and research activity at Jan Kazimierz University. In 1940 the university was renamed to Ivan Franko University and Roman Ingarden taught there until 1941. Eugenia Ginsberg-Blaustein, together with her husband Leopold Blaustein, belonged to the second generation of Twardowski's pupils (who were also Kazimierz Ajdukiewicz's pupils) and like her husband was under the influence of Ingarden's research. The Blausteins shared the tragic fate of many other Jews, and together with their son, they were executed by the Nazis.

The structure of my paper is as follows:

1. First, I shall give some basic definitions and theorems from Roman Ingarden's ontology.
2. Then, I shall use them in my analysis of mental content.
3. I shall finish my considerations with a brief conclusion, presenting the results which should contribute to the current research in the philosophy of mind.

2 Ingarden's Definitions and Theorems

I start my considerations with some chosen definitions and theorems from Ingarden's ontology. In some cases, I shall not present them in exactly the same formulations which they have in Ingarden's original text, nevertheless, the formulations, while sufficient for the needs of my analysis, will retain the senses intended by Ingarden.

In his *Controversy over the Existence of the World* [20, p. 109] Ingarden distinguishes four different pairs of opposite existential moments which are:

1. autonomy—heteronomy
2. originality—derivativeness
3. selfsufficiency—non-selfsufficiency
4. independence—dependence.

These existential moments of a given object always occur within the framework of modes of existence of this object and are distinguished only in abstraction. Starting his ontological considerations with the analysis of existential moments, Ingarden as a phenomenologist makes no claim about the existence of the objects considered, but merely attempts to discover which existential moments are proper for particular modes of existence, such as: absolute, real, ideal and intentional.[2]

I shall start with the definition of a self-existing object.[3] The notion of self-existence is treated by Ingarden as synonymous to 'existential autonomy' and such a translation is applied by the translator Arthur Szylewicz in the recent English edition of vol. I of *The Controversy over the Existence of the World* [20].

A definition of self-existence, i.e. existential autonomy

D1. Something is selfexisting, i.e. existentially autonomous, if it has its existential foundation within itself [20, p. 109].[4]

[2]The philosophers who work on Ingarden's ontology considers the problem of the interpretation of the modes of existence (sometimes attempting to interpret them improperly in the category of relations or properties). Ingarden was, however, clear in this issue, and his pupils and associates provided elucidative responses to those erroneous interpretations.

[3]This is an exact translation of the Polish notion 'samoistny'. This translation was proposed by Helen Michejda in her first English translation of some fragments of Ingarden's *Spór o Istnienie Świata*. Later, however, she changed it for 'existential autonomy' [18]. See the remark made by A. Szylewicz in footnote 247 of [20].

[4]Ingarden writes down his definitions as conditional statements. I am, however, more inclined to write them down in the form of equivalence using the sign 'iff' (if and only if) in place of 'if'. It seems to

It has its existential foundation within itself, if it is immanently determined within itself. This determination is thanks to its own essence (in Husserl's terminology *Eigenwesen*) [20, p. 115]. In Husserl's view, the essence of an object is its constitutive nature. Ingarden, taking this assumption from Husserl, adds—in agreement with his three-aspectual definition of an object—that the essence of an object consists of its properties together with formal and existential moments [17, p. 208]. Earlier in *Essentiale Fragen* [15] he defined the essence of an individual object only as the material (qualitative) equipment of this object. In consequence he propounds the following thesis:

T1. A selfexisting individual object is unambiguously completely determined in every aspect of its qualitative equipment.

Thanks to this completeness it is comprehensively separated from other entities, and as separated and total, it is formally perfectly closed in itself (in the particular case of a real object, it is spatially closed). Nevertheless, some selfexisting objects (real objects, for example) can enter into interaction with their environment thanks to their material (qualitative) properties.
A definition of non-self-existence, i.e. existential heteronomy

D2. Something is non-selfexisting, i.e. exists heteronomously, if it has its existential foundation 'outside of itself', i.e. in something else [20, p. 109f].

In consequence, it has such essential properties which have been given to it from outside. It means that it does not have them within itself. The above definition allows us to formulate the following thesis:

T2. Every property is existentially non-selfsufficient in regard to an object to which it is ascribed.

A definition of an existentially original object

D3. An object is existentially original if, in accordance with its essence, it cannot be produced by any other object [20, p. 118].

Thus, originality is ascribed only to an autonomous object the essence of which forces it into existence [20, p. 118]. But an autonomous object does not have to be existentially original (for example a real object is autonomous, but not original).
A definition of an existentially derivative object

D4. An object is existentially derivative if it can or does exist in virtue of having been produced by another object [20, p. 118].

An existentially derivative object can be either existentially autonomous or heteronomous. This is why there are different kinds of derivativeness. Each derivative object is existentially imperfect which is seen not only in derivativeness of its existence, but also in its

me more adequate for definitions if only 'something' is not understood in the narrow sense of an object as a thing. Ingarden, however, uses the notion 'object' where his translator Szylewicz uses the notion 'entity'. In D1 Ingarden in fact, uses the word 'something' (Szylewicz uses here the word 'entity'), but his comment given in a footnote makes it clear that Ingarden is aware of danger of its antinomy and treats 'something' merely as a helpful abbreviation. On the analysis of Ingarden's ontology see also [25].

fragility and brittleness of existence. As Ingarden says 'no derivative object can exist [...], but once it has originated (for whatever reason) and does exist, it can always cease to exist' because its own essence does not sustain it in existence [20, p. 144f].
A definition of an existentially selfsufficient object

D5. An object is existentially selfsufficient, if in accordance with its essence, it requires for its existence no other entity which would have to coexist with it within the unity of some whole [20, p. 147].

A definition of existential non-selfsufficiency

D6. Something is existentially non-selfsufficient if, as implied by its essence, its existence involves a necessary coexistence with some other entity [20, p. 147].

Non-selfsufficiency has different degrees, for example, the colour red as a feature of an individual red object (let's say the red colour of the rose in the vase on my table) can be distinguished from the moment 'redness' in the colour red. In first case, the red colour coexists with an object (this rose, for instance) as its bearer, in the second one (where non-sufficiency is of a higher degree) the moment 'redness' (beyond its bearer) coexists also with colouration within the whole of 'red colour'. Non-sufficiency—as is emphasized by Ingarden—is characterized by peculiar relativity [20, p. 149]. There are different kinds of existential non-sufficiency the variations of which are distinguished with regard to different criteria [20, p. 152].

Within the framework of existential selfsufficient objects, Ingarden distinguishes between existentially dependent and independent objects.

D7. An object is existentially dependent if in order to continue its existence it requires another selfsufficient object [20, p. 153].
D8. An existentially selfsufficient object is independent if, in virtue of its essence it requires no other selfsufficient object for its own existence. In this sense it is absolutely independent [20, p. 153].

The above definitions have their application in distinguishing among different modes of existence of objects. Because my further considerations will be focused on mental sphere with its intentional objects I shall refer also to Ingarden's definitions of pure intentional objects. Speaking about pure intentional objects Ingarden distinguishes between primary and secondary ones.

A definition of an intentional object

D9. An object is primary purely intentional if it has been directly or indirectly produced by someone's acts of consciousness in such a way that thanks to their immanent intentionality it has its source of existence and complete equipment (as it is, for example, in someone's imagination).
D10. An object is secondary (as derived) purely intentional if its intentionality is given to it (as for instance sense of utterances, linguistic expressions).

An intentional object has a two-sided formal structure, that is: (1) the content consists of the non-intuitive content which belongs to the act of supposition, and (2) the intentional structure. Ingarden (like Husserl as well as representatives of the Lvov School) makes also distinctions between judgments and suppositions in regard to the function of assertion. Judgments in strict sense are sentences which have the function of assertion, while suppositions lack this function. This division has its grounds in Platonic epistemological distinction between *episteme* and *doksa* (opinion).

I shall stop with these initial formulas without further developing Ingarden's exceptionally rich analysis, and I shall go to on to analyse the mental sphere of thoughts.

3 Dependence in the Mental Sphere of Thoughts

The subject of my analysis is the content of someone's beliefs, wishes, desires, etc. simply someone's thoughts, not in the Fregean sense, but rather in the epistemological-psychological approach, when thoughts occur in someone's process of thinking. At the beginning, it is useful to make some terminological remarks concerning the basic notions such as: judgment, proposition and basic assumptions. I define judgment in the standard sense of epistemology as the product of someone's cognitive process of judging, presented in a propositional structure expressing its content which is affirmed or rejected by a given cognitive subject. I mean here only assertoric judgements. Affirmation or rejection is treated as a certain kind of objective operation of the cognitive subject (not psychologically where it would be subjective). This is objective as a cognitive operation in regards to reference to independently existing reality (when, for instance, someone asserts that in 2016 Poland celebrates 1050th anniversary of its baptism, and this is a fact which took place independently of this judgment). Reference to independently existing reality allows the philosopher evaluate a judgment, which in this aspect is treated the same as proposition in logical semantics. It can be useful here to assume the Kantian approach according to which proposition (Germ. *Satz*) is semantic content of judgment (Germ. *Urteil*).[5] Although Kant's treatment of judgment is not limited to assertoric (theoretical) judgments I am close to his cognitivism according to which judgment is a specific kind of cognition (Germ. *Erkenntnis*).

I make a distinction between judgment and belief. I use the notion 'judgment' appealing to this objective operation leading to objective knowledge according to usual standard criteria.[6] In consequence of this, each (assertoric) judgment is true or false in the system of objective knowledge, and not only of someone's knowledge. Beliefs are typically elements of someone's knowledge and as such I admit the case when someone is convinced about the truth of someone's belief while it is false. In logical analysis, this is distinguished by the propositional attitude of having knowledge which is drawn down by the operator 'K' ('to know') and by the attitude of having a belief which in turn is drawn down by the operator 'B' ('to believe'). The philosophers of mind who espouse analytic philosophy respect this distinction without referring to the traditional epistemological concept of 'judgment', only to the logical notion of 'proposition'. The notion of 'judgment' is used by analytic philosophers, but in epistemology and in the philosophy of science, it is used mainly in reference to the issue of justification. As

[5]There is an extraordinarily rich literature on Kant's theory of judgments and numerous comments on his treatment of the relation between judgment and proposition (here in the sense of its content).

[6]I do not discuss here the problem of these criteria which—as it is known—belongs to the central topics of analytic epistemology.

it is known from philosophy, the issue of judgment has a long history, and convey lots of particular problems. At the beginning of the twentieth century, the concept of judgment analysed in connection with the concept of proposition was especially important in overcoming logical psychologism (which was observed in Husserl's *Logical Investigations* 1900/01 [11] and his *Formal and Transcendental Logic* 1929 [10]).

As my starting point, I assume simply that our thoughts have contents (which can be referred to the Brentanian tradition) and thoughts require a thinking subject (a Cartesian approach). However, my approach is not Cartesian because contrary to Cartesian internalism I take the externalist position according to which contents (in the sense of Germ. *Inhalt*) of our thoughts are acquired through interactions with external environment. It is my answer to the question 'what causes the content of our thoughts?' If I were asked whether the subject would have any thoughts without any interactions with its external environment, including other persons, I would reply 'no, it wouldn't'. It does not mean that I exclude any internal determination because in order to have thoughts the subject must have a proper 'cognitive equipment' (which is described by disciplines of cognitive science). However, I want to emphasize the distinction between searching for the explanation of having thoughts and for the explanation of their contents. The first one requires referring to the cognitive 'equipment', i.e. the faculties of the cognitive subject, while the second one requires mainly the research concerning the character of mental content (first of all its nature and structure). Hence, the explanation of having thoughts can be advanced on the grounds of philosophical epistemology and it is the aim for philosophy of mind and cognitive science. Especially important in this research both for philosophers and scientists, is the Kantian question 'how cognition is possible?', which can be reformulated here as 'how having (conceptual) thoughts is possible?', in other words 'what is required in order to have thoughts?', 'what conditions should be satisfied by the subject?'.[7] In the second issue concerning the mental content the explanation is searched for today also in the philosophy of mind, but in connection with research in cognitive semantics. In both kinds of issues I shall refer to Ingarden's research.

At the level of epistemological considerations in which I refer to the Brentanian tradition, we can say that thoughts are produced in acts of consciousness. In my considerations I take into account only such processes of thinking in which someone consciously holds certain beliefs, desires, makes some decisions, makes certain plans, etc., although they are connected with other beliefs, desires, etc. of a given person who does not need to be aware of them at present. I assume that beliefs, wishes, desires, etc. have propositional (i.e. conceptual) character. This means that the content of the mental state of believing, judging, desiring etc. is in a form of proposition (it is a propositional content). Immediately the problem raises whether a propositional content is mental, as I have already stated announcing the subject of my interest, which is the mental content, or whether it is abstract as proposition itself. It is an ontological question. Today both issues: the issue of mental content and the issue of proposition are controversial. In philosophical considerations, even the concept of proposition is not so clear as it is in formal semantics (where it is a primary bearer of truth-value) and its definition poses many problems and

[7] I purposefully use here the term 'cognition', not 'knowledge' in reference to the German word '*Erkenntnis*' although Kant's well-known question from his *Critique of Pure Reason* is usually quoted as 'What can I know?', and further 'What should I do?' and 'What may I hope?'.

provokes discussions even today (e.g. [13, 14]). In order to avoid confusion I reply that in my approach the propositional content is mental, but only in the sense that it is the content (just a mental content) of someone's mental state. Because I take into account only the structured content in a propositional form this means that also a process of thinking which is taken here into account is structured and consists of linguistic units.[8] Mental content is expressed in them in propositions. I do not prejudge here whether every process of thinking is linguistic, although I am interested only in this kind of processes. This is why I take into account conceptual thoughts which in the process of thinking have propositional forms.[9] I assume that propositions in their nature are primary logical items (well-formed both syntactically and semantically), thus mind-independent. For my analysis here, it is enough to say that propositionally structured mental content (i.e. in proposition) is shareable by different subjects as well as by the same person at different times. In virtue of their logical nature, propositions (as mind-independent) are treated as constituents of objective knowledge (without any cognitive subject, as it might be said in Popper's terms).

However, in my considerations I deal with propositional thoughts of cognitive subjects. A given cognitive subject entertains different attitudes to propositional content, such as: affirmation or rejection (for judgments), doubt (for problems), duty (for decisions for instance), etc. Someone's thoughts, considered here in propositional forms, can also be linguistically expressed by a given person in speech or in written language. A linguistic form gives them the attribute of inter-subjectivity, and in this sense—objectivity. Thoughts given in a propositional form (performed as judgments) develop certain states of affairs, appropriate to their contents. I shall refer here to Ingarden's theory, according to which propositional content develops a state of affairs, proper for this propositional content, which is a pure intentional equivalent, so-called *objectum formale* of a given judgment.[10] Thus, the intentional state of affairs is never selfexisting (autonomous), but heteronomous with regard to its judgment. If, however, a propositional content is also satisfied by a state of affairs which occurs independently of the occurrence of that judgment, then we have to do with an objective state of affairs, so-called *objectum materiale* of this judgment, which has not only a function of predication as a sentence, but also a function of assertion. Assertion is a proper function of judgment thanks to which a given judgment is true.[11] I

[8]I do not refer to any phenomenal content which is beyond my interest here.

[9]The issue of the relation between thought (or process of thinking) and language belongs to the crucial topics of contemporary debates in analytic philosophy, cognitive linguistics and psychology. In the Polish literature an outstanding monograph on this matter is the book by Arkadiusz Gut [8].

[10]The issue of states of affairs has been developed under the influence of philosophical ideas of G.W. Leibniz, B. Bolzano, G. Frege, H. Lotze, the Brentanists (as K. Stumpf, A. Marty), and especially of Husserl, who dealt with it not only in epistemology, but also in formal ontology. The concept of the state of affairs (facts) has played an important role in formal logic and also in the philosophy of logic, thanks to L. Wittgenstein and B. Russell. In the philosophy of logic the state of affairs is treated as an abstract object existing in possible worlds, identified with proposition or distinguished from it as its correlate, defined also as a possible object of entertainment or assertion. In *Logical Investigation* [11] Husserl proposed the concept of the state of affairs (Germ. *Sachverhalte*) as an objective correlate of judgment (Germ. *Urteil*). In the rich literature on this topic see the overview papers by Chisholm [3], Smith [24], and a monographic book by Armstrong [1].

[11]Ingarden distinguishes between different kinds of judgments according to their formal structure, which is most evident in his analysis concerning categorical judgments and conditional judgments. Only the

assume that also other kinds of thoughts (being someone's desires, decisions, etc.), if only they are given a propositional form, develop, thanks to their content, intentional states of affairs, appropriately to their formal structure.

With reference to psychological considerations, I assume the occurrence of propositional content also in the case of a child who still does not know any ethnic language, but if only it is able to say something, even only syllables (for example, 'ma-ma') in reference to something in a meaningful way, then such a phrase can be treated as an abbreviation of a sentence ('this is the mother') which expresses a proposition (as the meaning of this sentence). I take into account such a level of the cognitive development of the child at which it is able to have conceptual thoughts. I assume the same in the case of some impairment in verbal communication, for instance in Broca's (motor) aphasia when a person is not able to communicate verbally in a proper way, but can communicate by means of single words, which are meaningful for her and a receiver tries to recognize their meaning.

Further, I assume that someone's beliefs constitutes his (or her) knowledge. This knowledge, enriched by the subject's wishes, desires, decisions, etc., constitutes the subject's mental life. In our body of knowledge, we can have a great number of shared beliefs which function in our common communication or belong to objective scientific knowledge. We can also share some desires, wishes, have similar intentions, make the same or similar decisions in a given situation. Some of them are common, others are quite subjective. Each person has their own personal experience, their own mental life, their own unique personality and identity, but can develop only by interactions with other persons. What needs to be emphasized here is the difference between the psychological approach (or the approach of psychophysical epistemology) and the approach of phenomenological pure epistemology. In the latter (developing in the Cartesian and Kantian tradition) the cognitive subject is a pure subject, namely pure 'self' being merely a 'satisfier' of his own conscious acts. Such an approach to the cognitive subject is, however, controversial, even among phenomenologists themselves which is seen in discussions between Ingarden and Husserl (when Husserl in his *Ideas* [12] was in favour of the existence of pure 'self').

Now I turn to the basic ontological notions by means of which I shall further analyze mental content, which is treated here as belonging to someone's thought. In the approach analyzed, someone's thought has no guarantee of its existence in itself, but finally in acts of consciousness of the cognitive subject. At the epistemological level in the structure of consciousness, following Ingarden's analysis (based on Twardowski's distinction, made also by Husserl in his *Logical Investigations* [11]), I distinguish between act, its content and an external object, pointed by the content thanks to an intentional moment of a given

first ones satisfy the function of assertion. Beyond assertoric judgments Ingarden also intended to analyze modal (problematic and necessary) judgments. This kind of analysis could be interesting when asking a metaphilosophical question concerning the relationship between the epistemological and semantic approaches (here possible world semantics). The difference between the epistemological notion of 'judgment' in a philosophical theory of modal judgments and the logical notion 'proposition' is blurred when modality is ascribed to the linguistic unit or to the subject of judgments, i.e. to states of affairs, but it remains relevant if modality is treated in the Kantian sense and then it is associated with the attitude of judging subject. On the application of possible world semantics to Ingarden's ontology see [22].

act.[12] Thus, the mental content considered as the content of someone's conscious act is not selfexisting, but heteronomous with regard to this act. If we have to do with an act of judging, then its mental content constitutes the content of judgment as the product of this act. As it has been already said, the content of judgment develops an intentional state of affairs. But as it has been also said, judgment in its function of assertion maintains a state of affairs which exists independently in the domain in which a given judgment 'locates' it. It is the domain of real or ideal objects. According to Ingarden's ontology, a domain is 'natural' (autonomous) plurality of autonomous objects. As it is known, a large body of Ingarden's research pertains to the theory of literary work, in which he analyzes the so-called quasi-judgments (for example, 'Oedipus loves Jocasta', 'Macbeth murders the king Duncan') with their reference to intentional objects. Intentional objects, however, do not constitute any domain in the strict sense of Ingarden's definition of this term (nevertheless, quasi-judgments can be evaluated according to the world presented in fiction).

Further, it is worth to emphasize Ingarden's radically realistic approach, according to which no act of consciousness is able to 'burst into'—as he says metaphorically—the domain of real objects and to make any changes in them [16, p. 557], [19, p. 466]. There is no existential connection between them, only a purely mental assignment. Thus, both spheres are in the above ontic sense existentially selfsufficient (mutually separated). In this way he defines real transcendency between the domain of real objects and consciousness. But also the domain of ideal objects is selfsufficient, separated and independent from acts of consciousness. I do not claim that there are any ideal objects in the sense of Ingarden's ontology (i.e. ideas and individual ideal objects). Nevertheless, I maintain that Ingarden's definition of objective correlates of judgments referring to the ideal domain can be applied in the domain of science, where scientific language is used in mathematical modelling of different aspects of reality (its physical processes or events). These models, being highly abstract in their character, are idealizations of the examined reality.

In my considerations, as I have stressed, I take into account not only judgments (in the strict sense of 'judgment'), but all these thoughts which are meaningful and expressed in a propositional form. As expressed in a propositional form they are products of certain linguistic operations, roughly speaking—operations for sentence production (as used in Ingarden's terminology). As meaningful, they are in turn products of significant acts which ascribe meaning to them. Following Ingarden's theory, the meaning of a linguistic unit (in which I also include propositional thoughts) is a purely intentional product of object apprehension consisting of a direction indicator, existential moments of a given object (whether it is real or ideal or only intentional, for example), its existential position (i.e. affirmation or not affirmation in existence) and its formal and material content. Formal content determines the formal structure of an object (whether it is an individual, an event, a process or something else) as well as the logical and grammatical form of a given linguistic unit, and material content determines a qualitative characterization of a given object. Ingarden's theory of meaning like traditional theories starts with the characterization of nominal meaning (meaning of names) as selfsufficient (but not autonomous, in the sense of semiotics—not selfinforming) linguistic units. Contrary to

[12] I make only very simplified remarks on intentionality in reference to Ingarden's theory, which is much richer than Brentano's and even than Husserl's theory of intentionality. On certain comparative analysis of Ingarden's and Husserl's theories see: Artur Chrudzimski [4] among others.

Ingarden, my interest is more focused on the meaning of sentences (not names [21]) because in my view the basic structure of our thoughts is already propositional. In this point, I direct my considerations more towards the Fregean compositional semantics which in this aspect is closer to current approaches.

Looking, however, from the epistemological point of view, meaning is not only considered semantically, but as intentional it is prescribed to linguistic units in significant acts of consciousness. In this way the intentional correlates of meaning are heteronomous, existentially dependent and derivative; primarily from the meaning itself, and secondarily from given significant acts.

A reader of this paper can ask here whether 'meaning' (sense) is the same as 'mental content'? If it were the same, then meanings would be only something mental in our heads [5]. Because I take advantage of Ingarden's conception, again I refer to his view. Ingarden attempts to grasp all aspects of meaning, that is, both its stable element and its dynamicity which is an effect of all changes of language. This is why the meaning of linguistic units is characterized as a complex structure (which has already been shown by its constituents). This is not all, because the meaning of linguistic units also depends on their context, both linguistic and situational. From a psychological point of view, situations in which a cognitive subject finds themselves have influence on his (or her) mental states, not only cognitive acts, but also emotional or sensitive ones. The content of someone's mental state be expressed linguistically in an adequate linguistic form and choice of utterances. In a speech act also the way of uttering sounds, the timbre and tone of someone's speech and the behaviour accompanying the speech act have a role in expressing the content of someone's thoughts. In different situations someone can manifest or hide his or her mental states. Thus, the meaning which is associated with utterances is not the same as the mental content of someone's thoughts although we can take into account the same schema for their characterization. Nevertheless, in the cognitive function, judgments and sentences which are used to perform the content of thoughts are transcendent with regard to someone's experiences. Then also the mental content in its propositional form can remain the same in repeated states of judging at different times or by different persons. As such it becomes a certain stable element of meaning. Under Husserl's influence, in his early works Ingarden assumes the occurrence of the so-called ideal concepts. In spite of his later intentional conception of meaning, he maintains a certain stable element of meaning without which no communication would be possible. What is also very important, his intentionalism in reference to meaning does not imply any psychological conception of meaning, but on the contrary, it is important in argumentation against psychologism, as in the case of Husserl[13] and late Twardowski.[14]

[13] Husserl only in his early work, which is *Philosophie der Arithmetik* (1891) [9] manifested traditional psychologistic tendencies, which immediately met objections from Frege. But according to Dan Zahavi's recent view, Husserl's criticism of psychologism and his own antipsychologistic position was not motivated by Frege's objection, but was an effect of his studies on H. Lotze and A. Pfänder writings [27]. Husserl's conception of meaning as an ideal object, in turn, is interpreted in the sense of logic which is close to Frege's theory, and not in the ontological Platonic approach.

[14] It is, however, not quite clear whether Twardowski overcame psychologism—as Jan Woleński [26, p. 41] pointed out referring to Izydora Dąmbska who observed that Twardowski's approach to meaning is not quite autonomous in regards to experience [6]. Ginsberg-Blaustein appeals to Twardowski's analysis of existential dependence and independence, but only in question of metaphysical parts and the whole.

From an externalist point of view, I assume that the mental content of our thoughts is determined by objective states of affairs, situations, events or facts. But I also admit inner determination. At the internal level, I take into account biological factors of our organisms, which occur in various consciously experienced states of the organism and psychological states. In the first ones, I include, for instance being tired, thirsty, etc. If a person is aware of these states, and focuses on them, then she is in a state of sensitive inner perception (as Husserl and Ingarden maintain). Thus these states can also determine the mental content of such thoughts as 'I feel tired', 'I am thirsty' and so on. I do not use the notion 'determine' in a strict sense of determination because the mechanism for sentence formation belongs to cognitive operations and it is not the same mechanism, which is biologically activated when the organism is in the state of being tired, thirsty and so on. I rather want to say that the cognitive subject must have special faculties to perceive such states, to be aware of them, to distinguish among them, and to express them in a linguistic form. The inner state of the organism is here in a certain analogy to an external state of affairs to which someone refers when, for example, he or she thinks about something in their environment. In spite of the fact that the state of being thirsty, as a sensitive state, is intuitive (in the sense of Twardowski, Husserl, Ingarden), the cognitive subject is able to associate his or her conceptual thoughts with it. Thus, a cognitive subject who has among his or her faculties the faculty of producing conceptual thoughts is able to refer to this state of his or her organism in the propositional form of his or her thought.

In the second kind of states—as has already been said—I include the states of emotion, feelings, as well as earlier experiences. In this case Husserl and Ingarden, talk about non-sensitive inner perception, as for instance 'I am happy', 'I am depressed' and so on, and not about sensitive inner perception. Also in these cases, I take into account only the propositional mental content which, in fact, is contained in someone's thoughts associated with these states, although their content is non-intuitive (as it is stressed in phenomenological epistemology and Twardowski's works).

Beyond that, as in a standard semantic analysis, I distinguish between content and a bearer of content. I say that the bearer of mental content is someone's thought. But at the neuronal level the vehicle for this meaningful thought is a certain configuration of neural network. One of the most important aims of interdisciplinary research in cognitive science is just to resolve the Kantian problem by showing how all these levels are integrated in the cognitive subject in his or her mental life, if only restricted to having propositional thoughts. But this is a topic for another discussion.

4 Concluding Remarks

In this paper, which was inspired by Ginsberg-Blaustein's analysis of the concepts of existential dependence and independence, I took advantage of Ingarden's analysis in application to my considerations on mental contents occurring in the states of judging, believing, etc. For the purpose of my considerations, I established the basic terminology (to make a distinction between judgment, belief, proposition on the one hand, and conscious act, its mental content and its object—on the other hand), and I presented some epistemological and ontological assumptions of Ingarden's ontology. The results of my considerations can be briefly summarized as follows:

1. Referring to the analytical tradition of the Lvov School and Ingarden's phenomenological approach allows us to see better the ontological assumptions which are important in the analysis of mental content.
2. Mental content occurring in the states of judging, believing, desiring, etc., has a propositional character and as such it can be sharable by different subjects or by the same person at different times.
3. Following Ingarden's view, I claim that in spite of the fact that mental content is propositional, it is not the same thing as linguistic meaning.
4. Mental content is determined both externally (by external objects and states of affairs which are cognitively accessible to the cognitive subject) and intrinsically thanks to special cognitive faculties of the cognitive subject who, being in different states (not only cognitive, but also sensitive) is able to produce conceptual thoughts.

Acknowledgements My special thanks are due to Dr Tomasz Fojt for his proofreading and helpful suggestions and to Mateusz Klonowski for his technical support.

References

1. Armstrong, D.: A World of States of Affairs. Cambridge University Press, Cambridge (1997)
2. Blausteinowa z Ginsbergów, E.: W sprawie pojęć samoistności i niesamoistności. In: Księga Pamiątkowa Polskiego Towarzystwa Filozoficznego we Lwowie 12.II.1904–12.II.1929. Polskie Towarzystwo Filozoficzne, Lwów (1931) (English translation: Ginsberg-Blaustein, E.: On the concepts of existential dependence and independence. In: Smith, B. (ed.) Parts and Moments. Studies in Logic and Formal Ontology, pp. 265–287. Philosophia Verlag, Munich (1982). English translation P. Simons)
3. Chisholm, R.: States of affairs again. Noûs **5**, 179–189 (1971)
4. Chrudzimski, A.: Teoria intencjonalności Romana Ingardena [Roman Ingarden's theory of intentionality]. Edukacja Filozoficzna **25**, 249–262 (1998)
5. Chrudzimski, A.: Are meanings in the head? Ingarden's theory of meaning. J. Br. Soc. Phenomenol. **30**(31), 306–326 (1999)
6. Dąmbska, I.: Semiotyczne koncepcje w filozofii Twardowskiego [Semiotical conceptions in Twardowski's philosophy]. Ruch Filozoficzny **27**(1), 1–9 (1969)
7. Ginsberg-Blaustein, E.: On the concepts of existential dependence and independence. In: Smith, B. (ed.) Parts and Moments. Studies in Logic and Formal Ontology, pp. 265–287. Philosophia Verlag, Munich (1982). English translation P. Simons
8. Gut, A.: O Relacji Między Myśla a Językiem. Studium Krytyczne Stanowisk Utożsamiających Myśl z Językiem [On the Relationship between Thought and Language. A Critical Study of Approaches That Identifies Thought with Language]. KUL, Lublin (2009)
9. Husserl, E.: Philosophie der Arithmetik. Psychologische und Logische Untersuchungen. In: Eley, L. (ed.) Gesammelte Werke. Husserliana, vol. 12. M. Nijhoff, The Hague (1891/1970)
10. Husserl, E.: Formal and Transcendental Logic. M. Nijhoff, The Hague (1969). Transl. D. Cairns (original work published 1929)
11. Husserl, E.: Logical Investigations. Routledge, London (1970, repr. 2001). Trans. J.N. Findlay from the second German edition 1913
12. Husserl, E.: Ideas Pertaining to a Pure Phenomenology and to a Phenomenological Philosophy. First Book: General Introduction to Pure Phenomenology. M. Nijhoff, The Hague (1982). Trans. F. Kersten (original work published 1913)

13. Iacona, A.: Propositions. Name, Genoa (2002)
14. Iacona, A.: Are there propositions. Erkenntnis **58**, 325–351 (2003)
15. Ingarden, R.: Essentiale Fragen. Ein Beitrag zum Problem des Wesens. Jahrbuch für Philosophie und Phänomenologische Forschung **7**, 125–304 (1925) (Polish translation in: Ingarden, R.: Z Teorii Języka i Filozoficznych Podstaw Logiki [From the Theory of Language and Philosophical Foundations of Logic], pp. 327–482. Państwowe Wydawnictwo Naukowe [Polish Scientific Publishers], Warszawa (1972))
16. Ingarden, R.: Spór o Istnienie Świata [Controversy over the Existence of the World], vol. II, 2nd edn. Państwowe Wydawnictwo Naukowe, Warszawa (1961)
17. Ingarden, R.: Spór o Istnienie Świata [Controversy over the Existence of the World], vol. I, 2nd edn. Państwowe Wydawnictwo Naukowe, Warszawa (1962)
18. Ingarden, R.: Time and Modes of Being. Charles Thomas, Springfield, IL (1964). Partial transl. of Ingarden, R.: Essentiale Fragen. Ein Beitrag zum Problem des Wesens. Jahrbuch für Philosophie und Phänomenologische Forschung **7**, 125–304 (1925) by H. Michejda
19. Ingarden, R.: Z Teorii Języka i Filozoficznych Podstaw Logiki [From the Theory of Language and Philosophical Foundations of Logic]. Państwowe Wydawnictwo Naukowe [Polish Scientific Publishers], Warszawa (1972)
20. Ingarden, R.: Controversy Over the Existence of the World, vol. I. Peter Lang Edition, Frankfurt a. Main (2013). Translated and annotated by Arthur Szylewicz
21. Jadacki, J.J.: O poglądach Romana Ingardena na język. Stud. Semiotyczne **5**, 22–54 (1974). English edition: On Roman Ingarden's semiotic views. Analecta Husserliana **27**, 523–540 (1989)
22. Krysztofiak, W.: The platonic idealism controversy in light of possible worlds semantics. Axiomathes **17**(1), 75–90 (2007)
23. Magdziak, M.: A logical analysis of existential dependence and some other ontological concepts. A comments to some ideas of Eugenia Ginsberg-Blaustein. Axioms **5**(3), 4–10 (2016). Ed. Urszula Wybraniec-Skardowska
24. Smith, B.: Logic and the Sachverhalt. Monist **72**(1), 52–69 (1989)
25. Wachter von, D.: Roman Ingarden's ontology. Existential dependence, substances, ideas and other things empirists do not like. In: Chrudzimski, A. (ed.) Existence, Culture and Persons. The Ontology of Roman Ingarden, pp. 55–82. Ontos, Frankfurt a. Main (2005)
26. Woleński, J.: Filozoficzna Szkoła Lwowsko-Warszawska. Państwowe Wydawnictwo Naukowe, Warszawa (1985). English edition: Logic and Philosophy in the Lvov-Warsaw School. Kluver, Dordrecht (1989)
27. Zahavi, D.: Husserl's Phenomenology. Stanford University Press, Stanford (2002)

U. M. Żegleń (✉)
Department of Cognitive Science and Epistemology, Institute of Philosophy, Nicolaus Copernicus University, Toruń, Poland
e-mail: zeglen@umk.pl

Józef I.M. Bocheński

Korneliusz Policki

Abstract Bocheński's philosophical biography consists in his original attempt to join the method of the Lvov-Warsaw School with an open Thomism. Paradigm changes in Bocheński's philosophy were preceded by changes in science, politics, sociology as well as ecclesiology.

Keywords History of logic · Philosophical methodology · Thomism · Sovietology · Logic of religion · Logic of authority · Logic of the business enterprise

Mathematics Subject Classification (2000) Primary 01A70

1 Life

J.I.M. Bocheński was born August 30, 1902 in Czuszów in the vicinity of Cracow. In 1920 he began studies in law and economics in Lwów (Lemberg, Lvov), completing them at the University of Poznań. He entered the Dominican order in 1927, and a year later he began to study philosophy at the University of Fribourg (CH) where he was awarded the doctoral degree. Theological studies followed at the Angelicum in Rome and culminated as well in the doctoral degree in 1935. In the following year he acceded to the 'Habilitation' in philosophy at the University of Cracow and began teaching logic. Throughout the period of the World War II he interrupted his scholarly pursuits in order to devote himself to the struggle for freedom in his capacity as military chaplain.

Following the cessation of hostilities Bocheński is named to the chair of modern and contemporary philosophy in the University of Fribourg, a position he held until his retirement in 1972. In the period 1951–1952 he was dean of the Faculté des lettres and in the years 1964–1966 rector of the University. In the latter capacity Bocheński was active in expanding the university. Moreover, in 1958 Bocheński founded, in Fribourg, the Institute of East European Studies, and 4 years later he became, thanks to Konrad Adenauer's support, cofounder and director of the Ost-Kolleg in Köln.

On November 14, 1992 Bocheński was granted the order 'Polonia Restituta' on the recommendation of President Lech Wałęsa and his scholarly achievements earned him entry into the august society of the Polish Academy of Science. Bocheński's long and

Á. Garrido, U. Wybraniec-Skardowska (eds.), *The Lvov-Warsaw School. Past and Present*, Studies in Universal Logic,
https://doi.org/10.1007/978-3-319-65430-0_20

productive life came to end on February 8, 1995, in Fribourg. The urn containing his ashes is found in the Dominican sepulcher of the Albertinum.

The following statistics, derived from Bocheński's curriculum vitae, are revealing:

- he served in three military campaigns,
- he studied over a course of 13 years in four different university faculties,
- he earned two doctorates and the habilitation,
- he taught at least one semester in 12 universities in Europe, South America, North America, and Africa,
- he was named to professorships in the Angelicum, Rome, and the University of Fribourg,
- he was twice dean of the Faculté des lettres and later rector of the University of Fribourg,
- he was awarded five honorary doctorates,
- he published 44 books and 191 scholarly articles,
- he completed a journey round the world as pilot in command.

2 Main Research Interests

The scope of Bocheński's research and the range of his interests were very wide. His areas of research can be divided into fundamental directions to which he attributed priority and secondary interests that occupied him per accidens. The main areas are logic (the history of logic, Hindu logic, applied logic, the logic of religion, the logic of authority, the logic of the business enterprise) and philosophy (Thomism, analytic philosophy, sovietology).

Of the secondary interests mention should be made of political science, systems theory, philosophical psychology, economics, Egyptology, as well as sapiential philosophy.

Bocheński was a polyglot. He lectured and wrote in all major European languages, including Russian. He was at ease with classical languages: Greek, Latin, and Hebrew. He knew Sanskrit as well as Chinese. Given his historical interests he was adept at paleography.

2.1 The Main Writings

Bocheński's most important publications

1932 Die Lehre vom Ding an sich bei M. von Straszewski
1933 Elementalogica Graecae
1938 Nove lezioni di logicasimbolica
1939 De virtuti militari
1947 Petri Hispani Summuas Logicales
1947 La logique de Théophraste
1947 Europäische Philosophie der Gegenwart
1949 Précis de logique mathématique
1951 Ancient Formal Logic

1952 Der sowjetrussische dialektische Materialismus
1954 Die zeitgenössischen Denkmethoden
1956 Formale Logik
1958 Handbuch des Weltkommunismus (together with G. Niemeyer)
1959 Wege zum philosophischen Denken
1965 The Logic of Religion
1974 Guide to Marxist Philosophy (together with T.J. Blakeley)
1974 Was ist Autorität?/Qu'est-ce que l'autorité ?
1985 Zur Philosophie der industriellen Unternehmung
1987 Sto zabobonów
1987 Über den Sinn des Lebens und über die Philosophie
1988 Między logiką a wiarą z J.M. Bocheńskim rozmawia J. Parys/«Entre la logique et la foi»
1988 Autorität, Freiheit, Glaube
1992 Podręcznik mądrości tego świata
1993 Religia w Trylogii
1994 Sens życia
2003 Gottes Dasein und Wesen

3 Bocheński's Views

3.1 Logic and the History of Logic

Bocheński was a passionate devotee of logic. Thanks to his style of writing and thinking he is counted among the members of the Polish school of logic. He maintained lively contacts with many Polish logicians, including S. Leśniewski, L. Chwistek, K. Ajdukiewicz, T. Kotarbiński, A. Tarski, but in particular J. Łukasiewicz whom he counted as a friend.

How did Bocheński conceive logic? How many parts of logic did he distinguish? The following table renders Bocheński's understanding of the parts of logic.
Logic (in general)

(a) in the strict sense

- pure logic (formal, mathematical)
- applied logic
  - Methodology
  - Semiotics

    Syntactics
    Semantics
    Pragmatics

(b) in the broad sense

- history of logic
- philosophy of logic

Bocheński made substantive contributions to each of these areas of logic.

Logic can be understood from a subjective and objective perspective. From the subjective perspective logic, according to Bocheński, "[. . .] is the ability to move about, to engage in thinking at a very high level of abstraction." Understood objectively, logic is the description of objects, the description of things, it investigates the general structure of the world; it determines not only relations among sentences but first of all among things in general. According to Bocheński, formal logic is but ontology carried out axiomatically. After Aristotle's death a controversy arose concerning the question whether logic is mereos, that is, a part, or organon, that is, a tool for philosophy, or also pajdagogos, that is, a means of education. As regards their material objects, logic and ontology do not differ. As disciplines they are distinguished only by their methods, that is, their respective formal objects.

Bocheński often cited an example from Albert the Great to illustrate the meaning in which logic is a tool (organon): "It is certain that in his workshop the smith produces a hammer. And when the hammer is ready it serves in the same workshop as a tool with which to produce other tools." The same is true of logic. Logic is an autonomous science, but this special science is a necessary tool of the other sciences, in particular philosophy and theology.

Logic qua pajdagogos is nothing other than the ethics of thinking and speaking. It has major educational significance.

3.2 Philosophy

Bocheński ventured down the path of philosophy in the shadow of three professors whose names all begin with the letter 'Z'. All three exerted considerable influence on the young student Bocheński, in particular in regard to his style and way of philosophizing. His professor of mathematics at the gymnasium was Zygmunt Zawirski, an analytic philosopher associated with the "Lvov-Warsaw School" whose founder was Kazimierz Twardowski, a student of Franz Brentano. Zawirski would later be among the experts who evaluated Bocheński's habilitation presented to the Jagiellonian University in 1938.

The second professor whose influence played a formative role for Bocheński was Florian Znaniecki, a student of Henri Bergson, later the president of the American Association of Sociology. Znaniecki fascinated Bocheński; it was Znaniecki who at that time awakened Bocheński's enthusiasm for statistics and sociology.

Bocheński owed his passion for precision to the school of Czesław Znamierowski, the creator of the analytic philosophy of law. Philosophy is not to be likened to the painterly art where each artist creates a new work; quite to the contrary, the philosopher should analyze already existing 'paintings' and in this way enrich science. Indeed, Bocheński created no philosophical system, but he did enrich a number of philosophical domains thanks to several important and innovative contributions.

Among the latter the following are the most important:

1. Metaphilosophy Bocheński pioneered the first broad concept of analytic philosophy. He proposed as well the idea of the cyclical development of philosophy, the specific kind of development to which it is subject. He set out a clear definition of the concept of 'worldview' and argued strongly for the claim that a philosopher should not attend to issues having to do with worldviews.

2. Ontology Bocheński argued that formal logic is axiomatic ontology. He proposed a new formulation of moderate Platonism with regard to universals according to which what is at stake is basically a specific concept of identity. For example, "redness" refers to the identity of red objects. Further, he systematically worked out the concept of the formal system and its purpose. Early in his career, he carried out empirical research into the concept of causality.

3. Metaphysics Bocheński has reservations with regard to over hasty constructions of metaphysical systems. He published a program of investigation concerning God as well as a critical study of Aquinas' theodicy (I,2–11).

4. Epistemology Bocheński carried out a critical analysis of the argument in favor of skepticism and concluded that skepticism is a philosophical superstition. He likewise criticized contextualism. Further, he discussed certain aspects of medical methodology. But above all he was the first to advance an analysis of analogy understood as isomorphism with the means of mathematical logic.

5. Philosophy of Nature He advanced a naturalist perspective, claiming that for science there is no fundamental difference between man and other animals. He undertook as well an analysis of the concept of the meaning of life.

6. Social Philosophy He was the first to propose an extensive logical analysis of the concept of authority. He advanced a novel analysis of the commercial enterprise and the free society. In addition, he provided a formal logical analysis of the concept of responsibility.

7. Ethics Bocheński was most probably the first to write on the ethics of war. Further, he insisted on the separation of morality, ethics, sapiential wisdom, and different kinds of religious morality (fear of the Lord). He carried out a logical analysis of the controversy concerning animal experimentation.

8. Philosophy of Religion Bocheński proposed a novel definition of the philosophy of religion and he created a new scholarly domain, the logic of religion. He formulated the theory of the religious hypothesis in metanoia/repentance. He advanced as well the religious hypothesis distinguishing fundamental faith from religious systems as well as from the history of saints and all souls.

9. The Philosophy of Spirituality Bocheński distinguished fundamental visions [of the world] from worldviews. He carried out a still relevant analysis of the fundamental visions of the Middle Ages and the Renaissance. Catholicism and Islam are the constituents of the fundamental medieval vision. They are static and anthropocentric, inimical to the power of progress. Anthropocentrism remained present within the Renaissance vision, however the Renaissance believes in man's unlimited potential, in progress, science, and rationality generally. The criterion of these two visions should be "contemporary spiritual situation" (a concept borrowed from Jaspers).

10. Sovietology It was common in the West to believe that, in order to understand thought in the Soviet Union, it sufficed to read several key works of Karl Marx, a preconception that Bocheński rejected as erroneous. He distinguished marxology, leninology, and sovietology and argued that each of these areas of research is far too extensive for any one person to be able to encompass all of them. To be sure, a sovietologist would have to be well read in Marx and Lenin, however he need not and cannot be a specialist in these areas. The converse holds equally: a specialist in Marx is not for that reason a specialist in Leninism, and the latter specialist is not for all that a sovietologist. Each successive sovietological sub-discipline incorporated the core of the earlier one completing it in its specific manner all the way to Gorbachev's perestrojka. Bocheński never ceased to be irritated when he was identified as a marxologist, considering himself on the contrary to be a sovietologist.

11. Anthropology We find in Bocheński reflections on philosophical and theological anthropology, the latter particularly in his early career. His theological anthropology is situated between theism and soteriology. In his later development, Bocheński gave up hylomorphism in his conception of man in favor of systems theory and substituted meta-anthropology for anthropology. He was an opponent of scientific humanism, allowing however for religious and intuitive humanism. This went in pair with his naturalism and minimalist anthropology. He rejected ontological as well as methodological humanism. In the course of his development he gave up his early Thomist perspective on the concepts of freedom and responsibility in favor of an analysis applying the theory of logical relations. In like manner Bocheński set out a formalized proof of the existence of the human soul.

12. Axiology In his value theory Bocheński defended a moderate realism. The mode of existence of values is like that of universals. The foundation of values is a two-place logical relation between the subject and the object. In the set of theoretical values the relation to the good is of a contemplative—passive—nature. These are values that entail the ought of existence though not the ought of action. For example, aesthetic values: beauty, ugliness, elegance, coarseness, delicateness, etc. On the other hand, in the set of practical values the relation/attitude to the good is of an active nature. These values comprise both the ought of existence and the ought of action. Their character is appellative. The set of practical values includes two subsets: technical values and spiritual values. For technical values the object of the action is external to the agent (facere). The techniques of physical activity evince a conditional (hypothetical) character. As for spiritual values, in keeping with his Thomism Bocheński saw them as primarily moral values possessing a categorical, unconditional character. For these values the object of action (agere) is internal to the agent. These are the so-called techniques of spiritual activity. Thomist traditionally included religious values among the moral values. In his later analytic period, however, Bocheński follows Max Scheler in distinguishing religious values from moral values.

13. Therapeutic Philosophy Under the influence of Hindu philosophy with which he had come in contact, Bocheński, while studying this civilization's logic, took an interest in subjective philosophy. It is well known that the basic question of Hindu philosophy concerns freeing oneself from suffering. Even logic should be able to assist (16 categories of Hindu logic help in freeing oneself from suffering). Bocheński took up the question of suffering in several essays: "On suffering," "The Sin of sadness," "To the broken man," "The phenomenology of the abyss," "With regard to evil people," "The meaning of life," "A handbook of worldly wisdom." Suffering is at times so great that even religion cannot help the sufferer. Homo patiens is not entirely coextensive with animal rationale. Bocheński postulated the creation of a distinct interdisciplinary science of suffering.

3.3 Theology

Bocheński was likewise a theologian, of a Thomist and equally an analytic stripe. Wherever possible he applied logical means to theology and believed that the right logic for theology is many-valued logic. In his Logic of Religion he constructed a semantics of theological language. The logical definition of logic has the following form: "Theology can be described as the domain of investigation in which, along with other axioms, at least one sentence from the Credo is admitted and accepted exclusively by the faithful of the corresponding religion." Bocheński broadened this definition by means of a more liberal conception of the Credo. For example, the "Our Father" is the "Credo" of the so-called basic faith that is the common feature of all monotheistic religions.

Bocheński did not accept the fideist understanding of theology, credo qui absurdum, but rather the thesis "credo quia ratio."

Pope Jean Paul II drew on many of Bochenski's ideas in his encyclical 'Fides et ratio'.

With regard to theology Bocheński preferred the reductive model of theology. Methodologically, theology is similar to physics. The role of protocol sentences is played by those sentences of the faith which require clarification by means of theory and theological hypotheses.

4 Reactions

4.1 Political Reaction

Of all the members of the Lvov-Warsaw School Bocheński exerted the greatest political influence on the changes that occurred in Central Europe as well as on the rise of the Perestrojka in the Soviet Union.

Bocheński educated 2/3 of the philosophical sovietologists working in the United States.

Bocheński's sovietological studies contributed to the rise of Solidarity, the fall of the Berlin Wall and the overthrow of Communism.

He was a co-founder of the review 'Kultura' in Paris which exercised an enormous influence on the Polish intelligentsia.

Despite repressive means applied by the communist authorities in Poland, Bocheński's works published in his second period were read and discussed in workers' and academic milieus.

Thanks to Dominican pastoral centers Bocheński's works found their way into Catholic circles and contributed to the renewal of the Catholic Church following Vatican Two.

4.2 Scholarly Reaction

So far five symposia have been devoted, in Poland, to Bochenski's scholarly heritage and two in Switzerland. Many directions in Bocheńki's work continue to be discussed and developed.

Bocheński's Thomism is carried forward by Ignacy Dec, J. Krucina, K. Wolsza, and Czeslaw Glombik.

With regard to Sovietology, Marxism-Leninism as well as Russian philosophy, Bocheński's work has influenced Edward Świderski in Switzerland and in Poland Jan Parys.

Development of Bocheński's logical ideas has been the work of logicians in Warsaw—Edward Nieznański, K. Świętorzecka, Marek Porwolik, R. Tomanek, and Anna Brożek.

In the areas of history of logic and metalogic mention should be made of J. Angelelli, M. Brander, Hans Burckhard, Peter Rutz, and Dariusz Gabler.

The logic of religion has been pursued by Roger Pouivet, M. Kaemfert, E. Żabski, and Jan Woleński.

Bocheński's interest in phenomenology played a significant role in the work of Anna Teresa Tymieniecka and the meeting of analytic philosophy and phenomenology was pursued by Bocheński's former assistant and later successor at the University of Fribourg, Guido Küng.

Aspects of Bocheński's ethics are deployed in studies by J. Weinsenbeck, Helmut Fleischer, Czesław Porębski, and Korneliusz Policki.

4.3 The Educative-Didactic Effect

Bocheński directed more than 50 doctoral dissertations and an astounding number of master's and licentiate theses.

He published four textbooks of logic in which he presents, in a most accessible manner, recent research in logic. In the history of philosophy he published two textbooks that have been translated into several languages. Bocheński enjoyed considerable celebrity thanks to his popular introductions to philosophy, his guides to Marxism-Leninism as well as Thomism. Many of these works were likewise translated throughout the world.

Bocheński was also an excellent preacher and retreat master. His talks and sermons have been published following his death. Virtually up to his death he conducted a philosophical seminar in Polish named by him "Sempol."

5 Criticism

In Bocheński's later works, in particular in his second phase, a certain change of philosophical paradigm is noticeable, a certain 'perestrojka'.

What is evident is a passage from a maximalist to a minimalist philosophy, from metaphysics to ontology, from a synthetic to analytic philosophy, from humanism to naturalism, from ethics to metaethics, from hylomorphism to systems theory.

Throughout his life Bocheński remained a realist, a defender of objectivity, an optimist struggling with "philosophical superstitions."

Unfortunately, Bocheński remained indifferent to developments in the philosophy of culture, to questions of social ontology, and current studies in the philosophy of science as well as in the methods of the natural and human sciences.

6 Summary

Stanisław Lem's description best characterizes Bocheński's life and activity:

> He was a person out of the ordinary endowed with so many diverse talents. Born in 1902, a cavalry man during the Polish-Bolshevik conflict, he spent the Italian campaign during World War II at the side of Bishop Gawlina. He travelled all over the world, encircled the earth with all the kilometers of his travels, he was personally acquainted with the leading logicians and philosophers — all the while dressed in a monkish dress.

K. Policki (✉)
University of Fribourg, Fribourg, Switzerland
e-mail: korpol@bluewin.ch

J.M. Bocheński's Theory of Signs

Korneliusz Policki

Abstract The article concerns Bocheński's thus far unpublished considerations pertaining to signs that are among the materials in my archive. Of particular interest are Bocheński's reflections on the ontological, psychological, and epistemological foundations of the theory of signs. In his semiotics the concept of the sign is broadened to include application to the logic of authority, the concept of bi-directionality of human phenomena as well the theory of 'philosophical superstitions'.

Keywords Signs: material, fundamental, formal · The emitter and receiver of signs

Mathematics Subject Classification (2000) Primary 03B80

1 Introduction

In the domain of semiotics Bocheński succeeded in rendering a number of basic concepts more precise. My task in this paper is to present some of the results from within this domain of research found among thus far unpublished materials of Bocheński's archive. In constructing his sign theory Bocheński relies on Thomism as well as the work of 'Lwów-Warsaw School' (the concept of the functor).

2 Ontological Assumptions

The world consists of things (*substantiae*). We name as 'thing' any object that exists self-sufficiently, is not determined by some other object, and undergoes change. Things include, e.g. people, planets, tables and chairs, etc. These things bear certain properties. In contrast with a thing, a property does not exist self-sufficiently (it is an accident), it appears on the basis of and is determined by the thing. Certain properties differ ontically (phenomenologically) from the thing in relation to which they emerge. Independently of the cognizing subject there exists a difference between a given property and the thing

Á. Garrido, U. Wybraniec-Skardowska (eds.), *The Lvov-Warsaw School. Past and Present*, Studies in Universal Logic,
https://doi.org/10.1007/978-3-319-65430-0_21

whose property it is. These properties include, for example, a man's anger, his knowledge, the shape of a piece of turf, an electron's spin. Other properties are but certain aspects of the thing, ontically indistinguishable from the latter, even though the human intellect separates them fictively (abstractly) from their basis. An example would be, for example, a man's humanity: humanity is abstractable in a man as his property. In the thing itself humanity is a constitutive component of the man. It is a characteristic property and the sign of being human.

A property of a specific kind is existence. We name as "existence" that which accounts for the existence of a given object. Existence is a very specific property in that it does not change the makeup of the object in any way, but only transfers it from the sphere of possibility to that of reality. What remains in an object when we exclude its existence we name its 'makeup' (content, or *essentia* in the ontological sense of the term). For that reason let us speak of the content of the sign as well as of its existence. A sign is characterized by its phenomenological, ontic, metaphysical, subjective and objective foundations [3, p. 55].

However, existence is multifarious. First of all, there is real existence, designated empirically. This is how, for example, living people, galaxies, street signs, and railway semaphores, etc. exist. Secondly, there is the mode of existence of merely possible objects; were possible objects to be deprived of a mode of existence it would be difficult even to think about them. However, it is a fact that we do think about such objects, for instance when we create new signs. Thirdly, another mode of existence characterizes objects that cannot exist empirically but which we can, so to speak, "allow ourselves" in creating fictional entities, e.g. the square circle, the Sphinx, the wooden stone, etc. The upshot is that we need to distinguish at least four modes of existence: actual, potential, possible, and purely putative existence.

Ontological assumptions are essential for understanding the objectively real and ideal content of a sign. A sign refers to an objective content that can be material, real, and ideal. A theory of signs will vary as the ontology varies. Thus Tadeusz Kotarbiński's theory of signs differs from that of Bocheński.

Whatever has a mode of existence (actual, potential, putative, possible, real) we traditionally call a 'being'. Mt Blanc is a being as is the as yet non-existent but possible railway line and the square circle; but whereas Mt Blanc is an actual being, the future railway line is a potential being (though both Mt Blanc and the railway line are real beings) and the square circle is an purely intentional being (*ens rationis*).

3 Psychological Assumptions

According to Bocheński the course of the cognitional process begins with a simple perception (*simplex apprehensio*) that is followed secondly by the judgment. The perception is a certain sort of intellectual assimilation of the cognized object; it differs from a material assimilation in that there arises in the psyche of the knower a mental picture of the object of knowledge which is a subjective concept (*conceptus subiectivus*). But the simple perception never grasps the object as a whole; it is always the grasp of only one aspect of the object. What the simple perception grasps is a property or properties of the object. These are called the 'ideal' concept (*conceptus obiectivus*). Thus in the course of the cognitional process we attain the extra-mental object thanks to the operation of two intermediate factors: first by means of the "subjective concept" and then by means of what is immediately conceived in the "ideal concept".

The subjective concept as well as the objective and ideal concepts play important roles in our practices. For example, a road sign that has been willfully displaced or knocked down by vandals retains its normative force as an ideal concept acquired during the driver-training course despite the absence of an actual perception and a subjective concept.

The ideal concept is always an abstraction from actual existence; quite simply, it is eternal. We can construct subjective and ideal concepts, that is to say, come to know objects that do not exist in a given moment, and that even cannot exist in the external world. Even when the object of knowledge is existence itself we construct the concept of existence in thought.

The construction of judgments proceeds analogously. In a judgment the counterpart of the subjective concept is the subjective, intended sentence. This sentence should not be confused with a spoken or written sentence. The counterpart of the objective concept is the objective sentence, the so-called "ideal sentence" (*enuntiabile*). It is by way of these sentences that the judgment attains its object, the fact. The ideal sentence has the same characteristics as the ideal concept: it is objective but is abstracted from real existence [1, pp. 1–4].

Tadeusz Kotarbiński did not agree with Bocheński's theory: according to Kotarbinski, concepts and ideal sentences are hypostases. He held that only nouns exist with designata in the world of persons and material things [7, p. 15]. These nouns are material signs. For Bocheński, on the contrary, there exist (non-material) psychological constructs such as concepts and subjective sentences. They are, in other words, so-called mental signs.

4 The Concept of the Sign

For Bocheński, a sign has three aspects: a material, fundamental, and formal aspect. The material sign is the thing that serves as the means to know something. The fundamental sign is some property (typically a certain shape) or the motion of the thing that is directly significant. For example, in the cavalry whereas the raising of the sabre three times by the commander means "gallop", the sabre itself is a material sign, and its motion and rest is the fundamental sign. Finally, the formal sign is the relation that holds between the fundamental sign and what the sign expresses, means, and denotes. This relation is called the 'formal sign' for the reason that the form determining that the material sign becomes a sign in the first place is precisely the relation: pragmatic, semantic and connotative.

We distinguish between natural and conventional signs. The smoke from Bocheński's pipe is the natural sign that Bocheński is smoking a pipe. The aforementioned raising of the sabre is a conventional sign in the cavalry. Whereas smoke leads anyone familiar with the use of tobacco to conclude, without recourse to any convention, to the fact of Bocheński's pipe-smoking, the raising of the sabre by Bocheński's cavalryman would lead no one to the thought that major Bocheński wishes to transit to the gallop were there no convention to this effect among the troopers. For Bocheński conventional signs are especially interesting.

A conventional sign in the fundamental sense is a characteristic that, thanks to an agreement between at least two persons, allows one of them (the recipient) to recognize something distinct from the characteristic as a natural sign, and this on the basis of an agreement between the recipient and another person who is the emitter of the sign.

5 The Theory of Signs and Authority

Bocheński created the logic of authority [5]. He distinguished two kinds of authority. One kind is epistemic authority (based on knowledge), the other is deontic authority (based on power). Bocheński divides deontic authority into the authority of solidarity and punishment (sanction). All of these kinds of authority are applicable to the theory of signs, in particular to the theory of conventional signs. Of the many emitters of conventional signs authorities are an important case. Authorities ascribe the intentional as well as the associative meaning to various objects that thereby become signs. However, this alone does not suffice. In order for conventional signs to become effective deontic authority is required; in order to institute the sign, there needs to be a power. Deontic authority motivates, sanctions, and regulates the signs defined by epistemic authority. This deontic power acts functionally to ensure solidarity among those who respect road signs and to apply sanctions (mandates).

How are signs received by the user? Bocheński writes about the bi-directionality of cognitive phenomena [8, pp. 141–145] that is present at all levels of man's cognitive activities. It is called the law of action and reaction. In the epistemology of signs bi-directionality is especially evident. On one hand, we have the centripetal direction; the sign is the stimulus, the call, it 'imposes' itself, as it were 'calls to', 'enters into' the knower. On the other hand, there is the centrifugal reaction of the receiver of the sign ranging over not only imagination, judgment, and emotions, but likewise external behavior and reactions, that is, centrifugal endeavors.

6 The Theory of Signs and 'Philosophical Superstitions'

Bocheński created as well the theory of 'philosophical superstitions'. A philosophical superstition is a minor false world view, though regarded by some as true. An absolute philosophical superstition is a view that is at odds with logic, methodology, science as well as common sense. A relative philosophical superstition is a view that is relative to some belief system, e.g. for Christians the pagan Roman religion is a superstition and conversely; for Tadeusz Kotarbiński Bocheński's theory of signs is a superstition since it resorts to hypostases, mental signs, whereas for Bocheński Kotarbinski's reism and concretism are "philosophical errors".

Bocheński's "One hundred superstitions" has so far appeared only in Polish [6], though a German translation is in preparation. Bocheński's philosophical and logical theories can be applied likewise to the theory of signs. An example of superstitions related to the theory of signs is astrology (the theory of zodiac signs), numerology (the theory of numerical signs exerting effects on human life), and semantic anthropocentrism (only humans employ signs), the confusion of sign-constitutive functors and arguments, extreme positivism, nominalism in the theory of signs.

7 Conclusion

Bocheński's studies in the theory of signs testify to the growth and development of the philosophy of signs and semiotics. Progress in this domain of philosophy consists less in attaining non-ambiguous and unquestionable results and more in discovering new conditions and associations. For example, Bocheński's theory of signs can be broadened to include not only the logic of authority but also the logic of religion ("sacred signs") [2], something that Bocheński did in fact attempt in his unpublished "Was ich glaube?" in my archive.

References

1. Bocheński, J.M.: A. Church, N. Goodmann (eds.) The Problem of Universals. Notre Dame Press, Notre Dame (1956)
2. Bocheński, J.M.: The Logic of Religion. New York University Press, New York (1965)
3. Bocheński, J.M.: Die zeitgenösssichen Denkmethoden. UTB Francke, Bern (1973)
4. Bocheński, J.M. Grundriss der Logistik. Ferdinand Schönigh, Padeborn (1973)
5. Bocheński, J.M.: Was ist Autorität?, Einführung In die Logik der Autorität. Herder, Freiburg (1974)
6. Bocheński, J.M.: Sto zabobonów [One Hundred Superstitions]. Instytut Literacki, Paris (1987)
7. Kotarbiński, T.: Kurs logiki dla prawników [Logic Course for Jurists]. PWN, Warszawa (1961)
8. Policki, K: Filozofia człowieka we wczesnej twórczości J.I.M. Bocheńskiego [Philosophical Anthropology in the Early Works of J.I.M. Bocheński]. PWT, Wrocław (2005)

K. Policki (✉)
University of Fribourg, Fribourg, Switzerland
e-mail: korpol@bluewin.ch

Jan Salamucha (1903–1944)

Kordula Świętorzecka

Abstract Father Jan Salamucha (1903–1944) was one of the most prominent Polish Catholic intellectuals of the first half of the twentieth century. He was a victim of the World War II, killed in the Warsaw Uprising. We present his intensive academic research and mention his didactic, pastoral, and pro-independence activities.

Keywords Jan Salamucha · Cracow circle · Analytical Christian philosophy

Mathematics Subject Classification (2000) Primary 01A60; Secondary 03-03

1 Beginnings

Jan Salamucha was born on 10 June 1903 in Warsaw to the family of Stanisława (née Marciniak) and Andrzej Salamucha. He was the first of two children—he had a younger sister, Genowefa (born in 1909). The Salamucha family had a working-class background (the father was a worker at a foundry in Warsaw). From (approx.) 1914 to 1919, Jan Salamucha studied at the Stefan Chrupczałowski humanistic junior high school, and later attended the Metropolitan Seminary. During his studies at the Seminary, he took part in the Polish-Soviet war of 1920 (August–October) as a volunteer paramedic.

2 Education

Upon graduating from the Seminary—in 1924—he commenced studies at the Faculty of Theology at the University of Warsaw, the Philosophy Section, which he completed in 1926, being awarded the master's degree on the basis of his dissertation entitled *O kategorii πρoξ u Arystotelesa* (*About the πρoξ category according to Aristotle*). A year later, he defended his doctoral dissertation in the field of Christian philosophy, entitled *Teoria wynikania modalnego u Arystotelesa. Studium krytyczne* (*Aristotle's theory of modal inference. A critical study*) Both dissertations were supervised by Rev. Stanisław

Á. Garrido, U. Wybraniec-Skardowska (eds.), *The Lvov-Warsaw School. Past and Present*, Studies in Universal Logic,
https://doi.org/10.1007/978-3-319-65430-0_22

Kobyłecki. From 1927 to 1929, he continued studies at the Gregorian University in Rome, which resulted in his master's degree dissertation entitled *De deductione aqud Aristotelem et S. Thomam*. A Polish version of this dissertation, *Pojęcie dedukcji u Arystotelesa i św. Tomasza z Akwinu* (*The idea of deduction according to Aristotle and Saint Thomas Aquinas*) was the basis for Salamucha's habilitation at the Jagiellonian University in 1933. However, the Ministry of Religion and Public Education approved the habilitation as late as 1936. The first attempt at having the habilitation approved by the Ministry of Religion and Public Education in 1933 ended unsuccessfully in unclear circumstances of personal intrigue (associated with father Andrzej Krzesiński, who was conflicted with Salamucha's supervisor, Rev. Konstanty Michalski) and anonymous denunciation letters regarding Salamucha (about his political views and close contact with people suspected of involvement with freemasonry—Jan Łukasiewicz and Wacław Sierpiński). Although the rejection of his habilitation seriously contributed to Salamucha's inability to conduct academic and educational work, luckily it did not stop him from further academic activity.

Salamucha was a student of numerous prominent figures of the Catholic Church and the Polish academic circles in the pre-war period. At the Seminary, among his teachers were: Rev. Mieczysław Węglewicz and Rev. Stanisław Mystkowski. In the period of studies at the University of Warsaw, he was also taught by: Tadeusz Kotarbiński, Władysław Tatarkiewicz, Wiktor Wąsik (philosophy); Stanisław Leśniewski and (probably) Jan Łukasiewicz (logic); Stefan Mazurkiewicz, Stefan Staszewicz, (probably) Wacław Sierpiński (mathematics); Rev. Stanisław Kobyłecki (psychology); Wincenty Kwiatkowski (theology). At the Gregorian University, Salamucha was taught, among others, by Pierre Hoenen. Salamucha's post-doctoral dissertation was written under the auspices of Rev. Konstanty Michalski. Under the influence of such excellent teachers, during the Third Polish Philosophical Congress in Cracow in 1936, father Salamucha, together with Rev. Innocenty Maria Bocheński and Jan Franciszek Drewnowski, initiated a discussion which resulted in the creation of the Catholic variant of the Lvov-Warsaw School—the so-called Cracow Circle.

The academic and personal life of father Salamucha was also strongly influenced by his patron of many years, the Metropolitan of Cracow, Archbishop Adam Sapieha.

3 Didactics

Salamucha started his didactic work in 1929 at the Warsaw Seminary, where he taught philosophy until 1933 (e.g. in formal logic, methodology of sciences, theory of cognition, and psychology). Due to the commencement of post-doctoral degree studies at the Jagiellonian University, he vacated his position at the Seminary. Because of the problems related to the approval of habilitation, the second Department of Philosophy (planned previously by the Ministry of Religion and Public Education) at the Jagiellonian University, to be headed by Salamucha, was not opened, and he was only offered the position of freelance lecturer. In such circumstances, father Salamucha declined that offer and gave no lectures for almost a year (his post at the Warsaw Seminary was already taken). In the academic year 1934/1935, he was able to resume his lectures at the Seminary, and decided to give freelance lectures at the Jagiellonian University. In 1938, the authorities of the Jagiellonian University proposed that father Salamucha be appointed

Associate Professor of Christian Philosophy, a title which he was finally awarded in the same year by the President of the Republic of Poland, Ignacy Mościcki.

In the entire period of his pre-war didactic activity at the Jagiellonian University, Salamucha taught formal logic, the history of logic, the theory of cognition, theodicy, cosmology, and the history of ancient philosophy. It should also be noted that in that period, one of his students was the future Pope—John Paul II.

Immediately after the outbreak of war, Salamucha came to Warsaw, where he became the chaplain for the forces defending one of the districts of Warsaw (Bem's Forts). In recognition of his valour and courage, the authorities of the occupied Poland awarded him the Cross of Valour. In the same year, he travelled back to Cracow to continue his didactic activities. On 6 November 1939, together with a group of 182 lecturers of the Jagiellonian University, he was arrested and transported to the Sachsenhausen concentration camp, and later to Dachau (in the so-called Sonderaktion Krakau). Thanks to the intervention of Heinrich Sholz, the German authorities agreed to Salamucha's release and return to Cracow (1941). After a short period of convalescence, Salamucha went to Warsaw. There, he continued his academic and activities. He participated in clandestine seminars organised by Tatarkiewicz, Łukasiewicz and Adam Krokiewicz and his close friend Bolesław Sobociński; he also taught philosophy (metaphysics, logic) at the secret Warsaw Seminary and at the University of the Western Lands (operating until 1945). At that time, Salamucha taught e.g. Henryk Hiż and Andrzej Grzegorczyk; he also reviewed the post-doctoral dissertation of Stefan Świeżawski.

4 Priesthood

Because of his spiritual formation, in addition to the didactic and academic activities, Salamucha was simultaneously active in pastoral work. Immediately after the completion of studies in Rome (in 1929), he took the post of vicar at the St. Wojciech Church in Wiązowna in the Otwock region (near Warsaw), were he stayed for almost a year. During the break in the academic work, resulting from his troubles with the habilitation (1933–1934), he served as a vicar at the Parish of Our Lady of Loreto in Warsaw's Praga district, at the St. Florian's Church. As already mentioned, he was the chaplain at Bem's Forts, as well as one of those priests who attracted numerous members of the intelligentsia in Warsaw under occupation. The last place of his ministry was the parish of St. James in Warsaw's Ochota district.

5 Death

While serving as the vicar at the Church of St. James, Salamucha decided to become involved in the Warsaw Uprising—he was assigned to the District of Warsaw Home Army, and as a military chaplain for the units defending the Wawelska Redoubt in Warsaw, participated in the fighting and helped with the evacuation of insurgents and residents of the Staszic Colony (housed just near Wawelska Redoubt). He served as a priest until the end—staying with the wounded and those who did not manage to escape before the end of the insurgents' struggle—and supporting them spiritually. On 11 August 1944, during mass murder of the residents of the old Ochota district in Warsaw, he was killed by the SS RONA brigade—a Ukrainian unit of the Russkaya Osvoboditelnaya Narodnaya Armiya.

6 Intellectual Formation

Salamucha's intellectual output was undoubtedly conditioned by certain characteristics of his personality. He was exceptionally talented and hard-working, with a scientific mind and great sensitivity; courageous, steadfast, and able to make sacrifices for the greater good. Therefore, his intellectual interests were broad—he was drawn to the issues of classical metaphysics and theodicy, the history of philosophy (in particular: the history of logic), formal logic, as well as matters of ethical and social nature. On the other hand, Salamucha's intellectual formation was influenced by the interests and views of some of his teachers. Father Salamucha was influenced intellectually by such prominent members of the Lvov-Warsaw school as: Łukasiewicz, Leśniewski, Tatarkiewicz and Kotarbiński. As a Catholic philosopher, Salamucha implemented the basic postulates of this school on the basis of Christian philosophy. (He also inspired other members of the Cracow Circle: Bocheński, Drewnowski and their consultant on logic: Sobociński.) Thus, in line with those postulates, he researched the issues of classical philosophy and its history, but using precise tools of contemporary formal logic. The strict method used by Salamucha (following the lead of Łukasiewicz and Leśniewski), was to bring the philosophical deliberations closer to science, and to help capture the rationalism, realism, impartiality, and maximalism of the Catholic philosophy understood as a modern continuation of scholastic philosophy. The most impressive realisation of the methodological and philosophical programme by Salamucha is the logical reconstruction of the ex motu argument by St. Thomas Aquinas proving the existence of God [2]. In the history of human thought, Salamucha's formalisation was the first attempt at substantiating the basic thesis of theodicy on the existence of the Absolute, made with the use of contemporary tools of formal logic and the set theory. It was also the starting point for broader research based on formalised theodicy, conducted e.g. by: Bocheński, Johannes Bendiek, Francesca Rivetti Barbo, Edward Nieznański, Korneliusz Policki.

During the occupation, Salamucha became more interested in ethical and social matters. Consequently, he saw those issues from a philosopher's and analyst's point of view, paying great attention to precision, accuracy, and clarity of his expressions. He analysed concepts such as: *evil*, *suffering*, *love*, *coercion*; he also attempted to specify the notion of *Catholic ethics*.

Father Jan Salamucha was one of the most prominent Polish Catholic intellectuals of the first half of the twentieth century. He had little time to pursue the goals he set for himself—he was a victim of one of the most dramatic periods in the contemporary history of Poland—World War II. His intensive academic, didactic, pastoral, and pro-independence activities allow us to recreate the image of an exceptional man—a talented and creative philosopher, a wise teacher and a spiritual guide, a patriot and a hero. This picture would be incomplete, if we failed to mention that Salamucha was a faithful and honest friend, and a lover of nature (an enthusiastic mountaineer and rower) and of arts.

Selected Publications

Salamucha's academic achievements are presented on a full list of publications in: *Jan Salamucha. Wiedza i wiara. Wybrane pisma filozoficzne*, ed.: Jacek J. Jadacki, Kordula Świętorzecka. TN KUL, Lublin (1997); English edition: *Jan Salamucha, Knowledge and Faith. Collected Works*, ed.: Jacek J. Jadacki, Kordula Świętorzecka, A. Mickiewicz University Press, Poznań (2003)

We list selected papers:

1. Pojęcie dedukcji u Arystotelesa i św. Tomasza z Akwinu. Studium historyczno-krytyczne [The idea of deduction according to Aristotle and Saint Thomas Aquinas. A Historical and Critical Study]. Polish Theological Society (1930)
2. Dowód *ex motu* na istnienie Boga. Analiza logiczna argumentacji św. Tomasza z Akwinu. Collectanea Theologica, ch. 15, vol. 54, book 1–2, 53–92 (1934) (English translation by Tadeusz Gierymski, Marian Heitzman: The Proof *ex motu* for the Existence of God. Logical Analysis of St. Thomas' Arguments. The New Scholasicism **32**, No 3, 327–372 (1958). French summary: Bolesław Sobociński, Le preuve *ex motu* de l'existence de Dieu. Analyse logique de l'argumentation de St. Thomas d'Aquin. Studia Philosophica, vol. 1, 483–484 (1935))
3. Pojawienie się zagadnień antynomialnych na gruncie logiki średniowiecznej [The appearance of antinomial questions in medieval logic]. Przegląd Filozoficzny, ch. **40**, book 1, 68–89; book 3, 320–343 (1937)
4. Cardinale Thomaso de Vio Gaetano nel quatro centenario della sua morte [rev.]. Przegląd Filozoficzny, ch. **40**, book 4, 474–476 (1937)
5. Zestawienie scholastycznych narzędzi logicznych z narzędziami logistycznymi [A comparison of scholastic logical tools with logistic tools]. In: Myśl katolicka wobec logiki współczesnej, pp. 35–48 (1937) (French summary of professional accomplishments: ibid., 167–169)
6. O «mechanizacji» myślenia [On the «Mechanisation» of Thinking]. In: Myśl katolicka wobec logiki współczesnej, pp. 112–121 (1937) (French summary of professional accomplishments: De la mecanisation de la pensee. ibid., 182–186)
7. O możliwościach ścisłego formalizowania dziedziny pojęć analogicznych [Possibilities of a Precise Formalisation of the Domain of Analogical Notions]. In: Myśl katolicka wobec logiki współczesnej, pp. 122–153 (1937) (French summary of professional accomplishments: A propos de la formalisation des domaines des concepts analogiques. ibid., 186–193)
8. Logika stara i nowa w wiekach średnich [Old and New Logic in the Middle Ages] (summary of professional accomplishments). Ruch Filozoficzny **15**(3), 156b (1938)
9. Z dziejów nominalizmu średniowiecznego (From the History of Medieval Nominalism]. In: Czeżowski, T. (ed.) Christeria. Rozprawy filozoficzne złożone w darze Władysławowi Tatarkiewiczowi, pp. 207–238. PWN, Warsaw (1960) (Work completed in 1938)
10. Krytyka poznania [The Critic of Cognition]. Filozofia Nauki, ch. 3, No 1–2, 193–225 (1995) (Work from 1935)

K. Świętorzecka (✉)
Institute of Philosophy, Cardinal Stefan Wyszyński University in Warsaw, Warsaw, Poland
e-mail: k.swietorzecka@uksw.edu.pl

Struve and Biegański: Towards Modern Approach to Logic

Roman Murawski

Abstract The paper is devoted to two Polish philosophers and logicians Henryk Struve and Władysław Biegański. They represented the traditional pre-mathematical approach to logic and stood on the threshold of the new paradigm of logic as formal mathematical logic. Their views concerning logic and its philosophy as well as its relations to philosophy and mathematics will be analyzed. Those views will be compared with views of their contemporaries and with views of Jan Łukasiewicz—one of the main leaders of Warsaw School of Logic who represented the new paradigm.

Keywords Traditional logic · Modern logic · Paradigm in logic · Struve · Biegański · Psychologism

Mathematics Subject Classification (2000) Primary 03-03; Secondary 01A55, 01A60

1 Henryk Struve

1.1 Life

Henryk Struve (alias Florian Gąsiorowski) was born in Gąsiorów near Koło on 27 June 1840. He studied theology and then philosophy at the University of Tübingen. In 1862, he obtained his doctorate at the University of Jena. In the years 1863–1903, he was a professor of the Main School [Szkoła Główna] in Warsaw, then the (Russian) Imperial University of Warsaw. In order to maintain this post he had to defend his doctorate again in Moscow (1870). In 1903, he decided to stay with his daughter in England. He died in Eltham (Kent) on 16 May 1912. He dealt with philosophy and the history of philosophy. He called his philosophical system 'ideal realism'—it was an attempt at reconciling realism with idealism. He popularized the history of Polish philosophy in Germany and England. He wrote also many logic textbooks as well as a history of logic.

Á. Garrido, U. Wybraniec-Skardowska (eds.), *The Lvov-Warsaw School. Past and Present*, Studies in Universal Logic,
https://doi.org/10.1007/978-3-319-65430-0_23

1.2 Struve's Conception of Logic

Henryk Struve is regarded as one of the most important figures of Polish logic in the nineteenth century (cf. for example Woleński [37, p. 30]). In the interwar period he was frequently referred to but his works were not analyzed and reprinted. We are interested here in Struve's views on logic as a science and his conception of logic.

Struve's views on logic were presented in his fundamental work *Historya logiki jako teoryi poznania w Polsce* [History of Logic as the Theory of Knowledge in Poland] [32] from 1911 and in the textbook *Logika elementarna* [Elementary Logic] [31] from 1907 as well as in various papers (cf. for example [29]). There arise certain difficulties in reconstructing his views—they result from the fact that he wanted to create a coherent system of philosophy that would embrace all the traditional branches of philosophy. Consequently the borders between particular branches were flexible and imprecise. The principles of one division influence the foundations of the other and conversely. He balanced between materialism and idealism aiming at the golden mean. This was reflected also in his understanding of the object of logic.

At first he thought that the object of logic was principles and rules of thinking. In his talk given in 1863, inaugurating his lectures at the Main School, he said:

> Gentlemen! Logic is most generally the science of rational thinking, having thinking, its principles and rules as its object.[1] [28, Lecture 1]

However, he added that thinking is one of the powers of the soul; it is "an objective, neutral consideration of this world by the soul"[2] [28, p. 55]. Thus he introduces a psychological element, and indirectly—an ontological one. In fact he claims that the soul is the ideal embryo of the human being and "the limits of our being are the limits of our correct thinking"[3] [28, p. 35].

According to Struve logic concerns objective reality. However, it does not concern it directly. The mediator between logic and the world is the thought. Nevertheless, this does not lead to the thesis that thought reflects the logical structure of the world or to the thesis that the world has some logical structure at all.

Struve's earlier views were even more inclined towards psychologism. Initially, he claimed—as indicated above—that thinking is "an objective, neutral consideration of this world by the soul." He upholds this thesis in *Logika elementarna* [31], but here he separates logic from psychology, writing that logic deals with thinking as "an auxiliary mean to get to know the truth" whereas psychology is interested in emotional and volitional motives of cognition. Logic has both a descriptive and normative character and is to oversee the application of the established norms and thus to evaluate the degree of the truth of cognition.

The foundation of logic is philosophy, but also conversely: philosophy can be developed only on the foundation of logical laws. The title of the main analyzed work of Struve, *Historya logiki jako teoryi poznania w Polsce* [History of Logic as the Theory of

---

[1] 'Panowie! Logika jest to w najogólniejszem pojęciu nauka myślenia mająca myślenie, jego zasady i prawidła za przedmiot.'

[2] 'obiektywne, neutralne rozpatrywanie tego świata przez duszę.'

[3] 'granice naszego bytu są granicami naszego myślenia prawidłowego.'

Knowledge in Poland], may suggest that he identified logic with the theory of knowledge. In the first editions of *Logika elementarna* in Russian[4] he made no clear distinction between these two disciplines, but in the Polish version of the textbook [31] he wrote:

> While it is true that thinking is the main co-factor of cognitive activity but not the only one; it unites directly and constantly with the suitable expressions of emotion and will.[5] [31, p. 3]

The examination of emotions and will as well as their relationships with thinking belongs to the sphere of psychology whereas logic deals with thinking merely in one aspect, namely:

> As an auxiliary mean of getting to know the truth [. . .]. Simultaneously, logic is not satisfied with the real course of mental activity but seeks principles, i.e. laws and rules which one should follow as norms if one wants to get to know the truth as exactly as possible. This separate view on thinking gives logic the character of an independent science, which is strictly different from psychology, namely this part of logic that investigates thinking as well.[6] [31, p. 5]

In *Historya logiki* one finds the following words:

> [. . .] many separate the theory and criticism of cognition from logic as science dealing only with thinking. Despite that, the connection between the development of correct thinking as well as arriving at and getting to know the truth is so close from the psychological perspective that these mental activities cannot be separated.[7] [32, p. 1]

One can see here some traces of his discussions conducted with Kazimierz Twardowski and the Lvov-Warsaw School. On the one hand, one can notice a certain readiness to recognize the new understanding of logic and on the other hand, a desire to abide by his current understanding of logic.

Struve differentiates between objective and subjective truth. The former is an ideal that is independent of the human cognition and the latter is the reconstruction of the content of being, of what exists in reality, in the mind—done through correct thinking. Logic controls this reconstruction and thus through formal means it reaches the real being. Thus thinking has a reconstructive and not a creative character. In Struve's opinion three forms of logic should be distinguished: (1) formal, (2) metaphysical and (3) logic treated as the theory of knowledge. Formal logic considers the principles and laws of thinking regardless of its object. Metaphysical logic (developed by Plato, Neo-Platonists, Spinoza and the German idealists: Fichte, Schelling and Hegel) states that since thinking contains its object directly in itself we get to know the very objective reality knowing the principles and laws of

[4] *Elemientarnaja logika* was first published in 1874; there were altogether 14 Russian editions. It was the obligatory manual of logic in classical junior high school from the year 1874. Its Polish version appeared in 1907.

[5] 'Myślenie jest wprawdzie głównym, ale nie jedynym współczynnikiem czynności poznawczej; jednoczy się ono zarazem bezpośrednio i stale z odpowiednimi objawami uczucia i woli.'

[6] 'Jako środek pomocniczy poznania prawdy [. . .]. Przytem nie zadowala się logika danym faktycznym przebiegiem czynności myślowej, lecz odszukuje zasady, t.j. prawa i prawidła, któremi w myśleniu jako normami kierować się należy, chcąc dojść do możliwie ścisłego poznania prawdy. Ten odrębny pogląd na myślenie nadaje logice charakter samodzielnej nauki, ściśle różniącej się od psychologii, a mianowicie tej części jej, która bada również myślenie.'

[7] '[. . .] wielu odróżnia zarówno teorię, jak i krytykę poznania od logiki jako nauki samego tylko myślenia. Pomimo to łączność pomiędzy rozwojem prawidłowego myślenia a dochodzeniem i poznaniem prawdy tak jest ścisła ze stanowiska psychologicznego, że tych czynności umysłowych rozerwać nie podobna.'

thinking. Struve accepts neither the first nor the second conception of logic. He opts for the third solution, treating it as the golden mean. Thus he understands logic as the method of investigation and cognition of truth. Its task is to discover the principles according to which man reconstructs the structure of the real world in his mind. Naturally, Struve sees the difficulties connected with this view. In *Logika elementarna* he wrote:

> The difficulties of examining the relation [...] between thinking and the objective world are obvious and can be reduced mainly to the fact that we are not able to compare directly our images and concepts of objects and our views on them with the objects themselves. The question concerning the objective knowledge of truth could be solved in favour of thinking only when it turned out that the laws of our mind, and thus thinking, were fundamentally consistent with the laws of the objective being which is independent of us. [...] Nonetheless, showing the accordance between the laws of the mind and the laws of the objective being requires a series of critical investigations concerning the results of scientific studies.[8] [31, pp. 6–7]

Struve calls this logic "logic of ideal realism", uses to constitute the framework of a coherent system of philosophy to give a general outlook on the world. We should add that Struve is far from ascribing Messianic tendencies to logic (as Hoene-Wroński did—cf. [24, 25]). He opts for a balance between the knower and the known, seeing the role of emotions and will in cognition. He was interested in Leibniz's view to which he referred many times—the view that logic is abstracted from reality.

Struve begins his lecture on logic by giving images and concepts. Then he introduces judgments. By "image" he means a kind of representation of the object through its characteristics, and "concept" is a set of essential features. Moreover, images result from certain mental processes. It is the object that makes the mind create images.

Struve attached great importance to the teaching of logic. He thought that teaching how to think correctly is much more important than giving students concrete contents. Consequently, he placed a strong emphasis on the teaching of logical culture.

We have already shown that Struve's conception of logic places him between the old and new paradigm or rather even in the old paradigm. We have mentioned that he did not value the role and significance of symbolic and mathematized formal logic but he stressed psychological questions. Consequently, one should ask what made him not see the advantages of the new attitude. It seems that one of the reasons was the fact that Struve saw no cognitive value in pure form devoid of content (cf. [33]). According to his conception it is the object, i.e., external world that stimulates our thinking, that realizes its existence and the characteristics of objects, and then using logical methods it creates notions which in turn it uses, applying logical methods, to formulate judgments. Thus there can be no cognition without content. Another reason may be that he set a low valuation on the role and importance of mathematics. Trzcieniecka-Schneider even claims that Struve "did not understand mathematics, reducing it only to the techniques of operations on numbers"

[8] 'Trudności zbadania stosunku [...] myślenia do świata przedmiotowego są oczywiste i sprowadzają się głównie do tego, że nie jesteśmy w stanie porównywać bezpośrednio naszych wyobrażeń i pojęć o przedmiotach ani poglądów na nie z samymi przedmiotami. Kwestya przedmiotowego poznania prawdy mogłaby być rozwiązaną na jego korzyść dopiero wtedy, gdyby się okazało, że prawa naszego umysłu, a więc i myślenia, są zasadniczo zgodne z prawami niezależnego od nas bytu przedmiotowego. [...] Wykazanie atoli tej zgodności praw umysłu z prawami bytu przedmiotowego wymaga szeregu badań krytycznych nad wynikami dociekań naukowych.'

[33, p. 93]. In "Filozofia i wykształcenie filozoficzne" [Philosophy and philosophical education] he wrote:

> Mathematics considers only quantitative factors: [...] But in its activities the human mind is not limited only to quantitative factors but everywhere supplements quantity with quality. [...] One cannot say that truth contains more or fewer thoughts than falsity: [...] These are all qualitative differences and they cannot be defined quantitatively; they cannot be understood rightly and characterized closely from the mathematical point of view but they can only be understood from a more general standpoint, going much beyond the scope of the quantitative factors alone.[9] [30, pp. 156–157]

Struve also opposed the introduction of quantifiers. He permitted quantitative elements in logic only in the case of the conversion of judgments and in the square of opposition. However, he was not consistent in his views when analyzing the relations between the scopes of concepts using Euler diagrams he actually used arithmetic notation.

In his opinion formal logic "leads [...] only to the development of one-sided formalism without elevating the essential cognitive value of the relevant forms"[10] [31, p. IX] and "mathematical logic depends on the completely dogmatic transfer of the quantitative and formal principles to the mental area where quality and content are of primary importance"[11] (*ibid.*). He also claims that "reducing judgment to equation and basing conclusion on substance, i.e. substituting equivalents, does not correspond to the real variety of judgments and conclusions"[12] [31, p. X].

Struve's aversion towards mathematics and mathematical methods in logic was connected with his views on the function of language in logic and cognition as well as with his conception of truth. Since if—in accordance with Aristotle—truth is the conformity of thought to the content of propositions, conducting operations on propositions as symbols means losing sight of this property to some extent. In fact, the characteristic of being true does not refer to propositions but to their contents. Symbolically identical propositions can differ with respect to their contents. Thus the operations conducted on the symbols of propositions do not have much in common with establishing their truth. And yet logic leads to the truth about the real world.

The above analysis shows that Struve stood on the threshold of the new way of understanding and cultivating logic, combining the old and new paradigms. His position and the meaning of his work was characterized by Kazimierz Twardowski in the following way:

---

[9]'Matematyka rozpatruje wyłącznie czynniki ilościowe: [...] Tymczasem umysł ludzki w działalności swej nie jest bynajmniej ograniczonym samemi tylko pojęciami ilościowymi, lecz uzupełnia wszędzie ilość jakością. [...] Nie można powiedzieć, że prawda zawiera w sobie więcej lub mniej myśli niż fałsz: [...] To wszystko są różnice jakościowe, których ilościowo określić nie można, które należycie zrozumiane i bliżej scharakteryzowane być nie mogą z punktu widzenia matematycznego, lecz pojęte być mogą tylko ze stanowiska ogólniejszego, wynoszącego się wysoko ponad zakres samych tylko czynników ilościowych.'

[10]'doprowadza [...] tylko do rozwoju formalizmu jednostronnego bez podniesienia istotnej wartości poznawczej odnośnych form.'

[11]'logika matematyczna polega na zupełnie dogmatycznym przeniesieniu zasad ilościowych i formalnych na pole umysłowe, gdzie jakość i treść mają znaczenie pierwszorzędne.'

[12]'sprowadzanie sądu do równania oraz oparcie wniosku na substytucyi, czyli podstawianiu równoważników, nie odpowiada rzeczywistej rozmaitości ani sądów, ani wniosków.'

> So Struve was as if a link connecting this new period with the previous one. Between the generations of the Cieszkowskis, the Gołuchowskis, the Kremers, the Libelts, the Trentowskis and the contemporary generation there appears the distinguished figure of this thinker, writer, who saved from the past what was of lasting value, and he showed the workers of today's Polish philosophy the direction through his prudent, and devoid of all prejudices, opinion.[13] [34, p. 101]

2 Władysław Biegański

2.1 *Life*

Władysław Biegański was born in Grabów near Ostrzeszów on 28 April 1857. He studied medicine in Warsaw in 1875–1880. After having completed his residency in Russia (1881–1882) he continued studies in Berlin (1882–1883) and Prague (1883). Then he settled in Częstochowa where he practiced medicine (internal diseases, neurology) for over 30 years working in a hospital and as a physician for a factory and the railways. At the same time, he conducted research. His scientific interests included many medical disciplines as well as the philosophy of medicine, and in particular the methodological and ethical issues connected with it (cf. his works [4–6, 16]). Biegański represented the Polish school of the philosophy of medicine (cf. [17]). He was interested also in epistemology (cf. [9, 10, 14]). Biegański was involved also in social activities. He died in Częstochowa on 29 January 1917.

2.2 *Biegański's Views on Logic*

Biegański's true passion was in fact logic. As a student he listened to Henryk Struve's lectures. Besides his medical practice, he taught logic in local secondary schools for some time. In 1914 there was even an initiative to appoint Biegański as the professor of the Jagiellonian University Chair of Logic. It did not occur because of his poor health condition and the outbreak of the First World War.

In discussing Biegański's philosophical views on logic, we first should notice that he was neither a formal nor a mathematical logician. It can be said that he was a

[13] 'Był tedy Henryk Struve jakby ogniwem, łączącym ten okres nowy z poprzednim. Między pokoleniem Cieszkowskich, Gołuchowskich, Kremerów, Libeltów, Trentowskich (It is worth explaining who those persons were. August Count Cieszkowski (1814–1894)—Polish philosopher, economist and social and political activist; Józef Gołuchowski (1797–1858)—Polish philosopher, co-creator of the Polish Romanticist „national philosophy"; Józef Kremer (1806–1875)—Polish historian of art, a philosopher, an aesthetician and a psychologist; Karol Libelt (1807–1875)—Polish philosopher, writer, political and social activist, social worker and liberal, nationalist politician; Bronisław Trentowski (1808–1869)—Polish "Messianist" philosopher, pedagogist, journalist and Freemason, and the chief representative of the Polish Messianist „national philosophy" [my remark—R.M.].) a pokoleniem współczesnem widnieje czcigodna postać myśliciela, nauczyciela, pisarza, który z przeszłości ocalił to, co miało w niej wartość trwałą, a dzisiejszym na polu filozofii polskiej pracownikom wskazał drogę rozważnym i dalekim od wszelkiego uprzedzenia sądem.'

philosophical logician from the standpoint formulated by Łukasiewicz (cf. [36]). The latter characterized philosophical logic in the following way:

> If we use here the term "philosophical logic" we mean the complex of problems included in books written by philosophers, and the logic we were taught in secondary school. Philosophical logic is not a homogenous science; it contains various issues; in particular, it enters the field of psychology when it speaks not only about a proposition in a logical sense but also this psychological phenomenon, which corresponds with a proposition and which is called "judgment" or "conviction." [...] Philosophical logic also embraces some issues from the theory of knowledge, for example, the problem of what truth is or whether any criterion of truth exists.[14] [22, pp. 12–13]

It is immediately worth adding that Łukasiewicz himself did not value philosophical logic—he was of the opinion that the scope of problems it considers is not homogenous, and also that philosophical logic mixes logic with psychology. Moreover, both fields are different and use different research methods.[15]

How did Biegański understand logic? Let us quote two definitions of logic given by him. In *Zasady logiki ogólnej* [Principles of General Logic] he wrote:

> Logic is the science of the ways or norms of true cognition.[16] [7, p. 1]

In *Podręcznik logiki i metodologii ogólnej* [Manual of Logic and General Methodology] we find the following definition:

> We call logic the science of the norms and rules of true cognition.[17] [8, p. 3]

Therefore, according to Biegański the laws of logic concern the relationships of mental phenomena because of its aim, which is true cognition. Consequently, logic aims at investigating cognitive activities of the mind. At the same time Biegański claims that one should separate and distinguish between logic and the theory of knowledge on the one hand, and psychology on the other. The reason is that logic is a normative and applied science whereas both the theory of knowledge and psychology are theoretical. However, in practice Biegański—like other authors of his time—did not distinguish strictly between logical and genetic questions, investigating logical constructions both from the precisely logical and psychological points of view. Yet, it should be noted that in *Zasady* [7] Biegański suggests that his conception of logic makes him reject the division into formal and material truth whereas in *Podręcznik* [8] he regards this distinction as correct. He also adds that logic embraces the formal side of cognition.

Consider now his large (638 pages) monograph entitled *Teoria logiki* [Theory of Logic] [12] from 1912 which was an attempt to consider the foundations of logic comprehensively. This work presented general problems concerning logic, the study of

---

[14] 'Jeżeli używamy terminu logika filozoficzna, to chodzi nam o ten kompleks zagadnień, które znajdują się w książkach pisanych przez filozofów, o tę logikę, której uczyliśmy się w szkole średniej. Logika filozoficzna nie jest jednolitą nauką, zawiera w sobie zagadnienia rozmaitej treści; w szczególności wkracza w dziedzinę psychologii, gdy mówi nie tylko o zdaniu w sensie logicznym, ale także o tym zjawisku psychicznym, które odpowiada zdaniu, a które nazywa się „sądem" albo „przekonaniem". [...] W logice filozoficznej zawierają się również niektóre zagadnienia z teorii poznania, np. zagadnienie, co to jest prawda lub czy istnieje jakieś kryterium prawdy.'

[15] Cf. [35] or [26].

[16] 'Logika jest to nauka o sposobach albo normach poznania prawdziwego.'

[17] 'Logiką nazywamy naukę o normach i prawidłach poznania prawdziwego.'

concepts, the study of judgments, the study of argumentation and the study of induction. Every problem was considered in historical and comparative perspectives on the one hand and a systematic perspective on the other hand. Although the author focused on the views of the representatives of traditional logic, he also analyzed the algebra of logic. Biegański wrote there:

> The main aim of logic is to control argumentation. [...] Logic, as the science and art of argumentation, is an a priori science, i.e. science that draws its content not from experience and not from the facts given in experience, but from certain a priori presumptions and constructions.[18] [12, pp. 34–35]

One sees that logic appears here as a normative science. Therefore Biegański proposes to use the name "pragmatic logic". He separates logic from psychology, ontology and epistemology. The basis of logic is formed by axioms: "the most general laws which are directly obvious, i.e. requiring no proof" [12, p. 41]. The axioms are the laws of identity, contradiction, excluded middle and sufficient reason.

This understanding of logic as the art of argumentation can be found in his earlier treatise "Czem jest logika?" [What is Logic?] [11]. He wrote there:

> [...] logic does not reproduce the processes of thought and it does not aim at doing it at all. Therefore, the definition of logic as the science or art of thinking is actually devoid of any basis. [...] But the origin of logic shows that this ability [i.e. logic—remark is mine] is neither a science nor art of thinking, but was created by Plato and Aristotle as the art of argument. Such differences in views cause serious consequences. If logic is a science or even an art of thinking, it is or should be a branch of psychology; on the contrary, if it is only the art of argument, it becomes a separate science that is independent from psychology. Logic as the art of argument does not describe the ordinary course of thoughts, used in argumentation; it does not reproduce it; it does not find laws for it, laws expressing the mutual causal relationship of thoughts, but uses ideal constructions which serve to control the ways of argumentation and in this respect it is explicitly separated from psychology.[19] [11, p. 144]

As a consequence logic "must be of normative character" [11, p. 145]. Biegański explains this in the following way:

> The essence of argumentation consists in valuing. Looking for a proof of any proposition we always follow the question about its cognitive value.[20] [11, p. 145]

---

[18] 'Logika ma na celu głównie kontrolę dowodzenia. [...] Logika, jako nauka i sztuka dowodzenia jest nauką aprioryczną, tj. taką, która swoją treść czerpie nie z doświadczenia, nie z faktów w doświadczeniu nam danych, lecz z pewnych naprzód powziętych założeń i konstrukcji.

[19] '[...] logika nie odtwarza procesów myśli i nie ma wcale na celu tego zadania. To też określenie logiki jako nauki lub sztuki myślenia jest pozbawione właściwie wszelkiej podstawy. [...] Tymczasem geneza logiki wykazuje, że umiejętność ta [tzn. logika—my remark, R.M.] nie jest ani nauką, ani sztuką myślenia, lecz utworzona została przez Platona i Arystotelesa jako sztuka dowodzenia. Takie różnice w zapatrywaniach prowadzą za sobą poważne konsekwencje. Jeżeli logika jest nauką lub nawet sztuką myślenia, to w każdym razie jest lub powinna być działem psychologii, przeciwnie, jeżeli jest tylko sztuką dowodzenia, to staje się nauką odrębną, niezależną od psychologii. Logika jako sztuka dowodzenia nie opisuje zwykłego biegu myśli, stosowanego przy dowodzeniu, nie odtwarza go, nie wynajduje dla niego praw, wyrażających wzajemny związek przyczynowy myśli, lecz posługuje się konstrukcyami idealnymi, które służą dla kontroli sposobów dowodzenia i pod tym względem odgranicza się wyraźnie od psychologii.'

[20] 'Istota dowodzenia polega na wartościowaniu. Poszukując dowodu dla jakiegokolwiek zdania, kierujemy się zawsze pytaniem o jego wartości poznawczej.'

What is important here is the veracity of a proposition and not its meaning and content. "Every proof consists in stating the consistency between the content of the proposition and the principles, which we recognize as true, and it is in this consistency that the essence of truth lies" [11, p. 145].[21]

One should ask now what then is the relation between logic and psychology? It must be noted that Biegański stressed their autonomy. He wrote:

> Any direct [...] dependence here is out of the question. Nonetheless, psychological investigations are not completely meaningless to logic since they constitute an important control for logical constructions [...]. An ideal logical construction would be one that is the closest to the real course of thoughts, that completely guarantees to distinguish truth and is easy to apply. [...] Thus psychological investigations are undoubtedly of great importance for the development of logic because they can contribute to formulating new constructions which are the closest to the natural course of thoughts.[22] [11, pp. 147–148]

In this context the following question arises: what did Biegański mean by argumentation? In fact, one do not find by him any clear idea of inference. He neither used the concept of logical deduction nor distinguished between deductive and inductive reasoning. He states only that inference is based on the idea of necessity, that the principles of logic refer to the form and not the content of cognition. These ideas, however, are not clear and they are mixed. In the work "Sposobność logiczna w świetle algebry logiki" [Logical Modality in the Light of the Algebra of Logic] [13] he speaks about reliable and possible deduction, which is a misunderstanding.

Discussing Biegański's conception of logic one must add that his departure from psychologism was not definitive. In fact in *Podręcznik logiki ogólnej* [Manual of General Logic] [15] from 1916 he returned to psychologism. He wrote there:

> We call logic the science about the ways of controlling the truth of our cognitive thoughts.[23] [15, p. 1]

Biegański's conceptions concerning the foundations and philosophy of logic did not evoke much interest and his work was criticized. Let us tell for example about the reaction of Łukasiewicz. He published in the journal *Ruch Filozoficzny* the review of Biegański's work "Czem jest logika?" [11]. Łukasiewicz stressed there Biegański's departure from psychologism but noticed that it was not completely consistent. He also emphasized the fact that Biegański's conception of logic was too narrow—in fact he limited it to inference. In Łukaszewicz's opinion the object of logic should be reasoning in general, which should include non-deductive reasoning as well. He wrote:

> Logic does not only concern argumentation but reasoning in general, while using the term 'reasoning' as more general than 'argumentation' in accordance with Prof. Twardowski's view

[21] 'Każdy dowód polega na stwierdzeniu zgodności treści zdania z zasadami, które uznajemy za prawdziwe i w tej właśnie zgodności tkwi istota prawdy.'

[22] 'O bezpośredniej [...] zależności nie może tu być mowy. Pomimo to badania psychologiczne nie są zupełnie bez znaczenia dla logiki, stanowią bowiem bardzo ważną kontrolę dla konstrukcji logicznych. [...] Ideałem konstrukcyi logicznej byłaby taka, którąby się najbardziej zbliżała do rzeczywistego biegu myśli, dawała zupełną gwarancyę w odróżnianiu prawdy i była łatwa do stosowania. [...] To też badania psychologiczne mają niewątpliwie duże znaczenie w rozwoju logiki, gdyż mogą się przyczynić do wynalezienia konstrukcyi nowych, najbardziej zbliżonych do naturalnego biegu myśli.'

[23] 'Logiką nazywamy naukę o sposobach kontrolowania prawdy naszych myśli poznawczych.'

(cf. my dissertation *O twórczości w nauce* [About Creativity in Science], p. 8). Secondly, argumentation or reasoning is also thinking, and thus psychologism returns. I would agree to distinguish between logic as 'science' and 'art' but I would use different terms. Namely, I think that logic as a theoretical science investigates relations which formal propositions (e.g. S is P) maintain because of their truth or falsity, and makes laws of these relations (e.g. 'if it is true that S is M and M is P, it is also true that S is P'); as a practical science it applies these laws to solve tasks in the area of reasoning in general, for instance to introduce some conclusion, like in induction, to verify or prove some thesis, etc.[24] (cf. [21])

So we have here the distinction between theoretical and practical logic, which corresponds to the distinction between *logica docens* and *logica utens*. Łukasiewicz defines both in an anti-psychologistic way. Briefly speaking, he understands logic as argumentation theory, dividing reasoning into deductive and reductive; further dividing deductive reasoning into concluding and verifying, and reductive reasoning: proving and explicating (cf. [12]).[25]

We see that Biegański was involved only in the traditional paradigm of logic. However, it should be added that he saw the advantages of the new approach, in particular the values and advantages of the algebra of logic. In the introduction to the work "Sposobność logiczna w świetle algebry logiki" [13], in which he attempted to apply the algebra of logic to the theory of modal categories, he wrote:

Although logical calculus, called the algebra of logic or logistics, has not and cannot have a large practical application, considering the logical evaluation of our judgments and conclusions, it has undoubtedly important theoretical significance. [...] Yet, algebraic symbols, which we use in logical calculus, separate clearly the object of investigation from psychological factors and objective relations, and bring to light all the properties of pure logical relations. Therefore, the main value of the algebra of logic consists in the fact that using it we can explain more thoroughly and mark strictly the relations that are explained variously in school logic.[26] [13, p. 67]

It should be added however that those attempts of Biegański brought a poor result.

[24]'Logika tyczy się nie tylko dowodzenia, ale w ogóle rozumowania, przy czym zgodnie z prof. Twardowskim używam terminu "rozumowanie" jako ogólniejszego od "dowodzenia" (cf. rozprawę moją O twórczości w nauce, str. 8). Po wtóre, dowodzenie czy rozumowanie jest także myśleniem, a więc psychologizm powraca. Zgodziłbym się natomiast na odróżnienie logiki jako "nauki" i "sztuki", tylko użyłbym innych terminów. Sądzę mianowicie, że logika jako nauka teoretyczna bada stosunki, w jakich zdania formalne (np. S jest P) pozostają do siebie ze względu na swoją prawdziwość lub fałszywość, i ustanawia prawa tych stosunków (np. "jeśli prawdą jest, że S jest M i M jest P, to prawdą jest, że S jest P"); jako nauka praktyczna stosuje te prawa do rozwiązywania zadań z zakresu rozumowania w ogóle, np. do wyprowadzenia jakiejś konkluzji, jak we wnioskowaniu indukcyjnym, do sprawdzenia lub udowodnienia jakiejś tezy itp.'

[25]More on Łukasiewicz's views on logic can be found in [35] as well as in [26] and [27].

[26]'Rachunek logiczny, zwany algebrą logiki lub inaczej jeszcze logistyką, jakkolwiek nie ma i nie może mieć rozległego zastosowania praktycznego przy ocenie wartości logicznej naszych sadów i wniosków, posiada jednak niewątpliwie ważne teoretyczne znaczenie. [...] Tymczasem symbole algebraiczne, jakimi się w rachunku logicznym posługujemy, odrywają wyraźnie przedmiot badania zarówno od czynników psychicznych jako też od stosunków objektywnych i wydobywają na jaw wszystkie właściwości czystych stosunków logicznych. To też główna wartość algebry logiki polega na tem, że przy jej pośrednictwie możemy dokładniej wyjaśnić i ściślej wyznaczyć stosunki, które w logice szkolnej rozmaicie bywają tłomaczone.'

3 Conclusion

The above analyses lead to the conclusion that Struve and Biegański represented the traditional "pre-mathematical" approach to logic. In fact the logical culture was decisively "pre-mathematical" in Poland at the turn of the nineteenth and the twentieth centuries. This opinion is supported, for example, by the first edition of *Poradnik dla samouków* [A Guide for Autodidacts] which includes Adam Marburg's paper "Logika i teoria poznania" [Logic and the Theory of Knowledge] (cf. [23]), written in a rather old-fashioned manner. It was so despite the fact that the first signs of interest in mathematical logic appeared in Poland already in the 1880s. Let us mention here the treatise of Stanisław Piątkiewicz *Algebra w logice* [Algebra in Logic] published in 1888—cf. [1] and [2] as well as [3].

Struve's and Biegański's views on logic were shared by their contemporaries, for example by Władysław Kozłowski (1832–1899) and Władysław Mieczysław Kozłowski (1858–1935). The former characterized logic in his work *Logika elementarna* [Elementary Logic] in the following way:

> logic is the science about mental activities with the aid of which we reach truth and prove it.[27] [18, p. 1]

In turn, Władysław M. Kozłowski wrote in *Podstawy logiki* [The Foundations of Logic]:

> Logic is the science about the activities of the mind which seeks truth.[28] [19, p. 8]

The first chapter of his work was entitled "Thinking as Object of Logic" [19, p. 22]. This thought was repeated in the work *Krótki zarys logiki* [A Brief Outline of Logic] where it was claimed that logic is a normative science and its task is "to examine the ways leading the mind to truth" [20, p. 1].[29] However, it was stressed that logic:

> analyses mental operations conducted to reach the truth in a form that is so general that it could be applied to any content. It investigates its form, separating it completely from the content. Logic shares this property with mathematics [...]. [...] This formal character, common to logic and mathematics, made these sciences close in their attempts, which were less or more developed, and led to the creation of mathematical logic.[30] [20, pp. 8–9]

---

[27] Logika jest nauką o czynnościach umysłowych, za pomocą których dochodzimy prawdy i jej dowodzimy.

[28] Logika jest nauka o czynnościach umysłu poszukującego prawdy.

[29] 'badanie dróg, prowadzących rozum do prawdy.'

[30] bada operacye umysłowe, wykonywane w celu osiągnięcia prawdy w formie tak ogólnej, iżby mogły zastosować się do jakiejkolwiekbądź treści. Bada je ze stanowiska ich formy, odrywając się zupełnie od treści. Własność tę podziela z logiką matematyka [...]. [...] Ten formalny charakter, wspólny logice z matematyką, spowodował zbliżenie do siebie obu nauk w próbach mniej lub dalej posuniętych i znalazł wyraz w utworzeniu logiki matematycznej.

Finally, it was stated that logic can be defined "as the science about the forms of every ordered field of real or imaginary objects" [20, p. 9].[31]

In Poland the road to the new paradigm in logic, to the new understanding of it, was long and difficult. Only in the next generation of Polish scholars the new mathematical approach to logic can be seen.

References

1. Batóg, T.: Stanisław Piątkiewicz — pionier logiki matematycznej w Polsce. Kwartalnik Historii Nauki i Techniki **16**, 553–563 (1971)
2. Batóg, T.: Stanisław Piątkiewicz — pionier logiki matematycznej w Polsce. Z Dziejów Kultury i Literatury Ziemi Przemyskiej **II**, 325–330 (1973)
3. Batóg, T., Murawski, R.: Stanisław Piątkiewicz and the beginnings of mathematical logic in Poland. Hist. Math. **23**, 68–73 (1996)
4. Biegański, Wł.: Logika medycyny, czyli Zasady ogólnej metodologii nauk lekarskich. W. Drukarnia W. Kowalewskiego, Warszawa (1894) (2nd edn. (1908); German translation: Medizinische Logik: Kritik der ärztlichen Erkenntnis. C. Kabitzsch, Würzburg (1909))
5. Biegański, Wł.: Zagadnienia ogólne z teorii nauk lekarskich. Drukarnia Towarzystwa S. Orgelbranda Synów, Warszawa (1897)
6. Biegański, Wł.: Myśli i aforyzmy o etyce lekarskiej. W.L. Anczyc, Warszawa (1899)
7. Biegański, Wł.: Zasady logiki ogólnej. Kasa Józefa Mianowskiego, Warszawa (1903)
8. Biegański, Wł.: Podręcznik logiki i metodologii ogólnej dla szkół średnich i dla samouków. E. Wende, Warszawa; W.L. Anczyc, Lwów; H. Altenberg, Kraków (1907)
9. Biegański, Wł.: Wnioskowanie z analogii. Polskie Towarzystwo Filozoficzne, Lwów (1909)
10. Biegański, Wł.: Traktat o poznaniu i prawdzie. Zapomoga Kasy dla Osób Pracujących na Polu Naukowem im. Józefa Mianowskiego, Warszawa (1910)
11. Biegański, Wł.: Czem jest logika? Sprawozdania z Posiedzeń Towarzystwa Naukowego Warszawskiego, Wydział Nauk Antropologicznych, Społecznych, Historii i Filozofii, year **III**(6), 119–152 (1910)
12. Biegański, Wł.: Teoria logiki. E. Wende i S-ka, Warszawa (1912)
13. Biegański, Wł.: Sposobność logiczna w świetle algebry logiki. Sprawozdania z Posiedzeń Towarzystwa Naukowego Warszawskiego, Wydział Nauk Antropologicznych, Społecznych, Historii i Filozofii, year **V**(9), 67–79 (1912)
14. Biegański, Wł.: Teoria poznania ze stanowiska zasady celowości. Towarzystwo Naukowe Warszawskie, Warszawa (1915)
15. Biegański, Wł.: Podręcznik logiki ogólnej. E. Wende i spółka (Ludwik Fiszer), Warszawa; H. Altenberg, G. Seyfarth, E. Wende i S-ka, Lwów; M. Niemierkiewicz, Poznań; Ludwik Fiszer, Łódź (1916)
16. Biegański, Wł.: Etyka ogólna. Gebethner i Wolff, Warszawa (1918)
17. Borzym, S.: Poglądy epistemologiczne Władysława Biegańskiego. Filozofia Nauki **6**(3–4), 11–17 (1998)
18. Kozłowski, Wł.: Logika elementarna. Nakładem Towarzystwa Nauczycieli Szkół Wyższych, Lwów (1891)
19. Kozłowski, Wł.M.: Podstawy logiki, czyli Zasady nauk: wykład systematyczny dla szkół wyższych oraz dla samouctwa. Wydawnictwo M. Arcta, Warszawa (1916)
20. Kozłowski, Wł.M.: Krótki zarys logiki wraz z elementami ideografii logicznej. Wydawnictwo M. Arcta, Warszawa (1918)
21. Łukasiewicz, J.: W. Biegański, Czym jest logika? [review]. Ruch Filozoficzny **2**(8), 149a–149b (1912)

[31] 'jako naukę o formach każdej uporządkowanej dziedziny przedmiotów rzeczywistych lub urojonych.'

22. Łukasiewicz, J.: Elementy logiki matematycznej. Komisja Wydawnicza Koła Matematyczno-Fizycznego Słuchaczów Uniwersytetu Warszawskiego, Warszawa (1929). Ed. by M. Presburger (Reprint: Wydawnictwo Naukowe Uniwersytetu im. Adama Mickiewicza, Poznań (2008))
23. Marburg, A.: Logika i teoria poznania. In: Poradnik dla samouków. Wydawnictwo A. Heflera i St. Michalskiego, Warszawa (1902)
24. Murawski, R.: Genius or Madman? On the Life and Work of J.M. Hoene-Wroński. In: Więsław, W. (ed.) European Mathematics in the Last Centuries, pp. 77–86. Stefan Banach International Mathematical Center/Institute of Mathematics, Wrocław University, Wrocław (2005)
25. Murawski, R.: The Philosophy of Hoene-Wroński. Organon **35**, 143–150 (2006)
26. Murawski, R.: Filozofia matematyki i logiki w Polsce miedzywojennej. Monografie Fundacji na rzecz Nauki Polskiej, Wydawnictwo Naukowe Uniwersytetu Mikołaja Kopernika, Toruń (2011)
27. Murawski, R.: The Philosophy of Mathematics and Logic in the 1920s and 1930s in Poland. Birkhäuser Verlag, Basel (2014)
28. Struve, H.: Logika poprzedzona wstępem psychologicznym. Odczyt Uniwersytecki Dra Henryka Struve profesora Szkoły Głównej w Warszawie (1863). Lithograph
29. Struve, H.: Wykład systematyczny logiki czyli nauka dochodzenia i poznania prawdy, vol. 1: Część wstępna (Historya logiki u obcych i w Polsce). Warszawa (1868–1870)
30. Struve, H.: Filozofia i wykształcenie filozoficzne. In: Lubomirski, J.T., Stawiski, E., Plebański, J.K. (eds.) Encyklopedia wychowawcza, vol. IV. J. Drukarnia J. Sikorskiego, Warszawa (1885)
31. Struve, H.: Logika elementarna. Podręcznik dla szkół i samouków z dodaniem słownika terminów logicznych. Wydawnictwo M. Arcta, Warszawa (1907)
32. Struve, H.: Historya logiki jako teoryi poznania w Polsce. Printed by the author, Warszawa (1911)
33. Trzcieniecka-Schneider, I.: Logika Henryka Struvego. U progu nowego paradygmatu. Wydawnictwo Naukowe Uniwersytetu Pedagogicznego, Kraków (2010)
34. Twardowski, K.: Henryk Struve. Ruch Filozoficzny **2**(6) (1912)
35. Woleński, J.: Logic and Philosophy in the Lvov-Warsaw School. Kluwer Academic Publishers, Dordrecht (1989)
36. Woleński, J.: Władysław Biegański jako logik. Filozofia Nauki **6**(3–4), 19–26 (1998)
37. Woleński, J.: Mathematical logic in Warsaw: 1918–1939. In: Ehrenfeucht, A., Marek, V.W., Srebrny, M. (eds.) Andrzej Mostowski and Foundational Studies, pp. 30–46. IOS Press, Amsterdam (2008)

R. Murawski (✉)
Adam Mickiewicz University in Poznań, Faculty of Mathematics and Computer Science, Poznań, Poland
e-mail: rmur@amu.edu.pl

Part II
Warsaw School of Logic, Its Main Figures and Ideas: The Period of Prosperity

As it was mentioned in the previous part, the famous branch of LWS was Warsaw School of Logic (WSL). Its founders were Jan Łukasiewicz and Stanisław Leśniewski, who, after re-activation of the University of Warsaw, got professorships at the Faculty of Mathematics and Natural Sciences. The biographies of both were introduced in Part I. Here, in Part II, we present some papers connected with their achievements, as well as biographies of some important disciples and representatives of WSL, followed by the articles on their output.

Łukasiewicz and His Followers in Many-Valued Logic

Grzegorz Malinowski

Abstract The aim of this work is a concise introduction to the Łukasiewicz logical world: three and n-valued, n-natural or infinite, denumerable or of the power of continuum. We present Łukasiewicz inventory work and its rationale, the elaboration of original ideas and their technical complementation. Finally, we attempt to show the impact of Łukasiewicz conceptions, their development and directions of resulting applications.

Keywords Logical value · Many-valuedness · Matrix consequence · Matrix semantics · Functional completeness · Many-valued algebra · Probability · Fuzzy set

Mathematics Subject Classification (2000) Primary 03B50; Secondary 03A99

1 Łukasiewicz's Way to Many-Valuedness

The assumption stating that to every proposition it may be ascribed exactly one of the two logical values, *truth* or *falsity*, called the *principle of bivalence*, constitutes the basis of classical logic. It determines both the subject matter and the scope of applicability of the logic. While the roots of many-valued logics can be seen in Aristotle—with his famous concern for future contingents and the *sea-battle tomorrow*—and traced through the middle ages and the nineteenth century, the real 'era of many-valuedness' began in 1920 with the work of Łukasiewicz and Post. In this chapter we will look carefully at the first approach, which since its history started as early as in 1918 may be considered as the first modern approach to many-valuedness.

The actual introduction of a third logical value by Łukasiewicz [20], next to truth and falsity, was preceded by thorough philosophical studies. Their crowning achievement was a three-valued propositional calculus which, from the point of view of logic, represented a standard line of approach. However, in view of the surprisingly rich motivation substantiating the new logic and also the hopes it produced, its exceptional position has been maintained.

Á. Garrido, U. Wybraniec-Skardowska (eds.), *The Lvov-Warsaw School. Past and Present*, Studies in Universal Logic,
https://doi.org/10.1007/978-3-319-65430-0_24

1.1 Łukasiewicz and the Lvov–Warsaw School

Łukasiewicz's work, especially that done in 1906–1922, resulted in a significant philosophical context. The first and second decade of the 1920s witnessed the development of the so-called Lvov–Warsaw philosophical school of which Łukasiewicz was a co-originator and a prominent representative (see Woleński [49]).

It is an unenviable task to decide which of the debates carried out in the Lvov–Warsaw school directly contributed to the chief logical discovery of Łukasiewicz. According to several notes in his works one may maintain that the main source of Łukasiewicz's views on the logic may be subscribed to a discussion concerning the Brentano–Twardowski–Meinong general theory of objects. Created by Meinong towards the end of the nineteenth century, the theory of contradictory objects postulated the existence of objects having contradictory properties, such as, for example, squaring the circle. Meinong claimed that the non-existence of contradictory objects would result in the inability to utter any true proposition, notably a proposition stating that they are not objects. Łukasiewicz advocated the theory of contradictory objects. What is more, he even tended to think that non-contradictory objects do not exist. It is noteworthy that the turn of the nineteenth century witnessed Russell's discovery of a paradox in set theory with the comprehension axiom. This fact was acknowledged by Łukasiewicz who, consequently, attacked in [18] the logical principle of contradiction (see Sect. 1.2).

The important part of Łukasiewicz's studies was concerned with the problems of induction and the theory of probability. Especially, while dealing with the latter, he extricated himself from the 'embarrassing' principle of contradiction (see [19]) and classified as *undefinite* the propositions with free nominal variables. Subsequently, he assigned fractional 'logical' values to undefinite propositions indicating the proportions between the number of actual variable values verifying a proposition and the number of all possible values of that variable. Under this conception, 'logical values' exhibit the feature of relativity and depend on the set of individuals actually evaluated. So, for example, the value of the proposition '$x^2 = 1$' amounts to 2/3 in the set $\{-1, 0, 1\}$ and to 2/5 in the set $\{-2, -1, 0, 1, 2\}$. Obviously enough, infinite sets of individuals are not admitted and this immediately implies that Łukasiewicz's suggestion cannot be taken seriously within the theory of probability (some attempts to improve the idea are due to Zawirski [54]). However, independently of the justification for combining the problems of many-valuedness with probability the crucial fact is that already in 1913 Łukasiewicz had employed the concept of logical value in an unorthodox manner.

The studies that finally led Łukasiewicz to the construction of three-valued logic touched upon determinism, indeterminism and some related problems like the causality principle and modality (i.e. possibility and necessity; see [17, 18]). Some historians of logic suspect Łukasiewicz of being influenced by the discourse (in the Lvov–Warsaw school) about freedom and creativity. Its main thesis was formulated by Kotarbiński who in [15] suggested the need for revising the two-valued logic that seemed to interfere with the freedom of human thinking. Łukasiewicz was a fierce follower of indeterminism, which found its expression, among others, in the introduction of the third logical value, next to truth and falsity, to be assigned to non-determined propositions; specifically, to propositions describing casual future events, i.e. *future contingents*.

1.2 The Third Logical Value

The very earliest remarks about the three-valued propositional calculus can be found in the Farewell Lecture given by Łukasiewicz in the Assembly Hall of Warsaw University on the 7th of March, 1918. Next in the paper 'O logice trójwartościowej', [20], one may find an outline of the three-valued logic and a brief motivation for the new logical construction casting off the principle of bivalence. Łukasiewicz in [21] analyses the sentence 'I shall be in Warsaw at noon on 21 December of the next year' and argues that at the time of its utterance, it is neither true nor false, since otherwise fatalist conclusions about necessity or impossibility of contingent future events would follow, its value (truth or falsity) is not settled. Hence, the sentences of this kind pertain to a 'third' logical category.

Consequently, to the two classical values 0 and 1, Łukasiewicz added an intermediate logical value ½ interpreted as 'possibility' or 'indeterminacy'—the former of these options was subsequently repudiated by him under the influence of the studies on modality. First, Łukasiewicz extends the classical interpretation of implication ($\rightarrow$) and negation ($\neg$) connectives, which resulted in the putting forward of the following tables[1]:

α	$\neg\alpha$
0	1
½	½
1	0

$\rightarrow$	0	½	1
0	1	1	1
½	½	1	1
1	0	½	1

The remaining connectives of disjunction, conjunction and equivalence were later introduced through the sequence of the following definitions:

$$\alpha \vee \beta = (\alpha \rightarrow \beta) \rightarrow \beta$$
$$\alpha \wedge \beta = \neg(\neg\alpha \vee \neg\beta)$$
$$\alpha \equiv \beta = (\alpha \rightarrow \beta) \wedge (\beta \rightarrow \alpha).$$

Consequently, their tables are as follows:

$\vee$	0	½	1
0	0	½	1
½	½	½	1
1	1	1	1

$\wedge$	0	½	1
0	0	0	0
½	0	½	½
1	0	½	1

$\equiv$	0	½	1
0	1	½	0
½	½	1	½
1	0	½	1

[1]The truth-tables of binary connectives $*$ are viewed as follows: the value of α is placed in the first vertical line, the value of β in the first horizontal line and the value of $\alpha * \beta$ at the intersection of the two lines.

A *valuation* of the set of formulas, *For*, in the three-valued logic is any function v : $For \to \{0, {}^1\!/\!_2, 1\}$ 'compatible' with the above tables; a three-valued *tautology* is a formula which under any valuation v takes on the *designated value* 1. An instant reflection shows that any such formula is the classical tautology.

System Ł3 = $Taut_3$ of the three-valued propositional calculus differs radically from CPC. On the one hand some important laws of the classical logic are not tautologies of Ł3, as

(1) $p \vee \neg p$ (law of the excluded middle)
(2) $\neg(p \wedge \neg p)$ (principle of contradiction).

but on the other hand some classically contradictory formulae[2] are consistent in the logic of Łukasiewicz. An important example of such formula is

(3) $p \equiv \neg p$.

All these properties can be checked with the help of any valuation which yields ½ for p. Such a valuation associates ½ with (1), (2) and 1 with (3).

The thorough-going refutation of the law of the excluded middle and the principle of contradiction was intended, in Łukasiewicz's opinion, to codify the principles of indeterminism: 'both disjunction and conjunction of two possible propositions are possible propositions (and nothing else)'. Whereas the consistency of (3) supports the claim that Łukasiewicz logic is adjusted to the formalization of reasonings about contradictory objects. Also, as Łukasiewicz remarks, Russell's paradox of 'the set of all sets that are not their own elements' ceases to be an antinomy in Ł3. The *Russell's set* Z is defined as $Z = \{x : x \notin x\}$, and therefore the equivalence

$$Z \in Z \equiv Z \notin Z,$$

which is a substitution of (3) is not-contradictory, since putting for ½ p makes it true.

1.3 Modality, Axiomatization and Interpretation

Another Łukasiewicz's intention was to get formalization of modal functors of possibility M and necessity L. which would preserve the consistency of middle ages' intuitive theorems on modal propositions. Since expressing these functors within the truth-functional classical logic is impossible, Łukasiewicz took the three-valued logic as a new base. In 1921 Tarski gave simple definitions, using negation and implication, of the two connectives which meeting the Łukasiewicz's requirements[3]:

x	Mx
0	0
½	1
1	1

x	Lx
0	0
½	0
1	1

$$M\alpha =^{df} \neg\alpha \to \alpha$$

$$L\alpha =^{df} \neg M \neg\alpha = \neg(\alpha \to \neg\alpha)$$

[2]That is, formulae taking 0 at arbitrary logical valuation.

[3]See Łukasiewicz [21].

Using M, L and other Łukasiewicz connectives we get third modal connective "it is contingent that"or, "it is modally indifferent":

α	$I\alpha$
0	0
½	1
1	0

$$I\alpha = M\alpha \wedge \neg L\alpha$$

We may treat I as an operator distinguishing the intermediate logical value ½. Further, applying I allows the formulation within Ł3, of counterparts of the law of the excluded middle and the principle of contradiction:

$$p \vee Ip \vee \neg p$$

$$\neg(p \wedge \neg Ip \wedge \neg p)$$

rendering altogether that Łukasiewicz's logic is three-valued.

In spite of the promising combination of trivalence and modality the full elaboration of modal logic on the basis of the three-valued logic never succeeded (with the mere exception of algebraic constructions of Moisil), which was the result of the further Łukasiewicz's investigations on modal sentences (see Łukasiewicz [21]). Many years after Łukasiewicz comes back to the idea of construction of modal logic of possibility and necessity founded of the four element Boolean matrix, Łukasiewicz [22].

Through the law of excluded fourth and the extended contradiction principle, Ł3 expresses its three-valuedness. However, contrary to the classical case this expression it is limited since not all connectives described by $\{0, {}^1\!/\!_2, 1\}$–tables are definable through formulas. One important example is the constant connective T, such that $Tx = {}^1\!/\!_2$ for any $x \in \{0, {}^1\!/\!_2, 1\}$.

The axiomatization of Ł3 due to Wajsberg [47] is the first known axiomatization of a system of many-valued logic. Accepting the rules MP (detachment) and SUB (substitution) Wajsberg axiom system for $(\neg, \rightarrow)$-version of Łukasiewicz's three-valued propositional calculus is as follows:

W1. $p \rightarrow (q \rightarrow p)$
W2. $(p \rightarrow q) \rightarrow ((q \rightarrow r) \rightarrow (p \rightarrow r))$
W3. $(\neg p \rightarrow \neg q) \rightarrow (q \rightarrow p)$
W4. $((p \rightarrow \neg p) \rightarrow p) \rightarrow p$.

The result obviously applies to the whole Ł3 since the other Łukasiewicz connectives are definable using those of negation and implication. Słupecki [39] enriched the set of primitives by T and extending Wajsberg system with

W5. $Tp \rightarrow \neg Tp$
W6. $\neg Tp \rightarrow Tp$

got an axiom system for a functionally complete three-valued logic.

2 Matrix Semantics for n-Valued Logics

In 1922 Łukasiewicz generalized his three-valued logic and defined the family of many-valued logics, both finite and infinite-valued.[4] *Łukasiewicz n-valued matrix has* the form

$$M_n = (L_n, \neg, \rightarrow, \vee, \wedge, \equiv, \{1\}),$$

where

$$L_n = \begin{cases} \{0, \ ^1/_{n-1}, \ ^2/_{n-1}, \ \ldots, \ 1\} & \text{if } n \in N, n \geq 2 \\ \{s/w \ : \ 0 \leq s \leq w, \ s, \ w \in N \ \text{ and } \ w \neq 0\} & \text{if } n = \aleph_0 \\ [0, 1] & \text{if } n = \aleph_1. \end{cases}$$

and the functions are defined on L_n as follows:

(i) $\neg x = 1 - x$
$x \rightarrow y = min(1, 1 - x + y)$
(ii) $x \vee y = (x \rightarrow y) \rightarrow y = max(x, y)$
$x \wedge y = \neg(\neg x \vee \neg y) = min(x, y)$
$x \equiv y = (x \rightarrow y) \wedge (y \rightarrow x) = 1 - |x - y|.$

2.1 *Basic Matrix Properties and Consequence*

The introduction of new many-valued logics was not supported by any separate argumentation. Łukasiewicz merely underlined, that the generalization was correct since for $n = 3$ one gets exactly the matrix of his 1920' three-valued logic. The future history will, however, show that Łukasiewicz logics have nice properties, which locate them among the most important logical constructions.

First, the Łukasiewicz matrix M_2 coincides with the matrix of the classical logic. And, since the set $\{0, 1\}$ is closed with respect to all Łukasiewicz connectives A_2 is a subalgebra of any algebra $(L_n, \neg, \rightarrow, \vee, \wedge, \equiv)$ and M_2 is a *submatrix* of M_n. Therefore all tautologies of n-valued Łukasiewicz propositional calculi, $Taut_n$ are included in $Taut$, i.e. the set of tautologies of the classical logic.

$$Taut_n \subseteq Taut_2 = Taut.$$

What is more, the family of all finite Łukasiewicz matrices has more submatrix properties, between tautologies of finite matrices further inclusion relations hold. Actually, they are established by the following Lindenbaum condition[5]:

For finite $n, m \in N$, $Taut_n \subseteq Taut_m$ iff $m - 1$ is a *divisor* of $n - 1$.

---

[4]See Łukasiewicz [23, 1970, p. 140].

[5]See Łukasiewicz and Tarski [24].

It may be proved that the infinite-valued Łukasiewicz matrices produce the same set of tautologies, which is an intersection of all $Taut_n$:

$$Taut_{\aleph_0} = Taut_{\aleph_1} = \bigcap \{Taut_n) : n \geq 2, n \in N\}.$$

For any n as above, one may define the matrix consequence relation $\models_n$ putting for a set of formulas X and a formula α, $X \models_n \alpha$ iff for any valuation v of formulas in M_n, $v(\alpha) = 1$ whenever $v(X) \subseteq \{1\}$). Then, using the same argument one may extend onto finite n and m the Lindenbaum tautology inclusion result for matrix consequence relations $\models_n$:

$$\models_n \subseteq \models_m \text{ iff } m - 1 \text{ is a } divisor \text{ of } n - 1.$$

Completing the view for n infinite, we should say that though the matrices $M_{\aleph_0}$ and $M_{\aleph_1}$ define the same sets of tautologies, their consequence relations are different, more precisely,

$$\models_{M_{\aleph_1}} \subseteq \models_{M_{\aleph_0}}, \text{ but } \models_{M_{\aleph_1}} \neq \models_{M_{\aleph_0}}.$$

The first inclusion is easy: all valuations in C are valuations in $M_{\aleph_1}$. The reverse, however, due to the absence of irrational numbers in $L_{\aleph_0}$, which obviously are in $L_{\aleph_1}$, is not true, see Wójcicki [51].

The last two formulas are but some of the properties of many-valued consequence relation and operation. An ingenious general elaboration Tarski's idea of logical consequence [44] by Wójcicki and his collaborators, led to a general matrix approach in the theory of sentential calculi, see [50]. Łukasiewicz matrices and logics played a major role as an experimental background in the crucial period of this creation, see Wójcicki, Malinowski [52].

Łukasiewicz n-valued logics $Ł_n$ or, equivalently, their algebras are not functionally complete. All what was established for $n = 3$ applies for each finite n. First, no constant except 0 and 1 is definable in $(L_n, \neg, \rightarrow, \vee, \wedge, \equiv)$. Second, adding the constants to the stock of connectives makes this algebra functionally complete. And, since M_n is one generated, either by ${}^{1}/_{n-1}$ or by ${}^{n-2}/_{n-1}$, also adding only one of them do the job as well. McNaughton [25] proved an ingenious definability criterion, both finite and infinite, which shows the mathematical beauty of Łukasiewicz's logic constructions.

2.2 Axiomatizability

A proof that finite matrices are axiomatizable was given in Łukasiewicz and Tarski [24]. However, the problem of formulation of a concrete axiom system for finite Łukasiewicz logics for $n > 3$ remains open till 1952. Rosser and Turquette [37] are the authors of a general method of axiomatization of n-valued logics with connectives satisfying the so-called standard conditions. The method can be applied, among others, to $Ł_n$ since such connectives are either primitive or definable in Łukasiewicz finite matrices. Hence, for every n an axiomatization of Łukasiewicz's n-valued propositional calculus can be obtained. The axiomatization, however, becomes very complicated due to the high generality of the method given by Rosser and Turquette.

In 1930 Łukasiewicz conjectured that his $\aleph_0$-valued logic was axiomatizable (see [24]) and that the axiomatization of the infinite-valued propositional calculus together with MP and SUB was the following:

L1. $p \rightarrow (q \rightarrow p)$
L2. $(p \rightarrow q) \rightarrow ((q \rightarrow r) \rightarrow (p \rightarrow r))$
L3. $((p \rightarrow q) \rightarrow q) \rightarrow ((q \rightarrow p) \rightarrow p)$
L4. $(\neg p \rightarrow \neg q) \rightarrow (q \rightarrow p)$
L5. $((p \rightarrow q) \rightarrow (q \rightarrow p)) \rightarrow (q \rightarrow p)$.

Due to Łukasiewicz[6] this hypothesis was confirmed by Wajsberg in 1931. Next comes the reduction of the axiom set: Meredith [26] and Chang [2] independently showed that axiom L5 is dependent on the others. There are two main accessible completeness proofs of L1–L4 (with MP and SUB): based on syntactic methods and linear inequalities by Rose and Rosser [36], and purely algebraic—by Chang [4].

Chang's proof is based on properties of MV algebras, algebraic counterparts of the infinite-valued Łukasiewicz logic, defined and studied in [3].[7] The key role in the approach have additional binary connectives + and ·, which directly correspond to the algebraic operations of MV algebras are defined by

$$\alpha + \beta =^{df} \neg\alpha \rightarrow \beta, \text{ and } \alpha \cdot \beta =^{df} \neg(\alpha \rightarrow \neg\beta).$$

Several axiomatizations for finite-valued Łukasiewicz logics ($n > 3$) were obtained by way of extension of the axiom system L1–L4. Grigolia [10] employs multiplying use of the connectives + and ·. Let $k\alpha$ will be a replacement of the formula $\alpha + \alpha + \ldots + \alpha$ (k times) and α^k a replacement of the formula $\alpha \cdot \alpha \cdot \ldots \cdot \alpha$ (k times). Given a finite $n > 3$, Grigolia's axiom system for $Ł_n$ consists of the schemes of L1–L4 and

L_n5. $n\alpha \rightarrow (n-1)\alpha$
L_n6. $(n-1)((\neg\alpha)^j + (\alpha \cdot (j-1)\alpha))$,

where $1 < j < n-1$ and j does not divide $n-1$.

Tokarz [45] extension of L1–L4 is based on the characteristic functions of the set L_n in [0, 1] and the properties of the consequence relation of $M_{\aleph_0}$. The axiom set for a given n-valued Łukasiewicz logic, including $n = 2$, results from L1–L4 by adding a single special "disjunctive" axiom

$$p \vee \neg p \vee \delta_n^1(p) \vee \ldots \vee \delta_n^{n-2}(p),$$

[6] Łukasiewicz [23, 1970, p. 144]; no publication on the topic by Wajsberg exists.
[7] MV algebras are presented in Sect. 4.2.

where, for any k, $1 \leq k \leq n-1$, the algebraic operation $\delta_n^k(x)$ corresponding to a formula is the "characteristic function" of the logical value ${}^k/_{n-1}$ in the infinite Łukasiewicz matrix $M_{\aleph_0}$. $v(\delta_n^k(x))$ is 1 only when $v(x) = {}^k/_{n-1}$, and is 0 otherwise. Such a formula is definable in virtue of the McNaughton criterion in [25].

Another axiomatization of finite Łukasiewicz logics, offered by Tuziak [46], is formulated in the standard propositional language with sequences of ascending implications defined inductively by: $p \to^0 q = q$, $p \to^{k+1} q = p \to (p \to^k q)$. The axiom set for n-valued Łukasiewicz logic consists of ten formulas taken from the Hilbert-Bernays axiomatization of CPC and two axioms:

T1. $(p \to^n q) \to (p \to^{n-1} q)$
T2. $(p \equiv (p \to^{s-2} \neg p)) \to^{n-1} p$ for any $2 \leq s \leq n-1$ such that s is not a divisor of $n-1$.

3 Post Logics and Słupecki's Functional Completeness Programme

Post n-valued logical constructions in [32, 33] were mainly inspired by the well-known formalization of the classical logic of *Principia Mathematica* [48] and by the truth-table method of verification of logical laws. Contrary to Łukasiewicz logics all they are *functionally complete*, i.e. they have the property that their any (finite-valued) propositional connective can be represented or defined as a composition of their primitive connectives.[8]

3.1 Post Matrices and Functional Completeness

The basic many-valued Post constructions are connected with two primitives: cyclic negation ($\neg$) and disjunction ($\vee$). For any natural $n \geq 2$ Post builds an n-valued logical algebra on the linearly ordered set of objects

$$P_n = \{t_1, t_2, \ldots, t_n\}$$

($t_i < t_j$ iff $i < j$) equipped with two operations: unary *rotation* $\neg$ (*cyclic negation*) and binary disjunction $\vee$, defined in the following way:

$$\neg t_i = \begin{cases} t_{i+1} & \text{if } i \neq n \\ t_1 & \text{if } i = n \end{cases} \qquad t_i \vee t_j = t_{max(i,j)}.$$

[8]The very property applies to the logic algebras i.e. matrices without designated set of elements. Nb. the applications of Post construction are focused on algebras.

The disjunction function fixes on a natural and intuitive meaning of the disjunction connective, typical for most known many-valued constructions. In plain terms, the logical value of disjunctive proposition equals the greater of the values of its components. The function of cyclic rotation permutes, in some specified manner, the set P_n and the negation corresponding to it, the case $n = 2$ being excluded, is quite special: $\neg t_1 = t_2, \neg t_2 = t_3, \ldots, \neg t_{n-1} = t_n, \neg t_n = t_1$. It is just the fact of combining the latter with an appropriate binary function of algebra on P_n that warrants the functional completeness of that algebra, i.e. it ensures that by means of the primitive functions, every finite-argument function can be defined on P_n, including constant functions and hence the objects $t_1, t_2, \ldots, t_n$. For a given finite $n \geq 2$ the algebraic structure:

$$P_n = (\{t_1, t_2, \ldots, t_n\}, \ \neg, \ \vee)$$

will be called an *n-valued Post algebra.*

The matrix P_n naturally associated with the algebra P_n:

$$P_n = (\{t_1, t_2, \ldots, t_n\}, \ \neg, \ \vee, \ \{t_n\})$$

will, in what follows, be referred to as the (basic) *n-valued Post matrix*. It is easily seen that the two-valued Post matrix is isomorphic to the negation–disjunction matrix for the classical propositional calculus. To check it, one must replace t_1 in P_2 by the falsity (0) and t_2 by the symbol of truth (1). Simultaneously, however, the matrices P_n for $n > 2$ are totally incompatible with the mentioned classical matrix, which is the result of the non-standard mode of the negation connective. Hence for $n = 3$, for instance, t_3 could be the only counterpart of 'truth' with respect to the adopted interpretation of disjunction but then t_1 would have to correspond to 'falsity' as $\neg t_3 = t_1$, which should not take place because $\neg\neg t_3 = \neg t_1 = t_2 \neq t_3$. A contradiction.

3.2 Słupecki Class of Functionally Complete Finite-Valued Logics

The pioneering Post construction of algebras of logic emerged scholar's interest in formulating handy criteria which might determine whether or not a given algebra (or, logic) is functionally complete.

An early important result on the field a partial result on definability of one-argument operations:

(Picard [31]). In U_n all one-argument operations are definable whenever the functions $H, \ R, \ S$ are definable in it.[9]

One immediate application of this criterion led Słupecki to complete the three-element *Łukasiewicz algebra* $L_3 = (\{0, \ ½, \ 1\}, \ \neg, \ \rightarrow, \ \vee, \ \wedge, \ \equiv)$ with the operations determined by tables in Sect. 1.2 is an example of an incomplete algebra. Słupecki then remarked that the one-argument constant function $T : Tx = ½$ for any $x = \{0, \ ½, \ 1\}$

[9]See next page.

is not definable in terms of the basic operations. However, in virtue of Picard's criterion, adding T to the stock of functions of L_3 leads to the (three-element) functionally complete algebra (Słupecki [39]).

In the late 1930's Słupecki undertook a serious investigation of the problem of functional completeness. In the result, he provided one of the best appreciated criterion for all finite logics [40], constructed the largest possible class of functionally complete finite logics and gave a general method of their axiomatization, [41]:

(Słupecki [40]). An n-valued algebra U_n ($n \geq 2, n$ finite) is functionally complete if and only if in U_n there are definable:

1^0. All one-argument operations
2^0. At least one two-argument operation $f(x, y)$ whose range consists of all the values i for $1 \leq i \leq n$.

The Słupecki matrix S_{nk} ($1 \leq k < n$) with k-valued set of designated elements $\{1, 2, \ldots, k\}$ is of the form

$$S_{nk} = (\{1, 2, \ldots, n\}, \ \rightarrow, \ R, \ S, \ \{1, 2, \ldots, k\}),$$

where $\rightarrow$ is a binary (implication), and R, S unary operations defined in the following way:

$$x \rightarrow y = \begin{cases} y \text{ if } 1 \leq x \leq k \\ 1 \text{ if } k < x \leq n \end{cases}$$

$$R(x) = \begin{cases} x + 1 \text{ if } 1 \leq x \leq n - 1 \\ 1 \quad\quad \text{if } x = n \end{cases}$$

$$S(x) = \begin{cases} 2 \text{ if } x = 1 \\ 1 \text{ if } x = 2 \\ x \text{ if } 3 \leq x \leq n. \end{cases}$$

Functional completeness of each of these matrices results from the Picard's criterion: R and S are two necessary functions. To define the third, it suffices to put:

$$Hx = (x \rightarrow R(x \rightarrow x)) \rightarrow Sx \text{ for } k = 1, \quad \text{then } Hx = \begin{cases} 1 \text{ if } x = 2 \\ x \text{ if } x \neq 2 \end{cases}$$

$$Hx = R(x \rightarrow x) \rightarrow x \qquad \text{for } k > 1, \quad \text{then } Hx = \begin{cases} 1 \text{ if } x = k \\ x \text{ if } x \neq k. \end{cases}$$

Słupecki produced an effective proof of axiomatizability of $Taut_{S_{nk}}$ (any pair (n, k) as above) giving a long list of axioms formulated in terms of implication and special one-argument connectives defined through the superpositions of R, S, and H. The chief line of approach here is to make capital of the character of implication, which can be classically axiomatized using the detachment rule (MP). Słupecki extends MP onto the whole language, taking the Łukasiewicz formula $((p \rightarrow q) \rightarrow r) \rightarrow ((r \rightarrow p)$

$\to (s \to p))$ as the only axiom for implication, and provides an inductive, combinatorial proof of completeness.

The original $(\neg, \vee)$ systems of Post logic have not yet been axiomatized. However, the problem of their axiomatizability has for years been taken for granted, so in [41] Słupecki constructed the largest possible class of functionally complete finite logics and gave a general method of axiomatization. From this it evidently follows that Post logics are also axiomatizable although the problem of providing axioms for their original version still remains open.

4 Algebraic Interpretations of Łukasiewicz Logics

The attempts to obtain algebras which would play the same role for Łukasiewicz calculi as Boolean algebras do for the classical logic bore several constructions. The most important of them are: Moisil and Chang algebras.

4.1 Moisil Algebras

Moisil's constructions, dating back to the 1940s, are best adapted to the finite case; see [28]. These are algebras with operations corresponding to 'modal' connectives of Łukasiewicz logics[10]; more specifically, distributive lattices with Boolean-valued endomorphisms.

A structure

$$(L, \cup, \cap, N, s_1, s_2, \ldots, s_{n-1}, 0, 1)$$

is an n-valued *Moisil algebra* if the following conditions are satisfied:

M1. $(L, \cup, \cap)$ is a distributive lattice with 0 and 1.
M2. N is an involution of L (i.e. $NNx = x$) also satisfying: $N(x \cup y) = N(x) \cap N(y)$, $N(x \cap y) = N(x) \cup N(y)$.
M3. The elements of $\{s_k\}_{1 \leq k < n}$ are endomorphisms of $(L, \cup, \cap)$: $s_k(x \cup y) = s_k(x) \cup s_k(y)$ and $s_k(x \cap y) = s_k(x) \cap s_k(y)$, such that

   (i) $s_k(x) \leq s_{k+1}(x)$
   (ii) $s_k(s_t(x)) = s_t(x)$
   (iii) $s_k(N(x)) = Ns_{n-k}(x)$
   (iv) $Ns_k(x) \cup s_k(x) = 1$, $Ns_k(x) \cap s_k(x) = 0$
   (v) If $s_k(x) = s_k(y)$ for every k, then $x = y$.

The simplest example of n-valued Moisil algebra (given finite n) is the linearly ordered n-element algebra constructed on the base of Łukasiewicz matrix M_n:

$$M_n = (L_n, \cup, \cap, N, s_1, s_2, \ldots, s_{n-1}, 0, 1),$$

[10] Compare 1.3 for $n = 3$.

where $x \cup y = x \vee y,\ \ x \cap y = x \wedge y,\ \ N(x) = \neg x$ and

$$s_k(j/n-1) = \begin{cases} 0 \text{ when } 1 \le k \le n-j-1 \\ 1 \text{ when } n-j-1 < k \le n-1. \end{cases}$$

In virtue of McNaughton criterion [25] the operations s_k are definable in M_n. The formulae definining them by means of $\neg$ and $\rightarrow$ were given by Suchoń in [43]. It is interesting to notice, that $s_1,\ s_2,\ \ldots, s_{n-1}$ are the modal functors in n-valued Łukasiewicz logic. Thus, for $n = 3$, $s_1 = L$ and $s_2 = M$, compare Sect. 1.3.

Moisil called his structures *Łukasiewicz algebras*. In the course of time the term *Moisil algebras* came to be used because, for $n \ge 5$, Łukasiewicz implication was not definable in n-valued structures of this kind. Despite that fault, however, Moisil algebras proved to be the most fertile source of applications of Łukasiewicz logics.

4.2 Chang Algebras

The concept of MV algebra, tailored by Chang in [3], resorts to mathematical group theory. Consider an algebra

$$A = (A,\ +,\ \cdot,\ -,\ 0,\ 1),$$

where $+$ and $\cdot$ are binary operations, $-$ is a unary operation and, finally, 0 and 1 are two different constants in A, $0 \ne 1$. Let us introduce, moreover, the following notations:

$$x \cup y = x \cdot y^- + y \quad x \cap y = (x + y^-) \cdot y.$$

The algebra A so defined is called an *MV algebra* if the following conditions are satisfied:

C1.	$x + y = y + x$		
C2.	$x + (y + z) = (x + y) + z$	C1∗.	$x \cdot y = y \cdot x$
C3.	$x + x^- = 1$	C2∗.	$x \cdot (y \cdot z) = (x \cdot y) \cdot z$
C4.	$x + 1 = 1$	C3∗.	$x \cdot x^- = 0$
C5.	$x + 0 = x$	C4∗.	$x \cdot 0 = 0$
C6.	$(x + y)^- = x^- \cdot y^-$	C5∗.	$x \cdot 1 = x$
C7.	$(x^-)^- = x$	C6∗.	$(x \cdot y)^- = x^- + y^-$
C9.	$x \cup y = y \cup x$	C8.	$0^- = 1$
C10.	$x \cup (y \cup z) = (x \cup y) \cup z$	C9∗.	$x \cap y = y \cap x$
C11.	$x + (y \cap z) = (x + y) \cap (x + z)$	C10∗.	$x \cap (y \cap z) = (x \cap y) \cap z$
C11∗.	$x \cdot (y \cup z) = (x \cdot y) \cup (x \cdot z).$		

The simplest example of the MV algebra is an arbitrary (finite or not) Łukasiewicz matrix—in this case $x + y = \neg x \rightarrow y,\ x \cdot y = \neg(x \rightarrow \neg y)$, and $\cup, \cap, -$ are the functions of disjunction, conjunction and negation, respectively.

4.3 MV_n Algebras and Axiomatization of L_n

The problem of adapting MV algebras to finite cases (i.e. of introducing such modifications which would allow algebras characteristic for particular cases of Łukasiewicz finite-valued logics to be obtained) was taken up in 1973 by Grigolia (see [10]) who introduced the notion of MV$_n$ algebra. Adopting $k\alpha$ as the replacement of the formula $\alpha + \alpha + \ldots + \alpha$ (k times) and α^k for the formula $\alpha \cdot \alpha \cdot \ldots \cdot \alpha$ (k times)), we say that an MV algebra A is an *MV_n algebra* provided that the following conditions are satisfied:

C12. $(n-1)x + x = (n-1)x$ C12 $*$. $x^{n-1} \cdot x = x^{n-1}$

and for $n > 3$:

C13. $\{(jx) \cdot (x^- + ((j-1)x)^-\}^{n-1} = 0$

C13 $*$. $(n-1)\{x^j + (x^- \cdot (x^{j-1})^-)\} = 1$

where $1 < j < n-1$ and j does not divide $n-1$.

The common property of the two algebraic interpretations of Łukasiewicz logics is the way in which they are related to Boolean algebras. Every Moisil algebra as well as each MV algebra contains some Boolean subalgebra. The set of Boolean elements of an arbitrary n-valued Moisil algebra is equal to

$$\{s_k(x) : x \in L,\ 1 \leq k < n\},$$

and the set of Boolean elements of an MV algebra A to

$$\{x \in A : x + x = x\} \text{ (or, equivalently, to } \{x \in A : x \cdot x = x\}).$$

From among the purely algebraic results for Moisil and Chang algebras those of the greatest importance for logic are the representation theorems which state that any algebra is (isomorphic to) a subdirect product of a class of linear algebras, i.e.definitional variants of Łukasiewicz matrices. Employing MV algebras, in [4] Chang gave a purely algebraic proof of completeness of the $\aleph_0$-valued Łukasiewicz logic. Thus, L1–L4 together with MP and SUB create the axiomatization of Łukasiewicz's $\aleph_0$-valued propositional calculus.

Several axiomatizations of finite-valued Łukasiewicz logics ($n > 3$) were obtained by way of extension of L1–L4 (MP and SUB being assumed) with the sets of 'specific axioms' depending on n. Thus, for instance, Grigolia for the purpose of axiomatization of n-valued Łukasiewicz logics uses in [10] two additional binary connectives + and $\cdot$ defined by $\alpha + \beta = \neg\alpha \to \beta,\ \alpha \cdot \beta = \neg(\alpha \to \neg\beta)$ (in what follows we write $k\alpha$ as the replacement of the formula $\alpha + \alpha + \ldots + \alpha$ (k times) and α^k for the formula $\alpha \cdot \alpha \cdot \ldots \cdot \alpha$ (k times)). Given a finite $n > 3$, Grigolia's axiom system for L_n consists of the schemes of L1–L4 and, additionally, the axioms

L$_n$5. $n\alpha \to (n-1)\alpha$
L$_n$6. $(n-1)((\neg\alpha)^j + (\alpha \cdot (j-1)\alpha))$,

where $1 < j < n - 1$ and j does not divide $n - 1$. The completeness proof, received by an adaptation of the Chang's method, uses MV_n algebras.

Tokarz in [45] worked out a method of axiomatization of finite Łukasiewicz logics based on the characteristic functions of the set L_n (in [0,1]) and the properties of the matrix consequence of $M_{\aleph_0}$.

5 In Search for Interpretation of Łukasiewicz Logics

The source interpretation of the third logical value as 'possibility' or 'indeterminacy' was strongly criticized as soon as 1938 by Gonseth, who noticed (see [9]) that the connectives in Łukasiewicz logic are incompatible with the suggested ways of interpreting $\frac{1}{2}$. Accordingly, for two propositions α and $\neg\alpha$ if α is undetermined, so is $\neg\alpha$, and then, according to the table of conjunction, $\alpha \wedge \neg\alpha$ is undetermined, which contradicts the intuition since, independently of α's content, $\alpha \wedge \neg\alpha$ is false. A similar argumentation applies to the disjunction: take the formula $\alpha \vee \neg\alpha$ and put $\frac{1}{2}$. We then get $\frac{1}{2}$ as a result, while intuitively $\alpha \vee \neg\alpha$ should be true. As it can be easily observed, the definitiveness of the original interpretations of Łukasiewicz is caused by the fact that they neglect the mutual dependence of some 'possible' propositions.

The history of attempts to give a satisfactory interpretation of Łukasiewicz logics is fairly long. Below, we shall discuss some interesting undeterminacy and probability approaches.

5.1 Three-Valued Logic as a Base for Non-Boolean Undetermined Events

An interesting attempt to interpret Łukasiewicz logic intuitively was made by Słupecki [42]. By an 'interpretation' Słupecki means pointing out some definite language which describes the property of the determination of events in a three-valued manner reconciling Łukasiewicz's truth-tables.

The language considered by Słupecki is a set-theoretical union of a set S comprising propositions about events and a set S^* of propositions that do not state events. Both sets contain simple (atomic) propositions and compound ones formed by means of disjunction ($\vee$), conjunction ($\wedge$) and negation ($\neg$) connectives. Since S^* does not play a significant role in what follows, it is omitted. Słupecki supposes the set of events Z described by propositions of S to be closed under the operations of union ($\cup$), meet (or intersection) ($\cap$) and complementation ($-$) and, furthermore, the structure $Z = (Z, \cup, \cap, -)$ to be a Boolean algebra. There is a causality relation $\mapsto$ on Z ('$f_1 \mapsto f_2$' means 'the event f_1 is a cause of the event f_2') providing the assumption that

(P1) $f \mapsto f_1 \cup f_2$ iff $f \mapsto f_1$ or $f \mapsto f_2$
(P2) $f \mapsto f_1 \cap f_2$ iff $f \mapsto f_1$ and $f \mapsto f_2$
(P3) If $f \mapsto f_1$ for some f, then $f^+ \mapsto -f_1$ for no f^+
(P4) If $f_1 \mapsto f$, then $f_1 \cap f_2 \mapsto f$

for any $f, f_1, f_2 \in Z$. For the purpose of defining the property of determination, he then singles out a set of past and present events hereafter denoted by symbols $g, g_1, g_2, \ldots$, and puts

$$D(f) = df \text{ there is a } g \in Z \text{ such that } g \mapsto f,$$
$$\overline{D}(f) = df \text{ not } D(f) \text{ and not } D(-f).$$

The above expressions can be read as:

$D(f)$ – f is (at the present moment) determined,
$\overline{D}(f)$ – f is not determined (at the present moment).

The relationship between expressions of the language S and events is established by relation $*$ of describability ('$p * f$' means: (proposition) p describes (event) f) satisfying the following conditions

$$(*) \quad \begin{array}{l} p \vee p_1 * f \cup f_1 \text{ whenever } p * f \text{ and } p_1 * f_1 \\ p \wedge p_1 * f \cap f_1 \text{ whenever } p * f \text{ and } p_1 * f_1 \\ \text{when } p * f, \text{ then } \neg p * -f \end{array}$$

for any $p, p_1 \in S$. Moreover, Słupecki assumes that every proposition of S refers to some event of Z and he then defines the properties $1(p), 0(p)$ and $\frac{1}{2}(p)$: $1(p) = p$ is true, $0(p) = p$ is false, $\frac{1}{2}(p) = p$ has the 'third' logical value, as follows:

$$(\mathrm{DT}) \quad \begin{array}{l} \text{if } p * f, \text{ then } \{ 1(p) \text{ iff } D(f) \} \\ \text{if } p * f, \text{ then } \{ 0(p) \text{ iff } D(-f) \} \\ \text{if } p * f, \text{ then } \{ \frac{1}{2}(p) \text{ iff } \overline{D}(f) \}. \end{array}$$

Using (P1)–(P4) and $(*)$, it is easy to check that for $x \in \{0, \frac{1}{2}, 1\}$

$$x(p \vee q) = x(p) \vee x(q)$$
$$x(p \wedge q) = x(p) \wedge (q)$$
$$x(\neg p) = \neg x(p),$$

where $\vee$, $\wedge$ and $\neg$ appearing on the right-hand side are the connectives of the Łukasiewicz three-valued logic. Therefore, what Łukasiewicz proposed, referring to the property of determining propositions as a manner of interpretation of logical values, is, by (DT), to some extent justified.

Słupecki's interpretation, however, is not free of faults. First, it is partial, lacking the implication connective. Admittedly, Słupecki extends it onto the language with modal connectives M and L, and in this enriched language the implication of Łukasiewicz is definable. However, it certainly does not change anything since the interpretation of the implication obtained is fairly unintuitive. Secondly, more profound analysis of the whole construction reveals that the assumption concerning Z has to be modified: Nowak in [30] proved the formal correctness of the interpretation exclusively when Z is a de Morgan lattice and not a Boolean algebra. This result, nevertheless, does not depreciate Słupecki's proposal but, on the contrary, makes it still more noteworthy; three-valued logic can thus be interpreted as a set of propositions describing events which form a non-classical algebra. If so, then (DT) implies that the third value of Łukasiewicz, $\frac{1}{2}$, is assigned to propositions concerning non-Boolean, undetermined events.

5.2 Logical Probability

Mathematical probability is a measure of casual events and the probability calculus in its simplest form resembles many-valued logic. Therefore, the question of a connection between probability and many-valuedness emerges quite naturally. Łukasiewicz [19] invented a theory of *logical probability* where undetermined propositions (i.e. propositional functions) are associated with a fraction which is proportional to the number of variable values verifying the proposition and the number of all values of a given finite domain. The differentiating feature of this comprehended probability in comparison to mathematical probability is the fact that the former refers to propositions and not events. The continuators of Łukasiewicz's conception, Reichenbach and Zawirski among them, exerted much effort to create a many-valued logic within which logical probability could find a satisfactory interpretation (see e.g. [54, 55] and [34]).

The foundations of the Reichenbach–Zawirski conception are provided by the assumption that there is a function Pr ranging over the set of propositions of the considered propositional language, with values from the interval [0,1]. The basic characteristics of Pr is embodied in the following rules:

(P1) $0 \leq Pr(p) \leq 1$
(P2) $Pr(p \vee \neg p) = 1$
(P3) $Pr(p \vee q) = Pr(p) + Pr(q)$ if p and q are mutually exclusive (i.e. if $p \wedge q$ is an 'impossible' proposition or, equivalently, if $Pr(p \wedge q) = 0$)
(P4) $Pr(p) = Pr(q)$ when p and q are logically equivalent, i.e. if $p \wedge \neg q$ and $\neg p \wedge q$ are 'impossible'.

From the above, other properties of Pr known in the probability calculus can be inferred. Among them are:

(1) $Pr(p \wedge \neg p) = 0$
(2) $Pr(\neg p) = 1 - Pr(p)$
(3) $Pr(p \vee q) = Pr(p) + Pr(q) - Pr(p \wedge q)$.

Therefore, if we identify the logical value $w(p)$ of p with the measure of its probability $Pr(p)$, $w(p) = Pr(p)$, then the function w should satisfy the conditions:

($\neg$) $w(\neg p) = 1 - Pr(p)$
($\wedge$) $w(p \wedge q) = Pr(p) + Pr(q) - Pr(p \vee q)$
($\vee$) $w(p \vee q) = Pr(p) + Pr(q)$ when p and q are exclusive.

However, let us note that completion of ($\vee$) through a direct reference to Pr is impossible. Whenever propositions p and q are not exclusive, $w(p \vee q)$ is a certain number from [0,1] not less than $Pr(p)$, $Pr(q)$ and not greater than $Pr(p) + Pr(q)$. This number obviously depends on the 'content' of both propositions and not only on their logical values. What follows is a practical inability to satisfy the extensionality principle in the system whose semantics resorts to w valuations; values of some compound propositions formed by means of the disjunction connective are not uniquely determined by the values of the component propositions. As an illustration, consider the case when $w(p) = \frac{1}{2}(p)$.

Then we would have

$$\tfrac{1}{2}(p) \vee \tfrac{1}{2}(p) = Pr(p \vee \neg p) = 1 \qquad \text{(according to (P2))}$$

and

$$\tfrac{1}{2}(p) \vee \tfrac{1}{2}(p) = Pr(p \vee p) = 2Pr(p) - Pr(p) = \tfrac{1}{2}(p). \qquad \text{(due to (3))}$$

Consequently, logical probability must not be identified with logical values in any ordinary extensional many-valued logic.

It was, however, not until the 1970s that Giles worked out an operationalistic conception of subjective probability interpreted unquestionably in denumerable Łukasiewicz logic.

5.3 Operationalistic Conception of Subjective Probability

The researches of Giles directed towards finding a logic appropriate for the formalization of physical theories, including those with undetermined propositions, resulted in a very convincing philosophical interpretation of countable Łukasiewicz logic $L_{\aleph_0}$ combining the conceptions of logical many-valuedness and subjective probability (see Giles [8]). The main point of the approach consists in a *dispersive* physical interpretation of standard logic language: it is assumed that each prime proposition in a physical theory is associated through the rules of interpretation with a certain experimental procedure which ends in one of two possible outcomes, 'yes' or 'no'. The tangible meaning of a proposition of the language is then related to the persons using it (i.e. speakers or observers). In the case of prime propositions it is determined from the values of probability of success ascribed by users in respective experiments, whereas in the case of compound propositions it is determined from the rules of obligation incurred by whoever asserts them and formulated in the nomenclature of *dialogue logic* (see Lorenz [16]).

The starting assumptions of the construction are stated in two principles:

(1) Everyone who asserts a prime proposition A commits themselves to paying a certain sum of money, say \$1, when the experiment associated with A results in 'no'. Secondly, to assert a proposition is not the same thing as to assert the same proposition twice (in the same debate).
(2) Every speaker (i.e. language user) is able to ascribe to any prime A a real number $p(A) \in [0, 1]$, the so-called *subjective probability*, such that, given any $\varepsilon < 0$, they are willing: $(*)$ to assert A in return for a payment of $\$(1 - p(A) + \varepsilon)$; $(*)$ to assert $\neg A$ in return for a payment of $\$(p(A) + \varepsilon)$.

which yield that all prime propositions are probability definite for all speakers. The *risk value* of A for a given person is $< A > = 1 - p(A)$, denoting the expected loss if she asserts it.

The meaning of compound propositions is appointed by the rules of debate of two participants: a given person and their partner who can be a fate as well. The rules tend to reduce all assertions of such propositions to the assertions of sequences (sets) of prime propositions. Giles adopts the following rules (for any propositions p and q):

The rule concerning negation is less handy than the others but it can be 'standardized'

Assertion	Obligation (commitment)
$p \vee q$	*Undertaking to assert either p or q at one's own choice*
$p \wedge q$	*Undertaking to assert either p or q at the opponent's choice*
$p \rightarrow q$	*Agreement to assert q if the opponent will assert p*
$\neg p$	*Agreement to pay* \$1 *to opponent if they will assert p.*

under the assumption that language contains a constant F (a falsum) signifying any false proposition; next define $\neg p$ as $p \rightarrow$F.

Every *dialogue* runs as follows: one of the participants asserts a (compound) proposition, the other joins the game or not. Subsequently, the uttered propositions that undergo no decomposition in a given step are repeated, which especially concerns prime propositions. Every position of a dialogue game is of the form W, Z being the names of the participants, while $W_1, W_2, \ldots, W_k$ and $Z_1, Z_2,$

$$\begin{array}{rcl} W & : & Z \\ W_1, W_2, \ldots, W_k & : & Z_1, Z_2, \ldots, Z_m \end{array}$$

$\ldots, Z_m$ are sequences of propositions asserted by W and Z, or *tenets* of W and Z, respectively. A position with prime propositions exclusively is final. Once it is reached, the game counts as finished, appropriate experiments are carried out and the accounts are settled. The two examples illustrate the course of a dialogue:

$$\begin{array}{rcl} W & : & Z \\ \emptyset & : & A \rightarrow (B \rightarrow A) \\ A & : & B \rightarrow A \\ A, B & : & A \end{array} \qquad \begin{array}{rcl} W & : & Z \\ \emptyset & : & A \rightarrow B \vee A \\ A & : & B \vee A \end{array}$$

$$A : B \qquad\qquad A : A$$

In the first case it is W who takes the risk of paying \$1 when the test proves falsity of B. In the second case, two *strategies* are possible: one safe for both sides ($A : A$), the other ending with $A : B$; according to the rules of dialogue the choice lies with Z and it depends on Z's evaluation of the risk values of A and B.

The *risk value* of a tenet $(A_1, A_2, \ldots, A_m)$ is the sum of risks of its elements $< A_1, A_2, \ldots, A_m > = \Sigma < A_j >$, and that of final position for Z is the difference: $< A_1, A_2, \ldots, A_m : B_1, B_2, \ldots, B_n > = \Sigma < B_i > - \Sigma < A_j >$.

From a fundamental result of game theory stating that every game with perfect information has a 'saddle point', it follows (see Giles [8]) that each valuation of prime propositions (i.e. assigning to them risk values or, equivalently, subjective probability values) has a unique extension onto the whole language guaranteeing both participants no increase in the risk value of the initial position—or to put it differently: that *optimal strategy* exists. The extension in question is defined for any statements P and Q through the schemes:

$$< P \rightarrow Q > = sup\{0, < Q > - < P >\}$$
$$< P \vee Q > = inf\{< P >, < Q >\}$$
$$< P \wedge Q > = sup\{< P >, < Q >\}$$
$$< \neg P > = 1- < P >;$$

sup and inf are the symbols of the operations of lower upper bound and greater lower bound in [0,1], respectively.

The formulae of the considered language to which any valuation assigns non-positive risk value are referred to as *tautologies*; there are propositions, the utterance of which may lead only to (theoretically) not losing final positions. Using the equality $pr(P) = inf\{1, 1- < P >\}$ one can describe the property of being a tautology in terms of subjective probability. A simple calculation shows that:

$$pr(P \rightarrow Q) = inf\{1,\ 1 - pr(P) + pr(Q)\}$$
$$pr(P \vee Q) = sup\{pr(P),\ pr(Q)\}$$
$$pr(P \wedge Q) = inf\{pr(P),\ pr(Q)\}$$
$$pr(\neg P) = 1 - pr(P).$$

Thus, pr is a valuation of Łukasiewicz logic $L_{\aleph_0}$ (compare Sect. 2). In consequence, the set of tautologies of dialogue logic just constructed is equal to $Taut_{\aleph_0}$: only those formulae are tautologies whose probability amounts to 1 independently of the values assigned to prime propositions as its components.

6 On Applications and Influence of Łukasiewicz Work

The philosophically well grounded and mathematically mastered Łukasiewicz work brought to science important theoretical results and subscribed to the development of important logical strategies, like the matrix approach. Independently of that, however, some concrete, worthy applications of Łukasiewicz many-valued logics and algebras may be indicated. The section contains a short presentation of some of them, like the use logical matrices to demonstrate the independence of axioms, the way the Łukasiewicz infinite logic grounds the fuzzy sets and truth functional fuzzy logic and, finally, the application of Moisil's algebras in switching theory.

6.1 Independence of Axioms

The logical method of testing axioms independence using algebras and matrices is credited to Bernays and Łukasiewicz. To prove that an axiom system is independent one singles out a property, mostly validity, which is common to all axioms besides of one chosen and is inherited, via accepted rules of inference, by all theorems of the systems. The procedure being repeated as many times as there is the number of axioms in the system.

Consider $(\neg, \to)$–system of the classical propositional calculus originating with Łukasiewicz. Its axioms are:

(A1) $(\neg p \to p) \to p$
(A2) $p \to (\neg p \to q)$
(A3) $(p \to q) \to ((q \to r) \to (p \to r))$,

and the rules MP and SUB. Let now

$$M_{(A1)} = (\{0, 1\}, \neg_1, \to, \{1\}), \quad M_{(A2)} = (\{0, 1\}, \neg_2, \to, \{1\})$$

be matrices wherein the implication connective $\to$ is determined classically (by the well-known truth table), $\neg_1(0) = \neg_1(1) = 0$ and $\neg_2(0) = \neg_2(1) = 1$. Moreover let

$$M_{(A3)} = (\{0, {}^1\!/\!_2, 1\}, \neg, \to, \{1\})$$

be a matrix with the connectives defined by the tables:

α	$\neg\alpha$
0	1
$1/2$	$1/2$
1	1

$\to$	0	$1/2$	1
0	1	1	1
$1/2$	1	0	1
1	0	0	1

It is readily seen that the sets of Tautologies of the three matrices, $Taut_{(A1)}$, $Taut_{(A2)}$ and $Taut_{(A3)}$, are closed under (MP) and (SUB). Since

(1) A1 $\notin Taut_{(A1)}$, A2 $\in Taut_{(A1)}$ and A3 $\in Taut_{(A1)}$,
(2) A1 $\in Taut_{(A2)}$, A2 $\notin Taut_{(A2)}$ and A3 $\in Taut_{(A2)}$,
(3) A1 $\in Taut_{(A3)}$, A2 $\in Taut_{(A3)}$ and A3 $\notin Taut_{(A3)}$.

the axiom system (A1)–(A3) is independent.

The application of the method described is not limited to the logical calculi. The proofs of independence in set theory through the use of matrices built on the base of Boolean algebras were presented by Scott and Solovay, [38].

6.2 *Fuzzy Sets and Truth Functional Fuzzy Logic*

Everyday reasonings operate on imprecise concepts and are supported by approximate inferences. That makes the possibility of applying the apparatus of the classical logic to formalize them very limited. Among special tools extending the formalization power of the standard approach are fuzzy sets and fuzzy logics. Below, we outline fuzzy concepts since, due to their properties, the infinite-valued Łukasiewicz logics are the most often used as their bases.

Fuzzy Sets Zadeh in [53] defines a fuzzy set A of a given domain U as an abstract object characterized by generalized characteristic function U_A in the real set [0,1]: $U_A : U \to [0, 1]$. The values of U_A are interpreted as degrees of membership of elements of U to the fuzzy set A. The extreme values of this function, 0 and 1, denote respectively: not belonging to A and entire membership to A. Limiting the scope of U_A to $\{0, 1\}$ results in an "ordinary" characteristic function and in this each "classical" set is a special case of a fuzzy set.

Fuzzy sets are an instrument of modelling inexact predicates appearing in natural languages. Thus, for example the property of "being much greater than 1" defined on the set of real positives R^+ can be assigned to a fuzzy set W with a non-decreasing characteristic function $R^+ \to [0, 1]$ which meets the conditions like: $R_W(0) = 0$, $R_W(1) = 0$, $R_W(5) = 0.01$, $R_W(100) = 0.95$, $R_W(500) = 1$, etc. Certainly, in the above example only values $R_W(0)$, $R_W(1)$ unquestionable and the selection of other values is somehow arbitrary.

In the family $F(U)$ of fuzzy (sub)sets of a given domain the relation of inclusion reflexes the order between the reals:

$$A \subseteq B \quad \text{if and only if} \quad U_A(x) \leq U_B(x) \quad \text{for any} \quad a \in U,$$

and the counterparts of the operations of complement ($-$), union ($\cup$) and intersection ($\cap$) are set by:

$$U_{-A}(x) = 1 - U_A(x)$$

$$U_{A \cup B}(x) = max\{U_A(x), U_B(x)\}$$

$$U_{A \cap B}(x) = min\{U_A(x), U_B(x)\}.$$

Bellman and Giertz [1] showed that $U_{A \cup B}$ and $U_{A \cap B}$ are the unique non-decreasing continuous functions warranting both the compatibility of the construction with the standard algebra of sets and the fact that $(F(U), \cup, \cap, -)$ is a de Morgan lattice if and only if U_{-A} is defined as above. It is in order to notice that, in spite of the naturalness of the proposal, several studies admit as (more) helpful, fuzzy sets algebras defined otherwise.

The values of generalized characteristic functions may be identified with logical values of propositions of the form "x ∈ X", where ∈ is a "generalized" set-theoretical predicate. Subsequently, using logical constants of a base logic one may set the inclusion and the operations of fuzzy set algebra as:

$$A \subseteq B =_{df} \forall x(x \in A \rightarrow x \in B)$$
$$-A = \{x : \neg(x \in A)\}$$
$$A \cup B = \{x : x \in A \vee x \in B\}$$
$$A \cap B = \{x : x \in A \wedge x \in B\}.$$

For Zadeh's algebra the choice of a base logic is to great extent prejudiced: it (this logic) must be based on $\aleph_1$-element matrix, wherein negation is expressed by the function $1 - p$, disjunction and conjunction, respectively by: $max\{p, q\}$ and $min\{p, q\}$, and the universal quantifier as the greatest lower bound (inf). The function of implication is not uniquely determined. However, evidently it should meet the requirement:
(.) If $p \rightarrow q = 1$, then $p \leq q$.
Though Łukasiewicz's implication and, consequently, his $\aleph_1$-valued logic, have been the most intensely applied, similar connectives of other logics have been also taken into account. The commonly shared belief among scholars working on fuzziness, both theoreticians and practicians, is that only a concrete application of fuzzy sets algebra can decide about the choice of the base logic (see Gaines [7]).

Truth Functional Fuzzy Logics Algebraic and metamathematical studies of the infinite-valed Łukasiewicz logic, see Cignoli et al. [5], are among the most important issues of recent investigations. Somewhat related to these studies are activities delineating a class of propositional logics, called by Hájek [12] fuzzy logics (in a narrow) sense. The influence of fuzzy set theory initiated the study of a class of systems of many-valued logics, whose semantics is based on the real interval [0,1]. Several comparisons between the systems serving as a base for particular constructions directed the scholar's attention to, possibly idempotent, strong conjunctions connectives, whose corresponding truth functions were associative, commutative, non-decreasing and have 1 as its neutral (unit) element. Such functions were called *t-norms*. Accordingly, a binary function $*$ on [0,1] is a *t-norm* (triangular norm) if for any $x, y, z \in [0,1]$

$x * (y * z) = (x * y) * z$
$x * y = y * x$
if $x \leq y$, then $x * z \leq y * z$
$x * 1 = x$

Connectives corresponding to t-norms are conjunctions. Further to this one may also define *t-conorms* which serve as truth functions of disjunctions and, possibly, relate the two functions using appropriate negation. In Sect. 2 we have examples of both, the t-norm ($min(x, y)$ i.e. Łukasiewicz conjunction) and t-conorm ($max(x, y)$ i.e. Łukasiewicz disjunction).

Hájek [12] is the main study of logics defined by continuous t-norms (a t-norm is continuous is considered in the mathematical terms is continuous as a mapping). Among the important continuous t-norms are the following:

Łukasiewicz t-norm $x * y = max(0, x + y - 1)$,
Gödel t-norm $x * y = min(x, y)$
product t-norm $x * y = x \cdot y$;

these functions have been used in numerous applications of fuzzy set theory as well as fuzzy logics.

The connectives defined through the continuous t-norm conjunctions (continuity with respect to the left argument is sufficient) are special since there is algebraically nice procedure relating them with implications, having good metalogical properties. Any such implication $\rightarrow$ is defined as *residuum* of a given continuous t-norm $*$, i.e.

$$x \rightarrow y = max(z : x * z \leq y).$$

Hájek's basic fuzzy propositional logic, *BL-logic*, is the logic of continuous t-norms on [0,1]. The language of BL has the connectives of conjunction &, implication $\rightarrow$ and the constant $\perp$ of falsity. The semantics of BL is established by the function of t-norm, all other functions corresponding to the connectives are derived. A formula is a BL tautology if and only if under each valuation of propositional variables compatible with the functions of connectives it takes the value 1. Hájek's axiom system adequate for BL logic is the following:

(H1) $(\alpha \rightarrow \beta)\rightarrow ((\beta \rightarrow \gamma) \rightarrow (\alpha \rightarrow \gamma)$
(H2) $(\alpha\&\beta) \rightarrow \alpha)$
(H3) $(\alpha \&\beta) \rightarrow(\beta \& \alpha)$
(H4) $(\alpha \&(\alpha \rightarrow \beta))\rightarrow(\beta \& (\beta \rightarrow \alpha)$
(H5a) $(\alpha \rightarrow(\beta \rightarrow \gamma)) \rightarrow ((\alpha\&\beta)\rightarrow \gamma)$
(H5b) $((\alpha\&\beta)\rightarrow \gamma)\rightarrow (\alpha \rightarrow(\beta \rightarrow \gamma))$
(H6) $(\alpha \rightarrow(\beta \rightarrow \gamma))\rightarrow (((\beta \rightarrow \alpha)\rightarrow \gamma)\rightarrow \gamma)$
(H7) $\perp \rightarrow \alpha$

Any system of propositional logic determined by a t-norm in the way indicated may be received as a strengthening of BL. For instance, Łukasiewicz logic results from BL by addition the single one axiom schema:

(Ł) $\neg\neg\alpha \rightarrow \alpha$

BL extends to the *basic fuzzy predicate logic* in a standard way. Hájek [12] shows interesting features of t-norm based predicate calculi, see also Hájek et al. [13], Montagna [29].

6.3 *Moisil Algebras and Switching Theory*

Soon after the successful applications of the classical logic, Boolean algebras and other algebraic structures (e.g. groups) in switching theory, in the 1950s, the scholars centered the interests on the possibility of the use of many-valued logic algebras for similar purposes (see e.g. Epstein et al. [6]). These interests brought about the birth of several

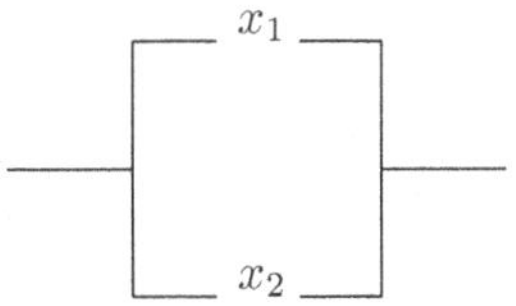

Fig. 1 Two oppositely oriented contacts

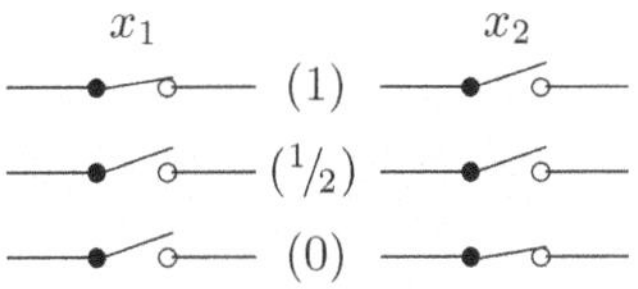

Fig. 2 The "real" switch-function

techniques of the analysis and synthesis of electronic circuits and relays based (mainly) on Moisil and Post algebras (see Rine [35]). Below, we confine ourselves to some remarks justifying the purposefulness of using many-valued algebras in switching and relay circuits theory. The most elementary composite of the traditional electronic circuit is a mechanical contact opening and closing some fragment of an electrical network. The switch over of contacts is affected mechanically or electromechanically (i.e. using relays). And, among the contacts of a given network one may find such pairs of contacts which according to the technical assumptions have to change their positions into complementary simultaneously. The simplest example of such situation is the gear of two oppositely oriented contacts x_1 and x_2 positioned in parallel branches of a circuit (see Fig. 1): x_1 is normally closed while x_2 normally open. When considering the ideal model of the circuit one assumes that both contacts react momentarily to an affection and thus stroking, as shown at Fig. 2, from the state (1) to (0).

Practically, however, it may happen that x_1 will open still before x_2 would be closed and, consequently, contrary to the technical presumptions the gear during a time moment will be open. That is just the reason for such a modelling in which the third state ($^1\!/_2$) (see Fig. 2) is considered; the table beforehand characterizes the "real" switch-function as a function of states and contact (1 inside the table denotes normal contact's state and 0 its denial). On the other hand, one also may read table treating x_1 and x_2 as (one-argument) functions of states i.e. of z, and their values as states as well putting: $x_1 = s_1(z)$, $x_2 = s_2(z)$.

z	x_1	x_2
1	1	1
$^1\!/_2$	0	1
0	0	0

Let us notice that then s_1 and s_2 are Moisil's operations on $\{0, {}^1\!/_2, 1\}$. Subsequently, to describe any network built of the contacts x_1, x_2 and their complements $\overline{x_1}$, $\overline{x_2}$ one should define binary operations $\cup$ and $\cap$ corresponding to two possible types of connections and unary operation N such that $\overline{x_i} = Nx_i$ $(i = 1, 2)$ and that $NNz = z$

for $z \in \{0, {}^1\!/\!_2, 1\}$. It appears that the most accurate ways of introducing these operations leads to the Moisil algebra $(\{0, {}^1\!/\!_2, 1\}, \cup, \cap, N, s_1, s_2)$.

A generalization of the outlined construction onto the case of any number of contacts similarly results in n-valued algebras. The algebraic treatment of switching devices aims at providing several techniques of the analysis, the synthesis and the minimalization of multiplex networks. The most important advantage of the many-valued approach is the possibility of elimination of possible switching disturbance through the algebraic synthesis of the networks, see e.g. Moisil [27].

Application of many-valued algebras is not limited to binary contacts. Investigations concerning multi-stable contacts and switches have also been undertaken. However, according to difficulties with technical realizations of devices working on voltage-mode and the progress of technology of binary highly integrated circuits these activities are not very common. Still, however, many-valued constructions attract attention of engineers. Thus, for instance, multiple values may be useful for describing transistors. Hayes [14], Hähnle and Kernig [11] give an example of such modelling MOS transistors. Due to a degradation of signals, a MOS transistor has different signal levels at source and terminals. Thus, a natural modelling leads to a many-valued logic, the values of which are organized in form of a lattice.

References

1. Bellman, R.E., Giertz, M.: On the analytic formalism of the theory of fuzzy sets. Inf. Sci. **5**, 149–156 (1973)
2. Chang, C.C.: Proof of an axiom of Łukasiewicz. Trans. Am. Math. Soc. **87**, 55–56 (1958)
3. Chang, C.C.: Algebraic analysis of many-valued logics. Trans. Am. Math. Soc. **88**, 467–490 (1958)
4. Chang, C.C.: A new proof of the completeness of the Łukasiewicz axioms. Trans. Am. Math. Soc. **93**, 74–80 (1959)
5. Cignoli, R.L.O., D'Ottaviano, I.M.L., Mundici, D.: Foundations of Many-Valued Reasoning. Trends in Logic: Studia Logica Library, vol. 7. Kluwer Academic Publishers, Dordrecht (1999)
6. Epstein, G., Frieder, G., Rine, D.C.: The development of multiple-valued logic as related to Computer Science. Computer **7**(9), 20–32 (1974)
7. Gaines, B.R.: Foundations of fuzzy reasoning. Int. J. Man Mach. Stud. **8**, 623–668 (1976)
8. Giles, R.: A non-classical logic for physics. Stud. Logica **33**, 397–416 (1974)
9. Gonseth, F. (ed.): Les entretiens de Zurich sur les fondements et la méthode des sciences mathématiques 6–9 décembre 1938. Zurich (1941)
10. Grigolia, R.: Algebraic analysis of Łukasiewicz–Tarski's n-valued logical systems. In: Wójcicki, R., Malinowski, G. (eds.) Selected papers on Łukasiewicz sentential calculi, pp. 81–92. Ossolineum, Wrocław (1977)
11. Hähnle, R., Kernig, W.: Verification of switch level designs with many-valued logic. In: Voronkov, A. (ed.) Proceedings LPAR'93, St. Petersburg, Russia. Lecture Notes on Computer Science, vol. 698, pp. 158–169. Springer, Berlin (1993)
12. Hájek, P.: Metamathematics of Fuzzy Logics. Trends in Logic: Studia Logica Library, vol. 4. Kluwer Academic Publishers, Dordrecht (1998)
13. Hájek, P., Paris J., Sheperdson, J.: The liar paradox and fuzzy logic. J. Symb. Log. **65**, 339–346 (2000)
14. Hayes, J.P.: Pseudo-Boolean logic circuits. IEEE Trans. Comput. **C-35**(7), 602–612 (1986)
15. Kotarbiński, T.: Zagadnienie istnienia przyszłości [The problem of existence of the future]. Przegląd Filozoficzny **VI.1** (1913)
16. Lorenz, K.: Arithmetik und Logik als Spiele. Kiel (1961)
17. Łukasiewicz, J.: Analiza i konstrukcja pojęcia przyczyny. Przegląd Filozoficzny **9**, 105–179 (1906)

18. Łukasiewicz, J.: O zasadzie sprzeczności u Arystotelesa. Studium krytyczne. Kraków (1910) (English tr.: On the principle of contradiction in Aristotle. Rev. Metaphys. **XXIV**, (1971))
19. Łukasiewicz, J.: Die logischen Grundlagen der Wahrscheinlichkeitsrechnung. Kraków (1913) (English tr.: Logical foundations of probability theory. In: Borkowski, L. (ed.) Selected Works, pp. 16–63. North-Holland, Amsterdam (1970)
20. Łukasiewicz, J.: O logice trójwartościowej. Ruch Filozoficzny **5**, 170–171 (1920) (English tr.: On three-valued logic. In: Borkowski, L. (ed.) Selected works, pp. 87–88. North-Holland, Amsterdam (1970))
21. Łukasiewicz, J.: Philosophische Bemerkungen zu mehrwertigen Systemen des Aussagenkalküls. Comptes rendus des séances de la Société des Sciences et des Lettres de Varsovie Cl. III **23**, 51–77 (1930) (English tr.: Philosophical remarks on many-valued systems of propositional logic. In: McCall, S. (ed.) Polish Logic 1920–1939, pp. 40–65. Clarendon Press, Oxford (1967))
22. Łukasiewicz, J.: A system of modal logic. J. Comput. Syst. **1**, 111–149 (1953)
23. Łukasiewicz, J.: Z zagadnień logiki i filozofii. Pisma wybrane. PWN, Warszawa (1961) (English tr.: Borkowski, L. (ed.): Selected Works. North-Holland, Amsterdam (1970))
24. Łukasiewicz, J., Tarski, A.: Untersuchungen über den Aussagenkalkül. Comptes rendus des séances de la Société des Sciences et des Lettres de Varsovie Cl. III **23**, 30–50 (1930)
25. McNaughton, R.: A theorem about infinite-valued sentential logic. J. Symb. Log. **16**, 1–13 (1951)
26. Meredith, C.A.: The dependence of an axiom of Łukasiewicz. Trans. Am. Math. Soc. **87**, 54 (1958)
27. Moisil, G.: Zastosowanie algebr Łukasiewicza do teorii układów przekaźnikowo-stykowych [Application of Łukasiewicz algebras to the study of relay-contact networks], vol. II, 1967 edn. Ossolineum, Wrocław (1958)
28. Moisil, G.: Essais sur les logiques non-chrisipiennes. Editions de l'Académie de la Republique Socialiste de Roumanie, Bucharest (1972)
29. Montagna, F.: An algebraic approach to propositional fuzzy logic. J. Log. Lang. Inf. **9**(1), 91–124 (2000). Special issue on many-valued Logics of Uncertainty. Mundici, D. (ed.)
30. Nowak, M.: O możliwości interpretowania trójwartościowej logiki Łukasiewicza metodą Słupeckiego [On the possibility of interpreting the three-valued Łukasiewicz logic using Słupecki's method]. Acta Universitatis Lodziensis, Folia Philosophica **5**, 3–13 (1988)
31. Picard, S.: Sur les fonctions définies dans les ensembles finis quelconques. Fundam. Math. **24**, 198–302 (1935)
32. Post, E.L.: Introduction to a general theory of elementary propositions. Bull. Am. Math. Soc. **26**, 437 (1920)
33. Post, E.L.: Introduction to a general theory of elementary propositions. Am. J. Math. **43**, 163–185 (1921)
34. Reichenbach, H.: Wahrscheinlichkeitslehre. Leiden (1935) (English tr.: The Theory of Probability. University of California Press, Berkeley (1949))
35. Rine, D.C. (ed.): Computer Science and Multiple-Valued Logic. Theory and Applications. North-Holland, Amsterdam (1977)
36. Rose, A., Rosser, J.B.: Fragments of many-valued statement calculi. Trans. Am. Math. Soc. **87**, 1–53 (1958)
37. Rosser, J.B.,Turquette, A.R.: Many-Valued Logics. North-Holland, Amsterdam (1952)
38. Scott, D., Solovay, R.: Boolean valued models for set theory. In: Proceedings of the American Mathematical Society Summer Institute on Axiomatic Set Theory 1967. University of California, Los Angeles. Proceedings of Symposia in Pure Mathematics, vol. 13 (1969)
39. Słupecki, J.: Der volle dreiwertige Aussagenkalkül. Comptes rendus des séances de la Société des Sciences et des Lettres de Varsovie Cl. III **29**, 9–11 (1936) (English tr.: The full three-valued propositional calculus. In: McCall, S. (ed.) Polish Logic 1920–1939, pp. 335–337. Clarendon Press, Oxford (1967))
40. Słupecki, J.: Kryterium pełności wielowartościowych systemów logiki zdań [A criterion of completeness of many-valued systems of propositional logic]. Comptes rendus des séances de la Société des Sciences et des Lettres de Varsovie Cl. III **32**, 102–109 (1939)
41. Słupecki, J.: Dowód aksjomatyzowalności pełnych systemów wielowartościowych rachunku zdań [Proof of the axiomatizability of full many-valued systems of propositional calculus]. Comptes rendus des séances de la Société des Sciences et des Lettres de Varsovie Cl. III **32**, 110–128 (1939)

42. Słupecki, J.: Próba intuicyjnej interpretacji logiki trójwartościowej Łukasiewicza [An attempt of intuitionistic interpretation of three-valued Łukasiewicz logic]. In: Rozprawy Logiczne. PWN, Warszawa (1964)
43. Suchoń, W.: Définition des founcteurs modaux de Moisil dans le calcul n-valent des propositions de Łukasiewicz avec implication et négation. Rep. Math. Log. **2**, 43–47 (1974)
44. Tarski, A.: O pojęciu wynikania logicznego [On the concept of logical consequence]. Przegląd Filozoficzny **39**, 58–68 (1936) (English tr. in: Tarski, A.: Logic, Semantics, Metamathematics: Papers from 1923 to 1938, pp. 409–420. Clarendon Press, Oxford (1956). Translated by J.H. Woodger)
45. Tokarz, M.: A method of axiomatization of Łukasiewicz logics. Bull. Sect. Log. **3**(2), 21–24 (1974)
46. Tuziak, R.: An axiomatization of the finitely-valued Łukasiewicz calculus. Stud. Logica **48**, 49–56 (1988)
47. Wajsberg, M.: Aksjomatyzacja trójwartościowego rachunku zdań. Comptes Rendus de la Société des Sciences et des Lettres de Varsovie Cl. III **24**, 126–148 (1931) (English tr.: Axiomatization of the three-valued propositional calculus. In: McCall, S. (ed.) Polish Logic 1920–1939, pp. 264–284. Clarendon Press, Oxford (1967))
48. Whitehead, A.N., Russell, B.: Principia Mathematica, vol. I. Cambridge University Press, Cambridge (1910)
49. Woleński, J.: Filozoficzna Szkoła Lwowsko–Warszawska. PWN, Warszawa (1985) (English tr.: Logic and philosophy in the Lvov–Warsaw School. Synthese Library, vol. 198. D. Reidel, Dordrecht (1989))
50. Wójcicki, R.: Theory of Logical Calculi. Basic Theory of Consequence Operations. Synthese Library, vol. 199. Kluwer Academic Publishers, Dordrecht (1988)
51. Wójcicki, R.: On matrix representations of the consequence operations of Łukasie-wicz's sentential calculi. In: Wójcicki, R. (ed.) Theory of Logical Calculi. Basic Theory of Consequence Operations. Synthese Library, vol. 199, pp. 101–111. Kluwer Academic Publishers, Dordrecht (1977/1988)
52. Wójcicki, R., Malinowski, G.: Selected papers on Łukasiewicz sentential calculi. Ossolineum, Wrocław (1977)
53. Zadeh, L.A.: Fuzzy sets. Inf. Control. **8**, 338–353 (1965)
54. Zawirski, Z.: Znaczenie logiki wielowartościowej i związek jej z rachunkiem prawdopodobieństwa [Significance of many-valued logic for cognition and its connection with the calculus of probability]. Przegląd Filozoficzny **37**, 393–398 (1934)
55. Zawirski, Z.: Stosunek logiki wielowartościowej do rachunku prawdopodobieństwa [Relation of many-valued logic to the calculus of probability]. Prace Komisji Filozoficznej Polskiego Towarzystwa Przyjaciół Nauk **4**, 155–240 (1934)

G. Malinowski (✉)
University of Łódź, Łódź, Poland
e-mail: gregmal@uni.lodz.pl

Tomorrow's Sea-Battle and the Beginning of Temporal Logic

Kazimierz Trzęsicki

Abstract Jan Łukasiewicz's consideration of the deterministic consequences of the law of the excluded middle and the principle of causality gave the incentive to the development of temporal logic. Formal reformulation of arguments in favour of determinism is possible in the language of temporal logic. Following Jan Łukasiewicz both the arguments, the argument from the law of the excluded middle and the argument from the principle of causality, will be discussed concomitantly.

Keywords Tomorrow's sea-battle · Temporal logic

Mathematics Subject Classification (2000) Primary 03B44; Secondary 03 A10

1 Introduction

Aristotle (384 BC–322 BC) in his famous tomorrow's sea-battle passage of *De Interpretatione* 9 19^a 30 stated the problem of the logical value of statements about future contingencies. The problem has been discussed since the age of Aristotle as an important philosophical question in connection with determinism. There are many solutions to it [1]. Diodorus Cronus from the Megarian school of philosophy is famous as the author of a version of the problem in his notorious Master Argument. In this subject some achievements are due to Avicenna (980–1037). His work influenced medieval logicians Albertus Magnus (1193/1206–1280) and William of Ockham (c. 1288–c. 1348). In the nineteenth century Charles Sanders Peirce (1839–1914) wrote that he did not share the common opinion that time is an extralogical matter. The deterministic consequence of the law of the excluded-middle (LEM) and the principle of causality (PC) were considered by Jan Łukasiewicz (1878–1956). It was, as writes Słupecki: [8, Introduction, p. vii]

> ... the problem in which Łukasiewicz was most interested almost all his life and which he strove to solve with extraordinary effort and passion was the problem of determinism. It inspired him with the most brilliant idea, that of many-valued logics.

To avoid fatalism, the deterministic (PRE-DET) consequence of LEM, he proposed abolishing the logical principle of bivalence (PB—every proposition is either true or

Á. Garrido, U. Wybraniec-Skardowska (eds.), *The Lvov-Warsaw School. Past and Present*, Studies in Universal Logic,
https://doi.org/10.1007/978-3-319-65430-0_25

false)—as Łukasiewicz called it [5]—and to introduce a new logical value [6, 7, 9].[1] He laid the ground for the historically first elaboration of three-valued logic [4]. The sentences about future contingencies would be neither true nor false; neither 1 nor 0. The logical value of such sentences would be intermediate. The inventor of temporal logic, Arthur Norman Prior (1914–1969), was concerned with the philosophical matters of free will and predestination. He learned logic from Polish logicians, in particular Jan Łukasiewicz (even using his parenthesis-free notation); in his early tense-logical considerations he referred to many-valued logic of Łukasiewicz [11–18]. Many-Valued approaches to temporal logic were also discussed by Rescher and Urquhart [19, pp. 213–237]. Though attempts at formalizing future contingents using many-valued logics were not successful, the idea of many valuedness in temporal logic is still fruitful in particular in *AI* [1, Łukasiewicz's contribution to temporal logic, 149–154] and model checking [3].

In Łukasiewicz's opinion [6, p. 126]:

> A trivalent system of logic ... differs from ordinary bivalent logic, that only one known so far, as much as non-Euclidean systems of geometry differ from Euclidean geometry.

For him [10, pp. 84–85] it contributes to the

> struggle for the liberation of the human spirit [from the] logical coercion given by Aristotle's science as a system of principles and theorems connected by logical relationships. [...] The creative mind revolts against this concept of science [...] A system of three-valued logic [...] destroys the former concept of science, based on necessity [...] Logic is a free product of man, like a work of art.

The thought that many-valuedness is connected with freedom is still vivid. For Karpenko [2, pp. 143–144]

> ... Łukasiewicz finite-valued logic, are not restrictions (as they are usually viewed), but in a sense, extensions of $\mathbf{C}_2$[the classical propositional logic]. The repercussions of such an extension are quite serious as well rather surprising. The extension of the very basic logical universe resulted in the logics of continual nature; in the possibility to characterize, structure, and describe classes of prime numbers. Are all of these required for the logical reasoning? On the other hand, the problem of fatalism and free will is also of continual nature, which usually goes unnoticed. As we wrote in this book for the refutation of the doctrine of logical fatalism, Łukasiewicz, without being aware of it, abandoned discreteness for continuity.

Referring to his book Karpenko even ask [2, p. 4]:

> Is there, however, any link between the doctrine of logical fatalism and prime numbers?

In contemporary temporal logic the discussion of PRE-DET is usually limited to logical PRE-DET. Łukasiewicz also considered the argument from PC, i.e. the principle that any temporal fact has its cause in another fact that occurs at an earlier moment. Usually the argument from PC in favour of PRE-DET is assumed as evident. Łukasiewicz questioned the argument and showed that both the arguments from LEM and PC are

---

[1] The article in question is a revised version of the address that Łukasiewicz delivered as a Rector during the inauguration of the academic year 1922–1923 of Warsaw University. Later on Łukasiewicz revised the address giving it the form of an article, without changing the essential claims and arguments. It was published for the first time in Polish in 1946 and also [6, pp. 114–126]. An English version of the paper is available in: [7, pp. 19–39]. It is worth noting that when the author delivered the talk initially, the theories and discoveries in the field of atomic physics that undermined determinism were still unknown.

independent of one another. Moreover, Łuksiewicz questioned what in temporal logic discussions is treated almost as dogma, that the past is completely determined (POST-DET): We should not treat the past differently from the future [8, p. 127]. He rejects the Latin saying "facta infecta fieri non possunt" that is, what once has happened cannot become not happened. In order to analyze POST-DET, the principle of effectivity (PE), as a principle symmetrical to PC, should be considered.

2 Language and Semantics

To use a formal logic to solve a philosophical problem, we have to have:

1. a formal language in that the problem can be expressed in an intuitively satisfactory way,
2. the logic should be neutral with respect to this problem, i.e. the formula that expresses it should not be a thesis of the logic (an analytical truth of the language).

The thesis of determinism as consisting of two theses, the thesis of pre-determinism PRE-DET and the thesis of post-determinism POST-DET, can be formulated as follows:
DET. If φ, then

- PRE-DET. at any earlier moment it was true that there would be φ
  and
- POST-DET. at any later moment it will be true that there was φ

The principle of causality says:

- PC. If φ occurs at t, then at t_1, some moment earlier than t, and at any moment between t and t_1 it was true that there will be φ.

The principle of effectivity, as symmetrical to PC, may be formulated as follows:

- PE. If φ occurs at t, then at t_1, some moment later than t, and at any moment between t and t_1 it will be true that there was φ.

The relation of causality is transitive [9, p. 28]:

> This means that for any facts ϕ, ψ, and χ, if ϕ is the cause of ψ and ψ is the cause of χ, then ϕ is the cause of χ.

Both the theses of pre- and post-determinism, PRE-DET and POST-DET, and both the principles of causality and effectivity, PC and PE, can be expressed in the language of temporal logic.

Let the language consist of:

1. $p_1, p_2, \ldots$—propositional letters,
2. a functionally complete set of classical propositional connectives,
3. temporal operators (past tense and future tense operators).

Let AP (Atomic Propositions) be the set of propositional letters. Formulas are defined in the usual way and will be denoted by Greek letters: φ, ψ, ..., if necessary with indices.

Let time be $\mathfrak{T} = \langle T, < \rangle$, where T is a non-empty set (of moments) and $<$ is a binary (earlier-later) relation on T. No conditions on $<$ are imposed.

The temporal world W consists of time $\langle T, < \rangle$ and facts that occur at elements of T. Let V be a function, valuation, that to each t ($\in T$) assigns a subset of AP, the set of propositional letters that are true at this point, $V : T \to 2^{AP}$.

Let temporal operators be defined in the usual way, i.e. as follows:

Definition 2.1 (G) $\langle T, <, V \rangle, t \models G\varphi$ iff for any $t_1, t < t_1 : \langle T, <, V \rangle, t_1 \models \varphi$.

Definition 2.2 (F) $\langle T, <, V \rangle, t \models F\varphi$ iff there is $t_1, t < t_1 : \langle T, <, V \rangle, t_1 \models \varphi$.

Definition 2.3 (H) $\langle T, <, V \rangle, t \models H\varphi$ iff for any $t_1, t_1 < t : \langle T, <, V \rangle, t_1 \models \varphi$.

Definition 2.4 (P) $\langle T, <, V \rangle, t \models P\varphi$ iff there is $t_1, t_1 < t : \langle T, <, V \rangle, t_1 \models \varphi$.

Operators G, H are dual to, respectively, F and P:

- $G\varphi \leftrightarrow \neg F \neg \varphi$,
- $H\varphi \leftrightarrow \neg P \neg \varphi$.

Let us call the language Priorean. Let K_t be the set of all formulas that are satisfied in any model of whatever are the set of moments of time T, the relation $<$ and the valuation V, $\varphi \in K_t$ iff for any $W : W \models \varphi$.

The operators are usually read as:

F — it will at some time be the case that ...
P — it has at some time been the case that ...
H — it has always been the case that ...
G — it will always be the case that ...

Formulas of the language are recursively defined by:

$$\varphi ::= p | \neg\varphi | \varphi \vee \varphi | \varphi \wedge \varphi | \varphi \to \varphi | F\varphi | P\varphi | G\varphi | H\varphi$$

In Priorean language PRE-DET and POST-DET are expressed as follows:

- PRE-DET. $\varphi \to HF\varphi$
- POST-DET. $\varphi \to GP\varphi$

Both the classes of formulas PRE-DET and POST-DET are theses of K_t [19, Chapter VI], the minimal logic of the Priorean language, i.e. they are propositions that are satisfied in any model independently of the property of time. Thus the Priorean language does not fulfil the condition of neutrality.

In the following two languages it will be considered in which:

- the rejection of both the theses PRE-DET and POST-DET will be possible, and
- the principles of PC and PE can be expressed, and
- LEM will hold.

3 Branching Time Logic

Since the thesis of determinism, DET, is a thesis of minimal K_t logic, a new language has to be defined. Ockamist and Peircean languages are well known proposals. In any case it is assumed that time is branching in the future with many possible time-lines (histories). There are many courses of events possible. Since we are interested in neutral language, i.e. a language such that no discussed thesis is a thesis of minimal logic of the language, we do allow also branching in the past.

A branch of $\mathfrak{T}$ is any maximal linearly ordered subset of T. Each branch represents a possible course of events. For a point t and branch b, if t is a member of b, we say that t lies on b or that b goes through t.

To say that at point t there will be φ means, that at any branch that goes through t there will be φ. Analogously the past tense operator is understood. Let $B(t)$ be the set of all branches that go through t.

Definition 3.1 ($F_{\Box}$) $\langle T, <, V\rangle,\ t \models F_{\Box}\varphi$ iff for any $b \in B(t)$ there is $t_1 \in b,\ t < t_1$: $\langle T, <, V\rangle,\ t_1 \models \varphi$.

Definition 3.2 ($P_{\Box}$) $\langle T, <, V\rangle,\ t \models P_{\Box}\varphi$ iff for any $b \in B(t)$ there is $t_1 \in b,\ t_1 < t$: $\langle T, <, V\rangle,\ t_1 \models \varphi$.

The formulations of definitions of G and H do not differ of respective definitions of the Priorean language.

The theses PRE-DET and POST-DET are expressible as:

- PRE-DET. $\varphi \rightarrow HF_{\Box}\varphi$
- POST-DET. $\varphi \rightarrow GP_{\Box}\varphi$

Neither $p \rightarrow HF_{\Box}p$ nor $p \rightarrow GP_{\Box}p$ are theses of the minimal logic of the language of branching time.

If PC is valid, then

- $\varphi \rightarrow P_{\Box}F_{\Box}\varphi$

should be valid.

If PE is valid, then

- $\varphi \rightarrow F_{\Box}P_{\Box}\varphi$

should be valid.

It could be remarked that neither formula expresses the fact that the cause/effect is in any "between" moment.

Neither $p \rightarrow P_{\Box}F_{\Box}p$ nor $p \rightarrow F_{\Box}P_{\Box}p$ are theses of the minimal logic of the language of branching time.

FS: for any t and for any $t_1,\ t_1 \leq t$:

1. there exists t_2 such that $t_2 < t_1$, and
2. for any t_3: if $t_2 < t_3$, then $t_3 < t_1$ or $t_1 \leq t_3 \leq t$ or $t < t_3$.

PS: for any t and for any $t_1,\ t \leq t_1$:

1. there exists t_2 such that $t_1 < t_2$, and
2. for any t_3: if $t_3 < t_2$, then $t_1 < t_3$ or $t \leq t_3 \leq t_1$ or $t_3 < t$.

The conditions FS and PS taken jointly characterize both directions open segment s. Openness means that for any $t \in s$:

1. there is $t_1 \in s$ such that $t_1 < t$, and
2. there is $t_1 \in s$ such that $t < t_1$.

If $<$ is dense, s can be finite, i.e.:

1. there is $t \in T$ such that for any $t_1 \in s : t < t_1$, and
2. there is $t \in T$ such that for any $t_1 \in s : t_1 < t$.

We maintain that PC holds if FS holds and PE holds if PS holds. It is true also when the relation of causality is transitive. Moreover, even if both the conditions are fulfilled, there are counter-models of PRE-DET and POST-DET. For any combination of PRE-DET, POST-DET, PC and PE there is a frame in which this combination is valid and for the other thesis or principle there is a counter-model. For example, for all PRE-DET, POST-DET, PC and PE there is a counter-model if time is branching in the past and in the future and the relation $<$ is discrete and irreflexive.

There are some objections to proposed semantics of branching time logic as a tool of formalizing arguments in favour of determinism. The relation of reachability of possible states of events is defined as the relation $<$ of earlier-later, a relation on the set of time points. Maybe it is supported by the idea of time as a result of change offered by the theory of the relativity of time. In the definition of the operators $F_\Box$ and $P_\Box$ any points of the branches are allowed. Due to the fact the formula $F\varphi$ rather expresses inevitability of φ than its determination. Analogously it is true in the case of $P_\Box$. It seems that time should be the same independently of branch.

4 Temporal Logic of Possible Worlds

Let us look for a solution such that the possibility of some event is referred to one and the same point of time, not—as it was the case in branching time logic—different moments are allowed. „There will be a sea-battle tomorrow" is true if it is determined that tomorrow there will be a sea-battle, i.e. in any possible course of events it will be tomorrow but not, e.g. in one possible course of events tomorrow, and another course the day after tomorrow.

The language of temporal logic of possible world not only will better expresses the idea of determination but it seems to be, in some sense, more typical for modal logics.

The relation of accessibility $\lhd$ will be a relation defined on sets of pairs consisting of a possible world and points of time. Possible worlds will not differ in time $\langle T, <\rangle$. They will differ only in valuation V.

Let $\mathfrak{W}(T, <) = \{\langle T, <, i\rangle : i \in I\}$. The relation $\lhd$ between (W, t) and (W_1, t_1) says that the moment $t_1 \in T$ of the world W_1 is reachable from a moment t of the world W. Formally:

$$\lhd \subseteq (W \times T) \times (W_1 \times T).$$

For intuitive reason it will be assumed that for any W, $W_1 \in \mathfrak{W}$, and for any $t, t_1 \in T$:

- $(W, t) \lhd (W, t)$, i.e. $\lhd$ is reflexive
- $(W, t) \lhd (W_1, t_1)$ only if $t \leq t_1$

Let us define the tense operators.

Definition 4.1 ($G_\lhd$) $W, t \models G_\lhd\varphi$ iff for any t_1, $t < t_1$, and for any (W_1, t_1), $(W, t) \lhd (W_1, t_1)$: $W_1, t_1 \models \varphi$.

Definition 4.2 ($F_\lhd$) $W, t \models F_\lhd\varphi$ iff there is t_1, $t < t_1$, such that for any (W_1, t_1), $(W, t) \lhd (W_1, t_1)$: $W_1, t_1 \models \varphi$.

Definition 4.3 ($H_\lhd$) $W, t \models H_\lhd\varphi$ iff for any t_1, $t_1 < t$, and for any (W_1, t_1), $(W, t) \lhd (W_1, t_1)$: $W_1, t_1 \models \varphi$.

Definition 4.4 ($P_\lhd$) $W, t \models P_\lhd\varphi$ iff there is t_1, $t_1 < t$, such that for any (W_1, t_1), $(W, t) \lhd (W_1, t_)$: $W_1, t_1 \models \varphi$.

The language is defined in the usual way. The theses of PRE-DET and POST-DET and the principles of PC and PE are expressible as:

- PRE-DET. $\varphi \rightarrow H_\lhd F_\lhd \varphi$
- POST-DET. $\varphi \rightarrow G_\lhd P_\lhd \varphi$
- PC. $\varphi \rightarrow P_\lhd F_\lhd \varphi$
- PE. $\varphi \rightarrow F_\lhd P_\lhd \varphi$

PC and PE are not theses of the minimal logic of the language. They hold if some conditions on the relation $\lhd$ are imposed.

FS1: for any W, t, t_1: if $(W, t_1) \lhd (W, t)$, then there exists

1. t_2 such that $t_2 < t_1$, and
2. for any W_1, t_3: if $(W, t_2) \lhd (W_1, t_3)$, and $t_3 \leq t$, then $W_1 = W$

PS1: for any W, t, t_1: if $(W, t) \lhd (W, t_1)$, then there exists

1. t_2 such that $t_1 < t_2$, and
2. for any W_1, t_3: if $(W, t_3) \lhd (W_1, t_2)$, and $t \leq t_3$, then $W_1 = W$

In the frame $\langle T, <, \lhd \rangle$ that fulfils the condition $FS1$, PC holds. If $PS1$ is fulfilled, PE is valid. In the case the frame fulfils both the conditions $FS1$ and $PS1$, PC as well PE are valid. In any case a construction of a counter-model for the theses PRE-DET or POST-DET is possible. Using discussed semantics it is possible—as it was in the case of branching time logic—to constructed frames in which any combination and only that of PRE-DET, POST-DET, PC, PE is valid.

5 Conclusion

Abolishing arguments in favour of determinism does not mean that the universe, and we as part of it, is not determined. Newtonian physics gives reasons for Laplacean determinism.

Quantum physics says that everything happens with probability. Though not all of the interpretations of quantum physics imply indeterministic 'choices' of events. A universe governed by deterministic laws is preferably for the sake of making predictions. To be free means to make decisions according to mind. To make rational decisions, knowledge about a state of affairs is needed. According to quantum theory any attempt to know the state of affairs is changing the state. Our acts have only probable effects. To be free we have to grasp the reality. Logic is the fundamental part of our consciousness. Maybe a candidate for a logic of free will is the quantum logic.

References

1. Hasle, P.F.V., Øhrstrøm, P.: Temporal Logic: From Ancient Ideas to Artificial Intelligence. Kluwer Academic Publishers, Dordrecht (1955)
2. Karpenko, A.S.: Łukasiewicz Logics and Prime Numbers. Luniver Press, Beckington (2006)
3. Lu, Z., Liu, J., Augusto, J.C., Wang, H.: A many-valued temporal logic and reasoning framework for decision making. In: Atlantis Computational Intelligence Systems. Computational Intelligence in Complex Decision Systems, vol. 2, pp. 125–146. Atlantis Press/World Scientific, Amsterdam/Paris (2010)
4. Łukasiewicz, J.: On three-valued logic. Ruch Filozoficzny **5**, 170–171 (1920). (English translation in Łukasiewicz, J.: In: Borkowski, L. (ed.) Selected Works. Polish Scientific Publishers/North-Holland Publishing Company, Warszawa/Amsterdam (1970))
5. Łukasiewicz, J.: Philosophische Bemerkungen zu mehrwertigen Systemem des Aussgankalkül. Sprawozdania Towarzystwa Naukowego Warszawskiego Wydział III **23**, 51–77 (1930)
6. Łukasiewicz, J.: O determiniźmie. In: Z zagadnień logiki i filozofii. Pisma wybrane, pp. 114–126. Państwowe Wydawnictwo Naukowe, Warszawa (1961)
7. Łukasiewicz, J.: On determinism. In: McCall, S. (ed.) Polish Logic 1920–1939, pp. 19–39. Oxford University Press, Oxford (1967)
8. Łukasiewicz, J.: Selected Works, Borkowski, L. (ed.). Polish Scientific Publishers/North-Holland Publishing Company, Warszawa/Amsterdam (1970)
9. Łukasiewicz, J.: On determinism. In: Borkowski, L. (ed.) Selected Work, pp. 110–128. North-Holland, Amsterdam (1970)
10. Łukasiewicz, J.: On three-valued logic. In: Borkowski, L. (ed.) Selected Work, pp. 87–88. North-Holland, Amsterdam (1970)
11. Prior, A.N.: The craft of formal logic. Unpublished manuscript, 806 pages (1951)
12. Prior, A.N.: In what sense is modal logic many-valued? Analysis **12**, 138–143 (1952)
13. Prior, A.N.: Łukasiewicz's symbolic logic. Australas. J. Philos. **30**, 121–130 (1952)
14. Prior, A.N.: Three-valued logic and future contingents. Philos. Q. **3**, 317–26 (1953)
15. Prior, A.N.: Many-valued modal systems. An intuitive approach. Philos. Rev. **63**, 626–630 (1954)
16. Prior, A.N.: Time and Modality. Oxford University Press, Oxford (1957)
17. Prior, A.N.: Past, Present and Future. Oxford University Press, Oxford (1967)
18. Prior, A.N.: Papers on Time and Tense. Oxford University Press, Oxford (1968)
19. Rescher, N., Urquhart, A.: Temporal Logic. Library of Exact Philosophy, vol. 3. Springer, Wien (1971)

K. Trzęsicki (✉)
Higher Vocational School in Suwałki, University of Białystok, Białystok, Poland
e-mail: kasimir@uwb.edu.pl

Leśniewski and Mereology

Peter Simons

Abstract This paper surveys mereology, the theory of parts and wholes, focussing on its origins in Leśniewski, and noting its intended employment as a surrogate for set theory. We examine parallel and independent work by Whitehead, Leonard and Goodman, and outline the subsequent adventures of mereology, both in its formal guises and in its now intensive application within philosophical ontology.

Keywords Mereology · Part · Class · Stanisław Leśniewski · Calculus of individuals

AMS Subject Classification (2000) Primary: 03B99, Secondary: 03A05, 01A99

1 How Leśniewski Came to Mereology

Philosophers have *used* the notion of *part* since the beginning of their subject: it is far too ubiquitous and important not to have entered their discussions. Plato employs it in the *Parmenides*, Aristotle anatomizes various meaning of the term in the *Metaphysics*, Euclid in the *Elements* defines a point as that which has no parts, and scholastic and modern philosophers used it also. One who did so explicitly and extensively was Franz Brentano. But the idea of a *formal* theory of part and whole did not occur to anyone until Brentano's student Edmund Husserl proposed it in his third Logical Investigation, "On the Theory of Wholes and Parts". Stanisław Leśniewski, who obtained his doctorate with another Brentano student, Kazimierz Twardowski, probably read Husserl's investigation. But the motivation for Leśniewski's mereology, to use the name he himself coined for the formal theory of part, whole and cognate concepts, was a different one. It came from the theory of sets or classes, and arose out of his endeavours to understand and neutralize Russell's Paradox.

Leśniewski first learnt about Russell's Paradox in 1911 ([1], p. 169, [2], p. 181) from the appendix to Jan Łukasiewicz's 1910 book *On the Principle of Contradiction in Aristotle*, where Łukasiewicz gives a short introduction to symbolic logic in Couturat's notation, and briefly outlines Russell's Paradox. That paradox, it will be recalled, concerns

Á. Garrido, U. Wybraniec-Skardowska (eds.), *The Lvov-Warsaw School. Past and Present*, Studies in Universal Logic,
https://doi.org/10.1007/978-3-319-65430-0_26

a certain class, the class of all classes that are not elements of themselves. Assuming there is such a class, if it is an element of itself, by its definition it is not an element of itself, but then by its definition it is an element of itself. Therefore it both is and is not an element of itself, which is a contradiction. So clearly there cannot be such a class. But the unreflective assumption about the existence of classes, namely that for any clearly statable condition there is a class of just the things satisfying that condition, called the Principle of Comprehension, seems to make it unproblematic that there is such a class: the condition is just (using a variable 'x' to mark the place for insertion of constants):

x is a class and x is not an element of x

and the class is then (using the standard brace notation to bind the free variable)

$\{x$: x is a class and x is not an element of $x\}$

Since some classes clearly are not elements of themselves, for example the class of all dogs is not a dog, there should be elements of such a class, so why not collect them all together?

When Leśniewski first read about the paradox, he was convinced, as initially was Russell and were others, that it was a silly minor problem that could easily be fixed. But as time went by, nearly everyone who dealt with it realised that solving the difficulty would take more work. Leśniewski worked for over a decade on it before he was satisfied that it could be avoided in a way that was both intuitively satisfying and rigorous. It clearly continued to fascinate him long afterwards, for he lectured on antinomies in 1934–35 ([3], p. 179) and when he died in 1939 he left behind him a large manuscript on antinomies, which was sadly destroyed in Bolesław Sobociński's apartment during the 1944 Warsaw Rising.

The paradox was not just a little local difficulty, because class theory, or set theory, as it came to be known, was widely held to be the most promising basis for a foundation of mathematics. When in 1902 Russell communicated his surprising result to Frege, whose logicist system of foundations was nearing completion, Frege recognised immediately that the antinomy undermined his whole system, which had been constructed to that date with unprecedented logical rigour. Frege quickly cobbled together a repair and published this as an afterword to the second volume (1903) of his *Grundgesetze der Arithmetik*, but at some point he must have realised that his repair blocked his derivation of arithmetic from logic, and gave up. Some time later, in 1938, Leśniewski showed that Frege's repair had the unacceptable consequence that there could not be more than one individual (cf. [4]). Russell, after much to-ing and fro-ing, came up with his theory of types as a way of maintaining consistency, while Ernst Zermelo developed axiomatic set theory in a way which presumptively avoided inconsistency while providing a strong (but no longer logicist) foundation for mathematics. Whitehead and Russell's three-volume *Principia Mathematica* carried the Russell theory through, but at a price: to derive basic arithmetic they had to assume the existence of an infinity of individuals, which certainly did not seem like a logical truth. Mathematicians shunned the complexities of the theory of types and, when they needed to, worked with Zermelo's much simpler set theory or one of its closely related descendants.

Leśniewski wanted to get to the bottom of the motivation driving Russell's Paradox, and to this end provided a first analysis in a paper published in 1914, "Is the Class of Classes not Subordinated to Themselves Subordinated to Itself?" [5] His provisional conclusion was that, if we understand the notion of class correctly, there is no paradox, because every class is subordinated to (is an element of) itself, so there is no Russell Class about which to pose the question. This paper used terms like 'class' and 'subordinated to' understood in an intuitive way. Clearly this did not satisfy Leśniewski, because his next publication, written in wartime Moscow and published in 1916, was "Foundations of the General Theory of Sets I" [6]. In it, he set forth his first axiomatic treatment of the concepts *part*, *ingredient*, *set*, *class*, *element*, *subset*, and others. This reaffirmed the 1914 conclusion on a much more satisfactory, axiomatic basis, and although he had yet to coin the term, it may count as his first treatment of axiomatic mereology, the theory of *part* and related concepts.

Over the next years, Leśniewski continued to develop and refine his views, modifying his axiom systems in the direction of greater elegance, but in no way changing his views on what classes or sets are. To avoid confusion of his system with those of the more prevalent, what he called "official", set theories, he renamed it 'mereology', from the Greek word *meros*, 'part'. He was able to show that there are several equivalent ways to axiomatize mereology, using different notions as primitive. Later his energies went into developing the logico-linguistic basis for mereology: first, the logic of names and predicates which he called 'ontology', and subsequently the logic of propositions, propositive functors and quantifiers that he called 'protothetic', which extended what Whitehead and Russell had called 'theory of deduction' and what is now called 'propositional logic'.

Appalled at the presentational sloppiness and use/mention conflations of *Principia Mathematica*, Leśniewski somewhat naively assumed that all symbolic logic had to be like that, and so preferred to develop mereology and ontology initially in a carefully restricted natural language idiom, a small fragment of his native Polish, supplemented by variables. In 1920 Leon Chwistek persuaded him that symbolization was advantageous and so Leśniewski set about formulating with unprecedented precision the axioms and rules of development of his systems, consistent with a nominalistic attitude to logical systems. He never published his mereology as a symbolic system, but some notes from his Warsaw lectures were published posthumously in 1988 in the collection *S. Leśniewski's Lecture Notes on Logic* under the title "Class Theory" and they give a flavour of how he presented his mature mereology to students (cf. [3], pp. 59–125).

2 Leśniewski's Understanding of Sets or Classes

Leśniewski took his understanding of sets or classes from the words of Georg Cantor, who wrote, "By a 'set' we understand any aggregation *M* of determined well-distinguished objects *m* of our intuition or thought (called the 'elements' of *M*), into a whole." It will be noted that Cantor speaks of aggregation into a whole (*Zusammenfassung zu einem Ganzen*). Both the idea of aggregation and that of a whole are resonant of mereology, so Leśniewski's way of taking Cantor's idea, which is admittedly expressed less than wholly perspicuously, is understandable. That way is to treat a set or class as a composite individual literally made up or composed of its elements. If *m* are my books, then their set *M*, or $\{m\}$, as Cantor sometimes writes it, is my library, understood as that single individual or whole which contains all my books as parts, and nothing else outside that whole.

Leśniewski referred to this as a class taken in the *collective* sense. There is another sense of class, taken in the *distributive* sense. For example, when we speak of the class of all my books, when understood in the distributive sense, this does not mean the composite individual made up of my books but simply my books, of which there are many. So the expression 'This paperback is an element of the class of all my books' can be understood in two different ways. In the distributive sense, it is simply an elaborate way of saying 'This paperback is one of my books', where the reference to a class drops out. The logical predicate 'is one of' is what Leśniewski came to symbolize by the Greek letter 'ε' and refer to as the functor of singular inclusion. It became the primitive predicate of his system of ontology. In many contexts, for example if we say 'Fido is one of (all) the dogs', it is simpler and more idiomatic to say 'Fido is a dog', and represent it semi-symbolically as 'Fido ε dog'. In Leśniewski's native Polish, which is not cluttered up by articles, the functor is rendered simply as 'jest': *Fido jest psem.*

Taken in the collective sense however, we cannot say that Fido is *one* of the class of all dogs, since there is only one such class and that would mean Fido was it and therefore the only dog. (Unless Fido has another dog as a proper part, which is conceivable but not how dogs generally are.) Rather we must say that he is an element, member, or, as Leśniewski is happy to say, that he is a *part* (*część*) of this class.

There are, it must be said right from the start, reasons to think that Leśniewski's treatment of classes is not exhaustive. Consider the following fallacious syllogisms:

Fido is one of the dogs
The dogs are a species
therefore
Fido is a species

or the following mathematical example

8 is one of the Pythagorean triple of numbers 6, 8, 10
The Pythagorean triple 6, 8, 10 is one of the Pythagorean triples containing 8
therefore
8 is one of the Pythagorean triples containing 8.

In each case, whether we understand the object of the sentence as a class in the distributive or in the collective sense, we get nonsense. If we understand 'the dogs' distributively, they cannot be one thing, a species, of which there are others, cats, horses etc., but equally if we understand 'the dogs' collectively, then not only is Fido not a species, though he is a member of a species, things other than dogs would have to be taken as members of the species, such as the class of all attached dogs' tails. What we have in each case is a class of several things being one of a class of several other classes: species, or Pythagorean triples. Neither of Leśniewski's ways of taking the notion of a class captures this feature, and that is why, when set theory became established as a framework for dealing with mathematical matters, his efforts were sidelined, since the representative and expressive power of set theory rests primarily on this ability to have classes of classes (of classes . . .) which do not collapse to the lowest level as they do on either of Leśniewski's understandings. Technically, it turns on the non-transitivity of the element relation. Whether Leśniewski understands '*A* is an element of the class of *a*s' to mean '*A* is one of the *a*s' or '*A* is a part of the (collective) class of the *a*s', the relational functor is transitive, whereas in standard set theory it is not, and that is why standard set theory can build hierarchies that Leśniewski cannot.

3 Terminology

We need to say something here about terminology, both Leśniewski's and ours, because several of the terms have been used differently by different people, and we are going to have to fix their use for later. We have seen that Leśniewski starts using terminology associated with standard set theory. Initially he called a collective class a multitude, multiplicity or set (*mnogość*). In this sense a set is one thing, an individual, something which is enshrined later in his axioms. A set of *m*s is a collective class of some *m*s, at least one, and perhaps, but not necessarily, all. The unique set composed of *all* the *m*s he calls the class (*klasa*) of all *m*s. Later Leśniewski dropped the term *mnogość* in favour of the more usual term for sets or collections in Polish, *zbiór*, but he continued to use *klasa* for the complete collective class of all things of a certain kind. Lejewski, who brought Leśniewski's ideas to England after the Second World War, used 'collection' for sets and 'complete collection' for classes. Later mereologists have used the terms 'sum' and 'fusion' for collective classes. This more modern usage derives from Leonard and Goodman, stressing the individuality of the result and moving away from Leśniewski's ancestral link to the theory of sets or classes. For the next section, we adopt Leśniewski's own terminology, in order to give its flavour, but subsequently we shall adopt the following terminology: a collective class of some but not necessarily all *a*s will be called a *fusion* of *a*s; a (in fact: the) collective class of all the *a*s will be called a (the) *sum* of *a*s.

Now for terminology about parts etc. Leśniewski used *część* ('part') for the asymmetric relation, which entails that no object is a part of itself. In English terminology, this is now usually called the *proper part* relation, and we will adopt this term. He used the term *ingredyens* ('ingredient') for the related improper part notion, according to which an object is an improper part of itself. Later he introduces the terms 'element' and 'subset' which turn out to be equivalent to 'ingredient', and in later expositions dropped 'ingredient' in favour of 'element'. We will stay closer to modern terminology and call this the *part* relation. Where we mean 'proper part' we will say so.

Other terms, such as 'overlap' and 'exterior' are less problematic and will be introduced as we come to them. None of this terminology is perfect, but once we understand closely enough what is meant, we can convert to symbols, where the meaning gets more precisely fixed by the axioms.

4 The Early System

In his 1916 monograph, Leśniewski takes the notion of proper part as primitive, and lays down four axioms, two of which require some definitions before their introduction. We will give them in a translation of his prose, and then in the following section give them a symbolic form. Our lettering does not exactly follow Leśniewski: he talks of 'object *P*', 'object P_1' and the like, using a capital letter '*P*' sometimes with numerical subscripts, when talking about an individual object (*przedmiot*) and 'objects *m*', ('*m*' as in '*mnogość*') using a lower-case letter, when talking about one or several individual objects. We will follow this distinction between upper-case and lower-case letters though it is a presentational device only and can be omitted without loss of meaning or precision. We recall that in 1916 Leśniewski used semi-prose only, no logical symbols.

Axiom I *If object A is a proper part of object B, then object B is not a proper part of object A.*

This is the asymmetry of proper parthood. Leśniewski's use of singular letters and the expression 'object *A*', 'object *B*' indicate that he intends that if *A* is a proper part of *B* then not only is the part *A* an individual (as signalled by its appearing as subject before 'is a'), but that the whole *B* is also an individual.

In his ontology, Leśniewski later made clear the difference between two ways to understand negative sentences of the form '*B* is not an *m*'. One is as a sentential negation, 'it is not the case that *B* is an *m*'. The other has a nominal negation, '*B* is a non-*m*'. In the latter case the presence of '*B*' before 'is a' clearly indicates that *B* is an individual. With sentential negation, the negative sentence is true if the subject term is empty or plural. The wording does not make it clear which is meant here, but nominal negation fits better.

Axiom II *If object A is a proper part of object B, and object B is a proper part of object C, then object A is a proper part of object C.*

This is the transitivity of proper parthood. Again it is clear that the three schematic terms are intended as singular denoting terms.

Definition I A *part* of *A* is (Def.) either *A* or a proper part of *A*.

Leśniewski actually expresses this definition as a metalinguistic description of how he intends to use the expression 'part' (*ingredyens*):

> I use the expression 'part of object *A*' to denote the same object *A* as well as every proper part of this object. ([1], p. 261; [2], p. 132).

Definition II A *set* of *m* is (Def.) an object *A*, all of whose parts have a part in common with an *m* which is part of that object *A*.

Employing, in advance of its definition, the idea of two objects overlapping when they have a common part, we can paraphrase this and make it a little easier to understand:

A set of *m* is an object *A*, all of whose parts overlap some *m* which is a part of *A*.

This rules out any object which has a part that is not part of any *m*, but it also means that any *m* that overlaps *A* is itself wholly within the set *A*, in other words, the set is composed of all parts of those *m* involved. But these need not be all the *m*. For example, if *m* is dogs, then one set is the whole composed of just three of these dogs, say Fido, Spot and Rover. Indeed, a limiting case of a set would be the set of just one dog, which is simply that dog itself. To get the biggest possible set, what Leśniewski calls the class of all *m*, he needs another definition:

Definition III A *class* of *m* is (Def.) an object *A* such that every *m* is a part of *A* and every part of *A* has a part which is also a part of some *m*.

Again, this is easier to understand if we employ the concept of overlapping, thus:

A class of *m* is an object *A* of which every *m* is a part, and of which every part overlaps some *m*.

Because of the requirement that every *m* be a part of the class, the class is a maximal set of *m*. Now Leśniewski is ready to state his final two axioms.

Axiom III *If some object is an m, then some object is a class of m.*

Axiom IV *If A is a class of m and B is a class of m, then A is B.*

The third axiom guarantees the existence of a class (sum) of all *m* provided at least one *m* exists, and the fourth axiom ensures the uniqueness of any such class. The uniqueness allows us to speak of '*the* class of *m*' when one exists. In fact, there fails to be a class of *m* only when there are no *m*. The final clause, '*A* is *B*', should be understood as meaning that *A* and *B* are the same object, so the 'is' is the 'is' of singular identity.

Leśniewski goes on to give a further six definitions and he states and proves 57 theorems using the primitive and defined notions. We give the definitions as they show something interesting.

Definition IV *A* is an *element* of *B* if and only if (Def.) for some *x*, *A* is an *x* and *B* is the class of *x*.

Here Leśniewski is introducing some standard set-theoretic terminology, alongside the terms 'set' and 'class'. But the term 'element' is redundant: the two theorems he proves after the definition show that object is an element of another if and only if it is a part of it.

Definition V *A* is a *subset* of *B* if and only if (Def.) every element of *A* is an element of *B*.

Once again it turns out that being a subset is coextensive with being a part (or an element). What I think Leśniewski is doing here is adopting standard set-theoretic terminology in order to hijack it and then show that it is not needed, that the original notions of part and class (sum) are sufficient. The aim here is then not logical elegance, as it would be later for Leśniewski: it is rather the annexation of set theory by mereology.

Definition VI *A* is a *proper subset* of *B* if and only if (Def.) *A* is a subset of *B* and *A* is not *B*.

Again this definition is unnecessary, because being a proper subset is coextensive with being a proper part.

Definition VII The *Universe* is the class of all objects.

Leśniewski goes on to state as a theorem:

Theorem XLIII *Some object is the class of non-contradictory objects.*

This class is the Universe (*wszechświat*) and it is unique. Leśniewski's result here smuggles in an unstated assumption: that there is at least one non-contradictory object. This is of course true, but it does not follow from his previous axioms, all of which would be true even if there were no objects. In his later work, Leśniewski would reject the unstated assumption here and ensure that none of his logical systems have as a theorem that there exists at least one object.

Definition VIII *A* is *exterior* to *B* if and only if (Def.) no object is a part of both object *A* and object *B*.

Note that this is not the logical contradictory of overlapping since either or both of the variables '*A*' and '*B*' might be plural or empty. But among existing individuals, two objects overlap if and only if they are not exterior to (disjoint from) one another.

Definition IX *A* is a *complement* of *B* with respect to *C* if and only if (Def.) *A* is the class of all parts of *C* that are exterior to *B*.

It is a quick consequence of this that

Theorem XLVIII *If A is a proper part of B, then some object is the complement of A with respect to B.*

This say that any proper part of a whole has a remainder, its complement, which disjointly exhausts the whole together with it. It is an instance of what is now called a *supplementation principle* ([7], p. 26).

5 Remarks About This System

While Leśniewski's meaning is clear, from the point of view of logic, the system is still imperfect. Apart from the fact that the underlying logics of connectives, quantifiers, names and functors had still to be developed, there are inelegancies of presentation. The definition of 'set' is not needed for the definition of 'class'. It is no doubt interesting to see that a class is a special case of a set, but the definition of 'class' does not use that of 'set', though it might have done. Perhaps in a first version, Leśniewski had defined 'class' in terms of 'set' but later found the definition of 'class' that he gave and used that but without removing the definition of 'set'. Leśniewski himself came to see the interpolation of definitions, those of 'part' and 'class', between the first two and the last two axioms, as an imperfection and preferred to look for an axiomatization in terms of the primitive alone, which he subsequently achieved. Also the fact that some of the variables are "unofficially" singular and occur in phrases in apposition like "object *A*" is not transparent logical practice. We will remedy this below in our symbolic reconstruction, drawing on his later work. Further, as we noted, Leśniewski deliberately introduces redundant terminology from set theory in order to try and demonstrate that set theory is using a lot of unnecessary terms to talk about something rather straightforward. So the 1916 paper sacrifices much logical elegance for the sake of propaganda.

Philosophically, there are two things to note. The first is that the first pair of axioms, governing the notion of a proper part, are unobjectionable. The second remark however is that the remaining two axioms, on the existence and uniqueness of sums, are far from anodyne. The third axiom entails the existence of arbitrary sums of objects, which is and remains one of the more hotly disputed propositions of mereology. We will not enter into the philosophical controversy here, but simply note that Leśniewski's strong commitment results principally from his keenness to use mereological concepts in providing a foundation for mathematics. The final axiom, in the presence of the others, ensures a form of mereological extensionality: objects with the same parts are identical. Because of the strength of these principles, Leśniewski's axiomatization does not really test the limits of the everyday notion of part: it goes all the way to the maximal extent. Theorem XLVIII above is much closer to analysing the essential characteristics of parthood, the existence of supplements of proper parts, but it is a mere consequence of the strong Axioms III and IV and so its conceptual importance gets lost in the big noise they make.

6 Symbolic Reconstruction

Here we draw on the symbolic developments Leśniewski was to introduce in his full logic comprising protothetic and ontology, in order to give symbolic equivalents of the prose axioms and definitions of 1916. For full explicitness some additional definitions are required beyond those Leśniewski gives.

In what follows Italic letters are used for variables, Roman letters for constants. The standard modern symbols for the logical constants, connectives and quantifiers are used, where it must be noted that quantification does not, for Leśniewski, carry ontological import. That is carried rather by the basic predicate of singular inclusion. We represent singular inclusion, 'is a' or 'is one of' by 'ε', the standard undefined constant of Leśniewski's ontology. A sentence of the form '$A \varepsilon m$' is true if and only if A is an existing individual object and it is one of the object or objects m. Universal quantifiers having a whole formula as their scope are omitted but should be understood as tacitly present: Leśniewski does not allow free variables into his stated theses. Quantifier scope is marked, as in Leśniewski, by upper corners. Unless otherwise indicated by parentheses or constrained by corners, the order of grouping by binary connectives, from loosest to tightest is: $\leftrightarrow$ $\rightarrow$ $\vee$ $\wedge$. Otherwise, the symbols are straightforward. We shall assume the axiom and rules of ontology in the background as to give them explicitly would bloat this article. We start with some definitions, where the definiendum is always on the left of the equivalence and the definiens on the right. As in Leśniewski these are to be understood as object-language equivalences rather than metalinguistic abbreviations.

D= $\quad A = B \leftrightarrow A \,\varepsilon\, B \wedge B \,\varepsilon\, A$ Singular identity

DV $\quad A \,\varepsilon\, \mathrm{V} \leftrightarrow A \,\varepsilon\, A$ Object

DN $\quad A \,\varepsilon\, \mathrm{N}(m) \leftrightarrow A \,\varepsilon\, A \wedge \sim(A \,\varepsilon\, m)$ Nominal negation

The undefined primitive functor is 'proper part'. We symbolize that A is a proper part of B using the name-forming functor of one nominal argument 'ppt' as follows: $A \,\varepsilon\, \mathrm{ppt}(B)$. Here now are Leśniewski's axioms and definitions:

A1 $\quad A \,\varepsilon\, \mathrm{ppt}(B) \rightarrow B \,\varepsilon\, \mathrm{N}(\mathrm{ppt}(A))$

A2 $\quad A \,\varepsilon\, \mathrm{ppt}(B) \wedge B \,\varepsilon\, \mathrm{ppt}(C) \rightarrow A \,\varepsilon\, \mathrm{ppt}(C)$

D1 $\quad A \,\varepsilon\, \mathrm{pt}(B) \leftrightarrow A = B \vee A \,\varepsilon\, \mathrm{ppt}(B)$ (Proper or improper) part

There is subtlety to this definition of 'part', as indeed of the two previous definitions, that is easy to overlook. Any expression of the form '$A \varepsilon m$' entails by the meaning of the singular inclusion functor the existence and singularity of A. Leśniewski covers this by talking of 'the object A' and the like. So a definition like this, which Leśniewski would later call an *ontological* definition, uses a variable, in this case 'A', in such a way that the left hand side (the definiendum) can only be true if this object A exists and is an individual. For this to be a satisfactory definition, that requirement has to be fulfilled by the right hand side (the definiens) as well. Now in this case it is, because the sentence '$A = B$' is true if and only if both names are singular and denote the same individual, as indicated in D=, and the second disjunct likewise requires A to be an existing individual. But not all of Leśniewski's verbal definitions make this equally clear, in particular that of fusion (our name from now on for 'set'):

D2 $A \,\varepsilon\, \mathrm{fu}(m) \leftrightarrow A \,\varepsilon\, \mathrm{V} \wedge \forall B \ulcorner B \,\varepsilon\, \mathrm{pt}(A) \rightarrow \exists C \ulcorner C \,\varepsilon\, m \wedge C \,\varepsilon\, \mathrm{pt}(A) \wedge \exists D \ulcorner D \,\varepsilon\, \mathrm{pt}(B) \wedge D \,\varepsilon\, \mathrm{pt}(C) \urcorner\urcorner\urcorner$ Fusion

Notice the need for the first conjunct in the definiens, since without it the right hand side could be vacuously true were the name 'A' to be empty or plural. The first conjunct rules these cases out and ensures that the definition is adequate. The import of this definition may not be immediately clear. To help make it clearer, let's introduce an auxiliary definition that Leśniewski was to add later, that of two objects overlapping. Two objects overlap when they have a common part:

Dov $A \,\varepsilon\, \mathrm{ov}(B) \leftrightarrow \exists C \ulcorner C \,\varepsilon\, \mathrm{pt}(A) \wedge C \,\varepsilon\, \mathrm{pt}(B) \urcorner$ Overlapping

We can then express the equivalence in D2 as
$A \,\varepsilon\, \mathrm{fu}(m) \leftrightarrow A \,\varepsilon\, \mathrm{V} \wedge \forall B \ulcorner B \,\varepsilon\, \mathrm{pt}(A) \rightarrow \exists C \ulcorner C \,\varepsilon\, m \wedge C \,\varepsilon\, \mathrm{pt}(A) \wedge B \,\varepsilon\, \mathrm{ov}(C) \urcorner\urcorner$

D3 $A \,\varepsilon\, \mathrm{Sm}(m) \leftrightarrow A \,\varepsilon\, \mathrm{V} \wedge \forall B \ulcorner B \,\varepsilon\, m \rightarrow B \,\varepsilon\, \mathrm{pt}(A) \urcorner \wedge \forall C \ulcorner C \,\varepsilon\, \mathrm{pt}(A) \rightarrow \exists D \ulcorner D \,\varepsilon\, \mathrm{pt}(C) \wedge \exists E \ulcorner E \,\varepsilon\, m \wedge D \,\varepsilon\, \mathrm{pt}(E) \urcorner\urcorner\urcorner$ Sum

Again we need to state explicitly in the definiens that A is an object. This too is easier to understand if we employ the concept of overlapping, so
$A \,\varepsilon\, \mathrm{Sm}(m) \leftrightarrow A \,\varepsilon\, \mathrm{V} \wedge \forall B \ulcorner B \,\varepsilon\, m \rightarrow B \,\varepsilon\, \mathrm{pt}(A) \urcorner \wedge \forall C \ulcorner C \,\varepsilon\, \mathrm{pt}(A) \rightarrow \exists D \ulcorner D \,\varepsilon\, m \wedge D \,\varepsilon\, \mathrm{ov}(C) \urcorner\urcorner$

A3 $\exists A \ulcorner A \,\varepsilon\, m \urcorner \rightarrow \exists B \ulcorner B \,\varepsilon\, \mathrm{Sm}(m) \urcorner$

A4 $A \,\varepsilon\, \mathrm{Sm}(m) \wedge B \,\varepsilon\, \mathrm{Sm}(m) \rightarrow A = B$

D4 $A \,\varepsilon\, \mathrm{el}(B) \leftrightarrow \exists x \ulcorner A \,\varepsilon\, x \wedge B \,\varepsilon\, \mathrm{Sm}(x) \urcorner$ Element

D5 $A \,\varepsilon\, \mathrm{sub}(B) \leftrightarrow A \,\varepsilon\, \mathrm{V} \wedge \forall C \ulcorner C \,\varepsilon\, \mathrm{el}(A) \rightarrow C \,\varepsilon\, \mathrm{el}(B) \urcorner$ Subset

As noted above, the three functors 'pt', 'el' and 'sub' are coextensive.
We now give all the other definitions Leśniewski uses:

D6 $A \,\varepsilon\, \mathrm{psub}(B) \leftrightarrow A \,\varepsilon\, \mathrm{sub}(B) \wedge \sim(A = B)$ Proper Subset

This is likewise coextensive with 'ppt'.

D7 $\mathrm{U} = \mathrm{Sm}(\mathrm{V})$ Universe

This is the form of the definition Leśniewski gives: note that it is an identity. In his mature work, nominal terms are introduced using singular inclusion, not identity, partly for symbolic parsimony, but mainly because existence conditions need to be given. Having a definition does not always mean something satisfies it, as in the case of a canonical empty term. The definition D7 would be false if nothing existed, unless the identity symbol were reinterpreted to be without existential import. Here then is an equivalent but unproblematic definition for the Universe, in that form:

D7* $A \, \varepsilon \, \mathrm{U}^* \leftrightarrow A \, \varepsilon \, \mathrm{V} \wedge \forall B \ulcorner B \, \varepsilon \, \mathrm{V} \rightarrow B \, \varepsilon \, \mathrm{pt}(A) \urcorner$

It is readily proved from the axioms and definitions that U* is unique, and that if U or U* exists, then U = U*. So if any object at all exists, the object U (= U*) exists.

D8 $A \, \varepsilon \, \mathrm{ex}(B) \leftrightarrow A \, \varepsilon \, \mathrm{V} \wedge B \, \varepsilon \, \mathrm{V} \wedge \sim\exists C \ulcorner C \, \varepsilon \, \mathrm{pt}(A) \wedge C \, \varepsilon \, \mathrm{pt}(B) \urcorner$ Exterior

Two objects are exterior to (disjoint from) one another if they both exist but have no common part.

Leśniewski's last definition is of the (mereological) complement of one object in another. To formulate it we need a way to yoke two names together conjunctively. Here is how:

D∩ $A \, \varepsilon \, m \cap n \leftrightarrow A \, \varepsilon \, m \wedge A \, \varepsilon \, n$ Nominal Conjunction

Now we can render

D9 $A \, \varepsilon \, \mathrm{Cpl}(B,C) \leftrightarrow B \, \varepsilon \, \mathrm{sub}(C) \wedge A \, \varepsilon \, \mathrm{Sm}(\mathrm{el}(C) \cap \mathrm{ex}(B))$ Complement

The complement of *B* in *C* is the sum of all parts of *C* that are disjoint from *C*'s part *B*. The supplementation theorem then comes out as

T48 $A \, \varepsilon \, \mathrm{ppt}(B) \rightarrow \exists C \ulcorner C \, \varepsilon \, \mathrm{Cpl}(A,B) \urcorner$

Since the complement is what is left when "subtracting" one part from a whole, another notation for 'Cpl(*A*,*B*)' could be '$B - A$'.

7 Later Improvements

We have dwelt at some length on Leśniewski's 1916 system for three reasons. The first is that it is his first published mereology, and so the first published mereology *überhaupt*. The second is that it is relatively underexposed in the secondary literature. The third is that its use of redundant definitions and set-theoretic terminology makes Leśniewski's motivation clear. He wishes not only to give a solid basis to his own views, but is interested in converting set theorists to his view by adopting their terminology but then interpreting it in his own way. It becomes clear from this that Leśniewski's discontent with "official" set theory, which later led to his falling out with several of his mathematical contemporaries, had roots going right back to his earliest work on foundational matters.

Despite intensive work on mereology as well as on his other systems, protothetic and ontology, the next time Leśniewski published on mereology was in 1927–1931, in a series of articles entitled *O podstawach matematyki* (On the Foundations of Mathematics), appearing in *Przegląd Filozoficzny* (cf. [1], pp. 295–468; [2], pp. 174–382). The articles are written in a somewhat unusual autobiographical manner, detailing the development of Leśniewski's own ideas in more or less chronological sequence. They contain a critical, destructive part, followed by a longer, constructive part, and finish with general remarks on singular propositions of the form '$A \, \varepsilon \, b$' which are relevant to ontology rather than mereology. In the critical part, Leśniewski disapproves strongly of the logical inexactness of Whitehead and Russell's *Principia Mathematica*, which he holds in much lower esteem than Frege's *Grundgesetze der Arithmetik*, the latter's inconsistency notwithstanding. But his chief criticism is reserved for standard systems of set theory, among whose proponents he singles out Zermelo, Hausdorff, Dedekind, Schröder, Sierpiński and Fraenkel. His main objections are that such systems "create", "invent" or "introduce" an empty set, an object consisting of or comprising no objects at all, and that they distinguish between a single object and the set whose sole element is that object, as we would now say, an object x and its singleton $\{x\}$. His own theory, he notes, admits no empty set and identifies an object with its singleton. His tone is scathingly ironic, and this no doubt made him no friends. In his critical remarks on Schröder's extensionalist conception of classes, Frege had criticised the idea of an empty class, as well as the distinction between an object and its singleton, as inappropriate for Schröder's conception, and Leśniewski quotes Frege with approval. But his own view is close to Schröder's, and Leśniewski professes inability to understand Frege's treatment of classes as the extensions of concepts, and so is thrown back on his own previous conception, which he claims to have held from the very beginning. Zermelo's principles of set construction are considered *ad hoc* by Leśniewski, simply blocking the paradoxes but without a satisfactory analysis of their source. Since many of Zermelo's constructive axioms closely follow the ideas of Cantor, this is a partly unfair accusation, but the Axiom of Separation, which is the principal block to paradox, does look like an artificially constrained version of the naïve Principle of Comprehension.

Leśniewski first reprises the 1916 ideas in a somewhat streamlined and improved form. He places the definition of sum (class) before that of fusion (set), which he now calls not *mnogość* but *zbiór* (collection); his theorems are developed in a slightly more streamlined way. He then adds further definitions and theorems going beyond the 1916 results until he has a total of 198 theorems, all presented and proved in much the same way as in 1916, without special symbols other than variables. These, he says, are results he obtained up to 1920.

Leśniewski soon found himself able to give axiomatic treatments of his theory, which from 1927 he took to calling 'mereology', without interpolating definitions. The first such treatment used 'ppt' as the primitive, but another used 'part', (now called 'element' rather than 'ingredient') and as this is slightly simpler, we give it here as Leśniewski formulated it, modulo changes of terminology and schematic variables.

Axiom (a) If *A* is a part of *B* and not (*B* is *A*) then *B* is not a part of *A*

Axiom (b) If *A* is a part of *B* and *B* is a part of *C*, then *A* is a part of *C*

Axiom (c) If every *a* is a part of *A* and a part of *B*, and for any *C*, if *C* is a part of *A* or a part of *B*, then some part of *C* is a part of some *a*, then *A* is *B*

Axiom (d) If some object is *a*, then for a certain *A*, ((for any *B*, if *B* is an *a*, then *B* is a part of *A*) and for any *C*, if *C* is a part of *A*, then some part of *B* is a part of some *a*).

The terms 'proper part' and 'class' (sum) are then defined as

Def (e) *A* is a proper part of *B* iff *A* is a part of *B* and not (*A* is the same object as *B*)

Def (f) *A* is the class of objects *a* iff (*A* is an object, for any *B*, if *B* is an *a*, then *B* is a part of *A*, and for any *C*, if *C* is a part of *A*, then some part of *B* is a part of some *a*)

Axiom (c) effectively asserts the uniqueness of sums, while Axiom (d) effectively asserts the existence of a sum of *as* if any *as* exist.

Leśniewski shows that this more logically elegant system is deductively equivalent to the original, and he then goes on to prove a grand total of 264 numbered theorems. But we notice here the beginnings of a trade-off between logical elegance on the one hand and perspicuity on the other. The axioms (c) and (d) are less easily motivated in the abstract than the original 1916 axioms which used intermediate definitions: they are as they are just in order to give the same results. Similar remarks apply to other systems with other primitives that Leśniewski developed, including one from 1921 based solely on 'exterior'. Leśniewski was able to state that quite a few terms could be taken as sole primitives, including the disjoint sum of two individuals $A + B$ and its generalization, the disjoint sum (*suma*) of individuals a, where

DDsm $\quad A \mathrel{\varepsilon} \mathrm{Dsm}(a) \leftrightarrow A \mathrel{\varepsilon} \mathrm{Sm}(a) \wedge \forall BC \ulcorner (B \mathrel{\varepsilon} a \wedge C \mathrel{\varepsilon} a) \rightarrow (B = C \vee B \mathrel{\varepsilon} \mathrm{ex}(C)) \urcorner$

By now, logical innovation had given way to logical perfectionism and the exploration of every last corner of possibility. Increasingly, the focus both in Leśniewski's own work and in that of his pupils and followers was on incremental refinements of the basic idea of the systems, rather than questioning, motivating or revising them.

8 Whitehead's Alternative Mereology

At almost exactly the same time as Leśniewski was formulating his mereology, a similar development was taking place in the work of Alfred North Whitehead. Whitehead needs no introduction, but the reason why he was interested in mereology does. Whitehead was of course Russell's teacher and his collaborator in the three-volumed *Principia Mathematica* (1910–13), the most extensive attempt to show that mathematics is derivable from logic, and the biggest compendium of results in mathematical logic to that date. In the three published volumes, Whitehead and Russell dealt with logical basics, including propositional logic and type theory, and showed how to formulate the basic notions of cardinal, ordinal and real-number arithmetic. Unlike Frege however, they thought the logicist approach could be applied to geometry, and they planned a fourth volume of *Principia* on geometry, which Whitehead was to author alone. Whitehead's approach to and work on geometry is highly complex, and underwent several transformations, which it is beyond the remit of this paper to outline. Suffice it to say that Whitehead had long been impressed by the geometric algebras of Hermann Günther Graßmann and William Rowan Hamilton, had studied modern axiomatics of affine and metric geometries with care, and written short textbooks on them. His inclinations to integrate geometry as the mathematical theory of space with the physics of the occupants of space and time were apparent as early as his 1906 Royal Society memoir 'On Mathematical Concepts of the Material World'. But it is clear that when working on the fourth volume of *Principia* (which for various reasons was never completed) Whitehead became interested in the part-whole relation. He was interested in providing a point-free account of geometry, and to this end developed, building on the earlier memoir, the idea of *extensive abstraction*, taking points, and other lower-dimensional elements, not to be existent items within his geometry but as special kinds of abstraction arrived at through sequences of successively finer approximations. In his 1916 paper "La théorie relationniste de l'espace" [8], based on a lecture given in Paris in 1914, he notes that 'part' can mean subset, a component ("heterogenous part") of a more complex whole, or a spatial part ("homogeneous part"). He defines a relation he calls 'inclusion' and shows it is reflexive and transitive, but his sketchy account falls short of providing a full mereology.

After the war however, Whitehead was more forthcoming. In his 1919 book *An Enquiry Concerning the Principles of Natural Knowledge* [9] he outlines, without pretence at a properly developed formal system, some principles governing a relation he calls 'extending over', defined on events ([9], pp. 101 ff.) *Extending over* is simply his name for the converse of the proper part relation. The term probably comes from the English translation of Graßmann's word *Ausdehnung*. Among the principles Whitehead supports are the asymmetry and transitivity of the proper part relation, a strong supplementation principle, and the notion of a disjoint sum, which he calls a partition of an object. But his mereology differs from that of Leśniewski in several ways. It is atomless and lacks a universe: every object both is and has a proper part. Further it is mereologically dense: if A is a proper part of B, there is a third object C which is a proper part of B and of which A is a proper part. The lack of a Universe is directly contrary to Leśniewski; atomlessness and density are independent of Leśniewski's system.

Whitehead's motivation for developing mereological notions is quite different from Leśniewski's: he is interested in providing an account of the spatiotemporal and geometrical characteristics of the universe in terms of the relations among its inhabitants, which are events. His treatment of mereology in *Enquiry* is done *en passant* and is formally offhand and imprecise. When in 1926 Tarski drew Leśniewski's attention to the similarities between their two theories, Leśniewski found little difficulty in detecting formal defects in Whitehead's exposition. He outlined his criticisms, along with copious quotations from Whitehead's English, in a very long note at the end of Chapter IV of 'Foundations'. Leśniewski is right to find fault, but the defects in Whitehead are quite easily remediable (cf. [10]), and Leśniewski characteristically fails to penetrate beyond the formal shortcomings and consider the interesting doctrinal differences and their motivation.

Whitehead developed a revised mereology in his 1929 *Process and Reality* [11], but it is there mixed with topological concepts, and deals directly with spacetime rather than with the events therein, underlining the geometric motivation of Whitehead's treatment. It is of course a huge pity that the work for Volume IV of *Principia* failed to survive, as it would have been developed to a much higher formal standard and would have provided a more robust comparison with Leśniewski's work. What we can say is that Whitehead and Leśniewski developed mereological theories at almost the same time, around 1914, though Leśniewski was first into print. Their motivations were very different, and that is why the details and implications of their theories differ, despite sharing a conceptual core.

9 The Polish Continuation

Leśniewski died a few short months before the Nazi invasion of Poland and the beginning of the Second World War. His widow gave his papers to Bolesław Sobociński to edit, but they were destroyed in 1944. Any results that Leśniewski possessed that had not been published either were lost or had to be reconstructed from the memory of those who had worked under him or heard him lecture in Warsaw. There were two principal contributors of this kind. The first was Sobociński, who emigrated to the United States and became professor at Notre Dame University, founding and editing the *Notre Dame Journal of Formal Logic*. Sobociński was a loyal adherent of Leśniewski and did not amend the master's overall conception. His principal contribution was to consider systems of atomistic mereology. In mereology, an *atom* is an object with no proper parts, and mereological atomism is the theory that every object is completely composed of atoms, or is a sum of atoms. The atomistic hypothesis is usually expressed as follows:

Dat $A \mathrel{\varepsilon} \mathrm{at} \leftrightarrow A \mathrel{\varepsilon} \mathrm{V} \wedge \sim\exists B \ulcorner B \mathrel{\varepsilon} \mathrm{ppt}(A)\urcorner$

Atm $\forall A \ulcorner A \mathrel{\varepsilon} \mathrm{V} \rightarrow \exists B \ulcorner B \mathrel{\varepsilon} \mathrm{at} \wedge B \mathrel{\varepsilon} \mathrm{pt}(A)\urcorner\urcorner$

Atomism, its contrary, atomlessness, and mixed intermediate positions are all logically independent of general mereology. Atomistic mereology, unlike its contraries, has finite models. If in a model there are n atoms, then the model has cardinality $2^n - 1$, since any non-empty collection of atoms has a sum. Sobociński showed how to axiomatize atomistic mereology in ways suited to its peculiarities. He also made several incremental improvements to axiom systems for general mereology.

The other adherent of Leśniewski who contributed to mereology was Czesław Lejewski, who though a student of classics had attended Leśniewski's advanced classes in the 1930s. After the war he settled in England, becoming Professor of Philosophy at the University of Manchester. Like Sobociński, he explored various alternative axiomatizations of mereology and made numerous incremental improvements. Among his novel contributions, one is worth mentioning here: he provided a syntactic consistency proof of mereology relative to protothetic (cf. [12]).

Tarski himself contributed to mereology in the 1920s. He showed that Axiom (a) of Leśniewski's *Przegląd* system is redundant: since an individual A is identical to the sum of its parts, we can use the definition and uniqueness of sum to show that if A is part of B and B is part of A, then $A = B$, which gives Axiom (a) by partial contraposition. He then, in an article published in French in 1929 [13], employed a mereology using singular variables only and a standard notion of class to provide a point-free formulation of the geometry of solids. Readers of Tarski in the English translation of this paper might be forgiven for thinking that the system of mereology Tarski there gives was in the original, but in fact it was added in the English translation. Arianna Betti has conjectured, plausibly, that Tarski might not have wished to become embroiled in a priority dispute with his former teacher, whose mature work on mereology had not yet been published, and also that the formulation of the system, using standard class theory, would have seriously displeased Leśniewski (cf. [14]).

Be that as it may, in the expanded English version of 1956, Tarski formulates mereology using two axioms and two definitions (of which the first is not used for the mereology as such but only for the ensuing geometry):

Definition I An individual *X* is called a *proper part* of an individual *Y* if *X* is a part of *Y* and *X* is not identical with *Y*.

Definition II An individual *X* is said to be *disjoint* from an individual *Y* if no individual *Z* is a part of both *X* and *Y*.

Definition III An individual *X* is called a *sum* of all elements of a class α of individuals if every element of α is a part of *X* and if no part of *X* is disjoint from all elements of α.

Postulate I If *X* is a part of *Y* and *Y* is a part of *Z*, then *X* is a part of *Z*.

Postulate II For every non-empty class α of individuals there exists exactly one individual *X* which is a sum of all elements of α.

This system forms the basis for an extremely elegant mereologically-based axiomatic treatment of a solid geometry without points, using the primitive geometric notion of sphere. From a mereological point of view, several remarks are in order. Tarski's elimination of Leśniewski's first axiom is indeed a simplification, as he stated, but the second Postulate is simply a telescoping of Leśniewski's third and fourth axioms into one, as revealed by the use of the expression 'exactly one', which conjoins the notions 'at least one' and 'at most one'. Further, the logical basis includes sets (classes) in the standard sense, and not in either Leśniewskian sense, though it could easily be transposed into Leśniewskian ontology. Finally, because Tarski is employing 'class' and 'element' in the standard sense, he cannot continue with Leśniewski's acquisitive mereological terminology, so 'sum' comes into play instead of 'class'. His effort thus lies partly outside Leśniewskian orthodoxy, unlike the contributions of Sobociński and Lejewski. Somewhat later, in 1937, Tarski used a system of mereology closely equivalent to this one in order to provide axiomatic backing for the English biologist Joseph Henry Woodger's use of mereology in his *The Axiomatic Method in Biology*, where it appears as Appendix E [15]. In a paper on extended Boolean algebras of 1935 [16], he noted the formal (we would now say: model-theoretic) similarity between mereology and extended Boolean algebras, that is, Boolean algebras with variable-binding meet and join (product and sum) operators. This consisted solely in the absence of a null element in mereology.

10 The Harvard Variant

Outside Poland, and the associated use of mereology in Woodger, formal systems of part and whole first emerged from Harvard, where Whitehead had been teaching since 1924. It was in 1940 that Henry S. Leonard and Nelson Goodman, both of whom studied there, published in *The Journal of Symbolic Logic* the paper that introduced formal part–whole theory to the majority of anglophone readers: 'The Calculus of Individuals and its Uses' [17]. Its prehistory is however surprisingly convoluted (cf. [18]) and goes back to around 1928, when Nelson Goodman submitted a Harvard Honors Dissertation (now seemingly lost) which made some use of part and whole. His friend Henry S. Leonard developed a formal calculus using mereological concepts in his 1930 PhD dissertation *Singular Terms*, written under Whitehead, and along with acknowledgement to his supervisor, Leonard notes that he discussed the subject extensively with Goodman. In 1935 they were alerted by Quine, who had met Leśniewski personally in 1933, that their work closely resembled the latter's mereology, of which at that time they had no knowledge. In 1936 they gave a joint talk to the Association for Symbolic Logic which developed into the 1940 paper. A year later, in 1941, Goodman defended his PhD thesis *A Study of Qualities*, written under C. I. Lewis, which also used their joint system. When Goodman came to publish a heavily revised version of the dissertation in 1951 under the title *The Structure of Appearance*, he had eliminated the modest amount of set theory that he and Leonard had used and was, like Leśniewski, resolutely opposed to set theory. We outline Leonard's system (quoting from [18], as it is unpublished) and that from the 1940 paper.

Leonard's *Singular Terms* is conceived as a parallel for singular terms of several sections of *Principia Mathematica* which deal with class theory: its formal sections are numbered like those of *PM*, and its results track those of Whitehead and Russell. The undefined primitive is the binary mereological sum operator, written '+', and introduced thus:

> By "$x + y$" we mean to describe that individual which arises from the most general togetherness of any two other individuals. (Quoted from [18].)

The nominal expression '$x + y$' is taken as a descriptive function in the sense of Whitehead and Russell, that is, a definite description of the form $\iota x \ulcorner \phi x \urcorner$. However, since the symbol '+' is primitive, the description cannot be analysed in the usual Russellian way, so all this means is that it is meaningful, firstly, to predicate singular existence of an expression '$x + y$' in the form $\mathrm{E}!x + y$, and secondly, to countenance the possibility that compound expressions employing '+' may sometimes be empty. In terms of '+', Leonard sets out numerous definitions, two of which are required for his statement of the postulates:

Definitions

*16.01 $x < y =_{\mathrm{Df.}} x + y = y$

*16.02 $x \circ y =_{\mathrm{Df.}} \exists z \ulcorner z < x \wedge z < y \urcorner$

*16.03 $xy =_{\mathrm{Df.}} \iota z \ulcorner z < x \wedge z < y \wedge \forall t \ulcorner t < x \wedge t < y \to t < z \urcorner \urcorner$

The first of these defines the part relation, the second overlapping, and the third the product or greatest common part of two individuals, which of course does not exist if they do not overlap.

Postulates

*16.1 $x + y = y + x$

*16.12 $\mathrm{E}!x + x \to x + x = x$

*16.14 $[\mathrm{E}!x + y \wedge \mathrm{E}!y + z \wedge \mathrm{E}!(x + y) + z \wedge \mathrm{E}!x + (y + z))] \to x + (y + z) = (x + y) + z$

*16.16 $x \circ y \to \mathrm{E}!xy$

*16.17 $x \circ (y + z) \to (x \circ y \vee x \circ z)$

*16.18 $x < (y + z) \to (x < y \vee x < z \vee x < xy + xz)$

It follows from these that any two individuals have a binary sum: $E!x + y$, and many other now familiar mereological theses also follow, but a general sum operator is missing.

By the time of the 1940 joint paper, this lacuna had been plugged, helped by Quine's allusion to Leśniewski and the appearance of Woodger's book with the Tarski appendix. The 1940 system uses a modest amount of set theory—just sets of individuals, as in Tarski. It is based on disjointness (discreteness, exteriority or non-overlapping) as primitive: that x is disjoint from y we write $x \mid y$. The relevant definitions and axioms are

I.01 $\quad x < y =_{\text{Df.}} \forall z \ulcorner z \mid y \to z \mid x \urcorner$

I.02 $\quad x \circ y =_{\text{Df.}} \exists z \ulcorner z < x \wedge z < y \urcorner$

I.03 $\quad x \operatorname{Fu} \alpha =_{\text{Df.}} \forall z \ulcorner z \mid x \leftrightarrow \forall y \ulcorner y \in \alpha \to z \mid y \urcorner \urcorner$

I.1 $\quad \exists x \ulcorner x \in \alpha \urcorner \to \exists y \ulcorner y \operatorname{Fu} \alpha \urcorner$

I.12 $\quad x < y \wedge y < x \to x = z$

I.13 $\quad x \circ y \leftrightarrow \sim(x \mid y)$

This is by no means a perspicuous axiom set, but the upshot is "ideologically" the same as in Leśniewski. The two-place predicate 'Fu' is worth a brief eludication. It can be read as 'fuses': the individual x fuses the class α when x is the sum of the elements of α. To have a singular term for this fusion Leonard and Goodman can use a definite description,

$$\iota x \ulcorner x \operatorname{Fu} \alpha \urcorner$$

which denotes an individual just in case α is non-empty (by I.1) and whose uniqueness is provable from the axioms and definitions.

11 Subsequent Developments

Here we have to be brief and incomplete, as the literature on mereology has exploded in recent years. For a long time, during the hegemony of ordinary-language philosophy through the 1940s–1970s, there was little interest in formal theories like mereology. Things changed only gradually at first. Two pieces commenting on Goodman appeared in the 1970s. Rolf Eberle's *Nominalistic Systems* [19] was based on his UCLA dissertation and looked at Goodman's nominalism and its use of mereology. Eberle was interested in building up to the full strength of Goodman's mereology in stages. Alfred Breitkopf's 1978 paper 'Axiomatisierung einiger Begriffe aus Nelson Goodmans *The Structure of Appearance*' [20] gives a precise axiomatization of the system behind Goodman's use of mereology in that book, and applies it to constructive matters of interest to Goodman: qualia, magnitudes of qualia, and categories. In 1979 David Bostock employed a mereological basis slightly weaker than full Leśniewskian strength to underpin a logicist analysis of the real numbers [21]. Peter Simons's 1982 essay 'The Formalisation of Husserl's Theory of Wholes and Parts' [22] attempted for the first time to provide a formal framework for discussing the non-extensional notions of founding, dependence and the like which occur in combination with mereological concepts in Husserl's 1901 investigation.

The first full monographic treatise on mereology was Simons's *Parts* [7], which appeared in 1987. Although the principal aim of that work was to show how mereology could be used as a tool in a range of metaphysical problems, as a preamble to such applications it contained a survey of mereology to that date and the first detailed comparison of Leśniewskian systems with those of Whitehead and of Leonard and Goodman. It developed a free logic-based mereology and also looked at extending mereology into the temporal and modal domains, the latter allowing discussion of mereological essentialism and of Husserl. In the light of criticisms of various mereological theses, such as the existence of sums of any collection of objects, it proposed isolating a core or *minimal* mereology which confined itself to analysing the concept of part and extracting the analytic principles governing it, so that the strength of additional principles may be more discerningly identified. To this end a set of core principles was presented based on a free logic and the relation of proper part ($\ll$) [7, p. 362]

Existence $x \ll y \rightarrow \mathrm{E!}x \wedge \mathrm{E!}y$

Asymmetry $x \ll y \rightarrow \sim(y \ll x)$

Transitivity $x \ll y \wedge y \ll z \rightarrow x \ll z$

Supplementation $x \ll y \rightarrow \exists z \ulcorner z \ll y \wedge z \mid x \urcorner$

The suggestion was that anything less than this fails to capture formally the notion of part, whereas anything more goes beyond this core and embodies substantive metaphysical hypotheses.

In the decades since then, mereology has developed in two directions. In formal terms, much greater attention has been paid to the strengths of mereological postulates lying between the analytic core and the full strength of systems of Leśniewski, Leonard and Goodman. The most thorough formal treatment is in Lothar Ridder's 2002 book *Mereologie* [23], which is punctilious in its proofs and detailed in its critical examinations of alternatives.

From a philosophical point of view, mereology has achieved hitherto unrivalled penetration of analytic metaphysics, being itself a subject of metaphysical scrutiny as well as a standard tool for use within ontology. Here we survey the principal areas of activity: composition, class theory, persistence, mereotopology, and vagueness, but for reasons of space cannot enter into associated controversies. A readable and comprehensive survey of both the formal and philosophical issues may be found in Achille Varzi's *Stanford Encyclopedia of Philosophy* article 'Mereology' [24], which also contains an extensive bibliography of recent writings.

The notion of *composition* is already present in mereology. It can take one of two forms. An object *X* is *simply* composed of objects *a* if *X* is, or (in case of non-identical objects with the same parts) mereologically coincides with, the mereological sum Sm(*a*). An object *X* is *disjointly* composed of objects *a* if each *a* is disjoint from every other, and they simply compose *X*, which is then Leśniewski's disjoint sum Dsm(*a*). In his book *Material Beings* of 1990 [25], Peter van Inwagen raised the question as to when a collection of several objects *a* disjointly compose another object. He called this the

Special Composition Question. His surprising and controversial answer was: only when X is an organism and a are all its atomic parts. This means that, strictly and philosophically speaking, the only composite objects that exist are organisms, so artefacts and other inanimate objects, and the non-atomic proper parts of organisms such as organs and cells, do not exist. The proposal contradicts standard mereology following Leśniewski, for which any collection of disjoint individuals has a sum which it accordingly disjointly composes. This position is now often called *compositional universalism*. If that represents one extreme, in the discussions following van Inwagen's proposal another extreme position has emerged, which is *compositional nihilism*, according to which disjoint composition *never* occurs. That may take one of two forms. Either there are many atoms, and nothing composed of them; or, the world is one huge atom, without proper parts. The former assumes, controversially, the truth of atomism, the latter is a form of monism. Attempts to find a stable and satisfying intermediate position more consonant with common sense than any of these theories have to date not been crowned with conspicuous success.

In 1991, David Lewis published his book *Parts of Classes* [26], in which he showed that if standard mereology is applied to classes in the sense of von Neuman–Bernays–Gödel set-and-class theory, with the subclass relation interpreted as part, then it yields the same objects as that theory, provided one can find a way to generate singletons so that a singleton $\{x\}$ is always different from its sole element x. Allowing this, and adding Lewis's identification of the null class with the mereological sum of all urelements, the construction works on those not unproblematic assumptions. In a sense it brings the development of mereology as an alternative to set theory unexpectedly full circle back to its origins, in a way which would no doubt have infuriated Leśniewski, since it leaves in abeyance the ontological status of singletons and the null set, precisely the issues which he identified as problematic in standard set theory.

In 1930, Leśniewski addressed the question whether the following three propositions are consistent (cf. [1], pp. 465 ff.; [2], pp. 379 ff.):
The Warsaw of 1830 is smaller than the Warsaw of 1930
The Warsaw of 1830 is Warsaw
The Warsaw of 1930 is Warsaw.

He argued that, if 'Warsaw' is not an empty name, and these are all true, and 'is' is governed by the axioms of ontology, then we can conclude the absurdity that

The Warsaw of 1930 is smaller than the Warsaw of 1930.

His way out of this was to treat Warsaw as a four-dimensional extended object, Warsaw from the beginning to the end of its existence, of which the Warsaw of 1830 and the Warsaw of 1930 were different temporal parts of lesser temporal extent than the whole, so that the second and third premises are false. Alternatively, if 'Warsaw' were a common name, the conclusion would not follow from the premises. Leśniewski preferred the first alternative, and so espoused a view of objects in time that has come to be called *four-dimensionalism*. His view is similar to those of other "hard-minded" logician–metaphysicians such as Bolzano, Carnap, Quine, and Lewis. There is an extensive modern debate about whether the account of change, as here adumbrated by Leśniewski, requires ordinary objects to be considered extended in time as well as in space. The principles of mereology are not in question in this debate, but the way in which mereological notions are best applied to concrete objects certainly is.

Another area to which mereological concepts have been applied is the structure of space and time and their relation to their occupants (cf. [27, 28]). One strand of this discussion takes off from Whitehead's treatment in *Process and Reality* of mereology in conjunction with topological notions, a combination that has come to be called 'mereotopology'. Another issue related to both this and the issue of the previous paragraph is whether there is one or more than one way in which the occupants of space and time occupy the regions they do occupy. Another is whether the parts of such regions correspond to parts of their objects, or whether it some objects may lack proper parts while occupying regions that do have proper parts, a view found in Whitehead but interestingly presaged in the pre-critical Kant. Such objects are generally known as *extended simples*.

An issue which for a long time remained in the background in ontology and in logic is the question of *vagueness*, that is, predications which are neither clearly true nor clearly false but which may occasion Sorites paradoxes. Standard bivalent logic rules out such cases or regards them as arising from epistemological limitations. Leśniewski, as a convinced adherent of bivalence, would no doubt have consigned the consideration of vagueness to the extra-logical dustbin of inexact vernacular language. However, if vagueness is taken seriously and not brushed under the carpet, it interacts in an interesting and not well understood way with mereology. Of ordinary objects, whether things or events, it is often unclear where they begin and end, or what their parts are, so there are cases where, of the following three statements

A exists
B exists
A is a proper part of B

the first two are unproblematically true, while the third is neither clearly true nor clearly false. This is despite the fact that the following mereological statement is clearly false:

$$B + A = B - A$$

whereas both

$$B = B + A$$

and

$$B = B - A$$

are neither clearly true nor clearly false. There are several competing theories of vagueness, and here is not the place to review them. Suffice it is to say that the application of these theories to mereology is an area which is at present relatively underdeveloped.

12 Prospects

Leśniewski's own mereology was, barring simplifications and variations, essentially completed no later than 1927, and he and his adherents did not deviate from it. Whitehead's mereology, developed for different purposes, was not published as a formal system and has attracted less attention. The Leonard–Goodman development shared many of Leśniewski's philosophical views and their mereology is "ideologically" equivalent to that of Leśniewski, differences of logical vehicle aside, so that this joint vision has come to be called *classical extensional mereology* [7, p. 37]. More recent developments have however questioned several of the assumptions behind this classical theory and investigated systems which have been ever more finely distinguished. This has been accompanied by hitherto unprecedented levels of use of mereology within philosophical ontology, as well as its extra-philosophical employment in database representations. The mereological breakdown of an object into its principal component parts is called, in database parlance, a 'meronomy', by analogy with 'taxonomy'. Classicists may rightly wince at the term 'partonomy' sometimes used instead. Not least because of this increased activity across several domains, and because of several unresolved questions in the applications, it is to be expected that mereology will continue for some time to come to provide material for philosophical controversy, and to remain an essential tool for philosophers and an object of metalogical research by logicians and mathematicians.

References

1. Leśniewski, S.: Pisma zebrane [Collected writings]. In: Jadacki, J.J. (ed.) vol. 2. Semper, Warsaw (2015)
2. Leśniewski, S.: Collected works. In: Surma, S.J. et al. (eds.) vol. 2. Kluwer/Polish Scientific Publishers, Dordrecht/Warsaw (1992)
3. Leśniewski, S.: Lesniewski's lecture notes in logic. In: Srzednicki, J.T.J., Stachniak, Z. (eds.) Kluwer, Dordrecht (1988)
4. Sobociński, B.: L'analyse de l'antinomie Russellienne par Leśniewski. Methodos **1**, 94–107, 220–228, 308–316 (1949); **2**, 237–257 (1950). English translation: Leśniewski's Analysis of Russell's Paradox. In: Srzednicki, J.T.J., Rickey, V.F. (eds.) Leśniewski's Systems: Ontology and Mereology, pp. 11–44. Nijhoff/Ossolineum, The Hague/Wrocław (1984)
5. Leśniewski, S.: Czy klasa klas nie podporządkowanych sobie jest podporządkowana sobie? Przegląd Filozoficzny. **17**, 63–75 (1914). Cf. [1, pp. 223–35, 2015]. Translation: Is the Class of Classes not Subordinated to Themselves, Subordinated to Itself? in [2, pp. 115–28]
6. Leśniewski, S.: Podstawy ogólnej teorii mnogości. I [No further parts appeared]. Moscow: Popławski (1916). (Prace Polskiego Koła Naukowego w Moskwie. Sekcya matematyczno-przyrodnicza, No. 2.) Cf. [1, pp. 256–94]. Translation: Foundations of the General Theory of Sets. I, in [2, pp. 129–73]
7. Simons, P.: Parts. A Study in Ontology. Clarendon, Oxford (1987)
8. Whitehead, A.N.: La théorie relationniste de l'espace. Rev. Metaphys. Morale. **23**, 423–454 (1916)
9. Whitehead, A.N.: An Enquiry Concerning the Principles of Natural Knowledge. Cambridge University Press, Cambridge (1919)
10. Simons, P.: Whitehead und die Mereologie. In: Hampe, M., Maaßen, H. (eds.) Die Gifford Lectures und ihre Deutung. Materialien zu Whiteheads Prozess and Realität, vol. 2, pp. 369–388. Suhrkamp, Frankfurt am Main (1991); English translation: Whitehead and mereology. In: Durand, G., Weber, M. (eds.) Les principes de la connaissance naturelle d'Alfred North Whitehead, pp. 215–233. Ontos, Frankfurt am Main (2007)

11. Whitehead, A.N.: Process and Reality. An Essay in Cosmology. Macmillan, New York/London (1929). Corrected edition, Macmillan, New York/London (1978)
12. Lejewski, C.: Consistency of mereology. J. Symb. Log. **34**, 321–328 (1969)
13. Tarski, A.: Les fondements de la géométrie du corps. Księga Pamiątkowa Pierwszego Polskiego Zjazdu Matematycznego, pp. 29–33. Supplement to Annales de la Société Polonaise de Mathématique, Kraków (1929). Extended translation: Foundations of the geometry of solids, in [29, pp. 24–29]
14. Betti, A.: Leśniewski, tarski and the axioms of mereology. In: Mulligan, K. et al. (eds.) The History and Philosophy of Polish Logic. Essays in Honour of Jan Woleński, pp. 242–258. Palgrave Macmillan, Basingstoke (2013)
15. Tarski, A.: Appendix E. In: Woodger, J.H. (ed.) The Axiomatic Method in Biology, pp. 161–172. Cambridge University Press, Cambridge (1937)
16. Tarski, A.: Zur Grundlegung der Booleschen Algebra I. Fund. Math. **24**, 177–98 (1935). Translation: On the Foundations of Boolean Algebra, in [14, pp. 320–41]
17. Leonard, H.S., Goodman, N.: The calculus of individuals and its uses. J. Symb. Log. **5**, 45–54 (1940)
18. Rossberg, M.: Leonard, Goodman, and the development of the calculus of individuals. In: Ernst, G., et al. (eds.) From Logic to Art: Themes from Nelson Goodman, pp. 51–69. Ontos, Frankfurt am Main (2009)
19. Eberle, R.: Nominalistic systems. Reidel, Dordrecht (1970)
20. Breitkopf, A.: Axiomatisierung einiger Begriffe aus Nelson Goodmans the structure of appearance. Erkenntnis. **12**, 229–247 (1978)
21. Bostock, D.: Logic and Arithmetic. Vol. 2. Rational and Irrational Numbers. Oxford, Clarendon (1979)
22. Simons, P.: The formalisation of Husserl's theory of wholes and parts. In: Smith, B. (ed.) Parts and Moments, pp. 113–159. Philosophia, Munich (1982)
23. Ridder, L.: Mereologie. Ein Beitrag zur Ontologie und Erkenntnistheorie. Klostermann, Frankfurt am Main (2002)
24. Varzi, A.: Mereology. In: Zalta, E.N. (ed.) The Stanford Encyclopedia of Philosophy. http://plato.stanford.edu/archives/spr2016/entries/mereology (2016)
25. van Inwagen, P.: Material Beings. Cornell University Press, Ithaca (1990)
26. Lewis, D.K.: Parts of Classes. Blackwell, Oxford (1991)
27. Casati, R., Varzi, A.: Parts and Places. The Structures of Spatial Representation. MIT Press, Cambridge (1999)
28. Kleinschmidt, S. (ed.): Mereology and Location. Oxford University Press, Oxford (2014)
29. Tarski, A.: Logic, Semantics, Metamathematics. Papers from 1923 to 1938. Translated by J. H. Woodger, 2nd ed. edited and introduced by J. Corcoran. Hackett, Philadelphia (1983) (1st edn. Oxford, Clarendon Press, 1956)

P. Simons (✉)
Department of Philosophy, Trinity College Dublin, Dublin, Ireland
e-mail: psimons@tcd.ie

Alfred Tarski (1901–1983)

Jan Woleński

Abstract This paper presents the life and work of Alfred Tarski, one of the most distinguished and influential logicians in the entire history.

Keywords Logic · Semantics · Truth · Metamathematics

Mathematics Subject Classification (2000) Primary 01A60

Alfred Tarski (AT hereafter) was born in Warszawa on January 14, 1901. His family was Jewish—Teitelbaum (Polish spelling: Tajtelbaum) was the original name of AT; Wacław, AT's brother was born 2 years later. Their parents, Ignacy Teitelbaum and Rosa Prussak, belonged to Jewish business families. Rosa's family was rich and involved in the textile industry in Łódź. The Teitelbaums were not particularly religious people, although decisively keeping Jewish identity; especially, they celebrated Jewish holidays. On the other hand, they wanted to be partially assimilated. The family lived outside Jewish settlements in Warszawa and spoke Polish. Consequently, both boys attended an elementary school with teaching in Polish or Russian (Warszawa was in the Russian Empire at that time). However, Alfred and Wacław also went to cheder (a Jewish school) where they studied Hebrew and Torah. Thus, AT was fairly familiar with Jewish culture and tradition. After years, Czesław Miłosz, AT's colleague in Berkeley, told the present author that when he translated a fragment of Torah into Polish, he was convinced that he did the first translation. However, AT explained to him that the entire Hebrew Bible was much earlier (in the years 1883–1914) translated into Polish by Izaak Cylkow. This story documents that AT was fairly versed in Jewish religious literature published in Poland. On the other hand, his general attitude toward religion was quite flexible. As I already noted, the Teitelbaums preserved Jewish tradition, but AT celebrated in his family home Easter and Christmas (the main Catholic holidays) as well. It can be considered as his early inclination toward Polish identification. This attitude did not save him against anti-Semitism. He and his brother frequently heard offending anti-Jewish remarks of their

Tarski's life is extensively described in [3] and [13]. According to the character of the present volume I concentrate on the years 1901–1939, that is, Tarski's life in Poland. His postwar curriculum vitae is treated less extensively. I mention dates of birth and death in the case of Polish people only.

Á. Garrido, U. Wybraniec-Skardowska (eds.), *The Lvov-Warsaw School. Past and Present*, Studies in Universal Logic,
https://doi.org/10.1007/978-3-319-65430-0_27

contemporaries. These unpleasant experiences from the youth made persistent traces in AT's consciousness.

AT entered high school in Warszawa. He studied languages (Latin, German, French), Polish literature, science, mathematics, religion and history. He ended high school with very good results in 1918. The high school years of AT happened to be very stormy from the political points of view. World War I began in 1914. German troops attacked Warszawa very soon and Russian army had to step back. What was extremely important for young Polish patriots (AT decisively belonged to them) consisted in great hopes that Poland would recover its independence lost at the end of the eighteenth century. It happened in the autumn of 1918, exactly on November 11. AT began his university study 4 weeks before this date; his matriculation (a celebrated nomination as a student) occurred on October 15. As far as the matter concerns his nationality, he considered himself as a Pole, although he did not deny his Jewish origin. As many others Jews living in Poland, AT considered full assimilation as the only possible defense against anti-Semitism. He stressed his "Polonization" by preferring the form Tajtelbaum over Teitelbaum; the latter looked too German for him. AT's political views were close to socialism at that time.

AT began his university education as a student of biology. Due to the difficult political situation in Poland (the problem of fighting for the borders of the country) the University of Warsaw was closed just after AT's matriculation. He was taken to army and served in a unit doing military logistic work. AT returned to his studies in 1919, but he decided to study mathematics. This change was related to his participation in a course in logic conducted by Stanisław Leśniewski, who had just started his teaching as the Professor of the Philosophy of Mathematics. Leśniewski informed his students on one open problems in set theory, concerning the concept of ordered set, and AT solved it. Leśniewski immediately suggested AT to switch from biology to mathematics, particularly mathematical logic. The result achieved by AT was published in his first scientific paper, published in *Przegląd Filozoficzny* (Philosophical Review) in 1921; the subtitle of this work has the subtitle "From the seminar of prof. Stanisław Leśniewski in the University of Warsaw". It was the very beginning of AT's spectacular scientific career as one of the greatest logicians in the entire history of logic.

Entering into the territory of logic, AT was probably not aware that he would inscribe himself to a fast growing superpower in mathematical logic and the foundations of mathematics. How did it happen that a country without a specially strong tradition in logic so quickly (during one generation in the years 1918–1939) became a real stronghold in this field? In particular, this qualification concerns the Warsaw School of Logic. This school was established by philosophers and mathematicians. In philosophy, everything goes to Kazimierz Twardowski, a student of Brentano and the founder of the Lvov-Warsaw School. Twardowski (1866–1938) was appointed professor of philosophy at Lvov University in 1895. He wanted to introduce Brentano's metaphilosophical program in Poland. In particular, he demanded clarity language and thought and believed in scientific philosophy. Following his teacher, Twardowski maintained that philosophical method is (or can be) exactly the same as that executed in so-called special sciences. Twardowski was not a logician and did not consider himself as such. On the other hand, his metaphilosophical views formed a very friendly environment for logic *sensu largo*, that is, covering formal logic (the term "mathematical logic" was rarely used at that time), semantics and the methodology of science. Twardowski lectured on elementary algebra of logic in the academic year 1899/1900; in fact, it was the first university course on this topic in Poland (more precisely in the part of Poland belonging to the Austro-Hungarian Empire). Jan Łukasiewicz (1878–1956) participated in this class and very soon became attracted by logic (originally, he studied law, but switched to philosophy under Twardowski's influence). Łukasiewicz began systematic courses in advanced algebra of logic and other logical topics. He trained many young philosophers with explicit interests in logic, including Kazimierz Ajdukiewicz (1890–1963), Tadeusz Czeżowski (1889–1981), Tadeusz Kotarbiński (1886–1981) and Zygmunt Zawirski (1882–1948); all of them also studied mathematics, mostly under Wacław Sierpiński (1882–1969), who acquainted his students with set theory. Stanisław Leśniewski (1886–1939) joined this circle in 1910. Although this group, the Lvov Collegium Philosophicum, as Leśniewski used to say, cannot be regarded as a logical school, logic played a distinguished role in this circle of scholars. Hence, Twardowski insisted that his students should know philosophical novelties, for instance, logical works of Gottlob Frege and Bertrand Russell were well-known in Lvov.

Warszawa appeared on the logical stage exactly in 1916, when the University of Warsaw was reopened; it was closed in 1831 and functioned in 1862–1869 as the Warsaw Main School. The academic staff was mainly imported from Lvov. Łukasiewicz was appointed professor of philosophy. He began lectures in logic and attracted many young mathematicians. When Poland recovered its independence in 1918, this also resulted in a great debate about the tasks and prospects of Polish science and culture (in fact, these discussions began about 1916). Scholars in every field discussed how to develop their disciplines in the new expected political situation and what should be done in order to catch up with world science. Particularly important was the discussion among mathematicians. In fact, it had already started in Lvov, but was rather as a private enterprise, involving Sierpiński and Zygmunt Janiszewski (1888–1920). Both were disappointed by a lack of a common language and interests among Polish mathematicians and both were convinced that set theory and topology should play a fundamental role in mathematics. The national discussion about science, its needs and perspectives, was a good occasion for manifesting views about the future of mathematics in Poland. Janiszewski became the main exponent of the project, later known as the Janiszewski program, which defined the ideological fundament of the Polish Mathematical School.

Roughly speaking, according to Janiszewski, Polish mathematicians should concentrate on chosen mathematical fields and work in one strong circle. The second point was very soon abandoned, but the first was adopted. Although Janiszewski did not mention any concrete topic to be cultivated in Poland, most Polish mathematicians understood it as favouring set theory, topology and their applications to other branches of mathematics. Janiszewski also postulated that Poland should have a special mathematical journal published in international languages. This idea found its realization in *Fundamenta Mathematicae* (the first volume appeared in 1920). Janiszewski's program attributed a great role to mathematical logic and the foundations of mathematics. The placement of logic and the foundations at the heart of mathematics required definite steps in the sphere of organization. The University of Warsaw had the Faculty of Mathematical and Natural Sciences. The Department of the Philosophy of Mathematics was very soon organized and Leśniewski became its head; Łukasiewicz left the University in 1918 in order to serve as the Minister of Religious Denominations and Education in the government under Ignacy Paderewski. He returned to the academic staff in 1919 and the University established for him a special position in philosophy at the Faculty of Mathematics and

Natural Sciences. Both professors began intensively teaching mathematical logic, mostly among mathematicians but also among philosophers; Ajdukiewicz taught at this faculty in the years 1926–1927. The first project of *Fundamenta Mathematicae* divided the journal into two series, one devoted to set theory, topology and their applications, and second to logic and the foundations. This project was finally abandoned, but the significance of mathematical logic in the eyes of the founders of the Polish mathematical school found its impressive manifestation in the composition of the Editorial Board of *Fundamenta*: Mazurkiewicz, Sierpiński, Leśniewski and Łukasiewicz. Logic was also popularized among students of philosophy by Kotarbiński.

Generally speaking, the logical circle in Warszawa, known as the Warsaw School of Logic, was a child of two movements, namely the Lvov-Warsaw Philosophical School and the Polish Mathematical School. Both determined the scientific environment in which AT grew as a logicians. In fact, he opened the list of young mathematicians and philosophers attracted by logic in Warszawa. This group included (in alphabetical order and covering the whole interwar period): Stanisław Jaśkowski (1906–1965), Adolf Lindenbaum (1904–1941?), Andrzej Mostowski (1913–1975), Moses Presburger (1904?–1943), Jerzy Słupecki (1904–1984), Bolesław Sobociński (1904–1980; a philosopher by training)) and Mordechaj Wajsberg (1902–1942?). The names of three other of logicians who graduated shortly before 1939 or studied during War World II and began their academic work after 1945 should be added to this list, namely Jan Kalicki (1922–1953; a mathematician), Czesław Lejewski (1913–2001; a classicist and philosopher) and Henryk Hiż (1917; a philosopher).

AT studied mathematics at the University of Warsaw from 1919 to 1923. He attended courses and seminars by (inter alia) Leśniewski (the foundations of mathematics), Kotarbiński (logic), Sierpiński (set theory), Mazurkiewicz (analysis), Kazimierz Kuratowski (topology) and Leon Petrażycki (sociology). AT also met his student fellows interested in logic, namely Lindenbaum and Wajsberg. The former became his close friend and collaborator; Bronisław Knaster (1893–1990), a mathematician was another close friend of AT. Working with Leśniewski, AT obtained important results concerning prothotetic (an extended propositional calculus), one of three logical systems constructed by Leśniewski. These results constituted AT's doctoral dissertation supervised by Leśniewski, defended in 1924. AT was the only person who did a doctorate under Leśniewski and the latter used to say that he had a 100% of genius doctoral students. AT's PhD dissertation was published in two papers which appeared in *Fundamenta Mathematicae*. In 1923, AT acted as the secretary of the logic section of the 1st Polish Philosophical Congress in Lvov. On that occasion he met Stefan Banach (1892–1945). At the same year they published (1892–1945) a famous paper on the paradoxical decomposition of a ball. This result, called the Banach-Tarski paradox shows some surprising consequences of the axiom of choice. In 1925, AT obtained his habilitation on the base of a dissertation on the concept of finite set and became the youngest docent (a scholar who had *veniam legendi* and thereby could lecture at university) in the entire history of mathematics in Poland. After his doctorate, AT was mostly involved in set theory. His deep and numerous results (partly achieved together with Lindenbaum) ensured him a distinguished place in the Polish Mathematical School.

AT all the time was thinking how to strengthen his Polonization. He decided to change his surname, following advices of Leśniewski and Łukasiewicz, AT and Wacław, his brother (he studied law), adopted the name Tarski. Firstly, it was added to Tajtelbaum. For instance, AT's mentioned paper published in *Przegląd Filozoficzny* is signed by Tajtelbaum-Tarski, but his PhD diploma is for Alfred Tarski. AT also converted to Catholicism. These moves were motivated by his intention to be recognized as a Pole. In 1929, AT married Maria Witkowska; they had two children; Jan born in 1934, and Ina born in 1938. Although changing name and converting to Catholicism helped to some extent, AT (and Wacław as well) were stigmatized as Jews by Polish anti-Semitic activists. In the late 1930s. a booklet appeared with a list of dangerous Jews, that is, such who changed names and converted. Both Alfred and Wacław were included into this infamous register. As far as the issue concerning AT's style of life, he belonged to a group of friends who met in cafes and used alcohol or even drugs; he was a heavy smoker until the end of his life. One of AT's friends deserves a special attention. It was Stanisław Ignacy Witkiewicz (1885–1939), called Witkacy. He was a writer, painter, philosopher and a very eccentric person. Witkacy made portraits of Alfred and Maria. The head of the former is presented inside spikes. A very accurate picture. AT and Maria frequently went to Zakopane, a very popular resort in the Tatra mountains, where they climbed.

The professional career of AT was not easy. Having *veniam legendi*; he could lecture and conduct seminars at university. He delivered many advanced courses, for instance, on set theory, methodology of deductive sciences, geometry or arithmetic of natural numbers, in order to mention a few. In 1929, he officially became an assistant of Łukasiewicz in the Department of Philosophy existing at the Faculty of Mathematics and Natural Sciences (this chair was especially established for Łukasiewicz). AT, in order to earn money needed for his family had to find a job outside the university. He worked as a teacher of mathematics in the Żeromski Secondary School in Warszawa and the National Pedagogical Institute, which organized training for teachers of mathematics; he had to resign from the latter for his Jewish origin. Perhaps AT's teaching of elementary geometry should be especially mentioned, because it resulted in his involvement into writing textbooks for schools and inventing problems. AT also published a textbook on mathematical logic (see [27]) . This small book was written for students of secondary schools, particularly interested in logic. It was translated into German in 1937 and, in a revised and extended version, into English in 1941 (see [28]), and reprinted many times (translations into other languages were published as well). When I told one of my American colleagues that the book in question was written for secondary schools, he replied with a surprise that it is too difficult for most students of American universities. In a guide for students of mathematics in Warsaw University published in 1926 after a list of textbooks of elementary logic, we find information that the 1st volume of *Principia Mathematica* is recommended of advanced students. These facts illustrate how logic was taught in Poland, particularly in Warszawa in the interwar period.

AT's academic ambitions went further than to occupy the position of a docent or assistant. He intended to be a university professor. The first opportunity appeared in 1928, when the Lvov University decided to establish the professorship in mathematical logic at the Faculty of Mathematics and Natural Sciences. Leon Chwistek (1884–1944) and AT were competitors. The former had a strong support of mathematicians, also Banach, who was in a difficult personal situation due to his mentioned work with AT. Philosophers, particularly Twardowski, acting in the name of Leśniewski and Łukasiewicz, preferred AT. The University invited Luitzen Brouwer, David Hilbert and

Bertrand Russell as referees. These names show how serious was this competition. It is not surprising for the great prestige of logic in the interwar period. In fact, Poland had five professorships in logic in the years 1918–1939. How many were there outside Poland? The answer is surprising: just one, in Münster, in Germany.

Chwistek won this rivalry. AT was convinced that his Jewishness played the decisive role in this issue. However, one should be very careful in evaluation of what was going on in the competition in question. Doubtless, Tarski's Jewish origin did not help. In fact, not many Jews became full professors in the interwar Poland, but some succeeded; for instance Hugo Steinhaus (1887–1972), a distinguished mathematician and very powerful professor in Lvov. Since his sister married Chwistek, Steinhaus' support for AT's rival was natural. On the other hand, Jewishness of Mrs. Chwistek did not do a favor to her husband, because anti-Semites did not like persons with Jewish spouses. Importantly, Chwistek was older, better known at that time and supported by Jagiellonian University in Kraków. According to Polish academic rules, universities had a right to opt for candidates for professorships. In his case, Warsaw University supported AT, but Cracow University voted for Chwistek. Last but not least, Russell wrote a well-know letter in which he recommended Chwistek (opinions of Brouwer and Hilbert are unknown) . Russell explicitly said that since he knew Chwistek and his work, the choice of Chwistek would be a good decision. However, he added that his recommendation was not based on a comparison of both rivals, because "The work of Mr. Tarski I do not at the moment remember and do not have access to at present". Ironically, Russell quoted AT's papers in the second edition of *Principia Mathematica.* According to recollections of some people (I heard this story from Hiż) , the second opportunity for AT to be appointed as professor appeared in Poznań in 1937, when Zawirski moved to Kraków; but Poznan University cancelled the professorship in logic, apparently to block Tarski as a Jew. This affair is not testified by existing documents, but if it actually happened, anti-Semtic attitude, strong in Poznan, played the decisive role in rejecting Tarski. Tarski as a docent could not supervise doctorates. However, he was the doctor father of Mostowski (1938)—Kuratowski acted as the official supervisor. There is also a very surprising story about Presburger. He proved the completeness of so-called Presburger arithmetic (theory of natural numbers with addition as the sole operation). Presburger asked AT whether this theorem is sufficient for obtaining the doctoral degree. The answer was entirely negative—AT considered this result as too trivial. He made a mistake, because Presburger's result is presently considered as a very serious achievement.

Tarski intensively worked in the 1930s. He continued his work in set theory, but was more and more involved in logic and metamathematics. In a series of papers, he defined several notions used by logicians (and mathematicians as well) rather in an intuitive way. For instance, he axiomatized the concept of logical consequence and deductive system. His most important work concerned the concept of truth. In 1930, AT delivered a talk about the concept of truth in deductive science. His famous monograph *Pojęcie prawdy w jęykach nauk dedukcyjnych* (The Concept of Truth in Languages of Deductive Sciences) appeared in 1933 (see [25]). Its German translation *Der Wahrheitsbegriff in den formalisierte Sprachen* (The Concept of Truth in Formalized Languages) was published in 1935 (see [26]) and English version in 1956 (see [29]). Since this idea is extensively presented in another paper in this volume, I only mention it in this place. At the moment let me add that the semantic definition of truth (it is, so to speak, the official label for AT's approach to truth) is (a) the most important result of AT; (b) one of the most important achievements in analytic philosophy; (c) the idea which originated model theory as one of the most important parts of contemporary logic; (d) the most important achievement in the entire history of Polish philosophy. Even if someone will say that there are no clear criteria of what belong to the most important philosophical achievements, sociological measures support the opinion expressed in (b) and (d).

In the interwar period, AT active participated in scientific life, mathematical and philosophical, in Poland, and on the international scale, in particular in all Polish Philosophical Congresses (Lvov—1923, Warszawa—1927, Kraków—1936), in many mathematical congresses (for example, the eight Mathematical Congress in Bologna, the one Congress of Mathematicians of Slavic countries), in the eight International Philosophical Congresses (Prague 1934) and in the 1st Congress for Scientific Philosophy (Paris 1935). AT's participation in the last event was remarkable. He, invited by Karl Menger, visited Vienna in 1930. Menger introduced AT to many philosophers, directly or indirectly associated with the Vienna Circle. In particular, AT met Rudolf Carnap and Kurt Gödel for the first time. In the next years, AT visited Vienna several times and had discussions with Viennese philosophers, particularly with Karl Popper. AT's reported his semantic ideas in discussions in Vienna. As it is known, the Vienna Circle was skeptical about semantics and its significance for philosophy. AT convinced Carnap and Popper (he was not a member of the Circle) to semantics. Carnap insisted that AT should deliver a talk on semantics at the Paris Congress in 1935. Tarski agreed, although he expected a criticism of semantics, and delivered one lecture on the foundations of semantics and the second on the concept of logical consequence. These talks were recognized as the most important scientific events of the Congress. The other paper about semantics was read by Maria Kokoszyńska (1905–1981), a close friend of AT. All these contributions caused a very hot discussion. The Poles (or the Polish camp as it was called) and Carnap definitely defended the role of semantics in philosophy, but Otto Neurath radically opposed to using semantic tools in philosophical analysis. He was afraid that semantics could introduce bad metaphysics into philosophy. The controversy over semantics was continued in the next years. Although Neurath (he died in 1945) never accepted semantics as a legitimate part of philosophy, the Polish camp finally won. Thus, AT became one of the most influential thinkers in contemporary philosophy. Even if someone were to say that contemporary anti-realistic semantics rejected the semantic theory of truth, it is still semantics.

In the 1930s several philosophers visited Poland, for instance, Carnap. Joseph Woodger, Willard van Orman Quine, and Heinrich Scholz. They appreciated very positively the philosophical situation in the country, particularly AT and his achievements. Scholz said once that he was surprised that AT had not been promoted to the professor position. A new possibility appeared after Leśniewski's death in May 1939. AT hoped to be his successor. Yet the political situation in Europe became more and more dangerous. Quine urged that AT should leave Poland. The Congress for Scientific Philosophy to be organized at the Harvard University, created a good opportunity. AT had mixed feelings. On the one hand, he liked to participate in scientific events, also because his ambitions were satisfied by his position as a great star in logic; but on the second hand, he wanted to be present in Warszawa, when the succession after Leśniewski would be decided; it was expected to happen in the autumn of 1939. Finally, he decided to go to USA. On August 11, AT embarked on a ship sailing to America. He took only a small suitcase, as he planned to return very soon. He landed in New York on August 24. The war began exactly one week later. Maria Tarski with two children remained in Poland. She left Warszawa and survived. Maria was Aryan, but Jan and Ina, the children, satisfied the condition of Jewishness introduced by the Nürnberg statutes. AT's parents were murdered by the Nazis in Auschwitz, and Wacław, his brother was killed in Warszawa just before the end of the war. If AT were to remain in Poland, his fate would probably be tragic. One of the chapters in the Feferman's book (see [3]) has "How the "Unity of Science" Saved Tarski's Life" as its title. Very proper, indeed. Several Polish logicians of Jewish origin, including Lindenbaum and Wajsberg, perished in the Holocaust.

AT participated in the Harvard Congress. However, the question "What then?" was urgent. Quine arranged a research appointment at Harvard for AT. He temporarily lectured at Columbia University for undergraduate students of Ernest Nagel. Then, he became a visiting professor at the City University of New York. Russell tried to help AT in getting a permanent position at Columbia, but these attempts were unsuccessful. Some support came from the Young Men's Hebrew Association in New York. The Guggenheim Fund gave a fellowship for AT; he used it to stay in Princeton where he met Gödel once again. Important things happened at Harvard in 1940–1941. A discussion group of logic was formed, including Carnap. Quine, AT and Russell as main participants (see [4]). Moreover, AT gave several talks in many places in the USA from the East Coast to the Midwest. Besides troubles with getting a permanent job, AT was terribly worried about his family in Poland. He was, of course, fully conscious of the danger for the life of Maria and children. From time to time, he had indirect news obtained with the help of Father Józef M. Bocheński (1902–1995) and Anders Wedberg. The problems with job ended in 1942, when the University of California at Berkeley offered a position for AT as a lecturer. AT became a USA citizen in 1945. In the early 1945, he received a letter from Maria informing him that she and the children survived. They arrived to Berkeley on January 6, 1946. This day ends AT's Odyssey from Warsaw to Berkeley.

AT became the full professor at the age of 45 in 1946. Perhaps it should be noted in the Guiness book as a world record in the category of eminent scientists waiting for the professorship. He was elected to the National Academy of Sciences, the Royal Netherlands Academy of Arts and Sciences and the British Academy, served as President of the Association of Symbolic Logic (1944–1946) and the International Union for the History and Philosophy of Science in 1956–1957; initiated the International Congresses of Logic, Methodology and Philosophy of Science (the 1st Congress took place in Stanford in 1960). He received honorary degrees from the Catholic University of Chile (1975), the University of Marseilles (1977) and the University of California (1982). He travelled all over the world and, as before the war, participated in many scientific congresses and conferences. He visited once Poland; it happened in 1959. He had a permanent sentiment to his first homeland, although he considered himself as American after 1945. When a conference was organized to celebrate AT's one hundredth anniversary in Warszawa, Leon Henkin wrote to me that Alfred's heart remained in this city for ever. The family spoke Polish in the daily life and his house in Berkeley was always opened to guests from Poland. AT, remembering his youth, was always very sensitive to all signs of anti-Semitism. He rejected an invitation to Poland in 1968, protesting against anti-Semitic campaign executed by Polish authorities at that time. On the other hand, AT supported the "Solidarity" movement in 1981–1982 and offered a considerable amount of money for the Kościuszko Fund, giving scholarships for Polish scientists. He maintained close contacts with Polish friends. AT's famous collection *Logic, Semantics, Metamathematics* (see [29]), published in 1956 has the dedication "To His Teacher Tadeusz Kotarbiński—the Author". AT had many distinguished teachers, but he decided to celebrate just Kotarbiński, as a pattern of humanity. And AT changed his earlier socialist political views to more social-democratic. Alfred Tarski died in Berkeley on October 26, 1983.

AT was a great teacher. Although he was very demanding and not always nice to his students, he attracted many people to logic. As a result, he supervised 24 doctoral dissertation. The list of his students includes (in the chronological order): Bjarni Jonsson, Louise Chin Lim, Julia Robinson, Wanda Szmielew, Frederick Thompson, Anne Morel, Robert Vaught, Cheng-Chung Chang, Solomon Feferman, Richard Montague, Jerome Keisler, Donald Monk, Haim Gaifman, William Hanf, John Doner, Robert Bratford, Haragauri Gupta, Donald Pigozzi, George McNaulty, Charles Martin, Roger Maddux, Benjamin Wells and Kan Ng. AT also directly influenced the work of such logicians as John MacKinsey, Dana Scott and Steven Givant. He created the California School of Logic, the most powerful logical circle in the USA. The number of mathematicians and philosophers indirectly influenced by AT and his ideas is enormous. His writings appeared at least in the following languages: Bulgarian, Czech, Dutch, English, French, German, Georgian, Hebrew, Hungarian, Italian, Polish, Portuguese, Romanian, Russian, Serbian-Croatian, Spanish and Swedish.

AT research in his American period is complex. Perhaps his contributions to the model theory are most important. AT and his students developed so-called Western (Californian) model theory. Its main idea consists in investigations of the relation between languages and mathematical structures. The former are regarded as the starting point. By contrast, the Eastern model theory, developed by Abraham Robinson, generalizes algebraic concepts. Roughly speaking, the former approach is more logical, but the latter—more mathematical, although these differences are rather vague at the present. Generally speaking, AT continued the ideology of the Warsaw School of Logic. Although he considered mathematical logic as a part of mathematics, he considered it as a relatively autonomous. In particular, he did not agree that logic is servant of mathematics. The most important feature of his approach to logic consisted in admission of all mathematically accepted methods, regardless whether they were constructive or not. For instance, AT had no scruples in using the axiom of choice. This attitude was very characteristic for the Polish Mathematical School. More specifically, philosophical controversies around this axiom are entirely independent of its mathematical applications. On the other hand, there is tension between a free use of infinistic methods by AT and his explicit predilections toward nominalism as a philosophical position rejecting abstract objects; he also accepted empiricism in epistemology. AT, asked how he reconciled his private philosophy with his mathematical practice, answered that he felt like a tortured nominalist. AT added that there exist various tale-stories, set theory belongs to this variety, but it is very useful. In his early years, AT understood logic relatively widely as covering set theory, but, according to his later views, he favored first-order, logic as *the* logic. As far as the issue concerns the plurality of logical systems, although AT obtained important results in non-classical logics, particularly, many-valued and intuitionistic, he preferred the classical system as *the* logic. On the other hand, he was ready to investigate every logical system provided that such a research was interesting from the mathematical point of view.

Bibliography

Selected Books on Tarski, His Background and His Ideas

1. Brożek, A., Chybińska, A., Jadacki, J.J., Woleński, J. (eds.): Tradition of the Lvov-Warsaw School. Ideas and Continuations. Brill/Rodopi, Leiden/Boston (2015)
2. Coniglione, F., Poli, R., Woleński, J. (eds.): Polish Scientific Philosophy: The Lvov-Warsaw School. Rodopi, Amsterdam (1993)
3. Feferman, A., Feferman, S.: Alfred Tarski: Life and Logic. Cambridge University Press, Cambridge (2004)
4. Frost-Arnold, G.: Carnap, Tarski and Quine at Harvard. Conversations on Logic, Mathematics and Science. Open Court, Chcicago (2013)
5. Gruber, M.: Alfred Tarski and the "Concept of Truth in Formalized Languages". A Running Commentary with Consideration of the Polish Original and the German Translation. Spinger, Heidelberg (2016)
6. Horsten, L.: The Tarskian Turn. Deflationism and Axiomatic Truth. The MIT Press, Cambridge (2011)
7. Jadacki, J.J.: Polish Analytical Philosophy: Studeis on Its Heritage, with the Appendix Containing the Bibliography of Polish Logic from the Second Half of the 14th Century to the 1st Half of the 20th Century. Semper, Warszawa (2009)
8. Jadacki, J.J. (ed.): Alfred Tarski: dedukcja i semantyka [Déduction et sémantique] [Deduction and Semantics]. Semper, Warszawa (2003)
9. Kijania-Placek, K., Woleński, J. (eds.): Lvov-Warsaw School and Contemporary Philosophy. Kluwer Academic Publishers, Dordrecht (1998)
10. Kuratowski, K.: Half Century of Polish Mathematics: Remembrances and Reflections. Pergamon Press, London (1980)
11. Lapointe, S., Marion, M., Miśkiewicz, W., Woleński, J.: The Golden Age of Polish Philosophy: Kazimierz Twardowski's Philosophical Legacy. Springer, Berlin (2009)
12. López-Escobar, E.G.K., Miraglia, F.: Definitions: The Primitive Concept of Logics or the Leśniewski-Tarski Legacy. Instytut Matematyczny PAN, Warszawa (2002)
13. McFarland, A., McFarland, J., Smith, T.J.: Alfred Tarski. Early Work in Poland – Geometry and Teaching. Birkhäuser, New York (2014)
14. Murawski, R.: The Philosophy of Mathematics and Logic in the 1920s and 1930s in Poland. Birkhäuser, Basel (2014)
15. Pantsar, M.: Truth, Proof and Gödelian Arguments: A Defence of Tarskian Truth in Mathematics. The University of Helsinki, Helsinki (2009)
16. Patterson, D.: Alfred Tarski. Philosophy of Language and Logic. Palgrave Macmillan, New York (2012)
17. Patterson, D. (ed.): New Essays on Tarski and Philosophy. Oxford University Press, Oxford (2008)
18. Rojszczak, A.: From the Act of Judging to the Sentence. The Problem of Truth Bearers from Bolzano to Tarski. Springer, Dordrecht (2005)
19. Skolimowski, J.: Polish Analytical Philosophy. A Survey and a Comparison with British Analytical Philosophy. Routledge and Kegan Paul, London (1967)
20. Szaniawski, K. (ed.): The Lvov-Warsaw School and the Vienna Circle. Kluwer Academic Publishers, Dordrecht (1989)
21. Twardowski, B., Woleński, J. (eds.): Sixty years of Tarski's Definition of Truth. Philed, Kraków (1994)
22. Wagon, S.: The Banach-Tarski Paradox. Cambridge Univerity Press, Cambridge (1985)
23. Woleński, J.: Logic and Philosophy in the Lvov-Warsaw School. Kluwer Academic Publishers, Dordrecht (1989)
24. Woleński, J., Köhler, E. (eds.): Aldred Tarski and the Vienna Circle. Austro-Polish Connections in Logical Empiricism. Kluwer Academic Publishers, Dodrecht (1999)

(The most complete bibliography works on Tarski is in McFarland, A., McFarland, J., Smith T. J; Alfred Tarski. Early Work in Poland – Geometry and Teaching (see above), pp. 398–471)

Selected Works of Tarski (Only Books)

25. Pojęcie prawdy w językach nauk dedukcyjnych [The Concept of Truth in Languages of Deductive Sciences]. Towarzystwo Naukowe Warszawskie, Warszawa (1933)
26. Der Wahrheitsbegriff in den formalisierten Sprachen. Studia Philosophica **I**, 261–405 (1935)
27. O logice matematycznej i metodzie dedukcyjnej [On Mathematical Logic and Deductive Method]. Książnica-Atlas, Lwów (1936)
28. Introduction to Logic and to the Methodology of Deductive Sciences. Oxford University Press, Oxford (1941). (4th ed. (1994))
29. Logic, Semantics, Metaamathematics. Papers from 1923 to 1938. Oxford University Press, Oxford (1956). (2nd edn., Hackett Publishing Company, Indianapolis (1983). Includes The Concept of Truth in Formalized Languages, 152–278)
30. Collected Papers, vols. 1–4. Birhäuser Verlag, Basel (1986)
(The most complete bibliography of Tarski is collected by S. Givant: Bibliography of Alfred Tarski. The Journal of Symbolic Logic **41**, 913–941 (1986))

J. Woleński (✉)
University of Information, Technology and Management Rzeszów, Rzeszów, Poland
e-mail: jan.wolenski@uj.edu.pl

Some Philosophical Aspects of Semantic Theory of Truth

Jan Woleński

Abstract The semantic theory of truth, formulated by Alfred Tarski in the 1939s, is primarily a mathematical theory. On the other hand, it also has a considerable philosophical content. This paper presents the second aspect of this theory. It can be shown that several traditional philosophical issues pertaining to the concept of truth can be illuminated by Tarski's account of truth. It concerns, for instance, the idea of correspondence, the relation of truth and logic, the problem of the relativity/absoluteness of truth, etc.

Keywords Satisfaction · Model · Truth-bearers · Truth-criterion · Correspondence

Mathematics Subject Classification (2000) Primary 03A05

The semantic theory of truth (**STT**), developed by Alfred Tarski, has two separate but interconnected aspects.[1] Firstly, **STT** is a formal mathematical theory of a central concept of model theory, one of the most important branch of mathematical logic. Secondly, **STT** is also a philosophical doctrine that elaborates the notion of truth investigated by philosophers since antiquity. As the title indicates, this paper focuses on the second issue, that is, **STT** as a philosophical theory. Due to their significance for philosophical analysis of the concept of truth, some formal questions must be taken into account too.

However, the fate of **STT** as a mathematical theory and as a philosophical doctrine is different. Consider the following prophecy [12, p. 135]:

> [...] you should ask yourself what your grandsons and granddaughters are likely to study when they settle down to their 'Logic for computing class' at 9.30 after school assembly. Will it be syllogisms? Just possibly it could be the difference between saturated objects and unsaturated concepts, though I doubt it. I put my money on Tarski's definition of truth for formalized languages. It has already reached the universal textbooks of logic programming, and another 10 years should see if safely into the sixth forms. This is a measure of how far Tarski has influenced the whole framework of logic.

[1] I prefer the label "semantic theory of truth" over "semantic definition of truth" or "semantic conception of truth", though I will use, mostly for stylistic reasons, the second name too. First of all, **STT** cannot be reduced to a definition, because it consists of a statement defining truth and many other assertion. On the other hand, the name "semantic conception of truth" is too vague.

Á. Garrido, U. Wybraniec-Skardowska (eds.), *The Lvov-Warsaw School. Past and Present*, Studies in Universal Logic,
https://doi.org/10.1007/978-3-319-65430-0_28

Clearly, Wilfrid Hodges in the quoted fragment says about **STT** as a mathematical theory. Independently whether he is right or not in his prophecy concerning the logical education of our grandsons and granddaughters, Tarski's truth-definition is en vogue amongst mathematical logicians and almost nobody denies its importance as an idea belonging to mathematical logic. If one wonders why I say "almost nobody", I recall what Alan Turing said once about **STT**: "Triviality can go no further" (see [31, p. 144]). Hao Wang grounds on this fact the following opinion: "There is a great difference of opinion on the importance of [Tarski's] contribution to this area [that is, the theory of truth—J. W.]" (ibidem, p. 144).[2] However, it is rather Wang's personal evaluation of the situation (I do not enter into his motives), and the real measure of the importance of Tarski's work as a mathematical enterprise is closer to Hodges' statement.

The fate of **STT** as a piece of philosophy is much more complex and there is really "a great difference of opinion on the importance of [Tarski's] contribution." It is, of course, no surprise, because being controversial is the mode of existence in the case of all interesting philosophical proposals, Tarski himself was convinced (at least, when he published his results about truth) that he contributed to philosophy [24, pp. 266–267][3]:

> Its [that is, 25—J. W.] central problem—the construction of the definition of true sentence and establishing the scientific foundations of the theory of truth—belongs to the theory of knowledge and forms one of the chief problems in this branch of philosophy. I therefore hope that this work will interest the student of the theory of knowledge above all that he will be able to analyse the results contained in it critically and to judge their value for further researches in this field, without allowing himself to be discouraged by the appearance of concepts methods used here, which in places have been difficult and have not hitherto been used in the field in which he works.

The moral to be derived from the end of the above passage is that according to Tarski, formal (mathematical aspects) aspects of **STT** are indispensable for its proper comprehension (see also later Tarski's papers, namely [28] and [31]).

Tarski's hopes about possible interests of philosophers in **STT** were correct. Most philosophers, who oriented philosophy toward logic and used logical tools in philosophical investigations, welcomed his ideas; In fact, it is difficult to find today a serious monograph concerning the concept of truth which would not refer to Tarski's truth-definition. The philosophical significance of the semantics theory of truth was recognized very soon. According to Alfred Ayer [2, p. 116]

> Philosophically the highlight of the Congress [of the Scientific Philosophy in Paris in 1935—J. W.] was the presentation by Tarski of a paper which summarized his theory of truth.

Note that Ayer did not belong to the protagonists of the semantic definitions of truth.

Three important philosophers, namely Kazimierz Ajdukiewicz, Rudolf Carnap and Karl Popper, radically changed their earlier views under Tarski's influence (see [33]). Ajdukiewicz abandoned radical conventionalism which was among others a theory of language and meaning [1, p. 315]:

[2] Jean-Yves Girard is another critic of **STT** as a mathematical theory. See [8], section 2.3 of this book has "Tarksism" as its title and contains rather nasty comments on Tarski's on pp. 36–37, 213, 491–496, 499–500. In fact, Girard's criticism is also directed against philosophy behind **STT**.

[3] Page-references are to translations or reprints, if they are mentioned in References at the end of the present paper.

> The objection [...] communicated to me by Tarski in a conversation [...] seems to show that the concept of meaning is not definable in purely syntactical terms without the use of semantic terms in the narrower sense.

A similar point was made by Carnap [4, p. X]:

> Tarski, both through his book, and in conversation, first called my attention to the fact that the formal method of syntax and semantics must be supplemented by semantical concepts, showing at the same time that these concepts can be defined by means not least exact than those of syntax. Thus the present book owes very much to Tarski, more indeed than to any other single influence.

Briefly, Carnap, under Tarski's influence (or better, above all under Tarski's influence) passed from philosophy as logical syntax to philosophy as exact semantic analysis. There is no exaggeration, if we say that Tarski essentially contributed to semantic revolution in philosophy (see [34]).

Popper recalls [18, p. 322]:

> [...] I met Tarski in July 1934 in Prague. It was early in 1935 that I met him again in Vienna in Karl Menger's Colloquium [...] It was in those days that I asked Tarski to explain me his theory of truth, and he did so in a lecture of perhaps 20 min on a bench (un unforgotten bench) in the *Volksgarten* in Vienna. He also allowed me to see the sequence of proofs sheets of the German translation of his great paper on the concept of truth, which were than just begin sent to him from [...] *Studia Philosophica*. No words can describe how much I learned from all this, and no words can express my gratitude for it. Although Tarski was only a little older than I, and although we were, in those days, on terms of considerable intimacy, I looked upon him as the one man whom I could truly regard as my teacher in philosophy, I have never learn so much from anybody else.

How Tarski's ideas influenced Popper? Generally speaking, Popper began to defend realism in his approach to science, because he came to the conclusion that **STT** renewed the classical idea that truth consisted in the conformity between propositions and the objective reality. Thus, Tarski essentially contributed to the development of scientific realism.

These three examples together with Ayer's general assessment are perhaps the most spectacular single traces of the Tarski's influence on philosophy consisting in the full acceptance of his ideas. However, the philosophical role of **STT** is by no means limited to such measures. Almost every book (introductory or advanced) in semantics, philosophy of language or the history of analytic philosophy mentions this theory. Almost every discussion of such topics as the definition of meaning, semantic realism or scientific realism uses Tarski's ideas or at least alludes to them. Several important views in contemporary philosophy employ **STT**, for example, Donald Davidson's theory of meaning (see [6]) as based on truth-conditions or various semantic theories of induction (Carnap and his followers). Tarski's theory was more or less modified, like in [14] (Saul Kripke) or [9] (Anil Gupta, Nuel Belnap or replaced by other constructions as in [11] (Jaakko Hintikka); both modifications and replacements refer to **STT** as the solid starting point. There is no exaggeration in the statement that every post-Tarskian theory of truth (at least in analytic philosophy) is propter-Tarskian.

STT is also strongly criticized. Of course, it is not surprising that most non-analytic philosophers simply ignore the semantic definition of truth. Others regard it as a typical degeneration of the logical or computational (whatever it means) mind. I will not comment on such criticisms. However, I would like to explain why a discussion between philosophers belonging to various metaphilosophical camps is a very delicate matter. The main problem is that metaphilosophical options largely decide about substantial solutions.

Thus, if someone says that truth is entirely outside logic or semantics and its problem must be located in philosophical anthropology, there is a very little chance for a fruitful discussion between such controversies and a philosopher who believes in philosophy based on logical analysis. As a logical philosopher I do not say that other philosophy is wrong and has no value; I only indicate that, perhaps except explaining fundamental misunderstandings and disagreements, there is not very much to discuss. A consequence of this attitude which I regard as rational is this: it sounds as a restriction of criticism of **STT** to those? which arose within the analytical camp.

Returning to the past, Carnap with an astonishment noted [5, p. 61]:

> When I met Tarski again in Vienna in the spring of 1935, I urged him to deliver a paper on semantics and on his definition of truth at the International Congress for Scientific Philosophy to be held in Paris in September. I told him that all those interested in scientific philosophy and the analysis of language would welcome this new instrument with enthusiasm, and would be eager to apply it in their own philosophical work. But Tarski was very sceptical. He thought that most philosophers, even those working in modern logic, would be not only indifferent, but hostile to the explication of the concept of truth. I promised to emphasize the importance of semantics in my paper and in the discussion at the Congress and he agreed to present the suggested paper.
> At the Congress it became clear [...] that Tarski's sceptical predictions had been right. To my surprise, there was vehement opposition to even on the side of our philosophical friends. [...]. Neurath believed that the semantical concept of truth could not be reconciled with a strictly empirical and anti-metaphysical point of view. Similar objections were raised in later publications by Felix Kaufmann and Hans Reichenbach.

The point is that objections raised by the enemies of **STT** belonging to logical empiricism denied any philosophical significance of it. The criticism noted by Carnap assumed a very concrete philosophical basis, namely rather radical logical empiricism as far as the matter concerned the empirical basis of knowledge. Moreover, Neurath argued that the semantic account of truth reintroduces a very bad metaphysics into philosophy.

However, several authors argued later for the philosophical sterility of **STT** from a quite general perspective. For instance, Max Black (see [3]) tried to show that Tarski's theory of truth, although correct from the purely logical point of view, is neutral in fact with respect to old philosophical controversies over the concept of truth. A very radical criticism against **STT** was raised by Hilary Putnam (see Putnam [19–21]) who maintained that this theory is completely mistaken, which—although does not cause troubles for mathematical logic, yet fatally deceives philosophers (see [35] for defending **STT** against Putnam's objections). Another criticism was advanced by from the point of view of anti-realism based on constructive (intuitionistic) logic, for instance. Michael Dummett (see [7]). According to objections forwarded by anti-realists of Dummett's brand, meaning-conditions should be defined as related to assertibility, but not as associated with truth in a sense of Tarskian semantics.

The above brief and selected survey focuses on positive as well as negative influences of Tarski's ideas. Both axes reportedly show that **STT** belong to the contemporary philosophical equipment, at least in the camp of analytic philosophy and those currents, even inside the continental school, provided that they are ready to discuss various issues with colleagues working with use of logical devices. Some philosophers try to continue Tarski's ideas and develop them according to new challenges, for example, advanced by anti-realism, whereas other are at least stimulated by **STT** in their philosophical investigations, even resulted in alternative semantic accounts. One should remember that, for instance, anti-realistic semantics is still a semantic theory. In fact, Tarski's talks (see [26, 27]) saved philosophical semantics in general, not only his version of it. Although in philosophy many unexpected happenings took place, the return to pre-semantic ear, for instance, to the dominance of the syntactic approach in the style of early logical empiricism seems very unlike. This general assessment of the role of **STT** in contemporary philosophy should be illustrated by more specified data. However, before focusing on a philosophical examination of **STT**, I will outline its formal shape.[4]

At the beginning of the present paper (see also footnote 1) I distinguished **STT** as a mathematical (logical) theory and a philosophical doctrine. I deliberately used the terms "theory" and "doctrine" as contrasts in their meaning. However, I abandoned this way speaking when I passed to a closer examination of tasks that any philosophical account (it is another convenient label) of truth has to confront (I also employ the expression "truth-theory" ("theory of truth") as referring to philosophical theories of truth. Now there appears a question of the ambiguity of the word "theory". Typically (at least in metamathematics) a theory is a set of sentences, in particular, definitions closed under a selected consequence operation. **STT** as a logical construction is a theory in this sense. It is based on definitions and lemmas which enable us to prove in the exact mathematical manner various properties of the set of true sentences, for instance, that this set forms maximally consistent set. It is also possible to embed **STT** into the weak second-order arithmetic with the axiom of arithmetic comprehension and assess its logical complexity (see [10]). **STT** as a philosophical theory belongs to different order of things. By a philosophical theory I understand a body of interconnected statements related to a set of philosophical and metaphilosophical constraints. For instance, materialism, idealism, rationalism or empiricism are such theories.

Some basic preliminary intuitions are as follows. Consider two stocks of ideas (for simplicity I limit informal as well as formal explanations for monadic formulas, that is, of the type $P(\ldots)$, where an individual variable or an individual name occurs in the place of dots occurring in the symbol $(\ldots)$); the letter **U** represents the assumed universe of discourse; it is convenient to claim that **U** is infinite):

(I) (General case): open formulas, satisfaction by an object from a given set **U**, non-satisfaction by an object from the complement of **U**;
(II) (Special case): closed formulas (sentences), a special case of satisfaction relatively to a given set of objects belonging to a given set **U**, satisfaction by no object from a given set **U**.

Informally speaking, open formulas are neither true nor false, but satisfied or non-satisfied by some objects. For instance, the number 2 satisfies the condition "x is a prime number", but the number 4 does not satisfy this condition. Yet we have an intimate connection of truth and satisfaction. Some substitutions convert open formulas into true sentences, but other ones—into falsehoods. This heuristics suggests to treat truth, resp. falsity, as

[4] My presentation of **STT** uses rather contemporary settings than original Tarski's version given in [24, 25] and [29]). The main issue concerns the explicit role of the concept of model and working with first-order languages.

a special case of satisfaction, resp. non-satisfaction. Since a (declarative) sentence is traditionally defined as a sentence which is true or false, we have a hint for defining truth.

Let *A* be a sentence (closed formula, that is not having free variables). Now, it is convenient to use sequences of objects, not objects as such. One can prove that *A* is either satisfied by all sequences of objects from **U** or is not satisfied by any sequences of objects (satisfied by no such sequence object). Tarski proposed to define truth as satisfaction by all sequences of objects and falsity as satisfaction by no sequence of objects. The above condition for truth is equivalent to saying that a sentence is true, if it is satisfied by at least one infinite sequence of objects or by the empty sequence of objects. Technically, since formulas can be of an arbitrary finite length, it is convenient to introduce infinite sequences of objects in order to obtain a general scheme for all possible syntactic cases. Moreover, because **U** is always associated with a structure of the type $\mathbf{M} = \langle \mathbf{U}, \mathbf{P} \rangle$, where **P** is collection of predicates (**M** is a model), we have the following definition :

(**TrDef**) A sentence *A* is true in a model **M** if and only if it is satisfied by all infinite sequences of objects from **U** (or at least by one such sequence or the empty sequence; otherwise *A* is false in **M**).

Thus, truth is defined in the above way is an outcome of an elegant analogy displayed by (II). In particular, (**TrDef**) satisfies Tarski's claim that a satisfactory definition of truth should logically entail the equivalence (at the moment I use its naive form)

(**TE**) A sentence *A* is true if and only if *A* (in symbols, $\mathbf{T}A \Leftrightarrow A$),

for any *A* belonging to a language **L**. This requirement is called the convention **T** (**CT**). (**TE**) is frequently called the **T**-scheme. (**CT**) establishes the condition of the material adequacy for a truth-definition, also for (**TrDef**).

The letter **L** in **CT** serves a new parameter in the entire construction. Hence (**TrDef**) should be completed as

(**TrDef***) A sentence *A* of a language **L** is true in a model **M** if and only if *A* is satisfied by all infinite sequences of objects from **U** (or at least one such sequence or the empty sequence; otherwise *A* is false in **M**).

The relativisation to **L** is associated with the semantic paradoxes, especially with the Liar antinomy. It is generated by (**TE**) by substituting "this sentence is false" for *A*. In order to solve the difficulty, (**TE**) is transformed into

(**TE***) $\mathrm{Tn}(A) \Leftrightarrow [A]$,

where n(*A*) is a name of the sentence and [*A*] refers to a retranslation of A into a metalanguage. Let **L** be German, but English serves as a metalanguage. We consider a German sentence "Schnee ist weiss" and say (it is a special case of (**TE***) that the sentence "Schnee is weiss" of German is true if and only if snow is white. The expression "Schnee ist weiss" is a name of the German sentence occurring inside quotes, but the right hand of the equivalence in question is a translation of the German sentence. Consequently, if A belongs to an object-language **L**, (**TE***) and (**TrDef***) must be formulated in the metalanguage **ML**. Generally speaking, a semantic theory for **L** should be formulated in **ML** in order to avoid semantic paradoxes. It is very important to see that (**TE**) is not a truth-definition. **CT** states that every instance of the **T**-scheme must be derivable from

(**TrDef***), but, except the case, when **L** is finite (languages are sets of sentences), we have no equivalence between (**TrDef***) and (**TE***).

To proceed more formally, consider a formalized language **L** for which truth (more precisely a set of true sentences) is defined. Due to arithmetization (or other similar technique), the syntax of **L** can be represented in **L** itself. However, the Tarski undefinability theorem (the set of true sentences of arithmetic of natural numbers in not arithmetically definable; **UT** for brevity)) shows that semantics of **L** is not fully expressible in **L** itself. In order to define semantic relations, we need to use **ML** which has a greater expressive power than **L**. Perhaps the most important observation is that **ML** remains partly informal. Even if we formalize **ML**, we must use **MML** and the story reappears. Generally speaking, the hierarchy $\mathbf{L_1}$, $\mathbf{L_2}$ (= $\mathbf{ML_1}$), $\mathbf{L_3}$ (=$\mathbf{ML_2}$), ..., $\mathbf{L_n}$ (=$\mathbf{ML_{n-1}}$), ... of languages has the following property: if $\mathbf{L_k}$ is formalized, $\mathbf{L_{k+1}}$ $(1 \leq k \leq n)$ has some informal features with respect to $\mathbf{L_{k+1}}$ (= $\mathbf{ML_k}$). Thus, there occurs a necessary connection between formal and informal aspects of **STT**. Yet one point requires an explanation. According to Tarski, the truth-definition for **L** generated by **STT** can be given in the syntax of **ML**. Does it mean that semantics is reducible to syntax? Not at all. Tarski worked in the paradigm of logic on which set theory belonged to logic. In fact, the definition of truth via satisfaction proceeds in the set-theoretical framework. Thus, we can say at most that defining semantic relation for **L** assumes a set theory. Putting this fact into the contemporary fashion, if **T** is a first-order theory (principally every deductive theory can be expressed in the first-order language), its semantics can be constructed in weak second-arithmetic with the arithmetic axiom of comprehension, employed as a metatheory **MT** of **T**. However, this does not mean that we reduce semantics of **T** to **T**-syntax. Otherwise speaking, if **T** is completely formalized, **MT** does not admit such a treatment. As it was pointed out above, it is possible to stay with first-order object languages and the weak-second order arithmetic as the metatheory.

What about constraints of a successful theory of truth? Omitting earlier proposals answering this question, I recall conditions stated by Bertrand Russell. According to him (see [23]), any theory pretending to be the satisfactory account of the concept of truth must conform to:

(A) the theory of truth must also explain the nature of falsehood;
(B) truth must be taken as a property of beliefs;
(C) truth is an external relation of belief to something existing outside them.

However, these conditions say too much on the one hand, but too little on the other. Too much, because (B) and (C) exclude some important ideas. The requirement (B) ignores other accounts of truth-bearers, but (C), selecting the correspondence theory as proper, rejects theories which consider truth as consisting in relations between judgments without making any reference to the external world; in particular, (C) excludes the coherence theory. On the other hand, Russell's constraints are too weak, if he intended to favour the correspondence theory. For example, if one says that truth consists in evidence of beliefs, one also offers an account which satisfies (C). In general, the Russellian conditions are unfair to the richness of problems usually investigated by truth-theories that are known from a very history of the subject.

Thus we need a more complex scheme of any philosophical truth-theory which intends to be historically faithful. Without ambition to completeness (the list is filtered by **STT**), I mention the following problems which should be touched by any philosophically reasonable. By being philosophically I do not mean 'correct', but 'deserving an attention in the world of philosophy') truth-theory in philosophy:

1. What are the bearers of truth?
2. What are initial intuitions associated with a given truth-definition?
3. How to define truth?
4. How is truth related to logic?
5. If we classify truth-bearers into true and false, is this division exhaustive and disjoint (are there values apart from truth and falsehood or perhaps are there truth-falsehood-value gaps)?
6. Is this division stable, that is, do at least some truth-bearers sometimes change their truth-values (briefly: is truth relative or absolute)?
7. What is a truth-criterion and what about the relations of truth-criteria and truth-definition?
8. What is the relation of a particular truth-theory to its rivals?
9. How a given truth-theory can be defended against various objections?
10. What is the relation of truth to other philosophical problems?

As we see, there is a lot to do for a theory of truth. I will try to show how **STT** of truth is related to these questions, or at least to some of them.

(Ad1) **STT** assumes that truth-bearers are sentences in the syntactic sense. We have several other possibilities (see [22] for a survey). Sentences? Propositions? Statements? Judgments? These entities can be either linguistic units or objects expressed by linguistic utterances. By contrast, concepts are not truth-bearers, contrary to Hegelians. To have a convenient label, we can say that according to **STT** entities qualified as true or false are of the propositional syntactic category. This ways of speaking has nothing to do with the question of the ontological nature of propositions, for instance, as abstract objects.

Tarski himself chose sentences as entities on which truth is predicated. But there is an additional very important point [25, pp. 166–167]:

> It remains perhaps to add that we are not interested here in 'formal' languages in sciences in one special sense of the word 'formal', namely sciences to the signs and expressions of which no material sense is attached. For such sciences the problem here discussed [the problem of truth] has no relevance, it is not even meaningful. We shall always ascribe quite concrete and, for us, intelligible meanings to the signs which occur in the language we shall consider. The expressions which we call sentences still remain sentences after the signs which they occur in have been translated into colloquial language.

Thus a language **L** for which **STT** applies is always interpreted, even if it is formalized. Consequently an interpretation of **L** always precedes definitions of semantic concepts including truth. Thus we arrived at the problem of how 'formal' is related to 'formalized'. The answer is that formal languages do not need to be equipped, contrary to formalized languages.

A common misunderstanding of Tarski's views consists in attributing to him the opinion that **STT** applies to formal languages only. This mistake neglects what Tarski explicitly explained (see essays in [16] for a discussion), that truth-bearers are correct syntactic units of the propositional category having meaning. It does not mean that Tarski's views about language and meaning have no weak points. In particular, he did not define the concept of meaning. In fact, he intentionally avoided this question and deliberately preferred to speak about interpreted languages as semantic items. For Tarski, the concept of language was clearer than the concept of meaning (see [17] for an extensive presentation of Tarski's philosophy of language). Yet **STT** does not seem to be especially dependent on a particular theory of meaning. Another controversial point concerns **STT** and natural language. Tarski regarded natural language as universal, and thereby generating semantic paradoxes. Roughly speaking, natural languages do not block self-referential use of semantic predicates. The Liar sentence "This sentence is false" illustrates this fact, as it attributes the property of being false to itself. In Tarski's view, dividing natural language into strata (levels), like the object language, the meta-language, the meta-meta-language, etc. is inconsistent with its naturalness and universality. On the other hand, it is quite legitimate to define truth for fragments of natural language (see [28]). Finally, Tarski's view on the relation of truth and meaning differs from that of Davidson. Whereas the latter maintained that truth-conditions define the meaning of sentences, the former considered meaning as prior to truth (see [38]).

(Ad2) Tarski always stressed that his definition follows intuitions of Aristotle. He was influenced by Stagirite himself and by his teachers (see [16, 17]). However, Tarski's statement in his particular writings on truth differ. In [24, p. 152] he says:

> [...] in this work I shall be concerned exclusively with grasping the intentions which are true, contained in the so-called *classical* conception of truth ('true—corresponding with reality), in contrast, for example, with the *utilitarian* conception ('true—in a certain respects useful').

Further (p. 155), he adds that

> true sentence is one which says that the state of affairs so and so, and the state of affairs is indeed so and so.

However, the Polish original text has no exact counterpart of the expression "state of affairs". On the other hand, English (but not German) translation quotes famous passage from the Aristotle's *Metaphysics*:

> To say of what is that it is not, or what is not that it is, is false; while to say that of what is that it is, or what is not that it is not, is true.

This quotation also appears in [28, p. 667] and is explained by statements "The truth of a sentence consists in the agreement with (or corresponding to) reality" and "A sentence is true if it designates an existing state of affairs". Tarski's comment (p. 267) is as follows:

> However all these formulations can lead to various misunderstandings, for none of them is sufficiently precise and clear (though this applies? much less to the original Aristotelian formulation); at any rate, none of them can be considered a satisfactory definition of truth.

In [30, pp. 402–403] the above formulations are repeated together with similar critical remarks. Tarski subsequently says (p. 403):

> The conception of truth which is found in its expression in the Aristotelian formulation (and in related formulations of some recent origin) is usually referred to as the *classical*, or *semantic*, conception of truth. By semantic, we mean the part of logic which, loosely speaking, discusses the relations between linguistic objects, e.g. sentences and what is expressed by these objects; the semantic character of the term "true" is clearly revealed by the explanation offered by Aristotle and by some formulations which will be given in our further discussion. One speaks sometimes of the correspondence theory of truth as the theory based upon the classical conception.

To sum up, Tarski, at the beginning, identified the classical and correspondence theory of truth, but later he expressed greater reservations with respect to explanations via expressions, like "agreement" or "correspondence" than to Aristotle's original formulation.

If we assume that **STT** follows Aristotle's intuitions, are they captured by (**TE***) of (**TrDef***)? The answer that the former seems fairly correct and justified by Tarski's own explanations. So the example 'the sentence "snow is white" is true if and only if snow is white', points out that because the sentence in question says that snow is white and it is so and so as this sentence says, it (the sentence) is true. What about the intuitive content of (**TrDef***)? We have two options; first, having some justifications in Tarski's explanations that is a mathematical trick, and second, that the official definition brings some intuitions. First of all, sequences of objects cannot be identified with facts. Moreover, the satisfaction by the empty sequence appears as an artificial construction (see [24, p. 195]). On the other hand, if the semantic truth-definition is a special case of the definition of satisfaction and the latter is based on explicit intuitions, it suggests that perhaps some intuitions are behind (**TrDef***) as well. I am inclined to take the last option; that whether an open formula is satisfied or not by an object depends of valuation of free variables. Such valuations are irrelevant in the case of sentences. Consequently every infinite sequence of objects can be ascribed to bound variables (note that individual constants can be eliminated by identity and existential quantification). The same can be expressed by saying that the empty sequence satisfies a sentence. What remains? The answer is that being true depends on how **L** is interpreted and, metaphorically speaking, how things are in **M** associated with **L**. And it precisely expresses what is established by the **T**-scheme. Informally speaking, truth depends on the domain which sentences to say about. (see [43]) to use the name "weak correspondence" or "semantic correspondence" in the case of (**TrDef***) and (**TE***) as something different from "strong correspondence", used, for example, by Russell in his definition of truth (the structure of a judgement or a proposition) which corresponds with the structure of a corresponding fact provided that this judgment is true. Thus I will consider **STT** as the classical truth-theory based on the weak concept of correspondence.

(Ad. 3) Tarski defined truth by a single formula (the definition satisfaction is recursive). He considered (see [26]) introducing truth by axioms, but rejected this possibility for philosophical reasons. More specifically, he was afraid of a criticism on the side of physicalism. This worry was associated with his scepticism mentioned by Carnap (see above). This motivation is presently completely historical. Tarski himself mentioned that taking all instances of **T**-scheme as axioms could be regarded as an axiomatization of the set of true sentences. Yet he was fully aware that such procedure would be trivial and leading to the infinite collection of axioms. Today, the axiomatization of the concept of truth is commonly applied (see [10, 13]) and also proposed in order to avoid semantic paradoxes. I will not enter into this issue.

Tarski's way has an important consequence because of his undefinability theorem. Assume, what is natural, that the collection **TRUTH** of all truths is infinite. By **UT** (see below), **TRUTH** is not definable by resources conceptually available within it. Yet saying that **TRUTH** exists appears to be philosophically tempted. The only way out admissible by set theory consists in conserving **TRUTH** as too big set (Zermelo-Fraenkel system), a class as distinct from sets (Bernays-Gödel-von Neumann) or a category. All these outcomes are formally correct but lead to not quite pleasant consequences, at least for philosophers who like having something to say on the set of all truths. However, set theory and **UT** seriously limit such theoretical ambitions. We can eventually say that **TRUTH** constitutes consistent deductive systems, which has no finite axiomatization, even by schemes. On the other hand, **TRUTH** is not compact, which means that although its every finite subset has a model, the collection of all truths has no model. Consequently, the collection **MOD** of all models is not a set as well. The medieval theory of transcendentals assumed that truth coincides with being and that *ens* (and other *transcendentalia*, for instance, *verum*, that is truth), *omnia genera transcendit*. In a sense, considerations about **TRUTH** and **MOD** (it can represent the being) justify the medieval intuition about the transcendentality of truth and being (see [40] for a more closer analysis). It is rather an unexpected application of **STT**.

Ad (4 and 5) Tarski proved that **STT** implies bivalence, which is the conjunction of metalogical principles of excluded middle and (non)contradiction. This means that this theory is inherently associated with classical logic. However, Tarski's proof is sometimes criticized contested as circular, as assuming classical logic in metatheory. It is possible to demonstrate that the above result can be constructively achieved (see [36]); the argument employs the fact that truth can be defined as satisfaction by the empty sequence). Now, the problem arises whether construction a la Tarski can be realized in the case of non-classical logic, in particularly, one rejecting the presumption that every sentence is either true of false. Many-valued logics and logics with truth-value gaps provide standard examples. Paraconsistent logic, intuitionistic logic or quantum logic are further examples. Clearly, more or less modified Tarskian ideas have applications for non-classical logics but according to my knowledge no general results are available. For instance, some constructions use partial models also for excluding paradoxes, but only special cases are available.

Consider **T** as a modality. We read **T***A* as "it is true that *A*". We have the following formulas $\mathbf{T}A$, $\mathbf{T}\neg A$, $\neg\mathbf{T}\neg A$, $\neg\mathbf{T}A$, $\mathbf{T}A \vee \mathbf{T}\neg A$, $\neg\mathbf{T}\neg A \wedge \neg\mathbf{T}A$. These formulas satisfy logical modal principles organized by a generalized logical square, for instance $\mathbf{T}A \Rightarrow \neg\mathbf{T}\neg A$. The conjunction $\neg\mathbf{T}\neg A \wedge \neg\mathbf{T}A$ is logically possible (consistent) and opens room for other (than truth and falsity) logical values or truth-value gaps. We have also the principle $\mathbf{T}A \Rightarrow A$, but its converse does not hold as a purely logical rule. Thus, we can add the formula $A \Rightarrow \mathbf{T}A$ and obtain the **T**-scheme as a new theorem, but it is not forced by logic. Adding the **T**-scheme results in cancelling $\neg\mathbf{T}\neg A \wedge \neg\mathbf{T}A$ as a possibility. Moreover, $\mathbf{T}A$, $\neg\mathbf{T}\neg A$, A as well as $\mathbf{T}\neg A$, $\neg\mathbf{T}A$, $\neg A$ become equivalent, and the same concerns $\mathbf{T}A \vee \mathbf{T}\neg A$, $\mathbf{T}A \vee \neg\mathbf{T}A$, $A \vee \neg A$. Since we can now interpret $\mathbf{T}\neg A$, $\neg\mathbf{T}A$, $\neg A$ as expressing "*A* is false", the principle of bivalence is valid in the logic of truth with the **T**-scheme. This reasoning shows that **T**-scheme favours classical logic. Since ¬(**TE***) ⇒ ¬(**TrDef***), rejecting **T**-scheme results in rejecting the semantic truth-definition. Yet, this conclusion does not preclude a revision of Tarski's definition for needs of particular logical systems. For instance, $\mathbf{T}A \Rightarrow A$ is a theorem of classical modal theory of truth, but a paraconsistent logician can accept its negation, that is the formula $\mathbf{T}A \wedge \neg A$, but its truth-condition requires a modification of (**TRDef***). Incidentally, modal analysis of truth shows that the **T**-scheme is not a logical tautology, because the formula $A \Rightarrow \mathbf{T}A$ is not logically valid. If one were to say that modal logic is not a logic in the proper sense, we could still point out that there is a difference in the status of both components of the **T**-scheme: the formula $\mathbf{T}A \Rightarrow A$ is a formal modal theorem, but its converse is not.

(Ad 6) The classical concept of truth is commonly considered as absolute, that is, if *A* is true such eternally (for ever) and sempiternally (since ever). On the other hand, (**TrDef***) indexes truth by **L** and **M**. Does this relativisation deprive truth of its absolute character? This question is connected with such issues as bivalence, logical determinism or many-valued logic. Without entering into details concerning this fairly complex stock of ideas, let me somehow dogmatically suggest (details are in [40, 43]) that we can model-theoretically prove that truth is eternal if and only if it is sempiternal. If so, the classical theory of truth in the semantic setting can be considered as associated with the absolute concept of truth. Even if this conclusion encounters reservations, the possibility of analysing the absolutism/relativism controversy within the philosophical theory of truth via (**TrDef***) is a remarkable fact.

(Ad 7) Clearly (**TrDef***) is a-criterial. This means that the definition in question does not generate any truth-criterion, though it says what truth is. If mathematics is taken into account, proof can be regarded as a measure of truth. However, there arises a problem. Let the symbol **Pr** denote the provability operator. By the Löb theorem, we have $\mathbf{Pr}A \Rightarrow A$, a theorem very similar to $\mathbf{T}A \Rightarrow A$. But, due to the first incompleteness theorem, the formula $A \Rightarrow \mathbf{Pr}A$ cannot be consistently added to the provability logic. Hence, there is no counterpart of (**TE***) with **Pr** instead **T**, that is, the scheme $\mathbf{Pr}A \Leftrightarrow A$ and we must conclude that proof is not a complete truth-criterion even in mathematics. This fact can motivate various ways out, for instance, modifying the concept of proof (every true mathematical assertion can be proved in a formal system; this assertion does not contradict the incompleteness theorem) or replacing truth by proof, eventually with additional constraints, for instance, that proofs must be constructive. However, such proposals are restricted to mathematics. Another suggestion is like this. Consider an open formula Px. It can be transformed to the true sentence Pa via substituting x by the term a denoting an object which has a property P. Another way is to go through from Px to $\exists x\, Px$. Both strategies require some empirical or deductive steps based on some criteria. And these criteria are truth-criteria. Generally speaking, truth-criteria consist of procedures which justify satisfaction of open formulas by some objects.[5] Note that the proposed criterion does not work for satisfaction by all sequence of objects.

(Ad 8) Tarski grew up in the tradition of division of truth-theories into the classical theory and so-called non-classical theories of truth, namely the evidence theory (A is true, if is evident), the coherence theory (A is true, if it can be embedded into a coherent system without destroying its coherence), the common agreement theory (A is true, if specialists agree about its correctness) and the utilitarian theory (A is true, if A is useful). The non-classical theories are criteria, because they appeal to procedures assuring that something is true.

Tarski himself mentioned the last definition (see above) and the coherence account (see [30, p. 403]). He considered them as lacking of precision and did not discuss them as serious alternatives for **STT**. However, it seems that the coherence theory can be discussed with help of some logical ideas. Let us agree that consistency is a component of coherence. Thus, we defined a system **S** as coherent if and only if **S** is consistent and satisfies some additional requirements, for instance is comprehensive, has an empirical support etc. Anyway, we have $\mathbf{S} \in \mathbf{COH} \Rightarrow \mathbf{S} \in \mathbf{CONS}$. It means that consistency is a necessary condition for coherence. The provability operator **Pr** satisfies the condition (*) $\mathbf{Pr}(A \Rightarrow B) \Rightarrow (\mathbf{Pr}A \Rightarrow \mathbf{Pr}B)$. By definition, $\mathbf{Pr}(\mathbf{S} \in \mathbf{COH} \Rightarrow \mathbf{S} \in \mathbf{CONS})$. This and the condition (*) gives $\mathbf{Pr}(\mathbf{S} \in \mathbf{COH}) \Rightarrow \mathbf{Pr}(\mathbf{S} \in \mathbf{CONS})$. By the rule contraposition, we obtain $\neg\mathbf{Pr}(\mathbf{S} \in \mathbf{CONS}) \Rightarrow \mathbf{Pr}(\mathbf{S} \in \mathbf{COH})$. Finally, the second incompleteness theorem, asserting

[5] This suggestion was made by Anna Kanik, a former student of mine.

that consistency is not generally provable, suggests that the coherence is not a general criterion of truth. Thus, at least under classical logic, there are truths, which satisfy (**TrDef***) but cannot be tested by coherence. On the other hand, the above argument does not suggest that there are truths not subjected to any justification. Yet, no absolutely universal criterion of truth based on exclusively deductive resources seems to be possible. By analogy, one can argue that any criterion of truth appealing to this or that kind of justification, works in the concrete circumstances and cannot define the concept of truth in its full generality. If so, **STT** (or the classical theory) is the only known account of truth, which is actually universal.

Another issue involving the relation between various truth-theories concerns substantial and minimalist account (see [15] for an analysis of conceptions of truth from the point of view of this axis). The latter approach (the redundancy theory, the deflationary theory, etc.) reduces the truth-definition to the **T**-scheme. Under this view, **STT** is a minimalist theory. Tarski himself (see [27, p. 682–683]) discussed this question. His counterexample was the sentence "All consequences of true sentences are true". To continue Tarski's analysis let us assume that the minimalist translation of the above assertion runs "for every *B*, if *A*, and *B* is a logical consequence of *A*, then *B*", we still need to explain the ground of truth in the case of universal closure (that is, sentences with the universal quantifier in front) and the meaning of the predicate "is a logical consequence". Thus, **T**-scheme does not justify to assert that all consequences of true sentences are true. There are much more complicated cases, for instance, the sentence "There exist true but not provable sentences", which seems to be not subjected to a minimalist translation. If so, **STT** is essentially richer than any minimalist theory of truth.

(Ad 9) I will address in this section three objections stated by Franz Brentano against the classical theory and try to show that **STT** meets them successfully (see [32]). Firstly, the concept of correspondence is obscure and cannot be explained satisfactorily. More precisely, in order to establish what a truth-bearer corresponds to reality, one must compare the former with the latter. But it is impossible, due to relata of such a comparison. However, this objection applies to the strong notion of correspondence, not to its weak form. The second objection is more serious. Assume that we define truth by a definition **D**. Yet **D** is a sentence. In order to have a good definition, **D** must be true. Now, the definition is either circular (if it uses itself) or falls into the *regressum ad infinitum*, because in order to formulate **D**, we must appeal to **D'** related to **D**, etc. Thirdly, the concept of correspondence does not explain truth of negative sentences.

How things are in **STT**, relatively to the second objection (see also [28, pp. 680–681])? Obviously, the answer depends on the relation of **L** (for which truth is defined) to **ML** (in which truth is defined). Tarski observed that the latter must be essentially richer than the former. Using a more contemporary way of speaking, it holds for semantics of **L**, **which** is not fully expressible in its syntax. Consequently, the expressive power of **ML** must be greater if the concept of truth is to be defined. In particular, **ML** has to have resources to define the concept of satisfaction. According to Tarski, (**TrDef***) is expressible in the syntax of **ML**. This explanation is slightly misleading, because the satisfaction is defined in set theory. The crucial point is that (**TrDef***) for the concepts of truth and of satisfaction is defined recursively. Although one should assume that the metatheory of truth is consistent, consistency is a syntactic property. Thus circularity does not occur in the entire procedure. The same concerns the *regressum ad infinitum*.

However, another interesting point appears. Clearly, if **L** is formalized, **ML** is not, at least not entirely. Consequently, the content of **ML** exceeds the content of **L**. One can observe that this phenomenon leads to the situation that semantic properties of poorer and thereby less problematic theories are defined in richer and thereby more problematic conceptual systems. In the case of **STT** this circumstance is somehow limited by the mentioned fact that the weak second-order arithmetic is sufficient for (**TrDef***), but the phenomenon in question is very intriguing from the philosophical point of view. The problem of negative sentences has a simple solution in **STT**, because they are true (or false) under the same definition as positive ones.

(Ad 10) According to Tarski [28, p. 686]:

> [...] We may accept the semantic theory of truth without giving up any epistemological attitude we may have had; we may remain naive realists, critical realists or idealists, empiricists or metaphysicians — whatever we were before. The semantic conception of truth is completely neutral toward all these issues.

These words seem to block any serious involvement of **STT** into traditional philosophical debates and controversies. Two remarks are in a row here. Firstly, if we look at the stock of terms used in the above passage, we can ask whether Tarski's evaluation might be generalized. My impression is that terms "naive realists", "critical realists", "idealists" or "empiricists" refer to views concerning the philosophy of perception; even "metaphysicians" can be taken as referring to philosophers speaking about the reality of what is perceived. Secondly, even if one claims that Tarski employed the labels in question as exemplifications of his general and somehow negative attitude to philosophy, we should ask whether he was right.

Independently of Tarski's intentions, it is easy to give an example of a philosophical problem closely related to **STT**, namely the semantic realism/semantic anti-realism debate. Generally speaking, (semantic) realists use **STT** but (semantic) anti-realists reject this account to truth (see [37]). This issue concerns the mutual relation of the condition of truth and condition of assertibility. Generally speaking, the realist says that the meaning of a sentence (**MS**) is given by its truth-conditions (**TC**), whereas the anti-realist argues that **MS** is given by its assertibility-conditions (**AC**). Thus we have two equalities:

(i) $\mathbf{MS} = \mathbf{TC}$;
(ii) $\mathbf{MS} = \mathbf{AC}$.

However, (i) and (ii) are still too vague. In fact, we should transform (i) and (ii) into

(iii) $(\mathbf{MS} = \mathbf{TC}) \wedge (\mathbf{TC} > \mathbf{AC})$;
(iv) $(\mathbf{MS} = \mathbf{AC}) \wedge (\mathbf{TC} = \mathbf{AC})$,

respectively. In fact, the realist says that truth-conditions exceed assertibility-conditions but the anti-realist identifies truth-conditions with the assertibility conditions.[6] How does **STT** work here? It justifies (iii), but refuses (iv). If, as Dummett maintains, the conditions of assertibility are governed by intuitionistic logic, it does not generate sufficient and necessary conditions for asserting any mathematical sentence. The point is that the

[6] I model these formulas on the earlier mentioned debate between Davidson (realism) and Dummett (anti-realism). My own view (also expressed in [38]) is that meaning of a sentence exceeds its truth-conditions. However, the points (iii)–(iv) suffice for further analysis.

incompleteness theorem constructively holds for the Heyting arithmetic, that is, the Peano arithmetic based on intuitionistic logic. If so, the anti-realist cannot say that there are true, but unprovable sentences, but the realist can do so by appealing to **STT**.

Although I accept the semantic theory of truth as *the* correct account of the concept of a true sentence, I am very far from saying that its philosophical uses are unproblematic and the only correct. For instance, my analysis of the issue of realism/anti-realism should be taken as an analysis of what follows, if **STT** is assumed. My main intention in the present paper consists in demonstrating that Tarski's semantic ideas are not philosophically sterile.

References

1. Ajdukiewicz, K.: Zagadnienie empiryzmu a koncepcja znaczenia [Empiricism and the Problem of Meaning]. Studia Filozoficzne **1**(36), 3–14 (1964) (Eng. tr. in: Ajdukiewicz, K., The Scientific World-Perspective and Other Essays, 1930–1963, pp. 306–319. D. Reidel, Dordrecht (1978))
2. Ayer, A.J.: A Part of My Life. Collins, London (1977)
3. Black, M.: The semantic definition of truth. Analysis **9**, 49–63 (1948) (Reprinted in: Black, M., Language and Philosophy, pp. 89–107. Cornell University Press, Ithaca (1949))
4. Carnap, R.: Introduction to Semantics. Harvard University Press, Cambridge (1942)
5. Carnap, R.: Intellectual autobiography. In: Schilpp, P. (ed.) The Philosophy of Rudolf Carnap, pp. 1–84. Open Court, LaSalle (1963)
6. Davidson, D.: Truth and Meaning. Synthese **17**, 304–323 (1967) (Repr. in: Davidson, D., Inquiries into Truth and Interpretation, pp. 65–75. Oxford University Press, Oxford (1984))
7. Dummett, M.: The Logical Basis of Metaphysics. Duckworth, London (1991)
8. Girard, J-Y.: The Blind Spot. Essays on Logic. European Mathematical Society, Zürich (2011)
9. Gupta, A., Belnap, N.: The Revision Theory of Truth. The MIT Press, Cambridge (1993)
10. Halbach, V.: Axiomatic Truth Theories. Cambridge University Press, Cambridge (2011)
11. Hintikka, J.: The Principles of Mathematics Revisited. Cambridge University Press, Cambridge (1996)
12. Hodges, W.: Truth in a structure. Proc. Aristot. Soc. **86**, 135–151 (1985–1986)
13. Horsten, L.: The Tarskian Turn: Deflationism and Axiomatic Truth. The MIT Press, Cambridge (2011)
14. Kripke, S.: An outline of a theory of truth. J. Philos. **72**, 690–716 (1975) (Repr. in: Kripke, S., Philosophical Troubles. Collected Papers, v. 1, pp. 75–98. Oxford University Press, Oxford (2011))
15. Künne, W.: Conceptions of Truth. Oxford University Press, Oxford (2005)
16. Murawski, R., Woleński, J.: Tarski and his precedessors on truth. In: Patterson, D. (ed.) New Essays on Tarski and Philosophy, pp. 21–43. Oxford University Press, Oxford (2008)
17. Patterson, D.: Alfred Tarski Philosophy of Language and Logic. Palgrave Macmillan, Hampshire (2018)
18. Popper, K.: Objective Knowledge: An Evolutionary Approach. Clarendon Press, Oxford (1972)
19. Putnam, H. (ed.): Do true assertions correspond to reality? In: Mind, Language and Reality. Philosophical Papers, vol. 2, pp. 70–84. Cambridge University Press, Cambridge (1975)
20. Putnam, H.: On truth. In: Cauman, L., Parsons, C, Schwartz, R. (eds.) How Many Questions? Essays in Honour of Sidney Morgenbesser, pp. 35–56. Hackett, Indianapolis (1983) (Repr. in: Putnam, H., World and Life, pp. 315–329. Harvard University Press, Cambridge (1994))
21. Putnam, H.: On comparison of something with something else. New Lit. Hist. **17**, 61–79 (1985) (Repr. in: Putnam, H., World and Life, pp. 330–350. Harvard University Press, Cambridge (1994))
22. Rojszczak, A.: From the Act of Judging to the Sentence. The Problem of Truth Bearers from Bolzano to Tarski. Springer, Dordrecht (2005)
23. Russell, B.: The Problems of Philosophy. Wiliam and Norgate, London (1912)

24. Tarski, A.: Pojęcie prawdy w językach nauk dedukcyjnych [The Concept of Truth in Languages of Deductive Sciences]. Towarzystwo Naukowe Warszawskie, Warszawa (1933)
25. Tarski, A.: Der Wahrheitsbegriff in den formalisierten Sprachen. Stud. Philos. **I**, 53–198 (1935)
26. Tarski, A.: Grundlegung der wissenschaftlichen Semantik. In: Actes du Congrès international de philosophie scientifique, Paris 1935, fasc. 3: Semantique, pp. 1–14. Herman, Paris (1936) (Eng. tr.: The Establishment of Scientific Semantics. In: Tarski, A., Logic, Semantics, Metamathematics. Papers of 1923 to 1938, pp. 401–408. Clarendon Press, Oxford (1956); 2nd ed., pp. 401–408. Hackett Publishing Company, Indianapolis (1983))
27. Tarski, A.: Über den Begriff der logischen Folgerung. In: Actes du Congrès international de philosophie scientifique, Paris 1935, fasc. 7: Logique, pp. 1–11. Herman, Paris (1936) (Eng. tr.: On the Concept of Logical Consequence. In: Tarski, A., Logic, Semantics, Metamathematics. Papers of 1923 to 1938, pp. 409–420. Clarendon Press, Oxford (1956); 2nd ed., pp. 409–420. Hackett Publishing Company, Indianapolis (1983)). See also [39, 41, 42]
28. Tarski, A.: The semantic conception of truth and the foundations of semantics. Philos. Phenomenol. Res. **4**, 341–395 (1944) (Repr. in: Tarski, A., Collected Papers, vol. 2, pp. 665–699. Birkhäuser, Basel (1986))
29. Tarski, A.: The concept of truth in formalized languages. In: Tarski, A. (ed.) Logic, Semantics, Metamathematics. Papers of 1923 to 1938, pp. 152–278. Clarendon Press, Oxford (1956) (2nd edn., pp. 152–278. Hackett Publishing Company, Indianapolis (1983))
30. Tarski, A.: Truth and proof. L'age de la Sci. **1**, 279–301 (1969) (Repr. in: Tarski, A., Collected Papers, v. 4, pp. 399–422. Basel: Birkhäuser (1986))
31. Wang, H.: Beyond Analytic Philosophy: Doing Justice to What We Know. The MIT Press, Cambridge (1986)
32. Woleński, J.: Brentano's criticism of the correspondence theory of truth and Tarski's semantic theory. Topoi **6**, 105–110 (1989) (Repr. in: Woleński, J., Essays in the History of Logic and Logical Philosophy, pp. 103–109. Jagiellonian University Press, Kraków (1999))
33. Woleński, J.: Semantic Revolution: Rudolf Carnap, Kurt Gödel, Alfred Tarski. In: Woleński, J., Köhler, E. (eds.) Alfred Tarski and the Vienna Circle, pp. 1–15. Kluwer Academic Publishers, Dordrecht (1999)
34. Woleński, J.: Semantic conception of truth as a philosophical theory. In: Peregrin, J. (ed.) Truth and Its Nature (If Any), pp. 51–65. Kluwer Academic Publishers, Dodrecht (1999)
35. Woleński, J.: In defence of the semantic definition of truth. Synthese **126**(1–2), 67–90 (2001)
36. Woleński, J.: Truth and satisfaction by the empty sequence. In: Rojszczak, A., Cachro, J., Kurczewski, G. (eds.) Philosophical Dimensions of Logic and Science. Selected Contributed Papers from the 11th International Congress of Logic, Methodology, and Philosophy of Science, Kraków, 1999, pp. 267–276. Kluwer Academic Publishers, Dordrecht (2003) (Repr. in: Woleński, J., Essays on Logic and Its Applications in Philosophy, pp. 17–24. Peter Lang, Frankfurt am Main (2011))
37. Woleński, J.: Logic, semantics and realism. In: Benmakhlouf, A. (ed.) Sémantique et épistemologie, pp. 135–148. Editions Le Fennec, Casablanca (2004) (Repr. in: Woleński, J., Essays on Logic and Its Applications in Philosophy, pp. 51–60. Peter Lang, Frankfurt am Main (2011))
38. Woleński, J.: Two critical contributions to the problem of truth and meaning. Polish J. Philos. **1**, 137–141 (2007) (Repr. in: Woleński, J., Essays on Logic and Its Applications in Philosophy, pp. 183–186. Peter Lang, Frankfurt am Main (2011))
39. Woleński, J.: Truth is eternal if and only if it is sempiternal. In: Tegtmeier, E. (ed.) Studies in the Philosophy of Herbert Hochberg, pp. 223–230. Ontos Verlag, Frankfurt am Main (2012)
40. Woleński, J.: Ens et Verum Concertuntur (Are Being and Truth Convertible)? A Contemporary Pesrpective. In: Dumitriu, M., Sandu, G. (eds.) Truth, pp. 75–83. Editura Universitătii din Bucureşti, Bucureşsti (2013)
41. Woleński, J.: An analysis of logical determinism. In: Lapointe, S. (ed.) Themes from Ontology, Mind and Logic. Present and Past. Essays in Honor of Peter Simons, pp. 423–442. Brill, Leiden (2015)
42. Woleński, J.: Formal and informal aspects of the semantic theory of truth. In: Kuźniar, A., Odrowąż-Sypniewska, J. (eds.) Uncovering Facts and Values. Studies in Contemporary Epistemology and Political Philosophy, pp. 56–66. Brill/Rodopi, Leiden (2016)
43. Woleński, J., Simons, P.: De Veritate: Austro-Polish contributions to the theory of truth from Brentano to Tarski. In: Szaniawski, K. (ed.) The Vienna Circle and the Lvov-Warsaw School, pp. 391–442. Kluwer Academic Publishers, Dordrecht (1989)

J. Woleński (✉)
University of Information, Technology and Management Rzeszów, Rzeszów, Poland
e-mail: jan.wolenski@uj.edu.pl

Tarski's Influence on Computer Science

Solomon Feferman

Abstract Alfred Tarski's influence on computer science was indirect but significant in a number of directions and was in certain respects fundamental. Here surveyed is Tarski's work on the decision procedure for algebra and geometry, the method of elimination of quantifiers, the semantics of formal languages, model-theoretic preservation theorems, and algebraic logic; various connections of each with computer science are taken up.

Keywords Tarski · Decision procedures · Quantifier elimination · Cylindrical algebraic decomposition (CAD) · Time complexity · Space complexity · Semantics for formal languages · Fixed point theorem · Finite model theory · Preservation theorems · Relational database theory · Algebraic logic · Relation algebras · Cylindric algebras

Mathematics Subject Classification (2000) Primary F.4.1; Secondary F.4.3

The following is the text of an invited lecture for the LICS 2005 meeting held in Chicago June 26–29, 2005.[1]

Almost exactly 8 years ago today, Anita Feferman gave a lecture for LICS 1997 at the University of Warsaw with the title, "The saga of Alfred Tarski: From Warsaw to Berkeley." Anita used the opportunity to tell various things we had learned about Tarski while working on our biography of him. We had no idea then how long it would take to finish that work; it was finally completed in 2004 and appeared in the fall of that

The author "Solomon Feferman" is deceased.

Reprinted from *Logical Methods in Computer Science*, vol. 2 (3:6), pp. 1–13, https://doi.org/10.2168/LMCS-2(3:6)2006.

This reprint for *Studies in Universal Logic* style in LaTeX is prepared by Zbigniew Bonikowski.

[1] I want to thank the organizers of LICS 2005 for inviting me to give this lecture and for suggesting the topic of Tarski's influence on computer science, a timely suggestion for several reasons. I appreciate the assistance of Deian Tabakov and Shawn Standefer in preparing the LaTeX version of this text. Except for the addition of references, footnotes, corrections of a few points and stylistic changes, the text is essentially as delivered. Subsequent to the lecture I received interesting comments from several colleagues that would have led me to expand on some of the topics as well as the list of references, had I had the time to do so.

Á. Garrido, U. Wybraniec-Skardowska (eds.), *The Lvov-Warsaw School. Past and Present*, Studies in Universal Logic,
https://doi.org/10.1007/978-3-319-65430-0_29

year under the title, *Alfred Tarski: Life and Logic* [14]. The saga that Anita recounted took Tarski from the beginning of the twentieth century with his birth to a middle-class Jewish family and upbringing in Warsaw, through his university studies and Ph.D. at the ripe young age of 23 and on to his rise as the premier logician in Poland in the 1930s and increasing visibility on the international scene—despite which he never succeeded in obtaining a chair as professor to match his achievements. The saga continued with Tarski coming to Harvard for a meeting in early September, 1939 when the Nazis invaded Poland on September 1st, at which point he was, in effect, stranded. Then, during the next few years he went from one temporary research or teaching position to another on the East Coast. He was finally offered a 1 year position in 1942 as Lecturer in Mathematics at the then far off University of California in Berkeley, with the suggestion that it might stretch into something longer. In fact, he not only succeeded in staying, but rose to the rank of Associate Professor by the end of the war and a year later was made Full Professor, thus finally obtaining the position he deserved. At Berkeley, Tarski built from scratch one of the world's leading centers in mathematical logic, and he remained there, working intensively with students, colleagues and visitors until his death in 1983.

Tarski became recognized as one of the most important logicians of the twentieth century through his many contributions to the areas of set theory, model theory, the semantics of formal languages, decidable theories and decision procedures, undecidable theories, universal algebra, axiomatics of geometry, and algebraic logic. What, in all that, are the connections with computer science? When Anita started working on the biography-which only later became a joint project-she asked me and some of my colleagues exactly that question, and my response was: none. In contrast to that-as she said at the conclusion of her Warsaw lecture-John Etchemendy (my colleague in Philosophy at Stanford, and now the Provost of the University) responded: "You see those big shiny Oracle towers on Highway 101? They would never have been built without Tarski's work on the recursive definitions of satisfaction and truth."[2] It took me a while to see in what sense that was right. Indeed-as I was to learn-there is much, much more to say about his influence on computer science, and that's the subject of my talk today. I owe a lot to a number of colleagues in the logic and computer science areas for pointing me in the right directions in which to pursue this and also for providing me with very helpful specific information.[3]

Alfred Tarski: Life and Logic was written for a general audience; the biographical material is interspersed with interludes that try to give a substantive yet accessible idea of Tarski's main accomplishments. Still, given the kind of book it is, we could not go into great detail about his achievements, and in particular could only touch on the relationship of his work to computer science. Before enlarging on that subject now, I want to tell a

[2]For those who may not know what the "big shiny Oracle towers" are, the reference is to the headquarters of Oracle Corporation on the Redwood Shores area of the San Francisco Peninsula. A duly shiny photograph of a few of these towers may be found at http://en.wikipedia.org/wiki/Image:OracleCorporationHQ.png.

[3]I am most indebted in this respect to Phokion Kolaitis. Besides him I have also received useful comments from Michael Beeson, Bruno Buchberger, George Collins, John Etchemendy, Donald Knuth, Janos Makowsky, Victor Marek, Ursula Martin, John Mitchell, Vaughan Pratt, Natarajan Shankar, and Adam Strzebonski. And finally, I would like to thank the two anonymous referees for a number of helpful corrections.

story that is in our biography [14, pp. 220–230], and is in many respects revelatory of his own attitude towards the connection.

I had the good fortune to be Tarski's student in the 1950s when he was beginning the systematic development of model theory and algebras of logic. In 1957, the year that I finished my Ph.D., a month-long Summer Institute in Symbolic Logic was held at Cornell University. That proved to be a legendary meeting; in the words of Anil Nerode: "There has been nothing else in logic remotely comparable." What the Cornell conference did was to bring together for the first time, leaders, up-and-coming researchers, and students in all the main areas of logic, namely model theory, set theory, recursion theory, and proof theory. Besides Tarski, the top people there—along with their coteries—were Alonzo Church, Stephen Kleene, Willard Quine, Barkley Rosser, and, in the next generation, Abraham Robinson and George Kreisel. The organization of the meeting itself had been inspired by the mathematician Paul Halmos, who, independently of Tarski, had developed another approach to the algebra of first-order logic. As Halmos wrote about it in his *Automathography* [23, p. 215]:

> There weren't many conferences, jamborees, colloquia in those days and the few that existed were treasured. . . . I decided it would be nice to have one in logic, particularly if it were at least partly algebraic.

And, "nice" it was.

In addition to the four main areas, the Cornell logic conference was the first to include many speakers from the emerging field of computer science, the theoretical foundations of which had been laid in the 1930s by Gödel, Church, Turing, Post, and Kleene. The connections between the theory and application of computation began toward the end of World War II when the first large scale electronic digital computers were built. At that point, for each kind of application, the hardware had to be programmed by hand, a long and arduous task. John von Neumann was instrumental in demonstrating how to circumvent that process by introducing the first form of software.

By 1957, companies such as IBM and Remington Rand were producing the first generation of commercial electronic computers, and the high-level programming language FORTRAN had become established as an industry standard. Some-but by no means all-logicians were quick to grasp the implications of these developments. At the Cornell meeting, Rosser gave a talk on the relation between Turing machines and actual computers; Church gave a series of talks on the logical synthesis of switching circuits for computer hardware; and Abraham Robinson spoke on theorem proving as done by man and machine. Among the younger contributors, Michael Rabin and Dana Scott spoke about finite automata, and Martin Davis talked about his implementation on the "Johnniac" computer (at the Institute for Advanced Study) of a decision procedure for the arithmetic of the integers under addition-a procedure that had been discovered in 1930 by Tarski's student Mojżesz Presburger in his Warsaw seminar.

On the industry side, IBM and some of the other companies employed a number of researchers with backgrounds in mathematics and logic, and these people turned out in large numbers at Cornell, both to listen and to speak. There were fifteen talks given by researchers from IBM, many of them demonstrating the utility of FORTRAN-like programs for solving problems of potential interest to logicians. In particular, the talk by George Collins—a former student of Rosser's—on the implementation of parts of Tarski's decision procedure for the algebra of real numbers on an IBM 704 should have caught

Tarski's attention because it suggested possible practical applications of his procedure. But a few years ago, when I asked Collins about Tarski's reaction to that, he said: "He didn't show any appreciation for my work, either then or later. I was somewhat surprised and disappointed." It is indeed surprising that-despite Tarski's own recognition of the importance and systematic pursuit of the decision problem for various algebraic theories, he did not evince the least bit of interest in the practical computational applications of those problems for which a decision procedure had been found. And, he didn't even seem to be interested either in the work of Rabin and Scott, both of whom were high on his list of favorites, Scott as a former student and collaborator, and Rabin as someone he wanted to bring to Berkeley.

Let's look at what Tarski and Collins were up to in more detail. At heart, Tarski's decision procedure rests on the solution of an algebraic problem for the reals, i.e., the ordered field $\langle \mathbb{R}, +, \times, <, 0, 1 \rangle$. Tarski's procedure uses the method of elimination of quantifiers to associate with each first-order formula of the language of the reals an equivalent quantifier-free formula; those without free variables are then easily decided. The procedure reduces to determining for each system P of polynomial equations and inequalities and one of its variables x, whether or not there exists a common real solution x; the answer is to be expressed in terms of the coefficients of the polynomials involved and the remaining variables. Algorithms for special cases of this problem go back through the history of algebra. Tarski's procedure generalizes one due to Sturm for computing the number of roots of a real polynomial in a given interval. On the face of it, Tarski's procedure is non-elementary in time complexity, i.e., greater than all finite towers of powers of 2, and so it was important for Collins to find a more feasible procedure than the one that he had talked about at the 1957 Cornell meeting. Further improvements on Tarski's procedure by Abraham Seidenberg and later by Paul Cohen didn't really help much in that respect. In the meantime, Collins was working on various aspects of computer algebra and in 1974 and 1975 he published [9, 10] a new method of doubly exponential upper bound complexity called Cylindrical Algebraic Decomposition (CAD). Incidentally, this was during a year that he was visiting Stanford University from Wisconsin.[4]

The most comprehensive source of information on the development of the CAD procedure and related work is [7].[5] Here are a few of the things I learned from that invaluable volume. In the first stage pursuant to quantifier elimination, the CAD algorithm takes all the polynomials in the matrix of a prenex formula φ with a total of m free and bound variables, and outputs a cell decomposition in $\mathbb{R}^m$, on each cell of which each of the given polynomials is sign invariant; furthermore, the cells are arranged in *cylinders*. The QE part of the algorithm uses the output of the CAD algorithm to determine which cells of the decomposition satisfy the matrix of φ in order to eliminate the bound variables. The first implementation of the CAD method was made in 1979–1980 by Collins' student Dennis S. Arnon. In 1991, Collins and another of his students, Hoon Hong, published a

[4]Collins reports [7, p. 86] a communication from Leonard Monk in 1974 stating that he and Bob Solovay had obtained a triply exponential upper bound decision procedure for real algebra, though not a quantifier elimination procedure. Fischer and Rabin say (*op. cit.*, p. 124) that Solovay found a doubly exponential upper bound, based on Monk's work.

[5]This includes a reprint of Tarski's "A decision method for elementary algebra and geometry" [39].

substantial improvement for various examples in practice, though not in complexity upper bound, requiring only *partial* CAD [11]. This was subsequently implemented by Hong under the acronym QEPCAD. The Caviness and Johnson volume presents a number of applications, including polynomial optimization, polynomial best approximation in lower degree (by $n-2$ degree polynomials), the topology of semi-algebraic sets, algebraic curve display, and robot motion planning. By the way, the system *Mathematica* implements another form of CAD, according to Adam Strzebonski of Wolfram Research, Inc.

To round out the complexity picture, Fischer and Rabin [17] gave an EXPTIME lower bound of the form 2^{cn} for deciding for sentences of length n whether or not they are true in the reals, no matter what algorithm is used; the same applies even with non-deterministic algorithms, such as via proof systems. They also showed that the cut-point by which EXPTIME sets in, i.e., the least n_0 such that for all inputs of length $n \geq n_0$, at least 2^{cn} steps are needed, is not larger than the length of the given algorithm or axiom system for proofs. Thus real algebra is definitely infeasible on sufficiently large, though not exceptionally large inputs. The applications mentioned above are in a gray area with relatively small numbers of variables, where feasibility in practice depends on the specific nature of the problems dealt with. As for space complexity, there is a PSPACE lower bound on the theory of the reals, as a consequence of a result of Stockmeyer's. Ben-Or, Kozen and Reif [4] established an EXPSPACE upper bound and conjectured that the set of true first-order sentences of the reals is EXPSPACE-complete. The exact time and space complexities of this set are to this date an open problem (Phokion Kolaitis, personal communication).[6]

Tarski's own route to the decision problem for the reals began in the mid-1920s with his development of an elegant first-order axiomatization of geometry [44]. One of his main goals was to prove the completeness of this axiomatization, and that led him to consider its interpretation in the first-order theory of the reals. Tarski recognized that the method of eliminating quantifiers that had been initiated by Leopold Löwenheim and then applied by Thoralf Skolem and C.H. Langford was—when it succeeded—a way of determining all the complete extensions of a first-order axiom system—and in particular of proving the completeness of complete systems. In the latter part of the twenties Tarski ran the "exercise sessions" for the seminar at Warsaw University led by the logic professor, Jan Łukasiewicz, and he used the opportunity to systematically pursue the method of elimination of quantifiers. As an "exercise", Tarski suggested to one of the students, Mojżesz Presburger, that he find an elimination-of-quantifiers procedure for the additive theory of natural numbers, i.e., for the structure $\langle \mathbb{N}, +, <, 0, 1 \rangle$. In that case, full quantifier-elimination is not possible, but can be carried out in a definitional extension of its language, obtained by adding as atomic formulas all those of the form $x \equiv y (\text{mod}\, m)$ for each $m = 2, 3, 4, \ldots$. Mathematically, the procedure comes down to solving a system of simultaneous congruences and thus the Chinese remainder theorem. Presburger's result served as his master's thesis in 1928 and it was published a year later [33]. This slim paper of nine pages was to be his sole work in logic; after that he went to work in the insurance

[6]Just minutes before my lecture, I learned from Prakash Panangaden that John Canny (U.C. Berkeley School of Engineering) proved [6] that the existential theory of the reals is in PSPACE.

industry. Some people think Presburger should have received the Ph.D. for that work, but it has to be admitted that its significance was not realized until much later.[7]

The set of first-order truths of the additive structure of natural numbers is called Presburger Arithmetic. As I mentioned earlier, Martin Davis presented his work on programming the Presburger procedure on the *Johnniac* at the Cornell conference in 1957. That was long before Fischer and Rabin [17] showed that there is a doubly-exponential time lower bound on any algorithmic procedure for Presburger Arithmetic, including non-deterministic ones. If Martin had known that, he might not even have tried, even with today's computers.[8] On the other hand, such lower bounds tell us little about the feasibility in practice of deciding relatively short statements. As to upper bounds, Presburger's own procedure is non-elementary; this was improved to triply-exponential by Derek Oppen [32]. A search on "Google Scholar" came up with a number of references to Presburger Arithmetic. Near the top are applications to the symbolic model checking of infinite state systems [5] and proving safety properties of infinite state systems [20]; further applications via combination decision procedures are indicated in [37].

Let's return to Tarski's own work on elimination of quantifiers for the elementary (i.e., first-order) theory of real numbers: although it was obtained by 1930 and he considered it to be one of his two most important results (the other being his theory of truth), it's surprising he didn't get around to preparing it for publication until 1939. That was under the title, "The completeness of elementary algebra and geometry"[9] for a new series on metamathematics planned by a Parisian publisher, but the actual publication was disrupted by the German invasion of France in 1940. As Tarski later wrote: "Two sets of page proofs which are in my possession seem to be the only material remainders of that venture." The next time he got around to working on its publication was in 1948 when his friend and colleague J.C.C. McKinsey was at the RAND Corporation in Santa Monica. My guess is that McKinsey suggested to his superiors that there would be potential value to applying Tarski's procedure to the computer calculation of optimal strategies in certain games. (Game theory was in those years a very popular subject at RAND.) However, any implementation would first require writing up its theoretical details in full. Working under Tarski's supervision, McKinsey took on the job, revising the 1939 manuscript in its entirety. That came out as a RAND Report under the new title "A decision procedure for elementary algebra and geometry" in 1948; it was finally brought out publicly 3 years later by UC Press as a second edition [39]. The change in title from 1939 to 1948/1951 corresponds to a change in aims, from completeness to decidability. (By the way, a lightly edited version of the 1939 page proofs eventually appeared under the original title in 1967 in France).

Though Tarski may not have been interested in actual computation at any time in this entire history, he *was* interested in mathematical applications of his procedure. In fact,

---

[7]A sad coda to this story is that Presburger, a Jew, perished in the Holocaust in 1943.

[8]Shankar [37] takes as an epigram a quote from Davis [12] re his experiment with Presburger Arithmetic: "Its great triumph was to prove that the sum of two even numbers is even." A second epigram from the same source, quoting Hao Wang, is that: "The most interesting lesson from these results is perhaps that even in a fairly rich domain, the theorems actually proved are mostly ones which call on a very small portion of the available resources ..."

[9]By the elementary theory of a structure, Tarski means the set of its first-order truths.

one of Tarski's strongest motivations throughout his career was to attract mathematicians to the results of work in logic, and he often did this by reformulating the results in a way that he thought would be more digestible by mathematicians. One side result he noticed about his elimination-of-quantifiers argument for the first-order theory of the real numbers is that every definable set has the form of a union of a finite number of intervals (not necessarily proper) with algebraic end-points. He used this to illustrate the general concept of definable set of elements in a structure. At the outset of his 1931 paper on definable sets of real numbers [42] he said that mathematicians in general don't like to deal with the notion of definability. One reason is that used informally it can lead to contradictions, like the paradox of Richard; that uses an enumeration in English (say) of all the real numbers definable in English, to define (in English) a real number not in that enumeration, by diagonalization. Another reason for mathematicians' aversion mentioned by Tarski is that mathematicians think the notion of definability is not really part of mathematics. In a way, he agrees, for he says that

> The problems of making [the meaning of definability] more precise, of removing the confusions and misunderstandings connected with it, and of establishing its fundamental properties belong to *another branch of science-metamathematics*. [Italics mine]

In fact, he says, he has "found a general method which allows us to construct a rigorous metamathematical definition of this notion".

But then

> by analyzing the definition thus obtained it proves to be possible ...to replace it by [one] formulated exclusively in *mathematical* terms. Under this new definition the notion of definability does not differ from other mathematical notions and need not arouse either fears or doubts; it can be discussed entirely within the domain of normal mathematical reasoning. [Italics mine]

The *metamathematical* explanation of definability in Tarski's 1931 paper is given in terms of the notion of satisfaction, whose definition is only indicated there. Under the *mathematical* definition, on the other hand, the definable sets and relations are simply those generated from certain primitive sets of finite sequences corresponding to the atomic formulas, by means of the Boolean operations and the operation of projection. Although the 1931 paper concentrates on the concept of *definability in a structure*, in a footnote to the metamathematical explanation it is stated that "an analogous method can be successfully applied to define other concepts in the field of metamathematics, e.g., that of *true sentence* ..."

Tarski later spelled this out in his famous 1935 paper "Der Wahrheitsbegriff in den formalisierten Sprachen" (The concept of truth in formalized languages, [41]).[10] Some regard this work as one of the most important instances of conceptual analysis in twentieth century logic, while others think he was merely belaboring the obvious. After all, logicians like Löwenheim, Skolem and Gödel had been confidently using the notions of satisfaction and truth in a structure in an informal sense for years before Tarski's work and an explicit definition was not deemed to be necessary, unlike, for example, the conceptual analysis of computability by Turing. I have to agree that there is some

[10]It was not until 1957, in a paper with Robert L. Vaught [45], that Tarski explicitly presented these notions as those of satisfaction and truth in a structure. See the discussion by Hodges [26] and Feferman [15] of the relationship of that to the "Wahrheitsbegriff" paper.

justice to the criticism since the definitions are practically forced on us, once one attends to providing them at all. But even if that's granted, Tarski's explication of these concepts has proved to be important as a paradigm for all the work in recent years on the semantics of a great variety of formal languages.

In particular, the influence of Tarski on the semantics of programming languages is so pervasive that to detail it would require an entire presentation in itself. Let me mention just one example, namely that of the semantics of the lambda calculus and its extensions via domain theory, as developed by Dana Scott and his followers. This happens to connect with the item in Tarski's list of publications that is most cited in the computer science literature, namely his lattice-theoretic fixpoint theorem [40], which is an elegant abstract formulation of the essential characteristic of definition by recursion.[11] There is also a significant personal connection: Scott began his studies in logic at Berkeley in the early 50s while still an undergraduate. His unusual abilities were soon recognized and he quickly moved on to graduate classes and seminars with Tarski and became part of the group that surrounded him, including me and Richard Montague; so it was at that time that we became friends. Scott was clearly in line to do a Ph. D. with Tarski, but they had a falling out for reasons explained in our biography of Tarski [14]. Upset by that, Scott left for Princeton where he finished with a Ph.D. under Alonzo Church. But it was not long before the relationship between them was mended to the point that Tarski could say to him, "I hope I can call you my student," and rightly so: not only did Scott's thesis deal with a problem that had been proposed by Tarski, but all of Scott's work is in the best Tarskian tradition of breadth, rigor, clarity of exposition and clarity of purpose. And, like Tarski, he prefers set-theoretic and algebraic methods, of which the domain-theoretic approach to the semantics of type-free functional programming languages is a perfect example. So Tarski's influence on computer science manifests itself here at just one remove, though of course Scott's contribution, beginning in 1976 [36] with the construction of a domain D isomorphic to $D \to D$, is completely novel.[12]

Satisfaction and truth in a structure are the basic notions of model theory, whose systematic development in the 1950s is initially largely due to Tarski and his school. The notions are relative to a formal language, which is usually taken to be first-order (FO), because of the many happy properties of FO logic such as that of compactness (cf. the texts by Chang and Keisler or Hodges). But other kinds of logics in which, e.g., compactness fails, turned out to be partially susceptible to useful model-theoretic methods as shown by the greatly varied contributions to the collection *Model-Theoretic Logics* [3]. Among these are the model-theory of infinitary languages as well as second-order and higher-order languages. For computer science, a great variety of finite structures, such as various classes of graphs, arise naturally, and it was discovered that a number of

[11] Tarski proved that every monotonic function over a complete lattice has a complete lattice of fixed points, and hence a least fixed point. This is a generalization of a much earlier joint result of Knaster and Tarski and so is sometimes referred to as the Knaster-Tarski theorem. A related result used in applications is that every continuous function on a complete lattice has a least fixed point; credit for it is unclear, and thus it is considered a "folk theorem". The history of these and other fixed point theorems relevant to computer science is surveyed in [29].

[12] Scott informed me that his use of lattice fixed points was initiated in the fall of 1969 in work with Christopher Strachey and exposed in many lectures in Oxford while on leave there. For further developments and a large bibliography see [21].

questions in complexity theory may be framed as questions in finite model theory (cf., e.g. [13]). In addition to first-order logic (FOL) and its finite variable fragments, other logics that have proved to be useful in finite model theory are finite-variable infinitary logics, monadic second-order logic (MSOL) and its fragments, and certain fixed point logics such as Datalog and least fixed-point logic LFP.

Recently there has been a surge of very interesting work on analogues in finite FO model theory to a class of general results called preservation theorems in classical FO model theory. The newest and most exciting of these is due to Ben Rossman [35]. So I can limit myself to explaining the general nature of the main results.

Given a relation R between structures and a sentence φ, we say that φ is preserved under R if whenever M satisfies φ and N is in the relation R to M, then N satisfies φ. The results from classical FO model theory characterize up to logical equivalence, the form of sentences preserved under various R. The most famous ones are the following, all from the 1950s.

- EPT (Łoś-Tarski). φ is preserved under extensions iff φ is equivalent to an existential sentence.
- OHPT (Lyndon). φ is preserved under onto-homomorphisms iff φ is equivalent to a positive sentence.
- HPT (Łoś-Tarski-Lyndon). φ is preserved under (into-)homomorphisms iff φ is equivalent to an existential positive sentence.

The finite analogues of these results are obtained by restricting to finite M and N. The 'if' directions of course hold in all the finite versions, but the 'only if' analogues of the Extension Preservation Theorem (EPT) and the Onto-Homomorphism Preservation Theorem (OHPT) are known to fail. In particular, the failure of the finite analogue of EPT is due to Bill Tait in 1959, who thereby disproved a conjecture of Scott and Suppes; Tait's result was rediscovered by Gurevich and Shelah in 1984. The failure of OHPT in the finite is due to Rosen [34].[13] What Rossman [35] has proved, surprisingly, is that HPT holds in the finite. There are interesting relations to Datalog programs, which are given by existential positive FO inductive definitions. A Datalog formula may thus be considered as an infinitary disjunction of existential positive FO formulas. Using a simple compactness argument, Rossman's result about HPT in the finite also implies the theorem of Ajtai and Gurevich [1] that on finite structures, if a Datalog sentence is equivalent to a FO sentence then it is equivalent to a single existential positive sentence, exactly those preserved by homomorphisms.[14] The failure of the Łoś-Tarski characterization in the finite shows that preservation theorems do not in general relativize from a class to a subclass. Thus it is also of interest to ask for which classes of finite structures HPT holds. This had already been investigated by Atserias, Dawar and Kolaitis [2] in which one of the main results is that HPT holds for every class of finite structures of bounded tree width; another result is that HPT holds for the class of all planar graphs.

[13]Lyndon's famous positivity theorem implies OHPT. Ajtai and Gurevich, and then Stolboushkin in a simpler way, proved failure in the finite of the positivity theorem, but their constructions did not prove failure in the finite of OHPT.

[14]According to Rossman, the implication was known to hold prior to his discovery.

At the Tarski Centenary Conference held in Warsaw in 2001, Johann A. Makowsky presented a survey of applications of another kind of preservation result that goes under the heading of the Feferman-Vaught Theorem [31]. What Vaught and I had shown in our joint 1959 paper [16] was that for a great variety of sum and product operations O on structures $M_i (i \in I)$, the first-order properties of

$$M = O\langle M_i | i \in I\rangle$$

are determined by the first order properties of each M_i together with the monadic second-order properties of a structure on the index set I. It follows that elementary equivalence between structures is preserved under such operations O. In later work, Läuchli, Gurevich and Shelah extended our reduction-to-factors theorem to monadic second-order properties. In his paper, Makowsky gives a unified presentation of this work with emphasis on its algorithmic applications, in particular to splitting theorems for graph polynomials. I'll have to leave it at that, since it would take too much time to try to go further into that here.

The final thing I want to tell something about is the connection of Tarski's ideas and work with database theory. Here it is not a matter of direct influence but rather of the pervasiveness of his approach to things, since the development of database theory apparently proceeded quite independently. Jan Van den Bussche has written an excellent survey [46] of the connections, which I urge you to read; here are a few of the high points. Codd [8] introduced a relational algebra for expressing a class of generic (i.e., isomorphism invariant) queries on databases; he also proved that the queries expressible in his relational algebra are exactly those that are domain independent and definable in FO logic. For those who know Tarski's work on relation algebra, cylindric algebras and algebraic logic more generally, the immediate question to raise is the nature of the connection.[15] (You can find a quick introduction to Tarski's work in this respect in Interlude VI of our biography [14].[16]) I view Tarski's work on algebraic logic as part of his general effort to reformulate logic in mathematical as opposed to metamathematical terms, in the hopes of thus making logic of greater mathematical interest. Tarski had done much work in the 1930s on Boolean algebras, of which algebras of sets and algebras of propositions (up to equivalence) are specific cases. Stone's representation theorem for Boolean algebras showed that every abstract BA is isomorphic to a concrete one in the sense of fields of sets, and in that sense the equational axioms of BA are complete. With Tarski's 1941 paper "On the calculus of relations" [38] he single-handedly revived and advanced the nineteenth century work on binary relations by Peirce and Schröder, and introduced an elegant finite equational axiom system for relation algebra, from which all known special cases of valid relational identities could be deduced. However, it was shown by Roger Lyndon in 1950 that there are non-representable relation algebras, so

[15] According to Van den Bussche (personal communication) the first people from the database community to recognize the connection between Codd's relational algebra and Tarski's cylindric algebras were Witold Lipski and his student Tomasz Imielinski, in a talk given at the very first edition of PODS (the ACM Symposium on Principles of Database Systems), held in Los Angeles, March 29–31, 1982. Their work was later published in Imielinski and Lipski [27].

[16] Some other applications to computer science — not discussed here — of Tarski's work on relation algebra are indicated on p. 339 and its footnote 4 of that interlude.

Tarski's axioms are not complete; later, Donald Monk proved (in 1964) that there is no finite axiomatization of the valid equations in the language of these algebras. This is related to the fact that what can be expressed in relation algebra is exactly what can be expressed about binary predicates in 3-variable FOL.[17] See [30] for more on the history of relation algebras.

Given the weakness of relation algebra, in the early 1950s Tarski introduced the idea of *cylindric algebras* (CAs) of dimension k for any $k \geq 2$, finite or infinite. (*NB*: Cylindric Algebras have nothing to do with Cylindrical Algebraic Decomposition.) In addition to the Boolean operations, these algebras use operations C_n of cylindrification for each $n < k$ and *diagonal* constants $d_{n,m}$ for each $n, m < k$. The concrete interpretations are given by fields of subsets of a k-ary space U^k, with the C_n interpreted as cylindrification along the nth axis, and the $d_{n,m}$ as the set of k-tuples in U for which the nth and mth terms are equal. Thus k-dimensional CAs abstract k-variable FOL with identity. The theory of CAs was extensively developed by students and colleagues of Tarski and the results are exposited in the volumes by Henkin, Monk and Tarski [24, 25]. It turns out that there are non-representable CAs for every $k \geq 2$, finite or infinite, but Henkin and Tarski showed that "locally finite" CAs are representable for every infinite k. A CA is called locally finite if for each element a of the algebra, $C_n(a) = a$ for all but a finite number of $n < k$. The local-finiteness condition corresponds to each formula in FOL having at most a finite number of free variables, and the representation theorem for infinite dimensional locally finite CAs corresponds to the completeness theorem for FOL with identity.

As Van den Bussche points out in [46], the language of ω-CAs provides an alternative to Codd's relational query language, and that of k-CAs for k finite is an alternative to queries definable in FOL with at most k distinct variables. But how does Codd's language match up with that of Relation Algebras (RAs)? In [43] it is shown that adjunction of a suitable "pairing axiom" to RA makes it as strong as FOL. It turns out that the corresponding idea has been developed in the case of database theory by Gyssens, Saxton and Van Gucht [22] using "tagging" operations, giving a form $RA^=$ that simulates Codd's relational algebra.[18]

So, does that justify John Etchemendy's statement that the shiny Oracle towers on Hwy 101 wouldn't be there without Tarski's recursive definition of satisfaction and truth? It would be more accurate to say that the Oracle towers wouldn't be there without the theoretical development of database theory, and *that* wouldn't be there without rethinking the model theory of first-order logic in relation-algebraic and/or cylindric-algebraic terms, and *that* wouldn't be there without Tarski's promotion of both model theory and algebraic logic. Does Larry Ellison know who Tarski is or anything about his work? At the time of my lecture, I wondered whether Ellison even knew who Codd was or the whole theoretical development of database theory, without which the Oracle towers would indeed not be there. I learned subsequently from Jan Van den Bussche that not only did Ellison know about Codd's work but he marks the reading of Codd's seminal paper as the starting point leading to the Oracle Corporation; cf. his biography given by the "Academy of Achievement" at http://www.achievement.org/autodoc/page/ell0bio-1/. Actually, Codd

[17] Cf. the papers [18, 19] dealing with expressibility/inexpressibility in Tarski's algebraic framework.

[18] Van den Bussche's article concludes with a survey of some interesting connections to constraint databases and geometric databases.

himself didn't refer to Tarski in his fundamental papers on database theory. But other workers in the subject, such as Imielinski and Lipski, and later, Kanellakis, did; they were well aware of the connection and brought explicit attention to it (cf., e.g. [28, p. 1085]). In whatever way the claim is formulated, *I* think it is fair to say that Tarski's ideas and the approaches he promoted are so pervasive that even if his influence in this and the various other areas of computer science about which I spoke was not direct it was there at the base, and—to mix a metaphor—it was there in the air, and so the nature and importance of his influence eminently deserves to be recognized.

References

1. Ajtai, M., Gurevich, Y.: Datalog vs. first-order logic. J. Comput. Syst. Sci. **49**, 562–588 (1994)
2. Atserias, A., Dawar, A., Kolaitis, P.G.: On preservation under homomorphisms and unions of conjunctive queries. In: 23rd ACM SIGMOD-SIGACT-SIGART Symposium on Principles of Database Systems (PODS), pp. 319–329. ACM (2004)
3. Barwise, J., Feferman, S. (eds.): Model-Theoretic Logics. Springer, Berlin (1985)
4. Ben-Or, M., Kozen, D., Reif, J.H.: The complexity of elementary algebra and geometry. J. Comput. Syst. Sci. **32**(2), 251–264 (1986)
5. Bultan, T., Gerber, R., Pugh, W.: Symbolic model checking of infinite state systems using Presburger arithmetic. In: CAV'97: Proceedings of the 9th International Conference on Computer Aided Verification, pp. 400–411. Springer, London (1997)
6. Canny, J.F.: Some algebraic and geometric computations in PSPACE. In: ACM Symposium on Theory of Computing, pp. 460–467. ACM, New York (1988)
7. Caviness, B.F., Johnson, J.R. (eds.): Quantifier Elimination and Cylindrical Algebraic Decomposition. Springer Series in Symbolic Computation. Springer, Berlin (1998)
8. Codd, E.F.: A relational model of data for large shared data banks. Commun. ACM **13**(6), 377–387 (1970)
9. Collins, G.E.: Quantifier elimination for real closed fields by cylindrical algebraic decomposition – preliminary report. SIGSAM Bull. **8**(3), 80–90 (1974)
10. Collins, G.E.: Quantifier elimination for real closed fields by cylindrical algebraic decomposition. In: Proceedings Second GI Conference on Automata Theory and Formal Languages. Lecture Notes in Computer Science, vol. 33, pp. 134–183. Springer, Berlin (1975) (Reprinted in: Caviness, B.F., Johnson, J.R. (eds.): Quantifier Elimination and Cylindrical Algebraic Decomposition. Springer Series in Symbolic Computation, pp. 85–121. Springer, Berlin (1998))
11. Collins, G.E., Hong, H.: Partial cylindrical algebraic decomposition for quantifier elimination. J. Symb. Comput. **12**(3), 299–328 (1991) (Reprinted in: (Reprinted in: Caviness, B.F., Johnson, J.R. (eds.): Quantifier Elimination and Cylindrical Algebraic Decomposition. Springer Series in Symbolic Computation, pp. 174–200. Springer, Berlin (1998))
12. Davis, M.: The prehistory and early history of automated deduction. In: Siekmann, J., Wrightson, G. (eds.) Automation of Reasoning 1: Classical Papers on Computational Logic 1957–1966, pp. 1–28. Springer, Berlin (1983)
13. Ebbinghaus, H.D., Flum, J.: Finite Model Theory, 2nd edn. Springer, Berlin (1991)
14. Feferman, A.B., Feferman, S.: Alfred Tarski: Life and Logic. Cambridge University Press, New York (2004)
15. Feferman, S.: Tarski's conceptual analysis of semantical notions. In: Benmakhlouf, A. (ed.) Sémantique et épistémologie, pp. 79–108. Editions Le Fennec, Casablanca [distrib. J. Vrin, Paris] (2004)
16. Feferman, S., Vaught, R.L.: The first order properties of products of algebraic systems. Fundam. Math. **47**, 57–103 (1959)
17. Fischer, M.J., Rabin, M.O.: Super-exponential complexity of Presburger arithmetic. In: Karp, R. (ed.) Complexity of Computation. SIAM-AMS Proceedings, vol. 7 , pp. 27–42. American Mathematical Society, Providence (1974) (Reprinted in: (Reprinted in: Caviness, B.F., Johnson, J.R.

(eds.): Quantifier Elimination and Cylindrical Algebraic Decomposition. Springer Series in Symbolic Computation, pp. 122–135. Springer, Berlin (1998))
18. Formisano, A., Omodeo, E.G., Policriti, A.: Three-variable statements of set-pairing. Theor. Comput. Sci. **322**(1), 147–173 (2004)
19. Formisano, A., Omodeo, E.G., Policriti, A.: The axiom of elementary sets on the edge of Peircean expressibility. J. Symb. Log. **70**, 953–968 (2005)
20. Fribourg, L., Olsén, H.: Proving safety properties of infinite state systems by compilation into Presburger arithmetic. In: Mazurkiewicz, A.W., Winkowski, J. (eds.) CONCUR. Lecture Notes in Computer Science, vol. 1243, pp. 213–227. Springer, Berlin (1997)
21. Gierz, G., Hoffman, K.H., Keimel, K., Lawson, J.D., Mislove, M., Scott, D.S.: Continuous Lattices and Domains. Encyclopedia of Mathematics and its Applications, vol. 93. Cambridge University Press, Cambridge (2003)
22. Gyssens, M., Saxton, L.V., Van Gucht, D.: Tagging as an alternative to object creation. In: Freytag, J.C., Maier, D., Vossen, G. (eds.) Query Processing for Advanced Database Systems, pp. 201–242. Kaufmann, San Mateo (1994)
23. Halmos, P.: I Want to be a Mathematician: An Automathography. Springer, Berlin (1985)
24. Henkin, L., Monk, D., Tarski, A.: Cylindric Algebras, vol. 1. North-Holland, Amsterdam (1971)
25. Henkin, L., Monk, D., Tarski, A.: Cylindric Algebras, vol. 2. North-Holland, Amsterdam (1971)
26. Hodges, W.: Truth in a structure. Proc. Aristot. Soc. **86**, 131–151 (1985–1986)
27. Imieliński, T., Lipski, W. Jr.: The relational model of data and cylindric algebras. J. Comput. Syst. Sci. **28**(1), 80–102 (1984)
28. Kanellakis, P.C.: Elements of relational database theory. In: van Leeuwen, J. (ed.) Handbook of Theoretical Computer Science, Volume B: Formal Models and Semantics (B), pp. 1073–1156. MIT Press, Cambridge (1990)
29. Lassez, J.L., Nguyen, V.L., Sonnenberg, E.A.: Fixed point theorems and semantics: a folk tale. Inf. Process. Lett. **14**, 112–116 (1982)
30. Maddux, R.D.: The origin of relation algebras in the development and axiomatization of the calculus of relations. Stud. Logica **50**(3–4), 421–455 (1991)
31. Makowsky, J.A.: Algorithmic uses of the Feferman-Vaught theorem. Ann. Pure Appl. Logic **126**, 159–213 (2004)
32. Oppen, D.C.: A $2^{2^{2^{pn}}}$ upper bound on the complexity of Presburger arithmetic. J. Comput. Syst. Sci. **16**, 323–332 (1978)
33. Presburger, M.: Über die Vollständigkeit eines gewissen Systems der Arithmetik ganzer Zahlen, in welchem die Addition als einzige Operation hervortritt. Comptes Rendus du I congrs des Mathmaticiens des Pays Slaves, 92–101 (1929)
34. Rosen, E.: Some aspects of model theory and finite structures. Bull. Symb. Log. **8**, 380–403 (2002)
35. Rossman, B.: Existential positive types and preservation under homomorphisisms. In: Panangaden, P. (ed.) Proceedings of the Twentieth Annual IEEE Symposium on Logic in Computer Science. LICS 2005, pp. 467–476. IEEE Computer Society Press, Washington (2005)
36. Scott, D.S.: Data types as lattices. SIAM J. Comput. **5**, 522–587 (1976)
37. Shankar, N.: Little engines of proof. In: Eriksson, L.-H., Lindsay, P. (eds.) FME 2002: Formal Methods – Getting IT Right, Copenhagen, pp. 1–20. Springer, Berlin (2002)
38. Tarski, A.: On the calculus of relations. J. Symb. Logic **6**, 73–89 (1941)
39. Tarski, A.: A Decision Method for Elementary Algebra and Geometry, 2nd edn. University of California Press, Berkeley (1951) (Reprinted in: Caviness, B.F., Johnson, J.R. (eds.): Quantifier Elimination and Cylindrical Algebraic Decomposition. Springer Series in Symbolic Computation, pp. 24–84. Springer, Berlin (1998))
40. Tarski, A.: A lattice-theoretical fixpoint theorem and its applications. Pac. J. Math. **5**, 285–309 (1955)
41. Tarski, A.: The concept of truth in formalized languages. In: Corcoran, J. (ed.) Logic, Semantics, Metamathematics. Papers from 1923 to 1938, 2nd edn., pp. 182–278. Oxford University Press, Oxford (1983) (original date (1935))
42. Tarski, A.: On definable sets of real numbers. In: Corcoran, J. (ed.) Logic, Semantics, Metamathematics. Papers from 1923 to 1938, 2nd edn., pp. 110–142. Oxford University Press, Oxford (1983) (Original date (1931))
43. Tarski, A., Givant, S.: A Formalization of Set Theory Without Variables. Number 41 in Colloquium Publications. American Mathematical Society, Providence (1987)

44. Tarski, A., Givant, S.: Tarski's system of geometry. Bull. Symb. Log. **5**, 175–214 (1999)
45. Tarski, A., Vaught, R.L.: Arithmetical extensions of relational systems. Compos. Math. **13**, 81–102 (1957)
46. Van den Bussche, J.: Applications of Alfred Tarski's Ideas in Database Theory. In: International Workshop on Computer Science Logic. Lecture Notes in Computer Science, vol. 2142, pp. 20–37. Springer, Berlin (2001)

The Absence of Multiple Universes of Discourse in the 1936 Tarski Consequence-Definition Paper

John Corcoran and José Miguel Sagüillo

Dedicated to Professor Roberto Torretti, philosopher of science, historian of mathematics, teacher, friend, collaborator—on his eightieth birthday.

Abstract This paper discusses the history of the confusion and controversies over whether the definition of consequence presented in the 11-page Tarski consequence-definition paper is based on a *monistic fixed-universe* framework—like *Begriffsschrift* and *Principia Mathematica*. Monistic fixed-universe frameworks, common in pre-WWII logic, keep the range of the individual variables fixed as 'the class of all individuals'. The contrary alternative is that the definition is predicated on a *pluralistic multiple-universe* framework—like the Gödel incompleteness paper. A pluralistic multiple-universe framework recognizes multiple universes of discourse serving as different ranges of the individual variables in different interpretations—as in post-WWII model theory. In the early 1960s, many logicians—mistakenly, as we show—held the 'contrary alternative' that Tarski had already adopted a Gödel-type, pluralistic, multiple-universe framework. We explain that Tarski had not yet shifted out of the monistic, Frege-Russell, fixed-universe paradigm. We further argue that between his *Principia*-influenced pre-WWII Warsaw period and his model-theoretic post-WWII Berkeley period, Tarski's philosophy underwent many other radical changes.

Keywords Tarski · Logical consequence · Universe of discourse · Monistic fixed-universe · Pluralistic multiple-universe

Mathematics Subject Classification (2000) Primary 03A05, 01A60; Secondary 03-03

Á. Garrido, U. Wybraniec-Skardowska (eds.), *The Lvov-Warsaw School. Past and Present*, Studies in Universal Logic,
https://doi.org/10.1007/978-3-319-65430-0_30

1 Introduction

In the long history of this text even what is obvious has often been overlooked. —Norman Kretzmann on a passage in the *Organon*, Buffalo (1972).

In his now-famous 11-page consequence-definition paper [63],[1] Tarski broke with various established traditions in several ways. He focused on a 'consequence' relation instead of a converse, i.e., an 'implication' relation: 'is a consequence of' replaces 'implies'. He further separated himself by taking consequence to be a relation of an individual to a 'class' instead of a relation of an individual to an individual.[2] His consequence cannot be expressed with a sentential connective. Even more, he chose his consequence relation to concern sentences of a given fully interpreted language[3] instead of 'propositional functions', uninterpreted 'formal sentences', or 'logical forms'—on one hand—or abstract 'propositions', 'judgments', 'statements', or 'thoughts'—on the other. Moreover, his consequence was metalinguistic relative to the object-language sentences it related. Thus, it could not be 'iterated'; it was impossible to say that one consequence sentence was a consequence of a class of object-language sentences—or conversely, that one object-language sentence was a consequence of a class of consequence sentences. Some of the 'paradoxes' were thus defused. For Tarski, one given sentence of a fully interpreted language is, or is not, a consequence of a given class of sentences of the same language.[4] Each language has its own consequence relation—just as in earlier Tarski works each language had its own rules of inference.

In addition for Tarski, the respective truth-values of the given sentence and those in the given class are irrelevant to determining consequence, except of course in case the given sentence is false and all those in the given class are true. For example, in some cases a given true sentence is not a consequence of a given class of true sentences; Tarski's consequence is not a material-consequence relation. And for Tarski, [sc. logical] consequence is formal in the traditional sense: if the conclusion of a given argument is a consequence of its premises, the same relation holds between the conclusion and the premises of any other argument in the same form.[5] Finally, Tarski avoided traditional logic's reference to premise-conclusion arguments: he left it entirely up to traditional

[1]Unless explicitly said otherwise, 'consequence' is used in a 'logical' sense so that the expression 'logical consequence' is redundant. Thus, we are not discussing the 'consequences' of events, of actions, or of inactions, etc.

[2]'Class' is the word Tarski used [63, pp. 414–419]. But, although others had distinguished 'classes' from multiplicities such as 'sets', 'aggregates', 'collections', 'extensions', and the like, no distinction is implied in [63] or elsewhere in his collection [69].

[3]One of the *History and Philosophy of Logic* referees required us to give more emphasis to this point than it had received in the submitted version. We are grateful that this needed improvement was brought to light; it helps us to highlight a major philosophical and technical change between Tarski's pre-WWII attitude and his post-WWII thinking. After the war uninterpreted constants became legitimate and were assigned an important role. Contrast [57].

[4]Tarski did not dwell on the philosophical ramifications of this 'relevance' or 'pertinence' requirement, which might have been foreshadowed by Aristotle. But others have noted its importance (Corcoran [8]).

[5]Tarski is not as explicit as one might wish. This point is not made in any single sentence though it can be gleaned from the paragraph that begins on p. 414. Incidentally, this is the only place in [69] that uses 'form' in the required sense of logical form. See Corcoran's [14] piece 'Logical form' in [1].

logicians to discern that an argument's being valid in their sense is its conclusion being a consequence of its premises in Tarski's terminology. They were left to infer that Tarski was defining argument validity: Tarski was silent on the connection between defining consequence and defining validity. Of course, breaking with one tradition is frequently keeping with some other tradition. In a sense our paper is about whether in his consequence-definition paper he had already broken with the Frege-Russell fixed-universe paradigm.

Although we discuss the details, background, and ramifications of Tarski's definition; as stated in the abstract, our central concern is the issue of whether his definition has a *monistic* framework that keeps the range of the individual variables fixed as 'the class of all individuals', or 'the actual universe of things' as in *Principia Mathematica*, for every interpretation, or whether—to the contrary—it has a *pluralistic* framework which, like that of the Gödel incompleteness paper [27, translation in 78], recognizes multiple universes of discourse serving as different ranges of the individual variables in different interpretations ('sequences' or 'models').

The 1936 consequence-definition paper did not become widely available until 1956 when J.H. Woodger translated it into English [64, 74]. Previously it had appeared in Polish and German versions [72, p. 409]. Church [6, p. 325], who refers only to the Polish and German versions, implied that Tarski [63] presupposed a Gödel-type changeable-universe framework. In the early 1960s, after the English had been available for years, nevertheless Corcoran and many other logicians still concurred with the Church interpretation. Now to the contrary, it is widely held that Tarski had not yet fully shifted out of the monistic Frege-Russell paradigm (Mancosu [41]). We too support the monistic, one-universe interpretation of the Tarski consequence-definition paper [63]. In other words, we argue for the absence of multiple universes of discourse in the Tarski consequence-definition paper [63].

After the issue of interpreting the 1936 paper came to Corcoran's attention in 1964, he reread the paper and felt forced to recognize his mistake. From then on his understanding has been that, contrary to Church and to his own former belief, Tarski [63] presupposes a monistic *Principia*-type fixed-universe framework within which the individual variables have a range which is fixed throughout and which does not change from interpretation to interpretation. The contrast between the two approaches to explicating 'logical truth' and 'logical consequence' is explained in Chapter 4 of Quine's *Philosophy of Logic* [47], where the older fixed-universe framework is preferred.

The issue of whether Tarski's paper [63] employs a fixed or a changeable range of individual variables is entirely separate from the question of whether Tarski in 1936 or before acknowledged the fact, already noted by Aristotle in *Posterior Analytics* (Bk. A, Ch. 10, esp. 76b10 ff.) that each science has its own separate domain of investigation, sometimes called its subject matter, or its genus (e.g., Sagüillo [51, p. 268]). As Sagüillo has amply established, there are many well-known formalizations of sciences in which the individual variables range over a universe of discourse that is wider than the domain of investigation of the science. One example is Tarski's own 1929 geometry of solids, where—as Tarski says—the term 'individual' is used in the same way as in *Principia Mathematica*. But the domain of investigation is not the range of the individual variables; rather it is 'the class of all [sc. geometrical] solids' [72, pp. 24–29].

However, the issue is related to, but not determined by, the question of whether Tarski was aware before 1936 of the De Morgan-Boole concept of *the universe of a given discourse* (Boole [3, p. 42]; Corcoran [15, p. 242], [16, p. 941]), and if aware, whether he accepted its theoretical importance in logic.

Underlying the historical and hermeneutic issue of how the 1936 paper is faithfully interpreted, there is a list of broader and more important historical and philosophical issues. Near the top of this list, we would put the question of whether, as De Morgan and Boole held, we have the capacity to discuss a limited domain of investigation without referring to things outside of the domain and in particular without referring to the entire universe of individuals. For example, is it *possible* to say that every person is mortal using a sentence such as 'for every person x, x is mortal' with the common noun 'person' indicating the range of the variable and without in any way referring to objects other than persons? Or to the contrary, in order to say that every person is mortal is it *necessary* to use a sentence such as 'for every [object] x, if x is a person, then x is mortal' with the 'universe of individuals' as the range of the variable? Do we really have to talk about *every* individual in order to make a statement about every person?[6]

The view held by Frege and Russell is that individual variables necessarily range over the entire universe of individuals and any restriction of subject is to be accomplished by means of a qualifying condition. For example, to be more explicit, on their view, to say that every person is mortal we would be required to use a sentence such as 'for every individual x, if x is a person, then x is mortal', where the common noun 'individual' indicates that the range of the variable is taken to be the class of all individuals. In practice, the quantifier phrase 'for every individual x' is often written elliptically as 'for every x' with the *range indicator* 'individual' to be 'understood'. Moreover, the entire quantifier phrase 'for every individual x as Tarski [66, p. 7] says, '[...] is often omitted and has to be inserted mentally': 'if x is a person, then x is mortal' is used to express the proposition 'for every individual x, if x is a person, then x is mortal'.[7] On this view, Aristotle's universal affirmative propositions cannot be expressed in a logically perfect language without using the truth-functional connective 'if-then' and using a variable that ranges over the entire universe of individuals.

Boole espoused the *Principle of Wholistic Reference* (PWR): every proposition refers to *its* entire universe of discourse, regardless of how limited the number of objects explicitly referred to by its non-logical concepts Corcoran [20]). Although within a given discourse, it is impossible to refer to objects outside of the universe of *that* discourse, as said, there are formalizations of sciences in which the individual variables range over a universe of discourse that is wider than the domain of investigation of the science. However, frequently but—as already said—not always, the universe of discourse of a proposition is not wider; it coincides with the domain of investigation of the science in which it occurs (Sagüillo [51]; Corcoran [15, 16]). In such cases, there need not be two different non-logical constants for the two different but coextensive concepts. But when there are two terms, the formalization often contains a 'theorem' to the effect

[6]According to Leonard Jacuzzo [36], whose 2005 dissertation research involved comparisons of dozens of introductory logic texts, most books he studied teach the affirmative answer to this question (personal communication). Our own less extensive experience confirms his sad finding.

[7]For more on these points see Gupta [31]; Tarski [66, Section 3]; Corcoran [17].

that every object in the universe of discourse is in the domain of investigation. Tarski [61, pp. 310–311] mentions a geometry containing a sentence 'which asserts that every *individual* [sc. in the universe of discourse] is a *point* [sc. member of the domain of investigation]'—emphasis added. Thus, the universe of discourse coincides with the domain of investigation. He also mentions an 'axiom system of arithmetic' that 'contains a sentence to the effect that every *individual* [sc. in the universe of discourse] is a *number* [sc. member of the domain of investigation]'—emphasis added.

Frege and Russell carried Boole's PWR one step further: by putting all of the domains into one all-encompassing cosmic universe of individuals over which every individual variable ranged. They and many of their followers replaced PWR by the *Principle of Cosmic Reference* (PCR)—although they did not, indeed could not, put it this way. The PCR is that every proposition refers to the entire, cosmic universe of all individuals. The hermeneutic question mentioned above is whether Tarski [63] subscribed to the Principle of Cosmic Reference.[8] Our opinion is that he did: in some passages he seems to think that the individual variables range over 'all possible objects' [63, p. 416]. He never made use of the De Morgan-Boole concept of universe of discourse, i.e., discourse universes, in the 400-odd pages of *Logic, Semantics, Metamathematics*. The expression 'universe of discourse' occurs there about a dozen times: never in the consequence-definition paper, never in the plural, and never in the sense of the range of the individual variables of the object language. Sometimes, it is used for the domain of investigation of a science [72, p. 28, pp. 135–140] and sometimes it is used for the carrier[9] of a structure such as a Boolean algebra [62, pp. 320, 322, 335, 347, 350, and 373]. Tarski's use of 'universe of discourse' is discussed further below.

[8]Space limitations preclude discussion of how Frege arrived at PCR (or a similar monistic universal variable-range view), why it was so widely adopted, how its conflict with pluralistic views escaped notice for so long, and when its soundness came under scrutiny.

[9]Today the set of elements of a Boolean algebra, or of any other algebraic structure such as a group or a ring, is called its *carrier*. The maximal element of the Boolean algebra is usually denoted by the digit '1' and is called its *unity*. If a given Boolean algebra is 'formalized' using a first-order language whose individual variables range over the carrier, then the carrier is the universe of discourse of the 'theory of the algebra'. But this can distort Boole's [3] viewpoint. Take Boole's 'universe of men (sc. humans)'. Boole used '1' to denote this class and '0' to denote the null class: two elements of the carrier of the corresponding Boolean algebra. In such cases, where a Boolean algebra of classes is under discussion, there are two things competing for the names 'universe', 'universe of discourse', 'universal class' and the like: the carrier and the carrier's unity. The carrier is often the powerset of the carrier's unity. In such cases, the carrier's unity is the union of the carrier. The Boolean tradition would incline towards using such terms for the unity. The modern abstract-algebra viewpoint that abstracts from the nature of the elements of an abstract algebra would incline toward using such terms for the carrier. Tarski called the carrier or the set of elements of a given Boolean algebra its universe of discourse on the first page of the article beginning on p. 347 of [72].

Even in the late 1970s and early 1980s, the issue of single versus multiple universes in Tarski [63] was still completely (and wrongly) settled in favor of the pluralistic interpretation in the minds of all logicians Corcoran was in contact with and who cared about it, except for the few he succeeded in convincing otherwise.[10] In fact, as far as we know, there was little or no discussion of this point until Corcoran brought it up in the winter of 1964 or the spring of 1965.

2 Corcoran's Awareness of the Issue in the 1960s and 1970s

> People who comb or concoct obscure sources in order to support usually lame or self-serving priority claims concerning well-known discoveries, theorems, 'theses', or conjectures have their priorities reversed. It is not the first person who discovers something that counts; it is the last person—the person who discovers it so that it never needs rediscovery. —Peter Freyd, 1968 Philadelphia Logic Colloquium.

For years, without having studied Tarski [63] very carefully, Corcoran was under the impression that it contained the pluralistic, multiple-universe conception of consequence, probably to some extent on the strength of Church's assertion [6, p. 325, fn 533]. In the fall of 1964, Corcoran reread the paper carefully and was surprised that his expectations were not fulfilled: he found that Tarski had a monistic fixed-universe framework. Still, Corcoran was not very interested in the issue; he thought that Church's mistake must have been caught by many others. He let the issue drop.

Corcoran traces the awakening of his interest in the issue to a conversation that semester with William Craig[11] in Craig's Dwinelle Hall office on the Berkeley campus of the University of California late one afternoon. He recalled casually mentioning to Craig that Church [6, p. 325, fn 533] was not entirely accurate in crediting Tarski [63] with explication of the concept of logical consequence Church used and that had been used by the American Postulate Theorists[12] including Church's teacher Oswald Veblen.

[10]To the best of Corcoran's knowledge, there is no chance that any unpublished writing by Tarski, either a passage in the Tarski-Corcoran correspondence (preserved at the Bancroft Library at U.C. Berkeley), or anywhere else, would suggest that Tarski [63] conceives of various models having various universes of discourse. In fact, according to Paolo Mancosu, evidence to the contrary is to be found not only in the Tarski-Corcoran correspondence (Mancosu [26 and 39, p. 451]) but also in Tarski's unpublished 1940 lecture 'On the completeness and categoricity of deductive systems' also in the Bancroft Library (Mancosu [41, pp. 754–756]).

[11]Even at that time, William Craig was a distinguished mathematical logician, former doctoral student of Quine at Harvard, Full Professor of Philosophy at Berkeley. He was a member of the UC Berkeley Group for Logic and Methodology, in which Tarski was still active. He was internationally known for what was then called the Craig Interpolation Lemma. The Craig Interpolation Theorem, as it is sometimes known today, is 'one of the basic results of the theory of models [2], almost on a par with, say, the compactness theorem' (Boolos et al. [4, p. 260]). An entire chapter of [4] is devoted to this theorem, which is to be distinguished from a less deep but equally famous result then called Craig's theorem, now sometimes the Craig Axiomatizability Lemma or the Craig Reaxiomatization Lemma (Boolos et al. [4, p. 198]).

[12]For more on the American Postulate Theorists see Scanlan's article 'Who were the American Postulate Theorists?' [53] and for the influence they had on Tarski see the same author's 'American Postulate Theorists and Alfred Tarski' [54].

The discrepancy was that the Church concept, contrary to Tarski's, involved changing universes of discourse. Corcoran was under the impression that it was common knowledge that the Tarski [63] definition did not Vary universes'. Corcoran was surprised that Craig had not heard of this rare inaccuracy in Church [6] and even more surprised to learn that Craig thought that Corcoran must be wrong.

Corcoran remembers excusing himself for a moment, going next-door to his office, and returning with a copy of *Logic, Semantics, Metamathematics*, which he handed to Craig asking him to show where Tarski says that different interpretations can have different universes of discourse, i.e., that the individual variables can have different ranges relative to different interpretations. As befits his dignity, Craig asked permission to read at his leisure and to continue the discussion the next day. The next day he came to Corcoran's office reporting he had reread the entire article and that, to his surprise, he found no mention of alternative universes of discourse in Tarski [63].

What Tarski [63] presents is a kind of updated and transformed version of the Russellian formal implication. In Russell's sense, *one* proposition *formally implies* a *second* if and only if the 'generalized conditional' of the former with the latter is true. By the generalized conditional of one proposition with another is meant the universal closure of the result of replacing all of the non-logical constants in the conditional by suitable variables (Russell [49, pp. 5, 11, 14, and 36–41]).[13] To use an adaptation of one of Russell's examples, the single proposition 'Socrates is human and everything which is human is mortal' formally implies the single proposition 'Socrates is mortal' if and only if the following single proposition is true: 'Given any individual x, any property P, and any property Q, if x is P and everything which is P is Q, then x is Q'. There is no place here for changing universes.[14]

As already noted, one of Tarski's innovations was his broadening the situation by taking the 'implicant' to be a class of propositions (interpreted sentences), possibly infinite, and thereby avoiding the conditional altogether in favor of a metalinguistic condition.

A person scrutinizing [63] for signs of interpretations having different variable ranges might first look for the words 'domain' and 'universe'; neither of them occur there. Moreover, the first full sentence on p. 416 criticizes the no-expressible-countermodels definition[15] on the ground that it implies 'the designations of all possible objects occurred

[13]We never use the expression 'formally implies' in Russell's sense without explicitly adding 'in Russell's sense' or an equivalent. In fact, as Nabrasa pointed out, few if any logicians do either. He reminded us that the fact that a person defines an expression in a certain sense, in and of itself, is no evidence that the person uses the expression in that sense (Frango Nabrasa, personal communication).

[14]This is yet another example of Russell's habit of using previously established terminology in a sense never before employed and without explaining or even alluding to the previous senses. His friend and colleague G. E. Moore criticized him for this in connection with using 'implies' in the sense of the truth-functional conditional. Previously no traditional logician would have said that 'Some animal is not a dog' logically implies 'Every oak is a tree'.

[15]See p. xxii of Corcoran's Editor's Introduction to Tarski [72]. A no-expressible-countermodels definition of consequence defines a premise-conclusion argument expressed in a given language to be valid if no argument in the same form expressible in the same language has all true premises and false conclusion. Such a definition is featured in Quine's *Philosophy of Logic* [47], where great care is taken to ensure that the language has the required 'richness', to use Tarski's expression [64, p. 416]. Even today, some authors claim, of course without giving any justification, that the no-expressible-countermodels

in the language in question', suggesting that he is thinking of languages whose individual variables are unrestricted in range, i.e., which have the cosmic universe of individuals as their range.[16] The fact that Corcoran's observation about the Church inaccuracy was a surprise to Craig highlighted the issue in Corcoran's mind. He recalls resolving to say something about it in print when appropriate opportunities arose.

Corcoran recalls discussing this with some of his friends, students, and teaching assistants then and over the next couple of years. He remembers Robert Barnes, Herbert Bohnert, Oswaldo Chateaubriand, William Frank, Edward Keenan, Ray Lucas, John Pollock, David Sherr, George Weaver, but there were probably some others. At the time he was in regular correspondence with Lucas (then at the University of Wisconsin) and with Bohnert (then at IBM Research). Chateaubriand, Pollock, and Barnes were in his 1964 'Philosophy of Mathematics' seminar. Frank, Keenan, Sherr, and Weaver attended Corcoran's 1966 graduate course 'Mathematical Logic', which had Church [6] as its only required text. In the late 1960s Corcoran was in constant contact with his friend and colleague Henry (or Henryk) Hiż, whose 'Reexamination of Tarski's semantics' [34] did not even touch this issue.[17]

Corcoran also mentioned Tarski's fixed-universe viewpoint, as it has come to be known,[18] in lectures at Georgia, Michigan, Pennsylvania, New York, Ontario, Quebec and Mexico over the next few years in connection with his work on argumentation in philosophy of logic and with his work on Aristotle. From 1971 to 1973, he circulated a series of typescripts that appeared as Corcoran [11] containing the following passage on pp. 126 and 127, bracketed matter added.

> Church [6, p. 325] attributes this [concept] [...] of logical consequence to Tarski [72, pp. 409–420], but Tarski's notion of true interpretation (model) seems too narrow [...] in that no mention of alternative universes of discourse is made or implied. In fact, the limited Tarskian notion seems to have been already known by Lewis and Langford [37, p. 342], to whom [...] I am indebted for the terms 'interpretation' and 'true interpretation', which seem heuristically superior to the Tarskian terms 'sequence' and 'model', the latter of which has engendered category mistakes—a 'model of a set of sentences' in the Tarskian sense is by no means a model, in any ordinary sense, of a set of sentences.

---

conception is *the classical definition of consequence*. E.g. in [4, p. 101] where 'argument' is elliptical for 'expressed argument', we find: 'Logic teaches that the premisses [...] (*logically*) *imply* or have as a (*logical*) *consequence* the conclusion [...], because [*sic*] in any argument of the same form, if the premisses are true, then the conclusion is true'.

[16]Tarski should not be construed as referring to possible objects as opposed to actual objects. By '*possible objects*' Tarski means 'objects'; the use of the modal adjective 'possible' is entirely empty—it is what is sometimes called redundant rhetoric, filler, or expletive. Tarski's usage of modal words is almost always, if not absolutely always, expletive, like putting 'absolutely' before 'always', 'entirely' before 'empty', 'no matter how small' after 'every real number' or 'if any' after 'all odd perfect numbers'. See Corcoran [18] and [19, p. 266]. Incidentally, Tarski may not be speaking very strictly in saying that designations of *all* objects are needed.

[17]Henry Hiż had been a fellow Quine PhD student at Harvard with William Craig and Robert McNaughton, supervisor of Corcoran's 1963 dissertation. McNaughton introduced Corcoran to Hiż in 1961 or 1962 and to Craig in 1963. Hiż helped with the second edition of *Logic, Semantics, Metamathematics* and with the Editor's Introduction [72, pp. viii, xxvii].

[18]See Sagüillo [52], but compare with Sagüillo [50, 51].

These points had been made in print earlier in 'Conceptual Structure of Classical Logic', (Corcoran [8, p. 43] quoted in Sagüillo [50, p. 238]), a paper written in 1969, and in 'A Mathematical Model of Aristotle's Syllogistic' (Corcoran [9]). Both papers were published in main-stream international journals with large circulations. To the best of our knowledge these points were never disputed by anyone at the time, or even in the 1980s. Scanlan and Shapiro [55, pp. 149–150] mention related events.

3 The Place of the Monistic-Pluralistic Distinction

Here and elsewhere we shall not obtain the best insight into things until we actually see them growing from the beginning. —Aristotle, *Politics*.

It would be interesting to know when the explicit observation that Tarski [63] holds to a monistic and not to a pluralistic multiple-universe viewpoint was first made in print. More generally, we can wonder when, after say 1972, it was first noted that Tarski [63] employed a fixed-universe framework. In the 1972 article Corcoran had more pressing issues to deal with and accordingly understated his observations[19] about [63]: 'There is room to doubt whether [63] permits changes in domain (universe of discourse)' (Corcoran [8, p. 43]). The question is not one of priority: the issue is so obvious that no one would want to claim credit for it. The question is when logicians came to feel that this obvious point needed to be made. Our evidence seems to indicate that Chapter 8 of the 1990 Etchemendy book made this explicit in reconsidering the status of pure cardinality sentences. Likewise, Chapter 2 of Simons [58] brought the issue up in reconsidering the criterion for logicality sketched in the still monistic setting of Tarski [73].

It would also be interesting to know when the true statement that Tarski [63] holds a monistic viewpoint was first explicitly *denied*. In our opinion, Church [6] makes no such denial. In fact, he gives no indication that he specifically considered the issue. He does not quote one sentence from the Tarski paper. It is likely that he did not deliberate on it and it seems certain that he would have retracted his statement given the chance. Our evidence seems to indicate that Gómez-Torrente [29 and 30] was in fact the first to explicitly *deny* that Tarski held a monistic view and to *claim* that Tarski held a pluralistic view in the 1936 consequence-definition paper (cf. Mancosu [40, pp. 463–468], and [41, pp. 751–752]).

As mentioned, Tarski published two versions of the consequence-definition paper in the year 1936, one in German and one in Polish.[20] The above remarks apply directly to the

[19]We should also point out that Corcoran is also to blame for a related misinterpretation of Tarski [63]. He said that Tarski substituted new non-logical constants in the manner of the American Postulate Theorists. Of course, Tarski substituted variables like Russell [49], not constants.

[20]The following background has been generously supplied by David Hitchcock (personal communication): 'Tarski wrote the paper in 1935. He delivered the German version at a conference in Paris in September 1935, and appears to have left a copy of the paper with the conference organizers for publication in the conference proceedings, which came out in 1936. The Polish version appeared in the first (January) issue of the 1936 volume of *Przegląd filozoficzny* (Polish for 'Philosophical Review', the leading Polish philosophy journal), and so must have been submitted by the end of 1935, allowing time for typesetting and correcting proofs. Thus, Tarski wrote the paper no later than 1935. It is likely that he did not write it earlier than 1935, since Carnap reports in his autobiography that Tarski visited Vienna in June

1956 English translation by Woodger of the 1936 German form of the Tarski consequence definition paper. From the 1960s though the 1990s, Corcoran had no motivation to consider the 1936 Polish form. In the intervening period Magda Stroińska and David Hitchcock of McMaster University have translated the Polish form into English and they have written extensive commentary accompanying their translation (Hitchcock and Stroińska [33]).

In comparing the 1936 Tarski conception of consequence as it appears in the Polish paper with modern conceptions, they wrote [33, p. 167]: 'Tarski [...] worked with formalized languages, in which [...] the domain is fixed.'

It is hard to believe that anyone who had read Tarski [63] and who was versed in the history of modern logic going back to Frege and Whitehead-Russell could find 'variable universes' in it. Most authors we know of who address the issue agree that the pluralistic multiple-universe viewpoint is nowhere to be found in Tarski [63]. For example, in his 1992 *Journal of Symbolic Logic* review of John Etchemendy's *The Concept of Logical Consequence* [25], Vann McGee wrote [42, p. 254]:

> Etchemendy emphasizes the divergence between Tarski's [1936] analysis of logical truth and what we may call the modified Tarski thesis, which identifies logical truth with truth in every model. In particular, Tarski's original analysis makes no provision for the special role of the universe of discourse [...]. It is unfortunate that Etchemendy focuses all his fire upon the original [1936] Tarski thesis, for it is the modified thesis which Tarski himself employed in his mature work that has won nearly universal acceptance.

By omitting acknowledgement of the De Morgan-Boole theory of discourse universes, Frege—and then Russell—might have virtually ruled it out for their more dedicated disciples, as is especially clear in the writings of van Heijenoort, Chateaubriand [5, pp. 174f], and others. In 1936, Tarski had not yet developed many of the aspects of his post-WWII, Berkeley-period, multiple-universe view that *rejected* features of the Frege-Russell philosophy, e.g., higher-order logic. As indicated in Sagüillo [50, p. 233], our present evidence suggests that Tarski's earliest explicit written endorsement of the pluralistic viewpoint is in his 1953 undecidability paper [68], published in [76]. However, according to Tarski's student Roger Maddux (personal communication), the *locus classicus* most often referred to in this connection is the 1957 Tarski-Vaught paper [77].

It is not the case, however, that Tarski [63] explicitly *rejected* the multiple-universe viewpoint. In fact, it shows no awareness of the issue. As said, the expression 'universe of discourse' does occur several times in Tarski [69]: but never in the consequence-definition paper, never in the plural, and never in a relevant sense. After discussing formalization of 'the arithmetic of real numbers' wherein the first-order variables range over *Rl* the set of real numbers, he 'generalized' his remarks by taking an arbitrary set instead. He wrote

---

1935 and that Carnap persuaded Tarski at that time to present his ideas on semantics at the September 1935 conference. Tarski had just finished translating his truth monograph into German (at the end of the historical note which Tarski added to his German translation of his truth monograph, there appears in Latin in italics, centered two lines below the end of the text the sentence

"Nachwort" allatum est die 13. Aprilis 1935.

—i.e. The 'afterword' was produced on the 13 April 1935.). The other paper that Tarski presented at the Paris conference is a kind of summary of the ideas in semantics in that work. Both papers in fact presuppose the concepts of the truth monograph.' For further information see Hitchcock-Stroińska [33, especially, pp. 155–158].

[69, p. 135]: 'The set *Rl* is now replaced by an arbitrary set *V* (the so-called universe of discourse or universal set)'. There is another passage in the 400-odd page Tarski [69]—footnote 2 of p. 310 of Article X (Tarski [61])—where Tarski implicitly recognizes the possibility of a formalized science whose domain is a proper subclass of what he calls 'the class of all individuals'. But, the expression 'universe of discourse' does not occur in this article and there is no special symbol for arbitrary universes of discourse. In fact, the symbol *V* is used for 'the class of all individuals'. Moreover, the first sentence of the footnote implies that Tarski thought it was 'customary' to discuss categoricity only in contexts where the individual variables were considered to range over all individuals and not merely over the genus or domain of objects relevant to the science being formalized.

We are not aware of any hint in any of Tarski's pre-WWII papers of appreciation of the De Morgan-Boole concept of the universe *of a discourse*. He never mentioned the fact that in a discourse the participants agree tacitly or explicitly to limit the subject-matter of that discourse, i.e., in modern terms, to restrict the range of the individual variables (as opposed to affixing restrictive relative clauses to the universal sentences leaving the variables to have unrestricted range).[21] It is true that Tarski uses the expression 'universe of discourse', e.g., in the 1935 Boolean algebras paper [62] written about the same time. However, here the 'universes of discourse', indicated by a non-logical constant *B* used in restrictive relative clauses, are the carriers of the algebras, not the range of the individual variables. On p. 199 of the 1933 truth-definition paper, Tarski shows acquaintance with a relativized notion of truth when he said the following.

> In the investigations which are in progress at the present day in the methodology of the deductive sciences (in particular in the work of the Göttingen school grouped around Hilbert) another concept of relative character plays a much greater part than the absolute concept of truth and includes it as a special case. This is the concept of *correct or true sentence in an individual domain a*.

Tarski never makes the point (made repeatedly by De Morgan and by Boole) that each discourse[22] or discussion has its limited 'universe' as its ultimate subject matter, a point that was central to Boole's theory of propositions (Corcoran [18] and [19, p. 275]). Nevertheless, by his repeated use of 'domain-dedicated' variables, Tarski's work exemplified Boole's theory that different discourses can have different domains and that it is not necessary to make statements about *every* thing. Among the domain-dedicated first-order, individual variables used in Tarski's pre-WWII papers, we find sentential variables,

---

[21] The De Morgan and Boole work discovered the concept of the universe *of a discourse* in the conceptual framework of the mathematics and science of their time. The role of universes of discourse persisted into the conceptual framework of Tarski's time. Tarski's work shows this role but it does not show awareness of that role nor does it show any appreciation of the De Morgan-Boole achievement. The fact that blood circulated in Plato's veins is no reason to credit him with discovering or knowing of blood circulation.

[22] The wisdom of Boole's choice of the word 'discourse' for a certain sort of extended exposition or discussion may be questioned and it is open to doubt whether the word had ever been used in his precise sense before. Others may have observed how much Boole enriched the English language by coining the phrase 'universe of discourse', but no one seems to have suggested that his use of 'discourse' may have been equally creative and meritorious. By the way, the word 'discourse' is more often used for a stretch of speech involving typically more than one sentence-like expression, e.g., a paragraph or an argumentation. It was used in this sense in Zellig Harris's Discourse Analysis Project, which is discussed in Corcoran's paper 'Discourse grammars and the structure of mathematical reasoning, Part I: Mathematical reasoning and the stratification of language' [7]. See also Corcoran [12].

string variables, and number variables not to mention the first-order object languages whose variables range over a universe of 'classes'. The most prominent of the latter, of course, is the object language whose truths are defined in the 1933 truth-definition paper [72, pp. 168ff.].

A no-countermodel view of logical consequence is implicit in the classic Padoa paper [43] on definitional and implicational independence. Some readers might be inclined to say that Padoa actually subscribed to the no-countermodels view of consequence. It is tempting to speculate that he would have stated it clearly except that he had not distinguished—on one hand—the *epistemic* relation of 'is-a-deduction-from' (intrinsically connected with deduction, the human capacity to reason logically) and—on the other—the *ontic* relation 'is-a-consequence-of' (extrinsically serving as the objective ground or standard of deduction). This distinction is not made despite the twin facts that Padoa refers to deduction as a human activity and that he takes the existence of a countermodel to be entirely objective, not dependent on human capabilities. The distinction is overlooked because a kind of absolute soundness and a kind of absolute completeness are presupposed [43, pp. 122–123]. Padoa's explicit statements imply, using the vocabulary of the English translation by van Heijenoort, that in order for one given proposition to be a logical consequence of a given set of propositions it is necessary and sufficient for no interpretation satisfying the set to satisfy the negation of the given proposition. Tarski was well acquainted with Padoa's thought. Moreover, although Padoa mentions individuals and variables, he never mentions change in the range of the individual variables. To all appearances, Padoa holds to a monistic fixed-universe viewpoint.

The topic of precursors of Tarski's 1936 consequence definition points to a remarkable difference between his attitude toward precursors taken in the 1936 consequence-definition paper and that taken by him in the earlier and more well-known 1933 truth-definition paper. In the earlier paper, Tarski emphasized an 'essential' similarity—perhaps identity—between the conception of truth Tarski characterized and that characterized in Aristotle's truth-definition. He quoted Aristotle: To say of what is that it is not, or what is not that it is, is false, while to say of what is that it is, or of what is not that it is that it is not, is true' [69, p. 155]. In the later paper, there is no reference to Aristotle; no comparison is made between Tarski's relation of consequence and anything Aristotle says about 'syllogistic consequence': the relation of the conclusion of a syllogism and its premises. There are two interesting points. First, in a course Tarski probably attended, Tarski's teacher Jan Łukasiewicz interpreted the syllogistic consequence to be something very close to a special case of Russellian formal implication (Łukasiewicz [38, pp. 103–112, especially p. 112]). Second, in *Prior Analytics* there is no succinct passage characterizing syllogistic consequence; there is nothing analogous to the pithy passage in *Metaphysics* characterizing truth. In fact, Aristotle seems to have deliberately avoided any verbal characterization of syllogistic consequence (Corcoran [13 and 21, pp. 151–153]).

4 The Origin of the Modern form of the Monistic Framework

> Every science which is ratiocinative or at all involves reasoning deals with causes and principles, exact or indeterminate; but all these sciences mark off some particular being—some genus, and inquire into this, but not into being simply or *qua* being. —Aristotle, *Metaphysics*, 1025b5–10.

Frege, in apparent willful ignorance of facts made clear by De Morgan and disseminated by Boole (Corcoran [16, 18, 19]), created the fictional framework of monistic fixed-universe logic. Cf. Peirce 1880 [45, p. 206f.]. Goldfarb [28, p. 351] says that for Frege as well as for Russell the ranges of the variables 'are fixed in advance once and for all' and the 'universe of discourse is always the universe, appropriately striated'. Peano (e.g., [44]) and then Whitehead-Russell (1910) follow Frege in this regard. They also follow Frege in showing no awareness of an alternative—even though all four of them repeatedly refer to [3], where we find the shift from—at first—using ' 1' for *the* universe to—later—using it as an indexical[23] that denotes in a given discourse the universe *of that discourse* (Boole [3, p. 54], Corcoran [18, p. 254]). To the best of our knowledge, the monistic fixed-universe viewpoint has never been accepted by any logicians who knew of the alternatives, except followers of Frege such as Jean van Heijenoort, Willard Quine [47], and those who find Quine's brand of wholistic naturalism attractive. Even the post-WWII, Berkeley-period, Tarski, to his credit, adopted the multiple-universe viewpoint—although the issue was never clear to him.

In pre-WWII logic and foundations, there were two philosophies living in tense but peaceful co-existence. On one hand, there were the *monists* who recognized one single fixed universe, who worked in a framework in which the universe of discourse, or range of individual variables, remained fixed as *the* class of individuals, i.e., who subscribed to the fixed-universe viewpoint. Many of the monists pursued what van Heijenoort has somewhat mysteriously called 'logic as language'. A few of them, also somewhat mysteriously, regarded metalanguage statements as a kind of inspired and revealing incoherence, as literally meaningless, or studiously avoided saying anything about their own metalanguage, which belonged to a different discourse having a different range for its individual variables. They included Boole (at least as late as 1848), the early Frege, Peano, Padoa, Russell (in many of his writings), the early Wittgenstein, Lewis, Carnap, pre-WWII Tarski, and others. As is to be expected, the monists misunderstood, ignored, or even denigrated the De Morgan-Boole doctrine of universe of discourse. For example, Lewis-Langford [37] uses the expression 'universe of discourse' twice in its 500-odd pages. Once it is used on p. 28 in connection with the unity in the Boole-Schroeder algebra (not the carrier) and once on p. 353, *in quotes* preceded by 'as it were', in connection with interpreting a monadic letter used in restrictive relative clauses—not for the range of the individual variables (which is never changed). Quine'1940 [46] never uses 'universe of discourse' although 'universe' and 'discourse' occur sparingly.

In addition, monists tended to be reductionists who were not comfortable with the idea of a plurality of separate ontologically independent mathematical sciences. Many of them

[23]By an indexical we mean a word such as 'I', 'you', 'here', 'this', and 'now' whose denotation changes according to the context of the speech act it is used in. See Corcoran [20, pp. 159–160] for a discussion of Boole's innovations involving the symbol '1'.

stretched reductionism to its logicist extreme of 'reducing' all mathematical sciences to logic. In some cases this meant exchanging an ontology of numbers, points, lines, solids, classes, vectors, etc. for an ontology that restricted literal existence to the individual entities over which the individual variables ranged, entities of the most unquestionably substantial—yet abstract—and thus the most mysterious status, knowledge of which was an article of faith. In this framework, it was natural to accept what has unfortunately come to be known as 'axioms of infinity', albeit with bad conscience in some cases. On the other hand, there were the pluralists who recognized with Boole a plurality of universes of discourse, who tended to construct formalizations of sciences having in each case its universe of discourse identified with the range of its individual variables, and who tended to avoid having a variable range over all individuals without exception [71, reprinted in 35]. These thinkers tended to accept with Aristotle a plurality of ontologically independent mathematical sciences [10]. Among those showing pluralistic tendencies, we find Dedekind, Poincaré, Hilbert, Veblen, Huntington, Gödel, and even Church—somewhat surprisingly given his deep appreciation of many aspects of monistic thinking in the work of Frege and Russell.[24] Hilbert's famous *Foundations of Geometry* [32], which was probably read by every logician that came later, was a stunning endorsement of pluralistic thinking: it took geometry to be an autonomous science and it had three sorts of domain-dedicated individual variables—one sort ranging over points, one ranging over lines, and one ranging over planes. Moreover, from Hilbert's consistency and independence results it is clear that his work presupposes a no-countermodels conception of consequence based on a pluralistic multiple-universe framework.

Within a monistic framework one natural way of *construing* logical consequence is a no-countermodels fixed-universe concept, one explication of which is given in the 1936 Tarski consequence-definition paper. Within a pluralistic framework one natural way of *construing* logical consequence[25] is a no-countermodels multiple-universe concept, one explication of which is given in the Church *Introduction to Mathematical Logic* [6]. In the pre-WWII period, the expression 'logical consequence' was ambiguous. Tarski [63] was mistaken when he repeatedly writes 'the concept of logical consequence', 'the concept of consequence', 'the common concept of consequence', 'the consequence relation', and so on, suggesting that, aside from a little vagueness, there is essentially only one concept expressed in the then-extant literature by the noun phrase 'logical consequence'. It is probably significant that Tarski does not mention Hilbert's *Foundations of Geometry* [32] in the 1936 consequence-definition paper, nor, for that matter, in the entire *Logic, Semantics, Metamathematics: Papers from 1923 to 1938*, nor even in the *Introduction to Logic and to the Methodology of Deductive Sciences* [66].[26] Nevertheless, as noted above, there is ample evidence that Tarski was aware of the pluralistic framework and

[24]Of course, Church's pluralistic tendencies might have been formed when he was studying with his teacher and dissertation supervisor Oswald Veblen, the American Postulate Theorist [53, 54], before studying Frege.

[25]The idea that logicians down through the history of logic were expressing a Tarskian no-countermodels concept by the phrase 'is a consequence of' is implausible to say the least. Tarski's stated goal was not to characterize the traditional concept but merely to define 'a new concept which coincided in extent with the common one' [63, p. 409].

[26]However, Hilbert's *Foundations of Geometry* [32] is mentioned in the second edition of *Logic, Semantics, Metamathematics* [72], but not by Tarski—it is on p. xvii of the Editor's Introduction.

its conception of multiple ranges for individual variables ([59], [72, pp. 199–209] to cite one prominent passage). Perhaps surprisingly, he nowhere shows awareness that in presenting the monistic fixed-universe conception he might be taken as rejecting the pluralistic multiple-universe conception. But, as far as we know there in not a scintilla of direct evidence relevant to determining his process of deliberation.

What difference does it make? If we define a tautology as a proposition that is a logical consequence of any and every set of propositions, then a tautology is a consequence both of a given proposition and of its negation, and it is a consequence of the null set. If we define a contradiction as a proposition that has as a consequence any and every proposition, then a contradiction has as a consequence the negation of each of its consequences, indeed its own negation. Under these two definitions, a large chasm between the two definitions of 'logical consequence' opens. In the case of the multiple-universe view the only pure cardinality propositions that are contradictory are those to the effect that the universe is empty and the only ones that are tautological are those to the effect that the universe is non-empty. By a pure cardinality proposition is meant one to the effect that, for a certain cardinal number, the universe does (or does not) have at least, exactly, or at most that number of members. In the case of the fixed-universe view, every pure cardinality proposition is contradictory or tautological, a result that is hard to swallow.

5 Concluding Remarks

> If you by your rules would measure what doth not with your rules agree, forgetting all your learning seek ye first what *its* rules may be. —Wagner, *Die Meistersinger*.

Why would Tarski base his consequence definition on the monistic fixed-universe framework? Before attempting to answer this question it is important to note that we are not asking why Tarski would make the monistic fixed-universe *choice*. We have no evidence that any such choice was ever made. The question we ask is how the monistic view came to play a foundation role in Tarski's consequence-definition paper [63]. Probably, the most important consideration is that in his earliest logic training and in his dissertation *Principia Mathematica* was authoritative if not scriptural.

Once the monistic view was in place what obstacles may have kept Tarski from questioning it? For one thing, Tarski was never bothered by the limiting cases of tautologies and contradictions, or by the question of what human faculty is needed to determine their truth or falsity. Moreover, Tarski never seemed to have appreciated the De Morgan-Boole discovery of independent discourses each with its own limited universe of discourse—a discovery that would come to demote what others had seen as 'the universe of individuals'—from an exalted and unique place, a *sui generis*, to being just one of many universes of discourse (and a most questionable one at that). Some scholars have argued that Tarski had more deliberate reasons for leaning toward a monistic view (Rodriguez-Consuegra [48], Corcoran [22], and Tarski [60]). Although we and many others believe Tarski did not question the monistic, fixed individual-variable range view, for completeness it should be said that some well-informed scholars think that Tarski had not only questioned it but actually rejected it. They think that its presence in the consequence-definition paper is to be attributed to space limitation, style, rhetoric, pedagogy, and other contingent considerations. Thus, they allow such contingencies to outweigh truth in Tarski's deliberations. For example, David Hitchcock wrote the following (personal communication).

> It seems quite puzzling that Tarski did not allow variation of the range of variables in his 1936 logical consequence paper, given (1) the extensive treatment in his truth monograph ([59], [72, pp. 199–209]) of the concept of a truth in a domain, (2) the fact that the logical consequence paper is an application of ideas from the truth monograph, and (3) the fact that the fixity of the domain overgenerates consequences by (for example) making the existence of at least two individuals a consequence of the existence of at least one individual. The following possible explanation occurs to us. The paragraph in brackets on p. 415 of Tarski [72] suggests that Tarski simplified the exposition of his ideas, as would be appropriate for an audience of philosophers who were not necessarily mathematically sophisticated. He may have consciously avoided exploration of the variable-domain alternative in his 1936 logical consequence paper [65], in order to keep things simple for his audience. As far as I can see, there is no mention of a domain-relative conception of truth in the other paper that Tarski presented at the September 1935 conference, The establishment of scientific semantics', evidently for the same reason of keeping things simple; see the remark at the bottom of page 406 of Tarski [72].

This is in keeping with what Hitchcock and Stroińska published [67 and 33, p. 170].

It is hard to imagine a motivation for promulgating revisionist history that would make Tarski's pre-WWII, Warsaw-period thinking agree with his (and the dominant) post-WWII thinking about logical consequence, unless it is simply an inability to accept the fact that even Tarski can change his mind. One dramatic example of the evolution of his thought is his change from—at first—working in higher-order frameworks before WWII to—then later—regarding them as wrong-headed: in the late 1970s he bluntly told Corcoran that the definitional equivalence results in Corcoran, Frank, and Maloney [23] were 'meaningless' because they concerned theories with second-order underlying logics (personal communication).[27] By 1936 Tarski had not yet fully shifted out of the monistic, Frege-Russell, fixed-universe paradigm that had been presupposed in his Warsaw training. Between his Warsaw period and his Berkeley period, Tarski's philosophy underwent many other radical changes [2]. Not only was higher-order logic banned and replaced by first-order logic, but also type theory gave way to set theory, categoricity was de-emphasized in favor of decidability and deductive completeness, and the prominence of propositional logic was weakened while that of equational logic strengthened (e.g., [56], [70] and Tarski-Givant [75]).

In one of his last sessions with Tarski, Corcoran asked him whether his research had caused him to change his mind over the years on fundamental issues. He said in immediate response that it would be sad if there were no cases of this—but then, after a long silence, he asked whether Corcoran had had anything in particular in mind. When Corcoran said no, Tarski said he would give it some thought. The topic never came up again (personal communication).

[27] Tarski's oral evaluation contrasts with the evaluation published by Haskell Curry in Mathematical Reviews [24].

Acknowledgements

> We call in others to aid us in deliberating on important questions—distrusting ourselves as not being equal to deciding. —Aristotle.

We started planning this paper in 2003 on a trip the two of us made to Lisbon from Santiago de Compostela for Ricardo Santos's Tarski Symposium. Earlier versions have been circulated (Mancosu [40, p. 576]). For bringing errors and omissions to our attention, for useful suggestions, and for other help, it is a pleasure to acknowledge the following scholars: O. Chateaubriand (Brazil), D. Hitchcock (Canada), R. Torretti (Chile), I. Grattan-Guinness (England), S. Nambiar (India), H. Masoud (Iran), R. Santos (Portugal), S. Read (Scotland), F. Rodriguez-Consuegra (Spain), J. Gasser (Switzerland), and, from USA, M. Brown, W. Goldfarb, R. Grandy, I. Hamid, L. Jacuzzo, C. Jongsma, E. Keenan, M. La Vine, R. Maddux, P. Mancosu, D. Merrill, J. Miller, M. Scanlan, S. Shapiro, J. Smith, J. Tarski, and G. Weaver. During the entire period in which this article was being written we have been in almost constant contact with our long-time friend Frango Nabrasa, whose incisive sarcasm and disarming skepticism dampened any excessive enthusiasm we might have had. David Hitchcock has been especially helpful: not only has he shared insights and given valuable criticisms, his published scholarship on this topic makes crucial, actually essential, contributions. We are also grateful to Volker Peckhaus and to two anonymous *History and Philosophy of Logic* referees for their useful suggestions and perceptive corrections.

References

1. Audi, R. (ed.): The Cambridge Dictionary of Philosophy. Cambridge University Press, Cambridge (1999)
2. Bays, T.: On Tarski on models. J. Symb. Log. **66**, 1701–1726 (2001)
3. Boole, G.: Laws of Thought. Macmillan, Cambridge (1854). Reprinted with introduction by J. Corcoran. Prometheus Books, Buffalo (2003)
4. Boolos, G., Burgess, J., Jeffrey, R.: Computability and Logic. Cambridge University Press, Cambridge (2002)
5. Chateaubriand, O.: Boole on reference and universes of discourse: reply to John Corcoran. Manuscrito **27**, 173–182 (2004)
6. Church, A.: Introduction to Mathematical Logic. Princeton University Press, Princeton (1956)
7. Corcoran, J.: Discourse grammars and the structure of mathematical reasoning, part I: mathematical reasoning and the stratification of language. J. Struct. Learn. **3**, 55–74 (1971)
8. Corcoran, J.: Conceptual structure of classical logic. Philos. Phenomenol. Res. **33**, 25–47 (1972)
9. Corcoran, J.: A mathematical model of Aristotle's Syllogistic. Archiv für Geschichte der Philosophie **55**, 191–219 (1973)
10. Corcoran, J. (ed.): Ancient Logic and Its Modern Interpretations. Kluwer, Dordrecht (1974)
11. Corcoran, J.: Aristotle's Natural Deduction System. In: Corcoran, J. (ed.) Ancient Logic and its Modern Interpretations, pp. 85–131. Kluwer, Dordrecht (1974)
12. Corcoran, J.: Argumentations and Logic. Argumentation **3**, 17–43 (1989)
13. Corcoran, J.: Review of Tarski 1986(2). Mathematical Reviews. 91h:01101-4 (1991)
14. Corcoran, J.: Logical form. In: Audi, R. (ed.) The Cambridge Dictionary of Philosophy, pp. 511–512. Cambridge University Press, Cambridge (1999)
15. Corcoran, J.: Domain. In: Audi, R. (ed.) The Cambridge Dictionary of Philosophy, p. 242. Cambridge University Press, Cambridge (1999)
16. Corcoran, J.: Universe of discourse. In: Audi, R. (ed.) The Cambridge Dictionary of Philosophy, p. 941. Cambridge University Press, Cambridge (1999)
17. Corcoran, J.: The logical form of quantifier phrases: quantifier-sortal-variable. Bull. Symb. Log. **5**, 418–419 (1999)

18. Corcoran, J.: Introduction. In: Boole, G. (ed.) Laws of Thought. Macmillan, Cambridge (1854). Reprinted with introduction by J. Corcoran. Prometheus Books, Buffalo (2003)
19. Corcoran, J.: Aristotle's prior analytics and Boole's laws of thought. Hist. Philos. Log. **24**, 261–288 (2003)
20. Corcoran, J.: The principle of wholistic reference. Manuscrito **27**, 159–171 (2004)
21. Corcoran, J.: Notes on the founding of logics and metalogic: Aristotle, Boole, and Tarski. In: Martínez, C., et al. (eds.) Current Issues in Logic/Temas Actuates de Lógica, pp. 143–176 (2007). Imprenta Univeridade Santiago de Compostela, Galicia (University of Santiago de Compostela Press)
22. Corcoran, J.: Review of: Sinaceur, H. B.: Tarski's practice and philosophy: between formalism and pragmatism. In: Lindström, S., Palmgren, E., Segerberg, K., Stoltenberg-Hansen, V. (eds.) Logicism, Intuitionism, and Formalism: What Has Become of Them? pp. 357–396. Springer, Dordrecht (2009). Mathematical Reviews. MR2509665 (2011b:03006)
23. Corcoran, J., Frank, W., Maloney, M.: String theory. J. Symb. Log. **39**, 625–637 (1974)
24. Curry, H.: Review of: Corcoran, J., Frank, W., Maloney, M.: String theory. J. Symb. Log. 39, 625–637 (1974). Mathematical Reviews. MR0398771 (53#2622) (1977)
25. Etchemendy, J.: The Concept of Logical Consequence. Harvard University Press, Cambridge (1990)
26. Ferreiros, J., Gray, J.: The Architecture of Modern Mathematics. Oxford University Press, Oxford (2006)
27. Gödel, K.: On Formally Undecidable Propositions. In: van Heijenoort, J. (ed.) From Frege to Gödel, pp. 596–616. Harvard University Press, Cambridge (1931/1967)
28. Goldfarb, W.: Logic in the twenties: the nature of the quantifier. J. Symb. Log. **44**, 351–368 (1979)
29. Gómez-Torrente, M.: Tarski on logical consequence. Notre Dame J. Formal Log. **37**, 125–151 (1996)
30. Gómez-Torrente, M.: Rereading Tarski on logical consequence. Rev. Symb. Log. **2**, 249–297 (2009)
31. Gupta, A.: The Logic of Common Nouns. Yale University Press, New Haven (1980)
32. Hilbert, D.: Foundations of Geometry. Open Court, La Salle (1899/1971)
33. Hitchcock, D., Stroińska, M.: Translators' introduction to Alfred Tarski's on the concept of following logically. Hist. Philo. Log. **23**, 155–196 (2002). See Tarski, A.: Über den Begriff der logischen Folgerung. Actes du Congrès International de Philosophie Scientifique, vol. 7 (Actualités Scientifiques et Industrielles, vol. 394), 1–11. Paris (1936)
34. Hiż, H.: Reexamination of Tarski's semantics. Ann. Pure Appl. Logic **126**, 39–48 (2004)
35. Hughes, R.: Philosophical Companion to First-order Logic. Hackett, Indianapolis (1993)
36. Jacuzzo, L.: The Pedagogy of Logic. PhD dissertation in Philosophy. University at Buffalo, Buffalo (2005)
37. Lewis, C. I., Langford, C. H.: Symbolic Logic. Century, New York (1932)
38. Łukasiewicz, J.: Elements of Mathematical Logic (1929). Wojtasiewicz, O. (trans.). Macmillan, New York (1963)
39. Mancosu, P.: Tarski on models and logical consequence. In: Ferreiros, J., Gray, J. (eds.) The Architecture of Modern Mathematics, pp. 209–237. Oxford University Press, Oxford (2006)
40. Mancosu, P.: The Adventure of Reason. Interplay between mathematical logic andphilosophy of mathematics: 1900–1940. Oxford University Press, Oxford (2010)
41. Mancosu, P.: Fixed- vs. variable-domain interpretations of Tarski's account of logical consequence. Philos. Compass. **5**, 745–759 (2010)
42. McGee, V.: Review of Etchemendy, J. The Concept of Logical Consequence. Harvard University Press, Cambridge (1990). J. Symb. Log. **51**, 254–255 (1992)
43. Padoa, A.: Essai d'une théorie algébrique des nombres entiers, précede d'une introduction logique à une théorie déductive quelconque [Logical introduction to any deductive theory]. Bibliothèque du Congrès international de philosophie, Paris. In: van Heijenoort, J. (ed.) From Frege to Gödel, pp. 118–123. Harvard University Press, Cambridge (1900/1967)
44. Peano, G.: Principles of mathematics presented by a new method. In: van Heijenoort, J. (ed.) From Frege to Gödel, pp. 83–97. Harvard University Press, Cambridge (1967/1889)
45. Peirce, C. S.: The Essential Peirce: Selected Philosophical Writings (1867–1893), vol. I. Houser, N., Kloesel, C. (eds.). Indiana University Press, Bloomington (1992)
46. Quine, W.: Mathematical Logic. Harvard University Press, Cambridge MA (1940). Reprinted (1951)

47. Quine, W.: Philosophy of Logic. Harvard University Press, Cambridge MA (1970/1986)
48. Rodriguez-Consuegra, F.: Tarski's intuitive notion of set. In: Sica, G. (ed.) Essays on the Foundations of Mathematics and Logic, pp. 227–266. Polimetrica, Monza (2005)
49. Russell, B.: The Principles of Mathematics. Cambridge University Press, Cambridge (1903)
50. Sagüillo, J.: Logical consequence revisited. Bull. Symb. Log. **3**, 216–241 (1997)
51. Sagüillo, J.: Domains of sciences, universe of discourse, and omega arguments. Hist. Philos. Log. **20**, 267–280 (1999)
52. Sagüillo, J.: Methodological practice and complementary concepts of logical consequence: Tarski's model-theoretic consequence and Corcoran's information-theoretic consequence. Hist. Philos. Log. **30**, 21–48 (2009)
53. Scanlan, M.: Who were the American postulate theorists? J. Symb. Log. **56**, 981–1002 (1991)
54. Scanlan, M.: American postulate theorists and Alfred Tarski. Hist. Philos. Log. **24**, 307–325 (2003)
55. Scanlan, M., Shapiro, S.: The work of John Corcoran. Hist. Philos. Log. **20**, 149–158 (1999)
56. Schmidt, H. A., et al. (eds.): Contributions to Mathematical Logic (Collog., Hannover, 1966). North-Holland, Amsterdam (1968)
57. Shapiro, S. (ed.): The Limits of Logic. Dartmouth Publishing company, Aldershot (1996)
58. Simons, P.: Philosophy of Logic in Central Europe from Bolzano to Tarski. Kluwer, Dordrecht (1992)
59. Tarski, A.: The concept of truth in formalized languages. In: Tarski, A. (ed.) Logic, Semantics, Metamathematics, 2nd edn., pp. 152–278. Introduced and edited by J. Corcoran. Hackett, Indianapolis (1933/1983)
60. Tarski, A.: Some observations on the concepts of ω-consistency and ω-completeness. In: Tarski, A. (ed.) Logic, Semantics, Metamathematics, 2nd edn. pp. 279–295. Introduced and edited by J. Corcoran. Hackett, Indianapolis (1933/1983)
61. Tarski, A.: Some Methodological Investigations on the Definability of Concepts. In: Tarski, A. (ed.) Logic, Semantics, Metamathematics, 2nd edn., pp. 296–319. Introduced and edited by J. Corcoran. Hackett, Indianapolis (1934/1983)
62. Tarski, A.: On the Foundations of Boolean algebra. In: Tarski, A. (ed.) Logic, Semantics, Metamathematics, 2nd edn., pp. 320–341. Introduced and edited by J. Corcoran. Hackett, Indianapolis (1935/1983)
63. Tarski, A.: Über den Begriff der logischen Folgerung. Actes du Congrès International de Philosophie Scientifique, vol. 7 (Actualités Scientifiques et Industrielles, vol. 394), 1–11. Paris (1936). Cited from English trans. in: Tarski, A.: Logic, Semantics, Metamathematics, 2nd edn., pp. 409–420. Introduced and edited by J. Corcoran. Hackett, Indianapolis (1983)
64. Tarski, A.: On the Concept of Logical Consequence (1936). English trans. (Woodger, J. H. (trans.)) in: Tarski, A.: Logic, Semantics, Metamathematics, 2nd edn., Introduced and edited by J. Corcoran. Hackett, Indianapolis (1983)
65. Tarski, A.: On the Concept of Following Logically (1936). English trans. (Stroińska, M., Hitchcock, D. (trans.)) in: History and Philosophy of Logic **23**, pp. 155–196
66. Tarski, A.: Introduction to Logic and to the Methodology of Deductive Sciences (1941). Helmer, O. (trans.). Oxford University Press, New York (1946)
67. Tarski, A.: Introduction to Logic and to the Methodology of Deductive Sciences (1941). Trans., ed. with preface and biographical sketch of the author by J. Tarski. Oxford University Press, New York (1994)
68. Tarski, A.: A general method in proofs of undecidability. In: Tarski, A., Mostowski, A., Robinson, R. (eds.), Undecidable Theories, pp. 3–35. North Holland, Amsterdam (1953)
69. Tarski, A.: Logic, Semantics, Metamathematics. Woodger, J. H. (trans.). Oxford University Press, Oxford (1956)
70. Tarski, A.: Equational Logic and Equational Theories of Algebras. In: Schmidt, H. A., et al. (eds.): Contributions to Mathematical Logic (Collog., Hannover, 1966), pp. 275–288. North-Holland, Amsterdam (1966/1968). Reprinted in: Tarski, A.: Collected Papers, Givant, S., McKenzie, R. (eds.), Vol. 4, pp. 347–362. Birkhäuser, Basel (1986)
71. Tarski, A.: Truth and proof. Sci. Am. (1969). Reprinted in: Hughes, R.: Philosophical Companion to First-order Logic. Hackett, Indianapolis (1993)
72. Tarski, A.: Logic, Semantics, Metamathematics, 2nd edn., Introduced and edited by J. Corcoran. Hackett, Indianapolis (1983)

73. Tarski, A.: What are logical notions? Hist. Philos. Log. **7**, 143–154 (1986). Reprinted in: Shapiro, S. (ed.): The Limits of Logic. Dartmouth Publishing company, Aldershot (1996)
74. Tarski, A.: On the Concept of Logical Consequence (1936). In Givant, S., McKenzie, R. (eds.) Collected Papers. Birkhäuser, Basel (1986)
75. Tarski, A., Givant, S.: A Formalization of Set Theory without Variables. American Mathematical Society, Providence (1987)
76. Tarski, A., Mostowski, A., Robinson, R.: Undecidable Theories. North Holland, Amsterdam (1953)
77. Tarski, A., Vaught, R.: Arithmetical extensions of relational systems. Compos. Math. **13**, 81–102 (1957)
78. van Heijenoort, J.: From Frege to Gödel. Harvard University Press, Cambridge (1967)

J. Corcoran (✉)
Department of Philosophy, University at Buffalo, Buffalo, NY, USA
e-mail: corcoran@buffalo.edu

J. M. Sagüillo
Department of Logic, University of Santiago de Compostela, Santiago, Spain
e-mail: josemiguel.saguillo@usc.es

Alfred Tarski: Auxiliary Notes on His Legacy

Jan Zygmunt

Abstract The purpose of this article is to highlight a selected few of Alfred Tarski's career achievements. The choice of these achievements is subjective. Section 1 is a general sketch of his life and work, emphasizing his role as researcher, teacher, organizer and founder of a scientific school. Section 2 discusses his contributions to set theory. Section 3 discusses his contributions to the foundations of geometry and to measure theory. Section 4 looks at his metamathematical work, and especially the decision problem for formalized theories. Section 5 is a selected bibliography to illustrate Sects. 1–4.

Keywords Alfred Tarski · Logic · Metamathematics · Set theory · Foundations of geometry

AMS Subject Classification (2000) Primary: 01A60, Secondary: 01A70, 01A72, 51-03, 54-03, 54A05

1 Life

Alfred Tarski was a logician and a mathematician. He exerted a significant influence on the whole of the twentieth century development of logic and the foundations of mathematics, and through his work in formal semantics and the foundations of logic also on epistemology, the methodology of science, and the philosophy of language. He was a famous representative of the Lwów-Warsaw school and founder of the Berkeley school of logic and the methodology of sciences. He left behind him a rich and wide

---

Translated and edited by Robert Purdy.

A major part of this chapter is an extension of the first four sections of: J. Zygmunt, *Alfred Tarski—logik i metamatematyk (Alfred Tarski—Logician and Metamathematician)*, in the volume: ***O przyrodzie i kulturze*, Studium Generale, Seminaria Interdyscyplinarne**, vol. XIII (2009), pp. 305–327, Wydawnictwo Uniwersytetu Wrocławskiego: Wrocław 2009. See also: (1) J. Zygmunt, *Szkic biograficzny (A Biographical Sketch)*, in [95m], pp. vii-xx; and (2) J. Zygmunt, *Alfred Tarski*, in ***Polska filozofia powojenna***, edited by Witold Mackiewicz, vol. 2, pp. 342–375, Agencja Wydawnicza Witmark: Warszawa 2001.

Á. Garrido, U. Wybraniec-Skardowska (eds.), *The Lvov-Warsaw School. Past and Present*, Studies in Universal Logic,
https://doi.org/10.1007/978-3-319-65430-0_31

scientific legacy in logic, metamathematics, semantics, set theory, the foundations of geometry, general algebra and algebraic logic (Boolean algebra, Boolean algebras with operators, relation algebras, cylindric algebras); nearly all of Tarski's output is influenced by the ideas of algebra. Sensitive to the idiomaticity of language, he wrote precisely and clearly, persuasively setting issues in historical and intuitive perspective. He was a charismatic teacher and lecturer. He recognized the role of aesthetics in science, and he drew attention to the "intrinsic charm and prettiness" of theoretical constructions. He was a fierce champion of academic freedom.

Tarski was born in Warsaw on January 14th, 1901, the eldest son of Ignacy aka Izaak Tajtelbaum (often spelled Teitelbaum) and Róża (née Prussak). He had a younger brother Wacław (1903–1944), who was a lawyer. Alfred died on October 27th, 1983, age 82, in Berkeley, California, U.S.A. His creative working life spanned more than 60 years, starting in 1921 and continuing in Warsaw up to August, 1939, then in the U.S.A., first on the East Coast, and then, from 1942 to the end of his life, at the University of California, Berkeley.

1.1 The Warsaw Years

From 1918 to 1924 Tarski was a student in the Faculty of Philosophy at the University of Warsaw. He studied logic, mathematics and philosophy. Despite the interruptions of the Soviet westward offensive of 1918–19 and the Polish–Soviet War of 1919–1921, which intermittently obliged the University to suspend its operations,[1] he managed to fit all his undergraduate and graduate course work, sitting his doctoral exams, and researching, writing and defending his dissertation into those six tumultuous years—while at the same time publishing four journal articles, and, from January of 1924, taking high school teacher training. His thesis advisor was Stanisław Leśniewski, and his Ph.D. thesis was titled *O wyrazie pierwotnym logistyki <On the Primitive Term of Logistic>*.

One year later he successfully defended his "habilitation" thesis, securing a "*veniam legendi*" in the philosophy of mathematics and becoming a "*docent*" of the University of Warsaw—licensed to solicit, accept and fulfill commissions to lecture in its name.[2] He gave "*ćwiczenia*" (tutorials) and "*wykłady zlecone*" (commissioned lectures) on set theory, the methodology of mathematics, the foundations of "school" geometry, and the

[1] Tarski's first university year, 1918–19, was a write-off. Classes were cancelled. Students and faculty signed up for military service. Stanisław Leśniewski, Stefan Mazurkiewicz and Wacław Sierpiński worked with the military on decoding Soviet communications. Tarski performed community service in lieu of military service.

In June, 1920, lectures were again cancelled and students and faculty again volunteered, Tadeusz Kotarbiński and Jan Łukasiewicz among them. This time Tarski served with a military supply and medical unit.

[2] Jacek Jadacki speculates that Tarski's habilitation thesis was the 50-page paper *Sur les ensembles finis*—i.e., [24c]. See J.J. Jadacki (ed.), ***Alfred Tarski*: *dedukcja i semantyka* (*déduction et semantique*)**, Wydawnictwo Naukowe Semper: Warszawa 2003, p. 117. If true, Tarski must have researched and written two dissertations simultaneously.

arithmetic of the natural numbers and the reals. On October 1st, 1929, he joined the University's payroll as a "*starszy asystent*" (senior assistant), and on October 1st, 1934, he was promoted to "*adiunkt*" (no exact English equivalent; higher than senior assistant) in Jan Łukasiewicz's Philosophy Seminar. He tried for the position of "*katedra*" (department chair) at the University Jana Kazimierza in Lwów in 1930, and at the University of Poznań in 1937, but in both cases without success.

From 1925 on, he also taught high school mathematics at *Gimnazjum im. Stefana Żeromskiego* in Warsaw. Throughout this period he kept up a frenetic pace of investigation into mathematical logic, semantics, set theory, measure theory, the foundations of geometry, and the teaching of logic and geometry. Arguably, the whole of his life-long output can be traced to these roots.

His work in sentential calculi, methodology of the deductive sciences, cardinal arithmetic, the Axiom of Choice, the definition of truth and more generally semantics brought him recognition and acclaim. During these years in Warsaw he published 16 abstracts and short notes, 62 longer papers, 3 reviews, 14 exercises[3] and 2 problems, 11 contributions to discussions, and 3 books...

- ***Pojęcie prawdy w językach nauk dedukcyjnych*** <*The Concept of Truth in the Languages of the Deductive Sciences*> (1933).
- ***O logice matematycznej i metodzie dedukcyjnej*** <*On Mathematical Logic and the Deductive Method*> (1936). This has been translated into 12 other languages. Over the years it has had four English-language editions.
- A high school geometry textbook.[4]

By 1930 he had already become a leading figure of the Warsaw school of logic and mathematics, collaborating with nearly all of its preeminent members, as well as members of the Lwów school. He co-authored publications with...

- Stefan Banach (on measure theory, decompositions of point sets in metric spaces, and the paradoxical decomposition of the solid sphere).
- Adolf Lindenbaum (on set theory, and the theory of definability).
- Jan Łukasiewicz (the landmark paper *Untersuchungen über den Aussagenkalkül* <*Investigations into the Sentential Calculus*>, 1930).
- Kazimierz Kuratowski (on projective sets—the *Tarski–Kuratowski algorithm*).
- Wacław Sierpiński (on inaccessible cardinal numbers).
- Andrzej Mostowski (on Boolean rings/algebras with ordered bases).

He participated actively in academic life both at home and abroad. He was a frequent speaker at meetings of the Warsaw Philosophic Society, the Lwów chapter of the Polish Philosophic Society, the Warsaw Mathematics Society, and the Warsaw Scientific Society. He took part in the Polish Philosophy Congresses of 1923, 1927 and 1936, and the Polish

[3]For English translations see Chapter 12, *Exercises Posed by Tarski*, in A. McFarland, J. McFarland, James T. Smith (eds.), ***Alfred Tarski: Early Work in Poland—Geometry and Teaching. With a Bibliographic Supplement***, Birkhäuser: New York 2014, pp. 243–272.

[4]***Geometria dla trzeciej klasy gimnazjalnej***, co-authored with Z. Chwiałkowski and W. Schayer. For an English translation see McFarland–McFarland–Smith [2014], pp. 273–318.

Mathematics Congresses of 1927, 1931 and 1937. He also took part in international mathematics and philosophy conferences in Bologna, Italy (1928), Warsaw (1929), Prague (1934), Paris (1935, 1937), and Amersfoort, the Netherlands (1938). He played a principal role in establishing cooperation and working contacts between the Viennese and Warsaw schools. Thanks to his visits to Vienna in 1930 and 1935, and his lectures *Grundlegung der wissenschaftlichen Semantik <On the Semantics of Science>* and *Über den Begriff der logischen Folgerung <On the Concept of Logical Consequence>* given at the first Paris Congress on Scientific Philosophy in 1935, Polish logic exerted a formative influence on the Vienna Circle.

1.2 America

On August 21st, 1939 Tarski traveled to the United States to address the 5th International Congress for the Unity of Science, being held September 3–9 at Harvard University in Cambridge, Massachusetts, and to give a lecture tour at several American universities. Owing to the German invasion of Poland on September 1st, 1939, the massive aerial bombardment of Warsaw that began on the same day, and the course of events that immediately followed, leading to the land siege of Warsaw from September 8th through the 28th, Tarski was obliged for his own good to remain in America.[5]

Until the summer of 1942 he stayed on the east coast, with temporary appointments at Harvard, CUNY, and the Institute for Advanced Study at Princeton. During this period he forged relationships with Paul Erdös (at Princeton) and J.C.C. McKinsey (in New York City, though not at CUNY—McKinsey was at NYU.) With Erdös he co-authored a paper on fields of sets and large cardinal numbers, published in 1943. With McKinsey he collaborated on algebraic aspects of general topology, and applications of topological methods in intuitionistic and modal logics; their results appeared much later in three jointly authored papers published in 1944–1948. He also renewed contacts with Kurt Gödel (at Princeton) and Rudolf Carnap and W.V.O. Quine (both at Harvard at that time). In 1941 he published what later turned out to be a groundbreaking work in algebraic logic, a paper titled *On the Calculus of Relations*.

In October, 1942, upon securing a lecturer position at the University of California, Berkeley, he relocated to the west coast. In 1948 he was promoted to full professor in the Department of Mathematics at Berkeley, a position which he held to the end of his life.[6]

His first 14 years at Berkeley, up to 1956, witnessed the realization/achievement of several major scholarly and organizational undertakings which together cemented Tarski's scientific reputation on a world scale. First, he completed and published the results of some important investigations he had begun before the war. These included . . .

[5] . . . leaving his wife and children in Warsaw. In fact he had little choice, as his ship's return sailing was cancelled.

[6] Although *emeritus* from 1968, he continued teaching until 1973, and continued supervising Ph.D. candidates right up until his death in 1983.

1. The paper *The Semantic Conception of Truth and the Foundation of Semantics* (1944)—a relatively non-technical, or "technical-lite", exposition of his thinking on semantics, addressed to the philosophical community.
2. The monograph ***A Decision Method for Elementary Algebra and Geometry*** (1st edition 1948, 2nd edition 1951), received to great acclaim by the mathematics community.
3. The twin books ***Cardinal Algebras*** (1949) and ***Ordinal Algebras*** (1956). Both books developed results that had been announced without proof in the 1926 omnibus paper *Communication sur les recherches de la théorie des ensembles*, co-authored by Tarski and Adolf Lindenbaum. J.C.C. McKinsey and two of Tarski's very first doctoral students at Berkeley, Louise H. Chin[7] and Bjarni Jónsson, played major roles in editing ***Cardinal Algebras***. ***Ordinal Algebras*** was the collective effort of Tarski, Bjarni Jónsson, and C.C. Chang.
4. The anthology ***Logic, Semantics, Metamathematics: Papers from 1928 to 1938*** (1956), presenting J.H. Woodger's English translations of 17 of Tarski's works from the 1920s and 1930s on logic, the theory of truth, and the methodology of the deductive sciences. This anthology made Tarski's pre-war output available to the wider world for the first time.

Second, he marked out new lines of inquiry to do with algebraic logic, decidability, and model theory. Algebraic logic quickly gave rise to Tarski and McKinsey's *The Algebra of Topology* (1944) and *On Closed Elements in Closure Algebras* (1946); Tarski and Jónsson's *Boolean Algebras with Operators* (Part 1, 1951; Part 2, 1952); and saw work begin on the algebraization of quantificational logic with the help of cylindric algebras. The foundations of the theory of cylindric algebras were worked out by Tarski and his students Louise H. Chin and Frederick B. Thompson in the years 1948–1952. Cylindric algebras were extensively researched at Berkeley, and indeed in centers of logical and algebraic research around the world, up to the 1990s.[8]

Tarski summarized his work on decidability in the short book ***Undecidable Theories*** (1953), written in collaboration with Andrzej Mostowski and Raphael M. Robinson. This book is widely considered a masterpiece of the literature on mathematical logic. In the articles *Some Notions and Methods on the Borderline of Algebra and Metamathematics* (1952) and *Contributions to the Theory of Models: I, II, III* (1954–1955), Tarski introduced the basic conceptual apparatus of model theory—a specialized area of set-theoretic semantics for formalized languages and theories—and he showed that model-theoretic techniques are productive in mathematics. He proved, specifically, that the class of representable relation algebras is an equational class, axiomatizable by a set of equations alone—a finding which spurred further investigation of the logical aspects of relation algebras. More generally, one of the tasks model theory was supposed to fulfill was to provide mathematically tractable definitions of metamathematical and metalogical

[7]The Mathematics Genealogy Project lists her as Louise Hoy Chin Lim.

[8]For a beautifully written exposition of the first stages of these investigations, and their prehistory, see [61a]. For later developments, see [71^{m}] and [85^{m}].

concepts, such as, for instance, "arithmetical class", definability, elementary equivalence, etc. Tarski also played an important role in the formulation of a general definition of "reduced product", and he demonstrated a range of interesting applications of reduced products. In the years 1957–1962 his students and colleagues C.C. Chang, J. Howard Keisler, Dana Stewart Scott, Thomas E. Frayne and Anne C. Morel obtained what are now recognized as classic results in model theory, by applying the methods and techniques of reduced products.

Tarski also invented predicate logic with infinitely long expressions (1957) and model theory for infinitary languages. This theory and its application to the study of inaccessible cardinals were developed and elaborated by Carol Ruth Karp, William Porter Hanf, and Dana Stewart Scott. Of these, Hanf was a student of Tarski, Scott a frequent collaborator, and Karp an avid disciple.

Third, he founded a strong center of logic and foundations of mathematics at Berkeley, which by the mid-1950s already enjoyed international renown. From 1952 to 1970 Tarski was director of a program called *Basic Research in the Foundations of Mathematics*. The program covered all the main areas of the foundations of mathematics, and the number of people taking part in it was impressive. The proceedings for the period July 1st, 1959 to June 30th, 1961 list the following topics: model theory, proof theory, infinitary logics, set theory and its foundations, general theory of algebraic and relational structures, algebraic structures in logic, and the foundations of geometry. At least 20 people were listed as participating in the program as researchers, among them the Poles Jerzy Łoś, Wanda Szmielew, Andrzej Ehrenfeucht, and Jan Mycielski.

In 1958 Tarski and his colleagues established *The Group in Logic and the Methodology of Science*, whose mission was, and remains to this day, to foster and promote interdisciplinary doctoral studies leading to a Ph.D. in logic and the philosophy of science. Its idea was (and is) above all about research in mathematical logic in the broadest sense, and its applications to information theory, computability theory, artificial intelligence, methodology of science, philosophy of science, and philosophy of language. Since 1989 the Group has sponsored *The Annual Alfred Tarski Lectures*. Eminent scholars are invited to give lectures on their current areas of interest.

With the aim of strengthening and broadening international collaboration Tarski organized two international symposia at Berkeley: the first, in 1957/58, devoted to the axiomatic method, with special emphasis on its applications to geometry and physics, and the second, in 1963, devoted to model theory. The fruits of these two conferences, beyond of course their stated aim of fostering international cooperation, which they roundly succeeded in achieving, were two substantial volumes of proceedings, co-edited by Tarski: ***The Axiomatic Method*** (1959) and ***The Theory of Models*** (1965). The papers they contain highlight multifarious, subtle and often surprising connections between the axiomatic method and model theory.

In subsequent years Tarski continued working in almost every area that had occupied him previously, but especially universal algebra, equational logic and cylindric algebras, and the foundations of geometry (see §3 below). His chapter zero to ***Cylindric Algebras. Part I*** (1971) is a beautiful monograph on general algebra. He devoted the last years of his life, with the help of Steven Givant, to the book ***A Formalization of Set Theory***

without Variables, a work which at last brought to fruition Tarski's life-long program of algebraizing logic and the foundations of set theory.

In 1956–1957 he was president of the International Union of the History and Philosophy of Science. Within the framework of the Union he set up a Division of Logic, Methodology, and Philosophy of Science, whose purpose was to organize international congresses on logic, methodology and philosophy of science. In setting it up Tarski was partly carrying out one of the pre-war aims of the Unity of Science movement.

Tarski guest lectured at numerous universities in America and internationally. He participated in a huge number of conferences, symposia and congresses. He loved traveling, and was curious about the world. He was a scholar who viewed the meaning of his work not as the mere achieving of results in the form of theorems or theories but as part of a communal effort toward the discovery of scientific truth. It was for this reason that he chose to announce so many of his results jointly with his colleagues and students. The Mathematics Genealogy Project shows him as having had 26 students,[9] among them five women. Almost all of them went on to occupy prominent positions in the world of learning (C.C. Chang, Solomon Feferman, Bjarni Jónsson, H. Jerome Keisler, Richard Montague, Andrzej Mostowski, Julia Robinson, Wanda Szmielew, Robert Vaught).

Tarski took a keen interest in cultural and academic life in post-war Poland. Many a Polish logician, philosopher and mathematician benefited from his hospitality and watchful professional care at Berkeley. He visited Poland several times, including...

- For the symposium *Metody infinitystyczne <Infinitistic Methods>*, held in Warsaw in September, 1959, at which he gave a talk *On Predicative Set Theory* (which was never published). On that occasion he also visited Wrocław and gave a lecture at a meeting of the Wrocław chapter of the Polish Philosophic Society titled *Czym są pojęcia logiczne? <What are Logical Notions?>* which was not published until after his death.
- For a Methodology Colloquium on the justification of assertions and decisions (in 1961).
- For a conference on general algebra (in September, 1964).

He was awarded three honorary doctorates: by Pontificia Universidad Católica de Chile (1974), Université d'Aix-Marseille II (1978), and the University of Calgary (1982). In 1966 he was recipient of the Alfred Jurzykowski Foundation's Millennium Award. In 1981 the University of California at Berkeley awarded him its Berkeley Citation.

He was a member of the United States National Academy of Science, the British Academy, and the Royal Netherlands Academy of Arts and Sciences. In 1945 he was appointed correspondent member of the Polish Academy of Learning.[10]

[9]The Mathematics Genealogy Project defines a "student of Tarski" as someone who was awarded a Ph.D. and whose dissertation listed Tarski as "Advisor 1" or "Advisor 2."

[10]In March, 1949, he was stripped of this distinction, as were other Polish scholars living abroad at that time, for failing to repatriate.

2 Set Theory

Set theory was one of Tarski's main research interests almost all his professional life,[11] and it was one of his favourite tools for obtaining results in metamathematics, universal algebra and infinitary logic. His first published work, [21], was a term paper he wrote for Leśniewski's seminar while just a third-year student at the University of Warsaw, analysing Cantor's notion of a well-ordered set. One of his last published works, a monograph written jointly with John E. Doner and Andrzej Mostowski which was published in 1978 3 years after Mostowski's death, was a metamathematical study of the elementary theory of well-ordering.[12] Among other things it established that this theory admits elimination of quantifiers, is axiomatizable by an infinite recursive axiom set, and consequently is decidable. Tarski also co-authored, with Richard M. Montague and Dana S. Scott, a manuscript for a planned book, to have been titled ***An Axiomatic Approach to Set Theory***.[13]

Tarski's set-theoretic works focused on three major areas: (a) general or "pure" set theory, (b) connections between set theory, measure theory and Boolean algebras, and (c) presenting aspects of set theory in abstract algebraic calculi. We cite five examples:

2.1 Cardinal Arithmetic

He developed the arithmetic of cardinal numbers, and he established recursive formulas for the exponentiation of alephs.[14]

2.2 Axiom of Choice

He studied the Axiom of Choice by looking for equivalents of it, weaker forms of it, and equivalents of its weaker forms, and mapping out logical relations between the Axiom of Choice and other sentences of set theory, such as the Generalized Continuum Hypothesis,

---

[11] He seemed to tire of it for a spell in 1936 when he wrote to Karl Popper, "*Ich arbeite an einer Monographie aus der Mengenlehre, aber es interessiert mich wenig: alte Sachen, mit denen ich mich schon seit Jahren nicht beschäftigt habe.*" It is not clear if "*schon seit Jahren nicht*" was truth or posturing. Or possibly he just meant he had lost interest in writing a survey of established results; he preferred to work on getting new results. Sadly, the monograph he was referring to, ***Theorie der eineinendeutigen Abbildungen***, which he was writing jointly with Adolf Lindenbaum and which was to have been "*ein großes mathematisches Buch*", paid with its life. No working drafts ever surfaced, as far as anyone today knows.

[12] Where "elementary theory of well-ordering" is understood as the set of all formulas of first-order predicate calculus that are true in every structure $<U, R>$, where U is a non-empty set and R is a (binary) relation which well-orders the set U.

[13] Montague died in 1971. For reasons which remain unclear, Scott and Tarski ceased work on the manuscript in 1972. It remains unpublished to this day.

[14] See [25] and [30f].

and sentences postulating the existence of very large cardinal numbers. Wacław Sierpiński had proposed such a research program in 1918, in his survey paper *L'axiome de M. Zermelo et son rôle dans la Théorie des Ensembles et Analyse*. From the late 1920s until the mid 1950s Tarski and Sierpiński collaborated and competed to populate the program with results.

Tarski established about 30 equivalents of the Axiom of Choice, many of which concerned operations on or relations between cardinal numbers.[15] One of the simplest he found was this proposition:

$\mathfrak{m} = \mathfrak{m} \cdot \mathfrak{m}$ *for every infinite cardinal number* $\mathfrak{m}$.

From time to time he posed open questions. In 1924 he asked whether "$\mathfrak{m} = 2 \cdot \mathfrak{m}$" was equivalent to the Axiom of Choice.[16] This was finally answered—in the negative, by Gershon Sageev—only in 1975. The solution had to wait for a new technique of model construction to be invented.

In [26a] Lindenbaum and Tarski stated without proof that the Generalized Continuum Hypothesis (in the version that speaks of "transfinite numbers" rather than alephs) implied the Axiom of Choice.[17] Proofs of this remarkable result were found by others only in 1947 (by Wacław Sierpiński) and 1954 (by Ernst Specker).

In 1939 he established that the Axiom of Choice is a consequence of [one version of] the assertion that inaccessible sets exist[18]:

For every set A there exists a set M with the following properties:

(i) *A is equipollent (equinumerous) to a subset of M*;
(ii) *the family of subsets of M which are not equipollent to M is equipollent to M*;
(iii) *there exists no set B such that the family of all subsets of B is equipollent to M*.

As mentioned above, he also studied weaker versions of the Axiom of Choice, such as the principle of dependent choice,[19] and the Boolean prime ideal theorem.[20]

2.3 Finite Sets

He gave both a philosophical and a formal analysis of the notion of "finite set" that avoided making any use of the Axiom of Choice or the axiom of infinity, and he showed how the arithmetic of natural numbers could be defined in purely set-theoretic terms if finiteness was taken to mean:

[15] See [24a], [26], [38d], [39b], [48b], [49], [54], [64^{a}b].

[16] In [49^{m}] cardinals satisfying "$\mathfrak{m} = 2 \cdot \mathfrak{m}$" were introduced on an abstract level as "idem-multiple" ($a + a = a$) elements of a cardinal algebra.

[17] Adolf Lindenbaum first posed it as an open question. In 1925 Lindenbaum and Tarski jointly proved it, and asserted it without proof in [26a], §1, page 314, theorem 94.

[18] See [39b].

[19] See [48b].

[20] See [54^{a}f], [54^{a}g], [54^{a}h].

The set A is finite iff every non-empty family of subsets of A contains a minimal element with respect to inclusion.[21]

He framed sentences respectively equivalent to the Axiom of Choice and to the Generalized Continuum Hypothesis in terms of finite sets as so defined.[22] In [65ᵃ] he announced several results obtainable in "weak" set theories—i.e., set theories without the Axiom of Choice—having to do with D-finite infinite sets and D-finite infinite cardinals.[23]

The existence of such sets and cardinals had been conjectured by Henri Lebesgue (1904) and by Russell and Whitehead ("mediate cardinals" in the terminology of ***Principia Mathematica***, *124·61). Tarski proved there was a set S of D-finite infinite cardinals that was isomorphic to the set $\mathbb{R}$ of real numbers, under the "natural" ordering of cardinals and reals. Issues of this kind could only have occurred to someone who cared about fundamental things; for whom the notion of a set still required critical analysis. His work on finiteness inspired many other authors. Andrzej Mostowski was the first, with a metamathematical treatise titled *On the Independence of Definitions of Finiteness in a System of Logic* (1938), followed by Azriel Lévy (1958), Arthur L. Rubin, Jean E. Rubin, Erik Ellentuck (1962, 1965, 1968), John K. Truss (1972, 1984) and Agatha C. Walczak-Typke (2005).

2.4 Theory of Large Cardinals

In a series of papers spanning a period of almost 35 years, from 1930 to 1964, he laid the foundations for the theory of large cardinal numbers, including . . .

- a new definition of strong inaccessibility, and characteristic properties of inaccessible cardinals[24];
- several formulations of axioms asserting the existence of inaccessible sets (one such axiom cited above)[25];
- characteristic properties of strongly compact, measurable, and weakly compact cardinals, and open questions surrounding them[26];

---

[21] See [24c]. The reader should bear in mind that "minimal element" and "least element" are different concepts. Notice that this definition is independent of the notion of a finite natural number.

[22] See [38c], page 163.

[23] A set is said to be *Dedekind-finite*, or *D-finite*, iff there is no bijection of the set onto a proper subset of itself (or equivalently, iff every one-to-one mapping of the set into itself is surjective.) A cardinal number is said to be a *Dedekind-finite cardinal*, or a *D-finite cardinal*, iff it is the cardinality of a D-finite set. In weak set theory—without the Axiom of Choice—it can be proved that if a set is finite by Tarski's definition then it is also D-finite, but it cannot be proved that if a set is D-finite then it is also finite by Tarski's definition.

In [49ᵐ] D-finite cardinals were introduced on an abstract level as *finite* elements of a cardinal algebra.

[24] See the joint paper with Wacław Sierpinski [30a].

[25] See [38a] and [39b].

[26] See [62]; the joint papers [43] and [61b] with Paul Erdös; and the joint paper [64] with H. Jerome Keisler.

- a research program to investigate interconnections between large cardinals and infinitary logic (some of Tarski's students—William Porter Hanf and H. Jerome Keisler among them—obtained important results in this area).

2.5 Algebraization of Set Theory

He algebraized key aspects of the general theory of sets, having to do with order types, relation types, cardinal and ordinal arithemetics and operations on infinite sequences. He summarized his work in these areas in the books: ***Cardinal Algebras*** [49^{m}], and ***Ordinal Algebras*** [56^{m}a]. Both books could trace their origins to the 1926 Lindenbaum–Tarski paper *Communication sur les recherches de la théorie des ensembles*, in which Lindenbaum's role had been immense. Tarski conceded that many results "were originally established by Lindenbaum" and were first stated without proof in [26]. He added that it was impossible to convey "an adequate idea of the extent of my indebtedness" to Lindenbaum.

Substantial parts of several classic texts on set theory are based on Tarski's results:

- W. Sierpiński's ***Zarys teorii mnogości***, third edition 1928; and ***Leçons sur les nombres transfinis***, 1928, second edition 1950 (cardinal arithmetic; and equivalents of the Axiom of Choice);
- Sierpiński's ***Cardinal and Ordinal Numbers***, 1958, second edition 1965 (equivalents of the Axiom of Choice and proofs of theorems from Tarski–Lindenbaum [26]);
- H. Bachmann's ***Transfinite Zahlen***, 1955, second edition 1967 (equivalents of the Axiom of Choice and inaccessible numbers);
- K. Kuratowski and A. Mostowski's ***Set Theory***, second edition 1976 (generalizations of Banach's formulation of the Cantor-Bernstein theorem, Tarski's recursive formulae for exponentiation of alephs; the number of prime ideals in the power-set algebra of an arbitrary set; basic cardinal equivalents of the Axiom of Choice; the exposition of higher types of inaccessible numbers).

Tarski almost singlehandedly steered the development of set theory: dictating what counted as set theory; what were its important questions, results, applications, methods and tools; where were its frontiers; what was worth working on, and why. In the words of Azriel Lévy[27]:

> Alfred Tarski started contributing to set theory at a time when the Zermelo-Fraenkel axiom system was not yet fully formulated, and as simple a concept as that of the inaccessible cardinal was not yet fully defined. At the end of Tarski's career the basic concepts of the three major areas and tools of modern axiomatic set theory, namely constructibility, large cardinals and forcing, were already clearly defined and were in the midst of rapid successful development. The role of Tarski in this development was somewhat similar to the role of Moses showing his people the way to the Promised Land and leading them along the way, while the actual entry into the Promised Land was done mostly by the next generation. The theory of large cardinals was started mostly by Tarski, and developed mostly by his school. The mathematical logicians of Tarski's school contributed

[27] A. Lévy, *Alfred Tarski's Work in Set Theory*, ***The Journal of Symbolic Logic***, vol. 53 (1988), pp. 2–6; p. 2.

much to the development of forcing, after its discovery by Paul Cohen, and to a lesser extent also to the development of the theory of constructibility, discovered by Kurt Gödel. As in other areas of logic and mathematics Tarski's contribution to set theory cannot be measured by his own results only; Tarski was a source of energy and inspiration to his pupils and collaborators, of which I was fortunate to be one, always confronting them with new problems and pushing them to gain new ground.

3 Geometry and Measure Theory

Tarski's work in geometry showcases his fascination with metamathematics, axiom systems and the axiomatic method, and it reveals his genius for developing innovative tools and techniques to extract new ore from old seams. He gave serious thought to the choice of primitive concepts, axiom sets and logical bases for various geometries—affine, hyperbolic, elliptical, Euclidean, projective—and, for each geometry, to the independence, economy and strength of its axioms, the independence, definability and predicativity of its notions, and the decidability and completeness of the system as a whole.[28] A volume he co-edited with Leon Henkin and Patrick Suppes, [59^{e}], ***The Axiomatic Method, with Special Reference to Geometry and Physics***, underscores geometry's pride of place by its title alone.

Tarski's earliest paper on geometry, [24b], *O równoważności wielokątów <On the Equivalence of Polygons>*, was of a different character. It was primarily a study of measure-theoretical properties of point sets in the Euclidean plane, $\mathbb{R}^2$. The paper is important from a historical perspective as it contains:

- One of the first formulations of the definition of equivalence by finite decompositions: "Two geometric figures (thus in particular, two polygons) are called equivalent when it is possible to divide them into the same finite number of respectively congruent *arbitrary* geometric figures not having *any* common points."
- A proof of the Wallace–Bolyai–Gerwien theorem: *two polygons are equivalent iff they have equal areas.* Tarski's proof relied essentially on Banach's theorem on additive measures on the plane.
- The first announcement of the result known as the Banach–Tarski Paradox: "Any two polyhedra are equivalent" or, in an alternative formulation: "Any cube can be divided into a finite number of parts (without common points), which can then be reassembled to form a cube with an edge twice as long".[29]
- The question, "C*zy koło i wielokąt o równych polach są równoważne*?" (Are a circle and a polygon with equal areas equivalent [by finite decompositions]?). This question has since come to be known as "the Tarski Circle Squaring Problem", ... a name with echoes of antiquity.[30]

[28] See [34], [56c], the joint paper [56b] with E.W. Beth, and the joint papers [65a] and [79] with L.W. Szczerba.

[29] The Banach–Tarski Paradox was fully articulated, and its proof elaborated, in [24d], a subsequent paper co-authored with Banach.

[30] Tarski's circle-squaring problem was finally answered in the affirmative in 1990 by Miklós Laczkovich. See: M. Laczkovich, *Equidecomposability and Discrepancy: a Solution to Tarski's Circle-squaring*

Jan Mycielski, an authority on paradoxical decompositions, writes:

> The Banach-Tarski paradox is not an inconsistency of mathematics, although it shows that 2 equals 1 in a certain sense. It is proved within the usual system of set theory (using the Axiom of Choice), and we have no reason to doubt the consistency of that system. But it destroys certain naive intuitions about point sets in three-dimensional space $\mathbb{R}^3$....
>
> This theorem is a striking demonstration that the unrestricted concept of a set of points has little to do with the idea of a physical body, and also that, to develop a reasonable theory of areas, volumes, etc., one must limit oneself to more special sets (e.g., Borel sets, or Lebesgue measurable sets).[31]

In [30] Tarski gave an effective proof—i.e., without using the Axiom of Choice—that for any infinite set E the following two statements were equivalent:

1. *There exists a finitely additive two-valued measure m on the family of all subsets of E such that $m(E) = 1$ and for every finite subset $F \subseteq E$, $m(F) = 0$.*
2. *There exists a maximal ideal in the field of all subsets of E.*

Then he proved that 2 was a theorem of general set theory with the Axiom of Choice added—i.e., that the Axiom of Choice implied 2. His demonstration of a maximal ideal relied on the well-ordering principle and induction on transfinite ordinals. He developed this same technique and line of reasoning further in the papers [39, 45] on ideals of various types—$\mathfrak{m}$-additive, prime, $\mathfrak{p}$-saturated—in complete fields of sets.[32]

In [29b] and [38g] Tarski explored logical connections between the existence of paradoxical decompositions and the non-existence of some invariant measures. Following von Neumann's terminology, for any arbitrary set E, any subset I of E, and any group G of transformations of E, a function m defined on the power set of E with values in the set of all nonnegative real numbers is said to be an $[E, I, G]$*-measure* iff m is additive, m is invariant under transformations in G, and $m(I) = 1$. Tarski's remarkable result was:

In order for an $[E, I, G]$-measure to exist, it is necessary and sufficient that there be no paradoxical decompositions of I relative to G.

For an algebraic treatment of this and related issues see [49^{m}], ***Cardinal Algebras***, theorems 14.13, 16.8, 16.12 and 16.13.

Tarski's particular achievement in the foundations of geometry was conceiving the idea of a system of geometry based only on first-order logic with identity (i.e., elementary logic) and completely free from any set theoretical assumptions (set theoretical notions, primitive terms, axioms, rules of inference), and showing that such a system was plausible. He called this system *elementary geometry*.

Problem, ***Journal für die Reine und Angewandte Mathematik***, vol. 404 (1990), pp. 77–117. Laczkovich proved that the circle could be decomposed into no more than 10^{50} different pieces, which could be rearranged to compose a square of equal area. He needed the Axiom of Choice to obtain his decomposition, which was highly non-constructive.

[31] J. Mycielski, Review of ***The Banach–Tarski Paradox*** by Stan Wagon, ***The American Mathematical Monthly***, vol. 94, no. 7, pp. 698–700. The quoted passages are from page 698.

[32] Though published 6 years apart, [39] and [45] were nominally parts I and II of the same two-part paper, and appeared in consecutive issues of ***Fundamenta Mathematicae***, vol. 32, pp. 45–63, and vol. 33, pp. 51–65. The journal's operations were interrupted by the Second World War.

Tarski's axiomatic system of two-dimensional elementary geometry, $\mathcal{E}_2$, has only two specific (or primitive) notions: a ternary predicate constant β for the relation of betweenness and a quaternary predicate constant δ for the equidistance relation. The formula $\beta(x,y,z)$ is read: *y lies between x and z*, while $\delta(x,y,z,u)$ is read: *x is as distant from y as z is from u*. Individual variables represent only points. By contrast, Hilbert's system of geometry has other primitive geometrical notions in addition to points, for instance lines and planes. Tarski limited himself to points by exploiting the fact that figures (configurations of points) dealt with in traditional geometry are uniquely determined by finite numbers of points.

His system $\mathcal{E}_2$ is based on twelve axioms and one axiom schema, each of which is framed solely in primitive and logical terms. Despite avoiding using defined terms, the axioms are relatively short and their meanings are intuitively clear. The axiom schema defines an infinite recursive set of axioms, and plays the role of Dedekind's continuity axiom—or more precisely, it corresponds to Dedekind's continuity axiom restricted to [all and only] sets which are first-order definable in the language of $\mathcal{E}_2$. Relaxing this restriction and going to a second-order language leads directly to the whole of Euclidean geometry. This means that Tarski did successfully separate elementary geometry from full geometry.

To define a model for $\mathcal{E}_2$ Tarski applied the apparatus of his formal semantics: in particular, his definition of truth. Namely, a relational system $\mathcal{M} = \langle A, B, D \rangle$ is a *model* of $\mathcal{E}_2$ iff...

(i) A is an arbitrary non-empty set, and B and D are respectively a ternary and a quaternary relation among elements of A; and
(ii) All the axioms of $\mathcal{E}_2$ hold in $\mathcal{M}$ if all the variables are assumed to range over elements of A, and the primitive constants β and δ are understood to denote the relations B and D, respectively.

The representation theorem reads:

A relational system $\mathcal{M}$ is a model of $\mathcal{E}_2$ iff $\mathcal{M}$ is isomorphic with the Cartesian space $\mathcal{C}_2(\mathcal{F})$ over some real closed field $\mathcal{F}$.

Thus the axiom set for $\mathcal{E}_2$ is sound and adequate. Moreover, since the set of its theorems is identical with the set of all true sentences in the Cartesian space $\mathcal{C}_2(\mathbb{R})$ over the field of real numbers, the system $\mathcal{E}_2$ is complete, and consequently decidable. And the so-called "Non-finitisability theorem" says that $\mathcal{E}_2$ is not finitely axiomatizable.

Tarski [59], *What is elementary geometry?*, was first published in [59e]: L. Henkin, P. Suppes and A. Tarski (eds.), ***The Axiomatic Method, with Special Reference to Geometry and Physics***, pp. 16–29. It is one of that volume's most highly polished gems. Perhaps it owes its lapidary sheen to 33 years of being turned over and over again in Tarski's great tumbler of a head, starting all the way back in 1926.

For a detailed account of the long and fascinating history of Tarski's thinking on his system of elementary geometry see [99].

4 Decidable and Undecidable Theories

Let's begin with a quotation:

> By a *decision procedure* for a given formalized theory T we understand a method which permits us to decide in each particular case whether a given sentence formulated in the symbolism of T can be proved by means of the devices available in T (or, more generally, can be recognized as valid in T). The *decision problem* for T is the problem of determining whether a decision procedure for T exists (and possibly of exhibiting such a procedure). A theory T is called *decidable* or *undecidable* according as the solution of the decision problem is positive or negative. As is well known, the decision problem is one of the central problems of contemporary metamathematics. Since only few theories turn out to be decidable, most endeavors are directed toward a negative solution.[33]

Although the search for particular decision *procedures* was plainly evident in Ernst Schröder's focus on the "*Lösungsprobleme*" (solution problems) for his relation algebras, an awareness that the decision problem as defined in [53^{m}] was a metamathematical issue in its own right emerged later, in the Hilbert school, with its work on the "*Entscheidungsproblem*" (decision-making problem) to do with the decidability of the "restricted functional calculus"—a particular version of first-order predicate logic. In 1936 Alonzo Church and Alan Turing published independent papers showing that, in general, first-order theories were undecidable. Some special cases of first-order theories, such as Presburger arithmetic, were shown to be decidable, but these were more the exception than the rule.

In his 1946 Princeton address[34] Tarski expressed the view that the decision problem was one of the central issues—if not *the* central issue—of metamathematics of the day, noting that, "Hilbert considered the main task of logic to be the construction of a symbolism for use in solving the general decision problem: this was the *raison d'être* of metamathematics." He presented various open questions in logic and mathematics, at that time including Cantor's continuum hypothesis and Hilbert's tenth problem. Then he surveyed contemporary work and known results on the decision problem and proposed a research program on it.

Broadly speaking, Tarski's contributions to the decision problem were:

1. Proving that certain important theories formalized in classical first-order predicate logic were decidable.
2. Laying the foundations for a general method for proofs of undecidability.
3. Proving the undecidability of certain important first-order and non-first-order theories.

There are several ways of proving theories decidable—model-theoretic techniques, syntactic methods, and others. One way is simply to reduce the decidability of the theory under question to that of a theory already known to be decidable, either by an

[33] ***Undecidable Theories*** by Alfred Tarski, in collaboration with Andrzej Mostowski and Raphael M. Robinson. North-Holland Publishing Co., Amsterdam, 1953. The quoted paragraph is from Chapter I, *A General Method in Proofs of Undecidability*, §I.1. Introduction, page 3.

[34] See H. Sinaceur (ed., with introduction), *Address at the Princeton University Bicentennial Conference on Problems of Mathematics, December 17–19, 1946, by Alfred Tarski*, **The Bulletin of Symbolic Logic**, vol. 6 (2000), pp. 1–44. See also *Odczyt Alfreda Tarskiego na Konferencji o Problemach Matematyki w Princeton, 17 grudnia 1946*, in [01^{m}], pp. 396–413.

embedding, an extension, or a representation. In proofs of decidability, the full-blown theory of recursive functions is often not required. It is enough just to recognize that certain procedures are algorithmic or effectively calculable.[35]

Historically one of the earliest techniques ever used for proving a theory decidable was the method of eliminating quantifiers... a syntactical "process" whose first appearance in the literature was in a paper by Leopold Löwenheim (1915),[36] and which seems to have been inspired by Schröder's work on the *Eliminationsproblem*. The scare quotes around "process" are intended as a warning that quantifier elimination is not a mechanical procedure. Doner and Hodges (1988) gives a beautifully readable précis of how it works[37]:

Let T be a first-order theory. We say that a set Φ of formulas is a *set of basic formulas for* T (or more briefly a *basic set*) if, using the axioms of T, one can prove that every formula ϕ of the language is equivalent to a Boolean combination ϕ^* of formulas in Φ which has only the same free variables as ϕ.

To analyze T by the method of quantifier elimination, one would look for a basic set for T. Every axiom system has at least one basic set, namely the set of all formulas of the language. But one would try to find a better basic set than that. There is no exact criterion for a "good" basic set, but one would hope for a basic set Φ with at least the following three properties. (1) It should be reasonably small and irredundant. (2) Every formula in Φ should have some straightforward mathematical meaning. (3) There should be an algorithm for reducing every formula ϕ to its corresponding ϕ^*.

(3) is precise. When we have it, we can claim to have effective quantifier elimination for T. (1) and (2) are more a matter of judgment.

In the best cases we have one thing more: (4) an algorithm which tells us, given any basic sentence Ψ, either that Ψ is provable or that it is refutable from T. Given (3) and (4), we have both a completeness proof and a decision procedure for the theory T.

To quote Chang and Keisler, "This method applies only to very special theories. Moreover, each time the method is applied to a new theory we must start from scratch... On the other hand, the method is extremely valuable when we want to beat a particular theory into the ground. When it can be carried out, the method of elimination of quantifiers

[35]For more comments on this see Tarski [48^{m}], note 10.

[36]See: L. Löwenheim, *Über Möglichkeiten im Relativkalkül <On Possibilities in the Calculus of Relatives>*, ***Mathematische Annalen***, vol. 76 (1915), pp. 447–470.

[37]The four paragraphs in the box are an extended quote from: John Doner and Wilfrid Hodges, *Alfred Tarski and Decidable Theories*, ***The Journal of Symbolic Logic***, Vol. 53, No. 1, (March, 1988), pp. 20–35. The text reproduced here is from page 24 of their article, where they present it in three paragraphs, not four.

gives a tremendous amount of information about a theory ... [it] may be thought of as a direct attack on a theory."[38]

Three things should be emphasized. First, quantifier elimination is not just, or even primarily, about proving a theory decidable. It is about discovery generally. Second, it can be applied to all formulas, not just to sentences. Free variables are welcome. Third, it is more art than science. But when it works, it works a treat.

Tarski encountered the method as a student and applied it in his seminar exercises when he was a *docent*. One of the exercises he set Mojżesz Presburger was to use the method of quantifier elimination to investigate the completeness of the arithmetic of integers with addition.[39]

Using the method of elimination of quantifiers, Tarski contributed to a deeper understanding of the first-order theories of dense orders and linear orders.[40] He proved, among other things, that every sentence in the elementary theory of dense order, T_{DO}, was deductively equivalent to a Boolean combination of the following two (basic) sentences: "There is no first element", and "There is no last element". Since the elimination of quantifiers was "effective", in Doner and Hodges's precise sense (3) above, T_{DO} turned out to be decidable. As a corollary, Tarski obtained a classification of all axiomatic extensions of T_{DO}, as well as a semantic characterization of all complete extensions of T_{DO}. There are four such complete extensions—the theories of the order types η, $\eta + 1$, $1 + \eta$, and $1 + \eta + 1$.[41]

Together with Andrzej Mostowski and John Elliott Doner, Tarski proved that the elementary theory of well-ordering was decidable, described the ordinals definable by first-order formulas, and gave a description and classification of all models of the theory.[42]

Again using quantifier elimination he obtained a series of decidability results on "elementary algebra and geometry" and "related systems".[43] It is worth noting a few

[38]See: C.C. Chang and H.J. Keisler, ***Model Theory***, North-Holland: Amsterdam 1973 (3rd edition, 1990), pages 49–60. This quotation is from page 49 in the 3rd edition.

[39]See: M. Presburger, *Über die Vollständigkeit eines gewissen Systems der Arithmetik ganzer Zahlen, in welchem die Addition als einzige Operation hervortritt*, ***Sprawozdanie z I Kongresu Matematików Krajów Słowiańskich, Warszawa 1929*** (=***Comptes-rendus du I Congrès des Mathématiciens des Pays Slaves***), Księżnica Atlas: Warszawa 1930, pp. 92–101 & p. 395; see p. 97, footnote 1.

See also: W. Hodges, *A Visit to Tarski's Seminar on Elimination of Quantifiers*, in J. van Benthem et al. (eds.), ***Proof, Computation and Agency***, **Synthese Library** 352, Springer Science+Business Media B.V. 2011; pp. 53–66.

[40]In [36d], §5, Tarski axiomatized these two theories, calling them, respectively, "*die elementare Theorie der dichten Anordnung*" (p. 290), and "*die elementare Theorie der isolierten Anordnung*" (p. 294). In footnote 1 on page 293 he asserted that his results on the theory of dense order supplemented those reported in Langford's 1927 paper. (Literally: "*Die unten angegebenen Tatsachen, die diese Theorie betreffen, bilden die Ergänzung der Ergebnisse Langfords*"). He used the phrase "*sukzessiven Elimination der Operatoren*" (successive elimination of quantifiers) in print for the first time on pp. 293 and 295... somewhat odd considering he had been promoting the method since 1926 or earlier.

[41]By "a theory of an order type α" is meant, the set of all first-order sentences true in any structure $<X,R>$, where the relation R orders the set X according to the type α.

[42]See [78]. Precise attribution of specific results is a long and complicated story. See especially the historical comments on page 1.

[43]These results were announced in a variety of published works, scattered over time, and in varying notations and terminologies. See: [31], [48^{m}], [48^{m}](1), [49^{a}d], and [67^{m}]. The last was originally intended for publication in 1939 but was interrupted by the Second World War.

of these results that are pertinent to the ordered field of real numbers. (For those related to geometry, see §3 above.)

Let $\mathcal{R} = \langle R, +, \cdot, >, 0, 1, -1\rangle$ be this field, where $+, \cdot, >, 0, 1$ and -1 have their usual meanings. The system of elementary algebra is by definition the set Th($\mathcal{R}$) of all first-order sentences formulable in the vocabulary of $\langle R, +, \cdot, >, 0, 1, -1\rangle$ which are *true* in $\mathcal{R}$. Tarski proved that the theory Th($\mathcal{R}$) admits effective elimination of quantifiers, and he described a decision procedure for it. Moreover, he showed that Th($\mathcal{R}$) is axiomatized by the first-order axioms for real closed fields and, in consequence, any two real closed fields are arithmetically indistinguishable (or, using Tarski's later terminology, elementarily equivalent). As a by-product, one gets the following important theorem:

- *a set of real numbers is first-order definable in the field* $\mathcal{R}$ iff *it is a sum of a finite number of intervals* (*open or closed, bounded or unbounded*) *with algebraic end points.*

In [48^{m}](1), p. 45, Tarski asked about the decidability of the elementary theory of the real field expanded by the "exponential with a fixed base, for example base 2", i.e., the theory Th ($\langle R, +, \cdot, >, 0, 1, -1, Exp\rangle$), where *Exp* is the unary operation given by the formula $y = 2^x$ for all $x \in R$. Commenting on this problem Tarski wrote, "The decision problem for the system just mentioned is of great theoretical and practical interest. But its solution seems to present considerable difficulties. These difficulties appear, however, to be of purely mathematical (not logical) nature: they arise from the fact that our knowledge of conditions for solvability of equations and inequalities in the enlarged system is far from adequate." In [67^{m}a], p. 38—an earlier version of [48^{m}], originally to have appeared in 1940, with an emphasis on completeness in its title—Tarski posed the problem of finding a complete set of axioms for the theory Th ($\langle R, +, \cdot, >, 0, 1, -1, Exp\rangle$), commenting on it in a similar vein: "The attempt to carry out [this task] is confronted by difficulties of a purely mathematical nature which nevertheless do not appear to raise any question of principle".[44]

Another question raised by Tarski in [48^{m}] was about the decidability of the theory of the structure Th $\langle R, +, \cdot, >, 0, 1, -1, Al\rangle$ where the relation $Al(x)$ means "x is an algebraic number". Abraham Robinson proved (1959) that one obtains a complete theory

[44]Tarski was perfectly right that difficulties resided in mathematics, and this has been confirmed by the further development of algebraic geometry. The importance of the two problems is discussed in: D.E. Marker, *Model Theory and Exponentiation*, **Notices of the American Mathematical Society**, vol. 43 (1996), pp. 753–759.

Tarski's theorem that the ordered field of real numbers (a first-order theory) admits quantifier elimination triggered a stream of research. Important contributions were made by: Abraham Seidenberg (1954), Abraham Robinson (1959 and 1971), Stanisław Łojasiewicz (1964–65), Paul Joseph Cohen (1969), Joseph R. Shoenfield (1971), George Edwin Collins (1982), and Helmut Wolter (1986). They are summarized in an expert presentation by Lou van den Dries: *Alfred Tarski's Elimination Theory for Real Closed Fields*, ***The Journal of Symbolic Logic***, vol. 53 (1988), pp. 7–19.

A comprehensive review by Charles I. Steinhorn of Alex J. Wilkie's paper *Model Completeness Results for Expansions of the Ordered Field of Real Numbers* in ***The Journal of Symbolic Logic***, vol. 64 (1999), pp. 910–913, contains a survey of research up to the end of the 1990s stimulated by Tarski's beautiful, influential, and far-reaching monograph [48^{m}], ***A Decision Method for Elementary Algebra and Geometry***. See also Bob F. Caviness and Jeremy Russell Johnson (eds.), ***Quantifier Elimination and Cylindrical Algebraic Decomposition***. Wien, New York: Springer 1998.

by adding to the axioms for the real closed ordered fields sentences which state that the elements of the model which satisfy *Al*(x) constitute a real closed proper subfield which is dense in the entire model.

Tarski once opined, "In fact, I am rather inclined to agree with those who maintain that the moments of greatest creative advancement in science frequently coincide with the introduction of new notions by means of definition".[45] As Feferman and Feferman observed, a touch swellheaded. But he knew he could get away with it. Tarski himself defined many notions that contributed enormously to advances in logic, metalogic and metamathematics. A general theory of undecidability could never have been created if Tarski had not given precise mathematical definitions of two concepts: (1) that of an essentially undecidable theory, and (2) of the interpretability of one theory in another.

The two concepts were introduced, and a theory of undecidability expounded, in the paper *A General Method in Proofs of Undecidability*, which constituted Chapter 1 of the book [53^{m}], ***Undecidable Theories***. Though authorship of the book as a whole was credited to Tarski in collaboration with Andrzej Mostowski and Raphael M. Robinson (and with some later contributions by Julia Robinson), this first part was penned by Tarski alone.

Theories considered in [53^{m}] were all formalized in a standard way within first-order predicate logic with identity (without predicate variables). The set of syntactically articulable formulas of the language of any theory was general recursive. A set of *valid sentences* was identified for each theory T, subject to the condition that the set be closed under consequence operations of predicate logic. T was said to be *axiomatizable* if every valid sentence could be logically derived from a fixed recursive set of axioms, and *finitely axiomatizable* if the set of axioms could be finite. If the set of valid sentences was recursive, T was said to be *decidable*; otherwise T was called *undecidable*.

A theory T_2 was said to be an *extension* of a theory T_1 (and T_1 a *subtheory* of T_2) if every valid sentence of T_1 was valid in T_2. Two theories T_1 and T_2 were said to be *compatible* if they had a common consistent extension—equivalent to saying that the union of T_1 and T_2 was consistent. And finally, a theory T was said to be *essentially undecidable*, if it was not only undecidable, but also every consistent extension of T with the same constants as T was undecidable.[46,47] Key properties of undecidability and essential undecidability were then presented in several theorems, among them the following:

- *Let* T_1 *and* T_2 *be two compatible theories such that every constant of* T_2 *is also a constant of* T_1. *If* T_2 *is finitely axiomatizable and essentially undecidable, then* T_1 *is undecidable, and so is every subtheory of* T_1 *which has the same constants as* T_1 (Theorem 6, page 18).

[45] See [44a], p. 359.

[46] Tarski's use of the word "essentially" here came from a distinction he drew between two different ways of extending a theory. An extension T_2 of T_1 was called *inessential,* if every constant of T_2 which did not occur in T_1 was an individual constant, and every valid sentence of T_2 was derivable from a set of valid sentences of T_1.

[47] By the work of Church and Rosser from the year 1936, elementary Peano arithmetic is essentially undecidable.

To exploit all these mechanisms, Tarski needed two more notions: (1) interpretability, and (2) relativization of quantifiers.

First, interpretability. Let T_1 and T_2 be two theories. Assume, first, that they have no non-logical (or "specific") constants in common. In this case, T_2 is said to be *interpretable* in T_1, if T_1 can be extended by adding to its set of valid sentences "possible" definitions of the non-logical constants of T_2 in such a way, that the resulting extension of T_1 turns out to be an extension of T_2 as well. Then, in the case when T_1 and T_2 do have some non-logical constants in common, first rewrite T_2 with new non-logical constants that do not occur in T_1, but without changing the syntactical structure of T_2 in any other respect. If the resulting theory $\mathsf{T}_2{}'$ is interpretable in T_1, then T_2 may be said to be interpretable in T_1 as well. And last, a theory T_2 is said to be *weakly interpretable* in T_1, if T_2 is interpretable in some consistent extension of T_1 with all the same constants (both logical and non-logical).

Now, relativization of quantifiers.[48] Let P be any unary predicate constant not in the language of T. For every formula Ψ of T in which P does not occur, replace all occurrences of $\forall x\Psi$ with $\forall x(Px \rightarrow \Psi)$, and replace all occurrences of $\exists x\Psi$ with $\exists x(Px \wedge \Psi)$. Call the result $\mathsf{T}^{(P)}$. Notice that this construction creates a one-to-one mapping from the formulas of T to the formulas of $\mathsf{T}^{(P)}$. Apply this mapping to the set of valid sentences in T, then apply standard consequence operations to the image-set to derive its logical closure in $\mathsf{T}^{(P)}$, and let that closure be the set of valid sentences of $\mathsf{T}^{(P)}$. Then $\mathsf{T}^{(P)}$ is said to be *obtained from* T *by relativization of the quantifiers of* T *to* P.

The requirement that P initially not occur in the language of T was, as things turned out, not logically necessary. But Tarski could not prove his main meta-theorem about $\mathsf{T}^{(P)}$ without making this assumption.

The transformation of T into $\mathsf{T}^{(P)}$ preserves some important properties of T in $\mathsf{T}^{(P)}$. For instance...

(1) $\mathsf{T}^{(P)}$ is axiomatizable iff T is axiomatizable.
(2) When only finitely many individual constants and operator symbols occur in T, $\mathsf{T}^{(P)}$ is finitely axiomatizable iff T is finitely axiomatizable.
(3) $\mathsf{T}^{(P)}$ is essentially undecidable iff T is essentially undecidable (Theorems 9 & 10).

By merging interpretability with relativization one obtains *relative interpretability*. A theory T_2 is said to be *relatively interpretable* [*/relatively weakly interpretable*] in a theory T_1 iff the correlated theory $\mathsf{T}_2{}^{(P)}$ (obtained by relativizing the quantifiers of T_2 to a predicate P) is interpretable [/weakly interpretable] in T_1 in the senses defined above.

To apply this apparatus, one needs a simple, finitely axiomatizable, essentially undecidable theory. The second paper in [53^{m}], *Undecidability and Essential Undecidability in Arithmetic*, written jointly by A. Mostowski, R.M. Robinson, and A. Tarski, constructs just such a theory. This is the famous theory Q, a subtheory of Peano arithmetic.

48 The general idea of relativization was introduced by Tarski in 1930. The idea of relativization of quantifiers was due to Adolf Lindenbaum and Tarski, and dated from 1935. See [56^{m}], p. 69, p. 314 (footnote 1) and p. 396. With Tarski as one of his doctoral advisors, Andrzej Mostowski used the method of relativization of quantifiers in his Ph.D. dissertation, which he published under the title *O niezależności definicji skończoności w systemie logiki* <*On the Independence of Definitions of Finiteness in a System of Logic*>, in ***Dodatek do Rocznika Polskiego Towarzystwa Matematycznego*** (Supplement to the ***Annales de la Société Polonaise de Mathématique***), Vol. 11 (1938), pp. 1–54.

It has seven simple axioms describing the successor function (a unary operation), multiplication and addition (binary operations), and the number zero. The authors show that Q is essentially undecidable, since any recursive function of a single variable is metamathematically definable in Q. The proof relies on Julia Robinson's characterization of recursive functions of a single variable, and a version of Tarski's theorem on undefinability of arithmetical truth, which together make it possible to avoid having to construct a special provability predicate.

In the third paper of [53^{m}], *Undecidability of the Elementary Theory of Groups*, written by Tarski alone, the headline result is proved by demonstrating the following...

(i) The complete theory $\mathrm{Th}(\langle I,+,\cdot\rangle)$ of integer arithmetic is undecidable, since there exists a natural relativization of Q that is interpretable in $\mathrm{Th}(\langle I,+,\cdot\rangle)$.
(ii) The theory $\mathrm{Th}(\langle I,+,\cdot\rangle)$ is itself interpretable in the elementary theory of groups in such a way, that a decision procedure for the theory of groups would yield a decision procedure for $\mathrm{Th}(\langle I,+,\cdot\rangle)$.

Other results obtained by the method of interpretation or relative interpretation include the undecidability of the elementary theory of lattices, of modular lattices, of modular lattices with complementation, and of rings. The method was also used in proving a weak system of set theory undecidable (see [52b], by Szmielew and Tarski). In the 1960s Tarski and Szczerba used the method of interpretation to investigate the decision problem in geometry.

Tarski's talk at the 1946 Princeton *Logic Conference on Problems of Mathematics* began with a short reflexion of a semantic nature:

> Now the word "problem" has two distinct senses: in one sense, a problem is a definite question like "Is such-and-such the case?"; in another sense, we mean by a problem something of a less determined nature—which could perhaps more properly be characterized as a task—such as "Construct something with such-and-such properties." It is in this second—more general, if you prefer—sense that I will call the attention of this assembly to some important unsolved problems in mathematical logic.

Tarski's later publications treated the decision problem as a problem in the second, more general sense, recasting it as a challenge to explore what decidability is really all about, what are the different kinds of decidability, and which kinds are worth investigating in more detail. He identified what he called *restricted* decision problems, and *second-degree* decision problems.[49]

By *restricted* decision problems Tarski meant, "problems of determining whether a set S of all valid sentences of a theory T *satisfying certain additional conditions* is recursive". He cited Hilbert's tenth problem, the word problem for groups, and other word problems as prime examples of problems of this type.[50] In [87^{m}], §5.5 & §8.5, he gave three more examples, and proved they were answerable in the negative[51]:

---

[49] Both terms were coined in [53^{m}] (see pp. 34–35).

[50] See [53^{m}], p. 35, and [68], p. 287.

[51] To obtain these proofs he applied a method that he had earlier set out, in [53^{m}], p. 22, footnote 17, which could be termed "generalized interpretation".

- Certain subsystems of classical propositional logic, finitely axiomatizable under the rules of substitution and detachment, are undecidable.
- The equational theory of relation algebras and the equational theory of representable relation algebras are undecidable.[52]
- The equational theory of omega-relation algebras is finitely based[53] and essentially undecidable.

By *second-degree* decision problems, Tarski understood a catch-all category of meta-questions about questions—a category more easily limned by "*e.g.*" than by "$=_{def.}$". He gave his first example at the Princeton address, citing the (at that time open) question "of finding a procedure to tell whether a given set of formulas is adequate as a set of axioms for the sentential calculus".[54] In [53^{m}], pp. 34–35, he considered "the problem of the existence of a method which would permit us in each particular case to decide whether or not a given theory is decidable". Narrowing this down to something more tractable, he asked if the family of all finite sets of sentences which are axiom sets for decidable first-order theories with standard formalizations is recursive.

Based on the provable existence of an essentially undecidable theory, and on specific properties of deducibility in theories with standard formalizations, Tarski concluded that these questions were answerable in the negative. Analogous reasoning gave him a negative answer to the question: Is the family of all finite consistent sets of sentences recursive?[55]

Abraham A. Fraenkel, Yehoshua Bar-Hillel, and Azriel Lévy surveyed a sweeping cast of second-degree decision problems in their ***Foundations of Set Theory***. They wrote[56]:

> The reader might have already asked himself whether there might not exist a decision method by which it could be effectively determined for every given formalized theory, whether it is decidable or not. However, a rather simple argument shows that, at any rate for finitely axiomatizable first-order theories, no such general method could possibly exist, hence that this second-degree decision problem, so to speak, is unsolvable. [The authors cite [53^{m}], p. 35]. Other higher-degree decision problems deal, for instance, with the existence of an effective procedure of deciding, for every given *presentation* of a certain algebraic structure, such as of a semi-group, of a semi-group with cancellation, or of a group, whether the structures defined by these presentations have certain algebraic properties such as cyclicity, finiteness, simplicity, decomposability into a finite product, etc.

In [68], on page 287, Tarski wrote, "Various notions, problems, and results discussed so far in this paper suggest in a natural way corresponding *decision problems*. These are

---

[52]For a class **K** of similar algebras (of a fixed similarity type τ), the equational theory of this class is defined as the set of all identities (in type τ) which are true or hold in every algebra belonging to the class **K**.

[53]In equational logic *finitely based* means axiomatizable by a finite number of equations (identities).

[54]That such a procedure does not exist was communicated in 1949 by Samuel Linial (Gulden) and Emil Leon Post. For a detailed presentation of their result, see Mary Katherine Yntema, *A detailed argument for the Post-Linial theorems*, ***Notre Dame Journal of Formal Logic***, vol. 5 (1964), pp. 37–51.

[55]See: George F. McNulty, *Alfred Tarski and Undecidable Theories*, ***The Journal of Symbolic Logic***, vol. 51 (1986), pp. 890–898.

[56]A. Fraenkel, Y. Bar-Hillel, A. Lévy, ***Foundations of Set Theory***, **Studies in Logic and the Foundations of Mathematics**, Vol. 67, Elsevier, Amsterdam 1958, (2nd edition 1973). The quotation is from the second edition, Chapter 5, §7, *The limitative theorems of Gödel, Tarski, Church and their generalizations*, page 320.

problems of the type: is a given set of equations, or of finite sets of equations, or of finite algebras, recursive? The meaning of the term *recursive* in these contexts is clear; finite algebras can be regarded as algebras whose universe consists of finitely many integers."

He set out examples of such decision problems on pp. 287–288. Let Σ range over finite subsets of equations of a fixed finite similarity type—i.e., equations in which only a finite number of operation symbols occur. Then the following six condition-schemata, with free variable Σ, can be postulated:

c1. Σ is a base for a given finite algebra $\mathcal{A}$.
c2. There is a finite algebra $\mathcal{A}$ for which Σ is a basis.
c3. For a given positive integer k, the equational theory generated by Σ has an independent base consisting of k equations.
c4. The equational theory generated by Σ is consistent.
c5. The equational theory generated by Σ is complete.
c6. The equational theory generated by Σ is decidable.

To each of the above six condition-schemata (c_i) there corresponds the decision problem (P_i): *Is the family of all* Σ *such that* Σ *has property* (c_i) *recursive*? Six parallel decision problems can be obtained when Σ is restricted to range over one-element sets of equations.

Also consider the following three condition-schemata on finite algebras, with free variable $\mathcal{A}$, and their related decision problems:

e1. $\mathcal{A}$ is finitely based (that is, the set of all equations true in $\mathcal{A}$ is axiomatizable by a finite set of equations).
e2. $\mathcal{A}$ is equationally complete.
e3. $\mathcal{A}$ has an independent base consisting of k elements, where $k \leq \aleph_0$.

Of these, the problem associated with condition (e1), known as *Tarski's finite basis problem*, turned out to be most difficult, and attempts to solve it profoundly influenced the development of universal algebra and computability theory for 30 years. It was only in 1996 that Ralph McKenzie announced the result that *the class of all finite algebras which are finitely based is not recursive*.[57]

Flashback and flashforward can be revealing devices not just in cinematography. Here is Tarski speaking in 1968[58]:

> There is much affinity between the formalisms of sentential calculus and equational logic; as a consequence various metalogical results established for sentential calculus can frequently be carried over to equational logic with appropriate changes in formulations and proofs.

Here are Łukasiewicz and Tarski speaking in 1930[59]:

> In conclusion we should like to add that, as the simplest deductive discipline, the sentential calculus is particularly suitable for metamathematical investigations. It is to be regarded as a laboratory in which metamathematical methods can be discovered and metamathematical concepts constructed which can then be carried over to more complicated mathematical systems.

---

[57] R. McKenzie, *Tarski's Finite Basis Problem is Undecidable*, ***International Journal of Algebra and Computation***, Vol. 6 (1996), pp. 49–104.

[58] See [68], p. 288.

[59] See [56^m], p. 59.

And here, through the medium of Steven Givant, is Tarski speaking from beyond the grave in 1987[60]:

> The above observations may seem somewhat paradoxical. The formalism $\mathcal{T}$ of two-valued sentential logic is usually regarded as the simplest and most trivial logical formalism, with an almost empty mathematical content. Nevertheless, the formalism $\mathcal{L}_r^\times$, so closely related to $\mathcal{T}$ in its syntactical part, presents an adequate basis for the development of set theory, which is, in a sense, the richest mathematical discipline; and even in its logical part $\mathcal{L}_r^\times$ embodies an interesting and far from trivial mathematical theory, namely the equational theory of relation algebras.
>
> One conclusion emerges from our discussion: the connection between the formal structure of the language and its intended semantical interpretation is much looser than we might be inclined to believe.
>
> It may be interesting to observe that the logic of $\mathcal{T}$ is well known to be decidable, while the logic of $\mathcal{L}_r^\times$ is undecidable... Thus, we have obtained an example of an undecidable subtheory of the two-valued sentential logic (in fact a subtheory based upon a finite set of axiom schemata). Also ... this subtheory can be supplemented by means of finitely many axioms to form an essentially undecidable theory $\boldsymbol{\Theta}$; this will be a theory in the same formalism (i.e., $\mathcal{L}_r^\times$) as sentential logic, but clearly not a subtheory of that logic.

5 Selected Works of Tarski

Listed below are all of Tarski's works cited or referred to in the present paper. Also listed are a few other works which, while not cited above, are central to his canon. The method and style of citing Tarski's works adhere as closely as possible to the conventions adopted in Steven R. Givant's *Bibliography of Alfred Tarski*, in ***The Journal of Symbolic Logic***, Vol. 51, No. 4 (Dec., 1986), pp. 913–941.

Givant's bibliography was updated, corrected, and translated into Polish by Jan Zygmunt in [95^{m}], pp. 333–372. Both were divided into the same ten sections: Papers, Abstracts, Monographs, Exercises and problems, Contributions to discussions, Reviews, Publication as editor, Project reports, Letters, and Appendix. Both give full information on later re-editions and translations of an item, as well as references to reviews in ***Mathematical Reviews*** and ***The Journal of Symbolic Logic***. The bibliography below is divided into only four sections: (A) Monographs, (B) Papers, (C) Abstracts, and (D) Publications as editor.

A more recently updated bibliography has been supplied by Andrew McFarland, Joanna McFarland and James T. Smith in [14^{m}]. See that volume's Ch. 16: *Posthumous Publications*, Ch. 17: *Biographical Studies*, Ch. 18: *Research surveys*, and especially its excellent *Bibliography*.

[60]See [87^{m}], pp. 167–168.

5.1 Monographs

[33^{m}] ***Pojęcie prawdy w językach nauk dedukcyjnych*. Prace Towarzystwa Naukowego Warszawskiego, Wydział III—Nauk Matematyczno-fizycznych**, nr. 34, Warszawa 1933, vii + 116 pp. + errata. (see [35b] for German translation.)

[36^{m}] ***O logice matematycznej i metodzie dedukcyjnej*. Bibljoteczka Matematyczna**, vol. 3–5, Książnica-Atlas, Lwów and Warszawa 1936, 167 pp.

(1) ***Einfürung in die mathematiche Logik und in die Methodologie der Mathematik*** Julius Springer Verlag, Vienna, 1937, x + 166 pp. (German translation of [36^{m}].)

[41^{m}] ***Introduction to Logic and to the Methodology of Deductive Sciences***. Oxford University Press, Oxford and New York 1941, xviii + 239 pp. [Fourth, enlarged edition of [41^{m}], edited by Jan Tarski, **Oxford Logic Guides**, vol. 24, Oxford University Press, New York – Oxford 1994, xxiv + 229 pp.]

[48^{m}] ***A Decision Method for Elementary Algebra and Geometry*** (prepared for publication by J.C.C. McKinsey) U.S. Air Force Project RAND, R-109, the RAND Corporation, Santa Monica, California 1948, iv + 60 pp.

(1) Second, revised edition of [48^{m}] (prepared for publication with the assistance of J.C.C. McKinsey), University of California Press, Berkeley and Los Angeles, California 1951, iii + 63 pp.

[49^{m}] ***Cardinal Algebras, With an Appendix: Cardinal Products of Isomorphism Types*** (by B. Jónsson and A. Tarski), Oxford University Press, Oxford and New York 1949, xii + 327 pp.

[53^{m}] ***Undecidable Theories*** (with A. Mostowski and R. M. Robinson), North-Holland Publishing Company, Amsterdam 1953, xii + 98 pp.

[56^{m}] ***Logic, Semantics, Metamathematics. Papers from 1923 to 1938*** (translated by J.H. Woodger), Clarendon Press, Oxford 1956, xiv + 471 pp.

(1) Second, revised edition, with editor's introduction and an analytic index (J. Corcoran, editor), Hackett Publishing Company, Indianapolis, Indiana 1983, xxx + 506 pp.

[56^{m}a] ***Ordinal Algebras***, with appendices: *Some Additional Theorems on Ordinal Algebras* (by C.C. Chang) and *A Unique Decomposition Theorem for Relational Addition* (by B. Jónsson), North-Holland Publishing Co., Amsterdam 1956, 133 pp.

[67^{m}a] ***The Completeness of Elementary Algebra and Geometry***, Institut Blaise Pascal, Paris 1967, iv + 50 pp. (A reprint from page proofs of the work scheduled to appear in 1940 in **Actualités Scientifiques et Industrielles**, Hermann & Cie, Paris. See [48^{m}].)

[71^{m}] ***Cylindric Algebras. Part I. With an Introductory Chapter: General Theory of Algebras*** (with L. Henkin and J.D. Monk), North-Holland Publishing Company, Amsterdam 1971, vi + 508 pp.

[83^{m}] ***Metamathematische Methoden in der Geometrie*** (with W. Schwabhäuser and W. Szmielew), Springer-Verlag, Berlin 1983, viii + 482 pp.

[85^{m}] ***Cylindric Algebras. Part II*** (with L. Henkin and J.D. Monk), North-Holland, Amsterdam 1985, vii + 302 pp.

[86^{m}] ***The Collected Works of Alfred Tarski***, vol. 1, ***1921–1934***; vol. 2, ***1935–1944***; vol. 3, ***1945–1957***; vol. 4, ***1958–1979*** (S.R. Givant and R.N. McKenzie, editors), Birkhäuser, Basel, Boston and Stuttgart 1986, xii + 658 pp. (vol. 1); xii + 699 pp. (vol. 2); xii + 682 pp. (vol. 3); xii + 757 pp. (vol. 4). [Volume 4 (pp. 739–757) contains a reprint of S. Givant, *Bibliography of Alfred Tarski*, ***JSL***, vol. 45 (1986), pp. 913–941]

[87^{m}] ***A Formalization of Set Theory without Variables***. (with S.R. Givant), **Colloquium Publications**, vol. 41, American Mathematical Society, Providence, Rhode Island 1987, xxi + 318 pp.

[95^{m}] ***Pisma logiczno-filozoficzne***, vol. 1: ***Prawda***. (Translated and annotated, with an introduction by Jan Zygmunt), **Biblioteka Współczesnych Filozofów**, Wydawnictwo Naukowe PWN, Warszawa 1995, xxiv + 390 pp. + errata.

[01^{m}] ***Pisma logiczno-filozoficzne***, vol. 2: ***Metalogika***. (Translated and annotated, with an intoduction by Jan Zygmunt), **Biblioteka Współczesnych Filozofów**, Wydawnictwo Naukowe PWN, Warszawa 2001, xiv + 516 pp.

[14^{m}] ***Alfred Tarski*: *Early Work in Poland—Geometry and Teaching*. *With a Bibliographic Supplement***. (A. McFarland, J. McFarland, James T. Smith, editors), Birkhäuser: New York 2014, xxiii + 499 pp.

5.2 Papers

[21] *Przyczynek do aksjomatyki zbioru dobrze uporządkowanego <A Contribution to the Axiomatics of Well-Ordered Sets>*, ***Przegląd Filozoficzny***, vol. 24 (1921), pp. 85–94.

[24a] *Sur quelques théorèmes qui équivalent à l'axiome du choix,* ***Fundamenta Mathematicae***, vol. 5 (1924), pp. 147–154.

[24b] *O równoważności wielokątów <On the Equivalence of Polygons>*, ***Przegląd Matematyczno-fizyczny***, vol. 2 (1924), pp. 47–60.

[24c] *Sur les ensembles finis.* ***Fundamenta Mathematicae***, vol. 6 (1924), pp. 45–95.

[24d] *Sur la décomposition des ensembles de points en parties respectivement congruentes* (with S. Banach), ***Fundamenta Mathematicae***, vol. 6 (1924), pp. 244–277.

[25] *Quelques théorèmes sur les alephs*, ***Fundamenta Mathematicae***, vol. 7 (1925), pp. 1–14.

[26] *Communication sur les recherches de la théorie des ensembles* (with A. Lindenbaum), ***Sprawozdania z Posiedzeń Towarzystwa Naukowego Warszawskiego, Wydział III Nauk Matematycznych i Przyrodniczych*** (= ***Comptes Rendus des Séances de la Société des Sciences et des Lettres de Varsovie, Classe III***), vol. 19 (1926), pp. 299–330.

[29a] *Geschichtliche Entwicklung und gegenwärtiger Zustand der Gleichmächtigkeitstheorie und der Kardinalzahlarithmetik,* ***Księga Pamiątkowa Pierwszego Polskiego Zjazdu Matematycznego*** (Supplement to ***Rocznik Polskiego Towarzystwa Matematycznego***), Kraków, 1929, pp. 48–54.

[29b] *Sur les fonctions additives dans les classes abstraites et leur application au problème de la mesure.* ***Sprawozdania z Posiedzeń Towarzystwa Naukowego Warszawskiego, Wydział III Nauk Matematycznych i Przyrodniczych*** (= ***Comptes Rendus des Séances de la Société des Sciences et des Lettres de Varsovie, Classe III***), vol. 22 (1929 – published 1930), pp. 114–117.

[30] *Une contribution à la théorie de la mesure*, ***Fundamenta Mathematicae***, vol. 15 (1930), pp. 42–50.

[30a] *Sur une propriété caractéristique des nombres inaccessibles* (with W. Sierpiński), ***Fundamenta Mathematicae***, vol. 5 (1930), pp. 292–300.

[30b] *Über Äquivalenz der Mengen in Bezug auf eine beliebige Klasse von Abbildungen*, ***Atti del Congresso Internazionale dei Matematici, Bologna, 3–10 settembre 1928***, vol. 6, Nicola Zanichelli, Bologna 1930, pp. 243–252.

[30c] *Über einige fundamentalen Begriffe der Metamathematik*, ***Sprawozdania z Posiedzeń Towarzystwa Naukowego Warszawskiego, Wydział III Nauk Matematyczno-fizycznych*** (=***Comptes Rendus des Séances de la Société des Sciences et des Lettres de Varsovie, Classe III***), vol. 23 (1930), pp. 22–29.

[30d] *Untersuchungen über den Aussagenkalkül* (with J. Łukasiewicz), ***Sprawozdania z Posiedzeń Towarzystwa Naukowego Warszawskiego, Wydział III Nauk Matematyczno-fizycznych*** (=***Comptes Rendus des Séances de la Société des Sciences et des Lettres de Varsovie, Classe III***), vol. 23 (1930), pp. 30–50.

[30e] *Fundamentale Begriffe der Methodologie der deduktiven Wissenschaften. I*, ***Monatshefte für Mathematik und Physik***, vol. 37 (1930), pp. 361–404.

[30f] *Sur les classes d'ensembles closes par rapport à certaines opérations élémentaires*, ***Fundamenta Mathematicae***, vol. 16 (1930), pp. 181–304.

[31] *Sur les ensembles définissables de nombres réels. I*, ***Fundamenta Mathematicae***, vol. 17 (1931), pp. 210–239.

[31a] *Les opérations logiques et les ensembles projectifs* (with C. Kuratowski), ***Fundamenta Mathematicae***, vol. 17 (1931), pp. 240–248.

[31b] *O stopniu równoważności wielokątów* <*On the Degree of Equivalence of Polygons*>, ***Młody Matematyk***, vol. 1 (Supplement to ***Parametr***, vol. 2) (1931), pp. 37–44.

[34] *Z badań metodologicznych nad definjowalnością terminów* <*Methodological Investigations into the Definability of Terms*>, ***Przegląd Filozoficzny***, vol. 37 (1934), pp. 438–460.

[35a] *Grundzüge des Systemenkalküls. Erster Teil*, ***Fundamenta Mathematicae***, vol. 25 (1935), pp. 503–526.

[35b] *Der Wahrheitsbegriff in den formalisierten Sprachen*, ***Studia Philosophica***, vol. 1 (1935), pp. 261–405. (German translation of [33^{m}] by L. Blaustein, with an appendix. Also printed separately in Lwów, 1935, 145 pp.)

[36] *O ugruntowaniu naukowej semantyki* <*On Foundations for a Semantics of Science*>, ***Przegląd Filozoficzny***, vol. 39 (1936), pp. 50–57.

[36a] *O pojęciu wynikania logicznego* <*On the Concept of Logical Consequence*>, ***Przegląd Filozoficzny***, vol. 39 (1936), pp. 58–68.

[36b] *Über die Beschränktheit der Ausdrucksmittel deduktiver Theorien* (with A. Lindenbaum), ***Ergebnisse eines Mathematischen Kolloquiums***, vol. 7 (1936), pp. 15–22.

[36d] *Grundzüge des Systemenkalküls. Zweiter Teil*, ***Fundamenta Mathematicae***, vol. 26 (1936), pp. 283–301.

[38a] *Über unerreichbare Kardinalzahlen*, ***Fundamenta Mathematicae***, vol. 30 (1938), pp. 68–89.

[38c] *Ein Überdeckungsästz für endliche Mengen*, ***Fundamenta Mathematicae***, vol. 30 (1938), pp. 156–163.

[38d] *Eine äquivalente Formulierung des Auswahlaxioms*, ***Fundamenta Mathematicae***, vol. 30 (1938), pp. 197–201.

[38g] *Algebraische Fassung des Maßproblems*, ***Fundamenta Mathematicae***, vol. 31 (1938), pp. 47–66.

[38h] *Der Aussagenkalkül und die Topologie*, ***Fundamenta Mathematicae***, vol. 31 (1938), pp. 103–134.

[39] *Ideale in vollständigen Mengenkörpern. I*, ***Fundamenta Mathematicae***, vol. 32 (1939), pp. 45–63.

[39a] *Boolesche Ringe mit geordneter Basis* (with A. Mostowski), ***Fundamenta Mathematicae***, vol. 32 (1939), pp. 69–86.

[39b] *On well-ordered subsets of any set*, ***Fundamenta Mathematicae***, vol. 32 (1939), pp. 176–183.

[41] *On the calculus of relations*, ***The Journal of Symbolic Logic***, vol. 6 (1941), pp. 73–89.

[43] *On families of mutually exclusive sets* (with Paul Erdös), ***Annals of Mathematics***, vol. 44 (1943), pp. 315–329.

[44a] *The semantic conception of truth and the foundations of semantics*, ***Philosophy and Phenomenological Research***, vol. 4 (1944), pp. 341–375.

[48b] *Axiomatic and algebraic aspects of two theorems on sums of cardinals*, ***Fundamenta Mathematicae***, vol. 35 (1948), pp. 79–104.

[49] *Cancellation laws in the arithmetic of cardinals*, ***Fundamenta Mathematicae***, vol. 36 (1949), pp. 77–92.

[52b] *Mutual interpretability of some essentially undecidable theories* (with W. Szmielew), ***Proceedings of the International Congress of Mathematicians. Cambridge, Massachusetts, U.S.A., August 30 – September 6, 1950***, vol. 1, American Mathematical Society, Providence, Rhode Island 1952, p. 734.

[54] *Theorems on the existence of successors of cardinals, and the axiom of choice*, ***Indagationes Mathematicae***, vol. 16 (1954), pp. 26–32.

[54a] *Contributions to the theory of models. I*, ***Indagationes Mathematicae***, vol. 16 (1954), pp. 572–581.

[54b] *Contributions to the theory of models. II*, ***Indagationes Mathematicae***, vol. 16 (1954), pp. 582–588.

[55] *Contributions to the theory of models. III*, ***Indagationes Mathematicae***, vol. 17 (1955), pp. 56–64.

[56b] *Equilaterality as the only primitive notion of Euclidean geometry* (with E.W. Beth), ***Indagationes Mathematicae***, vol. 18 (1956), pp. 462–467.

[56c] *A general theorem concerning primitive notions of Euclidean geometry*, ***Indagationes Mathematicae***, vol. 18 (1956), pp. 468–474.

[57a] *Arithmetical extensions of relational systems* (with R.L. Vaught), ***Compositio Mathematica***, vol. 13 (1957), pp. 81–102.

[59] *What is elementary geometry?* in [59^{e}], pp. 16–29.

[61a] *Cylindric algebras* (with L. Henkin), ***Lattice theory***, **Proceedings of Symposia in Pure Mathematics**, vol. 2 (R.P. Dilworth, editor), American Mathematical Society, Providence, Rhode Island, 1961, pp. 83–113.

[61b] *On some problems involving inaccessible cardinals* (with Paul Erdös), ***Essays on the Foundations of Mathematics*** (Y. Bar-Hillel, E.I.J. Poznański, M.O. Rabin and A. Robinson, editors), Magnus Press, Jerusalem 1961, pp. 50–82.

[62] *Some problems and results relevant to the foundations of set theory*, in [62^{e}], pp. 125–135.

[64] *From accessible to inaccessible cardinals. Results holding for all accessible cardinal numbers and the problem of their extension to inaccessible ones* (with H.J. Keisler), ***Fundamenta Mathematicae***, vol. 53 (1964), pp. 225–308.

[65] *A simplified formalization of predicate logic with identity*, ***Archiv für mathematische Logik und Grundlagenforschung***, vol. 7 (1965), pp. 61–79.

[65a] *Metamathematical properties of some affine geometries* (with L.W. Szczerba), **Logic, *Methodology and Philosophy of Science. Proceedings of the 1964 International Congress*** (Y. Bar-Hillel, editor), North-Holland Publishing Company, Amsterdam 1965, pp. 166–178.

[68] *Equational logic and equational theories of algebras*, ***Contributions to Mathematical Logic. Proceedings of the Logic Colloquium, Hannover 1966*** (H.A. Schmidt, K. Schütte and H.J. Thiele, editors), North-Holland Publishing Company, Amsterdam 1968, pp. 275–288.

[78] *The elementary theory of well-ordering—A metamathematical study* (with J. Doner and A. Mostowski), ***Logic Colloquium '77*** (A. MacIntyre, L. Pacholski and J. Paris, editors), North-Holland Publishing Company, Amsterdam 1978, pp. 1–54.

[79] *Metamathematical discussion of some affine geometries* (with L.W. Szczerba), ***Fundamenta Mathematicae***, vol. 104 (1979), pp. 155–192.

[86] *What are logical notions?* (S. Givant, editor), ***History and Philosophy of Logic***, vol. 7 (1986), pp. 143–154.

[99] *Tarski's system of geometry* (with S. Givant), ***Bulletin of Symbolic Logic***, vol. 5 (1999), pp. 175–214.

5.3 Abstracts

[49^{a}c] *Arithmetical classes and types of Boolean algebras*, ***Bulletin of the American Mathematical Society***, vol. 55 (1949), pp. 64, 1192.

[54^{a}f] *Prime ideal theorems for Boolean algebras and the axiom of choice*, ***Bulletin of the American Mathematical Society***, vol. 60 (1954), pp. 390–391.

[54^{a}g] *Prime ideal theorems for set algebras and ordering principles*, ***Bulletin of the American Mathematical Society***, vol. 60 (1954), p. 391.

[54^{a}h] *Prime ideal theorems for set algebras and the axiom of choice*, ***Bulletin of the American Mathematical Society***, vol. 60 (1954), p. 391.

[64^{a}b] *The comparability of cardinals and the axiom of choice*, ***Bulletin of the American Mathematical Society***, vol. 11 (1964), p. 578.

[65^{a}] *On the existence of large sets of Dedekind cardinals*, ***Notices of the American Mathematical Society***, vol. 12 (1965), p. 719.

5.4 *Publications as Editor*

[59^e] ***The Axiomatic Method, with Special Reference to Geometry and Physics*** (edited with L. Henkin and P. Suppes), North-Holland Publishing Company, Amsterdam 1959, xi + 488 pp.

[62^e] ***Logic, Methodology and Philosophy of Science. Proceedings of the 1960 International Congress*** (edited with E. Nagel and P. Suppes), Stanford University Press, Stanford, California 1962, ix + 661 pp.

[65^e] ***The Theory of Models. Proceedings of the 1963 International Symposium at Berkeley*** (edited with J.W. Addison and L. Henkin), North-Holland Publishing Company, Amsterdam 1965, xv + 494 pp.

J. Zygmunt (✉)
University of Wrocław, Wrocław, Polska

Stanisław Jaśkowski: Life and Work

Andrzej Indrzejczak

Abstract In this brief note we would like to outline the main events of life and the main achievements of Stanisław Jaśkowski (1906–1965) one of the important Polish logician and mathematician of the first half of twentieth century.

Keywords Jaśkowski · Natural deduction · Nonclassical logics · Decidability

Mathematics Subject Classification (2000) Primary 03F03; Secondary 01A60, 01A70

1 Life

Stanisław Jaśkowski was born on 22.04.1906 in Warsaw as a son of landowner Feliks Jaśkowski and Kazimiera Dzierzbicka. His family represented high standards of humanistic culture. In particular, his grandfather—Jan Nepomucen Jaśkowski—was a poet and a writer; his father was a musician. It was not surprising that his parents expected him to study on humanistic faculty. Despite family's expectations in 1924 Jaśkowski started studies in mathematic faculty of Warsaw University. As a result he became one of the representatives of Warsaw School of Logic. Among his teachers were Leśniewski, Tarski and Łukasiewicz, who had the strongest influence on his early development. In particular, Łukasiewicz had a direct impact on his first, and one of the most important logical contribution—the invention of natural deduction. It was Łukasiewicz who posed on his seminar a problem of providing a fully formal account of methods of proof applied by mathematicians in their practice. The first results of Jaśkowski concerning this problem were presented in 1926 on the seminar and then announced on the First Polish Mathematical Conference in Lwów in 1927 in: *Księga Pamiątkowa I Polskiego Zjazdu Matematycznego*, Uniwersytet Jagielloński, Kraków 1929 [8].

Because of the serious health problems with lungs Jaśkowski had a lengthy break in his education and scientific career. In 1929–1930 he was cured in Davos in Switzerland. After recovery he participated in the Second Polish Mathematical Conference in Vilnius in 1931 where he presented his first axiomatization of the geometry of solids based on the primitive notion of semispace. This work shows for the early origins of his second passion—foundations of mathematics.

Á. Garrido, U. Wybraniec-Skardowska (eds.), *The Lvov-Warsaw School. Past and Present*, Studies in Universal Logic,
https://doi.org/10.1007/978-3-319-65430-0_32

In 1932 Jaśkowski gained his doctor's degree under the supervision of Łukasiewicz. The thesis [9], devoted to the presentation of the first system of natural deduction, was printed in 1934 as the first volume of "Studia Logica", the new journal edited by Łukasiewicz. Strongly delayed publication of his first, and one of the most important, discoveries was very unfortunate since in the same year Gerhard Gentzen published the first part of his own version of natural deduction [7] and Jaśkowski lost the priority. Moreover, in contrast to Jaśkowski's paper, Gentzen's work was soon widely known to logical community and made Jaśkowski's contribution relatively unknown.

In 1935 Jaśkowski took part in the International Congress of Scientific Philosophy in Paris where he presented important results concerning adequate matrix characterization for intuitionistic logic. This event was very important for his scientific development but 1937 was particularly important for his private life. Jaśkowski married Aniela Holewińska (1905–1976) a student of mathematics in Warsaw University. In 1939 their daughter Anna was born.

Till the September 1939 Jaśkowski continued scientific work, mainly on modal functions and logical systems based on the notion of dependent variable. This last contribution was very often applied in his later publications as a kind of methodological basis for studies on paraconsistent and causal logics. The Second World War interrupted his work on habilitation. Due to his health problems he was not admitted to the regular army. Still he served as a volunteer in the defence of Warsaw; he gave his car to the disposal of the 151st Column of Heavy Trucks. During the Nazi occupation of Poland, he was living in his estate in Wolka near Rawa Mazowiecka and in Warsaw where he worked as a bookkeeper. He was also arrested for a few weeks in 1942. Almost all of his scientific manuscripts were lost during the Warsaw Uprising in 1944.

After the War Jaśkowski came back to scientific work. He was working for half a year as a lecturer in the newly founded Łódź University, then he moved to Toruń where he lived and worked from 1.10.1945 until his death in 1965. Even when he was proposed to move to Kraków to became the Chair of Logic Department in Jagiellonian University after Professor Zawirski's death in 1949, he decided to stay in Toruń. Over the last 20 years of life he played a very important role as a scientist, a teacher, and an organizer in the development of the new Nicolas Copernicus University in Toruń. Concerning the last issue we list some of the most important facts. In 1945 Jaśkowski organised the Department of Mathematical Logic and was its first Chair. Until 1965 he was the director of mathematical departments in Toruń University. In 1952–53 he organized the Faculty of Mathematics, Physics and Chemistry and was its first dean in 1953–54. In 1956–59 he was a deputy prorector for science and in 1959–62 the rector of Toruń University. One may also mention that he was a co-founder and the first president of Toruń division of the Polish Mathematical Society. Despite the numerous occupations in Toruń from 1950 he was also a member of the National Institute of Mathematics at the Polish Academy of Sciences.

He was also very active as a teacher. In the early years of its existence Toruń University had serious problems with completing qualified staff sufficient for providing all necessary courses. As a result, in addition to teaching mathematical logic Jaśkowski was forced to provide also courses in analysis, set theory, geometry, probability theory. In the last years of his life he was also strongly engaged in organization of computer laboratory; his last seminar was devoted to the theory of automated deduction.

In the meantime Jaśkowski finished his habilitation concerned with a new definition of real numbers under the supervision of Zygmunt Zawirski. The habilitation colloquium was conducted on 1.10.1945 and confirmed on 7.04.1946. He obtained the title of associate professor in July 1946 and in 1957 he was nominated a full professor.

One should also mention his efforts in preparation of a new modern programme of teaching mathematics in secondary school. The new syllabus was under a great influence of Jaśkowski's ideas of how to teach mathematics and was finally introduced in 1960s. Despite all these organizational and educational duties he was still active on the field of scientific research in logic and mathematics. Some of the most important achievements will be sketched in the next two sections.

Unexpectedly he fell ill with infectious jaundice in 1962 and this caused a lot of health complications. In consequence of postjaundice complications Jaśkowski died on 16.11.1965.

2 Works

The scientific activity of Stanisław Jaśkowski may be roughly divided into two overlapping fields: logic and mathematics. In what follows we briefly characterize his most important achievements. A deeper presentation of his scientific ideas may be found in [32, 34] and [33].

2.1 Logic

The invention of natural deduction systems, presented in [9], is the first of his logical achievements and it is often claimed as the most important contribution of Jaśkowski to logic. We do not attempt to describe his approach here since this is the subject of the special paper presented in this volume.

Jaśkowski was particularly active on the field of investigation on nonclassical logics. Not only he provided some results for already known logics like intuitionistic logic or modal logics. He was also the inventor of many new and important systems, in particular:

1. In his study on natural deduction [9] he constructed the first system of inclusive logic. This is a kind of first-order logic which admits models with empty domains in the semantics. Syntactically it is weaker than classical logic since some theses (valid for nonempty domains only) are excluded. Logics which are inclusive and additionally free, in the sense of having terms which do not denote existing things, are called universally free (see Bencivenga [1]). Such systems are very often treated as philosophically more neutral basis than classical logic and commonly applied for constructing intuitively sound modal first-order logics (see e.g. Garson [6]). It should be emphasized that the first recognized systems of inclusive and free logics were proposed much later in 1950s by Mostowski, Leblanc, Hintikka, to mention just a few scholars. The fact that Jaśkowski constructed the first inclusive logic (and implicitly

also the first universally free logic—see Bencivenga [2]) has gone unnoticed and, surprisingly, is still not very well known even for authors writing on Jaśkowski's achievements. We characterize briefly his approach to first order logic in the paper on natural deduction in this volume.
2. In the field of studies on intuitionistic propositional logic Jaśkowski provided not only a natural deduction system but also an adequate matrix characterization. In his famous result Gödel has shown that there is no adequate matrix for intuitionistic logic with finitely many values. Jaśkowski's study from 1930s shed a new light on this result by providing a recipe for construction of adequate matrix characterization of intuitionistic logic. The construction consists of the infinite sequence of finite matrices. It was presented on the International Congress of Scientific Philosophy in Paris and published in [10].
3. In [13] Jaśkowski provided a philosophical justification and a formal construction of discursive logic which was the first system of paraconsistent logic. In such logical systems the presence of contradictory statements does not lead to trivialization of the system, in contrast to classical logic where, by Duns' law, all propositions follow logically from contradiction. In his system D2, material implication is replaced with discursive implication which may be read: if the antecedent is possibly true, then the succedent is true (with possibility understood as in modal logic S5). This work was then extended in [21] where the notion of discursive conjunction was introduced. Jaśkowski developed discursive logic mainly as a tool for analysing the situation of contradictory views represented in discussion. Soon it appeared that many other serious motivations for developing such systems may be provided and it gave the impact for development of several paraconsistent logics.
4. Jaśkowski provided also the basis for the development of causal logic by means of dependent sentential variables, i.e. variables representing propositions whose truth depends on some arguments. The theory of dependent variables was first developed in [14] and applied to modal functions. In [24, 25] a theory of causal functions is introduced where three notions of causal implication are defined: factorial, efficient and definitive. This work of Jaśkowski is not widely known but the study on causal logic was continued by August Pieczkowski and Max Urchs.

His work on logic includes also some papers on classical logic and their fragments as well as on traditional logic. Jaśkowski [15] is devoted to axiomatization of classical logic, definition of ternary connectives and reversible substitution. References [30, 31] provide a decision procedure for some fragmentary propositional calculi. Finally [23] contains a new interpretation of Aristotelian syllogistics, where non-Aristotelian terms (empty and universal names) are admitted. In this study he proposed a way of interpretation of categorical statements in the first order-logic which keeps all classical laws as valid even when non-Aristotelian terms are substituted for variables.

2.2 Mathematics

As a mathematician Jaśkowski was mainly interested in investigations on the foundations of mathematics. In fact, the border between his logical and mathematical works is very light since he always insisted on using logical tools in mathematics. In particular, he was working on the notion of number, foundations of geometry and decidability problems where he obtained both positive and negative results. Namely, he proved in [22] a decidability of the elementary theory of Boolean rings and generalised the method to elementary additive Boolean algebra. Above we also mentioned his decidability result for parts of propositional logic [30, 31]. As for negative results, in [17] he proved that some interesting classes of formulas of theory of groups and topology are undecidable. He has shown that the theory of free groupoids is undecidable [28]. Also the abovementioned paper [22] contains a theorem of the undecidability of some class of equalities of Boolean algebras. Interesting results on the undecidability of some existential problems in certain system of differential equations are provided in [27]. All these studies show that Jaśkowski was able to apply logical tools to sophisticated mathematical problems of great importance.

The investigation on the notion of number was the basis for Jaśkowski's habilitation and was summarised in [16]. He has shown that integers and real numbers may be defined in terms of some operations on the classes of sets.

Jaśkowski was also engaged in developing the geometry of solids which avoids ideal objects like points and straight lines. He was strongly convinced that such an approach is better for applications in quantum physics. In [19, 20] Jaśkowski modified Tarski's approach to geometry of solids and introduced the axiomatics based on the notion of "semispace" as primitive term in [12, 18]. He was preparing a book on the foundations of geometries but his premature death interrupted this work.

Last but not least, Jaśkowski was very active in educational sphere. He wrote two popular books on the geometry of ornament [26, 29] and prepared a textbook which may be seen as the first course in logic based on the application of natural deduction [11]. The latter will be treated in detail in the next paper in this volume. As for his books on geometry it must be emphasized that they show a great educational talent of Jaśkowski. In a popular way, not demanding any mathematical knowledge from the readers, he presented many interesting issues from geometry illustrated with examples taken from science, history and art. He also shows how to apply an abstract mathematical theory of groups to a concrete problem like classification of ornaments.

He actively participated in the work on modernization of the program of teaching mathematics in secondary schools. The problem was widely discussed in several commissions organized by the Ministry of Education, the Institute of Pedagogics and the Polish Mathematical Society. Jaśkowski was very critical with respect to the present state of the art. In his opinion the content of the present programme corresponds to the level which was obtained in seventeenth century mathematics before the achievements of Newton and Leibniz. The outcome of his activity in this field is a series of papers and notes where Jaśkowski insisted that teaching mathematics should be based on the application of logical tools and closer to the application in science and technique. His postulates concerning the changes in teaching mathematics were used in the construction of the new syllabus which was put to work in 1960s.

3 Influence

Although Jaśkowski passed away so early and his list of publications is not particularly long[1] he had significant achievements in many fields: logic, mathematics, education and organization of science. The problem of his influence on the development of logic and mathematics is a complicated matter. In Poland Jaśkowski had a number of students (e.g. Lech Dubikajtis, Jerzy Kotas, August Pieczkowski) who continued his work and made at least some of his achievements better known. However, it should be underlined that his contribution to logic is very often not recognized. In addition to his premature death there were some other unfortunate circumstances connected with his scientific achievements. Almost all of his ideas and results become known to logical community in the World much later after they were introduced, sometimes long after his death. We briefly comment on three selected issues.

His invention of natural deduction was not only significantly delayed in print but even after 1934 it was not as widely known as Gentzen's approach. In fact nowadays many logicians writing on, or presenting some systems of natural deduction, even do not mention Jaśkowski. But in fact, his style of presenting proofs, as a series of subordinate derivations with several bookkeeping devices applied to separate the scope of assumptions is widely applied. We discuss these matters in the second paper, here we only mention that Jaśkowski introduced a graphical method of separating subproofs and a technique of prefixes. The former was then popularised by Fitch [4] and is nowadays commonly called Fitch-style natural deduction.[2] The latter may be treated as the first application of labels which is now widely used in logic to construct several kinds of proof systems.[3] Thus Jaśkowski, even if not explicitly mentioned, is the true father of numerous natural deduction systems presented in textbooks and applied in practice. Gentzen's approach, based on the application of trees as proof-structures, is rather connected with theoretical investigations in proof theory.

Similarly, his system of inclusive logic, also presented in [9], was recognized as the first system of this type much later after the advent of studies on free logics.[4]

The work on paraconsistent logics was also developed in early stages without any knowledge of Jaśkowski's work; his contribution was recognized much later. Still it seems that his invention of discursive logic is better known than any other of his achievements. In fact, both his papers on discursive logic were written in Polish and published in the journal of local character, so there was no chance to make them known to logical community immediately after publication. However, soon after the works of Jaśkowski, a serious investigations on paraconsistent logics was undertaken by a group of logicians from Latin America, in particular by Newton da Costa in Brazil and Florentio Asenjo from Argentina. Later, in 1960s, a similar research started in United States and in Australia (Michael Dunn, Rober Meyer, Richard Routley (Sylvan), Graham Priest and others), in strong connection with the investigation on relevant implication. Nowadays,

[1] The full list of Jaśkowski's publications has 48 items and may be found in [3].

[2] In fact Fitch mentioned Jaśkowski in the Preface to his textbook as a source of inspiration.

[3] See e.g. Gabbay's theory of labelled systems in [5].

[4] See historical remarks in [1].

paraconsistent logics form one of the most well known group of nonclassical logics with wide spectrum of applications. Fortunately enough Jaśkowski work in this field was continued by Jerzy Kotas who cooperated with Newton da Costa. It seems that this joint work made Jaśkowski's achievements widely known to the community of researchers in paraconsistent logic.

In general, the problem with the lack of propagation of Jaśkowski's achievements was mainly connected with the fact that his papers were published very often in the journal of local character, sometimes written in Polish. After the II World War Polish scholars had in general difficulties with dissemination of their ideas in the World due to the new geopolitical situation. It seems that Jaśkowski's case was not exceptional. Summing up, we can say that he had a strong influence on some domains of research in logic but usually it was not a direct impact. In most cases we may rather say about later recognition of his ideas.

References

1. Bencivenga, E.: Free logics. In: Gabbay, D., Guenthner, F. (eds.) Handbook of Philosophical Logic, vol. III, pp. 373–426. Reidel Publishing Company, Dordrecht (1986)
2. Bencivenga, E.: Jaśkowski's universally free logic. Stud. Logica **102**(6), 1095–1102 (2014)
3. Dubikajtis, L.: The life and works of Stanisław Jaśkowski. Stud. Logica **34**(2), 109–116 (1975)
4. Fitch, F.: Symbolic Logic. Ronald Press Co, New York (1952)
5. Gabbay, D.: LDS – Labelled Deductive Systems. Clarendon Press, Oxford (1996)
6. Garson, J.W.: Modal Logic for Philosophers. Cambridge University Press, Cambridge (2006)
7. Gentzen, G.: Untersuchungen über das Logische Schliessen. Mathematische Zeitschrift **39**, 176–210, 405–431 (1934)
8. Jaśkowski, S.: Teoria dedukcji oparta na dyrektywach zał ożeniowych [Theory of deduction based on suppositional directives]. In: Księga Pamiątkowa I Polskiego Zjazdu Matematycznego. Uniwersytet Jagielloński, Kraków (1929)
9. Jaśkowski, S.: On the rules of suppositions in formal logic. Stud. Logica **1**, 5–32 (1934) (Reprinted in: Storrs McCall (ed.) Polish Logic 1920–1939, pp. 232–258. Oxford (1967))
10. Jaśkowski, S.: Recherches sur le systéme de la logique intuitioniste. In: Actes du Congrés International de Philosophie Scientifique, pp. 58–61. Sorbonne, Paris (1936) (English translation in: Studia Logica **34**(2), pp. 117–120 (1975))
11. Jaśkowski, S.: Elementy logiki matematycznej i metodologii nauk ścisł ych [Elements of mathematical logic and methodology of formal sciences]. Lecture notes, Toruń (1947)
12. Jaśkowski, S.: O aksjomatyce geometrii brył [On the axiomatization of the geometry of solids]. The Report of the VI-th Congress of the Polish Mathematicians (1948)
13. Jaśkowski, S.: Rachunek zdań dla systemów dedukcyjnych sprzecznych [A propositional calculus for inconsistent deductive systems]. Studia Societatis Scientiarum Torunensis, Sec. A **1**, 57–77 (1948) (Reprinted in: Studia Logica **24**, pp. 143–157 (1969))
14. Jaśkowski, S.: Sur les variables propositionnelles dépendantes. Studia Societatis Scientiarum Torunensis, Sec. A **24**, 17–22 (1948)
15. Jaśkowski, S.: Trois contributions au calcul des propositions bivalent. Studia Societatis Scientiarum Torunensis, Sec. A **1**, 3–15 (1948) (English translation in: Studia Logica **34**(2), pp. 121–132 (1975))
16. Jaśkowski, S.: Sur certains groupes formés de classes d'ensambles et leur applications aux définitions des nombres. Studia Societatis Scientiarum Torunensis, Sec. A **1**, 23–35 (1948) (English translation in: Studia Logica **34**(2), pp. 132–144 (1975))
17. Jaśkowski, S.: Sur le probléme de decision de la topologie et de la theorie des groups. Colloq. Math. **1**, 176–178 (1948)

18. Jaśkowski, S.: Sur certains axiomes de la géométrie élémentaire. Ann. Polish Math. Soc. **21**, 349–350 (1948)
19. Jaśkowski, S.: Une modification des définitions fondamentales de la géométrie des corps de A. Tarski. Ann. Polish Math. Soc. **21**, 298–301 (1948)
20. Jaśkowski, S.: Geometria brył [The geometry of solids]. Matematyka **1**, 1–7 (1949)
21. Jaśkowski, S.: O koniunkcji dyskusyjnej w rachunku zdań dla systemów dedukcyjnych sprzecznych [On the discussive conjunction in the propositional calculus for inconsistent deductive systems]. Studia Societatis Scientiarum Torunensis **1**, 171–172 (1949) (Reprinted in: Logic and Philosophy **7**, pp. 57–59 (1999))
22. Jaśkowski, S.: Z badań nad rozstrzygalnością rozszerzonej algebry Boole'a [Investigations on decidability of extended Boolean algebra]. Casopis pro pestovani matematiky a fizyky, roc. **74**, 136–137 (1949)
23. Jaśkowski, S.: O interpretacji zdań kategorycznych Arystotelesa w rachunku predykatów [On interpretations of Aristotle's categorical sentences in predicate calculus]. Studia Societatis Scientiarum Torunensis, Sec. A **2**, 77–90 (1950)
24. Jaśkowski, S.: Interpretacje funkcji przyczynowych w rachunku zmiennej zdaniowej zależnej [Interpretations of causal functions in the calculus of dependent propositional variable]. Sprawozdania Toruńskiego Towarzystwa Naukowego **1–4**, 123–124 (1950)
25. Jaśkowski, S.: On the modal and causal functions in symbolic logic. Stud. Philos. **4**, 71–92 (1951)
26. Jaśkowski, S.: O symetrii w zdobnictwie i przyrodzie [On symmetry in decorative art and in nature]. Warszawa (1952)
27. Jaśkowski, S.: Example of class of systems of ordinary differential equations having no decision method for existence problems. Bulletin de l'Academie Polonaise des Sciences **2**(4), 153–155 (1954)
28. Jaśkowski, S.: Undecidability of first order sentences in the theory of free groupoids. Fundam. Math. **43**, 36–45 (1956)
29. Jaśkowski, S.: Matematyka ornamentu [The mathematics of ornament]. Warszawa (1957)
30. Jaśkowski, S.: Über Tautologien, in welchen keine Variable mehr als zweimal vorkommt. Zeitschrift für mathematische Logik und Grundlagen der Mathematik, Bd. **9**, 219–228 (1963)
31. Jaśkowski, S.: On formulas in which no individual variable occurs more than twice. J. Symb. Log. **31**(1), 1–6 (1966)
32. Kotas, J., Pieczkowski, A.: Scientific works of Stanisław Jaśkowski. Stud. Log. **21**, 7–16 (1967)
33. Pietka, D.: Stanisław Jaśkowski's logical investigations. Organon **37**(40), 39–69 (2008)
34. Woleński J.: Logic and Philosophy in the Lvov-Warsaw School. Dordrecht, Kluwer (1989)

A. Indrzejczak (✉)
Department of Logic, University of Łódź, Łódź, Poland
e-mail: indrzej@filozof.uni.lodz.pl

Stanisław Jaśkowski and Natural Deduction Systems

Andrzej Indrzejczak

Abstract In 1934 Stanisław Jaśkowski published his groundbreaking work on natural deduction. At the same year Gerhard Gentzen also published a work on the same topic. We aim at presenting (three versions) of Jaśkowski's system and provide a comparison with Gentzen's approach. We also try to outline the influence of Jaśkowski's approach on the later development of natural deduction systems.

Keywords Jaśkowski · Gentzen · Natural deduction · Inclusive logic

Mathematics Subject Classification (2000) Primary 03F03; Secondary 01A60, 01A70

1 Introduction

Stanisław Jaśkowski is one of the founders of modern systems of natural deduction (ND). He presented his system in 1934 as the first volume of the series STUDIA LOGICA initiated by J. Łukasiewicz.[1] In fact, ND systems were constructed independently by two logicians; the second was Gerhard Gentzen. It is a matter of coincidence that at the same year Gentzen started to publish his Habilitationschrift which appeared in two parts in *Mathematische Zeitschrift* [13]. The name ND is due to Gentzen—he has called his system Natürliche Kalkül. Jaśkowski used the term "composite system" in contrast to Hilbert axiomatic "simple system"; below we explain the sense of this term in Sect. 3.1.2. Despite the differences in both approaches ND systems were conceived as formal realizations of traditional means of proving theorems in mathematics, science and ordinary discourse. Since then, several variants of ND were devised and presented in hundreds of logic textbooks, giving an evidence that ND systems are commonly accepted as the most efficient way of teaching logic. Still, simplifying a bit, but truly indeed, we may say that everything so far constructed in the field of ND and the related systems is based, more or less directly, on the ideas developed by these two researchers.

[1]This series after the War was revitalised as the well known logical journal.

Á. Garrido, U. Wybraniec-Skardowska (eds.), *The Lvov-Warsaw School. Past and Present*, Studies in Universal Logic,
https://doi.org/10.1007/978-3-319-65430-0_33

However, Jaśkowski's role in this enterprise and influence of his results on later developments in the field are not very well recognized. It is a common practice that authors presenting natural deduction systems of one sort or another are mentioning only Gentzen as the inventor of this kind of proof systems. It is not at all surprising that Gentzen's work is widely known since it provides much better developed body of research of a great generality. He did not present just an ND system but also a sequent calculus and provided important theoretical results. His famous normalization theorem for ND was indirectly proved on the basis of equally famous cut elimination theorem holding for sequent calculus. As a byproduct of cut elimination he obtained consistency and decidability results for propositional classical and intuitionistic logic and a version of Herbrand theorem. These profound results of Gentzen are the cornerstones of modern proof theory and this is the main reason that rather modestly looking, relatively short paper of Jaśkowski seems to have gone unnoticed.

But is that true that Jaśkowski's work on ND was really unnoticed and has no impact on the development of research on ND? It is the aim of this paper to show that despite of the absence of Jaśkowski's name in many textbooks presenting ND systems, his solutions had real and strong influence on further research. Roughly, we can say that whereas Gentzen had an enormous impact on the development of theoretical investigations on proof theory, Jaśkowski greatly influenced the practical side of the story. Most of ND systems popularised by hundreds of logic textbooks are directly based on Jaśkowski's proposal, usually without the authors' awareness of the roots of their solutions. We will point out also some other contributions of Jaśkowski's paper, like e.g. introduction of inclusion logic which also passed unnoticed at the time of publication, and were rediscovered by other scholars in later years. Surprisingly the same situation was connected with his invention of the first systems of paraconsistent logics (discursive logics) in 1948. Systematic research on this kind of non-classical logics started in 1960s without the knowledge of Jaśkowski's results.

We start with a general remarks on ND,[2] then we provide a detailed description of Jaśkowski's work on ND and compare his solutions with Gentzen's approach. Finally we describe some types of ND which were based on Jaśkowski's solution.

2 Natural Deduction in General

It is a lot of proof systems in use which are called ND systems and sometimes they differ greatly at first sight. Accordingly it is hard to provide a precise definition of ND-systems that would be generally accepted. Some authors tend to use this term in a broad sense, so that it covers also Gentzen's sequent calculus and various forms of tableau systems. In fact, all these systems are in close relationship to each other but we prefer to use this notion in a narrower sense. There are at least three reasons to make such a choice. First, for Gentzen his sequent calculus was meant as a technical tool to prove some metatheorems on natural deduction, not as a kind of ND. Secondly, both for him and for Jaśkowski, ND was supposed to reconstruct, in a formally proper way, traditional ways of reasoning. It

[2] This part is an excerpt from more detailed considerations contained in my [16] and [17].

may be a matter of discussion if existing ND systems realize this task in a satisfying way, but certainly systems like tableaux or resolution are worse in this respect. Finally, taking a term ND in a wide sense would be a classifying operation of doubtful usefulness.

So what is ND in a narrow sense? In Pelletier [24] and Pelletier and Hazen [25] one may find a discussion of several definitions of ND and their inadequacy. Instead of precise definition we provide three criteria which should be satisfied for genuine ND systems:

- ND system allows for entering assumptions into a proof and also for eliminating them.
- ND system consists of rules; there are no (or, at least, very limited set of) axioms.
- ND system admits a lot of freedom in proof construction and possibility of applying several proof search strategies.

Some authors (c.f. [3] or [24, 25]) formulated additional conditions characterising ND, but in our opinion these three are essential. According to this loose characteristics ND system should be open for reasoning from arbitrary assumptions and for the application of different proof constructions. The user is free in constructing direct, indirect, or conditional proofs. He may build more complex formulae or decompose them, as respective introduction/elimination rules allow. Instead of using axioms, or already proved theses, he is rather encouraged to introduce assumptions and derive consequences from them. The presence of axioms is permitted but not essential since their role is taken over by the set of primitive rules. This flexibility of proof construction in ND is in striking contrast to other types of deductive systems usually based on one form of proof.

The above characteristics of ND is still very broad and it allows a lot of freedom in the selection of primitive rules, design of a proof or graphical devices used as bookkeeping devices for indicating the scope of an assumption. These are also important features which make a variety of ND systems presented in textbooks apparently different but are of no real importance in delimiting this class of proof systems. In particular, both Jaśkowski's and Gentzen's approaches were similar in the three points we mentioned although different in many other respects, and this should be treated as a decisive argument for such a characterization of ND.

Additionally, in the first ND systems proposed by Jaśkowski and Gentzen one can identify two types of rules which we will call *rules of inference* and *proof construction rules*. The former have the form $\Gamma \ / \ \varphi$; we read them as follows: if we have all formulae from Γ (premises) present in the derivation we can add φ (conclusion) to this derivation. By derivation we mean an attempted, i.e. unfinished proof. *Proof construction rules* are more complex. In general they allow us to build a proof, enter additional assumptions opening nested subderivations, and show under which conditions we may discharge these assumptions and close the respective subderivations. Typical proof construction rules are meant to formalize the old and well known proof techniques like conditional proof, indirect proof, proof by cases, etc.

Although much can be said about the prehistory of ND, 1934 is commonly accepted as the first year in the official history of such systems. In this year two groundbreaking papers of Jaśkowski [19] and Gentzen [13] were published. It should be of no surprise that the two logicians with no knowledge of each other's work, independently proposed quite different solutions to the same problem. The need for deduction systems of this sort was in the air. Hilbert's proof theory already offered high standards of precise formalization in terms of axiom systems but the process of actual deduction in Hilbert calculi is usually complicated and needs a lot of invention. Moreover, axiomatic proofs are lengthy, difficult to decipher, and far from informal proofs provided by mathematicians. In consequence, axiom systems, although theoretically satisfying, were considered by many researchers as practically inadequate and artificial. Hence, two goals were involved in this enterprise: a theoretical justification of traditional proof methods on the ground of modern logic, and a formally correct and practically useful system of deduction.

A closer look at the circumstances of Jaśkowski's discovery shows that he may be rightly treated as the first inventor of ND. He was influenced by Łukasiewicz, who posed on his Warsaw seminar in 1926 a problem: how to describe, in a formally proper way, proof methods applied in practice by mathematicians (cf. Woleński [32]). In response to Łukasiewicz problem, Jaśkowski, as a young student[3] presented a first solution in the same year to his tutor. Officially, his first results on ND were announced in 1927, at the First Polish Mathematical Congress in Lvov, mentioned in [18]. Unfortunately, Jaśkowski had a lengthy break in his research due to serious health problems. After recovery in 1932 Jaśkowski gained his doctor's degree under the supervision of Łukasiewicz on the basis of his work on ND. The thesis was eventually published as [19].

3 Jaśkowski's Research on ND

Usually two versions of ND are attributed to Jaśkowski, the first called by Pelletier [24] a graphical method and the second a bookkeeping method. We are going to show that it is reasonable to say that Jaśkowski provided three versions of ND, quite similar yet different in a significant way. His first version of ND system (graphical) was not published in 1920s, and we do not know exactly for what logics, in what languages, and by means of what rules, it was conducted. The only thing we know is the format of proof applied by Jaśkowski in the original version since he provided examples in the footnote to [19]. Yet this feature is important enough to treat this proposal as different from the one officially presented in [19]. The latter, called by Pelletier a bookkeeping method differs significantly at least with respect to proof layout. We will call it the second ND (or the official) system of Jaśkowski.

After the War, Jaśkowski published his lecture notes [20] on mathematical logic in 1947. His presentation of classical logic in the script is not axiomatic but based on the application of ND. It seems that it is the first educational application of ND in the World where adequate system of ND is consequently applied as a form of presentation of classical logic in a textbook. His treatment of ND in [20] is different in some respects from [19] so we feel justified in saying about the third version of Jaśkowski's ND. In what follows we describe in separate sections the second and the third version. Remarks on the first version will be added to the presentation of the second one, because of the lack of knowledge mentioned above.

[3]In 1926 he was 20 years old.

3.1 Jaśkowski's Official ND

Jaśkowski's dissertation is not very long. However, on the 27 pages he provided ND systems for the following logics:

1. positive propositional logic;
2. intuitionistic propositional logic in the version of Kolmogoroff [22];
3. classical propositional logic (CPL);
4. (classical) propositional logic with quantifiers;
5. (inclusive) first-order logic.

Jaśkowski is using a language with $\rightarrow$, $\neg$ and $\forall$ as primitives, and applies so called Polish notation (parentheses-free) due to Łukasiewicz. In what follows we will be using standard notation for better readability. We do not repeat also the original formulation of rules since it is strongly connected with Jaśkowski's way of displaying proofs in the system and we explain this issue below. Instead we apply some neutral (to the proof format) way of description and additionally apply $\perp$ (not used as primitive by Jaśkowski, but introduced for illustration) as as a metalinguistic sign of inconsistency.

3.1.1 Rules

He started with CPL, then he just get rid with negation and a suitable rule for it, corresponding to indirect proof technique. Next, intuitionistic logic is obtained by a slight modification of this rule. So what are the rules for CPL? There are four such rules:

Rule I allows for introduction of an assumption prefixed with the letter 'S' (for supposition) in any place of the proof, hence it is neither inference nor proof construction rule. Rule II–IV formalize (in that order) Conditional Proof, Modus Ponens and (the strong form of) Indirect Proof. So rules II and IV are proof construction rules and III is the only rule of inference in the system. In proof-theoretic formulation (and without specific Jaśkowski's devices) the rules may be described in the following way:

Rule II If $\Gamma, \varphi \vdash \psi$, then $\Gamma \vdash \varphi \rightarrow \psi$
Rule III $\varphi, \varphi \rightarrow \psi \;/\; \psi$
Rule IV If $\Gamma, \neg\varphi \vdash \perp$, then $\Gamma \vdash \varphi$

where Γ denotes a set (possibly empty) of other active assumptions. As we mentioned Jaśkowski proposed also modifications of his calculus leading to weaker (he call them incomplete) propositional logics, namely, he has observed that the last rule may be weakened:

Rule IVa If $\Gamma, \varphi \vdash \perp$, then $\Gamma \vdash \neg\varphi$

This form yields ND formalization of Kolmogoroff's version of intuitionistic logic, whereas deletion of any rule for $\neg$ captures positive logic of Hilbert. It should be underlined that the version of intuitionistic logic considered by Jaśkowski is weaker than the well known Heyting's formalization. In particular, the intuitionistic thesis $\neg p \rightarrow (p \rightarrow q)$ is not provable in his system (although $p \rightarrow (\neg p \rightarrow \neg q)$ is provable). Note that in Gentzen's system this weaker form of indirect proof is sufficient for obtaining Heyting's intuitionistic logic but Gentzen is using $\bot$ as a primitive constant ($\neg$ is definable) and a rule of trivialization $\bot$ / φ, so deduction of q from $\neg p$ and p is not a problem. Incidentally, Jaśkowski is also mentioning a theory obtained by addition of $\top$ and $\bot$; we will describe it in connection with the proof format. Also, by the end of his paper he considered proper rules for conjunction and proposed the obvious ones:

$$(\wedge I) \quad \varphi, \psi \,/\, \varphi \wedge \psi$$
$$(\wedge E) \quad \varphi \wedge \psi \,/\, \varphi \text{ and } \varphi \wedge \psi \,/\, \psi$$

Jaśkowski formulated also ND system for propositional logic with universal quantifier, called by him the extended theory of deduction. In such a system we can define $\bot$ as $\forall p, p$ which is in fact shown by Jaśkowski. He defines in an obvious way the notions of free (real) and bound (apparent) propositional variable and add to CPL two new rules which may be formally displayed as follows:

$$\text{Rule V} \quad \forall p\varphi \,/\, \varphi[p/\psi]$$
$$\text{Rule VI} \quad \text{If } \Gamma \vdash \varphi, \text{ then } \Gamma \vdash \forall p\varphi$$

In rule V $\varphi[p/\psi]$ denotes the operation of proper substitution of ψ for p in φ, which means that all occurrences of p which were bound in $\forall p\varphi$ are substituted by ψ and no propositional variable in ψ is bound in $\varphi[p/\psi]$. Rule VI has a side condition that p is not free in any active assumption in Γ. Although we have formulated it as a proof construction rule it may be also described as inference rule with side conditions since there is no subtraction from the set of active assumptions.

Finally Jaśkowski developed ND for first order logic (calculus of functions) but with explicit remark that it is weaker than classical version. Jaśkowski states that "whether individuals exist or not, it is better to solve this problem through other theories. We shall present therefore a system [...] where all theses will be satisfied in the null field of individuals" (p. 28). Thus he introduced the first system of inclusive logic, whereas the first recognized system of this type was presented in 1950s.[4]

Jaśkowski realised that in such a system free variables are in fact not variables but rather (nonlogical) constants whose existence we assume in the proof. Such variables are called valid in the part of a proof where we declared their existence. He did not introduce different kind of letters for denoting free variables (as Gentzen did) but instead he applied a special technique of explicit signification that some variable is held constant for the sake of proof. It is an additional rule which parallels the rule I. This rule VII allows to introduce a term supposition Tx for any variable not valid (so far) in this part of proof. Also rule I must be restricted; we can introduce as assumptions only such formulae which do not

[4]Cf. historical remarks in Bencivenga [4]; in [5] he also pointed out that Jaśkowski's system may be easily changed into a system of free logic.

contain free variables not valid in the respective part of proof. The remaining rules are variants of rules V and VI:

Rule Va	$Ty, \forall x\varphi \;/\; \varphi[x/y]$
Rule VIa	If $\Gamma, Tx \vdash \varphi$, then $\Gamma \vdash \forall p\varphi$

Note that $\varphi[x/y]$ denotes the proper substitution of y for x in φ but only if y is valid in respective part of a proof (Ty was introduced earlier by rule VII), hence in contrast to rule V this is a two-premise rule. Also VIa is a "real" proof construction rule since the term-assumption Tx ceased to be valid in the result of its application and it is deleted from the set of active assumptions (a respective subproof is discharged).

3.1.2 Proof Format

In the case of ND it is very important how we define a proof since uncontrolled introduction of assumptions without clear indication of their scope may lead to troubles very often met when students are taught to use ND.

Let us analyse the following "proof" apparently performed with the help of Jaśkowski's rules (with added rule for disjunction introduction ($\vee I$) in line 3):

1	$Sp \vee r \to q$	$R.I$
2	Sp	$R.I$
3	$p \vee r$	$2, \vee I$
4	q	$1, 3, R.III$
5	$p \to q$	$2, 4, R.II$
6	$(p \to q) \wedge p$	$2, 5, \wedge I$

But $(p \to q) \wedge p$ does not follow logically from $p \vee r \to q$. We cannot apply ($\wedge I$) correctly to formulae from lines 2 and 5 since assumption Sp is not active in this place. The application of rule II in line 5 discharged the assumption 2 and this formula cannot be further used in the proof. This problem is generally connected with the application of such rules as II or IV (and VIa) which are not simply inference rules but proof construction rules showing that if some proof is being constructed on the basis of some assumption, then, in the effect, we obtain another proof in which this assumption is not in force. Thus linear proof admitting additional assumptions and means for closing dependant parts of a proof is in fact not a simple sequence of formulae but rather a richer structure containing nested subderivations (subordinate proofs). To avoid scoping difficulties some devices must be used for separating the parts of proof which are in the scope of discharged assumption, hence not available.

All that we know about the first version of Jaśkowski's ND is that he provided a clear indication of the scope of every assumption introduced into a proof. His first original solution to the problem consisted in making boxes for each assumption and dependent part of a proof. Every introduction of an assumption was connected with starting a new box, and this assumption was always put as the first formula in it (he did not applied the prefix "S" for that). An application of any proof construction rule like II or IV was connected with closing a current box, and inferred formula was immediately written down as the next element of outer derivation. He also used an additional rule of repetition to shift a formula from outer open box to the inner; transition in the other direction was of course forbidden. Schematically, the application of both proof construction rules in his propositional system[5] looks like this:

$$
\begin{array}{c}
\Gamma \\
\boxed{\begin{array}{l} \varphi \\ \Gamma' \\ \vdots \\ \psi \end{array}} \\
\varphi \rightarrow \psi
\end{array}
\qquad\qquad
\begin{array}{c}
\Gamma \\
\boxed{\begin{array}{l} \neg\varphi \\ \Gamma' \\ \vdots \\ \bot \end{array}} \\
\varphi
\end{array}
$$

On the diagrams possibly empty $\Gamma' \subseteq \Gamma$ refers to formulae obtained by repetition.

In his official version Jaśkowski applied an apparatus of numeric prefixes, instead of boxes. These are finite sequences of natural numbers separated with dots and written before each formula in a proof (except of a thesis). Each time we enter an assumption we extend prefix with additional number (and a prefix S in front of an assumption.). Each application of rules II, IV (IVa) and VIa is connected with subtraction of the last number in the prefix. Application of inference rules, like II or V is admitted only if prefixes of their premises are initial parts, or are identical, to the prefix of the last formula in a proof (prefix of a conclusion must be identical to it). This way Jaśkowski avoided introduction of repetition as a rule.

Application of prefixes instead of boxes may seem as an editorial simplification but in fact it is connected with some philosophical motivations taken from Leśniewski, concerning dynamic nature of deductive system. Jaśkowski thought of prefixes as indicators of domains in which formulae with this prefix are valid. Thus formulae with empty prefixes were ordinary theses (valid in every domain), and those with nonempty prefixes are theses relative to some domain in which some suppositions are postulated as valid. Prefixes are then records of dependency of a formula on assumptions in the context of a proof or on the existence of some objects named by term-suppositions. Domain-validity is hereditary with respect to nested subdomains. Thus a formula e.g. $3.2.1.\varphi$ is valid not only in the domain 3.2.1. but also in 3.2.1.1., 3.2.1.5.2 but not in 3.2. or 3.3.1. Jaśkowski directly pointed out that "Every domain can be considered as a system having its own axioms and constants, though not every domain gives a complete system, much less an interesting one." (p. 14). As an example of "interesting" system he considered the one with added suppositions St (representing $\top$) and $S\neg u$ (interpreted as $\neg\bot$). As the first is introduced with prefix 7. and the second with prefix 7.1. we can consider every formula valid in the domain prefixed with 7.1. as a thesis of a system with $\top$ and $\neg\bot$ being additional axioms. Such a rationale behind using prefixes is also a reason for using a name "composite system" for ND, as it is "composed of many systems".[6]

[5]The two examples provided in the footnote in [19] show only propositional proofs.

[6]By the way, an innovation introduced by Jaśkowski (i.e. prefixes) may be classified in a different way; we may treat his second version as the first example of ND defined not on formulae but on labelled formulae.

Below we illustrate both versions with an example of a proof.

1	p	$R.I$
2	$\neg\neg\neg p$	$R.I$
3	p	$(1, rep.)$
4	$\neg\neg p$	$R.I.$
5	$\neg\neg\neg p$	$(2, rep.)$
6	$\neg p$	$4, 5, R.IV$
7	$\neg\neg p$	$3, 6, R.IV$
8	$p \rightarrow \neg\neg p$	$1, 7R.II$

1	$1.Sp$	$R.I$
2	$1.1.S\neg\neg\neg p$	$R.I$
3	$1.1.1.S\neg\neg p$	$R.I$
4	$1.1.\neg p$	$2, 3, R.IV$
5	$1.\neg\neg p$	$1, 4, R.IV$
6	$p \rightarrow \neg\neg p$	$1, 5, R.II$

Additionally we provide an example of proof of the first two theses in his system with quantifiers:

$cf1$	$1.S\forall xyAxy$	$R.1$
$cf2$	$1.1.Tz$	$R.VII$
$cf3$	$1.1.\forall yAzy$	$1, R.Va$
$cf4$	$1.1.Azz$	$3, R.Va$
$cf5$	$1.\forall zAzz$	$4, R.VIa$
$cf6$	$\forall xyAxy \rightarrow \forall zAzz$	$1, 5, R.II$
$cf7$	$1.1.1.Tv$	$R.VI1$
$cf8$	$1.1.1.Azv$	$3, 7, R.Va$
$cf9$	$1.1.\forall vAzv$	$7, 8, R.VIa$
$cf10$	$1.\forall zvAzv$	$2, 9, R.VIa$
$cf11$	$1.2.Tx$	$R.VII$
$cf12$	$1.2.1.Ty$	$R.VII$
$cf13$	$1.2.1.\forall vAyv$	$10, 12, R.Va$
$cf14$	$1.2.1.Ayx$	$11, 13, R.Va$
$cf15$	$1.2.\forall yAyx$	$12, 14, R.VIa$
$cf16$	$1.\forall xyAyx$	$11, 15, R.VIa$
$cf17$	$\forall xyAxy \rightarrow \forall xyAyx$	$1, 16, R.II$

Each line contains a successive thesis of a calculus of function (hence 'cf') valid in respective domain. Absolute theses are formulae cf 6 and cf 17. The example illustrates not only the application of rules for inclusive quantifiers but also a dynamics of the system. For example in line 7 a term assumption is introduced not with a prefix 2 but with a prefix 1.1.1, as a continuation of domain 1.1. In fact, a thesis cf 6 is not in itself very interesting and may be seen as an auxiliary result required for proving cf 17 (to be more precise lines 1–3 are necessary for a proof of cf 17.).

3.1.3 Adequacy

Demonstration of completeness for ND systems is not demanding. If we have at our disposal adequate axiomatic system it is enough to show that all axioms are provable and primitive rules may by simulated in ND. We can also directly prove completeness for ND with respect to semantics in the same way as it is done for axiomatic systems. Indirect results of the first kind are provided in [19] for all ND systems except his inclusive logics since there were neither axiomatic formulation of such a logic nor semantic one.

Usually the problem for ND systems with additional bookkeeping devices, is to prove their soundness, because all this additional machinery must be somehow "translated" either into semantics of suitable logic or into a simpler syntax of axiomatic system. Jaśkowski [19] established some standard form of soundness proof extensively used by many logicians in respective proofs for ND-systems. Shortly, for each prefixed formula we build its development, which is a descending implication with suppositions for each number in the prefix as antecedents and formula itself as the consequent. For example, the development of prefixed formula $i_1 \ldots .i_n.\varphi$ is $\psi_1 \rightarrow (\psi_2 \rightarrow \ldots(\psi_n \rightarrow \varphi)\ldots)$, where each ψ_k, $1 \leq k \leq n$ is an assumption introduced with addition of successive i_k to already existing prefix, i.e. we have $i_1 \ldots .i_k.S\psi_k$ above $i_1 \ldots .i_n.\varphi$ in the proof. Now we may either directly prove that the development of a formula in each line is semantically valid or that it is provable in respective axiomatic system. In the first case we proceed by showing that the first line of a proof is valid ($\varphi \rightarrow \varphi$) and that all rules expressed in terms of developments are validity preserving. In the second case we must prove that some formulae are theses of axiomatic systems. Specifically, Jaśkowski has proved for his rules I, II, III (and additionally IVa) that the development of a formula in each line is a thesis of an axiomatic system of positive (and intuitionistic logic). For CPL he proved a soundness of his ND directly whereas for the system with quantifiers the result is only pointed out. In case of his inclusive logic no such result was possible of course.

This manner of showing soundness for ND systems is commonly applied. There are many variants of this technique but essentially we proceed in such a demonstration by turning formulae of any proof into formulae or a kind of sequents (we add a record of active assumptions), and then by showing that (such modified) rules are validity preserving.[7]

3.2 *Natural Deduction in Jaśkowski's Lecture Notes*

In 1947 Jaśkowski published in mimeographed form his lecture notes "Elements of Mathematical Logic and Methodology of Deductive Sciences" in polish. The book consists of 105 pages and is of great importance for us since Jaśkowski decided to apply in it his natural deduction rules. It is perhaps the first logic textbook where natural deduction is uniformly used as a method for presentation of logic. It is used from the beginning for proving theorems of logic without any reference to axiomatic systems. Moreover, it

[7] Although this approach is by no means the only one possible. For example, in [16] we proposed a different general strategy of proving soundness for any ND system in Jaśkowski format.

is applied also in proofs of metalogical results and even truth-functional semantics is introduced via analysis of ND proofs of selected theses.

A system described in [20] deserves the separate treatment since it contains significant differences with the version from [19]. In [20] Jaśkowski presented:

1. classical propositional logic;
2. propositional logic with quantifiers;
3. classical first-order logic;
4. theory of identity in the second order language.

The most important changes in his later approach to ND are the following:

1. Richer language. Jaśkowski introduced $\wedge$, $\vee$, $\leftrightarrow$ and $\exists$ as primitive constants and defined introduction and elimination rules for all of them.
2. Omission of nonclassical logics. Only classical logic is presented in lecture notes. In particular, instead of inclusive logic there is a set of rules characterising classical quantifiers. Moreover, the rules are generalised for second order variables and theory of identity is expressed in the extended language.
3. Different style of layout for proofs.

3.2.1 Rules

Again the list of rules is opened by the rule for introducing assumptions: any formula may be added with a horizontal bracket above it as an assumption (no prefix 'S' is attached in front of). The rules for $\rightarrow$ and $\neg$ are without changes and the names for them are: $C1$ (implication introduction), $C2$ (MP) and $N1$ (negation elimination). For the remaining connectives we have the following inference rules:

$$
\begin{array}{ll}
(K1) & \varphi, \psi \ / \ \varphi \wedge \psi \\
(K2) & \varphi \wedge \psi \ / \ \varphi \text{ and } \varphi \wedge \psi \ / \ \psi \\
(A1) & \varphi \ / \ \varphi \vee \psi \text{ and } \psi \ / \ \varphi \vee \psi \\
(A2) & \varphi \vee \psi, \varphi \rightarrow \chi, \psi \rightarrow \chi \ / \ \chi \\
(E1) & \varphi \rightarrow \psi, \psi \rightarrow \varphi \ / \ \varphi \leftrightarrow \psi \\
(E2) & \varphi \leftrightarrow \psi, \varphi \ / \ \psi \text{ and } \varphi \leftrightarrow \psi, \psi \ / \ \varphi
\end{array}
$$

In case of propositional logic with quantifiers two rules for $\forall$ are the same as in [19] (now called ($\forall$1) and ($\forall$2)) but he added three more rules for $\exists$:

$$
\begin{array}{ll}
(\exists 1) & \varphi[p/\psi] \ / \ \exists p\varphi \\
(\exists 2) & \forall p(\varphi \rightarrow \psi), \exists p\varphi \ / \ \exists p\psi \\
(\exists 3) & \exists p\varphi \ / \ \varphi
\end{array}
$$

The last one is the rule of eliminating vacuous quantification since it has a side condition that p does not occur in φ. In ($\exists$1) we have of course a proper substitution of ψ for p in the premiss. Note that there is not one elimination rule for $\exists$ in the system. The possible effect is divided into two rule with ($\exists$2) being of rather mixed character.

The rules for quantifiers in first order logic are identical, the only difference is that individual variables are bounded instead of propositional ones. Hence in particular, introduction of $\forall$ is not a "real" proof construction rule in this system; we add $\forall x$ to some φ only after checking that x is not free in any active assumption. In contrast to rule VIa from [19] there is no closure of a subproof and the rule exactly parallels rule VI. Of course there is also no rule of introduction of term-assumptions in this system.

In both logics with quantifiers there is some innelegancy in the treatment of $\exists$. However, it works and we avoid the problems usually generated in other ND systems where some rule for $\exists$ elimination is postulated.[8] In order to show that the set of rules is complete it is enough to demonstrate as a thesis $\exists x Ax \leftrightarrow \neg\forall\neg Ax$. We will show it in the next section after characterising proof format.

Jaśkowski extended the application of his rules for quantifiers to second order logic, mainly to develop the theory of identity. In an informal way he describe first the conditions of proper substitution for predicate variables which is necessary for rules ($\forall$1) and ($\exists$1). Identity is first characterised by Leibniz condition $\forall A(Ax \leftrightarrow Ay)$ without explicit introduction of a new constant. Then, in the section on definitions, he formulates its definition $\forall xy(x = y \leftrightarrow \forall A(Ax \leftrightarrow Ay))$ and shows that by addition of it as a new assumption we can deduce all characteristic properties of identity. Finally, when discussing axiomatic systems, he provides also axiomatic characterization:

A1 $\forall x, x = x$
A2 $\forall A \forall xy(x = x \rightarrow (Ax \rightarrow Ay))$

In all these formulae A is a predicate variable.

3.2.2 Proof Format

Proofs in [20] are written in linear form but the system of prefixes is absent in this presentation. Instead Jaśkowski is using horizontal brackets over the assumption and under the last formula in a subproof. This solution looks like a simplified version of his first idea of using boxes. In fact, he tends to simplify other things as well; there is no rule of repetition, no prefixes indicating suppositions and even no numeration of lines. Perhaps resignation from prefixes is connected with resignation from Leśniewski's like treatment of deductive system as a dynamic body of theses valid in different domains. Jaśkowski in [20] just provided a series of separate proofs of theses.

[8] The history of successive versions of Copi's ND with numerous mistaken formulation of this rule is particularly instructive—see e.g. Annellis [1].

Below we provide a proof of the same propositional thesis which was displayed in Sect. 3.1.2. for illustration.

$$
\begin{array}{lll}
1 & \overbrace{p} & ass. \\
2 & \overbrace{\neg\neg\neg p} & ass. \\
3 & \overbrace{\neg\neg p} & ass. \\
4 & \underbrace{\neg p} & 2, 3, N1 \\
5 & \underbrace{\neg\neg p} & 1, 4, N1 \\
6 & p \rightarrow \neg\neg p & 1, 5, C1
\end{array}
$$

In order to show the difference between the inclusive rules from [19] and classical rules we provide two proofs of a thesis which is also valid in inclusive logic. For easier comparison we settle the first proof also in Jaśkowski's new (bracketing) style.

$$
\begin{array}{lll}
1 & \overbrace{\forall x(Ax \rightarrow Bx)} & R.I \\
2 & \overbrace{\forall x Ax} & R.I \\
3 & \overbrace{Ty} & R.VII. \\
4 & Ay \rightarrow By & 1, R.Va \\
5 & Ay & 2, R.Va \\
6 & \underbrace{By} & 4, 5, R.III \\
7 & \underbrace{\forall x Bx} & 3, 6, R.VIa \\
8 & \underbrace{\forall x Ax \rightarrow \forall x Bx} & 2, 7, R.II \\
9 & \forall x(Ax \rightarrow Bx) \rightarrow (\forall x Ax \rightarrow \forall x Bx) & 1, 8, R.II
\end{array}
$$

$$
\begin{array}{lll}
1 & \overbrace{\forall x(Ax \rightarrow Bx)} & ass. \\
2 & \overbrace{\forall x Ax} & ass. \\
3 & Ax \rightarrow Bx & 1, \forall 1 \\
4 & Ax & 2, \forall 1 \\
5 & Bx & 3, 4, C2 \\
6 & \underbrace{\forall x Bx} & 5, \forall 2 \\
7 & \underbrace{\forall x Ax \rightarrow \forall x Bx} & 2, 6, C1 \\
8 & \forall x(Ax \rightarrow Bx) \rightarrow (\forall x Ax \rightarrow \forall x Bx) & 1, 7, C1
\end{array}
$$

The selection of rules for $\exists$ may seem doubtful at first but, in contrast to some other solutions, it has some advantages. All the rules are simple in their form and normal in the sense that premisses logically imply conclusions. In Gentzen's approach the rule for elimination of $\exists$ is a proof construction rule introducing additional subproof. In systems where some inference rule of this kind is provided it is connected with some (sometimes considerably complicated) side conditions (like in Quine's natural deduction [31] or Słupecki and Borkowski's [6] solution).

In order to show that Jaśkowski's characterization of $\exists$ is sufficient it is enough to demonstrate that $\exists x Ax \leftrightarrow \neg\forall x\neg Ax$ is derivable (normality of rules yields soundness).

$$
\begin{array}{lll}
1 & \overbrace{Ax} & ass. \\
2 & \quad\overbrace{\neg\neg\forall x\neg Ax} & ass. \\
3 & \quad\forall x\neg Ax & 2, CPL \\
4 & \quad\underbrace{\neg Ax} & 3, \forall 1 \\
5 & \underbrace{\neg\forall x\neg Ax} & 1, 4, N1 \\
6 & Ax \rightarrow \neg\forall x\neg Ax & 1, 5, C1 \\
7 & \forall x(Ax \rightarrow \neg\forall x\neg Ax) & 6, \forall 2 \\
8 & \overbrace{\exists x Ax} & ass. \\
9 & \exists x\neg\forall x\neg Ax & 7, 8, \exists 2 \\
10 & \underbrace{\neg\forall x\neg Ax} & 9, \exists 3 \\
11 & \exists x Ax \rightarrow \neg\forall x\neg Ax & 8, 10, C1
\end{array}
$$

We first prove an auxiliary thesis (in line 7) which is then used as one of the premisses for the application of ($\exists 2$). Also ($\exists 3$) is used in line 10 to eliminate vacuous quantification.

$$
\begin{array}{lll}
1 & \overbrace{\neg\exists x Ax} & ass. \\
2 & \quad\overbrace{\neg\neg Ax} & ass. \\
3 & \quad Ax & 2, CPL \\
4 & \quad\underbrace{\exists x Ax} & 3, \exists 1 \\
5 & \quad\neg Ax & 1, 4, N1 \\
6 & \underbrace{\forall x\neg Ax} & 5, \forall 2 \\
7 & \neg\exists x Ax \rightarrow \forall x\neg Ax & 2, 6, C1 \\
8 & \neg\forall x\neg Ax \rightarrow \exists x Ax & 7, CPL
\end{array}
$$

In the proof of the converse we applied ($\exists 1$) in line 4 thus showing that all three rules for $\exists$ yield a complete characterization of $\exists$.

4 Other Approaches to ND

ND was not also independently proposed by Gentzen but his proposal is widely known, in contrast to Jaśkowski. Before we try to explain why Gentzen is better recognized as a father of ND we briefly describe the most important similarities and differences.[9]

When we consider the rules both approaches are very similar. Gentzen also considered classical and intuitionistic logic; the former also in first order case, the latter in Heyting's version (and not restricted to propositional part). In contrast to Jaśkowski he prefers to work with richer language, in particular because he was interested in the philosophical project of syntactic characterization of logical constants by means of rules later developed in inferentialist program. The main difference lies in the format of proof chosen by both authors. Gentzen defined proofs as trees labelled with formulae, where leafs are assumptions and the conclusion is put in the root. Transitions between nodes correspond to elementary inferences.

The distinction between ND-systems using tree- or linear-format of proof seems to be not very serious from the theoretical point of view. After all we can redefine every binary tree as a sequence. But in practice the difference is very important because in linear proof we are dealing with formulae, whereas in tree-proofs we are dealing with their concrete occurrences. Since we may use the same formula many times in linear proof, we are forced to introduce some devices for cancelling the part of a proof which is in the scope of an assumption already discharged. Otherwise we could "prove" everything, as was illustrated in Sect. 3.1.2.

Scoping difficulties do not occur in tree-proofs because we are operating not on formulae but on their single occurrences. Thus premises of any application of a rule must always be displayed directly over the conclusion. Consequently, we cannot use in a proof something which depends on discharged assumptions, because a part of a proof responsible for deduction of a formula must be reproduced above. So Gentzen did not need to bother about technical devices to block nonvalid deductions. Tree format requires less complicated machinery and is very good in representing ready proofs, because the structure of inferential dependencies is readably represented. Moreover, tree proofs are better for proving theoretical results in proof theory. Prawitz [26] proved the profound result on the existence of proofs in normal forms for many ND systems with tree-proofs (Gentzen did it indirectly by showing equivalence with cut-free sequent calculus). No wonder that in works concerned with theoretical investigations this format is very popular (good witness is Negri and von Plato [23].).

But the features of tree proofs that make them so attractive are also the source of some problems. Tree nicely shows the structure of a finished proof but it is hardly suitable for actual derivation. Mental process of proof construction has rather linear structure; we start with assumptions and deduce conclusions until we get the desired goal. Gentzen himself was well aware of this fact, when he wrote that "we are deviating somewhat from the analogy with actual reasoning. This is so, since in actual reasoning we necessarily have (1) a linear sequence of propositions due to the linear ordering of our utterances, and (2) we are accustomed to applying repeatedly a result once it has been obtained, whereas the

[9]More detailed comparison of both approaches may be found in Hazen and Pelletier [15].

tree form permits only of a single use of a derived formula." ([13] citation from [30, page 76]) According to Gentzen however, this form of representation is simpler and resulting deviations are inessential.

The choice of linear proof format has also some computational advantages; we can show that for each proof D in tree format we can provide a linear proof D' such that the length of D' is the same or smaller than the length of D. The converse does not hold because in tree proofs we work with occurrences of formulae when in linear proofs we work with formulae themselves. It forces us to repeat many times the same proof-trees if their starting assumptions are used several times. This also shows that linear proofs are better from the practical point of view.

It seems that Jaśkowski decided to use linear format as much closer to actual reasoning, and much more useful for actual proof-search. The applications of ND in logic textbooks are good witnesses of this choice; there is only a few such books using tree format. Majority of them use linear proofs and basically all of them are some variations on the first two solutions introduced by Jaśkowski.

Despite the apparent differences, one thing is common to all variants of Jaśkowski's ND—an essential idea of dividing a proof into separated and partially ordered subproofs. It appears as the most popular solution in hundreds of textbooks where ND-techniques are applied. His first version (graphical), although abandoned by the author himself, is much more popular nowadays. It has many variants but there is always some graphical device added to linear sequence of formulae in a proof. The original format of boxes was used by Kalish and Montague [21], but with some adjustments which make their system one of the most flexible in practise. In their system each box is preceded with so called 'show-line' which indicates the goal of deduction to be performed inside the box. After closing a box such a show-line is treated as a new formula in the proof. Simplified account, where each assumption is entered with the vertical line which continues until this subproof is in force, is due to Fitch [10], whereas popular system of Copi [8] applies vertical bracketing to closed subproofs. These solutions were repeated in hundreds of textbooks.

It should be noted also that this approach proved especially useful with respect to many nonclassical logics formalized via ND systems. Because parts of proof are separated graphically it is easy to distinguish between different types of subproofs and formulate several kinds of repetition rules with restrictions on the form of formulae which may be shifted to subproofs. One may find ND systems of this kind for modal logics (c.f. Fitting [11], Garson [12], Indrzejczak [16]), relevant logics (Anderson and Belnap [2]) and many others.

The second solution of Jaśkowski is not so popular in ND setting. Borkowski and Słupecki in their ND system from [6] followed this route but with significant simplifications. First of all they treat prefixes as just line-numbers of the proof. They also avoid a proliferation of prefixes since they do not introduce a new prefix for every assumption. Each thesis is analysed in terms of descending implication and all antecedents are written in the same proof level. The proof is ready if the succedent is deduced. For example, if we want to prove a thesis of the form $\varphi_1 \to (\varphi_2 \to \psi)$ we construct a proof looking like this:

$$\begin{array}{l} 1.\ \varphi_1\ ass. \\ 2.\ \varphi_2\ ass. \\ \quad\vdots \\ n.\ \psi \end{array}$$

instead of Jaśkowski's more complicated form:

$$\begin{array}{lll} 1 & 1.S\varphi_1 & R.I \\ 2 & 1.1.S\varphi_2 & R.I \\ & \vdots & \\ n & 1.k.\psi & \\ n+1 & 1.\varphi_2 \to \psi & 2, n, R.II \\ n+2 & \varphi_1 \to (\varphi_2 \to \psi) & 1, n+1, R.II \end{array}$$

Of course, if a thesis to be proved is not an implication we must start with indirect assumption and proceed with indirect proof, hence a rule for indirect proof is a primitive one. In fact it is also the only indispensable proof construction rule in the system since in the definition of a proof they allow for introduction of previously proved theses. In consequence, such rules like introduction of implication, or other based on the introduction of additional assumptions, are admissible in their system but in fact we can dispense with subproofs and additional column of numerals for their indication. The problem of elimination rule for ∃ is also solved in the original way in their system. They apply the inference rule which implicitly uses skolemization. One may find an extensive applications of their system to logic and set theory in many textbooks written in Polish as well as in English translation [28].

The system of Słupecki and Borkowski is rather not known outside Poland but it was also interestingly applied in the field of automated deduction. In 1970s Andrzej Trybulec started to develop an integrated framework for deduction of theorems in mathematical theories called MIZAR. It is basically a computer environment allowing formalization and proof-checking on the basis of rich library. In 1980s Professor Marciszewski initiated a research program concerned with the applicability of MIZAR to construct and to check formal proofs in Słupecki and Borkowski system. The program was developed by numerous scholars in many centers, for example in Opole (Wybraniec-Skardowska, Bryniarski) and Łódź (Malinowski, Nowak, Łukowski). It shows a great potential of MIZAR and natural deduction system in formalization of logic and formal theories. Proofs in MIZAR are constructed similarly as in Słupecki and Borkowski system although some additional devices for users-friendly presentation are added. The present library is based on axioms of set theory in the version due to Tarski-Grothendieck and includes over 1300 articles written by nearly three hundreds of researchers (see www.mizar.org).

There is a kind of ND systems which at first sight may be seen also as a simplification of Jaśkowski's second variant. I mean here a system of Suppes [29] where in each line of a proof we have added a set of numerals of all assumptions active for the formula in question. But the similarity to Jaśkowski's prefixes is apparent in this kind of ND. Lines in Suppes' ND correspond rather to sequents; a set of numerals is a shortcut for antecedent of a sequent. Such a simplification is possible if all rules operate only on succedents of sequents. We do not enter into the details of such solution but it should be noted that such kind of ND is rather a by-product of Gentzen's later paper [14].

Jaśkowski's third system was not known and it is hard to find similar solutions, except perhaps a system presented by Corcoran and Weaver [9]. Here proofs are written down horizontally with subproofs put in brackets. Thus our example proof in Corcoran's style looks like that:

$$[p[\neg\neg\neg p, p[\neg\neg p, \neg\neg\neg p]\neg p]\neg\neg p]p \to \neg\neg p$$

Our claim that Jaśkowski's lecture notes are perhaps the first consequent textbook application of ND requires some justification. Quine [31] claims that the first textbook applying ND is due to Cooley [7] and was printed in 1942, then reprinted in 1946. In fact, Cooley applies numerous inference rules throughout the book, however it may be disputable if it is ND system satisfying our three criteria. Conditional proofs based on additional assumptions are only described on pp. 126–140 but not used as the main form of presentation of logic. Moreover, Cooley did not apply any devices for separating subproofs and a rule for elimination of existential quantifier is stated without sufficient restrictions. Hence in our opinion it cannot be treated as a correct system of ND. It seems that the first textbook which consequently applies ND is that of Fitch [10] published in 1952. In Quine's [31] from 1950, ND is also introduced only in three sections as an illustration rather, not as the main proof system. Quine mentioned also some earlier mimeographed notes of himself and of Rosser which applied ND but I had no possibility to check them. A well known textbook of Rosser [27] is using axiomatic system and introduces additional ND-like rules only as a metalogical devices for simplification of axiomatic proofs.

We can conclude our considerations with the following remark concerning Jaśkowski and Gentzen. Both authors laid down the foundations for further investigations on ND but in a slightly different fashion. Jaśkowski seemed to be more concerned with practical aspects of deduction and his general approach, as well as his technical solutions, are of common classroom and textbook use. On the other hand, Gentzen was more theoretically oriented; his investigations led him to profound results in general proof theory.

This is my own evaluation of Jaśkowski's real influence on ND. It is based on the analysis of his texts and easily verifiable. But it should be contrasted with the real knowledge of his achievements and impact on ND. In the earliest applications of ND, like in Quine's or Fitch's book, the origins of the method are known and confirmed. For example, Fitch in foreword claimed that he is using the method of subordinate proofs since 1941 but both Gentzen and Jaśkowski are mentioned as the source of inspiration. Unfortunately, later authors often tend to say about Fitch's ND and forget about Jaśkowski.

References

1. Anellis, I.H.: Forty years of "unnatural" natural deduction and quantification. A history of first-order systems of natural deduction from Gentzen to Copi. Mod. Log. **2**(2), 113–152 (1991)
2. Anderson, A.R., Belnap, N.D.: Entailment: The Logic of Relevance and Necessity, vol. I. Princeton University Press, Princeton (1975)
3. Barth, E.M., Krabbe, E.C.: From Axiom to Dialogue. Walter de Gruyter, Berlin (1982)
4. Bencivenga, E.: Free logics. In: Gabbay, D., Guenthner, F. (eds.) Handbook of Philosophical Logic, vol. III, pp. 373–426. Reidel Publishing Company, Dordrecht (1986)
5. Bencivenga, E.: Jaśkowski's universally free logic. Stud. Logica **102**(6), 1095–1102 (2014)
6. Borkowski, L., Słupecki, J.: A Logical System based on rules and its applications in teaching Mathematical Logic. Stud. Logica **7**, 71–113 (1958)
7. Cooley, J.C.: A Primer of Formal Logic. Macmillan, New York (1942)
8. Copi, I.M.: Symbolic Logic. The Macmillan Company, New York (1954)
9. Corcoran, J., Weaver, G.: Logical consequence in modal logic: Natural Deduction in S5. Notre Dame J. Formal Log. **10**, 370–384 (1969)
10. Fitch, F.: Symbolic Logic. Ronald Press Company, New York (1952)
11. Fitting, M.: Proof Methods for Modal and Intuitionistic Logics. Reidel, Dordrecht (1983)
12. Garson, J.W.: Modal Logic for Philosophers. Cambridge University Press, Cambridge (2006)
13. Gentzen, G.: Untersuchungen über das Logische Schliessen. Math. Z. **39**, 176–210, 405–431 (1934)
14. Gentzen, G.: Die Widerspruchsfreiheit der reinen Zahlentheorie. Math. Ann. **112**, 493–565 (1936)
15. Hazen, A.P., Pelletier, F.J.: Gentzen and Jaśkowski Natural Deduction: Fundamentally Similar but Importantly Different. Stud. Logica **102**(6), 1103–1142 (2014)
16. Indrzejczak, A.: Natural Deduction, Hybrid Systems and Modal Logics. Springer, Berlin (2010)
17. Indrzejczak, A.: Natural deduction. In: Internet Encyclopedia of Philosophy. http://www.iep.utm.edu/nat-ded/
18. Jaśkowski, S.: Teoria dedukcji oparta na dyrektywach założeniowych [Theory of deduction based on suppositional directives]. In: Księga Pamiątkowa I Polskiego Zjazdu Matematycznego. Uniwersytet Jagielloński, Kraków (1929)
19. Jaśkowski, S.: On the rules of suppositions in formal logic. Stud. Logica **1**, 5–32 (1934) (Reprinted in: Storrs McCall (ed.) Polish Logic 1920–1939, pp. 232–258. Oxford (1967))
20. Jaśkowski, S.: Elementy logiki matematycznej i metodologii nauk ścisłych [Elements of mathematical logic and methodology of formal sciences]. Lecture notes, Toruń (1947)
21. Kalish, D., Montague, R.: Logic, Techniques of Formal Reasoning. Harcourt, Brace and World, New York (1964)
22. Kolmogoroff, A.: O principie tertium non datur. Matematicesi Sbornik **32**, 651
23. Negri, S., von Plato, J.: Structural Proof Theory. Cambridge University Press, Cambridge (2001)
24. Pelletier, F.J.: A Brief History of Natural Deduction. Hist. Philos. Log **20**, 1–31 (1999)
25. Pelletier, F.J., Hazen, A.P.: A history of natural deduction. In: Gabbay, D., Pelletier, F.J., Woods, E. (eds.) Handbook of the History of Logic, vol. 11, pp. 341–414 (2012)
26. Prawitz, D.: Natural Deduction. Almqvist and Wiksell, Stockholm (1965)
27. Rosser, J.B.: Logic for Mathematicians. McGraw Hill, New York (1953)
28. Słupecki, J., Borkowski, L.: Elements of Mathematical Logic and Set Theory. Pergamon Press, Oxford (1967)
29. Suppes, P.: Introduction to Logic. Van Nostrand, Princeton (1957)
30. Szabo, M.E.: The Collected Papers of Gerhard Gentzen. North-Holland, Amsterdam (1969)
31. Van Orman Quine, W.: Methods of Logic. Holt, Rinehart and Winston, New York (1950)
32. Woleński, J.: Logic and Philosophy in the Lvov-Warsaw School. Kluwer, Dordrecht (1989)

A. Indrzejczak (✉)
Department of Logic, University of Lodz, Lodz, Poland
e-mail: indrzej@filozof.uni.lodz.pl

Variations on Jaśkowski's Discursive Logic

Barbara Dunin-Kęplicz, Alina Powała, and Andrzej Szałas

Abstract Stanisław Jaśkowski, in his 1948–1949 papers on propositional calculus for contradictory deductive systems, proposed discursive logic D_2. The main motivation behind D_2 is the need to properly deal with contradictions that naturally appear in many areas of philosophy and discourse. The intuitive justification of this logic reflects knowledge fusion occurring when "the theses advanced by several participants in a discourse are combined into a single system." This point of view was seminal in the mid twentieth century and remains visionary nowadays.

In contemporary autonomous systems operating in dynamic, unpredictable information-rich environments, distributed reasoning routinely takes place. This explains the key role of knowledge fusion, among others, in Distributed Artificial Intelligence. Therefore, different types of modern knowledge and belief bases become primarily concerned with inconsistent or lacking information. This requirement leads to recent approaches to paraconsistent and paracomplete reasoning, where nonmonotonic techniques for disambiguating inconsistencies and completing missing knowledge can be applied.

In this chapter we remind Jaśkowski's seminal, pioneering work on paraconsistent reasoning and indicate some of its relations to contemporary research on reasoning in Distributed AI.

Keywords Discussive logic · Paraconsistent reasoning · Argumentation · Belief structures

Mathematics Subject Classification (2000) Primary 03B53; Secondary 03B50, 03B42, 68T27, 68T30

1 Prelude

Stanisław Jaśkowski introduced *discursive logic* D_2 (called also *discussive logic*) in his visionary papers [24, 25] (for their English versions see [26, 27]). It has been the first formal paraconsistent logic proposed in the literature and has opened a wide area of paraconsistent reasoning (for surveys see [2, 4, 34]). It also inspired many researchers

Á. Garrido, U. Wybraniec-Skardowska (eds.), *The Lvov-Warsaw School. Past and Present*, Studies in Universal Logic,
https://doi.org/10.1007/978-3-319-65430-0_34

who published many papers focusing solely on or directly motivated by D_2 (like, e.g., [3, 8, 9, 11, 31, 39]).

In defining D_2, Jaśkowski used S5 worlds to model sets of beliefs of the discussing participants. We say that a statement is a consequence of a discussion if it follows from at least one belief set (i.e., at least one S5 world). That way different participants may express contradictory statements α and $\neg\alpha$ while the conjunction $\alpha \wedge \neg\alpha$ cannot be derived.

In the current chapter we recall D_2 and indicate its connections to contemporary research on reasoning in many subareas of artificial intelligence. In particular, we provide a new formalization of Jaśkowski ideas in terms of belief structures introduced by Dunin-Kęplicz and Szałas in [17–19]. Belief structures are built over a four-valued logic of [40] with truth values $\mathfrak{t}$ (true), $\mathfrak{f}$ (false), $\mathfrak{i}$ (inconsistent) and $\mathfrak{u}$ (unknown). This new paraconsistent and paracomplete formalization provides a shift from the deductive perspective to belief bases perspective. While, in the former, reasoning depends on deriving conclusions valid in all models of premises—in the later, one derives conclusions valid in a single model representing the current state of the world. Of course, the formalization in belief structures is not equivalent to D_2 as it is well-known that there is no characterization of S5 in any finitely-valued logic [13].

In the formalism of belief structures, belief bases are understood as sets of worlds. However, these worlds can contain contradictory claims what makes them incompatible with modal worlds. Also, there is no need to use Kripke-like accessibility relation on worlds. Instead, we focus on epistemic profiles designed for reflecting the dynamics of belief formation and revision. The concept of epistemic profile embodies an individual's (alternatively called an agent) or group of individuals reasoning capabilities encompassing techniques suitable for different aspects of activities.

Arguably, Jaśkowski with his ideas addressing paraconsistent reasoning, especially in the context of discursive logics, has been much ahead of his times. To show the bridge between D_2 and contemporary research on belief bases, argumentation, knowledge representation, artificial intelligence, autonomous systems, etc., we define a new logic D_4. While formalizing Jaśkowski's ideas behind D_2, it also enjoys the following features:

- the formalization allows to distinguish among statements supplied by different participants of discussion;
- it provides tools for both paraconsistent and paracomplete reasoning, allowing for disambiguating of inconsistencies and completing missing knowledge in a nonmonotonic manner;
- it is computationally feasible: for implementation one can use 4QL, a rule language developed in [28, 29, 37].[1]

Our variations on Jaśkowski's ideas start with the current prelude. Next, in Sect. 2 the main theme, Jaśkowski's Discursive Logic D_2, is reminded. The "movement" (Sect. 3) presents the main ideas and definitions related to belief structures. Then, in Sects. 4–6 three variations on D_2 are composed. The first one (Sect. 4) presents D_4, a new four-valued formalization of Jaśkowski's intuitions behind D_2. The second variation (Sect. 5) relates D_2 and D_4 to contemporary research on dialogues. The last variation (Sect. 6) elaborates on connections of discursive logics to selected work on argumentation. Finally, the coda (Sect. 7) concludes our variations.

[1] For open-source interpreters of 4QL, see `4ql.org`.

2 Theme: Jaśkowski's Discursive Logic D_2

In his papers on D_2 [24–27], Jaśkowski addressed the following problem:

> [...] the problem of the logic of contradictory systems [inconsistent systems] is formulated here in the following manner: the task is to find a system of the sentential calculus which: 1) when applied to contradictory systems would not always entail their over-completeness; 2) would be rich enough to enable practical inference; 3) would have an intuitive justification.

For simplicity, as the underlying logic Jaśkowski has chosen propositional modal logic S5 with usual classical connectives $\neg, \wedge, \vee, \rightarrow, \equiv$ together with modalities $\Box$, $\Diamond$, and considered additional connectives:

- discussive implication: $p \rightarrow_d q \stackrel{\text{def}}{\equiv} (\Diamond p \rightarrow q)$;
- discussive equivalence: $p \equiv_d q \stackrel{\text{def}}{\equiv} (p \rightarrow_d q) \wedge_d (q \rightarrow_d p)$;
- discussive conjunction: $p \wedge_d q \stackrel{\text{def}}{\equiv} (p \wedge \Diamond q)$.

As summarized in [34],

> we think of each participant's belief set as the set of sentences true at a world in a S5 model M. Thus, a sentence α asserted by a participant in a discourse is interpreted as "it is possible that α" ($\Diamond\alpha$).

Let us now define the discursive consequence relation. For a similar formulation see, e.g., Example 24 of [7]. We shall need the following translation function from D_2 formulas into S5 formulas:

$$Tr(p) \stackrel{\text{def}}{=} p \text{ for } p \text{ being a propositional variable;}$$
$$Tr(\alpha \wedge_d \beta) \stackrel{\text{def}}{=} Tr(\alpha) \wedge \Diamond Tr(\beta);$$
$$Tr(\alpha \rightarrow_d \beta) \stackrel{\text{def}}{=} \Diamond Tr(\alpha) \rightarrow Tr(\beta);$$
$$Tr(\alpha \equiv_d \beta) \stackrel{\text{def}}{=} (\Diamond Tr(\alpha) \rightarrow Tr(\beta)) \wedge (\Diamond Tr(\beta) \rightarrow Tr(\alpha)).$$

We assume that Tr preserves all other connectives and, for a set of formulas F,

$$Tr(F) \stackrel{\text{def}}{=} \{Tr(\alpha) \mid \alpha \in F\} \text{ and } \Diamond F \stackrel{\text{def}}{=} \{\Diamond\alpha \mid \alpha \in F\}.$$

Definition 2.1 The *discursive consequence relation*, $\Vdash_{D_2}$ is defined by:

$$F \Vdash_{D_2} \alpha \text{ iff } \Diamond Tr(F) \Vdash_{S5} \Diamond Tr(\alpha), \tag{2.1}$$

where F is a set of formulas and α is a formula. ◁

We are now in position to recall Jaśkowski's motivations concerning discussive connectives $\rightarrow_d$, $\wedge_d$ and $\equiv_d$.

First, the motivation behind $\rightarrow_d$, as stated by Jaśkowski (see [26]), is the failure of modus ponens-based reasoning when traditional implication is used:

> If implication is interpreted so as it is done in two-valued logic, then out of the two theses one of which is $p \rightarrow q$ and thus states "it is possible that if p then q", and the other is p, and thus states "it is possible that p", it does not follow that "it is possible that q", so that the thesis q does not follow intuitively, as the rule of modus ponens requires.

> [. . .] This is why in the search for a "logic of discourse" the prime task is to choose such a function which, when applied to discursive theses, would play the role analogous to that which in ordinary systems is played by implication.

Indeed, from p (i.e., $\Diamond p$) together with $p \rightarrow_d q$ (i.e. $\Diamond p \rightarrow q$) we can deduce q (so $\Diamond q$, too).

The discussive conjunction and equivalence are motivated by the following important theorem (see [27]):

> Each thesis α of the two-valued classical calculus containing no other symbols than $\rightarrow$, $\equiv$, $\vee$ or $\wedge$ is transformed into thesis of the discussive calculus D_2 by replacing in α functors $\rightarrow$ by $\rightarrow_d$, $\equiv$ by $\equiv_d$, and $\wedge$ by $\wedge_d$, respectively.

Additionally, discussive conjunction maintains the *adjunction principle* according to which $p, \neg p \models p \wedge \neg p$. Namely, for $\wedge_d$ adjunction holds [25], since from $p, \neg p$ one can deduce $p \wedge_d \neg p$.

3 Movement: Belief Bases and Belief Structures

This section is based on [19]. However, for clarity, we restrict the presentation to propositional logic. We use the classical propositional syntax but the presented semantics substantially differs from the classical one. Namely,

- truth values $\mathfrak{t}, \mathfrak{i}, \mathfrak{u}, \mathfrak{f}$ (true, inconsistent, unknown, false) are explicitly present;
- the semantics is based on sets of literals rather than on valuations of propositional variables.

This allows one to deal with the lack of information as well as inconsistencies. The underlying semantics of propositional connectives is the one of [40]. It is summarized in Table 1. Observe that definitions of $\wedge$ and $\vee$ reflect minimum and maximum w.r.t. the ordering:

$$\mathfrak{f} < \mathfrak{u} < \mathfrak{i} < \mathfrak{t}, \tag{3.1}$$

as advocated, e.g., in [10, 28, 37, 40]. Such a truth ordering appears to be natural and reflecting intuitions of the classical two-valued logic. For example, a conjunction is true if all its operands are true, etc.

Table 1 Truth tables for $\wedge$, $\vee$, $\rightarrow$ and $\neg$ (see [28, 29, 40])

$\wedge$	f	u	i	t	$\vee$	f	u	i	t	$\rightarrow$	f	u	i	t		$\neg$
f	f	f	f	f	f	f	u	i	t	f	t	t	t	t	f	t
u	f	u	u	u	u	u	u	i	t	u	t	t	t	t	u	u
i	f	u	i	i	i	i	i	i	t	i	f	f	t	f	i	i
t	f	u	i	t	t	t	t	t	t	t	f	f	t	t	t	f

A *positive literal* is a propositional variable and a *negative literal* is a negated propositional variable.

Definition 3.1 The *truth value* of a literal ℓ w.r.t. a set of literals L, denoted by $\ell(L)$, is defined as follows:

$$\ell(L) \stackrel{\text{def}}{=} \begin{cases} \mathfrak{t} \text{ if } \ell \in \text{ L and } (\neg\ell) \notin \text{ L;} \\ \mathfrak{i} \text{ if } \ell \in \text{ L and } (\neg\ell) \in \text{ L;} \\ \mathfrak{u} \text{ if } \ell \notin \text{ L and } (\neg\ell) \notin \text{ L;} \\ \mathfrak{f} \text{ if } \ell \notin \text{ L and } (\neg\ell) \in \text{L.} \end{cases}$$ ◁

Definition 3.1 is extended to all propositional formulas in the standard way, using the semantics provided in Table 1.

If S is a set then by $\text{FIN}(S)$ we understand the set of all finite subsets of S. By $\mathbb{C}$ we denote the set of all finite sets of literals.

Definition 3.2 By a *belief base* we understand any finite set Δ of finite sets of literals, i.e., any finite set $\Delta \subseteq \mathbb{C}$. ◁

Note that such belief bases can be tractably implemented using the 4QL rule language [28, 29, 37]. They serve as basis for belief structures. Indeed, constituents and consequents being basic building blocks of belief structures are, in fact, belief bases in the sense of Definition 3.2.

By *information ordering* we understand the ordering on truth values shown in Fig. 1. This ordering reflects the process of gathering and fusing information. Starting from the lack of information, in the course of belief acquisition, evidence supporting or denying investigated hypotheses are collected. This finally permits one to decide about the truth value of the hypotheses.

Definition 3.3 Let Δ be a belief base and α be a formula. We define the *belief operator* by: $\text{Bel}_\Delta(\alpha) \stackrel{\text{def}}{\equiv} \text{LUB}\{\alpha(D) \mid D \in \Delta\}$, where LUB denotes the least upper bound w.r.t. the ordering shown in Fig. 1. ◁

For clarity let us indicate that:

$$\text{Bel}_\Delta(t) = t \text{ when } t \in \{\mathfrak{t}, \mathfrak{i}, \mathfrak{f}, \mathfrak{u}\}. \tag{3.2}$$

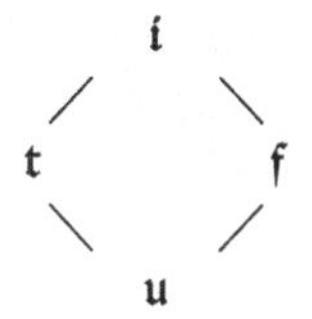

Fig. 1 Information ordering on truth values

Note that sets $D \in \Delta$ appearing in Definition 3.3 can be considered as four-valued worlds. Comparing to Kripke-like semantics for beliefs (see, e.g., [20]), at this point the main differences are:

- we do not require fixed, rigid structure connecting worlds via accessibility relations;
- we use four rather than two truth values.

We are now ready to define (indeterministic) belief structures, as in [19].[2] Belief structures consist of constituents and consequents: an agent starts with constituents, which are further transformed into consequents via the agent's or group's epistemic profile. While constituents contain initial, "raw" beliefs acquired by perception, expert-supplied knowledge, communication, discussion and other ways, consequents contain final, "mature" beliefs. In short, an epistemic profile encapsulates agents' or groups' reasoning capabilities, including methods of both disambiguation of inconsistencies and completing missing information.

Definition 3.4

- By a *constituent* we understand any set $C \in \mathbb{C}$;
- by an *indeterministic epistemic profile* we understand any function $\mathcal{E}$ of the sort $\text{FIN}(\mathbb{C}) \longrightarrow \text{FIN}(\mathbb{C})$;
- by an *indeterministic belief structure over an indeterministic epistemic profile* $\mathcal{E}$ we mean $\mathcal{B}^{\mathcal{E}} = \langle \mathcal{C}, \mathcal{F} \rangle$, where:
  - $\mathcal{C} \subseteq \mathbb{C}$ is a nonempty set of constituents;
  - $\mathcal{F} \stackrel{\text{def}}{=} \mathcal{E}(\mathcal{C})$ is the set of *consequents* of $\mathcal{B}^{\mathcal{E}}$. ◁

A formula is Bel ()-*free* if it does not contain belief operators. Let us emphasize that Bel ()-free formulas reflect properties of initial beliefs, being evaluated in constituents while the belief operator Bel () refers to consequents, so allows us to express properties of final beliefs, as stated in the following definition.

Definition 3.5 Let $\mathcal{B}_1^{\mathcal{E}} = \langle \mathcal{C}_1, \mathcal{F}_1 \rangle$ and $\mathcal{B}_2^{\mathcal{P}} = \langle \mathcal{C}_2, \mathcal{F}_2 \rangle$ be indeterministic belief structures. The semantics of formulas is defined by:

$$\alpha(\mathcal{B}_1^{\mathcal{E}}) \stackrel{\text{def}}{=} \begin{cases} \text{Bel}_{\mathcal{C}_1}(\alpha) & \text{when } \alpha \text{ is Bel}()\text{-free;} \\ \text{Bel}_{\mathcal{F}_2}(\beta) & \text{when } \alpha \text{ is of the form} \text{Bel}_{\mathcal{B}_2^{\mathcal{P}}}(\beta) \text{ and } \beta \text{ is Bel}()\text{-free,} \end{cases}$$

where $\text{Bel}_{\mathcal{C}_1,v}(\alpha)$ and $\text{Bel}_{\mathcal{F}_2,v}(\beta)$ are defined in Definition 3.3.[3] ◁

The above definition can be extended for all formulas by defining the semantics of connectives as in Sect. 3 and nested Bel () operators starting from the innermost ones.

[2]Note that epistemic profiles of [17, 18] are functions of the sort $\text{FIN}(\mathbb{C}) \longrightarrow \mathbb{C}$. That is, they basically are deterministic epistemic profiles with $\mathcal{F}$ consisting of one consequent.

[3]Note that, in the simplest case, $\mathcal{B}_1^{\mathcal{E}}$ and $\mathcal{B}_2^{\mathcal{E}}$ can be identical.

Recall after [19] that typical requirements as to belief operators are satisfied, where α is any formula and $\mathcal{B}^{\mathcal{E}}$ is any belief structure[4]:

$$\begin{aligned} &(\neg \mathrm{Bel}_{\mathcal{B}^{\mathcal{E}}}(\mathsf{f}))(\mathcal{B}^{\mathcal{E}}) = \mathsf{t} && \text{(consistency of beliefs)} \\ &(\mathrm{Bel}_{\mathcal{B}^{\mathcal{E}}}(\alpha) \rightarrow \mathrm{Bel}_{\mathcal{B}^{\mathcal{E}}}(\mathrm{Bel}_{\mathcal{B}^{\mathcal{E}}}(\alpha)))(\mathcal{B}^{\mathcal{E}}) = \mathsf{t} && \text{(positive introspection)} \\ &(\neg \mathrm{Bel}_{\mathcal{B}^{\mathcal{E}}}(\alpha) \rightarrow \mathrm{Bel}_{\mathcal{B}^{\mathcal{E}}}(\neg \mathrm{Bel}_{\mathcal{B}^{\mathcal{E}}}(\alpha)))(\mathcal{B}^{\mathcal{E}}) = \mathsf{t} && \text{(negative introspection)} \end{aligned}$$

4 Variations Part I: D_4—A New Framework for Discursive Logics

D_2 has a potential to be extended in many directions. In particular, the following aspects can be addressed.

- A participant in a discussion should be allowed to submit inconsistent statements, as advocated, among others, in [23]. Therefore, the relevant worlds should not exclude contradictory statements, as it happens in S5, so in D_2, too.
- In contemporary systems it is often important to distinguish among statements supplied by various participants of distributed reasoning and knowledge fusion. This aspect might be essential in formulating adequate strategies of disambiguation of inconsistencies.
- As the accessible information may be incomplete as well, to reflect this property not only paraconsistent but also paracomplete reasoning is often needed.

In order to formalize Jaśkowski's intuitions behind discursive logic while addressing the above aspects, one can use the framework of belief structures. Technically speaking, beliefs are represented as sets of literals constituting paraconsistent belief bases. Epistemic profiles are represented as specific rules operating on possibly complex belief structures in order to draw individual conclusions. Discursive reasoning can be used to define epistemic profiles of individuals and groups.

In order to define a logic D_4, let us first assume that discussion participants have, as a group, an associated belief structure, say $\mathcal{B}^{\mathcal{E}}$. Since $\mathrm{Bel}()$ corresponds to modal $\Box$, we define $\Diamond$ as usually:

$$\Diamond \alpha \stackrel{\text{def}}{\equiv} \neg \mathrm{Bel}_{\mathcal{B}^{\mathcal{E}}}(\neg \alpha), \tag{4.1}$$

and, consequently, modify translation Tr, in such a way that wherever $\Diamond$ occurs, it is replaced by $\neg \mathrm{Bel}()\neg$. We denote this modified translation by Tr_m.

To compute the consequences according to Definition 2.1, we have to evaluate the formula $\Diamond Tr(F) \Vdash_{S5} \Diamond Tr(\alpha)$. Since formulas involved are completely modalized, we

[4]Observe that the property of consistency of beliefs requires beliefs to exclude only falsity f. On the other hand, beliefs can contain contradictory claims.

use deduction theorem for S5 [43] and obtain that $\Diamond Tr(F) \Vdash_{S5} \Diamond Tr(\alpha)$ is equivalent to:

$$\Vdash_{S5} \Diamond Tr(F) \rightarrow \Diamond Tr(\alpha). \tag{4.2}$$

Now, rather than using S5, we use our formalization by evaluating the implication:

$$\Big(\bigwedge_{\phi \in F} \neg \mathrm{Bel}_{\mathcal{B}^{\mathcal{E}}}\big(\neg Tr_m(\phi)\big)\Big) \rightarrow \neg \mathrm{Bel}_{\mathcal{B}^{\mathcal{E}}}\big(\neg Tr_m(\alpha)\big). \tag{4.3}$$

To distinguish among different discussion participants, we consider operators $\Diamond_A$, where A is a discussion participant, rather than just $\Diamond$ as in the original D_2. This is a rather immediate extension of the method outlined above. Namely, the translation Tr_m should be applicable to modal operators $\Diamond_A$, so we replace such operators by $\neg \mathrm{Bel}_{\mathcal{B}^{\mathcal{E}}_A}(\neg \ldots)$, where $\mathcal{B}^{\mathcal{E}}_A$ is a belief structure associated with participant A. Now one can use (4.3) with such modified translation Tr_m.

5 Variations Part II: Relation to Dialogue

Complex communication patterns are essential in intelligent systems. Nowadays, rather than rigid communication protocols, more relaxed communication forms are developed. Indeed, communicative actions are "actions that change your mind" [38]. Taking a commonsense reasoning perspective calls for defeasible reasoning.

Contemporary approaches to communication in intelligent systems draw upon Walton and Krabbe's semi-formal theory of dialogue [42], adapting the normative models of human communication, including paradigmatic dialogue types like inquiry, information seeking, deliberation, persuasion and negotiation. See [5, 6, 12, 20, 30, 32, 33, 35] for investigations in argumentation-based dialogue, and [42] for the definitions of dialogue types. Each model of dialogue is defined by its initial situation, the participants' individual goals, and the aim of the dialogue as a whole (see Table 2).

Complex dialogues are composed with the use of *speech acts*—the basic building blocks of communication. Contemporary understanding of speech acts comes form the works of Austin and Searle [1, 36] including the most popular taxonomy of speech acts, identifying:

- *assertives*, committing to the truth of a proposition, e.g., stating;
- *directives*, which get the hearer to do something, e.g., asking;
- *commissives*, committing the speaker to some future action, e.g., promising;
- *expressives*, expressing a psychological state, e.g., thanking;
- *declaratives*, changing reality according to the proposition e.g., baptizing

Recently we developed a paraconsistent, paracomplete, dynamic and tractable formal model of communication including:

- a formal model of speech-acts and reasoning schemes [16, 21];
- formalization of *inquiry* as a dialogue type for *knowledge acquisition* [15];
- formalization of *persuasion* as a dialogue type for *conflict resolution* [14].

Table 2 Types of dialogue recalled from [41]

Type of dialogue	Initial situation	Participants' goal	Goal of dialogue
Persuasion	Conflict of opinions	Persuade other party	Resolve or clarify issue
Inquiry	Need to have proof	Find and verify evidence	Prove (disprove) hypothesis
Negotiation	Conflict of interests	Get what you most want	Reasonable settlement both can live with
Information seeking	Need information	Acquire or give information	Exchange information
Deliberation	Dilemma or practical choice	Coordinate goals and actions	Decide best available course of actions
Eristics	Personal conflict	Verbally hit out at opponent	Reveal deeper basis of conflict

Such a model of communication can be used to enrich D_2 by developing discussion patterns and related schemes.

The nature of multi-party inquiry and persuasion dialogues resembles distributed defeasible reasoning processes, especially collective problem solving. The complex logical architecture of both dialogue types permits to associate specific belief structures with each of them. Namely, the specific rules governing each dialogue type are included in the epistemic profile of a discussing group. Also specific methods for disambiguation of inconsistencies and information completion, specific for inquiry and persuasion are included in the involved epistemic profiles. Such an encapsulation of methods in epistemic profiles permits to effectively model and investigate different dialogue types indicated in Table 2. Technically, with each dialogue D, terminated or in progress, a specific epistemic profile and a belief structure $\mathcal{B}^{\mathcal{D}}$ is associated and one can use belief operators $\mathrm{Bel}_{\mathcal{B}^{\mathcal{D}}}()$ to formalize Jaśkowski's discursive connectives, as outlined in Sect. 4.

Using this framework, one can obtain a rich formalism, adjustable to a variety of dialogue types indicated in Table 2. Such a broader scope can still be rooted in D_2 or D_4, and deserves further investigations.[5]

6 Variations Part III: Relation to Argumentation

In realistic environments, heterogeneity of argumentation participants w.r.t. reasoning manifests itself in different conclusions drawn by participants even facing the same evidence. The notion of epistemic profile directly exposes this concept. In its abstract form, epistemic profile, being arbitrary function, conveys all reasoning capabilities of an argumentation participant. Due to this generic definition, also non-deductive reasoning methods like argumentation schemes, can be included as a part of epistemic profiles.

[5]Of course, one should take into considerations rich theories developed outside of logical formalisms, in particular in the case of negotiations.

Argumentation schemes, originating from legal argumentation, attempt to classify different types of everyday arguments, utilizing the ideas underlying nonmonotonic formalisms. Each scheme is accompanied by a set of critical questions, used to evaluate the argument. Although particular schemes may represent different types of reasoning (e.g., deduction, induction, abduction, presumption), in general they aim to model plausible, thus defeasible, reasoning.

In [16], paraconsistent argumentation schemes are modeled with the use of two dedicated sets of *premises* and *exceptions*. Intuitively, when all premises are present and none of the exceptions is present, the conclusion of the scheme can be drawn. To model such schemes, we consider three sets of ground literals: premises (P), exceptions (E) and conclusions (Con), together with a function $\mathcal{PAS}(\{P, E\}) = Con$, which represents the paraconsistent argumentation scheme. The set P contains *candidates for conclusion* of the scheme. They are obtained by means specific to every argumentation scheme. The elements of E are *triggers* that, when present, prevent drawing the respective candidate conclusion. Intuitively, a conclusion c cannot be obtained when the exceptions indicate $\neg c$. Ultimately, the conclusion of the scheme is obtained as follows. If there exists a candidate for a conclusion $c \in P$ (value of c is not $\mathbf{u}$), check whether there exists a trigger $\neg c \in E$ blocking this candidate (value of $\neg c$ is $\mathbf{t}$). If the trigger:

- does not exist, the candidate conclusion becomes the final scheme conclusion,
- exists, the scheme cannot be applied causing the value of $c \in Con$ to be $\mathbf{u}$.

In short, a conclusion c is established based on the supporting arguments given by the set P (i.e., $c(P, v) \neq \mathbf{u}$) and (lack of) rebutting triggers provided by the set E (i.e. $\neg c(E, v) \neq \mathbf{t}$).

The definition below presents the paraconsistent argumentation scheme as a partial function: a fragment of agent's epistemic profile that expresses agent's or group's argumentative skills.

Definition 6.1 Let P and E be two constituents, representing the set of premises and exceptions, respectively, and let $\mathcal{S} = \{P, E\} \subseteq \mathbb{C}$ be a nonempty set of constituents. Then, a *paraconsistent argumentation scheme* (over $\mathcal{S}$ and Con) is a partial function: $\mathcal{PAS} : \text{FIN}(\mathbb{C}) \to \mathbb{C}$ such that for $Con \stackrel{\text{def}}{=} \mathcal{PAS}(\{P, E\})$ and c being a literal, we have:

$$c(Con) \stackrel{\text{def}}{=} \begin{cases} \mathbf{t} \text{ iff } c(P) = \mathbf{t} \text{ and } \neg c(E) \neq \mathbf{t}; \\ \mathfrak{i} \text{ iff } c(P) = \mathfrak{i} \text{ and } \neg c(E) \neq \mathbf{t}; \\ \mathfrak{u} \text{ iff } c(P) = \mathfrak{u} \text{ or } \neg c(E) = \mathbf{t}; \\ \mathfrak{f} \text{ iff } c(P) = \mathfrak{f} \text{ and } \neg c(E) \neq \mathbf{t}. \end{cases}$$

By a *belief structure associated with* $\mathcal{PAS}$ we mean $\mathcal{B}^{\mathcal{PAS}} = \langle \mathcal{S}, \{Con\} \rangle$. ◁

Note that the belief structure $\mathcal{B}^{\mathcal{PAS}}$ in Definition 6.1 is, in fact, deterministic as the set of consequents contains only Con. This reflects the intuition that conclusions are determined, e.g., by applying belief operators. A more comprehensive theory of argumentation and communication founded on belief structures and 4QL, TALKLOG, is presented in [14–16, 21]. Observe that 4QL as the implementation tool guarantees the tractability of approach [28, 29, 37].

All and all, as in the case of dialogues, paraconsistent (and paracomplete) argumentation schemes can be viewed as a part of an agent's epistemic profile utilizing the notions of belief structures that can be directly translated into 4QL. Since with every paraconsistent argumentation schema $\mathcal{PAS}$ there is an associated belief structure $\mathcal{B}^{\mathcal{PAS}}$, one can consider belief operators $\text{Bel}_{\mathcal{B}^{\mathcal{PAS}}}()$ and other operators of Jaśkowski's discursive logic, as indicated in the end of Sect. 6. This framework, as in the case of dialogues, opens a wide spectrum of applications of D_2 and D_4 in modeling argumentation schemes and reasoning about them.

7 Coda

Jaśkowski;s discursive logic occupies a meaningful place in philosophical logic from the moment of its inauguration. Importantly, nowadays we observe an increased demand for paraconsistent logics, which is stimulated by the needs of complex, real world applications. As Dov Gabbay [22] noticed, "New logic areas have become established and the old areas were enriched and expanded". D_2 fits in perfectly with this current trend.

As expressed in Jaśkowski's motivations behind discursive logic, inconsistency should not immediately trivialize reasoning. This approach opens up the opportunity to continue inference even when some information sources deliver contradictory information. In real-world complex applications such a situation might be common for many practical reasons. Ultimately, the inconsistencies are typically being resolved according to a chosen strategy as to timing which, again, depends on the application in question. Apparently, various forms of defeasible reasoning are applicable in this context.

In the current paper, when defining D_4 we indicate a shift from modal perspective, with reasoning over arbitrary theories, to reasoning from knowledge bases. While modeling the world and reasoning usually ends up in models of high complexity, we generally have more humble expectations from contemporary intelligent systems. We, therefore, often lean to tailor the reasoning to rule-based approaches. Long investigations on complexity of reasoning, in particular in the field of descriptive complexity, provide us with a very good picture of what is and what is not tractable and supports this shift. Therefore, a knowledge base perspective on reasoning presented in this chapter is beneficial also from the complexity point of view.

Taking into account highly complex nature of environments real-world intelligent systems are embedded in, the use of paracomplete and paraconsistent reasoning methods proves invaluable. Also within that picture, Jaśkowski's ideas are viable and inspiring.

Acknowledgements Supported by the Polish National Science Centre grants 2015/19/B/ST6/02589 and 2015/17/N/ST6/03642.

References

1. Austin, J.L.: In: Urmson, J.O., Sbisa, M. (eds.) How to Do Things with Words, 2nd edn. Clarendon Press, Oxford (1975)
2. Béziau, J.-Y.: What is paraconsistent logic? In: Priest, G., Batens, D., Mortensen, C., vanBendegem, J.P. (eds.) Frontiers in Paraconsistent Logic, pp. 95–111. Research Studies Press, Philadelphia (2000)
3. Béziau, J.-Y.: The paraconsistent logic Z. A possible solution to Jaśkowski's problem. Log. Logical Philos. **15**(2), 99–111 (2006)
4. Béziau, J.-Y., Carnielli, W., Gabbay, D.M.: Handbook of Paraconsistency. College Publications, London (2007)
5. Black, E.: A generative framework for argumentation-based inquiry dialogues. PhD thesis, University College London (2007)
6. Bonzon, E., Maudet, N.: On the outcomes of multiparty persuasion. In: McBurney, P., Parsons, S., Rahwan, I. (eds.) Argumentation in Multi-Agent Systems. LNCS, vol. 7543, pp. 86–101. Springer, Berlin (2012)
7. Carnielli, W., Marcelo, E.C., Marcos, J.: Logics of formal inconsistency. In: Béziau, J.-Y., Carnielli, W., Gabbay, D.M. (eds.) Handbook of Paraconsistency. College Publications, London (2007)
8. Ciuciura, J.: Algebraization of Jaśkowski's paraconsistent logic D2. Stud. Log. Grammar Rhetoric **42**(1), 173–193 (2015)
9. da Costa, N.C.A., Francisco, D.A.: On Jaśkowski's discussive logics. Stud. Logica **54**(1), 33–60 (1995)
10. de Amo, S., Pais, M.S.: A paraconsistent logic approach for querying inconsistent databases. Int. J. Approx. Reason. **46**, 366–386 (2007)
11. de Moraes, L., Abe, J.: Some results on Jaśkowski's discursive logic. Log. Logical Philos. **9**, 25–33 (2004)
12. Dignum, F., Dunin-Kęplicz, B., Verbrugge, R.: Creating collective intention through dialogue. Log. J. IGPL **9**, 145–158 (2001)
13. Dugundji, J.: Note on a property of matrices for Lewis and Langford's calculi of propositions. J. Symb. Log. **12**(4), 150–151 (1940)
14. Dunin-Kęplicz, B., Strachocka, A.: Paraconsistent multi-party persuasion in TalkLOG. In: Proceedings of the PRIMA'15: Principles and Practice of Multi-Agent Systems – 18th International Conference. LNCS, vol. 9387, pp. 265–283. Springer, Cham (2015)
15. Dunin-Kęplicz, B., Strachocka, A.: Tractable inquiry in information-rich environments. In: Proceedings of the 24th International Joint Conference on Artificial Intelligence, IJCAI'15, pp. 53–60. AAAI Press, Menlo Park (2015)
16. Dunin-Kęplicz, B., Strachocka, A.: Paraconsistent argumentation schemes. Web Intell. **14**, 43–65 (2016)
17. Dunin-Kęplicz, B., Szałas, A.: Epistemic profiles and belief structures. In: Proc. KES-AMSTA 2012: Agents and Multi-agent Systems: Technologies and Applications. LNCS, vol. 7327, pp. 360–369. Springer, Berlin (2012)
18. Dunin-Kęplicz, B., Szałas, A.: Taming complex beliefs. Transactions on Computational Collective Intelligence XI. LNCS, vol. 8065, pp. 1–21. Springer, Berlin (2013)
19. Dunin-Kęplicz, B., Szałas, A.: Indeterministic belief structures. In: Jezic, G., Kusek, M., Lovrek, I., Howlett, J.R.J., Lakhmi, J.C. (eds.) Agent and Multi-Agent Systems: Technologies and Applications: Proc. 8th Int. Conf. KES-AMSTA, pp. 57–66. Springer, Berlin (2014)
20. Dunin-Kęplicz, B., Verbrugge, R.: Teamwork in Multi-Agent Systems. A Formal Approach. Wiley, Hoboken (2010)
21. Dunin-Kęplicz, B., Strachocka, A., Szałas, A., Verbrugge, R.: Paraconsistent semantics of speech acts. Neurocomputing **151**, 943–952 (2015)
22. Gabbay, D.M.: Preface to the second edition. In: Handbook of Philosophical Logic, vol. 4, pp. vii–ix. Kluwer Academic, Dordrecht (2002)
23. Gabbay, D.M., Hunter, A.: Making inconsistency respectable: a logical framework for inconsistency in reasoning, part I—a position paper. In: Jorrand, Ph., Kelemen, J. (eds.) Fundamentals of Artificial Intelligence Research: International Workshop FAIR'91, pp. 19–32. Springer, Berlin (1991)
24. Jaśkowski, S.: Rachunek zdań dla systemów dedukcyjnych sprzecznych. Stud. Soc. Sci. Torun. **5**, 55–77 (1948)

25. Jaśkowski, S.: O koniunkcji dyskusyjnej w rachunku zdań dla systemów dedukcyjnych sprzecznych. Stud. Soc. Sci. Torun. **8**, 171–172 (1949)
26. Jaśkowski, S.: Propositional calculus for contradictory deductive systems. Stud. Logica **24**, 143–157 (1969)
27. Jaśkowski, S.: On the discussive conjunction in the propositional calculus for inconsistent deductive systems. Log. Logical Philos. **7**, 57–59 (1999)
28. Małuszyński, J., Szałas, A.: Living with inconsistency and taming nonmonotonicity. In: de Moor, O., Gottlob, G., Furche, T., Sellers, A. (eds.) Datalog 2.0. LNCS, vol. 6702, pp. 384–398. Springer, Berlin (2011)
29. Małuszyński, J., Szałas, A.: Partiality and inconsistency in agents' belief bases. In: Barbucha, D., Le, M.T., Howlett, R.J., Jain, L.C. (eds.) KES-AMSTA. Frontiers in Artificial Intelligence and Applications, vol. 252, pp. 3–17. IOS Press, Amsterdam (2013)
30. McBurney, P., Hitchcock, D., Parsons, S.: The eightfold way of deliberation dialogue. Int. J. Intell. Syst. **22**(1), 95–132 (2007)
31. Nasieniewski, M., Pietruszczak, A.: On modal logics defining Jaśkowski-like discussive logics. In: Beziau, J.-Y., Chakraborty, M., Dutta, S. (eds.) New Directions in Paraconsistent Logic, pp. 213–228. Springer, Berlin (2015)
32. Parsons, S., McBurney, P.: Argumentation-based dialogues for agent coordination. Group Decis. Negot. **12**, 415–439 (2003)
33. Prakken, H.: Models of persuasion dialogue. In: Argumentation in Artificial Intelligence, pp. 281–300. Springer, Dordrecht (2009)
34. Priest, G., Tanaka, K., Weber, Z.: Paraconsistent logic. In: Zalta, E.N. (ed.) The Stanford Encyclopedia of Philosophy. Metaphysics Research Lab, Center for the Study of Language and Information, Stanford University (Spring 2015 Edition). http://plato.stanford.edu/archives/spr2015/entries/logic-paraconsistent/
35. Reed, C., Walton, D.: Towards a formal and implemented model of argumentation schemes in agent communication. Auton. Agent. Multi-Agent Syst. **11**(2), 173–188 (2005)
36. Searle, J.R., Vanderveken, D.: Foundations of Illocutionary Logic. Cambridge University Press, Cambridge (1985)
37. Szałas, A.: How an agent might think. Log. J. IGPL **21**(3), 515–535 (2013)
38. van Linder, B., van der Hoek, W., Meyer, J.-J.Ch.: Actions that make you change your mind. In: KI-95: Advances in Artificial Intelligence. LNCS, vol. 981, pp. 185–196. Springer, Berlin (1995)
39. Vasyukov, V.: A new axiomatization of Jaśkowski's discussive logic. Log. Logical Philos. **9**, 35–46 (2001)
40. Vitória, A., Małuszyński, J., Szałas, A.: Modeling and reasoning with paraconsistent rough sets. Fundam. Inform. **97**(4), 405–438 (2009)
41. Walton, D.: Types of dialogue and burdens of proof. In: Proceedings of COMA'10: Conference on Computational Models of Argument, pp. 13–24. IOS Press (2010)
42. Walton, D.N., Krabbe, E.C.W.: Commitment in Dialogue: Basic Concepts of Interpersonal Reasoning. State University of New York Press, Albany (1995)
43. Zeman, J.J.: The deduction theorem in S4, S4.2, and S5. Notre Dame J. Formal Log. **8**(1–2), 56–60 (1967)

B. Dunin-Kęplicz
Institute of Informatics, University of Warsaw, Warsaw, Poland
e-mail: keplicz@mimuw.edu.pl

A. Powała
(Formerly Alina Strachocka) Institute of Informatics, University of Warsaw, Warsaw, Poland
e-mail: alina.powala@mimuw.edu.pl

A. Szałas (✉)
Institute of Informatics, University of Warsaw, Warsaw, Poland

Department of Computer and Information Science, Linköping University, Linköping, Sweden
e-mail: andrzej.szalas@mimuw.edu.pl; andrzej.szalas@liu.se

Czesław Lejewski: Propagator of Lvov-Warsaw Ideas Abroad

Peter Simons

Abstract Czesław Lejewski studied in Warsaw before the Second World War, after which he settled in England and resumed an academic career, becoming Professor of Philosophy in Manchester. His writings, all articles, continue and extend the ideas of his teachers, especially Stanisław Leśniewski in logic and Tadeusz Kotarbiński in metaphysics.

Keywords Czesław Lejewski · Stanisław Leśniewski · Protothetic · Ontology · Mereology · Reism

Mathematics Subject Classification (2000) Primary 01A70; Secondary 03B05, 03B15, 03B65

1 Life

Czesław Lejewski was born in Minsk in the Russian Empire, on 14 April 1913. In 1920 his family moved to Lublin, where he attended *Gimnazjum*. From 1931 he studied Classics at Warsaw University, where he obtained a master's degree in 1936 with a dissertation on tropes in the sceptics. After military service he returned to the university in 1937 to study for a doctorate in Classics, concentrating on ancient logic. This interest drew him to courses and seminars on logic given by Jan Łukasiewicz and Stanislaw Leśniewski, and philosophy with Tadeusz Kotarbiński. His dissertation *De Aenesidemi Studiis Logicis* was examined and passed by the examiners, including Łukasiewicz, but he was unable to take his degree due to the outbreak of war. He was taken prisoner by the invading Soviets and spent 2 years in terrible conditions in Soviet labour camps, before joining the 2nd Polish Army Corps under General Władysław Anders after the Nazi invasion of the USSR. In 1942 he made his way by land and sea from Russia to Britain via Iran, Iraq, Palestine, Egypt, South Africa, South America, and the USA. Lejewski joined Polish military intelligence in London. After the war he taught English to Polish ex-servicemen in London, marrying an English woman in 1949, and remarrying in 1973 after the death of his first wife. He became a British citizen in 1955.

Á. Garrido, U. Wybraniec-Skardowska (eds.), *The Lvov-Warsaw School. Past and Present*, Studies in Universal Logic,
https://doi.org/10.1007/978-3-319-65430-0_35

The communist takeover in Poland decided Lejewski to make his home in England, and he took up doctoral studies again, under the supervision of Karl Popper at the London School of Economics, passing (with Łukasiewicz as examiner again, flown specially from Dublin) with a dissertation *Studies in the Logic of Propositions* in 1954. In 1956 he joined the Philosophy Department at the University of Manchester, where in 1966 he succeeded Arthur Prior as professor, and he remained there until his retirement in 1980. He was visiting professor at Notre Dame University (USA) in 1960–1961 and at Salzburg (Austria) in 1984. He never returned to his native land, even after the end of communist rule. After his retirement he lived quietly in Manchester, and died after a long illness on 9 July 2001. His scientific books and posthumous post-war papers are housed in the Special Collections Department of the library at the University of Leeds.

2 Main Works

Lejewski published only papers, of which probably the most influential was his 1954 article "Logic and Existence" [1], in which he opposed the standard understanding of quantification, coming from Tarski and Quine, with the Leśniewskian understanding in which he had been trained. The difference turns on the fact that in the standard interpretation, which Lejewski called *restricted*, only denoting names (terms) may be substituted for bound variables in the rules of inference, whereas in the unrestricted interpretation also non-denoting names could be so substituted. Another quite widely quoted piece is his 1958 article "On Leśniewski's Ontology" [2], which is the clearest exposition of this, Leśniewski's central logical system. To aid understanding of the system, which in contrast to standard predicate logic includes not only singular and empty terms but also plurally referring terms, Lejewski introduced modified Euler diagrams for representing logical relations among the extensions of terms. He also defined numerous ontological functors and showing that partial inclusion ("Some *a* are *b*") can serve as the sole primitive functor. A particular *tour de force* among Lejewski's historical articles is the long encyclopedia entry "History of Logic", written for the 15th edition of *Encyclopedia Britannica* (1975) [3], which gives evidence of Lejewski's remarkable knowledge of the history of logic.

3 Views

Lejewski's views are based on and extend the logical and philosophical views of his principal teachers: in logic, Stanisław Leśniewski; and in philosophy, Tadeusz Kotarbiński. His logical systems are all either reformulations, adaptations, or extensions of Leśniewski's ideas, while his fewer philosophical articles are either defences of Leśniewski against criticism, or comparisons of Leśniewski with other approaches, and in one case an elaboration of Kotarbiński's materialistic reism.

Lejewski's published papers fall into several thematically connected groups. One group expound or simplify Leśniewski's major logical systems of mereology, ontology and protothetic. A second group offer alternatives and extensions to Leśniewski's systems, often prompted by criticisms of the artificiality of the latter. A third group of articles reflect on the philosophical implications of logic from a Leśniewskian standpoint. A fourth group examine miscellaneous aspects of logic, while a fifth group cover topics in the history of ancient and contemporary logic. We treat the last two groups together.

Lejewski did not start to publish until he was forty, and despite his Polish origins, most of his published papers are in English or are translations from his English.

3.1 Exposition and Simplification of Leśniewski

Lejewski's early articles on mereology propose minor simplifications or new axiomatizations of Lesniewski's theory, based on single axioms. One article discusses atomless and atomistic extensions to mereology [4]; another shows that mereology is consistent relative to protothetic [5]. Lejewski modified Leśniewski's mereological terminology to make it more understandable for English speakers, replacing 'ingredient', 'part' and 'class' respectively by 'part', proper part' and 'complete collection'. In ontology, in addition to the expository 1958 article, Lejewski showed that Boolean Algebra in the axiomatization of Ernst Schröder could be understood as elementary ontology (quantifying only nominal variables), and with nominal definitions can be based on a single axiom for weak inclusion ("Any *a* are *b*") [6]. Lejewski also showed, in work parallel to that of Jerzy Słupecki, how to bridge the gap between Aristotle's syllogistic in Łukasiewicz's modernized form and the elementary Ontology of Leśniewski [7].

3.2 Alternatives and Extensions to Leśniewski

Whereas Leśniewski's Ontology treats all names, whether singular, empty or plural, as belonging to a single category, natural languages tend to use only singular proper names, which may be empty. To accommodate this, Lejewski proposed a system of what he called *non-reflexive identity*, employing only singular or empty names, and based on identity sentences $a=b$ which are only true if both a and b exist and are the same individual [8]. Another and more radical innovation by Lejewski follows an idea of Kazimierz Ajdukiewicz and introduces a special category of names for abstract classes, resulting in what Lejewski called a *bicategorial ontology* [9]. Lejewski offered this system in an ecumenical and ontologically neutral spirit, as offering a medium for those who believe in classes (sets) to be able to talk with a Leśniewskian accent about such classes, but he himself denied that there are abstract classes. In the 1982 article "Ontology: What Next?" [10], Lejewski goes beyond mereology to *chronology*, an ontological theory of temporal objects, based on two primitive notions, one topological, that of an object's being wholly earlier than another object, and one metrical, that of an object's having a shorter duration than another object. He envisaged further extensions beyond chronology, to *stereology*, a theory of objects extended in space as well as in time, and *kinematics*, a theory of objects in motion through space, but though also outlined in a descriptive piece [11], a formal development did not attain publication. It is clear that the extensions were conceived in a Leśniewskian spirit.

Leśniewski based protothetic, his extended propositional logic, on material equivalence and universal quantification, with definitions. In a part of his London dissertation, later extracted and published as a paper in 1958 [11], Lejewski showed that material implication, together with universal quantification and definitions, would also do the job. It is a measure of the remarkable consistency of Lejewski's work that over 30 years later, in 1989, he published another paper showing how to fully formalize the 1958 system using Leśniewskian terminological explanations [12].

3.3 Philosophical Reflections on Leśniewskian Logic

Lejewski not only expounded and extended Leśniewski's views, but defended them against all criticisms. This is apparent in the 1954 article on quantification. One frequently made criticism of Leśniewski's extreme extensionalism was that an extensional logic such as his cannot adequately handle belief sentences such as 'John believes that snow is white', and other intensional contexts. Lejewski's account of belief contexts [13] is close to that of Davidson's paratactic view. Lejewski also reflected on the relationship between natural language and the idealized languages of Leśniewski and other logicians: the clearest statement of his view that a mutual give and take is required between natural and artificial languages can be found in the 1979 article "Idealization of Natural Languages for the Purpose of Logic" [14]. Though it does not purport to discuss Leśniewski directly, Lejewski's 1976 article "Ontology and Logic" [15] expounds a conception of the relationship between those two disciplines which is purely Leśniewskian, denying that either logic as a science or quantification as a device carries any ontological import whatever.

3.4 Miscellaneous and Historical Pieces

Among Lejewski's other pieces is another on propositional calculus derived from his London dissertation [16], examining the groups of truth-functions that can be taken as severally independent and jointly sufficient for functional completeness. Lejewski honoured (and corrected) his London *Doktorvater* in the large commemorative volume on Sir Karl Popper with a paper, "Popper's Theory of Formal or Deductive Inference" [17], which sympathetically reconstructs and revises Popper's faulty attempts to reform deductive logic.

In history of logic, apart from the magnificent *Britannica* article, Lejewski wrote two papers on Theophrastus' concept of prosleptic syllogisms [18, 19], a short Italian memoir on Leśniewski and his systems [20], and an Arabic paper on Łukasiewicz. When the latter died in 1956, Lejewski, at the request of the widow Regina Łukasiewicza, performed the service of seeing the second edition (1957) of the classic monograph *Aristotle's Syllogistic* through the press, after Łukasiewicz had became too ill to complete the editing and proofreading himself. It was also through Lejewski that the bulk of Łukasiewicz's post-war letters and manuscripts came to be deposited at the John Rylands Library in Manchester.

Lejewski's most straightforwardly philosophical piece is his 1976 article "Outline of an Ontology" [21]. All that exists according to Lejewski are bulky and temporally extended material bodies, and nothing else. His was thus a four-dimensionalist materialist, a nominalist, and an atheist. The article denies that there are mereological atoms, objects without proper parts, where Kotarbiński, whose views Lejewski otherwise closely follows, had been agnostic about whether there are atoms or not.

Lejewski was also an assiduous reviewer of books and articles for various journals, concentrating naturally but not exclusively on works connected with Polish philosophy and logic.

4 Influence, Teaching, Personality

Because of the narrow scope of his interests, Lejewski's influence was confined to a small circle of logicians, mostly those interested in Leśniewski, such as his fellow Leśniewski student Bolesław Sobociński, the New Zealand logician Arthur Prior, the British philosopher Peter Geach, the historian of medieval logic Desmond Paul Henry, and in mereology, the present author. In retirement Lejewski was pleased to act as mentor to Manchester doctoral student Audoënus Le Blanc, whose work on mereology and protothetic he considered an advance on his own. There is however very little secondary literature on Lejewski's work, and he did not aspire to a wide following, accepting with equanimity the esoteric nature of his interests, but he inspired respect, most strikingly that of Prior, among those who shared his interests and his conviction of the importance of formal precision in philosophy.

As a teacher of undergraduates, Lejewski kept close to classical texts, from Aristotle to Russell, though in advanced seminars he would venture into more technical work, proving theorems on the board at a steady pace. His rather dry style and ontological asceticism were not popular with students accustomed to the headier delights of existentialism or Marxism, and his polite, somewhat old-fashioned reserve was considered aloof, but away from the lecture room he could be outgoing and amusing. Friends in the local community in Manchester knew him as 'Czek'. He had no children, but he and both his wives were fond of Shetland collie dogs, and were welcoming to visitors at their home in Cheadle Hulme.

Works by Czesław Lejewski Mentioned in the Paper

1. Logic and existence. Br. J. Philos. Sci. **5**, 1–16 (1954)
2. On Leśniewski's ontology. Ratio **1**, 150–176 (1958)
3. History of Logic. Encyclopaedia Britannica, vol. 23, 15th edn., pp. 234–250. Chicago (1974)
4. A contribution to the study of extended mereologies. Notre Dame J. Formal Log. **14**, 55–67 (1973)
5. Consistency of Leśniewski's mereology. J. Symb. Log. **34**, 321–328 (1969)
6. Studies in the axiomatic foundations of boolean algebra. Notre Dame J. Formal Log. **1**, 23–47, 91–106 (1960); **2**, 79–93 (1961)
7. Aristotle's syllogistic and its extensions. Synthese **15**, 125–154 (1963)

8. A theory of non-reflexive identity and its ontological ramifications. In: Weingartner, P. (ed.) Grundfragen der Wissenschaften und ihre Wurzeln in der Metaphysik, pp. 65–102. Pustet, Salzburg (1967)
9. A system of logic for bicategorial ontology. J. Philos. Log. **3**, 265–283 (1974)
10. Ontology: what next? In: Leinfellner, W., et al. (eds.) Language and Ontology. Proceedings of the 6th International Wittgenstein Symposium, pp. 173–185. Hölder-Pichler-Tempsky, Vienna (1982)
11. On implicational definitions. Stud. Logica **8**, 189–211 (1958)
12. Formalization of functionally complete propositional calculus with the functor of implication as the only primitive term. Stud. Logica **48**, 479–494 (1989)
13. Propositional attitudes and extensionality. In: Morscher, E., Neumaier, O., Zecha, G. (eds.) Philosophie als Wissenschaft, pp. 211–228. Comes, Piding (1981)
14. Idealization of ordinary language for the purposes of logic. In: Alleton, D.J., Carney, E., Holdcroft, D. (eds.) Function and Context in Linguistic Analysis, pp. 94–110. Cambridge University Press, Cambridge (1979)
15. Ontology and logic. In: Körner, S. (ed.) Philosophy of Logic, pp. 1–28. Blackwell, Oxford (1976)
16. A propositional calculus in which three mutually undefinable functors are used as primitive terms. Stud. Logica **22**, 17–50 (1968)
17. Popper's theory of formal or deductive inference. In: Schilpp, P.A. (ed.) The Philosophy of Karl Popper, pp. 632–670. Open Court, La Salle (1974)
18. On prosleptic syllogisms. Notre Dame J. Formal Log. **2**, 158–176 (1961)
19. On prosleptic premisses. Notre Dame J. Formal Log. **17**, 1–18 (1976)
20. Ricordando Stanislaw Lesniewski [Remembering Stanisław Leśniewski]. Quaderni del Centro Studi per la Filosofia Mitteleuropea **1**, 1–46 (1989)
21. Outline of an ontology. Bull. John Rylands Univ. Library Manchester **59**, 127–147 (1976)
22. Logic, ontology and metaphysics. In: Shanker, S.G. (ed.) Philosophy in Britain Today, pp. 171–197. Croom Helm, London (1986) (Polish translation: Logika, ontologia, metafizyka. Filozofia Nauki **1**, 15–35 (1993))

P. Simons (✉)
Department of Philosophy, Trinity College Dublin, Dublin, Ireland
e-mail: psimons@tcd.ie

Adolf Lindenbaum, Metric Spaces and Decompositions

Robert Purdy and Jan Zygmunt

Abstract This paper revisits the life of Adolf Lindenbaum in light of new research findings, then looks at two areas among many—metric spaces, and decompositions of point sets—where his work has been underappreciated.

Keywords Adolf Lindenbaum · Biography · Bibliography · Metric spaces · Monomorphy · Decomposition · Warsaw school of mathematics · Wacław Sierpiński

AMS Subject Classification (2000) Primary 01A60, Secondary 01A70, 01A72, 51-03, 54-03, 54A05

1 Introduction

"I think your reasoning is really interesting for its simplicity and effective character. I would just like to make one remark (which I've already mentioned to Erdős). As I recall, Adolf Lindenbaum had a more general result—a proof of the existence of $2^{\mathfrak{c}}$ sets not equivalent by countable decomposition in relation to an arbitrary family of bijective transformations—not necessarily a family of isometric transformations (and perhaps even a more general result for arbitrary cardinal numbers). I do not remember the proof at all, and also I do not remember whether the family of bijective transformations was subjected to some additional assumptions. I am under the impression (but can be completely mistaken) that Lindenbaum announced his result without proof either in an article in ***Fundamenta Mathematicae*** or in the reports of talks in the ***Annales de la Société Polonaise de Mathématique***. In any event it would be worthwhile to reconstruct and announce the result. Overall it seems to me that there is an obligation to mathematics and to the memory of Lindenbaum to encourage people to become acquainted with what Lindenbaum left behind in print and to publish proofs of results [that he] supplied without proof."

Dedicated to Mariusz Pandura, in gratitude for his tireless research efforts.

Á. Garrido, U. Wybraniec-Skardowska (eds.), *The Lvov-Warsaw School. Past and Present*, Studies in Universal Logic,
https://doi.org/10.1007/978-3-319-65430-0_36

The paragraph above is from a letter Alfred Tarski wrote to Wacław Sierpiński on October 30th, 1946,[1] discussing a pre-publication copy of Sierpiński [1947], a paper which strengthened a result from Erdős [1943]: using the axiom of choice, Erdős had proved (page 644) that for every cardinal $\mathfrak{m} < \mathfrak{c}$ there is a family of $2^{\mathfrak{c}}$ sets of real numbers, no two of which can be decomposed into $\mathfrak{m}$ disjoint mutually congruent subsets. Citing something he thought he recalled Tarski telling him, Erdős had credited Lindenbaum for announcing, without proof, a less general version of that theorem. Sierpiński [1947] obtained a more general version of Erdős's 1943 result, without using the axiom of choice. However, Erdős must have been misremembering what Tarski had told him. Tarski in fact believed that Lindenbaum had achieved an even more general result than Sierpiński's, much earlier, and without the axiom of choice.

So do we (see the concluding paragraphs of our §**4** below), though we cannot find any published reference to it apart from Tarski's 1946 letter to Sierpiński. Moreover, we agree wholeheartedly with Tarski's judgment that "there is an obligation to mathematics and to the memory of Lindenbaum to encourage people to become acquainted with what Lindenbaum left behind in print and to publish proofs of results [that he] supplied without proof."

It is the object of the present article to provide just such encouragement.

2 A Short Life

Lindenbaum, Adolf (1904–1941); Polish-Jewish mathematician and logician; *docent* of the University of Warsaw; member of the Warsaw school of mathematics and the Warsaw school of mathematical logic; early supporter of ***Fundamenta Mathematicae***; co-founder of the Polish Logical Society; Alfred Tarski's closest collaborator of the inter-war period; logical positivist and member of the Vienna Circle; member of the International Unity of Science movement; anti-war campaigner; Polish Communist Party activist; Holocaust victim (Figs. 1 and 2).

Fig. 1 Adolf Lindenbaum, September, 1922

[1] More precisely, it is from our translation of Tarski's letter. The original letter was in Polish. McFarland–McFarland–Smith [2014] offer their own translation on pp. 377–379, very close to ours, but with two material differences: "efficiency" instead of "effective character", and "two" instead of "$2^{\mathfrak{c}}$". We make some changes to parts of their footnote 82 (*ibid.*, page 378), while adopting other parts of it *verbatim*.

Fig. 2 Adolf Lindenbaum, November, 1927

Fig. 3 Stefanja Lindenbaum. Photograph from her university student "*indeks*", October, 1926

Adolf Lindenbaum was born 12 June 1904 in Warsaw, the son of Mowsza Henoch aka Maurycy Henryk (1878–1932) and Emilja née Krykus (1875–1939 or later). He had a younger sister Stefanja (Fig. 3), born 22 March 1908.[2]

Mowsza was a businessman. On Adolf's birth certificate he described himself as a "*прикащикъ*"—an accordion word that can mean shop clerk, sales assistant, steward, purser, branch manager, manager, director, superintendent, majordomo, or overseer. We surmise that Mowsza's father owned several businesses and put Mowsza in charge of one or more of them. Soon afterward Mowsza switched to describing himself as a "*kupiec*"—which simply means businessman. On a 1924 document he is named as one of the officers of the Jewish Businessmen's Mutual Assistance Society.

Adolf's sister Stefanja entered *pensja dla dziewcząt Pauliny Hewelkówny*[3] in 1917 and matriculated in 1926. She was accepted into the Faculty of Law at the University of Warsaw on 2 September 1926. She attended all three trimesters of the 1926/27 academic

[2]Some of this article draws on Marczewski–Mostowski [1971] and in several places is a straightforward translation of that dictionary entry—a debt which the present authors are keen to acknowledge up front. As well, some of this article overlaps with the paper Zygmunt–Purdy [2014], where readers will find a more detailed treatment of Lindenbaum's university student years, professional life and participation in congresses. That said, much of the present material is new: some of it even overturns parts of Zygmunt–Purdy [2014].

[3]A private school for girls, at that time on the corner of Marszałkowska and Sienkiewicza (ul. Marszałkowska 122). In 1919 the school was nationalized and renamed *Państwowe Gimnazjum Żeńskie im. Klementyny z Tańskich Hoffmanowej*.

year, but did not sit any exams, and she formally withdrew from the university on 31 August 1927.

There is some slight evidence hinting at two more Lindenbaums in the household—an S. Lindenbaum aka Z. Lindenbaum (possibly a younger sibling of Mowsza Henoch's), born 11 December 1886; and an M. Lindenbaum, born sometime in 1912—both of whom appear to have been registered at the family's home address in the 1930s and to have emigrated to England either during or shortly after the Second World War.[4]

The family's financial circumstances were boom and bust. Mowsza was in the movie distribution and movie-theater franchising, leasing and financing businesses. From 1926 he was co-owner and general manager of *Spółka Kinematograficzna* "*Kolos*" (in Warsaw) and "*Kolos Małopolski*" (in Kraków). In the later 1920s, Stefanja was brought into several of the businesses as a co-owner and board member, and Mowsza stepped down from some of their boards of directors in favour of his daughter. He ventured into movie production in 1931 with a production company called *Towarzystwo Kinematograficzne* "*Tempofilm*", with which he had at least one box-office success that we know of.

Then he ran into financial difficulties. "*Kolos Małopolski*" went into bankruptcy, destroying some 50,000 złotych of Mowsza's net worth.[5] He took out some very large loans, and was unable to pay them back. He was last seen alive on 27 December 1932. His body was recovered from the Vistula River next spring when the ice melted.

"*Tempofilm*" was legally dissolved on 26 January 1934. Some of the other businesses survived. Records show that Stefanja remained a shareholder throughout most of the 1930s. But shareholders outside the Lindenbaum family were also recorded, and it is not known if Stefanja held a controlling interest, or if the businesses were profitable after Mowsza's death, or if any of his erstwhile creditors had claims on the earnings.

There is no record of the elementary school Adolf Lindenbaum attended. In his first year of secondary school, 1914–1915, he attended *gimnazjum Rocha Kowalskiego*, and then from 1915 to 1922 *gimnazjum Michała Kreczmara*. In 1922 he entered the University of Warsaw, and on 22 June 1928 was awarded a Ph.D. for a thesis titled "*O własnościach metrycznych mnogości punktowych*" <On the metric properties of point sets>, written under the supervision of Wacław Sierpiński.

Lindenbaum's Ph.D. diploma reads, "*primum in mathematica, deinde in physica et in philosophia*," but in his bio-bibliography for ***Erkenntnis*** he added a telling qualification: "*Hauptfach–Mathematik*; *Nebenfächer–Philosophie*, *Experimentalphysik*." In other words, not simply physics, but experimental physics—involving measurement.

He took courses in descriptive geometry and projective geometry, dealing with transforms and invariants, and in mathematical astronomy and astrometry, Maxwell's

[4]At the present time this cannot be substantiated, as the Polish National Archives have "masked" the relevant records, and the U.K. National Archives have "closed" them for 100 years, on the grounds that they "contain sensitive personal information which would substantially distress or endanger a living person or his or her descendants". The "slight evidence" consists of scanned pages from the registry of residents of the building at ulica Złota 45 inadvertently revealing some details that are less-than-perfectly masked, and U.K. naturalization records HO 405/33558 and HO 405/33790.

[5]Although purchasing power comparisons across 80 years are dicey, he is thought to have lost between 350,000 and half a million in 2009/2010 PLN zł.

equations, Planck radiation and *Lichtquanta.*[6] He knew of Hendrik Lorentz's group-theoretic treatment of frame transforms and their invariants, and the role they played in the geometrization of space-time. He wrote his thesis in the wake of the Banach-Tarski paradox, which had possibly stirred some philosophical misgivings in him about measures in physics—Hausdorff, Dirac, Liouville-Hamiltonian, Lorentz. With a nod to the Vienna Circle and verificationism he took to using the term *metrologia* (metrology) in the titles of some of his courses. Together with Edward Szpilrajn (Marczewski) he co-authored an encyclopedia entry on measurement in geometry, aimed at a general audience; at the bottom of the entry the reader is urged to consult a Polish translation of Hermann von Helmholtz's "*Zählen und Messen*", which emphasizes the empirical nature of measurement.[7] Today the term "metrology" can be encountered in the literature on the Hausdorff dimension of space-time.

Lindenbaum was a keen participant in student academic organizations and societies. Through his papers, reviews, comments and personal contacts he exerted a strong influence on younger mathematicians. Together with Alfred Tarski he became an active contributor to two schools of scientific inquiry: the Warsaw school of mathematics, under the intellectual influence of Wacław Sierpiński and Stefan Mazurkiewicz, and the Warsaw school of mathematical logic, principally under the leadership of Jan Łukasiewicz and Stanisław Leśniewski. He also kept in close contact with the Lwów school of mathematics through Stefan Banach, Hugo Steinhaus and Stanisław Ulam, and two of his publications show him encouraging and rising to the defense of a doctoral student of Steinhaus called Sala Weinlös.

He published mainly in ***Fundamenta Mathematicae***. His first major paper, "*Contributions à l'étude de l'espace métrique. I.*", appeared in ***FM***, volume 8 (1926), pp. 209–222. Written while he was in his fourth year and loaded down with course work, it was a remarkable accomplishment. Later he used it as part of his doctoral dissertation.

His second major paper, "*Communication sur les recherches de la théorie des ensembles*," ***Sprawozdania z posiedzeń Towarzystwa Naukowego Warszawskiego, Wydział III***, volume 19 (1926), pp. 299–330, was co-authored with Alfred Tarski. It was a monumental work, setting out 30 pages of new results in general set theory and the arithmetic of transfinite ordinals all presented without proofs. Some were results obtained jointly by Lindenbaum and Mojżesz Dawid Kirszbraun, a classmate of his at *gimnazjum Kreczmara*. Most came from work Lindenbaum and Tarski had undertaken prior to 1926 on the theory of one-to-one transforms.

Despite the value and sheer multitude of his contributions to set theory, cardinal and ordinal arithmetic, the axiom of choice, the continuum hypothesis, theory of functions, measure theory, point set topology, geometry and real analysis, Lindenbaum's name continues to be associated principally with his work in mathematical logic, a field which in the 1920s was not yet widely developed.

---

[6]The courses "*Teorja Promieniowania*" and "*Promieniowanie i Kwanty*" were taught by Czesław Białobrzeski. Mathematical astronomy and astrometry were taught by Michał Kamieński.

[7]See [35], page 595.

His most important result in logic was his conjecture that any propositional calculus can be characterized by a denumerable (i.e., finite or at most countably infinite) matrix. He never published this conjecture. Its first appearance in print was by Łukasiewicz and Tarski in 1930. Jerzy Łoś first published a full proof of it in 1949. Lindenbaum conceived a method of constructing matrices by using the expressions of a propositional calculus (or more strictly speaking, equivalence classes of expressions) as elements of the matrix. For logicians in the 1920s, this idea was a revelation. Lindenbaum's method spawned waves of research, and it became generally accepted practice to refer to such algebras as Lindenbaum algebras.

His second most important result, widely known among logicians as "Lindenbaum's Lemma," was his theorem, framed in the terminology and concepts of Tarski's <*S*,*Cn*> methodology of deductive systems, that every *Cn*-consistent set of sentences in a language *S* can be extended to form a *Cn*-consistent and *Cn*-complete deductive system in *S*. Or, more loosely put, that every consistent theory, formulated in a suitable language and assuming a suitable underlying logic, has a complete (maximal) and consistent extension. Again, as with much of Lindenbaum's legacy, he never stated it in print. Its statement and proof were first published by Tarski (see, respectively, Tarski [1928] and Tarski [1930]), who scrupulously attributed both the idea and the proof to Lindenbaum. The "Lemma" (i.e., theorem) quickly became an essential tool in every logician's toolkit. Some writers have even ventured that Lindenbaum maximalization, with the notions of completeness and consistency defined as in Tarski [1928], is the only essential thing that all logics have in common.

Lindenbaum was the co-author (with Tarski) of a 1936 paper proving that all the logical notions of Russell and Whitehead's ***Principia Mathematica*** are invariant under one-to-one transformations (automorphisms) of the domain of discourse of the model onto itself: "*Über die Beschränktheit der Ausdrucksmittel deduktiver Theorien*"—a paper which anticipated and did most of the heavy lifting for Tarski's much later work on "What are Logical Notions?" The results they obtained in this paper had many and various applications, among other things to the foundations of geometry, to discussions of which mathematical concepts count as purely logical and which as specifically mathematical, to the study of the interdependence of primitive concepts in axiomatic systems, and ultimately to the study of the independence of the axiom of choice. Lindenbaum and Andrzej Mostowski documented their findings on this last problem in 1938 in the joint paper, "*Über die Unabhängigkeit des Auswahlaxioms und einiger seiner Folgerungen*."

Lindenbaum was an adept crossover artist, equally at home pursuing the program of the axiomatists (David Hilbert, Bertrand Russell, the Italians Giuseppe Peano, Alessandro Padoa and Mario Pieri) to reduce mathematics to language, truth and logic, and the program of the algebraists (quintessentially Tarski) to reduce language, truth and logic to mathematics. The same whiff of circularity, or more charitably coherentism, can be found today in the interplay between model theory and proof theory. In his published works and public lectures Lindenbaum concentrated on large themes, fundamental issues, general concepts and synoptic solutions. He sought, throughout all, to apply whatever means necessary to achieve the clearest possible understanding of the underlying reality of things.

Lindenbaum was an adherent of Logical Empiricism. In March, 1930, he spent time in Vienna where he met Rudolf Carnap, Herbert Feigl, Carl Hempel, Abraham

Fraenkel and Samuel and Lilian Broadwin. Later in the 1930s he participated in and contributed to the International Unity of Science movement. He played an active role in the movement's founding congresses in Prague (Aug 31st–Sept 1st, 1934) and Paris (Sept 15th–23rd, 1935), and he corresponded with Otto Neurath and Jørgen Jørgensen on detailed arrangements for its next two congresses, in Copenhagen (June 21st–26th, 1936), and again in Paris (July 29th–31st, 1937). Neurath twice solicited Lindenbaum's bio-bibliography, in 1930 and again in 1934, for inclusion in his surveys (published in ***Erkenntnis***) of who's who in the movement. In his article "After 6 years", ***Synthese***, vol. 5, no. 1/2 (May–June, 1946), pp. 77–82 [and date-lined Oxford, December 19th, 1945, 3 days before his death], Neurath states outright that Adolf Lindenbaum had been a member of the original Vienna Circle.

In a letter dated July 1st, 1935, Neurath invited Lindenbaum to speak on the subject of formal simplicity (*die formale Einfachheit*) at the 1935 Paris congress. Lindenbaum obliged with a lecture of the same title, which he gave in German on the morning of September 18th in Room 1 (*Vormittag, 18 September, Saal I*). Two days later, on the afternoon of September 20th, a debate took place in the same room on the question of standardizing logical symbolism (*Aussprache über Vereinheitlichung der logischen Symbolik*). The debate concluded by agreeing to establish a working committee charged with advising on and promoting the international standardization of logical symbolism. Lindenbaum was appointed to this committee.

Records show that Lindenbaum expected to take part in the 1937 Paris congress (July 29th–31st, 1937) in his capacity as a member of this committee, but at the last moment it emerged that he was unable to attend: he was denied a travel document to leave Poland. In place of attending in person, he sent a letter which was read aloud to the congress on the morning of July 30th, expressing his and other Polish logicians' concerns with the interim results of the committee ("*meine Meinung aussprechen, wobei ich im Voraus bemerken möchte, dass die polnischen Logiker verschiedentlich anderen Standpunkt einnehmen.*").

Lindenbaum is known to have been an *asystent* in Łukasiewicz's Philosophical Seminar in the faculty of mathematics and natural sciences from (at least) the fall of 1931. Samuel Eilenberg recalled him being in charge of the library at that time.[8] In 1934 he

[8]See Eilenberg [1993], page 1. Eilenberg writes:

> In the academic year 1930–31 I was a first-year student at the University of Warsaw, while Karol Borsuk was an assistant conducting exercises in real analysis. I was a member of a class which was huge but he soon started to notice me and we got involved in several conversations. In the spring of 1931 he received his doctorate and I attended the ceremony. At the same time I attended a course on set theory given by Docent Bronisław Knaster. There were two other students in the course, however, I was the only one who did all the homework. I struck up a friendship with Knaster which lasted as long as he did. Set theory naturally led to Topology, which in Warsaw meant strictly Set-theoretical Topology.
>
> I remember a curious incident. In the fall of 1931 I was browsing through the Mathematics Library and I came across a book entitled ***Topology*** by Solomon Lefschetz. I looked at the bibliography to see to what extent the "Polish School" was quoted. I found only one reference. It was a paper of Knaster, Kuratowski and Mazurkiewicz in volume 15 of ***Fundamenta Matematicae*** containing a combinatorial proof of the Brouwer Fixed Point Theorem. I was very surprised to find no other references and I conveyed my concern to Dr. Adolf Lindenbaum (an excellent logician) who was then the assistant in charge of the library. He told me that it was a terminological misunderstanding, that the book was not about Topology but about some sort of algebra.

successfully defended a habilitation thesis. Its title is lost, but there are strong indications it might have been the paper, "*Z teorii uporządkowania wielokrotnego*" <*Sur la théorie de l'ordre multiple*>, ***Wiadomości Matematyczne***, vol. 37 (1934), pp. 1–35, on an extension of Cantor's notion of multiply-ordered sets. On February 1st, 1935, he started lecturing as a *docent* of the University of Warsaw, and from the commencement of the 1935–1936 academic year he took up the position of *adiunkt* (assistant professor) in the Philosophical Seminar.

Lindenbaum's political sympathies were left-leaning, anti-fascist and anti-war, and some of his political activities were illegal for the time. He attended the World Congress Against War held in Amsterdam on 27th–29th August 1932. He belonged to the Polish Communist Party from at least the mid-1930s up until it was disbanded in 1938 by Stalin, and campaigned for it in intelligentsia circles. In 1936, as one of the "editors and co-workers" of ***Głos Współczesny***, he signed a petition to Professor Halvdan Koht of the Nobel Committee in Oslo, urging that Karl von Ossietzky, a German political journalist imprisoned by the Nazis, be awarded the Nobel Peace Prize.[9] Together with many writers and social activists of the day Lindenbaum added his name to an open letter "to the workers of Lwów" expressing solidarity with "the proletariat's protest against the bloody massacre [of April, 1936] of workers fighting for jobs, bread and freedom".[10]

Lindenbaum took a keen interest in pedagogy, and in the second half of the 1930s he delivered various lecture series to teachers' groups and organizations. He was also said to have been interested in art, literature, hiking and mountain climbing. One of his lecture series to the Polish Teachers Union included a lecture on the creative and "artistic" elements in mathematics. It seems he enjoyed teaching beyond the ranks of his own profession, and especially teaching other teachers. It appears, too, that he was an avid solver of newspaper chess problems: his name pops up time and again in the mid-1920s in ***ABC Nowiny Codzienne*** and ***Nasz Przegląd*** for having solved the previous issue's challenge puzzle. He even published a mathematical paper, "*Sur le «problème fondamental» du jeu d'échecs*," ***Rocznik Polskiego Towarzystwa Matematycznego*** (=***Annales de la Société Polonaise de Mathématique***), vol. 13 (année 1934, publ. 1935), pp. 124–125. At the same time it was rumored he liked "having a good time", being among people, frequenting cafés and partying. Interestingly though, in group photos he was never front row center. He seemed to prefer the third or fourth row, or somewhere over by a wall (Figs. 4 and 5).

Around the end of October or the beginning of November 1935 Lindenbaum married Janina Hosiasson (Fig. 6), an established philosopher of logic, fellow member of the Lwów-Warsaw school of mathematical logic, fellow graduate of the University of Warsaw, and four-and-a-half years older than him. She also was named by Neurath as

---

[9]The petition was issued in the name of the editorial staff, co-workers and "friends" of ***Głos Współczesny*** <*Contemporary Voice*>, a left-leaning monthly newspaper with national circulation, and was splashed across the front page of the March 1936 issue. The signatures of editorial staff and co-workers were grouped separately from those of "friends". Lindenbaum's signature was included among editorial staff and co-workers.

[10]Published in ***Lewar***, 15th May, 1936, no. 4, page 10. ***Lewar*** was a biweekly literary magazine sponsored and influenced by the Polish Communist Party from 1933 through 1936. Its name was a play on words, combining "leverage" and "leftist".

Fig. 4 First congress of mathematicians from Slavic countries, Warsaw, 23 September 1929. Adolf Lindenbaum, arms folded, wearing a blazer with a boutonnière in the lapel, is standing by the door

having been a member of the original Vienna Circle. During their engagement they stayed for 2 or 3 months with Adolf's mother and sister in the Lindenbaums' family home at Złota 45/4 before moving into their own apartment at Krasińskiego 16/34 on the 31st of October, 1935.

It was, according to Antoni Marianowicz (Kazimierz Jerzy Berman), a marriage of convenience for both of them.[11] Janina was in love with Antoni Ludwik Pański (1895–1942), philosopher, statistician, social activist and eldest son of the neurologist Aleksander Pański (Fig. 7). Janina and Antoni had been living together, on again off again, since Janina's late teens. During this time Antoni had a string of affairs with other women, and was briefly married (to a singer, Elza Aftergut), but his love for Janina was apparently "the real thing". Awkwardly, they were second cousins.[12] Marianowicz writes that their respective "aunts and uncles" pressured them to choose other marriage partners.

[11]See Marianowicz [1995], pp. 230–231, or Marianowicz [1995] (1), pp. 192–193.

[12]Janina's mother Zofja Hosiasson, née Feigenblatt, and Antoni's mother Róża Pańska, née Seidemann, were first cousins—(Antoni's maternal grandfather Adolf Seidemann and Janina's maternal grandmother Leona Feigenblatt, née Seidemann, were brother and sister).

Fig. 5 Meeting of the mathematics, physics and astronomy circles in Warsaw, 5 May 1932. Łukasiewicz, Leśniewski and Tarski are seated in the front row. Adolf Lindenbaum is standing farther right, one or two rows farther back, behind and looking out from between two gloved men in wool coats. It is our guess that in the very back row, plumb above the toe of Leśniewski's airborne boot, stands a 27-year-old Mojżesz Presburger

Marianowicz does not spell out why he thinks it was a marriage of convenience for Adolf. It is not clear if Adolf regarded the marriage the same way Janina did. To all outward appearances they cohabited just fine, and seemed to care for one another. Certainly they held each other's professional abilities in high regard. They were fired by the same progressive social ideals and political convictions. But was Janina the apple of Adolf's eye?... the "little man" (אישון עין)[13] of his eye? We are indebted to Arie Hinkis for his suggestion that Adolf was delicate, feminine, the exact opposite of Tarski; that he exhibited none of Tarski's machismo, competitiveness or ego; and that only Tarski could have "extracted" from the young Adolf the prodigious achievements of [26] and [26a]. Perhaps Lindenbaum was under social pressures of a different kind.

[13]The biblical Hebrew origin of today's expression (see: Deuteronomy 32:10, Psalms 17:8, Proverbs 7:2, Lamentations 2:18).

Fig. 6 Janina Hosiasson. Photograph from her university student "*indeks*", October, 1919

Fig. 7 The three Pański brothers, from left to right: Jerzy, Antoni and Wacław (Solski), *circa* 1920

On September 6th, 1939, Adolf and Janina abandoned their apartment and all their belongings and fled Warsaw on foot,[14] heading (Janina's letters to Otto Neurath and G.E. Moore suggest) either due east in the direction of Siedlce, or south-east in the direction of Dęblin. Janina writes to Neurath and Moore that progress on foot was slow and that the road was repeatedly strafed by Luftwaffe planes. A friend with a motorcycle encountered them on the road and gave Janina a ride on the saddle behind him, leaving Adolf to continue on foot. The motorcyclist managed to get Janina as far east as Rivne[15] where he left her before heading back alone in the direction of Warsaw. From Rivne, Janina made her way, catch as catch can, partly by train and partly by road, to Vilnius. There she eventually learned through friends and acquaintances that Adolf was in Białystok.[16]

[14]German forces invaded Poland on September 1st and within days German artillery shells were raining down on Warsaw. Janina and Adolf fled their home under fire.

[15]Polish: Równe; Ukrainian: Рівне; Russian: Ровно; Hebrew: רובנו; Yiddish: ראָוונע—a name that reverberates in Aliyah consciousness. According to Polish Wikipedia, in 1939 Rivne had a population of 41,500 persons, slightly more than half of whom (21,000) were Jews. See: https://pl.wikipedia.org/wiki/R%C3%B3wne.

[16]On September 17th Soviet forces entered Poland and on September 22nd Białystok came under Soviet occupation. We do not know if Lindenbaum was already in Białystok by then, or if he arrived after it was

Fig. 8 Members of the Department of Mathematics and Physics, Pedagogical Institute, Białystok, spring of 1941. Left to right: Stanisław Romanowski, Samuel Steckel, Adolf Lindenbaum, Salomon (Szlama) Lubelski, Henryk Ferencowicz, Edward Litwinowicz

Adolf and Janina tried writing to each other, without much success. Apparently, most of their letters to each other were not allowed to get through—confiscated by one postal authority or another. Janina visited Adolf in Białystok for a day, then she returned to Vilnius without him. Her letters to Neurath and Moore suggested that she and Adolf had "agreed to disagree" about where best to try to survive.[17]

Adolf found work in the Pedagogical Institute in Białystok as a mathematics teacher and the head of its mathematics department (Figs. 8 and 9).[18] Then on June 22nd, 1941, Germany invaded the Soviet Union, and on the same day, the Vilnius Uprising began.[19] Within days, German forces were in Białystok, and not much later in Vilnius. Sometime

---

in Soviet hands. It would have been a difference of only a few days. In either case it would have suited him, as he was attracted by Soviet communism—or at least, by his imagined picture of it.

[17]On September 19th, 1939, Soviet forces wrested the city of Wilno from Poland and on October 28th re-attached it to its ancestral home of Lithuania, upon which they bestowed (notional) "independent statehood", a dubious arrangement which lasted only until August 3rd, 1940.

[18]The Pedagogical Institute was a Soviet institution.

[19]In Zygmunt–Purdy [2014], §**1**, p. 299, we wrote that on June 22nd, 1941, Germany "declared war" on the Soviet Union. Germany of course did nothing of the sort. It simply invaded, without bothering to observe such niceties as telling anyone what it was doing. We thank Piotr Wojtylak for pointing this out.

Fig. 9 Street view of the Pedagogical Institute in Białystok, 1940

before the beginning of July, 1941, Adolf came to Vilnius and stayed, possibly for about 6 weeks, in a small satellite community on the eastern outskirts of the city called Pavilnys (Polish: Kolonia Wileńska).[20] By coincidence Pavilnys was where Anna Borkowska, aka Mother Bertranda, famously hid members of HaShomer HaTza'ir in her Dominican convent. However, there is no indication that Adolf had any knowledge of this.

Why he came when he did, and indeed at all, and why he chose to stay in Pavilnys, rather than with his wife in her apartment downtown,[21] remains unclear. Perhaps he was finally persuaded of the wisdom of trying to emigrate to the West, and hoped that Janina's

[20]We have this on the authority of Professor Bogusław Wolniewicz, who cites testimony of Professor Maria Renata Mayenowa (born Rachela Gurewicz), from a conversation he held with her on 26 April 1986. See Wolniewicz [2015].

[21]Remarks attributed to Oskar Lange suggest that Janina and Antoni Pański were living in the same apartment in Vilnius at that time, which might explain why Adolf chose to live elsewhere.

contacts in Vilnius might help him do so, but was wary of putting her in danger by openly associating with her. Sara Bender writes that under the Soviets there was a systematic campaign of destruction and arrests in Białystok from May to June, 1941, cut short only by the German invasion, and that a "fourth wave of arrests began on the night of June 20th, 1941, when members of the NKVD went from house to house with their lists, sending entire families, most of them Jewish, in cattle and freight trucks to the Soviet hinterland." Then on June 22nd, 1941, "the bombing of Białystok sowed panic in the city. As the Red Army began to flee, [. . .] anyone who could, fled with the Russians".[22]

Possibly he summoned his sister Stefanja to join him in Vilnius from wherever she had been hiding. Or possibly she was already staying with him in Białystok and simply came with him. We don't really know where she had been staying before this, or how or why she came to Vilnius when she did. All we know is that they both showed up together. Sometime before the middle of August, 1941, Adolf and Stefanja were arrested and shot. The timing of their arrests and murders—Adolf and Stefanja together—suggests that Stefanja was staying with Adolf in Pavilnys.[23]

There is some evidence to suggest that Janina finally did marry Antoni Pański . . . either immediately upon learning of Adolf's death, or sometime earlier. In September, 1941, both Antoni and Janina were arrested—he first, she a week later. She had two passports in her possession at the time of her arrest, one in the name of Janina Lindenbaumowa, the other in the name of Janina Pańska. Possibly Adolf had given her a divorce. Or she had committed bigamy. Or one or both of the passports, or supporting documents used to obtain them, had been forged.

Jerzy Dadaczyński writes that Janina applied for an American visa.[24] We know from American sources[25,26] that she repeatedly tried, to no avail, to be sponsored into the U.S. by the Rockefeller Foundation's New School for Social Research as a "refugee scholar". Rudolf Carnap, Oskar Lange, William Gruen, Ernest Nagel, Albert Hofstadter, Alfred

[22]See Bender [1997] (1), pp. 87–90.

[23]According to the *Dédicace*, ***Fundamenta Mathematicae***, vol. 33 (1945), p.V, Adolf Lindenbaum was shot in Naujoji Vilnia (Polish: Nowa Wilejka). This was where Soviet forces withdrew to on October 28th, 1939, after handing Vilnius over to a notionally independent Lithuania. Naujoji Vilnia is one train stop—4 km—east of Pavilnys/Kolonia Wileńska, and larger. However it must be underlined that by August, 1941, German forces were in control of both places. Stefanja and Adolf were not shot by the Soviets. They were shot either by Germans or by Lithuanian collaborators.

[24]Dadaczyński [2003].

[25]SUNY, University at Albany, Science Library 352, M.E. Grenander Department of Special Collections and Archives, German and Jewish Intellectual Émigré Collection (GER-017), Series 4: individual files from Else Staudinger, Director of the American Council for Émigrés in the Professions (ACEP), Box 3, folder 147.

[26]The Rockefeller Archive Center, the Rockefeller Foundation (RF) Archives collection, Record Group 2, RG2 1940, Series 200, Box 192, folders 1368 & 1369; and RG2 1941, Series 200, Box 212, folder 1487.

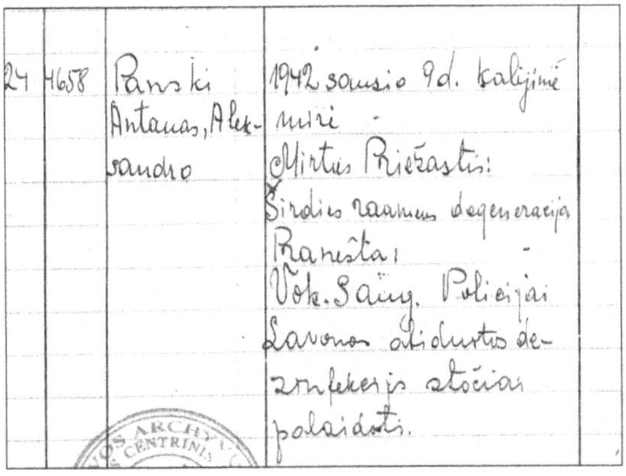
24 | 4658 | Panski Antanas, Aleksandro | 1942 sausio 9 d. kalėjime mirė - Mirties Priežastis: Širdies raumens degeneracija Raneštai - Vok. Saug. Policijai Lavonas atiduotas dezinfekcijos stočiai palaidoti.

24. Wpis do rejestru zgonu nr 4658 więźnia Antoniego Pańskiego zmarłego 9 stycznia 1942 r. w szpitalu więziennym

Fig. 10 Journal entry recording the death of prisoner #4658, Antoni Pański, on 9 January 1942 in the prison hospital. Reproduced from Monika Tomkiewicz's book, ***Zbrodnia w Ponarach 1941–1944*. Monografie Komisji Ścigania Zbrodni Przeciwko Narodowi Polskiemu**, vol. 43. Instytut Pamięci Narodowej: Warszawa 2008. In her book this is Fig. 24 of 51 figures on 17 unnumbered pages at the end of the book. Reproduced by kind permission of the author and Instytut Pamięci Narodowej

Tarski, Henry S. Leonard, Herbert Feigl, Mason W. Gross, G.E. Moore, C.A. Baylis, Carl Hempel, J.C.C. McKinsey, Sidney Hook, Willard V.O. Quine, Victor F. Lenzen and C.J. DuCasse all wrote letters to the Rockefeller Foundation supporting her applications.[27] In the end, the New School decided it was willing to accept her application to enter the U.S., but could not itself provide financial support for her, or recommend that the Rockefeller Foundation provide financial support. Dadaczyński also writes that "probably" her efforts to help her husband attracted the attention of the Lithuanian authorities that led to her arrest. What those efforts were, we do not know.

Antoni Pański died in Lukiškės prison[28] in Vilnius on January 9th, 1942 while under interrogation, probably tortured to death. His death certificate, issued by the prison hospital staff and handwritten in Lithuanian,[29] identifies the cause of death as "*Širdies raumens degeneracija*" <heart muscle degeneration>, which probably just meant his heart stopped beating (Fig. 10).

Janina's friends engineered a prison break for her, but she fluffed it. In April, 1942, after 7 months of imprisonment, she was taken to Paneriai (Polish: Ponary) and shot.

---

[27]Lord Russell, three of whose books Janina had translated, and himself an "economic immigrant" in the U.S. at that time, on a work visa, declined to support her, claiming that he did not recall her.

[28]Lithuanian: *Lukiškių tardymo izoliatorius kalėjimas*. Polish: *więzienie na Łukiszkach*, or simply *Łukiszki*. It was on the same street—Gedimino—as Janina's apartment.

[29]The prison and its hospital were under German direction and control but staffed by Lithuanians.

3 Metric Spaces[30]

Metric spaces interested Lindenbaum from early on, as can be seen from:

- his first published paper, [26], "*Contributions à l'étude de l'espace métrique, I*",[31] later incorporated into his doctoral dissertation
- §5: "*Théorie des ensembles des points*" of the paper [26a], "*Communication sur les recherches de la théorie des ensembles*", co-authored with Tarski. §5 sets out results pertaining to the decomposability of sets and their congruence in metric spaces, some of which results were obtained by Lindenbaum alone, some jointly with Tarski, and some jointly with M.D. Kirszbraun
- his doctoral dissertation, "*O własnościach metrycznych mnogości punktowych*" <On metric properties of point sets>, submitted in 1927 and defended in 1928 (though never published in its entirety)
- the short note [29^{a}a], lifted from his doctoral dissertation, summarizing his talk at the First Polish Mathematical Congress in Lwów in 1927
- the short note [31^{a}a], "*La projection comme transformation continue la plus générale*"
- the short note [31^{a}b], "*Sur les figures convexes*"
- the lecture "*Badania nad własnościami metrycznymi mnogości punktowych*", given at the Second Polish Mathematical Congress in Vilnius in 1931 (listed as [31^{1}a] in Zygmunt–Purdy [2014], p. 306)
- the paper [33a], "*Sur les ensembles localement dénombrables dans l'espace métrique*"

It seems certain Lindenbaum's interest in metric spaces was sparked most of all by Tarski[32] and also considerably by Banach and Kuratowski.[33] Lindenbaum had a ringside

[30] Metric spaces and decompositions are only two of many areas where Lindenbaum's work is underappreciated. The constraints of the present publication limit our treatment to these two areas. Future articles will consider his contributions to sentential logics, metalogic, the general theory of sets, and the independence of the axiom of choice.

[31] The Roman "I" in the title implies that a sequel was planned. In footnote 1, p. 214 (see also footnote 1, p. 218) Lindenbaum indicated what he intended the sequel to be about: it was going to be a study of the notion of the equivalence of point sets by decomposition, in the sense of Banach and Tarski (see also [26a], p. 327). In the event, no sequel was ever published. Nor was his doctoral dissertation ever published in its entirety, although both he and Sierpiński expected it to be: they each referred to it as "à paraître" (see [33a], p. 106, note 18; and Sierpiński [1936], p. 32). As there is no surviving copy in Warsaw University's archives, or anywhere else we know of, we are unsure precisely how his dissertation related to his publications. Many authors, including Lindenbaum himself, credited his dissertation for results that were never set out in any of his (other) publications (see, e.g., Aronszajn [1932], p. 99, note 12$''$; Kirszbraun [1934], p. 78, note 4, and p.102; Sierpiński [1936], p. 32; Lindenbaum [33a], p. 106, note 18).

[32] Lindenbaum and Tarski's collaboration on decomposability and congruence of point sets in Euclidean and general metric spaces began as early as 1923, when Lindenbaum was a freshman/sophomore (see [26a], p. 327).

[33] In [26], p. 210, one reads, "Je termine cette préface par remercier MM. Kuratowski et Tarski, qui ont bien voulu prendre intérêt à ces recherches: j'en ai profité beaucoup." Recognition of Kuratowski's impact on [26] is also expressed on p. 216 in a parenthetical suffix to the statement of Théorème 7 ("C'est M. Kuratowski qui a su généraliser de cette manière intéressante une idée de ma démonstration primitive du th. 8") and on p. 222, note 1 ("Le problème auquel le théorème (II) donne réponse m'a été posé par M. Kuratowski.")

seat at the gestation and birth of Tarski's "*O równoważności wielokątów*" <On the equivalence of polygons> and Banach and Tarski's famous so-called paradox, "*Sur la décomposition des ensembles de points en parties respectivement congruentes*", the latter of which explicitly mentioned a result obtained by Lindenbaum and thus constituted, in a "proxy" sort of way, Lindenbaum's first published result. (We cite this reference below.)

On the other hand Wacław Sierpiński, Lindenbaum's PhD thesis supervisor, probably did not exert a formative influence on Lindenbaum's interest in metric spaces... at least not at first. Only when as editor of ***Fundamenta Mathematicae*** he recognized good work in [26] and agreed to guide Lindenbaum's doctoral dissertation did Sierpiński begin to exert an influence on the direction of his pupil's inquiries. And in due course vice-versa: by [33a] and Sierpiński [1933], their mutual influence on each other started to show.

In the mid-1920s when Lindenbaum began investigating metric spaces the theory was still in its formative stages of development. Maurice Fréchet introduced the concept, using different terminology, in his doctoral dissertation in 1906. Felix Hausdorff, in his ***Grundzüge der Mengenlehre***,[34] was the first to use the expression *metrischer Raum* in place of Fréchet's terminology to denote such entities.[35] Lindenbaum's [26] can thus be seen as one of its earlier systematic treatments.

We present a selection of Lindenbaum's results from [26] and [33a]. To do this, we introduce some necessary terminology.

A space M is a collection of undefined entities called points, and a *metric space* in the sense of Fréchet and Hausdorff is a pair $<M, \rho>$ consisting of a space M and a real-valued non-negative function ρ on the cross product $M \times M$ satisfying the following conditions:

(M1) $\rho(x, y) = 0$ if and only if $x = y$ (law of coincidence)

(M2) $\rho(x, y) = \rho(y, x)$ (law of symmetry)

(M3) $\rho(x, y) + \rho(y, z) \geq \rho(x, z)$ (the triangle law)

The function ρ is called a *distance function*, and the number $\rho(x, y)$ is called the *distance* between points x and y. The triangle law (M3) is so called because of its formal expression of Euclid's Proposition I. 20: "Any two sides of a triangle are together greater than the third side." The three laws as formulated above are consistent and independent.

Lindenbaum observed (see [26], p. 211) that both the law of symmetry (M2) and the stipulation that ρ be non-negative can be derived, and thereby dispensed with, by formulating the triangle law *d'une façon plus avantageuse*. Thus he showed that two independent axioms sufficed[36,37]:

(ML1) $\rho(x, y) = 0$ if and only if $x = y$ (law of coincidence)

(ML3) $\rho(x, y) + \rho(x, z) \geq \rho(y, z)$ (modified triangle law).

---

[34]See Hausdorff [1914].

[35]The terms themselves—*l'espace métrique* and *metrischer Raum*—were not new. *L'espace métrique* even predated Fréchet's dissertation. See Couturat [1905], page 204; and Couturat [1905] (1), page 216. However, these earlier usages were not related to the theory of metric spaces discussed here.

[36]In the 1920s Stanisław Leśniewski and Jan Łukasiewicz, among others, advocated minimizing the number of axioms in formalized deductive systems. For more on this issue see, e.g., Sobociński [1955].

[37]He noted two other *avantageuses* modifications which could as well do the trick—"$\rho(x, z) + \rho(y, z) \geq \rho(x, y)$" and "$\rho(z, x) + \rho(y, z) \geq \rho(x, y)$"—the second of which he credited to Piotr Szymański. Garrett Birkhoff, citing Lindenbaum's [26], gave yet another modification (see Birkhoff [1944], p. 466): "$\rho(x, y) + \rho(y, z) \geq \rho(z, x)$". Birkhoff thought this "circularity postulate," as he called it, had "a clear intuitive content: if one journeys from p to q and then from q to r, the minimum energy required to get back to p is not more than that already expended."

More precisely, he showed that: *If* ρ *is a real-valued function defined on* $M \times M$ *and satisfying* (ML1) *and* (ML3), *then the laws* (M2) *and* (M3) *hold, and* ρ *takes only non-negative values, which is to say,* <*M*, ρ> *is a metric space in the usual sense.*

A metric space <*N*, σ> is said to be a *subspace* of a metric space <*M*, ρ> iff

$N \subseteq M$, and
σ is a restriction of ρ to $N \times N$.

It is obvious . . .

that every subset of *M* determines a *unique* restriction of ρ,
that any such restriction satisfies (ML1) and (ML3), and hence
that every subset of *M* determines a *unique* subspace of <*M*, ρ>.

Consequently it is common practice to speak of a subset $N \subseteq M$ as being a *subspace* of *M* (and *M* a *superspace* of *N*) without explicitly presenting them as ordered pairs <*M*, ρ> and <*N*, σ>, or mentioning their distance functions. Moreover, when their distance functions are explicitly mentioned, they are often presented using the same symbol to designate both the distance function on the superspace *M* and its restriction to the subspace *N*, provided this leads to no confusion.[38]

The relation of being a subspace is transitive. Hence, in regarding a subset *N* as a subspace of *M*, we need not consider whether the metric on *N* is inherited directly from *M*, or indirectly, from the metric on some intermediate subspace *Q*, where $N \subseteq Q \subseteq M$. Thus, if *N* is a subset of *M*, we may refer to *N* as either a subset or a subspace of *M*. The choice is one of emphasis only. If we refer to the *subspace N*, we are focusing our attention primarily on *N* itself, *qua* metric space, whereas if we refer to the *subset N*, we are considering *N*'s set-theoretic properties in relation to *M*.

[26] began by defining basic notions of point-set topology and metric spaces. It cited Hausdorff's 1914 classic ***Mengenlehre***, but in its choice of definitions, which Lindenbaum conceded "*diffèrent souvent de celles qu'on trouve ailleurs*" <often differ from those found elsewhere>, it was mainly motivated by a desire to avoid using the axiom of choice (AC).[39] This is particularly evident in its definitions of a closure operation and a compact set.

Namely, if *A* is a subset of *M*, then the *closure* $\overline{A}$ of *A* is the set $A \cup A'$, where A' is the derived set of *A* in the space <*M*, ρ>, i.e., the set of all accumulation points of *A* in <*M*, ρ>.[40] Hence the closure operation is not defined in the spirit of Fréchet as the set of all limit points of *A* (or points "adherent to" *A*).[41]

---

[38]A similar shorthand is frequently adopted in speaking of metric spaces of differing dimensions, where the space of lower dimension can be considered as embedded in the higher-dimensional space: in this case, too, their distance functions are often presented using the same symbol to designate both.

[39]This attitude toward the axiom of choice is stated explicitly in [26] on page 212, footnotes 1 and 3.

[40]Let $a \in M$ and $A \subseteq M$. Recall that *a* is an *accumulation point* of *A* in the metric space <*M*, ρ> iff every open sphere with centre *a* contains at least one point of *A* which is distinct from *a* (and consequently an infinite number of points of *A*). It is easy to see that *a*'s being (or not being) an accumulation point of *A* does not depend on the whole space *M*, but only on the subspace $A \cup \{a\}$.

[41]As a student of Sierpiński, Lindenbaum must certainly have known that the theorem "If *A* is closed ($=\overline{A}$), then *A* contains all its limit points" is provable without using the axiom of choice, whereas the proof of the converse implication needs AC. See Sierpiński [1918].

Lindenbaum defined compactness by means of the *Cantor condition*: a metric space $<M, \rho>$ is said to be *compact* iff, for every finite decreasing sequence $\{F_k\}$ of non-empty closed subsets of M, $\ldots F_{k+1} \subseteq F_k \subseteq \ldots \subseteq F_1 \subseteq M$, the intersection $\bigcap\{F_k\colon k < \infty\}$ is non-empty.[42] And he commented that an alternative definition might read as follows: a metric space $<M, \rho>$ is *compact* iff either M is finite or, for every infinite subset X of M, the derived set X' is non-empty (i.e., every divergent subset of X is finite).[43]

An *isometric transformation* or *isometry* between metric spaces $<M, \rho>$ and $<N, \sigma>$ is a surjective mapping f between the points of M and the points of N which preserves distance. Thus:

$$(*) \quad \sigma(f(x), f(y)) = \rho(x, y) \text{ for every pair } x, y \text{ of points in } M.$$

It is easy to see that any f satisfying the above conditions must be one-to-one. So, for any function $f{:}M \to N$, if f satisfies (*) then it is an isometry between $<M, \rho>$ and the subspace $<f(M), \sigma>$ of the space $<N, \sigma>$... called the *image space of M in N under f.* One also says that such an f is an *isometry of M into N.*

Metric spaces $<M, \rho>$ and $<N, \sigma>$ are said to be *isomorphic*, or *congruent*, or *superposable* (in symbols $<M, \rho> \cong <N, \sigma>$; or simply $M \cong N$), iff there exists an isometry between them.

Notice that any two subspaces A and B of the real line with the standard or "natural" distance function $\sigma(x, y) = |x - y|$ are superposable only by means of a translation or rotation. That is, if $<A, \rho> \cong <B, \rho>$ then the isometric transformation f establishing this congruence is either $f(x) = x + c$, or $f(x) = -x + c$, where c is a constant.[44]

An interesting theorem due to Sierpiński states: *Any linear set A contains no more than one point p such that* $A - \{p\} \cong A$. A corollary states: *In any non-empty linear set A there exists a point p such that A is <u>not</u> congruent to* $A - \{p\}$. See Sierpiński [1954], page 7.

The congruence relation $\cong$ is of course nothing more than the familiar concept of isomorphism as applied to the class of metric spaces. It is an equivalence relation on this class. But the class of metric spaces admits of another, more inclusive[45] equivalence relation, namely *homeomorphism*, which turns out to be much more important than simple isometry.

Aware of this, Lindenbaum devoted the second half of §3 of [26] to "the topological properties of congruence and the problem of extending a given congruence" (see p. 214), by which he apparently meant, laying some groundwork for relating congruence

---

[42] To be more precise, Lindenbaum defined what it means for an arbitrary subset $A \subseteq M$ to be *compact in a metric space* $<M, \rho>$. Then if A is *closed and compact in* $<M, \rho>$, then the *subspace* $<A, \rho>$ is compact. In general the assumption of closedness cannot be omitted.

[43] One can prove without AC that if a metric space is compact by the first definition, using the Cantor condition, then it is also compact by the second definition. The proof of the converse implication requires AC. In general topology, a topological Hausdorff space (a T_2 space) satisfying the Cantor condition, or equivalently the dual Borel condition for open sets, is called *countably compact*. In the class of metric spaces, compactness and countable compactness are equivalent.

[44] ... though there are plenty of non-standard distance functions for which this is not true (readers of a certain age may recall slide rules). Subspaces of the real line with the standard distance function are called *linear* sets (not to be confused with *linearly ordered* sets).

[45] An equivalence relation is said to be more (/less) inclusive iff the corresponding partition has coarser (/finer) granularity.

to topology. Then in §4 he turned to the concept of *monomorphism*. We employ our own numbering for the results of these last nine pages of [26], departing from Lindenbaum's original, slightly shambolic scheme:

Theorem 1 *Every isometry is a homeomorphism.*

Theorem 2 *If an isometry f maps a compact set A onto a compact set B, then there exists a unique isometry f^* of $\overline{A}$ onto $\overline{B}$ which is identical with f on A.*

Hausdorff [1914] defined a *totally bounded* metric space as: $<M, \rho>$ is called *totally bounded* iff, for every $\varepsilon > 0$, there is a finite subset $A \subseteq M$ such that for every $x \in M$ there is an $a \in A$ with $\rho(x, a) < \varepsilon$. This is equivalent to the condition that, for every $\varepsilon > 0$, M is the union of a finite number of open spheres (or balls) of radii $< \varepsilon$. Any compact metric space is totally bounded but not all totally bounded metric spaces are compact.

Theorem 3 *If $<M, \rho>$ is a totally bounded metric space, and $f: M \to M$ is an isometry of M into itself, then the image f(M) is dense in M,* i.e. $\overline{f(M)} = M$.[46]

To prove Theorem 3, Lindenbaum used the Dedekind chain method, which was used extensively in the 1920s by Sierpiński, Kuratowski and Banach.

In §4 of [26], a set is said to be *monomorphic* iff it is not congruent with any of its proper subsets. Lindenbaum defined it thus: "The set A is *monomorphic*, if the relations $B \cong A$ and $B \subseteq A$ occur only when $B = A$" (Def. 3, p. 217), and he added that a set is monomorphic if it is a minimal ("*irréductible*") element in the class of all sets on which it is superposable.

It follows that monomorphy (or non-monomorphy), although being a property of a set *qua* metric space, does not depend on the distance function used to construct a metric space out of the set; it inheres in the nature of the set itself, not in the nature of the distance function. Hence the definition should more correctly be worded thus: a metric space is monomorphic iff it is not congruent with any of its proper subspaces. Alternatively: iff all distance-preserving transformations of the space into itself are surjective.

Lindenbaum remarked that the following hold for Euclidean spaces:

(a) There exist non-monomorphic linear sets;
(b) Every bounded linear set is monomorphic;
(c) There exists a bounded plane set which is not monomorphic.

He then derived some sufficient conditions for a metric space to be monomorphic.[47] As mentioned earlier, we use a different numbering scheme from his:

Theorem 4 *If $<M, \rho>$ is a compact space, then it is monomorphic. Alternatively: Any closed and compact subset of a metric space is monomorphic. (Théorème 8 in his original numbering scheme)*[48]

[46]This readable formulation of Lindenbaum's Théorème 7 (p.216) is due to Ryszard Engelking [1989], p. 278.

[47]In his words, "*Le théorème 8, les corollaires 15 et 14, nous fourniront des conditions suffisantes, de plus en plus générales, pour qu'un ensemble compact soit monomorphe.*"

[48]Theorem 4 was generalized by Tarski as theorem **17**(T) in [26a], p. 329.

Theorem 5a *Every bounded subset of a compact metric space which is both F_σ and G_δ is monomorphic.*[49]

Theorem 5b *Every bounded set in an n-dimensional Euclidean*[50] *space which is both F_σ and G_δ is monomorphic.*[51]

Several theorems in §5 of [26a] are closely related to [26]: We mention three of them here: theorem **4**, due to Kirszbraun and Lindenbaum, and theorems **5**(*L*) and **14**(*L*), due to Lindenbaum alone.[52] The first two each give sufficient conditions for an *expanding*[53] mapping to be an isometry on a Euclidean space:

4: *Let $B \subset \mathbb{R}$ be a* <u>*bounded*</u> *linear space, and let $A \subseteq \mathbb{R}^n$ for $1 \leq n < \infty$ be a subspace of an n-dimensional Euclidean space. If $\delta(A) \geq \delta(B)$, that is, if the diameter of A is not less than the diameter of B,*[54] *and if $f{:}A \rightarrow B$ is an expanding, surjective mapping of A onto B, then f is an isometry between A and B, i.e., $A \cong B$.*

5(*L*): *Let $A \subset \mathbb{R}^n$ for $1 \leq n < \infty$ be a* <u>*bounded*</u> *subspace of an n-dimensional Euclidean space. Then any expanding mapping $f{:}A \rightarrow A$ of the subspace into itself is an isometry between A and $f(A)$, i.e., $A \cong f(A)$.*

The third one, theorem **14**(*L*), gives a necessary and sufficient condition for a set to be non-monomorphic:

14(*L*): *A set is not monomorphic iff it has a denumerable non-monomorphic subset.*

The first two of these theorems, particularly **5**(*L*), seem to have gone unnoticed, or been quickly forgotten, because Hans Freudenthal and Witold Hurewicz published a note in *Fundamenta Mathematicae* in 1936 proving a theorem very closely related to **5**(*L*) using exactly Lindenbaum's methods from [26].[55]

Indeed it can be argued that [26a] went generally unremarked before the Second World War. Stanisław Ruziewicz's review of it in ***JFM*** was perfunctory nearly to the point of dereliction of the reviewer's duty: scarcely 3½ or 4 lines of text, suggesting Ruziewicz himself had only skimmed the work, and offering no reasons why anyone else should

---

[49] In the Introduction to [26] Lindenbaum wrote, "*Au* §4 *j'examine la propriété singulière d'un ensemble de points* <u>*d'être superposable avec son vrai sous-ensemble*</u>. *On peut indiquer des ensembles plans bornés jouissant de cette propriété paradoxale, bien qu'ils ne puissent être* F_σ *et* G_δ *à la fois, ni linéaires; donc, à plus forte raison, ils ne sauraient être fermés, ni ouverts, cependant il y en a qui sont* F_σ *ou* G_δ. *Voilà le sujet principal, mais, à ce propos, j'étudie encore de plus près la notion (bien élémentaire) de congruence* (§3)." And in a footnote he explained, "*Un ensemble est* F_σ, *s'il est une somme dénombrable d'ensembles fermés; s'il est complémentaire d'un* F_σ *(c.-à-d.: produit dénombrable d'ensembles ouverts)—il est* G_δ."

[50] By "Euclidean" is meant, that the distance function is the "natural" or "standard" distance function on $\mathbb{R}^n$—i.e., the square root of the sum of the squares: $\sqrt{(x_2 - x_1)^2 + (y_2 - y_1)^2 + \ldots}$.

[51] Theorem 5b for $n = 2$ was generalized by Tarski as theorem **18**(*T*) in [26a], p. 330.

[52] See [26a], pp. 327–329.

[53] If $<M, \rho>$ and $<N, \sigma>$ are two metric spaces, then a mapping $f{:}M \rightarrow N$ fulfilling $\sigma(f(x), f(y)) \geq \rho(x, y)$ for all $x, y \in M$ is called *expanding*.

[54] For a given (fixed) distance function ρ, the *diameter* $\delta(X)$ of a set X is the farthest distance between any two points in X—i.e., $\delta(X) = \sup\{\rho(x, y): x, y \in X\}$. Since a distance function is by definition real valued, it is always possible to compare diameters of spaces of different dimension.

[55] See Freudenthal and Hurewicz [1936].

want to do even that much. Other than Ruziewicz's review, neither ***Zbl***, ***JFM*** nor ***JSL*** contains any reference to [26a] that would suggest it was studied, cited or worked on by others during the interwar period. This was possibly owing to where it appeared: in ***Sprawozdania z posiedzeń Towarzystwa Naukowego Warszawskiego*** <Minutes of the meetings of the Warsaw Society of Arts and Sciences> as opposed to a weighty mathematics journal. It was only during and (mostly) after the Second World War that, thanks to Sierpiński's diligently filling in almost all of its missing proofs, the work started receiving serious attention and garnering citations. Then in 1958 Sierpiński's monograph ***Cardinal and Ordinal Numbers*** capped these efforts and placed the value of [26a] beyond question.

It is instructive to view Lindenbaum's paper [33a] "*Sur les ensembles localement dénombrables dans l'espace métrique*" in the context of Sierpiński [1933] "*Sur les espaces métriques localement séparables*". Both papers are about local properties—then as now a subject of lively interest in topology. They exchanged ideas, shared preliminary drafts of their manuscripts,[56] and published their results in consecutive papers in the same issue of ***Fundamenta Mathematicae***.

Sierpiński's paper characterized locally denumerable[57] sets in a separable metric space. Alluding to this in [33a] at the bottom of page 101 Lindenbaum wrote, "Or, M. Sierpiński a posé la question *quels sont des ensembles localement dénombrables quand l'espace métrique n'est pas séparable.*" In fact Sierpiński's paper did not explicitly pose this question, only dangled it. Lindenbaum may have meant "m'a posé", i.e., in conversation, or in *marginalia* on a shared manuscript. In any event [33a] took up the question and developed new set-theoretical tools to answer it. "Voici une réponse:" he wrote.

Maurice Fréchet introduced the notion of a separable space in his doctoral dissertation and the importance of the concept was quickly recognized. The definition of a separable metric space can be expressed in various (equivalent) ways. For example, a metric space is said to be *separable* ...

iff it has a denumerable open base; or ...
iff any open covering of the space admits a denumerable subcovering.

A topological space X is called *separable* iff it contains a denumerable subset D which is dense in X, that is to say, for which $\overline{D} = X$.

A metric space $<M, \rho>$ is said to be *locally separable at a point* $x \in M$ iff there exists an open sphere centred on x which, *qua* subspace, is a separable space. Then a space $<M, \rho>$ is called *locally separable* (simpliciter) iff it is locally separable at every point $x \in M$.

The central result of Sierpiński's paper was its proof of the following "*Théorème: Pour qu'un espace métrique soit localement séparable, il faut et il suffit qu'il soit une somme*

[56]See [33a], p. 102, footnote 10, and p. 104, footnote 13; and Sierpiński [1933], p. 107, footnote 2.

[57]We use the word "denumerable" in the sense of "*at most* denumerable", i.e., either finite or at most countably infinite. We understand, for example, that a singleton is denumerable. So is the empty set.

disjointe d'ensembles ouverts séparables." <A metric space is locally separable iff it is a disjoint sum of open separable sets>.[58]

Lindenbaum framed his "*réponse*" in brand-new set-theoretical concepts. These are worth spelling out, as it is not generally realized how ground-breaking they were for 1933. Let $\mathbf{P}$ be a given class of sets, and let <M, ρ> be a given metric space. We will say that a subset $Z \subseteq M$ *is locally* $\mathbf{P}$ (or *has the property* $\mathbf{P}$ *locally*) *at a point* $z \in M$, iff there exists a real number $r > 0$, and a set $Y \in \mathbf{P}$, such that...

$$Y \cap S(z, r) = Z \cap S(z, r)$$

...where $S(z, r)$ is an open sphere of M centred on z with radius r.

Note that for any r satisfying the above condition there is a smaller one that does so too, and thus an infinitely descending sequence of them. This follows from a rudimentary property of open spheres, namely: that for every $p \in S(z, r)$ there is an r' such that $0 < r' < r$ and $S(p, r') \subseteq S(z, r)$.

Then we can say simply that Z *is locally* $\mathbf{P}$ (or *has the property* $\mathbf{P}$ *locally*) iff Z is locally $\mathbf{P}$ at z for all $z \in Z$. The class of all sets which are locally $\mathbf{P}$ will be denoted by $\boldsymbol{L}(\mathbf{P})$.[59]

We say that the set Z is *locally* $\mathbf{P}$ *in the restricted sense* iff Z is locally $\mathbf{P}$ at z for every $z \in M$. Note that, since $Z \subseteq M$, clearly if Z is locally $\mathbf{P}$ in the restricted sense then it is locally $\mathbf{P}$ simpliciter. The class of all sets which are locally $\mathbf{P}$ in the restricted sense will be denoted by $\boldsymbol{L}'(\mathbf{P})$.

The operators $\boldsymbol{L}'$ and $\boldsymbol{L}$ share basic properties with a closure operator; they are...

(1) extensive: $\mathbf{P} \subseteq \boldsymbol{L}'(\mathbf{P}) \subseteq \boldsymbol{L}(\mathbf{P})$
(2) isotone: $\mathbf{P} \subseteq \mathbf{Q}$ implies $\boldsymbol{L}(\mathbf{P}) \subseteq \boldsymbol{L}(\mathbf{Q})$ and $\boldsymbol{L}'(\mathbf{P}) \subseteq \boldsymbol{L}'(\mathbf{Q})$
(3) idempotent: $\boldsymbol{L}\boldsymbol{L}(\mathbf{P}) = \boldsymbol{L}(\mathbf{P})$ and $\boldsymbol{L}'\boldsymbol{L}'(\mathbf{P}) = \boldsymbol{L}'(\mathbf{P})$

As an exercise in cardinal arithmetic, one can estimate the cardinalities of $\boldsymbol{L}(\mathbf{P})$ and $\boldsymbol{L}'(\mathbf{P})$. Assume that the space <M, ρ> is infinite. Let $\mathfrak{m}$ be the cardinality of a dense set in M, and $\mathfrak{p}$ be the cardinality of the class $\mathbf{P}$. Then the class $\boldsymbol{L}'(\mathbf{P})$ is of cardinality $\leq \mathfrak{p}^{\mathfrak{m}}$. If $\mathfrak{p} > 1$, then the class $\boldsymbol{L}(\mathbf{P})$ is also of cardinality $\leq \mathfrak{p}^{\mathfrak{m}}$.

For a given cardinal number $\mathfrak{n}$, let $\mathbf{M}_{\mathfrak{n}}$ be the class of all sets of cardinality $< \mathfrak{n}$ which are contained in M. Thus, for example, $\mathbf{M}_2$ consists of the empty set and all singleton subsets of M, while $\mathbf{M}_{\aleph_0}$ is the class of all finite subsets of M.

A set Z is said to be *isolated* in a space <M, ρ> if Z and its derived set Z' are disjoint, i.e., if no accumulation point of Z is in Z. It is said to be *divergent* in the space <M, ρ> if it has no accumulation points in M, i.e., if its derived set $Z' = \varnothing$. Every divergent set is isolated; however the converse is not necessarily true.[60] It is easy to see that the elements

[58] Sierpiński admitted (p. 107, footnote 2) that Lindenbaum had pointed out to him that this theorem was "*implicitement contenu dans un théorème de M. Alexandroff* (***Math. Ann.*** 92, p. 299, *Fundamentalsatz* 2)", but that Alexandrov's proof was "*plus compliquée que la nôtre*".

[59] Always bearing in mind, of course, that this definition of $\boldsymbol{L}(\mathbf{P})$ is relative to the given metric space <M, ρ>.

[60] We allow ourselves to go off on a small sidetrack here. For readers who may be wondering if *isolated* and *scattered* are the same notion: no, they are not. Every isolated set is scattered, but in general, not all scattered sets are isolated.

of $\mathsf{L}(\mathsf{M}_2)$ are isolated sets, while $\mathsf{L}'(\mathsf{M}_2)$ consists of divergent sets. More interestingly, $\mathsf{L}(\mathsf{M}_{\aleph_0}) = \mathsf{L}(\mathsf{M}_2)$ and $\mathsf{L}'(\mathsf{M}_{\aleph_0}) = \mathsf{L}'(\mathsf{M}_2)$.

Since $\mathsf{M}_{\aleph_1}$ is the class of all denumerable subsets of M, then $\mathsf{L}(\mathsf{M}_{\aleph_1})$ is the class of all *locally denumerable* subsets of M. If M is a separable space, then by definition $\mathsf{L}(\mathsf{M}_{\aleph_1}) = \mathsf{L}'(\mathsf{M}_{\aleph_1}) = \mathsf{M}_{\aleph_1}$.

Lindenbaum's main theorems on locally denumerable sets were as follows[61]:

Theorem 6 *For a set Z to be locally denumerable, i.e., to belong to* $\mathsf{L}(\mathsf{M}_{\aleph_1})$, *it is necessary and sufficient that there exists a sequence of positive real numbers* $\{d_n\}$ *("distances") and a sequence of divergent sets* $\{Z_n\}$ *such that...*

6.1 $Z = \bigcup\{ Z_n : n < \infty\}$; *and...*

6.2 *if n and m are positive integers, and z an arbitrary element of* Z_n, *then for all points y of* Z_m, *with the possible exception of (at most) one point,* $\rho(x, y) \geq d_m$.

Then by using Sierpiński's result Lindenbaum reformulated the above Theorem 6 in terms of open sets as:

Theorem 7 *For a set Z to be locally denumerable it is necessary and sufficient that there exists a class* G *of open sets such that...*

7.1 *Z is contained in the union of the sets of* G;

7.2 *for every* $G \in \mathsf{G}$, *the set* $Z \cap G$ *is denumerable; and...*

7.3 *if* G_1 *and* G_2 *are distinct elements of* G, *then* $Z \cap G_1 \cap G_2$ *is empty.*

Lindenbaum remarked ([33a], p.106, footnote 18) that the above Theorems 6 and 7 (plus several others) remained valid even in a class of spaces—broader than the class of *metric* spaces strictly understood—where there may be no distance at all between distinct points of M, i.e., where ρ need satisfy only the conditions...

$$\text{(Mt 1.1)} \quad if\ x = y\ then\ \rho(x, y) = 0 \quad \text{(“half” the law of coincidence) ; and...}$$

$$\text{(Mt 2)} \quad \rho(x, y) \leq \rho(z, x) + \rho(y, z) \quad \big(\text{Szymański's modified triangle law}\big).$$

For a proof of this, he referred the reader to §22 of his doctoral thesis.[62]

This means that Lindenbaum was one of the first, along with E.W. Chittenden (1917) and W.A. Wilson (1931), to broaden or generalize the notion of metric space, and to use this generalized concept as a tool for solving topological problems.[63] Various kinds of generalized metric spaces were considered in later years, e.g., by Karl Menger (1935), Garrett Birkhoff (1936), and Hugo Ribeiro (1943), but none of these authors recognized or remarked on the fact that Lindenbaum had been there before them.

---

[61] Again, using our own numbering scheme, not Lindenbaum's original numbering.

[62] "*...la condition Mt 1.2 n'étant point essentielle [Cf. ma Thèse (Varsovie, 1927; à paraître), §22].*"

[63] Ryszard Engelking, in his treatise *General Topology*, makes extensive use of pseudometric spaces "as a convenient tool" for investigating a wide variety of topological spaces.

4 Decomposition of Point Sets, and Their Equivalence by Decomposition

Decomposition of point sets and their equivalence by decomposition, i.e., the congruence of their respective parts, was a subject of lively interest among the University of Warsaw's mathematicians in the early-to-mid-1920s, when Adolf Lindenbaum entered the university and began studying under them and working with them. Results in 1924 alone included Kuratowski's "*Une propriété des correspondances biunivoques*" <A property of bijections>, Banach's "*Un théorème sur les transformations biunivoques*" <A theorem on bijections>, Tarski's "*O równoważności wielokątów*" <On the equivalence of polygons>, and Banach and Tarski's famous paradox "*Sur la décomposition des ensembles de points en parties respectivement congruentes*" <On the decomposition of point sets into respectively congruent parts>. The first two established some eyebrow-raising facts about one-to-one mappings in purely abstract set-theoretical contexts. The second two were directly about equivalence by decomposition—of polygons in plane geometry, and of point sets in a Euclidean space of finite dimension. It was against this backdrop that Tarski and Lindenbaum set out their results on the theory of cardinal numbers and decompositions of abstract sets in [26a], "*Communication sur les recherches de la théorie des ensembles,*" §2: "*Propriétés des transformations univoques.*"

To say that a set A is *decomposed* into a family $\boldsymbol{F}$ of sets means that $\boldsymbol{F}$ is a *partition* on the set A, i.e., $\boldsymbol{F}$ is a family of non-empty disjoint subsets of A whose union $\cup\{X : X \in \boldsymbol{F}\}$ is all of A. If $\mathfrak{m}$ is the cardinality of $\boldsymbol{F}$, then A is said to be $\mathfrak{m}$-decomposed, or decomposed into $\mathfrak{m}$ parts. In the same way, A is said to be finitely decomposed, $\aleph_0$-decomposed, $\aleph_1$-decomposed, $2^{\aleph_0}$-decomposed, etc.

Stefan Mazurkiewicz and Wacław Sierpiński had whetted Warsaw's appetite 10 years earlier with their 1914 paper "*Sur un ensemble superposable avec chacune de ses deux parties*" <On a set congruent with each of its two parts> wherein they proved that there were nonempty sets A, A_1, $A_2 \subset \mathbb{R}^2$ such that $A = A_1 \cup A_2$, $A_1 \cap A_2 = \varnothing$, $A \cong A_1$, and $A \cong A_2$. What made their result so striking[64] was that the transforms they employed to superpose A_1 and A_2 on A were *rigid*, i.e., they preserved all the "standard" distance relationships on $\mathbb{R}^2$. One of their transforms was a simple rotation through an angle of 1 radian; the other was a straight-line translation over a distance of $+1$. And they had defined all three sets "effectively", which is to say, without resorting to the axiom of choice, transfinite induction, or the well-ordering theorem; all three sets were denumerable.

[64]On the most obvious level, of course, their result was another example of the paradox of infinity, namely, that the part could equal the whole—a family of paradoxes, actually, with an august lineage, from Zeno of Elea in the fifth century BC, through Galileo's ***Two New Sciences*** of 1638, to Bolzano's ***Paradoxes of the Infinite***, published posthumously in 1851. But after Cantor's work on cardinality, and certainly by 1914, mathematicians had gotten used to such paradoxes. The novelty of Mazurkiewicz and Sierpiński's result lay not in showing that the part could equal the whole (which by then was old news), but in the partition and the transformations they devised, which were truly novel, prefiguring and in a sense anticipating the isometry group $E(n)$ of Euclidean motions and the notion of $E(n)$-equidecomposability, and ultimately the definition of a *paradoxical set*. See also Hausdorff's 1914 paradoxical decomposition of the sphere.

Just as their result appeared, however, the First World War broke out, followed by the Soviet westward offensive of 1918–1919 and the Polish–Soviet War of 1919–1921, all of which combined to put a damper on Warsaw University's research activities.

In 1921 Stanisław Ruziewicz, working in Lwów, picked up where Mazurkiewicz and Sierpiński had left off, obtaining a related (though not fully analogous) result for a non-denumerable set in $\mathbb{R}^2$: "*Sur un ensemble non dénombrable de points, superposable avec les moitiés de sa partie aliquote*" <On a non-denumerable point set, congruent with halves of its proper subset>.

Using the axiom of choice Ruziewicz defined non-denumerable sets $A, B, C, D \subset \mathbb{R}^2$ such that $A = B \cup C \cup D$, $C \cap D = \varnothing$, and $A \cong C$ and $A \cong D$. The family $\boldsymbol{F} = \{B, C, D\}$ was not a strict partition of A, only a cover of A, as B was not necessarily disjoint from $C \cup D$. In fact the set B played a rather similar role to the number 0 in Mazurkiewicz and Sierpiński's 1914 proof.[65]

This result of Ruziewicz went some way toward answering (but stopped short of fully answering) a question that Hugo Steinhaus had earlier posed: *Does there exist an uncountable planar set which admits a 2-decomposition each of whose parts is congruent with the whole?*

Lindenbaum went the full distance and answered Steinhaus's question in the affirmative in the following theorems:

1. *If A is a linear set congruent with each of two subsets $B \subseteq A$ and $C \subseteq A$, then it is congruent with a subset $D \subseteq (B \cap C) \subseteq A$.*
   (See: [26a], p. 327, theorem **1**(*L*).)
2. *If A is a bounded planar set congruent with each of two subsets $B \subseteq A$ and $C \subseteq A$, then it is congruent with a subset $D \subseteq (B \cap C) \subseteq A$.*
   (*Ibid.*, theorem **2**(*L*).)

Corollary of 1 + 2: *No linear set, and no bounded planar set, can be decomposed into two parts each of which is congruent with the whole set.*
   (See: [26], p. 218, footnote 1.)
   Recall of course that Ruziewicz and Sierpiński [1914], and Hausdorff [1914], had shown there *do* exist unbounded planar sets, and sets on the surface of a 3-dimensional sphere, which can be so decomposed.

3. *For every cardinal number $\mathfrak{m} \leq 2^{\aleph_0}$, there exists an unbounded planar set which can be decomposed into $\mathfrak{m}$ parts each of which is congruent with the whole. A similar set can be constructed on the surface of a sphere.*
   (See: [26a], *loc. cit.*, theorem **3***(*L*).)

[65] Ruziewicz actually obtained this result some 2 years earlier, in the summer of 1919, and ran it past Sierpiński for publication in the first issue of ***Fundamenta Mathematicae***, i.e., the 1920 issue (Sierpiński was its founding editor). Sierpiński suggested a simplification, which Ruziewicz incorporated. The existence of the set B could be proved directly from Zermelo's axioms (including the axiom of choice), without any need of Zermelo's theorem on well-ordering, or transfinite numbers. Sierpiński also urged Ruziewicz to consult Hugo Steinhaus, who allegedly had an example of a non-denumerable planar set congruent to its halves. As it turned out, Steinhaus did not have such an example. See the letters from Sierpiński to Ruziewicz dated: 01 April 1919, 17 August 1919, 20 August 1919, and 19 April 1920, in Więsław [2004], pp. 141–143.

The above results nicely illustrated how Sierpiński's own later research was shaped by Lindenbaum's. The asterisk* by **3***(*L*) meant that Lindenbaum had used the axiom of choice in his (unpublished) proof. Twenty-one years later Sierpiński [1947b] proved Lindenbaum's third theorem without using the axiom of choice or any of its equivalents,[66] and in so doing answered Steinhaus's question in an effective manner: *There is a constructive proof that, yes, there does exist an uncountable planar set which admits a 2-decomposition each of whose parts is congruent with the whole.* Then Sierpiński went on to supply the missing proofs for 1, 2, and the Corollary of 1+2 above, which Lindenbaum had said he would deliver "*à plus tard*" but never got around to doing.

Many of the results in [26a] exploited and expanded on Stefan Banach's [1924] "*Un théorème sur les transformations biunivoques*" <A theorem on bijections>, applying it to the theory of functions, cardinal arithmetic and decompositions of point sets.[67] Banach's central theorem could be called "decompositions of pure sets", or the DPS theorem. It stated that:

(DPS) *For arbitrary sets A and B, if f is a one-to-one transformation of A onto a subset of B, and g is a one-to-one transformation of a subset of A onto all of B, then there exist decompositions of A and B:* $A = A_1 \cup A_2$, $B = B_1 \cup B_2$, $A_1 \cap A_2 = \varnothing = B_1 \cap B_2$, *such that...* $f(A_1) = B_1$ *and* $g(A_2) = B_2$.

Banach then set out a pair of useful properties which a 2-place relation between sets might possess. He called them property (α), and property (β):

(α) Whenever $A\ R\ B$, there exists a bijection $f{:}A \to B$ such that for every $X \subseteq A$, $X\,R\,f(X)$.

(β) If $A_1 \cap A_2 = \varnothing = B_1 \cap B_2$, and $A_1\ R\ B_1$, and $A_2\ R\ B_2$, then $(A_1 \cup A_2)\ R\ (B_1 \cup B_2)$.

And with these tools to hand, he proved the following two further theorems:

DPS2 *For a relation R with property* (α), *if A stands in relation R to some subset of B, and some subset of A stands in relation R to B, then there exist decompositions of A and B:* $A = A_1 \cup A_2$, $B = B_1 \cup B_2$, $A_1 \cap A_2 = \varnothing$, $B_1 \cap B_2 = \varnothing$, *such that* $A_1\ R\ B_1$ and $A_2\ R\ B_2$.

DPS3 *For a relation R with properties* (α) *and* (β), *if A stands in relation R to some subset of B, and some subset of A stands in relation R to B, then A R B.*

Some fundamental relations in set theory and topology turn out to possess properties (α) and (β), e.g., equipollence of pure sets, similarity of ordered sets, homeomorphism of topological spaces, and congruence of point sets. Properties (α) and (β) also crop up (indeed figure prominently) in work on "les types de dimensions", to use Fréchet's expression.[68]

[66] Sierpiński wrote: "*La démonstration de A. Lindenbaum n'a pas été publiée et elle m'est inconnue.*" (p. 9).

[67] See especially [26a], §2, "*Propriétés des transformations univoques*" <Properties of one-to-one functions>, pp. 316–318.

[68] See Arboleda [1981].

In [26a] Lindenbaum and Tarski used the terms "*relation transformante*" for a relation possessing property (α), and "*relation additive*" for a relation possessing property (β). Two of Lindenbaum's results from [26a] were[69] . . .

3(*L*) *If* $A \subseteq B \subseteq C$, $A_1 \subseteq C$, *and function* $f{:}A \to A_1$ *is surjective, then there exist four sets* B_1, *D*, D_1, *and E such that* . . .

(a) $A_1 \subseteq B_1 \subseteq C$,
(b) $B = D \cup E$ *and* $B_1 = D_1 \cup E$,
(c) $D \cap E = \varnothing$ *and* $D_1 \cap E = \varnothing$, *and*
(d) $f(D) = D_1$.

4(*L*) *If* $A \subseteq B \subseteq C$, $A \subseteq C_1$, *and function* $g{:}C_1 \to C$ *is surjective, then there exist four sets* B_1, *D*, *E*, *and* E_1 *such that* . . .

(a) $A \subseteq B_1 \subseteq C_1$,
(b) $B = D \cup E$ *and* $B_1 = D \cup E_1$,
(c) $D \cap E = \varnothing$ *and* $D \cap E_1 = \varnothing$, *and*
(d) $g(E_1) = E$.

The first of these, **3**(*L*), entailed Banach's DPS theorem, even though it did not require that the function *f* was one-to-one.

In [26a], §1, "*Théorie des nombres cardinaux*," Lindenbaum used the above results to help him prove **14**(*L*) and **14**(**bis**)[70] relating to the Cantor–Bernstein theorem, which provided Tarski with the machinery he was looking for to derive **15**(*T*).[71] The latter result has since come to be known as Tarski's Mean-value Theorem, and it has its counterpart in theorem **5**(*T*) of §2, p. 318, concerning decompositions.

Let *n* be a natural number. Point sets *A* and *B* in a metric space $<M, \rho>$ are said to be *equivalent by n-decomposition*—written $A \equiv_n B$—iff there exist two families of subsets, $\boldsymbol{F}_{\mathrm{A}} = \{A_1, A_2, \ldots, A_n\}$, and $\boldsymbol{F}_{\mathrm{B}} = \{B_1, B_2, \ldots, B_n\}$, such that . . .

$\boldsymbol{F}_{\mathrm{A}}$ is an *n*-decomposition of *A*,
$\boldsymbol{F}_{\mathrm{B}}$ is an *n*-decomposition of *B*, and
A_k is congruent to B_k, i.e., $A_k \cong B_k$, for all *k*: $1 \leq k \leq n$;

. . . and they are said to be *equivalent by finite decomposition*—written $A \equiv_f B$—iff there exists a natural number *n* for which $A \equiv_n B$. This can be extended in a natural way to *equivalence by* $\mathfrak{m}$*-decomposition*—written $A \equiv_{\mathfrak{m}} B$—where $\mathfrak{m}$ is an arbitrary transfinite cardinal number.

The above definitions—of equivalence by *n*-decomposition, by finite decomposition, and by $\mathfrak{m}$-decomposition—are from Banach and Tarski [1924], who established fundamental properties of these relations. Firstly, and most obviously, that for $n = 1$, $A \equiv_n B$ is simply an isometry $\cong$ in the space $<M, \rho>$. Secondly, that for all $m \geq n$, $A \equiv_n B$ implies $A \equiv_m B$. Thirdly, that for fixed *n*, the relation $\equiv_n$ is not transitive (simple counterexamples suffice to show this). Fourthly, and more interestingly, that equivalence

[69] See [26a], page 318.
[70] See pp. 302–303.
[71] *Ibid*, p. 303.

by *finite* decomposition—where you are free to choose a different n for each pair of sets—is transitive. Their proof of this used what they called a "double network" method. And since reflexivity and symmetry obviously hold, the relation $\equiv_f$ is an equivalence relation. Moreover, they showed that $\equiv_f$ has Banach properties ($\boldsymbol{\alpha}$) and ($\boldsymbol{\beta}$).

Lindenbaum and Tarski restated most of the above definitions and properties in [26a], along with several new findings (some joint, some by Tarski or Lindenbaum alone).[72] They demonstrated that:

(i) *The relation of equivalence by* $\mathfrak{m}$*-decomposition for* $\mathfrak{m} \geq \aleph_0$ *is completely additive. That is to say, if* $X_1, X_2, \ldots, X_n, \ldots$ *and* $Y_1, Y_2, \ldots, Y_n, \ldots$ *are two sequences of mutually disjoint sets, and* $X_k \equiv_{\mathfrak{m}} Y_k$ *for all naturals k, then* $(\bigcup_{k=1}^{k<\infty}\{X_k\}) \equiv_{\mathfrak{m}} (\bigcup_{k=1}^{k<\infty}\{Y_k\})$.
(See [26a], p. 328, theorem **6**.)

(ii) *If* $A \subseteq B \subseteq C$ *and* $A \equiv_n C$, *then* $A \equiv_{n+1} B \equiv_{n+1} C$.
(*Ibid.*, theorem **9**.)

Lindenbaum supplemented his result from [26], that every bounded linear set was monomorphic, with:

13(*L*). *There exists a bounded linear set A which has a proper subset B such that* $A \equiv_2 B$.
(*Ibid.*, page 329.)

He also considered combinatorial properties of congruence and decomposability, as in a result mentioned in [26], p. 218, footnote 2:

- *Let n be a natural number, and let A and B be subspaces of a metric space. If* $A \cong B$ *and* $A \cap B$ *contains fewer than* $\frac{n(n+1)}{2}$ *elements, then* $(A - B) \equiv_n (B - A)$.

This was proved for the first time only by Sierpiński [1954], pp. 110–113, who also showed, by means of a suitable counter-example on the straight line, that the number $\frac{n(n+1)}{2}$ cannot be any greater.

In his 1943 paper "Some remarks on set theory"[73] Paul Erdős recounted the following interesting story:

> Professor Tarski communicated to me the following result of Lindenbaum: There exist $2^{\mathfrak{c}}$ linear sets no two of which are countable equivalent [by decomposition]. This result was never published, and Tarski does not remember the details of the proof. I have succeeded in proving that if $\mathfrak{m}$ is any cardinal number $< \mathfrak{c}$, then there exist $2^{\mathfrak{c}}$ linear sets no two of which are $\mathfrak{m}$-equivalent. I do not know whether my proof differs from that of Lindenbaum, but I have thought it might be worth publishing, since the result has some interesting applications.[74]

Erdős's proof used the axiom of choice in an essential way. Enter Sierpiński, who 4 years later was able to prove a generalization—a strengthening—of the same theorem[75] without the axiom of choice, using the same von Neumann function that he and Ruziewicz had used in the 1930s... work which Lindenbaum had surely been aware of at the time. We have to conclude that, quite possibly, Sierpiński's proof merely recapitulated Lindenbaum's original.

---

[72] See [26a], §5, "*Théorie des ensembles de points*," p. 328.

[73] Erdős [1943].

[74] *Ibid.*, page 644.

[75] See: Sierpiński [1947].

From 1931 on, Lindenbaum contributed importantly to investigating Fréchet's "les types de dimensions" of topological spaces.[76]

5 Decompositions and Equivalence of Polygons in Elementary Geometry

Lindenbaum published only one short note in this area:

[37ªa] "*Sur l'équivalence de deux figures par décomposition en nombre fini de parties respectivement congruentes.*" ***Rocznik Polskiego Towarzystwa Matematycznego*** (=***Annales de la Société Polonaise de Mathématique***), vol. 16 (année 1937, publ. 1938), p. 197.

This is a (six-line) summary of a lecture given by Lindenbaum on 30 September 1937 to the Third Polish Mathematical Congress in Warsaw. His talk began by setting the problem in its historical perspective, then he outlined recent trends, which he called "quantitatives", then he worked through the proof of a theorem he and Zenon Waraszkiewicz had obtained in 1932, and in conclusion he presented some other related theorems without proofs. We state the Lindenbaum–Waraszkiewicz theorem below. Other details as to what he said in the talk we can only surmise.

Fortunately we know quite a bit about Lindenbaum's work in the theory of equivalence of polygons from two of Tarski's papers: [1931], "*O stopniu równoważności wielokątów*" <On the degree of equivalence of polygons>, and [1931/32], "*Uwagi o stopniu równoważności wielokątów*" <Remarks on the degree of equivalence of polygons>.

In elementary geometry, two polygons are said to be *equivalent* if it is possible to dissect them into the same finite number of respectively congruent polygons having no common *interior* points... though the parts will invariably share common *boundary* points: sides, or segments of sides, or vertices, etc.

This notion has a similar logical structure to the purely set-theoretical notion of equivalence by finite decomposability, $A \equiv_f B$, discussed above. One important difference between them is how they treat boundary points. Another is how they define congruence. The notion of congruent planar figures in elementary geometry is much narrower than the general idea of congruent metric spaces, and even differs in essential ways from the definition of superposable subsets of $\mathbb{R}^2$. Some of the other differences, however, are more superficial: in elementary geometry it is common to use the symbol '$\equiv$' without a subscript, finiteness being taken for granted, and to use verbs like 'divide', 'dissect', or 'cut' (as with scissors) instead of the septic 'decompose', which conjures images of bacterial decay.

To appreciate how big a difference common boundary points make, consider a square S with sides of length 1, and an isosceles right triangle T with sides of length 2, $\sqrt{2}$ and $\sqrt{2}$.

[76]See [31ª], [34ªa], [36ª]; see also Sierpiński's paper "Sur un problème concernant les types de dimensions" in ***Fundamenta Mathematicae***, vol. 19 (1932), pp. 65–71, wherein Sierpiński on pages 67–68 quotes verbatim Lindenbaum's two-page proof of a generalization—a strengthening—of one of Sierpiński's own results.

Both S and T can be cut exactly in half, into mirror-image pairs of isosceles right triangles with sides of length $\sqrt{2}$, 1, and 1. But neither of these divisions is a decomposition, in the set-theoretical sense of a *partition*, because the resulting halves are not disjoint: in both cases they share an edge. And it is not obvious at first sight how to decompose S and T into respectively disjoint parts such that $S \equiv_f T$. (Sierpiński [1954], p. 43, Theorem 15 offers one.)

The following two theorems about equivalence of polygons are provable in the usual axiomatic systems of elementary geometry:

- *If polygon V is a part of polygon W, then these polygons are not equivalent* (known as De Zolt's axiom); and ...
- *Polygons V and W are equivalent if and only if they have equal areas* (known as the Wallace–Bolyai–Gerwien theorem).

Tarski was the first to ask if these (or analogous) theorems remain true when equivalence is understood as set-theoretical equivalence $\equiv_f$ by finite decomposition into parts having no common points. Tarski [1924], "*O równoważności wielokątów*" <On the equivalence of polygons>, answered this question in the affirmative using Banach's measure theorem for bounded planar sets (see Banach [1923]).

It is clear that no dissection of equivalent polygons into congruent parts can be unique: two equivalent polygons can always be divided into congruent parts in various ways, with respect to both the form and the number of those parts. Hence the question arises: what is the smallest number of respectively congruent parts that two equivalent polygons can be divided into? For equivalent polygons, W and V, this smallest number is called their *degree of equivalence*, and is denoted by $\sigma(W, V)$.[77] (See Tarski [1931], pp. 37–38.)

Tarski attributed this definition to Lindenbaum in a footnote: "*O ile nam wiadomo, pojęcie to wprowadził Dr. Adolf Lindenbaum (Warszawa), który wraz z autorem artykułu ustalił pewne własności tego pojęcia*" <As far as we know, this notion was introduced by Dr. Adolf Lindenbaum (Warsaw) who together with the author of this article established some properties of the concept>. (See Tarski [1931], p. 38.) Oddly enough, none of the properties or results that he reports in that paper are attributed to Lindenbaum.

Tarski defined a function τ as follows: "Let Q be a square with edge a, and let P be a rectangle with edges $x \cdot a$ and ${}^{a}/_{x}$, where x is any positive real number. Polygons P and Q are obviously equivalent, and it is easy to see that their degree of equivalence is a function of x; we shall denote this function by the symbol $\tau(x)$. Thus $\tau(x) = \sigma(Q, P)$," ... and then he urged his colleagues to join him in investigating its properties.

He himself offered some general results, including upper bounds on $\tau(x)$ for certain values of x. For instance: $\tau(1^1/_3) \leq 3$; $\tau(2^1/_4) \leq 4$; $\tau(n) \leq n$ for all natural numbers n. This last inequality is easy to prove. A square with edge a can be dissected into n mutually congruent rectangular strips with edges a and ${}^{a}/_{n}$, from which a rectangle with edges $n \cdot a$ and ${}^{a}/_{n}$ can be assembled by arranging all the strips end to end.

Tarski conjectured that: (i) $\tau(n) = n$ for every natural number n; and asked whether: (ii) $\tau(x) \geq 3$ for every positive x different from ½, 1 and 2?

[77]The symbol "σ" was chosen because of the first letter of the Polish word "*stopień*", which means "degree".

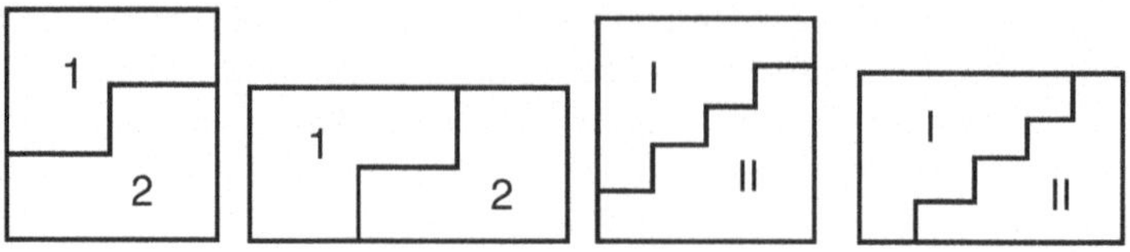

Fig. 11 Sierpiński's sketch, showing how Lindenbaum's examples work

In addition to Tarski himself, Henryk Moese, Adolf Lindenbaum, Bronisław Knaster and Zenon Waraszkiewicz investigated the function τ. Moese published a detailed proof of (i) in Moese [1932]. The others' results were published *en masse* in Tarski [1931/32], where it was noted that Zenon Waraszkiewicz was the first to come up with a proof of (i) but did not publish his proof at the time.

Lindenbaum noticed that, "all the results so far obtained for the function $\tau(x)$ still hold when, in the definition of this function, the square is replaced by any rectangle", that is, for a function $\tau_r(x) = \sigma(Q, P)$, where Q is a rectangle with edges a and b, and P is a rectangle with edges $x \cdot a$ and $^{b}/_{x}$. (See Tarski [1931/32], p. 313) And he also obtained the following:

- If $x = n + {}^{1}/_{p}$ where n and p are naturals ≥ 1, then $\tau(x) = n + 1$, i.e., $\tau(x)$ rounds x up.
- If $1 < x \leq 2$ then $\tau(x) = 2$ if and only if $x = 1 + {}^{1}/_{p}$ where p is a natural ≥ 1.

The second of these he obtained jointly with Zenon Waraszkiewicz, and its proof was—in Tarski's words—"somewhat complicated" and required "some subtle methods of reasoning." (See Tarski [1931/32], p. 312) It was this theorem and its proof that Lindenbaum delivered to the Third Polish Mathematical Congress in Warsaw on 30 September 1937. In [37^{a}a] it is formulated: *For non-congruent rectangles* $a_1 \times a_2$ *and* $b_1 \times b_2$ *each to be decomposed into 2 respectively congruent parts, it is necessary and sufficient that either* a_1/a_2 *or* b_1/b_2 *equal* $k + 1/k$ *for some whole positive k.*

At the end of Tarski [1931/32] there is a set of "exercises" for the ambitious reader. Presumably, all of the problems were known to their respective contributors to be solvable. Lindenbaum offered the following challenge[78]:

- Prove that, if W is a convex figure situated in the plane, then $s(W) \geq \frac{\delta(W)\cdot\omega(W)}{2}$, where $s(W)$, $\delta(W)$ and $\omega(W)$ are respectively the area, diameter and width of W. Moreover, in the above formula we will have equality if and only if W is a triangle.

Sierpiński, in a letter to Stanisław Ruziewicz dated Warszawa, 14/II/1932 r.,[79] wrote: "*Lindenbaum znalazł ciekawy przykład równoważności przez rozkład kwadratu i prostokąta,*" <Lindenbaum found an interesting example of equivalence by decomposition of a square and a rectangle>, and he drew a sketch in his letter to show Ruziewicz how it worked (Fig. 11).

It turned out, the decompositions Sierpiński sketched were crucial to the proof that Lindenbaum and Waraszkiewicz devised.

[78] *Ibid.*, page 314.

[79] See Więsław [2004], page 158.

Note. A deeper and broader presentation of the papers Tarski [1924], Banach–Tarski [1924], Tarski [1932], Tarski [1931/32] and Moese [1932], together with excellent English translations of them, can be found in the book McFarland–McFarland–Smith [2014]. For a survey of related results in the 70 years after Hausdorff's, Banach's and Tarski's works, see the fascinating monograph Wagon [1993].

6 Bibliography

6.1 Works by Adolf Lindenbaum

In Sect. 6.1 we select the main works of Adolf Lindenbaum. We follow the citation style of the Lindenbaum bibliography in Zygmunt–Purdy [2014]. A work is cited by a two-digit year in square brackets. Where several works appear in the same year, alphabetic suffixes designate their order of appearance. Citations of abstracts and short notes are distinguished by a superscript[a]; reviews by a superscript[r].

The Lindenbaum bibliography in Zygmunt–Purdy [2014] identified him as the author of seven reviews. That turned out to be a lamentable understatement. We are now able to attribute 22 reviews to him, all of them published in ***Zentralblatt für Mathematik***.

We also note that the Lindenbaum bibliography in Zygmunt–Purdy [2014] included a fourth sub-part, canvassing his public lectures, with citations distinguished by a superscript[l]. We omit that sub-part here, as we are currently reviewing its contents. We have learned that Lindenbaum delivered two lecture series, both titled „*O matematyce*" <On Mathematics>, to the *Instytut Wyższej Kultury Umysłowej Związku Nauczycielstwa Polskiego* <Institute of Advanced Learning of the Polish Teachers Union>: the first a 15-hour series over the month of July, 1938,[80] and the second a 20-hour series sometime between October 15th, 1938 and May 31st, 1939.[81] These presently await detailed documenting.

6.1.1 Papers

[26] Contributions à l'étude de l'espace métrique I. ***Fundamenta Mathematicae***, vol. 8 (1926), pp. 209–222. (The second part of [26] was announced on p. 214, note 1, but was never published. See [26a], p. 327, note 1, for a correction.) [JFM 52.0585.01 (E. Pannwitz)].

---

[80] Cited in ***Ruch Pedagogiczny***, rok 27, nr 6–10 (1938).

[81] An advance notice was given in ***Głos Nauczycielski***, nr 5, (25 September 1938), page 78. Some of the titles of individual lectures in the second series are rather philosophical in tone: *Co to są nauki scisłe* (*matematyczno-przyrodnicze*)? <What are exact sciences (mathematical-natural)?> *Jakie nauki należą do tej grupy*? <What sciences belong to this group?> *Trudności podziału nauk: przedmiot a metoda.* <Difficulties in the divisions of sciences: subject matter vs. method.> *Metoda dedukcyjna.* <The deductive method.> *Czy matematyka posiada przedmiot i jaki*? <Does mathematics have a subject, and if so, what?> *Rachmistrz-technik a matematyk-badacz.* <Number cruncher-technician versus mathematician-researcher.> *Element twórczy i "artystyczny" w matematyce.* <The creative and "artistic" element in mathematics.>

[26a] Communication sur les recherches de la théorie des ensembles (with A.Tarski). ***Sprawozdania z posiedzeń Towarzystwa Naukowego Warszawskiego, Wydział III Nauk Matematycznych i Przyrodniczych*** (=***Comptes rendus des séances de la Société des Sciences et des Lettres de Varsovie, Classe III***), vol. 19 (1926), pp. 299–330. [JFM 57.1330.02 (S. Ruziewicz)].
(1) Reprinted in: ***Alfred Tarski, Collected Papers. Volume 1, 1921–1934***, ed. by S.R. Givant and R. N. McKenzie, Birkhäuser Verlag: Basel 1986, pp. 173–204.

[30] Remarques sur une question de la méthode axiomatique. ***Fundamenta Mathematicae***, vol. 15 (1930), pp. 313–321. [JFM 56.0488.03 (A. Rosenthal)].

[30a] Sur les opérations d'addition et de multiplication dans les classes d'ensembles (with A. Koźniewski). ***Fundamenta Mathematicae***, vol. 15 (1930), pp. 342–355. [JFM 56.0084.02 (R. Baer)].

[31] Sur les ensembles ordonnés. ***Comptes rendus hebdomadaires des séances de l'Academie des Sciences***, Paris, vol. 192 (1931), pp. 1511–1514. [JFM 57.0091.03 (A. Fraenkel); Zbl 0002.18405 (A. Kolmogoroff)].

[33] Sur les ensembles dans lesquels toutes les équations d'une famille donnée ont un nombre de solutions fixé d'avance. ***Fundamenta Mathematicae***, vol. 20 (1933), pp. 1–29. Errata, p. 287. [JFM 59.0095.02 (W. Hurewicz); Zbl 0006.34001 (B. Knaster)].

[33a] Sur les ensembles localement dénombrables dans l'espace métrique. ***Fundamenta Mathematicae***, vol. 21 (1933), pp. 99–106; Errata p. 295. [JFM 59.0567.04 (G. Aumann); Zbl 0008.08806 (E. Čech)].

[33b] Sur les superpositions des fonctions représentables analytiquement. ***Comptes rendus hebdomadaires des séances de l'Académie des Sciences***, Paris, vol. 196 (1933), pp. 1455–1457. [JFM 59.0267.03 (A. Rosenthal); Zbl 0007.06005 (S. Saks)]. (For errata see [34a], p. 13, note 1.)

[34] Z teorii uporządkowania wielokrotnego <Sur la théorie de l'ordre multiple> (Polish with French summary). ***Wiadomości Matematyczne***, vol. 37 (1934), pp. 1–35. [JFM 60.0867.01 (S. Ruziewicz); Zbl 0009.30304 (Th. Motzkin)].

[34a] Sur les superpositions des fonctions représentables analytiquement. ***Fundamenta Mathematicae***, vol. 23 (1934), pp. 15–37; errata, *op. cit.*, p. 304. [JFM 60.0195.02 & JFM 60.0195.03 (A. Rosenthal)]; [Zbl 0009.30502 & Zbl 0010.01403 (S. Saks)].

[35] Miara w geometrji (with E. Szpilrajn) <Measure in geometry. Polish>. ***Świat i życie. Zarys encyklopedyczny współczesnej wiedzy i kultury***, vol. 3, Lwów–Warszawa 1935, pp. 586–595.

[36] Sur la simplicité formelle des notions. ***Actes du Congrès International de Philosophie Scientifique, Sorbonne, Paris 1935, VII Logique***, **Actualités scientifiques et industrielles**, vol. 394 (1936), Hermann & Cie: Paris 1936, pp. 29–38. [JFM 62.1050.01 (W. Ackermann); JSL 2 (1937), pp. 55–56 (S. C. Kleene)].

[36a] Über die Beschränktheit der Ausdrucksmittel deduktiver Theorien (with A. Tarski). ***Ergebnisse eines mathematischen Kolloquiums***, vol. 7 (1936), pp. 15–22. [JFM 62.0039.02 (F. Bachmann); Zbl 0014.38602 (A. Schmidt); JSL 1, p. 115–116 (B. Rosser)].

(1) On the limitations of the means of expression of deductive theories. In: A. Tarski, ***Logic, Semantics, Metamathematics. Papers from 1923–1938***. Translated by J.H. Woodger, Clarendon Press: Oxford 1956, pp. 384–392. (Revised English translation of Lindenbaum [36a].)
(2) Sur les limitations des moyens d'expression des théories déductives. In: A. Tarski, ***Logique, sémantique, métamathématique. 1923–1944***, vol. 2, **Philosophie pour l'Âge de la Science**, Librairie Armand Collin: Paris 1974, pp. 109–120. (Revised French translation of Lindenbaum [36a].)
(3) Reprinted in ***Alfred Tarski, Collected Papers. Volume 2, 1935–1944***, ed. by S.R. Givant and R. N. McKenzie, Birkhäuser Verlag: Basel 1986, pp. 205–212.
(4) O ograniczeniach środków wyrazu teorii dedukcyjnych. In: A. Tarski, ***Pisma logiczno-filozoficzne. Volume 2***, translated and annotated by Jan Zygmunt, Wydawnictwo Naukowe PWN: Warsaw 2001, pp. 147–157. (Polish translation of [36a](1).)

[38] Über die Unabhängigkeit des Auswahlaxioms und einiger seiner Folgerungen (with A. Mostowski). ***Sprawozdania z posiedzeń Towarzystwa Naukowego Warszawskiego, Wydział III Nauk Matematyczno-Fizycznych*** (=***Comptes rendus des séances de la Société des Sciences et des Lettres de Varsovie, Classe III***), vol. 31 (1938), pp. 27–32. [JFM 64.0932.01 (Th. Skolem); Zbl 0019.29502, p. 295 (A. Schmidt); JSL 4, pp. 30–31 (A. A. Fraenkel)].
(1) On the independence of the axiom of choice and some of its consequences. In: A. Mostowski, ***Foundational Studies. Selected Works***, vol. 2, edited by K. Kuratowski, W. Marek, L. Pacholski, H. Rasiowa, C. Ryll-Nardzewski, and P. Zbierski. **Studies in Logic and the Foundations of Mathematics**, vol. 93, North-Holland Publishing Company: Amsterdam 1979; and PWN—Polish Scientific Publishers: Warszawa 1979, pp. 70–74. (English translation of Lindenbaum [38] by M.J. Mączyński.) [Zbl 0425.01021 (E. Mendelson)].

6.1.2 Abstracts and Short Notes

[26[a]] Sur l'arithmétique des types ordinaux. ***Rocznik Polskiego Towarzystwa Matematycznego*** (=***Annales de la Société Polonaise de Mathématique***), vol. 5 (année 1926, publ. 1927), pp. 103–104. (Summary of §3 of [26a], presented at the meeting of the Polish Mathematical Society, Warsaw Section, on 23 April 1926.)

[26[a]a] Sur l'indépendence des notions primitives dans les systèmes mathématiques (with A. Tarski). ***Rocznik Polskiego Towarzystwa Matematycznego*** (=***Annales de la Société Polonaise de Mathématique***), vol. 5 (année 1926, publ. 1927), pp. 111–113. (Presented at the meeting of the Polish Mathematical Society, Warsaw Section, on 17 December 1926.)
(1) Reprinted in ***Alfred Tarski, Collected Papers. Volume 4, 1958–1979***, ed. by S.R. Givant and R.N. McKenzie, Birkhäuser Verlag, Basel 1986, pp. 538–540.

[27[a]] Sur quelques propriétés des fonctions de variable réelle. ***Rocznik Polskiego Towarzystwa Matematycznego*** (=***Annales de la Société Polonaise de Mathématique***), vol. 6 (année 1927, publ. 1928), pp. 129–130. [JFM 54.0298.01] (Presented at the meeting of the Polish Mathematical Society, Warsaw Section, on 24 October 1926.)

[29[a]] Méthodes mathématiques dans les recherches sur le système de la théorie de déduction. ***Księga Pamiątkowa Pierwszego Polskiego Zjazdu Matematycznego we Lwowie w 1927 roku*** (supplement to ***Annales de la Société Polonaise de Mathématique***), Kraków 1929, p. 36.

[29[a]a] O pewnych własnościach metrycznych mnogości punktowych.—Sur certaines propriétés métriques des ensembles de points. ***Księga Pamiątkowa Pierwszego Polskiego Zjazdu Matematycznego, Lwów, 7–10. IX. 1927*** (supplement to ***Annales de la Société Polonaise de Mathématique***), Kraków 1929, p. 96.

[31[a]] Sur un ensemble linéaire extrêmement non homogène par rapport aux transformations continues et sur le nombre des invariants de ces transformations. ***Rocznik Polskiego Towarzystwa Matematycznego*** (=***Annales de la Société Polonaise de Mathématique***), vol. 10 (année 1931, publ. 1932), pp. 113–114. [JFM 58.0648.12] (Presented at the meeting of the Polish Mathematical Society, Warsaw Section, on 16 January 1931.)

[31[a]a] La projection comme transformation continue la plus générale. ***Rocznik Polskiego Towarzystwa Matematycznego*** (=***Annales de la Société Polonaise de Mathématique***), vol. 10 (année 1931, publ. 1932), pp. 116–117. [JFM 58.0648.13] (Presented at the meeting of the Polish Mathematical Society, Warsaw Section, on 8 May 1931.)

[31[a]b] Sur les figures convexes. ***Rocznik Polskiego Towarzystwa Matematycznego*** (=***Annales de la Société Polonaise de Mathématique***), vol. 10 (année 1931, publ. 1932), pp. 117–118. [JFM 58. 0805.05] (Presented at the meeting of the Polish Mathematical Society, Warsaw Section, on 8 May 1931.)

[31[a]c] Sur les constructions non-effectives dans l'arithmétique élémentaire. ***Rocznik Polskiego Towarzystwa Matematycznego*** (=***Annales de la Société Polonaise de Mathématique***), vol. 10 (année 1931, publ. 1932), pp. 118–119. [JFM 58. 1001.14] (Presented at the meeting of the Polish Mathematical Society, Warsaw Section, on 8 May 1931.)

[34[a]] Sur le "problème fondamental" du jeu d'échecs. ***Rocznik Polskiego Towarzystwa Matematycznego*** (=***Annales de la Société Polonaise de Mathématique***), vol. 13 (année 1934, publ. 1935), pp. 124–125. (Presented at the meeting of the Polish Mathematical Society, Warsaw Section, on 24 February 1933.)

[34[a]a] Sur le nombre des invariants des familles de transformations arbitraires. ***Rocznik Polskiego Towarzystwa Matematycznego*** (=***Annales de la Société Polonaise de Mathématique***), vol. 13 (année 1934, publ. 1935), p. 131. (Presented at the meeting of the Polish Mathematical Society, Warsaw Section, on 19 January 1934.)

[34[a]b] Remarques sur le groupe des permutations de l'ensemble des nombres entiers. ***Rocznik Polskiego Towarzystwa Matematycznego*** (=***Annales de la Société Polonaise de Mathématique***), vol. 13 (année 1934, publ. 1935), p. 131. (Presented at the meeting of the Polish Mathematical Society, Warsaw Section, on 19 January 1934.)

[34[a]c] Sur les relations contenues dans les relations ordinales. ***Rocznik Polskiego Towarzystwa Matematycznego*** (=***Annales de la Société Polonaise de Mathématique***), vol. 13 (année 1934, publ. 1935), p. 132. (Presented at the meeting of the Polish Mathematical Society, Warsaw Section, on 19 January 1934.)

[36[a]] Sur le nombre des invariants des familles de transformations arbitraires, II. ***Rocznik Polskiego Towarzystwa Matematycznego*** (=***Annales de la Société Polonaise de Mathématique***), vol. 15 (année 1936, publ. 1937), p. 185. (Presented at the meeting of the Polish Mathematical Society, Warsaw Section, on 31 January 1936.)

[37[a]] Numérotage des types logiques. ***Rocznik Polskiego Towarzystwa Matematycznego*** (=***Annales de la Société Polonaise de Mathématique***), vol. 16 (année 1937, publ. 1938), p. 191. (Summary of a lecture given by Lindenbaum on 30 September 1937 to the Third Polish Mathematical Congress in Warsaw.)

[37[a]a] Sur l'équivalence de deux figures par décomposition en nombre fini de parties respectivement congruentes. ***Rocznik Polskiego Towarzystwa Matematycznego*** (=***Annales de la Société Polonaise de Mathématique***), vol. 16 (année 1937, publ. 1938), p. 197. (Summary of a lecture given by Lindenbaum on 30 September 1937 to the Third Polish Mathematical Congress in Warsaw.)

[38[a]] Sur les bases des familles de fonctions. ***Rocznik Polskiego Towarzystwa Matematycznego*** (=***Annales de la Société Polonaise de Mathématique***), vol. 17 (année 1938, publ. 1939), pp. 124–126.

6.1.3 Reviews

[36[r]] Tsao-Chen Tang, The theorem "p $\prec$ q. = .pq = p" and Huntington's relation between Lewis's strict implication and Boolean algebra. Bull. Am. Math. Soc. 42 (1936), 743–746. ***Zentralblatt für Mathematik***, vol. 15, no 4 (1936), p. 146. [Zbl 0015.14601].

[36[r]a] Tsao-Chen Tang, A paradox of Lewis's strict implication. Bull. Am. Math. Soc. 42 (1936), 707–709. ***Zentralblatt für Mathematik***, vol. 15, no. 4 (1936), p. 146. [Zbl 0015.14602].

[36[r]b] F. B. Fitch, A system of formal logic without an analogue to the Curry *W* operator. Jour. Symb. Log. 1 (1936), 92–100. ***Zentralblatt für Mathematik***, vol. 15, no. 6 (1936), p. 241. [Zbl 0015.24101].

[36[r]c] L. Kalmár, Zurückführung des Entscheidungsproblem auf des Fall von Formeln mit einer einzigen, binären, Funktionsvariablen. Compositio Math. 4 (1936), 137–144. ***Zentralblatt für Mathematik***, vol. 15, no. 8 (1936), pp. 338–339. [Zbl 0015.33804].

[37[r]] W. Hetper, Bases de la sémantique. (Polish) Wiadom. Mat. 43 (1937), 57–86. ***Zentralblatt für Mathematik***, vol. 16, no. 5 (1937), p. 193. [Zbl 0016.19302].

[37[r]a] Z. Zawirski, Über die Anwendung der mehrwertigen Logik in der empirischen Wissenschaft. Erkenntnis 6 (1937), 430–435. ***Zentralblatt für Mathematik***, vol. 16, no. 5 (1937), p. 195. [Zbl 0016.19502].

[37[r]b] Jiro Hirano, Zum Zerlegungssatz im erweiterten einstelligen Prädikatenkalkül. Proc. Phys.-Math. Soc. Japan, III, Ser. 19 (1937), 395–412. ***Zentralblatt für Mathematik***, vol. 16, no. 8 (1937), p. 337. [Zbl 0016.33702].

[38[r]] V. F. von Seckendorff, Beweis des Induktionsschlusses der natürlichen Zahlen aus der Dedekindschen Definition endlicher Mengen. Sitzungsber. Berlin. Math. Ges. 36 (1937), 16–24. ***Zentralblatt für Mathematik***, vol. 17, no. 4 (1938), p. 156. [Zbl 0017.15602].

[38ra] Carl G. Hempel, A purely topological form of non-Aristotelian logic. Jour. Symb. Log. 2 (1937), 97–112. ***Zentralblatt für Mathematik***, vol. 17, no. 6 (1938), p. 241. [Zbl 0017.24102].

[38rb] J.-L. Destouches, Groupe d'équivalence d'une théorie déductive. C. R. Acad. Sci., Paris 205 (1937), 725–727. ***Zentralblatt für Mathematik***, vol. 17, no. 6 (1938), p. 243. [Zbl 0017.24301].

[38rc] L. Kalmár, Zur Reduktion des Entscheidungsproblems. Norsk Mat. Tidsskr. 19 (1937), 121–130. ***Zentralblatt für Mathematik***, vol. 17, no. 8 (1938), p. 337. [Zbl 0017.33702].

[38rd] F. B. Fitch, Modal functions in two-valued logic. J. Symb. Log. 2 (1937), 125–128. ***Zentralblatt für Mathematik***, vol. 17, no. 8 (1938), pp. 337–338. [Zbl 0017.33703].

[38re] L. P. Gokieli, Über den Funktionsbegriff. Trav. Inst. Math. Tbilissi 2 (1937), 1–35. ***Zentralblatt für Mathematik***, vol. 18, no. 1 (1938), p. 1. [Zbl 0018.00102].

[38rf] W. Wilkosz, Sur la notion de l'équivalence des systèmes déductifs. Annales de la Soc. Polon. Math. 15 (1937), 161–164. ***Zentralblatt für Mathematik***, vol. 18, no. 1 (1938), pp. 2–3. [Zbl 0018.00204].

[38rg] Jiro Hirano, Einige Bemerkungen zum v. Neumannschen Axiomensystem der Mengenlehre. Proc. Phys.-Math. Soc. Japan, III. Ser. 19 (1937), 1027–1045. ***Zentralblatt für Mathematik***, vol. 18, no. 1 (1938), p. 3. [Zbl 0018.00301].

[38rh] L. Couffignal, Sur un problème d'analyse mécanique abstraite: la théorie de la réduction résulte de fonctions mécaniques. C. R. Acad. Sci., Paris 206 (1938), 1336–1338. ***Zentralblatt für Mathematik***, vol. 18, no. 8 (1938), pp. 338–339. [Zbl 0018.33805].

[38ri] W. Hetper, Le rôle des schémas indépendants dans le système de la sémantique élémentaire. Arch. Towarz. Nauk. Lwów 9 (1938), 253–263. ***Zentralblatt für Mathematik***, vol. 19, no. 4 (1938), p. 145. [Zbl 0019.14501].

[38rj] W. Hetper, Relations ancestrales dans le système de la sémantique. Arch. Towarz. Nauk. Lwów 9 (1938), 265–280. ***Zentralblatt für Mathematik***, vol. 19, no. 4 (1938), p. 145. [Zbl 0019.14502].

[38rk] L. Chwistek, Remarques critiques concernant la notion de la variable dans le système de la sémantique rationnelle. Arch. Towarz. Nauk. Lwów 9 (1939), 283–333. ***Zentralblatt für Matheatik***, vol. 19, no. 4 (1938), p. 145. [Zbl 0019.14503].

[38rl] L. Couffignal, Solution générale, par des moyens mécaniques, des problèmes fondamentaux de la logique déductive. C. R. Acad. Sci., Paris 206 (1938), 1529–1531. ***Zentralblatt für Mathematik***, vol. 19, no. 4 (1938), p. 145. [Zbl 0019.14504].

[38rm] L. Couffignal, Les opérations des mathématiques pures sont toutes des fonctions mécaniques. C. R. Acad. Sci., Paris 207 (1938), 20–22. ***Zentralblatt für Mathematik***, vol. 19, no. 4 (1938), p. 146. [Zbl 0019.14601].

[39r] J. J. Burckhardt, Zur Neubegründung der Mengenlehre. Jahresber. Dtsch. Math.-Ver. 48 (1938), 146–165. ***Zentralblatt für Mathematik***, vol. 19, no. 5 (1939), p. 201. [Zbl 0019.20101].

6.2 Works by Other Authors Referred to in This Paper

L.C. Arboleda

1981 Les Recherches de M. Fréchet, P. Alexandrov, W. Sierpiński et K. Kuratowski sur la théorie des types de dimensions et les débuts de la topologie générale. ***Archive for History of Exact Sciences***, vol. 24, no. 4 (1981), pp. 339–388.

N. Aronszajn

1932 Sur les invariants des transformations continues d'ensembles. ***Fundamenta Mathematicae***, vol. 19 (1932), pp. 92–142.

S. Banach

1923 Sur le problème de la mesure. ***Fundamenta Mathematicae***, vol. 4 (1923), pp. 7–33.

1924 Un théorème sur les transformations biunivoques. ***Fundamenta Mathematicae***, vol. 6 (1924), pp. 236–239.

S. Banach and A. Tarski

1924 Sur la décomposition des ensembles de points en parties respectivement congruentes. ***Fundamenta Mathematicae***, vol. 6 (1924), pp. 244–277.
(1) On decomposition of point sets into respectively congruent parts. In: McFarland–McFarland–Smith [2014], pp.95–123. (English translation by James T. Smith.)

S. Bender

1997 ***Mul mavet orev: Yehudei Bialystock be-milhemet ha-olam hasheniyah, 1939–1943***, Am Oved Publishers Ltd: Tel Aviv 1997.
(1) ***The Jews of Białystok during World War II and the Holocaust***, Brandeis University Press: Waltham, Massachusetts 2008. (English translation by Yaffa Murciano.)

G. Birkhoff

1944 Metric foundations of geometry, I. ***Transactions of the American Mathematical Society***, vol. 55, no. 3 (May, 1944), pp. 465–492.

L. Couturat

1905 ***Les principes des mathématiques. Avec un appendice sur la philosophie des mathématiques de Kant***, F. Alcan: Paris 1905.
(1) ***Die philosophischen Prinzipien der Mathematik***, Werner Klinkhardt: Wien–Leipzig 1908. (German translation by Carl Siegel.)

J. Dadaczyński

2003 Hosiasson-Lindenbaumowa, Janina. ***Powszechna Encyklopedia Filozofii***, vol. 4, Polskie Towarzystwo Tomasza z Akwinu: Lublin 2003, pp. 603–604.

S. Eilenberg

1993 Karol Borsuk—personal reminiscences, ***Topological Methods in Nonlinear Analysis***, vol. 1, no. 1 (1993), pp. 1–2.

R. Engelking

1989 ***General Topology.*** Revised and completed edition. Heldermann: Berlin 1989.

P. Erdős

1943 Some remarks on set theory. ***Annals of Mathematics, Series 2***, vol. 44 (1943), pp. 643–646.

H. Freudenthal and W. Hurewicz

1936 Dehnungen, Verkürzungen, Isometrien. ***Fundamenta Mathematicae***, vol. 26 (1936), pp. 120–122.

F. Hausdorff

1914 ***Grundzüge der Mengenlehre***. Veit & Comp: Leipzig 1914.

M.D. Kirszbraun

1934 Über die zusammenziehende und Lipschitzsche Transformationen. ***Fundamenta Mathematicae***, vol. **22** (1934), pp. 77–108.

J. Łoś

1949 ***O matrycach logicznych*** <On logical matrices. Polish>. **Prace Wrocławskiego Towarzystwa Naukowego**, Seria B, vol. 42, Wrocław 1949.

A. McFarland, J. McFarland, James T. Smith (eds.)

2014 ***Alfred Tarski***: ***Early Work in Poland—Geometry and Teaching. With a Bibliographic Supplement***. Birkhäuser: New York 2014.

E. Marczewski and A. Mostowski

1971 Lindenbaum Adolf (1904–1941). ***Polski Słownik Biograficzny*** <Polish Biographical Dictionary>, vol.17, Kraków 1971, pp. 364b–365b.

A. Marianowicz

1995 ***Życie surowo wzbronione***. Czytelnik: Warszawa 1995.
(1) ***Life Strictly Forbidden***. Valentine Mitchell: London 2004. (English translation by Alicja Nitecki.)

S. Mazurkiewicz and W. Sierpiński

1914 Sur un ensemble superposable avec chacune de ses deux parties. ***Comptes Rendus hebdomadaires des séances de l'Académie des Sciences***, Paris, vol. 158 (1914), pp. 618–619. (Stefan Mazurkiewicz uses the name Etienne Mazurkiewicz in French.)

H. Moese

1932 Przyczynek do problemu A. Tarskiego: "O stopniu równoważności wielokątów". ***Parametr***, vol. 2 (1931–1932), pp. 305–309.
(1) A Contribution to the problem of A. Tarski: "On the degree of equivalence of polygons", in: McFarland–McFarland–Smith [2014], pp. 145–151. (English translation by A. McFarland, J. McFarland and James T. Smith.)

S. Ruziewicz

1921 Sur un ensemble non dénombrable de points, superposable avec les moitiés de sa partie aliquote. ***Fundamenta Mathematicae***, vol. 2 (1921), pp. 4–7.

S. Ruziewicz and W. Sierpiński

1932 Sur un ensemble parfait qui a avec toute sa translation au plus un point commun. ***Fundamenta Mathematicae***, vol. 19 (1932), pp. 17–21.

W. Sierpiński

1918 L'axiome de M. Zermelo et son rôle dans la théorie des ensembles et l'analyse. ***Bulletin International de l'Académie des Sciences de Cracovie. Classe des Sciences Mathématiques et Naturelles. Serie A, Sciences Mathématiques, Comptes-Rendus des Séances de l'année 1918***, pp. 97–152.

1933 Sur les espaces métriques localement séparables. ***Fundamenta Mathematicae***, vol. 21 (1933), pp. 107–113.

1936 Sur une propriété du segment. ***Prace matematyczno-fizyczne***, vol. 43 (1936), pp. 25–32.

1947 Sur un théorème de A. Lindenbaum. ***Annals of Mathematics, Series 2***, vol. 8 (1947), pp. 641–642.

1947a Sur la congruence des ensembles de points et ses généralisations. ***Commentarii Mathematici Helvetici***, vol. 19 (1947), pp. 215–226.

1947b Sur un ensemble plan qui se décompose en $2^{\aleph_0}$ ensembles disjoints superposables avec lui. ***Fundamenta Mathematicae***, vol. 34 (1947), pp. 9–13.

1954 ***On the Congruence of Sets and their Equivalence by Finite Decomposition.*** **Lucknow University Studies**, no. XX. The Lucknow University: Lucknow 1954.

1958 ***Cardinal and Ordinal Numbers***. **Monografie Matematyczne**, vol. 34, PWN–Polish Scientific Publishers, Warszawa 1958.
(1) Second edition revised, 1965.

B. Sobociński

1956 On well constructed axiom systems. ***Rocznik VI Polskiego Towarzystwa Naukowego na Obczyźnie*** <Yearbook VI of the Polish Society of Arts and Sciences Abroad>, London 1955–56, pp. 54–65.

A. Tarski

1924 O równoważności wielokątów <On the equivalence of polygons. Polish with French summary>. ***Przegląd matematyczno-fizyczny***, vol. 2 (1924), pp. 47–60.
(1) On the equivalence of polygons, in: McFarland–McFarland–Smith [2014], pp. 79–91. (English translation by A. McFarland, J. McFarland and James T. Smith.)

1928 Remarques sur les notions fondamentales de la méthodologie des mathématiques. ***Rocznik Polskiego Towarzystwa Matematycznego*** (=***Annales de la Société Polonaise de Mathématique***), vol. 7 (année 1928, publ. 1929), pp. 270–272.
(1) Remarks on fundamental concepts of the methodology of mathematics. In J.-Y. Béziau (ed.), ***Universal Logic***: ***An Anthology***, Birkhauser 2012, pp. 67–68. (English translation by R. Purdy and J. Zygmunt.)

1930 Über Äquivalenz der Mengen in Bezug auf eine beliebige Klasse von Abbildungen. ***Atti del Congersso internationale di Matematici***, ***Bologna***, ***3–10 settembre 1928***, vol. 6, Nicola Zanichelli: Bologna 1930, pp.243–252.

1930a Über einige fundamentalen Begriffe der Metamathematik. ***Sprawozdania z posiedzeń Towarzystwa Naukowego Warszawskiego***, ***Wydział III Nauk Matematycznych i Przyrodniczych*** (=***Comptes rendus des séances de la Societé des Sciences et des Lettres de Varsovie***, ***Classe III***), vol. 23 (1930), pp. 22–29.
(1) On some fundamental concepts of metamathematics. In: Tarski [1956], pp. 30–37. (Revised English translation of Tarski [1930].)

1930b Fundamentale Begriffe der Methodologie der deduktiven Wissenschaften. I. ***Monatshefte für Mathematik und Physik***, vol. 37 (1930), pp. 361–404.
(1) Fundamental concepts of the methodology of the deductive sciences. In: Tarski [1956], pp. 60–109. (Revised English translation of Tarski [1930a].)

1931 O stopniu równoważności wielokątów <On the degree of equivalence of polygons. Polish>. ***Młody Matematyk***, vol. 1 (supplement to ***Parametr***, vol. 2) (1931), pp. 37–44. [Exercise 6 on p. 44 is due to A. L. = most probably Adolf Lindenbaum.]
(1) Reprinted, and English translation by I. Wirszup, in: ***Alfred Tarski***, ***Collected Papers***. ***Volume 1***, ***1921–1934***, ed. by S.R. Givant and R.N. McKenzie, Birkhäuser Verlag: Basel 1986, pp. 563–570, and pp. 573–580.
(2) On the degree of equivalence of polygons. In: McFarland–McFarland–Smith [2014], pp. 135–144. (English translation by A. McFarland, J. McFarland and James T. Smith.)

1931/32 Uwagi o stopniu równoważności wielokątów <Remarks on the degree of equivalence of polygons. Polish>. ***Parametr***, vol. 2 (1931–32), pp. 310–314.
(1) Reprinted, and English translation titled “Further remarks about the degree of equivalence of polygons” by I. Wirszup, in: ***Alfred Tarski***, ***Collected Papers***. ***Volume 1***, ***1921–1934***, ed. by S.R. Givant and R.N. McKenzie, Birkhäuser Verlag: Basel 1986, pp. 597–601, and pp. 605–611.

(2) Remarks on the degree of equivalence of polygons. In: McFarland–McFarland–Smith [2014], pp. 152–158. (English translation by A. McFarland, J. McFarland and James T. Smith.)

1936 Grundzüge des Systemenkalküls. Zweiter Teil. ***Fundamenta Mathematicae***, vol. 26 (1936), pp. 283–301.
(1) Foundations of the calculus of systems. In: Tarski [1956], pp. 342–383. (Revised English translation of Tarski [1936].)

1956 ***Logic, Semantics, Metamathematics. Papers from 1923 to 1938***. (Translated by J.H. Woodger.) Clarendon Press: Oxford 1956.
(1) Second, revised edition, with editor's introduction and an analytic index, J. Corcoran (ed.), Hackett Publishing Company: Indianapolis, Indiana 1983.

M. Tomkiewicz

2008 ***Zbrodnia w Ponarach 1941–1944***. **Monografie Komisji Ścigania Zbrodni Przeciwko Narodowi Polskiemu**, vol. 43. Instytut Pamięci Narodowej: Warszawa 2008.

S. Wagon

1993 ***The Banach–Tarski Paradox***. Paperback edition. Cambridge University Press: Cambridge 1993.

W. Więsław

2004 Listy Wacława Sierpińskiego do Stanisława Ruziewicza <Letters from Wacław Sierpiński to Stanisław Ruziewicz. Polish>. ***Wiadomości Matematyczne***, vol. 40 (2004), pp. 139–167.

B. Wolniewicz

2015 Nota do biogramu Lindenbauma <A note on the life of Lindenbaum. Polish>. ***Edukacja Filozoficzna***, vol. 60 (2015), pp. 163–164.

J. Zygmunt and R. Purdy

2014 Adolf Lindenbaum: Notes on his life, with bibliography and selected references. ***Logica Universalis***, vol. 8, no. 3–4 (2014), pp. 285–320.

7 Acknowledgments

7.1 Institutions, Digital Sources and Individuals

The authors are immensely grateful for library and archival services and invaluable advice and expertise provided by the following institutions, digital sources and individuals.

7.1.1 Universities

Archiwum Uniwersytetu Warszawskiego
Muzeum Uniwersytetu Warszawskiego; Jan Łukasiewicz fonds
Cambridge University Library
Bibliothek der Universität Konstanz
McMaster University Archives
University of Oxford, Bodleian Library
SUNY at Albany, M.E. Grenander Dept. of Special Collections & Archives
UC Berkeley, Bancroft Library
University of Pittsburgh, Hillman Library
Uniwersytet Wrocławski, Biblioteka Wydziału Nauk Społecznych

7.1.2 Other Archives

Archiwum Państwowe w Białymstoku
Archiwum Państwowe w Warszawie
Archiwum Akt Nowych
Archiwum Polskiej Akademii Nauk
Biblioteka Narodowa
Instytut Pamięci Narodowej
New York Public Library
The Rockefeller Archive Center
U.K. National Archives
Noord-Hollands Archief, Haarlem; *Wiener Kreis Archief* collection
Дзяржаўны архіў Гродзенскай вобласці (State Archives of Grodno Oblast)

7.1.3 Digital Sources

e-biblioteka Uniwersytetu Warszawskiego (e-bUW)
Jewish Records Indexing—Poland
Kujawsko-Pomorska Digital Library
Mazowiecka Biblioteka Cyfrowa
Narodowe Archiwum Cyfrowe
The Polish Digital Mathematics Library
Wielkopolska Biblioteka Cyfrowa
Yad Vashem Database

7.1.4 Individuals

Apart from persons mentioned in the text, particular thanks for their help in preparing this article are due to:

Bethany J. Antos—The Rockefeller Archive Center
Brigitta Arden—Hillman Library, University of Pittsburgh

Godelieve Bolten—Noord-Hollands Archief, Haarlem (*Wiener Kreis Archief*)
Danuta Bondaryk—Archiwum Państwowe w Warszawie
Frank Bowles—Cambridge University Library
Jacqueline Cox—University of Cambridge, University Archives
Jerzy Dadaczyński—Uniwersytet Papieski Jana Pawła II w Krakowie
Anna Dziedzic—Archiwum Uniwersytetu Warszawskiego
Renata Grzegórska—Archiwum Uniwersytetu Warszawskiego
Colin Harris—University of Oxford, Bodleian Library
Arie Hinkis—Cohn Institute, Tel Aviv University
Danuta Kamińska—Archiwum Państwowe w Warszawie
Brian Keough—SUNY at Albany, M.E. Grenander Dept of Special Collections & Archives
Marek Kietliński—Archiwum Państwowe w Białymstoku
Jean Lambert—CTC & Hughes Hall, University of Cambridge
Magdalena Masłowska—Archiwum Państwowe w Warszawie
Mariusz Pandura—Uniwersytet Wrocławski, Biblioteka Wydziału Nauk Społecznych
Brigitte Parakenings—Bibliothek der Universität Konstanz
Janusz Rudziński—Muzeum Uniwersytetu Warszawskiego
Wojciech Śleszyński—Uniwersytet w Białymstoku
Monika Tomkiewicz—Instytut Pamięci Narodowej. Oddział w Gdańsku
Piotr Wojtylak—Uniwersytet Opolski
Stephen Wordsworth—Council for At-Risk Academics, London South Bank University
Witold Żarnowski—Muzeum Niepodległości w Warszawie

7.2 Photograph Credits

7.2.1 Photographs of Adolf Lindenbaum (1922 and 1927), Janina Hosiasson (1919), and Stefanja Lindenbaum (1926) by kind permission of the University of Warsaw Archives.

7.2.2 Photograph of the three Pański brothers (*circa* 1920) by kind permission of Spółdzielnia Wydawnicza ‚Czytelnik" (Czytelnik Publishing House), Warsaw.

7.2.3 Photograph of the first congress of mathematicians from Slavic countries (1929) is reproduced from ***Comptes-Rendus du I Congrès des Mathématiciens des Pays Slaves, Warszawa 1929***, edited by Franciszek Leja, Książnica Atlas T.N.S.W., Warszawa 1930.

7.2.4 Photograph of the meeting of the mathematics, physics and astronomy circles (1932) by kind permission of the Museum of the University of Warsaw, Jan Łukasiewicz Fonds.

7.2.5 Photograph of the mathematics and physics faculty members of the Pedagogical Institute of Białystok (1941) by kind permission of the National Archives in Białystok.

7.2.6 Photograph of the façade of the Pedagogical Institute of Białystok (1940) by kind permission of the Byelorussian Telegraph Agency (УП „БЕЛТА").

R. Purdy
Tranby Avenue, Toronto, Ontario M5R 1N5, Canada

J. Zygmunt (✉)
University of Wrocław, Wrocław, Polska

Andrzej Mostowski: A Biographical Note

Marcin Mostowski

Abstract This is a short biography of the distinguished representative of Warsaw School of Logic and his activity in post-war Poland.

Keywords Andrzej Mostowski · Axiom of choice · Skolem arithmetic · Generalized quantifiers · Foundations of mathematics

Mathematics Subject Classification (2000) Primary 01A70, 01A60

1 Life

Andrzej Stanisław Mostowski was born in Lwów,[1] on the first November 1913. He died on the 22nd of August 1975 in Vancouver, British Columbia, Canada. His parents were Zofia Kramstyk and Stanisław Mostowski. In 1920 his family moved to Warsaw.

In 1938 Mostowski defended his PhD thesis *O niezależności definicji skończoności w systemie logiki* (On the Independence of Finiteness Definitions in a System of Logic) at Warsaw University under supervision of Kazimierz Kuratowski and Alfred Tarski.

Just after the war he obtained habilitation degree at Jagiellonian University in Kraków for the thesis *Axiom of choice for finite sets*. After graduation he got an university position in Łódź.[2] He moved to Warsaw in 1947, and then he created one of the most influential schools of mathematical logic in the world at the Warsaw University.

2 Main Works

Selected most important research papers:

The author "Marcin Mostowski" is deceased.

[1] In this time it was Austro–Hungarian part of Poland, then Poland, and nowadays Ukraine.

[2] In this time Warsaw was completely destroyed and many institutions were temporarily moved to Łódź.

Á. Garrido, U. Wybraniec-Skardowska (eds.), *The Lvov-Warsaw School. Past and Present*, Studies in Universal Logic,
https://doi.org/10.1007/978-3-319-65430-0_37

- *Über die Unabhängigkeit des Wohlordnungssatzes vom Ordnungsprinzip*, [2];
- *On direct products of theories*, [5];
- *On a generalization of quantifiers*, [7].

The most important research and survey books:

- *Sentences Undecidable in Formalized Arithmetic: An Exposition of the Theory of Kurt Godel*, [6],
- *Thirty years of foundational studies. Lectures on the development of mathematical logic and the study of the foundations of mathematics in 1930–1964*, [8]
- *Constructible sets with applications*, [9].

All of the mentioned works and many others are reprinted in [10] and [11]. He also published two important monographs *Logika Matematyczna* [3] and jointly with Kazimierz Kuratowski *Teoria Mnogości* [1]. They were very influential and useful as university handbooks.

3 Main Scientific Achievements

The best known theorem proved by Andrzej Mostowski is the following:

Theorem 1 (Mostowski Collapse Lemma, [4]) *Let us assume that a binary relation* $R \subseteq A^2$ *is well founded* ($\forall B \subseteq A(B \neq \emptyset \Rightarrow \exists x \in B(x \text{ is } R\text{-minimal in } B))$) *and extensional* ($\forall x, y \in A(x = y \equiv \forall z(zRx \equiv zRy))$)*, then there is unique transitive set* Z ($\forall x \in Z\; x \subseteq Z$) *and a bijection* $\pi : A \longrightarrow Z$ *such that*

$$\forall x, y \in A(xRy \equiv \pi(x) \in \pi(y)).$$

Other broadly known ideas of Andrzej Mostowski are:

- *Kleene–Mostowski hierarchy*, classification of arithmetical sets via quantifier prefixes required for defining them;
- the notion of *generalized quantifier*, formulated by him in [7].

However, some of less known results of Andrzej Mostowski are also very important and influential. Let us mention two of them.

One of a few hard important[3] results obtained before the second world war was the following:

Theorem 2 ([2]) *In Zermelo–Fraenkel set theory with atoms (ZFA) Zermelo theorem, saying that every set can be well ordered, is independent of Ordering Principle, saying that every set can be linearly ordered.*

The proof of this theorem was based on the method proposed by Abraham Fraenkel. Later on Andrzej Mostowski elaborated the method (see [9]). The basic construction is nowadays called *Fraenkel–Mostowski permutation models*.

[3]Let us observe that Andrzej Mostowski was 26 years old, when the result was published.

Another important work was the method of obtaining decidability results for theories of the product of two models having decidable theories, see [5]. As a corollary it gives elimination of quantifiers for arithmetic of multiplication of natural numbers. This is the first published proof of the theorem that first order arithmetic of natural numbers is decidable. The theorem is commonly attributed to Thoralf Skolem.

4 Influence of His Works

In mathematical logic ideas of Andrzej Mostowski influenced mainly foundations of set theory. He was also one of the important builders of model theory. Some of his ideas are also present in the recursion theory. However, he was thinking rather in terms of definability than computations.

In logic, in larger sense, Mostowski's most influential idea was the concept of generalized quantifiers. It is studied now not only by mathematicians, but also by philosophers, linguists, and computer scientists.

5 Students

He promoted the following PhD students: Zofia Adamowicz, Krzysztof Apt, Maciej Bryński, Andrzej Ehrenfeucht, Andrzej Grzegorczyk, Wojciech Guzicki, Andrzej Janiczak, Stanisław Krajewski, Michał Krynicki, Moshe Machover, Mihály Makkai, Wiktor Marek, Janusz Onyszkiewicz, Helena Rasiowa, Roman Sikorski, Kazimierz Wiśniewski, Andrzej Zarach, Pawel Zbierski.

Probably, the most influential in logic were Andrzej Ehrenfeucht, Andrzej Grzegorczyk, and Helena Rasiowa.

6 Last Words

Andrzej Mostowski founded one of the greatest schools of foundations of mathematics. How great and important he was becomes clear if on considers what happened later with his group at Warsaw University. It completely disappeared.

Acknowledgement This work was supported by the Polish National Science Centre [2013/11/B/HS1/04168].

References

1. Kuratowski, K., Mostowski, A.: Teoria mnogości: wraz ze wstępem do opisowej teorii mnogości. Monografie Matematyczne, 27. Państwowe Wydawnictwo Naukowe (PWN), Warszawa (1952)
2. Mostowski, A.: Über die Unabhängigkeit des Wohlordnungssatzes vom Ordnungsprinzip. Fundam. Math. **32**(1), 12–36 (1939)
3. Mostowski, A.: Logika Matematyczna. Monografie Matematyczne, 18. Państwowe Wydawnictwo Naukowe (PWN), Warszawa (1948)
4. Mostowski, A.: An undecidable arithmetical statement. Fundam. Math. **36**(1), 143–164 (1949)
5. Mostowski, A.: On direct products of theories. J. Symb. Log. **17**(1), 1–31 (1952)
6. Mostowski, A.: Sentences Undecidable in Formalized Arithmetic: An Exposition of the Theory of Kurt Godel. North-Holland, Amsterdam (1952)
7. Mostowski, A.: On a generalization of quantifiers. Fundam. Math. **44**(1), 12–36 (1957)
8. Mostowski, A.: Thirty years of foundational studies. Lectures on the development of mathematical logic and the study of the foundations of mathematics in 1930–1964. Acta Philosophica Fennica, Fasc. XVII. Barnes and Noble, Inc., New York (1966)
9. Mostowski, A.: Constructible sets with applications. Studies in Logic and the Foundations of Mathematics. North-Holland Publishing Co., Amsterdam; PWN-Polish Scientific Publishers, Warszawa (1969)
10. Mostowski, A.: Foundational Studies, Vol. I. PWN, North-Holland (1979)
11. Mostowski, A.: Foundational Studies, Vol. II. PWN, North-Holland (1979)

Foundations and Philosophy of Mathematics in Warsaw, the School of Andrzej Mostowski and Philosophy

Marcin Mostowski

Abstract The paper is a mathematical and philosophical essay devoted to mathematical logic school created and guided by Andrzej Mostowski. Firstly, we discuss some of the main—still actual—achievements of Andrzej Mostowski, then we discuss weak sides of his scientific project. They are computational and philosophical.

Scientific challenges of our times in logic are mainly computational. The Warsaw school of mathematical logic did not support this direction.

Partially because of the political situation in Poland public philosophical discussions were strongly influenced by hard politics, including personal politics in academic institutions. Therefore, many people think that isolation of the foundations of mathematics and philosophy was forced by the communist ideology. This impression is false. The abyss between philosophy and the foundations was basically independent of the political situation in Poland.

What can we learn from this experience?

Keywords Andrzej Mostowski · Axiom of Choice · Skolem arithmetic · Generalized quantifiers · Foundations of mathematics · Philosophy of mathematics

Mathematics Subject Classification (2000) Primary 00A30, 03A05, 00A35; Secondary 03E25, 03C80, 03F30

1 Introduction

This work can be divided into two parts. The first one is devoted to three selected topics of Andrzej Mostowski's work in logic. Undoubtedly, I could not cover all his important and influential works. The selection is partially personal and partially based on my opinion what was characteristic to his scientific interests.

The author "Marcin Mostowski" is deceased.

Á. Garrido, U. Wybraniec-Skardowska (eds.), *The Lvov-Warsaw School. Past and Present*, Studies in Universal Logic,
https://doi.org/10.1007/978-3-319-65430-0_38

The second part is a philosophical and historical essay devoted to relations of his school to philosophy and a new way of computational thinking. In comparison to the first part it is much more personal and informal.

2 Some Mathematical Works in the Foundations

We discuss here some selected ideas of Andrzej Mostowski.[1]

2.1 *Permutation Models*

One of the most fascinating idea in foundation of mathematics of twentieth century was undoubtedly the Axiom of Choice (AC).[2] AC says that for every family F of nonempty sets there is a choice function $f : F \longrightarrow \bigcup F$ such that $\forall A \in F\ f(A) \in A$. Zermelo invented it trying to prove his theorem, TZ: Every set can be well ordered. He proved TZ assuming AC. However, having TZ it is easy to prove AC. For any family F of nonempty sets, if we take any well ordering R on $\bigcup F$ then we can define a choice function taking $f(A) =$ the R-smallest element of A, for all $A \in F$.

The first question related to AC was whether it can be proved in Zermelo–Fraenkel set theory (ZF). A partial answer was known relatively early. However it was done only for the theory ZF extended by allowing elements not being sets, so called atoms, it is called Zermelo–Fraenkel set theory with atoms (ZFA).[3] Models of ZFA are of the form $V(A) = \bigcup_\alpha V_\alpha(A)$, where $V_0(A) = A$ and for all ordinal numbers α we take $V_{\alpha+1}(A) = P(V_\alpha(A)) \cup V_\alpha(A)$, for a limit ordinal λ we take $V_\lambda(A) = \bigcup_{\alpha<\lambda} V_\alpha(A)$. We assume that the set A is infinite.

Let us consider any permutation σ of the set A, that is a bijection $\sigma : A \longrightarrow A$. It can be extended on all sets from the model by taking $B^\sigma = \{x^\sigma : x \in B\}$. We say that B is stable for σ if $B^\sigma = B$. For any set $Z \subseteq A$ we say that a permutation σ fixes Z if $\sigma(a) = a$, for every $a \in Z$.

Now we define $S(A)$ a subuniverse of $V(A)$ taking all sets B from $V(A)$ such that there is a finite $Z \subseteq A$ such that for every permutation σ of the set A which fixes Z, B is stable for σ. Of course all pure sets, having no atoms in their transitive closures, are stable for any permutation of A, then all of them belong to $S(A)$. Additionally all finite and co-finite subsets of A belong to $S(A)$. On the other hand no infinite and co-infinite $B \subseteq A$ belong to $S(A)$. Because for any finite $Z \subseteq A$ we can find $a, b \in A - Z$ such that $a \in B$ and $b \notin B$, then we take a transposition σ exchanging a and b and not moving anything else. σ fixes Z and $B^\sigma \neq B$.

The next step is proving that $S(A)$ is a model for ZFA provided $V(A)$ is a model for ZFA. The class $S(A)$ would not contain any well ordering of A. Otherwise, having a well

[1] Some of the other topics are discussed in part 4 of this book.

[2] A very good monography of the topic can be found in [6].

[3] A good introduction to ZFA can be found in [5].

ordering R on A we can split A into two disjoint sets B containing even successors in the sense R and the remaining part containing odd successors.

In this way we gave a sketchy proof of the following:

Theorem 1 *If ZFA is consistent then neither AC nor TZ are provable in ZFA.*

By refinement of the method of Fraenkel, Andrzej Mostowski proved the following:

Theorem 2 ([9]) *In Zermelo–Fraenkel set theory with atoms (ZFA) Zermelo theorem, saying that every set can be well ordered, is independent of Ordering Principle, saying that every set can be linearly ordered.*

Later on he elaborated the method. His book [11] presents the results. Nowadays the basic construction is called *Fraenkel–Mostowski permutation models.*

These prewar ideas strongly influenced later research in the school of Andrzej Mostowski for many years.

2.2 Skolem Arithmetic and Direct Products of Theories

In 1952 Andrzej Mostowski published the paper "On direct products of theories" reprinted in [12]. It contains the first published proof of the theorem commonly attributed to Thoralf Skolem, saying that first order arithmetic of multiplication is complete and decidable.

Neither the idea nor the details of the proof are well known. In 1929 Mojżesz Presburger published his paper presenting his proof of decidability of first order arithmetic of addition [14]. The proof can be found in almost every handbook of mathematical logic.[4] It is one of the paradigmatic proofs by elimination of quantifiers.

The structure $(P^{<\omega}(N), \cup, -, \emptyset)$ of finite sets of natural numbers with union and difference has much more simpler first order theory $T_I = Th(P^{<\omega}(N), \cup, -, \emptyset)$. As a matter of fact this theory is simply the theory of nonprincipal maximal ideals in atomic infinite boolean algebras.[5]

Well known axioms of first order Peano arithmetic PA, with $0, S, +, \times$ as primitive notions, are the following:

(PA1)$\forall x\ (x = 0 \equiv \neg\exists y\ x = S(y)$,

(PA2)$\forall x, y\ (S(x) = S(y) \Rightarrow x = y)$,

(PA3)$\forall x\ x + 0 = x$,

(PA4)$\forall x, y\ x + S(y) = S(x + y)$,

(PA5)$\forall x\ x \times 0 = 0$,

(PA6)$\forall x, y\ x \times S(y) = (x \times y) + x$

and the induction axiom scheme, for each arithmetical formula $\varphi(x_1, \ldots, x_n, y)$:

(PA7$_\varphi$)$\forall x_1, \ldots, x_n\ (\varphi(x_1, \ldots, x_n, 0) \wedge\ \forall y\ (\varphi(x_1, \ldots, x_n, y) \Rightarrow \varphi(x_1, \ldots, x_n, S(y)))$
$\Rightarrow \forall y\ \varphi(x_1, \ldots, x_n, y))$.

---

[4]Professor Andrzej Grzegorczyk at least twice told me that Alfred Tarski decided that the result by Presbuger was too weak for PhD. Undoubtedly intending this terrible mistake as a lesson for future supervisors.

[5]A proper boolean algebra can be obtained by adding all the complements of the elements of an ideal.

The arithmetic of addition, called also Presburger arithmetic, T_P can be axiomatized by **(PA1)–(PA4)** and **(PA7$_\varphi$)** restricted to formulae with $0, S, +$ as only primitive notions.

The arithmetic of multiplication, called also Skolem arithmetic T_S, cannot be so easily extracted from axioms of PA. However it can be defined as the set of all first order consequences of PA which contain only multiplication and all quantifiers are of the form $\forall x \neq 0$ and $\exists x \neq 0$.[6] On the other hand, we define T_M as the first order theory of the structure $(N, \times)$. Of course $T_S \subseteq T_M$.

Theories T_I and T_P allow elimination of quantifiers. In the paper "On direct products of theories" Andrzej Mostowski proves that this two methods give elimination of quantifiers for T_M. His theorem is essentially more general. It gives elimination of quantifiers for T_S as a corollary. It was generalized in the work [4].

However we are interested mainly in properties of T_S. The method was elaborated for this case by Patrick Cegielski [1]. Cegielski gives also an axiomatic characterization of Skolem arithmetic, see also [15]. However the argument is still complicated.

The idea of the proof is based on so called Prime Factorization Theorem which says that for each integer $a > 0$ there are uniquely determined a set of prime divisors of a: $Supp(a) = \{q_1 < q_2 < \ldots < q_n\}$ and a sequence of positive integers $a_1, a_2, \ldots, a_n$ such that

$$a = q_1^{a_1} q_2^{a_2} \ldots q_n^{a_n}.$$

The result of multiplication of a by

$$b = r_1^{b_1} r_2^{b_2} \ldots r_m^{b_m}$$

is

$$c = s_1^{c_1} s_2^{c_2} \ldots s_k^{c_k},$$

where $Supp(b) = \{r_1 < r_2 < \ldots < r_m\}$, $Supp(c) = \{s_1 < s_2 < \ldots < s_k\}$, $Supp(c) = Supp(a) \cup Supp(b)$ and for $i = 1, 2, \ldots, k$ the exponent c_i is either the sum of exponents $a_{i'} + b_{i''}$, if $s_i = q_{i'} = r_{i''}$, or c_i is one of $a_{i'}$ or $b_{i''}$, if s_i divides only one of a, b.

Therefore each model $M = (U, \times^M)$ of T_M can be split into the model $(I^M, \cap, -, \emptyset)$ for T_I and the family of models $(M_p)_{p \in Primes^M}$ for T_P, where $I^M = \{Supp(a) \subseteq Primes^M : a \in U\}$ and $M_p = \{Component(p, a) : a \in U\}$, $Component(p, a)$ is the greatest power of the prime p dividing a.

Reception of the proof of Andrzej Mostowski and later refinements is very poor and in what follows we will discuss easier and less general argument for completeness and decidability of T_S. Because it is an axiomatic theory then it suffices to prove its completeness.

[6]Zero element in T_S is easy to define and inessential from the point of view of characterizing models of this theory.

Much easier proof was given by Nadel [13], who proved the completeness of Skolem arithmetic by using Ehrefeucht–Fraïssé games, [3].[7]

Theorem 3 *Any two models of T_S are elementary equivalent.*

The proof given by Nadel combines winning strategies of $\exists$-player for two models of T_I and two models of T_P for obtaining the same for two models of T_S. As a byproduct it gives that $T_M \subseteq T_S$.

This can be done simply by proving in PA proper statements justifying possibility of needed representation. We give two examples of them.

Lemma 1 (Prime Factorization Theorem: A Multiplicative Version) *In PA the following statement is provable:*

$$\forall x \forall y\ (x = y \equiv \forall p \in Primes\ \ Component(p, x) = Component(p, y)).$$

Because the statement uses only multiplicative language then it is also provable in T_S.

Lemma 2 (Selection: A Multiplicative Version) *For each arithmetical first order formula $\varphi(x_1, \ldots, x_n, y, z)$ The following statement is provable in PA:*

$$\forall x_1, \ldots, x_n\ \forall b\ (\forall p \in Supp(b) \exists a\ ((Pow(p, a) \wedge \varphi(x_1, \ldots, x_n, p, a)) \Rightarrow$$

$$\exists b'\ (Supp(b) = Supp(b') \wedge \forall p \in Supp(b')\ \varphi(x_1, \ldots, x_n, p, Component(p, b')))).$$

If additionally φ is in multiplicative language of T_S then the statement is also provable in T_S.

Proof Let us fix a formula $\varphi(x_1, \ldots, x_n, y, z)$. Then we prove the statement in PA by induction on the standard enumeration of primes $p_0, p_1, p_2, \ldots$. Firstly we take any $x_1, \ldots, x_n, b$.

We take as b_0 any element such that $Pow(p_0, b_0) \wedge \varphi(x_1, \ldots, x_n, p_0, b_0)$,[8] or $b_0 = 1$ if $\neg\ p_0 \mid b$. Let us assume that $\forall p \in Supp(b) \exists a\ ((Pow(p, a) \wedge \varphi(x_1, \ldots, x_n, p, a))$.

Let us assume that b_n is defined in such a way that $\forall i \leq n \forall p_i \in Supp(b')\ \varphi(x_1, \ldots, x_n, p, Component(p_i, b_n))$.

If $\neg\ p_{n+1} \mid b$ then we take $b_{n+1} = b_n$, otherwise—by the assumption—we have a such that $Pow(p_{n+1}, a) \wedge \varphi(x_1, \ldots, x_n, p_{n+1}, a)$. In this case we take $b_{n+1} = b_n \times a$. We stop at stage n when p_n is the greatest prime in $Supp(b)$, then we take $b' = b_n$. □

[7] The method is currently a standard one and can be found in many textbooks of mathematical logic. A good presentation of it can be found e.g. in [2].

[8] $Pow(q, d)$ means that d is a power of a prime q.

2.3 Generalized Quantifiers

Another very influential work by Andrzej Mostowski was presented in his paper "On a generalization of quantifiers"[10]. The paper does not contain any hard results, but it presents a new very influential idea of *generalized quantifiers*. Traditionally, in mathematics we have used two quantifiers: universal $\forall$ and existential $\exists$. In the Type Theory they have many interpretations depending on the types of variables they can bound. However, restricting our attention to first order—elementary language, they have unique interpretation.

2.3.1 Basic Idea

Let us consider a model M with the universe U. Interpretations of quantifiers $\forall$ and $\exists$ depend only on U. So we take

$$\forall_U = \{A \subseteq U : U - A = \emptyset\},$$

and

$$\exists_U = \{A \subseteq U : A \neq \emptyset\}.$$

Now for every formula $\varphi(x)$ with one free variable x we have:

$$M \models Qx\ \varphi(x) \text{ if and only if } \{a \in U : M \models \varphi(a)\} \in Q_U,$$

for any of quantifiers $Q = \forall, \exists$.

Are there more such quantifiers? Restricting our attention only to logical—topic independent—notions, there are much more than two. Logicality condition for Q says that for any bijection $f : U \longrightarrow W$ and for any $A \subseteq U$:

$$A \in Q_U \text{ if and only if } f(A) \in Q_W,$$

where $f(A) = \{f(a) : a \in A\}$. Quantifiers satisfying this condition are determined by classes of pairs of cardinal numbers, in the following sense $K_Q = \{(card(A), card(U - A)) : A \in Q_U\}$. For instance $K_\exists = \{(\kappa_1, \kappa_2) : \kappa_1 > 0\}$. On the other hand Q_U can be obtained back from K_Q as follows:

$$Q_U = \{A \subseteq U : (card(A), card(U - A)) \in K_Q\}.$$

The basic notion of a *generalized quantifier* is a quantifier in the above sense satisfying the logicality condition. They are called also *Mostowski quantifiers*

Andrzej Mostowski observed that logics with such quantifiers are hard to axiomatize. For instance the quantifier *there are only finitely many*—$\exists^{<\aleph_0}$, defined as

$$\exists^{<\aleph_0}{}_U = \{A \subseteq U : card(A) < \aleph_0\},$$

allows finite axiomatization of the standard model of natural numbers, what means that, by the Tarski undefinability of truth theorem, the set of theorems cannot be arithmetical, therefore it is not axiomatizable.

The axiomatization can be given by **(PA1)–(PA6)** and instead of scheme **(PA7$_\varphi$)** we take $\forall x\ \exists^{<\aleph_0} y\ \ y < x$.[9]

Therefore, the result obtained by H. Jerome Keisler [7] giving an example of axiomatizable logic was surprising. He gave a complete axiomatization for the logic with the quantifier $\exists^{>\aleph_0}$, defined as

$$\exists^{>\aleph_0}{}_U = \{A \subseteq U : card(A) > \aleph_0\}.$$

Let us consider the logic $FO(Q)$ first order logic with an additional quantifier Q interpreted as arbitrary Mostowski quantifier. It means that formulae are interpreted in models of the form (M, Q_U), where M is a usual model and U is its universe.

Theorem 4 (Per Lindström[10]) *The set of all universally valid arithmetical formulae in $FO(Q)$ is not arithmetical and therefore not axiomatizable.*

Proof Let φ be a conjunction of **(PA1)–(PA6)** and the statement

$$\psi\ =\ \forall x(Qy\ y < x \equiv \neg Qy\ y < S(x)).$$

Let us observe that ψ says that Q gives different truth values on initial segments determined by a and by $S(a)$, for all a, what is possible only when these segments are finite. Therefore for each arithmetical sentence ξ the conjunction $(\varphi \wedge \xi)$ is consistent in $FO(Q)$ if and only if ξ is true in the standard model of natural numbers.[11] □

2.3.2 Lindström's Generalization

Currently the term *generalized quantifiers* is used in the sense given by Per Lindström in [8]. For every finite sequence $\mathbf{t} = (t_1, \ldots, t_k)$ of positive integers, we define a generalized quantifier Q of type $\mathbf{t}$ as follows:

- On a syntactic level, for each formulae $\varphi_1, \ldots, \varphi_k$, we have a new formula $Q\mathbf{x}(\varphi_1(\mathbf{x}_1), \ldots, \varphi_k(\mathbf{x}_k))$, where $\mathbf{x} = x_1, \ldots, x_t$, $t = max(t_1, \ldots, t_k)$ and $\mathbf{x}_i = x_1, \ldots, x_{t_i}$, for $i = 1, \ldots, k$;
- On a semantic level, for each nonempty U the set Q_U contains k-tuples $(R_1, \ldots, R_k)$, where $R_i \subseteq U^{t_i}$, for $i = 1, \ldots, k$. Additionally we assume that *the logicalness condition* is satisfied, in the following sense: for each bijection $f : U \longrightarrow W$ and

[9]The ordering is defined by $x \leq y \equiv \exists z\ x + z = y$ and $x < y$ means $x \leq y \wedge x \neq y$.

[10]Personal communication at the joint meeting LMPhS and LC in Uppsala, 1991.

[11]The argument given here is mine, but Per Lindström gave an argument in a similar style in a conversation with Michał Krynicki and me (see footnote 10).

for each k-tuple $(R_1, \ldots, R_k)$:

$$(R_1, \ldots, R_k) \in Q_U \text{ if and only if } (f(R_1), \ldots, f(R_k)) \in Q_W.$$

For a model M with the universe U we define $M \models Q\mathbf{x}(\varphi_1(\mathbf{x}_1), \ldots, \varphi_k(\mathbf{x}_k))$ if and only if $(\varphi_1^{M,\mathbf{x}_1}, \ldots, \varphi_k^{M,\mathbf{x}_k}) \in Q_U$, where $\varphi_i^{M,\mathbf{x}_i} = \{(a_1, \ldots, a_{t_i}) \in U^{t_i} : M \models \varphi_i(a_1, \ldots, a_{t_i})\}$, for $i = 1, \ldots, k$.

Generalized quantifiers defined by Andrzej Mostowski are exactly Lindström's quantifiers of type (1).

3 Old School and Old Ideas

Here I am starting the esseistic part of the paper, changing to first person perspective (also grammatically) from, usual in science, plural of majesty style.

The great project of foundations of mathematics started as both philosophical and mathematical project. Some of the important researchers, as Bernard Bolzano and Bertrand Russell, were mainly philosophically motivated. Others, like Gottlob Frege and David Hilbert, were building on both philosophical and mathematical traditions, which were for them not clearly separated. It is particularly striking in the case of Hilbert. He has got his high position and influence in mathematics by purely mathematical works. However his research program of grounding foundations of mathematics by reduction to finitistic mathematics was formulated and justified in philosophical spirit. There were also other influential logicians motivated from both sides, let me mention: Rudolf Carnap, Willard Quine, Hilary Putnam, and Jon Barwise.

In Poland the most influential logicians of the first half of twentieth century, Jan Łukasiewicz and Alfred Tarski, started with philosophical problems. Jan Łukasiewicz started with the problem of justifying the basic logical laws and the problem of determinism. Alfred Tarski started with the problem of defining truth.

These were old ideas. New ideas disappeared in philosophy. Some people in Poland could think that the main reason was communistic dictature in years 1948–1989. Andrzej Mostowski openly claimed that any philosophical discussions should be removed from foundations of mathematics. The attitude can be justified by a political situation. Philosophical faculties in Poland were dominated by the communist party expositors.[12] All this unpleasant things did not touch mathematics. Stalin, who indirectly governed in Poland in 1945–1953, thought that mathematics and physics should be independent, everything else have to be penetrated by the communist party.

[12]I remember, I was a teenager, a comment of my father Andrzej Włodzimierz Mostowski about the book on philosophy and non-classical logics. He said: This is neither on philosophy nor on logic, this is about who should be relegated and who can keep his position.

It is worth to mention that this attitude was not accepted by Mostowski's oldest students, Andrzej Grzegorczyk and Helena Rasiowa. Grzegorczyk went to study philosophy in 1970s, and Rasiowa strongly supported joining philosophical and mathematical interests.[13] However for all later students the attitude was obvious and acceptable.

In mathematics, philosophy was replaced by bad philosophy and in philosophy philosophy was replaced by even worse philosophy.

What is important the same thing happened in many countries without communistic dictatorship. I think that the main reason is the idea of autonomy of faculties. Why should we confront relevance and importance of our results with people thinking in other way? Of course philosophy is loosing in this confrontation. However, mathematics is loosing as well.

4 The Old School and New Ideas: Computations

The other weakness of the school of Andrzej Mostowski was not absorbing the new idea of computability. Andrzej Mostowski thought of computability not in terms of algorithms, but in a sense of arithmetically definable sets. His only student really thinking in terms of computability was Andrzej Grzegorczyk. Reception of his ideas was very poor because he was far from the center. The center was Andrzej Mostowski. The department of Mathematical Logic guided by Helena Rasiowa reoriented to computer science in the late 1970s.[14]

In these times the old department of foundation of mathematics practically disappeared. The department of Helena Rasiowa was quickly developing and in majority passed to the institute of computer science preserving the only strong group in logic at mathematical faculty. It is symbolic that the main results establishing position of this group in the new faculty were results related to Büchi–Rabin automata—the topic which was introduced in Poland by Andrzej Włodzimerz Mostowski.[15]

5 What for Philosophy?

There is a common opinion in mathematics that a good achievement is just a difficult proof. Joining it with a view that axiomatically defined ZFC is a good basis for mathematics means that mathematics is a game for finding difficult proofs from ZF. In other words all mathematical questions are of the form: is φ provable from ZFC. More

---

[13] I was her student in 1982–1983.

[14] Andrzej Salwicki told that they did not know works by Andrzej Grzegorczyk. Only later on they recognized his works as relevant and important.

[15] He is my father, and younger cousin of Andrzej Mostowski. Similarly as sons of Andrzej Mostowski: Tadeusz and Jan, he lived in a shadow of Andrzej Mostowski. I was the the first person in the family who took the topics of Andrzej Mostowski.

difficult proof, better result. Of course we know that ZFC is not complete. Therefore, we allow questions of the form: is φ independent of ZFC.

I have asked a few mathematicians whether they would accept their activity as playing such a game. Nobody answered yes. Mathematicians lost their philosophical sensitivity. This is the only explanation of these contradictory views. They need serious philosophical thinking.

On the other hand, philosophy, without real interaction with current science starts to be infertile. It is going to problems from its history. So working on history of its history, and so on.

6 Who Is Your Master?

In the late 1970s, when I was a young student, Krzysztof Maurin[16] asked me "who is your master?". I was surprised, I did not know what to answer. After a while I answered "Jan Łukasiewicz". Łukasiewicz has died in 1956. After a few years I would answer "Alfred Tarski", but he died either. Never in my life I met anyone of them. After some time I realized that I could not honestly answer such a question. I would say that I learned from many people, frequently older ones, but I learned also a lot from younger people.

Later I was sometimes asked by my colleagues to give them a problem to work on. When I was young I would answered "if you do not know it then you are not ready to work in science". With time I have changed my mind. By having a good problem and a good support you get a few years in advance in your scientific carrier. This was probably what I lost in my life.

Summarizing, we need masters, more masters than one.

Acknowledgement This work was supported by the Polish National Science Centre [2013/11/B/HS1/04168].

References

1. Cegielski, P.: Théorie élémentaire de la multiplication des entiers naturels. In: Model Theory and Arithmetic, pp. 44–89. Springer, New York (1981)
2. Ebbinghaus, H.D., Flum, J., Thomas, W.: Mathematical Logic. Springer, Berlin (1994)
3. Ehrenfeucht, A.: An application of games to the completeness problem for formalized theories. Fundam. Math. **49**(2), 129–141 (1960)
4. Feferman, S., Vaught, R.L.: The first order properties of algebraic systems. Fundam. Math. **47**(1), 57–103 (1959)
5. Fraenkel, A., Bar-Hillel, Y., Levy, A.: Foundations of Set Theory, 2nd rev. edn. Studies in Logic and the Foundations of Mathematics. Elsevier, Amsterdam (1973)
6. Jech, T.J.: The Axiom of Choice. Studies in Logic and the Foundations of Mathematics. North-Holland/Elsevier, New York (1973)

[16]He was in this time a very eminent professor working in mathematical analysis and mathematical physics. He was also deeply interested in the philosophy of mathematics.

7. Keisler, H.J.: Logic with the quantifier "there exist uncountably many". Ann. Math. Log. **1**(1), 1–93 (1970)
8. Lindström, P.: First order predicate logic with generalized quantifiers. Theoria **32**(3), 186–195 (1966)
9. Mostowski, A.: Über die Unabhängigkeit des Wohlordnungssatzes vom Ordnungsprinzip. Fundam. Math. **32**(1), 12–36 (1939)
10. Mostowski, A.: On a generalization of quantifiers. Fundam. Math. **44**(1), 12–36 (1957)
11. Mostowski, A.: Constructible sets with applications. Studies in Logic and the Foundations of Mathematics. North-Holland Publishing Co., Amsterdam/PWN-Polish Scientific Publishers, Warszawa (1969)
12. Mostowski, A.: Foundational Studies, Vols. I and II. PWN, North-Holland (1979)
13. Nadel, M.E.: The completeness of Peano multiplication. Isr. J. Math. **39**(3), 225–233 (1981)
14. Presburger, M.: Über die Vollständigkeit eines gewissen Systems der Arithmetik ganzer Zahlen, in welchem die Addition als einzige Operation hervortritt. In: Comptes Rendus du I congres de Mathématiciens des Pays Slaves, pp. 92–101. Warszawa (1929)
15. Smorynski, C.: Logical Number Theory I; An Introduction. Springer, Berlin (1991)

Jerzy Słupecki (1904–1987)

Jan Woleński

Abstract This paper presents life and work in logic of Jerzy Słupecki, a distinguished Polish logician.

Keywords Propositional calculus · Many-valued logic · Aristotelian syllogistic

Mathematics Subject Classification (2000) Primary 01A60

Jerzy Słupecki (JS hereafter), a Polish mathematician and logician, was born in Harbin (Manjuria, at that time Russia, now China) on August 29, 1904. Stanisław Słupecki and Szczęsna Słupecki, née Jaszewska were his parents. The father was a military engineer in Russian army (in the rank of captain) and worked for a company constructing railways in the Far East. Most specifically, Stanisław Słupecki was employed in building the East China Railway. His salary was high and provided an income sufficient the family to live for living the family in very good material conditions. In particular, the Słupeckis had resources to organize a private education of JS and his older brother at home. In 1915, JS entered the secondary school in Tiumen (a city in the river-basin Obu in Western Siberia, fairly far away from Harbin). Due to stormy times after the October Revolution in Russia in 1917, the parents decided to bring back JS back to home in 1918. Since the living conditions considerably worsened in the entire Russia after the revolution, JS had to work in order to increase the income of the family. Moreover, he and his brother got tuberculosis, a very dangerous illness at that time.

The above circumstances strongly motivated the Słupeckis to go back to Poland. Stanisław Słupecki succeeded as the first member of the family; his wife and sons returned in 1921. JS continued his education in Warszawa. His 4 years in the secondary school in Tiumen were partially recognized in Poland in this sense that he had to pass examinations to be allowed to continue his education in Polish secondary schools. After successful examinations, JS entered the Tadeusz Rejtan State High School in Warszawa and graduated in 1926. In the same year, he began his academic studies (architecture) at the Warsaw Technical University (Politechnika Warszawska in Polish). He had to interrupt

This paper is based on a joint contribution of Jan Zygmunt and the present author (see [26]).

Á. Garrido, U. Wybraniec-Skardowska (eds.), *The Lvov-Warsaw School. Past and Present*, Studies in Universal Logic,
https://doi.org/10.1007/978-3-319-65430-0_39

studies because his health deteriorated due to relapse come back of tuberculosis. JS spend a year in a sanatorium in Otwock, near Warszawa.

After returning from Otwock to Warszawa, JS changed his interests and decided to study mathematics at the Warsaw University; in fact, mathematics was his beloved subject in the years spent in the Far East. Since his father died, JS had to interrupt studies once again and was forced to work in order to help his mother. He was appointed in the city office in Warszawa and in the Historical Museum. Although he tried tries to continue his mathematical studies, he could not do that regularly. This situation changed in 1932 and JS began to study in a normal way. Andrzej Mostowski, Stanisław Hartman and Julian Perkal, all students of mathematics, and Antoni Raabe, the studenta student of physics, were close friends of JS at that time (and later as well). They frequently met in Lourse's Cafe located at the heart of the capital of Poland on the famous Krakowskie Przedmieście Street. Due to his complex life story JS was considerably older than his fellow students; the difference in the age amounted to about 10 years. And he was an authority to his academic colleagues in many issues. The study years of JS were a difficult political time in Poland, particularly in Warszawa. Several rightist parties and organizations proclaimed and executed radically anti-Semitic policy, also at universities. In particular, they forced Jewish students to occupy special parts in the lecture rooms (so-called bench-ghetto) and claimed that Jews should be eliminated from the leadership in students' scientific clubs, including that acting at the activity in the Mathematical-Physical Faculty. JS decisively opposed these tendencies and defended his Jewish fellow students.

JS as a student became particularly interested in mathematical logic. The University of Warsaw was one of the most important centers in this field at that time. Jan Łukasiewicz and Stanisław Leśniewski acted as professors of logic and the foundations of mathematics (official denominations of their positions were different, but their scientific activities belonged to logic *sensu largo*), and Alfred Tarski, Adolf Lindenbaum, Mordechaj Wajsberg, Moses Presburger, Bolesław Sobociński and Stanisław Jaśkowski belonged to the most important students of both professors. These people formed the Warsaw School of Logic with Łukasiewicz, Leśniewski and Tarski as leaders. Łukasiewicz became the main mentor of JS and the supervisor of his master thesis concerning three-valued logic. JS graduated in 1935 and his diploma work won the Prize of the Council of the Faculty. Since the University had a limited number of positions for young scholars, JS could not pursue his academic career immediately after graduation. He was appointed as a teacher in Maria Curie-Skłodowska High School in Warszawa; he taught introduction to philosophy and mathematics. Yet he continued his scientific research and was considered as a member of the Warsaw School of Logic. In 1938 he obtained PhD on the base of dissertation on the axiomatization of many-valued system of logic (see [5]). Łukasiewicz was the supervisor and Wacław Sierpiński, a famous mathematician acted as one of referees. JS's PhD thesis contained new and original logical results and reaffirmed his reputation in the Warsaw Logical Group. At that time, Lourse's Café was not only a place for students-discussions, but also the locum of real scientific debates with participation of Łukasiewicz, Leśniewski (he also strongly influenced logical interests of JS) and Tarski. The relationships between JS and Łukasiewicz were fairly close. JS and his wife Stanisława, nee Izydorzak (they married in 1937 and had the daughter) spent the New Year's Eve 1938/1939 (the last one before the World War II) in Łukasiewicz's home in Warszawa.

JS lived in Warszawa almost the entire years 1939–1944. He participated in the clandestine education system and lectured in a private grammar school as well as at the underground Warsaw University. He taught together with Łukasiewicz, Mostowski and Sobociński; also Henryk Hiż joined this group. Moreover, JS was an active member of Polish military resistance movement (AK—Armia Krajowa, Home Army). In particular, he participated in the operation "Żegota", that is, helping Jews. In fact, thousands of Jews, mostly children, were saved by "Żegota". When the Warsaw Uprising began on August 2, 1944, JS took part in it as a soldier in Praga, the right-bank, relatively to the direction of the river Vistula, part of Warszawa. After the uprising, he was taken by Germans and transported to Germany for forced labor (he was released due to his bad health). JS was very proud of his war activities. In private talks (the present author had an opportunity to listen to his recollections) about his past, he stressed that he considered his teaching military service in the years 1939–1944 as something of the utmost importance.

After liberation of Warszawa in January 1945, JS worked for the Telecommunication School in Warszawa and the secondary school (Maria Curie-Skłodowska in which he was teaching before 1939). In 1945, Maria Curie-Skłodowska University was established and JS was appointed as Associate Professor of and the Head of the Department of Logic and the Foundations of Mathematics. This meant the real beginning of his academic career. In 1947, he obtained his habilitation on the base of the dissertation on Aristotle's syllogistic (see [7]); the procedure was completed at Jagiellonian University in Kraków. After obtaining the habilitation JS moved to Wrocław and became Assistant Professor of Logic and Methodology of Science at Wrocław University and Wrocław Polytechnic (both schools acted together; after their separation, JS remained at the former). In the years 1953–1955, he served as the Dean of the Faculty of Mathematics, Physics and Chemistry. JS was nominated as the full professor in 1962; he was the head of the department until his retirement in 1974.

JS worked not only for Wrocław University, but also for Higher Pedagogical School (a counterpart of a college in American education system; HPS hereafter) in the same city. This school was moved to Opole in 1954; JS was its rector in 1962–1966 and the professor until 1974. Another important position was held by JS in the Polish Academy of Sciences. In particular, he was associated with the Institute of Philosophy and Sociology of this academy since 1957. The Department of Applications of Logic was established as part of the Institute of Philosophy and Sociology and JS was its head in the years 1970–1974. He also served as the Editor-in-Chief of *Studia Logica* in 1964–1969. He helped in the editorial work in "Matematyka (Mathematics) and "Zeszyty Naukowe WSP w Opolu" (The Scientific Journal of the School of HPS in Opole). He was a member of Polish Mathematical Society, Polish Philosophical Society, Wrocław Scientific Society and Opole Society of the Friend of Science.

In spite of the fact that JS's position at Wrocław University was more prestigious, he considered Opole as the place of his main academic activity. It was caused by the situation of logic at Wrocław University, particularly among mathematicians. The first generation of mathematicians working at Wrocław University came from Lvov and inherited the style of doing mathematics represented by Lvov Mathematical School, in particular, by Stefan Banach and Hugo Steinhaus (the latter became professor in Wrocław). This group recognized mathematical logic as an important field, but secondary as compared with such branches of mathematics, like analysis (functional analysis, in particular), algebra, geometry or applied mathematics. On the other hand JS grew in Warszawa, in the environment in which mathematical logic and the foundations of mathematics belonged to the very heart of mathematics. This view was held by Łukasiewicz, Leśniewski and Tarski as well as accepted by Zygmunt Janiszewski and Wacław Sierpiński, the leaders of Warsaw Mathematical School. The Warsaw School's logical ideology regarded logic as an autonomous branch of mathematics and having its own scientific problems, unnecessarily connected with other mathematical issues. Moreover, for the matter concerning the meaning of the phrase "the foundations of mathematics", logicians and mathematicians working in Warszawa saw it as referring just to the foundations of mathematics as something fundamental or basic for the theoretical and practical aspects of mathematics as well as the practice of doing it in a proper manner, whereas the Lvov tradition saw the foundations of mathematics as one in the entire variety of mathematical disciplines. This view about mathematical logic and its place in mathematics was also continued by the generations of mathematicians working in Wrocław after 1945.

Consequently and somehow paradoxically, JS, influenced by his teachers, considered himself rather as a logician than a mathematician. Although he was respected in Wrocław for his achievements and personality (he was a very kind person, always ready to help other people), he had to feel alienated to some degree from the main stream of mathematical research conducted at Wrocław University. He did not have many students, at least not many became logicians in his sense of belonging to this profession. In Lublin, he trained two later distinguished scholars, namely Czesław Ryll-Nardzewski and Jerzy Łoś (both were his assistants), but both chose their own scientific way very soon and went in the direction of the foundations, not very much related to logic. Słupecki used to say that his two pupils did their own scientific "housekeeping" and although it sounded somehow sad in his mouth, he understood this situation. In Wrocław, he supervised PhD dissertations of Edward Bałuka, Ludwik Borkowski, Edmund Glibowski, Bogusław Iwanuś, Tadeusz Kubiński, Jerzy Nowak, Witold A. Pogorzelski or Juliusz Reichbach. Although most of them worked in logic and achieved fairly important results, no Wrocław School of Logic arose. Anyway, several circumstances decided that JS observed a very deep contrast between the status of logic in Warszawa (also after 1945) as compared with Wrocław.

Certainly, JS's deep intention was to repeat the Warsaw experience at least to some degree. Since HPS was a entirely new academic institution without any limit bounding the style of teaching and doing science, JS could start from scratch and try to realize his dream in cultivating logic. In the years 1954–1964, HPS had the Department of Mathematics without specified particular mathematical fields. JS succeeded in establishing the Department of the Foundations of Mathematics and began a very intensive teaching of logic. He trained, among others, Grzegorz Bryll, Katarzyna Hałkowska, Marian Maduch, Krystyna Piróg-Rzepecka Tadeusz Prucnal and Urszula Wybraniec-Skardowska. These logicians constituted the Slupecki School of Logic. Thus his dream was fulfilled. Moreover, JS's seminar conducted in Opole became very famous over the entire country and collected many logicians from other cities: Gliwice, Katowice, Kraków and Wrocław. In fact, JS strongly influenced investigations of many logicians working outside Opole, including (I omit earlier mentioned names) Paweł Bielak, Janusz Czelakowski, Jacek Kabziński, Grzegorz Malinowski, Stanisław J. Surma, Marek Tokarz, Ryszard Wójcicki, Andrzej Wroński and Jan Zygmunt. Every issue of the scientific journal of HPS had a considerable section of mathematical logic in which were published results obtained by JS, his students and participants of his seminars. JS co-authored many papers with his students. He also published textbook in mathematical logic and the foundations of mathematics.

JS died on January 15, 1987 in Wrocław. Although the day of his funeral ceremony was very cold, many people, friends, colleagues, an students, participated in it. It documented that JS was well remembered for his unusual personality. His death was symbolic for the history of Warsaw School of Logic. JS was the last logician who graduated in Warszawa before 1939 that is, in golden period (1918–1939) of the development of logic in Poland; Hiż and Lejewski, who studied with Łukasiewicz and Leśniewski, obtained their PhD degrees after the end of World War II.

The last section of this paper is devoted to a brief presentation of JS's logical research (I proceed by keeping the chronological order). JS was a very faithful inheritor of the tradition of Warsaw School of Logic. Although he considered Leśniewski as the most brilliant mind in this circle, Łukasiewicz was his beloved teacher (this statement is based on my private conversation with JS). JS continued Łukasiewicz's style of doing logic as well as investigated similar problems as his main teacher; these questions mostly concerned propositional calculi. As I already mentioned, first results of JS concerned many-valued logic, more specifically the three-valued system. He solved the problem of functional-completeness (or functional fullness) of three-valued logic (see [3, 4, 6]). The question concerns the choice of sentential functors (connectives) sufficient for defining all operations of a given calculus. For instance, negation and conjunction (disjunction, implication) allow to define all remained sentential functions in two-valued logic; the same role is played by two Sheffer connectives. On the other hand, negation and equivalence are not enough. It was known that full bases for two-valued logic are not sufficient for many-valued logic. JS introduced the T-functor such that, in the case of three-valued logic, the value $\mathrm{T}p = \frac{1}{2}$ for any value of p. Employing the T-operation. JS showed that the system $Ł_3$ (three-valued logic of Łukasiewicz) can be axiomatized by supplementing already known axioms by the formulas *CTpNTp* and *CNTpTp*. He also generalized his result on the fullness of many-valued logic by formulating the criterion (called the Słupecki criterion) universally applicable to the problem in question. Roughly speaking, this criterion points out that the functional fullness depends on the definability of all unary connectives and having a truth-table by at least one binary connective.

Aristotle's syllogistic was one of favorite topics of Łukasiewicz. JS joined this path of investigations even before 1939. One of open problems concerned decidability. Łukasiewicz himself obtained a partial result and observed that the final solution depends on introducing rejection rules next to assertion rules (likes modus ponens) usually used as inference rules in logical system. JS showed that a special form of such a (rejection) rule added to syllogistic results in decidability of this system (see [7]). According to Łukasiewicz himself, this result of JS should be recognized as the most important formal discovery concerning syllogistics after its construction by the Stagirite. The rejection function was intensively investigated by JS and his students from a more general point of view, that is, in the framework of general theory of consequence operation (see [17, 18]). In particular, Słupecki, Bryll and Wybraniec-Skardowska introduced the operation Cn^{-1}—the rejection consequence operation. We can say that a sentence A is rejected on the base of a set of X of sentences, if the falsity of X implies the falsity of A. Yet this definition is based on the classical consequence operation Cn introduced by

Tarski in 1930s. The question is how to define rejection directly, that is, not appealing to being false. Cn^{-1} realizes this task. Roughly speaking, a sentence A is rejected in the sense of Cn^{-1} on the basis of the set of sentences X if and only if some sentence B from X is deducible from A. The definition is conceived in such a way that the operation Cn^{-1} is a topological closure operator, because $Cn^{-1}\emptyset = \emptyset$ (in the case Cn as defined by Tarski, the set of logical consequences of the empty set is not empty). JS together with Pogorzelski investigated deductive systems based on non-classical logics. These studies generalized Tarski's works and showed how to defined consequence operations associated with logics other than classical two-valued system (see [12]). The same authors presented a new syntactic proof of the Gödel's completeness theorem (see [13]).

JS essentially contributed to popularization of two Leśniewski's logical systems, namely protothetics and ontology (see [8, 9]). His papers on these topics were based on lecture notes written by himself before 1939. On the other hand, he also generalized Leśniewski's mereology (see [11]). Other work of JS concerned a satisfactory intuitive interpretation of many-valued logic and modal logic based on many-valueness (see [14, 16]). Finally let me mention JS's writings on Łukasiewicz (see [10, 19]) and the entire Warsaw School of Logic (see [15, 18]). As a person who witnessed the glorious past of Polish logic, he formulated many very interesting remarks enlightening many issues in the history of contemporary logic. There is a problem with evaluating the logical work of JS. He is sometimes treated as a great loser of logic, because he worked inside an old-fashioned paradigm of logical investigations, dominant before 1939, but changed after World War II in the direction of the foundations of mathematics. This paradigm concentrated on logical calculi and was deeply associated with using the minimum mathematical methods. For example, logicians working in this style were not particularly interested in algebraization of logic. On the other hand, JS essentially developed the old paradigm and solved several fundamental problems generated by it.

References and Bibliography

A. Selected Works by Jerzy Słupecki

Books:

1. O dowodzie matematycznym [On Mathematical Proof]. PZWS, Warszawa (1962). With W.A. Pogorzelski
2. Elementy logiki matematycznej i teorii mnogości. PWN, Warszawa (1963). With L. Borkowski (Eng. Tr.: Elements of Mathematical Logic and Set Theory. Pergamon Press/PWN, Oxford (1967))

Papers:

3. Der volle dreiwertige Aussagenkalkül. Comptes Rendus de Séances de la Société de Sciences et de Lettres de Varsovie, Cl. III, **29**, 9–11 (1936) (Eng. tr. in: McCall, S. (ed.) Polish Logic 1920–1939, pp. 335–337. Oxford University Press, Oxford (1967))
4. Kryterium pełnosci wielowartościowych systemów rachunku zdań [A Criterion of Fullness of Many-Valued Systems of Propositional Logic]. Comptes Rendus de Séances de la Société de Sciences et de Lettres de Varsovie, Cl. III, **32**, 102–109 (1939) (Eng. tr.: Stud. Logica **30**, 153–157 (1972))
5. Dowód akjsomatyzowalności pełnych systemów wielowartościowych rachunków zdań [Proof of Axiomatiozability of Full Many-Valued Systems of Calculus of Propositions]. Comptes Rendus de Séances de la Société de Sciences et de Lettres de Varsovie, Cl. III (1939) (Eng. tr.: Stud. Logica **29**, 155–168 (1971))

6. Pełny trójwartościowy rachunek zdań [Full three-valued sentential calculus]. Ann. Univ. Mariae Curie-Skłodowska Sect. F, 193–208 (1946)
7. Z badań nad sylogistyką Arystotelesa. Travaux Comptes Rendus de Séances de la Société de Sciences et de Lettres de Wrocław, Wrocław (1948) (Eng. tr.: Stud. Philos. **4**, 275–300 (1949–1950))
8. St. Leśniewski's Protothetics. Stud. Logica **I**, 44–112 (1953)
9. St. Leśniewski's calculus of names. Stud. Logica **III**, 7–76 (1955) (Repr. in: Srzednicki, J. et allies (eds.) Leśniewski's Systems. Ontology and Mereology, pp. 59–122. Nijhoff, The Hague (1984))
10. The logical works of J. Łukasiewicz. Stud. Logica **8**, 7–56 (1958). With L. Borkowski
11. Toward a generalized mereology of Leśniewski. Stud. Logica **8**, 131–163 (1958)
12. Podstawowe własności systemów dedukcyjnych opartych na nieklasycznych logikach [Basic properties of deductive systems based on non-classical logics]. Part I, Stud. Logica **9**, 163–176 (1960); Part II, Stud. Logica **10**, 77–98 (1960). With W.A. Pogorzelski
13. A variant of the proof of completeness of the first-order functional calculus. Stud. Logica **12**, 125–134 (1961). With W.A. Pogorzelski
14. Próba intuicyjnej interpretacji logiki trójwartościowej Łukasiewicza [An attempt of an intuitive interpretation of Łukasiewicz's three-valued logic]. In: Rozprawy logiczne. Księga Pamiątkowa ku czci profesora Kazimierza Ajdukiewicza, pp. 185–191. PWN, Warszawa (1964)
15. Logic in Poland. In: Klibansky, R. (ed.) Contemporary Philosophy. A Survey I: Logic and Foundations of Mathematics, pp. 190–201. La Nuova Italia, Firenze (1968)
16. A generalization of modal logic. Stud. Logica **21**, 7–17 (1971)
17. A theory of rejected propositions. Part I, Stud. Logica **29**, 75–123 (1971); Part II, Stud. Logica **39**, 97–145 (1972). With G. Bryll and U. Wybraniec-Skardowska
18. Warszawska szkoła logiczna [The Warsaw School of Logic]. Wiadomości Matematyczne **15**, 65–72 (1972)
19. Jan Łukasiewicz. Wiadomości Matematyczne **13**, 73–78 (1972)

B. Selected Papers on Jerzy Słupecki

20. Bryll, G., Iwanuś, B., Piróg-Rzepecka, K.: Działalność naukowa prof. Jerzego Słupeckiego (w 76 rocznicę urodzin) [Scientific Activity of J. Słupecki (on the Occasion of His 76th Birth Anniversary)]. Zeszyty Naukowe WSP w Opolu, Matematyka **4**, 7–34 (1983)
21. Bryll, G., Hałkowska, K., Piróg-Rzepecka, K.: Działalność naukowa prof. Jerzego Słupeckiego w środowisku opolskim [Scientific Activity of Prof. Jerzy Słupecki in Opole], pp. 27–37. Zeszyty Naukowe WSP w Opolu, Matematyka 28 (1992)
22. Duda, R.: Słupecki Jerzy (1904–1987) – logik. In: Duda, R., Matematycy XIX and XX w. związani z Polską [Mathematicians of the 19th and 20th Century Associated with Poland], pp. 437–439. Wydawnictwo Uniwersytetu Wrocławskiego, Wrocław (2012)
23. Hartman, S.: Wspomnienia o Jerzym Słupeckim [Recollections on Jerzy Słupecki]. Zeszyty Naukowe WSP w Opolu, Matematyka **28**, 21–25 (1992)
24. Iwanuś, B., Piróg-Rzepecka, K.: Działalność naukowa prof. Jerzego Słupeckiego [Scientific Activity of Prof. Jerzy Słupecki]. Ruch Filozoficzny **45**(3), 227–239 (1988)
25. Woleński, J.: Jerzy Słupecki (1904–1987). Wiadomości Matematyczne **28**(2), 183–194 (1990)
26. Woleński, J., Zygmunt, J.: Jerzy Słupecki (1904–1987): life and work. Stud. Logica **XLVIII**, 401–411 (1989)
27. Zygmunt, J.: A bibliography of the published works of Jerzy Słupecki. Stud. Logica **XLVIII**, 413–421 (1989)

J. Woleński (✉)
University of Information, Technology and Management, Rzeszów, Poland
e-mail: jan.wolenski@uj.edu.pl

Rejection in Łukasiewicz's and Słupecki's Sense

Urszula Wybraniec-Skardowska

Abstract The idea of rejection originated by Aristotle. The notion of rejection was introduced into formal logic by Łukasiewicz. He applied it to complete syntactic characterization of deductive systems using an axiomatic method of rejection of propositions. The paper gives not only genesis, but also development and generalization of the notion of rejection. It also emphasizes the methodological approach to biaspectual axiomatic method of characterization of deductive systems as acceptance (asserted) systems and rejection (refutation) systems, introduced by Łukasiewicz and developed by his student Słupecki, the pioneers of the method, which becomes relevant in modern approaches to logic.

Keywords Aristotle's syllogistic · Genesis of rejection notion · Rejection in Łukasiewicz's sense · Słupecki's solution of decidability problem of Aristotle's syllogistic · Deductive systems · Axiomatic method of rejection · Refutation systems · Słupecki's rejection function · Generalization of rejection notion

Mathematics Subject Classification (2000) Primary 01A60, 03B22; Secondary 03B30

1 Introduction

The European logic as a science arose in fourth century BC in ancient Greece. Aristotle is thought to be the creator of logic. First of all, he is recognized as the creator of the first formal—logic system, deductive system, the so called Aristotle's syllogistic, which together with the theory of immediate reasoning (square of opposition, conversion, obversion, contraposition, inversion) is treated as *traditional logic* in a narrower meaning. It is the *logic of names*.

The idea of rejection of some sentences on the basis of others was originated by Aristotle, who in his systematic investigations regarding syllogistic forms not only proves the proper (true) forms but also rejects the false (invalid, erroneous) ones. Aristotle to reject some false syllogistic forms very often used examples, but he also used another way of rejecting false forms by reducing them to other ones, already rejected (see Łukasiewicz [22, 23]).

Á. Garrido, U. Wybraniec-Skardowska (eds.), *The Lvov-Warsaw School. Past and Present*, Studies in Universal Logic,
https://doi.org/10.1007/978-3-319-65430-0_40

The method of rejecting sentences always functioned in empirical sciences in connection with the procedure of refutation of hypotheses, and we can assume that it was also known to Stoics: because within the five 'unprovable', hypothetical syllogistic rules, their logic of sentences, formulated probably by Chrisippus, the rule of deduction called in Latin *modus tollendo tollens*, or *modus tollens* (If p, then q; not-q, so not-p) appears as the second one.

Although the method of rejection of sentences has always existed in empirical science, in relation to the procedure of refutation of hypothesis, the Aristotle's idea of rejection of some sentences on the basis of the others has never been properly understood by logicians or mathematicians, especially the ones convinced that rejection of closed sentences of the language of deductive system can be always replaced by introducing into such system the negation of such sentences.

The proper understanding of Aristotle's ideas and implementation of the concept of rejection into the formal researches on logical deductive systems (including on Aristotle's syllogistics) we owe to Jan Łukasiewicz—the co-founder of world famous Warsaw Logical School, which functioned in the interwar period (1918–1939). Łukasiewicz and his pupil—Jerzy Słupecki are classified as pioneers of meta-logical studies on the concept of rejection and related to it the notion of saturation (in Słupecki's terminology—*Ł-decidability*) of deductive systems.

The notion of rejection was introduced into formal logic by Łukasiewicz in his work "Logika dwuwartościowa" ("Two-valued logic") [20], in which, apart from the term "assertion" (introduced by Frege), Łukasiewicz introduces also the term "rejection". In adding "rejection" to "assertion" he, as he states himself, followed Brentano, but he did not mention any more about it. The notion of rejected proposition later played an important role in his research on Aristotle's syllogistic [22, 23], as well as in his metalogical studies of some propositional calculi [24, 25].

In that research Łukasiewicz makes use of the idea of rejection originated by Aristotle. Łukasiewicz applied it to complete a syntactical characterization of deductive systems using an axiomatic method of rejection, introduced by him into the formal logic in his paper on Aristotle's syllogistic [22], and then, after the war, in a monograph [23], which followed many years of research on Aristotelian logic, and which included the results presented in the paper prepared before the war.

As was pointed out by Łukasiewicz (see [23, p.67]),

- Aristotle, in his systematic investigations of syllogistic forms, not only proves the true ones but also shows that all the others are false, and must be rejected.

Further on (p.74), Łukasiewicz observes:

- Aristotle rejects invalid forms by exemplification through concrete terms. This procedure is logically correct, but it introduces into the systems terms and propositions not germane to it. There are, however, cases where he applies a more logical procedure, reducing one invalid form to another already rejected. On the basis of this remark, a rule of rejection could be stated corresponding to the rule of detachment by assertion; this can be regarded as the commencement of a new field of logical inquiries and of new problems that have to be solved.

- Modern formal logic, as far as I know—writes Łukasiewicz (see [23, p.71])—does not use 'rejection' as an operation opposed to Frege's 'assertion'. The rules of rejection are not yet known.

As a rule of rejection corresponding to the rule of detachment by assertion, Łukasiewicz adopts [22, 23] the following rule, which was anticipated by Aristotle:

- the rule of rejection by detachment:

  if the implication "If α, then β" is asserted, but its consequent β is rejected, then its antecedent α must be rejected, too.

As a rule of rejection corresponding to the rule of substitution for assertion Łukasiewicz adopts [22, 23] the following rule, which was unknown to Aristotle:

- the rule of rejection by substitution:

  if β is a substitution instance of α, and β is rejected, then α must be rejected, too.

Both rules enable us to reject some syllogistic forms, provided that some other forms have already been rejected.

As we mentioned above, Aristotle used the procedure of rejection of some forms by means of concrete terms, but such a procedure, though correct, introduces into logic terms and propositions that are not germane to it.

To avoid this difficulty, Łukasiewicz rejects some forms axiomatically, which leads him to biaspectual axiomatic characterization of deductive systems analyzed by him [22–25]. The idea of rejection was first used by Łukasiewicz for two-level syntactic description of Aristotle's axiomatic system of syllogistics: as a system with respect to acceptance (the first level) and as a system with respect to rejection (the second level). The sentence rejected from the system he understood as a false sentence, or a sentence which due to some reasons we cannot classify as a thesis of this system. The idea of rejection was used by Łukasiewicz while studying the decidability (saturation) of the system: each sentence, which is not the thesis of a decidable (saturated) system, is rejected.

Reconstruction of concepts of rejection and decidability (saturation) used by Łukasiewicz, was done by his pupil—Jerzy Słupecki [39, 40] (see also Słupecki et al. [45]). Słupecki modified, developed and later generalized the concept of rejected sentence, he also made its certain formalization. He also inspired systematic, formal studies on this and related concepts, and also initiated research on the decidability of many deductive systems.

In this paper[1] I am starting from Słupecki's and mine reconstruction of concepts rejection and decidability—the notions introduced and used by Łukasiewicz (Sect. 2).

Later on (Sect. 3), I am discussing the problem of decidability of Aristotle's syllogistic, set by Łukasiewicz, and I outline its solution given by Słupecki. To follow with, I am describing Słupecki's modification and generalization of the concept of rejection and also presenting the importance of Słupecki's research on syllogistics for contemporary metalogic (Sect. 4). Later on, I am showing the relation of idea of decidability in Łukasiewicz's sense with the common concept of decidability given by Słupecki, and I am discussing the decidability of more important logical systems (Sect. 5). Finally, I

[1] The paper is elaborated on the basis of my works [63–65] and Słupecki et al. [45].

am presenting Słupecki's generalization of the concept of rejection to the function of rejection, and its formalization in the theory of rejected sentences (Sect. 6). The final notes relate to presentation of different than Łukasiewicz's way of two-level formalization of deductive systems (Sect. 7).

2 Reconstruction of Concepts of Rejection and Decidability: The Notions Introduced and Used by Łukasiewicz

The notion of decidability of the deductive system that was used by Łukasiewicz in his research on Aristotle's syllogistic [22, 23] and systems of propositional calculi [24, 25] is based on the notion of a rejected sentence introduced by him.

The main idea of a syntactic biaspectual characterization of deductive systems in Łukasiewicz's sense is compatible with providing both:

- the axioms and inference rules for the given deductive system, which intuitively lead from some true formulas to true ones of this system

and

- the rejected axioms (treated as false formulas of this system) and rejection (refutation) rules of this system, which intuitively lead from some false formulas to false ones of this system.[2]

Łukasiewicz used the terms 'decidable system' and 'consistent system' in the meaning different from the one accepted in logic. Łukasiewicz does not give clear definitions of these terms, but the context points out that he used them in the following meaning:

- The system is *decidable* if every its expression which is not its thesis is rejected on the ground of finite number of axiomatically rejected expressions;
- The system is *consistent* if none of its thesis is rejected.

Łukasiewcz did not use the term 'decidable system' consequently. He also employed interchangeably the terms 'saturated system' or 'categorical system'.

In nomenclature introduced by Słupecki, decidability of a deductive system in Łukasewicz's sense was called Ł-*decidability* and its consistency was called Ł-*consistency*.

The meaning of the term 'decidable system' compatible with the understanding of the notion of a decidable system by Łukasiewicz gives the following definition:

- The deductive system determined by means of the ordered triple:

$$\langle F, A, R\rangle$$

[2]Such a syntactic formalization of some propositional calculi was also, probably independently, introduced by Rudolf Carnap [6, 7]; see Citkin [10].

where F is the set of all well-formed formulas of this system, A is the set of its axioms and R is the set of its primitive inference rules, is Ł-*decidable* if and only if there exist finite sets: the set A^{-1} of rejected axioms (included in F) and the set R^{-1} of primitive rejected rules, such that the following two conditions are satisfied:

$$\text{I. } T \cap T^{-1} = \emptyset \text{ and II. } T \cup T^{-1} = F,$$

where T is the set of all theses of the system, T^{-1} is the set of all rejected formulas (the smallest set including the set A^{-1} and closed with respect to every relation determined by rejected rules of the set R^{-1}).

The conditions I and II we call, respectively, Ł-*consistency* of the deductive system and Ł-*completeness* of the system.

Characterizing the deductive system by means of tuples

$$\langle F, A, R; A^{-1}, R^{-1}\rangle \text{ or } \langle F, T; T^{-1}\rangle,$$

the definition given above can be defined as follows:

The deductive system is Ł-*decidable* if and only if the following conditions are satisfied:

I. the set of all its theses (asserted expressions) determining the system is disjoined with the set of all its rejected formulas,
II. every propositional formula is either asserted or rejected.

3 The Problem of Decidability of Aristotle's Syllogistics, Set by Łukasiewicz, and Its Solution Given by Słupecki

3.1 Łukasiewicz's Biaspectual Formalization of Aristotelian Logic AS

On the first level **AS** is characterized as follows (cf. Łukasiewicz [21]):

The Vocabulary of ***AS***:

- constant symbols of classical logic *CL*, i.e., the connectives of *CL*,
- primitive terms of **AS**: constants a and i which are sentence forming functors of two-term arguments: 'all ... are ...' and 'some ... are ...',
- nominal variables: S, P, M, N, ...

Well-Formed Expressions of ***AS***:

- atomic affirmative expressions: formulas of the form of $S\ a\ P$ and $S\ i\ P$, which are read: 'all S are P', 'some S are P', respectively, compound expressions: formulas that are created from the atomic ones using the connectives of *CL*,
- the set F of all well-formed expressions—the smallest set of formulas including the atomic expressions and closed under the connectives of *CL*.

Defined Terms of ***AS***:

- the remaining constants of Aristotelian logic: *e* and *o*, i.e. the functors: 'no ... are ...' and 'some ... are not ...' which are defined as follows:

D1. $S\ e\ P \stackrel{\text{df}}{=} {\sim}S\ i\ P$,
D2. $S\ o\ P \stackrel{\text{df}}{=} {\sim}S\ a\ P$.

The negative expressions $S\ e\ P$ and $S\ o\ P$ are read, respectively: 'no *S* are *P*' and 'some *S* are not *P*'. Apart from the atomic expressions, we include them into the so called *simple expressions*.

Axioms of ***AS***:

$A^{+}1$. $S\ a\ S$,
$A^{+}2$. $S\ i\ S$,
$A^{+}3$. $M\ a\ P\ \wedge\ S\ a\ M \rightarrow S\ a\ P$ (*Barbara*),
$A^{+}4$. $M\ a\ P\ \wedge\ M\ i\ S \rightarrow S\ i\ P$ (*Datisi*).

Axioms $A^{+}1$ and $A^{+}2$ are the two laws of identity; Aristotle did not accept them.

Primitive Inference Rules for ***AS***:

rŁ^: *the rule of definitional replacement* (according to D1, $S\ e\ P$ may be everywhere replaced by ${\sim}S\ i\ P$, and, according to D2, $S\ o\ P$ may be everywhere replaced by ${\sim}S\ a\ P$);
$r^{+}1$: *the rule of detachment (modus ponens)* (if '$\alpha \rightarrow \beta$' and α are asserted expressions of a system, then β is an asserted expression);
$r^{+}2$: *the rule of substitution* (if α is an asserted expression of the system, then any expression produced from α by a valid substitution is also an asserted expression; the valid substitution is to put, for term-variables, other term-variables).

The schemes of the rules $r^{+}1$ and $r^{+}2$ are as follows:

$$r^{+}1:\ \frac{\vdash \alpha \rightarrow \beta \qquad \vdash \alpha}{\vdash \beta} \qquad\qquad r^{+}2:\ \frac{\vdash \alpha}{\vdash e(\alpha)}$$

The symbol '$\vdash$' is a sign of assertion introduced by Frege, whereas, the expression '$e(\alpha)$' denotes a substitution instance of α.

Characterization of the system **AS** on the second level consists in supplementing it with rejected axioms and rejection rules.

Łukasiewicz formulates the following rejected axioms and rejection rules:

Rejected Axioms of ***AS***:

$A^{-}1$. $P\ a\ M\ \wedge\ S\ a\ M \rightarrow S\ i\ P$,
$A^{-}2$. $P\ e\ M\ \wedge\ S\ e\ M \rightarrow S\ i\ P$.

Primitive Rejection Rules for ***AS***:

$r^{-}1$: the rule of rejection by detachment (reverse *modus ponens*),
$r^{-}2$: the rule of rejection by substitution.

The schemes of the rules are the following:

$$r^{-}1: \quad \frac{\vdash \alpha \to \beta \quad \dashv \beta}{\dashv \alpha} \qquad\qquad r^{-}2: \quad \frac{\dashv e(\alpha)}{\dashv \alpha}$$

The symbol '$\dashv$' is the sign of rejection.

The characterized system **AS** is determined by the following ordered 5-tuple, which may be called the *basis* of **AS**:

$$\langle F, A^{+}, R^{+}; A^{-}, R^{-} \rangle, \tag{B}$$

where F is the set of all well-formed formulas of this system; A^{+}—the set of its axioms; R^{+}—the set of its primitive inference rules; A^{-}—the set of its rejected axioms, and R^{-}—the set of its primitive rejection rules. The tuples

$$\langle F, A^{+}, R^{+} \rangle \text{ and } \langle F, A^{-}, R^{-} \rangle$$

determine, respectively, the set T^{+} of all *theses* of this system and the set T^{-} of all its *rejected formulas*.

The first tuple may be called the *assertion system* for **AS**, whereas, the second one may be called the *refutation system* for **AS**.

- T^{+} is the set of all well-formed formulas derivable from the set of theses of metalogically formulated *CL* and axioms of A^{+} by means of inference rules of R^{+}, while
- T^{-} is the set of all well-formed formulas derivable from the rejected axioms A^{-} by means of theses of T^{+} and rejection (refutation) rules of R^{-}. So

$$T^{+} = Cn^{+}(CL \cup A^{+}, R^{+}),$$

and T^{+} is the smallest set including $CL \cup A^{+}$ and closed under the inference rules of R^{+}.

$$T^{-} = Cn^{-}(T^{+} \cup A^{-}, R^{-}), \tag{D(Ł)}$$

and T^{-} is the smallest set including A^{-} and closed under the rejection rules of R^{-}. The set T^{-} is the *set of all rejected expressions of the system in Łukasiewicz's sense*.

To the set T^{+} there also belong all 24 valid syllogistic forms, the laws of logical square and the laws of conversion, and, to the set T^{-} of all rejected formulas, there belong all the remaining 232 invalid forms. However, it turned out that there exists such well-formed expression of **AS**, which is neither a thesis of this system nor a rejected expression of the set T^{-}. Such, for example, is the formula:

$$S\ i\ P \to ({\sim} S\ a\ P \wedge P\ a\ S). \tag{Fl}$$

In order to remove this difficulty, we could reject the expression (Fl) axiomatically. However, a question arises whether there exists some other formula of the same kind as (Fl), or, may be, an infinite number of such formulas, which can be called undecidable on the strength of our basis (B). Therefore, we may only claim that the following condition holds:

$$T^+ \cup T^- \subset F.$$

3.2 The Problem of Ł-Decidability of AS

The system **AS** whose basis is (B) analyzed by us, is not *saturated* or *decidable* in the sense that it is both 1° Ł-*consistent* and 2° Ł-*complete*, i.e.

$$1^\circ \quad T^+ \cap T^- = \emptyset \quad \text{and} \quad 2^\circ \quad T^+ \cup T^- = F.$$

As we know, a system satisfying both conditions was called by Słupecki an Ł-*decidable system* (see Sect. 2).

The problem concerning the finite Ł-decidability of Aristotelian syllogistic was raised by Łukasiewicz in December 1937, during his seminar on Mathematical Logic at the University of Warsaw.

Łukasiewicz presented the problem in the form of the following questions:

Q1. Are the axioms of A^+ for **AS** together with the inference rules of R^+ for **AS** sufficient to prove all true expressions of the **AS**?

Q2. Are the rules of rejection of $R^- = \{r^-1, r^-2\}$ for **AS** sufficient to reject all false expressions (every formula of F that is not a thesis of T^+), provided that a finite number of them are rejected axiomatically?

3.3 Słupecki's Solution of the Problem of Ł-Decidability of AS

Jerzy Słupecki, who participated in Łukasiewicz's seminar, solved in 1938 the problem providing a basis for which the system of Aristotelian syllogistic is Ł-decidable (see Łukasiewicz [22, 23]). His answer to the question Ql was positive; to the second one, negative.

Słupecki was able to prove that it is not possible to reject all the false expressions of **AS** by means of the rules r^-1 and r^-2, provided a finite number of them is rejected axiomatically.

This way Słupecki gave a negative answer to the question Q2:

(i) The system **AS** with the basis (B) is not Ł-complete with any finite set of rejected axioms of A^-.

Słupecki extended the system **AS**, adding to it a new rejection rule, called by Łukasiewicz *Słupecki's rule of rejection*. It is denoted by $r^{-}S$ and has the following scheme:

$$r^{-}S:\quad \frac{\dashv \alpha \to \gamma \qquad \dashv \beta \to \gamma}{\dashv \alpha \wedge \beta \to \gamma}$$

where α and β denote negative expressions in the form: $S\ e\ P$ or $S\ o\ P$ and γ denotes a simple expression or an implication the consequent of which is a simple expression and the antecedent, a conjunction of such expressions.

The Słupecki's rule says: *If the expression* γ *does not follow from any of two negative expressions then it does not follow from their conjunction.*

Słupecki's rule is closely related to the principle of traditional logic (*ex mere negative nihil sequitur*).

As was noticed by Łukasiewicz, having added Słupecki's rule, it is enough to adopt merely one rejected axiom, namely, $A^{-}1$.

Słupecki demonstrated that:

(ii) System of Aristotle's syllogistic determined by the following base:

$$\langle F, A^{+}, R^{+}; \{A^{-}1\}, R^{-} \cup \{r^{-}S\}\rangle \qquad \text{(BS)}$$

with the refutation system:

(iii) $$\langle F, \{A^{-}1\}, R^{-} \cup \{r^{-}S\}\rangle,$$

is Ł-decidability system, i.e.

$$1^{\circ}\quad T^{+} \cap T^{-S} = \emptyset \quad \text{and} \quad 2^{\circ}\quad T^{+} \cup T^{-S} = F, \text{ where}$$

$$T^{-S} = Cn^{-}(T^{+} \cup \{A^{-}1\}, R^{-} \cup \{r^{-}S\}), \qquad \text{D}(\underline{\text{Ł}})$$

is the *set of all rejected propositions*, i.e. the set of propositions derivable from the axiom $A^{-}1$ by means of the thesis of **AS** and Łukasiewicz's rejection rules and Słupecki's rejection rule.

And the problem of Ł-decidability of **AS** has been solved: any well-formed formula of **AS** is either a thesis or is a rejected formula of **AS**.

It is clear that Słupecki extended the notion of the rejected proposition used by Łukasiewicz, because:

$$T^{-} \subset T^{-S}. \qquad \text{D(Ł)} \subset \text{D}(\underline{\text{Ł}})$$

The results obtained by Słupecki were summarized by Łukasiewicz in his work [22] containing also the text of his paper on Aristotle's syllogistic.

The results of research of both Łukasiewicz and Słupecki were later, after the war, presented in detail in Łukasiewicz's monograph [23]. In both works Łukasiewicz expressed his high opinion of Słupecki's findings, which, in the words of Łukasiewicz [22] were

> organically united with researches of the author [...] the author regards as the most significant discovery made in the field of syllogistic since Aristotle.

3.4 *Słupecki's Definition of a Rejected Proposition*

Słupecki failed to publish his findings before the war. After the war ends, Słupecki published them in [38] and in a monograph [39]. In the monograph [39], in his proof of Ł-completeness (condition (ii), 2°), he also used a definition of the rejected proposition different than Łukasiewicz D(Ł), and additionally, he modified its extension D(Ł̲), adopted earlier by himself. Instead of Łukasiewicz's definition D(Ł), Słupecki adopts the following equivalent definition:
D(Sł). *A rejected proposition* on the ground of the basis

$$\langle F, A^+, R^+; A^-, \emptyset \rangle, \qquad (B \setminus R^-)$$

is such an expression for which there exists a rejected axiom of the set A^- *which is derivable from it and theses of the set* T^+ *by means of inference rules of* R^+.
Denoting a set of all rejected propositions in the sense of the definition D(Sł) by $Cn^{-1}(T^+ \cup A^-, R^+)$, we obtain the following symbolic notation of it:

$$\alpha \in Cn^{-1}(T^+ \cup A^-, R^+) \Leftrightarrow \exists \beta \in A^-(\beta \in Cn^+(T^+ \cup \{\alpha\}, R^+)), \qquad \text{D(Sł)}$$

where Cn^+ is a consequence operation with respect to the set T^+ of all theses of the system and its set of rules R^+.

The definition of a rejected proposition D(Sł) is closer to Aristotle's idea of refutation of syllogisms by means of reducing them to syllogisms rejected earlier.

We note (see D(Ł)) that:

$$T^- = Cn^-(T^+ \cup A^-, R^-) = Cn^{-1}(T^+ \cup A^-, R^+). \qquad \text{D(Ł)} \approx \text{D(Sł)}$$

Thus, the notions of rejected propositions, both the one used by Łukasiewicz and that introduced by Słupecki in the form of the definition D(Sł), are equivalent.

3.5 *Słupecki's Definition of an Extended Notion of Rejected Proposition*

We will reconstruct Słupecki's definition of an extended notion of the rejected proposition, equivalent to the definition D(Ł̲).

D($\underline{\text{Sł}}$). *A rejected proposition* on the ground of the basis

$$\langle F, A^+, R^+; \{A^-1\}, \{r^-S\}\rangle \tag{BSł}$$

is either a rejected axiom A^-1 *or a proposition rejected with respect to those rejected earlier* in the sense of D(Sł; A^-/X), *or an expression rejected on the basis of those rejected earlier, by means of using Słupecki's rule.*

Let $Cn'(T^+ \cup \{A^-1\}, R^+; \{r^-S\})$ be the set of all rejected propositions in the sense of D($\underline{\text{Sł}}$). We may note that

$$T^{-S} = Cn^-(T^+ \cup \{A^-1\}, R^- \cup \{r^-S\}) = Cn'(T^+ \cup \{A^-1\}, R^+; \{r^-S\}).$$

D($\underline{\text{Ł}}$) $\approx$ D($\underline{\text{Sł}}$)

3.6 Three Different Ways of Understanding the Notion of Rejected Proposition

It is easy to see that the given definitions of rejected propositions provide three different ways of understanding this notion:

$$\begin{array}{lccc}
\text{D(Sł)} \subset \text{D}(\underline{\text{Sł}}). & Cn^{-1}(T^+ \cup A^-, R^+) & \subset & Cn'(T^+ \cup \{A^-1\}, R^+; \{r^-S\}) \\
 & \| & & \| \\
 & T^- & & T^{-S} \\
 & \| & & \| \\
\text{D(Ł)} \subset \text{D}(\underline{\text{Ł}}). & Cn^-(T^+ \cup A^-, R^-) & \subset & Cn^-(T^+ \cup \{A^-1\}, R^- \cup \{r^-S\})
\end{array}$$

The first of them refers to Łukasiewicz's understanding of the rejected proposition (see D(Ł)) and to its strengthening given by Słupecki (see D($\underline{\text{Ł}}$)); the second and the third ones, to Słupecki's understanding of the rejected proposition (see D(Sł) and D($\underline{\text{Sł}}$). At the same time, the second one refers directly to Aristotle's method of rejection of syllogisms by reducing them to previously rejected syllogisms and makes it possible to simplify the procedure of rejection without supplementing the system with the rules of rejection, and the third one (see D($\underline{\text{Sł}}$)) is a combination of both former methods.

4 Notions of Rejection in a Deductive System and Notions of Ł-Decidability

4.1 Three Different Notions of Rejection

We are adapting the previous definitions of rejection for **AS** for any deductive system.

Let S be any deductive system with biaspectual formalization and with the basis

$$\langle F_S, A_S^+, R_S^+; A_S^-, R_S^- \cup R_S'\rangle \tag{B$_S$}$$

determined, respectively, by computable sets: the set F_S of all well-formed formulas, the set A_S^+ of axioms (asserted axioms), the set R_S^+ of inference rules, the set A_S^- of rejected axioms and by the set $R_S^- \cup R'_S$ of rejection rules, with an assumption that the sets R_S^+ and R_S^- are sets of mutual dual rules, while the set R'_S is a set of non-dual rejection rules.

Let T_S^+ be a set of all theses of S. Then, according to Łukasiewicz's conception, on the analogy to D(Ł̲),

the set ${}^1T_S^-$ *of all rejected propositions of* S, with respect to the set T_S^+ and the basis (B_S) (the refutation system $\langle F_S, A_S^-, R_S^- \cup R'_S\rangle$) is defined as follows:

$$^1T_S^- = Cn_S^-(T_S^+ \cup A_S^-, R_S^- \cup R'_S) \qquad \mathrm{D}_S(\underline{\text{Ł}})$$

And ${}^1T_S^-$ is *a set of all formulas with rejection proofs in Łukasiewicz's sense*, i.e. it is a set of all formulas derivable from rejected axioms of A_S^- by means of theses of T_S^+ and rejection rules of $R_S^- \cup R'_S$; i.e. ${}^1T_S^-$ is the smallest set including A_S^- and closed under the rejection rules of $R_S^- \cup R'_S$.

If $R'_S = \emptyset$, the basis (B_S) of the system S can be replaced by the basis

$$\langle F_S, A_S^+, R_S^+; A_S^-, \emptyset\rangle \qquad (\mathrm{B}_S \setminus R_S^-)$$

and *the set* ${}^2T_S^-$ *of all rejected propositions of* S with respect to the set T_S^+ and the basis $(\mathrm{B}_S \setminus R_S^-)$, on the analogy to D(Sł), is defined as follows:

$$^2T_S^- = Cn_S^{-1}(T_S^+ \cup A_S^-, R_S^+) \qquad \mathrm{D}_S(\text{Sł})$$

and ${}^2T_S^-$ is *a set of all propositions with rejection proofs in Słupecki's sense*, i.e. it is a set of all such formulas from which, and from theses of T_S^+, and by means of inference rules of R_S^+ a rejected axiom of A_S^- is derivable.

If $R'_S \neq \emptyset$, the basis (B_S) can be replaced by the basis

$$\langle F_S, A_S^+, R_S^+; A_S^-, R'_S\rangle \qquad (\underline{\mathrm{B}}_S)$$

and *the set* ${}^3T_S^-$ *of all rejected propositions of* S with respect to the set T_S^+ and the basis $(\underline{\mathrm{B}}_S)$, on the analogy to D(S̲ł̲), can be defined as follows:

$$^3T_S^- = Cn'_S(T_S^+ \cup A_S^-, R_S^+; R'_S) \qquad \mathrm{D}_S(\underline{\text{Sł}})$$

and the set ${}^3T_S^-$ is *a set of rejected propositions with rejected proofs in Słupecki-Łukasiewicz's sense*, i.e. it is a set of all such propositions every one of which is either:

1° a rejected axiom of A_S^-, or 2° a proposition rejected with respect to those rejected earlier in Słupecki's sense, or 3° a proposition rejected on the basis of those rejected earlier by means of rejection rules of R'_S.

Let us note that the given above definitions D_S(Ł̲), D_S(Sł) and D_S(S̲ł̲) are in some extent a simplification and require the precise definitions of the above-mentioned rejection proofs on the basis of any set $X \subseteq F_S$, with respect to the bases (B_S), $(\mathrm{B}_S \setminus R_S^-)$ and $(\underline{\mathrm{B}}_S)$ (see Słupecki [43], Słupecki and Bryll [44], Wybraniec-Skardowska [63, 64]).

Let us observe that among the set of rejected propositions defined above, the following relationships hold:

a. If $R'_S = \emptyset$ then ${}^2T_S^- = {}^1T_S^-$ b. ${}^2T_S^- \subseteq {}^1T_S^- = {}^3T_S^-$.

4.2 *Three Different Notions of Ł-Decidability; Decidability*

The three different definitions of the sets of rejected propositions of the system S entail three different definitions of Ł-decidability of this system.

Definition The system S is *Ł-decidable* if and only if, for some $i = 1, 2, 3$, S is ^{i}*Ł-decidable*, i.e. it satisfies the two following conditions:

$$1^\circ \quad T_S^+ \cap {}^iT_S^- = \emptyset, \qquad 2^\circ \quad T_S^+ \cup {}^iT_S^- = F_S.$$

The condition 1° is called ^{i}*Ł-consistence* condition and the condition 2° is called ^{i}*Ł-completeness* condition of the system S.

A question arises: What the relationship between Ł-decidability and decidability in the usual sense is? The answer was given by J. Słupecki.

Słupecki's Theorem [43]: *If the system S is Ł-decidable and any rejection rule, except for the rule of rejection by detachment, is computable, then the system S is decidable in the usual meaning.*[3]

It is easy to show that

Corollary *The system* **AS** *of Aristotle's syllogistic with the bases* (BSŁ) *is decidable.*

In the next section we will present Ł-decidable systems that are also decidable.

5 More Important Findings Concerning Ł-Decidability of Deductive Systems

The axiomatic rejection method introduced by Lukasiewicz to complete, biaspectual characterization of deductive systems (which was effectively used and developed by Słupecki) provided a broad response in literature after the Second World War. Łukasiewicz, already living in Dublin, uses such method in his research on intuitionistic logic [24], as well

[3]Słupecki's theorem is an immediate consequence of the following theorem of the theory recursion, which we quote from Grzegorczyk's book [14, p. 355 in the Eng. ed.]: *If the union of two recursively enumerable disjoint sets T and S is computable set, then the sets T and S are also computable.*

as in a four-valued modal system of propositional calculus [25], built by himself. At the same time, in Poland, further studies, inspired by Słupecki on Ł-decidability of deductive systems and the very notion of rejected proposition were taken up. In studies on Ł-decidability and providing complete refutation systems for logical systems (assertion systems) one of three methods, modeled on those described in the previous section, is usually used. In the next sections we will present a few, more important results, connected with this research.[4]

5.1 Calculi of Names

J. Słupecki's research on Aristotle's syllogistic was continued mainly by B. Iwanuś. Using the Słupecki-Łukasiewicz's methods, Iwanuś managed to prove Ł-decidability of a few systems of calculi of names.

- Iwanuś [17] gave a proof of Ł-decidability of the whole traditional calculus of names, i.e. the system of Aristotle's syllogistic enriched by nominal negation.
- Another interesting, though much later obtained result of Iwanuś's research [18] is a proof of Ł-decidability of the system of Aristotle's syllogistic built by Słupecki [38]. In this system, the two initial Łukasiewicz's axioms (laws of identity) of the system **AS** (those that are absent in Aristotle's logic) are replaced with the following axioms:

  $$S\ a\ P \Rightarrow S\ i\ P, \qquad S\ i\ P \Rightarrow P\ i\ S.$$

  In Słupecki's system, unlike in the system **AS**, it is permissible for variables to represent empty names. A complete refutation system for the system given by Iwanuś [18] is based on three rejected axioms and one rejected rule that is germane to the traditional calculus of names.
- Iwanuś also, in [16], gave a proof of Ł-decidability of a certain version of *elementary ontology*, distinguished from the system of *Leśniewski's Ontology* (both terms were invented by Słupecki in [41], who presented a system of calculus of names, based on Leśniewski's findings). In elementary ontology it is possible to interpret both the asserted system **AS** and other asserted syllogistic systems richer than **AS**, with nominal negation. Iwanuś gave a complete refutation system for the version of the system of elementary ontology; Iwanuś's refutation system consist of two independent rejected axioms and one non-Łukasiewicz's specific rejection rule.

[4]G. Bryll (as well as his book [2]) and T. Skura have been very helpful in verifying certain significant facts.

5.2 *Propositional Logic*

As we have already mentioned, Łukasiewicz used to apply the axiomatic method of rejection also to some systems of propositional calculus.

5.2.1 Classical Logic

Łukasiewicz mentioned in [25], that the classical propositional calculus with the inference rules $r^{+}1$ and $r^{+}2$ is Ł-decidable. A complete refutation system for it determines one rejected axiom, namely the sentential variable p, and two Łukasiewicz's rules: $r^{-}1$ and $r^{-}2$.

Let us note that the method of rejection of false formulas (i.e. non-theses) used by Łukasiewicz can be replaced by Słupecki's method, omitting Łukasiewicz's rules, i.e. applying the following principle: *If, from a formula of propositional calculus and the set of theses, it is possible to deduce, according to the rules* $r^{+}1$ *and* $r^{+}2$*, the rejected axiom* p*, then the formula is rejected* (see D(Sł)).

For the classical first-order calculus rejected axioms and rejected rules were formulated by T. Skura in [32]. Skura made use of the 'tableaux' method.

The rejection procedure accompanying the syntactic characterization of systems of propositional logic became a standard among logicians.

We will limit our presentation of results in the scope of the rejection procedure in the systems to a brief mention of only intuitionistic logic and extensions, modal logic and Łukasiewicz's logics.

5.2.2 Intuitionistic Logic and Extensions

In his research on intuitionistic propositional calculus, Łukasiewicz [24] advanced the hypothesis that it is Ł-decidable. Moreover, he supposed that a sole rejected axiom of it is a propositional variable, and that the rejection rules are: $r^{-}1$, $r^{-}2$, and one special rule (Gödel rule) which states that the disjunction $\alpha \vee \beta$ is rejected whenever so are α and β. Thanks to Kreisel-Putnam [19], we know that the rules proposed by Łukasiewicz do not suffice to reject all non-theses of the intuitionistic calculus. It is also known that there is no finite set of rejected axioms that, together with Łukasiewicz's rules, gives a complete refutation of intuitionistic system (cf. Maduch [28]).

Ł-decidability of the intuitionistic propositional logic was achieved by Scott [29] using a countable number of non-structural rejection rules. However, Scott's results, which were presented in *Summaries of Talks* at Cornell University, were inaccessible to Słupecki's circle in behind the Iron Curtain (then) Poland.

Independently of Scott's results, the proof of Ł-decidability of the intuitionistic propositional calculus was provided by Dutkiewicz [11]. In his approach, a complete refutation intuitionistic system is compounded of one axiom and three rejection rules: Łukasiewicz's rules, and a new, original rejection rule, which is, in fact, an infinite countable class of rejection rules of a common scheme. In his proof, Dutkiewicz uses the method of rejection modeled on the one applied by Łukasiewicz, as well as the method of Beth's semantic tableaux.

Another proof of Ł-decidability for the intuitionistic calculus was given by Skura [30, 36], who, while defining a complete intuitionistic refutation system, added to Łukasiewicz's rules a new rule, or rather, a class of structural rules of rejection, the number of which is infinite. Skura in [31], provided a complete refutation system for certain intermediate logics.

5.2.3 Modal Logic

The first research into a complete syntactic characterization of a modal propositional calculus was undertaken by Łukasiewicz [25]: he extends the four-valued modal system, built by himself, by two rejected axioms and his rejection rules, obtaining Ł-decidability of this system.

Afterwards, Słupecki initiated research on Ł-decidability of Lewis system S5. In his and Bryll's paper [44], the proof of Ł-decidability was achieved with an assumption of one rejected axiom (the prepositional variable p) and, apart from Łukasiewicz's rules, a class of rejection rules of the common scheme.

Skura [31, 32, 34] gave a simpler proof of Ł-decidability of Lewis system S5, assuming that the language of this system was supplemented with a symbol '$\bot$', i.e. *the constant of falsity*. Skura adopts this constant as the rejected axiom and extends the systems of Łukasiewicz's rules by: 1) a rule stating that: if formula $\Box p$ is rejected, the formula p is rejected, too; and 2) a class of structural rejection rules of the same scheme.

Skura in [31, 35, 36], using the algebraic method, also provided a complete refutation system for the logic S4 and for some of its extension (Grzegorczyk's logic).

A little earlier, Goranko [12, 13] formulated a complete refutation system for some normal modal propositional logics (including S4 and Grzegorczyk's logic) that are characterized by a class of finite trees. His refutation systems for these logics are based on the same rejected axiom (the constant '$\bot$'), Łukasiewicz's rules, and a class of non-structural rejected rules of the same scheme.

The method of constructing refutation systems corresponding to classes of finite models was used by Skura [31] for intermediate logics and, by Skura [33] and Goranko [13], for certain normal modal logics. Skura in [33], showed that the refutation systems can be useful in such cases when a given system of logic cannot be characterized by any class of finite models: there is a decidable modal logic without a finite model property that has a simple refutation system.

Tomasz Skura is regarded as an expert on the methods in refutation systems; his book [37] is devoted to refutation methods in modal propositional logics.

5.2.4 Łukasiewicz's Many-Valued Logics

Researches into Ł-decidability of Łukasiewicz's sentential calculus were conducted in the Opole circle of logical research, which, for many years, was led by Jerzy Słupecki. Bryll and Maduch [3], formulated a uniform method of rejection of formulas in an $n+1$-valued implicational, implicative-negative and definitionally complete Łukasiewicz's calculus. In these systems the same formula may be adopted as the sole rejected axiom[5]

$$C(Cp)^n q(Cp)^{n-1} q$$

(for $n = 1$ we get, in particular, the rejected axiom: $CCpqq$ in the classical implicational calculus). The only rejection rules are, here, Łukasiewicz's rules.

A complete refutation system for $\aleph_0$-valued Łukasiewicz's calculus was built by Skura [32] by extension of Łukasiewicz's rejection rules. Research into Ł-decidability of this system has been also conducted by Bryll [2]. His research was continued by R. Sochacki.

Sochacki [50] gave complete refutation systems for all invariant Łukasiewicz's many-valued logics (in which the rule of rejection by substitution was eliminated; see also Sochacki [48], Bryll and Sochacki [4, 5].

Sochacki also built refutation systems for selected many-valued logics: the k-valued logic of Sobociński and some systems of nonsense logic [49, 51].

5.3 *The Generalized Method of Natural Deduction*

The method of rejection introduced to metalogical investigations by Łukasiewicz in many cases can be replaced with the generalized method of natural deduction. The latter is applicable to all propositional calculi which have finite adequate matrices, as well as to the intuitionistic propositional calculus and the first-order predicate calculus, that is, to almost all logics discussed in this section.

The basis for such systems consists then of only assertion rules and rejection rules; the sets asserted and the rejected axioms are empty sets. The method used in the proofs is similar to Słupecki-Łukasiewicz's method of rejection, though they are apagogic proofs (by *reductio ad absurdum*).

This method refers to the 'tableaux' method. It is presented by Bryll [2], who in his studies refers to results obtained by Hintikka [15], Smullyan [47], Suchoń [54], Surma [55, 56] and Carnielli [8, 9].

6 Rejection Operation

As was noticed by Słupecki, the notion of rejected proposition is so general that it is most convenient to base studies concerning this notion on Tarski's theory of deductive systems, i.e. the axiomatic theory of consequence built by Alfred Tarski [57]. Let us recall that the only primitive notions of this theory are:

[5]It is saved in Łukasiewicz's, so called *Polish notation*.

- the set F of all propositions (well-formed formulas) of an arbitrary but fixed language of a given system and the consequence operation:

$$Cn^{+}: 2^{F} \to 2^{F};$$

- the symbol '$Cn^{+}(X)$' denotes the set of all consequences of the set $X \subseteq F$.

Properties of deductive systems characterized by the bases with refutation systems can be established by means of Tarski's theory of consequence enriched by the following definition of rejection operation:

$$Cn^{-1}: 2^{F} \to 2^{F},$$

determined by the consequence operation Cn^{+} (see D(Sł)): For any $X \subseteq F, \alpha \in F$

$$\alpha \in Cn^{-1}(X) \Leftrightarrow \exists \beta \in X(\beta \in Cn^{+}(\{\alpha\})) \qquad \text{MD(Sł)}$$

According to the definition MD(Sł): a proposition α is a *rejected proposition* on the ground of the set X if and only if at least one of the propositions of X is derivable from (is a consequence of) α; the symbol '$Cn^{-1}(X)$' denotes the set of all propositions rejected on the ground of propositions of X.

The following theorem helps to understand the intuitive sense of the definition MD(Sł):

$$\forall X \subseteq F(X \subseteq Y \Rightarrow Cn^{+}(X) \subseteq Y) \Rightarrow \forall X \subseteq F(X \subseteq Y' \Rightarrow Cn^{-1}(X) \subseteq Y').$$

If Y is the set of all true propositions of F (then the set Y' is the set of all false propositions of F), then, according to the above theorem, we can state that:

> *If consequence operation always leads from true propositions to true propositions, then a rejected operation always leads from false propositions to false propositions.*

So, the defined rejection operation (function) Cn^{-1} is a generalization of the notion of rejection introduced by Łukasiewicz.

The definition MD(Sł) of the function Cn^{-1} was formulated by Słupecki [42]. Słupecki proved that the function satisfies all axioms of Tarski's general theory of deductive systems [57] for the consequence Cn^{+} and that it is additive. Thus, the rejected operation is another consequence operation and is called a *rejection consequence*. It is clear that

> *Every deductive system with a bi-level formalization (with the assertion system and the refutation system) can be characterized by the basis:*
>
> $$\langle F, Cn^{+}, Cn^{-1} \rangle$$
>
> *and that the extension of Tarski's theory of the definition MD(Sł) describes every such system.*

This theory has been developed in the form of the *theory of rejected propositions* by Wybraniec-Skardowska [62], and later also by Bryll [1]. Their researches have been a continuation of investigations initiated by J. Słupecki and have been conducted under his supervision to be later presented in co-authored papers (see Słupecki et al. [45, 46]).

The theory of rejected propositions contains many significant theorems that are not counterparts of any theorem of Tarski's theory [58]. The following are examples of such theorems about the rejection operation Cn^{-1}:

- $Cn^{-1}(\emptyset) = \emptyset$ – *it is normal,*
- $Cn^{-1}(\bigcup\{X \subseteq G \colon G \subseteq F\}) = \bigcup\{Cn^{-1}(X) \colon X \subseteq G \subseteq F\}$ – *it is complete additive,*
- $Cn^{-1}(X) = \bigcup\{Cn^{-1}(\{\alpha\}) \colon \alpha \in X\}$ – *it is unit operation,*
- $\alpha \in Cn^{-1}(X) \Leftrightarrow \exists\beta \in X(Cn^{-1}(\{\alpha\} \subseteq Cn^{-1}(\{\beta\}))$ – *it is a unit consequence.*

In the theory the following formulation of the rule of rejection by detachment is valid:

- $c\,\alpha\beta \in Cn^{+}(X) \ \wedge \ \beta \in Cn^{-1}(Y \cup X) \Rightarrow \alpha \in Cn^{-1}(Y \cup X)$,

while in the Tarski's theory the formulation of rule of detachment has the form:

- $c\,\alpha\beta \in Cn^{+}(Y) \ \wedge \ \alpha \in Cn^{+}(Y \cup X) \Rightarrow \beta \in Cn^{+}(Y \cup X)$.

The set $Y \subseteq F$ can be understood as a set of axioms A^{+} or the set of theses T^{+} of a given deductive system.

7 Rejection Operation as a Primitive Notion

There is a possibility of the axiomatization of a theory of rejected propositions in a dual and equivalent way, i.e. assuming that a primitive notion is the rejection consequence Cn^{-1}, while Cn^{+} is defined operation.

Such dual theory of rejected propositions was formulated by Wybraniec-Skardowska [62] (see also [66, 67]). It can be understood as the theory describing the deductive systems with the basis

$$\langle F, Cn^{-1}, Cn^{+}\rangle.$$

A theory of rejected propositions was developed also to formalize some problems of methodology of empirical sciences, mainly by Bryll [1] (see also Słupecki et al. [46]).

The theories of rejected propositions have a natural interpretation, which was given by Staszek [53]: the set $Cn^{-1}(X)$ can be understood as a set of all rejection proofs on the ground of a proposition of the set X, in the sense relating to the nature of rejection proofs used by Słupecki in his researches on Aristotle's syllogistic.

In the theory of rejected propositions can be defined the notion of Ł-decidability.

7.1 *Rejection Operation as a Finitistic Consequence: Dual Consequences*

Rejection function Cn^{-1} can be generalized into a **dual, finitistic consequence** in the usual meaning. The notion of the dual consequence $\mathrm{d}Cn^{+}$ relating to the consequence

Cn^{+} was introduced by Wójcicki [61] by definition:

$$\alpha \in \mathrm{d}Cn^{+}(X) \Leftrightarrow \exists Y \subseteq X \wedge \mathrm{card}(Y) < \aleph_0(\bigcap\{Cn^{+}(\{\beta\}) \colon \beta \in Y\} \subseteq Cn^{+}(\{\alpha\})).$$

The dual consequence $\mathrm{d}Cn^{+}$ is stronger than the rejected consequence Cn^{-1} (i.e. $Cn^{-1} \leq \mathrm{d}Cn^{+}$), though the former is linked with the latter by a number of interesting relationships (see Spasowski [52]).
Facts: a. $\mathrm{d}Cn^{+1} = Cn^{-1}$ and b. $\mathrm{d}Cn^{-1} = Cn^{+1}$,
where the unit consequence Cn^{+1} is defined as follows:

$$\alpha \in Cn^{+1}(X) \Leftrightarrow \exists \beta \in X(\alpha \in Cn^{+}(\{\beta\}).$$

Certain studies of generalization of the notion of rejected expression in the form of a function of a consequence of rejection or a dual consequence are discussed by Wybraniec-Skardowska and Waldmajer [68].

The dual consequences Cn^{+} and $\mathrm{d}Cn^{+}$ as well as Cn^{+1} and Cn^{-1} can be used to study both true (asserted), and respectively, false (rejected) contents of a given theory. The fact was noticed by Woleński [60] in his studies relating to Popper's conception of a comparison of scientific theories by their contents.

References

1. Bryll, G.: Kilka uzupełnień teorii zdań odrzuconych [Some supplements to the theory of rejected sentences]. In: Wybraniec-Skardowska, U., Bryll, G.: Z badań nad teorią zdań odrzuconych [Investigations on the theory of rejected propositions]. Zeszyty Naukowe Wyższej Szkoły Pedagogicznej w Opolu, Matematyka, seria B: Studia i Monografie, no 22, Opole (1969), pp. 133–154
2. Bryll, G.: Metody odrzucania wyrażeń [The methods of the rejection of expressions]. Akademicka Oficyna Wydawnicza PLJ, Warszawa (1996)
3. Bryll, G., Maduch, M.: Aksjomaty odrzucone dla wielowartościowych logik Łukasiewicza [Rejected axioms for Łukasiewicz's many-valued logics]. Zeszyty Naukowe Wyższej Szkoły Pedagogicznej w Opolu, Matematyka **6**, 3–17 (1968)
4. Bryll, G., Sochacki, R.: Aksjomatyczne odrzucanie w inwariantnych rachunkach zdaniowych Łukasiewicza [Axiomatic rejection in invariant Łukasiewicz's propositional calculi]. Zeszyty Naukowe Uniwersytetu Opolskiego, Matematyka **29**, 29–37 (1995)
5. Bryll, G., Sochacki, R.: Elimination of the rejection rule by substituting in the three-valued Łukasiewicz calculus. Acta Universitatis Wratislaviensis, Logika **18**, 129–131 (1998)
6. Carnap, R.: Introduction to Semantics. Harvard University Press, Cambridge (1942)
7. Carnap, R.: Formalization of Logic. Harvard University Press, Cambridge (1943)
8. Carnielli, W.A.: Systematization of finite many-valued logics. J. Symb. Log. **52**(2), 473–493 (1987)
9. Carnielli, W.A.: On sequents and Tableaux for many-valued logics. J. Non-Classical Logics **8**(1), 59–76 (1991)
10. Citkin, A.: A meta-logic of inference rules. Syntax. Log. Logical Philos. **24**, 313–337 (2015)
11. Dutkiewicz, R.: The method of axiomatic rejection for the intuitionistic propositional logic. Stud. Logica **48**(4), 449–459 (1989)
12. Goranko, V.: Proving unprovability in some normal modal logics. Bull. Sect. Log. Polish Acad. Sci. **20**, 23–29 (1991)
13. Goranko, V.: Refutation systems in modal logic. Stud. Logica **53**, 299–324 (1994)
14. Grzegorczyk, A.: Zarys logiki matematycznej. PWN, Warszawa (1961) (English trans.: Outline of mathematical logic. R.Reidel Publishing Company, Dordrecht (1974))

15. Hintikka, J.: Form and content in quantification theory. Acta Philosophica Fennica **8**, 7–55 (1955)
16. Iwanuś, B.: On Leśniewski's elementary ontology. Stud. Logica **31**, 73–147 (1972)
17. Iwanuś, B.: Proof of decidability of the traditional calculus of names. Stud. Logica **32**, 73–90 (1973)
18. Iwanuś, B.: Z badań nad sylogistyką Arystotelesa [Investigations on Aristotle's syllogistic]. Zeszyty Naukowe Wyższej Szkoły Pedagogicznej w Opolu, Matematyka **28**, 41–56 (1992)
19. Kreisel, G., Putnam, H.: Unableitbarkeits-beweismethode für den intuitionistischen Aussagenkalkül. Archiv für Mathematische Logik und Grundlagenforschung **3**, 205–220 (1957)
20. Łukasiewicz, J.: Logika dwuwartościowa [Two-valued logic)] Przegląd Filozoficzny **23**, 189–205 (1921) (English trans. by Wojtasiewicz, O.A. in: Łukasiewicz, J.: Selected Works. North-Holland, Amsterdam (1970). Borkowski, L. (ed.), pp. 89–109)
21. Łukasiewicz, J.: Elementy logiki matematycznej [an authorized collection of Łukasiewicz's lectures]. ed. Presburger, M. (1929). (Second edition, PWN, Warszawa (1958); English trans.: Elements of Mathematical Logic. Pergamon Press, Oxford; The Macmillan Company, New York (1963))
22. Łukasiewicz, J.: O sylogistyce Arystotelesa [On Aristotle's Syllogistic]. Sprawozdania z Czynności i Posiedzeń Polskiej Akademii Umiejętności, no 44 (1939) (Reprinted in: Łukasiewicz, J.: Z zagadnień logiki i filozofii. Pisma wybrane [Problems in logic and philosophy. Selected papers]. PWN, Warszawa (1961). Słupecki, J. (ed., selection, introduction and footnotes), pp. 220–227)
23. Łukasiewicz, J.: Aristotle's Syllogistic from the Standpoint of Modern Formal Logic. Clarendon Press, Oxford (1951) (Second edition in 1955)
24. Łukasiewicz, J.: On the intuitionistic theory of deduction. Indag. Math. Ser. A **3**, 202–212 (1952) (Reprinted in: Łukasiewicz, J.: Selected Works. North-Holland, Amsterdam (1970). Borkowski, L. (ed.), pp. 325–340)
25. Łukasiewicz, J.: A system of modal logic. J. Comput. Syst. **1**(3), 111–149 (1953) (Reprinted in: Łukasiewicz, J.: Selected Works. North-Holland, Amsterdam (1970). Borkowski, L. (ed.), pp. 352–390)
26. Łukasiewicz, J.: Z zagadnień logiki i filozofii. Pisma wybrane [Problems in logic and philosophy. Selected papers]. PWN, Warszawa (1961). Słupecki, J. (ed., selection, introduction and footnotes)
27. Łukasiewicz, J.: Selected Works. North-Holland, Amsterdam (1970). Borkowski, L. (ed.)
28. Maduch, M.: O Łukasiewiczowskich regułach odrzucania [On Łukasiewicz's rules of rejection]. Zeszyty Naukowe Wyższej Szkoły Pedagogicznej w Opolu, Matematyka **13**, 115–122 (1973)
29. Scott, D.: Completeness proofs for intuitionistic sentential calculus. In: Summaries of talks presented at the Summer Institute of Symbolic Logic, Cornell University (1957) (Second edition by: Communications Research Div., pp. 231–242, Princeton (1960))
30. Skura, T.: A complete syntactic characterization of the intuitionistic logic. Rep. Math. Log. **23**, 75–80 (1989)
31. Skura, T.: Refutation calculi for certain intermediate propositional logics. Notre Dame J. Formal Log. **33**, 552–560 (1992)
32. Skura, T.: Some results concerning refutation procedures. Acta Universitatis Wratislaviensis, Logika **15**, 83–95 (1993)
33. Skura, T.: Syntactic refutation against finite models in modal logic. Notre Dame J. Formal Log. **35**, 573–582 (1994)
34. Skura, T.: A Łukasiewicz-style refutation system for the modal logic S5. J. Philos. Log. **24**, 573–582 (1995)
35. Skura, T.: Refutation and proofs in S4. In: Wansing, H. (ed.) Proof Theory of Modal Logic, pp. 45–51. Kluwer Academic, Dordrecht (1996)
36. Skura, T.: Aspects of refutation procedures in the intuitionistic logic and related modal systems. Acta Universitatis Wratislawiensis, Logika **20**, Wrocław (1999)
37. Skura, T.: Refutation Methods in Modal Propositional logic. Semper, Warszawa (2013)
38. Słupecki, J.: Uwagi o sylogistyce Arystotelesa [Remarks about Aristotle's syllogistic]. Annales Universitatis M. Curie-Skłodowska **I**(3), 187–191 (1946)
39. Słupecki, J.: Z badań nad sylogistyką Arystotelesa. Prace Wrocławskiego Towarzystwa Naukowego (B), **6**, Wrocław (1948) (English trans. in: Słupecki, J.: On Aristotelian syllogistic. Stud. Philos. **4**, 275–300 (1951))
40. Słupecki, J.: On Aristotelian syllogistic. Stud. Philos. **4**, 275–300 (1951)
41. Słupecki, J.: St. Leśniewski's calculus of names. Stud. Logica **3**, 7–76 (1955)

42. Słupecki, J.: Funkcja Łukasiewicza [The Łukasiewicz function]. Zeszyty Naukowe Uniwersytetu Wrocławskiego, seria B, **3**, 30–40 (1959)
43. Słupecki, J.: Ł-decidability and decidability. Bull. Sect. Log. Polish Acad. Sci. **1**(3), 38–43 (1972)
44. Słupecki, J., Bryll, G.: Proof of Ł-decidability of Lewis system S5. Stud. Logica **12**, 99–107 (1973)
45. Słupecki, J., Bryll, G., Wybraniec-Skardowska, U.: Theory of rejected propositions, part I. Stud. Logica **29**, 75–123 (1971)
46. Słupecki, J., Bryll, G., Wybraniec-Skardowska, U.: Theory of rejected propositions, part II. Stud. Logica **30**, 97–145 (1972)
47. Smullyan, R.M.: First Order Logic. Springer, Berlin (1968)
48. Sochacki, R.: Axiomatic rejection in the implicational-negational invariant sentential calculi of Łukasiewicz. Bull. Sect. Log. Polish Acad. Sci. **36**, 1–6 (2007)
49. Sochacki, R.: Rejected axioms for the "nonsense-logic" W and the k-valued logic of Sobociński. Log. Logical Philos. **17**, 321–327 (2008)
50. Sochacki, R.: Metody refutacyjne w badaniach nad systemami logicznymi [Refutation methods in research on logical systems]. Studia i Monografie, nr **445**, 1–122, Wydawnictwo Uniwersytetu Opolskiego, Opole (2010)
51. Sochacki, R.: Refutation systems for a system of nonsense-logic. Log. Logical Philos. **20**, 233–239 (2011)
52. Spasowski, M.: Some connections between Cn, Cn^{-1}, dCn. Bull. Sect. Log. Polish Acad. Sci. **2**(1), 53–56 (1973)
53. Staszek, W.: A certain interpretation of theory of rejected propositions. Stud. Logica **30**, 147–150 (1973)
54. Suchoń, W.: Smullyan's method of constructing Łukasiewicz's m-valued implicational-negational sentential calculus. In: Wójcicki, R., Malinowski, G. (eds.) Selected Papers on Łukasiewicz's Sentential Calculi, pp. 119–124. Ossolineum, Wrocław-Warszawa (1977)
55. Surma, S.J.: A method of constructing Łukasiewicz's algebras and its applications to a Gentzen-style characterization of finite logics. Rep. Math. Log. **2**, 49–54 (1974)
56. Surma, S.J.: An algorithm of axiomatizing every finite logic. Rep. Math. Log. **3**, 57–62 (1974)
57. Tarski, A.: Fundamentale Begriffe der Methodologie der deductiven Wissenschaften I. Monatshefte für Mathematik und Phisik **37**, 361–404, Leipzig (1930) (English trans.: Fundamental concepts of methodology of the deductive sciences. In: Tarski, A.: Logic, Semanics, Metamathematics. Papers from 1929 to 1938. Second edition editing and introduced by John Corcoran, Hackett Publishing Company, Indianapolis, Indiana (1983) (First edition published in 1956 by OUP; trans. by Woodger JH), pp. 60–109)
58. Tarski, A.: Über einige fundamentale Begriffe der Metamathematik. Comptes Rendus des séances de la Société des Sciences et des Lettres de Varsovie **23**, 22–29 (1930) (English trans.: On some fundamental concepts of metamathematics. In: Tarski, A.: Logic, Semanics, Metamathematics. Papers from 1929 to 1938. Second edition editing and introduced by John Corcoran, Hackett Publishing Company, Indianapolis, Indiana (1983) (First edition published in 1956 by OUP; trans. by Woodger JH), pp. 30–37)
59. Tarski, A.: Logic, Semanics, Metamathematics. Papers from 1929 to 1938. Second edition editing and introduced by John Corcoran, Hackett Publishing Company, Indianapolis, Indiana (1983) (First edition published in 1956 by OUP; trans. by Woodger JH)
60. Woleński, J.: On comparison of theories by their contents. Stud. Logica **48**, 617–622 (1989)
61. Wójcicki, R.: Dual counterparts of consequence operation. Bull. Sect. Log. Polish Acad. Sci. **2**(1), 54–57 (1973)
62. Wybraniec-Skardowska, U.: Teoria zdań odrzuconych [The theory of rejected propositions]. In: Wybraniec-Skardowska, U., Bryll, G.: Z badań nad teorią zdań odrzuconych [Investigations on the theory of rejected propositions]. Zeszyty Naukowe Wyższej Szkoły Pedagogicznej w Opolu, Matematyka, seria B: Studia i Monografie, no **22**, Opole (1969), pp. 5–131
63. Wybraniec-Skardowska, U.: Badania Jerzego Słupeckiego nad sylogistyką Arystotelesa i ich rezonans we współczesnej logice [Jerzy Słupecki's investigations on Aristotle's syllogistic ad their response in the contemporary logic]. Zeszyty Naukowe Wyższej Szkoły Inżynierskiej w Opolu, seria Matematyka **4**, 35–61 (1983)
64. Wybraniec-Skardowska, U.: On the notion and function of the rejection of propositions. Acta Universitatis Wratislaviensis, No 2754, Logica **23**, 179–202 (2005)

65. Wybraniec-Skardowska, U.: Rejection in traditional and modern logics. In: Proceedings of the 10th Panhellenic Logic Symposium, Samos, Greece, 11–15 June 2015, pp. 59–62 (2015)
66. Wybraniec-Skardowska, U.: On the mutual definability of the notions of entailment, rejection, and inconsistency. Axioms **5**, 15 (2016). https://doi.org/10.3390/axioms5020015
67. Wybraniec-Skardowska, U., Bryll, G.: Z badań nad teorią zdań odrzuconych [Investigations on the theory of rejected propositions]. Zeszyty Naukowe Wyższej Szkoły Pedagogicznej w Opolu, Matematyka, seria B: Studia i Monografie, no 22, Opole (1969)
68. Wybraniec-Skardowska, U., Waldmajer, J.: On pairs of dual consequence operations. Log. Univers. **5**, 177–203 (2011)

U. Wybraniec-Skardowska (✉)
Cardinal Stefan Wyszyński University in Warsaw, Warsaw, Poland
e-mail: skadowska@gmail.com

Bolesław Sobociński: The Ace of the Second Generation of the LWS

Kordula Świętorzecka

Abstract The here presented biography of Bolesław Sobociński (1906–1980) describes the history of his life, his patriotic activity and scientific results. Sobociński, as a pupil of Jan Łukasiewicz and Stanisław Leśniewski, was one of the main representatives of the second generation of the Lvov-Warsaw School, a member of the Warsaw School of Logic. He is known as the most influential popularizer of Leśniewski's systems of prothotetics, ontology and mereology, on the international stage, as well as the author of many important results in symbolic logic: many-valued logics, modal logics, as well in set theory and theory of algebras. The international community of logicians knows Sobociński as the creator and, for many years, editor of the prestigious scientific periodical *Notre Dame Journal of Formal Logic*. The material is divided in the following parts: (1) Life, (2) National service, (3) Academic career, (4) Interests and achievements, (5) Sobociński presonally, (6) Selected publications by Bolesław Sobociński.

Keywords Bolesław Sobociński · Warsaw School of Logic · Leśniewski's systems · Symbolic logic

Mathematics Subject Classification (2000) Primary 01A60; Secondary 03-03

In search of source materials for Bolesław Sobociński's biography, I came across the work entitled "Profesor Sobociński i kolega Bum" ["Professor Sobociński and colleague Bum"][1] by a Polish mathematician Krzysztof Tatarkiewicz (1923–2011) [18]. This text is exceptional, mainly because the author was able to capture the personality of Sobociński through a direct contact with him, through anecdotes heard from the members and supporters of the Lvov-Warsaw School, through the recollection and comments of his own father—Władysław Tatarkiewicz (1886–1980). It is especially worth noting how Tatarkiewicz made an attempt to refer to a particularly difficult chapter in Sobociński's life: his national service in the Second World War and in the early post-war period. The great historical value of Tatarkiewicz's paper stems from historical data and caution in passing judgment too quickly.

[1] English translations of Polish titles are put in square brackets.

Á. Garrido, U. Wybraniec-Skardowska (eds.), *The Lvov-Warsaw School. Past and Present*, Studies in Universal Logic,
https://doi.org/10.1007/978-3-319-65430-0_41

Sobociński's biography was written and published by K. Tatarkiewicz in Polish. The aim of this paper is to present one of the leading personalities of the second-generation of the Lvov-Warsaw School, and the Warsaw School of Logic more specifically, to a broader public, partly drawing on the text of this biography. We repeat a great deal of information after K. Tatarkiewicz, supplementing it, whenever possible, with data from other texts about Sobociński, and verifying the entire piece with our own research conducted in the Warsaw archives. We also add a commentary on the academic achievements of Sobociński.

1 Life

Bolesław Sobociński was born in St. Petersburg (Russian Empire) on 28.06.1906. He was the only child of Waleria, née Jasiewicz, and Antoni Sobociński. Waleria Sobocińska was the daughter of Katarzyna, née Misiewicz, and Karol Jasiewicz. She was born in Pułtusk (now Poland) on 31.05.1873 and died in Warsaw on 2.02.1942. Antoni Sobociński was the son of Bolesława, née Sokołowska, and Antoni Sobociński; his sister Wanda (married name Chawłowska) was born in Moscow around 1867 and died in Warsaw on 24.06.1934 [5]. It is known that Sobociński's parents married in 1905.

In 1916, Sobociński started attending the Catholic Gymnasium at the Saint Catherine of Alexandria Church in St. Petersburg. After 2 years, his parents decided to change his schooling to home education. In his 1926 biography from [1], Sobociński explains this change, stating "[...] in 1918, I was taken away by my parents due to the gymnasium's nationalisation and the introduction of the Bolshevik curriculum. Because of the conditions created by the Russian Revolution, I had to abandon regular education."*[2] (A deep dislike for communism in general, and not just Russian, was a distinctive characteristic of Sobociński's outlook in his adult life.)

Following the Riga treaty, the Sobocińskis were awarded Polish citizenship, and in October 1922 they came to Warsaw as repatriates. Here, Bolesław obtained his extramural baccalaureate (he started attending courses at the humanistic gymnasium in February 1923, and passed his maturity examination on 23.02.1926 [1]). In 1926, he started his studies in philosophy at the Warsaw University, where he obtained a Master of Arts degree (30.06.1930), a doctoral degree (30.06.1936), and successfully completed the habilitation (14.06.1939).

At the beginning of the war (around 6.09.1939), Sobociński fled from Warsaw and headed to the estate of W. Tatarkiewicz's mother-in-law, Ksawera Potworowska, located near Lublin (Radoryż). There he met Krzysztof Tatarkiewicz, who was sixteen at the time. Sobociński planned to reach the estate of the Skirmunt family (at that time in Poland, now Belarus), but he was stopped (by a "gang of peasants"—as he described the

[2] Translations of Polish fragments done by Daniela Szczygieł are marked with: *.

attackers) and taken to a town nearby (Motol).[3] Sobociński kept his acquaintance with the Skirmunts secret (the whole family was murdered in 1939 by the Bolsheviks); in this way, he saved his life and was able to leave Motol. He travelled to Vilnius first (at that time under Lithuanian). There, he gave a lecture entitled "O prototetyce prof. Leśniewskiego" ["On Professor Leśniewski's Prothotetics"] at a meeting of the Vilnius Philosophical Society (18.10.1939). He went to the estate belonging to his maternal relative, Karol Parczewski (1875–1957), where he stayed from December 1939 to mid-1941 (Stończe, Lithuania, at that time not occupied by the Soviets). In 1941, he returned to Warsaw. In February 1942, Sobociński's mother died. Until the outbreak of the Warsaw Uprising against Germans in 1944, he attended seminars at the underground Warsaw University, where he gave lectures in logic and edited the manuscripts by his master, Stanisław Leśniewski (deceased just before the war). He actively participated in the independence conspiracy within the structures of the underground National Armed Forces. Sobociński survived the Warsaw Uprising and he was a witness to the last hours of the life of his friend and companion in the underground fight, Jan Salamucha (1903–1944), who was killed in the uprising by SS-Sturmbrigade RONA[4] soldiers. After the fall of the uprising, K. Tatarkiewicz met Sobociński at a temporary camp near Warsaw (around 12 or 13 August 1944). It is also known that in 1944 Sobociński helped to organise the journey of the Łukasiewicz family to Switzerland. Together with Leśniewski, Jan Łukasiewicz (1878–1956) played a crucial role in his academic career. There are few known facts about Sobociński's life during the following 2 years, until September 1946. This lack of data is connected with his pro-independence activity. Sobociński held prominent functions within the underground National Armed Forces, and for the rest of his later life he kept absolute discretion regarding any issues related to this activity. Before the war (around 1923) he met his future wife Ewa, née Wrześniewska (1909–2005). Born in Warsaw, she graduated in Philosophy at the Warsaw University (she was a student of W. Tatarkiewicz). Later she was activist of the external women's Wiara i Wola [Faith and Will] structure cooperating with the Polish Organisation.[5] (Wrześniewska held a managerial function there [14, p. 251].)

After the war, in 1945, Sobociński was offered a position at the newly-established University in Łódź, which he declined. Two or three days before starting this job, he withdrew from public life. At the same time, he was wanted by the communistic Office of Public Security.

In September 1946, he secretly left Poland, together with Ewa Wrześniewska.[6] They travelled to Regensburg via Katowice and they eventually married in Germany.

Between 1946 and 1949, the Sobocińskis stayed in Brussels. During this time, Sobociński worked at the Polish Scientific Institute in Belgium. Between March 13 and

[3] We have some information about Sobociński's adventures at the beginning of the war, ended with his stay at the Stończe estate, through his direct testimony, which he gave in his letters to Father Bocheński, parts of which were published in [13, pp. 123–133] (we know their full versions from [6]). These events were also described by Tatarkiewicz in [18].

[4] RONA—Russkaya Osvoboditelnaya Narodnaya Armiya.

[5] The Polish Organisation was formed already in 1934 as a secret right-wing organisation and a continuation of the National Radical Camp, later (partly) included in the National Armed Forces.

[6] On Łukasiewicz's request, their travel was facilitated by Zbigniew Jordan (1911–1977).

30, 1948 they went to Freiburg (Switzerland) to see Father Józef M. Bocheński (1902–1995). Together with his wife, Sobociński visited the Łukasiewicz family in Dublin (July 1947). He also travelled to Paris.

In December 1949, the Sobocińskis moved to the States, where they remained for the rest of their lives. Thanks to the efforts of Father Bocheński, Sobociński was employed at the College of St. Thomas in St. Paul, USA (Minnesota). (He also availed himself the help of Marian Heitzman (1899–1964), already employed there.)

During the first period of their stay in the States, they were not entirely successful. Sobociński lost his job after the first semester. Due to financial constraints, Ewa Sobocińska took up a job as a typist. From c. 1951 to c. 1955, Sobociński got a permanent position at the Institute of Applied Logic in St. Paul. Ewa Sobocińska also became employed there as an IT engineer.

In 1955, the Sobocińskis received American citizenship, and a year later they moved to South Bend, USA (Indiana), where Sobociński became employed at the Faculty of Philosophy at the University of Notre Dame. Mrs Sobocińska was able to get a job at the same University, in the technical department. Due to financial stability they bought a house in South Bend [18, p. 141]. In 1972, Sobociński retired.

Bolesław Sobociński died on 21.11.1980, after a long illness. His wife Ewa died on 15.01.2005. Both of them were buried at the Cedar Grove university cemetery (St. Joseph County Indiana, USA).

The Sobocińskis had no children.

2 National Service

Sobociński was actively involved in the social and political life of pre-war Poland. In 1934, he became involved with the National Radical Camp (a fraction ABC), and the Polish Organisation.[7]

A reconstruction of Sobociński's patriotic activity and his ties with the Polish underground after the start of the war is clearly difficult: the information available is incomplete, and the fragments available were preserved despite the planned distortion of the image of right-wing pro-independence groups by the communist authorities. In 1949, Sobociński wrote to Bocheński: "It is obvious that my person there [in Warsaw] is completely unpalatable (no less than 7 trials ended with a death sentence—all executed, in which I played the role of the lead but absent actor and leader). [...] Matters are also complicated by some other concerns (the wish not to endanger those who survived) so that even anti-communist elements know nothing about me"* [13, p. 128].[8]

We will only outline this chapter of Sobociński's life, referring to information from [14, 15, 18]. As far as it was possible, we compared these datas with the documents available in the Polish Institute of National Remembrance [4].

[7]Cf. footnote 5.

[8]Former employee of the Office of Public Security—Lt.-Col. Różański, called Sobociński "a famous spy" in his report [7].

We know that after his return to Warsaw from the Stończe estate, in spring of 1941, Sobociński became involved in the activities of the Polish Organisation. He operated under the organisational nickname "Rawicz" (or "prof. Rawicz").[9] He joined the Christian-National Lizard Union (National Order) and led the 7th group responsible for secret service [14, p. 247]. In the spring of 1942, he became a member, and subsequently the chairman of the Executive Committee of the Polish Organisation, to which also Salamucha belonged. He performed important functions in the National Armed Forces: he was the head of the IVa Office of the Central Intelligence Service of the National Armed Forces Command. In 1943, he was a full member—together with Salamucha—of the highest level (A) of the Political Committee (the Polish Organisation leadership). In July 1944, he participated in the first secret meeting of A level members (the meeting was organised in his flat) during the occupation. At the meeting, an evacuation plan for the Polish Organisation leaders was also discussed. Sobociński and Salamucha decided to stay in Warsaw [14, pp. 329–331].[10] Members of the Political Committee met again in November 1943 in Kraków, to discuss the possibility to "establish tactical contact with the Germans"* [14, p. 347] and broadcast an anti-communist Polish radio station. The leaders of the Political Committee opposed this plan and resigned. We do not know Sobociński's position on this matter. However, we do know that in 1945, he was elected chairman of the Political Committee, and held this position until his departure in the autumn of 1946. (From January 1945 he was the political commissioner for the "South" and "West" areas of the National Armed Forces.) Thanks to his foresight, Sobociński avoided arrest by the Soviet authorities, and he managed to avoid the so-called Trial of the Sixteen. On 27.03.1946, delegates of the Polish government and the Home Army were deceitfully invited to take part in another round of talks about a "compromise" on the composition of "the temporary Government of National Unity" established under the Yalta agreement (1945), on which basis Poland fell under the Soviet control. The meeting (in Pruszków near Warsaw) was attended by the Polish Deputy Prime Minister J. S. Jankowski, last Commander-in-Chief of the Home Army, General L. Okulicki, and 14 other leaders of Underground Poland. Sobociński was also to attend that meeting, but having explored the site, he decided not to go and thus was not taken to Moscow for a staged Stalinist trial. Until September 1946, Sobociński went into hiding (he was wanted by the Office of Public Security) and in the end—after consulting with the leadership of the Internal Organisation—he made a decision to leave the country. We do not know the extent of Sobociński's political activity after his departure from Poland. His correspondence with Bocheński leads us to conclude that at least at the beginning of his stay in Brussels Sobociński was an energetic political activist. He wrote: "[...] I am of the opinion that the fuss from 1939 has not ended yet, we live in times of respite, short or long, and for this

[9]Among his relatives and friends he was called "Bum"; this nickname (used also in [4]) was supposedly given to Sobociński to commemorate a comment he made during one of the bombardments of Warsaw (according to Tatarkiewicz, Sobociński allegedly said "in a calm voice and with a characteristic intonation: What a great boom"* and he made his friends laugh [18, p. 136]. Sobociński stuttered and tried to "help" himself by using the word "bum"). Muszyński mentions some other nicknames used by Sobociński: "Professor", "Uncle" [14, p. 247].

[10]Tatarkiewicz denies the claim that during the uprising, Sobociński allegedly helped crack ciphers at the Home Army Command [18, p. 130].

reason, my presence in Europe will be necessary."* (letter of 16.12.1946 [6]).[11] However, we find no historical records describing his further political involvement.

3 Academic Career

Sobociński was awarded the degree of Doctor of Philosophy 30.06.1936. His doctoral thesis entitled "Aksjomatyzacja pewnych wielowartościowych systemów teorii dedukcji" ["Axiomatisation of certain many-valued systems of the theory of deduction"] was supervised by Łukasiewicz.

As already mentioned, in 1930 Sobociński was awarded an MA in Philosophy at the Faculty of Humanistics at the Warsaw University. His dissertation, entitled "Z badań nad teorią dedukcji" ["Investigations on the theory of deduction"*], was also supervised by Łukasiewicz. During his studies in philosophy, he was taught and examined, among others, by W. Tatarkiewicz (history of modern philosophy), T. Kotarbiński (logic), W. Witwicki (psychology), K. Ajdukiewicz (theory of cognition, Frege's logic), J. Łukasiewicz (Kant's philosophy, contemporary philosophy), W. Wąsik (Polish philosophy), W. Sierpiński (set theory), S. Leśniewski (introduction to mathematics, group theory) [3].

After completing his master's studies, Sobociński maintained and developed his interest in logic and became actively involved in publishing.

From 1931 (or 1932) and until the outbreak of the war, he worked as the secretary of *Przegląd Filozoficzny* (at that time edited by W. Tatarkiewicz; in 1939 Sobociński took up the position of editor while Tatarkiewicz became editor-in-chief).

In 1936, he became a member of the Interim Board of the Polish Association for Logic (together with A. Lindenbaum, A. Mostowski, and A. Tarski), and 2 years later, he started his work as an editorial secretary for the journal *Collectanea Logica* established on Łukasiewicz's initiative and published by the Polish Association for Logic. In the same year, he also became the secretary for *Organon* (founded in 1936 by S. Michalski as an international review and published by the Mianowski Institute for the Promotion of Science and Letters).[12]

In 1934, he attended a congress for the students of the new Faculty of Mathematics and Natural Sciences at the Warsaw University (where he later defended his doctorate).

Immediately after his doctoral defence, Sobociński began working there as the first titular assistant (1934 (36?)–1938), and later as a senior assistant (1938–1939) for the

[11] Maybe he knew about and took into account the possible implementation of the *Operation Unthinkable* by Churchill?

[12] Unfortunately, the proofs of almost all texts that were due to be published in these journals were burnt during the Warsaw Uprising (in the flats of Sobociński and W. Tatarkiewcz). The first issue of *Collectanea Logica*, printed in August 1938, did not come out because of the outbreak of the war, and was burnt together with the second volume prepared by Sobociński, in his apartment. In additions to the journals, Leśniewski's original manuscripts, as well as their editions done by Sobociński, were also burnt. According to Tatarkiewicz [18, p. 126], at that time a few dozen academic papers were lost (including 13 texts for *Collectanea*, 4 of which were authored by Sobociński).

Philosophy of Mathematics Seminars led by Leśniewski. Thus, he formally joined the Warsaw School of Logic.

During that period, Sobociński cooperated closely with Leśniewski, and maintained his interest in protothetics, ontology, and mereology for many years.

Leśniewski died on 13.05.1939, only a few weeks before Sobociński's habilitation.

On 14.06.1939, Sobociński passed habilitation colloquium at the Faculty of Mathematics and Natural Sciences of the Warsaw University presenting dissertation entitled "Z badań nad prototetyką" ["Research on Protothetics"]. The dissertation was reviewed by Łukasiewicz and S. Mazurkiewicz (the committee also comprised Białobrzeski [3]). The Polish Ministry of Religion and Public Education was unable to approve the habilitation before the outbreak of the war.

Understandably, the war also ruined the plans for Sobociński's academic career; however, it did not diminish his interest in academic matters.

During his stay at the Stończe estate (1939–1941), Sobociński conducted research in logic, and went back to academic research and teaching activities immediately after his return to Warsaw. As already mentioned, he participated in secret seminars; he also took over Łukasiewicz's lectures in the underground Warsaw University. At the same time, he took part in a seminar on the history of logic, conducted by Łukasiewicz and A. Krokiewicz. Together with Łukasiewicz, he was an academic supervisor for H. Rasiowa (her first master's thesis, written under the direction of Łukasiewicz and Sobociński, was burnt during the Warsaw Uprising). In the autumn of 1943, after Łukasiewicz's departure, Sobociński delivered lectures in logic for mathematicians (among his students were K. Szaniawski and K. Tatarkiewicz). Until the outbreak of the uprising, he worked on editing Leśniewski's notes given to him in 1939 by Leśniewski's widow. As mentioned earlier, the results of his work (approx. 1000 pages of text) and the original manuscripts by Leśniewski, were also burnt during the Uprising.

After the end of the war, it was possible to attempt to complete the habilitation procedure. On 14.09.1945, the Board of the Faculty of Mathematics and Natural Sciences of the Warsaw University applied to the Ministry of Education for approval of Sobociński's habilitation, which had taken place in 1938/39. The Ministry granted the application in May 1947, but Sobociński did not receive its official response—he had already gone abroad [3].

Also in 1945, Kotarbiński made efforts to secure for Sobociński the position of associate professor and head of the Department of Deductive Sciences at the Faculty of Mathematics and Natural Sciences of the Łódź University (at that time, Kotarbiński was Rector of the Łódź University [3]). On 15.09.1945, Sobociński accepted the post, but had to miss his first classes.[13] As we already know, due to the threat of arrest, he decided to leave Poland illegally. As late as 15.11.1946, Kotarbiński, as Rector of the Łódź University, informed the Ministry of Education that "the appointment of Dr Bolesław Sobociński associate professor of deductive sciences at the Faculty of Mathematics and Natural Sciences of the Łódź University is no longer valid"* [3]. The end of his formal

[13]According to K. Tatarkiewicz, who refers to information obtained from C. Lejewski, Sobociński was also to be employed at the Maria Curie-Skłodowska University (Lublin), and allegedly failed to assume this post in the same way. However, Tatarkiewicz could not find any documents in the archives of the Maria Curie-Skłodowska University to prove this [18, p. 132].

ties with the Warsaw University and the Łódź University can be dated, respectively, to 28.02.1946 (when of the Rector of the Warsaw University decided to cease salary payment [3]) and 5.12.1947 (when the Ministry of Education was notified by the Rector of the Łódź University that Sobociński "failed to report to assume duties and his place of stay is [...] unknown"* [3]).

Sobociński's departure from Warsaw to Brussels in 1946 is associated with more professional troubles. As already mentioned, Sobociński was employed by the Polish Scientific Institute in Belgium (headed by J.F. Drewnowski). He was given a research position after Łukasiewicz's retirement and was supposed to continue teaching classes in. The stay in Brussels was difficult for Sobociński: he struggled financially, feared another war turmoil,[14] was troubled by the lack of scientific contact with logicians and philosophers he had known previously.

The departure for the States in 1949 was to open up a new and better life, as well as more professional opportunities. With recommendations from Bocheński and Heitzman, Sobociński became employed at the College of St. Thomas, as previously mentioned, as a philosophy teacher. He taught Thomism, but his classes were not popular (according to Bocheński, Sobociński's diligence reflected in his rigorous reference to original source texts was not entirely appropriate, considering the rather basic level of the school [9, p. 120]). In the same year, Sobociński's lectures were suspended. Between 1951 and 1955, Sobociński acted as the Director of Research at the Institute of Applied Logic at St. Paul's, where he dealt with logical issues for the purposes of the computer industry. Thanks to Bocheński's help once again, in 1956 he was invited to an interview by the authorities of the Notre Dame University. According to Bocheński, because of a somewhat unfortunate course of the interview, Sobociński was offered the lowest possible position of Research Associate.[15] However, in 1961 (1958?), he was given the position of a Professor of Philosophy, and held this post until his retirement (1972). He gave lectures even in the autumn of 1979 (this date is provided by Rickey [8, p. 11]).

At Notre Dame, Sobociński was successful again with his research (which we will describe in the next section) and editorial efforts; he also proved to be an excellent teacher. In the early years of his work, he may not have been as popular with his students as Łukasiewicz, but at Notre Dame he was able to restore, to a certain degree, the atmosphere of the Warsaw School of Logic and encourage a big group of young logicians to become involved in academic work. Over the entire period of his career, he taught such courses as: symbolic logic, modal logic, combinatory logic, algebraic logic, metalogic, Leśniewski's systems, basic mathematics, logical basis of set theory, theory of cardinal and ordinal numbers (a detailed list of all subjects is provided in [8, pp. 9–10]). He successfully supervised fifteen Doctors of Philosophy (R.E. Clay, J.T. Canty, and V. F. Rickey among others); he also worked with other students on the results published by them between

[14]In his correspondence with Bocheński he mentioned anxiety in each letter, and repeatedly enquired about the possibility of leaving Belgium. In 1948, he asked Bocheński several times to help him to obtain a Spanish visa (due to concerns about the *Berlin Blockade* by the Soviets?). Eventually, he declared he wanted to leave Europe altogether [6].

[15]Bocheński recalls that during the meeting he had a problem with communication due to poor knowledge of English.

1972 and 1979. (A list of Sobociński's students and doctoral students is also provided in [8, pp. 7–9].)

In the States, he resumed his publishing activities. Between 1953 and 1954, he was the editor of *The Journal of Computing Systems*. In 1960, he founded one of the prestigious logic journals, the *Notre Dame Journal of Formal Logic* (NDJFL), and edited it until 1978. He published almost 70 papers in the NDJFL (the last one in 1979).

In his recollection of Sobociński, C. Lejewski states: "The academic achievements of Sobociński and his contributions in the areas of teaching and editing are clearly more than average"* [10, p. 401]. Let us now take a brief look at the main directions and results of his research.

4 Interests and Achievements

As Lejewski accurately remarks, Sobociński's intellectual inclinations combined the minds of his two great teachers: Jan Łukasiewicz and Stanisław Leśniewski [11]. His philosophical background and the "mathematical" way of thinking, were inherited by Sobociński from Łukasiewicz; likewise, a great share of his intellectual involvement was dedicated to the work of Leśniewski—a prominent formalist, who officially denied (almost) all existing philosophical perspectives, but created deeply philosophical systems: protothetics, ontology and mereology. Sobociński researched and improved these theories in the "mathematical" style of Łukasiewicz.

His early philosophical interests are known because of his collaboration with the Kraków Circle—the Catholic "branch" of the Lvov-Warsaw School, established in 1936 at the initiative of Bocheński and Salamucha (with support from Łukasiewicz). Bocheński considered Salamucha, Drewnowski, Sobociński and himself as key members of the association. The most prominent exponent of the Circle—whose aim was to improve methods of practising "Catholic" philosophy and theology—was Salamucha. Sobociński did not publish any papers concerning the topics researched by the school, and had an "advisory" role on the subject of logic. However, as evidenced by his correspondence with Father Bocheński, Sobociński also had his own, original philosophical reflections going beyond mere formalism, linked to the topics addressed by the Circle. In his letters, he discussed the issue of the existence of universals, and proposed their formalisation; he considered the possibility to formalise the concept of the Universe on the basis of mereology, and outlined the applications of mereological concepts to theological issues; he also wrote remarks on analogy [6]. Recently, there have been attempts at analysing and reconstructing some of these philosophical ideas of Sobociński: [12, 16], and also [17]. The main ideas of the Kraków Circle were still at play in a few activities undertaken jointly by Bocheński and Sobociński after the war, during their stay at Notre Dame. Here, together with theologian and philosopher Ivo Thomas (1912–1976),[16] they actively promoted the logical style of doing philosophy, and in 1956 their efforts led to a

[16]Ivo Thomas was invited to Notre Dame thanks to Bocheński's efforts; he collaborated in the creation of NDJFL, and attended Sobociński's seminars. Bocheński recollects that their team was called "Sobo, Ivo and Bobo" [9, p. 222].

meeting of some hundred logicians and mathematicians to discuss the status and nature of universals (among others, the discussion was attended by N. Goodman and A. Church).[17]

If we measure the creative contribution to a particular area of scientific research by the quantity and quality of publications, then we can safely say that Sobociński achieved remarkable results in logic. He published often, and his works were of recognised and substantial value. His achievements include almost 90 academic texts.

Lejewski classifies Sobociński's outputs into the following research areas: (1) symbolic logic: sentential systems and Leśniewski's systems, (2) border between logic and mathematics: set theory, axiomatics of algebraic systems, (3) history of logic [11]. This list can be supplemented by adding Sobociński's interest in history of contemporary Polish philosophy and philosophical didactics.

Before the war, Sobociński wrote several papers of a rather local nature. He published a guide to philosophical studies at the Warsaw University, a report on Polish publishing in the area of philosophy from 1918 to 1936, and a description of the current trends in Polish philosophy, after the 3rd Polish Philosophical Meeting in Kraków in 1936.

Sobociński's contribution to the history of logic includes two brief biographies of Łukasiewicz, a brief biography of Salamucha, and a text in Italian on the history of the Polish school of logic (from 1957).

The main core of Sobociński's achievements are in the area of symbolic logic and concern several issues in logic and mathematics.

Let us start with certain issues in sentential logics—the subject matter of Sobociński's master's, doctoral and post-doctoral dissertations. In his master's dissertation, Sobociński positively solved two proof-theoretic problems concerning classical sentential logic, first put by Łukasiewicz and Tarski (the original text can be found in [1]). He proved that: (i) there exists one sole organic axiom for sentential logic, made only of original symbols of implication and negation. (The axiom is organic, when none of its sub-formulae is a thesis of a given logic). Sobociński was the first to formulate such an axiom, which—using the Polish notation (without brackets)—comprised 139 symbols. Subsequently, he formulated further, shorter formulae of that type (with 47 and 27 symbols). Addressing Tarski's problem, he demonstrated that (ii) for each natural number n, classical sentential logic has a basis comprising exactly n elements. (A set of formulae Y is a basis for a set of formulae X, when Y is an independent set of formulae—i.e. where for any of its sub-sets Z, which has exactly the same consequences as Y, $Y = Z$—and when Y and X have the same consequences). It can be said that according to (ii), for any number n there is a complete set of n logically independent axioms for classical sentential logic. Research on the deductive minima of classical sentential logic was conducted by Sobociński mainly before the war (among other things, he found an axiomatic system for the implication-conjunction fragment of classical sentential logic). In his doctoral thesis, he examined selected n-valued logics with $n - 1$ distinguished values, for which he developed axiomatic systems. His post-doctoral thesis concerned Leśniewski's sentential system—protothetics, which he continued to study also after the war. He analysed its systems in the 1960s, identifying different possible sets of axioms (he found the shortest, sole axiom of protothetics with the only original symbol of equivalence). His results

[17] Sobociński did not participate in this meeting and we do not know to what extent he was involved in its organisation.

are considered to be crucial technical improvements of this theory. He also improved, axiomatised, and extended ontology and mereology. Thanks to him, Leśniewski's theories became known abroad. He also redefined other concepts by his master: he recreated Leśniewski's proposal for the solution of Russell's antinomy and promoted Leśniewski's ideas concerning the methodological correctness of axiomatic systems of deductive theories. Similarly to Leśniewski, he preferred classical logic to many-valued logics. He inspired many students with Leśniewski's ideas.

After the war, Sobociński became interested in modal logics, and achieved significant results in this field as well. He wrote almost 20 papers about various issues in sentential modalities. One of his first results was a proof that of modal systems T of R. Feys and M of G.H. von Wright are deductively equivalent. He also showed the same equivalence between systems M' and M'' of von Wright with $S4$ and $S5$ of C.I. Lewis. He analysed several axiom systems of Lewis and Feys. He established that system T has infinitely many non-equivalent modalities. He introduced a division of modal logics into regular (i.e. where the elimination of the modal symbols from any thesis yields a classical tautology), and irregular. He also proved that each of Lewis's modal systems containing T must be regular, and that there exist irregular extensions of system T^o of Feys-von Wright. He defined a certain family of modal irregular logics which are sublogics of system $S5$.

With respect to the last group of issues mentioned by Lejewski, we may notice that Sobociński was particularly interested in deductive dependencies between some key theorems of set theory. He researched logical relationships between selected formulae concerning cardinal numbers and the axiom of choice, generalised continuum hypothesis, and Cantor's hypothesis on alephs. He became interested in the theory of lattices in the early 1960s, and he maintained this interest—as in the case of the theory of algebras—until the end of his activity. He wrote papers on diverse axiom systems for different types of lattices and algebras as late as 1978 and 1979.

5 Sobociński Personality

We do not know whether Sobociński was a "mysterious" person, but he surely made surprisingly contrasting impressions under different circumstances and on different people. In a letter to Kazimierz Twardowski, Leśniewski described him as a "Very decent and well-educated person"* [20]. In same statements by Bocheński, K. Tatarkiewicz, and Lejewski, the reader will easily find the picture of a fragile man and slightly confused scholar ("eccentric") as well as a courageous and at the same time cautious WWII combatant (as evidenced by the events in Pruszków).

Sobociński was tall and slim (thin), had a "soft" eastern accent (he also stammered and had a lisp?) and had troubles with the English pronunciation (at least in the first period of his stay in America). To Bocheński, he seemed somewhat very "candid" (Bocheński described him as "naive"). Before the war he wore old-fashioned clothes, thus attracting attention (even during the war, he did not change his style and used to wear bow tie, a bowler hat, and an umbrella...).[18] He never had a driving licence. In his academic work, he combined research creativity, exceptional memory, determination to popularise the works of his master, efficiency in independent and original logical research, and care about teaching. In his political activity he was an efficient and discrete organiser of war secret services, as well as an industrious conspirator who bravely fought for independence.

Perhaps the multiple facets of Sobociński's personality were the result of the times in which he lived. It is true, however, that by combining them with natural talents, he gave witness to two cardinal virtues: bravery and wisdom. Thanks to them, he has a lasting place in the contemporary history of Poland, and has made a significant contribution to contemporary logic.

6 Selected Publications by Bolesław Sobociński

A complete list of publications, excluding reviews written by Sobociński over many years of his activity, is to be found in [19]. (Before 1939, Sobociński published 13 reviews, mainly in *Przegląd Filozoficzny*, but also in *Organon* and *Nauka Polska*; between 1938 and 1956, 11 reviews were published in the *Journal of Symbolic Logic* [8, pp. 5–7].) The list compiled by K. Tatarkiewicz was provided to him by Professor J. Woleński and contains a collection of articles published outside of NDJFL. The list was completed by Rickey [8]. Tatarkiewicz also enumerates all of Sobociński's papers published in the NDJFL, and several other articles published after the list prepared by Rickey.

We will not include Sobociński's publications in NDJFL. A full list is already available on the Leibniz—Zentrum für Informatik, Schloss Dagstuhl, Universität Trier: http://dblp.uni-trier.de/pers/hd/s/Sobocinski:Boleslaw.

Since 1957, Sobociński published only in English. Almost all earlier, significant works were translated or reconstructed in English. In line with [19], we present the articles in a chronological order, and English versions are followed by original Polish titles.

(1932) Z badań nad teorią dedukcji. [An Investigation on the Theory of Deduction] *Przegląd Filozoficzny* 35, 171–193 (master's degree thesis)

(1933) a. *Informator o studjach i egzaminach z nauk filozoficznych na Uniwersytecie Warszawskim.* [A Guidebook About Studies and Examinations from Philosophical Sciences at the Warsaw University] Warszawa

b. Polish original of: (1967) Successive Simplifications of the Axiom-System of Leśniewski's Ontology. In: McCall F. (ed.) *Polish Logic 1920–1939*, pp. 188–200. Oxford (O kolejnych uproszczeniach aksjomatyki Ontologji prof. St. Leśniewskiego. In: Księga Pamiątkowa ku uczczeniu 15-lecia pracy nauczycielskiej w Uniwersytecie Warszawskim prof. Tadeusza Kotarbińskiego. Warszawa)

[18]According to K. Tatarkiewicz: "[...] Sobociński's outfit drew attention. [...] in the interwar period, he would always wear a black suit with a waistcoat, a white shirt with a stiff collar, a black bow tie, and a bowler hat. Even before 1939 such attire would be worn only for formal occasions, and was almost never seen in the streets. But B. Sobociński started his journey around the country in this outfit—looking rather sensational... In addition to a suitcase, he always held a black umbrella."* [18, pp. 127–128].

(1935) Aksjomatyzacja implikacyjno-konjunkcyjnej teorii dedukcji. [Axiomatisation of the Implication-Conjunctive Theory of Deduction] *Przegląd Filozoficzny* 38, 85–95

(1936) a. Polskie wydawnictwa filozoficzne w latach 1918–1936. [Polish Philosophical Publications in the Years 1918-1936] *Nowa Książka* 3, 13–121

b. Tendencje rozwojowe współczesnej filozofii polskiej (Refleksje na marginesie III Polskiego Zjazdu Filozoficznego. Kraków 24–27.IX.1936). [Development Tendencies of Contemporary Polish Philosophy] *Nowa Książka* 3, 433–437

c. Aksjomatyzacja pewnych wielowartościowych systemów teorji dedukcji. [Axiomatisation of Certain Many-Valued Systems of the Theory of Deduction] Roczniki Prac Naukowych Zrzeszenia Asystentów Uniwersytetu Józefa Piłsudskiego w Warszawie. T.1, 399–419 (doctoral thesis)

(1939) a. Habilitation in Polish reconstructed as: (1949) An Investigation on Protothetics. Cahiers de l'Institut d'Études Polonaises en Belgique 5, V+44. In: McCall F. (ed.) *Polish Logic 1920–1939*, pp. 201–206. Oxford (Z badań nad prototetyką. *Collectanea Logica* 1, 171–177)

b. Polish text reconstructed as: (1953) Axiomatization of a Conjunctive-Negative Calculus of Propositions. *Journal of Computing Systems*, 229–242 (Aksjomatyzacja konjunkcyjno-negacyjnej teorii dedukcji. *Collectanea Logica* 1, 179–193)

(1949) French text translated as: (1984) Leśniewski's Analysis of Russell's Paradox. In: Srzednicki, J.T., Rickey, V.F. (eds.) *Leśniewski's Systems. Ontology and Mereology*, pp. 11–44. The Hague-Wrocław (L'analyse de l'antinomie russellenne par Leśniewski I, II, III, IV. *Methodos* 1, 94–100, 200–228, 308–316; 2 237–257)

(1953) a. Axiomatization of Partial System of Three-Value Calculus of Propositions. *Journal of Computing Systems* 1, 23–55

b. On a Universal Decision Element. *Journal of Computing Systems* 1, 71–80

c. Note on a Modal System of Feys-von Wright. *Journal of Computing Systems* 1, 171–178

(1954) Z badań nad aksjomatyką prototetyki Stanisława Leśniewskiego. [An Investigation on the Axiomatisation of Prototetics by Stanisław Leśniewski] *Yearbooks of the Polish Society of Arts and Sciences Abroad, for years 1953/1954*, 4, 18–20

(1955) a. Note on a Problem of Paul Bernays. *Journal of Symbolic Logic* 20, 109–114

b. Studies in Leśniewski's Mereology. *Yearbooks of the Polish Society of Arts and Sciences Abroad, for years 1954/1955*, 5, 34–65 (Reprinted in: Srzednicki, J.T., Rickey, V.F. (eds.) *Leśniewski's Systems. Ontology and Mereology*, pp. 217–227. The Hague-Wrocław)

(1956) a. On Well-constructed Axiom-Systems. *Yearbooks of the Polish Society of Arts and Sciences Abroad, for years 1955/1956*, 6, 54–65

b. In Memoriam Jan Łukasiewicz (1878–1956). *Philosophical Studies* 6, 3–49

(1957) a. Jan Łukasiewicz. *Yearbooks of the Polish Society of Arts and Sciences Abroad, for years 1956/1957*, 7

b. La Genesis de la Escuela Polaca de Logica. *Revista Orient Europe* 7, 25, 83–95

(1958) Jan Salamucha (1903–1944): A Biographical Note. *The New Scholasticism* 32, 327–333

(1995) Lata wojny... Listy do Józefa M. Bocheńskiego. [Years of War ... Letters to Józef M. Bocheński] In: Markiewicz, B., Jadacki, J. (eds.) Próg istnienia. Zdziesiątkowane pokolenie, pp. 123–128. Polskie Towarzystwo Filozoficzne, Warsaw

Acknowledgements Notes by Professor Jacek Jadacki [7] (and [20]) were an important source of information, and I wish to thank him for allowing me to use them. I wish to thank Professor V. Frederick Rickey for bringing amendments to some of the bibliographic details, as well Marek J. Minakowski, PhD, for helping me reconstruct the family connections of Bolesław Sobociński. The dates and historical facts were verified, as far as possible, in the Warsaw archives. Data from the archives of the Notre Dame University in South Bend are quoted from [18]. I am grateful to Mr Zbigniew Siemaszko for help in arranging information on Sobociński's political activity.

References

Archival Materials

1. Student files and a folder with the master's degree thesis along with examination reports, Archives of the Warsaw University, catalogue number RP21816, KEM 829
2. Employee records, Archives of the Warsaw University, catalogue number K996
3. Personal Files from the Ministry of Higher Education Institutions and Science, Archive of New Files, catalogue number MSW, 553
4. Files from the Polish Institute of National Remembrance, catalogue numbers: IPN BU 0203/2411, IPN Kr 074/216/1, IPN KR 074/216/2
5. Death records at the Archive of the Roman Catholic Diocese of Warszawa–Praga, entry on Wanda Chawłowska, reference number 493/1934
6. Letters by B. Sobociński to J.M. Bocheński. Copies of manuscripts
7. Archival materials of Professor Jacek Jadacki, with information obtained from Henryk Hiż and Witold Strawiński
8. Rickey, V.F.: Bibliography of Bolesław Sobociński, typescript from the collection of Professor J. Woleński, donated by K. Tatarkiewicz to the Archives of the Warsaw University (appended to the files [2]), p.12 (1981)

Others

9. Bocheński, J.: Wspomnienia [Memoirs], 2nd edn. Philed, Kraków (1994)
10. Lejewski, C.: Ś.P. Bolesław Sobociński. Znak, XXXVI, No (2–3), February–March, 351–352 Kraków (1984)
11. Lejewski, C.: Powojenna działalność polskich logików na Zachodzie [Post-war activity of Polish logicians in the West]. In: Strzałkowski, W. (ed.) Filozofia polska na obczyźnie [Polish philosophy abroad], pp. 161–175. Polskie Towarzystwo Naukowe na Obczyźnie, Londyn (1987)
12. Łyczak, M., Porwolik, M., Świętorzecka, K.: The Universe in Leśniewski's Mereology: Some Comments on Sobociński's Reflections. Axioms **5**(3), 23 (2016). https://doi.org/10.3390/axioms5030023 (open access http://www.mdpi.com/2075-1680/5/3/23)
13. Markiewicz, B., Jadacki, J. (eds.): Próg istnienia. Zdziesiątkowane pokolenie. [Threshold of existence. Decimated generation] Polskie Towarzystwo Filozoficzne, Warszawa (1995) (includes: Sobociński, B., Lata wojny... Listy do Józefa M. Bocheńskiego, 123–128)

14. Muszyński, W.J.: Duch młodych. Organizacja Polska i Obóz Narodowo-Radykalny w latach 1934–1944. Od studenckiej rewolty do konspiracji niepodległościowej [The Spirit of the Young: The Polish National Radical Camp in the Years 1934–1944. From Student Revolts to Conspiracy for Polish Independence]. Instytut Pamięci Narodowej, Warsaw (2011)
15. Siemaszko, Z.S.: Narodowe Siły Zbrojne [National Armed Forces]. London (1982) (3rd edn., Warsaw (1991))
16. Świętorzecka, K., Porwolik, M.: O pewnym przesądzie dotyczącym Uniwersaliów. Uwagi do sformalizowanego przez Bolesława Sobocińskiego argumentu na rzecz tezy o nieistnieniu powszechników podanego przez Stanisława Leśniewskiego [About a superstition concerning Universals. Comments on Bolesław Sobociński's formalised argument for the non-existence of Universals as presented by Stanisław Leśniewski*]. In: Brożek, A., Chybińska, A., Grygianiec, M., Tkaczyk, M. (eds.) Myśli o języku, nauce i wartościach. Seria druga. Profesorowi Jackowi Juliuszowi Jadackiemu w 70. rocznicę urodzin [Thoughts about language, science and values. Second series. To Professor Jacek Juliusz Jadacki on the 70th anniversary of his birth]. Semper, Warszawa (2016), 252–263
17. Świętorzecka, K., Porwolik, M.: Bolesław Sobociński on Universals. From Leśniewski's Nominalism to Meta-Conceptualism of Sobociński. In: Garrido, Á., Wybraniec-Skardowska, U. (eds.) The Lvov-Warsaw School. Past and Present, Studies in Universal Logic. https://doi.org/10.1007/978-3-319-65430-0_42
18. Tatarkiewicz, K.: Profesor Sobociński i kolega Bum [Professor Sobociński and colleague Bum]. Wiadomości Matematyczne Seria 2, **XXXIV**, 122–146 (1998)
19. Tatarkiewicz, K.: Logik i polityk (Bolesław Sobociński) [Logician and politician (Bolesław Sobociński)]. Materiały i Konferencje, Szczecin University, vol. 3, pp. 167–183 (1998)
20. Jadacki, J.: Stanisław Leśniewski: geniusz logiki [Stanisław Leśniewski: Genius of Logic]. Epigram, Warsaw (2016)

K. Świętorzecka (✉)
Institute of Philosophy, Cardinal Stefan Wyszyński University in Warsaw, Warsaw, Poland
e-mail: k.swietorzecka@uksw.edu.pl

Bolesław Sobociński on Universals

Leśniewski's Nominalism and Sobociński's Metaconceptualism

Kordula Świętorzecka and Marek Porwolik

Abstract The present paper proposes a comparative analysis of two standpoints on the existence and nature of universals hold by Stanisław Leśniewski and Bolesław Sobociński. We consider first the nominalistic argumentation of Leśniewski formalized by Sobocińki and described in the correspondence with J. M. Bocheński in 1956. Sobociński's formalization revealed a fundamental pragmatic weakness of the reconstructed argumentation which was also mentioned by Sobociński. He himself was aware of the difficulties connected with an adequate interpretation of the crucial axiom, whose acceptance Leśniewski imputed to supporters of all theories of universals. Finally, the problem of the existence and nature of universals was elaborated by Sobociński also in a separate typescript "Uwagi w sprawie powszechników" (Remarks on universals). The view formulated by Sobociński comes from a combination of the methodology of deductive systems and the conceptualist standpoint. From the philosophical perspective Sobociński's idea is both interesting and original, but it remained unknown to philosophers and logicians in general. For these reason we describe it and compare it with Leśniewski's approach. We use in this description epistemological notions of R. Suszko. Our analysis enables to speak about universals in sense of Leśniewski, which are described by some universal in sense of Sobociński.

Keywords Bolesław Sobociński · Stanisław Leśniewski · Universals · Nominalism · Conceptualism

Mathematics Subject Classification (2000) Primary 03A05; Secondary 01A60

Research was carried out within the project Logic, Concepts, and Communication (LogiCCom, IP-2014-09-9378) financed by Croatian Science Foundation (HRZZ).

Á. Garrido, U. Wybraniec-Skardowska (eds.), *The Lvov-Warsaw School. Past and Present*, Studies in Universal Logic,
https://doi.org/10.1007/978-3-319-65430-0_42

The medieval problem of universals was undertaken by members of the Lvov-Warsaw School.[1] A number of discussions concerning the nature and existence of general objects were conducted in the new, analytical style, sometimes utilizing formal methods. The main opponent of the views supporting the existence of universals was Stanislaw Leśniewski, one of the most committed nominalists of the school. What is interesting, his argumentations influenced the discussion mainly thanks to the information provided by his friend, Tadeusz Kotarbinski, who reported on Leśniewski's claims in his writings.[2] Regardless of these historical circumstances, it bears emphasizing that Leśniewski's theories constitute interesting material, which has already been formalized a few times. One of these formalizations will be the subject of our attention in the present paper. The author of the formalization to be discussed here, is Boleslaw Sobociński, a close co-worker and promoter of Leśniewski's achievements. Sobociński described his theory in the correspondence with Father Bocheński in 1956, which was published in 2004 in [19]. In this frame he wanted to reconstruct as closely as possible Leśniewski's argumentation of 1927, concerning the non-existence of universals [9]. Indeed, it can be claimed that this goal was achieved: Sobociński was a fine expert on all the systems created by Leśniewski, including his ontology, which he chose as the formal frame for his reconstruction. He was well acquainted with his master's philosophical views. As an exceptionally capable formalist, Sobociński helped Bocheński in some of his analyses, although he usually preferred to remain only a logical consultant, who kept a number of the discussed philosophical decisions in perspective.[3] Thus, it can be assumed that his formalization was actually meant to provide the addressee with the explanation of Leśniewski's line of thought. Independently of this direct goal, Sobociński's formalization revealed a fundamental pragmatic weakness of the reconstructed argumentation. Sobociński himself was aware of the difficulties connected with the adequate interpretation of the crucial axiom, whose acceptance Leśniewski imputed to supporters of all theories of universals. In a letter to Bocheński he tried to sketch out the possibility to avoid the problem, but he failed to develop his analysis. Finally, the problem of the existence and nature of universals proved to be important enough for him to elaborate on his opinion in the next letter to Bocheński, where he added an attachment titled "Uwagi w sprawie powszechników" (Remarks on universals). The view formulated by Sobociński comes from a combination of the methodology of deductive systems and the conceptualist standpoint. From the philosophical perspective Sobociński's idea is both interesting and original, but it remained unknown to philosophers and logicians in general. As far as we know, the reprint of Remarks in the original version was published only once in [19] and

---

[1]The historical presentation of all the discussion is shown by M. Grygianiec in [5].

[2]Such a role was played by the work "Sprawa istnienia przedmiotów idealnych" (The problem of the existence of ideal objects) [6] and by fragments devoted to Leśniewski's argumentations from "Elementy teorii poznania, logiki formalnej i metodologii nauk" (Elements of cognitive theory, formal logic, and methodology of science) [7], which elicited replies from R. Ingarden and K. Ajdukiewicz. A report on this discussion can be found in [5, pp. 110–113].

[3]Such intentions can be found in the text of Sobociński's letters. R. Murawski names Sobociński "an expert on logical problems" who has collaborated with all members of the Cracow Circle (J. M. Bocheński, J. Salamucha, F. Drewnowski). He participated in all of the meetings of the Circle, but he did not published any works related to the philosophical topics considered by the Circle [11, pp. 360–361].

it has not been subject to any analysis. For these reasons we have decided to present it in comparison with Sobociński's formalization of Leśniewski's nominalist argumentation.

1 Leśniewski's Nominalist Argumentation in Sobociński's Version

As it has already been noted, Sobociński formulated his formalization in 1956, but the fragments of correspondence which contained it were published almost 50 years later. The author of the publication, T. Waragai, received the manuscripts from Father Bocheński in 1982. However, in the literature we can find Sobociński's formalization published earlier. In 1997 it was reprinted by Woleński [20] and quoted from this source by Grygianiec in 2000 [5]. Additionally, also Urbaniak [18] referred to Gryganiec's presentation in his own paraphrase of Leśniewski's argumentation. Urbaniak used Sobociński's main specific axiom, but he did not follow the original structure of argumentation, and he also showed the possibility of simplifying Leśniewski's (and Sobociński's) reasoning. In 2016 our work was published [17], which we are going to refer to in the present paper.

Leśniewski expressed his claim that general objects do not exist several times, and he modified his justification. The first argument was published in 1911, and its amended version was published in 1913. In 1927 Leśniewski provided an argument based on the idea of the general object, which was similar to the one he used in his previous justifications, but this time his reasoning had a new, different structure. The source analyzed by Sobociński was the text of 1927.

The argumentation of 1927 had a broader range than the previous two arguments concerning the non-existence of general objects did (1911, 1913), and was supposed to represent Leśniewski's radical nominalist point of view. At that time, Leśniewski believed that only particulars exist in opposition to any non-individual objects, which also comprise "the so-called properties" and "the so-called relations" [9, p. 314]. He claimed that whoever has "a tendency to argue that "if X is a general object in relation to objects a, X is b, and Y is a, then Y must be b" ", he must accept that "that sentence involves the sentence "if there are at least two different a's, there is no general object in relation to objects a" ". The whole reasoning was formulated as a sequence of numbered sentences with a commentary on the inferential connections among them. Sobociński conducted a formal reconstruction of this text on the basis of Leśniewski's elementary ontology.

1.1 *Original Text*

Let us start with the presentation of manuscript reprinted in [19, pp. 211–214].

Sobociński uses two types of individual variables: $X, Y, \ldots; a, b, \ldots$ (symbol a is also used as an individual constant); two-argument predicate ε ($\ldots is \ldots$) in the sense assumed in Leśniewski's ontology; constant $op\alpha$, which appears in such contexts as $X\varepsilon op\alpha(a)$ (*X is a general object in relation to objects a*). We assume here standard notation of logical connectives, quantifiers, and brackets.

He provides the following definitions:

Def=. $\forall X, Y\ (X = Y \leftrightarrow X\varepsilon Y \wedge Y\varepsilon X)$
Def≠. $\forall X, Y\ (X \neq Y \leftrightarrow X\varepsilon X \wedge Y\varepsilon Y \wedge \neg(X = Y))$
Defι. $\forall X, Y\ (X\varepsilon\iota(Y) \leftrightarrow X\varepsilon X \wedge X = Y)$
(*Y is identical with individual X*)
Defδ. $\forall X, Y\ (X\varepsilon\delta(Y) \leftrightarrow X\varepsilon X \wedge X \neq Y)$
(*Y is different from individual X*)

and writes two ontological theses used in the argumentation:

(T1) $\forall X, Y\ (X\varepsilon Y \rightarrow X\varepsilon X)$
(T2) $\forall X \neg(X \neq X)$

Formalizations of the subsequent sentences from Leśniewski's argumentation in Sobociński's version are following[4]:
The empirical assumption:

(Z1) $\exists Z, Y(Z\varepsilon a \wedge Y\varepsilon a \wedge \neg(Z = Y))$, a is an auxiliary constant
there are at least two different [*objects*] *a* [the antecedent of the implication proved by Leśniewski]

The "philosophical"[5] *assumption*:

(A1) $\forall X, Y, Z(X\varepsilon opa(a) \wedge X\varepsilon Z \wedge Y\varepsilon a \rightarrow Y\varepsilon Z)$
1. if X is a general object in relation to objects a, X is Z and Y is a, then Y is Z

Now he formulates the following derivation:

(A2) $\forall X, Z\ (X\varepsilon opa(a) \wedge X \neq Z \wedge Z\varepsilon a \rightarrow Z \neq Z)$
2. if X is a general object in relation to objects a, X is different from Z, and Z is a, then Z is different from Z
(A3) $\forall X, Z, Y\ (X\varepsilon opa(a) \wedge X = Z \wedge Y\varepsilon a \rightarrow Y = Z)$
3. if X is a general object in relation to objects a, X is identical with Z, and Y is a, then Y is identical with Z
(A4) $\forall X, Z\ (X\varepsilon opa(a) \wedge Z\varepsilon a \rightarrow X = Z)$
4. X is a general object in relation to objects a, and Z is a, then X is identical with Z
(A5) $\forall X, Z, Y\ (X\varepsilon opa(a) \wedge Z\varepsilon a \wedge Y\varepsilon a \rightarrow X\varepsilon opa(a) \wedge X = Z \wedge Y\varepsilon a)$
5. if X is a general object in relation to objects a, Z is a, and Y is a, then (X is a general object in relation to objects a, X is identical with Z, and Y is a)
(A6) $\forall X, Z, Y\ (X\varepsilon opa(a) \wedge Z\varepsilon a \wedge Y\varepsilon a \rightarrow Y = Z)$
6. if X is a general object in relation to objects a, Z is a, and Y is a, then Y is identical with Z
(A7) $\forall X \neg(X\varepsilon opa(a))$
7. there is no general object in relation to objects a [the consequent of the implication proved by Leśniewski]

[4]Under the subsequent symbolic expressions we quote Leśniewski's original words from [9] (only variables are changed).

[5]Sobociński uses quotation marks here.

The proof is done in frame of natural deduction system. For auxiliary constant a it is assumed that formulas A1 and Z1 are fulfilled, and as a result: A7 is also fulfilled. Sobociński may have assumed rules that give the possibility to introduce or eliminate auxiliary constants, which are equivalent to those used by Słupecki in [13, pp. 19–25]. Słupecki's calculus fulfills the deduction theorem with restriction on the rule introducing $\forall$. Thus, we can say that Sobociński actually proved the following implication:

$TLS.$ $A1 \wedge Z1 \rightarrow A7$ (a is an individual variable)

or its general closure.

In fact, the subject of Leśniewski's original argumentation is the implication (or its general closure) formalized by TLS (or the general closure of TLS).[6]

Now, our task is to describe the theory which enables to express Sobociński's formalization.

1.2 The Formal Background of Sobociński's Formalization

In the letter to Bocheński we read:

> This formalization shows that Leśniewski's argumentations from that period of time complied with ontology rules, although the last one was constructed only in 1920.[7] [19, p. 214]

We assume that the formal basis of the analyzed argumentation is Leśniewski's ontology, or actually, its small fragment, which, following A. Pietruszczak, we call *little ontology* eO.

The vocabulary of eO comprises: individual variables: $u, v, w, x, y, z, \ldots$; predicate constant ε (*is*), logical connectives: $\neg, \wedge, \vee, \rightarrow, \leftrightarrow, \forall, \exists$, and parentheses: (,). The terms are individual variables. We construct the formulas in the standard way.

The axioms characterizing eO are:

(0) all theses of the classical propositional logic,
the formulas of schemas:

(1) $\forall x A(x) \rightarrow A(y/x)$

(2) $\forall x(A \rightarrow B) \rightarrow (A \rightarrow \forall x B)$, where $x \notin FV(A)$

(3) $\exists x A \leftrightarrow \neg\forall x \neg A$

and Leśniewski's axiom for ε:

AO $x \varepsilon y \leftrightarrow \exists z(z \varepsilon x) \wedge \forall z, u(z \varepsilon x \wedge u \varepsilon x \rightarrow z \varepsilon u) \wedge \forall z(z \varepsilon x \rightarrow z \varepsilon y)$

The primitive rules are **MP**: $\vdash A \rightarrow B, A \Longrightarrow \vdash B$; **Gen**: $\vdash A \Longrightarrow \vdash \forall x A$.

In our presentation we take the definitions of $=, \neq\, \iota$ rewritten in the introduced language and also

($\subset$) $x \subset y \leftrightarrow \forall z(z \varepsilon x \rightarrow z \varepsilon y)$ (subsumption)

($\angle$) $x \angle y \leftrightarrow \exists z z \varepsilon x \wedge \forall z(z \varepsilon x \rightarrow z \varepsilon y)$ (strong subsumption)

---

[6]The actual aim of the proof is discussed by K. Ajdukiewicz in his remarks on T. Kotarbiński's report on Leśniewski's idea [1, p. 227]. If someone accepts A1 (as an axiom), then the argument justifies implication $Z1 \rightarrow A7$, and not the thesis of the non-existence of universals.

[7]Sobociński might have confused the dates of the argumentations of 1913 and 1927.

We also use the following

Fact 1 In the eO theory and its extensions with the appropriate definitions from the above-mentioned ones the following formulas are provable:

(i) $x \varepsilon x \wedge y \varepsilon x \rightarrow x \varepsilon y$
(ii) $x \varepsilon y \wedge y \varepsilon z \rightarrow x \varepsilon z$
(iii) $x \varepsilon y \rightarrow x \varepsilon x$
(iv) $x \varepsilon x \rightarrow (x \varepsilon y \leftrightarrow x \subset y)$
(v) $x \angle y \wedge y \varepsilon z \rightarrow x \varepsilon z$

The proofs for (i)–(iii) and (v) can be formulated in the same way as in [8, p. 33]: cf. T6, T7, T14, for (v)—cf. T119 (p. 37); for (iv) cf. T13.2 in [13, p. 39].

The eO theory is weaker than Iwanuś's elementary ontology and than a stronger version of Leśniewski's EO ontology, where it is assumed that the set of theses comprises all substitutions of the equivalents of the definition schema:

($\star$) $\exists z \forall u (u \varepsilon z \leftrightarrow u \varepsilon x \wedge \phi)$,
where x is any variable different from z and ϕ is any formula in which z is not free.

Schema $\star$ allows us to introduce to EO definitions of name-forming functors and named constants, constructed in the way Leśniewski wanted:

def-f $\forall u (u \varepsilon f(x_1 \ldots x_n) \leftrightarrow u \varepsilon x_i \wedge \phi_f)$, where $x_i \in \{u, x_1, \ldots, x_n\}$

def-n $\forall u (u \varepsilon n \leftrightarrow u \varepsilon u \wedge \phi_n)$

In **def-f** and **def-n** on both sides of $\leftrightarrow$ there are the same free variables.

The relationships between eO, Iwanuś's ontology, and EO are described, for instance, in [12], and we use this source in our presentation.

We use eO_{gen} to mark the conservative extension of eO, created through adding one-argument function constant gen to the dictionary of eO, and through the respective extension of the set of terms and the set of formulas. Symbol gen is a counterpart of $op\alpha$ from Sobociński's formalism.

Symbol $T[F]$ is used for any extension of theory T with formula F.

First, we consider the equivalent of *"philosophical" assumption* $A1$:

$ALS.\ x \varepsilon gen(y) \wedge x \varepsilon u \wedge v \varepsilon y \rightarrow v \varepsilon u$

This implication follows directly from the substitution of **def-f**

$\forall u (u \varepsilon gen(x) \leftrightarrow u \varepsilon u \wedge \forall v (u \varepsilon v \leftrightarrow \forall z (z \varepsilon x \rightarrow z\ \varepsilon v))).$[8]

So, $\mathsf{eO}_{gen}[ALS]$ is consistent.

The crucial step in Leśniewski's argumentation is showing that just with the use of ontology theses and definitions the consequence of *"philosophical" assumption* $A1$ is that "if X is a general object in relation to objects a, Z is a, and Y is a, then Y is identical with Z". That consequence is expressed in implication $A6$ and now it is written

$NR.\ x \varepsilon gen(y) \wedge z \varepsilon y \wedge v \varepsilon y \rightarrow z = v.$

ALS implies formula NR in eO_{gen} with the definition of identity (=). In a similar way to Sobociński, who uses T1, we use thesis (iii) from Fact 1, but later on we significantly

[8]This formula is equivalent to the definition of universal assumed in Luschei's formalization of Leśniewski's argument (cf. [10, p. 309]).

shorten Sobociński's original version. We use Urbaniak's idea, in which the key role is played by thesis (i) [18, pp. 97–98]:

Fact 2 $\mathsf{eO}_{gen}[=] \vdash ALS \rightarrow NR$

Proof

1. $x\varepsilon gen(y) \wedge x\varepsilon x \wedge v\varepsilon y \rightarrow v\varepsilon x$	ALS
2. $x\varepsilon gen(y) \wedge x\varepsilon x \wedge v\varepsilon y \rightarrow v\varepsilon x \wedge x\varepsilon x$	1
3. $x\varepsilon gen(y) \wedge v\varepsilon y \rightarrow v\varepsilon x \wedge x\varepsilon x$	(iii), 2
4. $x\varepsilon gen(y) \wedge v\varepsilon y \rightarrow x\varepsilon v$	(i), 3
5. $x\varepsilon gen(y) \wedge x\varepsilon x \wedge z\varepsilon y \rightarrow z\varepsilon x$	ALS
6. $x\varepsilon gen(y) \wedge v\varepsilon y \wedge z\varepsilon y \rightarrow z\varepsilon v$	(iii), (i), 4, 5
7. $x\varepsilon gen(y) \wedge v\varepsilon y \wedge z\varepsilon y \rightarrow v\varepsilon z$	as: 1–6
8. $x\varepsilon gen(y) \wedge v\varepsilon y \wedge z\varepsilon y \rightarrow z\varepsilon v \wedge v\varepsilon z$	6, 7
9. $x\varepsilon gen(y) \wedge v\varepsilon y \wedge z\varepsilon y \rightarrow v = z$	8, =

Next, we add the definition of the diversity of individuals ($\neq$) to eO_{gen}, and following Sobociński's argumentation, we take into account the counterpart of implication $Z1 \rightarrow A7$:

$RNP.\ z\varepsilon y \wedge u\varepsilon y \wedge z \neq u \rightarrow \neg x\varepsilon gen(y)$

Using just the rules of classical logic we notice that $\mathsf{eO}_{gen}[=, \neq] \vdash NR \rightarrow RNP$ and therefore also the following:

Fact 3 $\mathsf{eO}_{gen}[=, \neq] \vdash ALS \rightarrow RNP$

By classical logic the implication $ALS \rightarrow RNP$ is an equivalent of TLS, proved by Sobociński.

Sobociński comments on the exposition given by Leśniewski in the following way:

> Leśniewski's argumentation does not lead to the conclusion that universals as such do not exist [...]. It only states that the theory of universals in which assumption A1 is used [...] is contradictory. [19, p. 215]

We can say that the main thesis of the theory of universals can be formulated in the following way:

$TP.\ \exists x \exists y(x\varepsilon gen(y) \wedge \exists z \exists u(z\varepsilon y \wedge u\varepsilon y \wedge z \neq u))$

In the closure of our reconstruction we can notice that:

Fact 4 $\mathsf{eO}_{gen}[=, \neq, ALS, TP]$ is a contradictory theory.

(Directly from the Fact 3.)

1.3 Universals in Theories with ALS

Now, we want to consider the consequences of accepting ALS in elementary ontology, and whether it can be assumed that any of these ideas corresponds to the attacked views of the supporters of universals. As we will see, ALS leads to a sort of collapse: it comes out that every universal representing any non-contradictory object is identical with it. In order to show this effect, we add the definition of the contradictory object to eO_{gen} as an axiom:

$(\bigwedge)\ x\varepsilon\bigwedge \leftrightarrow x\varepsilon x \wedge \neg(x\varepsilon x)$

Now, we can notice that:

Observation 1a. $\mathsf{eO}_{gen}[=, \subset, \angle, \bigwedge] \vdash \neg(y \subset \bigwedge) \rightarrow (ALS \rightarrow (x\varepsilon gen(y) \rightarrow x = y))$[9]

Proof

1. $x\varepsilon gen(y) \wedge x\varepsilon x \rightarrow \forall v(v\varepsilon y \rightarrow v\varepsilon x)$ — ALS
2. $\neg(y \subset \bigwedge) \wedge x\varepsilon gen(y) \wedge x\varepsilon x \rightarrow \forall v(v\varepsilon y \rightarrow v\varepsilon x) \wedge x\varepsilon x \wedge \neg(y \subset \bigwedge)$ — 1
3. $\neg(y \subset \bigwedge) \wedge x\varepsilon gen(y) \wedge x\varepsilon x \rightarrow \forall v(v\varepsilon y \rightarrow v\varepsilon x) \wedge x\varepsilon x \wedge \exists z(z\varepsilon y \wedge \neg z\varepsilon \bigwedge)$ — $\subset$, 2
4. $\neg(y \subset \bigwedge) \wedge x\varepsilon gen(y) \wedge x\varepsilon x \rightarrow y\angle x \wedge x\varepsilon x$ — $\angle$, 3
5. $\neg(y \subset \bigwedge) \wedge x\varepsilon gen(y) \wedge x\varepsilon x \rightarrow y\varepsilon x \wedge x\varepsilon x$ — (v), 4
6. $\neg(y \subset \bigwedge) \rightarrow (x\varepsilon gen(y) \rightarrow x = y)$ — (i), =, 5

Restriction $\neg(y \subset \bigwedge)$ is important because: $\mathsf{eO}_{gen}[=, \subset, \angle, \bigwedge] \nvdash ALS \rightarrow (x\varepsilon gen(y) \rightarrow x = y)$. It is sufficient to take into account any $\boldsymbol{\varepsilon}$-structure $< 2^{\{a\}}, \boldsymbol{gen}, \boldsymbol{\varepsilon} >$, in which $\boldsymbol{gen}$, $\boldsymbol{\varepsilon}$ are interpretations of symbols gen and ε respectively, where $\boldsymbol{\varepsilon} = \{< \{a\}, \{a\} >\}, \boldsymbol{gen}(\emptyset) = \boldsymbol{gen}(\{a\}) = \{a\}$. In this case, for the valuation, which gives $\{a\}, \emptyset$ to the variables x, y respectively, formula ALS is fulfilled because of the falsity of its antecedent ($< d, \emptyset > \notin \boldsymbol{\varepsilon}$ for every $d \in 2^{\{a\}}$), and it is not true that $x = y$ ($< \emptyset, \{a\} > \notin \boldsymbol{\varepsilon}$).

On the other hand, we can notice that our equivalent of the *"philosophical" assumption* is derivable from the implication, where it is said that being a universal of any object y is sufficient to be identical with it:

Observation 1b. $\mathsf{eO}_{gen}[=] \vdash (x\varepsilon gen(y) \rightarrow x = y) \rightarrow ALS$

Proof Let us assume that 1. $x\varepsilon gen(y) \rightarrow x = y$. From (ii) and (=) we get: 2. $x = y \rightarrow (x\varepsilon u \rightarrow y\varepsilon u)$, and from AO: 3. $x = y \rightarrow (x\varepsilon u \rightarrow \forall v(v\varepsilon y \rightarrow v\varepsilon u))$. From 1 and 3: $x\varepsilon gen(y) \wedge x\varepsilon u \wedge v\varepsilon y \rightarrow v\varepsilon u$.

Observations 1a and 1b allow to notice that:

Observation 2. Theories $\mathsf{eO}_{gen}[=, \subset, \bigwedge, \neg(y \subset \bigwedge) \rightarrow ALS]$ and $\mathsf{eO}_{gen}[=, \subset, \bigwedge, \neg(y \subset \bigwedge) \rightarrow (x\varepsilon gen(y) \rightarrow x = y)]$ are deductively equivalent.

A supporter of the theory of universals, who would accept ALS in its limitation to universals corresponding to nonempty objects, can equivalently accept a theory in which being a universal of any nonempty object means being identical with it. The problem is that for a supporter of universals, even for one who ascribes to universals the same status of individual objects as to objects for which they are universals, that latter claim is unacceptable and any dispute based on it (or equally on the basis of ALS limited to nonempty objects) is simply futile. This line of defense against Leśniewski's

[9]Let us notice that although the thesis of EO is $\neg\exists x(x\varepsilon\bigwedge)$, also $\exists x(x \subset \bigwedge)$. The fact that $x \subset \bigwedge$ is only equivalent to the fact that: $\neg\exists z(z\varepsilon x)$—we can say that in this situation x is *an empty object*.

argumentation is more effective than trying to modify the definition of a universal, which is criticized even by Sobociński himself:

> I do not know if there is any chance of weakening any premises of assumption $A1$, which would not lead to at least paradoxical conclusions. Adding the intuitive, as it appears, premise: $\forall v(v\varepsilon op\alpha(a) \rightarrow \neg(v\varepsilon a))$ (e.g. a universal of a cat is not a cat) results in a paradoxical thesis. [19, p. 215]

In fact, the assumption considered by Sobociński even strengthens Leśniewski's argument. As it can be easily shown, the assumption

$S.\ x\varepsilon gen(y) \rightarrow \neg x\varepsilon y$

immediately leads to

Observation 3. $\mathsf{eO}_{gen}[=, \angle, \bigwedge, S] \vdash \neg(y \subset \bigwedge) \rightarrow (ALS \rightarrow \neg x\varepsilon gen(y))$
The direct proof from Observation 1a and S.
This time being a universal cannot be reduced to identity, but it is an empty attribute (!).

The solution that does not fall into the described extremes and that accepts the existence of objects that are described in TP could consist of weakening ALS in such a way that we would limit attributes that are vested to universal x: x can have any attributes except *being x itself.*[10] This idea seems to be effective. If we assume:
$ALS^*.\ x\varepsilon gen(y) \rightarrow (\neg u\varepsilon x \wedge x\varepsilon u \wedge v\varepsilon y \rightarrow v\varepsilon u)$
instead of ALS, then we do not get TLS, which leads to a contradiction with TP. In fact:

Observation 4. $\mathsf{eO}_{gen}[=, \neq] \nvdash ALS^* \rightarrow RNP$

Proof Let us remember: $ALS^* \rightarrow RNP$ has the form:
$(x\varepsilon gen(y) \rightarrow (\neg u\varepsilon x \wedge x\varepsilon u \wedge v\varepsilon y \rightarrow v\varepsilon u)) \rightarrow (z\varepsilon y \wedge v'\varepsilon y \wedge z \neq v' \rightarrow \neg x\varepsilon gen(y))$.

We can consider $\boldsymbol{\varepsilon}$-structure $< 2^{\{a,b\}}, \boldsymbol{gen}, \boldsymbol{\varepsilon} >$ (where: $a \neq b$), where $\boldsymbol{gen}$, $\boldsymbol{\varepsilon}$ are interpretations of functors gen and ε respectively, such that: $\boldsymbol{\varepsilon} = \{< \{a\}, \{a\} >, < \{b\}, \{b\} >, < \{a\}, \{a, b\} >, < \{b\}, \{a, b\} >\}, \boldsymbol{gen}(\emptyset) = \boldsymbol{gen}(\{a\}) = \boldsymbol{gen}(\{b\}) = \emptyset$, $\boldsymbol{gen}(\{a, b\}) = \{a, b\}$. For the variables: x, y, z, v' we take the valuation that ascribes to them respectively: $\{a\}, \{a, b\}, \{a\}, \{b\}$. Then: (1) for every u, v-variant of our valuation ALS^* is true; (2) $z\varepsilon y \wedge v'\varepsilon y \wedge z \neq v'$ is true because $< \{a\}, \{a, b\} >, < \{b\}, \{a, b\} > \in \boldsymbol{\varepsilon}$ and (3) $\neg x\varepsilon gen(y)$ is false because $\boldsymbol{gen}(\{a, b\}) = \{a, b\}$.

However, we can say that this improvement of the idea of universals by means of weakening axiom ALS to ALS^* would not be satisfactory for numerous supporters of universals, including the author and the addressee of the letter we are studying. Bocheński, a moderate realist when it comes to universals, wrote about his point of view in the following way "Nominalists claim that in objects there are no universals, and that only general words exist; followers of Plato claim that universals exist in objects. Well, following Aristotle, I reject both views. I deny the existence of universals in objects, but I think that in those objects there is something real that allows us to create general notions" [4, p. 224]. Nevertheless, in his views, general notions are supposed to be "the content of properties" [2, p. 146], and those, in line with the Aristotelian-Thomistic thought, belong to a different ontic category than individuals. Bocheński analyzed the idea of universals twice in [2] and [3]. He expressed his formalizations in the language with two types of

[10] We owe this conception to M. Łyczak.

variables, and in this way he respected Ajdukiewicz's main objection to the nominalist view of universals shown in Kotarbiński's (and Leśniewski's) version. As Ajdukiewicz argues, the linking word "is" is used by Aristotle as a functor of a individual name and a predicative name, or of two predicative names, and the latter use, which is necessary to express the Aristotelian definition of a universal, is absent from Leśniewski's ontology [1, pp. 227–228]. Actually, elementary ontology is a theory of objects belonging to one category, its language contains only one type of variables, and symbol *gen* ($op\alpha$) creates terms belonging to the same category as individual variables. In a language constructed in this way, universals, if they exist, can only belong to the same category as individuals, for which they are universals, so they can only be something similar to Plato's ideas. Thus, if ontology is supposed to be a frame for discussion in the case of universals, then just because of its language the argument may concern some versions of Platonism [10, p. 23]. Sobociński's idea differs from the analyzed formalization in another way, and this is now the point of our interest.

2 Sobociński's Metaconceptualism

In a letter to Bocheński, Sobociński writes the following words:

> [...] the problem of universals does not belong to the area of logic (and to the deductive mathematical systems in general). This means that a system of logic (and mathematics), in whose assumptions there are these or other concepts concerning universals is not a scientific system. However, this does not mean that the problems of universals do not belong to the so-called science. [19, p. 209]

Then, he adds the following commentary on Leśniewski's reasoning:

> I claim that the problem of universals can be positively solved on a completely different ground. Namely: a universal is a deductive theory describing objects belonging to a particular class in an adequate way (but not in a specific way). [19, p. 215]

Actually Sobociński developed this idea in his Remarks added to the correspondence with Bocheński. In the present paper we describe this approach, complemented with a few our original comments. In our sketch we use elements of the semantic description of cognitive processes suggested by R. Suszko from [16]. We set together these two conceptions motivated by their concurrence.

2.1 The Original Text of Remarks

Sobociński divided the exposition of his conception into three parts without titles. In the first part, he presented his intuitions concerning the distinguished components of the cognitive process, which lead to knowledge expressed in certain types of axiomatic systems. He considered the core of each system of this type, that is a set of specified axioms, to be a universal (although sometimes the term "universal" was used by him to name the whole deductive system founded on such a set of axioms). This metatheoretical approach was explained in the second part of the analyzed text. Here the author presented also his own views on the existence of universals, which he described as being conceptualistic and he proposed a specific way of understanding both Platonic and Aristotelian realism. In the third part he provided some negative remarks: he gave examples of theories which, thanks to sufficient formal means, allow us to recreate the definitions of the original ideas determined in other systems in the axiomatic way. In this situation, the recreated idea cannot be a candidate to be a universal in the sense suggested by Sobociński.

Sobociński begins his consideration with the following view of the pre-theoretical phase of the cognitive process:

> The human mind only has the ability to perceive sensory data, as well as the ability to abstract and reason, which is its characteristics. Thus, our view of reality is determined by the properties of our senses and the way we reason. [19, p. 220]

The process of abstraction leads us to "distinguishing in our mind a distributive class [...] of objects that are of interest to us at the moment"[11] The distinguished classes can be *homogeneous* or *heterogeneous*. An example of the former type is a class of all people; the latter can be exemplified by a class of people and natural numbers. When we provide an axiomatic description for a homogeneous class, such a description can be considered as a universal:

> [...] the definition created through the postulates (with the assumption that it is intrinsically non-contradictory) of a class of some homogenous objects, and constructed by our mind from a specific point of view, can be, it appears to me, treated as a "universal" of these objects for the sake of our minds, so, as I think, it is an approach similar to conceptualism. [19, p. 223]

Thanks to the development of the epistemic subject's knowledge, universals can undergo changes:

> The axiomatic system, which is a more or less appropriate definition through postulates, because of its nature must be a system that deductively is highly complicated. Obviously, in most cases we only create a very imprecise and imperfect system of this kind. Improving our knowledge about objects of a given class means improving this system in us. [19, p. 223]

That *improving* is possible within the limits of non-contradiction: "Our only limitation in this respect is the intrinsic contradiction in the system", although the created systems (and the universals that generate them) can be (extrinsically) contradictory with one another, but at least they ought not to be subsystems of one another: "The distributive classes described in such a way are systems standing next to one another, but included in one another". Sobociński connects explications from *Platonic realism* with the problem of what else, apart from axioms, constitutes a component of our knowledge, resulting from the formulated deductive system. The full knowledge of this type is possessed only by God:

> If we had a precise and thorough knowledge concerning a specific deductive system which determines a class of some homogenous objects, then we had a precise description of those objects. Such a knowledge may be possessed only by the infinite mind, that is God, and I think that in this sense universals exist as ideal objects (Platonic realism). [19, p. 223]

[11] In the further part of the text he will add that the effect of abstraction depends also on "our knowledge and other external circumstances".

Some universals can be treated as corresponding to the theory called by Sobociński "Aristotelian realism". Whether a given universal is a universal understood in accordance with *Aristotelian realism* depends on the way individuals characterized by the universal exist. Although in general creating universals is not connected with the (real) existence of objects that they are supposed to refer to, certain universals cannot be universals in the sense of *Aristotelian realism* because their *realization* contradicts the structure of the universe planned by God:

> [...] it is easy for us to create in this way [that is, through determining non-contradictory axioms for a given set] concepts of objects that not only do not exist in reality, but that cannot even be realized, for instance, the concept of "Greek gods". Thus, I claim that we can even possess such a concept, but it is determined by such assumptions that are too weak to cause any intrinsic contradiction in the system. However, if we modify such a system by adding assumptions concerning the existence of God, then there is an intrinsic contradiction. In other words, in any concept of "Greek gods" the term "god" is used with so weak a meaning that it does not cause any contradiction in the system. Concepts of this kind cannot be realized (in my understanding of Aristotelian realism) because they might cause a contradiction in the universe. Concepts that are less roughly thought could be realized, as I think, if God wanted it and if they did not clash with his plan of the world, in which we dwell. [19, p. 224]

A system, whose axioms constitute a universal "must possess its own deductive basis, that is a system of logic", which is "an integral part of a given system". The choice of a particular logic depends on "the will and the starting point" of the epistemic subject.

Comparing universals must be done on the basis of one logic, which may differ from logics that function as the formal ground for these universals. Such an external logic—*a general logic*—ought to be "extensional, saturated with functors because of sentential functors,[12] and it should be compatible with properties of colloquial language. These assumptions mean that the logic must be the two-valued logic". General logic ought not to contain any existential assertions, and the assertions concerning the logic itself can only have the relative character (as it happens in the case of the set theory).

In the last part of Remarks Sobociński provides examples, which show that for him a universal is the way of describing certain concepts, and not concepts themselves. Real numbers can be defined on the ground of the theory of natural numbers (with the help of Cauchy sequences or Dedekind cuts), but this definition is not a universal. In this situation real numbers are *logical creations*, which differ from the categorial point of view from natural numbers, which are identified with the use of an object of the lowest rank, representing a universal, a natural number. The construction of a universal of a real number means the determination of the axiom of a deductive theory, in which axioms are true in the set of real numbers:

> I claim that the right idea of one or another arithmetic is based on its definition through postulates, among which there must be assumptions of one or another logic. This means that the concepts of natural numbers are given to us by logic and, for instance, Peano axioms. In a similar way we receive concepts of rational numbers, real numbers, complex numbers, etc. as well as other so-called mathematical objects. [19, p. 225]

[12]For any sentential functor, an expression created by this functor and any sentence, should also have a logical value.

Finally, Sobociński presents also his conceptualistic view on the existence of mathematical objects, which corresponds with his theory of universals:

> Thus, do numbers and other mathematical objects exist? Undoubtedly, they exist as concepts in our minds. In people who are not advanced at mathematics these concepts are imperfect [because they are described by relatively weak universals], but in people who are experienced in mathematics, they are much better developed [because their universals have been "improved"]. Do they exist in reality (in the sense of moderate realism)? They do not exist as real objects. However, in concepts (universals) of real objects, which, in line with the approach presented here, are deductive systems there is one or another mathematical theory. Thus, objects realized in accordance with a universal understood in this way fulfill, among others, the rules of this theory. For that reason we may use assertions belonging to this theory when we examine and describe these objects. [19, p. 225]

2.2 *Sobociński's Universals in the Frame of Suszko's Epistemic Approach*

Sobociński expressed his conceptualist standpoint in the frame of the metatheory of deductive systems and for that reason we call it "metaconceptualism". Obviously, this notion exceeds metalogic, because of the fact that some of its important fundamental assumptions concern human activities that are considered still in epistemology. The description of the pre-theoretical cognitive processes, whose results are deductive theories, was undertaken also by Suszko in [16] and this description is now used. Suszko meant his epistemological analysis to be an introduction to research into diachronic logic. That logic was supposed to help in the description of the changeability of knowledge that involves improving *the conceptual apparatus* associated with a chosen model or improving knowledge (with the use of a given *conceptual apparatus*).

Suszko's approach and Sobociński's theory turn out to be very similar when it comes to certain observations and solutions. For that reason we use distinctions suggested by Suszko in our explanations of Sobociński's conception.[13] Our proposal, just like Suszko's epistemological analysis, is only a partially formalized sketch, which we formulate in order to encourage our readers to develop further research, and which we do not consider to be a ready interpretation of Sobociński's views.

Suszko starts his reflection with combining certain a conception of cognitive acts and the model-theoretic view on the interpretation of languages and theories, which are results of those acts.[14] He restricts his investigations to theories expressed in standard formal languages, that is, such languages that can be described with Ajdukiewicz's method of semantic categories. In line with Suszko's conception, for every language it is possible to assign a family of models founded on any non-empty set of individuals, that differ among

---

[13]In 1946, only a few days before Sobociński left Poland forever, the young Suszko visited him in Warsaw. Many years later Suszko told Professor Mieczysław Omyła about that only meeting with Sobociński, saying that they discussed logic. We do not know what problems they discussed during that meeting. Although the similarities between the approaches described by us may be accidental, to us it seems to be interesting also in the context of that meeting.

[14]Suszko was involved in his research from the early 1950s. The text of 1966 [16] is a summary of this research. Technical details assumed in [16] which are assumed also here, are elaborated in [15].

one another when it view of their *characterization*, which contains properties and relations occurring between these individuals—the so-called *aspects* of individuals. Equivalently, having in mind a specific model, we may link with it a family of symbolic languages that can be used to describe that model. The language that is interpreted in our model in accordance with the *certain circumstances of the sense of that language* constitutes *the conceptual apparatus* for the selected model, and that model is called "the proper model" in those circumstances.

Developing the conceptual apparatus for the proper model is the *primary cognitive act*: "its course is obviously extremely complicated. [...] conducting the development of the conceptual apparatus for model M in a given group of people would not be possible unless that group of people were involved in a sort of direct contact with model M as part of their activity (sensory perception, practical activity)." [16, pp. 529–530]. Following Sobociński we would say that *the ability to perceive sensory data* precedes the ability of abstraction and this results in distinguishing the set of individuals, which we are going to describe in the frame of a formed deductive system. The effect of abstraction, about which Sobociński speaks, would be named by Suszko "the characterization of a given set of considered individuals" which constitute a given model.

Let us consider Sobociński's theory with reference to the systems expressed in such languages that are described by Suszko. The parameter of *circumstances*, which, according to Suszko, involves *the social group* that uses a particular language and *the totality of activities performed by the members of that group*, will be replaced with the relativization of the proper model to the defined epistemic subject x, who has a knowledge, which determines state i.

Following Suszko, we accept that

Definition 1 *A model considered by* x *in* i is an ordered pair:

$$M^i_{\mathsf{x}} =< U, C^i_{\mathsf{x}} >$$

where $U \neq \emptyset$ is a set of individuals and $C^i_{\mathsf{x}} = \{c_1, \ldots, c_n\}$ is a characterization of individuals from U considered by x in i.

Every model M^i_{x} is an effect of *an abstraction* undertaken by x.

Depending on a certain state of knowledge, x may investigate different characterizations.

All models founded on a given U, which may be considered by x form *epistemic space of* x:

Definition 2 Epistemic space of x is $\mathbb{M}_x = \{U\} \times \{C^i_{\mathsf{x}}\}_{i \in I}$.

(Perhaps, if x is a human being the family $\{C^i_{\mathsf{x}}\}_{i \in I}$ should be finite.)

Let us now consider some given element of $\mathbb{M}_{\mathsf{x}}$: $M_{\mathsf{x}} =< U, C_{\mathsf{x}} >$ with $C_{\mathsf{x}} = \{c_1, \ldots, c_m\}$ which is determined by the given stage of conceptual knowledge of x.

M_{x} determines a certain type of symbolic languages adequate to describe M_{x} which we call $\mathfrak{L}(M_{\mathsf{x}})$. Each of these languages contains specific terms (non-logical constants) which correspond to elements of C_{x}. Languages from $\mathfrak{L}(M_{\mathsf{x}})$ do not have more specific terms as is needed to describe elements of C_{x}.

We choose certain language $\mathcal{L} \in \mathfrak{L}(M_{\mathsf{x}})$ as the *conceptual apparatus* used by x to describe model M_{x}. We call M_{x} *a proper model of* $\mathcal{L}$. Now, epistemic subject x chooses certain logic L expressed in language $\mathcal{L}$. According to Sobociński, L may be non classical logic. We use the notation $\mathsf{L}[\mathbf{A}]$ to speak about extension of L by the set of formulas $\mathbf{A} \subseteq \mathcal{L}$. The set of all sentences of $\mathcal{L}$ true in model M_{x} we name like Suszko $Ver(\mathcal{L}, M_{\mathsf{x}})$.

Extensions of L are theories based on L. The act of forming a theory based on L is named by Suszko as "the secondary cognitive act" [16, p. 530]. At least some sets of formulas which may be added to L are considered by Sobociński as *universals*.

We take the set of all L-consistent theories which are adequate to M_{x}:

Definition 3 $\mathbb{T}(M_{\mathsf{x}}) = \{T : \exists_{\mathbf{A}}\, T = \mathsf{L}[\mathbf{A}] \text{ and } T \subseteq Ver(\mathcal{L}, M_{\mathsf{x}})\}$

We use primitive predicate $\mathsf{GEN}(\mathsf{x}, \mathbf{A})$ to be read: *set (of formulas)* $\mathbf{A}$ *is an universal for* x.

Let us propose the following few axioms characterizing universals in Sobociński's sense:

(A1) $\mathsf{GEN}(\mathbf{A}, \mathsf{x}) \Rightarrow \mathsf{L}[\mathbf{A}] \in \mathbb{T}(M_{\mathsf{x}})$ (*every universal is an axiomatization of some theory adequate for the proper model*)
(A2) $\exists_{\mathbf{A}}(\mathsf{GEN}(\mathbf{A}, \mathsf{x}))$ (*conceptualism*)
(A3) $\mathsf{GEN}(\mathbf{A}, \mathsf{x}) \Rightarrow \mathbf{A} \neq \emptyset$ (*universals are not empty*)
(A4) $\mathsf{GEN}(\mathbf{A}, \mathsf{x}) \Rightarrow \mathbf{A} \cap \mathsf{L} = \emptyset$ (*universals do not belong to logic*)

In order to express Sobociński's belief, which he calls "Platonic realism", we take the abbreviation $G\mathsf{x}$ to be read x *is God*.

Following Sobociński we say that God knows all consequences of any his universal:

(A5) $G\mathsf{x} \Rightarrow (\mathsf{GEN}(\mathbf{A}, \mathsf{x}) \Rightarrow \mathbf{A} = \mathsf{L}[\mathbf{A}] - \mathsf{L})$

Concerning the question of *improving* of universals we would say only that:

(A6) $\mathsf{GEN}(\mathbf{A}, \mathsf{x})$ and $\mathbf{A} \subseteq \mathbf{B}$ and $\mathbf{B} \subseteq Ver(\mathcal{L}, M_{\mathsf{x}})$ and $\mathbf{B} \cap \mathsf{L} = \emptyset \Rightarrow \mathsf{GEN}(\mathbf{B}, \mathsf{x})$

Let us note that the development of a given deductive system in order to *improve* a universal is connected only with adding axioms to a new theory, but not with enriching *the conceptual apparatus*. Changing the *conceptual apparatus* causes the need to redefine a selected model that we aim at and selecting a respectively changed language.

Our axioms allow us to formulate at least a few dependencies which we find in Sobociński's text:

(t1) $\mathsf{GEN}(\mathbf{A}, \mathsf{x}) \Rightarrow \mathbf{A}$ is L-consistent (A1)
(*universals are non-contradictory with their logic*)
(t2) not $(\mathsf{GEN}(\mathsf{L}, \mathsf{x}))$ (A4)
(*logic is not a universal*)
(t3) $\mathsf{GEN}(\mathbf{A}, \mathsf{x})$ and $\mathbf{A} \subseteq \mathbf{B} \Rightarrow \mathbf{B} \neq \mathsf{L}$ (A4)
(*improving of universals is done outside of logic*)
(t4) $\mathsf{GEN}(\mathbf{A}, \mathsf{x})$ and $\mathsf{GEN}(\mathbf{B}, \mathsf{x}) \Rightarrow \mathsf{GEN}(\mathbf{A} \cup \mathbf{B}, \mathsf{x})$ (A1, A6, Def3)
(*the sum of universals is also a universal*)
(t5) $\mathsf{GEN}(\mathbf{A}, \mathsf{x})$ and $\mathsf{GEN}(\mathbf{B} - \mathbf{A}, \mathsf{x}) \Rightarrow \mathsf{GEN}(\mathbf{B}, \mathsf{x})$ (A1, A6, Def3)
(*if* $\mathbf{B}$*-complement of universal* $\mathbf{A}$ *is a universal, then* $\mathbf{B}$ *is also an universal*)

We hope that Sobociński would accept the above considerations as a kind of sketch of *a general* theory of universals (obviously, requiring further complementation). Sobociński might expect us to include the conditions of adequacy and independence of axioms constituting each universal along with the condition of non-contradiction.[15] Sobociński lists these requirements, following Leśniewski, in [14] as the basis for all *well-constructed axiom systems*. Axiom systems ought to fulfill also other conditions, which were classified by Sobociński into the following groups: conditions concerning the primitive terms of axiom theories, specific conditions concerning groups of axioms from any theories and conditions that generate Leśniewski's systems or theories based on them. All additional requirements in his views result from these basic ones and their nature is rather *aesthetic*. We believe that a contemporary analysis of them might show that their nature is rather pragmatic, connected with the pragmatic criteria of choosing axioms for a particular theory. This question remains open.

The conditions of *well-constructed* axiom systems grouped by Sobociński only have the necessary character. If universals are supposed to be *well-constructed* axiom systems, these conditions only have such a meaning for them. What then decides that a given axiom system is a universal in Sobociński's understanding?

Let us analyze two of the already described theories.

In line with Fact 4, set $\{=, \neq, ALS, TP\}$ is non-consistent on the ground of eO_{gen} and as such it cannot be a universal for a given epistemic subject, whose cognitive apparatus is the language of system eO_{gen}, chosen by that subject to describe any model from his epistemic space, which is suitable for being an interpretation of that language. Next, set $\{ALS\}$ is consistent on the ground of eO_{gen} and it fulfills **all** the conditions of *well-constructed* axiom systems for theory $\mathsf{eO}_{gen}[ALS]$. Is $\{ALS\}$ that determines the notion of a universal in Leśniewski's understanding suitable to be a universal in Sobociński's understanding? Let us take into account epistemic subject S that uses system eO_{gen} enriched with equivalences $=, \subset, \angle, \bigwedge$ and ι, which he treats as theses of the logical nature. In line with Observation 1a, on the basis of a selected formal ground, subject S will be forced to accept that the notion of a universal in Leśniewski's understanding is essentially weaker than identity in the sense of ι, because in his theory ALS implies that $x\varepsilon gen(y) \rightarrow x\varepsilon\iota(y)$, if $\neg(y \subset \bigwedge)$. Next, let us assume that the right model for S is M^*_{S}—ordered pair consisted of a non-empty set U, founded on some set of objects Ind: $\mathsf{U} \subseteq 2^{Ind}$ (elements of set U are individuals in our model) and characteristic C^*_{S}, containing at least: relation $\boldsymbol{\varepsilon}$, a semantic correlate of predicate ε, $\emptyset$—the semantic correlate of constant $\bigwedge$, and functions: $\boldsymbol{gen}, \boldsymbol{\iota} : \mathsf{U} \rightarrow \mathsf{U}$, which are counterparts of constants gen and ι. In line with what is determined by the axioms of system $\mathsf{eO}_{gen}[=, \subset, \angle, \bigwedge, \iota]$ extended by ALS we notice that the function of being a universal in Leśniewski's understanding, that is $\boldsymbol{gen}$ assigns the following: (a) to every singleton the same singleton, (b) to every set bigger than a singleton it assigns $\emptyset$, and (c) to $\emptyset$ it assigns $\emptyset$ or any singleton. Only the second version of case (c) distinguishes function $\boldsymbol{gen}$ from identity $\boldsymbol{\iota}$ (!), which assigns the same singleton to every singleton, and $\emptyset$ to every non-singleton. It's hard to say if S would accept $\{ALS\}$ as a universal in Sobociński's understanding in this situation. It can

[15]The notion of the adequacy of axioms was distinguished by Sobociński from the notions of the adequacy and completeness of theories. We assume the following simplification of his explanation: (1) a set of formulas expressed in a language of a given theory is adequate when the set of all the theses of a given theory can be deduced from that set of formulas (cf. [14, pp. 56–57]); (2) independence is understood as such a property of a set of axioms that none of its elements can be deduced on the grounds of a specific theory from other axioms.

be said that there are no formal obstacles to do that, but it is hard to accept that being a universal in Leśniewski's understanding, which corresponds to function ***gen*** from C^*_{S} constitutes an interesting (not only from the philosophical point of view) *aspect* of objects from set U.

Acknowledgements We would like to express our gratitude to Professor M. Omyła for his advice concerning R. Suszko's epistemological analysis, and to M. Łyczak for valuable seminar discussions. Fragments of Sobociński's and Suszko's texts were translated by Marcin Kuczok.

References

1. Ajdukiewicz, K.: W sprawie uniwersaliów. [On universals] Przegląd Filozoficzny **37**, 219–234 (1934)
2. Bocheński, J.M.: Powszechniki jako treści cech w filozofii św. Tomasza z Akwinu. [Universals as the content of properties in st. Thomas Aquinas' philosophy] Przegląd Filozoficzny **41**, 136–149 (1938)
3. Bocheński, J.M.: The Problem of Universals. In: Bocheński, J.M., Church, A., Goodman, N. (eds.) The Problem of Universals, pp. 33–54. Notre Dame Press, Notre Dame (1956) (Polish translation: Zagadnienie powszechników. In: Parys, J. (ed.) Logika i filozofia. Wybór pism, pp. 79–105. PWN, Warszawa (1993). Trans. T. Baszniak)
4. Bocheński, J.M.: Wspomnienia. [Recollections] Philed, Kraków (1994)
5. Grygianiec, M.: Leśniewski przeciw powszechnikom. [Leśniewski against universals] Filozofia Nauki **3/4**, 109–125 (2000)
6. Kotarbiński, T.: Sprawa istnienia przedmiotów idealnych. [The Problem of the Existence of Ideal Objects] In: Księga pamiątkowa ku uczczeniu 25-letniej działalności nauczycielskiej na katedrze filozofii w Uniwersytecie Lwowskim Kazimierza Twardowskiego, pp. 149–170. Uniwersytet Lwowski, Lwów (1921)
7. Kotarbiński, T.: Elementy teorii poznania, logiki formalnej i metodologii nauk. [Elements of Epistemology, Formal Logic, and Methodology of Sciences] Ossolineum, Wrocław-Warszawa-Kraków (1961)
8. Leśniewski, S.: Definitions and theses of Leśniewski's Ontology. In: Srzednicki, J.T.J., Stachniak, Z. (eds.) S. Leśniewski's Lecture Notes in Logic, pp. 29–58. Kluwer Academic Publishers, Dordrecht/Boston/London (1988)
9. Leśniewski, S.: O podstawach matematyki [On the foundation of mathematics]. Przegląd Filozoficzny **30**, 164–206 (1927); In: Jadacki, J. (ed.) Leśniewski. Pisma zebrane [Leśniewski. Collected Works], vols. 1 and 2, pp. 295–337. Semper, Warszawa (2015)
10. Luschei, E.C.: The Logical Systems of Leśniewski. North-Holland Publishing Company, Amsterdam (1962)
11. Murawski, R.: Cracow circle and its philosophy of logic and mathematics. Axiomates **25**, 359–376 (2015)
12. Pietruszczak, A.: O teoriach pierwszego rzędu związanych z elementarnym fragmentem ontologii Leśniewskiego. [On first order theories connected to Leśniewski's fundamental fragment of ontology] In: Perzanowski, J., Pietruszczak, A. (eds.) Logika and Filozofia Logiczna. [Logic and Logical Philosophy] FLFL 1996–1998, pp. 127–168. Wydawnictwo UMK, Toruń (2000)
13. Słupecki, J.: S. Leśniewski's Calculus of Names. Stud. Logica **3**(1), 7–73 (1955)
14. Sobociński, B.: On well-constructed axiom-systems. Rocznik Polskiego Towarzystwa Naukowego na Obczyźnie **6**, 54–65 (1955/1956)
15. Suszko, R.: Syntactic structure and semantical reference I. and II. Stud. Logica **VIII**, 213–144 (1958), **IX**, 63–93 (1960)
16. Suszko, R.: Logika formalna a niektóre zagadnienia teorii poznania. [Formal logic and some aspects of epistemology] In: Pawłowski, T. (ed.) Logiczna teoria nauki. Wybór artykułów. [Logical Theory of Science. A Selection of Articles] PWN, Warszawa, 505–576 (1966)
17. Świętorzecka, K., Porwolik, M.: O pewnym przesądzie dotyczącym uniwersaliów. Uwagi do sformalizowanego przez Bolesława Sobocińskiego argumentu na rzecz tezy o nieistnieniu powszechników

podanego przez Stanisława Leśniewskiego [About a superstition concerning universals. Comments on Bolesław Sobociński's formalised argument for the non-existence of universals as presented by Stanisław Leśniewski]. In: Brożek, A., Chybińska, A., Grygianiec, M., Tkaczyk, M. (eds.) Myśli o języku, nauce i wartościach. Seria druga. Profesorowi Jackowi Juliuszowi Jadackiemu w 70. rocznicę urodzin [Thoughts about Language, Science and Values. Second series. To Professor Jacek Juliusz Jadacki on the 70th Anniversary of His Birth]. Semper, Warszawa 252–263 (2016)
18. Urbaniak, R.: Leśniewski's Systems of Logic and Foundations of Mathematics. Springer, Cham (2014)
19. Waragai, T.: Letters and typescripts send to Bocheński from Sobociński: Historical Documents about Leśniewski's refutation of universals. Reports of the Keio Institute of Cultural and Linguistic Studies **36**, 203–228 (2004)
20. Woleński, J.: Szkoła Lwowsko-Warszawska w polemikach. [The Lvov-Warsaw School in Polemics] Scholar, Warszawa (1997)

K. Świętorzecka • M. Porwolik (✉)
Institute of Philosophy, Cardinal Stefan Wyszyński University in Warsaw, Warsaw, Poland
e-mail: k.swietorzecka@uksw.edu.pl; m.porwolik@uksw.edu.pl

Many-Valued Logics in the Iberian Peninsula

Angel Garrido

To Prof. Dr. Solomon Marcus (1925–2016), who recently passed away.

Abstract The roots of the Lvov-Warsaw School (LWS, by acronym) can be traced back to Aristotle himself. But in later times we better put them into thinking GW Leibniz and who somehow inherited many of these ways of thinking, such as the philosopher and mathematician Bernhard Bolzano. Since he would pass the key figure of Franz Brentano, who had as one of his disciples to Kazimierz Twardowski, which starts with the brilliant Polish school of mathematics and philosophy dealt with. Among them, one of the most interesting thinkers must be Jan Lukasiewicz, the father of many-valued logic.

Jan Łukasiewicz (1878–1956) began teaching at the University of Lvov (now Lwiw; former Lemberg, but also Leópolis), and then at Warsaw, but after World War II must to continue in Dublin. Some questions may be very astonishing in the CV of Łukasiewicz. For instance, that a firstly Polish Minister of Education in Paderewski cabinet, into the new Polish Republic, and also Rector for two times at Warsaw University, was awarded with a Doctorate 'Honoris Causa' in spring 1936, at University of Münster, into the maximum of effervescency of Nazism in Germany. The explanation must be their good relation with a very good friend, the former theologian, and then logician, Heinrich Schölz, who was the first Chairman of Mathematical Logic in German universities.

Łukasiewicz firstly studied Law, and then Mathematics and Philosophy in Lvov (then Lemberg). His doctoral supervisor was Kazimierz Twardowski, and in 1902 he obtain his Ph. D. title with a very special mention: 'sub auspiciis Imperatoris' (i.e., under the auspices of the Kaiser). Also he received a doctorate ring with diamonds from the Kaiser of the Austro-Hungarian Empire, Franz Joseph I.

From 1902, Łukasiewicz was employed as a private teacher, and also as a desk in the Universitary Library of Lvov. So it was until 1904 when he obtained a scholarship to study abroad. He defends his 'Habilitationschrift' in 1906, entitled "Analysis and construction of the concept of cause". This permits to give university courses. His first lectures were on the *Algebra of Logic,* according to the recent translation to Polish of this book of the French logician Louis Couturat.

Between 1902 and 1906, Lukasiewicz continued his studies in the universities of Berlin and Leuwen (Lovaina). In 1906, by his 'Habilitationschrift', he obtain the qualification as

Á. Garrido, U. Wybraniec-Skardowska (eds.), *The Lvov-Warsaw School. Past and Present*, Studies in Universal Logic,
https://doi.org/10.1007/978-3-319-65430-0_43

university professor at Lvov. And the, in 1911, he was appointed as associate professor in his ‘alma mater’ (Lemberg).

Jan Łukasiewicz was also very active in historical research on logic, giving a new and up-to-date interpretation of Aristotle’s syllogism and of the Stoics’ propositional calculus. According to Scholz, the better pages on history of logic are due to him. And also, as Arianna Betti says, “Jan Lukasiewicz is first and foremost associated with the rejection of the Principle of Bivalence and the discovery of Many-Valued Logic.”

The discovery of MVL by Lukasiewicz was in 1918, a little earlier than Emil Leon Post. According to Jan Wolenski, “although Post’s remarks were parenthetical and extremely condensed, Lukasiewicz explained his intuitions and motivations carefully and at length. He was guided by considerations about future contingents and the concept of possibility”. So, he introduces, firstly, three-valued logic, then four-valued logic, generalized to logics with an arbitrary finite number of veritative values, and finally, to logics with a countably infinite-valued number of such values.

Very noteworthy is his treatment of the history of logic in the light of the new formal logic (then called Logistics). Thus, not only he addressed the issue of future contingents departing from Aristotle, but also put in value logic of the Stoics, at least so far taken. In fact, Heinrich Scholz said, rightly, that Lukasiewicz had written the most lucid pages on the history of logic.

Keywords Aristotle · G.W. Leibniz · Non-classical logics · Many-valued logics · Fuzzy logic · Fuzzy measures and integrals · Applied logic in the Iberian Peninsula

Mathematics Subject Classification (2000) Primary 01A20, 01A55, 01A60, 03A05, 03B50, 03B52, 03B60, 03B80, 03E72, 03E75, 03G20, 08E72

1 Introduction

As we know, logic is the study of the structure and principles of correct reasoning, and more specifically, attempts to establish the principles that guarantee the validity of deductive arguments [14, 15]. The central concept of validity is for logic, because when we affirm the validity of an argument are saying that it is impossible that its conclusion is false if its premises are true.

Propositions are descriptions of the world, that is, are affirmations or denials of events in various possible worlds, of which the “real world” is just one of them [20, 21, 30, 31]. There is a long philosophical tradition of distinguishing between truth necessary (a priori or “logical”) and facts “contingent” (a posteriori or “factual”).

Both have really led the two concepts of logical truth, without being opposed to each other, are quite different: the conception of truth as coherence, and the conception of truth as correspondence. According to the point of view of consistency, a proposition is true or false depending on their relationship with respect to a given set of propositions, because of the rules of that system. Under the terms of correspondence, a proposition is true or false, if it agrees with reality, that is, the fact referred to [22, 24].

To further enhance the complexity of the problem, not only analyze trueness or falsity of propositions, but also of theories, ideas and models [10, 19]. And so, we allow new and different conception of truth.

The basic idea underlying all these approaches is that of an intrinsic dichotomy between true and false. This opposition implies the validity of two fundamental laws of classical logic:

- Principle of excluded middle: Every proposition is true or false, and there is another possibility.
- Principle of non-contradiction: No statement is true and false simultaneously.

Such fundamental ideas produce some series of paradoxes and dissatisfaction that is based on the need to overcome this strict truth-bivalence of classical logic.

Searching for the origins could lead too far and eventually disperse, which, as we know is not very convenient for a job pretending to be research. So we will refer to these first signs that appear in the East (China, India...), and then we may analyze the problem of "future contingents", treated by Aristotle in *Peri hermeneias*.

> For a man may predict an event ten thousand years beforehand, and another may predict the reverse; that which was truly predicted at the moment in the past will of necessity take place in the fullness of time. (Aristotle, *Peri Hermeneias*, ch. 9)

About Future Contingent Propositions, we must remember that they are statements about states of affairs in the future that are neither necessarily true nor necessarily false. Suppose that a sea-battle will not be fought tomorrow. Then it was also true yesterday (and the week before, and last year) that it will not be fought, since any true statement about the case that will be was also true in the past. But all past truths are now necessary truths; therefore it is now necessarily true that the battle will not be fought, and thus the statement that it will be fought is necessarily false. Therefore it is not possible that the battle will be fought. In general, if something will not be the case, it is not possible for it to be the case [23, 29, 36].

As we know, although the starting point of Leibniz "calculus universalis" were Stagirite's theories, Leibniz ends to be dependent from the ideas of Aristotle, to finally develop its own axiomatic system, a more general type, based on applying the Combinatorial Instrument to syllogistic [18, 34].

That issue (Future Contingent's problem, with variations) would be then crucial in medieval times, as during the Scholasticism, with William of Ockham, and Duns Scotus, or Richard of Lavenham, among others, looked at from different point of views, for its relationships with Determinism and 'Divine Foreknowledge'. Then, this issue is taken up by Spanish Jesuit F. Luis de Molina (and the famous controversy 'De Auxiliis' maintained with the Dominican Fray Domingo Báñez), or Francisco Suarez, and even the great polymath G.W. Leibniz dedicated his time [15, 20].

The controversial "De Auxiliis" involves two key works: the *Concordia*, from the Jesuit Father Luis de Molina (1535–1600), and the *Apology*, from Fray Domingo Báñez (1528–1604), a Dominican School of Salamanca and San Esteban's Convent. In essence, it represented the possible antagonism between free will of humans and efficacy of Divine Grace. In short: Is Omniscience and Omnipotence compatible with the man's freedom? The discussion took a particularly interesting way during the Middle Age. In this period philosophy was interconnected with theology. And one of the most important theological issues was precisely the problem of future contingents, in its direct relationship with

Christian doctrine. According to this tradition, related with the Divine Foreknowledge. It includes knowledge of future possibilities to be made by human beings. But this assumption seems to lead to a simple argument. It leads from foreknowledge to the need of future events: now known as God and I will take the decision tomorrow, it's true that my choice of morning is given. My choice then, it seems necessary but not free. Therefore, there appears to be no basis for claim that we have freedom of choice among alternatives. The conclusion, however, would violate the idea of human freedom and of moral responsibility [16, 35, 37].

Even then there is a dark time for the logic, and reappearing in the nineteenth century, philosophers and mathematicians such as George Cantor, Augustus De Morgan and George Boole, Gottlob Frege, ... There was born the new set theory, now called "classic", but then also had terrible enemies, as the then almighty Leopold Kronecker, who from his professorship in Berlin did everything possible to hinder the work of Cantor, and the rise of those new ideas.

As Kluge said, Frege's logico-mathematical and philosophical speculations were not historically isolated phenomena that arose completely de novo, as it were like fulgurations of genius out of a conceptually unclouded sky. They were more like nodal points in a long series of speculative endeavors that began with people like Raymon Lull in the Middle Ages, continued through René Descartes, Athanasius Kircher, Jakob Böhme, and G.W. Leibniz, that drew on the thoughts of people like Giordano Bruno and Blaise Pascal, J.D. Gergonne and Thomas Hobbes, Pierre Gregoire and Bernhard Bolzano, and in turn constituted the basis of much contemporary thought—the works of Bertrand Russell and Rudolf Carnap, Edmund Husserl and Ludwig Wittgenstein, Alonzo Church, Strawson, and Willard van Orman Quine immediately come to mind.

Remember that, and according to SEP (Stanford Encyclopedia of Philosophy), Bernard Bolzano (1781–1848) was an outstanding mathematician and one of the greatest logicians or even (as some would have it) the greatest logician who lived in the long stretch of time between G.W. Leibniz and Gottlob Frege. As far as logic is concerned, Bolzano anticipated almost exactly 100 years before Alfred Tarski and Rudolf Carnap their semantic definitions of logical truth and logical consequence; and in mathematics he is not only known for his famous Paradoxes of the Infinite, but also for certain results that have become and still are standard in textbooks of mathematics such as the Bolzano-Weierstrass theorem. Bolzano also made important contributions to other fields of knowledge in and outside of philosophy. Due to the versatility of his talents and the various fields to which he made substantial contributions, Bolzano became one of the last great polymaths in the history of ideas.

2 Many-Valued Logics and the Lvov-Warsaw School

Parallel to this, there arises a new kind of thought and way of seeing must be the act of philosophizing: the Polish Lwow-Warsaw School (LWS, by acronym). This is happening like tributaries of a great river and sub-tributaries, departing from Leibniz, from masters to disciples [11, 18, 35, 38, 42, 43]. Start with the aforementioned Bernard

Bolzano, which influence-much about his intellectual heir, Franz Brentano. This, in turn, greatly influence on all his subsequent students. Among these disciples of Franz Brentano will be one that particularly interested us. This was the Polish philosopher Kazimierz Twardowski, who shared many characteristics with his teacher: love for precision and clarity of ideas, charisma among those who treated him, preference for the spoken to the written word, etc... From his chair in the city of Lvov spread many of the ideas of Franz Brentano, adding their own. Its members took the logical-philosophical and mathematical studies in Poland to the forefront of global world research. It was during the "interbellum", or period between the two World Wars, i.e. ranging from 1918–1939. Then, rouse the Diaspora, after the war and by the strong communist dictatorship. Jan Łukasiewicz introduced the three-valued logic and then generalized to the infinite-valued [22, 24]. He was the effective mentor of Alfred Tarski, whereas officially it was Stanisław Leśniewski.

The biographers of Tarski, Anita and Solomon Feferman, state that "along with his contemporary, Kurt Gödel, he changed the face of logic in the twentieth century, especially through his work on the concept of truth and the theory of models." Tarski had gone to the US to participate in a conference when Nazi troops invaded his native Poland and could not return to it. Over time, he created in California the most powerful logical school of his time; in fact, you can consider continuing the tradition inaugurated by the LWS, outside the continent in ruins (Europe). Its 'Semantic Theory of Truth' is one of the greatest achievements of the human thinking of all time.

Many notable names were among the members of this school of logic, but could cite [35, 37, 38] to:

- Jan Łukasiewicz,
- Stanisław Leśniewski,
- Kazimierz Ajdukiewicz,
- Tadeusz Kotarbiński,
- Mordechai Wajsberg,
- Alfred Tarski,
- Jerzy Słupecki,
- Andrzej Mostowski.

Also must be cited:

- Jan Woleński (as vindicator of the LWS' memory),
- Helena Rasiowa,
- Roman Sikorski,
- Zdzisław Pawlak,
- Andrzej Skowron,
- Roman Murawski,
- etc.

Among them, one of the most interesting must be Jan Łukasiewicz, the father of many-valued logics (MVLs, by acronym). Jan Łukasiewicz began teaching at the University of Lvov, and then at Warsaw, but after World War II he had to continue at the Royal Academy of Dublin, and then at Manchester.

At first, Jan Łukasiewicz introduced the three-valued logic and then generalized to the infinite-valued. That possibility modulation can be expressed by a membership function, with values which run over all real numbers of the unit real interval, [0, 1], instead of being reduced to the dichotomic {0, 1} of classical logic:

- True vs. False,
- 0 vs 1,
- White vs Black,
- etc.,

so, allowing the treatment of uncertainty and vagueness, important not only from the theoretical point of view, but from the applications.

The logical research of Łukasiewicz suffered a long slumber, until Zadeh, among others. The one who would see the potential utility in 1965, firstly obtaining a generalized version of the classical theory of sets, now denoted by FST, acronym of the so-called "Fuzzy Set Theory", and later, its application to logic, introducing the "Fuzzy Logic".

We must not forget that Zadeh, an engineer, knew Jan Łukasiewicz's research as explained by his colleague, the brilliant American logician Stephen Cole Kleene [21].

According to this theory, we have a transfer function derived from the characteristic function usually called the "membership function", which runs from the universe of discourse, U, until the unit closed interval of reals, which is [0, 1]. Not so in the sets "classic" or "crisp sets", where the range of the function is reduced to a set consisting of only two elements, namely was the {0, 1}. Therefore, fuzzy set theory is a generalization of classical set theory [14].

We may mention the uncertainty principle of quantum physics by Werner Heisenberg. The theory of "vague sets" (today, so-called Fuzzy Sets) proceeds from the quantum physicist and German philosopher Max Black (1937), also analyzes the problem of modeling "vagueness". He differs from Russell in that he proposes that traditional logic can be used by representing vagueness at an appropriate level of detail and suggests that Russell's definition of vagueness confuses vagueness with generality. He discusses vagueness of terms or symbols by using borderline cases where it is unclear whether the term can be used to describe the case. When discussing scientific measurement he points out "the indeterminacy which is characteristic in vagueness is present also in all scientific measurement".

An idea put forward by Black is the idea of a consistency profile or curve to enable some analysis of the ambiguity of a word or symbol. To the fuzzy logic researcher of today these curves bear a strong resemblance to the membership functions of (type-1)-fuzzy sets. Also may be considered the subsequent contribution of the Polish Jan Łukasiewicz (1878–1956).

So, they must have greatly influenced Lofti A. Zadeh (b. 1921) to publish his seminal paper in the journal *Information and Control*, and 3 years later (since 1968), the so-called "Fuzzy Algorithm" [39–41].

In 1923, the British logician Bertrand Russell wrote that all traditional logic habitually assumes that precise symbols are being employed. It is, therefore, not applicable to this terrestrial life but only to an imagined celestial existence.

And Lofti Asker Zadeh, says that according the complexity of a system increases, our ability to make precise and yet significant statements about its behavior diminishes until a threshold is reached beyond which precision and significance (or relevance) become almost mutually exclusive characteristics.

For such reasons—during the last decades—some very powerful 'Fuzzy Mathematics' has been developed, basically in Japan, but also in Europe, where these ideas came to fruition, creating a powerful technological "boom", with new techniques based on "fuzzy" concepts. This trend was particularly strong in Oriental countries, such as South Korea, China or India.

And much later these ideas, even more applications came to Western countries, both European and American, with brilliant studies both from a mathematical point of view and its philosophical implications, as always connected therewith. Very active research groups [15], where papers on Many-Valued Logic have been published, proceed currently from good European universities, for instance of:

- Warsaw,
- Prague,
- Ostrava,
- Vienna,
- Lisbon,
- Opole,
- Barcelona,
- Madrid,
- Toulouse,
- Pamplona,
- Granada,
- etc.

Today, some emerging countries, such as Brazil (Newton Da Costa or Jean-Yves Béziau), India or Turkey, are becoming related with the investigation of all these theories and associated methods, paraconsistent logics and so one [1–9].

From the above it follows that you may need a radical rethink of our classical concepts of truth and falsehood, replacing the concept of fuzziness (vagueness or fuzziness) as a result of which the truth or falsity are only extreme cases. By fuzziness we understand the fact that a proposition may be partially true and partially false simultaneously. A person is not just tall or short, but partially may participate in both features, so that only above and below certain heights it is necessary to call upper or lower bound, while in the intermediate zone of both heights exist as a graduation which is ceasing to be high. It seems intuitively clear that the concept of fuzziness is rooted in most of our ways of thinking and speaking [16]. Another separate issue is the valuation of that each individual granted such a fuzzy character (the glass half full or half empty), which depend on subjective psychological issues and are difficult to evaluate.

The fuzzy principle states that everything is a matter of degree. It will be its more famous "leitmotiv". All propositions acquire a truth value between one (true) and zero (false), inclusive. The allocation of these extreme values will only be given in the case of logical truths or falsehoods or strong inductions: "All men are mortal" can be an example of strong induction, since there is no counterexample.

The arguments for introducing the concept of fuzziness in logic have already been exposed, but it will be necessary to examine in detail some key aspects:

(a) The historical background and methodological concept.
(b) The possibility of building an infinite-valued formal language, and if so, try to define their properties and laws.
(c) The philosophical and practical consequences stemming from such introduction.

One of the most interesting cases in the history of AI is the country of Romania [25, 27, 28]. We have the greatest landmark in the person of the mathematician Grigore Constantin Moisil (1906–1973), who introduced Computer Science in the country; after he had left a very brilliant school of researchers from Romania devoted to mathematics and AI, many of them scattered around the world by the 'economic diaspora', after the Communist period. After World War II, Grigore C. Moisil started teaching Mathematical Logic and Algebra at Iasi and Bucharest, as he understood that the new emerging field of computers would have enormous repercussions for the social fabric of society. He continued working about the ideas of Shannon on Circuits, and some Łukasiewicz's fundamental advances on Many-Valued Logics, where the Fuzzy Logic eventually derived from.

The Łukasiewicz-Moisil Algebras (LMA) was created by G. Moisil as an algebraic counterpart for the many-valued logics of J. Łukasiewicz. They are an attempt to give semantic consistency to n-valued logics. This theory has developed to a considerable extent both as an algebraic theory of intrinsic interest and in view of its applications to logic and switching theory.

The study of LMA was followed by G. Georgescu and A. Iorgulescu, from Bucharest; also are very relevant C. Calude, G. Paun (membrane computing), and some others, in different areas. Also worthy of mentioning is the figure of Solomon Marcus (1925–2016), an inspired disciple of G.C. Moisil, because Marcus has made great contributions to many fields of Mathematics, such as Logic, Analysis, or Computational Linguistics, of which he is one of the founders and a principal contributor [25].

Antonio Monteiro (1907–1980), mathematician born in Portuguese Angola, showed that for every monadic Boolean-algebra we can construct a 3-valued Łukasiewicz-algebra, and that any 3-valued Łukasiewicz-algebra is isomorphic to a Łukasiewicz-algebra thus derived from a monadic Boolean-algebra. Roberto Cignoli says about it that since it was shown by Halmos that monadic Boolean-algebras are the algebraic counterparts of classical first order monadic calculus, Monteiro considered that the representation of 3-valued Łukasiewicz-algebras into monadic Boolean-algebras gives a proof of the consistency of 3-valued Łukasiewicz-logicas relative to classical logic. He showed that, from the algebraic point of view, the three-valued Łukasiewicz-logic stands in the same relation to constructive logic with strong negation as classical logic does to intuitionistic logic.

Of course there is an increasing production of publications on the area. But many of the best papers on Many-Valued Logics currently come from good European Universities and very active research groups. This is possible because very remarkable researchers on MVLs (in particular on Mathematical Fuzzy Logic) have created a solid and consistent basis for these theories. Such has been the case for Petr Hájek, from the Charles University (Prague), P. Cintula, Jan Pavelka, Líbor Behounek, or Vilem Novak, from Ostrava. They have powerful research groups, with publications which are among the most internationally valued in this field [16, 26].

And they are not alone, as in France we have the important task of dissemination and investigation of D. Dubois and H. Prade, Elie Sanchez (1944–2014), or B. Bouchon-Meunier. In Germany, H.-J. Zimmerman, or S. Gottwald (1943–2015). Also we find US researchers as Z. Wang, G. Klir or R.R. Yager; in Hungary (J. Fodor, ...), Canada (W. Pedrycz, ...), UK, Pays Bas (E. Kerre, B. de Baets, G. Cooman, M. De Cock, ...), Italy (G. Gerla, A. Di Nola, ...), Austria, Argentina (R. Cignoli), Brasil, Turkey, etc.

In Poland they follow the great tradition of the LWS of logic and mathematics, and with contributions to research the uncertainty topic through the Rough Sets, by Zdzisław Pawlak (1926–2006), and continued by Andrzej Skowron, among others [15, 38].

3 Reception of Many-valued Logics and Fuzzy Logic in the Iberian Peninsula

As remarkable precursor of Automatics, we must mention the mathematician and Spanish engineer Leonardo Torres Quevedo (1852–1936).

Also, in medieval times, Raymond Lully, and his famous book, *Ars Magna*.

One of the first Hispanic scholars giving notice of the new currents was Juan David Garcia Bacca, who in 1936 published his *Introduction to modern logic*, a work very praised by I. M. Bocheński and Heinrich Schölz.

Later, try so eminent teachers, among them Alfredo Deaño (editor of the Spanish translation of Łukasiewicz's selected papers), Miguel Sánchez-Mazas (interpreting in the more deep sense the logico-mathematical works and ideas of Gottfried Whilhelm Leibniz, as—for instance—the known "characteristica universalis"), or Manuel Sacristán (prosecuted in 'academia' due to its Marxist point of views), all them very often clashing against a very conservative and not so good innovative ideas [32, 33].

One very good initiative has been the creation in the old mining town of Mieres, and by the Government of Asturias, named the 'Research Center for Artificial Intelligence and Soft Computing', initially around someone as well-known as Enric Trillas, who can be considered the father of the introduction of Fuzzy Logic in the Spanish University curricula. This center has attracted many of the most famous international researchers, such as well-known Japanese Professor Michio Sugeno. His topics of research are very broad working, but revolve around fuzzy methods, as well as philosophical implications these carry.

Although I have left it for last, a name should not be omitted from those that appear only from time to time in Spain. I am referring to the Father Pablo Domínguez Prieto (1966–2009), Spanish philosopher and theologian [11–13], who wrote the first major book in Spain on the LWS, from his PhD thesis, at Madrid (1993). Such work is so-called *Indeterminación y Verdad. La polivalencia lógica en la Escuela de Lwow-Warsaw* (*Indeterminacy and Truth*), and was published in 1995, with a foreword by Arch. J.M. życiński, and showing a very strong influence by Jan Woleński.

Pablo can be considered as one of the Spanish precursors in the study of MVLs, from the philosophical and historical point of view.

Another interesting Spanish author who has been reporting these new streams of logic is Prof. Julián Velarde, with *Polyvalent Logic*, or his book *Formal Logic*, a volume II belonging to its *History of Logic*, all of them around the University of Oviedo and its service publications, or later, to the Editorial Pentalfa. Also of great interest may be his work *Gnoseology of Fuzzy Systems*, which analyzes the deep philosophical connections of these issues [34].

New research groups have been formed in recent times [17], as the Spanish institution CSIC (Consejo Superior de Investigaciones Científicas), centered in Barcelona, led by Lluis Godo and Francesc Esteva. Or the group that belongs to the UPNA (Public University of Navarra), headed by Humberto Bustince. Or in the University of Granada (lead by Miguel Delgado Calvo-Flores).

Observe that the impulse to the study of Mathematical Logic in Spanish universities came, among others, through the aforementioned Prof. Miguel Sánchez-Mazas, and also by Prof. Manuel Garrido.

In Portugal the origins of the study of AI are linked to the names of Luis Moniz Pereira, Helder Coelho and Fernando Pereira, who in 1973 created the LNEC, within which the following year formed a division of Computer Science [16].

In 1977 the programming language called DEC-10 Prolog is introduced, which Helder Coelho contributes greatly to propagate in Brazil.

In 1984 is created the Portuguese Association for AI (being APPIA, by acronym), which maintains its vitality with many publications, and also organizing congresses.

Currently, there exist in Portugal four basic areas of work, related respectively with:

- Learning;
- Knowledge Representation;
- Knowledge in general, and
- Applications.

The research on Logic and AI in Portugal maintains its vitality through good publications, and also organizing very important conferences, in Lisbon, Porto, etc.

4 Final Note

In early January, about 700 scientists specialized in technological areas signed an open letter, warning of the dangers of AI. There have been many reactions to this paper, with catastrophic scenarios described in science fiction that warn of a revolution of the machines and extinction of humanity. However, the current risks are different. For instance, the ability of processing huge amounts of data by computers, which can be released to those who are in control. Today, the main threat of artificial intelligence is a misuse of the abilities of the equipment, which can extract and analyze data in bulk.

The director of the Institute of Artificial Intelligence (IIA) of Barcelona, Ramón López de Mántaras, also adds autonomous weapons as a threat: “in order to have robots soldiers is very worrying, because for a robot it is almost impossible to distinguish between an innocent civilian and a fighter”.

“Today we can ensure that none of the current robots, nor any that will be in short, medium and even long term fairly, would be out of control”.

"The problem is not in technology itself, but in humanity. Is more likely to be the man with evil intentions, who may produce a very possible war between humans and machines", he qualifies.

The loss of workplaces is another hazard of AI.

According to López de Mántaras,

"So far the robots moved people from repetitive or dangerous task, but with advances in artificial intelligence begin to endanger related services sector; for example, jobs".

"Experts agree that education is the most important measure. We will look for other jobs where creativity is essential, and therefore will require investment in education to add art between engineering and mathematics".

Meanwhile, "it must be given much more importance to lifelong learning so that people can be recycled more easily and be able to change careers. At least so far, the technological changes, which have destroyed workplaces, have also created other ones instead."

Acknowledgements This work was supported by the MICINN's Research Project FFI2016-77574-P, and the Investigation Group of our Spanish University (UNED) that we belong to, into a section about 'Science and Technique', that belongs to the Project entitled *"Estudio sistemático de las lecturas heideggerianas de Jacques Derrida. Confluencias y divergencias"*. Being Principal Researcher (IP), Prof. Dr. Cristina De Peretti.

References

1. Belnap, N.D.: A useful four-valued logic. In: Dunn, J.M., Epstein, G., Wolf, R.G. (eds.) Modern Uses of Multiple-Valued Logic, pp. 8–37. Reidel, Dordrecht (1977)
2. Béziau, J.-Y.: Universal logic. In: Childers, T., Majer, O. (eds.) Logica'94: Proceedings of the 8th International Symposium, pp. 73–93. Czech Academy of Sciences, Prague (1994)
3. Béziau, J.-Y.: La véritable portée du théoréme de Lindenbaum-Asser. Log. Anal. **42**(167–168), 341–359 (1999)
4. Béziau, J.-Y.: From paraconsistent logic to universal logic. Sorites **12**, 5–32 (2001)
5. Béziau, J.-Y.: Sequents and bivaluations. Log. Anal. **44**(176), 373–394 (2001)
6. Béziau, J.-Y.: Many-valued and Kripke semantics. In: J. van Benthem et al. (eds.) The Age of Alternative Logics, pp. 89–91. Springer, Dordrecht (2006)
7. Béziau, J.-Y.: 13 questions about universal logic. Bull. Sect. Log. **35**(2–3), 133–150 (2006)
8. Béziau, J.-Y.: From consequence operator to universal logic: a survey of general abstract logic. Logica Universalis, pp. 3–17. Birkhäuser, Basel (2007)
9. Birkhoff, G.: Universal algebra. In: Proceedings of the First Canadian Mathematical Congress 1945, pp. 310–326. University of Toronto Press, Toronto (1946)
10. Da Costa, N., Béziau, J-Y.: Logiques Classiques et Non-Classiques: essais sur le fondements de la logique. Editora Masson et Cie, Paris (1997)
11. Domínguez Prieto, P.: Indeterminación y Verdad. La polivalencia lógica en la Escuela de Lvov-Varsovia. Nossa y Jara Editores, Móstoles (1995)
12. Domínguez Prieto, P.: Teoría del Contorno Lógico. Nossa y Jara Editores, Móstoles (1999)
13. Domínguez Prieto, P.: Lógica Modal y Ontología. Nossa y Jara Editores, Móstoles (2001)
14. Dubois, D., Prade, H.: Fuzzy Sets and Systems: Theory and Applications. Academic, New York, (1980)
15. Garrido, A.: Lógicas de nuestro tiempo. Madrid, Editorial Dykinson (2014)
16. Garrido, A.: Lógica Aplicada. Vaguedad e Incertidumbre. Madrid, Editorial Dykinson (2014)
17. Garrido, A.: The Challenges of Artificial Intelligence. Acta Universitatis Apulensis (AUA), Special Issue ICTAMI 2015, pp. 155–165. Alba Iulia (2015)
18. Garrido, A.: Leibniz y las Nuevas Lógicas. Editorial Comares, Granada (2016)

19. Goguen, J.A.: The logic of inexact concepts. Synthese **19**(3/4), 325–373 (1969)
20. Hájek, P.: Metamathematics of Fuzzy Logic. Kluwer, Boston/Dordrecht (1998)
21. Kleene, S.C.: Introduction to Metamathematics. North-Holland, Amsterdam/Van Nostrand, New York (1952)
22. Łukasiewicz, J.: Elements of Mathematical Logic. Państwowe Wydawnictwo Naukowe (PWN), Warsaw (1963)
23. Łukasiewicz, J.: Selected Works, L. Borkowski (ed.). North Holland, Amsterdam (1970)
24. Łukasiewicz, J.: Écrits logiques et philosophiques. Librairie Philosophique Jacques Vrin, Paris (2013)
25. Marcus, M., Marcinkiewicz, M.: Building a large annotated corpus of English: The Penn Treebank. Computat. Linguist. **19**(2), 313–330 (1993)
26. Mira, J., et al.: Aspectos Básicos de Inteligencia Artificial. UNED (1995)
27. Moisil, G.: Încercări vechi şi noi în logica neclasică. Bucuresti (1965)
28. Moisil, G.: Elemente de logică matematică şi teoria mulţimilor. Bucuresti (1968)
29. Murawski, R.: Essays in the Philosophy and History of Logic and Mathematics. Poznań Studies in the Philosophy of the Sciences and the Humanities. Rodopi (2010)
30. Rasiowa, H., Sikorski, R.: The Mathematics of Metamathematics. Polish Scientific, Warszawa (1963)
31. Rosser, J.B., Turquette, A.R.: Many-Valued Logics. North-Holland, Amsterdam (1952)
32. Sacristán, M.: Introducción a la Lógica y al Análisis Formal, Colección del Círculo de Lectores. Barcelona (1990)
33. Sánchez-Mazas, M.: Obras Escogidas. Servicio Editorial Universidad del País Vasco (UPV). En dos volúmenes. Edición de Javier de Lorenzo. Donostia/ San Sebastián (2002)
34. Velarde, J.: Leibniz y la lógica. Themata. Revista de Filosofía **29** (Treinta años de estudios leibnizianos), 217–231 (2002)
35. Woleński, J.: Logic and Philosophy in the Lvov-Warsaw School. Kluwer Academic Publishers, Dordrecht (1989)
36. Woleński, J.: Historico-Philosophical Essays, vol. I. Copernicus Center, Krakow (2013)
37. Wolenski, J., Zygmunt, J.: Jerzy Slupecki (1904–1987): Life and Work. Studia Logica: Int. J. Symb. Logic, 401–411 (1989)
38. Wybraniec-Skardowska, U.: Polish Logic. Some Lines from a Personal Perspective. Publications of Institute for Logic, Language and Computation, Amsterdam (2009)
39. Yager, R.R.: Fuzzy Sets and Applications: Selected Papers by L.A. Zadeh. Wiley, New York (1987). Yager et al. (eds.)
40. Zadeh, L.A.: Fuzzy sets. Inf. Control **8**, 338–355 (1965)
41. Zadeh, L.A.: Fuzzy Logic and approximate reasoning. Synthese **30**, 407–428 (1975)
42. Zimmermann, H.J.: Fuzzy Set Theory and Applications. Kluwer Academic Press, Boston (1991)
43. Zinoviev, A.A.: Philosophical Problems of Many-Valued Logic. Reidel Verlag, Dordrecht (1963)

A. Garrido (✉)
Department of Fundamental Mathematics, Faculty of Sciences of the National University of Distance Education (UNED), Madrid, Spain
e-mail: agarrido@mat.uned.es

Ontology of Logic and Mathematics in Lvov-Warsaw School

Roman Murawski

Abstract The aim of the paper is to consider ontological views connected with mathematics and logic of main representatives of Lvov-Warsaw School of Philosophy. In particular views of the following scholars will be presented and discussed: Jan Łukasiewicz, Stanisław Leśniewski, Alfred Tarski, Tadeusz Kotarbiński and Kazimierz Ajdukiewicz. We shall consider also views of Andrzej Mostowski who belonged to the second generation of the school as well as of Leon Chwistek who was not directly the member of this group but whose conceptions are of interest.

Keywords Philosophy of logic · Philosophy of mathematics · Ontology · Platonism · Nominalism · Intuitionism

Mathematics Subject Classification (2000) Primary 00A30; Secondary 03A05, 01A60

The aim of the paper is to describe ontological views concerning logic and mathematics of representatives of Lvov-Warsaw School of Philosophy.[1] As will be shown there was in fact no common conception. The element unifying members of the school was not a particular philosophical doctrine but rather the method of practicing and developing philosophy. I. Dąmbska charactered it in the following way:

> There was no common doctrine, no uniform view of the world shared by Lvov philosophers. What formed the foundation of the spiritual community of those scholars was not the content of conceptions but rather the way, the method of doing philosophy as well as the common scientific language. Thanks to that feature members of this group could be both spiritualists and materialists, nominalists and realists, logicians and psychologists, philosophers of nature and theoreticians of art.[2] [8, p. 17]

[1]On Lvov-Warsaw School in Philosophy see Woleński [38] and [39] as well as Murawski [35] and [36].

[2]Nie łączyła bowiem filozofów lwowskich jakaś wspólna doktryna, jakiś jednolity pogląd na świat. To, co stworzyło podstawę wspólnoty duchowej tych ludzi, to była nie treść nauki, tylko sposób, metoda filozofowania i wspólny język naukowy. Dlatego wyjść z tej szkoły mogli: spirytualiści i materialiści, nominaliści i realiści, logistycy i psychologowie, filozofowie przyrody i teoretycy sztuki.

Á. Garrido, U. Wybraniec-Skardowska (eds.), *The Lvov-Warsaw School. Past and Present*, Studies in Universal Logic,
https://doi.org/10.1007/978-3-319-65430-0_44

1 Jan Łukasiewicz: (Neo)Platonism

Let us start by considering views of Jan Łukasiewicz (1879–1956). One should begin by stressing his anti-psychological attitude. Psychologism claims that the objects investigated by logic and mathematics exist as psychological beings and are got to know like other psychological facts. This approach was popular in the philosophy of logic and mathematics at the end of the nineteenth century. It was criticised in particular by Frege, Husserl and Meinong. In the paper "Logika a psychologia" [Logic and psychology] (cf. [21]) Łukasiewicz formulated his arguments against psychologism. First he stated that laws of psychology are empirical and consequently only probable whereas laws of logic are certain. Laws of both those disciplines are also of different character: the laws of logic concern relations between the truth and falsity of judgements whereas the laws of psychology state the relations between psychological phenomena. Łukasiewicz concludes:

> Exposing the attitude of logic towards psychology can be to the advantage of both sciences. Logic will be cleared from the weeds of psychologism and empiricism, which choke its right development and the psychology of cognition will get rid of *a priori* traces which hid the light of the sincere splendour of its truths. Since one should remember that logic is an *a priori* science, like mathematics, whereas psychology, like any other natural science, is based and must be based on experience.[3]

Łukasiewicz stressed explicitly the apriorism of logic. In the paper "O twórczości w nauce" [On creativity in science] (cf. [25]) he wrote:

> Logic is an *a priori* science. Its theorems are true by virtue of definitions and axioms flowing from reason and not from experience. This science is a domain of pure mental creativity. [...] Logical and mathematical judgements are truths only in the world of ideal beings. We will never know whether some real objects correspond with these beings.
> *A priori constructions of the mind, being part of every synthesis, imbue the whole science with an ideal and creative element.*[4]

Łukasiewicz admitted that logic and mathematics have a nominalistic robe (see for example his paper "Logistyka a filozofia" [Logistics and philosophy], [23, p. 119]) but simultaneously he saw some difficulties in the nominalistic approach. An individual can create only a finite number of inscriptions. Hence a set of inscriptions is finite what would mean that the set of theses of logic and mathematics would be finite as well but "on this basis it would be as difficult to practise logistics, especially metalogistics, as to build

[3]Wyświetlenie stosunku logiki do psychologii przynieść może korzyści obu tym naukom. Logika oczyści się z chwastów psychologistycznych i empirystycznych, które tłumią jej prawidłowy rozwój, a psychologia poznania pozbędzie się naleciałości apriorycznych, spod których szczery blask jej prawd nie mógł jakoś dotąd zajaśnieć. Należy bowiem pamiętać, że logika jest nauką aprioryczną, tak jak matematyka, a psychologia, tak jak każda nauka przyrodnicza, opiera się i opierać się musi na doświadczeniu [21, p. 491].

[4]Logika jest nauką aprioryczną. Twierdzenia jej są prawdziwe na mocy określeń i pewników płynących z rozumu, nie z doświadczenia. Nauka ta jest dziedziną czystej twórczości myślowej. [...] Sądy logiczne i matematyczne są prawdami jedynie w świecie bytów idealnych. Czy bytom tym odpowiadają jakieś przedmioty rzeczywiste, o tym zapewne nigdy się nie dowiemy.
Aprioryczne konstrukcje umysłu, wchodząc w skład każdej syntezy, przepajają całą naukę pierwiastkiem idealnym i twórczym [22, pp. 13–14].

arithmetic on the assumption that the set of natural numbers is finite" [25, p. 224].[5] It would also lead to make logic dependent on certain empirical facts, i.e., on the existence of inscriptions, which is difficult to accept.

According to Łukasiewicz the nominalism of logic and mathematics is virtual. Moreover, logic was developed without solving the problem of its nominalism. In his article "Logistyka a filozofia" [Logistics and philosophy] he wrote:

> We have so far been little worried by these difficulties, and this is the strangest point. It was so probably because, while we use nominalistic terminology, we are not true nominalists but incline toward some unanalysed conceptualism or even idealism.[6] [25, p. 224]

Łukasiewicz himself thought that the objects that logic investigated existed only beyond the sphere of inscriptions. He did not develop some alternative to nominalism—he just formulated his personal view. But his view resulted from his personal religious convictions—influenced by these convictions Łukasiewicz opted for the Neoplatonic interpretation of logic. In the paper "W obronie logistyki" [In defence of logic] he wrote:

> In concluding these remarks I should like to outline an image which is connected with the most profound intuitions which I always experience in the face of logistic. That image will perhaps shed more light on the true background of that discipline, at least in my case, than all discursive description could. Now, whenever I work even on the least significant logistic problem, for instance, when I search for the shortest axiom of the implicational propositional calculus I always have the impression that I am facing a powerful, most coherent and most resistant structure. I sense that structure as if it were a concrete, tangible object, made of the hardest metal, a hundred times stronger than steel and concrete. I cannot change anything in it; I do not create anything of my own will, but by strenuous work I discover in it ever new details and arrive at unshakable and eternal truth. Where is and what is that ideal structure? A believer would say that it is in God and is His thought.[7] [25, p. 249]

Łukasiewicz stressed that this was his personal view. He was of the opinion that logic is neither called nor allowed to solve the eternal philosophical debate concerning universals.

[5]na takiej podstawie byłoby równie trudno uprawiać logistykę, a zwłaszcza metalogistykę, jak trudno byłoby zbudować arytmetykę na gruncie założenia, że zbiór liczb naturalnych jest skończony [23, p. 120].

[6]Mało dotychczas przejmowaliśmy się tymi trudnościami i to jest w tym wszystkim najdziwniejsze. Działo się to chyba dlatego, że używając terminologii nominalistycznej, nie jesteśmy naprawdę nominalistami, lecz hołdujemy jakiemuś nie zanalizowanemu konceptualizmowi czy nawet idealizmowi [23, p. 120].

[7]Chciałbym na zakończenie tych uwag nakreślić obraz związany z najgłębszymi intuicjami, jakie odczuwam zawsze wobec logistyki. Obraz ten rzuci może więcej światła na istotne podłoże, z jakiego przynajmniej u mnie wyrasta ta nauka niż wszelkie wywody dyskursywne. Otóż ilekroć zajmuję się najdrobniejszym nawet zagadnieniem logistycznym, szukając np. najkrótszego aksjomatu rachunku implikacyjnego, tylekroć mam wrażenie, że znajduję się wobec jakiejś potężnej, niesłychanie zwartej i niezmiernie odpornej konstrukcji. Konstrukcja ta działa na mnie jak jakiś konkretny dotykalny przedmiot, zrobiony z najtwardszego materiału, stokroć mocniejszego od betonu i stali. Nic w niej zmienić nie mogę, nic sam dowolnie nie tworzę, lecz w wytężonej pracy odkrywam w niej tylko coraz to nowe szczegóły, zdobywając prawdy niewzruszone i wieczne. Gdzie jest i czym jest ta idealna konstrukcja? Filozof wierzący powiedziałby, że jest w Bogu i jest myślą Jego [24, p. 165].

2 Stanisław Leśniewski: Nominalism

Nominalism mentioned above was the philosophical doctrine of another representative of Lvov-Warsaw School in Philosophy and, together with Łukasiewicz, the founder of Warsaw School of Logic, Stanisław Leśniewski (1886–1939). This doctrine had strong influence even on the contents as well as on the form of his logical constructions. His views Leśniewski called constructive nominalism.

Leśniewski treated language as a collection of concrete inscriptions and expressions of a language as finite sequences of signs. Two inscriptions of the same shape were treated by him as two separate, different inscriptions. In his opinion there only exist as many expressions as they have been written. One cannot speak of some potential existence of expressions. Consequently a given logical system contains only so many theorems as they have been written until a given moment, i.e., every logical system consists of only a finite number of theorems. Leśniewski did not allow the existence of any general objects, in particular of common properties of individual objects. Another consequence of Leśniewski's nominalism was the fact that two equivalent systems, for example the system of propositional calculus based on negation and implication and the system based on negation and disjunction as primitive connectives usually treated as two variants of the same logic should be treated now as two different systems. Leśniewski's systems are never something complete at a given moment.

Leśniewski connected the described view with the so-called intuitive formalism. According to it a language of logic—uniquely and completely codified—says always "something" and about "something". In the work "Grundzüge eines neuen System der Grundlagen der Mathematik" [17] he wrote:

> Having no predilection for various 'mathematical games' that consist in writing out according to one or another conventional rule various more or less picturesque formulae which need not be meaningful, or even—as some of the 'mathematical gamers' might prefer—which should necessarily be meaningless, I would not have taken the trouble to systematize and to often check quite scrupulously the directives of my system, had I not imputed to its theses a certain specific and completely determined sense, in virtue of which its axioms, definitions and final directives, have for me an irresistible intuitive validity.[8] [18, p. 487]

In the work "O podstawach matematyki" [On the foundations of mathematics] (cf. [16]) one reads:

> They encouraged the disappearance of the feeling for the distinction between the mathematical sciences, conceived as deductive theories, which serve to capture various realities of the world in the most exact laws possible, and such non-contradictory deductive systems, which indeed ensure the possibility of obtaining, on their basis, an abundance of ever new theorems, but which

[8]Da ich keine Vorliebe für verschiedene «Mathematikspiele» habe, welche darin bestehen, dass man nach diesen oder jenen konventionellen Regeln verschiedene mehr oder minder malerische Formeln aufschreibt, die nicht notwendig sinnvoll zu sein brauchen oder auch sogar, wie es einige der «Mathematikspiele» lieber haben möchten, notwendig sinnlos sein sollen,—hätte ich mir nicht die Möhe der Systematisierung und der vielmaligen skrupulösen Kontrollierung der Direktiven meines Systems gegeben, wenn ich nicht in die Thesen dieses Systems einen gewissen ganz bestimmten, eben diesen und nicht einen anderen, Sinn legen wörde, bei dem för mich die Axiome des Systems und die in den Direktiven zu diesem System kodifizierten Schluss- und Definitionsmethoden eine unwiderstehliche intuitive Geltung haben [17, p. 78].

> simultaneously distinguish themselves by the lack of any connection with reality of any intuitive, scientific value.[9] [18, pp. 177–178]

Leśniewski treated formal systems as a means to transmit certain information about the world and as a way to express what is intuitively true. This may seem not to be fully in accordance with his nominalism and radical formalism. However he did not consider those views as contradictory. In fact in "Grundzüge eines neuen System der Grundlagen der Mathematik" [17] he wrote:

> I see no contradiction, therefore, in saying that I advocate a rather radical 'formalism' in the construction of my system even though I am an obdurate 'intuitionist'. Having endeavoured to express some of my thoughts on various particular topics by representing them a s a series of propositions meaningful in various deductive theories, and to derive one proposition from others in a way that would harmonize with the way I finally considered intuitively binding ....[10] [18, p. 487]

For Leśniewski logic was the description of most general features of being (the same—under the influence of Leśniewski—was claimed by Kotarbiński). Hence it plays the role of a general theory of objects. This view was in accordance with the fact that Warsaw School of Logic rejected the so-called analytic interpretation of logic, i.e., the thesis that logic and mathematics are the set of tautologies that do not say anything about the world. Logic and mathematics were thought to refer to the formal aspects of reality. Add that Leśniewski rejected also the conventionalism in the style of Poincaré.

Leśniewski took a firm stand in the dispute concerning universals—he rejected the existence of any ideal and general objects. In the paper "Krytyka logicznej zasady wyłączonego środka" [Critique of the logical principle of excluded middle] (cf. [20]) he gave the proof of non-existence of such objects that became very popular in Poland. In the proof the concept of a feature as well as the principle of excluded middle and the principle of contradiction were used. It was quoted—with some modifications—by Kotarbiński in the paper "Sprawa istnienia przedmiotów idealnych" [The problem of existence of ideal objects] (cf. [11]) and repeated in his book *Elementy teorji poznania, logiki formalnej i metodologji nauk* [Elements of the theory of cognition, formal logic and methodology] (cf. [12]). It became one of the justifications of reism propagated by him. Leśniewski return to his proof in the work "O podstawach matematyki" [On the foundations of mathematics] (cf. [16, pp. 183–184]) where he gave a new version of it in

[9]Sprzyjało to zanikowi poczucia różnicy między naukami matematycznymi pojmowanymi jako teorie dedukcyjne, służące do ujęcia w prawa możliwie ścisłe różnorodnej rzeczywistości świata, a takimi niesprzecznymi systemami dedukcyjnymi, które zabezpieczają wprawdzie możność otrzymania na ich gruncie obfitości wciąż nowych twierdzeń, odznaczających się jednak jednocześnie brakiem jakichkolwiek łączących je z rzeczywistością walorów intuicyjno-naukowych [16, p. 166].

[10]Ich sähe keinen Widerspruch darin, wenn ich behaupten wollte, dass ich eben deshalb beim Aufbau meines Systems einen ziemlich radikalen «Formalismus» treibe, weil ich ein versteckter «Intuitionist» bin: indem ich mich beim Darstellen von verschiedenen deduktiven Theorien bemöhe, in einer Reihe sinnvolle Sätze eine Reihe von Gedanken auszudröcken, welche ich öber dieses oder jedes Thema hege, und die einen Sätze aus den anderen Sätzen auf eine Weise abzuleiten, die mit den Schlussweisen harmonisieren wörden, welche ich «intuitiv» als för mich bindend betrachte ... [17, p. 78].

which the concept of "feature" does not appear. The proof was preceded by the following explanations:

> At the time I wrote that passage [Leśniewski says about the appropriate fragment of his [19]—my remark - RM] I believed that there are in existence in this world so called features and so called relations, as two special kinds of objects, and I felt no scruples about using the expressions 'feature' and 'relations'. It is a long time since I believed in the cxistence of objects which are features, or in the existence of objects which are relations and now nothing induces me to believe in the existence of such objects [...] and in situations of a more 'delicate' character I do not use the expressions 'feature' and 'relation' without the application of various extensive precautions and circumlocutions. I also have no inclination at present—considering the possibility of various interpretational misunderstandings—to ascribe this or that opinion on the question of 'general objects' to the authors mentioned in the passage mentioned above.[11] [18, p. 198]

3 Alfred Tarski: Nominalism

A follower of nominalism was also Alfred Tarski (1901–1983). This pronominalistic attitude was the source of the fact that in the interwar period he treated language as a set of sentences understood in a strictly nominalistic way as physical objects. However his sympathies towards nominalism were in fact stronger. Mostowski wrote about this in the following way:

> Tarski, in oral discussions, has often indicated his sympathies with nominalism. While he never accepted the » reism« of Tadeusz Kotarbiński, he was certainly attracted to it in the early phase of his work. However, the set-theoretical methods that form the basis of his logical and mathematical studies compel him constantly to use the abstract and general notions that a nominalist seeks to avoid. In the absence of more extensive publications by Tarski on philosophical subjects, this conflict appears to have remained unresolved. [32, p. 81]

Tarski's pronominalistic attitude is confirmed in various sources. Firstly, it was Tarski's remark (preserved on a tape cassette) made during the symposium organised by the Association for Symbolic Logic and the American Philosophical Association, held in Chicago on 29th–30th April 1965, and dedicated to philosophical implications of Gödel's incompleteness theorem. Tarski said:

> I happen to be, you know, a much more extreme anti-Platonist. [...] However, I represent this very [c]rude, naive kind of anti-Platonism, one thing which I would describe as materialism, or nominalism with some materialistic taint, and it is very difficult for a man to live his whole life with this philosophical attitude, especially if he is a mathematician, especially if for some reasons he has a hobby which is called set theory. [9, p. 52]

[11] W czasie, gdy ustęp ten [chodzi tu o stosowny fragment pracy [19]—uwaga moja, R.M.] pisałem, wierzyłem, iż istnieją na świecie tzw. cechy i tzw. stosunki jako dwa specjalne rodzaje przedmiotów, i nie odczuwałem żadnych skrupułów przy posługiwaniu się wyrazami "cecha" i "stosunek". Obecnie nie wierzę już od dawna w istnienie przedmiotów będących cechami, ani też w istnienie przedmiotów będących stosunkami, nic mnie też nie skłania do wierzenia w istnienie takich przedmiotów [...], wyrazami zaś "cecha" i "stosunek" staram się w sytuacjach o cokolwiek "delikatniejszym" charakterze nie posługiwać bez daleko idących ostrożności i omówień. Nie mam dziś także skłonności—wobec możliwości rozmaitych nieporozumień interpretacyjnych—do przypisywania tych lub innych poglądów w sprawie „przedmiotów ogólnych" tym lub innym z autorów, wymienionym w ustępie wyżej przytoczonym [16, p. 183].

Fefermans' book [9] contains more similar words concerning Tarski himself or other people's opinions about Tarski. These opinions were expressed on Tarski's 70th birthday celebrations and remembered by Chihara, Chateaubriand and the Fefermans:

> I am a nominalist. This is a very deep conviction of mine. It is so deep, indeed, that even after my third reincarnation, I will still be a nominalist. [...] People have asked me, » How can you, a nominalist, do work in set theory and logic, which are theories about things you do not believe in?« [...] I believe that there is a value even in fairy tales.
> [I am] a tortured nominalist.
> Elsewhere Tarski has said more specifically that he subscribed to reism or concretism (a kind of physicalistic nominalism) of his teacher Tadeusz Kotarbiński. [9, p. 52]

Also Tarski's letter to Woodger, dated 21st November 1948, testifies to the importance he attached to nominalism:

> The problem of constructing nominalistic logic and mathematics has intensively interested me for many-many years. Mathematics—at least the so-called classical mathematics—is at present an indispensable tool for scientific research in empirical sciences. The main problem for me is whether this tool can be interpreted nominalistically or replaced by another nominalistic tool which should be adequate for the same purposes. [27, p. 147]

On many occasions Tarski stressed his sympathies towards Kotarbiński's reism and physicalism. He also translated into English (together with David Rynin) Kotarbiński's work "Zasadnicze myśli pansomatyzmu" [The Fundamental Ideas of Pansomatism] (cf. [13]). The translation was published in *Mind*, one of the most important English periodicals dedicated to philosophy. It was included in Tarski's *Collected Works* [37].[12]

More details about Tarski's sympathies and inclinations towards nominalism can be found in the recently discovered protocols of Carnap from the discussions conducted at Harvard in the academic year 1940/1941. Besides Carnap the other participants were Tarski and Quine as well as—occasionally—Russell.

In the protocol of 10th January 1941 Carnap wrote down the following remarks concerning nominalism and finitism:

> Tarski: I understand basically only languages which satisfy the following conditions:
>
> 1. Finite number of individuals;
> 2. Realistic (Kotarbinski): the individuals are physical things;
> 3. Non-platonic: there are only variables for individuals (things) not for universals (classes and so on) [26, p. 342].[13]

Mancosu notices [26, p. 343] a mistake: instead of 'realistic' it should be 'reistic,' which is confirmed by the reference to Kotarbiński.

Carnap's notes also contain the following exchange of views:

> I [Carnap]: Should we construct the language of science with or without types?
> He [Tarski]: Perhaps something else will emerge. One would hope and perhaps conjecture that the whole general set theory, however beautiful it is, will in the future disappear. With the higher types Platonism begins. The tendencies of Chwistek and others ('Nominalism') of speaking only

[12] It also testifies to Kotarbiński's strong influence on Tarski.

[13] Tarski: Ich verstehe im Grunde nur eine Sprache die folgende Bedingungen erföllt: [1] Finite Anzahl der Individuen; [2] Realistisch (Kotarbiński): Die Individuen sind physikalische Dinge; [3] Nicht-platonisch: Es kommen nur Variable för Individuen (Dinge) vor, nicht för Universalien (Klassen usw.).

of what can be named are healthy. The problem is only how to find a good implementation.[14] [26, p. 334]

Of special interest—in the context of the problem of passing from the systems of the theory of classes—is also Carnap's summary of his conversation with Tarski on 12th February 1941:

The Warsaw logicians, especially Leśniewski and Kotarbiński saw a system like PM (but with simple type theory) as the obvious system form. This restriction influenced strongly all the disciples; including Tarski until the 'Concept of Truth' (where the finiteness of the levels is implicitly assumed and neither transfinite types nor systems without types are taken into consideration; they are discussed only in the Postscript added later). Then Tarski realized that in set theory one uses with great success a different system form. So he eventually came to see this type-free system form as more natural and simpler.[15] [26, p. 335]

One should notice that Tarski's research practice, in particular his investigations concerning set theory or the theory of models, contradicted in fact his nominalism to a certain extent and would rather suggest that he was a follower of Platonism (this explains the question mark in the title of this section). This discrepancy can be explained by the spirit and ideological canon of the Polish School. According to them, research should not be limited by any *a priori* philosophical foundations and all correct methods should be allowed and applied.

4 Tadeusz Kotarbiński: Reism

Presenting philosophical views of Tarski we have mentioned Tadeusz Kotarbiński (1886–1981) and his doctrine of reism to which Tarski referred. Let us say now something more about this.

Reism is by Kotarbiński both a semantical and an ontological doctrine, moreover both levels are in a certain sense parallel. Kotarbiński admitted that developing reism he used some logical ideas of Leśniewski explained by the latter in his system of the calculus of names called ontology. In the Preface to *Elementy teorii poznania, logiki formalnej i metodologii nauk* [Elements of epistemology, formal logic and methodology of science]

[14]Ich: Sollen wir vielleicht die Sprache der Wissenschaften mit oder ohne Typen machen? Er: Vielleicht wird sich etwas ganz Anderes entwickeln. Es wäre zu wönschen und vielleicht zu vermuten, dass die ganze allgemeine Mengenlehre, so schön sie auch ist, in der Zukunft verschwinden wird. Mit den höheren Stufen fängt der Platonismus an. Die Tendenzen von Chwistek und anderen (» Nominalismus«), nur öber Bezeichenbaren zu sprechen, sind gesund. Problem nur, wie gute Durchföhrung zu finden.

[15]*Die Warschauer Logiker*, besonders Leśniewski und Kotarbiński, sahen ein System wie PM (aber mit einfacher Typentheorie) ganz selbstverständlich als die Systemform an. Diese Beschränkung wirkte stark suggestiv auf alle Schöler; auf T. selbst noch bis zu » Wahrheitsbegriff« (wo weder transfinite Stufen noch stufenloses System betrachtet wird, und Endlichkeit der Stufen stillschweigend vorausgesetzt wird, erst im später hinzugefögten Anhang werden sie besprochen). Dann aber sah T., dass in der Mengenlehre mit grossem Erfolg eine ganz andere Systemform verwendet wird. So kam er schliesslich dazu, diese stufenlose Systemform als natörlicher und einfacher zu sehen.

(cf. [14]) Kotarbiński wrote:

> Still, I have learnt most things from Prof. Dr Stanisław Leśniewski. I admit that in many places of the book. And they are the most important and clearest points. Besides I admit that all my thoughts are deeply saturated with the influences of that extraordinary mind whose precious gifts I have used, thanks to good luck, almost every day for a number of years. I am undoubtedly a disciple of my colleague Leśniewski whom here I thank cordially and respectfully for all that he has ever taught me.[16] [15, pp. 9–10]

Add that Leśniewski himself valued his collaboration with Kotarbiński. He admitted that he owed him a lot (see for example [36, p. 93]).

The source of Kotarbiński's reism were his doubts concerning the existence of properties and other ideal objects. He expressed them for the first time in his paper "Sprawa istnienia przedmiotów idealnych" [The problem of existence of ideal objects] (cf. [11]). He criticized there conceptions assuming the existence of ideal objects. He wrote that there were no foundations to assume the existence of such objects. He tried to show that there were no imaginary (only conceivable) objects, no mathematical objects; there were no types (universals), features, relations, intentional objects, thinking processes and psychological contents.

Reism was explained by Kotarbinski in his book *Elementy* [Elements] (cf. [12] and [14]) and in various papers. Reism in the ontological sense can be reduced to the following two theses: (1) every object is a thing, (2) no object is a state, a relation, a feature.[17] Kotarbiński assumes also that things are bodies, and thus extensive beings existing in time and space. Therefore, we are dealing with somatism strengthened to become pansomatism—there are only bodies. This distinguishes reism from other concretisms, for example from the concretism of Leibniz who towards the end of his life assumed that there were only concrete entities (note that this concretism was of spiritualistic nature because those concrete entities were spiritual monads). Reism can be seen as a certain interpretation of Leśniewski's ontology (the latter was not a reist although he was a nominalist).

Reism faces various difficulties when applied to logic and mathematics. Using the language of reism one can speak about sets in a distributive sense that is fundamental for set theory, on which in turn the whole building of mathematics is constructed, but only providing that those statements refer to the elements of these sets. Hence it allows us to develop the elementary algebra of sets but not to define, for instance the concept of finite or infinite set. However, it is not sufficient for mathematics. Leśniewski was aware of these difficulties and proposed to use the concept of a set in a collective sense (mereological)—such an approach does not allow realising all that mathematicians expect of set theory. It should be added that reism had numerous followers, the greatest one being

[16]Najwięcej wszelako nauczyłem się od prof. dra Stanisława Leśniewskiego. W wielu miejscach książki wyraźnie z tego zdaję sprawę. Ale to są punkty najważniejsze i najwyraźniejsze. Poza tym, przyznaję, cała myśl moja przesycona jest do głębi wpływami tego niezwykłego umysłu, z którego bezcennych darów los przychylny pozwolił mi przez szereg lat korzystać w obcowaniu niemal codziennym. Jestem niewątpliwie uczniem kolegi Leśniewskiego, któremu na tym miejscu serdecznie i z głębokim szacunkiem dziękuję za wszystko, czego mnie kiedykolwiek nauczył [14, pp. 9–10].

[17]A clear reference to the four categories proposed by W. Wundt can be seen here.

Alfred Tarski.[18] Furthermore, reism, thanks to its logical tools, allows achieving more than any other nominalism.

In the ontology of mathematics Kotarbiński proclaimed himself in favour of nominalism. In *Elementy* he wrote:

> In this variety of opinions, let us single out, and declare for, the position of nominalism. [...] no object is a number, and [...] neither arithmetic, nor the theory of numbers, nor—*a fortiori*—mathematics in general build statements which might strictly be called statements about numbers in the same sense in which zoology makes statements about animals.[19] [15, p. 317]

Mathematics speaks about all things—and hence, its universality.

Kotarbiński—firmly refuting the conception that mathematics investigates a certain world of ideal objects dependent on time, space and cognitive mind—did not follow any concrete conception. He stated that mathematics can be characterised in at least three ways:

(1) as the body of systems in which theorems are justified only in a deductive way and 'the theorems are formulated correctly as statements containing only the following types of signs—variables, connectives, what are called 'names of numbers", 'names of sets", 'names of figures", or terms defined by such signs, names of relations (such as 'greater than", 'equal to", etc,) and finally punctuation marks and signs informing about the role of the remaining signs' [15, p. 322][20]—mathematics thus understood embraces the whole formal logic (in its propositions these 'names' do not occur) and the so-called proper mathematics;
(2) as proper mathematics or mathematics in a narrower sense, which is characterised by the fact that those 'names' occur in its thesis;
(3) as a science that is characterised like proper mathematics but adding the condition that its propositions have the feature of apriority, i.e., its axioms are assigned the feature of obviousness, and justifying its theorems we do not refer to empirical data.

Add that according to Kotarbiński nominalism is consistent with the thesis about aprioristic character of mathematics a science.

---

[18] It is worth quoting the words of Andrzej Mostowski uttered after returning from a conference dedicated to the foundations of set theory: "Just imagine that there I sighed for reism. The presented conceptions resulted from so breakneck speculations, so unattainable for intuition and so incomprehensible that reism seemed to be an oasis where one can breathe fresh air" (cf. [10, p. 73]).

[19] W tym nadmiarze rozmaitych stanowisk niechaj nam wolno będzie wyróżnić stanowisko nominalizmu i przy nim się opowiedzieć. [...] żaden przedmiot nie jest liczbą i [...] ani arytmetyka, ani tzw. "teoria liczb", ani tym bardziej matematyka w ogóle nie budują zdań, które by można nazwać ściśle zdaniami o liczbach w tym sensie, w jakim np. zoologia mówi o zwierzętach [14, p. 373].

[20] których twierdzenia wypowiada się poprawnie w zdaniach, zawierających tylko następujące rodzaje znaków: symbole zmienne, spójniki, tzw. » nazwy liczb«, tzw. » nazwy zbiorów«, tzw. » nazwy figur«, lub terminy przez takie znaki zdefiniowane, dalej terminy stosunkowe, jak » większy«, » równy« itp., wreszcie znaki przestankowe oraz znaki informujące o roli pozostałych znaków [14, p. 379].

5 Leon Chwistek: Nominalism

Talking about nominalism one should mention also Leon Chwistek (1884–1944). Though he did not belong directly to Lvov-Warsaw School, he went always along his own paths being a "separate" scholar, nevertheless his conceptions were important.

Chwistek declared himself as a nominalist. According to him the subject of deductive sciences, hence also of mathematics, are expressions constructed in them according to accepted rules of construction. Hence the subject of mathematics are not ideal objects like points, lines, numbers or sets. Expressions being subjects of mathematics are physical objects given us in experience. They can be transformed according to accepted rules. In every system one accepts such rules as well as some expressions that play the role of axioms and that form the base on which theorems are deduced. Transformation rules and axioms are chosen in such a way that expressions could be interpreted as descriptions of considered states of affairs. To be able to apply deductive theories to specific sciences and generally to perceive concrete areas of reality, the elements of the latter should be schematised.

In particular geometry is—according to Chwistek an experimental discipline. In Chapter VIII of *Granice nauki* [Limits of Science] he wrote:

> Geometry is an experimental science. It depends upon the measurement of segments, angles, and areas. The Egyptians conceived it in this way and it has remained essentially the same up to this very day. Today what is generally regarded as geometry, i.e. what is included in textbooks, is the peculiar mixture of experimental geometry and the geometrical metaphysics which was inherited from the Greeks as Euclid's *Elements*.[21] [5, p. 170]

The rise of the systems of non-Euclidean geometry of Bolyai, Gauss and Lobachevsky in the nineteenth century—regarded by Chwistek as the most important achievement in exact sciences—abolished in his opinion Kant's idealism. These geometries showed that, for example, the concept of a straight line is not of an objective character, but depends on the accepted axioms. It may suggest that conventionalism is the proper philosophy for geometry. Indeed, in his first works, e.g. the paper "Trzy odczyty odnoszące się do pojęcia istnienia" [Three Talks Concerning the Concept of Existence] (see [3]), he states that the existence of systems of non-Euclidean geometry, which are consistent, refutes the thesis of the *a priori* character of geometry. It seems that he would tend to accept conventionalism, although he does not state this explicitly. However in *Granice nauki* [4] he explicitly and categorically rejected conventionalism claiming that geometry—similarly as all other fundamental experimental sciences—should be based on the theory of expressions. This is because conventionalism introduces hypothetical entities, as was the case in John Stuart Mill's works or later Poincaré's, a promoter of this direction. Chwistek wrote:

> It seems that it is impossible to attain a general concept of geometry without using formulae. It is therefore clear that the conception of geometry as the science of ideal spatial constructions

[21] Geometria jest nauką doświadczalną. Polega ona na mierzeniu odcinków, kątów i powierzchni. Tak pojmowali ją Egipcjanie i taką pozostała w istocie swojej do dzisiaj. To, co uważa się powszechnie za geometrię za naszych czasów, tj. to, o czym pisze się w podręcznikach, jest osobliwą mieszaniną geometrii doświadczalnej i metafizyki geometrycznej, którą pozostawili nam w spadku Grecy pod postacią elementów Euklidesa ([4, p. 190]; see also [6, p. 170]).

must be nullified .... To speak of different four-dimensional space-times it is necessary to employ five-dimensional spacetime. It is clear that all this has only as much meaning as do mathematical formulae.[22] [5, pp. 186–187]

In a similar way as geometry one should treat arithmetic, mathematical analysis and other mathematical theories obtaining in this way consequently their nominalistic interpretations.

6 Kazimierz Ajdukiewicz and Ontology of Mathematics

Let us turn back to Lvov-Warsaw School and consider ontological views of Kazimierz Ajdukiewicz (1890–1963). He considered the ontology of mathematics and logic in his *Habilitationsschrift* entitled *Z metodologii nauk dedukcyjnych* [From the Methodology of Deductive Sciences] from 1921 (cf. [1]) in which he discussed the problem of existence, in particular the problem what does it mean "to exist" in deductive sciences. He wrote there:

An analysis of meaning of the word 'exist' as used in deductive theories does not amount to the problem: what kind of existence is among the attributes of existing objects of deductive theories; our own position permits us to doubt whether any kind of being at all is among the attributes of these objects. Our problem then is not the question what kind of being is attributable to objects under discussion, but the question what is the meaning of the word 'exist' as used in deductive theories. It may be that it is being used quite erroneously and has nothing at all to do with existence.[23] [2, p. 34]

Ajdukiewicz argued in the considered work that the existence in deductive sciences cannot be identified with consistency and that consistency is neither sufficient nor necessary condition of existence. He claimed that the necessary conditions of existence are: (I) being included into the domain of the given theory, and (II) consistency:

My contention is, namely, that for an object p defined by $\Omega(p)$ to exist it is necessary that p be an element of the domain of the given theory, in other words that $\Omega(p)$ entailed $A(p)$ [...].
In order to exist an object must, therefore, satisfy another requirement—besides the above condition of being an element of the domain of the theory—scil. its definition must not have any consequences inconsistent with the consequences of $A(p)$. [...]

[22]Okazuje się, że dotarcie do ogólnego pojęcia geometrii bez formuł jest niemożliwe. Jasne jest, że idąc tą drogą, musimy dojść do unicestwienia geometrii jako nauki o idealnych utworach przestrzennych. [...] żeby mówić o różnych czterowymiarowych czasoprzestrzeniach, musimy się odwołać do czasoprzestrzeni pięciowymiarowej. Jest jasne, że wszystko to ma tyle sensu, ile zawierają go formuły matematyczne [4, pp. 186–187].

[23]Analiza znaczenia wyrazu "istnieć" w naukach dedukcyjnych nie jest zatem równoznaczna z zagadnieniem: jaki rodzaj istnienia przysługuje istniejącym przedmiotom nauk dedukcyjnych; problemat nasz pozwala nam w ogóle wątpić o tym, czy jakikolwiek rodzaj bytu przedmiotom tym przysługuje. Kwestią naszą zatem nie jest pytanie, co za rodzaj bytu mają przedmioty przez nas rozważane, ale co znaczy wyraz "istnieć" w naukach dedukcyjnych. Być może, że jest on całkiem mylnie używany i nie ma z istnieniem nic wspólnego [1, p. 46].

> Objects which do not satisfy either the first or the second requirement, do not exist, are nonexisting. From existing and nonexisting objects we ought to distinguish objects which are possible in the given theory.[24] [2, pp. 42–43]

Ajdukiewicz comes to the conclusion that if an object is to exist it must satisfy the requirements (I) and (II) as well as 'not restrict the domain of possible objects'[25] [2, p. 44] in the given theory. And he concludes his considerations in the following way:

> In the deductive sciences we do not speak of existence in absolute sense but only relatively to a given system. For there exist Euclidean straight lines and non-Euclidean straight lines; however, both cannot co-exist and their co-existence would be a consequence of their existence if this word were taken in either case in the absolute sense. We may only speak of existence in a system as we speak of inclusion in a domain. Nevertheless it is possible to construct a 'universe' consisting of the domains of several compatible theories, thus forming a system whose axioms would be all axioms of all compatible theories. We could then speak of absolute existence, not quite absolute, though, since it would be possible by choosing various theories, to construct many such 'universes,' self-compatible but mutually exclusive.[26] [2, p. 45]

7 Andrzej Mostowski: Nominalism, Reism, Constructivism

Let us finish our considerations by presenting ontological views of Andrzej Mostowski (1913–1975). He is usually treated as a representative of the second generation of Lvov-Warsaw School. He was disciple of Tarski, some influence of Alfred Lindenbaum on him can be seen.

Mostowski inherited from Tarski general philosophical views, in particular tendency towards empiricism and apparent respect for nominalism. His sympathies were also, as it seems, with Kotarbiński's reism, i.e., the view that there exist only individual physical things. However he avoided in his logical and mathematical works explicit philosophical declarations. Nevertheless there are some exceptions from that tendency.

---

[24] Twierdzę mianowicie, że koniecznym warunkiem na to, by przedmiot określony przez $\Omega(p)$ istniał, jest iżby przedmiot p należał do zakresu danej teorii, czyli iżby z $\Omega(p)$ wynikało $A(p)$ [...].

Musi tedy przedmiot na to, aby istniał, spełniać prócz pierwszego (wyżej wymienionego warunku zawierania się) warunek drugi, musi mianowicie jego określenie nie posiadać następstw sprzecznych z następstwami $A(p)$. [...]

Przedmioty, które nie czynią zadość pierwszemu albo drugiemu warunkowi, nie istnieją i są nieistniejące. Prócz przedmiotów istniejących i nieistniejących należy jeszcze rozróżnić, naszym zdaniem, przedmioty możliwe w danej teorii [1, pp. 59–60].

[25] nie ograniczał [on] zakresu przedmiotów możliwych [1, p. 62].

[26] O istnieniu bezwzględnym w naukach dedukcyjnych nie mówimy wcale. Zawsze tylko o istnieniu w pewnym systemie. Wszakże istnieją i proste euklidesowe, i nieeuklidesowe, obie nie mogą jednak współistnieć, a współistnienie ich byłoby konsekwencją ich istnienia, gdyby ten wyraz wziąć w odniesieniu do obu w tym samym sensie bezwzględnym. Można więc mówić tylko o istnieniu w pewnym systemie, podobnie jak o zawieraniu się tylko w pewnym zakresie. Niemniej jednak można utworzyć "uniwersum" z zakresów kilku zgodnych z sobą teorii, tworząc system, którego aksjomaty byłyby wszystkimi aksjomatami wszystkich teorii zgodnych. Można by wtedy mówić o istnieniu bezwzględnym, jakkolwiek niezupełnie bezwzględnym, bo można by, dobierając rozmaite teorie, potworzyć wiele takich "uniwersów" w sobie zgodnych, lecz między sobą wykluczających się [1, p. 63].

In the paper “A Classification of Logical Systems” [29] Mostowski assumes a certain philosophical presumption on the analysed logical systems. Declaring at the beginning of his work that “The subject itself as well as the method of its presentation will be of a mathematical rather than philosophical character” [29, p. 245] he openly states:

> Although our investigations will be purely formal we shall nevertheless accept a definitive philosophical point of view with respect to logical systems. We shall not consider logical systems as void schemata deprived of any interpretation. On the contrary we shall assume the objective existence of a kind of » mathematical reality« (e.g. of the set of all integers or the set of all real numbers). By objective existence we mean existence independently of all linguistic constructions. [29, pp. 246–247]

The task of logical systems is—according to Mostowski—just to describe that “mathematical reality”. Consequently, every sentence of logic is equipped with a certain meaning—it says that mathematical reality is entitled to have this or that property. The fact that there exist true sentences that are unprovable in a given system—what is a consequence of Gödel’s incompleteness theorems—can be explained by the bigger complexity of the properties of this “mathematical reality” than the complexity of the properties that can be deduced from axioms by the accepted inference rules. Mostowski concludes in a characteristic way:

> We do not intend to defend the philosophical correctness or even the philosophical acceptability of the point of view here described. It is evident that it is entirely opposite to the point of view of nominalism and related trends. [29, pp. 247]

One can easily see here certain tension between aforementioned inclinations towards nominalism and his concrete logical and mathematical investigations.

Philosophical questions, in particular ontological ones, appear in Mostowski’s papers devoted to set theory. Considering Gödel’s and Cohen’s results on consistency and independence of Axiom of Choice and Continuum Hypothesis Mostowski claimed that they can be treated as “one of the most important arguments against mathematical Platonism” [33, p. 176]. After Gödel’s and Cohen’s results it is possible to construct consistent but mutually inconsistent set theories. If such theories are constructed “we shall be forced to admit that in the match between Platonism and formalism the latter has again scored one point” [33, p. 182]. Since metamathematical results on set theory do not provide decisions concerning the way of existence of sets, and in general of objects of mathematics, and consequently the controversy between formalism and Platonism, Mostowski concludes in the paper “Sets” [34]:

> Whatever the final outcome of the fight between these two opposing trends will be, it is obvious that we should concentrate on the study of concepts which seem perfectly clear and perspicuous to us. [34, p. 28]

At another place however he wrote that “the ultimate formulation of axioms of set theory should be preceded by a discussion of the fundamental assumptions of this theory, including the constructive standpoint” [30, p. 20].

Mostowski had a keen interest in constructivism. However it should be added that he was more interested in its aims than in proposed solutions (see for example [31, p. 192]).

In the monograph *Logika matematyczna* [Mathematical Logic] [28] he wrote even:

> I am inclined to think that a satisfactory solution of the foundations of mathematics will happen on the way shown by constructivism or a similar direction. However, one cannot now write a textbook of logic on this basis.[27] [28, p. VI]

Mostowski saw advantages of constructivism in the fact that:

> [...] it wants to inquire into the nature of mathematical entities and to find a justification for the general laws which govern them, whereas Platonism takes these laws as granted without any further discussion. [31, p. 192]

According to Mostowski constructivistic conceptions are closer to nominalism than to Platonism. This implies that constructivism does not accept general concepts of mathematics as being given but attempts to construct them. "This leads to the result that one can identify mathematical concepts with their definitions" [31, p. 178]. In arithmetic constructivism allows us to give up assuming actual infinity or to use solutions requiring only the nominalist approach. Whereas one of the advantages of nominalism is that many important mathematical theories have been satisfactorily reconstructed on the nominalist basis, and these reconstructions have turned out to be equivalent to the classical theories.

Mostowski was aware of some limitations of constructive methods in mathematics and of the fact that they do not suffice (cf. for example [34, pp. 29–32]). Nevertheless he investigated principles of constructivism. He was of the opinion that sometimes constructivism is philosophically more satisfactory—this is the case of arithmetic or of applied mathematics where it seems to reveal new promising perspectives.

It should be noticed that Mostowski considered constructivism in a way connected with the classical point of view. Hence he was not connected with pure constructivism of Brouwer, Heyting and other intuitionists. In fact he represented rather certain combination of constructivism and set-theoretical program. This combination formed according to him the base on which the foundations of mathematics should be developed.

8 Conclusion

The described panorama of ontological views concerning logic and mathematics of representatives of Lvov-Warsaw School shows that one does not find here any dominating position, just opposite, there is a full range of positions from Platonism through constructivism till nominalism. What was the reason of that? First of all it should be stressed that philosophical views were formulated in Lvov-Warsaw School on the margin of proper logical or metamathematical investigations. One was convinced that mathematical and logical investigations should not bounded by any *a priori* philosophical assumptions. Mathematics and logic should be autonomous and neutral with respect to philosophy. Hence opinions about philosophical aspects of logic or mathematics were rather fragmentary and incomplete, they concerned first of all particular issues connected with problems actually studied. Formulated remarks were often simply comments to

[27] Jestem skłonny mniemać, że zadowalające rozstrzygnięcie zagadnienia podstaw matematyki nastąpi na drodze wskazanej przez konstruktywizm lub kierunek do niego zbliżony. Na tej jednak podstawie nie można by już teraz napisać podręcznika logiki.

concrete technical results from the foundations of mathematics or logic. Exceptions were here Leśniewski and Chwistek whose logical investigations were a consequence and result of some philosophical considerations.

Philosophical views were generally treated as private matters. They should not bound the research activity in logic, mathematics or the foundations of mathematics and during the very investigations of concrete mathematical or logical problems should be suspended. Moreover it happed that declared philosophical views were in fact not compatible with research practice. It can be easily seen for example by Tarski who declared himself as nominalist but simultaneously in his research practice used without any restrictions infinitistic methods far from what has been accepted by nominalists.

References

1. Ajdukiewicz, K.: Z metodologii nauk dedukcyjnych. Nakładem Polskiego Towarzystwa Filozoficznego, Lwów (English translation see Ajdukiewicz, K.: From the Methodology of the Deductive Sciences. Stud. Logica **19**, 9–45 (1966))
2. Ajdukiewicz, K.: From the Methodology of the Deductive Sciences. Stud. Logica **19**, 9–45 (1966)
3. Chwistek, L.: Trzy odczyty odnoszące się do pojęcia istnienia. Przegląd Filozoficzny **20**, 122–151 (1917). (Reprinted in: Chwistek, L.: Pisma filozoficzne i logiczne, Tom 1. Państwowe Wydawnictwo Naukowe, Warszawa (1961), pp. 3–29)
4. Chwistek, L.: Granice nauki. Zarys logiki i metodologii nauk ścisłych. Książnica-Atlas, Lwów-Warszawa (1935) (Reprinted in: Chwistek, L.: Pisma filozoficzne i logiczne, Tom 2. Państwowe Wydawnictwo Naukowe, Warszawa (1963), pp. 1–232; English translation see Chwistek, L.: The Limits of Science. Outline of Logic and of the Methodology of the Exact Sciences. Kegan Paul, Trench Trubner and Co, New York/London (1948). Translated by H.Ch. Brodie and A.P. Coleman (Reprinted by Routledge, London (2000)))
5. Chwistek, L.: The Limits of Science. Outline of Logic and of the Methodology of the Exact Sciences. Kegan Paul, Trench Trubner and Co, New York/London (1948). Translated by H.Ch. Brodie and A.P. Coleman (Reprinted by Routledge, London (2000))
6. Chwistek, L.: Pisma filozoficzne i logiczne, Tom 1. Państwowe Wydawnictwo Naukowe, Warszawa (1961)
7. Chwistek, L.: Pisma filozoficzne i logiczne, Tom 2. Państwowe Wydawnictwo Naukowe, Warszawa (1963)
8. Dąmbska I.: Czterdzieści lat filozofii we Lwowie. In: Pięćdziesiąt lat filozofii w Polsce. Przegląd Filozoficzny **44**, 14–25 (1948)
9. Feferman, A.B., Feferman, S.: Alfred Tarski. Life and Logic. Cambridge University Press, Cambridge (2004)
10. Kotarbińska, J.: Głos w dyskusji. Studia Filozoficzne **5** (1984)
11. Kotarbiński, T.: Sprawa istnienia przedmiotów idealnych. Przegląd Filozoficzny **23**, 149–170 (1920)
12. Kotarbiński, T.: Elementy teorji poznania, logiki formalnej i metodologji nauk. Wydawnictwo Zakładu Naukowego im. Ossolińskich, Lwów (1929)
13. Kotarbiński, T.: Zasadnicze myśli pansomatyzmu. Przegląd Filozoficzny **38**, 283–294 (1935)
14. Kotarbiński, T.: Elementy logiki, teorii poznania, logiki formalnej i metodologii nauk, 2nd edn. Ossolineum, Wrocław-Warszawa-Kraków (1961) (3rd edition: Ossolineum, Warszawa (1986); English translation see Kotarbiński, T.: Gnosiology: The Scientific Approach to the Theory of Knowledge. Pergamon, Oxford/Warszawa, 1966)
15. Kotarbiński, T.: Gnosiology: The Scientific Approach to the Theory of Knowledge. Pergamon, Oxford/Warszawa (1966)
16. Leśniewski, S.: O podstawach matematyki. Wstęp oraz §§1–3. Przegląd Filozoficzny **30**, 164–206 (1927)
17. Leśniewski, S.: Grundzüge ge eines neuen System der Grundlagen der Mathematik. Fundamenta Mathematicae **14**, 1–81 (1929)

18. Leśniewski, S.: Collected Works. PWN (Polish Scientific Publishers) and Kluwer Academic Publishers, Dordrecht-Boston-London (1992). Edited by S.J. Surma, J.T. Srzednicki, I. Barnett, V.F. Rickey
19. Leśniewski, S.: Czy prawda jest wieczna czy też wieczna i odwieczna?. Nowe Tory **18**, 493–528 (1913)
20. Leśniewski, S.: Krytyka logicznej zasady wyłączonego środka. Przegląd Filozoficzny **16**, 315–352 (1913) (Reprint (with certain omissions) in: Jadacki, J.J., Bigaj, T., Lissowska, A. (eds.) Co istnieje?, vol. 1, pp. 143–167. PETIT, Warszawa (1996))
21. Łukasiewicz, J.: Logika a psychologia. Przegląd Filozoficzny **10**, 489–491 (1907)
22. Łukasiewicz, J.: O twórczości w nauce. In: Księga pamiątkowa ku uczczeniu 250-tej rocznicy założenia Uniwersytetu Lwowskiego przez Króla Jana Kazimierza r. 1661, vol. 1, pp. 1–15. Uniwersytet Lwowski, Lwów (1912) (Second revised and enlarged edition in: Poradnik dla samouków. Wskazówki metodyczne dla studjujących poszczególne nauki. New edition, Vol. 1, pp. 15–39. Wydawnictwo A. Heflera i St. Michalskiego, Warszawa (1915)
23. Łukasiewicz, J.: Logistyka a filozofia. Przegląd Filozoficzny **39**, 115–131 (1936)
24. Łukasiewicz, J.: W obronie logistyki. In: Myśl katolicka wobec logiki współczesnej. Studia Gnesnensia **15**, 12–26, 159–165 (1937)
25. Łukasiewicz, J.: Selected Works. North-Holland Publishing Company, Amsterdam-London and PWN (Polish Scientific Publishers), Warszawa (1970). Edited by L. Borkowski
26. Mancosu, P.: Harvard 1940–1941: Tarski, Carnap and Quine on a finitistic language of mathematics for science. Hist. Philos. Log. **26**, 327–357 (2005)
27. Mancosu, P.: Tarski's Engagement with Philosophy. In: Lapointe, S., Woleński, J., Marion, M., Miskiewicz, W. (eds.) The Golden Age of Polish Philosophy. Kazimierz Twardowski's Philosophical Legacy, pp. 131–153. Springer, Dordrecht-Heidelberg-London-New York (2009)
28. Mostowski, A.: Logika matematyczna. Kurs uniwersytecki. Monografie Matematyczne, Warszawa-Wrocław (1948)
29. Mostowski, A.: A Classification of Logical Systems. Stud. Philos. **4**, 237–274 (1949–1950) (published in 1951) (Reprinted in: Murawski, R.: Filozofia matematyki i logiki w Polsce międzywojennej. Monografie Fundacji na rzecz Nauki Polskiej, Wydawnictwo Naukowe Uniwersytetu Mikołaja Kopernika, Toruń (2011), vol. II, pp. 154–191)
30. Mostowski, A.: The Present State of Investigations of the Foundations of Mathematics. Rozprawy Matematyczne **9**, 1–48 (1955). In collaboration with: A. Grzegorczyk, S. Jaśkowski, J. Łoś, S. Mazur, H. Rasiowa and R. Sikorski
31. Mostowski, A.: On Various Degrees of Constructivism. In: Heyting, A. (ed.) Contructivity in Mathematics. Proceedings of the Colloquium held in Amsterdam, 1957, pp. 178–194. North-Holland Publishing Company, Amsterdam (1959)
32. Mostowski, A.: Tarski Alfred. In: Edwards, P. (ed.) The Encyclopedia of Philosophy, vol. 8, pp. 77–81. The Macmillan Company and the Free Press, New York/Collie-Macmillan, London (1967)
33. Mostowski, A.: Niesprzeczność i niezależność hipotezy kontinuum. Roczniki Polskiego Towarzystwa Matematycznego, Seria II: Wiadomości Matematyczne **10**, 175–182 (1968)
34. Mostowski, A.: Sets. In: Scientific Thought. Some Underlying Concepts, Methods and Procedures, pp. 1–34. Mouton/UNESCO, The Hague (1972)
35. Murawski, R.: Filozofia matematyki i logiki w Polsce międzywojennej. Monografie Fundacji na rzecz Nauki Polskiej, Wydawnictwo Naukowe Uniwersytetu Mikołaja Kopernika, Toruń (2011)
36. Murawski, R.: The Philosophy of Mathematics and Logic in the 1920s and 1930s in Poland. Birkhäuser, Basel (2014)
37. Tarski, A.: Collected Papers, vols. 1–4. Birkhäuser, Basel-Boston-Stuttgart (1986). Eds. S.R. Givant and R.N. McKenzie
38. Woleński, J.: Filozoficzna szkoła lwowsko-warszawska. Państwowe Wydawnictwo Naukowe, Warszawa (1985)
39. Woleński, J.: Logic and Philosophy in the Lvov-Warsaw School. Kluwer Academic Publishers, Dordrecht-Boston-London (1989)

R. Murawski (✉)
Adam Mickiewicz University in Poznań, Faculty of Mathematics and Computer Science, Poznań, Poland
e-mail: rmur@amu.edu.pl

Part III
The War and Post-War Period

A View of Revival of Mathematical Logic in Warsaw, 1945–1975

Victor W. Marek

To the memories of my teachers: Andrzej Mostowski, Helena Rasiowa, and Andrzej Grzegorczyk

Abstract This is a (biased) account of the revival of Foundations of Mathematical Logic, and more generally Foundations of Mathematics after WWII. The perspective is limited to years 1945–1975. This coincided with the activities of Professor Andrzej Mostowski, the principal leader of the foundational research in Poland during the reported time. Moreover, in this text, we focus on the development of Foundations in Warsaw, which misses an important aspect of the foundational research in Poland of the time, specifically in Wrocław. Even the description of the Warsaw efforts is incomplete, not covering Foundations of Geometry. We focus on activities of three main researchers: Andrzej Mostowski, Helena Rasiowa and Andrzej Grzegorczyk.

Keywords Logic in Warsaw · Andrzej Mostowski · Helena Rasiowa · Andrzej Grzegorczyk

Mathematics Subject Classification (2000) Primary 01A60, 03-03

1 Introduction

World War II separates significant achievements of mathematical community of Poland during the so-called "Second Polish Republic" (1918–1939) and very different "Polish Peoples Republic" that was created after liberating Poland from German occupation. The war resulted in significant territorial changes. In effect, Poland was moved some 200 miles to the west, loosing significant part of its Eastern part and gaining territories to the West of the pre-war borders that had previously been German. Poland lost two significant academic centers: Lwów, (now Lviv in Ukraine) and Wilno (now Vilnius in Lithuania). Instead, Poland gained important centers of science: Wrocław (previously German Breslau) as well as Gdańsk (previously a free city). Significant part of Prussia became part of Polish territory.

Á. Garrido, U. Wybraniec-Skardowska (eds.), *The Lvov-Warsaw School. Past and Present*, Studies in Universal Logic,
https://doi.org/10.1007/978-3-319-65430-0_45

Science in Poland, and in particular Mathematics (thus also Mathematical Logic) suffered greatly as a result of German occupation and, more generally, World War II. The losses were both due to extermination of intellectuals by Germans (especially of those with Jewish roots) and emigration. Logicians such as Adolf Lindenbaum, Mordechaj Wajsberg, Mojzesz Presburger, and others were killed by Germans. Several important logicians, in particular Alfred Tarski, Jan Łukasiewicz, Bolesław Sobociński and others left Poland and emigrated as the result of the war and continued their work in foreign institutions.

There was a conscious effort to re-create Jan Kazimierz University of Lwów in Wrocław, and to some extent, Lwów mathematics, with Hugo Steinhaus continuing Lwów traditions. Likely the result of WWII, the death of Stefan Banach in 1945, in a relatively early age of 53, deprived Poland of one of its great mathematicians open to many foundational challenges.

2 The Beginnings

It is very hard to imagine the amount of destruction both of Warsaw and Wrocław as the result of WWII. The western part of Warsaw (where the Warsaw University was, and is now) was destroyed in 95%. This author, born in the second half of WWII, remembers well the sea of ruins in parts of Warsaw. This was the result of a systematic destruction of that part of Warsaw by Germans after the fall of Warsaw Uprising (August–October 1944.) Wrocław, converted by Nazis into a fortress, was similarly destroyed. Both places were, essentially, human deserts.

During the WWII, there were several underground Universities in Warsaw, functioning in spite of prohibition by the occupiers of both secondary and university education. At a significant risk, the principal Warsaw mathematicians (Sierpiński, Kuratowski, and others) were conducting lectures and doing research. All of them left Warsaw with the rest of the population when Germans started their destruction of the city.

Among the other mathematicians active in the underground Warsaw University was Andrzej Mostowski, a student of Lindenbaum, Tarski, Gödel and Kuratowski. Mostowski conducted an active research during WWII (a story of his "black notebook" containing the results of his research and the alternative he faced as he was leaving Warsaw in 1944 is told in several publications devoted to his life). Mostowski, along with the rest of population of Warsaw had to leave the town. Eventually, he found his way to Kraków. In 1945, Mostowski received his *habilitation* degree at Jagiellonian University. Mostowski returned to Warsaw in 1946, and was associated with the revived Warsaw University till the end of his life in 1975.

The faculty of Warsaw University with interests in widely understood Logic and Foundations, Kazimierz Kuratowski and Waclaw Sierpiński, while still active, had principal interests in a variety of areas, with Kuratowski mainly interested in Topology, and Sierpiński in Number Theory and Combinatorial Set Theory. Both of them had residual interest in Foundations, but it was not the principal area of their research. In this situation, the fundamental role in rebuilding foundational research in Poland fell on Mostowski, who, eventually, educated a large group of students who helped him in returning Warsaw to one of the prime centers in foundational research. Already in 1946, both Helena Rasiowa and Wanda Szmielew started their research in logic (both studied in the underground Warsaw University during German occupation). Eventually, both Professor Rasiowa and Professor Szmielew developed interests in areas not directly related to Professor Mostowski research (with Rasiowa mainly interested in Algebraic Foundations of Logic, and Szmielew working in Foundations of Geometry). Both Rasiowa and Szmielew educated groups of logicians following the directions of research outlined above. Additionally, Professor Rasiowa interests were later influenced by Computer Science applications of Logic. The results of Professor Szmielew initially developed during German occupation and then continued during her stay at the University of California, Berkeley, 1949–1950, (decidability of the theory of Abelian groups) were instantly recognized as a major result. Even earlier, 1948–1949, Professor Mostowski spent a year as a visitor at the Institute of Advanced Studies, in Princeton, NY. Mostowski's relation with Gödel (since Mostowski's extended visit in Vienna in 1936) played an important role in Mostowski's research in several areas of Foundations, specifically in the metamathematics of set theory and related formalisms.

3 Andrzej Mostowski and His Students

The most amazing aspect of Mostowski's activities during the first 15 years after WWII is the breadth of his interests as reflected in his publications. He found significant results in Recursion Theory, Set Theory, Proof Theory, and Model Theory. This needs to be related to the fact that the notebooks that he wrote during WWII were lost (Germans allowed inhabitants of Warsaw expelled from the city to take one kilogram of items, and he (as stated to this author) chose a kilogram of bread). Reconstruction of the results and their proofs took time, and not all was remembered. Still, results on independence of various set-theoretic statements, results on existence of models of arithmetical theories, results on the arithmetical hierarchy, on proof theory of intuitionistic logic and results on the incompleteness of Peano arithmetic and its extensions (culminating in the monograph with R. Robinson and A. Tarski) all followed in quick succession. The handbook on Mathematical Logic (in Polish, 1948) was a powerful tool in educating generations of Polish logicians.[1] Almost every year in the 1950s Mostowski was publishing work that was widely read, commented, and even more importantly further expanded by mathematicians all over the world. It is enough to mention his work on the direct product of theories (significantly expanded by Solomon Feferman and Robert Vaught), generalized quantifiers (which pioneered the area of significant importance both in mathematical logic and (much later) computer science, model theory of second order arithmetic (together with Andrzej Grzegorczyk and Czesław Ryll-Nardzewski), and one of most important tools of model theory—existence of models with indiscernibles (joint work with Andrzej Ehrenfeucht). This created the opportunity for many individuals to study with Professor Mostowski.[2]

[1] I believe that one of the articles in the present volume discusses this monograph.

[2] The author of the present article was Mostowski's Ph.D. student in 1964–1968.

It is safe to say that by the end of 1950s, Warsaw, and more generally, Poland, returned to its place as one of most important centers of research in the Foundations of Mathematics. Moreover, by that time, a cadre of researchers both within direct circle of Mostowski, as well as circles of his former students, such as Andrzej Grzegorczyk, Helena Rasiowa and Wanda Szmielew was active in Warsaw. Each of these individuals (and, of course, Professor Mostowski himself) directed a seminar at Warsaw University. For that reason, the access to the knowledge in the areas of Foundations was easy both for students and faculty.

The unusually wide interests of Mostowski continued into 1960s. The breakthrough in the area of independence proofs in Set Theory due to Paul J. Cohen using the technique of forcing influenced Mostowski and his group of collaborators. Mostowski devoted to this topic a monograph that presented Cohen arguments in a new perspective. Even more importantly, in a series of lectures at Helsinki (subsequently published as a monograph), Mostowski presented a unified perspective of Foundations of Mathematics of the middle of 1960s. It established him as one of the leaders and visionaries of the area.

Continuing exploration of other areas of foundations, models of second order arithmetic, impredicative theories of classes, abstract model theory, weak systems of arithmetic and other current areas of research, Mostowski constantly expanded the perspective of research in Warsaw. A number of scientific meetings in Warsaw, culminating with the Foundations of Mathematics Semester at the Banach Center in 1973, firmly established Warsaw as a place worth visiting. During 1960s and first half of 1970, there was a constant stream of visiting logicians benefiting of Mostowski's hospitality and the cutting edge of foundational research in Warsaw.

Professor Mostowski's death in 1975, at age 62, was a significant blow to foundational work in Warsaw, and more generally, in Poland. Additionally, later on, important political disturbances, including martial law, resulted in further losses in the foundational research in Warsaw.

A most comprehensive and reasoned bibliography of Professor Andrzej Mostowski is provided by Jan Zygmunt [12] and supersedes earlier efforts by students of Mostowski. The bibliography lists over 110 papers and their translations including papers published posthumously) as well as over 15 monographs and handbooks. Together with a large number of reviews of papers, this bibliography presents the enormous body of highest-class research and other contributions to Foundations.

At present, Professor Mostowski has 18 direct descendants in the Mathematical Genealogy Database as well as 300 indirect descendants. Since some Mostowski's students and their descendants moved to Computer Science and Philosophy, the number of indirect descendants is, clearly, bigger.

4 Helena Rasiowa and Her Group

Professor Helena Rasiowa started her Mathematics studies in 1938. The World War II, with its consequences for Polish citizens (with dual occupation at first, and then suffering under German occupation) limited opportunities for individuals interested in higher education to "underground University". Rasiowa was studying in that institution, and wrote her M.Sc. thesis under the direction of Łukasiewicz and Sobociński. She described to her younger collaborators (this author included) the horrible experience of being buried under ruins of a house bombed during the Warsaw Uprising of 1944. As the rest of the population of Warsaw she was expelled of the city, and survived as a secondary school teacher. Andrzej Mostowski brought her back to the Warsaw University where she started her Ph.D. studies under his guidance. Her M.Sc. thesis was reconstructed (recall Mostowski's experience mentioned above). Following Mostowski's work on algebraic methods of logic (the area 100 years old at the time, already), Rasiowa developed algebraic techniques for modal and intuitionistic logics. Soon, jointly with Roman Sikorski, she developed a comprehensive theory of algebraic methods in logic. Several papers in this area resulted in habilitation at the Mathematical Institute of the Academy of Sciences. The essence of that theory consisted in expanding the notion of semantics to one where values form a Boolean Algebra (or Heyting algebra, in intuitionistic case). A fundamental result of this area, often called Rasiowa-Sikorski Lemma proves the existence of filters closed under denumerable family of meets. This result provides an alternative means to prove completeness theorem for predicate calculus and is Baire Theorem in disguise. It also explains why Cohen's technique of forcing works (although Rasiowa and Sikorski did not take an opportunity to provide this, maybe the most spectacular, application of their techniques in their famous monograph "Mathematics of Metamathematics").

The work on the algebraic methods in logic continued throughout the entire further life of Rasiowa. She created a large group of collaborators who worked on algebraic methods in a variety of logics, including logics that stemmed of application of formal methods in Computer Science.

Professor Rasiowa was very active in the science administration, both at Warsaw University (she served as a Dean of Mathematics and Physics, later Mathematics, Mechanics and Computer Science for many years) and international scientific institutions. She helped to establish *Studia Logica* as a leading journal on the borderline of Mathematics and Philosophy, and *Fundamenta Informaticae* (together with Zdzisław Pawlak).

Beginning in 1960s Professor Rasiowa promoted foundational issues in Computer Science; her contributions to the logic of programs (algorithmic logic) formed a widely-read text published by the Institute of Computer Science of Polish Academy of Sciences.

Professor Rasiowa supervised 19 PhD dissertations. At present she has 12 direct descendants in the Mathematical Genealogy Database as well as 93 indirect descendants. Since some (in fact most) of her students and their descendants moved to Computer Science, the number of indirect descendants is, clearly, bigger.

The paper by Bartol, Orłowska and Skowron includes the full list of Rasiowa's work.

5 Andrzej Grzegorczyk

Of the four main leaders of mathematical logic, and more generally, foundations of mathematics in post 1945 Poland, the youngest researcher was Professor Andrzej Grzegorczyk (1922–2014).

Professor Grzegorczyk studied (like Wanda Szmielew and Helena Rasiowa) at underground universities in Warsaw where he studied chemistry, physics and philosophy. He completed his studies in Cracow, immediately after the war; his MA thesis generalized Leśniewski's system to higher types. His interests at the time were motivated by philosophical questions—and involvement in widely understood philosophy continued throughout his entire scientific career. After WWII, Grzegorczyk turned to mathematical logic, with his Ph.D. (written under the direction of Andrzej Mostowski) devoted to formal topology. Several important contributions to Foundations of Mathematics followed. Of these, the hierarchy of primitive-recursive functions is commonly called *Grzegorczyk hierarchy*. The paper on the hierarchy can be treated as one of the first research papers on computational complexity of recursive functions. In 2003 Grzegorczyk proposed an new proof for Gödel's undecidability result. The proof omits arithmetization. Grzegorczyk's proof is much simpler than the original Gödel's proof. Other important contributions involved modal logic (with the logic commonly called *Grz*, Leśniewski's mereology and λ-calculus). Joint work with Mostowski and Ryll-Nardzewski on ω-models of second-order arithmetic established fundamental properties of these structures, creating foundations for significant body of work in Warsaw.

Professor Grzegorczyk, like many philosophically-oriented logicians was concerned with the issues of syntax and definability of fundamental notions. This, to some extent, resulted in tensions between Professor Grzegorczyk and other researchers in foundations, esp. mathematicians. The point was that mathematicians proved correctly results over the millennia, and these results, and more importantly, their proofs are still valid. This results in a certain attitude among mathematicians ("this is philosophy, not mathematics") which alienates some philosophical thinkers. Eventually, Professor Grzegorczyk left mathematics (at least formally) transferring to the Institute of Philosophy and Sociology of the Academy of Sciences.

Of the researchers of foundations of older generation, Professor Grzegorczyk was the most concerned with the questions of ethics. A rigid set of values and opinions, often going "against the grain" distinguished Grzegorczyk among his peers. During the "real socialism" he was willing to protest against injustices (this author benefited from Grzegorczyk's help and encouragement when subjected to repressions), but was willing to speak against popular opinions when he felt that these bring more harm than societal benefit. As a result, Grzegorczyk stood out among many, but "not with many".

The complex nature of the research of Professor Grzegorczyk (firmly placed on the intersection of Mathematical Logic and Philosophy) resulted in some controversies, esp. from the mathematical side and critiques of his fundamental contribution to Foundations of Mathematics, *Outline of Mathematical Logic*.

Mathematical Genealogy Projects lists one direct descendant of Professor Grzegorczyk and 6 indirect descendants. Since the unique descendant listed in that project is a Computer Scientist, the number of descendants is, likely, bigger. The influence of Professor Grzegorzyk on the foundational research in Philosophy was also significant.

More details and references to Grzegorczyk's detailed bibliography can be found in the paper by Krajewski in this volume.

6 Conclusions

The picture of revival of the Warsaw School of Foundations after the WWII that we present in this notes, raises a variety of issues. The first, and the most important one refers to the role of an individual in the scientific process. Certainly, the developments in Warsaw after 1945, were dominated by the contributions and the personality of Professor Andrzej Mostowski. The two other individuals treated in this text, Professor Helena Rasiowa and Professor Andrzej Grzegorczyk were able to contribute to the area of Foundations, but in the reported period, practically all foundational research in Warsaw (possibly with the exception of the work by Professor Wanda Szmielew group) was originally motivated by Professor Mostowski. He not only conducted his own research and motivated his collaborators, but also initiated research of both Professors Rasiowa and Grzegorczyk.

This raises the question of dependence of such scientific events (revival of a scientific school) on the individuals. Of course, it is difficult to hypothesise, but what would happen if Professor Mostowski did perish during the German occupation? Would the Warsaw school revive?

While one may wonder if such questions are meaningful, the many disasters during the second half of twentieth century and also more recent violent developments in various parts of the world raise the issue on the dependence of scientific "schools" on the availability and contributions of individuals. Maybe the events and developments of the Warsaw School of Foundations tells us something and suggest blueprint for revival in analogous situations.

Acknowledgements The author is grateful to Andrzej Jankowski, Stanisław Krajewski, Marcin Mostowski and Andrzej Skowron for their corrections and other helpful suggestions.

References

Andrzej Mostowski

1. Axiom of choice for finite sets. Fundam. Math. **33**, 137–168 (1945)
2. On definable sets of positive integers. Fundam. Math. **35**, 81–112 (1947)
3. Mathematical logic, An Academic Course. Mathematical Monographs, vol. 18. Warsaw-Wrocław (1948) (in Polish)
4. Sentences undecidable in formalized arithmetic, An exposition of the theory of Kurt Gödel. Studies in Logic and the Foundations of Mathematics, North-Holland Publishing, Amsterdam (1952)
5. (with Kuratowski, K.) Set theory. Mathematical Monographs, vol. 27. Polish Scientific Publishers, Warsaw-Wrocław (1952)
6. (with Robinson, R.M. and Tarski, A.) Undecidability and essential undecidability in arithmetic. In: Tarski, A., Mostowski, A., Robinson, R.M.: Undecidable theories. Studies in Logic and the Foundations of Mathematics, Part II. North-Holland Publishing, Amsterdam (1953)
7. (with Ehrenfeucht, A.) Models of axiomatic theories admitting automorphism. Fundam. Math. **43**, 50–68 (1956)
8. On a generalization of quantifiers. Fundam. Math. **44**, 12–36 (1957)
9. (with Grzegorczyk, A. and Ryll-Nardzewski, Cz.) The classical and the ω-complete arithmetic. J. Symb. Log. **23**, 188–206 (1958)

10. Thirty years of foundational studies. Lectures on the development of mathematical logic and the study of the foundations of mathematics in 1930–1964. Acta Philosophica Fennica **17**, 1–180 (1965)
11. Constructible sets with applications. Studies in Logic and the Foundations of Mathematics. Polish Scientific Publishers, Warsaw and North-Holland Publishing, Amsterdam (1969)
12. Zygmunt, J.: Bibliography of Andrzej Mostowski. In: Ehrenfeucht, A., Marek, V.W., Srebrny, M.: Andrzej Mostowski and Foundational Studies. IOS Press, Amsterdam (2007)

Helena Rasiowa

1. (with Sikorski, R.) A proof of the completeness theorem of Gödel. Fundam. Math. **37**, 193–200 (1950)
2. (with Sikorski, R.) Mathematics of metamathematics. Wydawnictwo Naukowe PWN, Warsaw (1963)
3. An algebraic approach to non-classical logics. Studies in Logic and the Foundations of Mathematics, vol. 78, Wydawnictwo Naukowe PWN/North-Holland, Warszawa/Amsterdam (1974)
4. Algebraic models of logics. Warsaw University (2001)
5. Bartol, W., Orłowska, E., Skowron, A.: Helena Rasiowa, 1917–1994. Modern Logic **5**(3), 231–247 (July 1995)
6. Jankowski, A., Skowron, A.: Helena Rasiowa (1917–1994). In: Á. Garrido and U. Wybraniec-Skardowska (eds.) The Lvov-Warsaw School. Past and Present, Studies in Universal Logic. https://doi.org/10.1007/978-3-319-65430-0_48

Andrzej Grzegorczyk

1. Some Classes of Recursive Functions. Rozprawy Matematyczne Nr IV, Instytut Matematyczny PAN, Warsaw (1953)
2. Zarys logiki matematycznej. Biblioteka Matematyczna tom 20, PWN, Warsaw (1961) (English version: An Outline of Mathematical Logic. Reidel-Holland (1974))
3. Axiomatizability of geometry without points. Synthèse **XII**(2/3), 228–235 (1960)
4. A theory without recursive models. Bulletin de l'Académie Polonaise des Sciences, Série Math. **10**, 63–69 (1962)
5. A philosophically plausible formal interpretation of intuitionistic logic. Indagationes Mathematicae **XXVI**, 596–601 (1964)
6. Krajewski, S.: Andrzej Grzegorczyk, a logician par excellence. In: Á. Garrido and U. Wybraniec-Skardowska (eds.) The Lvov-Warsaw School. Past and Present, Studies in Universal Logic. https://doi.org/10.1007/978-3-319-65430-0_50

V. W. Marek (✉)
Department of Computer Science, University of Kentucky, Lexington, KY, USA
e-mail: marek@cs.uky.edu

Andrzej Mostowski and the Notion of a Model

Wilfrid Hodges

Abstract Model theory became an independent discipline within logic during the first half of the 1950s. Andrzej Mostowski made several distinctive contributions to this development through papers of his. Also his 1948 textbook of logic covers material in the foundations of model theory, and in 1966 he published a survey book with several chapters on model theory. We examine his choice of technical terms and concepts during this period, and we discuss a criticism made by Abraham Robinson of the coverage of model theory in the 1966 book. On this basis we draw some conclusions about Mostowski's aims and attitudes, which were often different from those of other pioneers in the field.

Keywords Model · Interpretation · Metamathematics · Mostowski · Tarski · Abraham Robinson

Mathematics Subject Classification (2000) Primary 03A05, 03C99

1 The Emergence of a New Discipline

The official birthday of model theory has to be the publication of Alfred Tarski's paper [47] in 1954, which began:

> Within the last years a new branch of metamathematics has been developing. It is called the *theory of models* and can be regarded as a part of the semantics of formalized theories. [47, p. 572] (1)

But we can see propaganda for the new discipline, as yet unnamed, 4 years earlier in the addresses by Abraham Robinson [36] and Tarski [45] to the International Congress of Mathematicians in 1950. One theme of the propaganda was that the new discipline would provide new tools that algebraists and other mathematicians could use within their own

Á. Garrido, U. Wybraniec-Skardowska (eds.), *The Lvov-Warsaw School. Past and Present*, Studies in Universal Logic,
https://doi.org/10.1007/978-3-319-65430-0_46

disciplines. Thus Robinson:

> ... contemporary symbolic logic can produce useful tools—though by no means omnipotent ones—for the development of actual mathematics, more particularly for the development of algebra and, it would appear, algebraic geometry. (Robinson [36, p. 694]) (2)

Likewise Tarski [45, p. 717] spoke of applications 'which may be of general interest to mathematicians and especially to algebraists'.

The new discipline absorbed various earlier pieces of work by Veblen, Löwenheim, Skolem, Gödel, Mal'tsev and others, going back to the beginning of the twentieth century. Some of the early contributors to the new discipline were seen at the time, and are still seen, as primarily model theorists (at least within logic—some like Mal'tsev and Robinson had mathematical interests outside logic). Besides Mal'tsev and Robinson this group includes Henkin, Vaught and Fraïssé. But Andrzej Mostowski, though he was a major figure in early model theory, was never primarily a model theorist. Like Feferman (his junior by 15 years), Mostowski had interests that ranged across the whole of logic, including set theory and the study of Gödel's incompleteness theorem. For example one of his papers with 'models' in the title [30, 1953] is mainly quoted for its proofs that Zermelo-Fraenkel set theory and first-order Peano arithmetic are not finitely axiomatisable.

The creation of a new mathematical discipline is always a challenge for historians of mathematics, to understand where the new discipline came from, what were the forces that drove it in the direction that it took, and what the creators of the discipline understood themselves to be doing.

As far as I know, Mostowski never published a reflective account of the aims of model theory, or of his own aims in this area. Of course we can infer something about his aims from the problems that he chose to work on. But in this paper I explore two other routes into his thinking. The first is his choice of terminology—not just the words that he chose to use, but the concepts that he chose to give names to. The second is an issue raised by Robinson, that Mostowski's historical book [33] (1966) might not be entirely objective in its account of model theory. We will assess Robinson's criticism; to the extent that it is sound, it throws some light on Mostowski's understanding of what model theory is.

My wife and I met Mostowski at a party at Richard Montague's house in Los Angeles in summer 1967; we spent some time talking with him and found him genial and friendly. My wife recalls that we met him again a few days later at a picnic on the Dockweiler State Beach, and on that occasion he burned his feet on the hot sand. I was sorry I never got to know him better.

I warmly thank the editors of this volume for their kind invitation to submit this chapter, Jan Woleński for suggesting Mostowski as a topic, and Barbara Bogacka for checking my translations of some of Mostowski's Polish. Of course none of these people are responsible for my errors. A copy of Mostowski's book [28] came into my hands too long ago for me to remember where it came from; my apologies to anybody I should be thanking for it.

2 Mostowski's Writings in Model Theory

We briefly review those parts of Mostowski's work that relate most directly to model theory.

There are three papers that had a major impact on model theory. The first is the paper [29] (1952) 'On direct products of theories', which analysed the sentences true in a cartesian power in terms of the sentences true in the factors. The paper [10] of Feferman and Vaught owes its existence largely to Feferman's study of ideas in [29] ([10, p. 58], [8, p. 37])—though Feferman may have learned them directly from Tarski rather than through reading the paper. (Possibly Tarski's reluctance to accept Feferman's early work on this topic as of dissertation standard was based on a concern about whether Feferman had added enough to what was already in [29]; see p. 211 in [9].) The second is the paper [7] (1956) 'Models of axiomatic theories admitting automorphisms', written jointly with Mostowski's student Andrzej Ehrenfeucht. This paper introduced the notion of indiscernibles, which quickly became an indispensable tool of both model theory and set theory. The third is the paper [32] (1957) 'On a generalization of quantifiers'; this paper created one of the main strands of research in generalised model theory, with applications spilling over into linguistics [34].

Mostowski's earlier paper [27] (1947) 'On absolute properties of relations' seems to have had little impact, because it was written around a rather obscure and specialised question. But the paper contains a version of Henkin models of second-order model theory, 3 years before Henkin published his own account; and on his page 34 Mostowski comes close to formulating the notion of an elementary extension, 9 years before Tarski and Vaught [49] published the definitive definition.

The paper [31] (1956) 'Concerning a problem of H. Scholz' was one of the earliest papers in a difficult area of work relating computation theory to the model theory of finite structures. See [6] for a recent assessment of the field and Mostowski's contribution to it.

The paper [5] (1978) 'The elementary theory of well-ordering—a metamathematical study' was a reconstruction by Doner and Tarski, after Mostowski's death, of work that Mostowski did with Tarski around 1940, applying the (syntactic) method of elimination of quantifiers to the first-order theory of well-orderings. The paper itself is not model-theoretic, but like other early work in quantifier elimination, its results were of great interest to model theorists. Mostowski counted Tarski as the actual supervisor of his doctoral thesis (1938) in set theory, though formally Kuratowski was the supervisor since Tarski was not a professor at the time (Krajewski and Srebrny [22, p. 5]).

The paper [30] (1953) 'On models of axiomatic systems' applies the model-theoretic notion of satisfaction. We will study its use of the word 'model'.

Mostowski also wrote two books which reviewed mathematical logic as a whole. One of these was his Polish textbook [28] *Logika Matematyczna: Kurs Uniwersytecki* of 1948, which competes with Kleene's more advanced *Introduction to Metamathematics* for the role of the last major pre-model-theoretic textbook of logic. The other was a historical survey *Thirty Years of Foundational Studies* [33, 1966]. The second of these books has two chapters on 'theory of models', and both books contain material close to the foundations of model theory.

3 Background and Notation

We will be discussing the formalisation of several informal notions. It will be helpful to have a convention for distinguishing the informal versions from the formal ones. I will distinguish the formal versions by writing them with small capitals. Thus we talk informally of assigning 'interpretations' to symbols, but Tarski's formal version of this notion is called an 'INTERPRETATION'.

Several important papers in foundations of mathematics in the period 1910–1940 have the following setting, which for convenience we can call the 'archetypal pattern'. A formal language L is described, and explanations are given for the meanings of the symbols of the language. Some of the symbols have logical meanings, for example conjunction. Other symbols, called the 'nonlogical symbols' or the 'primitives', have meanings that come from the topic under discussion. The meanings of the nonlogical symbols add up to a description of a structure M with a given universe or domain of elements; in a modern usage we call M the 'standard interpretation' of the language L or the theory T. A set T of sentences of L is presented as sentences that are true of M under the given explanation of their meanings. The paper studies the structure M by analysing logical properties of the set T. Often T, called a 'theory', is presented as the main topic of study. Probably the best-known example of this general format is Gödel's paper [11] on the incompleteness of first-order Peano arithmetic.

This format came under strain as scholars asked new questions. Two particular areas of strain are worth noting at once. The first is that in the archetypal pattern, the nonlogical symbols of the language L have a fixed set of meanings that determine a particular structure. But sometimes one wants to talk about two or more structures that make the same sentences true. So it became necessary to have a way of detaching the fixed meanings from the nonlogical symbols.

The second area of strain was that logicians became increasingly interested in the justification of metatheoretical arguments. So these arguments should be formalised, and some axioms for them set down. But then there was a question of what to formalise, and in what language.

One example of these strains is the paper of Löwenheim [23], which shows that if a first-order sentence is true in some structure, then it is true in some structure with at most denumerably many elements. In today's terminology, Löwenheim takes a structure M and constructs a substructure N of M that satisfies some of the same sentences of L as M did. But Löwenheim's proof sometimes descends into obscurity, because he has no explicit notion of a substructure; see Badesa [1], particularly his sections 6.1 and 6.2, for documentation of this.

The word 'model', in a sense relevant to model theory, begins to appear in German mathematical writing of the mid 1920s, in order to handle the situation where a structure M is introduced and then a second structure N is constructed from M. The second structure is called a model. One of the earliest examples of this usage is Hermann Weyl's discussion [54] (written in 1925) of geometric 'models' for proofs of consistency. For example on his pages 30 and 31 we read of a 'Modell' of Lobachevsky's geometry within Euclid's geometry; here Euclid's geometry forms the standard interpretation of the primitives 'point', 'line' etc., and the model is the nonstandard interpretation of these expressions that proves the consistency of Lobachevsky's axioms.

Another example of this usage, again from 1925, is in von Neumann's paper [53] on 'models' of set theory. Von Neumann supposes that we have a set—call it T—of axioms for set theory, and he shows how to construct, within the universe of sets described by T, a denumerable 'model' (his word) of the axioms. In order to carry out the construction, he describes a second system of axioms—call it U—and claims 'there obviously exists a smallest' system of sets Σ' satisfying U. To support the claim, he describes a procedure for 'constructing' the system Σ' in infinitely many steps (p. 407f in van Heijenoort [52]).

If this procedure is to be justified in a formal system, what formal system should be used? Could it be the axiom system T presented by von Neumann himself in [53]? Or should it be some other more powerful system? And exactly what calculations need to be represented in the formal system? Besides formalising the construction of Σ', do we also need to formalise a proof that a system of sets satisfying U also satisfies T? And if we do, would it be enough to show how to formalise this claim separately for each axiom, or must we have a single formal proof covering all of them?

The moral of the von Neumann example is that a piece of metatheory may have different formalisations, not all equivalent. We will see in Sect. 6 below that even when we know what parts of the metatheory we want to formalise, there may be more than one way of choosing concepts to do the required job.

4 Some of Tarski's Proposals

The second quarter of the twentieth century saw attempts by various people, both to apply the archetypal pattern to a wider range of problems, and to improve the formalisation of the pattern. Tarski made a number of contributions. In [41] (1933) he showed how to define, within the setting of the archetypal pattern, the notion 'ϕ is a true sentence of L'. This notion makes sense, given that in the archetypal pattern the expressions of L all have appropriate fixed meanings, so it is determinate whether any given sentence of L is true or not. This definition was Tarski's original 'truth definition'. Along the way, Tarski also defined 'satisfaction' in the following sense. Suppose $\phi(\bar{x})$ is a formula of L with free variables $\bar{x}$, and we assign meanings to the variables in $\bar{x}$. Then it makes sense to ask whether the assigned meanings make $\phi(\bar{x})$ true, in other words, whether they 'satisfy' $\phi(\bar{x})$. Tarski analysed what set-theoretic content the assigned meanings would need to have in order for us to give a formal definition of SATISFACTION; the resulting definition defines when an assignment of set-theoretic objects to the free variables counts as satisfying the formula. To formalise the definition, Tarski introduced a second theory $T^\star$ (a 'metatheory'), which would contain an exact copy of T but also enough set theory to formalise the syntax of L and carry out some definitions by induction on the complexity of formulas of L.

This was useful work in itself, but no help for dealing with the problem of alternative structures that make the same set of sentences true. In [17] I documented how progressive advances in the aims of metamathematics forced Tarski to adapt his truth definition step by step, until eventually he had the model-theoretic form which he published with Vaught in [49]. Already in 1933 Tarski could handle the case of two structures, one a substructure of the other, so that he was equipped to formalise Löwenheim's argument discussed above.

But this case is in a way degenerate, because the relations etc. of the substructure agree with those of the larger structure, so that all that is needed to specify the substructure is a formula expressing its domain.

In 1936 [44] Tarski adapted the truth definition to allow new meanings to be assigned to the nonlogical symbols of L. His idea was to consider an assignment α of appropriate set-theoretic objects to the nonlogical symbols of L, and a sentence ϕ of L. He would replace the nonlogical symbols in ϕ by distinct variables, thus getting a formula ψ. The assignment α was defined to be a MODEL of ϕ if it satisfied ψ, where the assignment is carried over from the nonlogical symbols to the variables put in place of them. For the particular purposes of the paper [44], Tarski wanted to talk about all possible assignments of meanings, so that the MODEL could in fact be exactly the same as the original assignment of meanings to nonlogical symbols in the theory T. But otherwise Tarski followed Weyl and von Neumann in using the expression MODEL for a new assignment of meanings.

The model-theoretic truth definition of Tarski and Vaught [49], which we are told was already available by 1952 or 1953 [49, p. 82, footnote], dropped this rigmarole of replacing the symbols by new variables, and assigned the meanings (or rather the set-theoretic objects representing them) directly to the nonlogical symbols. It was no longer assumed that the nonlogical symbols came with preassigned meanings.

In [51] Tarski defined a notion that he called INTERPRETATION. (The book went through several revisions. In the 1994 edition the definition appears on page 114 in §37, but in the earliest editions the definition is in §33.) It relates two theories, say T and $T^\star$ in languages L and $L^\star$ respectively. Like a MODEL of T, an INTERPRETATION of T involves an assignment β to the nonlogical symbols of L. But instead of assigning set-theoretic objects that convey meanings, β assigns to each nonlogical symbol of L an expression of $L^\star$. For each formula ϕ of L we construct a formula ϕ^β of $L^\star$ by replacing each nonlogical symbol of L by the expression assigned to it by β. We call β an INTERPRETATION of T in $T^\star$ if for each sentence ϕ of T, ϕ^β is provable from $T^\star$.

Tarski gives some simple examples of INTERPRETATIONS. He takes T to be a theory expressing that $\cong$ is an equivalence relation on the set S. Then for example let $T^\star$ be a theory of the arithmetic of rational numbers, with a symbol Q for the rational numbers and a symbol Z for the integers. We can write a formula $x \equiv y$ of $L^\star$ which expresses that x and y are in Q and the difference $x - y$ is in Z. Then let β be the assignment that assigns Q to S and $\equiv$ to $\cong$. Assuming that $T^\star$ is strong enough to allow us to prove that $\equiv$ is an equivalence relation on Q, the assignment β is an INTERPRETATION of T in $T^\star$. One can also construct geometric examples that interpret hyperbolic geometry in euclidean geometry, representing the Klein-Beltrami model of hyperbolic geometry as an INTERPRETATION in Tarski's sense.

In Tarski's terminology, INTERPRETATIONS and MODELS are really quite different kinds of thing. What they have in common is that they both consist of assignments of things to the nonlogical symbols of L. But the things assigned, and the condition for the assignment to be a MODEL or an INTERPRETATION, are quite different. The notion of an INTERPRETATION is purely syntactic: the assignment assigns strings of symbols to symbols, and the condition for the assignment to be an INTERPRETATION is that certain things are formally provable from a given theory. By contrast for a MODEL the assignment assigns set-theoretic objects, and the condition for the assignment to be a MODEL involves the notion of satisfaction; it requires that certain things are *true*, not that they are *provable*.

As we see from the example of the Klein-Beltrami model, the names 'interpretation' and 'model' have sometimes been applied to the same things. On p. 114f of the 1994 edition of [51] Tarski makes some remarks that seem to be intended to show how an INTERPRETATION in his sense could sometimes be regarded as a MODEL in his sense. He notes that if $L^\star$ has a standard interpretation, then the expressions assigned by β all have meanings determined by this interpretation, and we can think of β as assigning these meanings rather than the expressions. Then *if the sentences of* $T^\star$ *are true for the standard interpretation*, anything provable from $T^\star$ will be true too, and one can infer that the assignment of meanings (rather than expressions) is a MODEL of T in $T^\star$. These remarks are correct, but Tarski may have created some confusion by making them. In general MODEL and INTERPRETATION are different notions, and neither is a special case of the other. The jump that Tarski describes from INTERPRETATION to MODEL is not just a change of viewpoint; it needs a substantial mathematical proof.

In [46] (1953), p. 20f, Tarski gives another definition of INTERPRETATION, which agrees with the one above but removes the assumption that the theories have standard interpretations. Also instead of physically altering the formula ϕ to ϕ^β, Tarski achieves the same effect by adding the assignment β in the form of explicit definitions of the nonlogical symbols of L in the theory $T^\star$. With this new definition of INTERPRETATION Tarski's attempt above to bring MODELS and INTERPRETATIONS together loses its purchase. (Strictly the new definition is of 'INTERPRETABLE IN', but Tarski still speaks of an 'INTERPRETATION', as in the footnote on his p. 22.)

5 'Theories'

With this much background in place, we can begin to look at Mostowski's papers, starting with [29] 'On direct products of theories'. This paper considers a structure A that is a cartesian power B^I of a structure B. (Mostowski also considers a variant of cartesian power.) Mostowski asks how the first-order theory of A can be calculated from information about what sentences are true in B.

Mostowski introduces the symbol T for the set of first-order sentences true in the structure B, and he writes T^I (where I is the index set of the power) for the set of first-order sentences true in A. The sets T and T^I are called 'theories'. This notation 'T^I' implies that T^I depends only on T and I. But this creates a problem: might T^I not also depend on B? For example T might also be the set of sentences true in some other structure B'; how do we know that $(B')^I$ satisfies the same sentences as B^I?

In fact one of the main results of the paper is that we do know exactly this: T^I is determined by T and I, or even (as Mostowski points out) by T and the cardinality of I. But it seems as if Mostowski has begged the question by assuming this result when he sets up his terminology.

Closer inspection shows that no question is begged, but Mostowski is using an archaic terminology. At the beginning of the paper he tells us

> Elementary mathematical theories are always concerned with certain functions defined in a set **I** (called the *universe of discourse* of the the–ory) and certain relations with the common domain **I**. [29, p. 1] (3)

In other words, a theory always has a standard interpretation. What Mostowski means when he talks of 'products of theories' is exactly the same as what the modern reader would express by talking of products of structures. Mostowski is working in the archetypal pattern.

Of course Mostowski can choose to restrict the word 'theory' to theories that come with a certain standard interpretation, provided that he makes it clear that he is doing this (as in fact he does). But how sensible is this, at a date when other researchers in the area are freely talking about 'all the models of a set of axioms' (as e.g. in Robinson [35] (1951), p. 36)? Mostowski's use of terminology suggests that he is simply not aware that other researchers are using model theory as a framework for studying axiomatic classes in algebra or other parts of pure mathematics.

A glance at what Mostowski says about 'elementary theories' in his textbook [28] of 1948 gives no reassurance at all. Under the head 'elementary theory of groups' Mostowski writes:

> The specific constants [i.e. nonlogical symbols] of this theory (apart from the signs of equality and inequality) are L and σ. The constant L is of type ($\star$), so it is the name of a set which in this theory we call the *group*. The constant σ is of type $(\star, \star, \star)$, which means that it is the name of a ternary relation ... [28, p. 234] (4)

He really does seem to be telling us that for purposes of logic there just is one group!

It seems to me that we should draw the conclusion that at least by December 1949—the date when [29] was submitted—Mostowski had not bought into the propaganda that Robinson and Tarski would present for model theory at the 1950 International Congress. At that date it was not one of his aims to provide tools for algebraists. In fact no move in that direction appears in any of his later model-theoretic papers either. To my eye, none of them contain anything that invites the description 'application of model theory to algebra'.

When he came to write the historical book [33], Mostowski did find himself reviewing other logicians' work in this area, and he seems to have adopted a more relaxed notion of 'theory'. But in [33] he gives no definition of 'theory', or any indication of how it differs from 'axiomatic theory' or from 'set of sentences', or any sign that he is now using it in a different meaning from his earlier papers.

6 'Models'

In his textbook [28] of 1948 and his paper [30] 'On models of axiomatic theories' (published in 1953) Mostowski introduces five different definitions of MODEL. For convenience we can call the two definitions in [28] definitions (48:1) and (48:2), and the three definitions in [30] definitions (53:1), (53:2) and (53:3). In brief they are as follows.

(48:1): This definition is on page 270 of [28]. It defines 'MODEL of T in $T^\star$', and it agrees with Tarski's definition of 'INTERPRETATION of T in $T^\star$ as in [51].

(48:2): This is on page 356 of [28]. It defines 'SEMANTIC MODEL of T', and it agrees with Tarski's definition of 'MODEL of T' in [44].

(53:1): This definition is on page 136 of [30]. It defines 'MODEL OF THE FIRST KIND of T in $T^\star$'. The definition is the same as (48:1), except that Mostowski requires $T^\star$ to contain enough set theory to code up the syntax of T, so that the requirements 'ϕ^β is provable in $T^\star$' take the strong form of saying that in $T^\star$ we can define ϕ^β from ϕ and β, and $T^\star$ proves a statement expressing that for every ϕ in T, ϕ^β is true.

(53:2): This definition is on page 142 of [30]. It defines 'MODEL OF THE SECOND KIND of T in $T^\star$'. It relates to (48:2) in the same way as (53:1) relates to (48:1); in other words, Mostowski requires that $T^\star$ contains enough notions—and in particular enough set theory—to allow us to define satisfaction of formulas of L, and then prove that every axiom of T is true under the given assignment. (Strictly Mostowski says that formulas expressing the truth 'hold', not that they are provable from $T^\star$. But then his argument uses the assumption that 'the existence of a real model of the second kind is provable' in $T^\star$, so the effect is the same.)

(53:3): This definition is on page 149 of [30]. It defines 'MODEL OF THE THIRD KIND of T in $T^\star$' like (53:2), but using arithmetic instead of set theory.

Let me make a few remarks first about (48:2) and (53:2), which correspond to Tarski's MODEL.

In [28] Mostowski proves Gödel's completeness theorem, using the notion of SATISFACTION but without mentioning models. For example he says that every first-order sentence is either satisfiable in the natural numbers or formally refutable (this paraphrases Theorem 7 on his p. 353). Then he moves on to the downward Löwenheim-Skolem, and this is where he introduces the definition (48:2). In fact the form of the Löwenheim-Skolem theorem that he proves is that a satisfiable theory has an at most denumerable model, and he proves it by deducing it from the form of Gödel's theorem that says that a syntactically consistent theory is satisfiable in the natural numbers. So his version of the downward Löwenheim-Skolem theorem needs no new concepts beyond the ones that he has already used for Gödel's theorem. Then why does he choose this place to introduce the notion of SEMANTIC MODEL? One possible guess is that he has at the back of his mind a form of the Löwenheim-Skolem theorem which says that a theory which is true in a standard interpretation has an at most denumerable model; in general the denumerable model will not be the model given by the standard interpretation, so we are changing the interpretations of the nonlogical symbols. If this is right, then Mostowski is harking back to the 1920s notion of 'model'.

In (53:2) Mostowski arranges the formal definition of satisfaction in $T^\star$ in such a way that there is no mention of replacing nonlogical symbols by variables. In effect, Mostowski anticipates the Tarski-Vaught definition of satisfaction in [49]. But this fact is well hidden in the technical details, and quite possibly none of Mostowski's readers realised that he had cleared away the syntactic complexities of Tarski's definition in [44].

Mostowski's aim in (53:2) was to prove a mathematical fact about Zermelo-Fraenkel set theory, taking this set theory to be $T^\star$. The proof proceeds by taking the language $L^\star$ of $T^\star$ as an object language, not as a metalanguage for expressing metamathematical properties of L. This quietly undermines Tarski's efforts to keep mathematics and metamathematics distinct. A few years later Robinson created nonstandard analysis in a similar way, by treating the language in which we do mathematics as one of the objects that we handle in the mathematics that we do; there is a breakdown of levels. Mathematicians appreciate this sort of move, which discovers new mathematical facts by looking at familiar things from an unfamiliar point of view.

Tarski certainly appreciated good mathematics, but it's hard to imagine moves like these of Mostowski and Robinson coming from Tarski himself. Feferman [9, p. 223] makes the interesting remark that Robinson 'had a certain looseness of presentation that annoyed Tarski'. It never struck me that Robinson was a careless mathematician, and I wonder if the remark has to do with Robinson's willingness to ignore the 'right' way of looking at things, as for example in nonstandard analysis. If that be so, then Mostowski's definition (53:2) is loose in much the same way.

Tarski by contrast had a programme to tidy up metamathematics by giving formally correct definitions of the needed concepts, and it was his view that the question 'Which are the needed concepts?' has an objectively right answer. In a telling passage near the beginning of [40, p. 112] he says:

> In geometry it was a question of making precise the spatial intuitions acquired empirically in everyday life, intuitions which are vague and confused by their very nature. Here [i.e. in metamathematics as opposed to geometry] we have to deal with intuitions more clear and conscious, those of a logical nature relating to another domain of science, metamathematics. To the geometers the necessity presented itself of choosing one of several incompatible meanings, but here arbitrariness in establishing the content of the term in question is reduced almost to zero. (5)

This view is likely to appeal to philosophers who regard conceptual analysis as one of the basic tools of mathematical foundations—a view encouraged by Frege's analysis of number and Turing's analysis of computability. But it's also a view that stands in the way of the kinds of conceptual sleight of hand that Mostowski and Robinson exploited. In this respect Mostowski and Robinson were the mathematicians and Tarski was the philosopher.

Turning to (48:1) and (53:1), an obvious question is why Mostowski blurs the distinction between syntax and semantics by using the word MODEL in these two cases. Again we see Mostowski disregarding distinctions that Tarski put in place. But there are some further things to be said about these cases. I have discussed and documented Tarski's position on the issues elsewhere (chiefly [18], [19] and [20]), so I beg leave to give the conclusions rather than the supporting evidence.

Mostowski notes in [30] that:

> Models of the first kind [i.e. for (53:1)] are the ones with which one has to do in the usual proofs of consistency and of independence of axiomatic systems [30, p. 138]. (6)

This comment closely matches the applications that he gives in [28] for models of the kind (48:1); these are the main contents of his Chapter XI on 'Methodological questions'. One case that he discusses in Chapter XI is Padoa's method for proving independence of concepts within a theory.

In 1900 Padoa gave a loose description of a procedure for showing that in a formal theory T with nonlogical symbols $R_0, \ldots, R_n$, the concept expressed by R_0 is not definable in terms of the concepts expressed by $R_1, \ldots, R_n$. The procedure was to give two different interpretations of T which agree on all of $R_1, \ldots, R_n$ but disagree on R_0.

Suppose we want to make Padoa's method formally correct. How should we proceed? For example in terms of the notions introduced by Tarski, should we treat the 'interpretations' of T as INTERPRETATIONS or as MODELS? When Tarski considered the question in the 1920s and 1930s [42], he came down on the side of INTERPRETATIONS. In his reading, Padoa's procedure was to give a second theory $T^\star$ and two INTERPRETATIONS β, γ of T in $T^\star$, such that β and γ assign the same expressions to $R_1, \ldots, R_n$, but it is a theorem of $T^\star$ that the expressions assigned to R_0 by β and γ are not equivalent. On this account, Padoa's method is purely syntactic. There is no reference to MODELS or SATISFACTION anywhere in it. Mostowski [28, pp. 283–291] follows this account.

Today I think most logicians would say that Tarski captured the essential mathematical content of Padoa's method, but he threw away the intuition behind it. That intuition is better captured by formalising Padoa's procedure in terms of MODELS, as we normally do today. The issue came to a head in 1953 when Beth [2] proved that in first-order logic Padoa's condition is both necessary and sufficient for R_0 to be not definable from $R_1, \ldots, R_n$ in T. Beth expressed the condition model-theoretically, though for technical reasons he used cut-free derivations in his proof. When he sent his proof to Tarski, Tarski responded through his student Feferman that Beth had misconstrued a syntactic theorem as a semantic one, and he should rewrite so as to remove the reference to models. (Feferman later thought he remembered saying the opposite to this. But the correspondence is available in Van Ulsen's doctoral thesis; I quote the relevant passage in [20].)

To return to the obvious question mentioned earlier: why did Mostowski blur the distinction between syntax and semantics by using 'model' for Tarski's INTERPRETATIONS? One naturally asks why these five notions are all called 'model'. In [28] Mostowski says nothing to answer this question, but a footnote on page 356 does call attention to the clash of terminology. In the footnote Mostowski says

> We need to distinguish the concept of model defined thus from the concept of a model of one theory in another, as defined in Chapter XI (§2 p. 270). This is the reason for using here the term *semantic model.* (7)

The puzzled reader might well insist that it would be a better reason for *not* using the expression 'model' for both these notions.

The case of Padoa's method discussed above suggests a reason for using 'model' in all these cases, namely that Mostowski realised that Tarski's INTERPRETATIONS and Tarski's MODELS were to some extent solutions of the same problems. They could both be used to formalise earlier informal metamathematical discussions involving variation of interpretations of symbols. So Mostowski could naturally see all his definitions of MODEL as related tools in a general logical toolkit. If this is correct, then Mostowski chooses his terminology more on the basis of possible mathematical applications than on the basis of conceptual analysis. Here again, comparing him with Tarski, Mostowski is the more typical mathematician.

There is a feature of [30] that supports this reading. As he introduces each of his three definitions of MODEL, Mostowski lists some 'general facts' about it. The lists are given in similar formats for the three kinds of MODEL, 'for comparison with other notions of model'. (For (58:1) this is on p. 138f, for (58:2) on p. 142 and for (58:3) on p. 208.) The effect is as if Mostowski is providing a set of tools together with notes on where the tools can appropriately be used.

Mostowski is probably responsible for one more use of the word 'model', namely its use to mean 'structure'. This use appears already in [27] (1947) where he speaks of an 'absolute model' without any reference to any theory that it is a model of. In [33] he often uses 'model' for 'structure'—which puts an extra burden of interpretation on the reader who has to work out whether or not he means 'model of' some salient theory. In mitigation it should be added that when he wrote [27] the word 'structure' was not yet in use among model theorists; at that date one usually said 'system', which is even more open to confusion. Among model theorists the word 'structure' came into common use in the late 1950s, probably under the influence of Robinson and Bourbaki. But whatever the merits of this use of the word 'model', it once more shows Mostowski disregarding Tarski's careful analysis of concepts.

7 Thirty Years of Foundational Studies

In 1964 Mostowski gave a series of lectures 'on the development of mathematical logic and of the study of foundations of mathematics in the years 1930–1964' at a summer school in Vaasa, Finland. Two years later he published a revised version of these lectures [33]. He remarks in his Foreword that he hopes to convey 'some of the enthusiasm with which I witnessed the creation of theories reported on in the following pages'. The lectures certainly live up to that hope.

They can also serve another purpose for today's reader. Round about 1960 a consensus was forming that mathematical logic should be classified under the main heads of Proof Theory, Set Theory, Model Theory and Recursion Theory. For most of Mostowski's career there was no such classification. He draws the divisions in quite different places, and as a result he makes connections between different areas of logic in ways that a modern reader may find fresh and stimulating. Just to pick one example at random, the chapter on 'Semantics' begins with the truth definition and finishes with speedup theorems in proof theory.

About his own contributions, Mostowski's account in this book is modest to a fault. His paper with Ehrenfeucht on indiscernibles [7] is in the bibliography but not mentioned in the text; the papers [29] and [32] are not even in the bibliography. He mentions his paper [30] (at [33, p. 142]), but only to record that its proof that Zermelo-Fraenkel set theory is not finitely axiomatisable contained a mistake, and the result should be credited to Richard Montague [26]. (I haven't been able to find out what mistake he has in mind, or whether it is significant.)

8 Robinson's Complaint

Joseph Dauben in his biography of Abraham Robinson [4] records that Robinson was severely critical of the treatment of model theory in Mostowski's book [33]. The criticism was made privately in a letter to Gerald Sacks. Robinson wrote:

> This term ['model theory'] was indeed coined by Tarski in the early fifties and this is where Mostowski in his "Thirty years of Foundational Studies" (according to which I apparently started my career in 1963) places the beginning of the subject.
> However, if you were to look at my "On the Metamathematics of Algebra" you will find that it contains not only algebraic applications but also the general framework of model theory (e.g. the general scheme of classes of sentences *versus* classes of models). At the same time I do not wish to belittle Henkin's influence on later developments. (8)
> In any case, I am not surprised to observe, again and again, that Tarski has trained his students (and that includes Mostowski) to see history in the way he wants them to. ([4, p. 450f], quoting a letter to Sacks dated 8 June 1972)

Dauben [4, p. 449] describes this as an 'uncharacteristic letter'. Certainly it's uncomfortable. But I was glad that Dauben included it, because it does represent a side of Robinson that I and other people witnessed in the early 1970s. Possibly it was an early sign of the illness that took him soon afterwards.

Let it be said straight away that Robinson's specific complaint about [33]—that Mostowski places the start of Robinson's career in 1963—is not true. The first of Robinson's papers listed in Mostowski's bibliography is [37] from 1955. On his pages 125f Mostowski describes the contents of this paper, noting that one of Robinson's 'most important applications' of these results, in a paper of 1959, 'could hardly be obtained' by Tarski's own preferred methods.

But it is true that Mostowski's book shows no awareness that Robinson made any contributions before 1955. These contributions include his address [36] to the International Congress of Mathematicians in 1950, reporting the main results of his PhD thesis [35] submitted in 1949. Robinson had submitted a paper to the Congress, but Tarski as chair of the Logic section had intervened to elevate Robinson to an invited lecturer; in order to do this Tarski had had to argue for allowing four invited papers in logic, as opposed to the three allowed to other subjects [4, p. 170]. Mostowski must have been aware that Robinson was an invited speaker at the International Congress, since he was present at the Congress himself [4, p. 171]; but apparently Robinson's paper didn't register with him.

In fact Mostowski does mention in [33] two of the innovations that appear in Robinson's PhD thesis, but he attributes neither of them to Robinson.

One of these innovations is the use of Steinitz's theorem to prove the completeness of the theory of algebraically closed fields of a given characteristic ([33, p. 123], [35, p. 60]). This was one of the more startling ideas that Robinson presented to the International Congress in 1950 [36]. Robinson showed that any two algebraically closed fields of the same characteristic are respectively elementarily equivalent to fields that have transcendence degree $\aleph_0$ and hence are isomorphic by Steinitz's theorem. Vaught later gave a simpler and more general form ('Vaught's test') to this argument by pointing out that every denumerable first-order theory that is categorical in some uncountable cardinality and has no finite models is complete—an adjustment that Robinson himself praised for its 'remarkable efficacy' [38, p. 11]. Robinson's result could be recovered by deducing from Steinitz's theorem that the theory of algebraically closed fields of a given characteristic is categorical in every uncountable cardinality. Mostowski attributes Vaught's test to Vaught but gives no attribution for the use of Steinitz.

With hindsight we can see that Robinson's use of Steinitz's theorem was a major step forwards in model theory. It showed that model theorists could apply algebraic embedding or isomorphism results to prove facts about first-order definability. This very quickly became, and remains still, one of the main themes of the subject. But it took some time for the novelty to be appreciated. For example Henkin, in his review of [35] in 1952, mentioned Robinson's result that the theory of algebraically closed fields of a given characteristic is complete, but added the dubious claim that

> the techniques underlying these derivations ... have been obtained earlier by others. [12, p. 206] (9)

As in Sect. 9 below, Henkin might have been better advised not to write reviews of publications that he regarded as in competition with his own work.

The second innovation that appears in [35] and is mentioned by Mostowski without an attribution to Robinson is the notion of the formally defined class of all models of a given set of first-order sentences. Mostowski [33, p. 119] writes this class as $E(X)$, where X is the set of sentences. People who know Tarski's work on the truth definition might reasonably expect that the notion $E(X)$ appears in papers of Tarski. But in fact Tarski had considerable misgivings about using this notion in mathematics, as opposed to having it available as an informal notion of metamathematics. It doesn't appear at all in [47] where we would certainly expect to find it; and in [45] it appears only for the case where X is a single sentence (p. 710). As far as I know, free-wheeling mathematical use of the notion

$E(X)$ is found first in Robinson's PhD thesis [35] (see his p. 36f), which is a passage that Robinson himself refers to in the letter quoted by Dauben. (See [21] end of §2 for some further discussion of this point. There may also be earlier mathematical arguments that explicitly use this notion in papers of Mal'tsev or the doctoral thesis of Henkin; I have not checked these in detail.)

In practice historical surveys always leave out something, and there are plenty of other things that Mostowski could have mentioned but didn't. I would just say here that his lack of interest in the applications of model theory to algebra and other disciplines of pure mathematics made it less likely that he would appreciate the significance of Robinson's use of Steinitz's theorem. Likewise a relative lack of interest in questions of conceptual analysis would make it less likely that he would appreciate the fine details of Tarski's views on metamathematical definitions. There is not the slightest reason to attribute either of these features of [33] to how Tarski 'trained his students' (as Robinson's letter suggests).

9 Mostowski's Attributions in General

An unspoken implication of Robinson's complaint is that Mostowski attributed too much to Tarski. Mostowski does say that

> The systematic development of model theory was initiated by Tarski in the early fifties' [33, p. 119] (10)

This is true in the sense that Tarski pointed the efforts of leading members of the Berkeley group into this area of logic during the 1950s. He did this partly by pressing some questions (e.g. can elementary equivalence be defined without reference to satisfaction?), and partly by giving basic definitions (e.g. truth in a model, elementary extension). Tarski himself proved very few mathematical results of model theory—far fewer than Robinson. But the one result that is regularly credited to him, namely the Łoś-Tarski theorem on formulas preserved in substructures, was proved in the early 1950s [48, 1954], and it certainly set a trend.

On p. 121 of [33] Mostowski attributes to Tarski [45] (1950) the notions of submodel and extension. This is correct in the sense that Tarski included these notions in a list of basic notions in [45]. But the notions were already in wide use, not least in Garrett Birkhoff's notion of subalgebras [3, 1935]. Earlier than Tarski's paper, Robinson had given careful definitions of substructures and extensions of structures in his PhD thesis [35, pp. 65–68]. But very likely Mostowski was unaware of the contents of [35].

Mostowski also attributes the notion of elementary equivalence to Tarski [45]. This is one place where Mostowski gives Tarski less credit than Tarski is entitled to: Tarski had already defined elementary equivalence in [43] (1936). Tarski implies on p. 283 of [50] that he had a 'correct and precise' definition of the notion as early as 1930. What Tarski claims in [45] is a mathematical (as opposed to metamathematical) definition of elementary equivalence; today this distinction is probably invisible to most logicians.

One other case is worth mentioning, because it shows Mostowski dealing with an attribution that was sensitive for some people. In [13] Henkin credited Mal'tsev [25] with giving the first completeness proof for first-order logic with arbitrarily many symbols. In Henkin's later paper [14] about the history of Henkin's own proof of this result, Mal'tsev doesn't even get a mention. What happened between these two papers of Henkin was that Henkin and Mostowski together wrote a review [15] of work of Mal'tsev, in which they commented on the proof of the completeness theorem in [25]. Mal'tsev uses the Skolem normal forms ϕ' of sentences ϕ, more precisely the normal forms that add new relation symbols. The two reviewers comment that Mal'tsev uses the fact that every model of ϕ expands to a model of ϕ'; Skolem himself claimed only that if ϕ has a model then so does ϕ'. They remark 'This stronger result does not seem to be formulated explicitly in the literature, although it can be discerned by a careful reading of the usual proofs of Skolem's theorem' [15, p. 56f]. For what it's worth, when I included this result of Skolem in my Model Theory text [16, Theorem 2.6.5, p. 63] I included the stronger statement as part of the theorem; it never occurred to me that it wasn't obvious from the proof.

Gaps like this in proofs are an embarrassment. I think usually we size up whether we believe the author knows what he or she is doing, and if the answer is Yes then we credit the theorem—though we may still point out the gap. This is exactly how Mostowski handles the issue in [33]:

> The possibility of applying the completeness theorem to such problems was first pointed out by Malcev ... who also published the first proof of the theorem independent of the cardinality of [the set of symbols]. (Malcev's proof was not entirely correct but his mistake can easily be corrected.) [33, p. 56] (11)

Mostowski's statement is faultless.

10 Conclusion

Andrzej Mostowski saw model theory mainly as a source of tools for answering questions raised in metamathematics, not as a source of new tools for other mathematical disciplines such as algebra. In this respect his view was quite different from those expressed by Robinson and Tarski in their 1950 manifestos.

As a corollary, Mostowski was insensitive to some features that were characteristic of model theory from 1950 onwards. These include the use of embeddings as a tool for analysing definability properties of structures. Tarski shared this insensitivity (cf. [21] §3(a)), and it set Mostowski and Tarski apart from Robinson who pioneered this use of embeddings.

But at the same time, Mostowski and Robinson were alike in that their work was driven by the need to find mathematical tools to solve certain problems, and not by any programme of conceptual analysis such as we find in Tarski. In fact Mostowski's use of the word 'model' shows little regard for Tarski's conceptual concerns.

Robinson saw the account of model theory in [33] as playing down Robinson's own contributions to the subject. There is truth in his criticism—though he exaggerated the point. But the main cause seems to have been Mostowski's own lack of interest in the more algebraic aspects of the subject. Where Mostowski had reason to take an interest in the historical details, he comes across as careful and scrupulously fair. Possibly he gave too much credit to Tarski; but if he did, he was certainly not the first or last student to exaggerate the achievements of his own doctoral supervisor.

Very few new mathematical disciplines are created by a single person, and certainly model theory was not one of them. But without Mostowski's contributions, model theory could easily have evolved in a different direction from the one that it took. As often happens in mathematics, some of his best ideas were adopted and used in ways that he could hardly have anticipated; he played a much more significant role than is suggested by his own modest account in [33].

References

1. Badesa, C.: The Birth of Model Theory: Löwenheim's Theorem in the Frame of the Theory of Relatives. Princeton University Press, Princeton (2004)
2. Beth, E.W.: On Padoa's method in the theory of definition. Ned. Akad. Wet. Proc. Ser. A **56** (= Indagationes Mathematicae **15**), 330–339 (1953)
3. Birkhoff, G.: On the structure of abstract algebras. Proc. Camb. Philol. Soc. **31**, 433–454 (1935)
4. Dauben, J.W.: Abraham Robinson: The Creation of Nonstandard Analysis, A Personal and Mathematical Odyssey. Princeton University Press, Princeton (1995)
5. Doner, J.E., Mostowski, A., Tarski, A.: The elementary theory of well-ordering—a metamathematical study. In: Macintyre, A., Pacholski, L., Paris, J. (eds.) Logic Colloquium '77, pp. 1–54. North-Holland, Amsterdam (1978)
6. Durand, A., Jones, N.D., Makowsky, J.A., More, M.: Fifty years of the spectrum problem: survey and new results. Bull. Symb. Log. **18**(4), 505–553 (2012)
7. Ehrenfeucht, A., Mostowski, A.: Models of axiomatic theories admitting automorphisms. Fundam. Math. **43**, 50–68 (1956)
8. Feferman, S.: A fortuitous year with Leon Henkin. In: Manzano, M., Sain, I., Alonso, E. (eds.) The Life and Work of Leon Henkin, pp. 35–40. Birkhäuser, Heidelberg (2014)
9. Feferman, A.B., Feferman, S.: Alfred Tarski: Life and Logic. Cambridge University Press, Cambridge (2004)
10. Feferman, S., Vaught, R.L.: The first order properties of products of algebraic systems. Fundam. Math. **47**, 57–103 (1959)
11. Gödel, K.: Über formal unentscheidbare Sätze der Principia Mathematica und verwandter Systeme I. Monatsh. Math. Phys. **38**, 173–198 (1931)
12. Henkin, L.: Review: A. Robinson, On the metamathematics of algebra. J. Symb. Log. **17**, 205–207 (1952)
13. Henkin, L.: Some interconnections between modern algebra and mathematical logic. Trans. Am. Math. Soc. **74**(3), 410–427 (1953)
14. Henkin, L.: The discovery of my completeness proofs. Bull. Symb. Log. **2**(2), 127–158 (1996)
15. Henkin, L., Mostowski, A.: Review: A. Mal'cev, on a general method for obtaining local theorems in group theory. J. Symb. Log. **24**, 55–57 (1959)
16. Hodges, W.: Model Theory. Cambridge University Press, Cambridge (1993)
17. Hodges, W.: What languages have Tarski truth definitions? Ann. Pure Appl. Logic **126**, 93–113 (2004)
18. Hodges, W.: Tarski's theory of definitions. In: Patterson, D.E. (ed.) New Essays on Tarski and Philosophy, pp. 94–132. Oxford University Press, Oxford (2008)
19. Hodges, W.: Tarski on Padoa's method: a test case for understanding logicians of other traditions. In: Chakraborti, M.K., et al. (eds.) Logic, Navya-Nyāya and Applications: Homage to Bimal Krishna Matilal, pp. 155–169. College Publications, London (2008)
20. Hodges, W.: From choosing elements to choosing concepts: the evolution of Feferman's work in model theory. In: Jäger, G., Sieg, W. (eds.) Feferman on Foundations. Outstanding Contributions to Logic. Springer, Switzerland (2018)

21. Hodges, W.: A short history of model theory. Appendix in: Button, T. and Walsh, S. Philosophy and Model Theory. Oxford University Press, Oxford, pp. 439–476 (2018)
22. Krajewski, S., Srebrny, M.: On the Life and Work of Andrzej Mostowski (1913–1975). In: Ehrenfeucht, A., Marek, V.W., Srebrny, M. (eds.) Andrzej Mostowski and Foundational Studies, pp. 3–14. IOS Press, Amsterdam (2008)
23. Löwenheim, L.: Über Möglichkeiten im Relativkalkül. Math. Ann. **76**, 447–470 (1915) (Tr. as: On possibilities in the calculus of relatives. In: Van Heijenoort, J.: From Frege to Gödel. Harvard University Press, Cambridge (1967), pp. 228–251)
24. Mal'cev (= Mal'tsev), A.I.: The Metamathematics of Algebraic Systems, Collected Papers: 1936–1967. North-Holland, Amsterdam (1971) (Tr. Wells, B.F. III)
25. Mal'tsev, A.I.: Untersuchungen aus dem Gebiete der mathematischen Logik. Matematicheskie Sb. **1**(43), 323–336 (1936) (Tr. as: Investigations in the realm of mathematical logic. In: Mal'cev (= Mal'tsev), A.I.: The Metamathematics of Algebraic Systems, Collected Papers: 1936–1967. North-Holland, Amsterdam (1971) (Tr. Wells, B.F. III), pp. 1–14)
26. Montague, R.: Semantical closure and non-finite axiomatizability I. In: Infinitistic Methods (editors unnamed), pp. 45–69. Państwowe Wydawnictwo Naukowe, Warsaw (1961)
27. Mostowski, A.: On absolute properties of relations. J. Symb. Log. **12**, 33–42 (1947)
28. Mostowski, A.: Logika Matematyczna: Kurs Uniwersytecki. Monografie Matematyczne, vol. 18. Uniwersytet of Warsaw, Warsaw (1948)
29. Mostowski, A.: On direct products of theories. J. Symb. Log. **17**(1), 1–31 (1952)
30. Mostowski, A.: On models of axiomatic systems. Fundam. Math. **39**, 133–158 (1953)
31. Mostowski, A.: Concerning a problem of H. Scholz. Zeitschrift für Mathematische Logik und Grundlagen der Mathematik **2**, 210–214 (1956)
32. Mostowski, A.: On a generalization of quantifiers. Fundam. Math. **44**, 12–36 (1957)
33. Mostowski, A.: Thirty Years of Foundational Studies: Lectures on the Development of Mathematical Logic and the Study of the Foundations of Mathematics in 1930–1964. Blackwell, Oxford (1966)
34. Peters, S., Westerståhl, D.: Quantifiers in Language and Logic. Clarendon Press, Oxford (2006)
35. Robinson, A.: On the Metamathematics of Algebra. North-Holland, Amsterdam (1951) (This is an almost exact transcript of Robinson's PhD thesis, Birkbeck College London, 1949)
36. Robinson, A.: On the application of symbolic logic to algebra. In: Proceedings of the International Congress of Mathematicians, Cambridge, Massachusetts 1950, vol. 1, pp. 686–694. American Mathematical Society, Providence (1952) (Reprinted in: Robinson, A.: Selected Papers of Abraham Robinson, Vol. 1: Model Theory and Algebra. Keisler, H.J. et al. (eds.) North-Holland, Amsterdam (1979), pp. 3–11)
37. Robinson, A.: Ordered structures and related concepts. In: Heyting, A. (ed.) Mathematical Interpretation of Formal Systems, pp. 51–56. North-Holland, Amsterdam (1955)
38. Robinson, A.: Complete Theories. North-Holland, Amsterdam (1956)
39. Robinson, A.: Selected Papers of Abraham Robinson, Vol. 1: Model Theory and Algebra. Keisler, H.J. et al. (eds.) North-Holland, Amsterdam (1979)
40. Tarski, A.: Sur les ensembles définissables de nombres réels I. Fundam. Math. **17**, 210–239 (1931)
41. Tarski, A.: Pojęcie prawdy w językach nauk dedukcyjnych. Prace Towarzystwa Naukowego Warszawskiego, Wydział III Nauk Matematyczno-Fizycznych **34**, Warsaw (1933) (Revised and expanded version tr. as: The Concept of Truth in Formalized Languages. In: Tarski, A: Logic, Semantics, Metamathematics, 2nd edn. Hackett, Indianapolis (1983) (Tr. Woodger, J.H. (ed.) and introduced by Corcoran, J.), pp. 152–278)
42. Tarski, A.: Einige methodologische Untersuchungen über die Definierbarkeit der Begriffe. Erkenntnis **5**, 80–100 (1935) (Tr. as: Some methodological investigations on the definability of concepts. In: Tarski, A: Logic, Semantics, Metamathematics, 2nd edn. Hackett, Indianapolis (1983) (Tr. Woodger, J.H. (ed.) and introduced by Corcoran, J.), pp. 296–319)
43. Tarski, A.: Grundzüge des Systemenkalküls. Fundam. Math. **25**, 503–526 (1935) and **26**, 283–301 (1936) (Tr. as: Foundations of the calculus of systems. In: Tarski, A: Logic, Semantics, Metamathematics, 2nd edn. Hackett, Indianapolis (1983) (Tr. Woodger, J.H. (ed.) and introduced by Corcoran, J.), pp. 342–383)
44. Tarski, A.: O pojęciu wynikania logicznego. Przegląd Filozoficzny **39**, 58–68 (1936) (Tr. as: On the concept of logical consequence. In: Tarski, A: Logic, Semantics, Metamathematics, 2nd edn. Hackett, Indianapolis (1983) (Tr. Woodger, J.H. (ed.) and introduced by Corcoran, J.), pp. 409–420)

45. Tarski, A.: Some notions and methods on the borderline of algebra and metamathematics. In: Proceedings of International Congress of Mathematicians, Cambridge Massachusetts 1950, vol. 1, pp. 705–720. American Mathematical Society, Providence (1952)
46. Tarski, A., Mostowski, A., Robinson, R.M.: Undecidable Theories. North-Holland, Amsterdam (1953)
47. Tarski, A.: Contributions to the theory of models I. Indag. Math. **16**, 572–581 (1954)
48. Tarski, A.: Contributions to the theory of models II. Indag. Math. **16**, 582–588 (1954)
49. Tarski, A., Vaught, R.L.: Arithmetical extensions of relational systems. Compos. Math. **13**, 81–102 (1956)
50. Tarski, A: Logic, Semantics, Metamathematics, 2nd edn. Hackett, Indianapolis (1983) (Tr. Woodger, J.H. (ed.) and introduced by Corcoran, J.)
51. Tarski, A.: Introduction to Logic and to the Methodology of Deductive Sciences. Oxford University Press, New York (1994) (Expanded and revised translation of Polish original, 1936)
52. Van Heijenoort, J.: From Frege to Gödel. Harvard University Press, Cambridge (1967)
53. Von Neumann, J.: Eine Axiomatisierung der Mengenlehre. J. Reine Angew. Math. **154**, 219–240 (1925). Correction *ibid.* **155**, 128 (1925) (Tr. as: An axiomatization of set theory. In: Van Heijenoort, J.: From Frege to Gödel. Harvard University Press, Cambridge (1967), pp. 393–413)
54. Weyl, H.: Riemanns geometrische Ideen, ihre Auswirkung und ihre Verknüpfung mit der Gruppentheorie. Springer, Berlin (1988) (Written 1925)

W. Hodges (✉)
Herons Brook, Sticklepath, Okehampton, UK
e-mail: wilfrid.hodges@btinternet.com

All Quantifiers Versus the Quantifier All

Stanisław Krajewski

Abstract In courses of logic for general students the general and existential quantifiers are the only ones distinguished from among all possible quantifier expressions of the natural language. One can argue that other quantifiers deserve mention, even though there are good reason for emphasizing the familiar ones: namely, they are the simplest, the universal quantifier is a counterpart of the operation of generalizing, the number of nested quantifiers is a good measure of logical complexity, and the expressive power of the general quantifier and its dual is considerable.

Yet, even in the teaching about these two simplest quantifiers it has not been resolved how to indicate the realm to which a given quantifier refers. The methods range from the Fregean assumption that they refer to the totality of objects in the world to the restricted quantifiers to many sorted logic. It turns out that these approaches are not fully equivalent, because the sorts are usually assumed to be nonempty, which results in a problem similar to the well-known issue with non-emptiness of names in syllogistics.

Logicians have studied various generalized quantifiers. It is, however, unclear how to treat the quantifier "many" and similar heavily context-dependent ones. They are not invariant under isomorphisms so no purely logical or mathematical treatment seems applicable. How else can one characterize the context-independent quantifiers among all possible quantifiers corresponding to quantifier expressions in natural language? The following thesis on quantifiers is proposed:

(Principal Thesis) Context-independence = definability in terms of the universal quantifier.

This thesis provides an additional reason for distinguishing the universal quantifier from among all other quantifiers: it suffices for defining all context-independent ones

Keywords Quantifier · Generalized quantifiers · Quantifier's range · Definability · Context-independence

Mathematics Subject Classification (2000) Primary 03C80; Secondary 00A30, 00A35, 03A05, 03B10, 03B65

Á. Garrido, U. Wybraniec-Skardowska (eds.), *The Lvov-Warsaw School. Past and Present*, Studies in Universal Logic,
https://doi.org/10.1007/978-3-319-65430-0_47

1 The Universal Quantifier and Its Dual

The introduction of quantifiers to logical systems constituted an essential progress with respect to the calculus of Boolean connectives—even though the propositional calculus was finally formulated as a system at about the same time as was the predicate (functional) calculus. The wider system, including the quantifiers "for all" ($\forall$) and "there exists" ($\exists$), was also a far reaching strengthening of syllogistics, the celebrated system that up to the nineteenth century was seen as the core of logic, and that can be seen as an ancient form of a fragment of the predicate calculus.

In logic, from Aristotle to Frege to mid-twentieth century predicate logic, only two quantifiers were incorporated: the general and the existential. They are still the only ones taught in general logic courses. Because in classical logic $\exists$ is the dual of $\forall$, that is, $\exists = \neg\forall\neg$, we can say that only the universal quantifier is added. (In some non-classical logic, e.g., the intuitionistic logic, we need to retain the two quantifiers.)

From a certain natural point of view, namely the approach based on the linguistic realities, it is not clear why the general and existential quantifiers are the only concepts distinguished from among all possible quantifier expressions of the natural language. In natural language there are dozens quantifier expressions, that is, expressions that state or estimate the number of objects of a certain kind, or the size of a collection, or compare sizes, etc. They include phrases like 'all', 'always', 'nowhere', 'almost never', 'most', 'infinitely many', 'many', 'from time to time', 'a few', 'quite a few", 'several', 'just one', 'at least one', 'an overwhelming part of', 'as many as', 'roughly as many as', and many, many more, including statements like "More girls study programming than boys learn boxing." In mathematics, some other quantifier expressions are used, for example 'there are finitely many', 'there are uncountably many', 'the set of ... is dense in ...', and the phrases like 'almost all', 'a negligible amount' are given various precise meanings in specific mathematical theories.

For a long time logic did not recognize the rich realm of quantifiers or at least did not perceive it as belonging to the domain of logic. What could be the reason for the distinguished role of the familiar quantifiers? Let us try to argue from a logical perspective.

First, simplicity. 'All things' corresponds to the full set—either of all things or of all things in our universe of discourse. 'At least one thing' corresponds to the notion of non-empty set, or is the negation of being the empty set.

The two standard quantifiers are the simplest ones. At the same time, and this is the second reason, we can see the general quantifier as an abstract counterpart of the operation of generalization, our important mental faculty. (Existential quantifier is, as mentioned above, its dual.) This operation can be seen as basic: in Richard Epstein's *Critical thinking* [1] only the operation of generalization is mentioned, the quantifiers are not.

The third and fourth reasons are given after the following Digression and a fifth reason emerges at the end of the paper.

2 A Digression on the Range of the Familiar Quantifiers

Whereas the quantifiers ∀ and ∃ are familiar now, their use causes some problems that are not trivial, especially in the educational context, when general students are taught. One problem is: what is the range of quantifiers, the realm to which a given quantifier refers? Another problem appears when a standard answer is given to the first one: how to indicate the range?

Frege introduced quantifiers in his framework, in which everything, literally: *every thing*, was included. For Frege, and similarly for his successors like Russell, (early) Wittgenstein or Quine, there is only one value-range for quantifiers, namely "all the actually existing individuals" (cf. Hintikka [7, p. 30] and Peters-Westerståhl [17, p. 40]) or even "all the conceivably existing individuals" (as in Russell [18]). That assumption seems, however, unsatisfactory. There are at least two main reasons for dissatisfaction. First, it seems that in order to apply this approach the world must be perceived as a collection of things. In particular, 'always' is expressible by ∀ only if time is seen as composed of things such as moments or segments. In addition, only timeless relations are naturally dealt with (cf. Epstein [2] and [3], Appendix A). Second, to generalize over everything seems odd. In practice, we almost always mean a specified limited range. Today we rarely share Frege's ontology, but we all continue to use his formalism (in a modified form, of course). The contemporary prevailing approach to this formalism is, however, vastly different from his. The world is complex, and, usually, in a given moment we consider only some objects, that is, we specify fragments of the whole world. We have overwhelmingly adopted the model theoretic approach: models vary, and generality means "for all elements of an intended range." The formalism remains but its interpretation is different: logic is no more about "the world" but rather about various "possible worlds", or models. It was Tarski who helped convince logicians to study truth in models. In Hintikka [7] it is also stressed that this new, model-theoretic approach overcomes the difficulty inherent in Frege's approach, namely, how to identify the basic simple things, the "urindividuals".

Having agreed on the limited range, we need to express that in the symbolism we use to deal with quantifiers. There are two traditional ways of expressing the restriction on the range of variables used inside (logical) formalism. One is the use of variables of different sorts. The other is the use of restricted, or relativized, quantifiers. As is well known, it is easy to express relativized quantifiers by unrestricted ones: for any predicates A, B, C

$$(\forall x)_{C(x)}A(x) \equiv (\forall x)(C(x) \rightarrow A(x)), \qquad (\exists x)_{C(x)}A(x) \equiv (\exists x)(C(x) \wedge A(x)).$$

In terms of a variable t, assumed to satisfy $C(t)$, we simply have $(\forall t)A(t)$ and $(\exists t)A(t)$.

Is there a difference between the two methods? Of course, the approach is different: the restrictions can be seen as imposed from outside the system in the case of the language with different sorts of variables, while they are an optional part within the system in the case of relativisation. Still, at the first glance it may seem that they are formally equivalent. But not quite—there is a subtle difference. We normally assume that all sorts are nonempty—similarly to the assumption that the universe is nonempty, or that in each model the universe of the model is nonempty. This assumption is not made about the predicate $C(x)$.

Sometimes the possible emptiness of C is harmless. Explicit restrictions on the quantifiers preserve the validity of de Morgan's laws:

$$\neg(\forall x)_{C(x)}A(x) \equiv (\exists x)_{C(x)}\neg A(x), \qquad \neg(\exists x)_{C(x)}A(x) \equiv (\forall x)_{(C(x)}\neg A(x).$$

What about the other tautologies that are so useful in manipulating quantifier prefixes? Using different sorts brings no harm. In contrast to that, relativisation may cause a problem! It seems to me that while the story with (some) Aristotelian syllogisms being valid only under the assumption that all the terms are non-empty is very well known, similar limitations concerning relativized quantifiers are not generally known. It is easy to see (as was remarked in Krajewski [11]) that the following theorem holds.

Remark on relativized quantifiers:

The following formulas are valid under relativisation to an arbitrary C (we assume that in A the variable x is not free):

$$(\forall x)(A \vee B(x)) \equiv A \vee (\forall x)B(x),$$
$$(\forall x)(A \rightarrow B(x)) \equiv A \rightarrow (\forall x)B(x),$$
$$(\exists x)(A \wedge B(x)) \equiv A \wedge (\exists x)B(x),$$
$$(\forall x)(B(x) \rightarrow A) \equiv (\exists x)B(x) \rightarrow A.$$

The following formulas remain valid only when relativized to nonempty C (we assume that in A the variable x is not free):

$$(\forall x)(A \wedge B(x)) \equiv A \wedge (\forall x)B(x),$$
$$(\exists x)(A \vee B(x)) \equiv A \vee (\exists x)B(x),$$
$$(\exists x)(A \rightarrow B(x)) \equiv A \rightarrow (\exists x)B(x),$$
$$(\exists x)(B(x) \rightarrow A) \equiv (\forall x)B(x) \rightarrow A.$$

In other words the relativisations of the above tautologies are not valid, they are sometimes false when C is interpreted as an empty set, but the following formulas are valid:

$$(\exists x)C(x) \rightarrow [(\forall x)_{C(x)}(A \wedge B(x)) \equiv A \wedge (\forall x)_{C(x)}B(x)],$$
$$(\exists x)C(x) \rightarrow [(\exists x)_{C(x)}(A \vee B(x)) \equiv A \vee (\exists x)_{C(x)}B(x)],$$
$$(\exists x)C(x) \rightarrow [(\exists x)(A \rightarrow B(x)) \equiv A \rightarrow (\exists x)B(x)],$$
$$(\exists x)C(x) \rightarrow [(\exists x)_{C(x)}(B(x) \rightarrow A) \equiv (\forall x)_{C(x)}B(x) \rightarrow A].$$

3 The Power of ∀ and ∃

The third reason for the distinguishing of ∀ and ∃ from among all possible quantifiers has to do with logical complexity. The number of nested quantifiers is a good indicator of logical complexity. The quantifiers ∀ and ∃ provide a great measure of complexity if the number of alternating nested quantifiers is counted. The realization of this possibility gave rise to the Kleene-Mostowski hierarchy, classifying the sets obtained from recursive sets by a series of projections and complements. (See Kleene [10], Mostowski [15].) Then other similar growing chains of ever more complicated objects were established, e.g., the analytic hierarchy. From such a perspective these simple familiar quantifiers look like anything but trivial. It is also of interest that neither Aristotle nor other premodern logicians considered nested quantifiers. If teaching about quantifiers is limited to formulas, and especially tautologies, with one quantifier or at most two, as is still done in coursed for general students, the matter looks rather trivial and it remains unclear why the quantifiers are needed. The power of quantifiers, even the simplest ones, is seen only when several are combined. This brings us to the next reason for the distinguishing of $\forall$ and $\exists$.

The fourth reason emerges when one realizes that these quantifiers bring much more expressive power than it would seem at first. When the standard additional machinery available in logic is employed many new quantifiers can be defined. Some of them can be easily defined within first order logic, for instance the numerical quantifiers: "there are exactly n", in short $\exists^{!n}$, "there are more than n", in short $\exists^{>n}$, and their combinations (like "there are three or four"), etc.

In higher order logics and in set theory many more quantifiers can be defined. Definitions in mathematics are expressed in a technical language of a given branch, but logicians have been able to express these definitions in the language of logic. Thus, for instance, "there are infinitely many" cannot be defined in the first order logic, but can be defined in the second order logic. The Henkin quantifier, the first example of a branching quantifier, namely "for every x there exists y, and independently of that for every z there exists t such that $R(x, y, z, t)$", also goes beyond first order logic (see Henkin [6] and Krynicki et al. [13]), even though it reflects such a way of using the familiar quantifier expressions corresponding to $\forall$ and $\exists$ that can be found in natural language; this quantifier is easily defined in second order logic: "there exist functions f, g such that for every x and for every z $R(x, f(x), z, g(z))$". The phrase "there are uncountably many" also defines a quantifier but it makes sense only in reference to a background set theory. It was unexpected that this quantifier can be recursively axiomatized. (See Keisler [8].) There are many more examples of mathematical quantifiers. They suggested to mathematical logicians the concept of a "generalized quantifier".

4 Generalized Quantifiers in Logic

Generalized quantifiers were introduced to logic by Mostowski [16]. The formula $(Qx)\varphi(x)$ is satisfied in a model M=$(M, \ldots)$ iff the set $\{a\colon \mathrm{M}\models \varphi[a]\}$ belongs to the family of subsets of M that serves as the interpretation of Q. (Thus $\forall$ is interpreted as $\{M\}$ and $\exists$ as the family of all non-empty subsets of M.)

This notion was useful but was not sufficient for many formulations that are used in natural language. Mostowski quantifiers are all of type <1>. The sentence "More girls study programming than boys learn boxing" cannot be analyzed in logic with type <1> quantifiers only. A more general definition was introduced by Lindström [14] who allowed quantifiers of an arbitrary type $<n_1,\ldots,n_k>$ that bind more variables and apply to several formulas, and in a model M are interpreted as relations between subsets of M (in the case of monadic quantifiers of type <1,1,...,1>) or, more generally, relations between relations on M.

Both Mostowski and Lindström were mathematicians so they made an important assumption which obviously seemed necessary to them: they consider only the quantifiers that are invariant with respect to isomorphism. Formally, if $M \cong M'$ then

$$M \models (Qx_1, \ldots, x_k)(\varphi_1, \ldots, \varphi_n) \text{ iff } M' \models (Qx_1, \ldots, x_k)(\varphi_1, \ldots, \varphi_n).$$

The assumption in the case of monadic quantifiers amounts to the fact that only the size of the sets defined by the quantified formulas matters (cf. Peters-Westerståhl [17] or Westerståhl [20]). The assumption that logic should be completely topic-neutral constitutes the reason for admitting into logic only the quantifiers invariant under isomorphism. Other mathematical properties can be defined by isomorphism-preserving quantifiers. Yet they are not sufficient for some quantifiers commonly used in natural language.

It is clear that logic is poorly equipped, if at all, to deal with many from among the quantifier expressions listed above. For example, the concept "many" is different from the more logical quantifiers and seems hardly definable in general since its meaning depends on the situation in which the term is used. It is context-dependent. Peters and Westerståhl call it "strongly" context-dependent and some authors call it intensional. (See Peters-Westerståhl [17, p. 213].) To evaluate a sentence with such a context-dependent quantifier we need an appropriate understanding of the world, or at least of the appropriate fragment of the world. Logic itself is not sufficient. To know whether it is true or not that *many* women at my university are pregnant or that *many* have been in the Himalayas, we need to know how many women of a given age are, on average, pregnant, and how many go to the Himalayas.

It is similarly with quantifier expressions like "a few", "several", "a huge number", "rarely", "often", etc., and even more obviously, with "surprisingly many", "almost everyone", "virtually nowhere", etc.

I believe that it should always be taught in general logic courses that many concepts can be defined by the simplest quantifiers, but at the same time the student should be made aware that many natural language constructions cannot. This seems to be generally ignored by teachers of logic. For example, in the otherwise comprehensive textbook by Andrzej Grzegorczyk [5] the other quantifiers are not even mentioned.

This postulate seems to be loosely connected to a remark by Gödel, possibly his only recorded statement on the present topic. According to Wang [19, p. 266], Gödel said: "Even though predicate logic is distinguished there are also other notions, such as *many, most, some* (in the sense of plurality), and *necessity*."

Despite the initial impression that the quantifier "many" is not definable, one could try to define it formally, or to model it, by adding a variable σ and defining "many" as more numerous than (the interpretation of) σ. This new variable can be either a numerical one, interpreted as a cardinal number, or a set variable, interpreted as a certain set S. Then "many x's (satisfying φ)" is defined as having more members than S, or as the requirement that the cardinality of the set of the values of x that satisfy the interpretation of φ is larger than the cardinality of S. The set S depends on the context; it is chosen specifically for each interpretation.

The problem with this attempt is that the definition of "a few" is the same, only with "<" instead of ">". And the phrase "more than a few" is formalized exactly as is "quite a few" and "many". And do we normally identify "many" with "more than a few"? Hardly.

The above remarks should be easy to understand, but an example can still be helpful. A certain number, say 7, can play the role of both delimitations in the same discourse. For example, if exactly 7 students among the 20 students in my Warsaw university class have read more than ten books in their lifetime and 7 are pregnant, I would say that it is true that "a few read books" and "many are pregnant". (Incidentally, I believe that there could exist schools somewhere in the world in which 7 pregnant among 20 students would be seen as "few", and 7 readers among them would be considered "many".)

So everything depends on the context and introducing σ is of no help. Only the context counts.

We can still maintain that the logical content is better explained, when this formalization is made. The important feature—and a problem from a normal logical perspective—is that "many" defined as "more than σ" is not invariant with respect to isomorphisms. To continue our example, $(G, P) \cong (G, R)$, where G is the class, P is the set of pregnant students in the class, R is the set of book readers in the class, but the sentence "there are many x that φ " is true in one and false in the other interpretation.

Dealing with context-dependent quantifiers one can wonder how many contexts there are. Infinitely many? This seems probable, at least in the case of a quantifier "many". Is this the reason we are unable to pin them down? If only finitely many contexts were possible, a fixed number, then perhaps we could give a definition by listing all the cases. Notice that if a finite but practically unmanageable number of contexts has to be taken into account then the quantifier is still undefinable by us. However, a sufficiently strong intelligence, or even robot, could perhaps do that. The problem is analogous to the problem whether a computer can handle the natural language. The hopes of some early pioneers of Artificial Intelligence that computers would speak as humans were naive. Yet in restricted settings, where contexts can be comprehensively listed, it is perfectly possible to have computers "speak."

5 Characterizing Context-Independent Quantifiers

It seems that context-independence means that any extralogical terms referring to some specific fragments of the world are irrelevant for the understanding of the formula. The topic covered in the statement is of no consequence, only logic counts. Thus

(1) Context-independent quantifiers = Topic neutral quantifiers

We can still maintain that a definition of (a quantifier) being context-independent is needed. This is clear in specific cases, but can a general definition be given? What is needed is a criterion—indeed, a context-independent criterion—for context-independence of quantifiers (or perhaps even more generally, context-independence as such). The idea is, of course, quite simple: there is no need for any specific knowledge about the world. Yet, one could say, to understand the Magidor-Malitz quantifier one certainly needs some non-trivial knowledge. It is, however, a purely logical knowledge (in the broad meaning of logic), different from the knowledge of the features of the world, physical or social, that are relevant for the specific situation. We might even try to say that what is needed for understanding the context-independent quantifiers is the familiarity with merely the necessary features of the world. One could ignore its contingent aspects.

It has been noticed above that when linguistic quantifier expressions are reconstructed within logic the requirement of context-independence is formulated as invariance with respect to isomorphisms. Thus, we get another thesis:

(2) Context-independent quantifiers = Quantifiers invariant under isomorphisms

Before another thesis proposing a characterization of context-independence of quantifiers is attempted let us consider the meaning of being a thesis in this context. Church's Thesis is the best known example of a thesis identifying a formal concept with an intuitive one. The mathematical concept of recursive function is identified with the intuitive concept of effectively computable function. For a long time, the general conviction was that such a thesis can be justified by various arguments, but there is no way to prove its correctness because the intuitive concept is too vague to be part of a proof. However in recent decades there have been various attempts (in particular by Robin Gandy, Wilfried Sieg, Yuri Gurevich) to prove the identification. Namely, a proper analysis of the intuitive concept of computability can provide principles that make possible a demonstration that a function satisfying them must be recursive. There are more examples of similar theses, for instance "the Cantor-Dedekind thesis" that real numbers are defined by the appropriate set theoretic constructions. (For a discussion of Church's Thesis and the other examples as well as references to literature see, e.g., Krajewski [12].)

In the case studied in the present paper, it is the context-independence applied to quantifiers that is the intuitive notion we want to characterize.

In addition to topic-neutrality and invariance under isomorphisms we can try look at the ways the quantifier can be defined. It seems that whatever definition is formulated it cannot be expressed without taking some specific logic into account. This is because quantifiers are logical objects. They function inside a logical framework. On the other hand, it would be hard to emphasize the logical nature but ignore any specific logic. The way out of the dilemma can be as follows: the defining property is assumed to make sense in whatever logic it is formulated. For instance, the phrase "$\varphi(x, y)$ defines a well-ordering" defines a type <2> quantifier, whether in second order logic or in set theory.

Any quantifier Q can give rise to a "logic" L(Q). Then Q is trivially definable in this logic. To avoid this triviality, let us call a logic *basic* if it is first order, second order, nth order, type theory or set theory. Hence the following thesis

(3) A quantifier is context-independent iff it is definable in some basic logic.

Because the common part of all such logics, as far as quantification is concerned, is the universal quantifier $\forall$, we can reformulate the thesis as

(3') A quantifier is context-independent iff it is $\forall$-definable in some (basic) logic.

Since we admit definability either in first order or second order or higher order logic or in (formalized) set theory, and the general quantifier appears in each of these logics we can say in short:

(4) A quantifier is context-independent iff it is definable in terms of $\forall$,

or briefly,

(Principal Thesis) Context-independence = definability in terms of $\forall$.

It is seen that the position of the general quantifier, or rather of our two familiar quantifiers, ∀ and ∃, is vindicated. This is the fifth—in addition to simplicity, the faculty of generalization, the measuring of complexity, and the expressive power—and rather unexpected reason for distinguishing ∀: in the presence of the appropriate amount of logical machinery but with no generalized quantifiers ∀ suffices to define all context-independent quantifiers. Thus the power of the universal and existential quantifiers is claimed to be even stronger than it seemed on the basis of the definability of so many quantifiers by ∀. According to the Principal Thesis, the power of ∀, at least in relation to quantifiers, extends to the whole realm of context-independence.

Let us repeat that in each basic logic the universal quantifier is included, so definability in terms of ∀ is really the same as definability in (predicate) logic. The Principal Thesis says that not only definability of generalized quantifiers in terms of ∀ gives context-independence but also that context-independent generalized quantifiers are so definable. Each specific example of a quantifier has been defined (in the proper logic) in terms of ∀; the Thesis states the generalization to all possible quantifiers.

6 Formalism-Free Definition of ∀-Definable Quantifiers?

There exists an alternative way of looking at the Principal Thesis. If we agree to it then we can treat the Thesis as the proposal to characterize the definability of quantifiers with the use of ∀ (in basic logics) as context-independence, that is, a feature formulated without the recourse to a specific syntactic machinery used in definitions. This brings to mind the problem of formalism-free characterization of concepts.

The issue of "formalism freeness" has been introduced by Gödel who commented on the fact that all formal definitions of computable functions give the same class of functions. Therefore, even though each definition requires some specific formalism, we have been able to isolate an important class of functions in a formalism-free manner. Computability is formalism-free, and Gödel [4] proposed to look for a similar grasping of definability and other notions. Some developments in mathematical logic, notably work done in model theory by Shelah and Zilber, can be seen as going in this direction—see Kennedy [9]. According to Shelah, in model theory, conceived as a tower, "the higher floors do not have formulas or anything syntactical at all." (Kennedy [9, p. 355])

In our case, the Principal Thesis gives the formalism-free characterization of the class of quantifiers definable in some standard (predicate) logic, that is, using some formalism. In each of these logics we have the quantifier ∀, so one can say that this is the class of quantifiers definable in logic by ∀. The class consists of context-independent quantifiers. This characterization is formalism-free.

References

1. Epstein, R.L.: Critical Thinking. Wadsworth/Thomson Learning, Belmont (1998)
2. Epstein, R.L.: The metaphysical basis of logic. Manuscrito **XXII**(2), 133–148 (1999)

3. Epstein, R.L.: Predicate Logic. The Semantic Foundations of Logic. Wadsworth/Thomson Learning, Belmont (2001)
4. Gödel, K.: Remarks before the Princeton bicentennial conference on problems in mathematics. In: Feferman, S., et al. (eds.) Kurt Gödel: Collected Works, Vol. II: Publications 1937–1958, pp. 150–153. Oxford University Press, Oxford (1990)
5. Grzegorczyk, A.: An Outline of Mathematical Logic. Reidel, Holland (1974)
6. Henkin, L.: Some remarks on infinitely long formulas. Infinitistic Methods, pp. 167–183. Pergamon Press, Oxford (1961)
7. Hintikka, J.: Selected Papers, vol. 2. Lingua Universalis vs. Calculus Ratiocinator: An Ultimate Presupposition of Twentieth-Century Philosophy. Kluwer, Dordrecht (1997)
8. Keisler, H.J.: Logic with the quantifier "there exist uncountably many". Ann. Math. Log. **1**, 1–93 (1970)
9. Kennedy, J.: On formalism freeness: implementing Gödel's 1946 Princeton bicentennial lecture. Bull. Symb. Log. **19**(3), 391–393 (2013)
10. Kleene, S.C.: Recursive predicates and quantifiers. Trans. AMS **53**, 41–73 (1943)
11. Krajewski, S.: Remark on quantifiers and their relativisations. Acta Univ. Wratislaviensis **23**, Logika, 37–41 (2005)
12. Krajewski, S.: Is Church's Thesis Unique? In: Olszewski, A., Brożek, B., Urbańczyk, P. (eds.) Church's Thesis. Logic, Mind and Nature, pp. 113–135. Copernicus Center Press, Kraków (2014)
13. Krynicki, M., Mostowski, M., Szczerba, L. (eds.): Quantifiers: Logic, Models and Computation, vols. I and II. Kluwer, Dordrecht (1995)
14. Lindström, P.: First order predicate logic with generalized quantifiers. Theoria **32**(3), 186–195 (1966)
15. Mostowski, A.: On definable sets of positive integers. Fundam. Math. **34**, 81–112 (1947)
16. Mostowski, A.: On a generalization of quantifiers. Fundam. Math. **44**, 12–36 (1957)
17. Peters, S., Westerståhl, D.: Quantifiers in Language and Logic. Clarendon Press, Oxford (2006)
18. Russell, B.: The Principles of Mathematics. Allen and Unwin, London (1903)
19. Wang, H.: A Logical Journey. From Gödel to Philosophy. MIT Press, Cambridge (1996)
20. Westerståhl, D.: Generalized quantifiers. In: Stanford Encyclopedia of Philosophy. http://plato.stanford.edu/entries/generalized-quantifiers/

S. Krajewski (✉)
Institute of Philosophy, University of Warsaw, Warsaw, Poland
e-mail: stankrajewski@uw.edu.pl

Helena Rasiowa (1917–1994)

Andrzej Jankowski and Andrzej Skowron

The only way to rectify our reasoning is to make them as tangible as those of the Mathematicians, so that we can find our error at a glance, and when there are disputes among persons, we can simply say: Let us calculate, without further ado, to see who is right.

G.W. Leibniz [6]

Abstract This is a biogram of Professor Helena Rasiowa (1917–1994) one of the leading representatives of logicians from Warsaw. She was not only the great scientist but also the great human being. Rasiowa influenced numerous researchers from all over the world, especially by her results in algebraic logic, as well as by her great contribution for the mathematical community in all respect. She is also co-founder of the Pawlak-Rasiowa School of Artificial Intelligence (AI) (Jankowski and Skowron, Andrzej Mostowski and Foundational Studies, pp 106–143, 2008).

Keywords Logic in Warsaw · Helena Rasiowa · Pawlak-Rasiowa School of Artificial Intelligence (AI) · Metamathematics · Algebraic methods in logic · Foundations of computer science · Logic in computer science and in AI

Mathematics Subject Classification (2000) Primary 03Fxx, 03Gxx, 03Bxx, 06Dxx, 06Exx, 68Txx; Secondary 03G27, 03G05, 03G05, 03B35, 03B45, 03B50, 03B55, 03B60, 03B70, 06D15, 06D20, 06D25, 06E15, 68T15, 68T27, 68T37, 97P20

Helena Rasiowa was born on June 23, 1917. During her school days and also later, she attended piano classes at the Fryderyk Chopin Academy of Music. In 1938, she was enrolled at the University of Warsaw (Faculty of Mathematics and Science). Among her classmates there were well-known personalities like Wanda Szmielew and Roman Sikorski who, after 1945, significantly contributed to the development of mathematics and logic in Poland. She had the opportunity to attend lectures of renowned mathematicians such as Wacław Sierpiński, Kazimierz Kuratowski, Jan Łukasiewicz, and Karol Borsuk.

In 1939, the World War II broke out, and Rasiowa had to interrupt her studies. She established contact with her professors of mathematics and physics at the University of Warsaw and resumed her studies at the underground university in German-occupied

Á. Garrido, U. Wybraniec-Skardowska (eds.), *The Lvov-Warsaw School. Past and Present*, Studies in Universal Logic,
https://doi.org/10.1007/978-3-319-65430-0_48

Warsaw functioning at significant risk in spite of prohibition by the German occupiers of both secondary and university education. By July 1944, she passed all mandatory exams required for the title of Master of Science in mathematics and prepared her MSc thesis devoted to mathematical logic under the academic supervision of Jan Łukasiewicz. The manuscript of her dissertation was burned during the Warsaw Uprising of 1944. Helena Rasiowa and her mother were miraculously saved after spending numerous days in different cellars, covered by the ruins of houses in the Old City in Warsaw that were destroyed during the bombing. The two women left Warsaw and survived thanks to the help of other members of their family.

When the war ended, Rasiowa rewrote her MSc thesis, and in 1945 she graduated from the University of Warsaw with MSc in mathematics. From September 1945 to November 1946, she worked at the Wojciech Górski's secondary school in Warsaw, pursuing her graduate studies at the Section of the Philosophy of Mathematics, supervised by Andrzej Mostowski. On November 1, 1946, she became the assistant lecturer at the University of Warsaw, where she worked for the rest of her life. During her work at the University of Warsaw, she was granted various academic degrees and titles, and she gradually promoted to academic positions like assistant professor, associate professor, and then full professor. Her first publications, *Axiomatisation d'un système partiel de la théorie de la déduction* and *Sur certaines matrices logiques*, came out in 1947 [7]. In 1950, she received doctorate in mathematics and science based on her PhD thesis, *Algebraic treatment of the functional calculi of Lewis and Heyting*, under the academic supervision of Andrzej Mostowski. The thesis reflects the main theme of her future research on algebraic methods in logic. In 1956 Rasiowa received her academic degree *Doctor of Science in Mathematics* (equivalent to habilitation today) at the Institute of Mathematics of the Polish Academy of Sciences. For the habilitation degree she submitted two papers, viz., *Algebraic models of axiomatic theories* and *Constructive theories*. Together these papers formed a dissertation entitled *Algebraic models of elementary theories and their applications*. Between 1954 and 1957 she held a post of Associate Professor there, jointly with an analogous position at the University, and became Professor in 1957 and subsequently Full Professor in 1967. During that time, she wrote about 20 different papers coauthored by Roman Sikorski, Andrzej Mostowski, Andrzej Białynicki-Birula, and Jerzy Łoś. In 1957, she was granted the title of a Professor by the Central Certifying Board and in 1967, the title of a Full Professor by the Council of State.

As an academic, Helena Rasiowa represented, with pride and dignity, the Polish school of logic, which was popularized during the interwar period by such scientists as Jan Łukasiewicz, Stanisław Leśniewski, Kazimierz Ajdukiewicz, and Alfred Tarski. Her research from the 1950s (including her PhD dissertation, habilitation thesis and numerous other publications, co-authored by other well-known scientists with whom she worked at that time) led to a deeper understanding of the possibilities related to the use of algebraic and topological methods in logic. These concerned both the classical and non-classical predicate calculus (including intuitionistic logic, modal logics, and many-valued logics). In the academic year of 1949/1950, Helena Rasiowa participated in the lectures by Stanisław Mazur, devoted to "constructive" analysis. At that time, she became familiarized with the results of the research on "constructive" models of computation in analysis. That research was initiated between 1936 and 1939 as part of the collaboration between Stefan Banach and Stanisław Mazur. Later on, Helena Rasiowa and Andrzej Grzegorczyk [4] published lecture notes out of that lectures.

The interest in algebraic methods of logic (including their applications in theorem proving) and models of computations for identifying constructive fragments of mathematical theories were naturally connected with the research of logical foundations of the emerging computer science and studies in the field of AI (e.g., methods of automatic theorem proving). As a result in 1960s, a group of researchers around Professor Rasiowa started conducting studies in such fields as automatic theorem proving, algorithmic logic and, later in 1970s, program logic, approximate reasoning, and other domains of foundations of computer science. For approximately 30 years, that is till the end of her life, Professor Rasiowa continued that research together with her students and collaborators. Her work helped to open many important projects related to applications of logic in the foundations of computer science. Moreover, together with Zdzisław Pawlak, she founded a scientific journal, known as *Fundamenta Informaticae*, devoted to the foundations of computer science. From the very beginnings of that journal in 1977 till the end of her life, she was its editor-in-chief. Thanks to her unimaginably hard scientific and organizational work, *Fundamenta Informaticae* managed to gain international renown. She was an active Collecting Editor with *Studia Logica* (since 1974) and Associate Editor of the *Journal of Approximate Reasoning* (since 1986).

Her devotion to work bore fruits in the fields of AI and computer science, resulting in the creation of an entire school of students and collaborators, who still continue the research initiated by Rasiowa and Pawlak. The school is sometimes referred to as the Polish Pawlak-Rasiowa School of AI [5].

The establishment and development of the School emphasize considerable contributions of Professor Rasiowa in the development of research devoted to the application of mathematical logic in Computer Science and AI. Much earlier than other mathematicians of her generation, she realized that computer science may be a great inspiration for logic. She also paid attention to the need of research on logical foundations of computer science by actively participating in their development. By 1960s she envisioned the significance of logic for the development of theoretical computer science and AI. She was aware of the facts that numerous scientific problems of theoretical computer science may be solved by methods developed in logic, and logical research in computer science may constitute an important source of inspiration for the development of logic. Helena Rasiowa inspired to conduct intensive, fruitful research on program logics, methods of reasoning with incomplete information and logical calculi for AI systems. As the head of the Section of Mathematical Logic at the Institute of Mathematics (University of Warsaw), she conducted seminars and research projects devoted to these fields.

It is worth recalling that one of the first great scientists to propose the research on relationships between algebraic systems and logical systems was George Boole (who referred to the ideas formulated by, e.g., Leibniz). The works by Boole and his successors led to the development of the concept of Boolean algebra and helped to establish its relationship with the classical predicate calculus. This relationship is naturally based on an algebraic construction introduced by Lindenbaum and Tarski, based on the equivalence classes generated by formulas. The Stone's representation theorem applied to the Lindenbaum–Tarski algebra contributed to the development of research on topological properties of models for logic.

The second very important algebraic construction, which defines the relationships between logic and algebra, was initiated by the research of Łukasiewicz and Post. This approach treats formulas as algebraic functions over particular algebras. Key algebraic methods in the research concerning intuitionistic logic and modal logic were developed by Marshall Stone, Alfred Tarski, and John McKinsey. Helena Rasiowa became actively involved in this research area. Together with Roman Sikorski, she developed the first algebraic proof of the Gödel's completeness theorem for the classical predicate calculus. Next, she used an algebraic method to prove analogous theorems for the predicate calculus of intuitionistic logic and modal logics. This enabled her to obtain algebraic proofs of a number of other important theorems concerning both classical logic and numerous non-classical logics. Since that moment, numerous logicians and mathematicians in the world have started to use analogous algebraic methods in their research on logical systems.

The first monograph by Helena Rasiowa, *The Mathematics of Metamathematics*, which she co-authored with Roman Sikorski, was published in 1963 [10]. It contains a systematic and comprehensive overview of the results related to the algebras of logical calculi (including classical, intuitionistic, modal, and positive logic). The monograph exerted a profound influence on numerous logicians in the world. Let us quote professor Melvin Fitting [3]:

> [...] *the ideas were of a sort I had never come across before, and I was enchanted. This way of using algebra, producing known results in classical logic, then applying similar techniques to non-classical logics to get new results\it all seemed like magic. Profound results fell out so effortlessly, it seemed. How could one read this book and remain unaffected?*

In the Rasiowa–Sikorski monograph we may find, e.g., a very interesting proof of the Rasiowa–Sikorski lemma, based on a popular (among the representatives of the Polish School of Mathematics) method of proving existential theorems by means of the Baire category theorem. This method was used by such well-known mathematicians as Stefan Banach, Hugo Steinhaus, and Stanisław Mazur. At the same time, we become familiarized with numerous interesting topological properties of models for the classical predicate calculus. Numerous generalizations of the Rasiowa–Sikorski lemma to other algebraic structures (e.g., Post's algebras, certain types of distributive lattices in the form of the Rauszer–Sabalski lemma) were created. Moreover, there appeared versions of the proof of the Rasiowa–Sikorski lemma that did not used the Baire category theorem.

In *The Mathematics of Metamathematics*, algebraic models for the predicate calculus, introduced by Rasiowa and Sikorski, also are presented. These include Boolean models for the classical predicate calculus and Heyting models for the intuitionistic predicate calculus. Dana Scott and Robert Solovay used the concept of Boolean models to simplify a well-known Cohen's proof of the independence of the axiom of choice, along with Cantor's continuum hypothesis. In this simplification, the key role is played by both Boolean models and the Rasiowa–Sikorski lemma itself [2]. In 1973, Denis Higgs showed that there are some strong relationships between algebraic semantics of Rasiowa and Sikorski and the semantics based on topos of sheaves [2]. Later, these relationships were independently developed by Dana Scott and his students. This led to further modifications of the Cohen's method and its generalizations in topos theory.

Another well-known monograph of Helena Rasiowa, *Algebraic Approach to Non Classical Logics* [8], presents algebraic theories for a wide class of non-classical logics. It encompasses positive and classical implicational logics, logics weaker than the positive implicational logic, minimal logic, positive logic with semi-negation, constructive logics with strong negations and Post's logics. The aim of the book was to identify a class of logics that is as broad as possible and for which a general algebraic theory could be formulated. The developed theory consists of tools for proving significant theorems about the logics from this class simultaneously for the entire class, and not—as it was treated earlier—separately for each logic.

Helena Rasiowa is the author of about one hundred scientific works and the academic textbook, viz., *The Introduction to Contemporary Mathematics*, which gained popularity both in Poland and abroad. Her works and monographs helped to educate whole generations of logicians in Poland and abroad. In different countries, one can meet scientists who after studying the monographs and works of Professor Helena Rasiowa proudly call themselves her students, despite the fact that they have never met her in person. She educated many generations of students and young academic researchers and was an academic supervisor of nearly 20 PhDs, of which several obtained titles of professors. This includes: Michael Bleicher (1961, Tulane University), Vladimir G. Kirin (1966, University of Zagreb), Andrzej Salwicki (1969), Nguen Cat Ho (1971, Vietnam Academy of Sciences), Cecylia Rauszer (1971), Ewa Orłowska (1971), Grażyna Mirkowska (1972), Maria Semeniuk-Polkowska (1972), Wiktor Bartol (1973), Antoni Kreczmar (1973), Jerzy Tiuryn (1975), Lech Banachowski (1975), Anita Wasilewska (1975), Michał Krynicki (1976), Bolesław Szymański (1976), Dimiter Vakarelov (1977, Sofia University), Bogdan Sabalski (1977), Halina Przymusińska (1979), and Leszek Rudak (1986).

The extremely high position of Helena Rasiowa within the international scientific environment resulted in numerous invitations from the best universities and research centers in Europe and both Americas. She gave lectures, including frequent plenary lectures, during numerous international congresses and conferences. She also chaired many scientific sessions.

During her last 2 years of life, Helena Rasiowa, not succumbing to the aggravating disease, continued working on her new monograph, *Algebraic models of logics*, of which she is the author of eight chapters [9]. In June 1994, she started the preparation of a special edition of *Studia Logica*, entitled *Reasoning with Incomplete Information*, and, in particular, prepared the list of its topics.

The readers are referred to [1] for additional details on the scientific achievements of Helena Rasiowa, in particular for the full list of her publications.

It should be emphasized that despite of her significant scientific activity, Professor Rasiowa did not refuse taking various roles that were important for the scientific and academic life. The university was her second home. She was a person of great perseverance. In retrospect, it is amazing to realize how she was able to meet all these challenges.

From 1955 to 1958, she was the Director of the Extramural Studies at the Institute of Mathematics (University of Warsaw). From 1964 to 1970, she headed the Section of the Foundations of Mathematics at the University of Warsaw and from 1970 till the end of her life, she was the head of the Section of Mathematical Logic at the same university. For more than 15 years she was the Dean of the Faculty of Mathematics, Computer Science and Mechanics (1958–1960, 1962–1966, 1968–1978). She also represented the Faculty Council of the Faculty of Mathematics, Computer Science and Mechanics to the Senate of the University of Warsaw. Helena Rasiowa actively participated in the scientific life and performed a number of important functions in academic organizations and associations both in Poland and abroad. From 1961 to 1968, she was a Scientific Secretary of the Institute of Mathematics of Polish Academy of Sciences, and later, of its board member. From 1970 to 1972 she was the member of the Committee of Mathematical Assessors at the Central Council of Higher Education and from 1968 to 1972 she was a Chairperson of the Committee of Mathematical Assessors for Teacher's Colleges. Professor Rasiowa was active in the Polish Mathematical Society (PMS). In PMS, she was a member of the board in the Warsaw Chapter, responsible for the popularization of mathematics and training of teachers. She was appointed as the President of the Warsaw Chapter of PMS twice (1957–1958, 1963–1964) and she was also the Secretary of the Society (1955–1956) and its Vice-President (1958) as well. In the Association for Symbolic Logic, she was a Council member (1958–1960) and a member of the Executive Committee for Foreign Affairs (1972–1974). In 1972, she was an Alternate Assessor (1972–1975) and Assessor (1975–1979) in the Division of Logic, Methodology and Philosophy of Science of the International Union of History and Philosophy of Sciences. For several years, she was the Chairperson of the Scientific Council of Computational Center by the Polish Academy of Sciences and later of the Institute of the Foundations of Computer Science of the Academy, as well.

Thanks to her active participation, efforts aimed at the creation of the *Polish Association for Logic and Philosophy of Science* were undertaken between 1980–1981. This association is the continuation of the Polish Logic Society, founded on April 22, 1936 by Jan Łukasiewicz and Alfred Tarski. The creation of the society, whose aim was to "practice and nurture logic and methodology, along with their history, didactics and possible application," was connected with a great development of logic in the interwar Poland. The Polish Logical Society was the second organization of this type in the world (after the Association for Symbolic Logic in the USA). After the war, the society was not reactivated. The efforts aimed at the creation of a similar society in 1980–1981 turned out futile as the authorities refused to register it. During the IX International Congress of Logic and Philosophy of Science in Uppsala in 1991, a group of Polish congress participants met to initiate the establishment of the association. On December 10, 1991, a constituent meeting of the Polish Association for Logic and Philosophy of Science took place.

It is possible to identify numerous areas of contemporary research, which are directly or indirectly inspired by the works of Helena Rasiowa. These include algebraic methods in logic, deductive systems like tableau, mathematical foundations of computer science, and artificial intelligence (including approximate reasoning and interactive granular computations).

Helena Rasiowa was an extremely hard-working, friendly and cheerful person, who has been always kind and sensitive to human problems with great intelligence. She was always ready to help, not only as a scientist, but also as a human being in real-life problems. In each, even most difficult situation, she tried to do as much good as possible—this was her life motto as an academic teacher, devoted wholeheartedly to logic and Polish science.

She was able to share the love for her children, Krystyna Kijewska and Zbigniew Raś, with the love for mathematics.

Professor Helena Rasiowa was a Franciscan tertiary, a fact which, during her life, was known only to some of her loved ones. The faith combined with a passion for mathematics, enabled her to survive the most difficult moments, like the war or the very last stage of her life, when the incurable disease began to destroy her. This was a great lesson for her students about how to behave in the most difficult situations.

Acknowledgements Authors are grateful to Victor Marek and Soma Dutta for their corrections and other helpful suggestions.

References

1. Bartol, W., Orłowska, E., Skowron, A.: Helena Rasiowa, 1917–1994. Modern Log. Int. J. Hist. Math. Log. Set Theory Found. Math. **5**(3), 231–248 (1995)
2. Bell, J.L.: Set Theory Boolean-Valued Models and Independence Proofs. Oxford Logic Guides, vol. 47. Oxford University Press, Oxford (2011)
3. Fitting, M.: A tribute to professor Helena Rasiowa. In: Orlowska, E. (ed.) Logic at Work. Essays Dedicated to the Memory of Helena Rasiowa, pp. v–viii. Physica, Heidelberg (1998)
4. Grzegorczyk, A., Rasiowa, H.: Computable analysis. Rozprawy Matematyczne, vol. 33, pp. 1–111. PWN, Warsaw (1963)
5. Jankowski, A., Skowron, A.: Logic for artificial intelligence: the Rasiowa–Pawlak school perspective. In: Ehrenfeucht, A., Marek, V., Srebrny M. (eds.) Andrzej Mostowski and Foundational Studies, pp. 106–143. IOS Press, Amsterdam (2008)
6. Leibniz, G.W.: Dissertatio de Arte Combinatoria. Leipzig (1666)
7. Rasiowa, H.: Axiomatisation d'un système partiel de la théorie de la déduction and Sur certaines matrices logiques. C. R. Soc. Sci. Lett. Varsovie, C1 **III**, 22–37 (1947)
8. Rasiowa, H.: Algebraic Approach to Non Classical Logics. Studies in Logic and the Foundations of Mathematics, vol. 78. North Holland, Amsterdam (1974)
9. Rasiowa, H.: Algebraic Models of Logics. Warsaw University, Warsaw (2001)
10. Rasiowa, H., Sikorski, R.: The Mathematics of Metamathematics. Monografie Matematyczne, vol. 41, 1st edn. (1963), 2nd edn. (1968), 3rd edn. (1970). PWN, Warsaw

A. Jankowski
The Dziubański Foundation of Knowledge Technology, Warsaw, Poland

A. Skowron (✉)
Institute of Mathematics, University of Warsaw, Warsaw, Poland

Systems Research Institute, Polish Academy of Sciences, Warsaw, Poland
e-mail: skowron@mimuw.edu.pl

Post Algebras in the Work of Helena Rasiowa

Ewa Orłowska

Abstract A survey of some classes of Post algebras is given including the class of plain semi-Post algebras, Post algebras of order m, m>1, as its particular instance, Post algebras of order ω^+, and Post algebras of order $\omega + \omega^*$. Representation theorems for each of the classes are given. Some examples of the algebras in the classes are constructed.

Keywords Ordered algebraic structures · Lattice · Boolean algebra · Post algebra · Representation theorem · Many-valued logic

Mathematics Subject Classification (2000) Primary 03G20, 06D25

Emil Post doctoral dissertation [24] (see also Post [25]) provided a description of an n-valued, $n > 1$, functionally complete algebra. The notion of Post algebra was introduced in Rosenbloom [43]. Then was the paper by Wade [53] and in Rousseau [44, 45] an equivalent formulation of Post algebra was given which became a starting point for an extensive research. Various generalizations of Post algebras inspired by computer science have been proposed. In 1970-ies Helena Rasiowa and Tadeusz Traczyk run a seminar on Post algebras in the Department of Mathematics of the University of Warsaw, gathering the participants both from the Polish universities as well as guests and PhD students from abroad. This paper is a survey of major classes of Post algebras which were the subject of research of Helena Rasiowa at that time and were studied by the participants of the seminar. The present paper is based on the earlier paper included in:

Iturrioz, L., Orłowska E., Turunen E. (eds.): Atlas of Many-valued Logics. Mathematics Report 75 of the Department of Information Technology of the Tampere University of Technology, 2000.

The Atlas was elaborated by the participants of the COST Action 15 "Many-valued Logics for Computer Science Applications", 1994–1999.

Á. Garrido, U. Wybraniec-Skardowska (eds.), *The Lvov-Warsaw School. Past and Present*, Studies in Universal Logic,
https://doi.org/10.1007/978-3-319-65430-0_49

1 Plain Semi-Post Algebras

These algebras were introduced and investigated in Cat Ho [1], Cat Ho and Rasiowa [2–4]. Let $(T, \leqslant)$ be a poset. A subset s of T is an ideal provided that $s \neq \emptyset$ and for all $t \in s$ and $w \in T$, if $w \leqslant t$ then $w \in s$. Let ET be the set of ideals of T together with the empty set $\emptyset$. Clearly, $T \in ET$. It is known that any $s \in ET$ is of the form $s = \bigcup\{s(t) : s(t) \subseteq s\}$, where $s(t) = \{w \in ET : w \leqslant t\}$. The system $(ET, \subseteq)$ is a complete lattice, where join and meet are set-theoretical union and intersection, respectively.

An abstract algebra

(P) $\mathbf{P} = (P, \cup, \cap, \rightarrow, \neg, \{d_t : t \in T\}, \{e_s : s \in ET\})$

where $\cup, \cap, \rightarrow$ are 2-argument operations, $\neg$, d_t for $t \in T$ are unary operations and e_s for $s \in ET$ are 0-argument operations (constants) is a plain semi-Post algebra (psP-algebra) of type T provided that the following conditions are satisfied:

(p0) $(P, \cup, \cap, \rightarrow, \neg)$ is a Heyting algebra with the zero element $\mathbf{0} = e_\emptyset$ and the unit element $\mathbf{1} = e_T$,

for any $a, b \in P$

(p1) $d_t(a \cup b) = d_t a \cup d_t b$,
(p2) $d_t(a \cap b) = d_t a \cap d_t b$,
(p3) $d_w d_t a = d_t a$,
(p4) $d_t e_s = \mathbf{1}$ if $t \in s$, otherwise $d_t e_s = \mathbf{0}$,
(p5) $d_t a \cup \neg d_t a = \mathbf{1}$,
(p6) $a = \bigcup\{e_{s(t)} \cap d_t a : t \in T\}$ where $\bigcup$ is the least upper bound in P.

Let $\mathbf{P} = (P, \cup, \cap, \rightarrow, \neg, \{d_t : t \in T\}, \{e_s : s \in ET\})$ be a psP-algebra of type $(T, \leqslant)$. By B_P we denote the set of elements of P of the form $d_t a$, $t \in T$.

Proposition 1.1

(a) *B_P is closed under the operations $\cup, \cap, \rightarrow, \neg$ of P.*
(b) *The algebra $\mathbf{B}_P = (B_P, \cup, \cap, \rightarrow, \neg, \mathbf{1}, \mathbf{0})$ is a Boolean algebra.*

Let C_P be the set of all complemented elements in the distributive lattice $(P, \cup, \cap)$. Then

Proposition 1.2

(a) *C_P is closed under the operations $\cup, \cap, \rightarrow, \neg$ of P.*
(b) *The algebra $\mathbf{C}_p = (C_P, \cup, \cap, \rightarrow, \neg, \mathbf{1}, \mathbf{0})$ is a Boolean algebra.*
(c) *For every $a \in C_P$, $d_t \neg a = \neg d_t a$, $t \in T$.*

Note that B_P and C_P do not always equal. Consider a poset $(T, \leqslant)$ such that $T = \{a, b, c\}$ and $\leqslant = \{(b, a)\}$. Then $ET = \{\emptyset, \{b\}, \{c\}, \{b, c\}, \{a, b\}, T\}$, $B_P = \{\emptyset, T\}$, and $C_P = \{\emptyset, T, \{c\}, \{a, b\}\}$.

Proposition 1.3 (Epstein Lemma) *For any set $\{a_j : j \in J\}$ of elements in P it holds*

(a) $a = \bigcup^P\{a_j : j \in J\}$ *iff for every* $t \in T$ $d_t a = \bigcup^{B_P}\{d_t aj : j \in J\}$,
(b) $a = \bigcap^P\{a_j : j \in J\}$ *iff for every* $t \in T$ $d_t a = \bigcap^{B_P}\{d_t aj : j \in J\}$,

where $\bigcup^P$, $\bigcap^P$, $\bigcup^{B_P}$, $\bigcap^{B_P}$ denote infinite joins and meets in the algebras $\mathbf{P}$ and $\mathbf{B}_P$, respectively.

Proposition 1.4

(a) $d_t(a \to c) = \bigcap\{d_w a \to d_w c : w \leqslant t\}$
(b) $d_t \neg a = \bigcap\{\neg d_w a : w \leqslant t\}$
(c) $d_w a \leqslant d_t a$ *whenever* $w \leqslant t$, *for any* $w, t \in T$
(d) $a \leqslant b$ *iff* $d_t a \leqslant d_t b$ *for all* $t \in T$
(e) $e_w \leqslant e_t$ *iff* $w \subseteq t$, *for any* $w, t \in ET$

It follows that every psP-algebra of type $(T, \leqslant)$ uniquely determines a set of infinite meets of P

$$M(P) = \left\{\bigcap\{d_w a \to d_w c : w \leqslant\} : t \in T\right\}.$$

Observe that for any sets $s', s'' \in ET$ there exists the relative pseudo-complement $s' \to s''$ defined by

$$s' \to s'' = \bigcup\{s \in ET : s' \cap a \subseteq s''\}$$

and the pseudo-complement $\neg s'$ defined by

$$\neg s' = s' \to \emptyset = \bigcup\{s \in ET : s' \cap s = \emptyset\}.$$

Clearly, $s' \to s'', \neg s' \in T$.

Proposition 1.5

(a) *For any poset* $(T, \leqslant)$, *the system* $(ET, \cup, \cap, \to, \neg, T, \emptyset)$, *where* $\cup$, $\cap$ *are set-theoretical operations of union and intersection, respectively, and* $\to$, $\neg$ *are defined as above, is a Heyting algebra with the unit element* T *and zero element* $\emptyset$.
(b) *Given a psP-algebra* $\mathbf{P} = (P, \cup, \cap, \to, \neg, \{d_t : t \in T\}, \{e_s : s \in ET\})$, *let* $EP = \{e_s : s \in ET\}$. *Then* $(EP, \leqslant)$ *is a poset isomorphic to* $(ET, \subseteq)$.

Condition (b) follows from Proposition 1.4(e).

Example 1.1 An important example of a psP-algebra is the following algebra, referred to as a basic psP-algebra:

$$(ET, \cup, \cap, \to, \neg, \{d_t : t \in T\}, \{e_s : s \in ET\})$$

where $(ET, \cup, \cap, \to, \neg)$ is the Heyting algebra defined above and the operations $d_t, t \in T$, and $e_s, s \in ET$, are defined by:

$$e_s = s, \text{ in particular } e_\emptyset = \emptyset \text{ and } e_\emptyset = T,$$

$$d_t s = T \text{ if } t \in s, \text{ otherwise } d_t s = \emptyset.$$

Proposition 1.6 *The basic psP-algebra is functionally complete, that is any n-argument operation* $f : ET^n \to ET$, $n = 0, 1, \ldots$, *is definable with the operations of this algebra.*

Given a Boolean algebra $\mathbf{B} = (B, \cup, \cap, \to, \neg, \mathbf{1}_B, \mathbf{0}_B)$ and a poset $(T, \leqslant)$ by a descending T-sequence of elements of B we mean an indexed family $(b_t)_{t \in T}$ of elements of B such that $w \leqslant t$ in T implies $b_t \leqslant b_w$ in B (for the sake of simplicity we denote the Boolean ordering of B with the same symbol). We say that B and T satisfy condition (erpc) of existence of relative pseudo-complement if

(erpc) For any two descending T-sequences $b = (b_t)_{t \in T}$, $c = (c_t)_{t \in T}$ of elements of B there exists $\bigcap^B \{b_w \to c_w : w \leqslant t\}$ for all $t \in T$.

Example 1.2 We present a psP-algebra $\mathbf{P}_T(\mathbf{B})$ of type T determined by a Boolean algebra $\mathbf{B} = (B, \cup, \cap, \to, -, \mathbf{1}_B, \mathbf{0}_B)$ such that $\mathbf{B}$ and T satisfy condition (erpc). The universe $P(B)$ of $\mathbf{P}_T(\mathbf{B})$ is the set of all descending T-sequences of elements of B. We define a partial ordering $\leqslant$ on $P(B)$ as follows. Let $b = (b_t)_{t \in T}$ and $c = (c_t)_{t \in T}$ be any elements of $P(B)$. Then

$$b \leqslant c \text{ in } P(B) \text{ iff } b_t \leqslant c_t \text{ in } B \text{ for all } t \in T.$$

The system $(P(B), \leqslant)$ is a lattice with join and meet defined by

$$b \cup c = (b_t \cup c_t)_{t \in T}, \quad b \cap c = (b_t \cap c_t)_{t \in T}.$$

Since B and T satisfy (erpc), for any b, c in $P(B)$ there exists the relative pseudo-complement $b \to c$ and

$$b \to c = (x_t)_{t \in T}, \text{ where } x_t = \bigcap \{b_w \to c_w : w \leqslant t\}.$$

For every $s \in ET$ we define

$$e_s = (x_t)_{t \in T}, \text{ where } x_t = \mathbf{1}_B \text{ if } t \in s, \text{ otherwise } x_t = \mathbf{0}_B.$$

Moreover, we put

$$d_w b = (x_t)_{t \in T}, \text{ where } x_t = b_w \text{ for every } t \in T,$$

$$\neg b = (x_t)_{t \in T}, \text{ where } x_t = \bigcap \{-b_w : w \leqslant t\}.$$

It is easy to verify that the algebra

$$\mathbf{P}_T(\mathbf{B}) = (P(B), \cup, \cap, \to, \neg, \{d_t : t \in T\}, \{e_s : s \in ET\})$$

defined above is a psP-algebra of type $(T, \leqslant)$.

Proposition 1.7 *Let a Boolean algebra* $\mathbf{B}$ *and a poset* $(T, \leqslant)$ *satisfying (erpc) be given. Let* $\mathbf{P}$ *be the algebra* $\mathbf{P}_T(\mathbf{B})$ *defined as in Example 1.2. Then the algebra* $\mathbf{B}_P$ *(see Proposition 1.4) is isomorphic to* $\mathbf{B}$.

Example 1.3 A particular instance of the algebra defined in Example 1.2 is a set algebra obtained by taking the field of all subsets of a set as the respective Boolean algebra. Let U be a nonempty set and let $B(U)$ be the field of all subsets of U. We have $\mathbf{1}_{B(U)} = U$ and $\mathbf{0}_{B(U)} = \emptyset$. For any poset $(T, \leqslant)$, $B(U)$ and T satisfy condition (erpc). Let $P(B(U))$ be the set of all descending T-sequences of sets from $B(U)$. The ordering on $P(B(U))$ is the set inclusion. The algebra $\mathbf{P}_T(\mathbf{B}(U)) = (P(B(U)), \cup, \cap, \rightarrow, \neg, \{d_t : t \in T\}, \{e_s : s \in ET\})$ defined as in Example 1.2 is a psP-algebra of type $(T, \leqslant)$. The infinite joins in the axiom (p6) are set unions.

Proposition 1.8 (Representation Theorem) *Let* $\mathbf{P} = (P, \cup, \cap, \rightarrow, \neg, \{d_t : t \in T\}, \{e_s : s \in ET\})$ *be a psP-algebra of type* $(T, \leqslant)$. *If T is denumerable and either well-founded or the set $M(P)$ is denumerable (in particular if P is denumerable), then for any denumerable set Q of infinite joins and meets in P there exists the field $B(U)$ of all subsets of a nonempty set U and a monomorphism h from* $\mathbf{P}$ *into* $\mathbf{P}_T(\mathbf{B}(U))$ *preserving all the operations in Q.*

2 Post Algebras of Order m

The first axiom system for the algebras characterising Post's m-valued logics, for a finite m greater than 2, was presented in Rosenbloom [43]. He called them Post algebras. The axiomatisation was then simplified in Epstein [9] and Traczyk [48]. Traczyk proved the equational definability of the class of Post algebras. Over the years the theory of Post algebras and several generalizations of these algebras have been developed. Here we define Post algebras of order m as a particular case of psP-algebras.

Let $(T_m, \leqslant)$ be a poset such that $T_m = \{1, \ldots, m-1\}$, where m is a natural number greater than 2, and $\leqslant$ is a natural ordering in T_m. Then $ET_m = \{\emptyset, s(1), \ldots, s(m-1)\}$, where $s(t) = \{w \in T_m : w \leqslant t\}$. Clearly, $(ET_m, \subseteq)$ is isomorphic to $\{0, 1, \ldots, m-1\}$ with the natural ordering. Hence, we can identify these two posets and assume that constants e_s are indexed with elements from $\{0, 1, \ldots, m-1\}$.

By a Post algebra of order m we mean a psP-algebra of type $(T_m, \leqslant)$.

It can be easily shown that this definition is equivalent to the standard definition of Rousseau [44, 45].

Example 2.1 A classical example of a Post algebra of order m is an m-element Post algebra such that $P = \{e_0, \ldots, e_{m-1}\}$, and for $i, j \in \{0, 1, \ldots, m-1\}$ the operations in P are defined as follows:

(ex1) $e_i \cup e_j = e_{\max(i,j)}$,
(ex2) $e_i \cap e_j = e_{\min(i,j)}$,
(ex3) $e_i \rightarrow e_j = \mathbf{1}$ if $i \leqslant j$, otherwise $e_i \rightarrow e_j = e_j$,
(ex4) $\neg e_i = e_i \rightarrow \mathbf{0}$,
(ex5) $d_i e_j = \mathbf{1}$ if $i \leqslant j$, otherwise $d_i e_j = \mathbf{0}$.

Proposition 2.1

(a) $e_0 \to a = e_{m-1}$,
(b) $e_s \to a = \bigcup\{d_t a \cap e_t : t \leqslant s\} \cup d_s a$,
(c) $e_{m-1} \to a = a$,
(d) $a \to e_s = e_s \cup \neg d_{s+1} a$, *for* $s = 0, \ldots, m-2$,
(e) $a \to e_{m-1} = e_{m-1}$.

We define disjoint operations c_s for $s \in \{0, 1, \ldots, m-1\}$ as follows:

(c1) $c_0 a = \neg d_1 a = \neg a$,
(c2) $c_s a = d_s a \cap \neg d_{s+1} a$ for $s \in T \setminus \{m-1\}$,
(c3) $c_{m-1} a = d_{m-1} a$.

We clearly have

$$c_s a \cap c_t a = e_o \text{ for } s \neq t.$$

Any element a of P has the following disjoint representation:

(c4) $a = \bigcup\{c_t a \cap e_t : t \in T\}$.

Theorems analogous to Propositions 1.1–1.8 hold and constructions from Examples 1.2 and 1.3 carry over to the case of type $(T_m, \leqslant)$. Algebras presented in Example 1.1 can be identified with those defined in Example 2.1. Moreover, the algebras $\mathbf{B}_P$ and $\mathbf{C}_P$ corresponding to a Post algebra of order m coincide.

The m-valued Post logic is a propositional logic with binary connectives $\wedge$, $\vee$, $\to$, unary connectives $\neg$, D_t for $t \in T$, and propositional constants E_s for $s \in \{0, 1, \ldots, m-l\}$. The algebraic semantics for the logic is determined in the standard way by the class of Post algebras of order m. A Hilbert-style axiomatisation of m-valued Post logic and its completeness with respect to the algebraic semantics is presented in Rasiowa [30]. The main results on m-valued Post logic include: Model existence theorem (Rasiowa [31]), Craig interpolation theorem (Rasiowa [31]), Herbrand theorem (Perkowska [23]). Applications of the m-valued Post logic are concerned with the theory of programming. An algorithmic logic based on m-valued Post logic is developed in Perkowska [23].

Post algebras and logics of any finite type $(T, \leqslant)$ are considered in Nour [16]. They are also treated in Konikowska, Morgan and Orłowska [13].

3 Post Algebras of Order ω^+

Let $(T_\omega, \leqslant)$ be a poset such that $T_\omega = \omega$ is the set of natural numbers and $\leqslant$ is the natural ordering of natural numbers. Then $ET_\omega = \{\emptyset, s(1), s(2), \ldots, T_\omega\}$. Clearly, $(ET_\omega, \subseteq)$ is isomorphic to $\{0, 1, 2, \ldots, \omega\}$ with the natural ordering. Hence, we can identify these two posets and assume that constants e_s are indexed with elements from $\{0, 1, 2, \ldots, \omega\}$.

A Post algebra of order ω^+ is a psP-algebra of type $(T_\omega, \leqslant)$.

An example of a Post algebra of order ω^+ can be defined in a way similar to that developed in Example 2.1.

Theorems analogous to Propositions 1.1–1.8 hold and constructions from examples 1.2 and 1.3 carry over to the case of type $(T_\omega, \leqslant)$. Moreover, the algebras $\mathbf{B}_P$ and $\mathbf{C}_P$ corresponding to a Post algebra of order ω^+ coincide.

Representation theory for Post algebras of order ω^+ has been also developed in Maksimova and Vakarelov [14], Rasiowa [39].

A Hilbert-style axiomatisation of ω^+-valued Post predicate logic and its completeness with respect to the algebraic semantics is presented in Rasiowa [34]. The other results on ω^+-valued Post logic include: Kripke style semantics (Maksimova and Vakarelov [15], Vakarelov [52]), Herbrand theorem and a resolution-style proof system (Orłowska [17–19]), relational semantics and a relational proof system (Orłowska [21]).

Applications of the ω^+-valued Post logic are concerned with the theory of programming. An algorithmic logic based on ω^+-valued Post logic is developed and investigated in Rasiowa [36–38].

4 Post Algebras of Order $\omega + \omega^*$

These algebras are introduced and investigated in Epstein and Rasiowa [11, 12]. Let $T = \{1, 2, \ldots, -2, -1\}$ and $E = \{0, 1, 2, \ldots, -2, -1\}$. A Post algebra of order $\omega + \omega^*$ is an algebra of the form (P) in Sect. 1 satisfying axioms (p0)–(p6), where in (p0) $\mathbf{0} = e_0$ and $\mathbf{1} = e_{-1}$, and the following

(p7) $d_1 a = d_{-1} a \cup \bigcup\{d_s a \cap \neg d_{s+1} a : 1 \leqslant s \leqslant -1\}$ pivot elimination axiom
(p8) $(a \to b) \cup (b \to a) = 1$.

The axiom (p7) says that an element e such that $e_t \leqslant e$ $(d_t e = \mathbf{1})$ for all positive t and $e < e_t$ $(d_t e = 0)$ for all negative t does not exist.

Propositions analogous to Propositions 1.1–1.4 hold for Post algebras of order $\omega + \omega^*$. Moreover, the algebras $\mathbf{B}_P$ and $\mathbf{C}_P$ corresponding to a Post algebra of order $\omega + \omega^*$ coincide.

Example 4.1 A most natural example of a Post algebra of order $\omega + \omega^*$ is a linear Post algebra of order $\omega + \omega^*$ defined as follows:

$P = \{e_s : s \in E\}$, and the operations in P are defined with

conditions analogous to (ex1)–(ex5) from Example 2.1.

Disjoint operations in Post algebras of order $\omega + \omega^*$ can be defined with conditions analogous to (cl), (c2), (c3) from Sect. 2 by replacing $m - 1$ with -1. Then any element a of P has a disjoint representation given by condition (c4).

In Post algebras of order $\omega + \omega^*$ one can define arithmetic-like operations in the following way.

The successor sa (the predecessor pa) of an element a of P is an element given by the following disjoint representation

$$sa = \bigcup\{c_t a \cap e_{t+1} : t \in E\}$$

$$pa = \bigcup\{c_t a \cap e_{t-1} : t \in E\}$$

provided that either of these exist.

The inverse $-a$ of an element a is given by the disjoint representation

$$-a = \bigcup\{c_t a \cap e_{-1} : t \in E\}$$

provided that it exists.

Addition and multiplication operations have disjoint representations as follows

$$a + b = \bigcup\{c_t(a + b) \cap e_t : t \in E\}$$

where for each $t \in T$ the infinite join $c_t(a + b) = \bigcup\{c_i a \cap c_j b : i + j = t\}$ exists,

$$a \cdot b = \bigcup\{c_t(a \cdot b) \cap e_t : t \in E\}$$

where for each t there is the finite join $c_t(a \cdot b) = \bigcup\{c_i a \cap c_j b : ij = t\}$.

Proposition 4.1 *A Post algebra of order $\omega + \omega^*$ with inverse, addition and multiplication is a commutative ring with unit, where the ring zero is e_0 and the ring unit is e_1.*

These rings have the characteristic 0.

For a descending T-sequence $X = (X_t)_{t \in T}$ of sets from the field $B(U)$ of all subsets of a nonempty set U we define

$$X^+ = \bigcap\{X_t : t \text{ positive}\}$$

$$X^- = \bigcup\{X_t : t \text{ negative}\}.$$

It can be shown that the algebra of descending T-sequences $X = (X_t)_{t \in T}$ of sets from $B(U)$ such that $X^+ = X^-$, with the operations defined as in Example 1.2, is a Post algebra of order $\omega + \omega^*$.

Representation theorem for Post algebras of order $\omega + \omega^*$ has the following form.

Proposition 4.2 (Representation Theorem) *For every denumerable Post algebra* $\mathbf{P}$ *of order $\omega + \omega^*$ there is a monomorphism h of* $\mathbf{P}$ *into a Post set algebra of order $\omega + \omega^*$ whose elements are descending T-sequences $X = (X_t)_{t \in T}$ of sets from the field $B\{U)$ of all subsets of a nonempty set U such that $X^+ = X^-$. Moreover, h preserves a given denumerable set Q of infinite joins and meets of* $\mathbf{P}$.

Applications of the logic are concerned with approximation reasoning. An approximation reasoning to recognise a subset S of a nonempty universe U is understood as a process of gradual approximating S by

$$\text{subsets of } U \; S \subseteq S_1 \subseteq S_2 \subseteq \ldots \text{ which cover } S$$

$$\text{and subsets } \ldots \subseteq S_{-2} \subseteq S_{-1} \subseteq S \text{ which are contained in } S.$$

Then the approximations of set S are defined as follows:

$$S^+ = \bigcap \{S_t : t \text{ positive}\}$$

$$S^- = \bigcup \{S_t : t \text{ negative}\}.$$

In Epstein and Rasiowa [12] a characterisation of sets S such that $S^+ = S^-$ is given.

Post algebras of order ϑ, where ϑ is an arbitrary ordinal number are introduced and investigated in Przymusinska [26–29].

Acknowledgment Support from the National Science Centre project DEC-2011/02/A/HS1/00395 is gratefully acknowledged.

References

1. Cat-Ho, N.: Generalized post algebras and their applications to some infinitary many-valued logics. Diss. Math. **57**, 1–76 (1973)
2. Cat-Ho, N., Rasiowa, H.: Semi-Post algebras. Stud. Logica. **46**(2), 147–158 (1987)
3. Cat-Ho, N., Rasiowa, H.: Plain semi-Post algebras as a poset-based generalization of Post algebras and their representability. Stud. Logica. **48**(4), 509–530 (1989)
4. Cat-Ho, N., Rasiowa, H.: LT-fuzzy logics. In: Zadeh, L.A., Kacprzyk, J. (eds.) Fuzzy Logic for Management of Uncertainty, pp. 121–139. Wiley, New York (1992)
5. Chang, C.C., Horn, A.: Prime ideals characterisation of generalized Post algebras. Proc. Symp. Pure Math. **2**, 43–48 (1961)
6. Dwinger, Ph.: Notes on Post algebras I, II. Indag. Math. **28**, 462–478 (1966)
7. Dwinger, Ph.: Generalized Post algebras. Bull. Acad. Pol. Sci. Math. **16**, 559–563 (1968)
8. Dwinger, Ph.: A survey of the theory of Post algebras and their generalizations. In: Dunn, J.M., Epstein, G. (eds.) Modern Uses of Multiple-Valued Logic, pp. 53–75. Reidel, Dordrecht (1977)
9. Epstein, G.: The lattice theory of Post algebras. Trans. Am. Math. Soc. **95**, 300–317 (1960)
10. Epstein, G., Horn, A.: P-algebras. An abstraction from Post algebras. Algebra Univers. **4**, 195–206 (1974)
11. Epstein, G., Rasiowa, H.: Theory and uses of Post algebras of order $\omega + \omega^*$, Part I. In: Proceedings of the 20th International Symposium on Multiple-Valued Logic, Charlotte, NC, pp. 42–47 (1990)
12. Epstein, G., Rasiowa, H.: Theory and uses of Post algebras of order $\omega + \omega^*$, Part II. In: Proceedings of the 21th International Symposium on Multiple-Valued Logic, Victoria, pp. 248–254 (1991)
13. Konikowska, B., Morgan, Ch., Orłowska, E.: Relational formalisation of arbitrary finite valued logics. Log. J. Interest Group Pure Appl. Logic **6**(5), 755–774 (1998)
14. Maksimova, L., Vakarelov, D.: Representation theorems for generalized Post algebras of order ω^+. Bull. Acad. Pol. Sci. Math. **22**(8), 757–764 (1974)
15. Maksimova, L., Vakarelov, D.: Semantics for ω^+-valued predicate calculi. Bull. Acad. Pol. Sci. Math. **22**(8), 765–771 (1974)

16. Nour, A.: Etude de modeles logiques. Extensions de la logique intuitionniste. PhD dissertation, University Joseph Fourier, Grenoble (1997)
17. Orłowska, E.: The Herbrand theorem for ω^+-valued logic. Bull. Acad. Pol. Sci. Math. **25**(11), 1069–1071 (1977)
18. Orłowska, E.: The resolution principle for ω^+-valued logic. Fundamenta Informaticae **2**, 1–15 (1978)
19. Orłowska, E.: Resolution systems and their applications II. Fundamenta Informaticae **3**, 333–362 (1980)
20. Orłowska, E.: Mechanical proof methods for Post logics. Log. Anal. **110–111**, 173–192 (1985)
21. Orłowska, E.: Post relation algebras and their proof system. In: Proceedings of the 21th International Symposium on Multiple-Valued Logic, Victoria, pp. 298–305 (1991)
22. Perkowska, E.: Herbrand theorem for theories based on m-valued logics. Bull. Acad. Pol. Sci. Math. **19**, 893–899 (1971)
23. Perkowska, E.: On algorithmic m-valued logics. Bull. Acad. Pol. Sci. Math. **20**, 717–719 (1972)
24. Post, E.: Introduction to a general theory of elementary propositions. Bull. Am. Math. Soc. **26** (1920)
25. Post, E.: Introduction to a general theory of elementary propositions. Am. J. Math. **43**, 163–185 (1921)
26. Przymusińska, H.: Craig interpolation theorem and Hanf number for ϑ-valued infinitary predicate calculi. Bull. Acad. Pol. Sci. Math. **28**, 207–211 (1980)
27. Przymusińska, H.: Gentzen type semantics for ϑ-valued infinitary predicate calculi. Bull. Acad. Pol. Sci. Math. **28**, 203–206 (1980)
28. Przymusińska, H.: On ϑ-valued infinitary predicate calculi. Bull. Acad. Pol. Sci. Math. **28**, 193–198 (1980)
29. Przymusińska, H.: Some results concerning ϑ-valued infinitary predicate calculi. Bull. Acad. Pol. Sci. Math. **28**, 199–202 (1980)
30. Rasiowa, H.: A theorem on existence of prime filters in Post algebras and the completeness theorem for some many-valued predicate calculi. Bull. Acad. Pol. Sci. Math. **17**, 347–354 (1969)
31. Rasiowa, H.: Ultraproducts of m-valued models and a generalization of the Löwenheim-Skolem-Gödel-Malcev theorem for theories based on m-valued logics. Bull. Acad. Pol. Sci. Math. **18**, 415–420 (1970)
32. Rasiowa, H.: The Craig interpolation theorem for m-valued predicate calculi. Bull. Acad. Pol. Sci. Math. **20**, 341–346 (1972)
33. Rasiowa, H.: Formalised ω^+-valued algorithmic systems. Bull. Acad. Pol. Sci. Math. **21**, 559–565 (1973)
34. Rasiowa, H.: On generalized Post algebras of order ω^+ and ω^+-valued predicate calculi. Bull. Acad. Pol. Sci. Math. **21**, 209–219 (1973)
35. Rasiowa, H.: An Algebraic Approach to Non-classical Logics. North Holland and Polish Scientific Publishers, Amsterdam, Warsaw (1974)
36. Rasiowa, H.: Extended ω^+-valued algorithmic logic (a formalised theory of programs with recursive procedures). Bull. Acad. Pol. Sci. Math. **22**, 605–610 (1974)
37. Rasiowa, H.: Many-valued algorithmic logic as a tool to investigate programs. In: Epstein, G., Dunn, J.M. (eds.) Modern Uses of Multiple–Valued Logic, pp.77–102. Reidel, Dordrecht (1977)
38. Rasiowa, H.: Algorithmic logic. Multiple-valued extension. Stud. Logica. **38**(4), 317–325 (1979)
39. Rasiowa, H.: Topological representations of Post algebras of order ω^+ and open theories based on ω^+-valued Post logic. Stud. Logica. **44**(4), 353–368 (1985)
40. Rasiowa, H.: Towards fuzzy logic. In: Zadeh, L.A., Kacprzyk, J. (eds.) Fuzzy Logic for Management of Uncertainty, pp. 5–25. Wiley, New York (1992)
41. Rasiowa, H.: Axiomatization and completeness of uncountably-valued approximation logic. Stud. Logica. **53**(1), 137–160 (1994)
42. Rasiowa, H., Epstein, G.: Approximation reasoning and Scott's information systems. In: Proceedings of the 2nd International Symposium on Methodologies for Intelligent Systems, North Holland, pp. 33–42 (1987)
43. Rosenbloom, P.C.: Post algebras, I. Postulates and general theory. Am. J. Math. **64**, 167–188 (1942)
44. Rousseau, G.: Logical systems with finitely truth values. Bull. Acad. Pol. Sci. Math. **17**, 189–194 (1969)
45. Rousseau, G.: Post algebras and pseudo-Post algebras. Fundam. Math. **67**, 133–145 (1970)
46. Saloni, Z.: Gentzen rules for the m-valued logic. Bull. Acad. Pol. Sci. Math. **20**, 819–826 (1972)

47. Traczyk, T.: Some theorems on independence in Post algebras. Bull. Acad. Pol. Sci. Math. **10**, 3–8 (1963)
48. Traczyk, T.: An equational definition of a class of Post algebras. Bull. Acad. Pol. Sci. Math. **12**, 147–149 (1964)
49. Traczyk, T.: On Post algebras with uncountable chain of constants. Algebras and homomorphisms. Bull. Acad. Pol. Sci. Math. **15**, 673–680 (1967)
50. Traczyk, T.: Prime ideals in generalized Post algebras. Bull. Acad. Pol. Sci. Math. **16**, 369–373 (1968)
51. Urquhart, A.: An interpretation of many-valued logic. Z. Math. Logik Grundlagen der Mathematik **19**, 111–114 (1973)
52. Vakarelov, D.: Lattices related to Post algebras and their applications to some logical systems. Stud. Logica. **36**, 89–107 (1977)
53. Wade, L.I.: Post algebras and rings. Duke Math. J. **12**, 389–395 (1945)
54. Zabel, N.: New techniques for automated deduction in many-valued first order, finite and infinite logics. PhD thesis, Institut National Polytechnique de Grenoble (1993)

E. Orłowska (✉)
National Institute of Telecommunications, Warszawa, Poland
e-mail: E.Orlowska@itl.waw.pl

Andrzej Grzegorczyk, a Logician Par Excellence

Stanisław Krajewski

Abstract A short biography of Andrzej Grzegorczyk (1922–2014) is presented, listing his main accomplishments in logic, mentioning his philosophical views, followed by a list of all his books and a selection of main papers. (This paper is based on the biographies listed in Part III of the references section.)

Keywords Andrzej Grzegorczyk · Logic · Undecidability · Rationalism · Anti-psychologism

Mathematics Subject Classification (2000) Primary 01A70

1 Life

Andrzej Grzegorczyk was born in Warsaw on August 22, 1922, as the only son of Piotr, a historian of Polish literature, and Zofia who was a physician. He obtained his high school certificate in 1940, already during World War II. He studied physics and philosophy at the clandestine Polish university in Warsaw. He was first attracted to logic before the war when he heard on the radio a lecture of Jan Łukasiewicz on Stoic logic.

After the war Grzegorczyk lived in Cracow, where he obtained MA in philosophy, and in 1946 he returned to Warsaw, where he became an assistant to Władysław Tatarkiewicz and a secretary of *Przegląd Filozoficzny* (Philosophical Review). He began an intensive study of logic and the foundations of mathematics, in part because this was the safest area of philosophy during the Communist rule. In 1950, he obtained PhD at the University of Warsaw. His dissertation, *On Topological Spaces in Topologies without Points*, was written under the supervision of Andrzej Mostowski. He was admitted to the Institute of Mathematics of the Polish Academy of Sciences, where after Habilitation he became a "docent" in 1953, an associate professor in 1961, and a full professor in 1972. Until 1968, he also lectured at the University of Warsaw. In 1974, Grzegorczyk moved to the Institute of Philosophy of the Polish Academy of Sciences, and became the head of its Ethics Group in 1982. He retired in 1990. From 1999 to 2003 he served as the chairman of the Committee of Philosophy of the Polish Academy of Sciences. It is worth mentioning

Á. Garrido, U. Wybraniec-Skardowska (eds.), *The Lvov-Warsaw School. Past and Present*, Studies in Universal Logic,
https://doi.org/10.1007/978-3-319-65430-0_50

that after a series of lectures at the University of Amsterdam in 1965, he was offered a permanent position there, but he refused since he did not want to leave Poland.

Often involved in organizing scholarly activities, he headed the 1973 Logical Semester that inaugurated the International Mathematical Center (the Banach Center) of the Polish Academy of Sciences. At that time it provided a rare occasion for extensive contacts of mathematicians from the Soviet bloc with those from the West. Cooperation with Russia and Ukraine was important for Grzegorczyk, particularly after 1990. In 1995–1997 he led a special project, "One Hundred Years of the Lvov-Warsaw School", which culminated in conferences in Lvov and Warsaw. Grzegorczyk participated in many other conferences in logic and philosophy. He served as an assessor in the Executive Committee of the International Union of History and Philosophy of Science, the Division of Logic, Methodology and Philosophy of Science, and in 1979 he was elected to the Institut International de Philosophie. He received two honorary doctorates: from University of Clermont-Ferrand (2010) and from Jagiellonian University in Kraków (2013). He died in 2014.

In 1953 Andrzej Grzegorczyk married Renata Majewska, who would become a professor of Polish language and literature at the University of Warsaw. They have two children and six grandchildren.

In addition to his scholarly activities, Grzegorczyk was also a writer and was involved in public activities. During the Communist period, he signed some petitions protesting against limitations of freedom. Ha had life-long interest in the ideology of non-violence, a method of fighting without violence. Quite early he showed serious interest in environmental issues: in the 1970s, he popularized the warnings of the Club of Rome.

Grzegorczyk was a devout Roman Catholic, but he was a very independent Christian and was often critical of the official Church policies. He was also open to other faiths and churches, especially the Russian Orthodox Christianity. What is even more important, he was on good terms with intellectuals of various ideological convictions, religious and atheist, pro-Communist and anti-Communist.

2 Logical Accomplishments

Grzegorczyk's best known papers belong to logic and the foundations of mathematics. One of his early papers, the interpretation of Lesniewski's ontology as Boolean algebra without **0** (see [P5]), suggested, according to him, that Leśniewski's calculus of names added nothing interesting to logic.

In widely quoted 1953 paper *Some Classes of Recursive Functions* [2], he described and investigated classes of recursive functions obtainable by superposition, restricted recursion and the operation of restricted minimum (*f* is defined form *g* and *h* by: *f(n) = the smallest x<h(n) such that g(n,x)=0*) from some initial functions containing addition, in the next step multiplication, then raising to power, then superpowering, etc. This leads to a subrecursive hierarchy, that is, the so-called Grzegorczyk hierarchy. It forms a strictly increasing infinite sequence of classes of functions such that its union is equal to the class of primitive recursive functions. The third class of the Grzegorczyk hierarchy is identical with the class of elementary functions, definable as the smallest class of functions containing addition and subtraction, and closed with respect to superposition as well as restricted summation and restricted multiplication.

Grzegorczyk continued the investigations in computable analysis that had been initiated by Banach and Mazur (see [P4, P6, P7, P9]). In particular, he co-edited (with Rasiowa) Mazur's posthumously published book *Computable Analysis*. Grzegorczyk also gave various definitions of computable real numbers. In this way he wanted to apply the concept of effectiveness used in the arithmetic of natural numbers in mathematical analysis. He also studied (see [P16]) computable functionals of higher types.

In 1958 Grzegorczyk co-authored (with Mostowski and Ryll-Nardzewski) a fundamental paper "The classical and the omega-complete arithmetic" ([P10], also [P12]) about second-order arithmetic, which treats it as formalized in first-order logic, but makes it possible to speak about natural numbers as well as their sets. Due to the introduction of the omega-rule, the Π^1_1-relations are representable in it.

Working on computability, decidability and undecidability, Grzegorczyk considered various proofs of undecidability using recursively enumerable sets which are not recursive (see [P8]), and proved that elementary topological algebra, that is, Boolean algebra with closure, is undecidable, because arithmetic is interpretable in it (see [P3]). He also showed (see [P14]) that the calculus of combinators has no recursive model. His most recent contribution, "Undecidability without arithmetization" (2004, see [P26]), deals with the theory of concatenation of texts (conceived as a primitive notion), introduced by Tarski in the 1920s. The theory of concatenation was shown by Grzegorczyk to be undecidable; therefore, it can replace arithmetization in metamathematics. Computability is replaced here by the effective recognizability of texts. The arithmetical relativization of quantifiers can be replaced by the relativization of expressions to subexpressions of a given expression.

Grzegorczyk, like other Polish logicians and mathematicians, for example, Tarski, Mostowski, Rasiowa, Sikorski, was interested in intuitionistic logic and constructive mathematics. Whereas in his work on constructivism he looked at constructive mathematics from the point of view of classical methods, he believed that intuitionists may "perhaps penetrate more deeply into the meaning of logical concepts than classical logic does." Grzegorczyk's semantics for intuitionistic logic introduced in "A philosophically plausible formal interpretation of intuitionistic logic" (1964) (see [P15, P17]) is similar to the much better known construction by Kripke. It is based on the phrase "in my inquiry, I am forced to assert the sentence A at the moment t," and the notion of strong assertion, that is, an assertion that may not be abandoned at a later moment.

A modal interpretation of Grzegorczyk's semantics for intuitionism leads to the system named by Boolos S4.Grz, that is, the modal system S4 plus the formula

$$\Box(\Box(A \Rightarrow \Box A) \Rightarrow A) \Rightarrow A,$$

called Grzegorczyk's axiom. This system has applications in the provability logic.

Grzegorczyk considered logic as belonging to the foundations of mathematics. In this field he worked according to Tarski's program: every formal method is admissible in foundational research, independently of whether it is finitist, constructive or completely infnitistic.

3 Views

Among his ten philosophical books there is a study *Logic—a Human Affair* (1997), and (all in Polish): *Schemata and the Human Being* (1963), *Philosophy in the Time of Challenge* (1979), *Ethics in Internal Experience* (1989), *Life as a Challenge. Introduction to Rationalistic Philosophy* (1993), *Europe, Discovering the Sense of Existence* (2001), as well as a book of short literary forms *Moral Stories* (1986). He also published over 60 papers on mathematical and philosophical logic and over a hundred articles on ethics, religion, social issues.

Grzegorczyk approved Łukasiewicz's statement that logic is the morality of speech and thought. Logic was conceived by Grzegorczyk in a broad sense as including semiotics and the methodology of science. According to him, logic in this sense constitutes a basic ingredient of European rationalism (see [18]). Since his youth he was convinced that everything can be expressed precisely and logically. He believed that European rationalism brought an exceptional intellectual and cultural success, and therefore must be protected. Teaching logic and proliferating its standards is the way.

For Grzegorczyk, logic is a human affair. He maintained that semantic relations are always of someone and for someone. This leads to the acceptance of psychologism in logic. He also claimed that anti-psychologism inevitably leads to idealism. Grzegorczyk argued that paradoxes should not be interpreted as showing that our language is inconsistent, but rather that our concepts and conceptual systems are limited. For example, the Grelling antinomy shows that there exists a set of expressions which cannot be correctly named (see [P22]) and the Liar paradox demonstrates the existence of problems about which nobody can think consistently, sincerely, and in a fully conscious way. In this approach, semantic antinomies appear as laws of thinking (see [17]).

Grzegorczyk also studied axiomatic geometry based on the concept of solid (see [P13]). He continued some works by Tarski, but he also motivated his work by referring to reism, a view formulated by Tadeusz Kotarbiński who claimed that individual spatio-temporal concrete things are the only elements of the furniture of the world. Grzegorczyk was acquainted with reism by Henryk Hiż, his clandestine teacher of logic. Grzegorczyk was always strongly attracted to this approach (see [P1, P2, P19, P24]), and believed that reism is in principle a good ontology for natural science. On the other hand, he clearly saw difficulties and limitations of reism in mathematics, particularly in set theory since it makes talking about infinite sets problematic.

Grzegorczyk was a Christian and believed that Christianity represents the values of European rationalism. According to him, the history of Christianity shows the sense of the world by revealing the realm of the sacred and transcendence. He also admitted the presence of the highest values in other religions or ideologies. Humanity is not reducible to biological facts. European rationalism is open to moral values. Jesus's testimony represents the highest pattern of morality. At the same time, Grzegorczyk was ready to (verbally) accept the need for some use of force in order to promote rational standards.

Grzegorczyk was strongly attracted to the attitude of radical non-violence, as represented by Mahatma Gandhi or Martin Luther King. He helped organize symposia on this issue. He was uncompromising in recommending compromise in all conflicts. We should talk with everyone, he claimed, also with a terrorist. He was also one of the first people in Poland accepting ecological challenges; he proposed limitations on consumption and opposed waste, although such appeals sounded highly abstract in Poland of the 1970s.

He considered moral issues and social challenges to be so important in the contemporary world that he criticized as immoral the focus, so common among the brightest minds, on abstract intellectual questions which can have no impact on social problems. His move from the Institute of Mathematics to the Institute of Philosophy and Sociology was partly motivated by this frustration. He felt he participated in the malaise. After all, he avowed that for him mathematical problems in formal logic were an "obsession, addiction, narcotic."

He believed that we, humans, need solidarity of all people, independently of where they live. He even appealed to the UN in the 1970s proposing to establish a law that everybody is entitled to help every person who is in a worse situation, wherever the other person lives. In particular, richer people or societies should help poorer ones, the stronger should protect the less powerful, the better educated should do something for the less educated, etc.

Grzegorczyk's religiosity and his understanding of religion provided an extension or correction to his views on the supremacy of logic in all reasoning, including the moral debates. In 1974 he wrote a poem in which it was stated that logic is powerless in dealing with religious matters: "I do not refute those who say You don't exist. I agree with the friends who say that Your concept is contradictory... only those have a pure concept of God who don't have it at all. ..."

4 Influence

Grzegorczyk will remain known due to his logical achievements: the Grzegorczyk hierarchy, results about undecidability, second-order arithmetic, the S4Grz system of modal logic, semantics for intuitionistic logic, his concatenation theory. Also his course in mathematical logic *An Outline of Mathematical Logic* (in Polish 1961, in English 1974, see [6, 8]) was for a long time the only detailed textbook in Polish in which logical calculus, model theory and recursion theory were presented in a balanced fashion, and influenced many students of mathematics and philosophy in Poland. His textbook on the foundations of arithmetic *An Outline of Theoretical Arithmetic* (1971, in Polish, see [7]), in which he simultaneously developed arithmetic and exposed its logical basis, was much less influential. He popularized recursive functions and the problems of computability and decidability not only in Poland but also in France (see [3, 4]).

Grzegorczyk was productive but he was always working alone as it was never easy to work with him. He established no school, and hardly had students. He supervised only two PhD dissertations, one in logic and one in ethics. The one in logic was mine, but as a matter of fact I received little from him; I owed my logical education primarily to Andrzej Mostowski.

Grzegorczyk was not a member of any ideological group, and in no circle was he considered fully "one of our guys". At the same time he was widely respected for being absolutely honest and sincere. Whereas he was recognized as an important logician, Grzegorczyk's philosophical and axiological views did not become important or influential. His other works were rarely noticed. There was a notable exception: in 1987 he obtained a literary prize for the book "Moral stories".

In addition to particular formal achievements two general main messages are expressed in Grzegorczyk's work: first, the utmost importance of logical thinking, second, the all-human solidarity. I believe that for all who knew him he can easily remain the best available personification of logic, or the individual embodying the ideal type of a logician. This is a rather ambivalent tribute, even among logicians.

References

In Part I all Grzegorczyk's books are listed, and in Part II a selection of his articles. A fairly comprehensive list is available in the paper A1 mentioned below.

Books

1. Logika popularna [Popular Logic, in Polish]. PWN, Warsaw (1955). (2nd ed. (1958); 3rd ed. (1961); Czech trans. (1957); Russian trans. (1965))
2. Some Classes of Recursive Functions. Rozprawy Matematyczne Nr IV. Instytut Matematyczny PAN, Warsaw (1953)
3. Zagadnienia rozstrzygalności [Problems of Decidability, in Polish]. PWN, Warsaw (1957)
4. Fonctions Récursives. Gauthier-Villars, Paris (1961)
5. Schematy i człowiek [Schemata and Human Being, in Polish]. Biblioteka Więzi. Znak, Kraków (1963)
6. Zarys logiki matematycznej [An Outline of Mathematical Logic, in Polish]. Biblioteka Matematyczna tom 20. PWN, Warsaw (1961) (subsequent improved editions up to 6th ed. (1984))
7. Zarys arytmetyki teoretycznej [An Outline of Theoretical Arithmetic, in Polish]. Biblioteka Matematyczna tom 37. PWN, Warsaw (1971) (2nd ed. (1983))
8. An Outline of Mathematical Logic. Reidel, Dordrecht (1974). English version of Zarys logiki matematycznej [An Outline of Mathematical Logic, in Polish]. Biblioteka Matematyczna tom 20. PWN, Warsaw (1961) (subsequent improved editions up to 6th ed. (1984))
9. Filozofia czasu próby [Philosophy in the Time of Challenge, in Polish]. Édition du Dialogue, Paris (1979) (2nd ed.: PAX, Warsaw (1984))
10. Próba treściowego opisu świata wartości i jej etyczne konsekwencje [An Attempt of a Contetual Description of the World of Values, and Its Ethical Consequences, in Polish]. Zakład Psychologii PAN, Ossolineum, Wrocław (1983)
11. Moralitety [Moral Stories, in Polish]. PAX, Warsaw (1986)
12. Etyka w doświadczeniu wewnętrznym [Ethics in Internal Experience, in Polish]. PAX, Warsaw (1989)
13. Mała propedeutyka filozofii naukowej [Small Introduction to Scientific Philosophy, in Polish] PAX, Warsaw (1989)
14. Życie jako wyzwanie. Wprowadzenie w filozofię racjonalistyczną [Life as a Challenge. Introduction in Rationalistic Philosophy]. IFiS PAN, Warsaw (1993) (2nd ed. (1995))
15. Життя як виклик [Life as a Challenge, in Ukrainian, with modifications]. Scholar, Warsaw—Lvov (1997). Transl. and commentaries by Борис Домбровский, Олег Гірний
16. Жизнь как вызов [Life as a Challenge, in Russian, with modifications]. Вузовская Книга, Moscow (2000). Transl. by Виктор Макаренко
17. Logic—a Human Affair. Scholar, Warsaw (1997)

18. Europa—odkrywanie sensu istnienia [Europe, Dicovering the Sense of Existence, in Polish]. Fundacja Akademii Teologii Katolickiej dla Wspierania Rozwoju Chrześcijańskich Nauk Społecznych i Społecznej Gospodarki Rynkowej, Warsaw (2001)
19. Psychiczna osobliwośćczłowieka [Psychological Peculiarity of Man, in Polish] Scholar, Warsaw (2003)

Selected Papers and Articles

P1. Próba ugruntowania semantyki języka opisowego. Przegląd Filozoficzny **44**, 330–371 (1949)
P2. The pragmatic foundations of semantics. Synthèse **8**, 300–324 (1950–1951)
P3. Undecidability of some topological theories. Fundam. Math. **38**, 137–152 (1951)
P4. Elementary definable analysis. Fundam. Math. **41**, 311–338 (1954)
P5. The systems of Leśniewski in relation to contemporary logical research. Stud. Logica. **3**, 77–97 (1955)
P6. Computable functionals. Fundam. Math. **42**, 168–202 (1955)
P7. On the definition of computable functionals. Fundam. Math. **42**, 232–239 (1955)
P8. Some proofs of undecidability of arithmetic. Fundam. Math. **43**, 166–177 (1956)
P9. On the definitions of computable real continuous functions. Fundam. Math. **44**, 61–71 (1957)
P10. Mostowski, A., Ryll-Nardzewski, C.: The classical and the omega-complete arithmetic. J. Symb. Log. **23**, 188–206 (1958)
P11. Some approaches to constructive analysis. In: Constructivity in Mathematics, pp. 43–61. North Holland, Amsterdam (1959)
P12. Mostowski, A., Ryll-Nardzewski, C.: Definability of sets in models of axiomatic theorems. Bulletin de l'Académie Polonaise des Sciences, Série Math. **9**, 163–167 (1961)
P13. Axiomatizability of geometry without points. Synthèse **XII**(2/3), 228–235 (1960)
P14. A theory without recursive models. Bulletin de l'Académie Polonaise des Sciences, Série Math. **10**, 63–69 (1962)
P15. A philosophically plausible formal interpretation of intuitionistic logic. Indag. Math. **XXVI**, 596–601 (1964)
P16. Recursive objects in all finite types. Fundam. Math. **54**, 73–93 (1964)
P17. Assertions depending on time and corresponding logical calculi. Compos. Math. **20**, 83–87 (1968)
P18. The moral basis for peace: the absolute value of the human individual. Dialect. Hum. **1**(1), 19–28 (1974)
P19. On certain formal consequences of reism. Dialect. Hum. **1**, 73–80 (1977)
P20. Classical, relativistic and constructivist ways of asserting theorems. In: Woleński, J. (ed.) Philosophical Logic in Poland, pp. 19–31. Kluwer, Dordrecht (1994)
P21. Non-violence: wychowanie do negocjacji, demokracji i współistnienia. In: Edukacja wobec wyzwań XXI wieku, pp. 57–92. Komitet Prognoz Polska 2000 PAN, Warszawa (1996)
P22. The paradox of Grelling and Nelson presented as a veridical observation concerning naming. In: Kijania-Placek, K. (ed.) The Lvov-Warsaw School and Contemporary Philosophy, pp. 183–190. Kluwer, Dordrecht (1998)
P23. Is Antipsychologism Still Tenable? In: Woleński, J., Köhler, E. (eds.) Alfred Tarski and the Vienna Circle, pp. 109–114. Kluwer, Dordrecht (1999)
P24. Reizm i analiza logiczna (Spojrzenie w epoce postmodernizmu). In: Polska Filozofia Analityczna, pp. 31–41. Toruń, Poland (1999)
P25. The Vocation of Europe. Dialogue and Universalism **IX**(5–6), 11–41 (1999)
P26. Undecidability without Arithmetization. Stud. Logica. **79**(2), 163–230 (2005)

Selected Articles on Andrzej Grzegorczyk

A1. Krajewski, S.: Andrzej Grzegorczyk (1922–2014) [in Polish]. Stud. Semiotyczne **XXVIII–XXIX**, 63–88 (2015)
A2. Krajewski, S., Woleński, J.: Andrzej Grzegorczyk: Logic and Philosophy. Fundam. Inf. **81**(1–3), 1–10 (2007)

S. Krajewski (✉)
Institute of Philosophy, University of Warsaw, Warsaw, Poland
e-mail: stankrajewski@uw.edu.pl

A Mystery of Grzegorczyk's Logic of Descriptions

Joanna Golińska-Pilarek and Taneli Huuskonen

> *It is important to approach the deeds of previous generations with the right attitude, accepting the fact that their actions were an actual step of their development. Regardless of any alternative ways in which reality could have evolved, we must not belittle their thoughts and actions, but rather augment them, keeping in mind all that our ancestors could not have taken into account. So, let us not claim that we are building a new philosophy on new principles, as we are not essentially wiser than our predecessors, but let us utilize the whole record of the past, carefully considering their whole evolution (biological and cultural). Their experiences form the basis of our growth.*
>
> Andrzej Grzegorczyk

Abstract In 2011, Andrzej Grzegorczyk formulated *Logic of Descriptions* (LD), a new logical system in which the classical equivalence has been replaced with the descriptive equivalence. Two sentences are descriptively equivalent whenever they describe the same state of affairs. Grzegorczyk's logic LD is built from the ground up by revising the axioms of classical propositional logic and rejecting those that do not correspond to the intended interpretation of the descriptive equivalence as the connective expressing equimeaning relations between sentences. Grzegorczyk's last paper, which introduced in detail philosophical motivations of LD and its axiomatization, has become an inspiration for investigating the properties of LD and its various modifications. In this paper we present the basics of Grzegorczyk's logic LD and then we survey the recent results on LD that have shed light on mysterious properties of the Grzegorczyk's descriptive equivalence connective.

Keywords Logical connectives · Paradoxes of classical logic · Descriptive equivalence · Equimeaning · Grzegorczyk's logic of descriptions

Mathematics Subject Classification (2000) Primary 03A05; Secondary 03-02, 03B60

Á. Garrido, U. Wybraniec-Skardowska (eds.), *The Lvov-Warsaw School. Past and Present*, Studies in Universal Logic,
https://doi.org/10.1007/978-3-319-65430-0_51

1 Introduction

Andrzej Grzegorczyk devoted the last 5 years of his life to the study of a new logical system avoiding weaknesses and paradoxes of the classical logical connectives. His main goal was to build a logic determined by the logical structure and properties of descriptions of the world that people use in everyday life. Grzegorczyk emphasized very strongly the extremely positive role played by formal and metamathematical studies in the development of science in the twentieth century. He appreciated the particularly powerful impact of classical logic on the development of information technology and artificial intelligence. Yet at the same time, he felt that the time had come to reject the current paradigm and to propose another way of describing the meaning of logical concepts, not yet used by researchers, and based on the analysis of the role of human language in its entirety to the challenge of the human condition.

The main philosophical assumption of Grzegorczyk's standpoint was that in the human description of the cognized world's phenomena, the roles of negation, conjunction, and disjunction differ significantly from those of implication and equivalence. Negation, conjunction, and disjunction are very primitive and have clear intuitive descriptive meaning, while the classical implication and equivalence are derivative and have no intuitively plausible sense. Furthermore, it is exactly implication and equivalence that are responsible for some paradoxical laws of classical logic, such as "false implies everything", "truth is implied by anything" and "all true sentences are logically equivalent to each other".

As a consequence, states Grzegorczyk, we are forced to accept that among all the logical connectives exactly negation, conjunction, and disjunction—together with the *equimeaning connective* (or *descriptive equivalence*) expressing the assertion that two descriptions have the same meaning—are well suited as the primitive concepts of a new logic. Indeed, those four connectives are crucial and necessary for descriptive practice, that is, for the way people actually describe reality. As *descriptions* and *descriptive equivalences* among them have became crucial for Grzegorczyk's approach, he called his new logical system the *Logic of Descriptions,* or LD for short.

The first exposition of Grzegorczyk's new logic, its philosophical motivations and assumptions was published in 2011 in the paper [5], cf. [6]. In the paper, Grzegorczyk proposed a number of axioms and rules that the equimeaning connective (descriptive equivalence), denoted by $\equiv$, should satisfy and he posed a number of open problems, in particular whether the new connective $\equiv$ is different than the classical equivalence. Grzegorczyk's study on the logic of descriptions was quickly joined by other researchers, in particular the authors of this paper. In 2012 in the paper [3], we published the first results on the formal properties of LD, showing that the descriptive equivalence connective is essentially different than the classical one and the logic itself is indeed new. Further results are presented in the forthcoming paper [4].

Andrzej Grzegorczyk was informed about our results, he actively participated in seminars and meetings devoted to the logic of descriptions, where we had lively discussions about Grzegorczyk's motivations and the adequacy of his axioms. Toward the end of his life, Andrzej Grzegorczyk was working on a deeper exposition of the whole philosophical system underlying the logic LD. He was writing a book concerned with a complete philosophical theory of acquisition of logical connectives, making use of results from empirical sciences and linguistics. Unfortunately, professor Grzegorczyk did not finish this work before his death in 2014.

The aim of the present paper is to survey the current research results on Grzegorczyk's Logic of Descriptions LD and some of its variants. The paper consists of three sections and conclusions. In Sect. 2, we present the basics of the logic LD, its language, axiomatization and semantics. Section 3 is concerned with the exposition of the formal properties of LD, in particular those that we consider to be the most amazing and striking. In Sect. 4 we discuss some extensions and modifications of the logic LD. The paper ends with the conclusions section, in which we list a few open problems, among others.

2 The Logic of Descriptions LD

The logic LD is a propositional logic. The vocabulary of LD consists of the following pairwise disjoint sets of symbols:

- $\mathbb{V} = \{p_0, p_1, p_2, \ldots\}$—an infinite countable set of propositional variables,
- $\{\neg, \wedge, \vee, \equiv\}$—propositional operations of negation $\neg$, conjunction $\wedge$, disjunction $\vee$, and descriptive equivalence $\equiv$.

The set of LD is defined in a standard way as the smallest set that contains all the propositional variables and is closed under the propositional operations of LD. We will use the following four shorthand notations for LD-formulas:

- $(p \to q) \stackrel{\text{df}}{=} (\neg p \vee q)$ (classical implication)
- $(p \leftrightarrow q) \stackrel{\text{df}}{=} (p \to q) \wedge (q \to p)$ (classical equivalence)
- $(p \Rightarrow q) \stackrel{\text{df}}{=} (p \equiv (p \wedge q))$ (descriptive implication)
- $(p \Leftrightarrow q) \stackrel{\text{df}}{=} [(p \Rightarrow q) \wedge (q \Rightarrow p)]$ (quasi descriptive equivalence)

The logic LD has been defined by a Hilbert-style axiomatization given in [5]. It consists of 17 axioms and 4 rules of inference. The axioms are as follows:

(Ax0) $\neg(p \wedge \neg p)$
(Ax1) $p \equiv p$
(Ax2) $\neg\neg p \equiv p$
(Ax3) $p \equiv (p \wedge p)$
(Ax4) $p \equiv (p \vee p)$
(Ax5) $(p \wedge q) \equiv (q \wedge p)$
(Ax6) $(p \vee q) \equiv (q \vee p)$
(Ax7) $(p \wedge (q \wedge r)) \equiv ((p \wedge q) \wedge r)$
(Ax8) $(p \vee (q \vee r)) \equiv ((p \vee q) \vee r)$

(Ax9) $(p \wedge (q \vee r)) \equiv ((p \wedge q) \vee (p \wedge r))$
(Ax10) $(p \vee (q \wedge r)) \equiv ((p \vee q) \wedge (p \vee r))$
(Ax11) $\neg(p \wedge q) \equiv (\neg p \vee \neg q)$
(Ax12) $\neg(p \vee q) \equiv (\neg p \wedge \neg q)$
(Ax13) $(p \equiv q) \equiv (q \equiv p)$
(Ax14) $(p \equiv q) \equiv (\neg p \equiv \neg q)$
(Ax15) $(p \equiv q) \Rightarrow ((p \equiv r) \equiv (q \equiv r))$
(Ax16) $(p \equiv q) \Rightarrow ((p \wedge r) \equiv (q \wedge r))$
(Ax17) $(p \equiv q) \Rightarrow ((p \vee r) \equiv (q \vee r))$

The axioms express basic properties of conjunction, disjunction and negation with respect to the descriptive equivalence. Note that only one axiom does not involve the connective $\equiv$, namely the axiom (Ax0), which reflects a kind of consistency of the logic. The intended meanings of the other axioms are:

- Descriptive equivalence $\equiv$ represents an equivalence relation: (Ax1), (Ax13), and (Ax15).
- Extensionality property of $\equiv$: (Ax14) through (Ax17).
- Idempotency of conjunction and disjunction: (Ax3) and (Ax4).
- Commutativity of conjunction and disjunction: (Ax5) and (Ax6).
- Associativity of conjunction and disjunction: (Ax7) and (Ax8).
- Distributivity of conjunction over disjunction: (Ax9).
- Distributivity of disjunction over conjunction: (Ax10).
- Involution of negation: (Ax2).
- De Morgan laws for negation: (Ax11) and (Ax12).

The rules have the following forms:

$$\text{(MPE)} \quad \frac{\varphi \equiv \psi, \varphi}{\psi} \qquad \text{(Sub)} \quad \frac{\varphi(p_0, \ldots, p_n)}{\varphi(p_0/\psi_0, \ldots, p_n/\psi_n)}$$

$$(\wedge_1) \quad \frac{\varphi, \psi}{\varphi \wedge \psi} \qquad (\wedge_2) \quad \frac{\varphi \wedge \psi}{\varphi, \psi}$$

where $\varphi, \psi, \psi_0, \ldots, \psi_n$ are any LD-formulas and $p_0, \ldots, p_n$ are propositional variables with the additional restriction that the rule (Sub) applies only to axioms. The rules (Sub), $(\wedge_1)$, $(\wedge_2)$ are standard in classical logic. However, it should be emphasized that instead of the classical Modus Ponens rule, the logic LD has a similar inference rule but with respect to the descriptive equivalence operator. Note also that LD does not include any rule for introduction or elimination of disjunction and negation.

The provability of a formula is defined in LD in a standard way. Thus, a formula φ is said to be *provable in* LD ($\vdash \varphi$ for short) whenever there exists a finite sequence $\varphi_1, \ldots, \varphi_n$ of LD-formulas, $n \geq 1$, such that $\varphi_n = \varphi$ and each φ_i, $i \in \{1, \ldots, n\}$, is an axiom or follows from earlier formulas in the sequence by one of the rules of inference. If X is any set of LD-formulas, then φ is said to be LD-*provable from* X ($X \vdash \varphi$ for short) whenever there exists a finite sequence $\varphi_1, \ldots, \varphi_n$ of LD-formulas, $n \geq 1$, such that $\varphi_n = \varphi$ and for each $i \in \{1, \ldots, n\}$, φ_i is an axiom or $\varphi_i \in X$ or φ_i follows from earlier formulas in the sequence by one of the rules of inference.

Observe that if we interpret $\equiv$ as the classical equivalence, then the LD-axioms are classical tautologies and all the rules preserve classical validity. It means that one of the possible models of LD is the two-element Boolean algebra of classical propositional logic. Therefore, $p \equiv \neg p$ is not provable in LD, and hence the logic LD is consistent in the sense that it does not entail all formulas. Moreover, as shown in [4], the logic LD is actually paraconsistent, that is, even a contradiction in LD does not entail all formulas. The proof of this fact uses semantics for LD.

The first sound and complete semantics for LD was introduced in [3]. Then, after some weakening and improvement of the LD-models, the strong soundness and completeness of LD with respect to the class of paraconsistent LD-models was proved in the paper [4]. Generally, models of the logic LD are based on the so-called *Grzegorczyk algebras* satisfying some further conditions. Now, following presentations given in [3] and [4], we present the definition of LD-models in detail.

A structure $(U, \oplus, \otimes)$ is said to be a *distributive bisemilattice* whenever the following hold, for all $a, b, c \in U$ and for any $\odot \in \{\otimes, \oplus\}$:

- $a \odot b = b \odot a$, (commutativity of $\otimes, \oplus$)
- $a \odot (b \odot c) = (a \odot b) \odot c$, (associativity of $\otimes, \oplus$)
- $a \odot a = a$, (idempotency of $\otimes, \oplus$)
- $a \oplus (b \otimes c) = (a \oplus b) \otimes (a \oplus c)$, (distributivity of $\otimes$ over $\oplus$)
- $a \otimes (b \oplus c) = (a \otimes b) \oplus (a \otimes c)$. (distributivity of $\oplus$ over $\otimes$)

A *de Morgan bisemilattice* is a structure $(U, \sim, \oplus, \otimes)$ such that $(U, \oplus, \otimes)$ is a distributive bisemilattice and for all $a, b \in U$, the following hold:

- $\sim\sim a = a$,
- $\sim(a \oplus b) = \sim a \otimes \sim b$.

A *Grzegorczyk algebra* is a structure $(U, \sim, \oplus, \otimes, \circ)$ such that $(U, \sim, \oplus, \otimes)$ is a de Morgan bisemilattice and for all $a, b, c \in U$, the following hold:

- $a \circ b = b \circ a$,
- $a \circ b = \sim a \circ \sim b$,
- $a \circ b = (a \circ b) \otimes ((a \circ c) \circ (b \circ c))$,
- $a \circ b = (a \circ b) \otimes ((a \oplus c) \circ (b \oplus c))$,
- $a \circ b = (a \circ b) \otimes ((a \otimes c) \circ (b \otimes c))$.

Fact 2.1 *A structure $(U, \sim, \oplus, \otimes, \circ)$ is a Grzegorczyk algebra if and only if the following conditions hold, for all $a, b, c \in U$:*

(LD1) $a \circ b = b \circ a$,
(LD2) $a \circ b = (a \circ b) \otimes ((a \circ c) \circ (b \circ c))$,
(LD3) $a \circ b = \sim a \circ \sim b$,
(LD4) $a \circ b = (a \circ b) \otimes ((a \oplus c) \circ (b \oplus c))$,
(LD5) $a \circ b = (a \circ b) \otimes ((a \otimes c) \circ (b \otimes c))$,
(LD6) $a \oplus b = b \oplus a$,
(LD7) $a \oplus (b \oplus c) = (a \oplus b) \oplus c$,
(LD8) $a \oplus a = a$,
(LD9) $a \otimes b = b \otimes a$,
(LD10) $a \otimes (b \otimes c) = (a \otimes b) \otimes c$,

(LD11) $a \otimes a = a$,
(LD12) $a \otimes (b \oplus c) = (a \otimes b) \oplus (a \otimes c)$,
(LD13) $a \oplus (b \otimes c) = (a \oplus b) \otimes (a \oplus c)$,
(LD14) $\sim(a \oplus b) = \sim a \otimes \sim b$,
(LD15) $\sim(a \otimes b) = \sim a \oplus \sim b$,
(LD16) $\sim\sim a = a$.

It is worth to emphasize the following fact:

Fact 2.2 *Boolean algebras, Kleene algebras, and de Morgan algebras are Grzegorczyk algebras.*

The converse of the above does not hold. The class of Grzegorczyk algebras contains subclasses that form bases for semantics of various non-classical logics of different types. This shows how different from other well known non-classical logics the logic LD is. Grzegorczyk algebras are a base for structures of LD.

Definition 2.3 A *paraconsistent* LD*-structure* is a structure of the form

$$\langle U, \sim, \oplus, \otimes, \circ, D\rangle, \text{ where}$$

- U, D are non-empty sets such that $D \subsetneq U$,
- $\langle U, \sim, \oplus, \otimes, \circ\rangle$ is a Grzegorczyk algebra,
- for all $a, b \in U$, the following hold:
  - $a \otimes b \in D$ if and only if $a \in D$ and $b \in D$,
  - $a \circ b \in D$ if and only if $a = b$,
  - $\sim(a \otimes \sim a) \in D$.

A paraconsistent LD-structure is called a *classical* LD*-structure* if there is no $a \in U$ such that $a \otimes \sim a \in D$. In what follows, paraconsistent LD-structures will be also referred to as LD-structures or LD-models.

Definition 2.4 Let $\mathcal{M} = \langle U, \sim, \oplus, \otimes, \circ, D\rangle$ be an LD-structure. A *valuation* on $\mathcal{M}$ is any mapping $v : \mathbb{V} \to U$ such that for all LD-formulas φ and ψ:

- $v(\neg\varphi) = \sim v(\varphi)$,
- $v(\varphi \wedge \psi) = v(\varphi) \otimes v(\psi)$,
- $v(\varphi \vee \psi) = v(\varphi) \oplus v(\psi)$,
- $v(\varphi \equiv \psi) = v(\varphi) \circ v(\psi)$.

A formula φ is said to be *satisfied in* $\mathcal{M}$ *by a valuation* v if and only if $v(\varphi) \in D$. It is *true in* $\mathcal{M}$ whenever it is satisfied in $\mathcal{M}$ by all the valuations on $\mathcal{M}$, and it is LD*-valid* if it is true in all paraconsistent LD-structures.

Definition 2.5 Let X and φ be a set of LD-formulas and a single LD-formula, respectively. The formula φ is a *semantic consequence* of X, denoted by $X \models_{\mathsf{LD}} \varphi$, if for every paraconsistent LD-structure $\mathcal{M}$ and every valuation v in $\mathcal{M}$ such that $\mathcal{M}, v \models X$, it holds that $\mathcal{M}, v \models \varphi$.

In [3] soundness and completeness with respect to the classical LD-structures is proved, while in [4] it has been proved that LD is strongly sound and complete with respect to the paraconsistent LD-structures. Hence, the following holds:

Theorem 2.6 (Soundness and Completeness of LD) *For every LD-formula φ the following conditions are equivalent:*

1. *φ is LD-provable.*
2. *φ is LD-valid.*
3. *φ is true in all classical LD-structures.*

Theorem 2.7 *Let X and φ be a set of LD-formulas and a single LD-formula, respectively. Then, the following conditions are equivalent:*

1. $X \vdash_{\mathsf{LD}} \varphi$.
2. $X \models_{\mathsf{LD}} \varphi$.

The above theorem provides a way of proving that the logic LD is paraconsistent in the following sense:

Definition 2.8 A logic is *paraconsistent* (or *contradiction-tolerant*) iff there are formulas φ, ψ such that $\varphi \wedge \neg\varphi \nvdash \psi$.

In [4] the following theorem is proved:

Theorem 2.9 *The logic LD is paraconsistent. In particular, $p \wedge \neg p \nvdash_{\mathsf{LD}} q$.*

Proof Let $\mathcal{M} = (U, \sim, \oplus, \otimes, \circ, D)$ be a paraconsistent LD-structure defined as:

$$U = \{0, 1, 2\},$$

$$\sim a = 2 - a,$$

$$a \oplus b = \max(a, b),$$

$$a \otimes b = \min(a, b),$$

$$a \circ b = \begin{cases} 1, & \text{if } a = b, \\ 0, & \text{otherwise;} \end{cases}$$

$$D = \{1, 2\}.$$

Now, let $v(p) = 1$, $v(q) = 0$. Then, it is easy to show that $v(p \wedge \neg p) = 1 \in D$, but $v(q) = 0 \notin D$. Hence, the logic LD is paraconsistent. □

The result on the semantics shows that LD has much in common with the non-Fregean logic SCI introduced by Suszko in [7], cf. [8]. Indeed, the logic LD can be seen as *non-Fregean* in the sense that it rejects the main assumption of classical Fregean logic, according to which sentences with the same truth value have the same denotations. Although the logics LD and SCI share the language and have similar philosophical motivations, they differ considerably in the formalization and they are *different logics*, as we will see.

3 Properties of LD

The first results on provable and unprovable formulas presented in [3] have shown in particular that: (1) the logic LD is new, in the sense that its class of valid formulas cannot be identified with valid formulas of any other well known non-classical logic, (2) the descriptive equivalence and implication connectives are different from the classical ones, (3) neither the absorption nor the boundedness laws are provable in LD.

In [3] the following has been proved:

Theorem 3.1 *The following formulas are not provable in* LD*:*

1. $(\varphi \equiv \psi) \leftrightarrow (\varphi \leftrightarrow \psi)$
2. $(\varphi \Rightarrow \psi) \leftrightarrow (\varphi \rightarrow \psi)$
3. $(\varphi \equiv \varphi) \equiv (\psi \equiv \psi)$
4. $(\varphi \vee (\varphi \wedge \psi)) \equiv \varphi$
5. $(\varphi \wedge (\varphi \vee \psi)) \equiv \varphi$
6. $(\varphi \vee \neg\varphi) \equiv (\psi \vee \neg\psi)$
7. $(\varphi \wedge \neg\varphi) \Rightarrow \psi$
8. $\varphi \Rightarrow (\psi \vee \neg\psi)$
9. $\neg(\varphi \equiv \neg\varphi)$

Proof By way of example, we will show that the absorption laws, i.e., formulas 4 and 5 of the theorem, are not provable in LD.

Let $\mathcal{M} = (U, \sim, \oplus, \otimes, \circ, D)$ be a paraconsistent LD-structure such that $U = \{0, 1, 2, 3\}$, $D = \{2, 3\}$ and the operations are defined as follows:

$\sim$	0	1	2	3
	3	2	1	0

$\circ$	0	1	2	3
0	3	0	0	0
1	0	3	0	0
2	0	0	3	0
3	0	0	0	3

$\otimes$	0	1	2	3
0	0	0	0	0
1	0	1	1	0
2	0	1	2	3
3	0	0	3	3

$\oplus$	0	1	2	3
0	0	0	3	3
1	0	1	2	3
2	3	2	2	3
3	3	3	3	3

It can be verified that the above tables indeed define a Grzegorczyk algebra. Then, the absorption laws do not hold, as $1 \oplus (1 \otimes 0) = 1 \otimes (1 \oplus 0) = 0 \neq 1$. Therefore, the formulas $(p \vee (p \wedge q)) \equiv p$ and $(p \wedge (p \vee q)) \equiv p$ are not satisfied in $\mathcal{M}$ by a valuation v such that $v(p) = 1$ and $v(q) = 0$, which by Theorem 2.6 means that these formulas are not provable in LD. Proofs for other formulas can be found in [3]. □

Observe the following particular consequences of Theorem 3.1:

- The descriptive equivalence and implication are different from the classical ones (unprovability of formulas 1 and 2).
- True descriptive equivalences may be not identical (unprovability of formula 3).
- The absorption laws are not theorems of LD (unprovability of formulas 4 and 5).
- True sentences are not necessarily descriptively equivalent (unprovability of formula 6).
- False does not imply everything (unprovability of formula 7).
- Not everything implies the truth (unprovability of formula 8).
- Negation and disjunction do not behave in a classical way, which is clearly manifested by the unprovability of formula 9. Indeed, $\neg(p \equiv \neg p)$ is not a theorem of LD, while it can be easily observed that the formula $p \equiv \neg p$ is not satisfiable in any LD-model and, on the other hand, the formula $\neg(p \equiv \neg p) \vee (p \equiv \neg p)$ is valid in LD.

This shows that LD can be seen as a logic that avoids the classical paradoxes of implication. On the other hand, LD shares many of the classical laws.

Theorem 3.2 *The following formulas are* LD*-valid:*

1. $\varphi \vee \neg\varphi$
2. $(\varphi \wedge (\varphi \rightarrow \psi)) \rightarrow \psi$[1]
3. $(\varphi \equiv \psi) \Rightarrow (\varphi \Leftrightarrow \psi)$

Moreover, $(\varphi \Leftrightarrow \psi) \vdash_{\mathsf{LD}} (\varphi \equiv \psi)$ *and the following rules are admissible in* LD*:*

$$\text{(tran)}\ \frac{\varphi \equiv \psi,\ \psi \equiv \theta}{\varphi \equiv \theta} \qquad (\vee)\ \frac{\varphi}{\varphi \vee \theta}$$

The proofs of the above properties can be found in [3] and [4].

The LD-validity of the formula $(\varphi \equiv \psi) \Rightarrow (\varphi \Leftrightarrow \psi)$ and the fact that $(\varphi \Leftrightarrow \psi) \vdash_{\mathsf{LD}} (\varphi \equiv \psi)$ connect the descriptive equivalence $\equiv$ with the quasi descriptive equivalence operator $\Leftrightarrow$. Recall that it is defined as follows:

$(p \Leftrightarrow q) \stackrel{\text{df}}{=} [(p \Rightarrow q) \wedge (q \Rightarrow p)]$

As in classical logic, it could be expected that the descriptive equivalence can be expressed as the conjunction of two implications, which would mean that the descriptive and quasi descriptive equivalences are indistinguishable in LD. However, as shown in [4] it does not hold, since the following theorem is true:

Theorem 3.3 *The following formula is not* LD*-valid:*

$$(p \equiv q) \equiv (p \Leftrightarrow q).$$

[1] Recall that $\varphi \rightarrow \psi$ is short for $\neg\varphi \vee \psi$.

In [4] the following specific formulas have been discussed:

$$[(p \equiv q) \wedge (q \equiv r)] \Rightarrow (p \equiv r) \quad \text{(AxT)}$$
$$((p \equiv q) \wedge p) \equiv ((p \equiv q) \wedge q) \quad \text{(AxD)}$$

The formula (AxT) expresses a strong form of transitivity of $\equiv$. The formula (AxD) is called the *Delusion Axiom* in reference to an example used to illustrate its intuitive content. Consider someone who believes that Vladimir Putin and Donald Trump are the same person, just wearing different masks. Such a deluded person should accept that the following two statements say the same:

(α) *Putin is the president of Russia*
(β) *Trump is the president of Russia*

Consequently, she would accept that the following statements are descriptively equivalent:

1. $\alpha \equiv \beta$ and α.
2. $\alpha \equiv \beta$ and β.

The acceptance of the identity of these claims above means evaluating a conjunction involving a delusion of identity as if the identity were true, which is expressed by axiom (AxD).

In [4] is has been shown that neither (AxT) nor (AxD) is a theorem of LD.

Theorem 3.4 *The following formulas are not provable in* LD*:*

1. $[(p \equiv q) \wedge (q \equiv r)] \Rightarrow (p \equiv r)$
2. $((p \equiv q) \wedge p) \equiv ((p \equiv q) \wedge q)$

There has been some controversy over the last three axioms of LD:

(Ax15) $(p \equiv q) \Rightarrow ((p \equiv r) \equiv (q \equiv r))$
(Ax16) $(p \equiv q) \Rightarrow ((p \wedge r) \equiv (q \wedge r))$
(Ax17) $(p \equiv q) \Rightarrow ((p \vee r) \equiv (q \vee r))$

One of their intended meanings is to express the extensionality principle for descriptions:

> Sentences that have equal meaning (have the same content or are the same descriptions) are interchangeable in all possible contexts.

Axioms (Ax15), (Ax16), (Ax17) seem to express this property in a stronger form. Their adequacy has been deeply discussed in [2] and the conclusion presented there is rather negative. Instead a new very weak *Minimal Grzegorczyk Logic* MGL has been introduced with a rule expressing the extensionality principle instead of the axioms. However, this account goes beyond the scope of this paper, so the interested reader should consult [2].

On the other hand, in the paper [4] other ways of expressing the extensionality principle have been explored. We start this discussion with the formal definitions.

Definition 3.5 The *Weak Extensionality Principle* (WEP) is the following metarule:

$$\frac{\varphi \equiv \psi}{\vartheta(p/\varphi) \equiv \vartheta(p/\psi)},$$

where φ, ψ, and ϑ are arbitrary formulas.

Definition 3.6 The *Strong Extensionality Principle* (SEP) is the claim

$$\vdash (\varphi \equiv \psi) \Rightarrow (\vartheta(p/\varphi) \equiv \vartheta(p/\psi)),$$

where φ and ψ are arbitrary formulas, and $\vartheta(p)$ is a formula in which p actually occurs.

Definition 3.7 The *Grzegorczyk Extensionality Principle* (GEP) is the statement

$$\vdash (\bar{\varphi} \equiv \bar{\psi}) \Rightarrow (\vartheta(p_1/\varphi_1, \ldots, p_n/\varphi_n) \equiv \vartheta(p_1/\psi_1, \ldots, p_n/\psi_n)),$$

where $\varphi_1, \ldots, \varphi_n, \psi_1, \ldots, \psi_n$ are arbitrary formulas, ϑ is a formula whose propositional variables are contained in $\{p_1, \ldots, p_n\}$, and $\bar{\varphi} \equiv \bar{\psi}$ is short for $(\varphi_1 \equiv \psi_1) \wedge \ldots \wedge (\varphi_n \equiv \psi_n)$.

Observe that SEP implies the WEP and the GEP implies the WEP for a given logic. In [4] the following have been proved

Theorem 3.8

1. *Neither the SEP nor the GEP holds for* LD.
2. *The WEP holds for* LD.

In the next section we will discuss extensions of LD with axioms (AxT) and (AxD) as well as some other modifications of LD, together with their properties and mutual relationships.

4 Extensions and Modifications of LD

In the previous section we discussed the most important properties of LD. In particular, we mentioned that the transitivity axiom (AxT) and the delusion axiom (AxD) are not provable in LD. Hence, it was quite natural to consider extensions of LD with these axioms, which were deeply studied in the paper [4].

Logics obtained by adding (AxT) and (AxD) to the axiomatization of LD are denoted by LDT and LDD, respectively. The notions of provability in LDT and LDD are defined in a similar way as in LD. Models of LDT and LDD are defined as paraconsistent LD-models that satisfy the semantic counterparts of axioms (AxT) and (AxD), respectively. Then, the notions of satisfaction, truth and validity in LDT and LDD are defined in a standard way.

Theorem 4.1

1. $\vdash_{\mathsf{LDD}}$ (AxT).
2. $\nvdash_{\mathsf{LDT}}$ (AxD).
3. $\vdash_{\mathsf{LDT}} (p \equiv q) \equiv (p \Leftrightarrow q)$.

The above theorem is proved in [4]. Note that the last formula is not provable in LD, while it is in LDT. Thus, LDT seems to be a better candidate than LD to compare Grzegorczyk's approach with various other non-classical logics, which are usually defined mainly in terms of implication.

In [4] the following is proved:

Theorem 4.2

1. $\nvdash_{\mathsf{LD}} p \Rightarrow (p \vee q)$.
2. $\nvdash_{\mathsf{LDT}} p \Rightarrow (p \vee q)$.
3. $\nvdash_{\mathsf{LDD}} p \Rightarrow (p \vee q)$.

The above theorem allows us to show that LD, LDT, and LDD are different from several non-classical logics. Recall that $p \rightarrow (p \vee q)$ is provable in intuitionistic logic and relevance logics T, E, R, EM, and RM, see e.g.,[1]. Hence, if we identify the descriptive implication $\Rightarrow$ with the implication of the other logic, then LD, LDT, and LDD are different from any of the aforementioned non-classical logics. Moreover, the descriptive equivalence $\equiv$ cannot be identified with a necessary equivalence in any class of Kripke frames since $\Box(p \leftrightarrow (p \wedge (p \vee q)))$ is true in all frames.

Definition 4.3 Let L and L′ be logics. A logic L is said to be:

- *not stronger than* L′, $\mathsf{L} \leq \mathsf{L}'$ for short, whenever all formulas valid in L are valid in L′,
- *equal with* L′, $\mathsf{L} = \mathsf{L}'$ for short, $\mathsf{L} \leq \mathsf{L}'$ and $\mathsf{L}' \leq \mathsf{L}$,
- *weaker than* L′, $\mathsf{L} < \mathsf{L}'$ for short, whenever $\mathsf{L} \leq \mathsf{L}'$ and $\mathsf{L} \neq \mathsf{L}'$,
- *uncomparable with* L′ whenever $\mathsf{L} \nleq \mathsf{L}'$ and $\mathsf{L}' \nleq \mathsf{L}$.

Comparing provable formulas in the logics LD, LDT, and LDD, we can state the following:

Fact 4.4 $\mathsf{LD} < \mathsf{LDT} < \mathsf{LDD}$.

As noted in the previous section, the question whether the last three axioms of LD adequately express the extensionality principle has caused some controversy. This has led to some modifications of LD in which these three axioms have been substituted with alternative forms. The most important ones, explored in the paper [4], are the logics LE and LDS obtained by replacing the axioms (Ax15), (Ax16), (Ax17) with the following formulas, respectively:

$$(\mathsf{Ax15})_{\mathsf{LE}} \quad ((p \equiv q) \wedge (p \equiv r)) \equiv ((p \equiv q) \wedge (q \equiv r))$$
$$(\mathsf{Ax16})_{\mathsf{LE}} \quad ((p \equiv q) \wedge (p \wedge r)) \equiv ((p \equiv q) \wedge (q \wedge r))$$
$$(\mathsf{Ax17})_{\mathsf{LE}} \quad ((p \equiv q) \wedge (p \vee r)) \equiv ((p \equiv q) \wedge (q \vee r))$$

$$(\mathsf{Ax15})_{\mathsf{LDS}} \quad ((p \equiv q) \wedge (r \equiv s)) \Rightarrow ((p \equiv r) \equiv (q \equiv s))$$
$$(\mathsf{Ax16})_{\mathsf{LDS}} \quad ((p \equiv q) \wedge (r \equiv s)) \Rightarrow ((p \wedge r) \equiv (q \wedge s))$$
$$(\mathsf{Ax17})_{\mathsf{LDS}} \quad ((p \equiv q) \wedge (r \equiv s)) \Rightarrow ((p \vee r) \equiv (q \vee s))$$

Table 1 The semilattice of Grzegorczyk's logics of descriptions

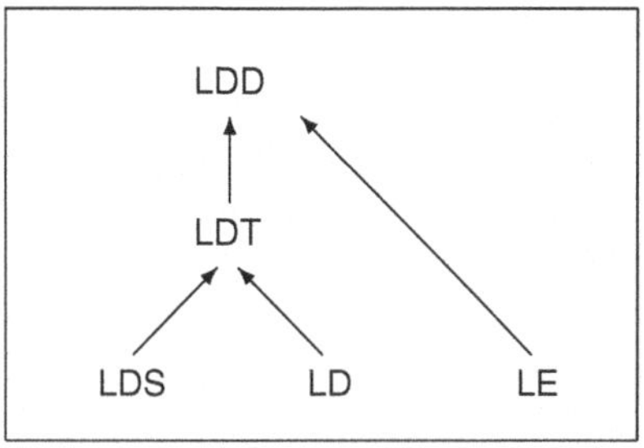

Table 2 Extensionality principles in Grzegorczyk's logics of descriptions

	LD	LDT	LDD	LE	LDS
WEP	+	+	+	−	+
SEP	−	+	+	−	−
GEP	−	+	+	−	+

The axioms for LDS are essentially copied from Suszko's Sentential Calculus of Identity, substituting the descriptive implication for the classical one.

Below we list some properties of the logics under consideration. Their proofs can be found in [4].

Theorem 4.5

1. *The formula* $(p \equiv q) \equiv ((p \equiv q) \wedge (q \equiv q))$ *is provable in* LE, *but not in* LD.
2. LE $\leq$ LDD.
3. LE $\not\leq$ LDT.
4. LE *is uncomparable with* LD *and* LDT.
5. LDS *is paraconsistent.*
6. LDS *and* LD *are uncomparable.*
7. LDS $<$ LDT.
8. *The GEP holds for* LDS.
9. *The SEP holds for* LDT.
10. *The WEP does not hold for* LE.

All these results lead to the following:

Theorem 4.6

1. LD $<$ LDT $<$ LDD
2. LDS $<$ LDT $<$ LDD
3. LE $<$ LDD

Hence, we have the following picture of dependencies among the logics LD, LDT, LDD, LE, LDS (Table 1):
We end this section with two summary tables (Table 2):

Table 3 Validity of some special axioms

	(AxT)	(AxD)	$(p \equiv q) \equiv (p \Leftrightarrow q)$	LD	LE	LDS
LD	−	−	−	+	−	−
LDT	+	−	+	+	−	+
LDD	+	+	+	+	+	+
LE	−	?	?	−	+	−
LDS	−	−	?	−	−	+

5 Conclusions

We have discussed the logic LD of descriptions together with its extensions LDT and LDD as well as the modifications LE and LDS. We have presented a strongly sound and complete semantics for LD, and used it to show that LD is a paraconsistent logic. Then, we have listed examples of classical laws which are provable and not provable in LD. We have reported what extensionality properties the logic LD has. Finally, we have presented formal properties of extended and modified versions of LD and stated the relationships among them (Table 3).

Still there are many open problems concerning the logics in question. Below we list a few of them. In what follows, L stand for any of LD, LE, LDS, LDT, or LDD.

1. Is L decidable? If so, what is its complexity?
2. Is L equivalent to a previously known paraconsistent logic? If not, how does it relate to them?
3. How is a logic obtained by replacing $\equiv$ with $\Leftrightarrow$ in L related to L?
4. The last three axioms of LE express a property that resembles extensionality. Can this property and its relationship with actual extensionality be formulated in an informal, intuitive way?

References

1. Anderson, A.R., Belnap, N.D.: Entailment. The Logic of Relevance and Necessity, vol. I. Princeton University Press, Princeton (1975)
2. Golińska-Pilarek, J.: On the minimal non-Fregean Grzegorczyk's logic. Stud. Logica. **104**(2), 209–234 (2015)
3. Golińska-Pilarek, J., Huuskonen, T.: Logic of descriptions. A new approach to the foundations of mathematics and science. Stud. Log. Grammar Rhetor. **27**(40), 63–94 (2012)
4. Grzegorczyk's non-Fregean logics and their formal properties. In: Urbaniak, R., Payette, G. (eds.) Applications of Formal Philosophy: The Road Less Travalled, Logic, Argumentation & Reasoning, vol. 14. Springer (2017)
5. Grzegorczyk, A.: Filozofia logiki i formalna LOGIKA NIESYMPLIFIKACYJNA [Philosophy of Logic and Perceptive Equivalence]. Zagadnienia Naukoznawstwa **4**(190), 445–450 (2011)
6. Grzegorczyk, A.: Errata to Grzegorczyk, A.: Filozofia logiki i formalna LOGIKA NIESYMPLIFIKACYJNA [Philosophy of Logic and Perceptive Equivalence]. Zagadnienia Naukoznawstwa **4**(190), 445–450 (2011). Zagadnienia Naukoznawstwa **4**(190) (2011)

7. Suszko, R.: Non-Fregean logic and theories. Analele Universitatii Bucuresti, Acta. Logica. **9**, 105–125 (1968)
8. Suszko, R.: Semantics for the sentential calculus with identity. Stud. Logica. **28**, 77–81 (1971)

J. Golińska-Pilarek (✉) • T. Huuskonen
Institute of Philosophy, University of Warsaw, Warsaw, Poland
e-mail: j.golinska@uw.edu.pl

Roman Suszko: Logician and Philosopher

Mieczysław Omyła

Abstract The paper presents basic information about life and main scientific achievements of Roman Suszko (1919–1979). He is known as one of the first logicians who applied the model theory to non-mathematical problems, particularly to study development of knowledge. The article is divided into five sections: 1. Life, 2. Main papers, 3. Views, 4. Influences, 5. Summary.

Keywords Roman Suszko · Logical material · Diachronic logic · Non-Fregean logic

Mathematics Subject Classification (2000) Primary 01A60; Secondary 03-03

1 Life

Roman Suszko was born on 9th of November 1919 in Podobora near Cieszyn in Poland (at present—Czech Republic).

Roman Suszko started his studies at Poznań University in 1937. In 1939, with outbreak of Second World War, he was displaced to Cracow, where he spent the whole of the occupation period; he worked as a night guard, telephonist and—at the same time—he attended the clandestine study classes of physics, mathematics and philosophy under the guidance of professors of Jagiellonian University. In 1945 he obtained a title of Master of Arts in Philosophy for the work *The achievements of polish logic* promoted by Professor Zygmunt Zawirski.

In 1946–1953 he worked at the Chair of the Theory and Methodology of Science, Faculty of Mathematical-Natural Sciences, University of Poznań. In 1948 he was awarded a title doctor for the thesis *On Normal Systems and Few Questions of Basic Logic* written under the supervision of Professor Kazimierz Ajdukiewicz. The thesis was published in the form of two articles *On Analytical Axioms and Logical Rules* and *From the Theory of Definitions*. In 1951 he was awarded a postdoctoral degree in logic also on the Faculty of Mathematical-Natural Sciences, University of Poznań for the dissertation *Canonic Axiomatic systems*, Studia Philosophica IV, Poznań 1951. In 1953 he moved to Warsaw and started to work at the Faculty of Philosophy at the University of Warsaw. From the very beginning of his work in Warsaw he had close contact with the Polish Academy

Á. Garrido, U. Wybraniec-Skardowska (eds.), *The Lvov-Warsaw School. Past and Present*, Studies in Universal Logic,
https://doi.org/10.1007/978-3-319-65430-0_52

of Sciences, at the beginning only with the Group of Algebra headed by Jerzy Łoś at the Institute of Mathematics, and then with the Group of Logic headed by Kazimierz Ajdukiewicz at the Institute of Philosophy and Sociology PAN.

In 1959 he was awarded a title of associate professor. In 1960 he was elected to the position of a Dean of the Faculty of Philosophy at the Warsaw University. In 1960 he moved to the Institute of Philosophy and Sociology of PAN, where in 1966–1969 he was a head of the Group of Logic Institute of Philosophy and Sociology Polish Academy of Sciences. In 1967–1969, 1970–1973 he worked as a professor at the Stevens Institute of Technology in Hoboken, New Jersey, USA. After returning from the USA in 1973 till his death he worked at the Group of Logic Institute Philosophy and Sociology Polish Academy of Sciences. Besides scientific and teaching activity Suszko played also important role as a publisher: he was one of the founders of Studia Logica—journal on logic with international coverage—and till the end he was a member of its Programming Board.

He died on cancer on 3 June 1979 in Warsaw.

2 Main Papers

Roman Suszko's works were published in prestigious international journals of mathematics, logic and philosophy, such as: "Fundamenta Mathematicae", "Journal Symbolic Logic", "Colloquium Mathematicum", "Synthese", "Theoria", "Logique et Analyse", "Archiv für anathematise Logik und Grundlagen-Forschung ", "Studia Logica" and others.

From the scientific achievement of Suszko we select those publications, which influenced greatly the choice scientific interest and investigation of many logicians, philosophers and mathematicians, both in Poland and abroad.

[1] Canonic axiomatic systems. Studia Philosophica **IV**, 301–330 (1951)
[2] On the extending of models. II: Common extensions (together with J. Łoś). Fund. Math. **XLII**, 343–347 (1955)
[3] Formalna teoria wartości logicznych I [A formal theory of logical values I]. Studia Logica **VI**, 145–237 (1957)
[4] On the extending of models. IV: Infinite sums of models (with J. Łoś). Fund. Math. **XLIV/1**, 52–60 (1957)
[5] Remarks on sentential logics (with J. Łoś). Indagationes Mathematicae **20**, 177–183 (1958)
[6] Syntactic structure and semantical reference II. Studia Logica **IX**, 63–91 (1960)
[7] Wykłady z logiki formalnej. Część I [Lectures on formal logic]. PWN, Warsaw (1965). Ed. B. Stanosz
[8] Formal logic and the development of knowledge. In: Problems in the Philosophy of Sciences, International Colloquium London, pp. 210–222. North-Holland, Amsterdam (1965)
[9] Ontology in the Tractatus of L. Wittgenstein. Notre Dame Journal of Formal Logic **9**, 7–33 (1968)

[10] Non-Fregean logic and theories. Analele Universitatii Bucuresti, Acta Logica **11**, 105–125 (Russian translation of that work is in the collection of articles Neklassičeskaja logika, Moscov 1970)
[11] Identity connective and modality. Studia Logica **XXVII**, 7–39 (1971)
[12] Abstracts Logics (with D.J. Brown). Dissertationes Mathematicae **CII**, PWN, Warsaw (1973)
[13] Abolition of the Fregean Axiom. Lecture Notes in Mathematics **453**, 169–239 (1975)
[14] The Fregean axiom and Polish mathematical logic in the 1920s. Studia Logica **36**, 376–380 (1977)
[15] The reification of situations. In: Woleński, J. (ed.) Philosophical Logic in Poland, pp. 247–270. Kluwer Academic Publishers (1994). Translation from Polish by T. Stazeski

Suszko's scientific work is very diverse and strictly connected with his philosophical views, and above all—with his views on philosophy of language and of logic.

In his doctoral thesis Suszko considered with so called "problem of logic without axioms". He refers—from one side—to the theory of language presented by K. Ajdukiewicz in *O znaczeniu wyrażeń* (On the meaning expressions), according to which one can among the semantic rules of a language distinguish: axiomatic, deductive and empirical rules, and—from other side—he refers to works written by Jaśkowski and Gentzen, which dealt with replacing logical axioms with sets of inference rule.

According to Suszko the rules offered by them are not "proper inference rules", because they let to infer infinitely many logical theorems without any assumptions. According to Ajdukiewicz they are at the same time axiomatic inference rules.

Another important thread in Suszko's work was the problem of connections between logical matrixes and sentential calculi. This subject appear again in many Suszko's works written at different age, devoted to both abstract and specific logics. In the work *Remarks on sentential logics* Łoś and Suszko formulated the general notion of strong adequacy of a matrix with respect to abstract sentential calculi and formulated general conditions of existence of such a matrix for any calculus. The work is still quoted in almost all publications on sentential calculus. Suszko in cooperation with J. Łoś wrote certain number of works on models for first-order predicate languages. The results obtained in those works Suszko used later in his research on formal aspects of development of knowledge. According to Suszko certain epistemological problems, which were investigated traditionally in an intuitive way, contemporary logic can search in a strict and formal way. It is possible thanks to the fact that logical semantics—and especially the model theory—is a branch of logic. In Suszko's opinion logic is strictly connected with epistemology, and he considered the mathematical model theory as a formal part of the theory of knowledge. Application of logic to the research on development of knowledge was called by Suszko "diachronic logic". He devoted to the diachronic logic certain number of publications.

Introducing non-Fregean logic was considered by Suszko as his most important achievement. The name "non-Fregean logic" is derived from the fact that Suszko rejected Frege's assumption, according to which all true sentences have one common semantic correlate, just as all false sentences have one common semantic correlate. Those assumption Suszko called "Frege's axiom" and more strictly—"semantic version of

Frege's axiom". The common correlate of true sentences is usually interpreted as: truth, being or universe of a model, which means: all objects that satisfy any true sentence. Analogically the common correlate of all false sentences is usually interpreted as: false, non-being, all objects that belong to the universe of model and that satisfy any false sentence. The term non-Fregean logic was introduced by Suszko in [10].

The "non-Fregean period" (1966–1979) was the most creative period of Suszko's life. He wrote 36 scientific works and promoted seven doctors. However for him it was also the time of tragedy, because during this period his 15-years-old son was killed in an accident and his 24-years-old stepson died in tragic circumstances.

3 Views

In Roman Suszko's logical writings there are many remarks and reflections on the idea of logic which is closely related to his work in formal logic.

According to Suszko the subject of logical investigations are any conceptual constructions that came into being as the result of the cognition of the world. The totality of those constructions is called by Suszko *logical material*. The state of logical studies at any time is determined by the logical material available as well as the research tools at the hand. Among various research tools used for studying logical material Suszko—in agreement with the trends of contemporary logic—gives priority to mathematical instruments, especially he prefers the set-theoretical, algebraic and topological methods.

According to Suszko logic—and especially semantics—is not the science devoid of assumptions but, on the contrary, logic has at its fundament certain general and schematic knowledge of reality, such as: there are objects, which has properties and which are related one with another; there are certain states of affairs and some of them occur, while other do not occur etc. In Suszko's opinion logic gains those knowledge from the set theory. Since the set theory contains certain ontological presumptions of logic, Suszko used to call this theory "formal ontology". What is essential in this view, is that there are certain general, structural, and—at the same time—formal properties of the world to be investigated by ontology.

Suszko took the view, that formal languages being designed and studied by logic are not free creations, but abstracts derived from common languages and languages of particular sciences. They may also constitute hypothetical presumptions of those languages. We investigate the consequences of those presumptions within logic. "Science, cognitive process and natural language, which play essential role in this process, are frame of reference for formal logical research." wrote Suszko in his *Formal theory of logical values I*. According to Suszko logic investigates language by means of all formal tools, i.e. by means of any mathematical method available.

Suszko distinguished two basic categories of expressions: sentences and names. All other categories have no any independent meaning but are designed for construction of more complex expressions. Those auxiliary expressions were called by Suszko "formators"; there are formators that do not bind variables (e.g. connectives, predicates) and formators that bind variables (such as quantifiers and description operators). Suszko was convinced that there is certain logico-philosophical parallelism, and namely: names correspond to objects; sentences correspond to situations; and formators correspond to functions. That is why Suszko used to quote J. Bocheński: “Syntax mirrors ontology”.

Suszko believed that there are certain semantical principles which connect the syntactical construction rules with theirs object references. Those principles allows us to introduce the notion of a model for any language which is formalized in a standard way. The philosophical views that inspired his logical work are here presented.

4 Influences

Roman Suszko had great influence over the choice of investigation problems and over the way of scientific investigations carried out by many philosophers, logicians and mathematicians, both in Poland and abroad.

Joint work Suszko’s and Łoś’s [5] was continued by R. Wójcicki and his school. Also the works [2], [4], [8] served as a source of inspiration for the logicians who applied the model theory to the methodology of empirical sciences.

Suszko’s reflections on the connections between natural languages and formal languages were continued among others by: B. Stanosz, A. Nowaczyk, U. Wybraniec-Skardowska, J. Pogonowski. Those reflections were continued in such works as:

Stanosz, B., Nowaczyk, A.: Logiczne podstawy języka (in Polish) [Logical Foundations of language]. Ossolineum (1976)

Wybraniec-Skardowska, U.: On language adequacy. Studies in Logic, Grammar and Rhetoric **40(53)**. In: Hensel, M., Poczobut, R. (eds.) Cognitive Systems, pp. 257–292 (2015)

Wybraniec-Skardowska, U.: On the Denotation of Quantifiers. In: Omyła, M. (ed.) Idee logiczne Romana Suszki (in Polish) [Logical ideas of Roman Suszko], pp. 89–119. Warsaw (2001)

In turn work [6] devoted to the parallelism between language and reality, and work [3] in which Suszko ascribes to sentences logical values different than true and false, serves as a reference base for G. Malinowski, M. Omyła, B. Wolniewicz, A. Wójtowicz. Suszko inspired with non-Fregean logic and abstract logic logicians in USA working at the Stevens Institute of Technology: S.L. Bloom, D.J. Brown, J.D. Kagan, A. Michaels, R. Quackenbush and others, and in Poland: W. Dzik, G. Malinowski, M. Omyła, J. Pogonowski, J. Zygmunt. Non-Fregean logic is also subject of study for younger generation logicians: T. Huuskonen, J. Golińska-Pilarek, P. Łukowski, A. Wójtowicz, J. Wesserling. With the course of years it was turned out that the abstract logic which was grown out from the considerations on characteristics of non–Fregean logic and on its place in the spectrum of logical calculi found its application in universal algebra. J. Czelakowski, W. Dzik, W. Dziobiak, A. Wroński and others—in Poland—and W. Blok, D. Pigozzi, J.M. Font—abroad—are working on abstract logics inspired by non-Fregean logic.

After Suszko's death three monographs at least were published which related meaningfully to his ideas and which—one can say—fulfill his program:

Omyła, M.: Zarys logiki niefregowskiej (in Polish) [Outline Non-Fregean Logic]. Warsaw (1986)
Wójtowicz, A.: Znaczenie nazw a znaczenie zdań, w obronie ontologii sytuacji (in Polish) [Meaning name and meaning sentences]. Semper, Warsaw (2007)
Dzik, W.: Unification Types in Logic. Silesian University, Katowice (2007)

Very good discussion of the work [13] constitutes the review:

Malinowski, G., Zygmunt, J.: R. Suszko "Abolition of the Fregean Axiom". Erkenntnis **12**, 369–380 (1978)

Full bibliography of Suszko and its description can be found in:

Omyła, M., Zygmunt, J.: Roman Suszko (1919–1979): Bibliography of the Published Work with an Outline of His Logical Investigations. Studia Logica **43**, 421–441 (1984)

Comparison between diachronic logic and non-Fregean logic can be found in the work:

Omyła, M.: Roman Suszko—from diachronic logic to non-Fregean logic. In: Krajewski, W. (ed.) Polish Philosophers of Science and Nature in the 20th Century, pp. 153–161. Amsterdam – New York (2001)

5 Summary

Suszko's logical ideas were a source of inspiration in the choice of problems and in the ways of carrying out logical investigations in the second half of twentieth century and in first decades of twenty-first century, both in Poland and in the world.

The majority of polish logicians of the second half of twentieth century remained under his scientific influence and his irresistible personal charm. Almost on all scientific conferences devoted to logic taking place in Poland, beginning from Suszko's death till today, both speakers and commentators refer to Suszko's ideas as great inspiration for other logicians.

M. Omyła (✉)
Faculty of Law and Administration, Cardinal Stefan Wyszyński University in Warsaw, Warszawa, Poland
e-mail: m.omyla@uw.edu.pl

From Formal Theory of Knowledge to Non-Fregean Logic

Mieczysław Omyła

Abstract The terms: 'diachronic logic' and 'non-Fregean logic' we owe to Roman Suszko. He called 'diachronic logic' an application of classical logic to study of the development of knowledge. But Non-Fregean logic is a logical calculus obtained from the classical logic by adding identity connective and axioms for it. The main goal of the paper is to proof that the non-Fregean logic is a continuation of diachronic logic. The article is divided into following parts: 1. Diachronic logic, 2. Non-Fregan logic, which contain 2.1. Introduction, 2.2. Axiomatic form of non-Fregean Logic, 2.3. Properties of non-Fregean logic, and Bibliography.

Keywords Diachronic logic · Development of knowledge · Non-Fregean logic

Mathematics Subject Classification (2000) Primary 03B60, 03B65, 03B99

Roman Suszko was logician who united in their research the mathematical form of logic with its philosophical content. One of his work starts with the following epigraph: "Abstract mathematics may be a thorough philosophy". This epigraph may be interpreted in various ways, but the most natural is that we can solve certain philosophical problems in mathematical way, by creating at the beginning their formal representation, and next—by solving formal problems related with them. Solving those related formal problem constitutes at the same time looking for an answer to the initial philosophical question. In practice the formulation of a given philosophical problem in a formally strictly and adequate way is a considerable achievement as such. As a matter of fact we restrict ourselves to creating certain pattern or formal model which has a proper philosophical interpretation.

During 1957–1968 Suszko devoted certain number of works to the formal analysis of development of knowledge; this analysis was done with the aid of the models theory for the classical predicate logic calculus.

Suszko called the research on development of knowledge "diachronic logic".

Hence diachronic logic is application of the classical models theory for first-order predicate languages to research development of knowledge.

Suszko used here eloquent terminology offered earlier by Ferdinand Saussure in the monography *Cours Lingistique Gènèrale*, where it was offered to distinguish synchronic

Á. Garrido, U. Wybraniec-Skardowska (eds.), *The Lvov-Warsaw School. Past and Present*, Studies in Universal Logic,
https://doi.org/10.1007/978-3-319-65430-0_53

linguistic from a diachronic one. Suszko defined "synchronic logic" as syntax and semantics of formal languages where classical logic is in force, and "diachronic logic"—as application of synchronic logic to research on development of knowledge. He devoted to those issues several articles in 1957–1968 [7, 9].

In 1968–1979 Suszko worked extremely intensely on non-Fregean logic, which he himself invented. This logic was inspired by Wittgenstein's *Tractatus Logico-Philosophicus*. Suszko invented this logic because he thought that the ontology of Tractatus—considered as certain philosophical theory—goes beyond the formal means that are available in the framework of the classical first-order predicate calculus.

In the present work I am trying to show, that research on non-Fregean logic is continuation of earlier research on diachronic logic.

1 Diachronic Logic

A lot of effort was devoted in the 1950s of twentieth century to problems on the border between formal logic and theory of knowledge. In Suszko's opinion the contemporary logic can describe certain aspects of development of knowledge in a precise way and can throw new light on traditional epistemological questions.

Theory of knowledge explores the epistemological opposition $< S, O >$ where S is a subject of knowledge and O is an object of knowledge. Suszko represented the subject of knowledge S by the sequent

(*) $$(L, Cn, A, T),$$

where:

(i) L is a formalized version of language used by the subject of knowledge,
(ii) Cn is operation of consequence defined on language L,
(iii) A is a set of analytical axioms formulated in language L,
(iv) T is set of sentences of the language L accepted by the subject of knowledge S.

Every sentence of language L represents certain thought, and every predicate of language L represents certain notion at the disposal of subject S. Operation of consequence Cn defined on language L, is the totality of logical thinking rules. Set of analytical axioms A contains logical and extra-logical thinking principles, which are at the disposal of subject. As far as the set of accepted sentences T is concerned we assume that $Cn(T) \neq T$, because subject of knowledge usually does not know all logical consequences accepted by himself and does not know all of his own assumptions. It is assumed also that $T - Cn(A) \neq \emptyset$, i.e. that the subject accepts certain sentences that are not true in an analytical way.

In order to study in a formal way the epistemological opposition $< S, O >$, the reality which is the object of knowledge has also to be defined in a strict way. For a finite subject the world as a the whole reality R is never an object of knowledge; at every moment t subject of knowledge sees only certain fragment of the world, which at the moment t becomes an object of knowledge O_t for subject S. Suszko calls the period during which given subject studies one and the same fragment of the world "epoch in development of knowledge". If during the given period t certain fragment of knowledge constitutes an object of knowledge, subject attaches to this world-fragment certain language in order to speak about it. Speaking loosely, the object of discursive knowledge O on a given stage of development of knowledge constitutes such a fragment of reality R, that may be caught by the net of notions being at the disposal of the subject, i.e. such a fragment, which serves as a model for language L used by the subject. For the sake of simplicity Suszko assumes that the language used by subject of knowledge is certain first-order predicate language L, and subject of knowledge is represented by certain intended model of language, it means by the structure of the following type:

$$M = < U, d(C_1), d(C_2), \ldots >,$$

where: U is universe of language L i.e. it is a set, from which the nominal variables of the language take their values, and $d(C_k)$ is denotation of the extra-logical constant C_k.

Selected objects from the set U, or certain sets of objects which represent adequate properties of objects and relations and relations between objects or—possibly—functions defined on the set of objects U serve as denotations of extra-logical constants of the given language. Therefore epistemological opposition
$< S, O >$ is represented by Suszko as the following:

(**) $$< (L, Cn, A, T), M >,$$

where (L, Cn, A, T) represents a subject of knowledge and model M of the language L constitutes a formal representation of object of knowledge. One of the consequences of the assumption, that the subject is equipped with the language L, which is—with the first-order predicate language is the following: we represent the object of knowledge with the intended model of the language.

For a given epistemological opposition we mark the set of true sentences of the language L in model M as $Ver(L, M)$. Let us notice that we have here to do with a double relativisation of the notion of truth: to the language and to the model.

Because the language L is a language defined by the structure of the model M, we can mark this set shorter with the symbol $Ver(M)$.

Research on the formal epistemological opposition (**) is a matter of synchronic logic. In turn research on changes of this opposition in time belongs to diachronic logic.

Development of knowledge consists—according to apt formulation of Suszko—in gaining more and more amount of truths about wider and wider object of knowledge. Within the framework of diachronic logic a development of knowledge is represented by transformation of epistemological oppositions

$$< (L, Cn, A, T), M > / < (L^*, Cn^*, A^*, T^*), M^* >$$

in such a way, that the set of sentences which are simultaneously true and which are accepted by the subject at the next stage of knowledge contains the set of sentences which were true and were accepted by the subject at the previous knowledge stage, i.e.

$$T \cap Ver(M) \subseteq T^* \cap Ver(M^*).$$

Presenting the basic ideas of diachronic logic Suszko used eloquent terminology, which—among others—was a result of modification of Ajdukiewicz's terminology used in works Ajdukiewicz [1, 2].

Let us present briefly those terminology. During every given period of development of knowledge t subject of knowledge sees certain fragment of the really existing reality R, which is called by Suszko "the world-layer in period t" and which is represented in the system of diachronic logic by certain model M_t. The period during which the subject sees one and the same layer of the world is called by Suszko an "epoch in development of knowledge". While initial research on object of knowledge, i.e. of a given world-layer, subject attaches the language which fits for speaking about this fragment of the world. Language L_t tailored to speak about the world-layer in epoch t is called by Suszko "conceptual apparatus in epoch t".

Let $< (L_t, Cn, A_t, T_t), M_t >$ be the epistemological opposition in a given epoch t. Suszko uses further the following terminology: the set of sentences T, accepted by the subject of knowledge is called by him the picture of the world in epoch t and in turn the set $T_t \cap Ver(M_t)$ is a true fragment of world-picture in epoch t, or "real world-knowledge in epoch t". Suszko calls the set $Cn(T_t)$ a "world-perspective in epoch t" or "potential world-knowledge in epoch t" and the set of sentences $Cn(T_t) \cap Ver(M_t)$ constitutes than "true fragment of world-perspective in epoch t".

Suszko distinguishes two main types of knowledge development:

(1) evolutionary, by which the object of knowledge does not change, what means that the subject of knowledge sees the same world-layer and the syntactic structure of the language remains also unchanged.
(2) revolutionary, by which the object of knowledge does change.

The evolutionary process of knowledge consists above all in the situation where the sequence of the sets of sentences accepted by subject of knowledge $T, T^*, T^{**}, \ldots$ contains more and more true sentences about the same object of knowledge. It happens that during the evolutionary development of knowledge the extra-logical principles of thinking change, i.e. set of axioms A becomes set of axioms A^*. Suszko considers following two cases of that kind:

(i) $A \neq A^*$ and $Cn(A) = Cn(A^*)$,
(ii) $A \subseteq A^*$ and $Cn(A) \neq Cn(A^*)$.

In the case (i) the systematization of axioms takes place. In the case (ii) we have to do with reinforcement of axioms. The case (ii) embraces the following sub-case:

$$(L, Cn, A, T)/(L, Cn, A^*, T^*)/(L, Cn, A^{**}, T^{**})$$

which consists in fact that certain sentence α the given language L, is initially not accepted, i.e. $\alpha \notin T$ and the development of knowledge goes in such a way, that at the beginning α becomes a non-analytical theorem, $\alpha \in (T^* - Cn(A^*))$, and at the next stage of knowledge this sentence becomes one of the analytical sentences of given language, i.e. $\alpha \in Cn(A^{**})$; according to Suszko this kind of development of knowledge was noticed by conventionalists. Evolutionary development of knowledge corresponds with what T.S. Kuhn calls in his book *The structure of Scientific Revolutions*, Chicago, 1962 "normal stage of science development".

In turn revolutionary development of knowledge consists in the situation, in which the object of knowledge—represented here by the model of language—change. It means that the change M/M^* takes place. Suszko shows two kinds of such a change:

(2a) The universe of objects U does not change, but new properties of objects and new relations between them are discovered, and—as a result—new model M^* is built, which constitutes an extension of model M

$$M =< U, R_1, R_2, \ldots, R_n > /M^* =< U, R_1, R_2, \ldots, R_n, Q_1, Q_2, \ldots, Q_m >$$

Such an extension of knowledge object causes an extension of the language L to language L^*, which corresponds to the model M^*. In this case

$$L \cap Ver(L^*, M^*) = Ver(L, M).$$

This condition means, that all sentences of language L which are true in model M remain to be true also in model M^*. Additionally there are in model M^* also true sentences, that were impossible to formulate in language L:

$$Ver(L^*, M^*) - Ver(L, M) \neq \emptyset$$

(2b) Universe of model $M =< U, R_1, R_2, \ldots, R_n >$ becomes extended and as a result the object of knowledge becomes $M^* =< U^*, R_1^*, R_2^*, \ldots, R_n^* >$, where $U \subseteq U^*$ and $U \neq U^*$, and relations R_i^* for $i = 1, 2, \ldots, n$ are extensions of—accordingly—relations R_i. New objects become then known. Model M constitutes sub-model of model M^*, it may happen that $Ver(L, M) \neq Ver(L, M^*)$ and certain sentences that are true in model M may not be true in model M^*.

Suszko considered the following as very important: the theory of models can be applied to formulate precisely and to study everything what traditional theory of knowledge had studied just intuitively. Suszko was fascinated by the possibility of precise grasping the traditional issues, although he remained skeptical about the possibility to achieve in such a way any more important results.

Suszko's works on diachronic logic constitute one of the first in the world attempts to apply the theory of models to extra-mathematical questions, and especially—to philosophical problems. His works initiated applications of the theory of models to methodological research carried out in Poland. Those research flourished in Poland especially in 1960s and 1970s of twentieth century.

2 Non-Fregean Logic

2.1 Introduction

During "diachronic logic" period Suszko assumed that subject of knowledge S is equipped with language L, which was the language of the classical predicate calculus.

One of the consequences of that assumption was that object of knowledge M is a model the language L, i.e. it is a structure of the type: $(U, R_1, R_2, \ldots, R_n)$. From philosophical point of view it means that the world is considered as universe of objects which inhere properties and stay in certain relations. Names of the language L refer to objects which are elements of the universe of the model M. One-place predicates refer to properties of objects and many-place predicates refer to relations occurring between objects. The question arises: what sentences of language L in model M refer to?

Answers this question are in papers Suszko [6, 8]. He introduces there the notion of generalized denotation for sentence formulas. And namely: any sentence formula $\alpha(x)$ of language L refers in a model M to all of those objects which satisfy this formula in the model M, i.e. $\{x \in U : \alpha(x)\}$. If the formula $\alpha(x)$ is a sentence, it means, if there are no free variables in the formula, then whether every object of the universe satisfies it or no object in the universe satisfies it. Hence there are in model M only two generalized denotations of sentences and all true sentences have the same generalized denotation and all false sentences have one common generalized denotation. The generalized denotation of true sentence in model M constitutes universe of model U, and the generalized denotation of false sentence is empty set. Because the sum total of generalized denotations of sentences in model M consists of two elements, both sentence variables and quantifiers binding those variables are redundant. What is important from our point of view, Suszko almost from the beginning of his scientific carrier assigned to sentences not only logical values (truth and false) but also semantic correlates, which he called "generalized denotations of sentences".

If set of generalized denotations of sentences equipped in set-theoretical operations corresponding with logical connective, then algebra semantic correlate sentences will be isomorphic with two-elements algebra of logical values.

Because of that isomorphism, Frege could supposed that sentences are names of special objects called "logical values of sentences".

Under the influence of *Tractatus* Suszko modified his view and started to consider situation presented in a sentence as semantic correlate of this sentence. Since Wittgenstein wrote in Tractatus: 4.03 [...] *A proposition communicates a situation to us, and so it must be essentially connected with the situation. And the connection is precisely that it is its logical picture.*

Besides Suszko was convinced that logic should not impose any quantitative restraints on the universe of semantic correlates, except the one: there are at least two correlates of sentences, because correlate of any true sentence is different from a correlate of false sentence.

The name "non-Fregean logic" is justified by the fact that in this logic there no theorems asserting how many semantic correlates of sentences there can be.

At the base of non-Fregean logic lies also convictions, that syntactic categories of linguistic expressions should conform to ontological categories of theirs semantic correlates. On behalf of that conformity Suszko postulated—after Wittgenstein—that situations stated by the sentences constitute semantic correlates of these sentences, and sentential variables take their values from the universe of all situations correlated with a given language.

Syntax and semantics of non-Fregean logic displays a logic-philosophical parallelism between language and reality: we have names, functors and quantifiers in language and—objects, situations and functions in the reality; and functors and quantifiers refer to certain kind of functions.

However one can't conclude from above, that every object in a given universe of our discourse has a name, but one can conclude that every object may be a value of certain name variable. Similarly not every situation or a state of affairs occurring between objects of our discourse's universe may be described with sentences of our language, but if there are in the language sentential variables, then every of those situations may be a value of certain sentential variable. Let us notice that sentential variable differ fundamentally from other kinds of variables, because they are at the same time sentential formulas, and therefore they enter into logical connections with the rest of sentences and sentential formulas of a given language. Because of that the logical consequence influences the interpretation of sentential formulas.

2.2 *Axiomatic Form of Non-Fregean Logic*

To speak in formal way about the structure of universe of situations and universe of objects, Suszko introduced to literature of logic languages which he called W-languages (in honor of L. Wittgenstein). In the alphabet of these languages, there are:

(1) two kinds variables: sentential variables: $p, q, r, \ldots$, and nominal variables: $x, y, z, \ldots$; (2) truth-functional connectives: $\neg$ (negation), $\wedge$ (conjunction), $\vee$ (disjunction), $\leftrightarrow$ (implication), $\leftrightarrow$ (equivalence); (3) predicate-letters: $P_1, P_2, \ldots, P_n$; (4) function-symbols: $F_1, F_2, \ldots, F_m$; (5) symbols identity: identity connective and identity predicate which both symbolized by the sign "$\equiv$"; (6) quantifiers: $\forall$, $\exists$ binding both kinds of variables.

Each of the quantifiers may bind both sentential variable or nominal one, depending on which variable follows directly after it. Analogous the context uniquely determines whether we have to do with identity connective or identity predicate since the expression $x \equiv p$ is not a formula of the language discussed. A detailed description of the syntax of the W-kind languages has been presented in the papers: Bloom [3], Suszko [10, 12] Operation Cn on L is generated by the Modus Ponens rule and the schemas of logical axioms. To describe the consequence Cn in W-languages the following notations are introduced:

Letters: $v, w, v_1, w_1, v_2, w_2, \ldots$ will be metalanguage variables denoted depending on the context, either sentential variables or the nominal variables. By the letters: $\alpha, \beta, \gamma, \ldots$, will be denote any sentential formulas, by the letters: $\zeta, \xi, \tau, \ldots$, we denote any nominal formulas, and finally: $\phi, \varphi, \psi, \ldots$, denote sentential formulas or nominal ones, depending on the context. Symbols $\alpha[v/\phi]$ denoted result substitution in formula $\alpha(v)$ for free variable v the expression ϕ.

The result of proceeding the formula α by any finite number of universal quantifiers, i.e. $\forall v_1 \forall v_2 \ldots \forall v_n \alpha$, where $n \geq 0$ is called generalization of the formula α. For any set of sentential formulas X, by $Gen(X)$ will be denoted the set of all generalizations of formulas in the set X.

The formulas of the form $\phi \equiv \varphi$, are called equations.

The structural version non-Fregean logic in W-language L is introduced by accepting logical axioms and the only inference rule Modus Ponens.

The logical axioms are those formulas which are generalizations of any formula of the following sorts:

(**A1**) Axiom Schemata for truth-functional connective (they are classical)

(**A2**) Axiom Schemata for quantifiers

(i) $\forall v\alpha \to \alpha[v/\phi]$
(ii) $\alpha \to \forall v\alpha$ (if v is not free in α)
(iii) $\forall v(\alpha \to \beta) \to (\forall v\alpha \to \forall v\beta)$
(iv) $\exists v\alpha \leftrightarrow \neg\forall v\neg\alpha$

(**A3**) Axiom Schemata for identity connective and predicate:

(**A3.1**) Congruence axioms. All formulas of the form:

(i) $\varphi \equiv \phi$ (when ϕ, φ vary in at most bound variables)
for every functor Ψ we accept the invariance axiom:
(ii) $\varphi_1 \equiv \phi_1 \wedge \varphi_2 \equiv \phi_2 \wedge \ldots \wedge \varphi_n \equiv \phi_n \to \Psi(\varphi_1, \varphi_2, \ldots, \varphi_n) \equiv \Psi(\phi_1, \phi_2, \ldots, \phi_n)$
(iii) $\forall v(\alpha \leftrightarrow \beta) \to (Qv\alpha \equiv Qv\beta)$, where $Q = \forall, \exists$.

(**A3.2**) Special axiom for identity:
$\varphi \equiv \phi \to (\alpha[v/\varphi]) \to \alpha[v/\phi])$

The set logical axioms **LA** is the sum of three sets: **A1**, **A2**, **A3** i.e. **LA** $=$ **A1** $\cup$ **A2** $\cup$ **A3**.

A set of all the sentential formulas which are derivable from any set X and from logical axioms in any finite number of steps through the application of MP rule is called theory and is denoted by $Cn(X)$.

A formula α is called logical theorem non-Fregean logic iff $\alpha \in Cn(\emptyset)$.

2.3 Properties of Non-Fregean Logic

If in logical theorems of non-Fregean logic we replace at every place the sign "$\equiv$" with sign "$\leftrightarrow$", then we receive theorems of classical logic. It means that the non-Fregean logic constitutes a generalisation of classical logic and the classical logic constitutes a reinforcement of non-Fregean logic.

The sentence:

(AF) $$\forall p\forall q[(p \equiv q) \equiv (p \leftrightarrow q)]$$

Suszko called "ontological version of Frege's axiom". In non-Fregean theories, in which (AF) is a theorem the connectives "$\equiv$" and "$\leftrightarrow$" are indistinguishable. The non-Fregean theorie in which (AF) is a theorem are classical theories expressed in non-Fregean language. We can also derive classical logic by addition to its theorems the—seemingly weaker than (AF)—axiom:

$$\forall p\forall q[(p \leftrightarrow q) \to (p \equiv q)]$$

From the philosophical point of view the most important properties of non-Fregean logic is its logical bivalence and extentionality. The logical bivalence of non-Fregean logic finds expression in the following theorem of that logic:

$$\forall p \forall q \forall r[(p \leftrightarrow q) \vee (q \leftrightarrow r) \vee (p \leftrightarrow r)]$$

In turn the extentionality of this logic finds its expression in the fact, that schemas (2) and:

$$\varphi \equiv \phi \rightarrow (\alpha[v/\varphi] \equiv \alpha[v/\phi])$$

$$\varphi \equiv \phi \rightarrow (\alpha[v/\varphi] \rightarrow \alpha[v/\phi])$$

are schemas of logical theorems. These schemas state that expressions that have the same semantical correlates are mutually interchangeable in all sentential contexts without accordingly changing semantic correlate of those contexts (salva identitate) and without changing logical values of those contexts (salva veritate).

To logical theorems of non-Fregean logic belong theorems:

$$\exists x(x \equiv x), \qquad \exists p \exists q \neg(p \equiv q)$$

which state accordingly that the universe of objects is non-empty and that universe of situations contains at least two elements. To logical theorems of non-Fregean logic though do not belong any conditions, that limit "from above" the number of objects and situations in universe, what means that for every natural number n the following formulas are not logical theorems:

$$(x_1 \equiv x_2) \vee (x_1 \equiv x_3) \vee \ldots \vee (x_{n-1} \equiv x_n)$$

$$(p_1 \equiv p_2) \vee (p_1 \equiv p_3) \vee \ldots \vee (p_{n-1} \equiv p_n)$$

Non-Fregean logic—as every calculus—can be developed without any philosophical presumptions. Nonetheless for this logic the ontology of situations contained in *Tractatus* served as a fundament. This logic presumes the ontology, according to which there exist objects, situations and functions. Suszko introduced attention to the fact that the division all of beings into objects, situations and functions has logical character, i.e. it results the fact, that we describe the world with languages in which we have names, predicates, connectives and quantifiers.

Suszko introduced non-Fregean logic because he was convinced that there exist in the world certain beings and aspects of beings, which can be properly told about with the aid of sentential variables. In other words, the ontology that underlies non-Fregean logic contains the view, that it is not enough to consider the world as the universe of objects only, but we have to consider it also as the universe of possibilities, among which some of them become realized, i.e.—there are facts. In [11] Suszko wrote: "*...perceiving an object x consists of perceiving at least one situation that x is so-an-so*".

It is worth stressing that the importance of non-Fregean logic goes beyond its ontological applications. Since this logic has initiated research on abstract logics. Those logics constitutes new chapter in application of algebra to logic. This problems are discussed among others in monographs Czelakowski [4], Dzik [5].

References

1. Ajdukiewicz, K.: Das Weltbild Und die Begriffsapparatur. Erkenntnis **IV**, 259–287 (1934)
2. Ajdukiewicz, K.: The scientific world perspective. Erkenntnis **V**, 22–30 (1935)
3. Bloom, S.L.: A completeness theorem for "theories of kind W". Stud. Logica **27**, 43–55 (1971)
4. Czelakowski, J.: Protoalgebraic Logics. Kluwer, Dordrecht (2001)
5. Dzik, W.: Unification Types in Logic. Silesian University Press, Katowice (2007)
6. Suszko, R.: Syntactic structure and semantical reference II. Stud. Logica **9**, 63–91 (1960)
7. Suszko, R.: An abstract scheme of development of knowledge. In: Actes X Congress International d'Histoire des Sciences Varsovie 1965, pp. 52–55. Ossolineum, Wrocław (1967)
8. Suszko, R.: An essay in the formal theory of extension and of intension. Stud. Logica **20**, 7–36 (1967)
9. Suszko, R.: Formal logic and the development of knowledge. In: Colloquium in Philosophy of Science. London 1965, pp. 210–222. North Holland, Amsterdam (1968)
10. Suszko, R.: Ontology in the Tractatus of L. Wittgenstein. Notre Dame J. Form. Log. **9**, 97–121 (1968)
11. Suszko, R.: Non-Fregean logic and theories. Analele Universitatii Bucuresti. Acta Logica **11**, 7–33 (1968)
12. Suszko, R.: Identity connective and modality. Stud. Logica **27**, 7–39 (1971)

M. Omyła (✉)
Faculty of Law and Administration, Cardinal Stefan Wyszyński University in Warsaw, Warsaw, Poland
e-mail: m.omyla@uw.edu.pl

Categories of First-Order Quantifiers

Urszula Wybraniec-Skardowska

Abstract One well known problem regarding quantifiers, in particular the 1st-order quantifiers, is connected with their syntactic categories and denotations. The unsatisfactory efforts to establish the syntactic and ontological categories of quantifiers in formalized first-order languages can be solved by means of the so called *principle of categorial compatibility* formulated by Roman Suszko, referring to some innovative ideas of Gottlob Frege and visible in syntactic and semantic compatibility of language expressions. In the paper the principle is introduced for *categorial languages* generated by the Ajdukiewicz's classical categorial grammar. The 1st-order quantifiers are typically ambiguous. Every 1st-order quantifier of the type $k > 0$ is treated as a two-argument functor-function defined on the variable standing at this quantifier and its scope (the sentential function with exactly k free variables, including the variable bound by this quantifier); a binary function defined on denotations of its two arguments is its denotation. Denotations of sentential functions, and hence also quantifiers, are defined separately in Fregean and in situational semantics. They belong to the ontological categories that correspond to the syntactic categories of these sentential functions and the considered quantifiers. The main result of the paper is a solution of the problem of categories of the 1st-order quantifiers based on the principle of categorial compatibility.

Keywords 1st-order quantifiers · Categorial languages · Syntactic categories · Denotation · Ontological categories · Denotational semantics · Compositionality · Categorial compatibility

Mathematics Subject Classification (2000) Primary 03C07, 03H05; Secondary 03A99

The main problems and their solutions connected with the subject of this paper were presented at the Logic Colloquium'15 held in Helsinki on August 3–8, 2015, co-located with the 15th International Congress of Logic, Methodology and Philosophy of Science, CLMPS (see my abstract [35]; see also [36]).

Á. Garrido, U. Wybraniec-Skardowska (eds.), *The Lvov-Warsaw School. Past and Present*, Studies in Universal Logic,
https://doi.org/10.1007/978-3-319-65430-0_54

1 Introduction

Around 1879 Frege and—independently—Charles Sanders Peirce developed a way to extend sentential logic by introducing symbols representing *determiners*, such as 'all', 'some', 'no', 'every', 'any', and so on.

Frege and Peirce used two symbols: the *universal quantifier* (which we will write $\forall$) corresponding roughly to the English 'all', 'every' and 'each' and the *existential quantifier* (which we will write $\exists$) corresponding to the English 'some', 'a', 'an'.

In this paper we will consider only standard, Fregean quantifiers $\forall$ and $\exists$ of the 1st-order as individual variable-binding operators. They are used in formulas of predicate logic of the 1st-order and in formalized languages of elementary theories based on this logic. Their syntactic role and semantic references, i.e. denotation, extension, created some problems that have not been satisfactorily solved yet.

In the next part (Sect. 2) I shall partially explicate the problem of quantifiers. In Sect. 4, I'll outline some intuitive foundations of my theory of categorial languages which gives the formal direction for justification of my solution of the problem of quantifiers. It corresponds to the principle (*CC*) of categorial compatibility based on some Frege's ideas and was formulated by Roman Suszko [24]. The solution of the problem is presented in Sect. 5.

2 Problem of Quantifiers

The problem of quantifiers is connected with the difficulty pertaining to establishing their syntactic and semantic categories.

Leśniewski's theory of semantic/syntactic categories [13, 14], which was improved by Ajdukiewicz [1] by introducing categorial indices, does not, obviously, solve this problem, which limits the universal character of the theory.

Leśniewski's hierarchy of semantic/syntactic category does not include any variable-binding operators. Leśniewski, in his protothetics and ontology systems, allows only one operator—the universal quantifier, noting it as parentheses, Ajdukiewicz, on the other hand, indicates the difficulty of assigning to quantifiers the index *s/s* or *s/ns*.

Assigning to them the index *s/s*, i.e. the category of sentence-forming functors of one-sentence argument, would mean that the quantifiers belong to the same category as one-argument connectives, and assigning to them the index *s/ns* of sentence-forming functors of one-name and one-sentence arguments would mean that we include them into the same category as some expressions of indirect speech, e.g. 'think that', 'know that', etc.

It has been suggested that the categorial grammar, which Bar-Hillel derived from Ajdukiewicz's version of the theory of semantic/syntactic categories, does not satisfactorily account for the role of bound variables and operators binding them.

Suszko [22, 24] assigns to them the index *s//s/n*, and thus the index of sentence-forming functor of the argument, which is a one-argument predicate. In this way, the index, for example in the sentence '$\forall x(x\, \mathit{flows})$' pertains to the entire quantifier-variable pattern '$\forall x(x \ldots)$' (see Simons [20]) which corresponds to English word 'everything' (see also Cresswell [4], Simons [21]).

Suszko and many other researchers of language syntax treat quantifiers as expressions independent of the quantifier variable. Generally, researchers avoid bound variables in attempting to solve the problem, for example by means of combinators (Curry [5, 6], Curry and Feys [7], see e.g. Simons [20]).

But earlier, Suszko stated that mounting variable-binding operators into a syntactic scheme requires general principles other than the theory of syntactic/semantic categories.

The principle (*CC*) of categorial compatibility is one such principle. It allows us to assign to every expression of a formalized 1st-order language, which possesses an index symbolizing a syntactic category, a denotation whose ontological category (relative to the universe U of a given model of the language) is indicated by the same index.

Suszko assumes that

- the denotation of the entire expression $\forall x(e(x))$, where $e(x)$ is a sentential function with the free variable x, is ether the logical value 1 (of truth) or the logical value 0 (of falsity) which belong to the ontological category with the index s, and
- the denotation of the universal quantifier $\forall$ is the function of generalization which has the value 1 in only one case, if its argument is the universe U.

The function of generalization belongs to the ontological category with the index *s//s/n* because its arguments are any sets belonging to the family $P(U)$ included into the ontological categories with the index *s/n*. In this way the principle (*CC*) holds although the principle of syntactic connection (*SC*) does not hold because no index is assigned to quantifier variable x, and the scope of the quantifier $\forall$ (here $e(x)$) is not one-argument predicate of the syntactic category with the index *s/n*.

In the next parts of this paper I explicate both the principle (*SC*) of syntactic connection and the principle (*CC*) of categorial compatibility on the basis of my theory of categorial languages [30–34] which allows us to give some solutions to the problem of quantifiers.

The essence of the approach proposed here is considering them to be typical syntactic notions: functors-functions mapping language expressions into language expressions that correspond to some functions on extralinguistic objects—on denotations of arguments of these functors.

Let us note that a standard background for research in the field of mentioned quantifiers assumes treating them as some functions or relations on extralinguistic objects, mostly functions with index *t//t/e* (cf. Mostowski [18], Lindström [15], Montague [16, 17], Nowaczyk [19], van Benthem [27, 28], van Benthem and Westerståhl [29]).

3 Some Intuitive Foundations of the Theory of Categorial Languages

3.1 Main Ideas of Formalization of Categorial Language

In the paper, formal-logical considerations relate to syntax and extensional semantics of any language L characterized *categorically*:

- in the spirit of some ideas of Husserl [12] and Leśniewski–Ajdukiewicz's theory of syntactic/semantic category (see Leśniewski [13, 14], Ajdukiewicz [1, 2]),

- in accordance with Frege's ontological canons [10],
- in accordance with Bocheński's motto [3]: *syntax mirrors ontology*, and
- some ideas of Suszko [22–25]: *language should be a linguistic scheme of ontological reality and simultaneously a tool of its cognition.*

The paper includes developing and some explications of these authors' ideas. It also presents, in a synthetic form, some ideas presented in my papers published in [30–34].

Language L is there defined, if the set S of all *well-formed expressions* (briefly *wfes*) is determined. These expressions must satisfy requirements of categorial syntax and categorial semantics.

3.2 Categorial Syntax

The categorial syntax of L is connected with generating the set S by the classical categorial grammar and belonging *wfes* of S to appropriate syntactic/semantic categories.

A characteristic feature of categorial syntax is that each composed *wfe* of the set S has a *functor-argument* structure, in this sense that, in accordance with the principle originated by Frege [8], it is possible to distinguish in it its constituent called the *main functor*, and the other constituents—called *arguments* of that functor, yet each constituent of the *wfe* has a determined syntactic category.

If e is a functor-argument *wfe* of S, f is its main functor and $e_1, e_2, \ldots, e_n$ its subsequent arguments then e can be written in the functional-argument form:

$$e = f(e_1, e_2, \ldots, e_n). \tag{e}$$

In categorial approach to the language L, syntactic categories of *wfes* of L are determined by attributing to them, like their expressions, categorial indices of a certain set I. To every *wfe* e of the set S is unambiguously assigned a categorial index (type) $i_S(e)$ of the set I; *wfes* belonging to the same syntactic category *CATa* have the same categorial index a.

Categorial indices were introduced by Ajdukiewicz [1] into logical semiotics with the aim to determine the syntactic role of expressions and to examine their syntactic connection, in compliance with the principle of syntactic connection (*SC*) discussed below.

The set S of all *wfes* of L is then intuitively defined as the smallest set including the vocabulary of L and closed with respect to the principle (*SC*), which in free formulation says that

(*SC*) *The categorial index of the main functor of each functor-argument expression of the language L is formed out of the categorial index of the expression which the functor forms together with its arguments, as well as out of the subsequent indices of arguments of this functor.*

In the formal definition of the set S it is required that each functor-argument constituent of the given expression should satisfy the principle (*SC*).

If the functor-argument expression $e = f(e_1, e_2, \ldots, e_n)$ is a *wfe* (it belongs to the set S), then in accordance to the principle of syntactic connection (*SC*) the index of its main functor f formed from the index a of e and successive indices $a_1, a_2, \ldots, a_n$ of successive arguments $e_1, e_2, \ldots, e_n$ of the functor f, can be written in the following quasi-fractional form:

$$i_S(f) = i_S(e)/i_S(e_1)i_S(e_2)\ldots i_S(e_n) = a/a_1a_2\ldots a_n. \qquad (i_S)$$

3.2.1 An Algebraic Structure of Categorial Language

In categorial language L we can distinguish two sets: the set B of all basic *wfes* of S and the set F of all functors of S such that

$$S = B \cup F \text{ and } B \cap F = \emptyset,$$

where functors of the set F differ from basic expressions of B that they have indices formed from simpler ones. If the functor f has the functoral index of the form ($\boldsymbol{i_S}$), i.e. the index of the form $a/a_1a_2\ldots a_n$ then it belongs to the syntactic category $CATa/a_1a_2\ldots a_n$ and so to the category of functors forming expressions with the index a if their arguments are n expressions with successive indices $a_1, a_2, \ldots, a_n$. So the functor f can be treated as the following partial function defined on *wfes* of S:

$$f : CATa_1 \times CATa_2 \times \ldots \times CATa_n \rightarrow CATa$$

mapping of *wfes* from Cartesian product of syntactic categories $CATa_1, CATa_2, \ldots, CATa_n$ into the category $CATa$. Then we have

$$f \in CATa/a_1a_2\ldots a_n = CATa^{CATa_1 \times CATa_2 \times \ldots \times CATa_n}. \qquad (CAT_f)$$

In this way we simultaneously can regard the categorial language L as an algebraic structure $\boldsymbol{L}$, partial algebra with the carrier S and the set $Fo \subseteq F$ of partial functions on S (simple functors of L):

$$\boldsymbol{L} = \langle S, Fo\rangle .$$

3.3 *Categorial Semantics*

Categorial extensional semantics is connected with *denotations* of *wfes* of S and with their belonging to an appropriate semantic extensional category. Each constituent of the composed *wfe* has determined a semantic extensional category and also a *denotation*, and thus—an *ontological category* (the *category of ontological objects*). *Denotations* (*extensions*) of *wfes* of L are sets of *object references* (*references*) of *wfes* of L, objects of the cognized reality, e.g.: individuals, sets of individuals, states of affairs, operation on the indicated objects, and the like.

We will concentrate only on referential relationships between expressions of L and reality to which they refer. We enrich the categorial grammar generating L by the *denotation operation* δ regarded as its semantic component. The denotation operation δ assigns to every *wfe* of the set S an object of ontological reality *ONT* describing by the language L—its denotation belonging to an ontological category. So

$$\delta : S \to ONT, \tag{δ}$$

where *ONT* is the sum of all ontological categories corresponding to *wfes* of S.

According to some innovative ideas of Frege [9, 10], Bocheński's (his famous motto: syntax mirrors ontology) and Suszko [22–24] who anticipated the research in categorial semantics and was the first to use categorial indices as a tool for coordination of expressions and their references, extralinguistic objects, the mutual dependence of syntactic and semantic formal description of L should be considered by keeping the *principle* (*CC*) *of categorial compatibility*, based on the compatibility of the syntactic category of each language expression of L with the ontological category assigned to its denotation. The principle (*CC*) of syntactic and semantic, i.e. also ontological categorial compatibility in Suszko's formulation can be given by keeping for any *wfe* e of categorial language L the relationship:

$$e \in CAT\iota \quad \text{iff} \quad \delta(e) \in ONT\iota, \tag{CC}$$

where $CAT\iota$ and $ONT\iota$ are: the syntactic category and the ontological category, respectively, with the same categorial index ι, and δ is the operation of denotation.

From the principle (*CC*) it follows that for any $e = f(e_1, e_2, \ldots, e_n) \in S$ with the main functor-function $f \in CATa/a_1a_2\ldots a_n$ satisfying the condition (CAT_f) the following conditions are satisfied:

$$\delta(f) \in ONTa/a_1a_2\ldots a_n = ONTa^{ONTa_1 \times ONTa_2 \times \ldots \times ONTa_n} \tag{ONTf}$$

and

$$\delta(f(e_1, e_2, \ldots, e_n)) = \delta(f)(\delta(e_1), \delta(e_2), \ldots, \delta(e_n)). \tag{PCD}$$

The condition (*ONTf*) states that the denotation (object reference) of the main functor of the composed *wfe* e of the set S is the set-theoretical function mapping the Cartesian product of ontological categories $ONTa_1 \times ONTa_2 \times \ldots \times ONTa_n$ into the ontological category *ONTa* and it is defined by means of the condition (*PCD*) connected with some Frege's ideas and called the principle of compositionality of denotation.

3.3.1 An Algebraic Ontological Structure Corresponding to the Partial Algebra *L*

The operation δ assigns the following ontological structure $\boldsymbol{R_L}$ of a reality corresponding to language L to the algebraic structure $\boldsymbol{L}$:

$$\boldsymbol{R_L} = \langle ONT, ONT_{Fo} \rangle ,$$

where ONT_{Fo} is the sum of all ontological categories corresponding to all functors of the set *Fo*. The structure $\boldsymbol{R_L}$ is a partial algebra similar to the algebra $\boldsymbol{L}$ and the principle (*PCD*) is simultaneously the condition of homomorphism of the algebra $\boldsymbol{L}$ into the algebra $\boldsymbol{R_L}$, i.e.

$$\delta : \langle S, Fo \rangle \underset{hom}{\longrightarrow} \langle ONT, ONT_{Fo} \rangle \,.$$

A ***model of language*** L is the structure of homomorphic images of components of $\boldsymbol{L}$, i.e. the substructure $\boldsymbol{M_L} = \langle \delta(S), \delta(Fo) \rangle$ of the structure $\boldsymbol{R_L}$.

If we distinguish in the set B of basic *wfes* of S the category *CATs* of all sentences of language L, then the notion of *truthfulness* of any sentence $e \in CATs$ in the model $\boldsymbol{M_L}$ is defined as follows:

$$e \text{ is a } \textit{true sentence} \text{ in the model } \boldsymbol{M_L} \quad \text{iff} \quad \delta(e) \in T, \qquad (T)$$

where T is primitive notion of the considered theory intuitively understood either as the singleton with the true value (in Fregean semantics) or as the set of all states of affairs that take place (in situational semantics).

4 The Solution of the Problem of Quantifiers of 1st-Order

The unsatisfactory efforts to establish, in the sense of the principle (*CC*) of categorial compatibility, the category of quantifiers in formalized 1st-order languages can be solved by means of notions and statements of the above outlined theory of categorial languages.

Let L_1 be any 1st-order formalized language. Let us treat any standard quantifier of L_1 as a context-dependent functor of two arguments:

1. a quantifier variable (the variable accompanying this quantifier) and
2. its scope, i.e. a sentential function including as a free variable the same variable as the quantifier variable.

4.1 Different Types of the 1st-Order Quantifiers and Their Syntactic Categories

A standard, the 1st-order quantifier is a functor forming a new sentential function (in particular a sentence of L_1) in which there occur one free variable less than in the scope of this quantifier (the variable bound by the quantifier). As such a functor, a quantifier can be treated as a set-theoretical function relative to the number of free individual variables occurring in its scope. So, we should not speak of one existential $\exists$ or one universal quantifier $\forall$ but about different types of such quantifiers depending of the number of free variables in their scope. We will use numerical superscripts in order to point out these different types of quantifiers.

Let

- *Var* be the set of all individual variables for L_1, with categorial index n_1;
- $S = S_0$—the set of all its sentences, with the categorial index s;
- $S_k (k \geq 1)$—the set of all sentential functions in which exactly k free variables occur, with the index s_k.

For example, if $\alpha(x_1, x_2, x_3) \in S_3$, where $x_1, x_2, x_3 \in Var$, then the expressions:

$$\forall^3 x_2 \alpha(x_1, x_2, x_3) \in S_2,$$
$$\exists^2 x_3 \forall^3 x_2 \alpha(x_1, x_2, x_3) \in S_1,$$
$$\forall^1 x_1 \exists^2 x_3 \forall^3 x_2 \alpha(x_1, x_2, x_3) \in S_0$$

and quantifiers $\forall^3$, $\exists^2$, $\forall^1$ belong to different syntactic categories with indices $s_2/n_1 s_3$, $s_1/n_1 s_2$, $s/n_1 s_1$, respectively.

More generally, the quantifiers $\forall^k$ and $\exists^k$ $(k \geq 1)$ are treated as the functors-functions:

$$\forall^k, \exists^k : Var \times S_k \rightarrow S_{k-1} \qquad (S_0 = S).$$

Thus, in accordance to (*CATf*), for $k > 0$ we have

$$(CAT\forall^k, \exists^k) \qquad \forall^k, \exists^k \in CATs_{k-1}/n_1 s_k \qquad (s_0 = s),$$

and the principle of syntactic connection (*SC*) for them is satisfied.

Their denotations and ontological categories should be defined in such a way as to satisfied the principle (*CC*) of categorial compatibility (their denotations should belong to the ontological category $ONTs_{k-1}/n_1 s_k$) and the principle (*PCD*) of compositionality of denotation.

Let the denotation operation for the language L_1 be the function d in Fregean, standard semantics and the function $\underline{d}$ in the situational, non-standard semantics:

$$d, \underline{d} : S(L_1) \rightarrow ONT(L_1)$$

mapping the set $S(L_1)$ of all *wfes* of L_1 into the set $ONT(L_1)$ which is the sum of all ontological categories in the ontological structure $\boldsymbol{R}_{L_1}$.

We will give here two possible solutions of denotations of quantifiers of the 1st-order taking into account two different ways of understanding of the denotation of sentences and sentential functions presented below.

4.2 Denotations of 1st-Order Quantifiers and Their Ontological Categories

4.2.1 Fregean Semantics

We assume that if U is the universe of individuals in an established model $\boldsymbol{M}_{L_1}$ of L_1, 1 is the value of truth, 0—the value of falsity then

$$\begin{aligned} &d(x) \in \{U\} = ONTn_1 && \text{for any } x \in CATn_1 = Var; \\ &d(p) \in \{0, 1\} = ONTs && \text{for any } p \in CATs = S; \\ &d(sf) \in 2^{U^k} = ONTs_k && \text{for any } sf \in CATs_k = S_k (k \geq 1) \end{aligned}$$

and for any $x_1, x_2, \ldots, x_k \in Var$ and for any $sf = \alpha(x_1, x_2, \ldots, x_k) \in S_k$

$$\begin{aligned} &d(\alpha(x_1, x_2, \ldots, x_k)) = \\ &\quad \{(u_1, u_2, \ldots, u_k) \in U^k \mid d(\alpha^o(x_1/u_1, x_2/u_2, \ldots, x_k/u_k)) = 1\}, \end{aligned}$$

where $\alpha^o(x_1/u_1, x_2/u_2, \ldots, x_k/u_k)$ is a sentence which we get from sentential function sf by replacement of its all free variables $x_1, x_2, \ldots, x_k$ of Var by suitable individual names of individuals $u_1, u_2, \ldots, u_k$ of the universe U, i.e. the denotation of sf is the set of all k-tuples from U^k which satisfy this sentential function.

Denotation for the quantifier $\forall^k$ of the type $k(k \geq 1)$ is defined by induction as follows:

(a) for $k = 1$ and any $\alpha(x) \in S_1$

$$d(\forall^1 x\alpha(x)) = d(\forall^1)(d(x)), d(\alpha(x))) = \begin{cases} 1, & d(x) = U = d(\alpha(x)) \\ 0, & d(x) = U \neq d(\alpha(x)); \end{cases}.$$

According to (a) the quantifier sentence obtained from any sentential function $\alpha(x)$ by preceding it with the universal quantifier $\forall^1$ is a true sentence in the established model $\boldsymbol{M}_{L_1}$ of L_1 with the universe of individuals U *iff* every object of the universe U satisfies the $\alpha(x)$ which is the scope of $\forall^1$.

(b) for $k = j + 1 (j > 0)$ and any $\alpha(x_1, x_2, \ldots, x, \ldots, x_{j+1}) \in S_{j+1}$

$$\begin{aligned} &d(\forall^{j+1} x\alpha(x_1, x_2, \ldots, x, \ldots, x_{j+1})) = \\ &\quad = d(\forall^{j+1})(d(x), d(\alpha(x_1, x_2, \ldots, x, \ldots, x_{j+1}))) = \\ &\quad = \{(u_1, u_2, \ldots, u_{j+1}) \in U^j \mid d(\alpha^o(x_1/u_1, x_2/u_2, \ldots, x/u, \ldots, x_{j+1}/u_{j+1})) = 1 \\ &\qquad \text{for each } u \in U\}. \end{aligned}$$

According to (b) the denotation of the sentential function $sf_{k-1} \in S_{k-1}$ obtained from the sentential function $\alpha(x_1, x_2, \ldots, x, \ldots, x_{j+1}) \in S_k (k > 1)$ by binding the variable x by the universal quantifier $\forall^k (k = j + 1 > 1)$ is the set of all $j = (k - 1)$-tuples $(u_1, u_2, \ldots, u_{k-1})$ of individuals of U such that all sentences obtained by the substitution of all j free variables in sf_{k-1}, respectively, by names of individuals of these tuples and names of any individuals of U representing x are true; in other words the denotation of sf_{k-1} is the set of all such $(k-1)$-tuples $(u_1, u_2, \ldots, u_{k-1})$ of individuals of U that for any individual u of U k-tuples $(u_1, u_2, \ldots, u, \ldots, u_{k-1})$ build from them satisfy the scope $\alpha(x_1, x_2, \ldots, x, \ldots, x_{j+1})$ of the quantifier $\forall^k$.

Thus for any $k \geq 1$

$$d(\forall^k) \in ONTs_{k-1}/n_1 s_k = ONTs_{k-1}^{ONTn_1 \times ONTs_k}.$$

Similarly for $d(\exists^k)$:

(a) for $k = 1$ and any $\alpha(x) \in S_1$

$$d(\exists^1 x\alpha(x)) = d(\exists^1)(d(x)), d(\alpha(x))) = \begin{cases} 1, & d(x) \cap d(\alpha(x)) \neq \emptyset \\ 0, & d(x) \cap d(\alpha(x)) = \emptyset; \end{cases}.$$

(b) for $k = j + 1 (j > 0)$ and any $\alpha(x_1, x_2, \ldots, x, \ldots, x_{j+1}) \in S_{j+1}$

$$\begin{aligned} &d(\exists^{j+1} x\alpha(x_1, x_2, \ldots, x, \ldots, x_{j+1})) = \\ &= d(\exists^{j+1})(d(x), d(\alpha(x_1, x_2, \ldots, x, \ldots, x_{j+1}))) = \\ &= \{(u_1, u_2, \ldots, u_{j+1}) \in U^j \mid d(\alpha^o(x_1/u_1, x_2/u_2, \ldots, x/u, \ldots, x_{j+1}/u_{j+1})) = 1 \\ &\quad \text{for some } u \in U\}. \end{aligned}$$

According to (a) the quantifier sentence obtained from any sentential function $\alpha(x)$ by preceding it with the existential quantifier $\exists^1$ is true sentence in the established model $\boldsymbol{M}_{L_1}$ of L_1 with the universe of individuals U *iff* at least one object of the universe U satisfies the $\alpha(x)$ which is the scope of $\exists^1$.

According to (b) the denotation of the sentential function $sf_{k-1} \in S_{k-1}$ obtained from the sentential function $\alpha(x_1, x_2, \ldots, x, \ldots, x_{j+1}) \in S_k (k > 1)$ by binding the variable x by the existential quantifier $\exists^k (k = j + 1 > 1)$ is the set of all $j = (k-1)$-tuples $(u_1, u_2, \ldots, u_{k-1})$ of individuals of U such that all sentences obtained by the substitution of all j free variables in sf_{k-1}, respectively, by names of individuals of these tuples and the substitution some individual name of u for x are true; in other words the denotation of sf_{k-1} is the set of all such $(k-1)$-tuples $(u_1, u_2, \ldots, u_{k-1})$ of individuals of U that for some individual u of U k-tuples $(u_1, u_2, \ldots, u, \ldots, u_{k-1})$ build from them satisfy the scope $\alpha(x_1, x_2, \ldots, x, \ldots, x_{j+1})$ of the quantifier $\exists^k$.

Thus, for any $k \geq 1$

$$d(\exists^k) \in ONTs_{k-1}/n_1 s_k = ONTs_{k-1}^{ONTn_1 \times ONTs_k}.$$

Moreover, the principle (CC) is also valid for $\forall^k$ and $\exists^k$ in situational semantics.

4.2.2 Situational Semantics

In situational semantic we assume that

$$\underline{d}(x) \in \{U\} = ONTn_1 \quad \text{for any } x \in CATn_1 = Var;$$
$$\underline{d}(p) \in \{St\} = ONTs \quad \text{for any } p \in CATs = S,$$

where St is the set of all states of affairs, $St = T \cup F, T \cap F = \emptyset$ and T is the nonempty set of all states of affairs that take place and F—the nonempty set of remaining states of affairs. $St_k \subset St$ is the set of states of affairs with k individuals.

$$\underline{d}(sf) \in 2^{St_k} = ONTs_k \quad \text{for any } sf \in CATs_k = S_k$$

and for any $x_1, x_2, \ldots, x_k \in Var$ and for any $sf = \alpha(x_1, x_2, \ldots, x_k) \in S_k$

$$\underline{d}(\alpha(x_1, x_2, \ldots, x_k)) =$$
$$\{s \in St_k \mid s = \underline{d}(\alpha^0(x_1/u_1, x_2/u_2, \ldots, x_k/u_k)) \text{ for any } (u_1, u_2, \ldots, u_k) \in U^k\}.$$

So, if the denotation operation is understood here as the operation $\underline{d}$ then the denotations of sentences are states of affairs and the denotation of any sentential function is the set of all states of affairs that are denotations all sentences represented by the sentential function.

Denotation for the quantifier $\forall^k$ is defined by induction as follows:

(a) for $k = 1$ and any $\alpha(x) \in S_1$

$$\underline{d}(\forall^1 x\alpha(x)) = \underline{d}(\forall^1)(\underline{d}(x)), \underline{d}(\alpha(x))) \in T \text{ iff } \underline{d}(\alpha^o(x/u)) \in T \text{ for each } u \in U;$$

(b) for $k = j + 1 (j > 0)$ and any $\alpha(x_1, x_2, \ldots, x, \ldots, x_{j+1}) \in S_{j+1}$

$$\begin{aligned}&\underline{d}(\forall^{j+1} x\alpha(x_1, x_2, \ldots, x, \ldots, x_{j+1})) =\\ &= \underline{d}(\forall^{j+1})(\underline{d}(x), \underline{d}(\alpha(x_1, x_2, \ldots, x, \ldots, x_{j+1}))) =\\ &= \{s \in St \mid s = \underline{d}(\alpha^o(x_1/u_1, x_2/u_2, \ldots, x/u, \ldots, x_k/u_k))\\ &\quad \text{for each } u \in U, \text{ any } (u_1, u_2, \ldots, u_{j+1}) \in U^j\}.\end{aligned}$$

According to (a) the quantifier sentence obtained from any sentential function $\alpha(x)$ by preceding it with the universal quantifier $\forall^1$ is a true sentence in an established model $\boldsymbol{M}_{L_1}$ of the language L_1 with the universe of individuals U *iff* every sentence representing this sentential function is true (because their denotations are states of affairs that take place).

According to (b) the denotation of sentential function $sf_{k-1} \in S_{k-1}$ obtained from the sentential function $\alpha(x_1, x_2, \ldots, x, \ldots, x_{j+1}) \in S_k (k > 1)$ by binding the variable x by the universal quantifier $\forall^k$ is the set of all denotations of sentences (intuitively—the set of all states of affairs describing by these sentences) which can be obtained from sf_{k-1} by replacing all free variables in it with individual names of any individuals of U; in other words, it is the set of all denotations of sentences (all states of affairs) which can be obtained from $\alpha(x_1, x_2, \ldots, x, \ldots, x_{j+1})$ by replacement for the variable x binding by $\forall^k$ individual names of any individual of U (of the denotation of this variable) and for remaining variables in it also individual names of any individuals of U.

Thus, for any $k \geq 1$

$$\underline{d}(\forall^k) \in ONTs_{k-1}/n_1 s_k = ONTs_{k-1}^{ONTn_1 \times ONTs_k}.$$

Similarly for $\underline{d}(\exists^k)$:

(a) for $k = 1$ and any $\alpha(x) \in S_1$

$$\underline{d}(\exists^1 x\alpha(x)) = \underline{d}(\exists^1)(\underline{d}(x)), \underline{d}(\alpha(x))) \in T \quad \text{iff} \quad T \cap \underline{d}(\alpha(x)) \neq \emptyset$$

(b) for $k = j + 1 (j > 0)$ and any $\alpha(x_1, x_2, \ldots, x, \ldots, x_{j+1}) \in S_{j+1}$

$$\begin{aligned} &\underline{d}(\exists^{j+1} x\alpha(x_1, x_2, \ldots, x, \ldots, x_{j+1})) = \\ &\quad = \underline{d}(\exists^{j+1})(\underline{d}(x), \underline{d}(\alpha(x_1, x_2, \ldots, x, \ldots, x_{j+1}))) = \\ &= \quad \{s \in St \mid s = \underline{d}(\alpha^o(x_1/u_1, x_2/u_2, \ldots, x/u, \ldots, x_{j+1}/u_{j+1})) \\ &\quad\quad \text{for some } u \in U, \text{ any } (u_1, u_2, \ldots, u_{j+1}) \in U^j\}. \end{aligned}$$

Thus, for any $k \geq 1$

$$\underline{d}(\exists^k) \in ONTs_{k-1}/n_1 s_k = ONTs_{k-1}^{ONTn_1 \times ONTs_k}.$$

4.3 The Syntactic and Semantic Compatibility of Quantifiers

In our categorial approach to syntax and semantics of the 1st-order formalized language L_1 its quantifiers have been treated as context-dependent two-argument functors-functions of different categorial types $k > 0$ (defined on the set *Var* of all its individual variables and the set of all its sentential functions S_k with exactly k free variables) and with values in the set of sentential functions S_{k-1} possessing one free variable less or, in particular, in the set of sentences S:

$$\forall^k, \exists^k : Var \times S_k \to S_{k-1} \qquad (S_0 = S).$$

Thus, according to the condition (*CATf*), quantifiers $\forall^k$, $\exists^k$ belong to syntactic categories:

$$(CAT\forall^k, \exists^k) \qquad \forall^k, \exists^k \in CATs_{k-1}/n_1 s_k = CATs_{k-1}^{CATn_1 \times CATs_k} \qquad (s_0 = s),$$

and it means that they satisfy the principle (*SC*) of syntactic connection.

It was also shown that for the denotation operations:

$$d, \underline{d} : S(L_1) \to ONT(L_1)$$

their denotations, according to the condition (*ONTf*), belong to ontological categories:

$$d(\forall^k), \underline{d}(\forall^k), d(\exists^k), \underline{d}(\exists^k) \in ONTs_{k-1}/n_1 s_k = ONTs_{k-1}^{ONTn_1 \times ONTs_k}. \qquad (ONT\forall^k, \exists^k)$$

5 Conclusions

From the conditions ($CAT\forall^k, \exists^k$) and ($ONT\forall^k, \exists^k$) follow the following conclusions:

1. *the 1st-order quantifiers* $\forall^k, \exists^k (k > 0)$ *satisfy the principle of syntactic connection (SC) and the principle of categorial compatibility (CC)*
   and
2. *the problem of standard quantifiers is solved by employing the conceptual apparatus and statements of the outlined theory of categorial languages.*

It should also be noted that

3. *in languages with other operators biding variables the problem of their denotations can be solved in an analogous way,*
   but
4. *for branching quantifiers used in Independence-Friendly logic (see Hintikka* [11]) *the outlined here denotational (compositional) semantics does not work.*

However,

5. *according to Frege's ideas, the proposed categorial approach to language syntax and semantics can be developed in the same spirit for formalized languages of higher order than 1.*
6. *the proposed approach to semantics of the 1st-order formalized languages of differ from the standard in the Tarski's approach* [26] *and other improved versions; first of all it refers to the concept of denotation of any language expression instead to the concept of satisfaction—the crucial ancillary notion in the definition of truth; this notion may be omitted in the definition of the concept of a true sentence and probably replaced by the notion of denotation.*

References

1. Ajdukiewicz, K.: Die syntaktische Konnexität. Stud. Philos. **1**, 1–27 (1935) [English translation: Syntactic connection. In: McCall, S. (ed.) Polish Logic 1920–1939, pp. 202–231. Clarendon Press, Oxford (1967)]
2. Ajdukiewicz, K.: Związki składniowe między członami zdań oznajmujących [Syntactical relations between constituents of declarative sentences]. Stud. Filozoficzne **6**(21), 73–86 (1960) [First presented in English at the International Linguistic Symposium in Erfurt, 27 September–2 October 1958]
3. Bocheński, J.M.: On the syntactical categories. New Scholasticism **23**, 257–280 (1949)
4. Cresswell, M.J.: Logics and Languages. Mathuen and Co. Ltd, London (1973)
5. Curry, H.B.: Grundlagen der kombinatorischen Logik. Am. J. Math. **52**, 509–536, 789–834 (1930)
6. Curry, H.B.: Some aspects of grammatical structure. In: Jakobson, R. (ed.) Structure of Language and Its Mathematical Aspects, vol. 12, pp. 57–68. AMS, Providence, RI (1961)

7. Curry, H.B., Feys, R.: Combinatory Logic, vol. 1. North Holland, Amsterdam (1958)
8. Frege, G.: Begriffschrift, eine der arithmetischen nachbildete Formalsprache des reinen Denkens. Halle (1879) [English translation in: Geach, P.T., Black, M. (eds.) Translations from the Philosophical Writings of Gottlob Frege. Blackwell, Oxford (1970)]
9. Frege, G.: Die Grudlagen der Arithmetik: eine logische-mathematische Untersuchungen über den Begriff der Zahl. Breslau (1884) [English translation: The Foundations of Arithmetic: A Logico-Mathematical Enquiry into the Concept of Number. Blackwell, Oxford (1950)]
10. Frege, G.: Über Sinn und Bedeutung. Zeitschrift für Philosophie und pilosophishe Kritik **100**, 25–50 (1892) [English translation in: Feigel, H., Sellars, W. (eds.) Readings in Philosophical Analysis. Appleton-Century-Crofts, New York (1949), and also In: Beaney, B. (ed.) The Frege Reader, pp. 151–171. Blackwell, Oxford (1997)]
11. Hintikka, J.: Principles of Mathematics Revisited. Cambridge University Press, Cambridge (1996)
12. Husserl, E.: Logische Untersuchungen. Vol. I, Halle (1900), Vol. II, Halle (1901)
13. Leśniewski, S.: Grundzüge eines neuen Systems der Grundlagen der Mathematik. Fundam. Math. **14**, 1–81 (1929)
14. Leśniewski, S.: Über die Grundlagen der Ontologie. Comptes rendus des séances de la Société des Sciences et des Lettres de Varsovie, Classe II **23**, 111–132 (1930)
15. Lindström, P.: First-order predicate logic with generalized quantifiers. Theoria **32**, 186–195 (1966)
16. Montague, R.: Universal grammar. Theoria **36**, 373–398 (1970)
17. Montague, R.: Formal Philosophy: Selected Papers of Richard Montague. Yale University Press, New Haven, CT (1974). Ed. and introd. R.H. Thomason
18. Mostowski, A.: On generalization of quantifiers. Fundam. Math. **44**, 12–36 (1957)
19. Nowaczyk, A.: Categorial languages and variable-binding operators. Stud. Logica **37**, 27–39 (1978)
20. Simons, P.: Combinators and categorial grammar. Notre Dame J. Form. Log. **30** (2), 241–261 (1989)
21. Simons, P.: Languages with variable-binding operators: categorial syntax and combinatorial semantics. In: Jadacki, J., Paśniczek, J. (eds.) The Lvov-Warsaw School—The New Generation. Poznań Studies in the Philosophy of Sciences and Humanities, vol. 89, pp. 239–268. Rodopi, Amsterdam/New York (2006)
22. Suszko, R.: Syntactic structure and semantical reference, Part I. Stud. Logica **8**, 213–144 (1958)
23. Suszko, R.: Syntactic structure and semantical reference, Part II. Stud. Logica **9**, 63–93 (1960)
24. Suszko, R.: O kategoriach syntaktycznych i denotacjach wyrażeń w językach sformalizowanych [On syntactic categories and denotation of expressions in formalized languages]. In: Rozprawy logiczne [Logical Dissertations to the Memory of Kazimierz Ajdukiewicz], pp. 193–204. PWN, Warsaw (1964)
25. Suszko, R.: Ontology in the tractatus of L. Wittgenstein. Notre Dame J. Form. Log. **9**, 7–33 (1968)
26. Tarski, A.: The semantic notion of truth: and the foundations of semantics. Philos. Phenomenol. Res. **4**(3), 341–376 (1944)
27. van Benthem, J.: Essays in Logical Semantics. Reidel, Dordrecht (1986)
28. van Benthem, J.: Quantifiers in the world of types. In: van der Does, J., van Eijck, J. (eds.) Quantifiers, Logic and Language, pp. 47–61. Stanford University, Stanford, CA (1996)
29. van Benthem, J., Westerståhl, D.: Directions in generalized quantifier theory. Stud. Logica **53**, 389–419 (1995)
30. Wybraniec-Skardowska, U.: Theory of Language Syntax. Categorial Approach. Kluwer Academic, Dordrecht/Boston/London (1991)
31. Wybraniec-Skardowska, U.: Logical and philosophical ideas in certain approaches to language. Synthese **116**(2), 231–277 (1998)
32. Wybraniec-Skardowska, U.: On denotations of quantifiers. In: Omyła, M. (ed.) Logical Ideas of Roman Suszko, pp. 89–119. Faculty of Philosophy and Sociology of Warsaw University, Warsaw (2001)
33. Wybraniec-Skardowska, U.: On the formalization of classical categorial grammar. In: Jadacki, J., Paśniczek, J. (eds.) The Lvov-Warsaw School—The New Generation. Poznań Studies in the Philosophy of Sciences and Humanities, vol. 89, pp. 269–288. Rodopi, Amsterdam/New York (2006)
34. Wybraniec-Skardowska, U.: On language adequacy. Stud. Log. Grammar Rhetor. **40** (53), 257–292 (2015)

35. Wybraniec-Skardowska, U.: Categories of first order quantifiers. Bull. Symb. Log. **22**(3), 427–428 (2015). doi: 10.1017/bsl.2016.22
36. Wybraniec-Skardowska, U.: Categorial compatibility of the 1st-order quantifiers. Non Class. Log. Theory Appl. **8**, 134–140 (2016)

U. Wybraniec-Skardowska (✉)
Cardinal Stefan Wyszyński University in Warsaw, Warsaw, Poland
e-mail: uws@uni.opole.pl

The Lvov-Warsaw School: A True Mythology

Jean-Yves Beziau

Dedicated to Jan Zygmunt, my host during my first stay in Poland, Wrocław, 1992–93

Abstract I discuss various aspects of the Lvov-Warsaw School: its past, present and future; its location, evolution, mathematics; the variety of its members. I develop this analysis on the basis of my 25-year experience with Poland.

Keywords Lvov-Warsaw School · Twardowski · Leśniewski · Łukasiewicz · Tarski · Polish logic · Metalogic · Methodology of deductive sciences · Consequence · Universal logic

Mathematics Subject Classification (2000) Primary 01A72; Secondary 0303, 03A05, 03B22, 03B45; 03B50; 03B53; 03C95

This paper is a mix between personal recollections and logico-philosophical reflections about the glorious Lvov-Warsaw School. When supervising the edition and production of the present book in the series *Studies in Universal Logic* I started to think about my experience with Poland and I realized how much I had been connected to this tradition from the very start of my research up to the development of my present and future projects. So I decided to develop the topic, to look at all the archives I have, and to investigate more about the Lvov-Warsaw School.[1]

---

[1] I have consulted many books and papers, among them: 4–6, 31, 43, 48, 49, 55, 57, 62, 65–67, 75, 77, 89, 92, 93, 95, 99, 102 and others also quoted in the list of references of this paper, which gathers the main works on the topic.

Á. Garrido, U. Wybraniec-Skardowska (eds.), *The Lvov-Warsaw School. Past and Present*, Studies in Universal Logic,
https://doi.org/10.1007/978-3-319-65430-0_55

1 From Bahía Blanca to Wrocław

1.1 Meeting with Stan Surma at the End of the World

It is like a fair trail: in August 1992 I took a train from Bahía Blanca, Argentina to Wrocław, Poland. It was an all night train and by no chance I was seated next to Stan Surma and his son Charles (Fig. 1). We said good bye in Buenos Aires this August 1992 and we were to meet again only in August 2007, 15th years later, for the *2nd World Congress on Universal Logic* (UNILOG) I organized in Xi'an, China. In the mean time I developed the Universal Logic project, defending a PhD in Paris (July 1995), organizing the 1st UNILOG in Montreux (March 2005) and launching the journal *Logica Universalis* (January 2007), a project rooted in Poland, considering in particular that I coined the expression "Universal Logic" when in Wrocław in 1992–93.

Stan Surma was the head of the logic group at the Jagiellonian University in Kraków during the sixties, many of his students became later main logicians in Poland: in particular Grzegorz Malinowski, Jan Woleński, Jan Zygmunt, Andrzej Wroński, Jerzy Perzanowski. He founded there in 1965 the journal *Prace z Logiki* which in 1973 became *Reports on Mathematical Logic*. He then escaped communism going first to Black Africa where he was teaching mathematics in the jungle, then Australia, then New Zealand (In Xi'an, his son Charles told me this story and we also produced an interview/film with

Fig. 1 JYB with Stan and Charles Surma at the 2nd UNILOG in Xi'an, 2007

Stan). When in New Zealand he edited *The Collected Works of Leśniewski* (cf. [50]). A very important book to maintain alive the Lvov-Warsaw tradition, since Leśniewski is one of the main figures of this school.

Still in Kraków, Surma "originated an ambitious program of reconstruction of various results obtained in past in logic but presented in an incomplete or inaccurate way. Surma's team was able to offer a complete and rigorous setting for some of Lindenbaum's, Post's, Wajsberg's and others' results." [84, p. 106]. April 27–29, 1973 he organized in Kraków the XIX edition of the conference of the history of logic with subject the scientific achievement of Mordchaj Wajsberg (published in the *Bulletin of the Section of Logic* the same year, see [69]). Surma also published the collected works of Wajsberg [81], and later on a paper entitled "The logical work of Mordachj Wajsberg" [71]. Surma wrote also several papers on Lindenbaum (see [70] and [68]).

But Surma's work did not restrict to this historical/editorial line of work, he pursued logical research in the very spirit of the Polish tradition. At the *9th Latin American Symposium on Mathematical Logic* which took place August 10–14, 1992 in Bahía Blanca he gave the talk "Alternatives to the consequence theoretic approach to metalogic" (the corresponding paper was published in the IX SLALM proceedings [72]). This work was much in the spirit of what I was working on for my PhD at this time. When I created the journal *Logica Universalis*, his paper "A Galois connection" was included in the first issue [73].

1.2 The Polish Brazilian Connection

I left Paris in August 1991, where I was doing a PhD in mathematical logic, to work 1 year with Newton da Costa in São Paulo, Brazil. I was first interested in the work of da Costa through paraconsistent logic, the topic of my Master in mathematical logic [8]. But when I came to Brazil we were working on what da Costa calls the *Theory of Valuation*, a general theory of logics based on bivaluations, that he started to develop after having provided non truth-functional bivalent semantics for his systems of paraconsistent logic and other non-classical logics. This is what I presented at the IXth SLAM in Bahía Blanca.[2]

In view of extending and generalizing this general theory of logic I started to use the expression "abstract logic". There were two reasons do to so: it fits the spirit of modern mathematics, a general theory of logical systems would be a theory of logical structures, and a good name for a logical structure would be "abstract logic" in the same way as "abstract algebra". The second point is more philosophical, emphasizing abstraction of the objects. From this perspective my project was at this time to have a PhD with title *From Formal Logic to Abstract Logic*.

Then, I discovered the work of Suszko at the library of the Department of Mathematics of the University of São Paulo. Suszko had himself used the expression "abstract logic".[3] I started to seriously think of going to Poland to know more about all this. I talked with da

[2] A systematic paper on that was published in *Logique et Analyse* [32] and a more philosophical note [33], related to the talk I presented in Bahía Blanca was published in the same proceedings as Surma's paper; see also Loparic and da Costa's paper reprinted in the anthology of universal logic [21] with my comments.

[3] See about this, one of Suszko's papers reprinted in the anthology of universal logic [21], with comments by Ramon Jansana and my paper "La logique abstraite au sein de la mathématique moderne" [9].

Costa and he encouraged me to do so. He had himself developed strong links with Poland, having been there several times in the 1960s and 1970s and inviting Polish logicians to come to Brazil: Lech Dubikajtis, Jerzy Kotas, Grzegorz Malinowski, Jerzy Perzanowski. And Ryszard Wójcicki, who wrote his book *Lectures on propositional calculi* [82] there:

> Although the work on this book was concluded only after my return to Poland, the substantial part of it have been done during my stay in Brazil sponsored by FAPESP (Fundação de Amparo a Pesquisa do Estado de Sao Paulo, Brazil, grant no 80/1188–8). I benefited a great deal and in various ways from the opportunity to have scientific context with my Brazilian colleagues and friends. My greatest debt has been to Prof. Ayda I. Arruda, at that time the Director of Instituto de Matemática e Estatística e Ciência de Computacão, Universidade Estadual de Campinas both for the care she took for creating me excellent conditions for work and for her keen and penetrating interest in the ideas I discussed in my lectures. Also I own a special debt to Prof. Newton C. A. da Costa for his invitation to Instituto de Matemática e Estatística, Universidade de São Paulo and stimulating discussions we held, and to Prof. Elias Alves for his introducing me to people from the Logical Center of UNICAMP and his assisting me on many occasions.

This book is a overall presentation and synthesis of the general approach to logical systems mainly develop by Polish logicians, which is sometimes qualified as *Polish Logic*. Wójcicki later wrote a second version of this book [83], with more details, which can be considered as the Bible of Polish logic, internationally published and promoted by Kluwer (now Springer). I myself prefer the style of the first edition, published by Ossolineum in Poland, which is nowadays difficult to find.

Newton da Costa himself was born in Curitiba, South of Brazil, which is the second most important Polish city in the world outside of Poland, after Chicago. And Brazil also has the second largest community of Polish immigrants after the USA, nearly two millions. The Polish colony was started by Sebastian Woś (1844–1933) from the city of Siolkowice (region of Opole). Then, between the two world wars, there was an important second wave of Polish immigration to Brazil, in particular many Jews. Famous among them is Leopoldo Nachbin (1922–1993), considered as the most important Brazilian mathematician, funding member of the IMPA (Institute for Pure and Applied Mathematics) in Rio de Janeiro, whose PhD was directed by Laurent Schwartz.

1.3 Polish Surroundings

Besides the logical attraction to Poland I was also interested to go to Poland for various reasons. When I was a student in Paris in the 1980s, two Polish artists were very popular: Roman Polanski and Witold Gombrowicz. I saw many movies by Polanski in particular *Repulsion* (1965) with Catehrine Deneuve and *Le Locataire* (1976) with Isabelle Adjani. I also saw him in a theatre on a one man show, interpreting *The Metamorphosis* by Franz Kafka. From Gombrowicz I read in that order three books: *Cosmos* (1965), *Ferdydurke* (1937), *Possessed* (1939) and was quite impressed. I saw the movie adaptation of *Ferdydurke* directed by Jerzy Skolimowski, a quite good adaptation (never easy to make a good movie about a literature masterpiece). Skolimowski is a famous Polish filmaker who has collaborated with Polanski in the seminal *Knife in the water* (1962), both being from the famous Łódź's film school. I also saw Skolimowski's very good movie, *Torrents of Spring* (1989), adaptation of a novel by Turgenev, with Nastassia Kinski, the daughter of the legendary actor Klaus Kinksi, also of Polish origin. There are several connections between Polish logicians and movies. Lindenbaum's father was a film producer (see

[104]) and Roman Suszko appeared in a mythical post-war Polish movie *A trip down the river* (1970) directed by Marek Piwowski (original title in Polish: *Rejs*).

Gombrowicz was exiled in Argentina and became famous when the French publisher Maurice Nadeau (1911–2013) got interested in his work. René Goscinny (1926–1977) is another Polish guy connecting Poland, Argentina and France. Of Polish Jewish origin, born in France, he spent his youth in Buenos Aires, moved back to France after WWII and became a cultural icon through comic books: *Astérix* of course, but also *Lucky Luke* and *Iznoud* (my favorite one). With *Astérix*, Goscinny promoted the Gauls as a founding cultural myth of French identity. It became a symbol of French culture in the same way as Mickey Mouse is a symbol of American culture. This myth is a mix of many things in particular figures imported from Argentina (cf. *Patoruzú*). Goscinny was really good at puns, in particular multilinguistic puns, therefore *intraduisible*... The name of his main hero *Astérix* has a double or triple meaning, one of them being *king of the stars*. The French mathematical society created in 1973 a journal called *Astérisque* (Fig. 2), punny name referring to this symbolic figure in the continuity of the Bourbachic funny naming tradition, rich of meaning and quite spiritual in the French way ("avoir de l'esprit")—one of a few foreign members of Bourbaki was Samuel Eilenberg (co funder of category theory with MacLane), a Polish Jew.

At the Sorbonne I was a student of Sarah Kofman, also from Polish Jewish origin. Her father was a Rabbi who was deported during WWII and was beaten to death in Auschwitz, because he didn't want to work during the Sabbath, by a Jewish kapo, who later on became a successful merchant in Paris. She wrote a very nice book about Plato: *Comment s'en sortir*, literally: *How to get out?* This is a book about the notion of "aporia" (Greek etymology: without a path). She pointed out that in the dialogues, when facing an "aporia", a deadlock, by using rational thinking, dialectics, then Plato uses the myth to go out, telling some stories. She showed very well that at the heart of Plato's philosophy there is a constant mix between *logos* and *muthos*, a true mytho-logy. Unfortunately, she thought, the mythical dimension was washed away in Aristotle's philosophy. She was one of my favorite teachers and I did a Master thesis with her on Plato's cave [7].[4]

On my family side, one of my mother's uncles emigrated from Marocco to Argentina in the 1950s and his three daughters all married Polish emigrants there. Moreover the sister of my mother married a guy of Polish origin, Jean Dybowsky, descendant of the famous

Fig. 2 Astérix and mathematics

[4]She later on committed suicide and I dedicated to her a paper I wrote on suicide [11]. When in Poland I visited Auschwitz and this was a terrific experience.

African explorer bearing the same name, giving birth to three neo-Polish cousins. Finally, my father told me several times that my mother (of Swiss citizenship) had some Polish Jewish ascendancy. I don't know if it is true and I never inquired to check this information, because on the one hand if it is true, it is a logical truth, a tautology, since good logicians are all Polish Jews, on the other hand I don't feel the necessity to identify myself with a specific community, subculture, or abelian subgroup. To simply be a rational animal is already quite complicated!

Anyway, for all these reasons, I had some sympathy for Poland. And after going from the 1st World (France) to the 3rd World (Brazil), I was curious to discover the 2nd World, which was a mystery for me. I remember that during an exchange in high school in France I spent 2 weeks in Bayreuth in 1979 and one of the attractions was to go to watch the iron curtain (Czech border). On the other side it looked really like another world. I knew very few things about Poland. In the train from Bahía Blanca to Buenos Aires Stan Surma draw me a logic map of Poland. As the erudite reader may know it is not possible to go all the way by train from Argentina to Poland. I took a bus from Buenos Aires to São Paulo, then a plane to Paris and then again a train to Poland. I stopped a few days in Brazil to meet again Newton da Costa and in France to visit my family.

I was going to Poland within an exchange program between France and Poland. The French Ministry of Foreign Affairs sent me a train ticket and I took a train from Paris to Warsaw on October 1st, 1992 (Fig. 3). It was the train Paris-Moscow, a 48 h trip, Warsaw being at the middle of the way, a 24 h trip. I was in a compartment with a friendly Polish professor. Arriving in Warsaw there was nobody from the French Embassy waiting for me as promised (Fig. 4). I didn't know where to go. I had lots of baggages, few money and didn't speak Polish. Delikatnej sytuacji ...

Fig. 3 JYB and grandma Alice, Oct 1 1992, France: departure to Poland

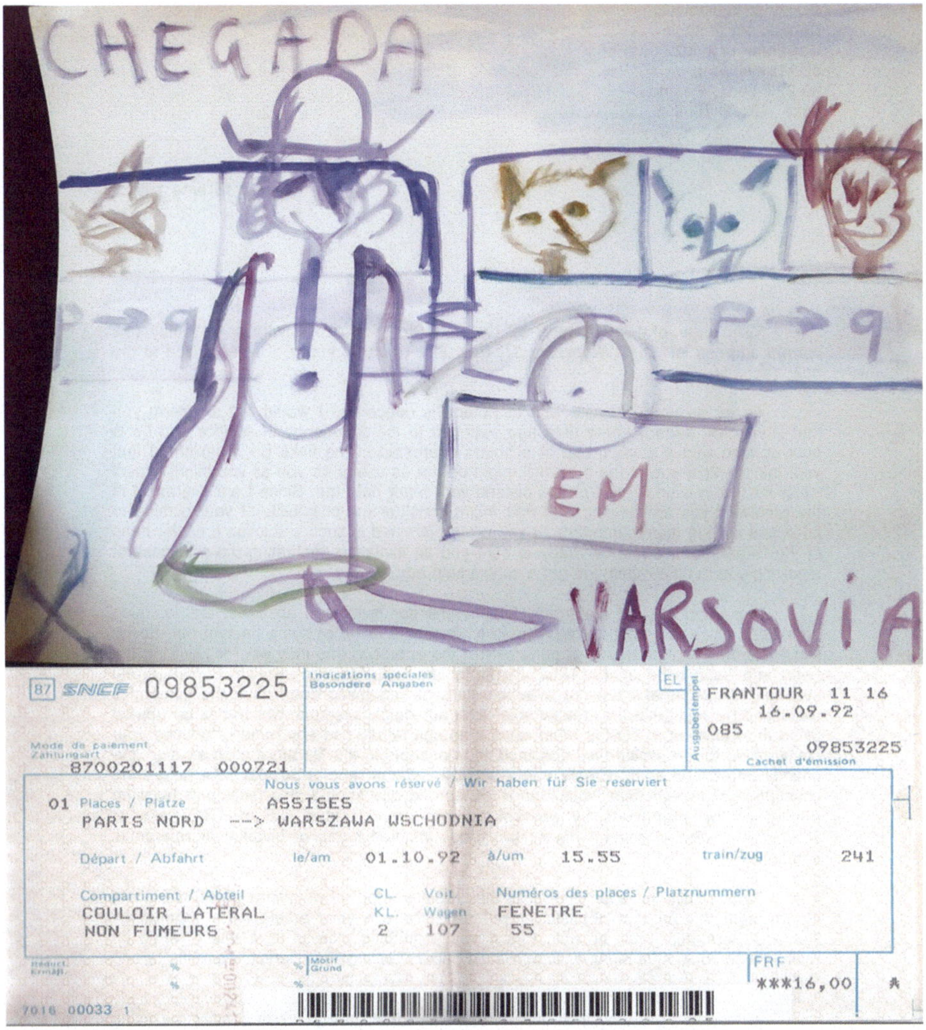

Fig. 4 Logical arrival in Warsaw

I met by chance on the platform a French girl who was in the same situation as me and the Polish professor helped us to call the French Embassy. It was a Friday, late afternoon. They told us they right now could not help us, but that we could use the diplomatic visitor flat at the Embassy and that we will talk on Monday. The flat was OK but with only one bed so we had to sleep on the same bed. And during the weekend I walked around Warsaw with this girl, I think her name was "Sylvie". Monday morning we had a meeting with the staff at the French embassy and they put us on a train to Wrocław. For this girl this was normal since she was supposed to study at the Art School there. In my case this was quite a surprise because I was prepared to go to Łódź. When in Brazil I had written to Wójcicki, he didn't reply to me, but after some months I received a letter from Grzegorz Malinowski inviting me to join his group in Łódź (Fig. 5).

Grzegorz Malinowski
Department of Logic
University of Łódź,
ul. Matejki 34 a
90-237 Łódź, Poland
Fax: 48 42 783958

Nantes, 15 june 1992.

Dear Mr. Beziau,

Your letter of 13 February, 92 reached Prof. R.Wójcicki with some delay (his current address is: UPT Warszawa 37, P.O. Box 61) and, finally, it was passed to me only recently.

As far as your possible visit to Poland is concerned I would like to inform you that R.Wójcicki already some time ago switched to the philosophy of science and he is now working with a small group of scholars interested in the field. So, I am afraid that your joining his group at the moment might be not as useful for you as you might expect - just for that reason Prof. Wójcicki passed your letter onto me. Since I am interested in the problems you are dealing with one might consider the possibility of your coming to Łódź and staying there some time. If you were interested in such a solution I, as the head of the Department, would be ready to sent you an invitation confirming the readiness of receiving you in case when you get a scholarship from France.

I passed through the article "Recherche sur la logique abstraite: Les logiques normales" which you sent us rather quickly. It seems to be of some interest but, taking into account the high degree of generality of the approach one may ask for justifying the need and usefulness of it. Thus, in particular: what kind of *non-structural* formal systems (*logics*) that have not available kind of semantics does have the semantics discussed by you, what are "reasonable" abstract logical systems, which may be treated using the framework. On the other hand, one can hardly find any remarks relating your construction to the similar well accomodated concepts in the literature such as e.g. the notions introduced some twenty years ago by S.Bloom and R.Suszko. As far as the conception of two-element valuations is concerned one should also receive at least an outline of the treatments by several authors among them D.Scott, R.Suszko et R.Wojcicki, which in several places provided justified forms of bivalent or referential semantics.

Since I have intension to discuss the problem of logical valuations of N.C. da Costa on my seminar, on that occasion I will pass through your article more thoroughly. Perhaps thereafter I will be able to say you something more: I think that it would be useful if you could sent me your unpublished paper "Les logiques paraconsistantes Ci and C1".

In Nantes I will stay until July 1, 92. After that date you can contact me using my constant address (as above).

With best regards,

Sincerely yours

Grzegorz Malinowski

Grzegorz Malinowski

PS. Please convey my best wishes and regards to Prof. Newton da Costa.

Fig. 5 Malinoswki's letter inviting me to join his logic group in Łódź

But for some reasons the French Ministry of Foreign Affairs organized my coming to Worcław. That's how I arrived in this town I knew nothing about. On the platform was waiting for me Jan Zygmunt who drove me in his little Fiat to a University House in Plac Grunwaldzki (cf.[39]).

2 The Atopicity of the Lvov-Warsaw School

The expression "Lvov-Warsaw School" has become canonical. It is rather a proper name than a definite description. But, like many proper names, it is also improper. It can reasonably serve as an identification device, we roughly know what we are talking about, like when we use names such as "Poland", "Truth" or "Alfred Tarski". But this does not mean we know exactly what it is. Identification does not confer identity. Maybe Poland and Truth do not have a proper identity. The identity of Alfred Tarski is more palpable, tangible, at least imaginable, paintable, photographical, not (yet!) completely mythical. The consistency of a phenomenon can be tested by trying to answer the five basic questions: when, where, what, why, how? Let's see if LWS passes the test…

I will use "LWS" as an acronym/abbreviation for the Lvov-Warsaw School. Sometimes people use "LWs" or "LW-s". I prefer to capitalize the School, as they do in Vienna with the Circle. Note also that in the canonical expression "Lvov-Warsaw School", "Lvov" has become a standard English spelling for the name of this city, but the tendency is nowadays to write "Lviv".

2.1 A School Without a Location

I progressively discovered that I arrived not only in Wrocław but also in Lvov and Breslau. Before WWII the city of Wrocław was named "Breslau" and was in Germany. The Friedriech Wilhelm University of Breslau was an important University frequented by famous people like the Nobel prizes Erwin Schrödinger, Max Bohr or Friedrich Bergius. The book by Frege *Die Grundlagen der Arithmetik, eine logisch-mathematische Untersuchung über den Begriff der Zahl* was published in Breslau in 1884 by W. Koebner.

After WWII "Breslau" became "Wrocław" and the surrounding region became part of Poland. On the other hand the city of Lvov and the surrounding region, which was part of Poland before WWII, became part of Soviet Union. Poland was displaced on the West (Fig. 6). This displacement was not purely physical, but also humanistical. Poland, like many countries, has a variable geometry…

Wrocław is a sort of transposition of Lvov. Polish people from Lvov moved to Wrocław. German people of Breslau fled or were expelled "home", i.e. in what would be called "DDR", *Deutsche Demokratische Republik*, now out of the map, like Prussia, Atlantis or Gaul.

Wrocław University became a major university of Poland after WWII, Wrocław a major student town, and also one of the most prosperous Polish towns, in particular due to his President (or Mayor) Rafał Dutkiewicz, who did study logic at Wrocław University and that I shortly met during my stay—he was not yet famous (Fig. 7).

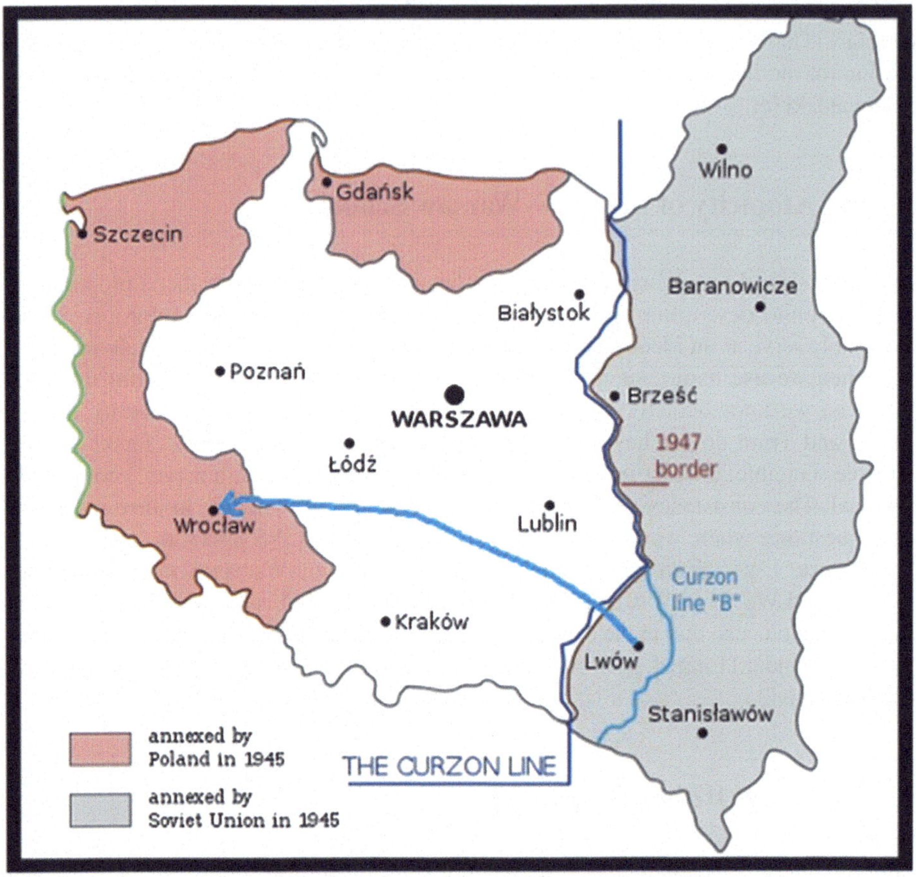

Fig. 6 After WWII Poland goes West—Lvov moves to Wrocław

Fig. 7 Rafał Dutkiewicz logically successful mayor of Wrocław

A symbol of the transposition of Lvov to Wrocław is the Ossolineum foundation. This foundation was created by the Count Józef Maksymilian Ossoliński (1748–1826) in Lvov. I used to have lunch at the canteen of the Ossolineum foundation which was one block from the Department of Logic in the same street, Szewska street. Ossolineum was during the communist time the main publisher of the Polish Academy of Science, publishing many books of logic and the journal *Studia Logica*. When I arrived there everything was changing. Wójcicki made a deal in 1992 with Kluwer (which later on was incorporated into Springer) and Jan Zygmunt was the editor-in-chief of *Studia Logica*. I went to the Ossolineum office in Wrocław asking if it was possible to buy old issues of the journal. I succeeded to buy a good number of issues which were in the cellar for a price as low as 2 or 3 cents each.

Wrocław became after WWII one of the most important logic centers in Poland. Wójcicki describes the situation as follows:

> Wrocław was another important center of postwar logic. The most prominent figure in foundations of mathematics has been Czeslaw Ryll-Nardzewski. He grouped around himself several very talented young peoples (L. Pacholski, B. Węglorz, A. Wojciechowska, and others) but that group was much more oriented towards foundational than purely logical problems. Also Jerzy Łoś, known for a number of outstanding theoretical results, notably his famous ultraproduct theorem, started his academic career in Wrocław. Another eminent Wrocław logician was Jerzy Słupecki. His most prominent collaborators were Ludwik Borkowski, Witold A. Pogorzelski and Boguslaw Iwanuś. ... my own academic biography starts in Wrocław too. [84, pp. 503–504].

Considering all these aspects we can maybe talk about a Lvov-Warsaw-Wrocław school of logic. But on the one hand Wrocław has not been the only and main center of logic after WWII in Poland and on the other hand there was a serious break in the development of logic in Poland after WWII, in particular with the emigration of Alfred Tarski to the USA where he founded a center for logic in Berkeley (California) who became during 25 years the most important logic center in the world. Considering this move, one could also talk of a Lvov-Warsaw-Berkeley school of logic. But this also would be controversial because few people of the "original" school followed Tarski in California—one of them was the young Kalicki who tragically died there in a car accident (see [101]).

Słupecki tried without success to develop a logic school in Wrocław, as explained by Woleński and Zygmunt:

> Since the beginning of his activities in Wrocław, Slupecki supervised graduate students and doctors, evaluated doctoral and habilitation theses and influenced mature logicians. This circle was formed by Edward Baluka, Ludwik Borkowski, Edmund Glibowski, Boguslaw Iwanuś, Tadeusz Kubinski, Jerzy Nowak, Witold A. Pogorzelski, Juliusz Reichbach and others, but it was not a school as Slupecki would have envisaged it, and such a school that could not develop in Wrocław mathematical circles, of which he was fully aware. [98]

My host in Wrocław, Jan Zygmunt, was the director for the Department of Logic, which full and official name was *Katedra Logiki i Metodology Nauk*. It literally means *Chair of Logic and Methodology of Science*. It was part of the Institute of Philosophy located in a building which was very close to the main building of the university where there was the beautiful "Aula Leopolodina" and a nice cafeteria to which Jan Zygmunt used to frequently invite me to chat having a tea.

At the department I had a big office room. On the wall it was written: "Alfred Tarski and Dana Scott were there in 1955". I don't know the exact impact of this on my mind, but I spent lots of time reading all the papers of Dana Scott on rules and general logic and wrote in January 1993 the paper "Rules, derived rules, permissible rules and the various types of systems of deduction" [12]. Later on I decided to include one of these papers by Scott, "Complenetess and axiomatizibility in many-valued logic", in the anthology of universal logic and asked Lloyd Humberstone, one of his former students to write the presentation of it (see [21]). This room was the one of Tomasz Skura, who at this time was in an exchange program in Konstanz, Germany invited by Andre Fuhrmann (who later on moved to Brazil and became my friend). Other permanent members of the department were Tomasz Furmanowski and Jacek Hawranek. I also met at this time the East-German logician Max Urchs who was partially employed at the department. The present director of the department, Marek Magdziak, was at this time doing his PhD there. There was a regular seminar to which I presented two talks (as recalled by Zygmunt [39]) and to which was coming Juliusz Reichbach, a retired logician, who had done some important work on completeness in the 1950s (cf. [60, 61]) and after that emigrated to Israel but didn't succeed to adapt and came back to Poland.

Reichbach was living in the same building as me on Plac Grunwaldzki. It was a Stalinist building called "the house of science" at the corner of Plac Grunwaldzki and Curie-Skłolodowskiej street. I used to bump into him on the nearby supermarket where he was buying vodka. One day I invited him to have a tea in my flat but when he saw my paintings on the wall (Fig. 8) he was quite afraid and decided not to enter. I had made only few painting when I was a child in Corsica and then in high school. For some reasons I started to paint quite a lot when I was living in Wrocław. In some sense this was quite natural if we think of someone like Leon Chwistek who was both a famous logician and painter.

I remember that I bought some tubes of paint on the Russian market just below my flat. At this time the (ex) Soviet Army was leaving the country after many years of occupation selling all they had, from watches to planes. After some years the department of logic was

Fig. 8 One of the paintings I did when in Wrocław

Fig. 9 Back to Wrocław in 1998: JYB and Valdimir Vasyukov

moved from the town center to the ex-Soviet Military based. Visiting Wrocław again in 1998 I had the opportunity to see the place. At this time the Russian logician Vladimir Vasyukov was also there. He was a good friend of Jan Zygmunt and also became a good friend of mine (he invited me to Moscow in 2001 and I went again to Russia in 2003, 2009, 2012, 2016). The father of Vasyukov had been a military officer in activity in Wrocław and Vladimir spent some years of his life there when he was a child and spoke good Polish (Fig. 9).

1992–93 was a great time in Wrocław, it was a transition period. After the liberation of Poland people were optimistic, new bars and restaurants were opening every day. There were lots of foreigners: business men coming to invest, visitors at the university and tourists. At the same time the way of life and style of life of the communist time was still very present. Moreover communism, in particular due to isolation, had frozen things, so it was like a travel in time, back to the 1960s or 1950s and even the 1930s.

Plac Grunwaldzki was used by the Germans at the end of the war as an airfield: most of the buildings were demolished for that and furthermore the region was heavily bombarded by the Russians. After the war it was developed in a "modern" way—after my departure it was further modernized with a big shopping center (Fig. 10). The department of mathematics was located at the entrance of Plac Grunwaldzki, near the bridge. I used to go there at the library, where there was a mix of books from the University of Breslau and the University of Lvov. There were some logicians there, like Węglorz, but I hardly met/knew him. This was in fact symptomatic of the separation between logic and mathematics in Poland which, according to the legend, goes back to a dispute between Sierpiński and Leśniewski. Also, though the department of logic was in the institute of

Fig. 10 Evolution of Plac Grunwaldzki

philosophy, we nearly had no contact with the other parts of the institute, still under heavy influence of Marxism.

During my first stay in Poland I visited also some other towns, in particular Łódź and Kraków which were two other important logic centers in Poland at this time, with important departments of logic. This was not the case of Warsaw. I went there only a few times to solve some questions with the French Embassy.

2.2 A School Without a Topic

How can we characterize the topic of the Lvov-Warsaw School? The Vienna Circle (VC) is strongly associated with "Logical empiricism" and "Logical positivism". It would be difficult to find an expression that could systematically be associated with the name "Lvov-Warsaw School".

There is a famous paper by Ajdukiewicz entitled "Logistic anti-irrationalism" [3]. So one may want to put the equation:

$$\text{LWS} = \text{Logical Anti-Irrationalism}$$

that would be parallel to the equation:

$$\text{VC} = \text{Logical Positivism}$$

But the expression "anti-irrationalism" is too much idiosyncratic and its meaning is not completely clear.[5] It is a bit confuse for a school wanting to promote conceptual clarification. And we may wonder why being so negative, with a double negation. Why not being purely positive? But then we would have something like "Logical Rationalism" which is rather pleonastic.

There is another expression, also used by Ajdukiewicz: "Methodology of deductive sciences". This expression is the title of his habilitation thesis defended in 1921 in Lvov [1], and was then systematically used and promoted by Alfred Tarski, in Polish, German, English and French, in one way or another:

- "Fundamentale Begriffe der Methodologie der deduktiven Wissenschaften. I." (paper published in 1930)
- *O Logice Matematycznej i Metodzie Dedukcyjnej* (book published in Lvov-Warsaw 1936 [76])
- Sur la méthode deductive (talk presented at the 9th International congress of philosophy in Paris in 1937, cf. [78])
- *Introduction to logic and the methodology of deductive sciences* (book published in Oxford in 1941. English version of [76])

The interesting thing about *Methodology of Deductive Sciences*, hereafter MDS, is that it keeps a philosophical perspective which disappears in the case of expressions like "Metalogic" or "Polish logic". The word *Metalogic* was mainly promoted in Poland, in particular it was used by Wajsberg. It is a very important word that rightly characterizes the Polish perspective: on the one hand the distinction between two levels, on the other hand the generally beyond it; it is not only the study of mathematical reasoning, but the deductive reasoning of any science. Tarski was also much interested in biology and physics. It is of course inspired by the *Metamathematics* of Hilbert and there is a direct connection since Ajdukiewicz visited Göttingen. However Polish logicians went much higher at the metalevel, because firstly, as it is well known, and contrarily to Hilbert, they were promoting no limitation at the metalevel, and secondly they had more imagination which led to the idea of arithmetization of syntax, crucial to Gödel's theorem. This was mainly promoted by Łukasiewicz in the early twenties as recalled by Tarski who explained that to Gödel during a visit in Vienna (cf. [80]).

---

[5] See our recent paper "Is logical relativity irrational?" in a special issue of *Studia Metodologiczne* dedicated to Ajdukiewicz [25].

Before WWII were already created some Departments of Logic and Methodology of Science in Poland and this expanded after WWII. Nowadays there are many such departments in Poland, most of them in philosophy institutes, but not always. Can we consider that

$$\text{LWS} = \text{MDS}$$

is a key equation explaining everything? This would be the easiest solution, but a simplification. First it important to notice that between the two wars, LWS was not a tiny part of the university but the dominating part. Łukasiewicz was twice the rector of the University of Warsaw. Good also to remember that Twardowski was rector of the University of Lvov and that Ajdukiewicz, who married his daughter after WWII, became the rector of the university of Poznań (Jaśkowski was also a rector after WWII: rector of Nicolas Copernicus University in Toruń).

Few years ago the University of Warsaw created a symbolic monument with four statues: those of Twardowski, Łukasiewicz, Tarski and Leśniewski. This means that these four men are considered as the most important figures of the history of the University of Warsaw (Fig. 11).

On the one hand we could say that at some point MDS was dominating the University of Warsaw and that it is not anymore the case. Or on the other hand we could say that LWS was much broader than MDS, which is true in some sense if we consider that LWS was an nebulous group of philosophers, mathematicians, linguists and artists.

But looking at the Fig. 11 we see another paradox: if we put aside Twardowski, the bridge between Lvov and Warsaw and considered as the founding father of the LWS (the figure of a father is typically mythical, before and after Freud), the three other men are mainly known as logicians. So why not putting the following equation

$$\text{LWS} = \text{Polish Logic?}$$

Here is what Jan Zygmunt wrote about this expression:

> The term "Polish logic" was coined by McCall to signal the important contributions to modern logic by logicians from Poland between the wars. There were several centres of research, of which

Fig. 11 Monument at Warsaw University with statues of Twardowski, Łukasiewicz, Tarski and Leśniewski

> the Warsaw school, which grew out of the earlier Lvov-Warsaw philosophical movement, was the most significant. Its development was closely connected with the Warsaw school of mathematics, which gave it its characteristic mathematical bent...
> The main centres of logic research between these dates were Cracow, Lwów, Poznań, Wilno and, most importantly, Warsaw. The Warsaw school of logic, founded by Łukasiewicz and Leśniewski, began as an intellectual offshoot of the Lvov-Warsaw philosophical movement, but quickly eclipsed it in both quantity and quality of research. The development of the Warsaw school of logic was closely connected with that of the Warsaw school of mathematics. In 1918 one could already speak of Warsaw as a fairly strong centre of research in set theory and topology under the direction of Janiszewski, Mazurkiewicz and Sierpiński. The two schools shared organizational structures, swapped directors and collaborated on many academic initiatives. This accelerated the development of both schools and deepened their research in key areas. In later years Kuratowski, Lindenbaum, Tarski and finally Mostowski made significant contributions to both schools. [103, p. 6634]

But on the other side there were also philosophers, as described by Wójcicki;

> One may also point out that an interwar (1920–1939) formation widely known as the "Polish School of Logic" overlapped in a very substantial way with another formation known as the "Lwow-Warsaw School of Philosophy". The members of the latter not only considered logic to be the main tool of philosophical analyses but they often contributed to logical investigations by themselves. [84, p. 498].

So at the end we can have the following picture: a group of three men—Lesńiewski, Łukasiewicz and Tarski—mainly logicians with on one side some mathematicians, on the other side some philosophers. In the French version of Wikipedia, the following is said about Leśniewski : "With Jan Łukasiewicz and Alfred Tarski, who was his only doctor, he formed a troika which during the decades 1920 and 1930 made the University of Warsaw one of the most important research centers of mathematical logic in the world." This troika scheme is confirmed by the entry on Łukasiewicz in the English version of Wikipedia: "He remained a professor at the University of Warsaw from 1920 until 1939 when the family house was destroyed by German bombs and the university was closed under German occupation. He had been a rector of the university twice. In this period Łukasiewicz and Stanisław Leśniewski founded the Lvov-Warsaw school of logic which was later made internationally famous by Alfred Tarski who had been Leśniewski's student." And Woleński provided a general picture as follows:

> Alfred Tarski (1901–1983), later recognized as one of the greatest logicians of all times, decided to specialize in logic (he graduated and obtained his PhD under Leśniewski) and became the third pillar of the Warsaw School of Logic (WSL for brevity). In the 1920s and early 1930s these Big Three were joined by (I list them in alphabetical order) Stanislaw Jaśkowski (1906–1965), Adolf Lindenbaum (1904–1941?) (the question marks in the dates indicate that exact data are uncertain or even unknown), Andrzej Mostowski (1913-1975), Mojżesz Presburger (1904–1943?), Jerzy Słupecki (1904–1987), Boleslaw Sobociński (1906–1980) and Mordechaj Wajsberg (1902–1943?). All of them except for Sobociński were mathematicians by training. ... The WSL as a working group had eleven members at its peak, that is, around 1937. Is this large or small? Of course, everything depends on the point of reference. Evaluating from the contemporary point of view, about a dozen people working together in logic is perhaps not so many. However, if one looks at this group from a broader international perspective, one should remember that no other place in the world in which logic was actively done had even one third of this amount. Thus, at the time Warsaw was the place most populated by professional logicians in the world. ... The Warsaw logical community was much larger than the WSL sensu stricto. Some mathematicians, already mentioned above, like Kuratowski, should be included. [94, p. 35]

Woleński also note that: "According to Łukasiewicz, logic is an autonomous subject which is subordinated neither to philosophy nor to mathematics. On this view, logic is no servant of any other science." [91, p. 377] So it makes sense to consider that logic is the heart of the LWS even if we don't use the expression "Lvov-Warsaw School of Logic" to make clear that LWS was not reduced to logic or to at least logic reducely considered.

Polish logic can be understood as literally logic made in Poland. But can we way that there is (or there was) a special way of doing logic in Poland? One may think of some peculiarity such as Polish notation, the search for the unique axiom or the focus on propositional logic. But, at the middle of all this, there is the theory of consequence developed by Tarski, as emphasized by Jan Zygmunt:

> A metamathematical theme was present in Polish logic from the early 1920s, beginning with Ajdukiewicz (1921). Tarski's efforts allowed investigations of the sentential calculi to be carried out within an explicitly metamathematical framework. Tarski generalized this framework to a mathematical theory of two primitive concepts (sentence and consequence) which he called 'the methodology of the deductive sciences'. Within this theory he was able to provide a conceptual apparatus for investigating deductive systems. (Lindenbaum contributed much to this work, including the widely known Lindenbaum maximality lemma.) [103, p. 6636]

and Ryszard Wójcicki:

> Note that already in the early thirties, Tarski, at that time a young docent at Warsaw University, formed his theory of deductive systems. It was intended to be the most general theory of logical reasoning. The central notion of Tarski's theory is a consequence operation, thus an operation which applied to any set of premises yields the set of all the conclusions derivable from (or, under semantic interpretation of the consequence operation entailed by) those premises. Since, depending on one's logical preferences, logical validity (and hence both derivability and entailment) can be understood in different ways, Tarski focused his attention on those characteristics of the consequence operation which are by and large independent of our logical preferences. Curiously enough the theory of deductive systems, known also as the theory of the consequence operation, has attracted little attention outside of Tarski's native country. In this way the theory of the consequence operation became a Polish specialty, not to say idiosyncrasy, often treated as some strange, not to say redundant, province of logical investigations. Only now, after nonmonotonic logic has been invented, the significance of Tarski's idea of studying the consequence operation in an abstract way, independent of any rules of inference derived from a specific system of logic, has become obvious. [84, p. 498].

This theory can be called *Polish logic* because it was developed mainly in Poland and it is not well-known outside of Poland. Tarski himself when he moved to California developed model theory because he had the idea that it was a more general framework and didn't work anymore directly or explicitly on the theory of consequence operator. Results about this theory can be found in the two books of Wójcicki [82, 83] and on the book *Completeness theory for propositional logics* by Pogorzelski and Wojtylak, originally published in 1982 by the Silesian University of Katowice, that we re-edited in 2008 in the book series *Studies in Universal Logic* [56]. A seminal work in this line is the monograph by Łoś, originally written in Polish [51], translated in English by Robert Purdy, but not yet published.

2.3 A School Which Is Not a School

The Aula Leopolidna of the University of Wrocław is nowadays considered as the baroque jewel of Poland. "Baroque" is a name of an artistic movement 1590–1725. The name was given only afterwards by the Swiss historian Jacob Burckardt (1897–1897) in his book *Der Cicerone Eine Anleitung zum Genu der Kunstwerke Italiens* published in 1855 (see [29]). The people who had developed the baroque style as a reaction to protestant's sobriety, didn't identify as a movement and didn't self-baptize them as "baroque".

Burckardt created the Baroque movement, he did not create the word. Jean-Jacques Rousseau notably used this word in his *Dictionary of Music*:

> Baroque : Une musique baroque est celle dont l'harmonie est confuse, chargée de modulations et dissonances, le chant dur et peu naturel, l'intonation difficile, et le mouvement contraint. Il y a bien de l'apparence que ce terme vient du baroco des logiciens. (A Baroque, or rough music, is that, whose harmony is confused, filled with modulations and dissonances, its notes hard and unnatural, the intonation difficult, and the movement constrained. It appears evidently that this term must be derived from the Baroco of the logicians, William Waring's translation). [63]

In syllogistic, Baroco is the name of the fourth mode of the second figure, an example of which would be:

All good logicians are Polish
Some humans are not Polish
Therefore some humans are not good logicians

Strange indeed! Maybe the Lvov-Warsaw School, by its heterogeneity and diversity, can be considered as baroque. Considering self-consciousness and self-baptism, the situation is not so baroque because LWS presented itself as such: at the event organized in 1935 in Paris by Louis Rougier, Ajdukiewicz, in his introductory speech used this expression:

> Presque tous les membres polonais de notre Congrès sont disciples de l'école connue sous le nom d'école de Lwôw et de Varsovie. (Most of the Polish members of our congress are disciples of a school known as school of Lwôw and Warsaw.) [2]

It was later on used again for example by Zbigniev Jordan [47], but the full canonization is certainly due to the works and activities of Jan Woleński (in particular: [86–88]).

With LWS we don't have something as explicit as with the Bourbaki's group, the Surrealist movement or the Vienne Circle. In the two last cases, besides the self-baptism, there were also manifestos. Chapman wrote the following about the Vienna circle:

> Philosophers are not generally known for being team players. Philosophical ideas may sometimes be attributed jointly to two or more thinkers, but it often turns out that these people disagreed on fundamental issues, or that they worked in separate countries or even separate centuries... The Vienna Circle was unusual. It consisted of a large but identifiable group of philosophers who met regularly, collaborated on work of mutual interest, and largely agreed on their conclusions. They even acknowledged their group identity by coining the term *The Vienna Circle*, or *Der Wiener Kreis*, in the title of a collaborative manifesto. [30, p. 7]

Let us compare this with what Surma wrote about LWS:

> A distinctive feature of the school and one of the secrets of its success was the spirit of teamwork. The mutual collaboration among the members was so close and intimate that it is often hard to decide who should be credited with which particular results. [71, p. 102]

But it is more connected with the school of Polish logic. There is a joke, pushing to the extreme this empathy, claiming that all important results proved by Tarski at this time are due to Lindenbaum. What is interesting is that this group of people were able to work together without sharing the same views. Lindenbaum was strongly communist (see [96, 104]), in a rather fanatic and irrational way, this was not the case of Tarski or Łukasiewicz.

Woleński wrote that: "Lindenbaum, Presburger, Tarski and Wajsberg were Jews, but Leśniewski and Sobociński (later also Łukasiewicz) were strongly antisemitic." [94, p. 38] As it is known Tarski changed his name and did as much as possible to hidden his Jewish identity to escape persecution but was inclined to consider Jews has a superior race (see [38]).

In Wikipedia, we find the following description:

> A school of thought (or intellectual tradition) is a collection or group of people who share common characteristics of opinion or outlook of a philosophy, discipline, belief, social movement, economics, cultural movement, or art movement...
> Schools are often named after their founders such as the "Rinzai school" of Zen named after Linji Yixuan and the Asharite school of early Muslim philosophy named after Abu l'Hasan al-Ashari. They are often also named after their places of origin, such as the Ionian School of philosophy that originated in Ionia and the Chicago school of architecture that originated in Chicago, Illinois and the Prague School of linguistics, named after a linguistic circle found in Prague, or Tartu-Moscow Semiotic School whose representatives lived in Tartu and Moscow. (Wikipedia, School of thought)

The name "Lvov-Warsaw School" is related to a name of a (bi)location, not a person. Although, according to the myth Twardowski is the father of LWS, nobody would use the expression "Twardowski School" instead of or as a synonym to LWS. When the name of a school is attached to a location, this means that some people gathered in a place, but shared something which is not limited to the place. In the case of LWS it is clear that a group of people gathered in Warsaw (Lvov in the shadow) and shared something which was part of a general rational movement but whose specific identity is connected to logic, however difficult to characterize.

On the other hand the situation of the group that Tarski developed in California is much simpler. It can be qualified in various ways: Tarski's school, Berkeley's school of model theory, etc. There is a main figure and a main topic. But funny enough, since 1957, it is called *Group in Logic and the Methodology of Science*...

3 The Future of the Lvov-Warsaw School

3.1 No Future

One may wonder what the future of the Lvov-Warsaw school is. A simple and direct reply would be: No Future. We can say that this school ended at the time of the second world war with:

- the death of Leśniewski (natural death few months before the war),
- the departure of Tarski to United States (just before the war),
- the destruction of Warsaw by the Nazi Germany,

- the invasion of Lvov by the Soviets,
- the departure of Łukasiewicz to Ireland (after the war).

These are five main central symbolic points with many corollaries and side-effects: The killing of Lindenbaum, Wajsberg and Presburger. The suicide of Witkiewicz. The departure of Bocheński, Sobociński, Kalicki, etc.

WWII officially started with the invasion of Poland by the Nazi Germany on September 1st 1939. On the other side Poland was invaded by the Soviets on September 17, 1939. During WWII Poland became a German Nazi death camp. Poland was chosen by the Nazis to develop the main extermination camps, the most famous being Auschwitz-Birkenau. The population of Poland before WWII was about 35 millions of people, after, less than 24. It is the country with the highest percentage of its citizens that died during WWII. It is difficult to imagine the survival of any school under such circumstances.

Moreover, after the war, Poland was dominated by the Soviet Union during about 40 years. Poland became free again only with the third republic in 1989, result of the Solidarność movement which played an important role in the final collapse of the Soviet Union in 1991. When I arrived in Wrocław in October 1992 the Soviet Army was still there. Things have been changing slowly: the 1st May of 2004 Poland entered the European Union. I was in Wrocław at this time, after taking part to a meeting of logic in Karpacz, and there was a great party on the main square. But 13 years after, for better or for worse, Poland is still not part of the eurozone.

The Soviet built the impressive tower dominating the city of Warsaw named *Joseph Stalin Palace of Culture and Science* (Fig. 12), nicknamed the *8th sister* because of its similarity with the seven sister Stalinist skyscrapers in Moscow, being the second highest one after the main building of Moscow State University. In this tower there was in

Fig. 12 Stalin power in Poland: the palace of culture and science

particular the department of mathematics of the University of Warsaw, where Andrzej Skowron, as he recently told me, attended a talk by Newton da Costa in the 1970s.

The result of the Soviet domination over the development of logic in Poland after WWII was on the one hand isolation from Western Europe and United States, on the other hand the domination of the Russian school. The best Polish students were sent to Moscow, this was the case of Surma who went to study with Kolgomorov. Philosophy in the Soviet sphere was strongly ideologically oriented towards Marxism. This certainly broke the interaction between logic and philosophy and in Poland logicians concentrated on more technical matters: algebraic methods applied to the study of non-classical logics (Rasiowa, see [59]), a general theory of logic (Suszko's abstract logic, see [45]), foundations of mathematics (in particular Mostowski, see [54]).[6]

The Lvov-Warsaw School is very much linked to the 2nd Republic of Poland which lasted from 1918 to 1939, a 20 year period, described as follows by Jan Zygmunt:

> In 1918, after 120 years of subjugation by foreign powers, Poland reappeared on the map of Europe as a free and independent nation-state. In 1939 the outbreak of the Second World War reversed these brief gains, bringing down the curtain on the Second Republic and inflicting massive damage on its economy, culture and learning. [103, p. 6634].

During this interwar period grew a milieu, a society, a world that was almost completely destroyed, didn't re-appear and never will. In this period logic developed, within a cocktail of philosophy, art and mathematics, an extraordinary situation which does not exist anymore anywhere in the world. This happened in Poland at the beginning of the twentieth century and also in Austria. Maybe it would make sense to speak about a Vienna-Lvov-Warsaw school, considering in particular that Twardowski studied in Vienna and that the interaction between logicians in Vienna and Warsaw was very important during the 1930s. As we already said Tarski went to Vienna and met Gödel (see [38] and [80]). The two schools met in Paris in 1935 at the congress organized by Louis Rougier (cf. [25]). Quine came to Europe and symbolically visited Vienna and Warsaw (see [36] and [58]).

Such atmosphere does not exist nowadays in other places in the world, in particular not in United States, where in some sense the center of civilization was displaced after WWII. Tarski created a school of logic in Berkeley, which became from the 1950s until his death the main school of logic in the world, but it turned out to be mainly a school of model theory centered on mathematics. At the start Tarski tried to develop applications of the axiomatic method to other sciences, in particular to physics. He organized, jointly with Leon Henkin and Patrick Suppes, an important event in December 1957 at Berkeley (cf. [44]). But there was no real follow up. Nowadays the main group of people working on the logic of physics, especially relativity theory, is in Hungary, a group formed by Istvan Németi.

Other Polish logicians were spread around the world: Sobociński in Notre Dame (USA), Bocheński in Fribourg (Switzerland), Łukasiewiz in Dublin (Ireland) [64], Surma in Auckland (New Zealand). Although these people were quite active, for example Sobociński created *Notre Dame Journal of Logic*, no one, excepted Tarski, created a group

---

[6]Stan Surma recalls this time in a film we did with him in Xi'an. In Warsaw there were important seminars directed by Andrzej Mostowski, Helena Rasiowa and Zdisław Pawlak. To know more about that period, see [37], [85], [90].

Fig. 13 USA won the cold war: Warsaw becoming as beautiful as New York

or a school of logic. The relations between Polish logicians outside of Poland were not developed in a systematic way. Moreover the relation between these expatriate logicians with those inside Poland were severely restricted due to the cold war and the iron curtain. So we cannot talk of a Polish school of logic around the world after WWII.

One may think that the end of Soviet Union has led to a true liberation of Poland, but this is a fairly naive perspective symbolically illustrated by the statue of liberty at the entrance of New York (Fig. 13). People say that during the Soviet period in Poland, they had money but there was nothing to buy, after this period there are many things to buy but they don't have money to buy them. You can see a Ferrari in a shopping center but it is a pleasure only for the eyes. One can also say more generally that during the Soviet time there was no choice, in the post-Soviet time, there is the freedom to become a slave (Fig. 14).

It is interesting to remember the etymology of the word "school", *skhole* in Greek: "spare time, leisure, rest ease; idleness; that in which leisure is employed; learned discussion" (*Online Etymology Dictionary*.) In the open market society, a researcher has to become a business man, i.e. someone who is always *busy*, in particular fundraising is considered as a major quality and request for a researcher. There cannot therefore be schools in the original sense, like was the Lvov-Warsaw School.

Fig. 14 After 1991 Poland went further West

3.2 *Logic in Poland 1992–2017*

Since the end of the Soviet Union things are changing in Poland. I will describe the situation from the point of view of my own experience having visited Poland 11 times in 25 years, my first visit being a 14 month stay at the Department of Logic of the University of Wrocław in 1992–93 and my latest visit being a visit in Warsaw in June 2017 for the 2nd World Congress on Logic and Religion I was co-organizing.

From 1995 to 1999 I was working at the National Laboratory for Scientific Computing (LNCC) in Rio de Janeiro. During this period I came three times to Poland for three different conferences:

- April 1997, Karpacz, 2nd Conference Applications of Logic in Philosophy and the Foundations of Mathematics
- July 1998, Toruń, Stanislaw Jaśkowski Memorial Symposium
- August 1999, Kraków, 11th LMPS (Logic, Methodology and Philosophy of Science) congress.

The series of conferences "Applications of Logic in Philosophy and the Foundations of Mathematics" (ALPFM) was launched in 1996 and it happens every year since then. It generally takes place in a very nice village named Karpacz (now in Szklarska Poręba) in the South West of Poland. It has been organized by Jan Zgymunt, Piotr Wojtylak, Janusz

Fig. 15 JYB with Jan Zygmunt in 1997 in Karpacz

Czelakowski, Marcin Selinger, and Tomasz Połacik. I was invited to the first edition but was not able to take part to it. Two neo-Brazilian friends of mine went there: Marcelo Tsuji and David Miller. In 1997 I presented at the second edition the talk "Universal Logic in Perspective" (Fig. 15).

July 15–18, 1998 was organized in Toruń a conference commemorating 50 years of the publication of the paper by Jaśkowski on discussive logic (cf. [46]). It was a kind of follow up of the first world congress on paraconsistency organized in Ghent (Belgium) in 1997. Jaśkowski is considered as one of the two forerunners of paraconsistent logic (the other one is the Russian logician Vasiliev). Generally Łukasiewicz is not considered as a forerunner, although he wrote a seminal book on the principle of contradiction in Aristotle [52], severely criticizing arguments of Aristotle defending this principle. But Łukasiewicz didn't present a paraconsistent system of logic. He famously presented a three-valued system of logic [53] (later on criticized by Suszko [74]). Funny enough the matrices of this system were later used by Asenjo, da Costa and D'Ottaviano and Priest, to develop paraconsistent systems of logic.

The work of Jaśkowski on discussive logic has been promoted outside of Poland mainly by Newton da Costa and his pupils, in particular Lafayette de Moraes (see e.g. [34]). For this reason, at the occasion of this memorial conference was attributed to da Costa Nicholas Copernicus University's medal of merit. Da Costa was not able to come. He was represented by his colleague Itala D'Ottaviano, with whom he developed a logic called "J3"—the "J" referring to Jaśkowski, the "3" to three-valuedness. It is a paraconsistent logic trying to modelize Jaśkowski's ideas using three-valued logical

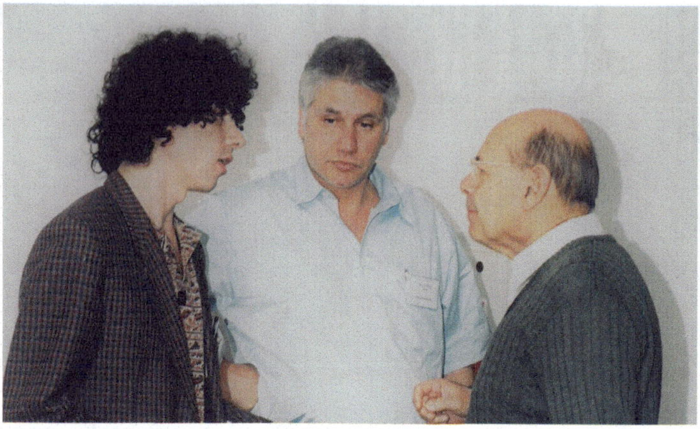

Fig. 16 Jaśkowski Memorial Symposium, Nicholas Copernicus University, Toruń, 1998. JYB, Arthur Buchsbaum, Lafayette de Moraes

matrices (see [35]). I myself presented at this meeting a talk entitled "The Paraconsistent Logic Z" [17], presenting yet another modelization, closer to Jaśkowski's ideas, based on modalities, inspired by a discussion I had with Arthur Buchsbaum [18] (another Brazilian logician, who also spent some times at the University of Wrocław through the connection I established with Jan Zygmunt) (Fig. 16).

In 1999 I went to Poland for the 11th LMPS congress which took place in Kraków August 20–26. It was my first participation to this series of events—after that I went to the 12th edition in Oviedo (Spain) in 2003, the 13th edition in Beijing (China) in 2007 and the 15th edition in 2015 in Helsinki (Finland). The edition in Kraków was by far the best. In particular because it was the more logical one. This series of event, mainly initiated and promoted by Tarski (cf. [38]), has become more and more oriented towards history and philosophy of science, most of the time without any serious logical basis. The event in Kraków was masterly organized by Jan Woleński. Stanislas Lem (1921–2006), the science fiction writer, author of *Solaris* (with was famously adapted in cinema by Tarkowsky in 1972) was invited to give a general audience talk. I met there the late Hartley Salter, with whom I had discussions which lead to my work on the square of opposition (many papers and a series of world events, see [22, 26]). I presented at this LMPS meeting a talk entitled "The Philosophical Import of Polish Logic" (cf. [13]) in which I emphasize the import of three logical notions which were clarified within the Polish school of logic: structurality, truth-functionality, extensionality.

In 2000 and 2001 I was at Stanford University working with Patrick Suppes and didn't come to Poland. From 2002 to 2008 I was working in Switzerland and during this time I also came three times to Poland for three conferences:

- May 2003, Karpacz, 8th ALPFM conference
- April 2004, Karpacz, 9th ALPFM conference
- March 2006, Zakopane, Applications of Algebra to Logic and Informatics X.

I was invited to work at the Institute of Logic of the University of Neuchâtel in Switzerland by Denis Miéville. Neuchâtel is where Jean Piaget was born and his father was the first rector of the university. Piaget taught in Paris and then built a school and research center in Geneva. One of his students, Jean-Blaise Grize, was from the region of Neuchâtel and developed logic and semiology there. He became rector of the University of Neuchâtel and created an institute of logic and a semiological research center. Miéville was his main student and he became head of the institute of logic and later on also rector of the university. Grize had interest for many things but Miéville main interest was about Leśniewski. He published many papers and book on Leśniewski and became one of the best specialist of the Polish bear followed by his student Pierre Joray. Knowing my interest for Polish logic, Miéville invited me to work with them in Neuchâtel.

In 2003 I went to Karpacz with Pierre Joray and Nadine Gessler, another student of Miéville. I took part to the 8th edition (May 6–10, 2003) of the ALPFM conference in Karpacz, together with Pierre and Nadine, presenting the talk "A New Four-Valued Approach to Modal Logic" (later publisher as [20]). The following year (2004) the event took place April, 26–30, I came again, with Pierre Joray, and presented the talk "Does logic need axioms?" (see my later paper [19]).

With Joray and Miéville we also went to a meeting in Nancy (November 21–22, 2003) organized by Roger Pouivet on Philosophy in Poland 1918–1939. A book was subsequently published including my paper about Tarski's axioms for the consequence operator [16]. I met there for the first time Katarzyna Gan-Krzywoszyńska, who later on was my partner for the organization of congresses on universal logic and the square of opposition.

Sandra Lapointe visited us in Neuchâtel and then invited us to a meeting she organized September 23–26, 2004, in Montréal, Canada: *Logic, Ontology, Aesthetics—The Golden Age of Polish Philosophy* where I presented a lecture entitled "Tarski on Consequence and Consequence" discussing the difference and relation between Tarski's operator of consequence and the notion of consequence he introduced in his famous paper "On the concept of logical consequence" [77].

I had heard about Zakopane, in particular because of Witkiewicz, but never visited the town, so I was glad to receive an invitation by Joanna Grygiel to come to the 10th edition of the workshop *Applications of Algebra to Logic and Informatics* which took place there March 6–12, 2006. I presented a talk entitled "Universal Algebra and Universal Logic". David Makinson also was there. There was an interesting excursion organized at night in the mountain on a sledge driven by horses on the snow (Fig. 17).

Since 2010 I am back to Rio de Janeiro as professor of logic of the department of philosophy of the University of Brazil. And since then I also came three times to Poland:

- February 2012, Poznań
- September 2016, Łódź, 8th International Conference—Non-Classical Logics Theory and Applications
- June 2017, Warsaw, 2nd World Congress on Logic and Religion.

In January 2012 I was part of an exchange program with the University of Munich with Matthias Schirn and I decided to come to visit Katarzyna in Poznań. I gave at Adam Mickiewicz University a talk on the square of opposition and Katarzyna showed me the place where Suszko was living (Figs. 18 and 19). Suszko's father was the rector of the University there after Ajdukiewicz.

In 2016 I was invited to take part to the *8th International Conference—Non-Classical Logics Theory and Applications*, which took place September 5–7, 2016 in Łódź,

Fig. 17 In Zakopane in 2006, night excursion. Applications of Algebra to Logic and Informatics

Poland. This is a series of events which alternatively takes place in Łódź and Toruń, two of the most active centers of logic in Poland since WWII. I presented there the talk "Paraconsistent Logic from A to Z". The pun is as follows: "A" is the first letter of "Angers" the place where I discovered paraconsistent logic (see [18]) and "Z" the last letter of "Łódź" where I was giving the talk and at the same time the name of a paraconsistent logic I developed and that I named like that because it is based on possible worlds and "Z" is the last letter of "Leibniz". The organizer of the event, Andrzej Indrzejczak, organized for us a special visit to Łódź Jewish cemetery which was very impressive, with a guide telling us all about the story of Łódź's ghetto.

During this visit I also did a stop in Warsaw in preparation of the 2nd world congress on logic and religion that we organized there in June 2017. After a first successful edition I organized in Brazil with my colleague Ricardo Silvestre in 2015 (cf. [28]), we received the proposal by Marcin Trepczyński to organize a second edition at the University of Warsaw. In September 2016 Marcin showed me the projected place of the congress and presented me to Stanisław Krajewski who was the main organizer of the meeting together with Marcin and Piotr Balcerowicz. The event was a great success with invited speakers such as Dov Gabbay, Michal Heller (Templeton Prize), Saul Kripke (I succeed to convince him to come for his first visit to Poland), Laurent Lafforgue (Fields medal), and Jan Woleński. There were many participants from all over the world that enjoyed very much Warsaw and the friendly atmosphere of the meeting (Fig. 20).

Fig. 18 JYB in front of Suszko's house in Poznań

Fig. 19 JYB in Poznań with Piotr Leśniewski

Fig. 20 The 2nd World Congress on Logic and Religion. University of Warsaw, June 2017

3.3 The Universal Logic Project

Since now more than 25 years I am developing the Universal Logic project. It is deeply connected with Polish logic and the Lvov-Warsaw School. Here are five basic steps corresponding to the development of the project:

- 1993, Wrocław, choice of the expression "Universal Logic"
- 1995, Paris, PhD *Recherches sur La Logique Universelle*
- 2005, Montreux, *1st World Congress on Universal Logic* (UNILOG)
- 2007, Neuchâtel-Basel, launching of the journal *Logica Universalis* and the book series *Studies in Universal Logic*
- 2018, Vichy, 6th UNILOG, World Logic Prizes, including *Alfred Tarski's Prize*

I chose the expression "Universal Logic" when in Wrocław in 1993. I have explained the choice and meaning of this expression in some other papers (see in particular [14], [15], [23] and [24]), so I will not enter in details here, just focusing on the essential. The expression "Universal Algebra" was promoted (not invented) by Garrett Birkhoff. The central feature of Birkhoff's approach is the abandonment of any axiom. I was much inspired by this to develop a general theory of logics, generalizing Tarski's theory of consequence operator, going one more step into abstraction, throwing out the axioms. I crystallized this into the expression *Axiomatic emptiness* (see [19] and my paper [16] discussing Tarski's axioms).[7]

I presented then a lecture entitled "Universal Logic" at the Logica'94 meeting in Czech Republic and defended in July 1995 a PhD in mathematical logic at the University of Paris 7 bearing this expression in the title [10]. Jan Zygmunt came to Paris and was a member of the Jury. I remember that he stayed at the house of a Polish friend of us, Richard Zuber, Rue Mouffetard. After that I spent time in Rio de Janeiro and California, in particular 2 years at Stanford University with Patrick Suppes.

I was able to systematically start to develop the Universal Logic project when I was in Switzerland during the period 2002–2008. I submitted a research project on universal logic to the Swiss Science Foundation (SNF) which was accepted. After organizing a small workshop at the Universality of Neuchâtel in 2004 I started to work on the preparation of the *1st UNILOG—World Congress and School on Universal Logic* in Montreux in Spring 2005 (Fig. 21, Fig. 22). The SNF was at this time giving special support for people from Eastern Europe in particular students so I decided to invite Katarzyna Gan-Krzywoszyńska from Poznań that I had known in Nancy in 2004 and she accepted to help us in the organization of the event. She was especially dedicated and efficient and she subsequently took part to the organization of the further editions

Fig. 21 UNILOG'2005 Montreux—train trip to the Marmot's Paradise, Katarzyna Gan-Krzywoszyńska with Andrei Rodin

[7]I recently defended the idea that anti-classical logic, the complementary of classical logic, can be considered as a logic in view in particular of the theory of refutation initiated by Łukasiewicz. Anti-classical logic is a concrete example of a logic obeying none of Tarski's axioms; See [27].

Fig. 22 UNILOG'2005, Montreux—excursion on the Lake. Urszula Wybraniec-Skardowska, Janusz Czelakowski, Saul Kripke, Musa Akrami, Joanna Grygiel, João Marcos

of UNILOG: the 2nd UNILOG in Xi'an, China in 2007, the 3rd UNILOG in Lisbon, Portugal in 2010, the 4th UNILOG in Rio de Janeiro, Brazil in 2013 and the 5th UNILOG in Istanbul in 2015. She also helped me to develop other projects such as the congresses on the square of opposition. We jointly organized a congress on analogy in Puebla Mexico in 2015. She invited me to write a paper on a special issue of the journal *Studia Metodologiczne* dedicated to Kazimierz Ajdukiewicz [25] and prepared the paper "Personal recollections about JYB by Newton da Costa and others" published in the second volume of the Festschrift for my 50th birthday [39].

At all UNILOGs we had important participations of Polish logicians: Janusz Czelakowski (invited speaker at UNILOG'2005), Piotr Wojtylak (tutorial on consequence operator at UNILOG'2005) and Joanna Grygiel (tutorial on universal algebra for logics at UNILOG'2005), Stan Surma and Jan Woleński (invited speakers at UNILOG'2007), Andrzej Wiśniewski (tutorial on erotetic logics at UNILOG'2013) and Beata Konikowska (invited speakers at UNILOG'2013), Roman Murawski (invited speaker at UNILOG'2015), Andrzej Indrzejczak (tutorial on cut-elimination at UNILOG'2015).

Urszula Wybraniec-Skardowska, one of the two editors of the present book, took part to the 1st UNILOG in Montreux and to successive editions, organizing in particular a workshop on logic and linguistics at UNILOG'2013 together with Marcos Lopes. As a result they published a special issue of the *Journal of Logic, Language and Information* (volume 23, issue 3, 2014). Urszula also took part to the SQUARE events, in particular the 2014 edition in the Vatican [100].

The forthcoming edition of UNILOG will happen in Vichy in June 2018. The Polish logician Jerzy Tomasik (who studied logic in Wrocław) was working at the University Blaise Pascal in Clermont-Ferrand. He knew the late Marcel Guillaume, also from this University, a friend of Newton da Costa, who helped him to publish his work on paraconsistent logic in France in the 1960s and was therefore seminal in the world promotion of paraconsistent logic (see [40–42]). Tomasik was at the UNILOG'2013 in Rio and during a visit in Vichy I told him about my project to organize the 6th edition there (He himself organized in Clermont-Ferrand the *Logic Colloquium* in 1994). He then presented me to Christophe Rey working at the Vichy Campus of the now unified new university Clermont Auvergne (UCA), with whom I am now working on the organization of the 6th UNILOG.

At this edition of UNILOG there will be a session entitled *World Logic Prizes*. In 2014 I decided to create in Brazil, the *Newton da Costa Logic Prize*. It is a prize open to any researcher working in logic in Brazil. People have to submit a non published paper. The winner won publication of his paper in *Logica Universalis* and participation to an international logic event. The first edition of the prize was organized in 2015 and the winner, Rodrigo Freire, was invited to take part to UNILOG'2015 in Istanbul. I then decided to work on the promotion of similar prizes in many countries. At UNILOG'2018 (Fig. 23) we will have about 15 countries presenting a winner, in particular the winner of the Polish Prize of Logic that I suggested to name *Alfred Tarski Logic Prize*.

The present book is published in Birkhäuser's book series *Studies in Universal Logic* (SUL). I was particularly glad to develop with Birkhäuser a journal and a book series on

Fig. 23 UNILOG'2018

universal logic, because the collected papers of Tarski were published by Birkhäuser. It is a four volume book which was released in 1986 [79]. Since many years it is out of print. We are now working on the re-edition of this book in the SUL series with an additional fifth volume including letters and unpublished works of Tarski.

Acknowledgements Thanks to Jan Zygmunt, Newton da Costa, Jan Woleński, Roman Murawski, Katarzyna Gan Krzywoszyńska, Pierre Cartier, Suasan Gómez, Vladimir Vsyukov, Urszula Wybraniec-Skardowska and Rober Purdy for information, photos, comments and remarks.

References

1. Ajdukiewicz, K.: Z metodologii nauk dedukcyjnych, Lvov (1921)
2. Ajdukiewicz, K.: Allocution d'ouverture VI. In: Actes du congrès international de philosophie scientifique, Sorbonne, Paris, 1935 Volume I, pp. 19–21. Hermann, Paris (1936)
3. Ajdukiewicz, K.: Logistyczny antyirracjonalizm w Polsce. Przeglad Filozoficzny, vol. 37, pp. 399–408 [English translation: Logistic Anti-Irrationalism in Poland. Poznań Studies in the Philosophy of the Sciences and the Humanities, vol. 74, pp. 241–250. Rodopi, Amsterdam (2001)]
4. Balcerzyk, S., Bartol, W., Orłowska, E., Wieczorek, A., Wojciechowska-Waszkiewicz, A.: Jerzy Łoś 1920–1998. Stud. Logica **65**, 301–314 (2000)
5. Batóg, T.: Ajdukiewicz and the development of formal logic. In: Sinisi, V., Woleński, J. (eds.) The Heritage of Kazimierz Ajdukiewicz. Poznań Studies in the Philosophy of Sciences and the Humanities, vol. 40, pp. 53–67. Rodopi, Amsterdam (1995)
6. Batóg, T., Łuszczewska-Romahnowa, S.: Logic and methodology of science. In: Krajewski, W. (ed.) Polish Philosophers of Science and Nature in the 20th Century. Poznań Studies in the Philosophy of Sciences and the Humanities, vol. 74, pp. 53–67. Rodopi, Amsterdam (2001)
7. Beziau, J.-Y.: D'une caverne à l'autre. Master Thesis, Department of Philosophy, University Panthón Sorbonne (Paris 1), Paris (1988)
8. Beziau, J.-Y.: La logique paraconsistante C1 de Newton da Costa. Master Thesis, Department of Mathematics, University Denis Diderot (Paris 7), Paris (1990)
9. Beziau, J.-Y.: La logique abstraite au sein de la mathématique moderne. Ruch Filozoficzny **50**, 289–293 (1993)
10. Beziau, J.-Y.: Recherches sur la logique universelle. PhD Thesis, Department of Mathematics, University Denis Diderot (Paris 7), Paris (1995)
11. Beziau, J.-Y.: O suicídio segundo Arthur Schopenhauer. Discurso **28**, 127–143 (1997)
12. Beziau, J.-Y.: Rules, derived rules, permissible rules and the various types of systems of deduction. In: Hauesler, E.H., Pereira, L.C. (eds.) Proof, Types and Categories, pp. 159–184. PUC, Rio de Janeiro (1999)
13. Beziau, J.-Y.: The philosophical import of Polish logic. In: Talasiewicz, M. (ed.) Methodology and Philosophy of Science at Warsaw University, pp. 109–124. Semper, Warsaw (2002)
14. Beziau, J.-Y.: From consequence operator to universal logic: a survey of general abstract logic. In: Logica Universalis: Towards a General Theory of Logic, pp. 3–17. Birkhäuser, Basel (2005)
15. Beziau, J.-Y.: 13 Questions about universal logic. Bull. Sect. Log. **35**, 133–150 (2006)
16. Beziau, J.-Y.: Les axiomes de Tarski. In: Pouivet, R., Rebuschi, M. (eds.) La philosophie en Pologne 1918–1939, pp. 135–149. Vrin, Paris (2006)
17. Beziau, J.-Y.: The paraconsistent logic Z - A possible solution to Jaśkowski's problem. Log. Log. Philos. **15**, 99–111 (2006)
18. Beziau, J.-Y.: Adventures in the paraconsistent jungle. In: Beziau, J.-Y., Carnielli, W.A., Gabbay, D.M. (eds.) Handbook of Paraconsistency, pp. 63–80. College Publication, London (2007)
19. Beziau, J.-Y.: What is a logic? - Towards axiomatic emptiness. Log. Investig. **16**, 272–279 (2010)
20. Beziau, J.-Y.: A new four-valued approach to modal logic. Log. Anal. **54**, 109–121 (2011)
21. Beziau, J.-Y. (ed.): Universal Logic: An Anthology. Birkhäuser, Basel (2012)

22. Beziau, J.-Y.: The new rising of the square of opposition. In: Beziau, J.-Y., Jacquette, D. (eds.) Around and Beyond the Square of Opposition, pp. 6–24. Birkhäuser, Basel (2012)
23. Beziau, J.-Y.: The relativity and universality of logic. Synthese. Special Issue Istvan Németi 70th Birthday **192**, 1939–1954 (2015)
24. Beziau, J.-Y.: Logical autobiography 50. In: Koslow, A., Buchsbaum, A. (eds.) The Road to Universal Logic: Festschrift for the 50th Birthday of Jean-Yves Beziau Volume II, pp. 19–104. Birkhäuser, Basel (2015)
25. Beziau, J.-Y.: Is logical relativity irrational? Stud. Metodologiczne. Special Issue Kazimierz Ajdukiewicz **35**, 63–78 (2015)
26. Beziau, J.-Y.: Hartley Slater and false contradictions. South Am. J. Log. **6**, 101–107 (2016)
27. Beziau, J.-Y., Buchsbaum, A.: Let us be antilogical: anti-classical logic as a logic. In: Moktefi, A., Moretti, A., Schang, F. (eds.) Soyons Logiques/Let Us Be Logical, pp. 1–10. College Publication, London (2016)
28. Beziau, J.-Y., Silvestre, R.: Logic and religion. Log. Univers. **11**, 1–12 (2017)
29. Burckhardt, J.: Der Cicerone Eine Anleitung zum Genu der Kunstwerke Italiens. Schweighauser, Basel (1855)
30. Chapman, S.: Language and Empiricism - After the Vienna Circle. Palgrave Macmillan, Basingstoke (2018)
31. Coniglione, F., Poli, R., Woleński, J. (eds.): Polish Scientific Philosophy: The Lvov-Warsaw School. Rodopi, Amsterdam (1993)
32. da Costa, N.C.A., Beziau, J.-Y.: La théorie de la valuation. Log. Anal. **146**, 95–117 (1994)
33. da Costa, N.C.A., Beziau, J.-Y.: La théorie de la valuation en question. In: Abad, M. (ed.) Proceedings of the Ninth Latin American Symposium on Mathematical Logic, pp. 95–104. Universidad del Sur, Baíha Blanca (1994)
34. de Moraes, L.: Sôbre a lógica discurisva de Jaśkowski. Master Thesis, University of São Paulo, São Paulo (1970)
35. D'Ottaviano, I.M.L., da Costa, N.C.A.: Sur un problème de Jaśkowski. C. R. Acad Sci. Paris **270**, 1349–1353 (1970)
36. Dresner, E.: Quines philosophy of language and Polish logic. Hist. Philos. Log. **20**, 79–96 (1999)
37. Ehrenfeucht, A., Marek, W., Srebrny, M. (eds.): Andrzej Mostowski and Foundational Studies. IOS Press, Amsterdam (2008)
38. Feferman, S., Feferman, A.: Alfred Tarski, Life and Work. Cambridge University Press, Cambridge (2004)
39. Gan-Krzywoszyńska, K.: Personal recollections about JYB by Newton da Costa and others. In: Koslow, A., Buchsbaum, A. (eds.) The Road to Universal Logic: Festschrift for the 50th Birthday of Jean-Yves Beziau Volume II, pp. 1–18. Birkhäuser, Basel (2015)
40. Guillaume, M.: La logique mathématique en sa jeunesse. In: Pier, J.-P. (ed.) Development of Mathematics 1900–1950, pp. 185–367. Birkhäuser, Basel (1994)
41. Guillaume, M.: Regard en arrière sur quinze annèes de coopération douce avec lécole brésilienne de logique paraconsistante. Log. Aanal. **39**, 5–14 (1996)
42. Guillaume, M.: La logique mathématique en France entre les deux guerres mondiales: Quelques repères. Rev. Hist. Sci. **62**, 177–219 (2009)
43. Hendricks, V.F., Malinowski, J.: 50 years of studia logica: editorial introduction. In: Trends in Logic 50 Years of Studia Logica, pp. 1–10. Springer, Dordrecht (2003)
44. Henkin, L., Suppes, P., Tarski, A. (eds.): The axiomatic method, with special reference to geometry and physics. In: Proceedings of an International Symposium Held at the University of California, Berkeley, December 26, 1957–January 4, 1958. North-Holland, Amsterdam (1959)
45. Jansana, R.: Bloom, Brown and Suszko's work on abstract logics. In: Beziau, J.-Y. (ed.) Universal Logic: An Anthology, pp. 251–256. Birkhäuser, Basel (2012)
46. Jáskowski, S.: Rachunek zdan dla systemów dedukcyjnych sprzecznych. Stud. Soc. Scientiarum Torunensis Sect. A **I**(5), 55–77 (1948)
47. Jordan, Z.: The development of mathematical logic and of logical positivism in Poland between the two wars. In: Polish Science and Learning, vol. 6. Oxford Oniversity Press, Oxford (1945)
48. Kijania-Placek, K., Woleński, J. (eds.): The Lvov-Warsaw School and Contemporary Philosophy. Springer, Dordrecht (1998)

49. Lapointe, A., Woleński, J., Marion, M., Miskiewicz, W.: The Golden Age of Polish Philosophy - Kazimierz Twardowski's Philosophical Legacy. Springer, Dordrecht (2009)
50. Leśniewski, S.: In: Surma, S.J., Srzednicki, J.T., Barnett, D.I., Rickey, V.F. (eds.) Collected Works, 2 volumes. Nijhoff - Kluwer, Amsterdam; Polish Academy of Science, Warsaw (1992)
51. Łoś, J.: O matrycach logicznych'. Prace Wrocławskiego Towarzystwa Naukowego Seria B **19**, 1–41 (1949)
52. Łukasiewicz, J.: O zasadzie sprzeczności u Arystotelesa. Akademia Umiejętniści, Kraków (1910)
53. Łukasiewicz, J.: O logice trójwartościowej Ruch Filozoficny **5**, 170–171 (1920)
54. Mostowski, A.: Foundational Studies, Selected Works, vols. 1 and 2. North-Holland, Amsterdam (1979)
55. Murawski, R.: The Philosophy of Mathematics and Logic in the 1920s and 1930s in Poland. Birkhäuser, Basel (2014)
56. Pogorzelski, W., Wojtylak, P.: Completeness Theory for Propositional Logics. Birkhäuser, Basel (2008)
57. Pouivet, R., Rebuschi, M. (eds.): La philosophie en Pologne 1918–1939. Vrin, Paris (2006)
58. Quine, W.V.O.: The Time of My Life. MIT Press, Cambridge (1985)
59. Rasiowa, H.: An Algebraic Approach to Non-Classical Logics. North-Holland, Amsterdam (1974)
60. Reichbach, J.: Completeness of the functional calculus of first-order. Stud. Logica **2**, 245–250 (1955)
61. Reichbach, J.: On the first-order functional calculus and the truncation of models. Stud. Logica **7**, 181–220 (1958)
62. Rescher, N.: The Berlin school of logical empiricism and its legacy. Erkenntnis **64**, 281–304 (2006)
63. Rousseau, J.-J.: Dictionnaire de la musique. Marc Michel Rey, Amsterdam (1768) [English translation by William Waring, London, 1779]
64. Schmidt Am Busch, H.C., Wehmeier, K.F.: On the relations between Heinrich Scholz and Jan Łukasiewicz. Hist. Philos. Log. **28**, 67–81 (2007)
65. Sinisi, V., Woleński, J. (eds.): The Heritage of Kazimierz Ajdukiewicz. Poznań Studies in the Philosophy of Sciences and the Humanities, vol. 40. Rodopi, Amsterdam (1995)
66. Skolimowski, H.: Polish Analytical Philosophy. Routledge and Kegan Paul, London (1967)
67. Sobociński, B.: In Memoriam Jan Łukasiewicz (1878–1956). Philos. Stud. **6**, 349 (1956)
68. Surma, S.: Studies in the History of Mathematical Logic. Ossolineum, Wrocław (1973)
69. Surma, S. (ed.): Special issue on Mordchaj Wajsberg. Bull. Sect. Log. **2**, 91–96 (1973)
70. Surma, S.: On the origin and subsequent applications of the concept of the Lindenbaum algebra. In: Cohen, L.J., Łoś, J., Pfeiffer, H., Podewski, K.-P. (eds.) Logic, Methodology and Philosophy of Science, VI (Hannover, 1979), pp. 719–734. North-Holland, Amsterdam (1982)
71. Surma, S.: The logical work of Mordchaj Wajsberg. In: Srzednicki, J. (ed.) Initiatives in Logic, pp. 101–115. Matrinus Nijhoff, Dordrecht (1987)
72. Surma, S.: Alternatives to the consequence theoretic approach to metalogic. In: Abad, M. (ed.) Proceedings of the Ninth Latin American Symposium on Mathematical Logic, pp. 1–30. Universidad del Sur, Baha Blanca (1994)
73. Surma, S.: A Galois connection. Logica Univers. **1**, 209–219 (2007)
74. Suszko, R.: The Fregean Axiom and Polish mathematical logic in the 1920s. Stud. Logica **36**, 377–380 (1978)
75. Szaniawski, K. (ed.): The Vienna Circle and the Lvov-Warsaw School. Nijhoff, Dordrecht (1988)
76. Tarski, A.: O Logice Matematycznej i Metodzie Dedukcyjnej, Atlas, Lvov-Warsaw (1936). Eng. ed., Introduction to Logic and to the Methodology of Deductive Sciences. Oxford University Press, Oxford (1941)
77. Tarski, A.: O pojciu wynikania logicznego. Przegld Filozoficzny **39**, 58–68 (1936)
78. Tarski, A.: Sur la méthode déductive. Travaux du IXe Congrès International de Philosophie, vol. 6, Actualités Scientifiques et Industrielles, vol. 535, pp. 95–103. Hermann & Cie, Paris (1937)
79. Tarski, A.: In: Givant, S.R., McKenzie, R.N. (eds.) Collected Papers - 4 Volumes. Birkhäuser, Basel (1986)
80. Tarski, A.: Drei Briefe an Otto Neurath. Grazer Philosophische Studien **43**, 1–32 (1992)
81. Wajsberg, M.: Logical Works (translated and edited by S.J. Surma). Ossolineun, Wrocław (1977)
82. Wójcicki, R.: Lectures on Propositional Calculi. Ossolineum, Wrocław (1984)
83. Wójcicki, R.: Theory of Logical Calculi. Basic Theory of Consequence Operations. Kluwer, Dordrecht (1988)

84. Wójcicki, R.: The postwar panorama of logic in Poland. In: dalla Chiara, M.L., et al. (eds.) Logic and Scientific Methods, pp. 497–508. Kluwer, Dordrecht (1997)
85. Wójcicki, R., in collaboration with Zygmunt, J.: Polish logic in postwar period. In: Trends in Logic 50 Years of Studia Logica, pp. 11–33. Springer, Dordrecht (2003)
86. Woleński, J.: Filozoficzna szkoła lwowsko-warszawska. Państwowe Wydawnictwo Naukowe, Warszawa (1985)
87. Woleński, J.: Filozofia szkoły lwowsko-warszawskiej. Uniwersyte Wrocławski, Wrocław (1986)
88. Woleński, J.: Logic and Philosophy in the Lvov-Warsaw. Kluwer, Dordrecht (1989)
89. Woleński, J.: Polish logic. Log. J. IGPL **2**, 93–100 (1992)
90. Woleński, J.: Philosophy inside communism: the case of Poland. Stud. Sov. Thought **43**, 93–100 (1992)
91. Woleński, J.: Mathematical logic in Poland 1900 1939: people, circles, institutions, ideas. Mod. Log. **5**, 363–405 (1995)
92. Woleński, J.: Essays in the History of Logic and Logical Philosophy. Jagiellonian University Press, Kraków (1999)
93. Woleński, J.: The achievements of Polish School of Logic. In: Baldwin, Th. (ed.) The Cambridge History of Philosophy 1870–1945, pp. 401–416. Cambridge University Press, Cambridge (2003)
94. Woleński, J.: Mathematical logic in Warsaw: 1918–1939. In: Ehrenfeucht, A., Marek, W., Srebrny, M. (eds.) Andrzej Mostowski and Foundational Studies, pp. 30–46. IOS Press, Amsterdam (2008)
95. Woleński, J.: Kazimierz Twardowski and the development of philosophy of science in Poland. In: Galavotti, M.C., Nemeth, E., Stadler, F. (eds.) European Philosophy of Science: Philosophy of Science in Europe and the Viennese Heritage, pp. 173–182. Springer, Cham (2014)
96. Woleński, J.: Adolf Lindenbaum. In: Internet Encyclopedia of Philosophy (2017)
97. Woleński, J., Köhler, E.: Alfred Tarski and the Vienna Circle - Austro-Polish Connections in Logical Empiricism. Kluwer, Dordrecht (1999)
98. Woleński, J., Zygmunt, J.: Jerzy Słupecki (1904–1987): life and work. Stud. Logica **48**, 401–411 (1989)
99. Wybraniec-Skardowska, U.: Polish Logic. Some Lines from a Personal Perspective. Publications of Institute for Logic, Language and Computation, Amsterdam (2009)
100. Wybraniec-Skardowska, U.: Logical squares for classical logic sentences. Logica Univers. **10**, 293–312 (2016)
101. Zygmunt, J.: The logical investigations of Jan Kalicki. Hist. Philos. Log. **2**, 41–53 (1981)
102. Zygmunt, J.: Mojżesz Presburger: life and work. Hist. Philos. Log. **12**, 221–223 (1991)
103. Zygmunt, J.: Polish logic. In: Routledge Encyclopedia of Philosophy, pp. 6634–6639. Routledge, London/New York (1998)
104. Zygmunt, J., Purdy, R.: Adolf Lindenbaum: notes on his life, with bibliography and selected references. Logica Univers. **8**, 285–320 (2014)

J.-Y. Beziau (✉)
Brazilian Research Council, University of Brazil, Rio de Janeiro, Brazil

Ecole Normale Supérieure, Paris, France
e-mail: jyb@ufrj.br

The manufacturer's authorised representative in the EU is Springer Nature Customer Service Centre GmbH, Europaplatz 3, 69115 Heidelberg, Germany. If you have any concerns regarding our products, please contact ProductSafety@springernature.com

Printed and bound by CPI Group (UK) Ltd, Croydon, CR0 4YY
30/09/2026
02225654-0019